Porous Carbons:

Syntheses and Applications

多孔碳的合成与应用

Michio Inagaki　Hiroyuki Itoi　Feiyu Kang

[日]稻垣道夫　[日]糸井弘行　康飞宇　著

清华大学出版社
北　京

内 容 简 介

本书分为5章，第1章为绪论，讨论了孔结构在碳材料中的重要性；第2章为多孔碳的合成、制备与改性，介绍了微孔碳、介孔碳、大孔碳的设计和控制；第3章为多孔碳在能量存储和转化中的应用，涉及各类电池和超级电容器；第4章为多孔碳的环境应用，介绍了多孔碳在吸附、净化和催化方面的应用；第5章对多孔碳的发展进行了展望。

ISBN 978-0-12-822115-0

图书在版编目（CIP）数据

多孔碳的合成与应用 = Porous Carbons : Syntheses and Applications : 英文 /（日）稻垣道夫，（日）糸井弘行，康飞宇著.—北京：清华大学出版社，2023.8

ISBN 978-7-302-64424-8

Ⅰ. ①多… Ⅱ. ①稻… ②糸…③康… Ⅲ. ①多孔碳－合成－英文 Ⅳ. ①TM242

中国国家版本馆CIP 数据核字(2023)第153293字

责任编辑：戚 亚 黎 强
封面设计：刘艳芝
责任印制：宋 林

出版发行：清华大学出版社
网 址：http://www.tup.com.cn, http://www.wqbook.com
地 址：北京清华大学学研大厦 A 座 邮 编：100084
社 总 机：010-83470000 邮 购：010-62786544
投稿与读者服务：010-62776969，c-service@tup.tsinghua.edu.cn
质量反馈：010-62772015，zhiliang@tup.tsinghua.edu.cn
印 装 者：涿州汇美亿浓印刷有限公司
经 销：全国新华书店
开 本：153mm ×228mm **印 张：**54.5
版 次：2023 年8 月第1 版 **印 次：**2023 年 8 月第1次印刷
定 价：399.00 元

产品编号：090428-01

Preface

One of characteristics of carbon materials is the formation of pores with a wide range of sizes and morphologies, from micropores with sizes of less than 1 nm to macropores with sizes of more than 50 μm, and from channel-like pores with homogeneous diameters in porous carbons and carbon nanotubes to round pores in porous carbons and fullerene cages, including irregular-shaped pores in polycrystalline carbon materials. Large quantities and rapid adsorption of different gas and liquid molecules have been utilized for the remediation of our circumstances since prehistorical times and nowadays for storage of foreign atoms and ions for energy storage and conversion. In addition, porous carbons can keep various functional groups on their surfaces, which can enable the storage of electrochemically active materials, such as Li^+, Na^+, etc., and the removal of the environmental pollutants, such as Pb^{2+}, Hg^0, etc., through chemical interaction.

The authors have published three books on carbon materials in a series entitled *Materials Science and Engineering of Carbons: Fundamentals, Advanced Materials Science and Engineering of Carbons*, and *Materials Science and Engineering of Carbons: Characterization*. A book focused on graphene and its related materials was also published by the same publishers, entitled *Graphene: Preparations, Properties, Applications and Prospects*. **Porous carbons** are among the most important themes in discussing their of the fabrication and applications of carbon materials in these books. In the first book (*Fundamentals*), their fabrication (synthesis and preparation) processes are described mainly on the bases of fabrication techniques, such as activation, etc., together with their applications in specific fields, such as energy storage and environment remediation. In the second book (*Advanced*), the focus is on novel fabrication techniques, such as template carbonization, carbon nanofibers via electrospinning and carbon foams, and applications in electrochemical capacitors, lithium-ion batteries, and adsorption. The characterization techniques of carbon materials are discussed on gas adsorption techniques for the evaluation of pore structures and the determination of electrochemical performances of various carbon materials in the third book (*Characterization*). In the fourth book (*Graphene*), graphene and its related materials are discussed, and the importance of the surface functional groups on reduced graphene oxides is demonstrated to develop new applications.

The present book, *Porous Carbons: Syntheses and Applications*, is focused on the syntheses and applications of porous carbons, the pore structures of which depend strongly on the precursors and synthesis conditions, temperatures and durations of carbonization and activation, activation reagents, templates, etc. The syntheses are explained by dividing the synthesis techniques into activation, template-assisted carbonization, and precursor-designing on the resultant microporous, mesoporous, and macroporous carbons. In the section on macroporous carbons, so-called carbon foams composed from exfoliated graphite and carbon nanofibers are included. The applications are discussed by dividing them into two chapters, one on the applications for energy storage and conversion, and another on the applications for environment remediation. In the applications for energy storage and conversion, rechargeable batteries (lithium-ion, sodium-ion and potassium-ion batteries, lithium-sulfur and lithium-oxygen batteries) and electric double-layer capacitors (supercapacitors), together with their hybridization, and fuel cells are discussed, in addition to the storage of energy sources (hydrogen and methane) and thermal energy. The applications for environment remediation are discussed by dividing them into adsorption, gas separation, capacitive deionization, electromagnetic interference shielding, and sensing. The section on adsorption is composed of adsorptive removal of environment pollutants (including inorganic and organic species), water vapor adsorption, CO_2 capture, metal ions trapping, and oil sorption. The section on sensing is discussed by separating sensors into chemical sensors, mechanical sensors (strain sensors), and biosensors. In the final chapter, concluding remarks are presented on syntheses and applications by emphasizing the pore structure of the carbon and the presence of surface functional groups. In addition, the possibility of some porous carbons, including carbon nanotubes and fullerenes, presenting constraint and reaction spaces is proposed by showing some experimental results.

It would be a source of great pleasure for the authors if the content of this book was to provide useful information to readers, to enable readers to gain a thorough understanding of the syntheses and applications of porous carbons and also to form new ideas on the carbon materials. For readers' convenience, it is recommended to consult the three books mentioned above which are published by Tsinghua University Press and Elsevier. These books will supply the fundamental knowledge on carbon materials and provide a broad understanding of the range of topics discussed in this book.

Acknowledgments

The authors would like to express their sincere thanks to the people who kindly provided the data and figures for this book, their origins being mentioned in the caption of figures and tables as reference numbers, with the names of contributing persons and journals published being presented in the list of the references. The authors also thank all of those people at Elsevier and also at Tsinghua University Press who helped in the publication of this book.

Contents

CHAPTER 1
Introduction

1.1 Carbon materials

Carbon materials are one of most important and essential materials for realizing and maintaining a low-carbon society. Various carbon materials are currently being investigated, such as the electrode materials of lithium-ion rechargeable batteries, electrochemical capacitors (supercapacitors), fuel cells, and solar batteries, which are important devices for creating, saving, and storing sustainable energies. Also, they serve as materials to contribute to environmental remediation, such as adsorbents for pollutants, sensor elements for toxic gases, and electrodes for capacitive desalination of water. Carbon nanotubes and graphenes have contributed greatly to accelerating the development of nanodevices. Some carbon materials are considered to be promising materials for biomedical applications, mainly because of their biocompatibility, such as drug-delivery carriers, cell culturing beds, and biosensors. In these carbon materials, pores with various sizes and morphologies play important roles in the improvement and realization of functional materials and their applications.

Before discussing porous carbons, the fundamentals of carbon materials science are briefly explained here, emphasizing the diversity of carbon materials in structure and textures in various scales. On the detailed explanation and discussion on carbon materials, the readers of this book are referred to fundamental books written by the current authors (M.I. and F.K.) [1,2].

1.1.1 Classification of carbon materials

Numerous kinds of carbon materials have been synthesized and widely used in many fields of industry. These carbon materials are proposed to be classified on the basis of chemical bonds of constituent carbon atoms using sp^3, sp^2, and sp hybrid orbitals [3]. The sp^2 hybrid bonding of carbon atoms results in two structures: flat layers composed of six-membered carbon rings, which have so far been represented by graphite but now are typified by graphene, and curved layers created by introducing five-membered carbon rings into six-membered rings, as occurs in fullerenes. Carbon layers composed of sp^2 orbitals, both flat and curved, are intrinsically anisotropic and have π-electron clouds on both sides of the layer, and these anisotropic layers create the broad diversities in the structure and properties of the carbon materials. Carbon nanotubes can be placed between fullerene and graphene, because the tips of the tube include five-membered rings (fullerene-like) and its wall is composed of six-membered rings (graphene-like) though it is rolled up. A classification of carbon materials based on hybrid bonds is presented, together with the diversities of the materials, in Fig. 1.1.

Due to the anisotropic nature and the presence of π-electron clouds in the carbon materials composed mainly of sp^2 hybrid orbitals, the number of carbon layers stacked in parallel has a strong influence on their properties. The importance of the number of stacked layers has been pointed out on carbon nanotubes and fullerenes, and now on graphene. In addition to the layer number stacked, the stacking regularity of layers and the size of anisotropic layers widen the diversity of carbon materials. An infinite number of large-sized layers stacked with regularity has been called

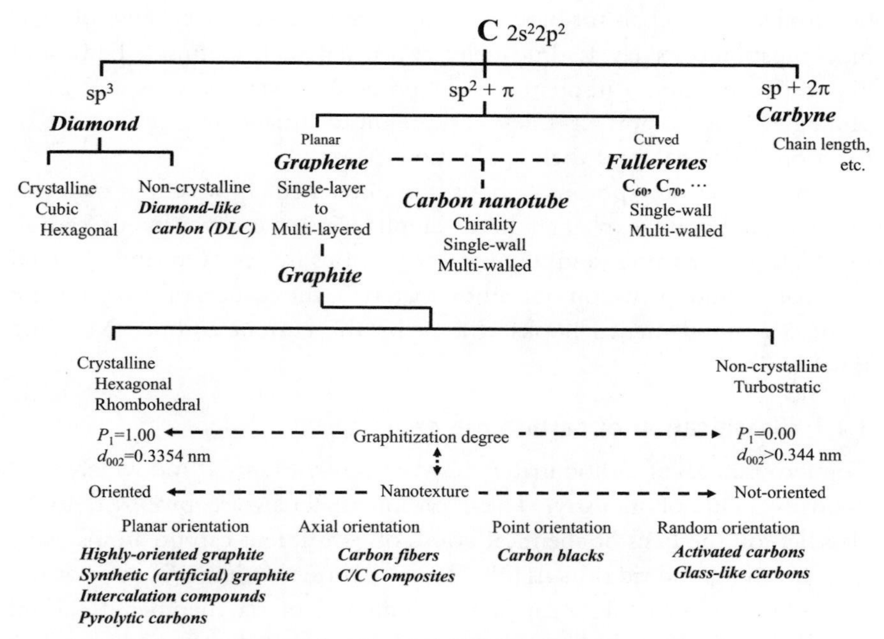

Figure 1.1 Classification and diversity of carbon materials.

graphite, which is crystalline and strongly anisotropic in structure, and as a consequence anisotropic in properties. In contrast, random aggregation of the units of irregularly stacked small layers results in so-called amorphous carbon, which is isotropic in properties with high mechanical strength and high hardness. In between crystalline graphite and noncrystalline amorphous carbon, various graphite-related materials having structures with different degrees of graphitization (proportion of regular stacking) and layer sizes have been produced in industries and used as important industrial materials.

1.1.2 Structure and nanotexture of carbon materials

Most carbon materials are composed of flat layers using sp^2 hybrid orbitals and consist of small units of layers stacked in parallel, which are called basic structural unit (BSU) or crystallite, as shown by the lattice fringe TEM image on an aggregation of some BSUs and the schematic illustration on a BSU in Fig. 1.2A and B, respectively. In the unit, two kinds of stacking regularity of layers coexist: random and regular stacking. The latter is graphitic stacking (usually written as AB stacking) with interlayer spacing d_{002} of 0.3354 nm (the same as natural graphite), while the former is stacking with a slightly larger d_{002} as 0.342 nm, called turbostratic stacking. These BSUs are strongly anisotropic in bonding nature, with strong covalent bonding using sp^2 hybrid orbital along the layer and weak van der Waals bonding due to the interaction between π-electron clouds of the layers.

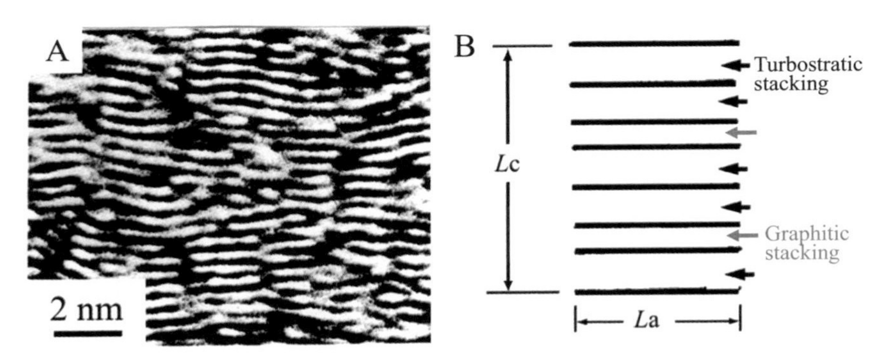

Figure 1.2 Basic structural unit (BSU) of carbon materials: (A) lattice fringe image and (B) schematic illustration of a unit.

The aggregation of these anisotropic nano-sized BSUs gives different textures to the particles, due to the different schema of preferred orientation of anisotropic BSUs, i.e., planar, axial, point, and random orientations, as illustrated together with some representative carbon materials [3,4] in Fig. 1.3. The aggregation of BSUs in different schema is called nanotexture. By planar orientation, films and platelets of highly oriented graphite are produced and some coke particles are principally composed by the planar orientation of BSUs. By axial orientation, fibrous carbon materials are produced from different precursors, such as carbon nanotubes and vapor-grown carbon fibers with a coaxial mode of orientation, and some mesophase-pitch-based carbon fibers with radial mode. By point orientation, a variety of carbon spheres are produced, as represented by various-sized fullerene particles and different nano-sized carbon blacks with concentric mode and mesophase spheres with radial mode. The particles composed of these oriented nanotextures are still anisotropic. In addition, random aggregation of small BSUs occurs in so-called glass-like carbon (glassy carbon), of which the particles are isotropic in nature.

1.1.3 Carbonization and graphitization

Most carbon materials used in industry are produced from organic precursors, such as pitches, biomasses, and organic polymers, via heat treatment at high temperatures in inert atmosphere, pyrolysis, carbonization, and

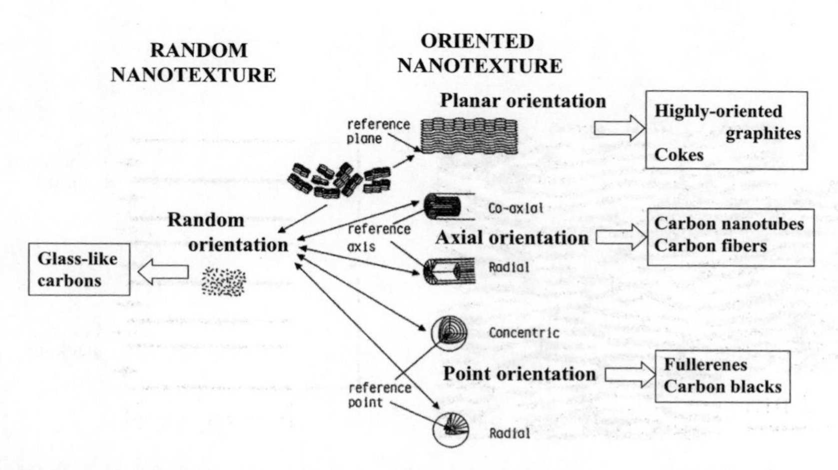

Figure 1.3 Nanotextures in carbon materials on the bases of preferred orientation of BSUs.

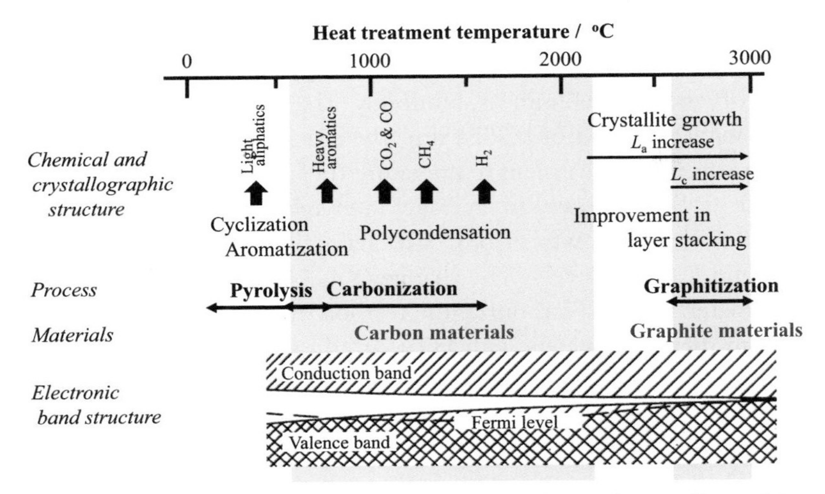

Figure 1.4 Schematic illustration of the changes in chemical, crystallographic, and electronic band structures with heat treatment temperature for an organic carbon precursor.

graphitization processes. In Fig. 1.4, changes in chemical, crystallographic, and electronic band structures with heat treatment temperature are illustrated.

The carbonization process proceeds after pyrolysis of organic precursors at the temperature range from 800°C up to 2000°C, in which the BSUs are formed and their basic aggregation scheme (nanotexture) is established through polycondensation of six-membered carbon rings, accompanying the emission of foreign atoms, oxygen, hydrogen, and nitrogen, as gases. This process is the most important in the production of various carbon and graphite materials, because the nanotexture of most carbon materials is established during this process and the nanotexture governs the development of the crystalline structure in carbon materials during the following graphitization process. During pyrolysis and carbonization processes, a high amount of shrinkage occurs due to the large amount and rapid emission of gas species, associated with cracking in the resultant carbon particles in many cases. Therefore, the process of carbonization is usually performed separately from that of graphitization.

Above 2000°C, the change in crystalline structure, in other words, the development of graphite structure, occurs mainly. The development of the graphite structure is evaluated by different techniques, including X-ray diffraction (XRD), electromagnetic property measurements, Raman spectroscopy, high-resolution transmission electron microscopy, etc. In BSUs formed during carbonization, turbostratic stacking with an interlayer spacing of about 0.342 nm is randomly changed to graphitic regular stacking with a spacing of 0.3354 nm (the spacing in the graphite crystal) with increasing heat treatment temperature (HTT) above 2000°C, which is evaluated as the decrease in average interlayer spacing d_{002} by XRD, associated with the growth of BSU sizes (crystallite sizes) along the a and c axes, L_a and L_c, respectively. The change in d_{002} with HTT depends greatly on the materials after carbonization (carbon materials). In Fig. 1.5, the changes in these parameters with HTT are shown for various carbon materials. In a needle-like coke with planar orientation scheme d_{002} decreases quickly to approach the value of graphite crystal (0.3354 nm) and L_c and L_a grow rapidly. In glass-like carbon with a random orientation scheme, in contrast, almost no decrease in d_{002} and no appreciable growth in L_c and L_a even after 3000°C treatment are observed, i.e., there is no development of graphite structure. Carbon blacks with a point orientation scheme exhibit intermediate behaviors, large-sized thermal black showing more structure improvement than small-sized furnace black.

Diffraction peaks of XRD for carbon materials are classified into three groups, 00*l*, hk0, and hk*l*, mainly due to the strong anisotropy of BSUs and the coexistence of two interlayer spacings, as shown in Fig. 1.2B. Diffraction peaks with the indices 00*l* give averaged interlayer spacing d_{002}, which decreases gradually with increasing HTT from more than 0.344 nm for small and highly defective layers to 0.3354 nm for graphitic stacking through about 0.342 nm for turbostratic stacking, in other words, with improving crystallinity of carbon, as shown in Fig. 1.5A. In turbostratic stacking of layers, there is no three-dimensional regularity in stacking, i.e., no l-index is defined. Therefore, their diffraction peaks are expressed as hk for carbon materials mainly consisting of turbostratic stacking of layers, although they are indexed as hk0 for the graphitic structure with three-dimensional regular stacking. Regular and random stackings, graphitic and turbostratic, are clearly demonstrated in diffraction profiles of hk0 and hk peaks, as 100 and 10 peaks. The 10 peak for turbostratic structure shows a characteristic asymmetric profile and is modulated by the appearing 101 peak, together with its sharpening and improving in symmetry with increasing HTT. The 100 peak for a graphitic structure is sharp and symmetrical, associated with the 101 peak due to the formation of three-dimensional stacking regularity. The peaks with the indices of hk*l* are caused by the graphite structure and so the 112 peak is often selected as an

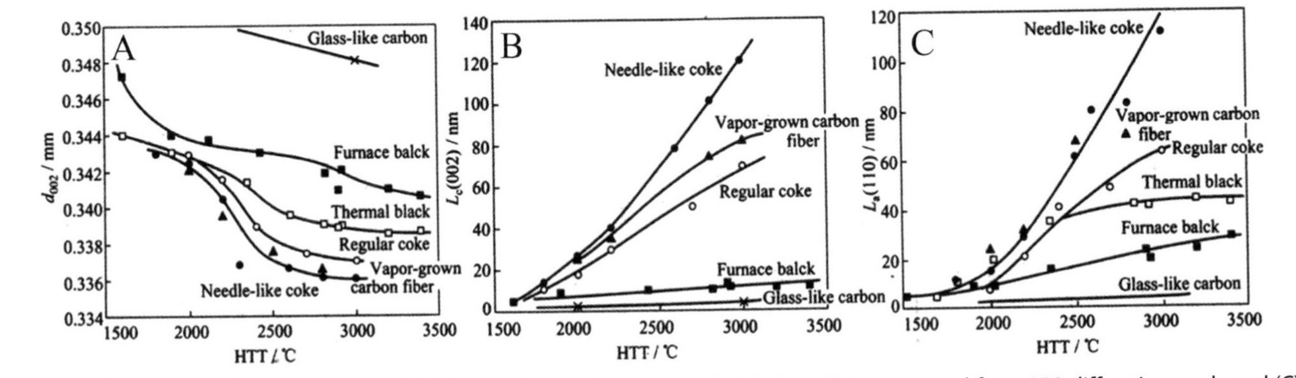

Figure 1.5 Changes in XRD parameters with HTT on various carbon materials: (A) d_{002}, (B) L_c measured from 002 diffraction peak, and (C) L_a from 110 peak.

indication of the formation of graphite structure, because the 112 peak is not overlapped with other peaks although the 101 peak is often overlapped with the 100 peak. Sharpening of these diffraction peaks, 002, 004, 100, 110, and 112, with increasing HTT provides the information on the growth of crystallite sizes, average size along the c-axis (L_c), that along the a-axis (L_a), as shown in Fig. 1.5B and C, respectively, and the size of three-dimensional graphite crystallites.

Since most carbon particles are anisotropic, except glass-like carbon, their aggregation into a block gives different textures in a larger scale, which may be called microtexture and macrotexture. One of the examples of macrotexture is illustrated for the carbon fiber-reinforced composites in Fig. 1.6. These large-scale textures must be controlled for practical applications, although the evaluation technique for these large-scale textures has not been established yet.

1.2 Pores in carbon materials

All carbon materials contain pores, with the exception of highly oriented graphite and graphite crystal, because they are polycrystalline products of thermal decomposition of organic precursors, such as various resins and pitches. During their pyrolysis and carbonization, a large amount of decomposition gases is formed in a wide range of temperatures, as shown in Fig. 1.4. Since the gas evolution behavior of organic precursors is strongly dependent on both the precursors used and the heating conditions, such as

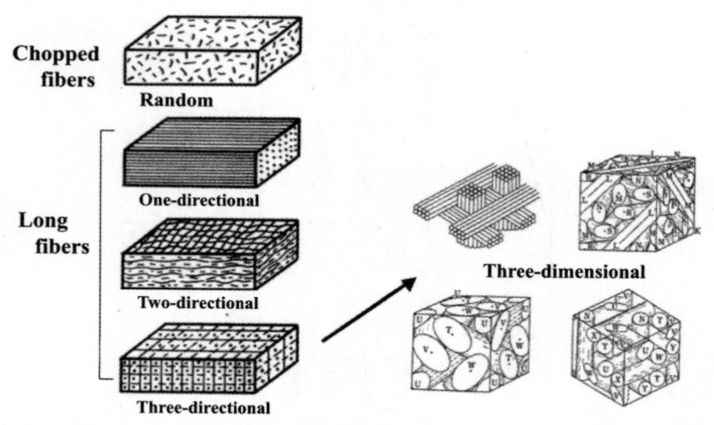

Figure 1.6 Different schema of the arrangements of carbon fibers for reinforcing the composites.

Table 1.1 Classification of pores in solid materials.

(1) Based on the sizes		
	Micropores <2 nm	Ultramicropores < 0.7 nm Supermicropores 0.7−2 nm
	Mesopores 2−50 nm Macropores >50 nm	
(2) Based on the origins		
	Intraparticle pores	Intrinsic intraparticle pores Extrinsic intraparticle pores
	Interparticle pores	Rigid interparticle pores Flexible interparticle pores
(3) Based on the states		
	Open pores Closed pores (latent pores)	

heating rate, pressure, etc., carbon materials contain a large amount of pores consisting of a wide range of sizes and morphologies. The pores in solids, including carbon materials, are classified on the bases of sizes, origins, and states, as summarized in Table 1.1.

The classification of pores based on their sizes was proposed by IUPAC (International Union for Pure and Applied Chemistry). As illustrated in Fig. 1.7, pores are usually classified into three types: macropores (>50 nm), mesopores (2−50 nm), and micropores (<2 nm) [5,6]. Micropores are further divided into supermicropores with a size of 0.7−2 nm and ultramicropores of less than 0.7 nm. The terms "nanopores" and "nanoporous carbons" are often used to show the presence of micropores and mesopores in the carbon.

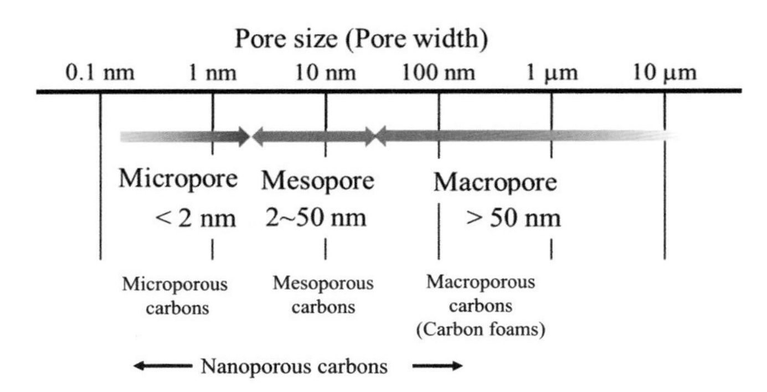

Figure 1.7 Classification of pores in carbon materials on the bases of sizes.

Pores are also classified into two types, intraparticle and interparticle pores, on the bases of their origins. The intraparticle pores are further classified into two subtypes, intrinsic and extrinsic. The former owes its origin to the crystal structure, and a typical example is the pores in zeolite crystals. The graphite gallery between neighboring hexagonal carbon layers can be a slit-shaped pore with a width of 0.3354 nm, which can accept various atoms, ions, and even molecules, with the insertion of foreign species into interlayer spaces of graphite often being called "intercalation." Therefore, the graphite gallery is an intrinsic intraparticle pore. Into the graphite gallery, which is widened by the intercalation of Cs ions (binary intercalation compound), hydrogen and some hydrocarbon molecules can be intercalated to form ternary intercalation compounds, with this intercalated gallery being an extrinsic intraparticle pore. In most activated carbons, a large amount of pores with various sizes of nanometer scale, micropores and mesopores, are formed because of random orientation of crystallites, which are rigid interparticle pores. In exfoliated graphite, which consists of worm-like particles, large pores in and among worm-like particles are formed, which are flexible intraparticle and interparticle pores, respectively. These pores can be easily deformed by compression and by sorption of heavy oil.

In addition, pores are classified on the bases of their states, either open or closed. In order to identify the pores by gas adsorption, they must be exposed to the adsorbate gas, in other words, the pores must be open for adsorbate gas molecules and have sufficient size to accommodate gas molecules. If some pores are too small to accept absorbate gas molecules, they cannot be recognized as pores by the gas adsorption technique. These pores are often called latent pores to differentiate them from structurally closed pores. Closed pores are not necessarily small in size.

Porous carbon materials are often classified into microporous, mesoporous, and macroporous carbons, whether the principal pores are micropores, mesopores, or macropores, as indicated in Fig. 1.7. Macroporous carbons are often called carbon foams, and sometimes carbon sponges. However, it has to be pointed out that these pores coexisted in most of the carbons without any intention. For example, micropores exist in the walls of mesopores and also between two mesopores as the windows to interconnect two mesopores. In most carbon foams, micropores, and even mesopores, often exist in the walls of macropores. Although the words "nanopores" and "nanoporous carbons" are used in some of the literature without definitions, they are used to indicate pores having nanoscale sizes in most cases, i.e., micropores and mesopores including small-sized macropores, as shown on the last line in Fig. 1.7. Recently hierarchical pore structures have attracted attention, mainly because the coexistence of micropores, mesopores, and macropores is desired for some applications of porous carbons.

In this book, the classification of porous carbons into four classes is employed: microporous carbons, mesoporous carbons, macroporous carbons (or carbon foams), and hierarchically porous carbons. The last class, hierarchically porous carbons, is limited to the carbons in which three kinds of pores, micropores, mesopores, and macropores, intentionally coexist and their structures (i.e., size and volume) are independently controlled, because most porous carbons are more or less hierarchically porous.

1.3 Identification and evaluation of pores in carbons

Pores in carbon materials have been identified and evaluated by different techniques depending mostly on their size. The techniques to evaluate pores in carbon materials are summarized in Fig. 1.8. Pores with nanometer sizes, i.e., micropores and mesopores, are identified by the analyses of gas adsorption–desorption isotherms mostly using nitrogen gas at 77 K. The fundamental theories, instruments, measurement practices, analysis procedures, and many results obtained so far by gas adsorption measurements have been reviewed in different publications [7]. X-ray small-angle

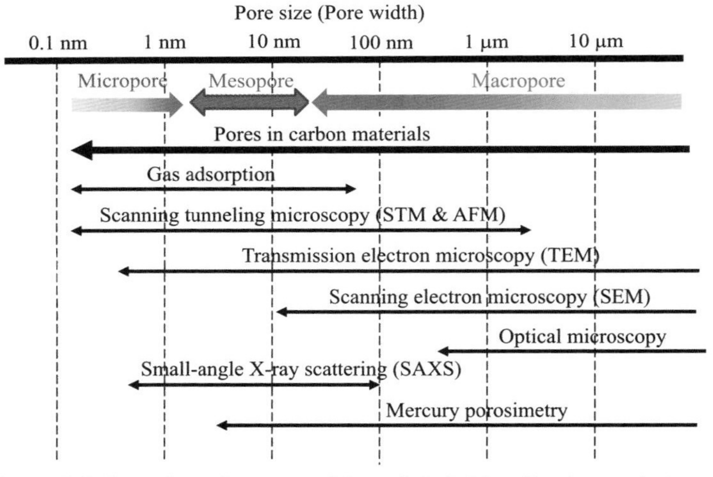

Figure 1.8 Pores in carbon materials and their identification techniques.

scattering has an advantage to identify the latent pores, including closed pores, though gas adsorption can detect only open pores which can accept gas molecules. For macropores, mercury porosimetry has been frequently applied [7]. Identification of intrinsic intraparticle pores, including interlayer space between hexagonal carbon layers in the case of carbon materials, is carried out by X-ray diffraction (XRD), while extrinsic intraparticle and rigid interparticle pores, including latent pores, are identified by small-angle X-ray scattering (SAXS). Direct observation of pores on the surface of carbon materials has been reported by using microscopy techniques coupled with an image-processing technique, e.g., scanning tunneling microscopy/atomic force microscopy (STM/AFM) with transmission electron microscopy (TEM) for micropores and mesopores, and scanning electron microscopy (SEM) with optical microscopy for macropores.

On the bases of these characterization techniques, the explanations focusing on the key issues for the measurements and analyses were presented in a separate book [8].

1.3.1 Gas adsorption

The pore structure of solids has usually been measured using physical adsorption of various gases. For the convenience of the measurements, nitrogen adsorption has often been used at a liquid nitrogen temperature of 77 K [9,10]. There are many reviews and books on pore structure determination by nitrogen gas adsorption focusing on carbon materials [7].

The isotherms of adsorption and desorption were classified by their shapes into six types by IUPAC (International Union of Pure and Applied Chemistry), as shown in Fig. 1.9A [5]. The type I isotherm is typical for microporous solids, especially for activated carbons, where micropore filling occurs at a low relative pressure P/P_0, and adsorption completes below 0.5 in most cases. In Fig. 1.10A, N_2 adsorption−desorption isotherms for some microporous carbons are shown, where an abrupt increase in N_2 adsorption at very low P/P_0 (below 0.05) is characteristic for the presence of micropores. The type IV isotherm is typical for mesoporous carbons, being characterized by a hysteresis, which is classified by IUPAC into four kinds from H1 to H4, as shown in Fig. 1.9B. Fig. 1.10B shows representative N_2 isotherms with clear H2 hysteresis and an abrupt increase in N_2 adsorption at low P/P_0, suggesting the coexistence of micropores with mesopores. As shown in Fig. 1.10C, isotherms of some macroporous carbons (carbon foams) show very small adsorption of N_2 even at a high P/P_0 of 0.8 and an

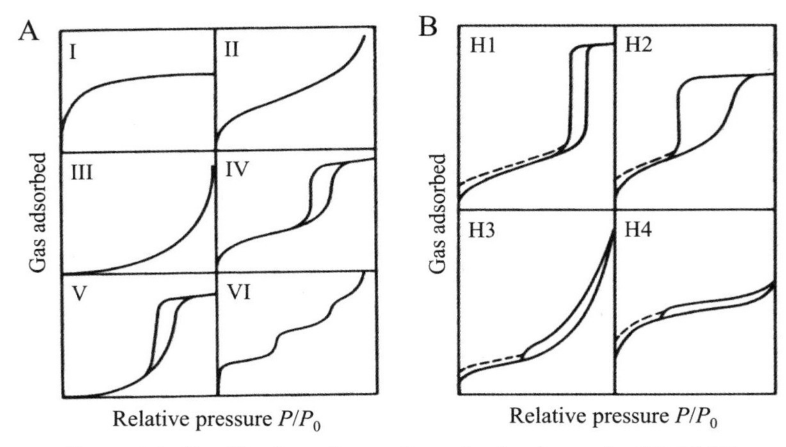

Figure 1.9 Classification of gas adsorption isotherms by IUPAC [5].

abrupt increase in N_2 adsorption above P/P_0 of 0.9. In Fig. 1.10C, pore size distribution curves are inserted, to show the existence of macropores with about 50 nm sizes.

For the analysis of these adsorption–desorption isotherms, various methods have been proposed: BET (Brunauer–Emmett–Teller) method, α_s plot, BJH (Barrett–Joyner–Halenda) method, DR (Dubinin–Radushkevich) plot, DH (Horvath–Kawazoe) method, t plot, and DFT (density functional theory) method, together with theoretical calculation based mainly on the Grand Canonical Monte-Carlo method. They have been applied on the isotherms measured on various carbon materials.

The BET method [11] is commonly used for carbon materials by determining the surface area S_{BET}, which is convenient to compare the pore structures among different carbon materials. S_{BET} is calculated from the slope and the intercept of the linear relation of $1/\{W[(P_0/P)-1]\}$ against P/P_0 (BET plot), where W is the amount of adsorbed gas at P/P_0. Porous carbons containing mesopores together with micropores, which are rather common in porous carbons, show composite isotherms of types I and IV. They show an abrupt increase in N_2 adsorption at low P/P_0 below 0.1 and hysteresis at P/P_0 above 0.5, providing certain errors in surface area calculated by the BET method. S_{BET} is frequently applied to all kinds of carbon materials to give a relative measure of their pore structures.

An α_s plot [12,13] is based on the comparison of isotherms between the sample and the reference (nonporous solid) and is often used for carbon materials. A typical plot with analysis procedures is schematically shown in

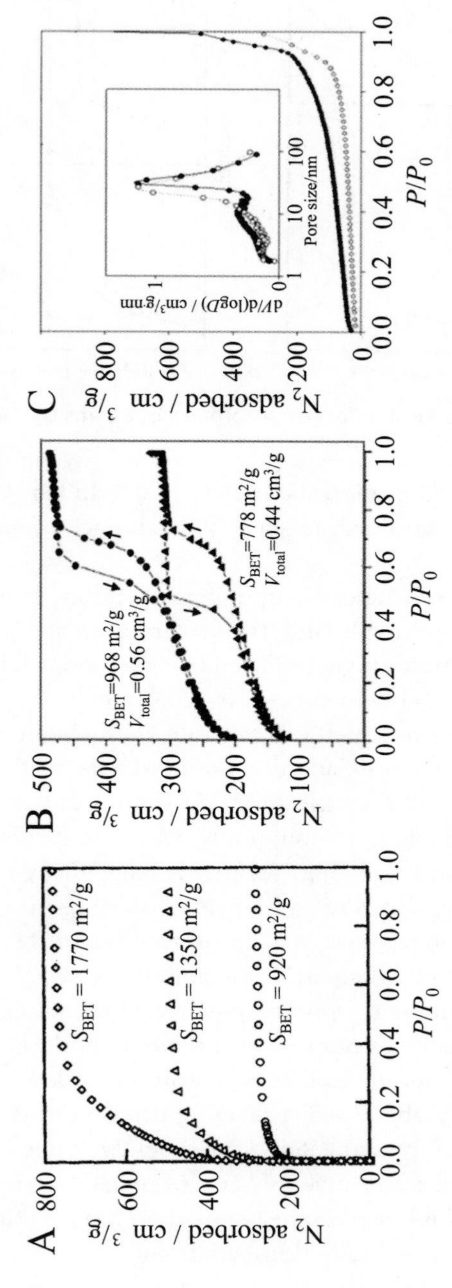

Figure 1.10 Representative N_2 adsorption-desorption isotherms for (A) microporous, (B) mesoporous, and (C) macroporous carbons.

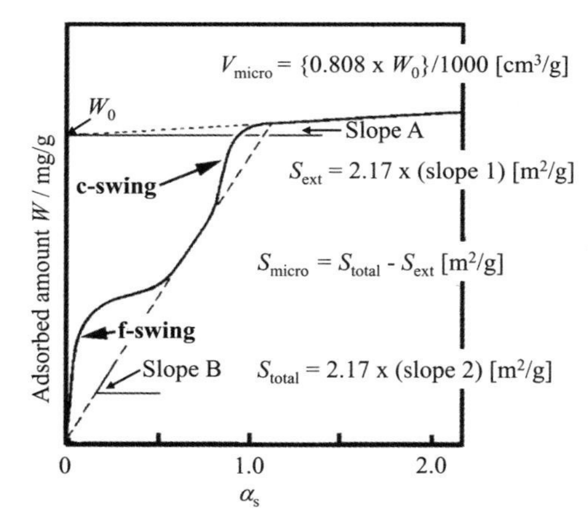

Figure 1.11 Illustration of α_s plot of N_2 adsorption at 77 K with the calculation procedure for the pore structure parameters.

Fig. 1.11: W in the unit of mg/g is plotted against α_s, which is nitrogen adsorption of the reference at each P/P_0 after normalizing at P/P_0 of 0.4. Often it shows the deviations upward from the line passing through the origin, f-swing (filling swing) below α_s of 0.5, and c-swing (cooperative filling) over α_s of 0.5—1.0. The f-swing and c-swing are mainly due to the strong adsorption into micropores and quasi-micropore condensation, respectively. External surface area S_{ext} is calculated from the slope A and total surface area S_{total} from the slope B in α_s plot using the respective equations in Fig. 1.11, and microporous surface area S_{micro} is obtained as a balance. The values of surface area thus calculated are those subtracted from the effect of strong potential field in micropores and so-called SPE surface areas (SPE: subtracting pore effect). From the intercept W_0, micropore volume V_{micro} can be obtained. These pore structure parameters are the reliable information on microporous and mesoporous carbons. In t plot analysis, the parameter t is the thickness of adsorbed layer on the reference at the corresponding P/P_0.

Analysis of isotherms by the BJH method [14] is used to determine the size distribution of mesopores assuming that micropores are absent from the sample and only mesopores are present with homogeneous morphology. However, the BJH method has frequently been applied to activated carbons containing a large amount of micropores and to calculate V_{micro} as well as mesopore volume V_{meso}.

In the DR method [15], Gaussian-type distribution of pore sizes is assumed. For four activated carbons having different pore structures, N_2 adsorption–desorption isotherms observed, their DR plots, and α_s plots are shown in Fig. 1.12A–C, respectively [16]. DR plots show a linear portion at low P/P_0 range, revealing the presence of micropores, of which the back extrapolation gives V_{micro}. In Table 1.2, V_{micro} values calculated from DR and α_s plots, V_{micro}(DR) and $V_{micro}(\alpha_s)$, are listed together with S_{BET}. The activated carbons having high S_{BET}, AX21 and JF517, show a marked deviation from the linearity at P/P_0 below 0.05 and give quite different values of V_{micro} calculated from DR and α_s plots, suggesting that these carbons have a wide range of micropore sizes and that the evaluation of V_{micro} via the DR plot seems not to be appropriate.

The HK method [17] assumes the presence of slit-shaped micropores between graphite layers to calculate the size distribution of micropores by using semiempirical formulae. The DFT method is based on the simulation of adsorption isotherms by theoretical calculation and has the advantage of obtaining the pore size distribution in a wide range of sizes from micropores to mesopores. The results of the DFT analysis depend strongly on the calculation parameters, such as interaction parameter between gas molecules and that between gases and solids.

The analysis of pore structure in carbon materials by CO_2 adsorption around room temperature was recommended to complement the characterization by N_2 adsorption at 77 K [18,19]. The studies using CO_2 adsorption at 298 K up to 4 MPa on activated carbon fibers [18] and at 273 K down to subatmospheric pressure on five activated carbons having different pore structures [19] demonstrated that it has a special relevance for the characterization of activated carbons with narrow micropores which are not accessible to N_2 at 77 K. In addition, it is more convenient to use the temperature 0°C (273 K) as the adsorption temperature instead of 298 K because of high uncertainty of the density of the adsorbed CO_2 at high temperature. In Table 1.3, V_{micro} determined by the DR method using N_2 at 77 K, $V_{micro}(N_2)$, is compared to that using CO_2 at 273 K, $V_{micro}(CO_2)$, on carbon materials with high V_{micro}, except nonactivated carbon fibers (CF) [20]. On nonactivated CF, no micropores are detected by N_2 adsorption but some micropores are detected by CO_2 adsorption, demonstrating CO_2 is adsorbed into the pores which N_2 cannot get into (the pores having a size less than 0.4 nm). $V_{micro}(CO_2)$ gives smaller values than $V_{micro}(N_2)$, probably because CO_2 is adsorbed selectively into narrow micropores and $V_{micro}(N_2)$ includes large micropores (supermicropores).

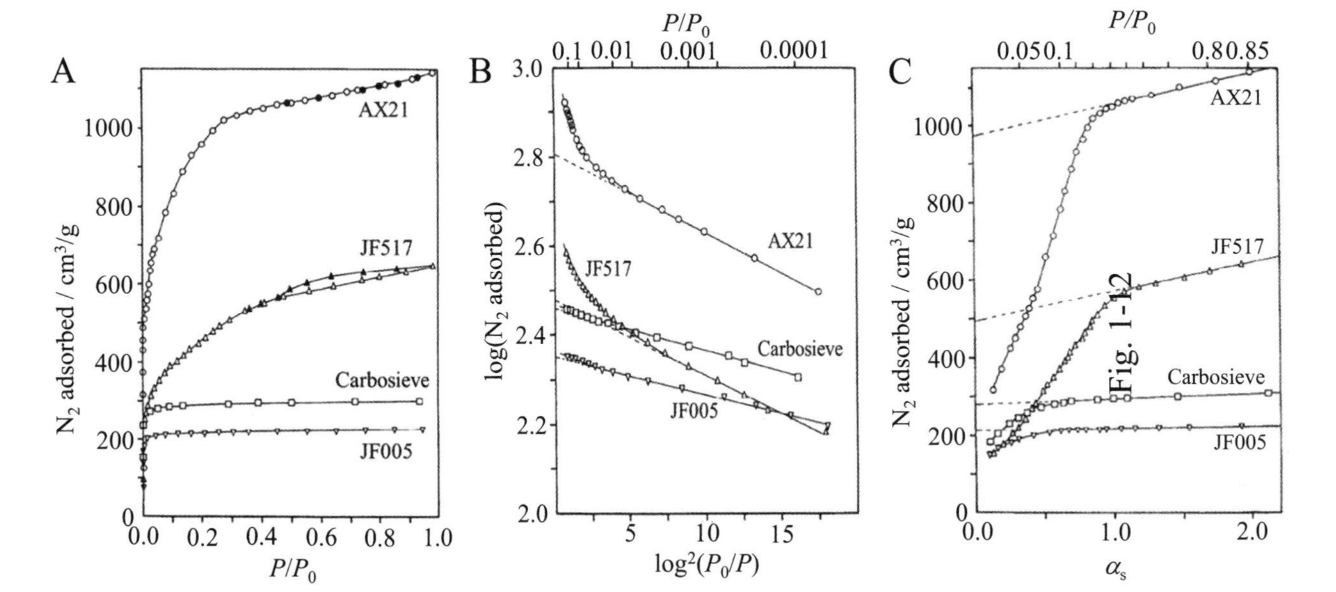

Figure 1.12 Comparison between DR and α_s plot analyses on activated carbons; (A) N_2 adsorption–desorption isotherms measured at 77 K, (B) DR plots, and (C) α_s plots [16].

Table 1.2 V_{micro} calculated from α_s and DR plots [16].

Porous carbon	S_{BET} (m^2/g)	V_{micro} (cm^3/g)	
		From α_s plot	From DR plot
AX21	3393	1.52	1.00
JF517	1657	0.76	0.47
Carbosieve	1179	0.43	0.45
JF005	882	0.31	0.35

Table 1.3 Micropore volumes V_{micro} (cm^3/g) measured using N_2 at 77 K and CO_2 at 273 K on carbon materials [19].

Carbon material	S_{BET} (m^2/g)	$V_{micro}(N_2)$ (cm^3/g)	$V_{micro}(CO_2)$ (cm^3/g)	Mean pore size (nm)	He density (g/cm^3)
CF	0	0	0.18	0.41	1.70
ACFC26	1079	0.39	0.39	0.75	1.70
ACFC50	1738	0.78	0.57	1.36	1.85
KUAI	1058	0.51	0.50	0.66	1.73
A20	2206	0.79	0.43	1.46	1.95
Fabric	1407	0.65	0.54	1.46	1.70
AC35	1204	0.54	0.40	1.54	1.83
AX21	2575	0.86	0.65	1.48	1.70

1.3.2 Mercury porosimetry

Pore structures of macroporous solids were often studied using a mercury porosimeter. For macroporous carbons, particularly carbon foams derived from exfoliated graphite (EG) and reduced graphene oxide (rGO), however, the application of a mercury porosimeter needs special care to prevent collapsing of fragile foams by heavy mercury. A new U-type dilatometer was proposed [21], as illustrated by comparing with the conventional N-type dilatometer in Fig. 1.13A, and pore size distributions measured by these two dilatometers are shown in Fig. 1.13B. By using a conventional N-type dilatometer, only pores having sizes up to 60 μm are detected in an EG foam, whereas the pores larger than 60 μm, which are reasonably supposed to be the interparticle pores, are collapsed during the intrusion of mercury and are not detected. By using a U-type dilatometer, the pores having much larger sizes up to 600 μm can be reproducibly detected. It has to be pointed out, however, that there is a certain risk of overlooking larger macropores, more than a few hundred, even using this U-type dilatometer.

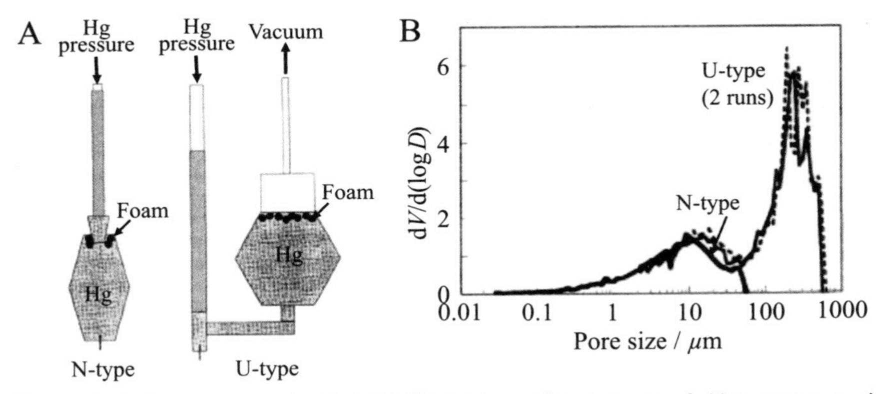

Figure 1.13 Mercury porosimetry: (A) illustrations of two types of dilatometers and (B) pore size distributions of an exfoliated graphite foam measured by two dilatometers [21].

1.3.3 Microscopy techniques and image processing

To identify the pores in carbon materials, various microscopic techniques are applied, as shown in Fig. 1.8, from nano-sized pores by tunneling microscopic techniques (STM and AFM) to millimeter-sized pores by an optical microscopy technique. Most microscopic data are presented as images in different scales. In order to obtain the quantitative data from these images, the assistance of image processing techniques is essential.

Image processing techniques have successfully been applied to SEM micrographs for quantitative characterization of pore structure in exfoliated graphite (EG) [22]. In most cases, either the bulk density or the exfoliation volume has been evaluated for EGs, which have low bulk density and are very fragile. EG consists of at least three kinds of pores, i.e., pores inside the work-like particles, crevice-like pores on the worm-like particles (these two are intraparticle pores), and pores formed by complicated entanglement of these fragile worm-like particles (interparticle pores). The techniques to prepare the fractured surface of these fragile foams, including worm-like particles of EG, were proposed: a simple cleavage of worm-like particles [23] and an impregnation of paraffin followed by cutting [24]. With the assistance of image processing on the cross-sections, quantitative characterization of pore structure became possible and a series of works has been performed [22]. In Fig. 1.14A and B, an SEM image of a cross-section prepared by cleaving a worm-like particle of EG and the distribution of cross-sectional area of pores inside the particles determined from more than

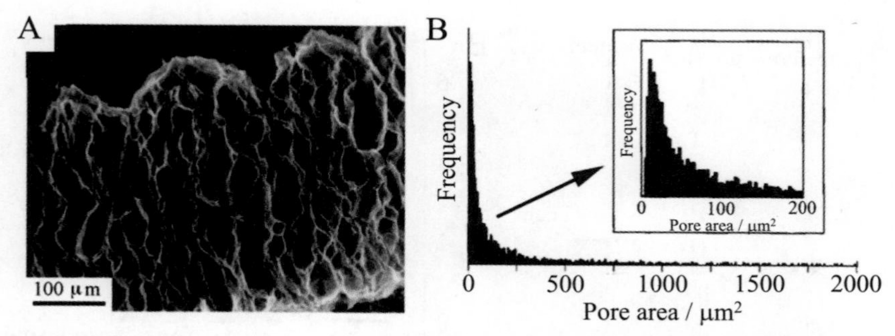

Figure 1.14 (A) SEM image of the cross-section of a worm-like particle and (B) distribution of the cross-sectional area of the pores [22].

7000 pores are shown, respectively. One of the advantages of this technique is to be able to observe large size pores, which cannot be identified by gas adsorption and mercury porosimetry.

STM observation is possible to observe not only pores of very small sizes but also the shape and fractal dimension along the pore surfaces through image processing, although only the entrances of the pores are observed and a large number of observations are necessary. STM analysis was performed on the surfaces of glass-like carbon spheres carbonized at different atmospheres, including oxidizing atmosphere [25,26]. An STM image on the surface of a carbon sphere and an example of a contour map around a pore are shown in Fig. 1.15A and B, respectively. From the observations on a large number of pores, a pore size distribution is determined, as shown in Fig. 1.15C. In Table 1.4, the results obtained from STM image analysis are compared with those from gas adsorption. The results of the STM characterization show quite good correspondence to those of the gas adsorption analysis; the density of micropores with a size of 0.5−1.5 nm (number per 1 μm^2 area of the surface) measured on four different spheres corresponds to the BET surface area (S_{BET}) determined by N_2 adsorption at 77 K.

In transmission electron micrographs taken on a thin section of carbon materials with sufficiently high magnification, pores look white because the electron beam passes through and pore walls look black because of scattering of the electron beam. The quantitative analysis of these micrographs with the aid of image processing gives information on the pore size distribution and also smoothness of pore walls (fractal dimension). Detailed studies have been carried out mainly on activated carbon fibers (ACFs) [27,28]. A TEM micrograph (bright-field image) is converted to a power spectrum as a curve showing a change in brightness with distance, which is

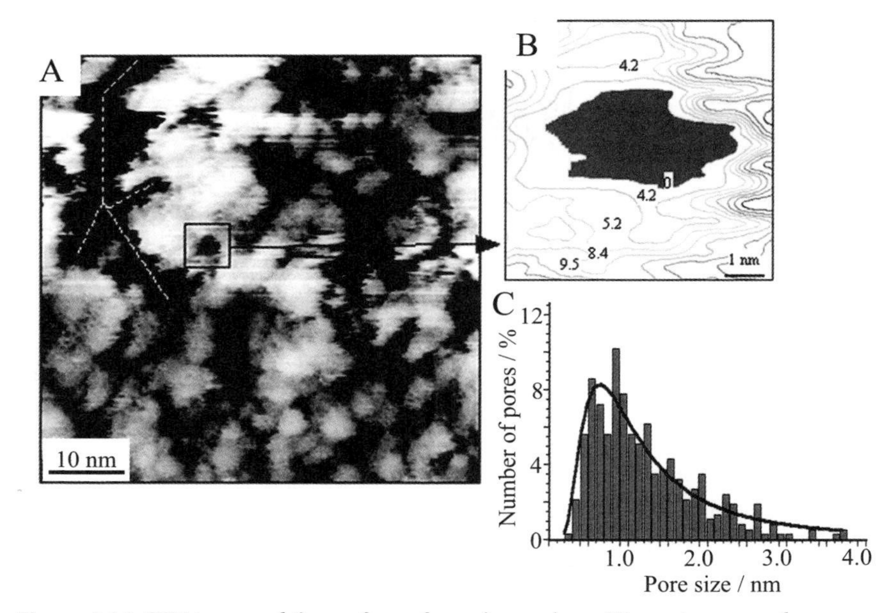

Figure 1.15 STM image of the surface of a carbon sphere (A), contour map for a pore (B), and pore size distribution (C) [26].

Table 1.4 The results of STM image analysis in comparison with gas adsorption results [26].

Sample	Pores with a size of 0.5–1.5 nm analyzed by STM		Gas adsorption analysis	
	Number in 1 μm²	Ratio to B	S_{BET} (m²/g)	Ratio to B
A	75	0.04	2	0.06
B	1788	1.0	307	1.0
C	4078	2.3	993	3.2
D	6322	3.5	1193	3.9

considered to reveal the pore size distribution. The area under the power spectrum curve corresponds to the relative pore volume. An original bright-field image exhibits a good correspondence to the bimodal image obtained by image processing on various carbons. In Fig. 1.16, the power spectrum obtained from TEM image is shown with pore size distribution determined by gas adsorption for three ACFs with different S_{BET}. Since the power spectrum is expressed in reciprocal space, the distance in real space indicated on the abscissa increases on the left-hand side. Therefore, pore size distributions are plotted in the same manner. Taking into account that

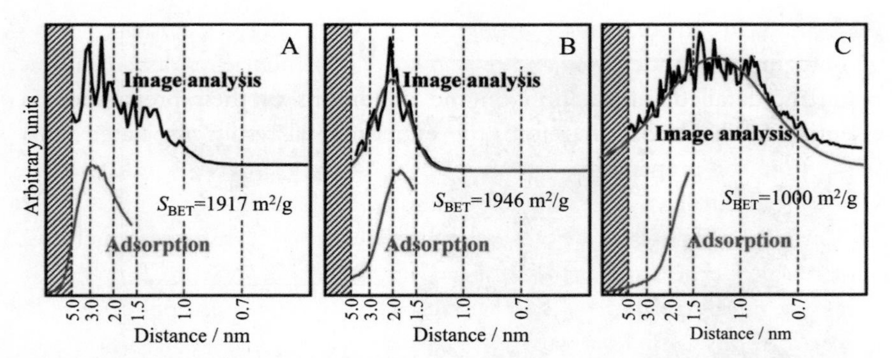

Figure 1.16 Power spectrum determined by TEM analysis and pore size distribution determined by gas adsorption for three activated carbon fibers [28].

the magnification of TEM observation for this analysis does not give the information on the distance more than 5 nm, relatively good correspondence was obtained between the power spectrum from the TEM observation and pore size distribution from N_2 gas adsorption. The distribution estimated from the TEM image is a little broader than that from gas adsorption, which is supposed to be due to the fact that three-dimensional averaging is performed in the former, but the minimum value of pore parameters is detected in the latter.

Optical microscopic images of isotropic high-density graphite blocks were analyzed by image processing and provided quantitative information on their macropores, which made discussion on the dependences of various properties on pore structure possible [29].

1.4 Purposes and construction of this book

The purpose of this book is to provide a basic and thorough understanding of the synthesis, structure, properties, and applications of porous carbon materials. As briefly explained above, carbon materials have a wide range of diversity in structure, textures, and properties, which mainly depend on the precursors and their conversion conditions to carbon materials, i.e., temperature, heating rate, atmosphere, pressure, etc. Even if a precursor, for example a pitch, is selected, the product obtained by pyrolysis, carbonization with a gentle heating rate, associated with flowing at the stage of liquid has graphitizing nature, such as needle-like cokes, which is used in graphite electrode production in industry, while the same pitch pyrolyzed with rapid heating is less graphitizing as a conventional coke. If its pyrolysis is performed in an oxidative atmosphere, the resulting carbon is non-graphitizing. To understand and discuss the experimental results on carbon materials, detailed information on the conditions of their preparation is essential. In this book, therefore, the experimental results associated with the keys to the preparation conditions of the carbon materials published in various journals have been collected as much as possible.

In this book, porous carbons are explained by their syntheses, structures, textures, properties, and applications. Their structures, textures, and properties are tightly related to their applications. The field of applications of porous carbons is divided into two areas: energy storage/conversion and environment remediation. In these applications, carbon nanotubes and nanofibers, as well as reduced graphene oxides, are included for comparison with porous carbons.

This book consists of the following five chapters.

In Chapter 1 **Introduction**, a brief explanation of carbon materials is presented with a classification based on the chemical bonds between carbon atoms, sp, sp^2, and sp^3. The carbon materials with sp^2 hybrid bonding are explained on the basis of nanotextures to show the position of porous carbons in whole carbon materials. In addition, a brief explanation of the identification and evaluation of pores is presented, although for detailed theories and procedures of these techniques readers are referred to specific books.

In Chapter 2 **Syntheses of porous carbons**, the procedure for synthesis of porous carbons is summarized by dividing them into microporous, mesoporous, and macroporous carbons, with a separate section on hierarchically porous carbons which have attracted attention recently. They are explained on the basis of synthesis procedures, i.e., activation, template-assisted carbonization, precursor design including polymer blend, molecular design, organic-frameworks, carbon aerogels, and ionic liquids. For the synthesis of macroporous carbons, carbonization with blowing and exfoliation of graphite is described, in addition to template-assisted and precursor design processes.

In Chapter 3 **Porous carbons for energy storage and conversion**, the applications of porous carbons are described by dividing them into rechargeable batteries, supercapacitors, hybrid cells, fuel cells, hydrogen storage, and methane storage. In the section on rechargeable batteries, lithium-ion, lithium-sulfur, and sodium-ion batteries are included. In the section on hybrid cells, the cells composed of an electrode based on the Faradaic reaction and another electrode based on adsorption are explained.

Hydrogen and methane are important future energy sources and thus their storage is discussed. In addition, the storage of thermal energy using porous carbons is also summarized.

In Chapter 4 **Porous carbons for environment remediation**, the applications of porous carbons mainly for environment remediation fields are summarized by dividing them into adsorption, gas separation, capacitive deionization, electromagnetic interference shielding, and sensing. In these application fields, not only porous carbons, but also other carbon materials, such as nanotubes and graphene-related materials, are important, and an explanation of these carbon materials is included. In the section on adsorption, adsorption removals of various pollutants are explained, followed by descriptions of focused adsorbates, i.e., water vapor, carbon dioxide, heavy metals, and oils. In the section on sensing, the applications for chemical, mechanical, and biomedical sensors are summarized.

In Chapter 5 **Concluding remarks and prospects**, the synthesis processes of porous carbons are summarized by reviewing briefly their history, and the applications of porous carbons are discussed by focusing on the contributions of pores in carbon materials. As prospects for porous carbons, the new possibility of pores as a constraint and reaction space in carbon materials is discussed by referring to various experimental results.

1.5 Abbreviations of technical terms employed

Technical terms used frequently in this book are abbreviated in order to save space. They are summarized in Table 1.5. Expressions of polyimides and ionic liquids are summarized in Fig. 4.89 and Table 2.21 to explain their application for gas separation and their carbonization, respectively.

Table 1.5 Abbreviations employed in this book.

Carbon materials
AC: Activated carbon
ACC: Activated carbon fiber cloth
ACF: Activated carbon fiber
CB: Carbon black
CDC: Carbide-derived carbon
CF: Carbon fiber
CNF: Carbon nanofiber
CNT: Carbon nanotube
DWCNT: Double-walled carbon nanotube
EG: Exfoliated graphite
GC: Glassy carbon (glass-like carbon)

(*Continued*)

Table 1.5 Abbreviations employed in this book.—cont'd

GO: Graphene oxide
HOPG: Highly oriented pyrolytic graphite
MCMB: Mesocarbon microbeads
MSC: Molecular sieve carbon
MWCNT: Multi-walled carbon nanotube
OMC: Ordered mesoporous carbon
rGO: Reduced graphene oxide
SWCNH: Single-wall carbon nanohorn
SWCNT: Single-wall carbon nanotube
VGCF: Vapor-grown carbon fiber
ZTC: Zeolite-templated carbon

Preparation procedures

CVD: Chemical vapor deposition
CVI: Chemical vapor infiltration

Polymers

AA: Ascorbic acid
CMC: Carboxymethyl cellulose
COF: Covalent organic framework
DA: Dopamine
EDA: Ethylenediamine
EDTA: Ethylenediaminetetraacetic acid
FA: Furfuryl alcohol
IL: Ionic liquid
MOF: Metal-organic-framework
PAA: Polyamic acid
PAH: Polycyclic aromatic hydrocarbons
PAN: Polyacrylonitrile
PANI: Polyaniline
PDMS: Polydimethylsiloxane
PEDOT: poly(3,4-ethylenedioxy-thiophene)
PEG: poly(ethylene glycol)
PEI: Polyetherimide
PET: Poly(ethylene terephthalate)
PF: Phenol-formaldehyde
PhF: Phloroglucinol/formaldehyde
PI: Polyimide
PMMA: Polymethylmethacrylate
POD: Polydopamine
PPO: poly(phenylene oxide)
PPY: Polypyrene
PPy: Polypyrrole

(*Continued*)

Table 1.5 Abbreviations employed in this book.—cont'd

PS: Polystyrene
PSS: Poly(styrene sulfonate)
PTFE: Polytetrafluoroethylene,
PVA: poly(vinyl alcohol)
PVDF: poly(vinylidenefluoride)
PVP: poly(vinylpyrrolidone)
RF: Resolcinol-formaldehyde
TEOS: Tetraethylorthosilicate
UA: Uric acid
VOC: Volatile organic compounds
ZIF: Zn-MOF
Characterization techniques
CV: Cyclic voltammogram
DTA: Differential thermal analysis
DTG: Differential thermal gravimetry
EDS: Energy dispersive X-ray spectroscopy
EXAFS: Extended X-ray absorption fine structure
SEM: Scanning electron microscope
TEM: Transmission electron microscope
TG: Thermal gravimetry
TPD: Temperature-programmed desorption
STEM: Scanning transmission electron microscopy
XAS: X-ray absorption spectroscopy
XPS: X-ray photoelectron spectroscopy
XRD: X-ray diffraction method
Characterization parameters
GF: Gauge factor
LOD: Limit of detection (detection limit)
Procedures for isotherm analysis
BET: Brunauer–Emmett–Teller method
BJH: Barrett–Joyner–Halenda method
DFT: Density functional theory
DH: Horvath–Kawazoe method
DR: Dubinin–Radushkevich plot
Pore structure parameters
S_{BET}: Surface area determined by BET method
S_{meso}: Surface area due to mesopores

(*Continued*)

Table 1.5 Abbreviations employed in this book.—cont'd

S_{micro} and $S_{micro}(N_2)$: Surface area due to micropores measured by using N_2 at 77 K, $S_{ultra} + S_{super}$
$S_{micro}(CO_2)$: Surface area due to narrow micropores, measured by using CO_2 at 273 K
S_{super}: Surface area due to supermicropores
S_{total}: Total surface area
S_{ultra}: Surface area due to ultramicropores
V_{total}: Total pore volume
V_{micro}: Volume of micropores
V_{meso}: Volume of mesopores
V_{narrow}: Volume of narrow micropores measured using CO_2
V_{super}: Volume of supermicropores
V_{ultra}: Volume of ultramicropores
Electrolytes
$EMIMBF_4$: 1-Ethyl-3-methyl-imidazolium tetrafluoroborate
LiTFSI: Lithium bis(trifluoromethane-sulfonyl)imide
NaTFSI: Sodium bis(trifluoro-methanesulfonyl)imide
$TBAPF_6$: Tetra-*n*-butyl-ammonium hexafluoro-phosphate
$TEABF_4$ or Et_4NBF_4: Tetraethylammonium tetrafluoroborate
$TEMABF_4$: Triethylmethyl-ammonium tetrafluoroborate
$TEMAPF_6$: Triethylmethyl-ammonium hexafluophosphate
PMPyrr-TFSI: 1-Methyl-1-propylpyrrolidinium TFSI
Solvents
AN: Acetonitrile
BC: Butylene carbonate
DEC: Diethyl carbonate
DMAc: di-methylacetamide
DMC: Dimethyl carbonate
DME: 1,2-Dimethoxyethane(ethylene glycol dimethyl ether)
DMF: N,N′-dimethylformamide
DMSO: Dimethyl sulfoxide
EC: Ethylene carbonate
EMC: ethyl methyl carbonate
GBL: γ-butyrolactone
NMP: *n*-methyl-2-pyrrolidone
PC: Propylene carbonate
DOL: 1,3-Dioxolane
PBS: Phosphate buffer solution
THF: Tetrahydrofrane

(*Continued*)

Table 1.5 Abbreviations employed in this book.—cont'd

di-EGDME: di(ethylene glycol) dimethyl ether tetra-EGDME or TEGDME:tetra(ethylene glycol)dimethyl ether tri-EGDME: tri(ethylene glycol)dimethyl ether
Li compounds
LCO: $LiCoO_2$ LFP: $LiFePO_4$ LMO: $LiMn_2O_4$ LTO: $Li_4Ti_5O_{12}$ LVP: α-$Li_3V_2(PO_4)$

References

[1] Inagaki M, Kang F. Materials science and engineering of carbon: fundamentals. 2nd ed. Tsinghua Univ. Press and Elsevier; 2014.
[2] Inagaki M, Kang F, Toyoda M, Konno H. Advanced materials science and engineering of carbon. Tsinghua Univ. Press and Elsevier; 2013.
[3] Inagaki M. New carbons - control of structure and functions. Elsevier; 2000.
[4] Inagaki M. Microtexture of carbon materials. TANSO 1985;122:114–21 [in Japanese].
[5] Sing KSW, Everett DH, Haul RAW, Moscou L, Pierotti RA, Rouquérol J, Siemieniewska T, IUPAC recommendations 1984. Reporting physisorption data for gas solid systems with special reference to the determination of surface area and porosity. Pure Appl Chem 1985;57:603–19.
[6] Thommes M, Kaneko K, Neimark AV, Olivier JP, Rodriguez-Reinoso F, Rouquerol J, Sing KSW. Physisorption of gases, with special reference to the evaluation of surface area and pore size distribution. Pure Appl Chem 2015;87(9–10):1051–69.
[7] Patrick JW, editor. Porosity in carbons: characterization and applications. London: Edward Arnold; 1995.
[8] Inagaki M, Kang F, editors. Materials science and engineering of carbon: characterization. Tsinghua Univ. Press and Elsevier; 2017.
[9] Lowell S, Shields JE, Thomas MA, Thommes M. Characterization of porous solids and powders: surface area, pore size and density. Kluwer Academic Puiblisher; 2004.
[10] Rouquerol F, Rouqerol J, Sing KSW, Llewellyn P, Maurin G. Adsorption by powders and porous solids Principles, methodology and applications. 2nd ed. Elsevier; 2014.
[11] Brunauer S, Emmett PH, Teller E. Adsorption of gases in multi-molecular layers. J Am Chem Soc 1938;60:309–19.
[12] Kaneko K, Ishii C. Superhigh surface area determination of microporous solids. Colloid Surf 1992;67:203–12.
[13] Kaneko K, Ishii C, Kanoh H, Hanzawa Y, Setoyama N, Suzuki T. Characterization of porous carbons with high resolution α_s-analysis and low temperatures magnetic susceptibility. Adv Colloid Interface Sci 1998;76–77:295–320.
[14] Barrett EP, Joyner LG, Halenda PP. The determination of pore volume and area distributions in porous substances. I. Computations from nitrogen isotherms. J Am Chem Soc 1951;73:373–80.
[15] Dubinin MM. The potential theory of adsorption of gases and vapors for adsorbents with energetically nonuniform surfaces. Chem Rev 1960;60:235–41.

[16] Carrott PJM, Roberts RA, Sing KSW. Adsorption of nitrogen by porous and non-porous carbons. Carbon 1987;25:59–68.
[17] Horvath G, Kawazoe K. Method for the calculation of effective pore size distribution in molecular sieve carbon. J Chem Eng Jpn 1983;16:470–5.
[18] Cazorla-Amoros D, Alcañiz-Monge J, Linares-Solano A. Characterization of activated carbon fibers by CO_2 adsorption. Langmuir 1996;12:2820–4.
[19] Cazorla-Amoros D, Alcañiz-Monge J, de la Casa-Lillo MA, Linares-Solano A. CO_2 as an adsorptive to characterize carbon molecular sieves and activated carbons. Langmuir 1998;14:4589–96.
[20] Lozano-Castello D, Cazorla-Amoros D, Linares-Solano A, Quinn DF. Micropore size distributions of activated carbons and carbon molecular sieves assessed by high-pressure methane and carbon dioxide adsorption isotherms. J Phys Chem B 2002;106:9372–9.
[21] Nishi Y, Iwashita N, Inagaki M. Evaluation of pore structure of exfoliated graphite by mercury porosimeter. TANSO 2002;201:31–4 [In Japanese].
[22] Inagaki M, Suwa T. Pore structure analysis of exfoliated graphite using image processing of scanning electron micrographas. Carbon 2001;39:915–20.
[23] Inagaki M, Saji N, Zheng Y-P, Kang F, Toyoda M. Pore development during exfoliation of natural graphite. TANSO 2004;215:258–64.
[24] Zheng YP, Wang HN, Kang FY, Wang L-N, Inagaki M. Sorption capacity of exfoliated graphite for oils-sorption in and among worm-like particles. Carbon 2004;42:2603–7.
[25] Vignal V, Morawski AW, Konno H, Inagaki M. Quantitative assessment of pores in oxidized carbon spheres using scanning tunneling microscopy. J Mater Res 1999;14:102–12.
[26] Inagaki M, Vignal V, Konno H, Morawski AW. Effects of carbonization atmosphere and subsequent oxidation on pore structure of carbon spheres observed by scanning tunneling microscopy. J Mater Res 1999;14:3152–7.
[27] Oshida K, Kogiso K, Matsubayashi K, Takeuchi K, Kobayashi S, Endo M, Dresselhaus MS, Dresselhaus G. Analysis of pore structure of activated carbon fibers using high resolution transmission electron microscopy and image processing. J Mater Res 1995;10:2507–17.
[28] Endo M, Furuta T, Minoura F, Kim C, Oshida K, Dresselhaus G, Dresselhaus MS. Visualized observation of pores in activated carbon fibers by HRTEM and combined image processor. Supramol Sci 1998;5:261–6.
[29] Oshida K, Ekinaga N, Endo M, Inagaki M. Pore analysis of isotropic graphite using image processing of optical micrographs. TANSO 1996;173:142–7 [in Japanese].

CHAPTER 2

Syntheses of porous carbons

Nano-sized pores are created in carbon materials and their structures are modified mainly by oxidation under controlled conditions; the process involves the oxidative gasification of precursor carbons to CO and CO_2 and is called "activation." The products are called activated carbons, having played important roles since prehistoric times and they are now becoming increasingly important materials in various technological fields. Different activation processes have been employed using air, steam, $ZnCl_2$, and KOH as activating reagents in order to develop micropores in carbon materials, in particular KOH activation, resulting in a very high surface area of more than 3000 m^2/g. Since the 1990s, various techniques to control the pore structure in carbon materials without an activation process have also been proposed, mainly with the aim of more precise control of pore structure and avoidance of sacrificial loss of carbon atoms; template-assisted carbonization using zeolite, mesoporous silicas, MgO, etc. Blending of precursor polymers was successfully employed to control the pore structure in carbons (polymer-blending). In addition, proper selection of the precursor organics for carbonization was also proposed, such as defluorination of fluorine-containing organics and gelation of organics by selecting appropriate precursors.

In this chapter, porous carbons are divided into microporous, mesoporous, and macroporous carbons, based on the principal pores, although these three kinds of pores coexist in most practical porous carbons. The techniques for the creation and control of pore structure in carbon materials are reviewed on the basis of the classification of the fundamental processes into activation, template-assisted carbonization, and precursor design. The "activation" process has been employed for the synthesis of so-called activated carbons since prehistoric times and is divided into physical activation and chemical activation. Since activated carbon fibers have advantages for some application fields, they are discussed in a separate section. The "template-assisted carbonization" process is reviewed on the bases of template materials, which are different from the desired micropores, mesopores, and macropores. The "precursor design" process is further divided into polymer-blend, molecular design, and process design. In the "polymer-blend" process, two kinds of polymers—matrix-forming polymer and pore-forming polymer (labile polymer)—are intentionally mixed before carbonization. The selection of a labile polymer is important to control the pore structure. For macroporous carbons, some labile polymers are selected to blow the matrix-forming polymer by evolving gaseous species from pyrolysis products. In the "molecular design" process, precursor molecules are designed to have some pending functional groups on the carbon framework, the former giving pores and the latter resulting in the matrix of porous carbons. The "process design" includes the gelation of precursor polymers before carbonization for the creation of mesopores and the pyrolysis of carbon precursors under pressure for creating macropores in carbons. The concept for this chapter is summarized in Table 2.1.

2.1 Microporous carbons

2.1.1 Activation

2.1.1.1 Physical activation

The principal reaction in activation of carbons is oxidation and consequently it is always accompanied by weight loss of matrix carbon, with weight loss (burn-off) being often used as a parameter to characterize the activation degree.

Oxidation of carbonaceous materials was studied as a fundamental process for coal gasification using various oxidation reagents. In Fig. 2.1A, burn-off of the carbon (char) prepared from coal at 1000°C is plotted against the oxidation (activation) time at 405°C in O_2/N_2 mixed gases with different O_2 contents [1], revealing that burn-off (in other words, activation degree) depends strongly on the O_2 content in the atmospheric gas. It has been shown that these burn-off data could be unified to a single curve using a normalized time scale $t/t_{0.5}$, where $t_{0.5}$ is the time to reach 50% burn-off, as shown in Fig. 2.1B. Unification curves are discussed experimentally and theoretically [2,3].

The pore growth process through O_2 activation was also demonstrated using glass-like carbon spheres [4–6]. Glass-like carbon spheres prepared from resole-type phenol-formaldehyde resin at 1000°C were selected, where activation by oxygen in dry air started from the surfaces of spheres because of their gas impermeability. Pore development in the carbons with activation is understood through the master curves for each pore parameter, which was obtained by shifting the experimental points at different

Table 2.1 Synthesis techniques of porous carbons.

Fundamental process		Microporous carbons	Mesoporous carbons	Macroporous carbons	Hierarchically porous carbons
Activation		Physical activation (O_2, CO_2, steam) chemical activation (KOH, $ZnCl_2$, etc.)	Physical and chemical activation with additives	–	–
Template-assisted carbonization		Zeolites Other hard templates	Silicas (mesoporous and colloidal silicas) Magnesium oxide Other hard templates	Organic foams	Dual assistance (template and labile polymer blend)
Precursor design	Polymer blending	Blending with labile polymers (PVP, PMMA, etc.)	Blending with block copolymers	Blending with blowing agents Labile polymer particles	
	Molecular design	Labile functional groups Defluorination Porous-organic frameworks Metal carbides	Metal-organic frameworks	Exfoliation of graphite	–
	Process design	–	Carbon aerogels	Pyrolysis under hydrothermal conditions	Hydrothermal treatment

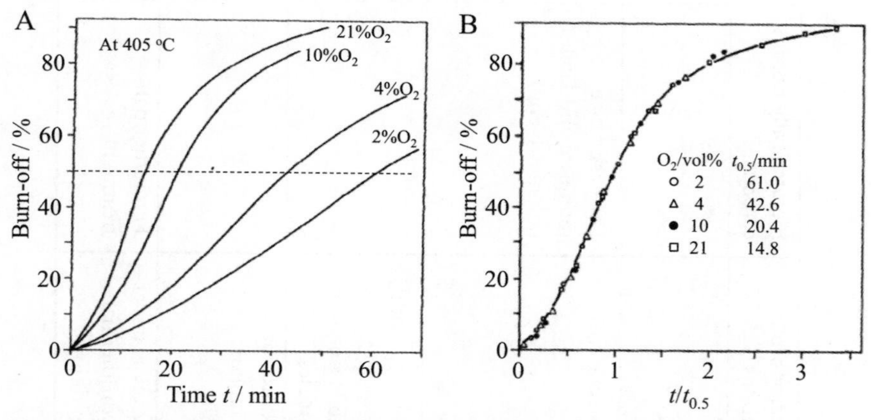

Figure 2.1 Burn-off in O_2/N_2 mixed gases with different O_2 contents for a carbon prepared from coal at 1000°C: (A) burn-off versus oxidation time t at 405°C and (B) burn-off versus normalized parameter $t/t_{0.5}$ [1].

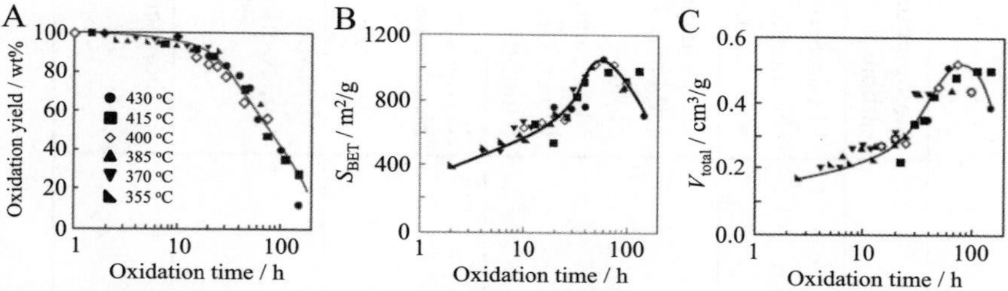

Figure 2.2 Master curves for O_2 activation of glass-like carbon spheres: (A) oxidation yield, (B) BET surface area S_{BET}, and (C) total pore volume V_{total} [6].

activation temperatures along the oxidation time axis in logarithmic scale to be fitted with the experimental points observed at the reference temperature. In Fig. 2.2, the master curves obtained at the reference temperature of 400°C on glass-like carbon spheres are shown for oxidation yield, S_{BET} and V_{total}. Those for S_{micro}, S_{meso}, V_{micro}, V_{ultra}, V_{super}, and V_{meso} were also obtained successfully [6]. On the bases of these master curves for pore structure parameters, O_2 activation of the glass-like carbon spheres is discussed. In Fig. 2.3, the changes in pore volumes for ultramicropores, supermicropores, and mesopores (V_{ultra}, V_{super}, and V_{meso}, respectively) together with micropore volume ($\boldsymbol{V_{micro} = V_{ultra} + V_{super}}$) and the corresponding SEM images of the carbon sphere are shown as a function of the activation time at 400°C. At the start of activation, i.e., up to 10 h at 400°C, the main process is the formation of ultramicropores. From 10 to

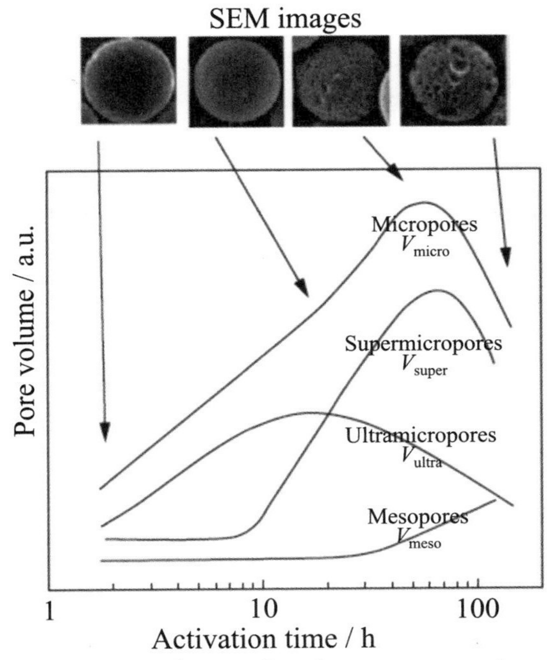

Figure 2.3 Changes in pore volumes, ultramicropores, supermicropores, and mesopores together with the sum of ultramicropores and supermicropores (micropores), with activation time at 400°C. The corresponding SEM images of the carbon spheres are also shown [6].

60 h, the relative amount of ultramicropores decreases, while the volumes of supermicropores and mesopores increase with increasing oxidation time, suggesting the transformation of ultramicropores to supermicropores and then mesopores. Above 65 h, the volume of micropores decreases rapidly associated with a slight increase in the mesopore volume, thereby resulting in a decrease in the surface areas measured by the different methods employed. The principal process in the formation of micropores was supposed to be the opening of closed pores intrinsically created in the matrix carbon during carbonization [7].

The carbons, which were prepared from commercially available Saran (copolymer of vinylidene and vinyl chlorides) by carbonization at 600°C, followed by treating in H_2 at 700°C to remove chlorine complexes, had very high surface areas of up to 3500 m^2/g after steam activation [8]. The Saran-derived carbon activated at 950°C for 135 min with a yield of 15 wt.% exhibited a surface area of 3500 m^2/g from N_2 adsorption isotherm and 3330 m^2/g from ethyl chloride adsorption isotherm.

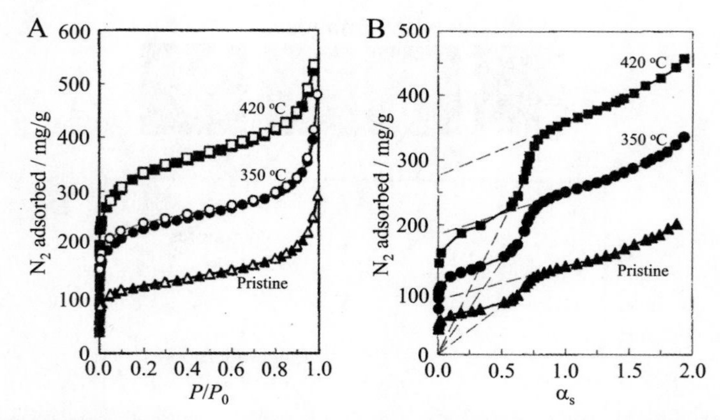

Figure 2.4 O_2 activation of single-walled carbon nanohorns at 350 and 420°C: (A) N_2 adsorption-desorption isotherms and (B) their α_s-plots [10].

A marked increase of V_{micro} in single-walled carbon nanohorns (SWCNHs) was achieved in a flow of pure O_2 at 350 and 420°C for 10 min [9]. In Fig. 2.4, N_2 adsorption–desorption isotherms and their α_s plots for activated SWCNHs are shown with those of the pristine SWCNHs. V_{micro} increased to 0.23 and 0.34 cm^3/g by oxidation at 350 and 420°C, respectively, from 0.11 cm^3/g for the pristine SWCNHs, although the increase in V_{meso} was negligible. As a consequence, the activated SWCNHs had V_{micro} comparable to V_{meso}, although the pristine one was mesopore rich. Mechanical treatment by compression of SWCNHs was reported to be effective in increasing V_{micro} and achieving a high CH_4 storage capacity [10,11].

An activation process using CO_2 gas was investigated on carbons obtained by carbonization of a bituminous coal and a coconut-shell char at 900°C in an N_2 atmosphere, followed by deashing with HF and diluted HCl [12]. Activation was performed in a CO_2 flow of 300 mL/min at 900°C for 1–15 h using a rotary furnace. Relations between burn-off during activation, resultant S_{BET}, and averaged interlayer spacing d_{002} are shown in Fig. 2.5A–C. Burn-off during activation increases linearly with time for both series of carbons and S_{BET} increases with increasing burn-off almost linearly (Fig. 2.5A and B). N_2 adsorption–desorption isotherms for short-time activation were of type I, indicating the predominant formation of micropores, and those for long-time activation suggested the development of mesopores. However, d_{002} kept the initial values up to around 500 m^2/g, i.e., around 20 wt.% of burn-off, and then increased gradually

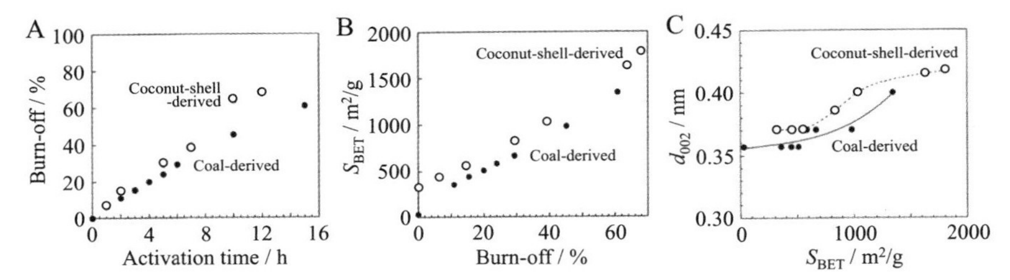

Figure 2.5 Activation of the carbons derived from coal and coconut-shell using CO_2: (A) burn-off versus activation time, (B) BET surface area S_{BET} versus burn-off, and (C) interlayer spacing d_{002} versus S_{BET} [12].

with continuing activation, suggesting degradation of the structure of parallel stacking of carbon layers (Fig. 2.5C). The stacking degradation of the crystallite on the carbons activated for a long time was also concluded from the analysis of 002 profiles. These experimental results resulted in the conclusion that CO_2 activation proceeds in two steps: (1) carbon atoms out of the crystallites are preferentially oxidized and removed to create micropores among crystallites, and (2) these crystallites are gasified from their edges, causing degradation of parallel stacking and resulting in enlargement of the ultramicropores to supermicropores and then mesopores.

On a carbon derived from furfuryl alcohol (FA), changes in pore structure with CO_2 activation at 800°C were studied by adding metal oxides [13,14]. On the FA-derived carbons without additives, marked pore development was observed at the start of activation, which was supposed to be due to the opening of closed pores in the matrix carbon and the widening of existing pores [13]. Marked acceleration of activation, i.e., micropore development, was observed by adding a small amount of Fe and Ni (atomic ratio of 10^{-3}) through the impregnation of $FeCl_3$ and $Ni(NO_3)_2$ [14]. Acceleration of activation for various chars (wood sawdust, coconut shells, and coal) was also reported on Fe_2O_3, NiO, Mo_2O_3, and Na_2CO_3 [15].

Pretreatment using $HClO_4$ before CO_2 activation was shown to be effective for developing micropores in anthracite [16]. Before pyrolysis at 1000°C, a demineralized anthracite was pretreated either by $Mg(ClO_4)_2$ at 450°C or by $HClO_4$ through a step-by-step increase of temperature from 120 to 160°C over 2.5 h with or without water washing. A series of pretreated anthracites (listed in Fig. 2.6A) were subjected to CO_2 activation by rapid heating up to 850°C and keeping for different times (5–24 h) to obtain 10%–80% burn-offs. S_{BET} of the activated anthracites with different pretreatments are plotted against burn-off in Fig. 2.6A, revealing that the

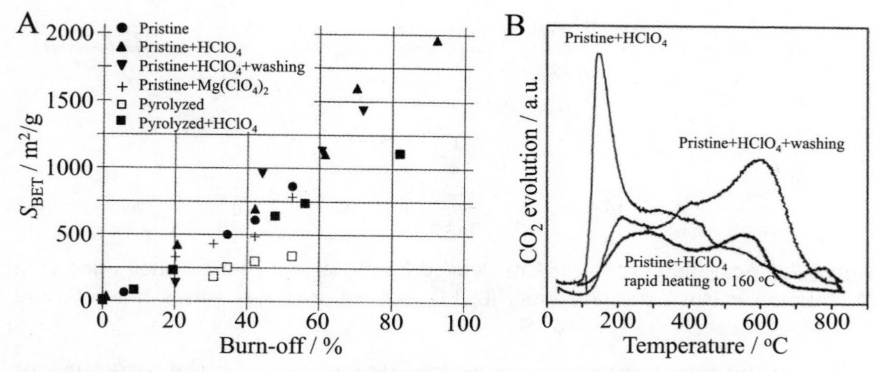

Figure 2.6 CO_2 activation of anthracites pretreated with $HClO_4$, $Mg(ClO_4)_2$, and pyrolysis: (A) S_{BET} versus burn-off and (B) CO_2 evolution spectra of $HClO_4$-pretreated anthracites before activation [16].

pretreatment using $HClO_4$ leads to high S_{BET}. After activation for 24 h, the anthracite pretreated by $HClO_4$ (Pristine + $HClO_4$) exhibited S_{BET} of 1600 m^2/g with V_{micro} of 0.61 cm^3/g and V_{meso} of 0.30 cm^3/g, and the anthracite washed after $HClO_4$ treatment (Pristine + $HClO_4$ + washing) gave similar S_{BET}, with these parameters being much higher than those of the pristine anthracite (Pristine) of 865 m^2/g, 0.54 cm^3/g, and 0.11 cm^3/g, respectively. $Mg(ClO_4)_2$ pretreatment did not assist CO_2 activation of anthracite. CO_2 evolution with temperature was followed by mass spectroscopy on $HClO_4$-pretreatment anthracites before activation, as shown in Fig. 2.6B. The $HClO_4$-pretreated anthracite with step-by-step heating (Pristine + $HClO_4$) shows a strong peak at 150—200°C, which can be ascribed to the low-temperature oxidation of carbon by incorporation of $HClO_4$, while that subjected to water washing (Pristine + $HClO_4$ + washing) evolves CO_2 in a broad range of temperatures. Meanwhile, the $HClO_4$-pretreated anthracite through rapid heating up to 160°C followed by holding for 4 h gives a small amount of CO_2 evolution because there was almost no free $HClO_4$ in the carbon and the chlorine atom of $HClO_4$ was bonded to aromatic carbons. These results suggested that ClO_4^- impregnation can modify the texture of anthracite particles and oxidized (activated) them at low temperatures, and consequently can assist CO_2 activation.

Thermal heating for activation is commonly used for the preparation of activated carbon (AC), although it has some disadvantages, such as the possibility of inhomogeneous heating between the surface and the interior of a particle, high cost, long time, etc. Microwave irradiation attracted attention as an alternative technique for heating [17]. Microwave

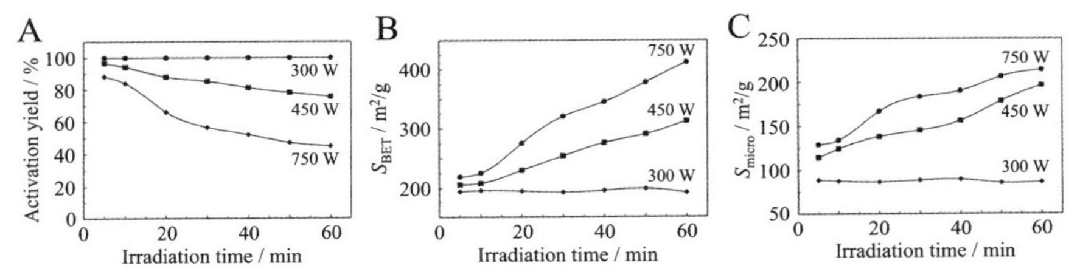

Figure 2.7 Microwave-assisted CO_2 activation of oil-palm-stone chars: (A) activation yield, (B) BET surface area S_{BET}, and (C) microporous surface area S_{micro} [18].

irradiation, instead of thermal heating, during the activation by CO_2 was applied to different carbon precursors. In Fig. 2.7, the effects of microwave power on the yield, S_{BET}, and S_{micro} are shown for oil-palm-stone chars [18]. The results reveal that a marked decrease in activation yield, i.e., an increase in burn-off, and an increase in S_{BET} can be obtained at high power irradiation, such as 750 W, due to the marked development of micropores. Irradiation at 750 W was resulted in an abrupt increase in the temperature of the char particles up to around 350°C, with no difference in temperature being expected between the surface and interior of the particles.

Steam activation at 800°C was applied to EG composites with the carbon derived from furfuryl alcohol (FA) [19]. The composites were fabricated by impregnation of FA into uniaxially compressed EG (bulk density of 0.0026 g/cm^3), followed by polymerization of FA and carbonization at 550°C. In Fig. 2.8A and B, the isotherms measured using N_2 at 77K and CO_2 at 273K, respectively, are shown on the activated composites with different burn-offs and are compared with the compressed EG and the composites carbonized at 550 and 800°C without steam. All isotherms measured using N_2 for the composites are type I with slight hysteresis, and N_2 adsorption is enhanced with increasing carbonization temperature and

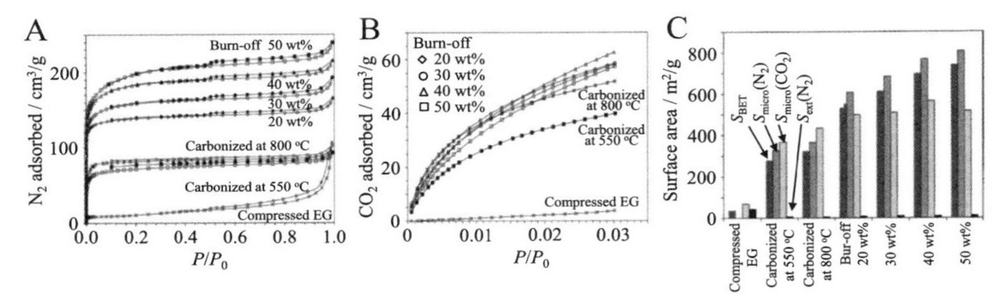

Figure 2.8 Steam activation of the composites of EG with furfuryl alcohol-derived carbon: (A) N_2 adsorption–desorption isotherms, (B) CO_2 adsorption isotherms, and (C) comparison of surface areas measured using N_2 and CO_2 [19].

burn-off with activation at 800°C (Fig. 2.8A). An abrupt increase in N_2 adsorption at low P/P_0 shows a marked increase in V_{micro}. CO_2 adsorption also increases with carbonization; the composite carbonized at 800°C adsorbing more CO_2 than that carbonized at 550°C (Fig. 2.8B), although N_2 adsorption is not as different in these two composites, suggesting the development of smaller micropores (confirmed from CO_2 adsorption) by carbonization. For the composites activated at 800°C with different burn-offs, CO_2 adsorption at low P/P_0 is even lower than that for the composite carbonized at 800°C without activation, indicating that the smaller micropores (measured by CO_2) are affected slightly by activation, whereas the larger micropores (measured by N_2) are markedly increased by activation. In Fig. 2.8C, a series of surface areas measured by N_2 and CO_2 are compared. The composites carbonized at 550 and 800°C are composed predominantly of the microporous surface owing to smaller micropores, and they are widened upon activation. The microporous surface area obtained by N_2 adsorption [$S_{micro}(N_2)$] becomes larger than that obtained by CO_2 for the activated composites [$S_{micro}(CO_2]$, the difference increasing as the burn-off increases. The external surface area $S_{ext}(N_2)$ due to a nonmicroporous surface area is generally negligible, except for the original compressed EG.

Activation processes using CO_2 and steam were compared at 890°C under a pressure of 0.1 MPa at a flow rate of 80 mL/min on isotropic carbon fibers carbonized at 1000°C [20]. For steam activation, a steam/N_2 mixture (l/l by volume) was used. As shown in Fig. 2.9A, steam has greater reactivity than CO_2, even though steam pressure is half of CO_2 pressure. In Fig. 2.9B, $S_{micro}(N_2)$ and $S_{micro}(CO_2)$ are compared for these activated carbon fibers (ACFs) with different burn-offs together with tensile strength

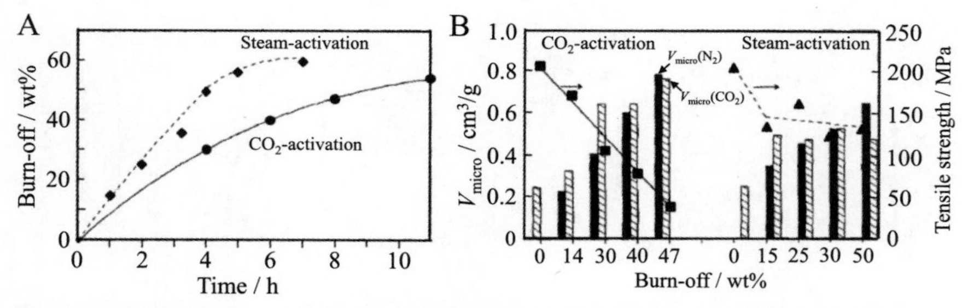

Figure 2.9 Comparison of two activation processes, steam and CO_2 activation, on carbon fibers: (A) burn-off and (B) microporous surface areas measured by N_2 and CO_2 [$V_{micro}(N_2)$ and $V_{micro}(CO_2)$, respectively] and tensile strength [20].

of fibers. At the low burn-offs, $S_{micro}(CO_2)$ is higher than $S_{micro}(N_2)$, suggesting the formation of narrow micropores is mainly occurring at the initial stage of activation. $S_{micro}(N_2)$ of the CO_2- and steam-activated fibers increases with increasing burn-off, and is lower for the CO_2-activated fibers than for the steam-activated ones at 14% burn-off, whereas the opposite happens in the high burn-off region. However, $S_{micro}(CO_2)$ of the CO_2-activated fibers increases with increasing burn-off, while that of the steam-activated ones increases up to 15% burn-off and remains almost constant above 15%. These results conclude that steam activation causes burn-off near the external surface of the fiber, and, in contrast, CO_2 activation takes place not on the external surface but within the fibers. CO_2 activation not only generates micropores but also deepens porosity inside the fibers, thereby decreasing the tensile strength of the fiber.

2.1.1.2 Chemical activation

Various activation reagents, such as $ZnCl_2$, KOH, NaOH, etc., have been employed to develop micropores in carbon materials.

KOH is frequently used as an activation reagent, mainly because it gives a very high surface area, as high as 3600 m^2/g. Various carbon precursors, such as petroleum pitch, cokes with different textures, wood charcoals, and coffee bean char, were mixed with KOH and heat treated at 550—900°C for 20 min to 4 h, followed by washing with water [21]. Most cokes gave S_{BET} of 1600—2700 m^2/g with burn-off of 28%—54% under the condition of 80 wt.% KOH addition at 800°C for 2 h, while metallurgical coke carbonized above 1200°C and calcined cokes were unreactive to KOH. KOH activation was examined for the carbons prepared from eucalyptus wood sawdust via flash heating. The sawdust was heated for a short time (5—10 min) at relatively low temperature (400°C) under a flow of air and then reacted with KOH at KOH/carbon weight ratios of 2/1—4/1 at temperatures of 600,700, and 800°C [22]. A high KOH/C ratio and high activation temperature resulted in a highly developed micropore structure, as shown by the adsorption—desorption isotherm in Fig. 2.10A; activation at 800°C with a KOH/C ratio of 4/1 delivers S_{total} of 2610 m^2/g, S_{micro} of 1892 m^2/g, V_{total} of 1.15 cm^3/g, and V_{micro} of 0.74 cm^3/g. Activation conditions also influence the stability of the activated carbon at high temperature, as shown by the TG curves in air flow in Fig. 2.10B.

Microwave irradiation was effective for enhancing the KOH activation of cotton stalk and compared with K_2CO_3 activation [23]. The mixtures of cotton stalk with KOH and K_2CO_3 at the weight ratios, activation reagent

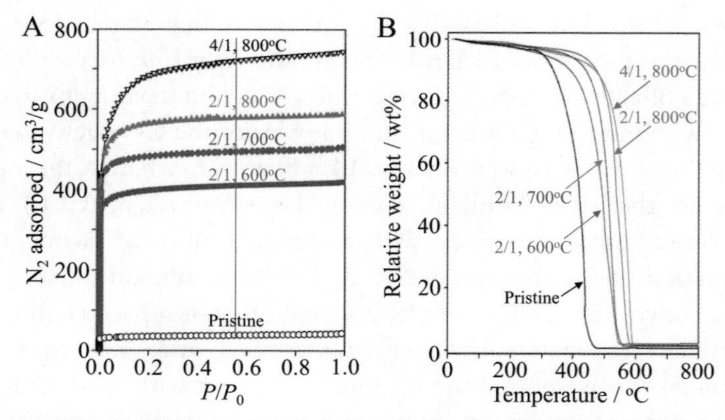

Figure 2.10 KOH activation of sawdust-derived carbons at different temperatures with KOH/C ratios of 2/1–4/1: (A) N_2 adsorption–desorption isotherms and (B) TG curves in air flow [22].

to stalk, of 0.6 and 0.8 were irradiated by microwave at 680 W power for 10 min and 660 W for 8 min, respectively. The resultant KOH-activated carbon exhibited S_{BET} of 729 m^2/g and V_{total} of 0.38 cm^3/g including V_{micro} of 0.26 cm^3/g, while the K_2CO_3-activated carbon exhibited 621 m^2/g, 0.38 cm^3/g, and 0.11 cm^3/g, respectively, revealing that KOH is more effective in creating micropores. When $ZnCl_2$ was used as an activation reagent at a ratio of 1.6, the activated carbon obtained by microwave irradiation with 560 W for 6 min was mesoporous, V_{total} of 0.63 cm^3/g including V_{micro} of 0.08 cm^3/g [24]. Microwave heating was compared with conventional heating using mesocarbon **microbeads (MCMBs);** with a particle size of about 20 μm) impregnated by KOH via its solution with a KOH/MCMB weight ratio of 3/1–9/1 [25]. The carbon obtained with a ratio of 8/1 exhibited the highest S_{BET} and V_{total}, 4106 m^2/g and 2.36 cm^3/g, by microwave irradiation with 3 kW power for 20 min and 3977 m^2/g and 2.03 cm^3/g by conventional heating at 850°C for 60 min, respectively. The microwave heating process shortened the activation time considerably.

By using potassium oxalate, $K_2(CO_2)_2$, highly microporous carbons were obtained from α-D-glucose through mixing in powders with a weight ratio $K_2(CO_2)_2$/glucose of 3.6 at 800–850°C for 1–5 h in N_2 flow, S_{BET} of 1270–1690 m^2/g, and V_{total} of 0.50–0.72 with V_{micro} of 0.49–0.67 cm^3/g (percentage of micropores of 93%–98%) [26]. $K_2(CO_2)_2$ is more effective for the development of micropores in carbon matrix than KOH and is less corrosive than KOH. The addition of melamine into glucose (melamine/glucose weight ratio of 1−3) resulted in micro-/mesoporous N-doped carbons with S_{BET} of 3460 m^2/g and V_{total} of 2.72 cm^3/g with V_{micro} of 1.00 cm^3/g (micropore percentage of 37%) and N content of 2.7 wt.%.

NaOH was also used as an activation reagent [27]. NaOH was mixed with a raw anthracite (without precarbonization) at different NaOH/anthracite weight ratios of 2/1, 3/1, and 4/1 via either mechanical powder mixing or mixing in aqueous solution of NaOH, and then heated at 750°C for 2 h in N_2 flow, followed by washing with deionized water and 0.5 M HCl. N_2 adsorption−desorption isotherms are shown for the carbons prepared via powder and solution mixing as a function of mixing ratio in Fig. 2.11A and B, respectively. The isotherms show three steps with the increase in P/P_0, steep increases at less than 0.02 owing to the adsorption into small micropores, gradual increases without any noticeable hysteresis owing to the progressive filling of large micropores and mesopores, and finally abrupt increases at P/P_0 near 1.0 because of active capillary condensation. The activation of the powder mixtures gives higher S_{BET}, V_{total}, and fraction of mesopore volume compared to those of the solution mixtures: 2063 m^2/g, 1.61 cm^3/g, and 49% for the powder mixture with a mixing ratio of 4/1, and 1763 m^2/g, 1.13 cm^3/g, and 32% for the solution mixture with the same mixing ratio.

Chemical and physical activation processes were compared using KOH and steam on fractionated pitches [28]. A coal tar pitch with a softening point of 274°C was fractionated into toluene soluble, pyridine soluble, quinoline soluble, and quinoline insoluble fractions (TS, PS, QS, and QI, respectively). Activation was performed with a KOH/pitch weight ratio of

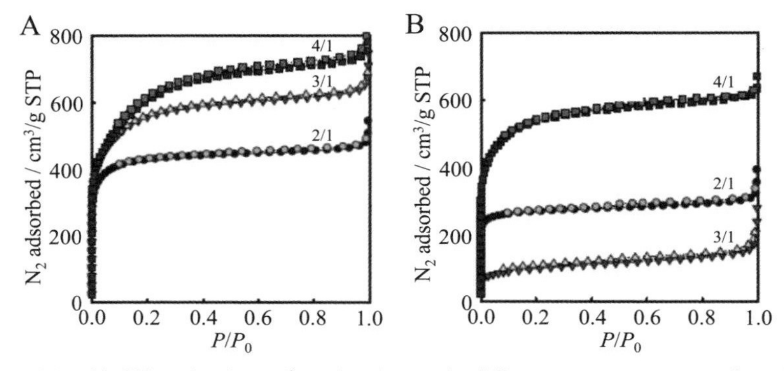

Figure 2.11 NaOH activation of anthracite with different mixing ratios of NaOH/C: (A) powder mixing and (B) solution mixing [27].

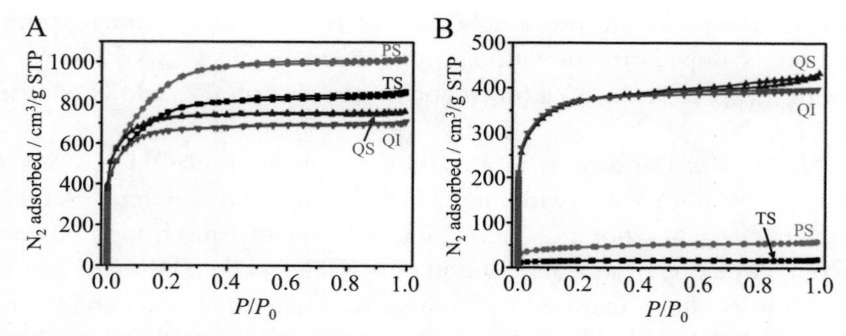

Figure 2.12 Adsorption–desorption isotherms of the fractionated pitches: (A) activated with KOH and (B) activated with steam [28].

4 at 750°C for 1 h and by steam at 900°C for 1 h. N_2 adsorption–desorption isotherms are shown in Fig. 2.12 and the determined pore parameters are listed in Table 2.2. The experimental results clearly show that KOH activation is more favorable to obtain microporous carbons from both light and heavy pitch fractions (from TS to QI fractions) than steam activation, and that the light fractions (TS and PS) give slightly more developed pore structure than the heavy fractions (QS and QI) by KOH activation. Meanwhile, steam activation led to poorer pore development for the former fractions in comparison with the latter fractions. Chemical and physical activation processes were compared also using KOH and CO_2 on phenol-formaldehyde resin [29]. At similar porosity levels the carbon yield during CO_2 activation was lower than that with KOH activation. Typical activation processes, physical activation using CO_2, and chemical activations using $ZnCl_2$, H_3PO_4, and KOH were compared on rGO monoliths made from GO colloids by the unidirectional freeze-drying method [30]. KOH activation was shown to be the most effective for the activation of rGO, of which S_{BET} reached 2200 m^2/g.

Table 2.2 Pore parameters of the fractionated pitches activated by KOH and steam [28].

Sample	Activation	Yield (%)	S_{BET} (m^2/g)	S_{micro} (m^2/g)	V_{total} (cm^3/g)	V_{micro} (cm^3/g)
TS	KOH	23.5	2732	706	1.315	0.270
PS		28.2	2981	721	1.521	0.331
QS		56.5	2513	1367	1.178	0.606
QI		69.3	2284	1498	1.106	0.732
TS	Steam	39.9	52	32	0.030	0.014
PS		35.8	158	76	0.091	0.034
QS		34.4	1288	692	0.671	0.312
QI		35.0	1271	781	0.616	0.355

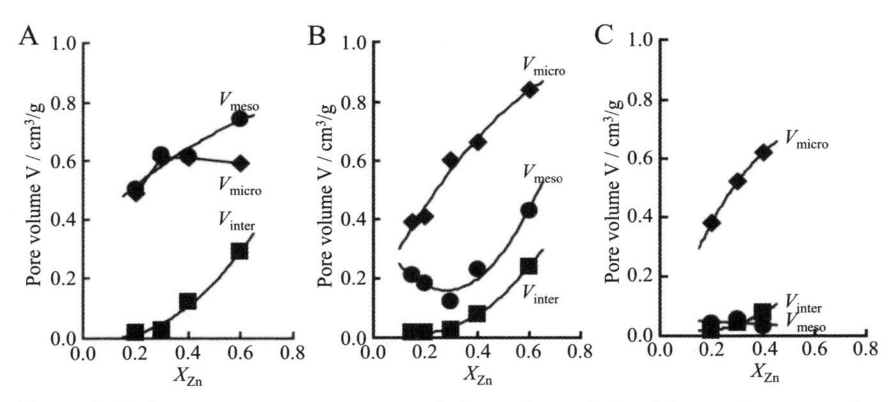

Figure 2.13 Pore structure parameters of the carbons derived from olive stones by $ZnCl_2$ activation: (A) the granules, (B) the discs pressed at 150°C, and (C) the discs pressed at 300°C [31].

Chemical activation of olive stones was carried out using $ZnCl_2$ [31]. $ZnCl_2$ was impregnated into the grains via its aqueous solutions with different concentrations (19–48 wt.%) at 85°C for 7 h. After the evaporation of water, the impregnated granules were hot-pressed into a disc (2 cm in diameter and 1 cm in thickness) under 130 MPa at either 150 or 300°C, followed by heat treatment at 500°C in N_2 flow. Fig. 2.13 shows the evolutions of V_{micro}, V_{meso}, and interparticle spaces, V_{inter}, with impregnation ratio X_{Zn} ($ZnCl_2$/stone by weight) for the granules and discs. Granular olive stones show increasing V_{meso} with X_{Zn}, but the V_{micro} levels off at X_{Zn} of 0.3. On the other hand, the discs show marked increase of V_{micro} with increasing X_{Zn}. In the disc pressed at 150°C, mesopore formation is accelerated with increasing X_{Zn}, although no marked development of mesopores is observed in the disc prepared at 300°C.

H_3PO_4 activation was also performed on olive stones by impregnation via aqueous solutions with different concentrations (29–52 wt.%) and hot-pressing at 100°C under 130 MPa, followed by heat treatment at 450°C in N_2 flow [32]. The impregnation ratio H_3PO_4/stone, X_P, was controlled to be 0.21–0.42 by changing the concentration of H_3PO_4. Micropore volume V_{micro} was calculated by application of the Dubinin–Radushkevich (DR) equation and mesopore volume V_{meso} was calculated by $V_{total} - V_{micro}$. In addition, the volume between carbon particles V_{inter} was estimated. In Fig. 2.14A, V_{micro}, V_{meso}, and V_{inter} are plotted against X_P. Up to X_P of 0.35, V_{micro} increases considerably with increasing X_P, associated with a decrease in V_{inter}, while at X_P of 0.42 marked increases in V_{meso} and V_{inter} are accompanied by a decrease in V_{micro}. The hot-pressing

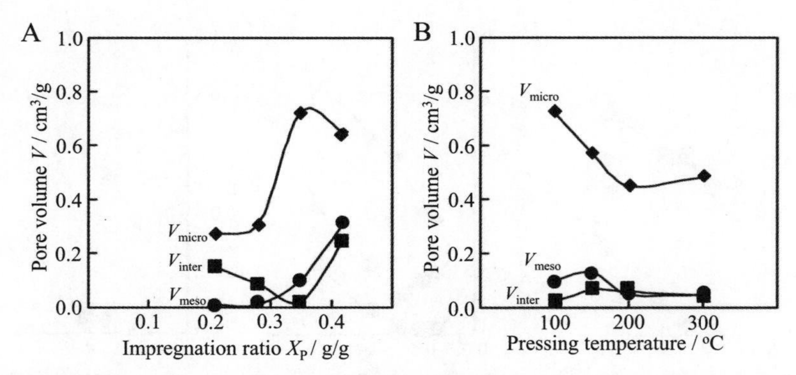

Figure 2.14 V_{micro}, V_{meso}, and V_{inter} of H_3PO_4-activated carbon discs: changes with (A) H_3PO_4 impregnation ratio and (B) pressing temperature on the disc with X_P of 0.35 [32].

condition influences micropore formation, as shown in the plot of pore structure parameters against pressing temperature in Fig. 2.14B, indicating that V_{micro} decreases with increasing pressing temperature. The disc with X_P of 0.35 prepared at 100°C delivers the highest V_{micro} and low V_{inter}, along with a high CH_4 storage capacity of 131 v/v.

2.1.1.3 Activated carbon fibers

Activated carbon fibers (ACFs) have the advantage of fast adsorption rate because most micropores are exposed on the physical surface of the fibers, in other words, the adsorbates are directly adsorbed into micropores without preliminary diffusion through macropores and/or mesopores from the physical surface of the grains of AC, as shown schematically in Fig. 2.15.

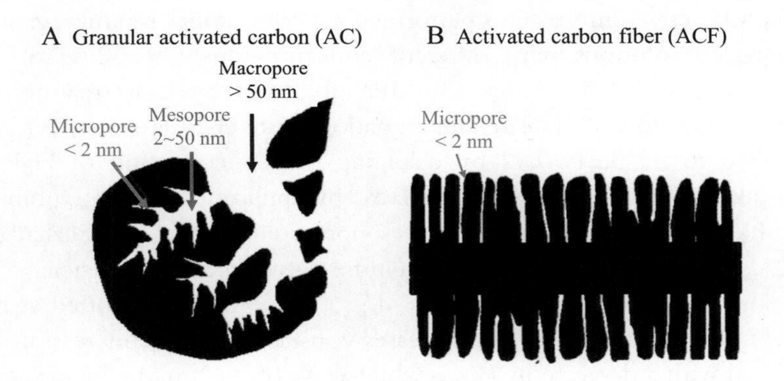

Figure 2.15 Schematic illustrations for the pore structures of activated carbons: (A) granular AC and (B) ACF.

In addition, ACFs have the advantage of ease of handling over granular and pelletized ACs and they are easily formed into webs or sheets. ACF production started to pyrolyze and activate viscose textiles, as reviewed in 1994 by focusing on the activity in Japan [33]. Now, most ACFs are produced from isotropic pitches by melt spinning and subsequent activation.

Pore structures of three commercially available ACFs, which were derived from different precursors, cellulose, pitch, and polyacrylonitrile (PAN), were studied by gas adsorption–desorption methods using N_2 and various hydrocarbons and by small-angle X-ray scattering [34–39]. N_2 adsorption isotherms and their α_s-plots are shown for three ACFs in Fig. 2.16A and B, respectively. The isotherms for three ACFs are classified into type I and almost complete adsorption below P/P_0 of 0.2, suggesting that the ACFs used are highly microporous [37]. However, the features of the isotherms at low P/P_0 are significantly different; the pitch-based ACF has the steepest uptake of N_2, while the cellulose-based one demonstrates gradual uptake. The α_s-plots for the cellulose- and PAN-based ACFs demonstrate pronounced upward deviations from linearity below α_s of 0.4, unlike that for the pitch-based ACF and activated carbon (AC), suggesting that the micropores of the pitch-based ACF and AC are narrower than those of the other ACFs. In Table 2.3, the pore structure parameters evaluated by t- and α_s-plots are compared. Both analyses give similar results: large V_{micro} and very small external surface area (S_{ext}), in other words, predominant micropore surface area (S_{micro}). Based on the measurements of

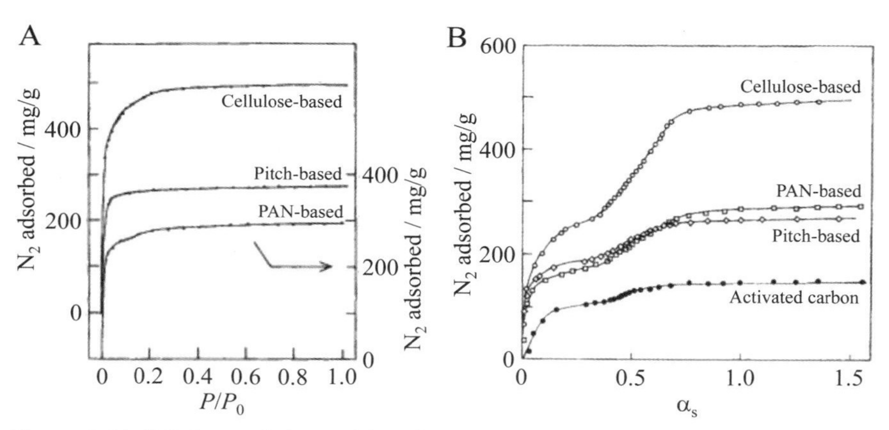

Figure 2.16 Cellulose-, pitch-, and PAN-based ACFs: (A) N_2 adsorption isotherms and (B) α_s-plots shown alongside the result of an AC [37].

Table 2.3 Parameters on microporosities evaluated using *t*- and α_s-plots [37].

Activated carbon fibers (ACFs)	***t*-plot**			**α_s-plot**	
	S_{total} (m^2/g)	**S_{ext} (m^2/g)**	**V_{micro} (cm^3/g)**	**S_{ext} (m^2/g)**	**V_{micro} (cm^3/g)**
Cellulose-based ACF	1410	22	0.590	54	0.606
PAN-based ACF	850	7.8	0.354	53	0.344
Pitch-based ACF	846	2.6	0.330	22	0.324
Activated carbon (AC)	468	4.7	0.186	19	0.173

N_2 adsorption—desorption isotherms, NMR spectra, and scanning tunneling micrographs on commercial pitch and PAN-based ACFs, a microdomain model was proposed [40]. ACF was composed of spherical microdomains where micropores were generated on each microdomain and mesopores were formed between the microdomains. The adsorption of N_2 into the micropores of these ACFs was shown to consist of multistages using the Dubinin—Radushkevich (DR) plot of gravimetric N_2 adsorption isotherms, as illustrated schematically in Fig. 2.17A [34]; nitrogen filling in the ultramicropores (nitrogen bilayer-sized pores), monolayer adsorption on the pore walls of supermicropores (nitrogen three- and/or four-layer-sized pores) and then cooperative filling in the monolayer-coated super-micropores, as shown as L-, M-, and H-regions, respectively. The detailed DR plots on pitch-based ACFs are shown in Fig. 2.17B, revealing that the ACF activated at low temperature (100°C) has three regions, while the L-region is missing on those activated at temperatures above 300°C, in

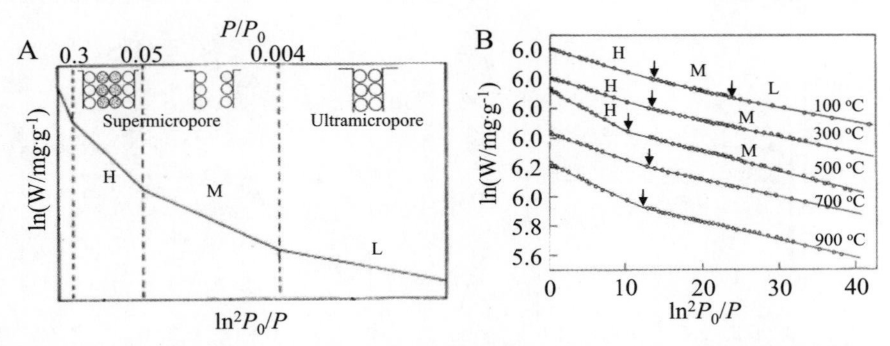

Figure 2.17 Detailed DR plots: (A) schematic plot with models for micropore filling and (B) plots of N_2 adsorption for pitch-based ACFs with different oxidation temperatures [34].

Table 2.4 S_{total}, V_{micro}, and mean micropore size of ACFs determined by α_s-plot analysis [42].

ACF	α_s plot		
	S_{total} (m^2/g)	V_{micro} (mL/g)	Mean pore size (nm)
A5	920	0.34	0.75
A10	1160	0.48	0.83
A15	1350	0.61	0.91
A20	1770	0.96	1.09

other words, ultramicropores are not detected by N_2 adsorption [34]. DR plot analysis was also applied to the ACFs using He adsorption isotherms at 4.2K [41]. V_{micro} values evaluated by He adsorption were higher than those obtained from the N_2 isotherms by 20%–50%, suggesting the presence of ultramicropores which could not be assessed by N_2 molecules at 77K.

A series of pitch-based ACFs with different activation degrees, A5, A10, A15, and A20, were also studied by α_s-plot of the N_2 adsorption isotherm [42]. The α_s-plots for the ACFs exhibited a clear upward deviation in the low α_s range below 0.2 (*f*-swing), and more significant deviation with increasing activation degree. For highly activated ACFs, A15 and A20, a slight deviation was also noticed above α_s of 0.5 (*c*-swing), suggesting the growth of the micropores to larger sizes. Total surface area (S_{total}), V_{micro}, and mean micropore size were determined using α_s-plot, as shown in Table 2.4. With increasing activation degree, the development and widening of micropores is clearly seen from the increases in S_{total}, V_{micro}, and mean pore size.

Commercially available poly(*p*-phenylene terephthalamide) pulp (Kevlar pulp) was carbonized at 900°C and activated at 750 and 800°C in CO_2 [43]. The pulps activated at 800°C up to 65% burn-off exhibited S_{BET} of 986 m^2/g and V_{micro} of 0.43 cm^3/g including V_{ultra} of 0.28 cm^3/g. Another aramid fiber, Nomex poly (*m*-phenylene isophthalamide), was carbonized at 850 and 900°C, followed by activation in CO_2 at a temperature 50°C lower [44]. The ACF obtained by the 850°C-activation exhibited S_{BET} of 1545 m^2/g and V_{micro} of 0.68 cm^3/g including V_{ultra} of 0.58 cm^3/g. The experimental results showed that the ACFs prepared from Nomex exhibited more uniform and narrower porosity than those prepared from Kevlar. Steam activation at 780°C was also effective for developing micropores in Nomex-derived ACFs; the carbon with burn-off of 63% exhibited S_{BET} of 1580 m^2/g and V_{micro} of 0.68 cm^3/g including V_{ultra} of 0.11 cm^3/g [45]. Activation of Nomex-derived carbon fibers using

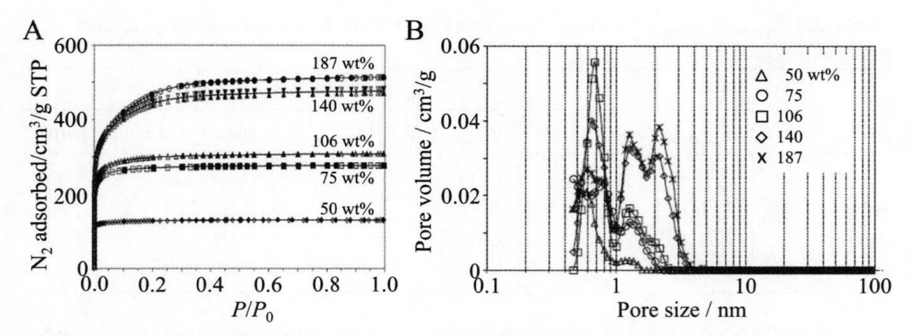

Figure 2.18 ACFs derived from Nomex fibers by H_3PO_4 activation at 700°C: (A) N_2 adsorption–desorption isotherms and (B) pore size distributions as a function of burn-off [47].

H_3PO_4 was effective to develop supermicropores in ACFs [46,47]. Nomex fibers were impregnated in H_3PO_4 aqueous solutions with different H_3PO_4/fiber weight ratio of 0.5–1.87. After evaporation of water the resultant mixtures were heat-treated at 700°C in Ar [47]. The resultant ACFs were microporous and their pore volume depends strongly on burn-off, as shown by the N_2 isotherms and pore size distribution in Fig. 2.18A and B, respectively. Ultramicropores are developed at the start of activation (low burn-off), while larger micropores (supermicropores) are formed with increasing burn-off. The ACF obtained using an H_3PO_4 impregnation ratio of 1.87 with 45% burn-off exhibited S_{BET} of 1688 m^2/g and V_{micro} of 0.74 cm^3/g including V_{ultra} of 0.03 cm^3/g, mainly supermicropores being formed. Nomex-derived ACFs activated using the ratio of 0.07 at 850°C were additionally activated by CO_2 at 800°C at up to 72% burn-off, resulting in enhanced S_{BET} of 2408 m^2/g and V_{micro} of 1.00 cm^3/g including V_{ultra} of 0.12 cm^3/g [46].

Commercial poly(vinyl alcohol) (PVA) fibers (softening point of 215–224°C) were heated up to 220°C in air under tension (stabilization). The stabilized fibers were heated at 250°C after impregnation of either diammonium phosphate (DAP) or a mixture of ammonium sulfate (AS) with DAP for dehydration and then heated up to 300°C in air for pyrolysis, followed by carbonization at 800–1000°C in N_2 flow and activation at the same temperature in either CO_2 flow, a steam/N_2 gas flow, or static air for 1 h to obtain ACFs [48]. The N_2 adsorption–desorption isotherms and pore size distribution are shown for the resultant ACFs in Fig. 2.19A and B, respectively. The ACF activated by CO_2 for 1 h exhibits the isotherm of typical type I as well as those activated by steam and air, and shows S_{BET} of 995 m^2/g and V_{total} of 0.424 cm^3/g including V_{micro} of 0.398 cm^3/g. However, the pore size distribution for the CO_2-activated ACF shows a sharp and strong peak at around 0.8 nm, in contrast to those for the ACFs

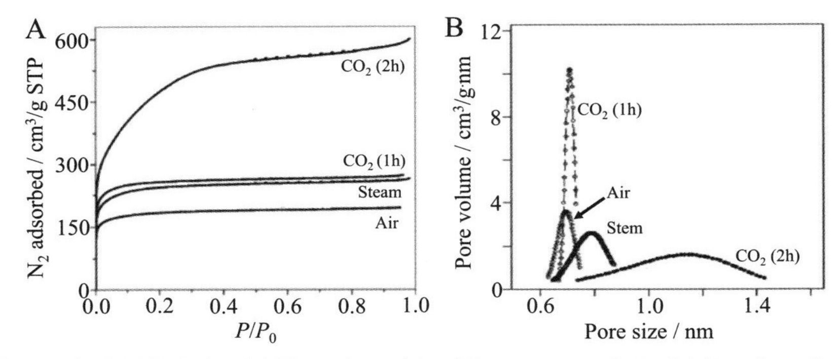

Figure 2.19 PVA-derived ACFs activated by CO_2, steam, and air: (A) N_2 adsorption–desorption isotherms and (B) pore size distribution in the micropore range [48].

activated by steam and air. By extending the activation time to 2 h in CO_2, S_{BET} and V_{total} markedly increased to 2128 m^2/g and 0.901 cm^3/g, but the increases in these parameters were due to the increase in larger micropores as shown by the broad pore size distribution in Fig. 2.19B. ACFs were produced from a pitch derived from poly(vinyl chloride) by melt-spinning, stabilization at 320°C in air, carbonization at 900°C, and then activation at 900°C in steam [49]. The development of pore structure depended strongly on the activation time; S_{BET} of 2096 m^2/g and V_{total} of 1.34 cm^3/g were obtained after 90 min activation. Activation of natural fibers (jute and coconut fibers) was performed using CO_2 and H_3PO_4 at 900°C [50]. The ACF prepared from coconut fibers by CO_2 activation delivered S_{BET} of 1303 m^2/g with V_{micro} of 0.536 cm^3/g and V_{meso} of 0.089 cm^3/g, and that by H_3PO_4 activation delivered 1088 m^2/g, 0.473 cm^3/g, and 0.132 cm^3/g, respectively.

Blending of PAN with a pitch improved the electrospinnability to make the fiber diameter thinner, down to about 750 nm, of which steam activation resulted in high microporosity [51]. As shown in the N_2 adsorption–desorption isotherms in Fig. 2.20, the fibers activated at 700 and 800°C deliver V_{total} of 0.324 cm^3/g with 79.5% micropores and V_{total} of 0.767 cm^3/g with 71.1% micropores, respectively, while activation at 900°C achieves a high V_{total} of 1.114 cm^3/g but microporosity decreases to 49%, increasing the mesoporosity.

2.1.2 Template-assisted carbonization

2.1.2.1 *Zeolites*

Zeolites are microporous aluminosilicate minerals with the framework consisting of tetrahedra of AlO_4 and SiO_4, and are widely used as

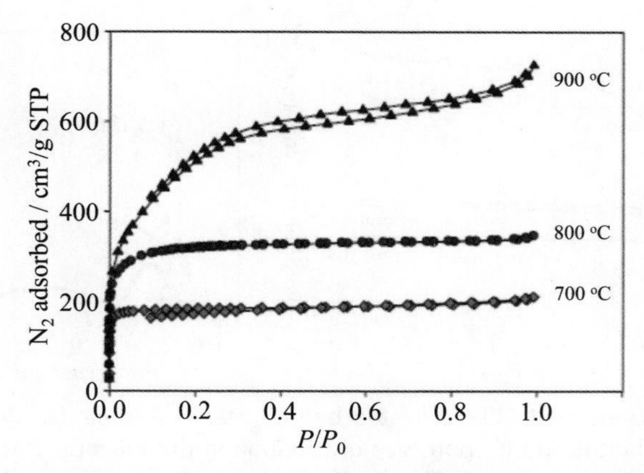

Figure 2.20 N_2 adsorption—desorption isotherms of electrospun PAN-blended pitch-based carbon fibers after steam activation at different temperatures [51].

adsorbents. Three types of zeolites, zeolites Y, β, and L, which have different frameworks and consequently different pore interconnections, have been used for template carbonization [52—55].

Filling of carbon precursors into the micropores of zeolite templates is an important process for highly developed pore structures in the resultant carbons. A two-step process was proposed to fill completely the micropores of the template zeolite using two carbon precursors, i.e., furfuryl alcohol (FA) and propylene. FA was firstly impregnated into an Na-form zeolite Y template under reduced pressure at room temperature and polymerized to form poly(furfuryl alcohol) (PFA) in the zeolite channels at 150°C. The PFA/zeolite composite was then heated to 700°C in N_2 flow, and the resultant composite (carbon-filled zeolite) was further impregnated with carbon by the chemical vapor infiltration (CVI) of propylene gas at 700 and 800°C, followed by washing with HF and HCl solutions to dissolve out the zeolite [56,57]. The carbon obtained by CVI at 700°C delivered S_{BET} of 3600 m^2/g, V_{micro} of 1.5 cm^3/g, and negligibly small V_{meso} [57]. The micropores in the resultant carbons (zeolite-templated carbons, ZTCs) are supposed to be regularly aligned because they show a sharp peak at around 6° in 2θ in the XRD pattern and a highly ordered lattice fringes in TEM image, as shown in Fig. 2.21, indicating structural regularity with a periodicity of about 1.4 nm replicated from the template zeolite [57]. Detailed studies have shown that the coupling of impregnation of a polymer, such as PFA, with CVI of an organic gas, such as propylene, and heat treatment at a high temperature such as 900°C were essential to obtain highly ordered microporous carbons [58]. Through this two-step filling, zeolite channels

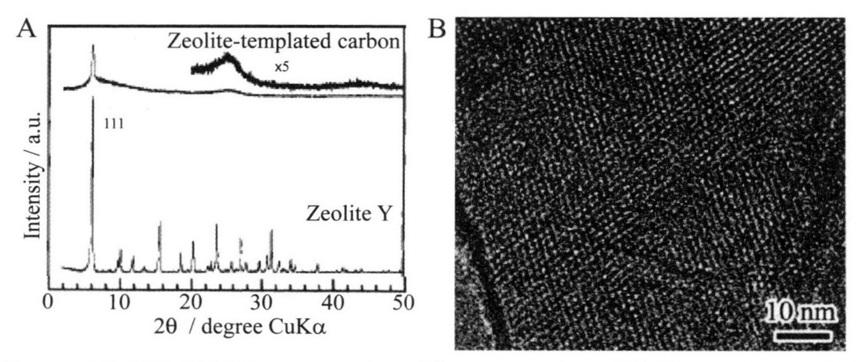

Figure 2.21 ZTC: (A) XRD pattern at low diffraction angle and (B) high-resolution TEM image [57].

were almost completely filled with carbon, forming a rigid carbon framework sufficient to be retained even after dissolving the zeolite framework. It could have as high S_{micro} as 3700 m^2/g, V_{micro} as 1.8 cm^3/g, and sharp pore size distribution in the range of 1.0–1.5 nm [59]. The ZTC prepared by CVI of acetylene (Ac) at 600°C, instead of liquid FA impregnation process, and further CVI of polypropylene (PP) at 700°C (Ac-CVI + PP-CVI), followed by heat treatment at 900°C in N_2 exhibited high S_{BET}, which was a little lower than the two-step filling of FA-impregnation and PP-CVI (FA-impregnation + PP-CVI) [60]. In Fig. 2.22A, its N_2 adsorption–desorption isotherm is shown together with those for the carbons prepared by another two-step process (Ac-CVI at 600°C + Ac-CVI at 700°C) and by a single CVI process of Ac at 600°C (Ac-CVI). Their pore parameters are summarized in Table 2.5. The microporosities of the carbons through

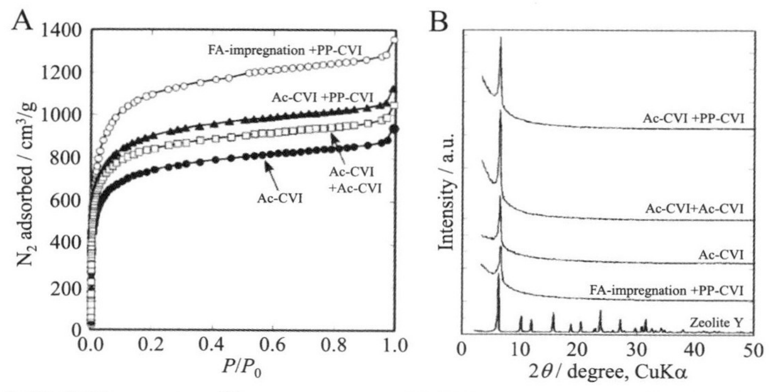

Figure 2.22 ZTC two-step filling processes: (A) XRD patterns and (B) N_2 adsorption–desorption isotherms [60].

Table 2.5 ZTCs prepared through different filling processes [60].

Filling process	Carbon content (wt.%)	S_{BET} (m^2/g)	V_{micro} (cm^3/g)	V_{meso} (cm^3/g)
Ac-CVI at 600°C	25.9	2760	1.12	0.22
Ac-CVI at 600°C + Ac-CVI at 700°C	16.9	3170	1.34	0.16
Ac-CVI at 600°C + PP-impregnation at 700°C	26.3	3370	1.37	0.26
FA-impregnation + PP-CVI at 700°C	22.3	4080	1.76	0.20
No filling (zeolite Y)	0.0	870	0.33	0.04

two-step CVI filling processes are comparable with each other, while a higher S_{BET} and large V_{micro} being possible to be obtained by combining FA-impregnation and PP-CVI. In all cases, long-range structural regularity was achieved, as shown by their XRD profiles in Fig. 2.22B. The time extension for Ac-CVI at 800°C influenced the microporosity of the resultant carbons; extension to 2 h resulted in S_{BET} of 3310 m^2/g and V_{micro} of 1.26 cm^3/g, whereas 1 h filling resulted in 2880 m^2/g and 1.06 cm^3/g, respectively [61]. However, a further extension to 3 h resulted in 2260 m^2/g and 0.91 cm^3/g, respectively. Two ZTC powders, Ac-CVI and FA-impregnation + PP-CVI, could be pelletized without any binder under conditions of 300°C and 147 MPa, with their bulk densities being 0.7–0.9 g/cm^3, whereas no pellets could be obtained from commercial ACs under the same conditions [62]. With this densification, the volumetric surface area increased to 1100–1300 m^2/cm^3 and the average micropore size decreased with increasing pelletizing pressure, suggesting the possibility of pore size control by pressure.

The ZTCs synthesized by impregnation of FA and CVI of propylene at 700°C using zeolite Y were modified by the following methods: heat treatment at 400°C to eliminate acid anhydride groups and functionalization with the $-B(OH)_2$ group through reaction of dimethylamine borane [63]. N-doped ZTC was also synthesized by changing the reactant for CVI to acetonitrile. The structural framework of ZTC was retained after heat treatment at 400°C, (S_{BET} of 3590 m^2/g and V_{total} of 1.65 cm^3/g), as shown by the N_2 isotherm and pore size distribution in Fig. 2.23A and B, respectively. ZTC has abundant edge sites and dimethylamine borane substitute edge hydrogen atoms with $-B(OH)_2$ without substantial change in structural framework, associated with a slight decrease in S_{BET} to

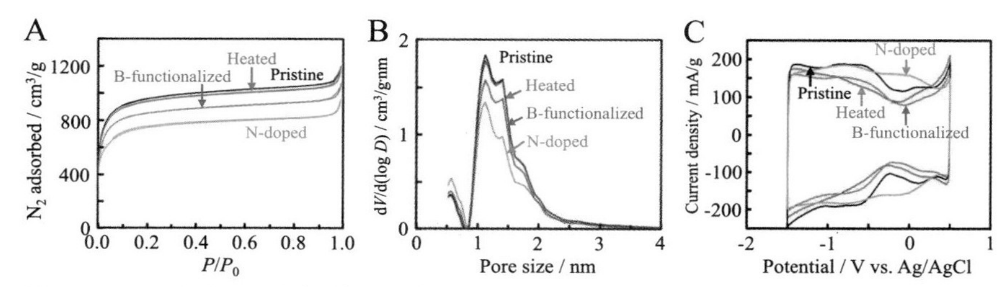

Figure 2.23 ZTCs modified by N-doping, B-functionalization, and heat treatment at 400°C: (A) N_2 adsorption—desorption isotherms, (B) pore-size distributions, and (C) CV curves at a scan rate of 1 mV/s in 1 M $TEABF_4$/PC at 25°C [63].

3280 m^2/g and V_{total} to 1.50 cm^3/g. N-doping restores the structural framework of ZTC, although it exhibits slightly lower S_{BET} of 2780 m^2/g and V_{total} of 1.31 cm^3/g. These modifications of the ZTC have certain effects on their supercapacitor performances in 1 M $TEABF_4$/PC at 25°C, as shown by their cyclic voltammograms (CVs) in Fig. 2.23C. The capacitances measured at 50 mA/g by galvanostatic charge—discharge analysis were 175, 151, 146, and 180 F/g for the pristine, heated, B-functionalized, and N-doped ZTCs, respectively. These results suggest that functional groups (acid anhydride) and N-doping induce pseudocapacitance, whereas -$B(OH)_2$ group makes no contribution to the capacitance enhancement.

Impregnation, polymerization, pyrolysis, and carbonization in the pores of different zeolites without CVI were performed using acrylonitrile, FA, vinyl acetate, phenol-formaldehyde, and pyrene (toluene solution) [64—66], however high surface areas greater than 2000 m^2/g were not obtained. Carbon filling of zeolite channels was also examined only by CVI of propylene and porous carbons with S_{BET} of 1380—2200 m^2/g were obtained, consisting of both micropores and mesopores [67,68].

The structure model of the carbon prepared by a two-step filling process using Na-form zeolite Y was proposed on the bases of the experimental results by various techniques [69], as shown in Fig. 2.24. The ZTC is composed of the assembly of nonstacked and nanometer-sized graphene fragments, and these graphene sheets are curved like buckybowls due to the steric hindrance of the template nanochannels. It contains some oxygen-containing functional groups bound to the edges of the graphene layers. The presence of curved graphene layers in ZTCs was thought to be responsible for their high-temperature ferromagnetism [70]. These curved graphene layers and ordered array of micropores were confirmed to be preserved **at** up to 380°C by X-ray diffraction and Raman spectroscopy [71]. A slightly modified model was recently proposed by the same authors [72].

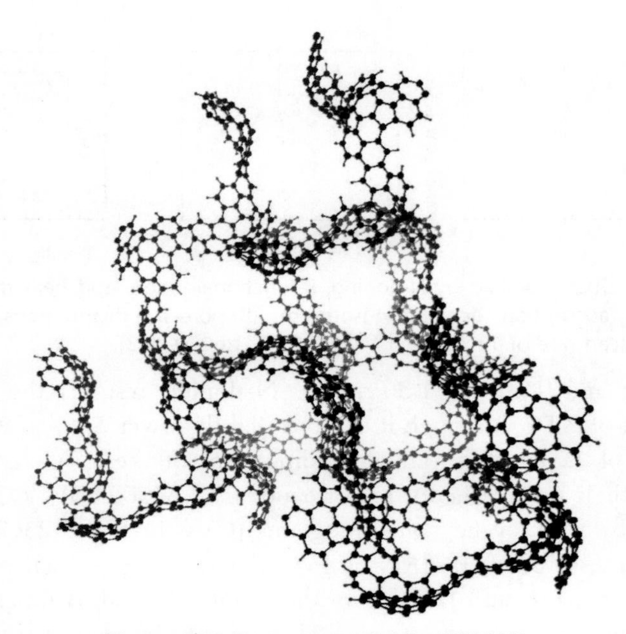

Figure 2.24 Structure model for ZTC [69].

By using NH_4-form zeolite Y, microporous carbons with nitrogen-containing functional groups were obtained by the impregnation of phenol-formaldehyde and carbonization above 900°C [73]. S_{BET} increased with increasing heat-treatment temperature, from more than 1750 m^2/g after 900°C treatment to 3681 m^2/g after 1100°C treatment. The particle size of the template zeolite was shown to have an influence on the pore structure of the resultant carbons: small particles of zeolite gave carbon with a slightly higher S_{BET}, V_{total}, and carbonization yield than those prepared using large zeolite particles [74]. A laboratory-prepared zeolite EMC-2, which had interconnected cages along the *a*-axis and straight channels along the *c*-axis, could be replicated by two-step filling with FA and propylene, resulting in carbon with highly ordered micropores and a high S_{BET} of 4000 m^2/g and V_{micro} of 1.8 cm^3/g [75], as shown in Fig. 2.25.

ZTCs were synthesized using two zeolites, zeolite Y and 13X, via acetonitrile and ethylene CVI at temperatures of 550–1000°C [76]. N_2 adsorption–desorption isotherms for the carbons prepared using zeolite 13X with acetonitrile, zeolite Y with acetonitrile, and zeolite Y with ethylene are shown in Fig. 2.26A–C, respectively. When acetonitrile CVI was performed at 850°C, zeolite Y gave slightly higher S_{BET} than zeolite 13X; with S_{BET} of 1918 and 1589 m^2/g, and S_{micro}, 662 and 194 m^2/g for

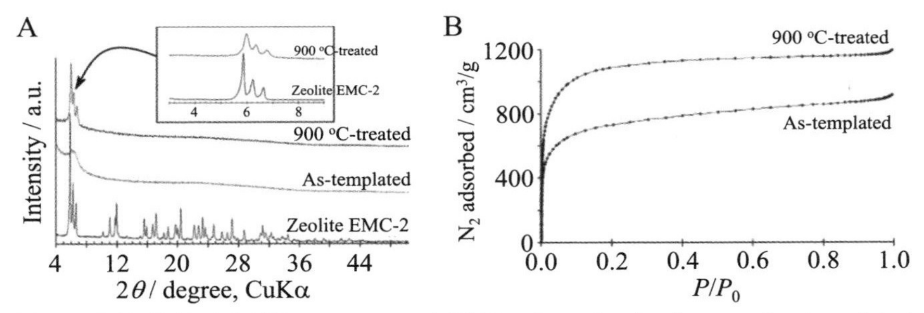

Figure 2.25 ZTCs: (A) XRD patterns and (B) N_2 adsorption isotherms of as-templated and 900°C-treated carbons in comparison with zeolite template [75].

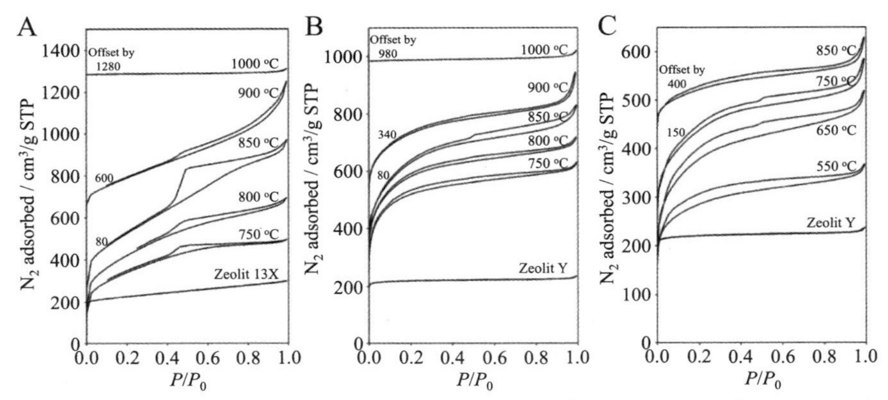

Figure 2.26 N_2 adsorption–desorption isotherms of ZTCs: (A) zeolite 13X with acetonitrile-CVD, (B) zeolite Y with acetonitrile-CVD, and (C) zeolite Y with ethylene-CVD [76].

zeolite Y and 13X, respectively. When ethylene was used, zeolite Y gave S_{BET} of 1307 m^2/g and S_{micro} of 533 m^2/g. The synthesized carbons were studied on hydrogen adsorption capacity to discuss the effect of N-doping in the adsorbent carbons.

2.1.2.2 Other hard templates

The Li-form of taeniolite was intercalated with either hydroxyaluminum [$Al_2(OH)_5Cl$] or hydroxyaluminum-zirconium ($Al_{1.2}Zr_{0.3}Cl$) and then saturated with a benzene solution of 80% furfuryl alcohol (FA), followed by polymerization of FA and carbonization at 700°C [77]. The resultant carbons were microporous, depending strongly on the water content of the inorganic matrix, and showed molecular sieving properties. Porous clay heterostructures prepared from natural montmorillonite by intercalation of

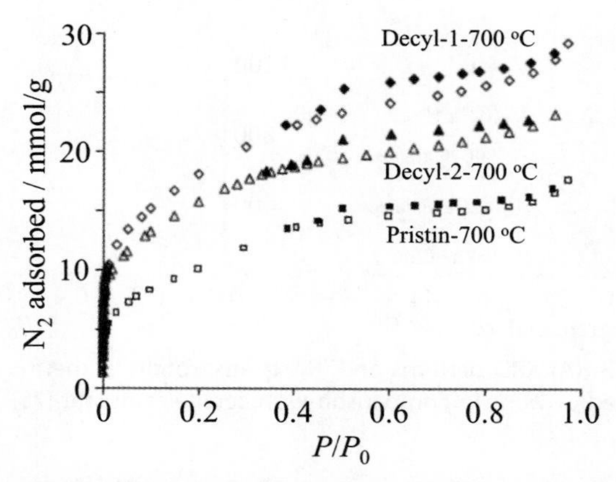

Figure 2.27 N_2 adsorption–desorption isotherms of the carbons prepared from furfuryl alcohol using a porous clay heterostructure template at 700°C. Sample code means the surfactant, FA-impregnation time, and carbonization temperature [78].

different surfactants, octyl-, decyl-, and dodecyl-amines, and tetraethyl orthosilicate (TEOS), were also used as a template to prepare nanoporous carbons [78]. A carbon precursor FA was impregnated into the template, polymerized at 95°C, and carbonized at 500–800°C, followed by removal of the template with 10% HF. The N_2 adsorption–desorption isotherms for the porous carbons obtained at 700°C are shown in Fig. 2.27. The pore structure of the resultant carbons consisted of pores with widths of about 2 nm and large mesopores (measured by α_s plot), the former being thought to be due to the inner structure of the template clays and the latter to aggregation of the clay particles. The carbon obtained using decyl-amine and carbonizing at 700°C (Decyl-1-700°C) gave the highest S_{BET} of 1469 m^2/g and V_{micro} of 0.68 cm^3/g. The carbon subjected to two impregnation/polymerization cycles (Decyl-2-700°C) exhibited lower S_{BET} and V_{micro} than Decyl-1-700°C.

A microporous carbon monolith was prepared by carbonization of the mixture of cyclodextrin and tetramethyl orthosilicate with sulfuric acid catalyst, which gave S_{BET} of 1970 m^2/g, V_{total} of 1.0 cm^3/g, and micropore width centered at 1.6 nm [79].

2.1.3 Precursor design

2.1.3.1 Polymer blending

2.1.3.1.1 Polyvinylpyrrolidone

Carbon molecular sieve membranes were prepared by blending polyvinylpyrrolidone (PVP) with different molecular weights (M_w) and polyimide (PI), which is synthesized from benzophenone tetracarboxylic dianhydride (BTDA) and 4,4-oxydianiline (ODA) [80,81]. In the dimethyl sulfoxide (DMSO) solution, BTDA and ODA were mixed to prepare a solution of 20 wt.% polyamic acid (PAA), followed by the addition of PVP. The PVP/PAA solution was cast onto a glass plate and then thermally imidized at 250°C to obtain thin films of PVP/PI polymer blend with a thickness of about 40 μm. In Fig. 2.28A, thermogravimetric (TG) curves for the PI films containing 5 and 10 wt.% PVP are compared with those of a PI film without PVP blending [80]. The TG curves for the blends show

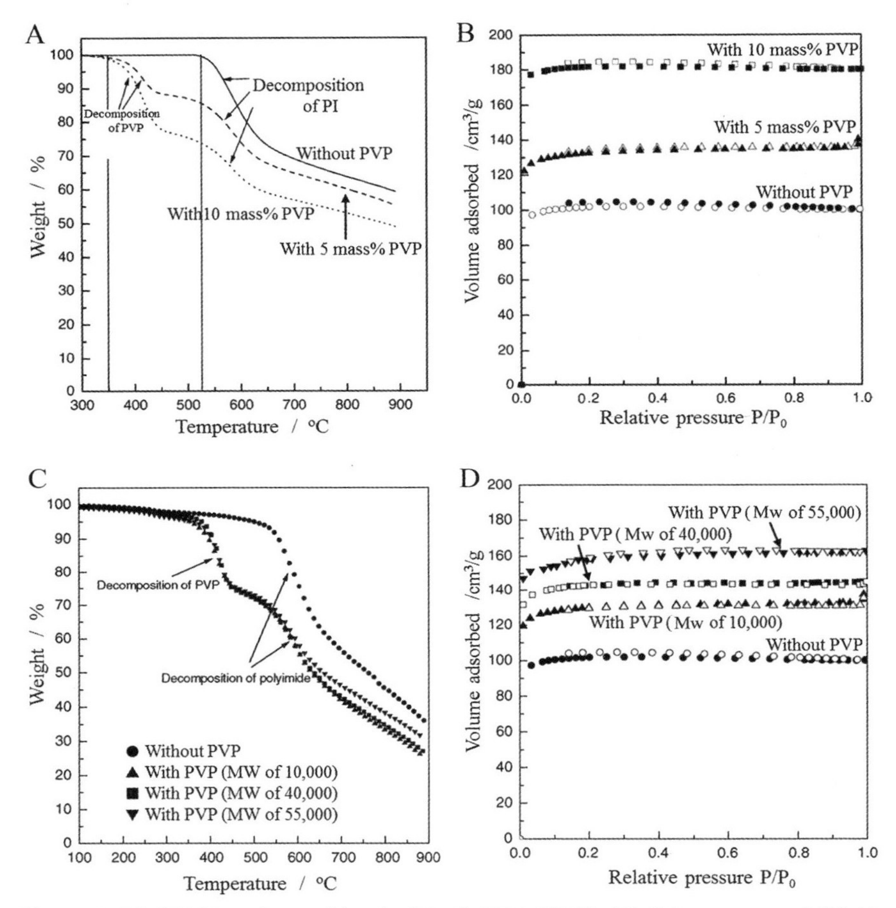

Figure 2.28 PVP/PI polymer blends (PI of BTDA/ODA): (A) TG curves and (B) N_2 adsorption–desorption isotherms for blends with different PVP contents, and (C) TG curves and (D) N_2 adsorption–desorption isotherms for blends with PVP having different M_w's [80,81].

well-separated two-step decompositions, the first and second being due to the decomposition of PVP and PI, respectively. In Fig. 2.28B, N_2 adsorption—desorption isotherms at 77K are shown for the films after carbonization at 700°C. In Fig. 2.28C and D, TG curves and N_2 isotherms are shown for blends with 5 wt.% PVP with different M_w's [81]. No temperature shift for the decomposition of PVP and PI was observed in the TG curves for the blends regardless of the M_w's of PVP. The N_2 adsorption—desorption isotherms in Fig. 2.28B and D suggest that the blending of PVP works effectively to increase micropores in the carbon films, V_{micro} increasing with increasing amount and M_w of PVP. The TG curves of the blends suggest no interaction between the matrix-forming PI and labile PVP. The imidization process of PAA to PI was indicated to be important to prepare carbon membrane from the materials composing different molecular structures of PAAs [82].

Microporous carbon fibers were prepared from the blend of PI (PMDA/ODA) with PVP at PVP/PI weight ratios of 0.2—2 by electrospinning and carbonizing at 900°C after imidization at 300°C [83]. Pyrolysis of the blends occurred in two steps, pyrolysis of PVP at around 400°C and that of PI at around 550°C. The micropore content increased with increasing PVP content up to the PVP/PI ratio of 1.0, and then decreased. The micropores in the resultant carbons consisted mainly of ultramicropores (<0.7 nm size) and the volume V_{ultra} increased from 0.14 cm^3/g for the carbon prepared without PVP to 0.228 cm^3/g for the carbon prepared from the blend with a PVP/PI ratio of 0.1. Hollow carbon nanofibers consisting of microporous walls were prepared using the coaxial electrospinning technique [84]. Styrene-acrylonitrile copolymer and PVP/PI (PMDA/ODA) blend (PVP/PI = 0.2) were used as the core and shell of nanofibers, through imidization at 300°C and carbonization at 900°C. The effectiveness of the hollow and microporous texture of carbon nanofibers as supercapacitor electrodes was discussed on the bases of the measurements of capacitive behavior in 6 M KOH electrolyte.

The hollow fibers were prepared by wet spinning of a polymer blend of poly(ethylene imide) (PEI) with PVP (17 mass% PEI and 13 mass% PVP blended in NMP) and formed into the membranes (webs) [85]. The membranes were carbonized at 800°C to obtain nanoporous membranes. The TG curves of PVP and PEI in Fig. 2.29A show marked contrast in pyrolysis behavior. PVP decomposes abruptly above 370°C and gives a very small amount of carbon residues above 500°C (about 4 wt.%), while PEI decomposes in the range of 475—640°C and gives a relatively high carbon

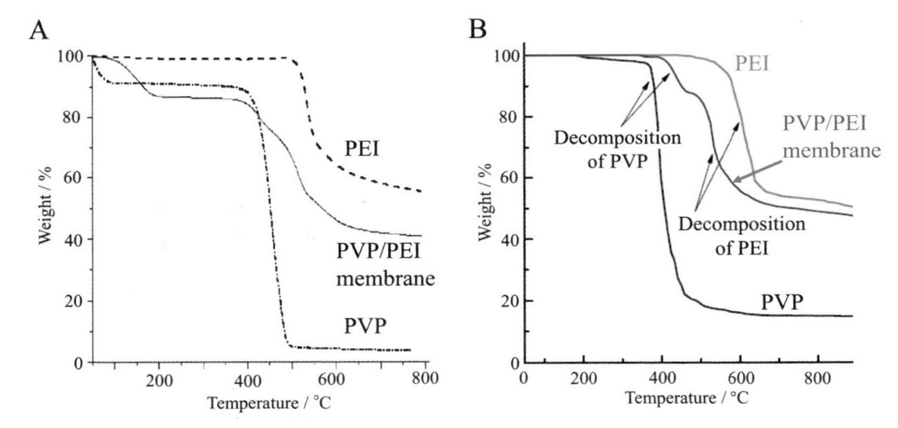

Figure 2.29 TG curves for PEI, PVP, and the membrane prepared from their blend with a heating rate of 10°C/min: (A) the PVP/PEI blend membrane without stabilization [85] and (B) the membrane stabilized at 300°C in air in advance [86].

residue (about 55 wt.%). The TG curve of the polymer blend membrane shows weight loss in a broad range of temperature after the desorption of water below 200°C, it being recognized that the decomposition of PVP occurs before that of PEI. Gas permselectivity of carbon membranes was improved by blending PVP with PEI. The membrane of PVP/PEI blend stabilized at 300°C for 30 min in air in advance, however, it demonstrates a slightly different aspect of the TG curve [86]. As shown by comparing with pure PVP and PEI in Fig. 2.29B, the TG curve for the polymer blend membrane shows two distinct weight losses due to PVP and PEI decomposition, slight shifting of the decomposition of PVP to a slightly higher temperature than pure PVP and that of PEI to a much lower temperature than pure PEI.

The polymeric membrane of the blend of poly(phenylene oxide) (PPO) with PVP was formed on the surface of a tubular macroporous α-alumina ceramic support (with an average pore size of 100 nm and porosity of 41.0%) and carbonized at varying temperatures (500–800°C) [87]. The polymer blend was prepared by mixing chloroform solutions of 2.4% PPO and 0.6% PVP with different M_w's of 10,000–90,000 (10–90K). PVP was rapidly degraded above 380°C and the degradation of PPO also occurred rapidly at around 420°C (slightly higher than for PVP) with a slightly higher carbon yield than PVP. The blends with PVP/PPO showed higher decomposition temperature than PVP. The micropore volume and surface area of the blend carbonized at 700°C, which were determined by CO_2

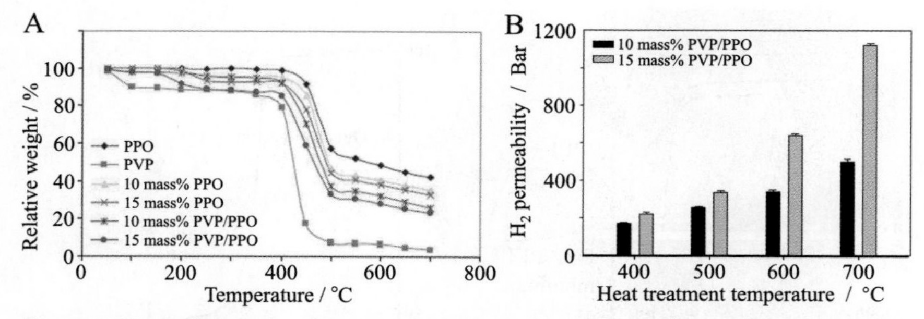

Figure 2.30 Carbon membranes prepared on the inner surface of a macroporous alumina tube by casting chloroform solution of 10 and 15 wt.% PPO, PVP, and a PVP/PPO blend: (A) TG curves in comparison with PPO and PVP resins and (B) H_2 permeability [88].

adsorption measurement at 25°C, were 0.14–0.21 cm^3/g and 188–380 m^2/g, respectively. The pyrolysis behavior of the PVP/PPO polymer blend (PVP content of 20 wt.%) spin-coated on the inner surface of a macroporous alumina tube was studied by focusing on the effect of the concentration of the blend in a chloroform solution (10 and 15 wt.%) [88]. The blend concentration in the solution did not affect the pyrolysis behavior, as shown in Fig. 2.30A, but enhanced H_2 permeability of the carbon membrane heat-treated at high temperatures, as shown in Fig. 2.30B. The H_2 permeability of the membrane prepared from a 15 wt.% solution is higher than that from a 10 wt.% solution and much higher than that from a 15 wt.% solution of PPO without PVP. The blending of PVP and solution concentration had a strong effect on the selectivity for gas pairs, such as H_2/N_2 and H_2/CH_4. Unfortunately, the pore structure in carbon membranes was not presented.

The DMF solution of the polymer blend of PVP/PAN at the weight ratio of 1/1 was electrospun to fibers with about 170 nm diameter and oxidized at 250°C, followed by leaching-out of PVP under supercritical condition at 250°C and 8 MPa, and then being carbonized at 800°C [89]. The matrix PAN with M_w of 150,000 and PVP with M_w of 1,300,000 were selected. Although the surface of the fibers became rough after leaching (Fig. 2.31A), it was still nonporous, as can be seen from the adsorption–desorption isotherm in Fig. 2.31C. After carbonization at 800°C, micropores were formed in the fiber (Fig. 2.31C) with macropores, albeit with shrinkage (Fig. 2.31B).

2.1.3.1.2 Poly(methyl methacrylate)

Polymer blends of phenol formaldehyde (PF) and poly(methyl methacrylate) (PMMA) were synthesized in DMF solution at 80°C from PF

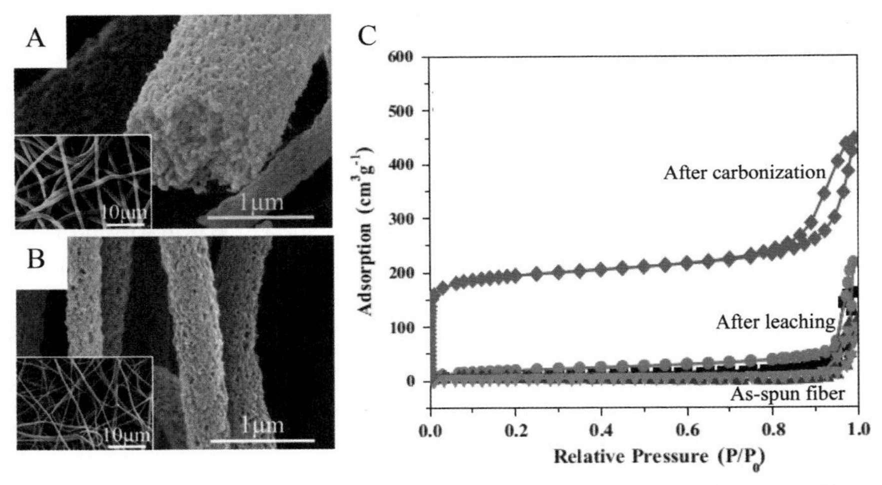

Figure 2.31 Fibers prepared from PVP/PAN blend: (A) SEM image of PVP/PAN fibers after leaching under supercritical condition, (B) SEM image of PVP/PAN fibers after carbonization at 800°C, and (C) N_2 adsorption–desorption isotherms [89].

prepolymer resins, methyl methacrylate (MMA), and ethylene glycol dimethacrylate as well as 2,20-azobisisobutyronitrile by polymerization, followed by curing of PF prepolymer and evaporation of solvent DMF at 160°C in Ar and carbonization at 800°C in N_2 [90]. Solution mixing of two component polymers was shown to be effective for the micropore development, as shown in Fig. 2.32. The carbon derived from this polymer blend delivered V_{total} of 0.50 cm^3/g with V_{micro} of 0.31 cm^3/g, although

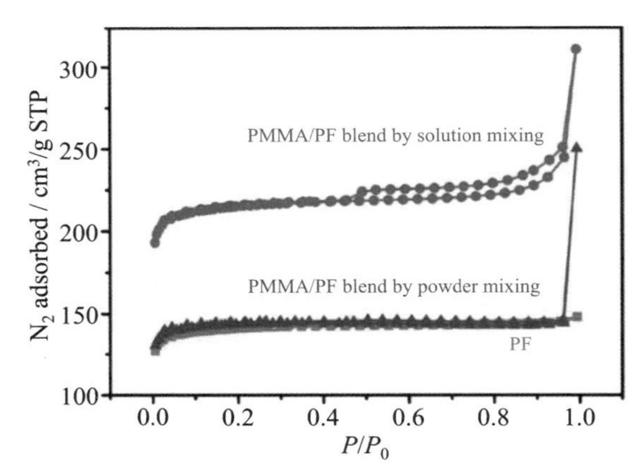

Figure 2.32 N_2 adsorption–desorption isotherms for the carbons derived from PMMA/PF blends via solution and powder mixtures [90].

that derived from the powder mixture of PF and PMMA delivered 0.42 with 0.22 cm^3/g, respectively.

Polymer blends of PF with PMMA were prepared by mixing of resole- and novolac-type phenol resins in a 60 wt.% methanol solution with PMMA in acetone solution, followed by removing the solvents in vacuum at room temperature, where hexamethylenetetramine (cross-linking agent) was added for the solution of novolac-type resin and the composition of the blends was PMMA/PF ratio of 37/63 by weight [91]. By dissolution of PMMA from the compressed mixtures after curing at 180°C, the porous PFs were obtained to be carbonized at 500–700°C in N_2 flow. The resultant PF-derived carbons were microporous, as shown by the N_2 adsorption–desorption isotherms for the resol- and novolac-type PFs in Fig. 2.33A and B, respectively. Blending of PMMA into PF was shown to be effective for increasing micropores, as well as mesopores, in the resultant carbons. For PF, *m*-phthalic acid (PA) and trimesic acid (TMA) were reported to work as micropore-forming agents when the mixtures of PF with either PA or TMA (at a ratio of 2/1 by weight) were directly carbonized at 1000°C [92].

Porous carbons with a broad pore size distribution at around 2 nm were fabricated from the mixture of PAN with PMMA (PMMA/PAN ratio of 3/1 by weight) in DMF by exchanging the solvent for water, followed by drying, stabilizing at 280°C in Ar, and carbonizing at 900°C [93]. The resultant porous carbons could be loaded by S through a chemical deposition reaction between $(S_2O_3)^{2-}$ and HCOOH, the loaded S reaching 53.7 wt.% after removing S deposited on the surface of the carbon particles by heating at 160°C.

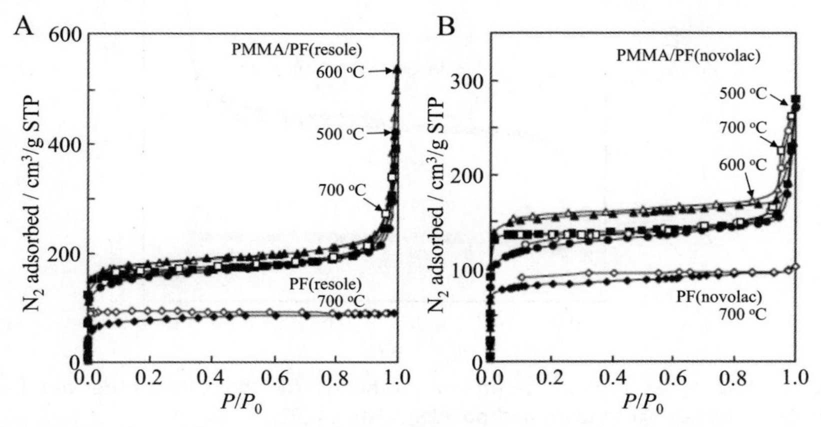

Figure 2.33 Microporous carbon prepared from polymer blends of phenol resins with PMMA: (A) resole-type phenol and (B) novolac-type phenol resins [91].

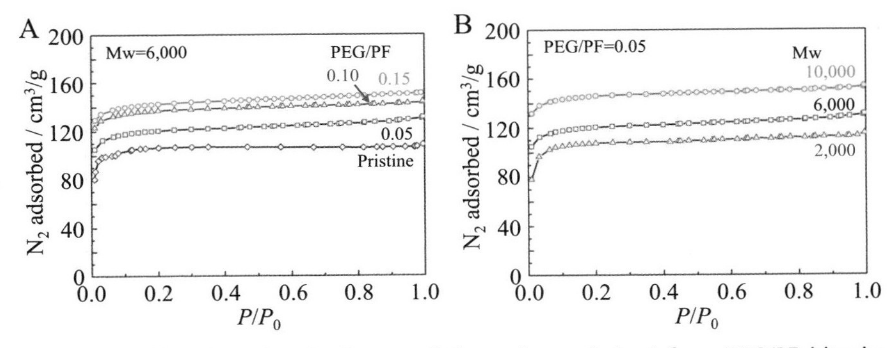

Figure 2.34 N_2 adsorption isotherms of the carbons derived from PEG/PF blends: (A) different contents of PFG with M_w of 6000 and (B) PEG with different M_w's with a constant PEG content [94].

2.1.3.1.3 Poly(ethylene glycol)

Polymer blends of novolac-type PF with poly(ethylene glycol) (PEG) via its 40 wt.% DMF solution gave microporous carbons after carbonization at 800°C in N_2 flow [94]. Blending of PF with PEGs of different molecular weights (M_w) was performed in PEG/PF weight ratios of 0.05–0.15 and sprayed onto a support to form membranes, followed by carbonization. The resultant carbon membranes are microporous, their N_2 adsorption isotherms being typical type I, as shown in Fig. 2.34. The carbon obtained from the blend with PEG/PF ratio of 0.05 using PEG with a M_w of 10,000 delivered V_{total} of 0.24 cm^3/g with V_{micro} of 0.23 cm^3/g.

2.1.3.1.4 Poly(ethylene oxide)

Blending of poly(ethylene oxide) (PEO) with PAN was effective in creating micropores in the resultant carbon. Blends in different ratios (PEO/PAN of 0/1, 1/2, 1/1, and 2/1 by weight) were prepared in a DMF solution and subjected to electrospinning [95]. The resultant fibers were stabilized at 280°C in O_2 and then carbonized at 1000°C. The fibers prepared from the blend with a PEO/PAN ratio of 2/1 had a highly developed microporous structure, S_{BET} of 2054 m^2/g, and V_{total} of 0.94 cm^3/g with V_{micro} of 0.77 cm^3/g, although those derived from PAN without PEO showed S_{BET} of 107 m^2/g and V_{total} of 0.09 cm^3/g.

2.1.3.1.5 Poly(vinyl butyral)

The porous carbon fibers were prepared from the polymer blends of novolac-type PF with poly(viny1 butyral) (PVB) in methanol by

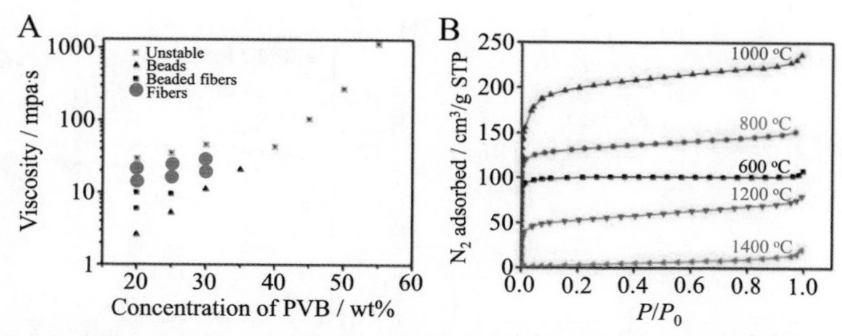

Figure 2.35 Electrospinning conditions for the polymer blends of PF with PVB (A) and N_2 adsorption–desorption isotherms for the carbon fibers prepared from the PVB/PF blend at different temperatures (B) [98].

melt-spinning at 180–190°C in a specially designed stabilization solution and subsequent carbonization at 900°C in N_2 [96]. The resultant carbon fibers were microporous, having a broad pore size distribution with a peak at about 0.6 nm. Novolac-type PF fibers were electrospun from a 30 wt.% methanol solution with 0.9 wt.% PVB and cured in a solution of formaldehyde and HCl, followed by carbonization at 800°C [97]. The resultant carbon fiber was microporous, showing a typical type I N_2 adsorption–desorption isotherm, high S_{BET} of 827 m^2/g, and high V_{micro} of 0.9 cm^3/g. For electrospinning of resole-type PF fibers, however, the addition of high-molecular-weight PVB was essential for the stable preparation of bead-free resole fabrics [98]. The regions of viscosity and PVB concentration of the precursor solution are significantly limited for stable spinning of fibers without beads, as shown in Fig. 2.35A. The resultant carbon fibers are microporous, as shown in the N_2 adsorption–desorption isotherms in Fig. 2.35B. The carbon fibers obtained at 1000°C gave S_{BET} of 858 m^2/g and V_{total} of 0.365 cm^3/g with V_{micro} of 0.326 cm^3/g.

2.1.3.1.6 Pitches

Polymer blends of polyacrylonitrile (PAN) with a pitch were prepared by mixing a DMF solution of PAN with a THF solution of pitch at a pitch/PAN weight ratio of 30/70, which was electrospun to produce the fiber webs [99]. The resultant webs were stabilized in air and then carbonized at 1000°C in N_2. The pitch used was fabricated from a heavy residual oil that is a by-product formed during naphtha cracking and had a M_w of 2380. The resultant nanofibers had average diameters of 150 nm with a core–shell texture, as shown in Fig. 2.36A and B, respectively, in which the carbon derived from low-molecular-weight pitch is preferentially located in the

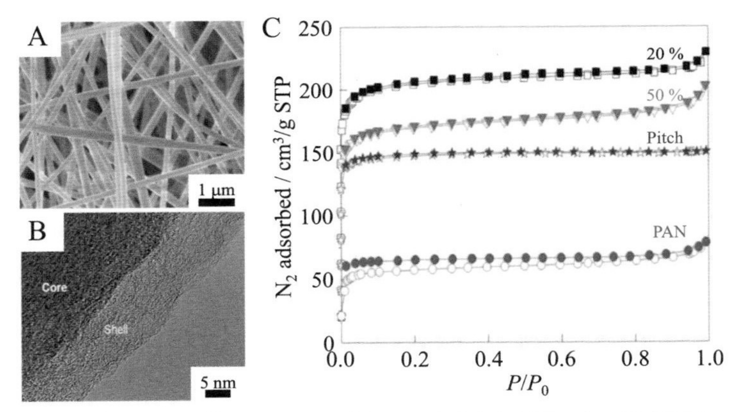

Figure 2.36 Carbon nanofibers prepared from pitch/PAN blends: (A) SEM image, (B) TEM image, and (C) N_2 adsorption–desorption isotherms of carbon fibers prepared using 20% and 50% THF solutions of pitch in comparison with carbon fibers from the pristine pitch and PAN [99].

shell, while that from PAN is in the core. The nanofibers are microporous as shown by their N_2 adsorption–desorption isotherms in Fig. 2.36C. The nanofibers prepared using 20 wt.% THF solution of the pitch delivered S_{BET} of 966 m^2/g and V_{total} of 0.379 cm^3/g.

Polymer blends of a coal tar pitch with either PAN, PVP, or PVPox (PVP thermally degraded at 300°C in air) at different weight ratios were prepared by dispersing polymer particles in liquid pitch at 250°C and then carbonized at 520°C, followed by steam activation at 800°C [100–102]. The activated carbon fibers prepared from the blend at a pitch/PAN ratio of 9/1 delivered V_{total} of 0.285 cm^3/g with V_{micro} of 0.207 cm^3/g and that from the blend of pitch/PVPox in 1/1 delivered 0.380 cm^3/g with 0.306 cm^3/g, although the pitch itself gave 0.112 cm^3/g with 0.086 cm^3/g, respectively [102].

2.1.3.2 Molecular design

2.1.3.2.1 Labile functional groups

Aromatic polyimides (PIs), which are synthesized from anhydride with diamine, have a variety of repeating units of imide framework with different labile functional groups (pendant groups) and have been used as the precursors for porous carbon membranes [82]. PIs with no labile functional groups were successfully used as the precursors to highly crystalline graphite films [103].

Aromatic polyimides were synthesized from a dianhydride, 3,3′,4,4′-biphenyltetra-carboxylic dianhydride (*s*-BPDA) and different aromatic diamines [104]. The repeating units of each dianhydride and diamine are shown in Fig. 2.37A, and the reaction scheme is shown using a couple of *s*-BPDA and TFMB in Fig. 2.37B to obtained PI membrane by casting on a PET film and following imidization at 300°C. The synthesized PIs are composed of the imide frameworks with different numbers of phenyl rings with the pendant groups of $-CF_3$ and $-CH_3$ (labile functional groups). They are characterized by phenyl carbon atom concentration in the repeating unit of the polyimides, M_{phenyl}/M_{unit}, ranging from 48.7% for the polyimide *s*-BPDA/Bis-A-AF to 71.3% for *s*-BPDA/FDA. PI membranes with thickness of about 40 μm were carbonized at 1000°C in N_2 flow by sandwiching between Al_2O_3 thin plates. As shown by the N_2 adsorption–desorption isotherms in Fig. 2.38A, carbonized PI membranes are microporous, typical type I isotherms with a sharp increase in N_2 adsorption at very low P/P_0, and strongly dependent on PI, in other words, on the chemical structure of the precursor PI. S_{micro} calculated as a balance between S_{total} and S_{ext} determined by α_s plots is plotted against M_{phenyl}/M_{unit} in Fig. 2.38B, revealing a strong dependence of S_{micro} on M_{phenyl}/M_{unit}. The results in Fig. 2.38B also suggest a strong effect of the presence of the pendant group $-CF_3$ in the repeating units on S_{micro}. The carbon membranes prepared from *s*-BPDA/Bis-A-AF and *s*-BPDA/TFMB deliver S_{micro} of 1169 and 1405 m^2/g, respectively, by simple carbonization without any activation process despite their low M_{phenyl}/M_{unit} values being less than 50%.

Polyimides containing pendant $-CF_3$ group gave microporous carbon membranes by single-step carbonization at 600°C [105]. In Fig. 2.39, S_{micro} determined from α_s plots is plotted against F content in the repeating units. The precursor polyimide containing 31.3 wt.% F gives S_{micro} of about 1340 m^2/g and V_{micro} of 0.44 cm^3/g without any activation process, where the micropore size is about 0.55 nm. These microporous carbon membranes adsorbed a large amount of water vapor [106].

By changing the polymerization conditions, solvents for PAA and its concentration, the morphology of the particles of a PI, benzidine and 3,3′,4,4′-benzophenone tetracarboxylic dianhydride (BZD/BTDA), can be easily controlled to be flower-, disk-, lantern-, and sphere-like, and could be carbonized by retaining the morphology with only slight shrinkage [107]. Carbonization of PIs was performed at 900°C for 1 h in N_2, and then in NH_3 at the same temperature for another 1 h. The carbon particles with

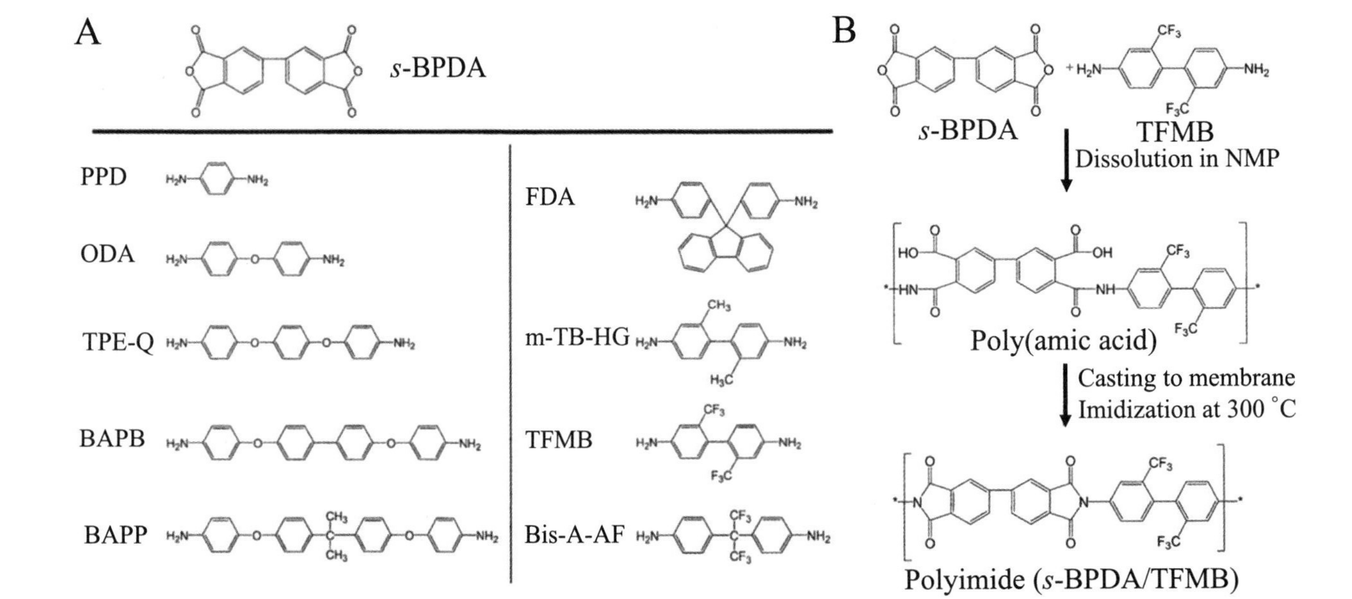

Figure 2.37 (A) Various polyimides by coupling an anhydride (*s*-BPDA) with different amines and (B) reaction scheme by employing dianhydride *s*-BPDA and aromatic diamine TFMB to synthesize polyimide film (*s*-BPDA/TFMB) [104].

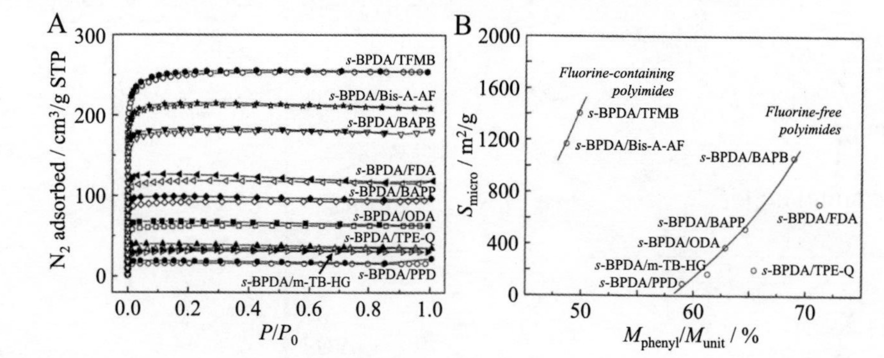

Figure 2.38 Carbon films prepared from the aromatic polyimides with different M_{phenyl}/M_{unit} values: (A) N_2 adsorption–desorption isotherms and (B) dependences of S_{micro} on M_{phenyl}/M_{unit} [104].

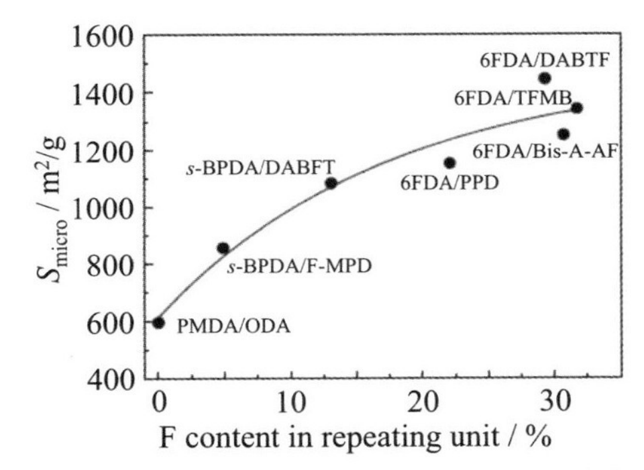

Figure 2.39 Microporous surface area (S_{micro}) as a function of fluorine content in the repeating unit of the precursor polyimides [105].

flower-, disk-, and lantern-like morphologies were synthesized using different solvents with different concentrations of PAA, DMF with a concentration of 81.8 mg/mL, NMP at 9.1 mg/mL, and NMP at 81.8 mg/mL, respectively. The resultant carbons with flower-, disk-, lantern-, and sphere-like morphologies are porous, having S_{BET} of 1375, 1250, 1187, and 983 m^2/g, respectively (Fig. 2.40A). The pore size distributions of these carbons show the peaks centering at 1.6, 1.9, and 3.9 nm, as shown in Fig. 2.40B. They contained approximately 3 at.% N with mainly graphitic and pyridinic configurations. These carbons, particularly flower-like carbon, delivered high oxygen reduction reaction (ORR) activities.

Highly microporous carbons were obtained from the PI synthesized from 6FDA and 3,30-dihydroxybenzidine (HAB) by carbonization at 1000°C in Ar [108]. By blending PI with poly(ethylene oxide) (PEO), mesopores could be introduced into the resultant carbons, depending on

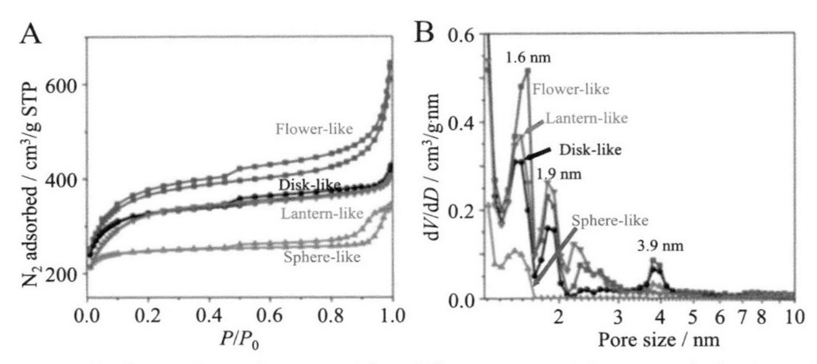

Figure 2.40 PI-derived carbons with different particle morphologies: (A) N_2 adsorption–desorption isotherms and (B) pore size distributions [107].

the PEO content. The carbon obtained without PEO delivered S_{BET} of 2467 m^2/g and S_{micro} of 2309 m^2/g, while the carbon obtained by mixing 20 wt.% PEO delivered 1827 and 1339 m^2/g, respectively.

2.1.3.2.2 Defluorination

The decomposition of polytetrafluoroethylene (PTFE) was found to occur by contacting with lithium amalgam at ambient temperature to form highly dispersed amorphous carbon with LiF nanoparticles [109]. This reaction between PTFE and 0.9 at% Li-amalgam was understood to proceed electrochemical defluorination of PTFE cathode coupled with amalgam anode (electrochemical reduction). The mixtures of LiF with highly dispersed amorphous carbon were prepared through defluorination of PTFE at a temperature of 100°C [110]. The resultant carbons after rinsing LiF in water had a bulk density of 0.123–0.143 g/cm^3 and their surface areas calculated from Ar adsorption isotherms at 77K reached 3300–4150 m^2/g, the isotherms showing no hysteresis. Three procedures for isolation of carbon were employed for a mixture of carbon and LiF which was prepared by defluorination of PTFE with 0.04 wt.% Li-amalgam at 100°C [111]. Carbon **A** was recovered from the mixture by complete washing out of LiF with 0.005 M HCl and with water at room temperature, and then drying in a vacuum at 300°C. Carbon **B** was obtained by heating carbon **A** at 950°C in He for 1 h. Carbon **C** was recovered from the mixture by heat treatment at **950°C** for 1 h in He and then washin out of LiF. Elemental composition, immersion density measured using toluene, and pore structure parameters measured from N_2 adsorption isotherms are shown for these three carbons, **A**, **B** and **C**, in Table 2.6. Polymeric carbon chains were formed by electrochemical reduction of linear PTFE molecules with Li-amalgam at

Table 2.6 Elemental composition, density, and pore structure parameters of carbons isolated using different procedures [111].

Isolated carbons	Elemental composition (wt.%)				Density[a] (g/cm^3)	S_{BET} (m^2/g)	V_{total} (cm^3/g)
	C	H	O	Ashes			
A	81.2	1.02	17.5	0.3	2.002	2940	1.62
B	97.5	0.41	1.7	0.4	2.114	2370	1.33
C	96.7	0.15	2.8	0.3	2.106	2170	1.54

A, washing with HCl and water at room temperature, followed by drying at 300°C in vacuum; **B**, heating the carbon **A** at 950°C in He; **C**, Heating at 950°C in He and then sieving to exclude large LiF, followed by the same procedure as for **A**.
[a]Immersion density (toluene).

100°C, which were encapsulated in LiF sheaths. After dissolving LiF in water at room temperature, carbon chains are bound to form skeletons including near 20 wt.% oxygen, which are hindered from coalescing with each other, resulting in the retention of porous texture of carbon even after 950°C annealing (in the cases of carbon **A** and subsequent carbon **B**). By annealed at 950°C before carbon isolation (the case of carbon **C**), however, LiF melted and coagulated to large spherical particles (the melting point of LiF is 842°C), the carbon chains have more opportunities to form large skeletons and to grow larger crystallites, which lowers the possibility of oxygen contamination during the isolation procedure (low O and H contents). On carbon **C**, marked hysteresis in the N_2 adsorption–desorption isotherm was observed, revealing the formation of mesopores, whereas carbons **A** and **B** showed no hysteresis. The formation of mesopores in carbon **C** may make S_{BET} lower.

PTFE film (100 μm thick) was compressed with Li metal foil (200 μm thick) under a pressure of 4 MPa in Ar for defluorination and then an excess Li metal was removed by rinsing with methanol, followed by annealing at 700°C to enhance the development of crystallinity of carbon and finally by extracting LiF with diluted HCl [112], as illustrated in Fig. 2.41A. The resultant carbon film was porous, S_{BET} reaching 1045 m^2/g. In Fig. 2.41B,

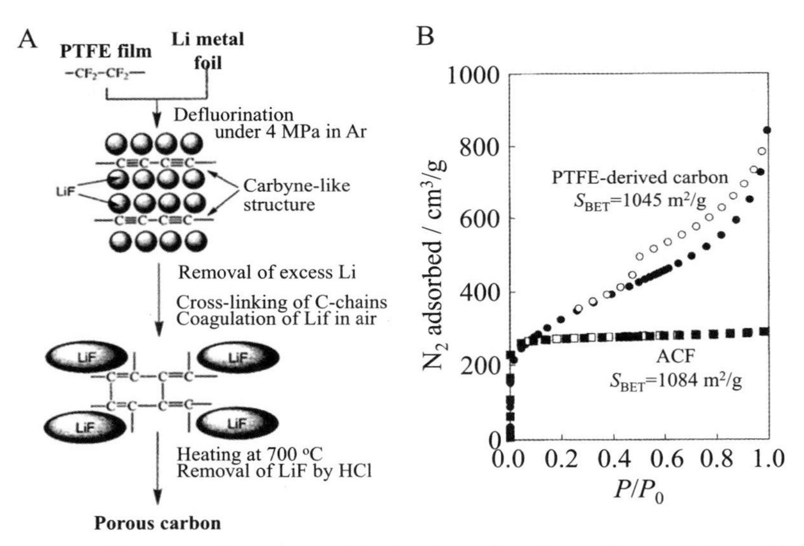

Figure 2.41 Porous carbon by defluorination of polytetrafluoroethylene (PTFE): (A) scheme of preparation and (B) N_2 adsorption–desorption isotherm in comparison with ACF [112].

the N_2 adsorption–desorption isotherm for the resultant porous carbon is compared with that for a phenol-based ACF having a similar S_{BET} (1084 m^2/g). The comparison between these two porous carbons suggests that the resultant PTFE-derived porous carbon has a relatively large amount of mesopores despite the amount of micropores being almost the same as in the ACF. The additionally formed mesopores are thought to be due to LiF particles grown during the annealing at 700°C. This result suggests that the annealing process governs the mesopore formation in the resultant carbon, in other words, the size and volume of mesopores can be tuned by the annealing process.

Microporous carbon fibers and films were fabricated by dehydrofluorination of poly(vinylidene fluoride) (PVDF) [113]. PVDF fibers (7–10 μm diameter) were prepared from bicomponent fibers consisting of PVDF core fibers and polystyrene (PS) matrix, which were melt-spun at 290°C, by soaking in THF at 50°C to dissolve the matrix PS. PVDF films were prepared by hot-pressing at 200°C. Dehydrofluorination of these PVDF fibers and films was performed using 1,8-diazabicyclo[5,4,0]undec-7-ene (DBU) in its 1.0 M solution of DMF/ethanol (**2/3** by volume) at 70°C. The resultant dehydrofluorinated fibers and films were carbonized at different temperatures in N_2 and then activated by CO_2 at 850°C. In Fig. 2.42A, the weight loss of the weight loss of the film, ΔW_{HT}, with different dehydrofluorination degrees ΔW_{HF} during heat treatment (carbonization and following CO_2-activation) is plotted against the carbonization temperature. In the temperature range of 300–400°C, rapid

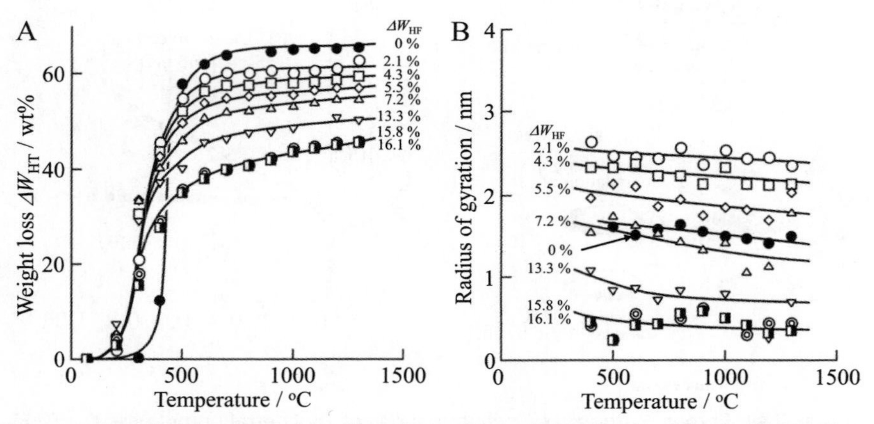

Figure 2.42 Microporous carbon films derived from PVDF film by dehydrofluorination: carbonization temperature dependences of (A) weight loss during carbonization and following activation ΔW_{HT} and (B) radius of gyration measured on the resultant porous carbon films for the films with different dehydrofluorination degrees ΔW_{HF} [113].

decomposition occurs as a steep increase in the weight loss ΔW_{HT} and pores are developed in the carbon. The pore size depends strongly on ΔW_{HF}, as shown by the gyration radius (pore radius) measured by small-angle X-ray diffraction against temperature in Fig. 2.41B; the pore size at 1300°C decreases from about 2.3 nm for the film with ΔW_{HF} of 2.1% to 0.3 nm for the film dehydrofluorinated completely (ΔW_{HF} above 15%). On PVDF fibers, almost the same results were obtained, although the pore sizes on the fibers with ΔW_{HF} of 7.2%–15% were slightly larger.

PVDF fiber webs were prepared from the DMF/water mixed solution with 6 wt.% PEO by electrospinning [114]. The resultant PVDF fiber webs were soaked in a mixture of DMF and methanol (9:1 in volume) with different amounts of 1,8-diazabicyclo[5.4.0]undec-7-ene (DBU) at 50°C for several hours to be dehydrofluorinated and stabilized. The dehydrofluorinated fibers were then carbonized at 1000°C in N_2. PVDF fibers before dehydrofluorination were melted by heating to elevated temperatures resulting in carbon chars. By dehydrofluorination in a solution of DBU/fiber of 1, the fiber morphology is retained, while the interior of the fibers is not porous. Dehydrofluorination in the solution with DBU/fiber of 10, i.e., more severe dehydrofluorination, resulted in pore development in the interior of the fibers by retaining fibrous morphology.

2.1.3.2.3 Porous organic frameworks

A porous organic framework (POF) was carbonized at different temperatures in the range of 350–450°C to improve the efficiency of CO_2 capture [115]. By heat treatment at 350–450°C, the amount of N_2 adsorbed decreases markedly with increasing heat treatment temperature (HTT) by retaining the type I isotherm character, as shown in Fig. 2.43; S_{BET}

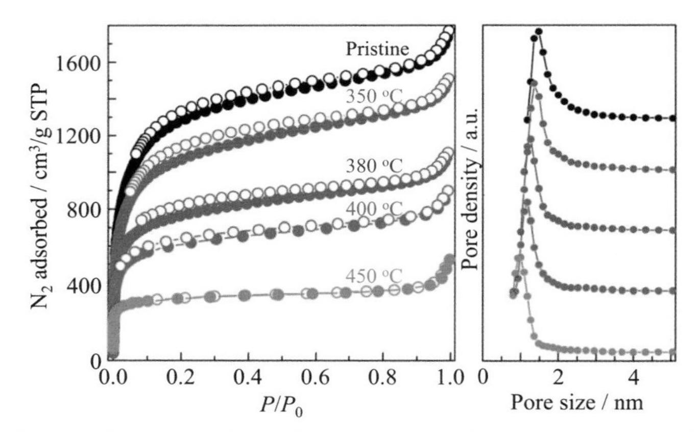

Figure 2.43 N_2 adsorption–desorption isotherms of a porous organic framework (POF) carbonized at different temperatures with pore size distributions [115].

decreased from 5300 to 1191 m^2/g by 450°C treatment. With increasing HTT, the micropore size decreases from 1.48 nm for the pristine POF to 1.00 nm for the 450°C-treated one, as shown in Fig. 2.43B, and V_{total} decreases from 2.43 to 0.53 cm^3/g. CO_2 uptake increased from 46 cm^3/g for the pristine to 100 cm^3/g after the 450°C treatment. A POF synthesized from 1,3,5-trihydroxybenzene using $FeCl_3$ catalyst was carbonized at 600–800°C for 2 h in an N_2 flow after mixing with KOH via its solution at a KOH/POF weight ratio of 4/1 [116]. The resultant carbons were highly microporous, consisting mainly of ultramicropores. The carbon obtained at 700°C delivered S_{BET} of 2363 m^2/g with S_{micro} of 2155 m^2/g and V_{total} of 1.09 cm^3/g and exhibited high CO_2 uptake of 6.3 mmol/g at 0°C.

A POF was mixed with FA after grafting sulfonic acid, which served as the catalytic site for the polymerization of FA, and then carbonized [117,118]. The synthesis procedures are illustrated for the preparation of an N-doped microporous carbon at 1000°C in Fig. 2.44. The resultant carbon contained 5.6 wt.% N (about 3 at% N) and relatively high oxygen reduction reaction (ORR) activity in oxygen-saturated 0.1 M KOH [117]. By carbonization at 900°C, S_{BET} decreased to 1174 m^2/g and the micropore size to 0.54 nm, in comparison with 4246 m^2/g and 1.45 nm for the pristine POF [118]. A POF synthesized from a 1,4-dioxane solution of 1,3,5-triformylbenzene and *p*-phenylenediamine was carbonized at 800°C in N_2, resulting in N-doped microporous carbon having S_{BET} of 525 m^2/g with V_{total} of 0.67 cm^3/g [119]. The resultant carbons are composed of hollow spheres with diameters of about 140 nm, the same morphology as the pristine POF framework, and their N-content decreased from 14.3% in the pristine to 4.7% after 800°C treatment. Cobalt-loaded N-doped microporous carbons were prepared by carbonization of Co-containing covalent organic frameworks (COFs) synthesized from the mixture of 1, 3, 5-tris(40-hydroxy-50-formylphenyl)benzene, 1, 2-diaminobenzene, and cobalt acetate in DMF at 700–900°C [120]. The resultant carbon delivered ORR activity comparable to commercially available Pt/C catalyst.

π-Conjugated microporous polymers (CMPs, a class of POFs) were synthesized by a condensation reaction of 1,2,4,5-benzenetetramine with triquinoyl hydrate at different temperatures of 300–500°C to obtain highly porous N-doped carbonaceous polymers (aza-CMPs) [121]. The resultant aza-CMPs are microporous, as shown in the N_2 adsorption–desorption isotherms in Fig. 2.45A. With increasing heat treatment temperature, S_{BET} increases quickly up to 400°C and tends to saturate (Fig. 2.45B), S_{BET} of 1227 m^2/g, V_{total} of 0.54 cm^3/g, and micropore size of about 0.85 nm after 500°C treatment.

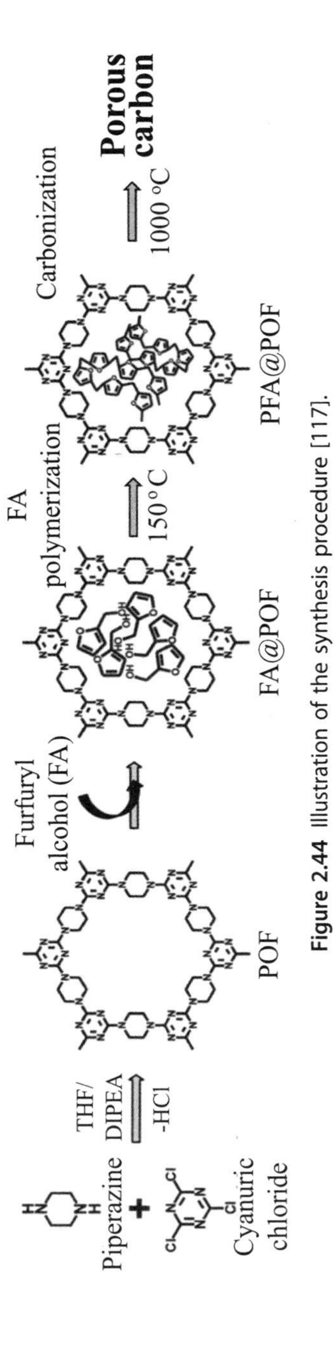

Figure 2.44 Illustration of the synthesis procedure [117].

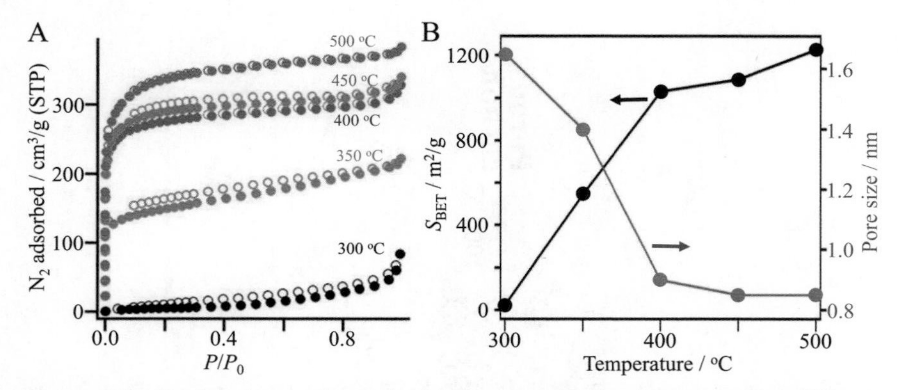

Figure 2.45 Aza-fused π-conjugated microporous polymers heat treated at different temperatures (300–500°C): (A) N_2 adsorption–desorption isotherms and (B) relations of S_{BET} and pore size to heat treatment temperature [121].

The rigid conjugated polymer networks synthesized from 1,3,5-tris-2′-biphenylbenzene (1 and 2 in Fig. 2.46A) as a carbon-rich unit by conjugating with a cross-linker (3 or 4 in Fig. 2.46A) were obtained as nanofibers [122]. These nanofibers could be carbonized at 600–800°C by keeping their fibrous morphology, as shown in the example in Fig. 2.46C. The resultant carbon nanofibers from the polymers composed of different combinations of units (A–D) are also microporous (Fig. 2.46B), with S_{BET} reaching about 900 m^2/g.

2.1.3.2.4 Metal carbides

Microporous carbons were prepared by chlorination of metal carbides, metal atoms being excluded as chlorides to leave micropores in the matrix carbon. The formation of a carbon layer on the surface of SiC particles of 1 μm size was observed when they were heat-treated at 1000°C in a flow of the mixture of H_2 with 3.5% Cl_2 [123]. Carbon films formed on the SiC particles were confirmed to be disordered carbon and calculated to have S_{BET} of about 1010 m^2/g. The formation of these porous amorphous carbon films on the SiC particles was supposed to occur through the exclusion of Si atoms as mainly $SiCl_4$ vapor. Aiming at the study of the structure and properties of the carbon films formed on SiC, sintered α-SiC disks (16 mm in diameter and 1 mm thick) and CVD synthesized β-SiC were heat-treated in a flow of gas mixtures of Ar, H_2, and Cl_2 at 1000°C [124]. Carbon films produced in pure Cl_2 demonstrated low hardness (1.8 GPa) and Young's modulus (18 GPa), while those produced in Cl_2/H_2 mixed gas had a microhardness up to 50 GPa and Young's

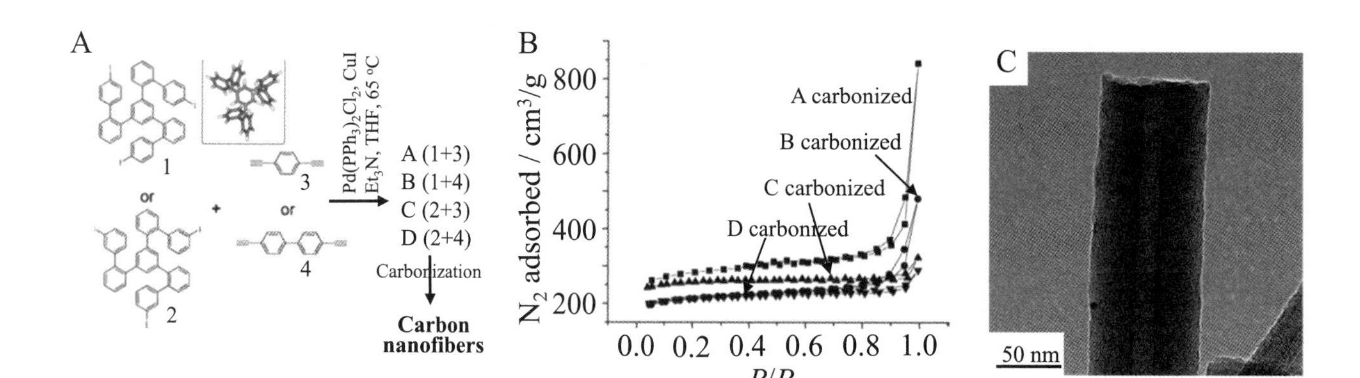

Figure 2.46 Microporous carbon nanofibers prepared through rigid conjugated polymer networks: (A) preparation process with 3D view (inset) of a polymer composed from 1 and 2 units, (B) N_2 adsorption–desorption isotherms of the carbonized polymers, (C) TEM image [122].

modulus up to 800 GPa, although no experimental results on pore structure of the resultant carbon films were presented.

Pore sizes in the carbons prepared from SiC, TiC, and Mo_2C by complete removal of metals in a Cl_2 flow at 700−1000°C were estimated to be 0.8−2 nm by small-angle X-ray scattering measurements, the pore sizes of the Mo_2C-derived carbons being approximately twice as large as those of SiC-derived ones [125]. A self-organization of the carbon framework left after chlorination of SiC was theoretically predicted to be able to form nanopores [126]. The dependence of pore sizes on the chlorination conditions was studied on SiC-derived carbons by different techniques: small-angle X-ray scattering, methyl chloride (CH_3Cl) adsorption, and Ar adsorption [127]. The pore size depends on the chlorination temperature, the mean size increasing from 0.5 nm at 300°C to 0.8 nm at 1200°C, as shown in Fig. 2.47. No mesopores or macropores were detected at up to 500°C, and a small volume of mesopores appeared above 700°C. A decrease in S_{micro} and an increase in S_{meso} above 800°C were observed on TiC and ZrC [128]. Pore sizes depended strongly on the kind of metal carbides as well as their chlorination temperatures [129]. In Fig. 2.48, pore size distributions are shown for the carbons derived from various metal carbides. Pore development behaviors in carbons derived from B_4C, TiC, ZrC, and β-SiC were studied through analyses of Ar adsorption−desorption isotherms and small-angle X-ray scattering [130]. The results are in very good qualitative agreement between the two techniques.

On Ti_2AlC-derived carbons, a marked dependence of pore size on chlorination temperature is observed, whereas there is a small dependence

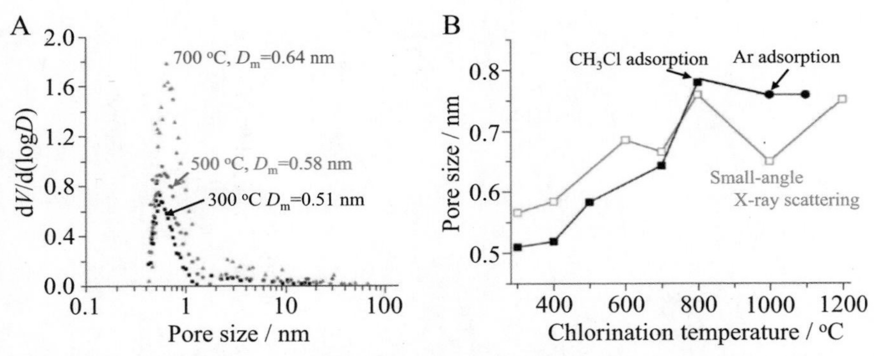

Figure 2.47 Pore sizes of SiC-derived carbons at different chlorination temperatures: (A) pore size distributions measured by CH_3Cl and (B) chlorination temperature dependences of mean pore size measured by small-angle X-ray scattering, CH_3Cl adsorption, and Ar adsorption [127].

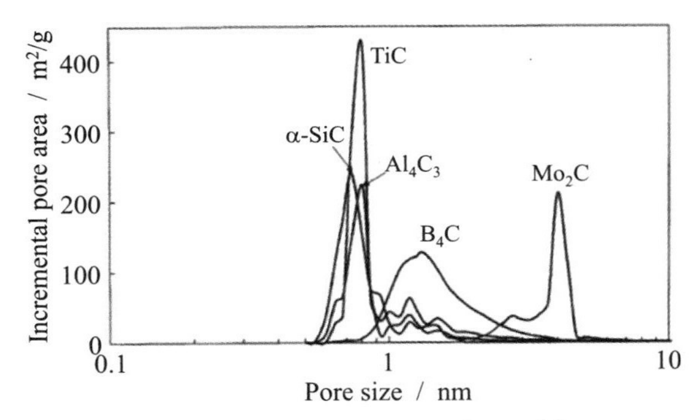

Figure 2.48 Pore size distributions of the carbons derived from various metal carbides [129].

on the B_4C-derived carbon [131], as shown in Fig. 2.49A. Marked dependences of S_{BET} and EDLC capacitance on chlorination temperature are observed for Tl_2AlC-derived carbons, giving the highest values at 1000°C, as shown in Fig. 2.49B. For B_4C-derived carbons, however, small dependences of S_{BET} and capacitance on chlorination temperature are observed, giving the highest values at 800–1000°C, as shown in Fig. 2.49C. Capacitance changes with chlorination temperature are almost parallel to S_{BET}.

The importance of chlorination temperature was shown on the pore structure of ZrC-derived carbon [132]. Changes in XRD profile, pore volumes of V_{total}, V_{micro} and V_{meso}, and S_{BET} with chlorination temperature are shown in Fig. 2.50A–C, respectively. Chlorination of ZrC seems to be completed up to 300°C from the XRD profile change with chlorination temperature, the resultant carbons being amorphous (Fig. 2.50A). The pore structure in the carbons develops with increasing temperature above 400°C, the increases in S_{BET} and V_{total} being almost linear with increasing temperature (Fig. 2.50B and C). V_{micro} tends to level off above 600°C, while V_{meso} increases, and consequently V_{total} increases with increasing temperature. The sizes of micropores formed in ZrC-derived carbons were about 0.74–0.83 nm up to 800°C and then pores were widened up with a broad distribution, with the average pore size being 1.41 nm for the carbon prepared at 1200°C.

Microporous structure and crystallinity of B_4C-derived carbons were studied as a function of the chlorination temperature [133]. The microporous structure was evaluated by calculating S_{BET} from Ar and N_2 adsorption isotherms at 77K and crystallinity was evaluated from full-width

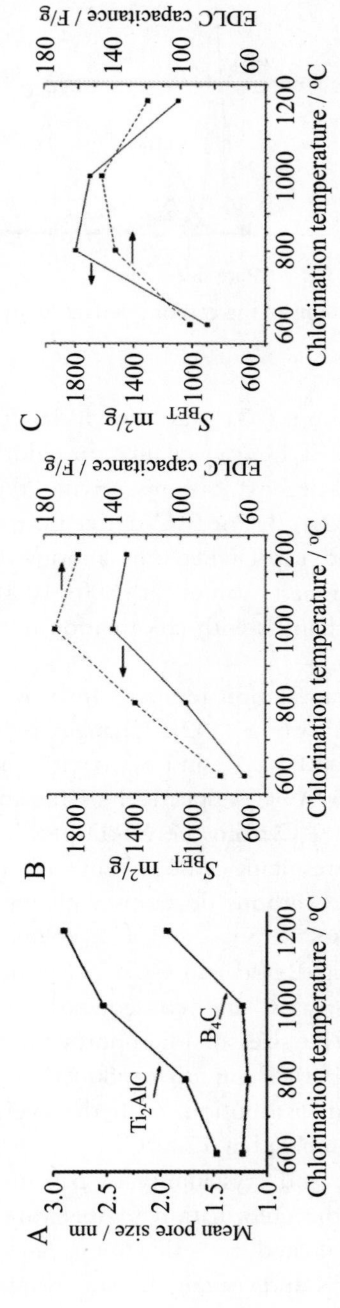

Figure 2.49 Ti_2AlC- and B_4C-derived carbons; chlorination temperature dependences of (A) mean pore sizes, (B) S_{BET} and EDLC capacitance for Ti_2AlC-derived carbon, and (C) those for B_4C-derived carbon [131].

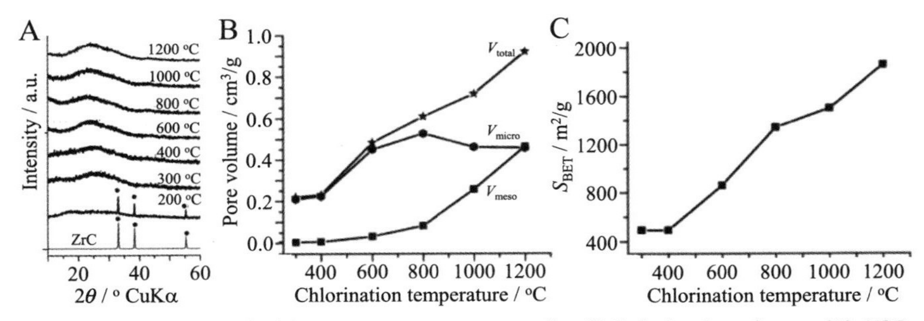

Figure 2.50 Effect of chlorination temperature for ZrC-derived carbons: (A) XRD, (B) pore volumes V_{total}, V_{micro}, and V_{meso}, and (C) S_{BET} [132].

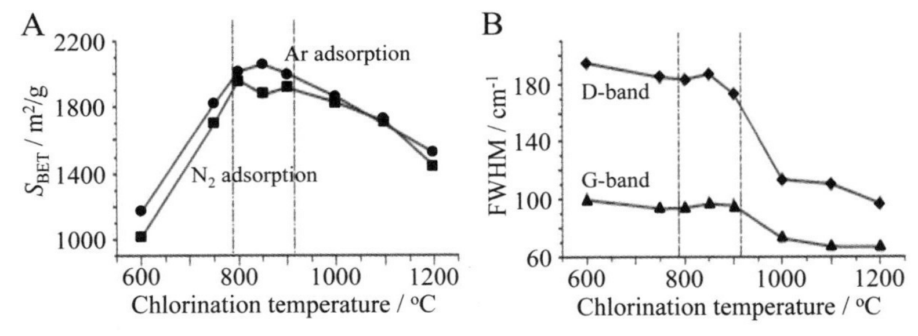

Figure 2.51 B_4C-derived carbons; Chlorination temperature dependences of (A) S_{BET} measured by Ar and N_2 adsorption and (B) FWHMs of G- and D-bands of Raman spectra [133].

at half-maximum intensity (FWHM) of G- and D-bands in Raman spectra. As shown in Fig. 2.51, S_{BET} values calculated from Ar and N_2 adsorptions are very similar, increasing from about 1100 m^2/g at 600°C to 2000 m^2/g at 800—900°C, and then decreasing gradually down to about 1500 m^2/g at 1200°C. The crystallinity of the carbon is considered to be improved above 900°C because FWHM values for D- and G-bands decrease markedly, more markedly for D-band. Chlorination at temperatures higher than 1300°C was shown to introduce marked structure changes in the resultant porous carbons derived from B_4C [134]. The parameters of the products after chlorination at 600—1800°C are shown in Table 2.7. By chlorination at temperatures above 1300°C, each parameter changes drastically from that chlorinated below 1300°C; S_{BET} drops down from more than 1000 m^2/g to less than 300 m^2/g and V_{micro} also decreases from about 0.50 cm^3/g to less than 0.10 cm^3/g, whereas V_{meso} remained almost the same value up to 1500°C but drops down also at 1800°C. Interlayer spacing d_{002} decreases

Table 2.7 Parameters of the porous carbons derived from B_4C by chlorination [134].

Chlorination temperature (°C)	Density by He (g/cm³)	S_{BET} (m²/g)	V_{micro} (cm³/g)	V_{meso} (cm³/g)	d_{002} (nm)
600	2.24	2208	0.93	0.15	0.370
800	2.23	2557	0.71	0.30	0.365
1000	2.19	1272	0.46	0.57	0.360
1300	2.21	288	0.10	0.36	0.345
1500	2.01	165	0.03	0.49	0.341
1800	1.80	36	0.01	0.19	0.336

gradually with an increase in chlorination temperature, and is close to the graphite value (0.3354 nm) after 1800°C.

Chlorination of iron carbide (Fe_3C) resulted in graphite, not porous amorphous carbons [135]. Chlorination was performed at different temperatures between 400–1200°C for 3 h in a Cl_2 flow. The product after chlorination at 1200°C showed the particle morphology of hexagonal platelets, a sharp G-band in Raman spectrum and no D-band, interlayer spacing d_{002} of 0.333 nm associated with three-dimensional diffraction lines of 101 and 112. The carbons obtained at low temperatures of 600 and 800°C were expected to have poor crystallinity because of the appearance of a strong D-band in the Raman spectrum, but have a relatively sharp 002 diffraction peak of XRD giving d_{002} of 0.334 nm. The addition of a small amount of $FeCl_3$, as well as $CoCl_2$ and $NiCl_2$, to TiC influenced the chlorination behavior [136]. TiC itself delivers highly porous carbon by chlorination at 400–1200°C, as shown by plotting of S_{BET} and adsorption capacity of benzene as a measure of pore volume in Fig. 2.52A; S_{BET} increases from 1180 m²/g by chlorination at 400°C to 1580 m²/g at 800°C.

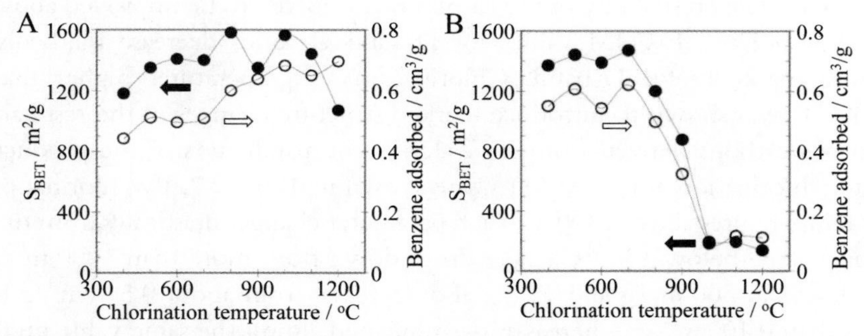

Figure 2.52 Changes in S_{BET} and benzene adsorption of the TiC-derived carbons with chlorination temperature: (A) TiC and (B) TiC mixed with transition metal chlorides [136].

The addition of a small amount of transition metal chlorides to TiC, however, leads to marked decreases in S_{BET} and benzene adsorption, as shown in Fig. 2.52B. The carbons obtained at temperatures above 1000°C had well-organized carbon structure, even large-sized graphitic crystallites being observed under high-resolution TEM observation.

Electrospinning of THF solution of polycarbomethylsilane and following pyrolysis at 800–900°C produced fibrous SiC, which could be converted to porous carbon fibers through chlorination at 850°C [137]. The resultant carbon fibers have very high S_{BET} and pore volume: S_{BET} of 2700–3100 m^2/g and V_{total} of 1.03–1.66 cm^3/g.

2.2 Mesoporous carbons

2.2.1 Activation

The activation process is efficient for creating micropores in carbons. To enlarge the created micropores to mesopore sizes, however, a long time is needed and it is associated with an increase in burn-off, in other words, a reduction in the activation yield. Therefore, it is preferable to use some additives to catalyze and accelerate the oxidation reaction.

Mesoporous carbons were prepared by mixing organo-rare-earth metal complexes, such as Ln-cyclopentadienyl and Ln-acetylacetonate, Ln(acac) (Ln: Y, La, Nd, Sm, Gd, Th, Er, Yb, and Lu), into petroleum pitches (softening point of 85°C) in THF, followed by mixing with a powdered pitch (softening point of 280°C), kneading at 360°C, air oxidation at 360°C, and finally steam activation at 930°C [138]. The resultant carbons with metal content of 2.5 wt.% provided a very high mesopore ratio of more than 70%, with an average size of about 5–6 nm. Mesopore-rich activated carbon fibers were prepared by melt-spinning of the mixture of novolac-type phenol resin with cobalt acetylacetonate Co(acac). followed by carbonization at 900°C in N_2 and activation in steam at 750–900°C [139]. Cobalt accelerated the activation process markedly at 800°C, where no difference in activation rate was observed between the fibers containing 38 and 100 ppm Co, as shown in Fig. 2.53A. Co in carbon fibers accelerates mesopore formation during the activation process. S_{meso} is shown by plotting against burn-off in Fig. 2.53B, revealing a predominant increase in S_{meso} in the fibers containing 38 ppm Co. The fibers containing 38 ppm Co have S_{BET} of 837 m^2/g, V_{micro} of 0.28 cm^3/g, and V_{meso} of 0.34 cm^3/g after the activation of up to 38% burn-off. Meanwhile, the pristine fibers without Co additives were microporous, S_{BET} of 1450 m^2/g with V_{micro} of

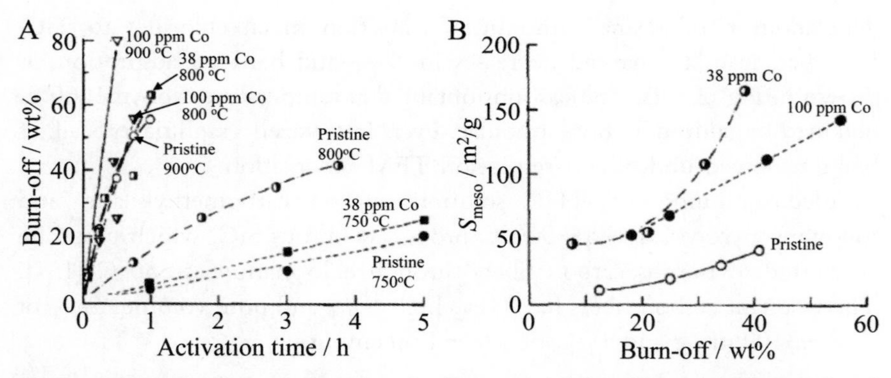

Figure 2.53 Activation of phenol resin fibers containing 38 and 100 ppm Co at different temperatures: (A) burn-off versus activation time and (B) S_{meso} versus burn-off [139].

0.57 cm^3/g and V_{meso} of 0.02 cm^3/g after 34% burn-off. Mesoporous carbon fibers prepared from phenol resin mixed with Co(acac) and 0.22 wt.% Ag exhibited antibacterial activity, even after reduction of Ag to a negligibly small amount [140].

Carbon fibers were prepared from a mixture of a pitch with Y-acetylacetonate [Y(acac)$_2$] by melt-spinning at its softening point (around 256°C) through small holes (diameters of 20–30 μm), stabilization at 360°C in air, and then carbonization/activation in steam-saturated N_2 flow at 875°C [141]. The ACFs prepared from the mixture with 1.0 wt.% Y(acac)$_2$ delivered relatively high S_{BET} of 1090 m^2/g with S_{meso} of 759 m^2/g (mesopore ratio of 70%). Other metal-acetylacetonates, TiO(acac)$_2$ and Al(acac)$_2$, could give S_{BET} comparable to Y(acac)$_2$, but much lower S_{meso}, suggesting these metallic species can accelerate the micropore formation but are not effective for mesopore development. The ACFs prepared from 0.3 wt.% Y(acac)$_2$ delivered a high mesopore percentage of 80.8% (pore size of 4.38 nm and S_{BET} of 1468 m^2/g) [142], which had high adsorption capacity of large-sized organic molecules, such as humic acid, γ-cyclodextrin, and vitamin B_{12}. In contrast, TiO(acac)$_2$ was reported to work effectively to create mesopores in the carbon matrix, similar to Y(acac)$_2$, on activated carbon granules prepared from three kinds of coal [143].

Viscose rayon fibers were impregnated by 3 w/v% boric acid with different amounts of NaCl in deionized water, followed by carbonization in N_2 flow at 850°C and then activation in CO_2 flow at 850°C [144]. N_2 adsorption–desorption isotherms of the resultant ACFs demonstrated the formation of microporous structures together with the creation of mesopores in the ACFs. The pristine ACFs without additives exhibited high S_{BET} of 1600 m^2/g, while S_{ext} was only 62 m^2/g, suggesting the presence of a negligibly small amount of mesopores. However, the ACFs obtained by adding 3 w/v% boric acid with 1.0 w/v% NaCl gave S_{BET} of 1370 m^2/g, a little lower than the pristine one, while S_{ext} was 207 m^2/g. Different alkali chlorides, LiCl, KCl, RbCl, and CsCl, could give S_{BET} of about 1500 m^2/g and S_{ext} of 122–196 m^2/g to the ACFs. Micropores in the resultant ACFs were analyzed using α_s plots, but no detailed analysis on mesopores formed in these ACFs was reported. From the same rayon clothes, mesoporous ACFs were prepared by impregnation of various phosphates, NaH_2PO_4, Na_2HPO_4, and Na_3PO_4, with a mixture of 3% $AlCl_3$, 3% $ZnCl_2$, and 3% NH_4Cl (in w/v%) through their aqueous solutions, followed by carbonization in N_2 and then activation in CO_2 [145]. The mixtures of metal chlorides could improve the carbonization yield of the viscose rayon chars, resulting in ACF cloths with improved mechanical strength [146]. Adsorption isotherms are shown on the ACFs using 10% NH_2PO_4 as a function of burn-off in Fig. 2.54A, revealing the development of a hysteresis loop with increasing burn-off. The addition of chlorides to the mixture containing NaH_2PO_4 helps the pore development in ACFs, as shown in Fig. 2.54B. At a high percentage burn-off, N_2 adsorption increases markedly at higher P/P_0, accompanied by a change in

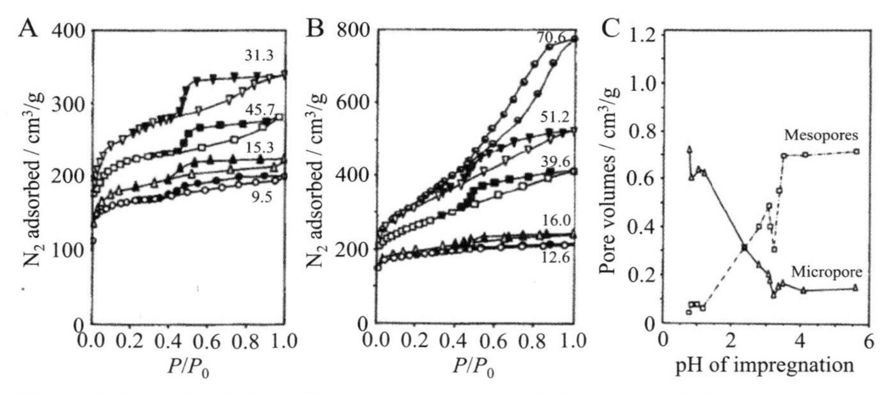

Figure 2.54 ACFs derived from viscose rayon clothes through impregnation of phosphate, carbonization in N_2, and activation in CO_2 at 850°C: (A) N_2 adsorption–desorption isotherms for the ACFs prepared by impregnation of 10 w/v% NaH_2PO_4 with different burn-offs, (B) those for the ACFs by impregnation of 5 w/v% NaH_2PO_4 with 3 w/v% $AlCl_3$, $ZnCl_2$, and NH_4Cl with different burn-offs, and (C) V_{micro} and V_{meso} versus pH of impregnation solution on the ACFs by impregnation of Na_3PO_4 and H_3PO_4 at burn-off of 52.8% [145].

the shape of the hysteresis loop. In the case of the mixture of Na_3PO_4 and H_3PO_4, the pH of the impregnating solution influences markedly on the pore structure in ACFs, as shown in Fig. 2.54C. With increasing pH, V_{micro} decreases markedly, associated with an increase in V_{meso} and mesopores are predominant above pH 3, whereas both pore volumes become constant above pH 4.

Mixing of H_3PO_4 with a lignosulfate was also effective for developing micropores and mesopores simultaneously [147]. Commercial lignosulfate, which had a degree of polymerization of 1650, was impregnated with H_3PO_4 through its 60% solution at a H_3PO_4/lignosulfate weight ratio of 1, followed by carbonization at a temperature of 400–1000°C in Ar flow. N_2 adsorption–desorption isotherms and pore size distributions are shown in Fig. 2.55A and B, respectively, in comparison with the pristine lignosulfate carbonized at 800°C (without H_3PO_4 impregnation). Pore development in the carbon depends strongly on the carbonization temperature. The carbon prepared at 1000°C in the presence of H_3PO_4 delivered S_{BET} of 1370 m^2/g and V_{total} of 0.97 cm^3/g including V_{meso} of 0.56 cm^3/g, whereas the pristine lignosulfate gave nonporous carbon, as shown in Fig. 2.55. The resultant H_3PO_4-activated carbon contained micropores with sizes of 0.9–1.3 and 2.3 nm, together with mesopores with sizes in the range of 6–20 nm. For the impregnation of H_3PO_4 solution, its drop-by-drop addition was recommended to produce swelling for incipient wetness with the carbon precursor solid (apple pulp) and to obtain efficient activation [148]. This incipient wetness impregnation was compared to hydrothermal

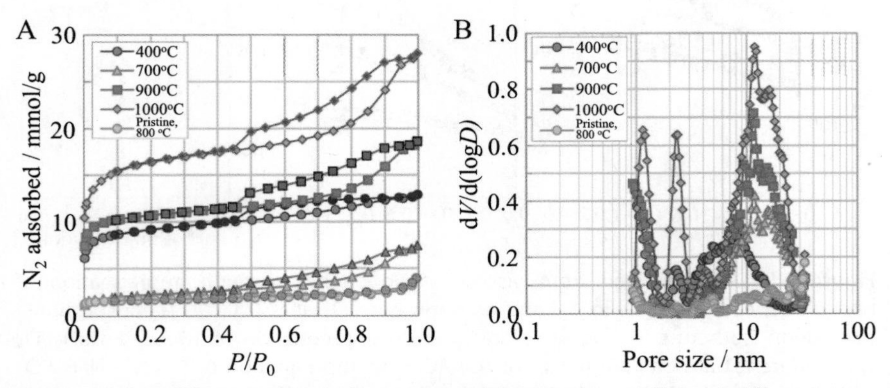

Figure 2.55 H_3PO_4-activated carbon prepared from a lignosulfate at different temperatures: (A) N_2 adsorption–desorption isotherms and (B) pore size distributions [147].

impregnation using biomass (coconut fibers and banana pseudostem) [149]. The former process was more effective than the latter process in developing mesopores; the former resulted in S_{BET} of 2179 m^2/g, V_{micro} of 0.86 cm^3/g, and V_{meso} of 0.85 cm^3/g whereas the latter resulted in 1578 m^2/g, 0.67 cm^3/g, and 0.39 cm^3/g, respectively, on the coconut fibers.

Composted spent coffee grounds provided mesopore-rich activated carbons via steam activation without any additives [150]. Spent coffee grounds (SCGs) were composted by mixing with lactic acid for 60 days (coffee grounds composted, CGCs). CGCs and SCGs were carbonized at 600°C in N_2 flow and then activated in steam flow at 800°C for 20–50 min. The CGC activated for 40 min gave a more developed pore structure with higher activation yield than the pristine SCG activated for 50 min, as shown in Fig. 2.56; the former delivered S_{BET} of 1181 m^2/g and V_{total} of 0.79 cm^3/g with 9.6 wt.% yield, while the latter delivered 1199 m^2/g and 0.99 cm^3/g with 12.2 wt.%, respectively.

The microwave irradiation method was compared with the conventional heating method using lotus stalks, which were crushed to a particle size of 0.5–1 mm [151]. The lotus stalks were impregnated by H_3PO_4 (H_3PO_4/stalks weight ratio of 2/1) and then irradiated by microwave (MW) at a power of 700 W for 15 min. In Table 2.8, pore structure parameters of the MW-assisted activated carbons are compared with the carbon activated by conventional heating at 450°C for 1 h [152]. MW irradiation assists effectively in activation by H_3PO_4 to increase S_{ext} and V_{ext} due to the increase in mesopores. Drying before MW irradiation was experimentally shown to be important for the development of mesopores

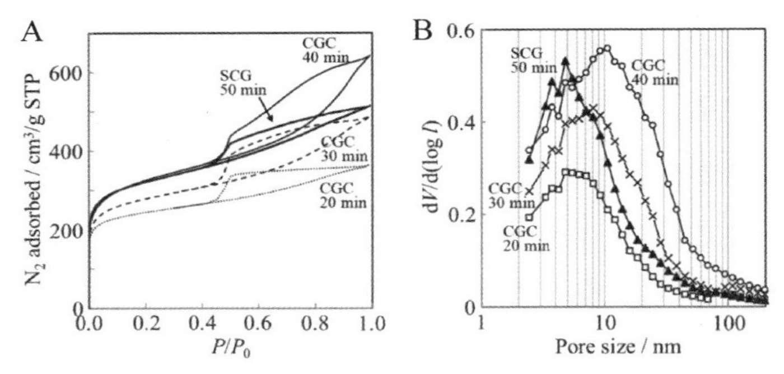

Figure 2.56 Coffee-derived activated carbon by steam activation: (A) N_2 adsorption–desorption curves and (B) mesopore size distribution for coffee grounds composted (CGCs) for different activation times in comparison with pristine spent coffee grounds (SCGs) activated for 50 min [150].

Table 2.8 Pore structure parameters of activated carbons from lotus stalks impregnated by H_3PO_4 [151].

	S_{BET} (m^2/g)	S_{micro} (m^2/g)	S_{ext} (m^2/g)	V_{total} (cm^3/g)	V_{micro} (cm^3/g)	V_{ext} (cm^3/g)	Yield (%)
Thermal activation	1220	503	717	1.191	0.286	0.905	56.3
Microwave activation	1431	454	928	1.337	0.306	1.031	40.1

on sawdust [153]. The sawdust was dried at 60–285°C for 1 h after the impregnation of H_3PO_4 solution and then subjected to MW irradiation of at 560 W and 2.45 GHz. The resultant activated carbon with predrying at 250°C gave the highest S_{BET} of 1509 m^2/g and V_{total} of 1.26 cm^3/g including V_{meso} of 1.05 cm^3/g, whereas that with 285°C predrying gave 1253 m^2/g, 0.96 cm^3/g, and 0.76 cm^3/g, and that without predrying gave 903 m^2/g, 0.84 cm^3/g, and 0.70 cm^3/g, respectively. When $ZnCl_2$ was coupled with MW irradiation (560 W power), mesoporous activated carbons were obtained from cotton stalks: V_{total} of 0.63 cm^3/g with a small V_{micro} of 0.08 cm^3/g [24]. The effects of MW irradiation for the production of activated carbons were reviewed by focusing on the biomass [17].

2.2.2 Template-assisted carbonization

2.2.2.1 Silicas

Ordered mesoporous structures of silicas, which were formed through self-assembly of surfactants and named as MCM-48, MCM-41, SBA-1, and SBA-15, etc., were successfully inherited to carbons to form ordered mesoporous carbons via template-assisted carbonization. In addition, colloidal silicas composed of nanospheres with homogeneous sizes were also used as templates for mesoporous carbons, even with ordered pore structure. In Table 2.9, pore structures in the carbons prepared using various silica templates are summarized. Reviews focusing on silica-templated carbonization have been published also [154–156].

2.2.2.1.1 Mesoporous silicas

The pores and/or channels in the mesoporous silicas are replicated in the carbons by either impregnation or CVI of a carbon precursor, followed by carbonization and removal of the templates by HF. The pore symmetry in these silica-templated carbons is measured by powder XRD pattern at very low diffraction angles, as shown on the silica templates and the carbons

Table 2.9 Silica template and resultant carbons.

Silica template			Mesoporous carbons synthesized			
Notation	Symmetry	Modification	Notation	Precursor	Process	Pore structure (some examples)
MCM-48	Cubic Uniform, ordered, interconnected channels		CMK-1	Sucrose, FA,	Impregnation	Ordered $S_{BET} = 970 \sim 1700\ m^2/g$ $V_{meso} = 1\ cm^3/g$
				Propylene	CVI	
		Alumination	CMK-4	PF Acetylene	Impregnation CVI	
		Silylation		Divinylbenzene		
		Calcination		Sucrose	Impregnation	
MCM-41	Hexagonal Uniform channels without interconnection			Sucrose, FA	Impregnation	Ordered $S_{BET} = 1100\ m^2/g$
			Alumination	PF	Impregnation	
			Leaf-like	Sucrose, FA	Impregnation	
SBA-15	Hexagonal Uniform channels interconnected		CMK-3	Sucrose, FA	Impregnation	2D-hexagonal $S_{BET} = 750 \sim 1800\ m^2/g$ $V_{meso} = 1\ cm^3/g$
		Calcined		Sucrose	Impregnation	Random
MSU-H	Hexagonal Ordered			Sucrose, FA	Impregnation	$S_{BET} = 1200\ m^2/g$ $V_{total} = 1.3\ cm^3/g$
Colloidal silica	Monodispersed spheres			RF, PAN, FA	Mixing	$S_{BET} = 1200 \sim 1500\ m^2/g$ $V_{meso} = 1 \sim 3.6\ cm^3/g$
		Stabilization				
		Sintered	Inverse opal	Phenol resin	Impregnation	3D periodical alignment

CVI, chemical vapor infiltration; *FA*, furfuryl alcohol; *PF*, phenol-formaldehyde resin; *RF*, resorcinol-formaldehyde resin; S_{BET}, BET surface area; V_{meso}, mesopore volume; V_{total}, total pore volume.

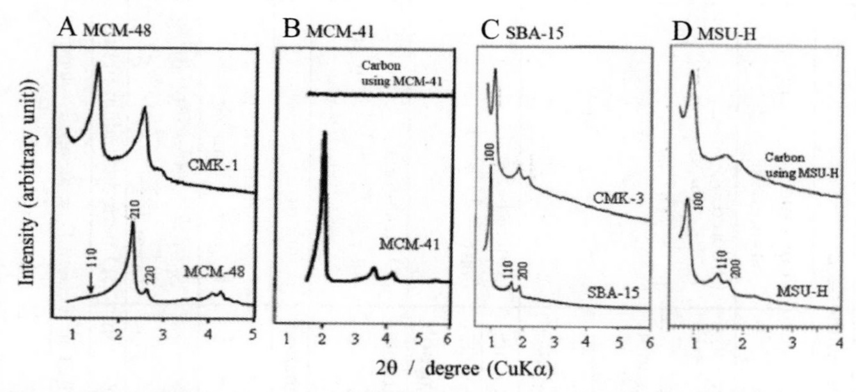

Figure 2.57 XRD patterns at low diffraction angles for the template silicas and the resultant carbons.

synthesized using each template in Fig. 2.57. Pore symmetries of most silica templates were replicated in the carbons, although no symmetry replication was achieved on MCM-41 and lower symmetry was obtained in some cases using MCM-48.

MCM-48 was impregnated by sucrose with sulfuric acid through their aqueous solution and carbonized at 800–1100°C, followed by dissolution of the template silica in aqueous solution containing NaOH and ethanol to isolate the carbon in powder [157]. The SEM and TEM images of the resultant carbon are shown in Fig. 2.58. The resultant carbon, called CMK-1, had S_{BET} of 1380 m^2/g and ordered mesopores with sizes of 3.0 nm, together with micropores of 0.5–0.8 nm size: V_{meso} of 1.1 cm^3/g and V_{micro} of 0.3 cm^3/g. Ordered pore structure is clearly seen by TEM

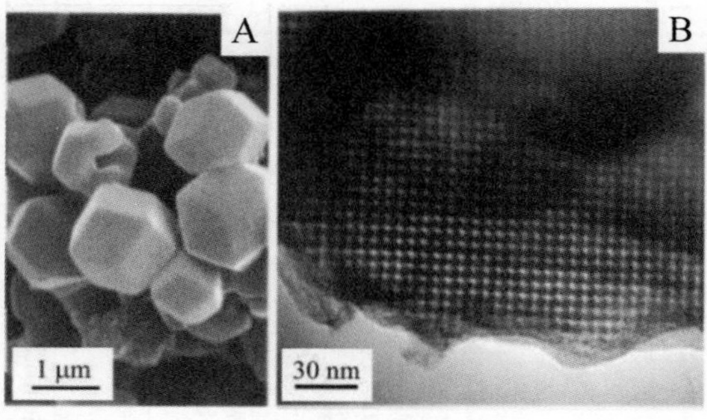

Figure 2.58 Ordered mesoporous carbon prepared using MCM-48 template from sucrose: (A) SEM and (B) TEM images [157].

(Fig. 2.58B) and is also confirmed by the diffraction peaks at 1.6 and 2.7° in 2θ (CuKα), although the symmetry of the carbon is different from that of the template MCM-48 (Ia3d) (Fig. 2.57A). Sugars, glucose, and xylose were successfully used as carbon precursors. In order to use phenol-formaldehyde resin as a carbon precursor, aluminum needed to be implanted onto the wall of mesopores of MCM-48 to generate strong acid catalytic sites on the channel wall for polymerizing the resin (Al-implantation or alumination) [158]. Silylation of the pore surfaces of MCM-48 using trimethylsilyl chloride was effective in getting highly ordered mesopores in the resultant carbons from divinylbenzene as a carbon precursor [159]. Pore structure symmetry Ia3d in the template MCM-48 was able to be replicated in the carbon by the repetition of impregnation/drying steps of sucrose solution [160]. By applying CVI of acetylene at 800°C to fill the pores of the template MCM-48, the symmetry of the template could be preserved in the resultant carbons even after carbonization at 900°C and removing the template (designated as CMK-4) [161]. In order to synthesize ordered mesoporous carbons, the optimization of the amount of sulfuric acid and the carbonization at temperatures above 600°C were essential for complete filling of the pores of the template by the repeated impregnation/carbonization of sucrose [162].

Mesoporous MCM-48-type silicas with different pore sizes were synthesized using alkyltrimethylammonium/cosurfactant mixtures, of which mesopore volume and size in the resultant silicas were changed using alkyltrimethylammonium with different alkyl chain lengths consisting of different n carbon atoms ($n = 12$, 14, 16, 18, and 20) [163]. Synthesized MCM-48-type silicas were used as templates for preparing CMK-1-type ordered mesoporous carbons by impregnation of sucrose with sulfuric acid, followed by carbonization at 900°C. As shown, N_2 adsorption–desorption isotherms of the templates and the resultant carbons in Fig. 2.59A and B, respectively, the mesoporous structure in the template is well replicated in the resultant carbon and the amount of N_2 adsorbed (i.e., pore volume) depends strongly on the alkyl chain length n used. The MCM-48 with n of 12 and 20 had V_{total} of 1.02 and 1.46 cm^3/g, respectively, and the resultant carbon from these MCM-48 had V_{total} of 1.12 and 1.23 cm^3/g, respectively. The symmetry of pore structure in the carbon depends on the n-number, as shown in Fig. 2.59C.

The carbons prepared using MCM-41 as a template contained disordered micropores, probably due to collapsing the template framework upon its removal, as shown in its XRD in Fig. 2.57B, and consisted of carbon

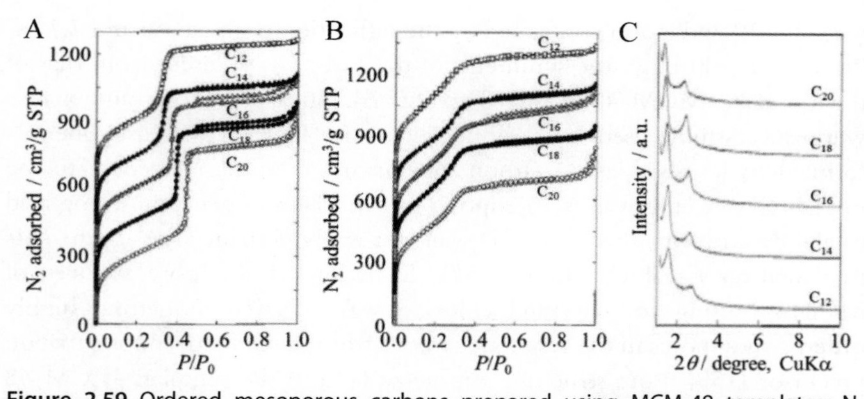

Figure 2.59 Ordered mesoporous carbons prepared using MCM-48 templates: N_2 adsorption–desorption isotherms of (A) MCM-48 templates synthesized using different *n*-number of alkyl chains and (B) resultant carbons prepared using these MCM-48, and (C) XRD patterns for the carbons [163].

nanowires separated randomly [163]. The resultant carbon had S_{BET} of 1170 m^2/g and V_{total} of 0.67 cm^3/g. Self-supported carbon nanowire arrays were prepared using MCM-41, which was subjected to microwave treatment to remove the remaining organics [164]. By using Al-implanted MCM-41 with leaf-like morphology, carbon nanowires with diameters of 4–5 nm and packed side-by-side were synthesized from FA.

Carbons with ordered mesopores in the same symmetry as the template SBA-15, as shown in the XRD patterns in Fig. 2.57C, were obtained by impregnation of sucrose via its aqueous solution containing sulfuric acid, followed by carbonization at 900°C, were named as CMK-3 [165]. As shown in the TEM image in Fig. 2.60A, the resultant carbon contains hollow channels with diameters of about 4.5 nm, of which the N_2 adsorption–desorption isotherm (Fig. 2.60B) indicates the coexistence of

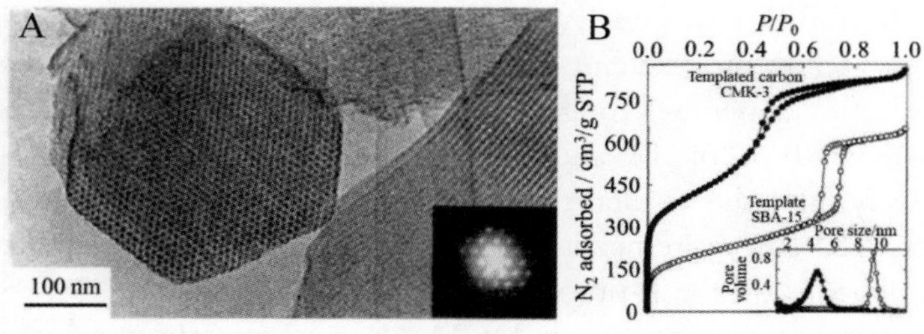

Figure 2.60 The carbon (CMK-3): (A) lattice fringe image and electron diffraction pattern (inset) and (B) N_2 adsorption–desorption isotherms with pore size distribution (inset) in comparison with the template SBA-15 [165].

mesopores with sizes of around 4 nm together with micropores, V_{total} reaching 1.3 cm^3/g with S_{BET} of 1520 m^2/g. To obtain ordered mesoporous carbons, the calcination temperature of the sucrose-impregnated template SBA-15 had to be below 880°C, because SBA-15 was calcined at 970°C and gave disordered pore arrangement above 880°C [166]. The size of mesopores in the carbon increased from 2.2 to 3.3 nm with increasing thickness of the wall of template SBA-15 from 1.4 to 2.2 nm, which was controlled by changing the ratio of hexadecyltrimethylammonium bromide/polyoxyethylene hexadecyl ether-type surfactants [167]. The use of SBA-15 particles with rod-like morphology delivered carbon CMK-3 rods, with S_{BET} of 1823 m^2/g, V_{total} of 2.23 cm^3/g, and mesopore width of 5.8 nm [168], of which mesopore channels are aligned parallel along with the rod length. Compression of a powder mixture of the SBA-15/PFA composite and NaCl with the particle size of about 0.2 μm particles at 1000 kg/m^3 resulted in a CMK-3 monolith with mesopores of 3.7 nm size and macropores of 550 nm size [169]. Through CVI of acetonitrile into the mesopores of SBA-15 at 950–1100°C, ordered mesoporous N-doped carbons were prepared [170].

An increase in the synthetic temperature of SBA-15 from 90 to 150°C resulted in a change in the mesopore size from 7.2 to 11.3 nm without a change in its structural order. By using these SBA-15 silicas as templates, mesoporous carbons were prepared at 800°C in N_2 through impregnation of FA [171]. Their N_2 adsorption–desorption isotherms and pore size distributions are shown in Fig. 2.61A and B, respectively. The resultant

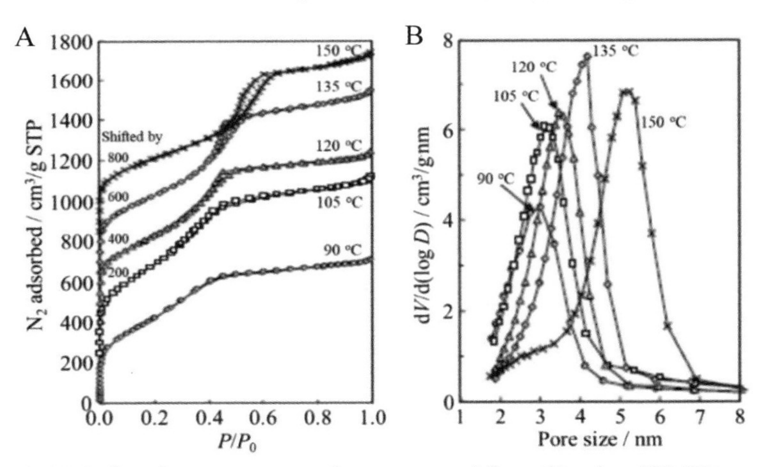

Figure 2.61 Ordered mesoporous carbons prepared from FA using SBA-15 templates synthesized at different temperatures: (A) N_2 adsorption–desorption isotherms and (B) pore size distribution as a function of temperature of the template synthesis [171].

carbons replicated the pore characteristics of the template. The mesopore size increases with increasing synthesis temperature of SBA-15, in other words, with increasing pore size of SBA-15, as shown in Fig. 2.61B. However, the pore parameters, S_{BET}, V_{total}, and V_{meso}, did not show marked dependences on synthesis temperature of the templates, in the ranges of 1450–1790 m^2/g, 1.10–1.47 cm^3/g, and 0.95–1.33 cm^3/g, respectively. When the impregnation of FA and carbonization was repeated twice, pore parameters decreased, S_{BET} from 1790 to 1280 m^2/g and V_{meso} from 1.31 to 0.85 cm^3/g, due to the disappearance and/or closing of the mesopores formed within the carbon rods by the second cycle.

The mesoporous silica MSU-H, which had a porous framework similar to SBA-15, was also used as a template for preparing mesoporous carbons [172]. The pore structure symmetry in the template MSU-H was preserved in the resultant carbon with low shrinkage (Fig. 2.57D). The carbon synthesized from sucrose at 900°C had a pore size distribution showing a sharp peak at 3.9 nm. The carbons obtained from FA in the presence of para-toluene sulfonic acid at 800°C had mesopore sizes from 3 to 14 nm [173]. By controlling the amount of carbon precursor impregnated, mesoporous carbons with unimodal (2.9 nm size) and bimodal (2.9 and 14 nm sizes) pore structures can be prepared from FA [173,174] and also from phenol resin [174]. N_2 adsorption–desorption isotherms and pore size distributions are shown for the carbons prepared from FA in Fig. 2.62.

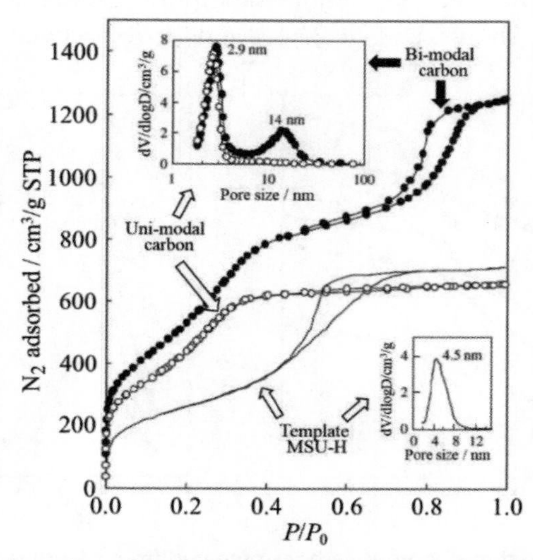

Figure 2.62 N_2 adsorption–desorption isotherms and pore-size distributions (insets) for mesoporous carbons with unimodal and bimodal pore size distributions and the template MSU-H [173].

The carbons consisting of unimodal pores with sizes of about 2.9 nm were formed by the repetition of impregnation and carbonization twice. The complementary mesopores with 14 nm size resulted from the coalescence of unfilled pores during removal of silica walls due to incomplete impregnation of carbon precursors. The size of mesopores in carbon was controlled by infiltration of the mixture of sucrose with different amounts of boric acid (0—25 mol%) from 3.8 to 10.5 nm, resulting in decreases of S_{BET} and V_{total} from 1337 to 848 m^2/g and from 1.60 to 1.25 cm^3/g, respectively [175]. Boric acid added to sucrose may change to boron oxide and borosilicate during carbonization and may result in an increase in mesopore size in the carbon. Another hexagonal mesoporous silica (HMS) was also used as template after alumination in ethanol solution of $AlCl_3$ at room temperature and the resultant mesoporous carbon had S_{BET} of 1056 m^2/g and V_{total} of 0.69 cm^3/g [176]. On the mixture of sucrose with MSU-H, carbonization by microwave irradiation was compared with carbonization by conventional thermal heating [177]. Microwave irradiation was advantageous for shortening the treatment time, microwave irradiation for 10 min resulting in a comparable pore structure in the carbon obtained by conventional heating at 900°C for 12 h.

Various aromatic compounds, such as acenaphthene, acenaphthylene, indene, indan, and substituted naphthalenes, were used as carbon precursors for the ordered mesoporous silica templates, such as MCM-48 and SBA-15, but their impregnation and carbonization above 750°C had to be done in an autoclave [178]. After further heat treatment under vacuum at 900°C, their XRDs showed strong 002 and unsymmetrical 10 diffraction peaks of carbon, indicating turbostratic stacking of carbon layers. A petroleum pitch with a low softening point between 114 and 122°C was impregnated into MCM-48 and SBA-15 templates at 302°C under atmospheric pressure and carbonized at 950°C in Ar [179]. The resultant carbons exhibited S_{BET} of 954 m^2/g and V_{total} of 0.6 cm^3/g using MCM-48, and 923 m^2/g and 0.6 cm^3/g using SBA-15, respectively.

The surface of the templates, SBA-15, FDU-1, and silica colloids, was modified by 3-(chlorodimethylsilyl)propyl 2-bromoisobutyrate and dimethylchlorosilane [180]. Mesoporous carbons were prepared using the modified templates and acrylonitrile through stabilization at 300°C in air and carbonization at 800°C in N_2 flow. The resultant carbons had a large pore volume (1.5—1.8 cm^3/g) because the surface modification enlarged the wall thickness of the template and the replicated carbons consequently

Table 2.10 Properties of ordered mesoporous carbons prepared by templating and high-temperature heating [181,182].

Carbon precursor	Template silica	HTT (°C)	S_{BET} (m^2/g)	V_{total} (cm^3/g)	d_{002} (nm)	σ (S/cm)
PVC	SBA-15	800	930	1.09	0.351	0.3
	SBA-15	2300	260	0.34	0.342	4.2
	MSU-1	800	950	1.60	0.354	—
FA	SBA-15	2300	1790	1.43	—	0.003

had large mesopore sizes. The carbons prepared using ordered mesoporous silicas (SBA-15 and FDU-1) exhibited narrow pore size distributions (centered at around 8 and 13 nm, respectively), whereas the carbon prepared using a disordered silica gel template had a broad pore size distribution.

Mesoporous carbons were prepared using silica templates, SBA-15 and MSU-H, and carbon precursors, poly(vinyl chloride) (PVC) and FA. By heat treatment at 2300°C in Ar [181,182]. It is known that PVC gives graphitizing carbon, while FA gives nongraphitizing carbon. The pore and structure parameters are listed in Table 2.10. The electrical conductivity (σ) of the PVC-derived carbon is two orders of magnitude higher than the FA-derived carbon, whereas the latter exhibits much higher pore structure parameters than the former. The 2300°C treatment of the former resulted in one order of magnitude increase in σ and a slight decrease in d_{002}, and a marked reduction in pore parameters, S_{BET} and V_{total}. On the FA-derived carbon, on the other hand, no marked increase in σ is observed, although the porous characteristics are kept even after 2300°C treatment. To use FA and furfural as carbon precursors with silica templates, such as SBA-15, acidification of the wall surfaces of silica templates by impregnation of phosphoric acid or sulfuric acid prior to their use was recommended to achieve highly faithful replication of the mesoporous structure of silica templates [183].

Ordered mesoporous carbons, CMK-3 and CMK-1, were prepared from sucrose using SBA-15 and MCM-48, respectively, and activated by CO_2 at 950°C to increase the number of micropores [184]. CO_2 activation was more effective on CMK-3 than CMK-1 and V_{micro} increased with increasing activation time. The pristine CMK-3 exhibited S_{BET} of 984 m^2/g, V_{total} of 1.09 cm^3/g, and V_{micro} of 0.37 cm^3/g, while that after CO_2 activation delivered 2749 m^2/g, 2.09 cm^3/g, and 0.96 cm^3/g, respectively. Mesoporous carbons prepared from sucrose using SBA-15 were doped by

Table 2.11 Pore structure parameters and N content of the mesoporous carbons doped by N by ammoxidation [185].

Ammoxidation	S_{BET} (m^2/g)	V_{total} (cm^3/g)	N-content (wt.%)
No ammoxidation	1271	1.3	—
Before carbonization	1106	1.0	8
After carbonization	1218	1.24	1.6

N via ammoxidation at 300°C for 5 h under a flow of ammonia/air (1/1 by volume) either before or after carbonization at 850°C in N_2 [185]. As shown in Table 2.11, the ammoxidation after carbonization deteriorates the pore structure condition only slightly and doped N only by a small amount (1.6 wt.%). On the other hand, ammoxidation before carbonization increased the amount of doped N up to 8 wt.% without a substantial decrease in the pore structure parameters. The configuration of N atoms doped into the carbon was different, depending on the order of the ammoxidation before or after carbonization: more than 60% N was pyrrolic in the latter but three configurations of pyrrolic, pyridinic, and graphitic were equal in the former. The EDLC capacitance for these carbons in 1 M H_2SO_4 electrolyte was strongly influenced by the order of N-doping.

Ordered mesoporous carbons (OMCs) were also prepared from sucrose using hexagonal mesoporous silicas (HMS) with a wormhole structure at 900°C [186]. HMSs were synthesized from TEOS using a surfactant $C_nH_{2n+1}NH_2$ with different *n*-values by calcining at 650°C in air. Pore characteristics of HMSs synthesized using the surfactant with *n*-values of 8, 10, 12, and 16 (different chain lengths of the surfactants) and the resultant carbons are summarized in Table 2.12. The resultant carbons, OMC-8 to OMC-16, are characterized by a bimodal pore structure.

Mesocellular silica foams having a wide range of pore sizes (24—42 nm) and window sizes (9—19 nm) were used as templates to synthesize mesoporous carbons from sucrose [187]. Impregnation of sucrose was performed through its aqueous solution with a small amount of H_2SO_4 twice and carbonized at 900°C in Ar flow. The resultant carbons had S_{BET} of 950—640 m^2/g with V_{total} of 0.56—1.8 cm^3/g, depending strongly on the pore structure of the template silicas. The carbons synthesized by the same process had spherical pores with diameters of 31 nm and window sizes of 22 nm and used as the cathode material of Li—O_2 batteries [188].

Table 2.12 Characteristics of HMSs with different pore sizes and OMCs prepared using HMSs as templates [186].

Silica	HMS-8	HMS-10	HMS-12	HMS-16
Surfactant	$C_8H_{17}NH_2$	$C_{10}H_{21}NH_2$	$C_{12}H_{25}NH_2$	$C_{16}H_{33}NH_2$
Expected micelle size (nm)	2.0	2.5	3.0	4.0
S_{BET} (m^2/g)	910	1270	1370	1280
Pore size (nm)	1.5	1.7	2.2	3.1
Wall thickness (nm)	2.4	2.3	2.1	2.2
V_{total} (cm^3/g)	1.21	1.34	1.09	1.83
Carbon (OMC)	**OMC-8**	**OMC-10**	**MC12**	**OMC-16**
S_{BET} (m^2/g)	980	1140	1240	1650
Pore diameter (nm)	1.8 and 3.9	1.7 and *a*	1.8 and 3.8	1.6 and 3.3

a, large pore was not easily separable for OMC-10.

2.2.2.1.2 Colloidal silicas

Colloidal silicas were also used as templates to synthesize porous carbons. A porous carbon with homogeneous-size spherical pores with 3D periodical alignment was synthesized from a phenolic resin with a silica template, which was prepared by sintering of colloidal SiO_2 spheres with uniform diameters in the range of 150–300 nm, followed by heat treatment at 1000°C [189]. SEM images of the template (opal) and the templated carbon are shown with a structural model of the carbon (inverse opal) in Fig. 2.63. Similar porous carbons with inverse opal structures were prepared by CVI of propylene into an opal slab at 800°C. The addition of transition metals (Ni and Fe) via their nitrate into novolac-type PF improved the electrical conductivity of the resultant mesoporous carbon prepared by templating silica xerogels [190]. Their S_{BET} and V_{total} became slightly smaller than those of the mesoporous carbon without transition metals and no differences in structural parameter (d_{002}) were detected, as listed in Table 2.13.

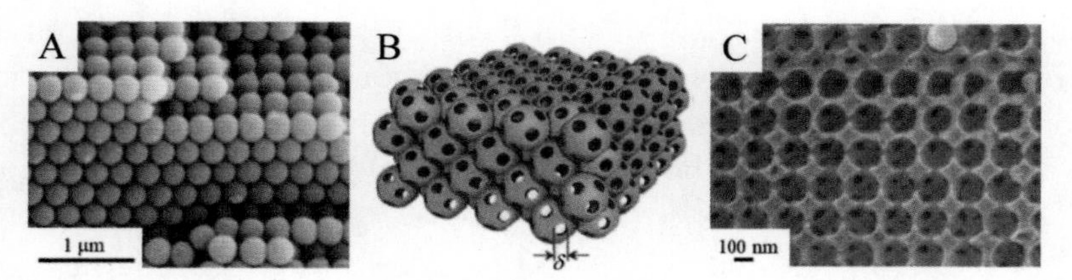

Figure 2.63 Opal and inverse opal structures. (A) SEM of a porous SiO_2 opal, (B) a structural model for an inverse opal, and (C) SEM image of the resultant carbon. δ is the diameter of the interconnections (windows) between void spaces [189].

Table 2.13 Electrical conductivity (σ) of mesoporous carbons prepared at 800°C by using silica xerogels and novolac-type phenol resins with and without the addition of Ni and Fe nitrates (c.3 mmol/g) [190].

Carbon precursor	S_{BET} (m^2/g)	V_{total} (cm^3/g)	d_{002} (nm)	σ (S/cm)
PF	1210	1.70	—	0.19
PF with $Ni(NO_3)_2$	1040	1.43	0.342	3.4
PF with $Fe(NO_3)_3$	1010	1.40	0.342	2.5

The carbons with bimodal and fully interconnected porosity were synthesized by impregnating FA into meso-/macroporous silica monoliths, followed by carbonization at 800°C [191]. They had mesopores of 2—4 nm and macropores of 0.5—30 μm, which were independently adjusted by varying the synthesis conditions of silica template. Colloidal silica spheres with 8 nm diameter and elongated silica particles (5—20 nm in diameter and 40—300 nm in length) were used as the templates by mixing with RF in aqueous solution to obtain mesoporous carbons with S_{BET} of 600—900 m^2/g and V_{total} of 1.0—2.0 cm^3/g [192].

A mesophase pitch was mixed with commercial colloidal silicas in ethanol and kept at 260°C for a short time (30 min) to obtain mesoporous carbon [193]. The temperature was slightly higher than the softening point of the pitch and was necessary to penetrate colloidal particles into the pitch before carbonization at 900°C. The silica with a surface area of about 230 m^2/g resulted in carbon with about **13 nm** mesopores and V_{total} of about 0.9 cm^3/g, and silica with 135 m^2/g resulted in the formation of mesopores of about 24 nm and V_{total} of about 0.9—1.6 cm^3/g. The mesoporous carbons prepared from a mesophase pitch using colloidal silica

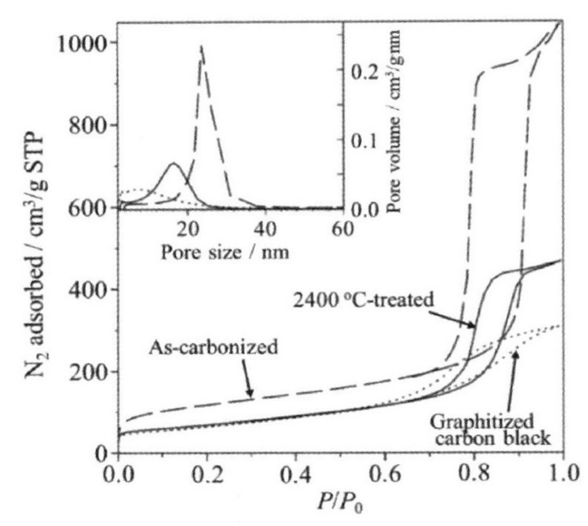

Figure 2.64 N_2 adsorption—desorption isotherms and pore size distributions (inset) of silica-templated carbons, as-carbonized (900°C) and 2400°C-treated, in comparison with a graphitized carbon black [194].

were subjected to high-temperature treatment at 2400°C [194]. N_2 adsorption–desorption isotherm and pore size distribution are shown for as-carbonized and 2400°C-treated carbons in Fig. 2.64. Although S_{BET}, V_{total}, and pore size were reduced by about a half after 2400°C treatment, from 425 to 239 m^2/g, from 1.57 to 0.72 cm^3/g, and from 23.6 to 16.5 nm, respectively, H1 type hysteresis was retained.

A mixture of silica sol stabilized by cetyltrimethylammonium bromide with RF was carbonized at 850°C [195,196]. The resultant carbon was mesoporous, with S_{BET} of 1512 m^2/g, V_{meso} of 3.6 cm^3/g, and about 12 nm sized mesopores. Through a sol–gel process using TEOS in the presence of FA and subsequent carbonization at 800°C, carbon with S_{BET} of 1170 m^2/g and V_{total} of 1.27 cm^3/g was obtained [197]. The pore structure was composed of 67% mesopores centering at 4 nm size by inheriting silica nanoparticles formed from silane. A silica-templated carbon film with a size larger than 15×25 mm^2 was prepared by spin-coating of an acidic aqueous solution of sucrose and TEOS onto silicon wafers, followed by carbonization at 400°C for 4 h and dissolution of silica template with HF [198]. The resultant carbon film had S_{BET} of 2600 m^2/g and V_{total} of 1.4 cm^3/g, consisting mainly of mesopores with the size centered at about 2.4 nm.

Carbon films prepared from resorcinol with oxalic acid using colloidal silicas with sizes of 22 and 80 nm as templates are compared with that using mesoporous silica SBA-15 [199,200]. N_2 adsorption–desorption isotherms of the carbons are shown together with size distributions of mesopores in Fig. 2.65. Silica colloids with sizes of 22 and 80 nm afforded carbons consisting of pores of the size resembling the diameter of silica colloids with

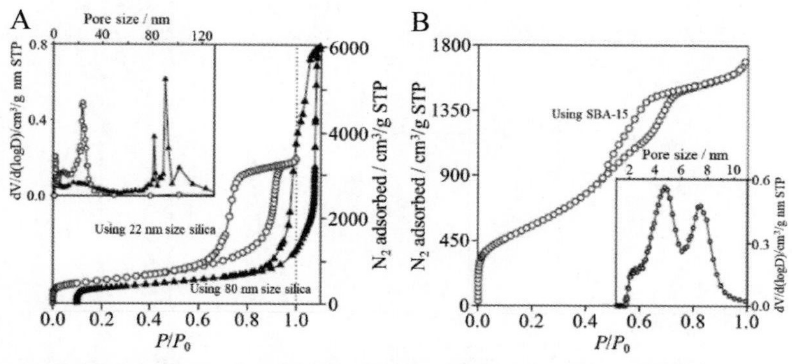

Figure 2.65 N_2 adsorption–desorption isotherms for silica-templated carbons: (A) colloidal silica templates with 22 and 80 nm diameters were used and (B) mesoporous silica SBA-15 was used [200].

extremely large pore volume (V_{total} of 5.1 and 9.1 cm^3/g, respectively), relatively large surface area (S_{BET} of 1900 and 1420 m^2/g, respectively), and thin walls (2.23 and 2.78 nm, respectively) (Fig. 2.65A). However, SBA-15 provided carbon with relatively large pore volume (2.6 cm^3/g), but it was smaller than the carbons provided by colloids, high surface area as 2000 m^2/g and bimodal pore size distribution centered at 4.5 and 7 nm (Fig. 2.65B), with the small mesopores (4.5 nm) replicating the silica template and the large pores (7 nm) being created by corruption during the dissolution of silica template. From the mixtures of polystyrene (PS) latex and colloidal silica with a volume ratio of 74/26, macroporous carbons with walls containing mesopores were prepared by heat treatment at 1000°C in Ar flow [201]. Three kinds of PS latex have uniform diameters of 450, 204, and 112 nm, and four kinds of colloidal silica have diameters of 4–6, 10–20, 40–50, and 70–100 nm. In the course of heating, PS was melted and penetrated into the spaces between the colloidal silica spheres below 300°C, with further carbonization at 1000°C to provide a very thin carbon layer on the colloidal silica, leaving macropores corresponding to the original PS particle size. The mixture of the PS with size 204 nm and silica spheres of 4–6 nm delivered carbon composed of macropores with about 190 nm size and mesopores of about 2 nm size with the highest S_{BET} of 1500 m^2/g including S_{micro} of 123 m^2/g. In these carbons, the sizes of macropores and mesopores were controllable by PS latex and silica colloid independently.

Mesocellular carbon foams with bimodal mesopore structure with sizes of 3.5 nm and larger than 20 nm, were prepared by selecting mesocellular aluminosilicates with various cell and window sizes as the templates and PF as the carbon precursor [202]. Mesoporous carbon hollow spheres were synthesized from an ethanol/water/NH_3 solution of resorcinol, formaldehyde, and tetrapropylorthosilicate (TPOS) by carbonization at 700°C for 5 h in N_2, followed by removal of silica by 5 wt.% HF [203]. The pores in the resultant carbons were tunable from micropores to 14 nm by controlling the process. TPOS was selected because of its slower hydrolysis and condensation than TEOS, and was favorable for the process control. By chlorination of ordered mesoporous SiC, which was synthesized through impregnation of SBA-15 by polycarbosilane, followed by heat treatment at 1000–1300°C, micro-/mesoporous carbons were obtained [204]. The resultant carbon chlorinated at 1000°C delivered S_{BET} of 2819 m^2/g and V_{total} of 2.26 cm^3/g including V_{micro} of 0.51 cm^3/g (supposedly V_{meso} reaching 1.7 cm^3/g).

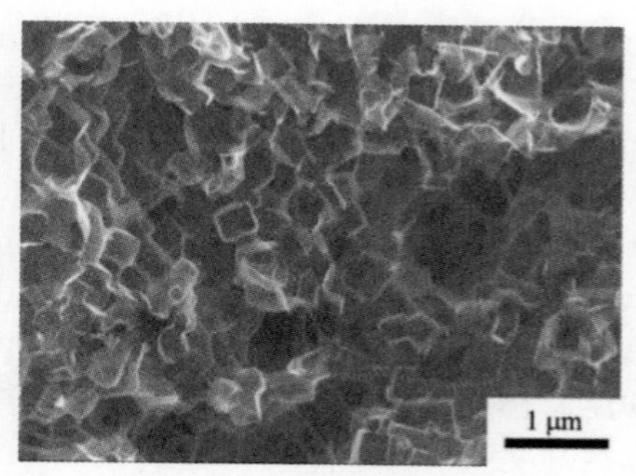

Figure 2.66 SEM image of the fractured surface of MgO-templated carbon [208].

2.2.2.2 Magnesium oxide

Porous carbons with a large amount of mesopores were prepared using magnesium oxide (MgO) particles as a template and the results were reviewed by focusing on the procedure of pore control [205,206] and on the applications in relation to their pore structure [207,208]. They are now in the market [209]. A mixture of a MgO precursor, which gave nano-sized MgO particles after its pyrolysis, with a carbon precursor was heat-treated at 900°C for 1 h in an inert atmosphere. From carbon-coated MgO particles thus obtained, template MgO was dissolved out using a diluted acid at room temperature to isolate the carbon formed. MgO was selected as a template mainly because of its chemical and thermal stability, i.e., no structural or compositional changes, no reaction with carbon at least up to the carbonization temperature of carbon precursors, and easy dissolution in a diluted acidic solution. In Fig. 2.66, an SEM image is shown of the porous carbon synthesized using a reagent grade MgO powder consisting of large cubic particles, revealing exact replication of the size and morphology of the template.

Different MgO precursors were used, magnesium acetate $Mg(CH_3COO)_2$, citrate $Mg_3(C_6H_5O_7)_2$, gluconate $Mg(C_{11}H_{22}O_{14})$, hydroxide $Mg(OH)_2$, and hydroxy-carbonate $3MgCO_3 \cdot Mg(OH)_2$. Poly(vinyl alcohol) (PVA) was often used as a carbon precursor. A coal tar pitch, hydroxyl propyl cellulose (HPC), poly(ethylene terephtharate) (PET), poly(amic acid) consisting of pyromellitic dianhydride and 4,4′-oxydianiline (PMDA/ODA), poly(vinyl pyrrolidone) (PVP), poly(-acrylamide) (PAA), and trimethylolmelamine (TMM) were also used. Mixing of two precursors was performed at different ratios either in powder (powder mixing) or solution (solution mixing). The mixing ratio was calculated by the MgO weight in the MgO precursor to the weight of the

Table 2.14 Surface areas of the MgO-templated carbons prepared.

Mixing process	Precursors	Mixing ratio in MgO/precursor	S_{BET} (m^2/g)	α_s plot analysis		
				S_{total}	S_{micro}	S_{ext}
Powder mixing	MgO/ PVA	7/3	920	959	822	137
		5/5	789	803	647	156
		3/7	546	567	457	110
	MgO/ HPC	7/3	741	723	102	621
		5/5	382	396	60	336
		3/7	249	264	40	224
	MgO/ PET	7/3	794	810	164	646
		5/5	701	724	167	557
		3/7	645	667	159	508
Powder mixing	Mg acetate/ PVA	7/3	1080	961	451	510
		5/5	886	878	466	412
		3/7	579	602	283	319
Solution mixing		7/3	1800	1788	87	1701
		5/5	980	966	65	901
		2/8	289	312	26	286
Powder mixing	Mg citrate/ PVA	7/3	1545	1459	121	1338
		5/5	1423	1346	7	1339
		3/7	1154	1102	53	1049
Solution mixing		5/5	1351	1253	2	1251
		3/7	1085	1055	97	958

carbon precursor. After carbonization at 900°C, the products prepared from most of the mixtures at MgO/carbon precursor ratios larger than 5/5 were obtained as a powder without marked aggregation of particles. No white particles were found even under high magnification, revealing that all MgO particles are coated by carbon.

Surface areas, S_{BET}, S_{total}, S_{micro}, and S_{ext}, determined by α_s plot analysis of N_2 adsorption isotherm at 77K are summarized in Table 2.14 for the carbons prepared from different combinations of MgO and carbon precursors. In most systems, S_{ext}, which is mainly due to mesopores, is predominant and depends greatly on the mixing ratio (MgO/carbon precursor), mixing method (either powder or solution mixing), and also MgO precursor.

The selection of MgO precursor is very important to obtain mesopores in the resultant carbons. In Fig. 2.67A and C, the changes to surface areas, S_{total}, S_{meso}, and S_{micro}, with mixing ratio of MgO/PVA are shown for Mg citrate and gluconate as MgO precursor, respectively [210,211]. For Mg citrate/PVA mixtures, S_{meso} increases with increasing MgO/PVA ratio,

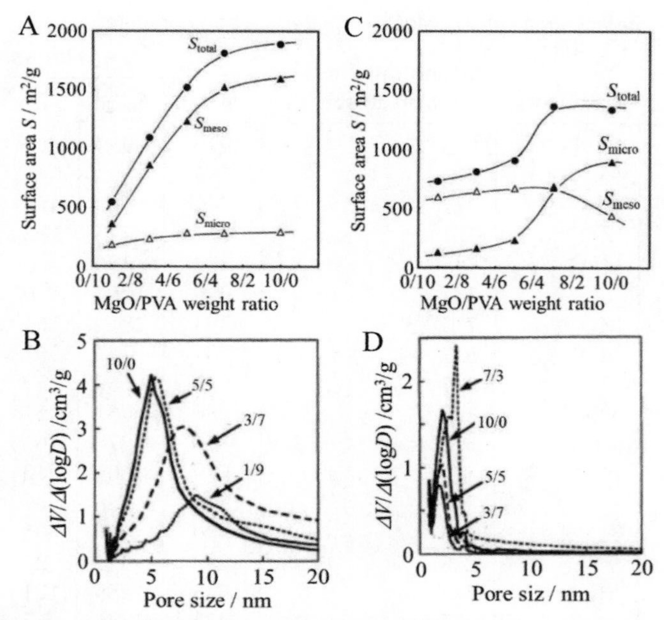

Figure 2.67 Porous carbons prepared in an Mg citrate/PVA system (A, B) and Mg gluconate/PVA system (C, D); changes in surface areas, S_{total}, S_{meso}, and S_{micro}, with a mixing ratio of MgO/PVA ratio (A, C) and pore size distributions with different MgO/PVA ratios (B, D) [211].

although S_{micro} increases slightly, i.e., mesopores are predominantly formed in this system. For Mg gluconate/PVA mixtures, however, relatively high S_{meso} of about 600 m^2/g is kept with small S_{micro} of about 200 m^2/g up to the MgO/PVA ratio of 5/5. S_{meso} then decreases and S_{micro} increases rapidly with increasing ratio above 5/5. At an MgO/PVA ratio of 5/5, S_{meso} is comparable with S_{micro}. When commercial MgO powder with a particle size of about 100 nm was used with PVA in the ratio of 7/3, microporous carbon with S_{micro} of about 880 m^2/g was obtained, although S_{meso} was about 130 m^2/g, whereas S_{meso} reached 1500 m^2/g in the systems using Mg citrate [212]. These experimental results reveal that mesopores are formed by replicating MgO particles and micropores are formed in the walls of mesopores. The decrease in S_{meso} at high MgO/PVA ratios in the system using Mg gluconate seems to be caused by the thickening of pore walls because of the high carbonization yield of gluconate.

MgO precursors also have a strong influence on mesopore sizes. In Fig. 2.67B and D, pore size distributions are shown as a function of the MgO/PVA ratio in two systems, Mg-citrate/PVA and Mg-gluconate/

PVA, respectively. Mesopores formed in the carbon derived from Mg gluconate and PVA exhibited a sharp distribution around 2–4 nm size, while the carbons derived from Mg-citrate and PVA system exhibited much broader distribution above 5 nm. Mg-citrate and Mg-gluconate delivered mesoporous carbon without PVA addition, and had sizes of about 5 and 2 nm, respectively. MgO crystals prepared from Mg-citrate and Mg-gluconate by pyrolysis at a temperature below 250°C were determined as 4 and 2 nm by XRD, corresponding to the mesopore sizes in the resultant carbons after dissolution of MgO particles. A mixture of Mg-citrate and Mg-gluconate in equal weights gave a carbon consisting of bimodal mesopores centered at 2 and 10 nm sizes [211]. By using MgO templates with sizes of 40 and 150 nm, meso-/macroporous carbons were fabricated, and were successfully applied for direct electron transfer enzymatic electrodes [213].

The mixing process also influenced the pore structure of the resultant carbons. In Fig. 2.68A and B, two processes, mixing in powder and in aqueous solution, are compared for the Mg-acetate/PVA system by S_{BET} [214]. The dependences of S_{BET} on MgO/PVA ratio are very similar by showing a maximum at the mixing ratio of 7/3, but the maximum values of S_{BET} are quite different, about 1900 m^2/g for solution mixing and about 1100 m^2/g for powder mixing. Pore composition was also different in the carbons prepared in these two systems, V_{meso} reaching 2.2 cm^3/g and a sharp peak in pore size distribution at around 13 nm for the solution mixing, but 0.37 cm^3/g and a very broad size distribution above 30 nm for the powder mixing. A slightly larger carbonization yield is obtained by the solution mixing than by the powder mixing. For Mg citrate/PVA and Mg

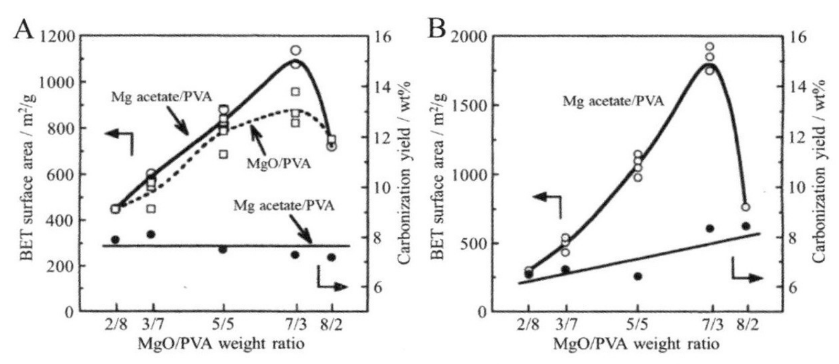

Figure 2.68 Changes in S_{BET} and yield of the resultant carbons with MgO/PVA mixing ratio in the system of Mg acetate/PVA: (A) powder mixing and (B) solution mixing. In powder mixing, S_{BET} change in the system of commercial MgO with PVA is shown for the comparison [214].

gluconate/PVA systems, however, no marked differences in surface areas and pore size distributions were observed between the powder and solution mixing.

Different carbon precursors were employed to synthesize mesoporous carbons by the MgO template method, which governed mainly the yield of the carbon. A coal tar pitch with a softening point of 85.2°C was used as a carbon precursor to obtain mesoporous carbons [215]. The Mg-acetate/pitch system gave carbons with mesopores centered at around 13 nm and the Mg-citrate/pitch system resulted in mesopores centered at around 5 nm, even though two precursors are mixed in powder. High wettability of pitches to the MgO surface seems to be the main reason for the sharp pore size distribution. Because of the high carbon yield of the pitch relative to PVA, the resultant carbons had thick pore walls and consequently contained micropores evaluated by S_{micro} as about 200 m^2/g. Amphiphilic carbonaceous materials derived from a coal tar pitch were also used by coupling with Mg citrate [216]. The carbonization process of the precursor mixture was studied by using a temperature-programmed desorption (TPD) technique. The mixtures of Mg-citrate with novolac-type PF with mixing ratios of 2/8–8/2 by weight were carbonized at 900°C in N_2 [217]. The resultant carbons were mesoporous; S_{BET} of 1037–1920 m^2/g with S_{meso} of 894–1569 m^2/g with increasing MgO/PF ratio. When poly(ethylene terephthalate) (PET) was used as a carbon precursor, dispersion of MgO precursor particles in the carbon precursor through the repeated fusion and crushing were needed to obtain a high surface area in the resultant carbons [218]. By mixing PET with either $MgCO_3$ or $Mg(OH)_2$ in MgO-precursor/PET of 3/7 and carbonization at 850°C after stabilization at 265°C in air, mesoporous carbons were obtained, which delivered S_{BET} of 1772 m^2/g and V_{total} of 3.35 cm^3/g including V_{micro} of 0.69 cm^3/g using $MgCO_3$ [219]. N-doped mesoporous carbons were synthesized using N-containing carbon precursors, poly(vinylpyrrolidone) (PVP), polyacrylamide (PAA), and trimethylolamine (TMM) with Mg-acetate [200]. Not only N and C contents but also pore structure parameters depend strongly on the carbon precursor and HTT, as shown in Fig. 2.69. PVP gave carbon with a high S_{BET} of about 2000 m^2/g, most of which was due to micropores, but relatively low N content. On the other hand, TMM gave carbon with S_{BET} of around 1000 m^2/g, but high N content.

The template MgO was experimentally demonstrated to be recycled. Acetic and citric acids were selected to dissolve out MgO from carbon-coated MgO. The recovered Mg-acetate and Mg-citrate aqueous

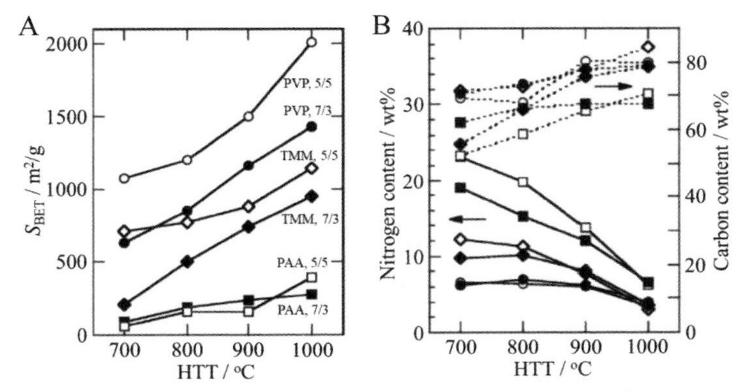

Figure 2.69 N-doped mesoporous carbons from the mixtures of Mg-acetate with N-containing carbon precursors (PVP, PAA, and TMM): (A) S_{BET} and (B) N and C contents against HTT [220].

Table 2.15 Recycling of MgO [206].

	Using acetic acid		Using citric acid			
	Mg acetate/PVA system MgO/ PVA = 5/5		Mg citrate/PVA system MgO/ PVA = 5/5		Mg citrate itself	
Cycle No.	S_{BET} (m²/g)	Carbon yield (mass%)	S_{BET} (m²/g)	Carbon yield (mass%)	S_{BET} (m²/g)	Carbon yield (mass%)
1	1210	9.8	1402	26.8	1680	9.7
2	1185	9.6	1468	25.2	1590	11.0
3	1262	9.3	1423	25.4	1621	10.1
4	1203	10.1	1415	26.1	1689	9.6
5	1249	9.1	1481	25.4	1575	10.5

solutions were mixed with PVA again and subjected to carbonization. In Table 2.15, S_{BET} and carbon yield are listed on the carbons obtained in each cycle. At least up to the fifth cycle, almost the same S_{BET} and carbon yield are obtained, revealing that MgO can be recycled in almost 100%, by supplying a carbon precursor in every cycle.

Needle-like $Mg(OH)_2$ particles were used as the precursor of the MgO template by coupling with various carbon precursors, starch [221], RF resin [222], and a coal tar pitch [223]. $Mg(OH)_2$ nanoparticles with needle-like morphology with diameters of 4−10 nm and lengths of 50−100 nm were transformed to MgO at 300−400°C by retaining their needle-like morphological feature. The needle-like morphology of these MgO

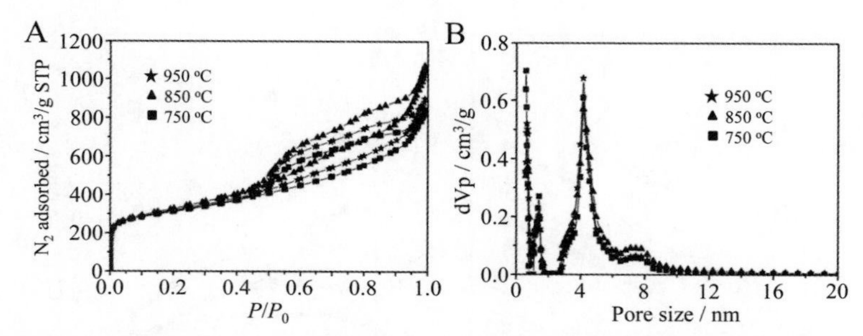

Figure 2.70 MgO-templated mesoporous carbons prepared from a PF resin with $Mg(OH)_2$ at different carbonization temperatures: (A) N_2 adsorption–desorption isotherms and (B) pore size distributions [222].

particles was replicated as hollow channels observed under TEM. N_2 adsorption–desorption isotherms and pore size distributions are shown on mesoporous carbons prepared from the mixture of PF and $Mg(OH)_2$ [222] in Fig. 2.70A and B, respectively, revealing the formation of a large amount of mesopores with diameters of about 4 nm. The carbon prepared at 950°C delivered V_{total} of 1.62 cm^3/g including V_{micro} of 0.80 cm^3/g. On the mixture of a coal tar pitch with $Mg(OH)_2$, stabilization in air before carbonization resulted in a larger pore volume than without stabilization; the carbon prepared from a mixture of $Mg(OH)_2$/pitch of 6.5/3.5 by weight via air stabilization and following carbonization at 950°C delivered V_{total} of 1.33 cm^3/g [223].

A powder mixture of sodium carboxymethyl cellulose (NaCMC), magnesium acetate [$Mg(OAc)_2 \cdot 4H_2O$], and zinc acetate [$Zn(OAc)_2 \cdot 2H_2O$] in a weight ratio of 1/5/0.5 was heat-treated at 1000°C in Ar flow, followed by rinsing with dilute HCl and water [224]. The resultant carbon delivered a high S_{BET} of 1596 m^2/g and a high V_{total} of 5.93 cm^3/g, whereas S_{micro} and V_{micro} were relatively small, 366 m^2/g, and 0.22 cm^3/g, respectively, suggesting the predominant contributions of mesopores in this carbon. During carbonization, Zn(OAc) decomposed to ZnO above 350°C, then reduced to metallic Zn by the reaction with carbon above 800°C and finally vaporized out at 1000°C.

The presence of MgO in carbon precursors was experimentally shown to influence strongly the yield of carbon [225]. Polypropylene (PP), low-density and high-density polyethylene (HDPE and LDPE, respectively), polystyrene (PS), polyethylene terephthalate (PET), and polyvinyl chloride (PVC) were mixed with MgO in the ratio of 1/1–8/1 by weight and then carbonized in an autoclave at 500°C in N_2. The carbon yield from the mixtures of PP, HDPE, LDPE, and PS with MgO in 6/1 ratio was in the range from 27% for PS to 24% for HDPE, although these polymers gave negligibly small carbon yields without MgO. On the other hand, the influence of MgO existence during carbonization was not marked on PVC and PET. The pyrolysis of PVC generated 38.5 wt.% of HCl to react with MgO to form $MgCl_2$.

MgO particles formed from $MgCl_2$ during carbonization with a carbon precursor were reported to work as a template for micropores in carbon. Porous carbon nanofibers were prepared by electrospinning of the DMF solution of PAN and $MgCl_2$ in different ratios, followed by stabilization at 250°C, carbonization at 1050°C, and dissolution of MgO with 1 M H_2SO_4 [226]. $MgCl_2$ changed to MgOHCl during the stabilization process and to MgO during carbonization. The nanofibers prepared are microporous, with S_{BET} of 800 m^2/g and V_{micro} of 3.32 cm^3/g. Nonwoven webs of porous carbon nanofibers were fabricated by electrospinning of the DMF solution of a PAN/pitch (3/7 by weight) with MgO, followed by stabilization at 300°C in air and carbonization at 1050°C in N_2 flow [227]. Loading of MgO nanoparticles onto the carbon nanofibers was reported to enhance the development of micropores in the fibers [228]. Nanofibers prepared from a polymer blend of a polyimide PI (PMDA/ODA) with PVP by electrospinning and imidization at 300°C were immersed into 1 M $Mg(NO_3)_2$ aqueous solution at 60°C after plasma surface treatment, followed by carbonization at 900°C with heating rates of 3 and 10°C/min and washing with 5 M HCl to remove MgO formed during carbonization. The presence of MgO nanoparticles on the PVP/PI nanofibers promotes the development of micropores in the fibers, as shown in the N_2 adsorption–desorption isotherms in Fig. 2.71A. The nanofibers obtained

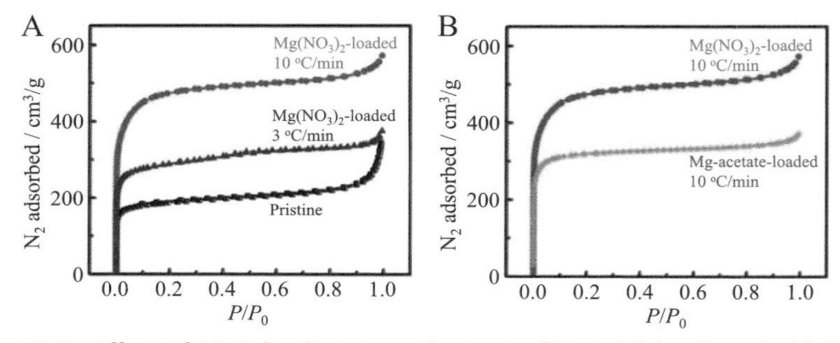

Figure 2.71 Effect of MgO loading on carbon nanofibers: (A) loading via $Mg(NO_3)_2$ with different heating rates and (B) comparison with loading of Mg-acetate [228].

by loading $Mg(NO_3)_2$ with the $10^{\circ}C/min$ heating rate delivered S_{BET} of $1836\ m^2/g$ and S_{micro} of $1762\ m^2/g$, whereas the pristine nanofibers without $Mg(NO_3)_2$ loading delivered 724 and $688\ m^2/g$, respectively. Loading of $Mg(CH_3COO)_2$ (Mg-acetate) is less effective, as shown in Fig. 2.71B. Not only MgO but also nitrogen oxides formed from $Mg(NO_3)_2$ during the carbonization process may assist in the formation (or opening) of micropores in the nanofibers.

In a CO_2 closed system of the mixture of MgO and Mg in powder (MgO/Mg of 8 by weight), the following reaction could self-propagate by an initial thermal stimulus,

$$Mg + CO_2 = 2MgO + C,$$

and deposit mesoporous carbon onto MgO particles [229]. This process was a self-propagating high-temperature synthesis (SHS) and both MgO as-mixed and newly formed by the above reaction worked as a template (coupling of SHS with MgO templating). The resultant carbon after dissolution of MgO showed a **type IV** N_2 adsorption–desorption isotherm with H3 hysteresis, of which S_{BET} was measured as $709\ m^2/g$.

Mesoporous carbons were synthesized from Ba citrate by carbonization at 700°C in N_2 and compared with those synthesized from Mg citrate [230]. BaO-templated carbon had S_{BET} of $955\ m^2/g$ with V_{meso} of $3.09\ cm^3/g$ due to large mesopores (average pore size of 13.3 nm) and low V_{micro} of $0.08\ cm^3/g$, whereas MgO-templated carbon had S_{BET} of $2322\ m^2/g$ with V_{meso} of $0.85\ cm^3/g$ due to small mesopores (average pores size of 2.3 nm) and V_{micro} of $0.48\ cm^3/g$.

2.2.2.3 Eutectic metal salts

Eutectic metal salts, $LiCl/ZnCl_2$ (23 mol% LiCl, T_m of 292°C), NaCl/ $ZnCl_2$ (42 mol% NaCl, T_m of 270°C), and $KCl/ZnCl_2$ (51 mol% KCl, T_m of 230°C), were used as templates to synthesize porous carbons, mainly mesoporous (salt templating). The mixture of glucose with different amounts of eutectic salt in water was carbonized under hydrothermal conditions at 180°C, followed by drying in a vacuum [231]. To synthesize N-doped porous carbons, 2-pyrrol-carboxyaldehyde (PCA) was added to the glucose (PCA/glucose of 1/6 by weight). The resultant carbons were composed of carbon particles with sizes in the 10 nm range, which were interconnected to give a porous, aerogel-like texture with S_{BET} of $400-650\ m^2/g$. By using the same eutectic salts, N-doped porous carbons were synthesized from ionic liquids (ILs), Bmp-dca and Emim-dca, as

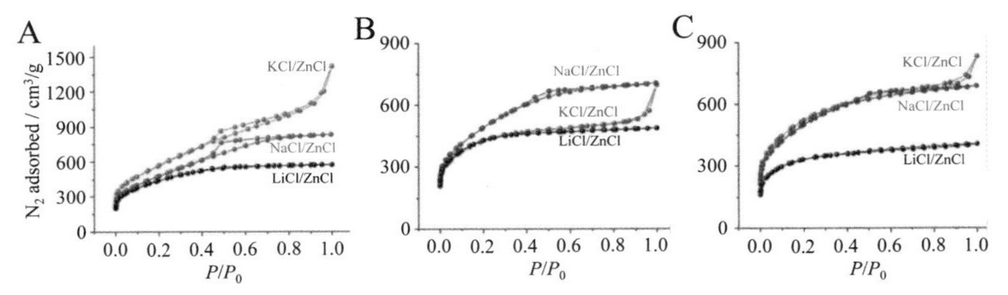

Figure 2.72 N_2 adsorption–desorption isotherms for the porous carbons prepared from ionic liquids by salt templating: (A) Bmp-dca, (B) Emin-dca, and (C) Emin-tcb [232].

carbon precursors at 1000°C, and N/B-co-doped carbon from Emim-tcb at 1400°C in N_2 flow [232]. The residual salts were washed out with water for several hours. The N_2 adsorption–desorption isotherms are shown on the carbons prepared from the mixtures with the eutectic salt/IL ratio of 3 by weight in Fig. 2.72. For the carbons templated with NaCl/$ZnCl_2$ their adsorption isotherms show a further uptake of N_2 in the medium P/P_0 region as well as a small hysteresis, suggesting the formation of supermicropores and small mesopores. For the carbons templated with KCl/$ZnCl_2$ having the lowest T_m, an additional uptake in the high P/P_0 is observed, showing the formation of macropores. The resultant carbons exhibited high S_{BET} of 1500–2000 m^2/g and V_{total} of 0.57–1.70 cm^3/g. The N-doped carbon prepared from Bmp-dca with KCl/$ZnCl_2$ gave the highest S_{BET} and V_{total} with the highest $V_{micro}(N_2)$ of 1.14 cm^3/g and V_{meso} of 0.56 cm^3/g (Fig. 2.72A). The N/B-co-doped carbons exhibited S_{BET} of 1100–1800 m^2/g and V_{total} of 0.57–1.04 cm^3/g with a low V_{meso} (Fig. 2.72C).

The main advantages of this salt templating technique are (1) the formation of a homogeneous starting aqueous solution of carbon precursor (glucose and ILs) and template (eutectic salts), which can be easily shaped and processed, (2) simple aqueous removal of the template after carbonization, and (3) the template salt can be recovered for recycling use. The technique was reviewed for the synthesis of functionalized carbons from ILs and polymerized ILs [233].

2.2.2.4 Other hard templates

A process using $Ni(OH)_2$ template with phenol in an ethanol solution was proposed to prepare mesoporous carbons which were subjected to study on supercapacitor performances [234]. Also, nanoparticles of TiO_2 were used as a template to obtain carbon hollow spheres [235]. In these cases, HCl and

HF had to be used to dissolve the templates (NiO and TiO_2). Ba-citrate gave porous carbons containing both micropores and mesopores, but much lower S_{BET} than the carbon prepared from Mg-citrate [230,236]. Ca-acetate was used as the precursor of $CaCO_3$ template for melamine-formaldehyde carbonization by coupling with KOH activation to fabricate the micro-/mesoporous carbons, giving S_{BET} of 1525 m^2/g, V_{total} of 1.59 cm^3/g, with V_{meso} of 1.12 cm^3/g [237].

The ZnO template employed for the carbonization of petroleum pitch (softening point of about 120°C) with activation agent, either K_2CO_3 or KOH, resulted in micro-/mesoporous carbons [238]. The mixture of ZnO/pitch/K_2CO_3 of 4/1/0.5 (by weight) delivered S_{BET} of 1170 m^2/g with S_{micro} of 397 m^2/g and S_{meso} of 773 m^2/g and that of ZnO/pitch/KOH of 4/1/1 delivered 1979 with 725 and 1254 m^2/g, respectively. Mesoporous N-doped carbons were obtained from Zn- and Ca-citrates by heating at 800°C for 1 h in N_2, followed by washing with diluted HCl [239]. The resultant carbons were mixed with melamine and heat-treated at 800°C for another 1 h to dope nitrogen. Metal oxides formed by decomposition of metal citrates were supposed to work as templates for mesopore formation, although the authors described it as a "template-free approach", and the citrate worked as a carbon precursor. N_2 adsorption–desorption isotherms and pore size distributions of the resultant carbons are shown in Fig. 2.73A and B, respectively. ZnO-templated carbon possesses bimodal mesopores, centering at 3.2 and 10.5 nm, and delivers S_{BET} of 1190 m^2/g, whereas a CaO-templated one composed from mesopores centered at 10.5 nm with S_{BET} of 1350 m^2/g. Both ZnO- and CaO-templated mesoporous carbons contained relatively high contents of

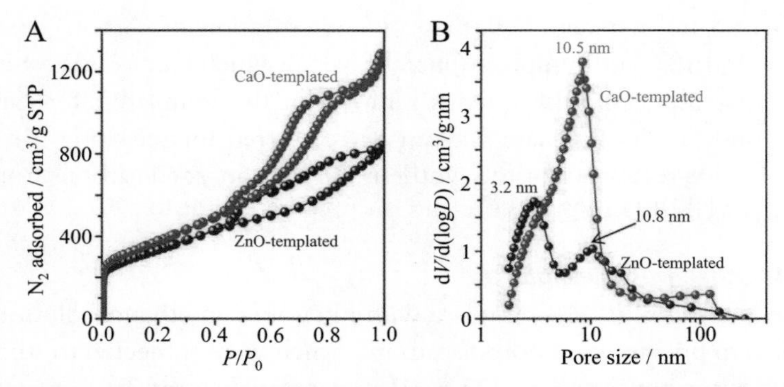

Figure 2.73 ZnO- and CaO-templated N-doped carbons: (A) N_2 adsorption–desorption isotherms and (B) pore size distributions [239].

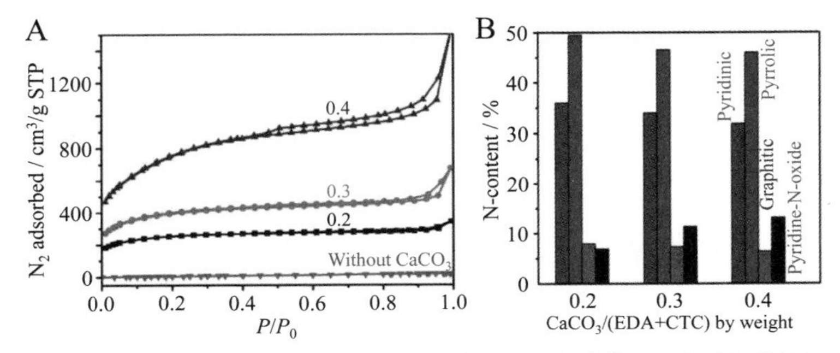

Figure 2.74 $CaCO_3$-templated mesoporous carbons with different $CaCO_3$/(EDA + CTC) weight ratios (0.2–0.4): (A) N_2 adsorption–desorption isotherms and (B) N configurations in the carbons [240].

nitrogen, 8.5 and 9.2 wt.%, respectively, and as a consequence high-performance oxygen reduction reaction (ORR) activity, almost comparable to the commercial catalyst Pt/C.

Hydrophobic $CaCO_3$ nanoparticles (about 50 nm diameter) were used as a template for carbonization at 700°C of the mixture ethylenediamine (EDA) and carbon tetrachloride (CTC), where the weight ratio of EDA to CTC was 0.4 and that of $CaCO_3$/(EDA + CTC) was 0.2–0.4, to synthesize N-doped porous carbons [240]. As shown in Fig. 2.74A, pore structure in the resultant carbons was developed with an increase in the template content, the carbon obtained using $CaCO_3$/(EDA + CTC) of 0.4 exhibiting S_{BET} of 2535 m^2/g with S_{meso} of 2090 m^2/g and V_{meso} of 1.40 cm^3/g with V_{micro} of 0.24 cm^3/g. The resultant carbons contain a relatively large amount of N, 16.1–9.5 wt.% in mostly pyrrolic and pyridinic configurations, as shown in Fig. 2.74B. $NaHCO_3$ worked as template by carbonization with methyl cellulose, urea, and KOH at 800°C in N_2 to obtain N-doped mesoporous carbons [241]. The resultant carbon had S_{BET} of 1748 m^2/g with V_{total} of 1.47 cm^3/g including 68% mesopores. The direct pyrolysis of Ca-citrate powder at 700–1000°C resulted in mesoporous carbons and the recovered Ca species could be recycled as template [242]. The mesoporous carbon prepared at 1000°C had S_{BET} of 1275 m^2/g with S_{ext} of 1140 m^2/g and the carbon prepared using recycled Ca-citrate had almost the same pore structure having S_{BET} of 1446 and S_{ext} of 1113 m^2/g.

γ-Alumina Al_2O_3 nanoparticles were used as template for the synthesis of sponge-like mesopore frameworks (mesoporous sponges) by CVD of CH_4 in 20 vol% CH_4/N_2 gas at 900°C [243]. The resultant sponges, as-deposited and annealed at 1800°C in vacuum, are compared with highly

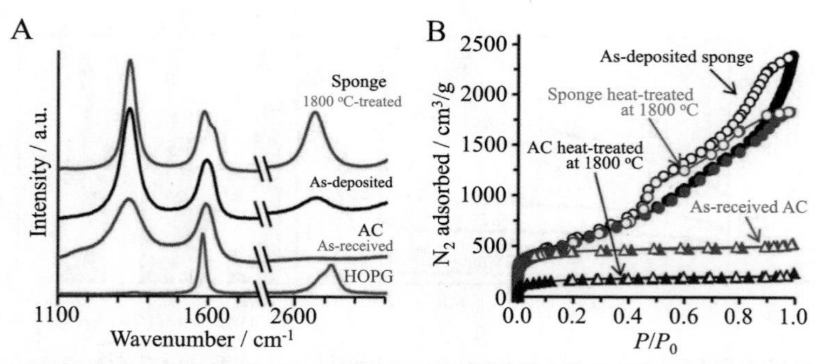

Figure 2.75 Al_2O_3-templated mesoporous sponges as-deposited and 1800°C-treated: (A) Raman spectrum and (B) N_2 adsorption−desorption isotherm in comparison with HOPG, an activated carbon (AC) [243].

oriented pyrolytic graphite (HOPG) and an as-received and 1800°C-annealed ACs on Raman spectra and N_2 adsorption−desorption isotherms in Fig. 2.75. The sponges were composed from mesopores with mean size of 5.8 nm, of which the walls were supposed to consist of thin graphene sheets. The sponge after 1800°C annealing had S_{BET} of 1940 m^2/g and V_{total} of 2.79 cm^3/g with a relatively low V_{micro} of 0.47 cm^3/g. The sponge was mechanically tough and extremely elastic, mesopores (5.8 nm size) being possible to be reversibly compressed down to 0.7 nm by applying mechanical force and had high electrical conductivity of about 7 S/cm under 86 MPa pressure, probably owing to the framework of thin graphene sheets.

Rice husks gave mesoporous carbons by carbonization and then removal of mineral residues (mainly SiO_2) [244]. Three step procedures were employed for carbonization, pyrolysis either in a fluidized bed at 450−600°C or in N_2 atmosphere at 400−700°C, carbonization and reacting silica with K_2CO_3 and Na_2CO_3 at 750−1000°C, and then removal of soluble potassium and sodium silicates by washing. Potassium and sodium silicates formed from mineral residues in the precursor husks by the reaction with K_2CO_3 and Na_2CO_3 during carbonization might work as templates for mesopores. The effect of carbonization temperature is shown by the N_2 adsorption−desorption isotherms in Fig. 2.76. The 900°C-treated carbon exhibited the highest S_{BET} of about 1600 m^2/g with S_{meso} of 934 m^2/g and V_{total} of 1.44 cm^3/g with V_{micro} of 0.35 cm^3/g. The fraction of mesopores from the total pore volume was about 76%.

Mesoporous ribbon-shaped carbon nanofibers were prepared by electrospinning of an ethanol solution of PF resin with $Co(NO_3)_2 \cdot 6H_2O$ (mass

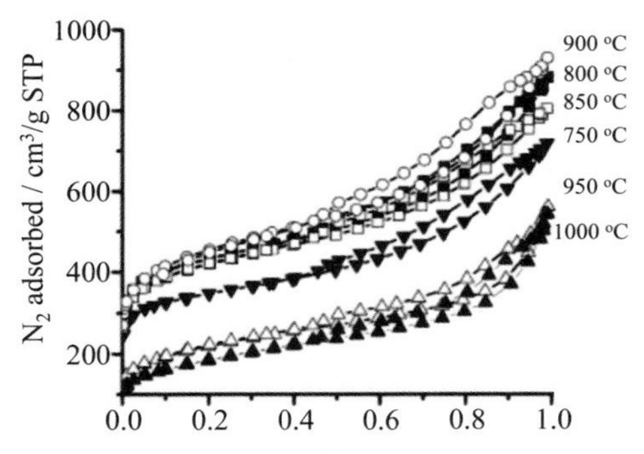

Figure 2.76 Mesoporous carbons prepared from rice husks by pyrolysis at 500°C in a fluidized bed, followed by carbonization at different temperatures [244].

ratio to phenolic resin of 0.5, 1, and 1.5) and PVP, followed by carbonization at 800°C and washing in 6 M HCl and deionized water [245]. The Co particles having sizes of 5–10 nm formed from $Co(NO_3)_2$ worked as the template for mesopores and PVP as that for micropores. The resultant nanofibers derived from the mass ratio of 1.5 exhibited V_{total} of 0.575 cm^3/g including V_{micro} and V_{meso} of 0.093 and 0.482 cm^3/g, respectively.

Metal-containing porous clay heterostructures (PCHs) were used as templates for the carbonization of sucrose at 900°C [246]. PCHs containing Fe, Zn, and Cu with and without modification by TEOS were mixed with sucrose in water containing H_2SO_4, followed by calcination at 160°C and carbonization at 900°C. The resultant carbons after removing the residues of the templates were mesoporous, for example, S_{BET} of 492 m^2/g with V_{total} of 0.541 cm^3/g including V_{micro} of 0.214 cm^3/g for the carbon synthesized using Fe-containing PCHs. Mesoporous carbons were prepared from FA and PCHs, which were synthesized by calcination of the mixture of a clay with TEOS and different neutral amine at 650°C [78]. PCHs contain large micropores and small mesopores were controlled by changing amines with different numbers of carbon atoms, i.e., octylamine, decylamine, and dodecylamine, which were coded as PCH8, PCH10, and PCH12, respectively. The pore structure of the PCH templates themselves and the carbons prepared from the mixtures of PCHs with FA at different temperatures were evaluated by S_{BET}, V_{total}, and V_{micro}, as listed in Table 2.16 together with (V_{total} − V_{micro}), which seems to be mainly due to small mesopores. The template PCHs themselves are rich in micropores, PCH10 having the highest V_{micro}. However, certain amounts of mesopores are created in the carbons prepared from FA using these PCHs as templated.

Table 2.16 Pore structure parameters of porous clay heterostructures (PCHs) and the carbons prepared from furfuryl alcohol (FA) using PCH templates [78].

PCH	Carbonization temperature (°C)	S_{BET} (m^2/g)	V_{micro} (cm^3/g)	$V_{total} - V_{micro}$ (cm^3/g)
PCH8	—	752	0.30	0.13
PCH8 + FA	700°C	1035	0.36	0.47
PCH10	—	818	0.45	0.14
PCH10 + FA	500°C	769	0.28	0.27
	600°C	1054	0.47	0.26
	700°C	1469	0.68	0.31
	800°C	1239	0.60	0.32
PCH12	—	917	0.38	0.13
PCH12 + FA	700°C	1202	0.55	0.28

Carbonization temperature influences micropore formation; the development in mesopores associating with micropores is obtained at 700°C.

Natural clay, halloysite [$Al_2Si_2O_5(OH)_4 \cdot 2H_2O$], was used as template to synthesize N-doped mesoporous carbons [247]. The mixture of halloysite powder with glucose and urea (glucose/urea of 1/1 by weight) in an aqueous solution was precarbonized under hydrothermal condition at 200°C, followed by carbonizing at 1000°C in N_2 and by dissolving out of the template using 4 M ammonium hydrogen difluoride. When furfural ($C_5H_4O_2$) was used as a carbon precursor in its ethanol solution, halloysite powder needed to be acidified using H_2SO_4 in advance. In Fig. 2.77, N_2 adsorption–desorption isotherms and pore size distributions are compared between the carbons derived from glucose and furfural, revealing that both yield type IV isotherms with H3 hysteresis loops ($P/P_0 > 0.4$), whereas N_2 adsorption volumes are very different. The furfural-derived carbon contains a large amount of mesopores at around 3.6 nm, together with micropores at around 1.5 nm (Fig. 2.77B), while the glucose-derived carbon has only

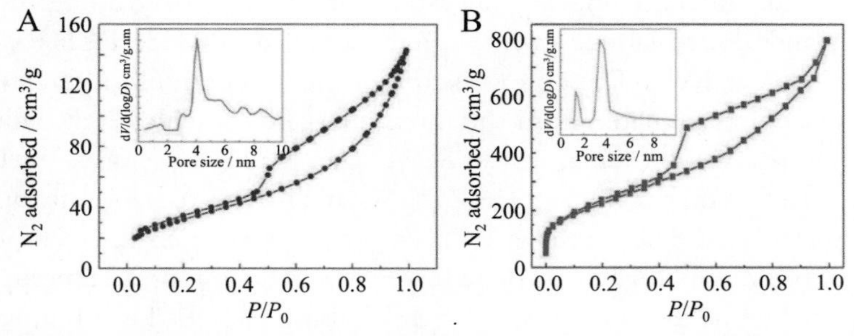

Figure 2.77 N_2 adsorption–desorption isotherms with pore size distributions (insets) of mesoporous carbons prepared using halloysite template from (A) glucose and (B) furfural [247].

mesopores at around 4.1 nm; the former has S_{BET} of 824 m^2/g and V_{total} of 1.23 cm^3/g and the latter 118 m^2/g and 0.22 cm^3/g, respectively.

2.2.3 Precursor design

2.2.3.1 Polymer blends

2.2.3.1.1 Block copolymers

Polymer blends of phenolic resins with block copolymers were successfully employed to fabricate mesoporous carbons. As the matrix-forming polymers, resol- and novolac-type phenol-formaldehyde (PF), resolcinol-formaldehyde (RF), and phloroglucinol-formaldehyde (PhF) were used. As labile polymers, triblock copolymers, poly(ethylene oxide)-*block*-poly(propylene oxide)-*block*-poly(ethylene oxide), either F127 (PEO_{106}-PPO_{70}-PEO_{106}), F108 (PEO_{132}-PPO_{50}-PEO_{132}), or P123 (PEO_{20}-PPO_{70}-PEO_{20}) were often used. Highly ordered mesoporous polymers could be synthesized from phenol with F127, for example, in a solution via an evaporation-induced self-assembly method (EISA), followed by thermal polymerization and carbonization at high temperatures, resulting in ordered mesoporous carbons. The advantage of this process is the direct use of the self-assembly of the surfactant block copolymers as templates, although the same surfactants were used for the preparation of mesoporous silicas. Since block copolymers act as a sacrificial template for pore formation in the polymers and consequently in the resultant carbons, this process is often classified into the template carbonization method and block copolymers are called soft templates in contrast to hard templates, such as zeolites, silicas, MgO, etc. [248]. However, it is also classified into one of polymer blend techniques [249]. Here, it is classified into a polymer blend because all additives blended with carbon precursors are decomposed to gases and no additional processes to remove residues are needed.

Mixture of a commercial triblock copolymer (Pluronic F127), resorcinol (R), triethyl orthoacetate (EOA), and formaldehyde (F) in a water/ethanol/HCl mixed solution was spin-coated on a silicon substrate and then carbonized up to 800°C to obtain mesoporous carbon films, where R, F, and EOA worked as carbon precursors [250]. The periodic structure of RF/EOA and F127 was established in the original organic films due to EISA with a spacing of 9.2 nm and was retained even after carbonization at 800°C, although the spacing was shortened to 7.0 nm. The surfactant F127 was decomposed up to 400°C to leave mesopores. The carbon films prepared at 800°C had S_{BET} of 1354 m^2/g, V_{total} of 0.743 cm^3/g, and mesopore sizes of around 5.9 nm. When a water/ethanol solution of

1,5-dihydroxynaphthalene with formaldehyde/F127 ratio of 1/0.003 was used, carbon having channels with diameters of 6.5 nm was obtained, the channels being arranged perpendicular to the film surface with hexagonal symmetry [251]. The self-assembly of resorcinol/formaldehyde and F127 under highly acidic conditions (1.5 M HCl) afforded the carbons a highly ordered hexagonal mesostructure with S_{BET} of about 600 m^2/g, V_{total} of about 0.60 cm^3/g, and channel diameter of 6.3 nm [252]. By changing the concentration of F127 in the starting solution, a carbon with hexagonal arrays of channels (phenol/F127 = 1/0.012) and one with cubic arrays of mesopores (phenol/F127 = 1/0.005−0.006) could be prepared [253]. The former consisted of channels with diameters of 2.9 nm, S_{BET} of 968 m^2/g, and V_{total} of 0.56 cm^3/g, and the latter consisted of mesopores with size of about 3.7 nm, S_{BET} of 778 m^2/g, and V_{total} of 0.44 cm^3/g. As shown in Fig. 2.78, N_2 adsorption−desorption isotherms for two mesoporous materials, polymers and carbons, are type IV with a pronounced hysteresis and have sharp pore size distribution. By coupling phloroglucinol/formaldehyde with F127, mesoporous carbons were prepared in monolith, fiber, and film morphologies much more quickly and under much milder conditions than when using phenol/ and resorcinol/formaldehydes [254]. From the mixture of PhF with F127 through a dual-phase separation process, macroporous carbon with mesoporous walls was prepared, where the macropore sizes were tunable without alteration of mesopore sizes [255]. From the composite prepared by deposition of benzyl alcohol at 300°C into a film of F127 on a silicon substrate, ultrathin carbon films with thickness of about 15 nm were synthesized [256]. In the resultant carbon films, the

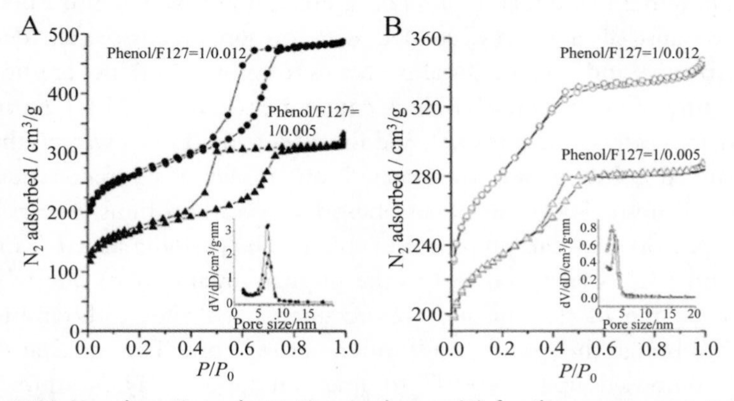

Figure 2.78 N_2 adsorption−desorption isotherms (A) for the mesoporous polymers and (B) for the resultant carbons with different phenol/F127 ratios and their pore size distribution (insets) [253].

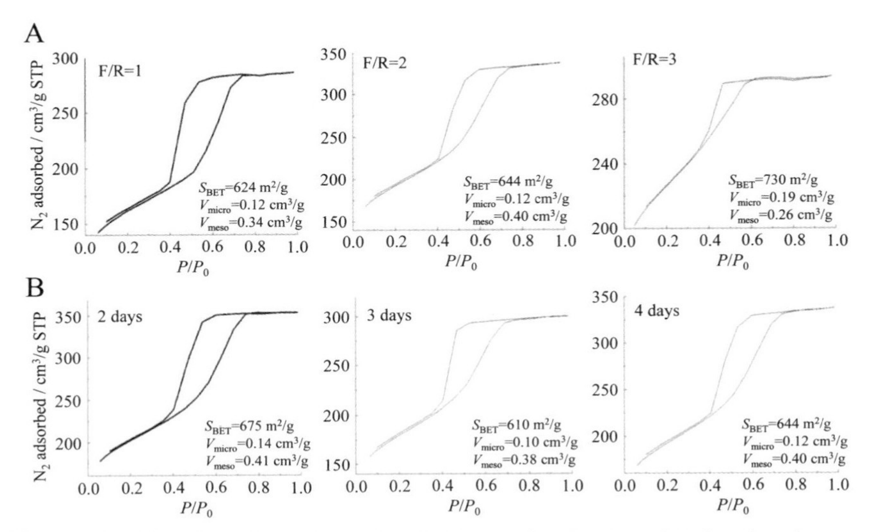

Figure 2.79 N_2 adsorption–desorption isotherms for the templated ordered mesoporous carbons under hydrothermal conditions: (A) effect of F/R molar ratios at a hydrothermal treatment time of 4 days and (B) effect of hydrothermal treatment time at an F/R ratio of 2 [257].

channels with diameter of 9.4 nm were running along the film surface and parallel with each other.

Ordered mesoporous carbons were synthesized from RF resin with F123 under hydrothermal conditions [257]. Effects of precursor composition expressed by the molar ratio of F/R and hydrothermal treatment time on the pore structure of the final carbons are shown by N_2 adsorption–desorption isotherms with some pore parameters in Fig. 2.79. To obtain a highly developed mesoporous structure, there is an optimum condition on the ratio of F/R (Fig. 2.79A); V_{meso} is the highest at F/R of 2. By increasing F/R to 3, V_{micro} tends to increase associated with the decrease in V_{meso}. The hydrothermal treatment time does not have a marked effect on the pore structure of the resultant carbons (Fig. 2.79B). Ordered mesoporous carbons were synthesized under hydrothermal treatment by replacing formaldehyde with hexamine [258].

Resins of RF and PhF were successfully used as carbon precursors for the process using F127 [259]. The precipitation of PF from the blend of phenol and formaldehyde with F127 in an ethanol/water mixed solution under a low concentration of HCl (10^{-2} M) at room temperature was so slow that the resulting polymer was not porous. However, the precipitation of RF occurred quicker than with PF, and the resulting carbon monoliths

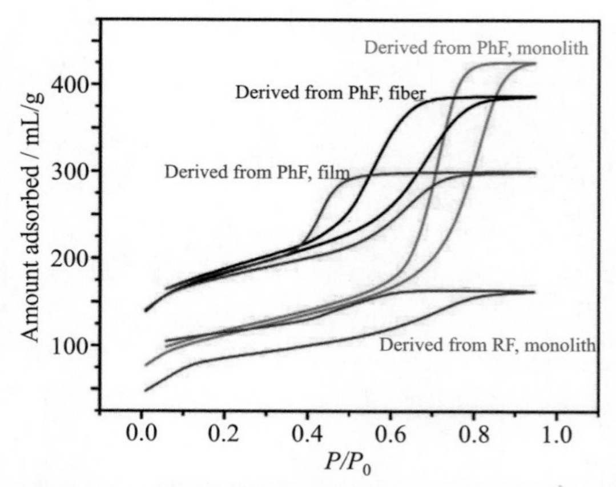

Figure 2.80 N_2 adsorption–desorption isotherms of the carbons with various morphologies derived from the blends of PhF and RF with F127 [259].

were mesoporous, giving S_{BET} of 288 m^2/g, as in the N_2 adsorption–desorption isotherms shown in Fig. 2.80. On the other hand, PhF was polymerized much faster than either PF or RF, and could give porous carbons of different morphologies, monolith, fiber, and film, which contained micropores together with mesopores, as suggested by Fig. 2.80. The differences in precipitation behavior and pore formation effectiveness were explained by the formation of hydrogen-bonding networks of ethylene oxide chains in F127 with phenol, resolcinol, and phloroglucinol.

By using another triblock copolymer Pluronic P123 with PF in an aqueous solution, mesopores with bicontinuous cubic structure were obtained in the resultant carbon [260]. Ordered mesoporous carbons were prepared through a basic aqueous solution of PF and triblock copolymer templates F127 and P123 [261]. A triblock copolymer $(PPO)_{15}$-$(PEO)_{22}$-$(PPO)_{15}$ was also used as a template by coupling with resorcinol, triethyl orthoacetate, and formaldehyde to synthesize a porous carbon with S_{BET} of 667 m^2/g and S_{ext} of 194 m^2/g with a mesopore percentage of 71% [262]. Other triblock copolymers such as L64 [$(EO)_{13}$-$(PO)_{30}$-$(EO)_{13}$], and Brij76 [$C_{18}(EO)_{10}$] were also used as templates for producing mesoporous carbons by coupling with PF and compared with F127 and P123 copolymers [263]. Brij76 resulted in carbon with S_{BET} of 1442 m^2/g and V_{total} of 1.38 cm^3/g with 2.3 nm-sized mesopores, and L64 in S_{BET} of 1898 m^2/g and V_{total} of 1.34 cm^3/g with 2.0 nm-sized mesopores. A triblock copolymer of acrylonitrile (AN) with η-butyl acrylate (BA), $(AN)_{45}$-$(BA)_{530}$-$(AN)_{45}$, itself could be converted to carbon on a cleaved mica or silicon wafer after

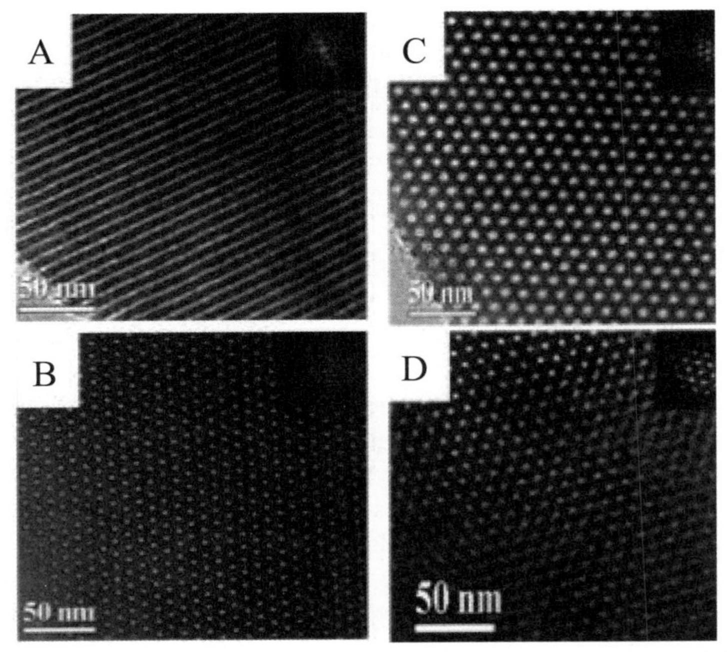

Figure 2.81 TEM images for the resultant ordered mesoporous carbons with hexagonal symmetry (A, B) and cubic symmetry (C, D). Images (A) and (C) are taken on cross-sections perpendicular to those of (B) and (D), respectively [265].

stabilization at 200–230°C, but no information on pore structure was presented [264].

Different pore structures in carbon materials, two-dimensional hexagonal, three-dimensional bicontinuous, body-centered cubic, and lamellar, were obtained by simply adjusting the ratio of phenol to template surfactant [265]. The compositions in a range of phenol/F127 ratio of 1/0.010–0.015 (by mol) and phenol/P123 of 1/0.007–0.016 gave carbons with channels with diameters of about 2.8 nm in a hexagonal arrangement even after carbonization at 1200°C, as shown in the TEM images of two perpendicular cross-sections in Fig. 2.81A and B. From the compositions in a range of phenol/F127 of 1/0.003–0.008 and phenol/F108 of 1/0.005–0.010, carbons having mesopores with mean diameters of 3.8 nm arranged in cubic symmetry were synthesized, as shown in Fig. 2.81C and D. Its N_2 adsorption–desorption isotherm was type IV, with clear hysteresis at P/P_0 of around 0.5. The carbon, in which bicontinuous mesopores with mean size of 2.3 nm were arranged in cubic symmetry, was obtained from the compositions in a very narrow range of phenol/P123 of 1/0.018–0.019.

By changing the phenol/P123 ratio to 1/0.022−0.027, a lamellar mesoporous structure was formed. A similar effect of mixing ratio of carbon precursor/F127 was reported, together with the results on the mixture of F127 and P123 [266].

The stability of hexagonal symmetry arrays of the channels in the film of PF and triblock copolymer composites and that after removal of template at 350°C was discussed taking into account the constraint due to wetting of thin film with the substrate and the contraction stress during template removal [267]. The removal of the template at 350°C from the composite film templated by F127 changed the pore structure from two-dimensional hexagonal symmetry to disordered structure, but the films templated by P123, which had shorter PEO segments than F127, could give well-ordered channels even after the removal of the template. The composition of phenol/P123 to give ordered channels was markedly narrowed for thin films in comparison to bulk powders, the range of 1/0.010−0.013 for the film but 1/0.008−0.016 for the powder. The frameworks in these mesoporous carbons were shown to be relatively rigid [252], even though S_{BET} and V_{total} decreased gradually with increasing heat treatment temperature, as shown in the N_2 adsorption−desorption isotherms and pore size distribution in Fig. 2.82A and B. S_{BET} and V_{total} were measured as 607 m^2/g and 0.58 cm^3/g, respectively, after 850°C treatment and 230 m^2/g and 0.30 cm^3/g after 2600°C.

An ethanol/water solution of phenol resin with F127 was coated onto the surface of a porous α-Al_2O_3 tubular support (average porosity of 40%)

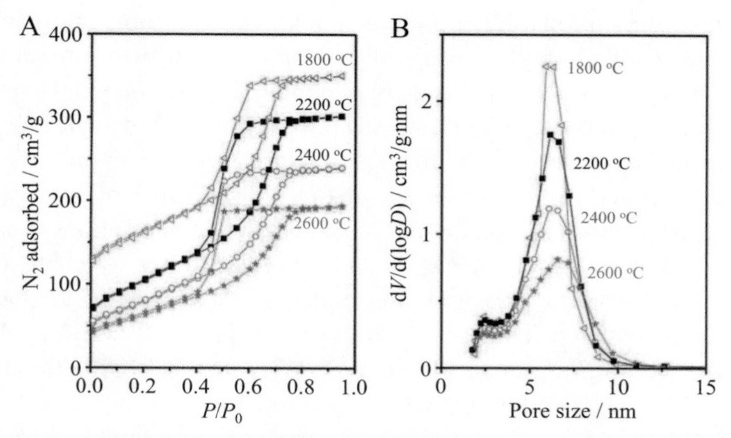

Figure 2.82 Effect of high-temperature treatment on mesoporous carbons prepared by using RF with F127: (A) adsorption−desorption isotherms (the isotherm for the 1800°C-treated one was shifted up by 50 cm^3/g) and (B) pore size distributions [252].

by dipping and then carbonized at 600°C in N_2 [268]. Carbon membrane formed on the Al_2O_3 tube walls had S_{BET} of 670 m^2/g, V_{total} of 0.58 cm^3/g, and mesopores with sizes of about 4.2 nm. The Al_2O_3/C composites thus prepared exhibited good permeation properties, high hydrothermal stability, and high alkaline resistance. Ordered mesoporous TiC/C composites were prepared by the same procedure using resol-type PF, titanium citrate, and F127 in ethanol/water solution, followed by carbothermal reduction at 1000°C [269]. Crystalline TiC particles with sizes of 4–7 nm were confined to the carbon pores and could enhance the oxidation resistance of carbon open frameworks.

From a mixture of a diblock copolymer, poly(styrene)-*b*-poly(4-vinylpridine) (PS–P4VP) as a template and RF as a carbon precursor, carbon films with hexagonal arrays of channels perpendicular to the film surface were synthesized [270,271]. A solution of PS-P4VP and resorcinol was cast onto a silica substrate to form a film, in which most of the resorcinol molecules were located in the P4VP domain due to the hydrogen-bond association between the basic P4VP blocks and the acidic resorcinol monomers [270]. By controlled evaporation of the solvent DMF in DMF/benzene vapor at 80°C, a highly ordered nanostructure of the film was obtained, where the PS domain became the cylinders directed perpendicular to the film surface. This film was exposed to formaldehyde vapor to form a highly cross-linked phenol resin in the P4VP domain and then carbonized to 800°C. The channels were formed perpendicularly in the carbon film, of which the diameter was 33.7 nm and the wall thickness was 9.0 nm. The resultant mesoporous carbons exhibited high adsorption capacity for methylene blue in water [271]. Mesoporous carbon thus obtained could be activated by using KOH to create micropores with only a slight sacrifice of mesopores. Impregnation of KOH through its aqueous solution into the carbon and heat-treatment at 700°C for 45 min resulted in increases in S_{BET} from 500 to 900 m^2/g and V_{total} from 0.70 to 0.87 cm^3/g, where V_{micro} increased from 0.04 to 0.22 cm^3/g and V_{meso} decreased from 0.66 to 0.64 cm^3/g [272]. KOH activation of the mesoporous carbons at 800°C increased S_{BET} from 694 to 1685 m^2/g and V_{total} from 0.54 to 0.94 cm^3/g, mainly due to the increase in micropores [273].

Carbons/silica composites were prepared from mixtures of resol-type PF, TEOS, and F127 (0.16–2.0, 2.08, and 1.0–2.3) by carbonization at 900°C in N_2 flow, from which either ordered mesoporous carbons by dissolution of SiO_2 with 10 wt.% HF or ordered mesoporous silica by burn out at 550°C in air were obtained [274]. The resultant carbons delivered

V_{total} of 0.73–2.0 cm^3/g and high S_{BET} of 1270–2470 m^2/g, and were composed of large mesopore sizes of about 6.7 nm.

2.2.3.1.2 Poly(ethylene glycol)

Mesoporous carbon membranes were prepared from polymer blends of one of PI (PMDA/ODA) with poly(ethylene glycol) (PEG) mixed in N,N-dimethylacetamide (DMAc) solution by casting onto a glass substrate, imidization at 200°C, and carbonization at 600°C [275]. PEGs with M_n of 2000, 10,000, and 20,000 and a blending ratio of PEG/PI of 0.85/1 and 1.7/1 by weight were used. TG curves are shown for PI, PEG, and their blend with a ratio of 0.85/1 in Fig. 2.83A. No carbon residue was yielded above 400°C from PEGs with different M_n's. The TG curve for the blend shows clearly that PEG decomposes as a first step, followed by the PI pyrolysis above 500°C. The N_2 adsorption–desorption isotherms in Fig. 2.83B demonstrate the formation of mesopores by blending PEG, accompanied by a slight increase in micropores. The volume of mesopores became larger with increasing PEG content, from 0.04 cm^3/g for PI (without PEG) to 0.35 cm^3/g for the blend with 1/6.8, while keeping almost the same micropore volume of 0.19 cm^3/g. M_n of PEG has only a small effect on the pore structure of the resultant carbon membranes.

2.2.3.1.3 Poly(methyl methacrylate)

Hollow carbon fibers were prepared from polymer blends of PAN with poly(methyl methacrylate) (PMMA) in different weight ratios (PMMA/PAN of 1/9, 3/7, and 5/5) in DMF at 60°C, followed by electrospinning, stabilization at 280°C in air and carbonization at 1000°C [276]. The

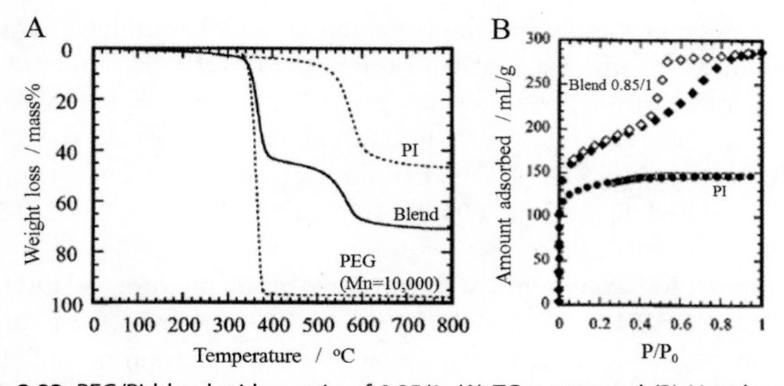

Figure 2.83 PEG/PI blend with a ratio of 0.85/1: (A) TG curves and (B) N_2 adsorption–desorption isotherms [275].

electrospun organic nanofibers exhibited long fibrous morphology with smooth outer surface and homogeneous diameters in the range of 200–400 nm, which consisted of two phases: the discontinuous and long rod-like PMMA phase at the core and the continuous PAN phase at the sheath. During heat-treatment up to 1000°C, the elongated PMMA phase decomposed without carbon residue and the PAN phase was easily transformed into carbon, resulting in continuous hollow carbon fibers. The pore volume of the walls of the carbon fibers increased with increasing PMMA content, V_{total} of 0.50 cm^3/g for the fibers from 1/9 mixture to 0.82 cm^3/g for those from 5/5 mixture, where mainly V_{meso} increased from 0.18 to 0.47 cm^3/g, although V_{micro} was kept at almost the same (0.32–0.35 cm^3/g).

Microemulsion latex of PMMA was added into an aqueous solution of resorcinol and formalin with the catalyst $NaHCO_3$ [277]. After 4 days of gelation at 85°C, the water in the resultant aquagels was exchanged with methanol to dissolved out PMMA colloidal particles. After complete removal of PMMA, the gels were dried by heating up to 60°C and then carbonized at 800°C in N_2. With increasing PMMA content, the S_{BET} increased from 356 to 765 m^2/g thanks to the increase in V_{meso} from 0.164 to 0.434 cm^3/g. The PMMA particles with size of 25 nm could leave mesopores with size of about 5 nm.

The TiO_2-loaded mesoporous carbon fiber webs were fabricated from the DMF solution of the polymer blends of PAN copolymer fibril (93.0 wt.% acrylonitrile, 5.3 wt.% methylacrylate, and 1.7 wt.% itaconic acid) and PMMA with $TiO(OAc)_2$ by electrospinning, stabilization at 280°C in air, and then carbonization at 600°C in N_2 [278]. TiO_2 particles formed in the carbon fibers had an anatase-type structure of which particle sizes were about 13 nm from XRD analysis. The fiber webs prepared from the blend with PMMA/PAN weight ratio of 1/3 exhibited high-rate performance in lithium-ion batteries.

2.2.3.1.4 Poly(vinyl butyral)

Mesoporous carbon fibers were prepared at 180–190°C by melt-spinning of the blend of novolac-type PF with poly(vinyl butyral) (PVB) (PVB/PF of 1/1 by weight in methanol) followed by stabilization in a solution of formaldehyde with HCl and carbonization at 900°C in N_2 [96]. In Fig. 2.84, TG curves for the precursors, PF and PVB, are compared with their 1/1 blend. PVB pyrolyzes abruptly before 400°C, before pyrolysis of matrix PF at 400–500°C, leaving mesopores with a broad size distribution,

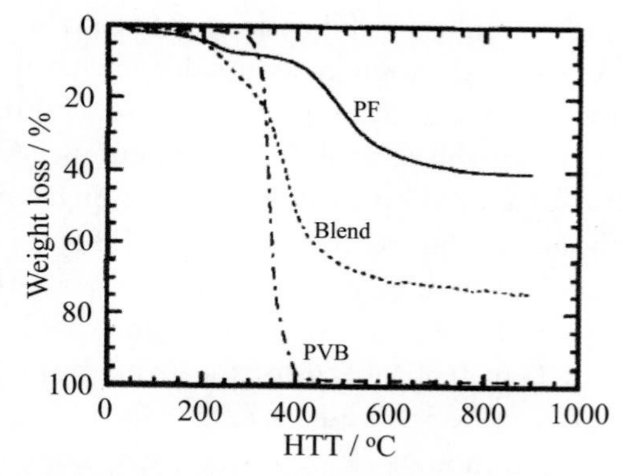

Figure 2.84 TG curves of PF, PVB, and PVB/PF blend (PVB/PF = 1/1) [96].

smaller than 20 nm. By mixing Pt acetylacetonate into the blend, mesoporous carbon fibers having S_{BET} of 390 m^2/g and loaded by 0.2 wt.% Pt nanoparticles were prepared [279].

2.2.3.1.5 Melamine

N-doped carbon aerogels were prepared from the polymer blends of melamine/RF by carbonization at 750°C in an N_2 atmosphere, followed by activation at 750°C in CO_2 [280]. Carbon monolith prepared from the blend with a melamine/RF weight ratio of 2 was mesoporous, V_{meso} of 0.56 cm^3/g in V_{total} of 0.80 cm^3/g, while that from the blend with 0.5 was microporous, V_{micro} of 0.20 cm^3/g in V_{total} of 0.27 cm^3/g, with both carbons having S_{BET} of around 700 m^2/g.

2.2.3.2 Metal organic and covalent organic frameworks

Metal organic frameworks (MOFs) and covalent organic frameworks [COFs, or porous-organic frameworks (POFs)] have intrinsically nano-scaled cavities and open channels, and so are used as templates to synthesize various porous and composite materials, nanostructured metal oxides, metal oxide composites, porous carbons, etc. [281]. They offer potential as a template, as well as carbon precursor, to synthesize porous carbons. One of the MOFs, MOF-5 [$Zn_4O(OOCC_6H_4COO)_3$], which had a three-dimensional intersecting channel system (diameter of 1.8 nm), was exposed to FA vapor at 150°C, with FA being polymerized in the channels of MOF-5, and then carbonized at 1000°C in Ar [282]. Template MOF-5 decomposed at 425–525°C and left ZnO, which was reduced to metallic Zn at around 800°C and then vaporized above its boiling point of

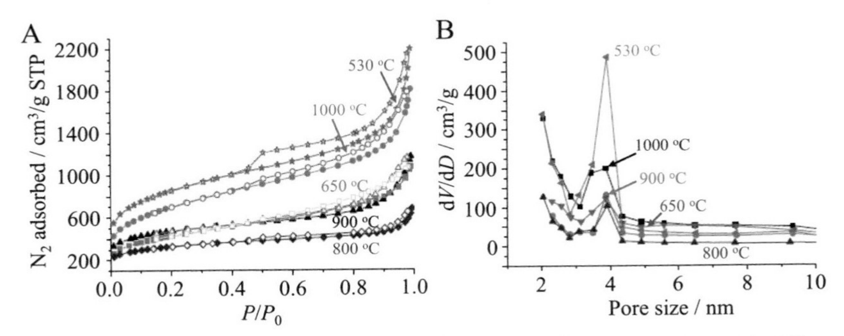

Figure 2.85 Mesoporous carbons prepared from FA/MOF-5 composite with different heat-treatment temperatures: (A) N_2 adsorption–desorption isotherms and (B) pore size distributions [283].

908°C, leaving porous carbon. The resultant carbon was mesoporous and had a high S_{BET} and V_{total} of 2872 m^2/g and 2.06 cm^3/g, respectively, giving a ratio of macropore/mesopore/micropore of about 1/15/5. A strong effect of carbonization temperature was observed on the FA/MOF-5 composite, showing that a temperature of 1000°C was needed to obtain a high surface area and consequently high performance in supercapacitors [283]. The N_2 adsorption–desorption isotherms and pore size distributions of the composites heat-treated at different temperatures from 530 to 1000°C are shown in Fig. 2.85A and B, respectively. Principal pores are mesopores centered at about 4 nm size. The changes in isotherm and pore size distribution (Fig. 2.85) suggest that carbonization of the composite occurs above 900°C after the decomposition of the framework of MOF-5 above 530°C, followed by vaporization of Zn above 900°C. The carbon prepared at 1000°C had S_{BET} of 2524 m^2/g and V_{total} of 2.44 cm^3/g, including V_{meso} of 2.05 cm^3/g and V_{micro} of 0.38 cm^3/g, whereas the carbon prepared at 900°C had 1647 m^2/g, 1.57, 1.11, and 0.26 cm^3/g, respectively. By using glycerol as the carbon precursor, mesoporous carbons with worm-like particle morphology were obtained [284], of which S_{BET} and V_{total} reached 2587 m^2/g and 3.14 cm^3/g, respectively, and pore size distribution exhibited peaks centered at 2.6, 5.5, and 12.4 nm. Porous carbons were prepared from MOF-5 itself, PF/MOF-5 composite, and a composite of carbon tetrachloride (CTC) and ethylenediamine (EDA) mixture with MOF-5 at 900°C [285]. The N_2 adsorption–desorption isotherms are shown for the carbons as-prepared and those after KOH activation in Fig. 2.86A and B, respectively. MOF-5 itself could leave a carbon with S_{BET} of 1812 m^2/g and V_{total} of 2.87 cm^3/g with V_{micro}/V_{total}

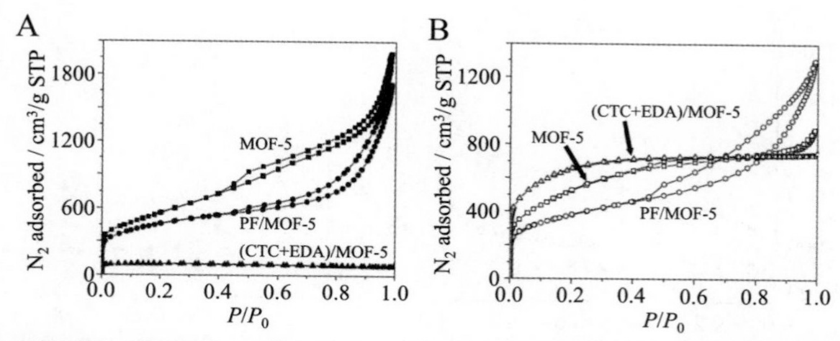

Figure 2.86 N_2 adsorption–desorption isotherms of porous carbons prepared from MOF-5, PF/MOF-5, and (CTC + EDA)/MOF-5 composites at 900°C: (A) as-prepared and (B) after KOH-activation at 700°C [285].

of 32%, and PF/MOF-5 composite resulted in a carbon with S_{BET} of 1534 m^2/g and V_{total} of 2.49 cm^3/g with V_{micro}/V_{total} of 36%. Although the carbon prepared from (CTC + EDA)/MOF-5 composite exhibited very low S_{BET} at 384 m^2/g, it gave microporous carbon with high S_{BET} of 2222 m^2/g and V_{total} of 1.14 cm^3/g with V_{micro}/V_{total} of 89% after KOH activation. For two other carbons derived from MOF-5 itself and PF/MOF-5 composite, KOH activation was not effective in improving their pore structures. By gelation of the mixture of RF and MOF-5 at 80°C, followed by carbonization at 950°C, a porous carbon with S_{BET} of 2368 m^2/g was obtained [286]. S-doped mesoporous carbons were prepared by carbonization of S-infiltrated MOF-5 at 950°C [287]. The resultant S-doped carbon contained 2.5 at% S and had S_{BET} of 573 m^2/g, which was much lower than that of precursor MOF-5 (2050 m^2/g) with a broad pore size distribution from 3 to 20 nm. MOF-5 without any additional carbon precursor gave porous carbons of cube-shaped particles with sizes of 5–20 μm by carbonization at 1000°C in Ar, of which S_{BET} was 2316 m^2/g, consisting of micropores with sizes of about 1 nm and mesopores of about 4 nm size [288].

From a zeolite-type MOF [$Zn(C_4H_5N_2)_2$, ZIF-8] as a template, which acted also as a carbon precursor, and FA as an additional carbon precursor, porous carbon was obtained at 800 and 1000°C [289]. N_2 adsorption–desorption isotherms are shown for the carbons prepared at 800 and 1000°C in Fig. 2.87. The carbon prepared at 1000°C delivered very high S_{BET} at 3405 m^2/g and V_{total} of 2.58 cm^3/g, whereas that at 800°C gave 2169 m^2/g and 1.50 cm^3/g, respectively. Sucrose, melamine, urea, and xylitol were impregnated as an additional carbon precursor for ZIF-8 and then carbonized at 950°C [290]. The resultant carbons had relatively high

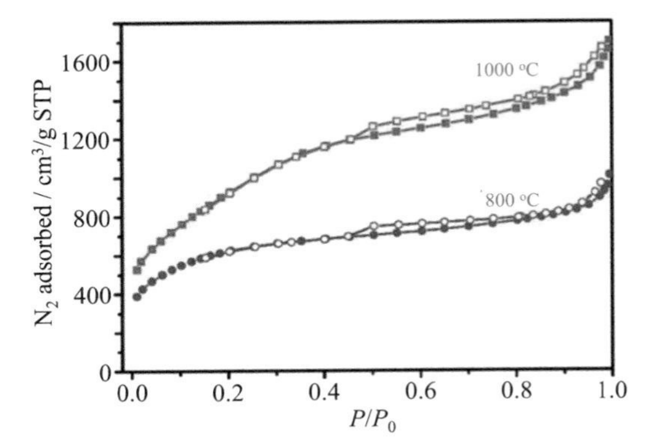

Figure 2.87 N_2 adsorption–desorption isotherms of the porous carbons prepared from a zeolite-type MOF (ZIF-8) with FA at 800 and 1000°C [289].

Table 2.17 N-doped porous carbons derived from ZIF-8 [290].

Additional precursor	S_{BET} (m^2/g)	V_{total} (cm^3/g)	N content (wt.%)	Capacitance[a] (F/g)
Non (ZIF-8)	603	0.30	2.99	213
Melamine	1038	0.58	2.98	237
Urea	2159	1.35	4.51	227
Xylitol	1470	0.68	4.06	248
Sucrose	934	0.47	4.5	286

[a]In 6 M KOH electrolyte with a current density of 0.1 A/g.

S_{BET} and V_{total} with a high N content, as summarized in Table 2.17. The carbon obtained by sucrose impregnation gave relatively high N content, comparable to those using urea, and excellent capacitive performance, which was thought to be due to the effective protection of nitrogen loss from the ZIF backbone because of the melting and polymerization of sucrose on the ZIF surface during carbonization.

ZIF-7 with the composition of $C_{14}H_{16}N_4O_3Zn$ was also used for the synthesis of N-doped porous carbons by mixing with an additional carbon precursor, glucose, through either aqueous solution or powder mixing and carbonizing at 950°C in Ar [291]. ZIF-7 itself also gives microporous carbon and glucose itself gives a nonporous solid, whereas the carbons resulting from their mixtures contained a certain amounts of mesopores, in addition to micropores. Mixing of glucose with ZIF-7 is effective for increasing micropores, particularly via solution mixing.

POFs (or COFs) were also employed as the precursors for porous carbons. Powders of a POF synthesized from 1,3,5-triformylphloroglucinol (TP) and *p*-phenylenediamine (PA) (TPPA-POF) were carbonized at

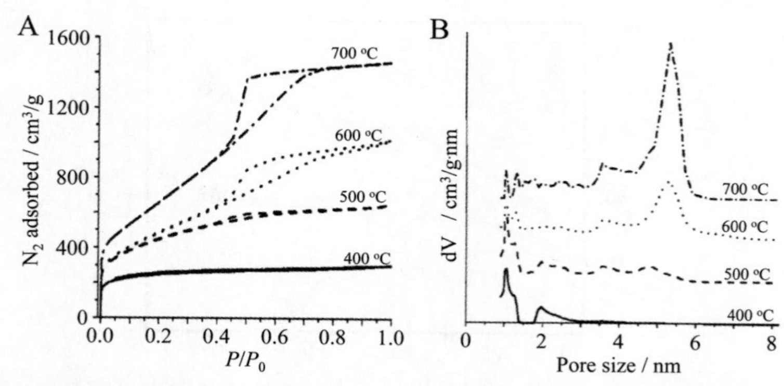

Figure 2.88 Polyaryltriazine frameworks synthesized at different temperatures: (A) N_2 adsorption–desorption isotherms and (B) pore size distributions [293].

different temperatures (500, 600, and 700°C) in N_2 flow [292]. N_2 adsorption–desorption isotherms showed typical type IV curves, indicating the existence of micropores and mesopores and giving S_{BET} of 380–482 m^2/g with V_{total} of 0.10 cm^3/g.

Nitrogen-rich organic frameworks were synthesized from 1,4-dicyanobenzene by mixing with $ZnCl_2$ (polyaryltriazine frameworks) and then carbonized at different temperatures between 400–700°C [293]. As shown in Fig. 2.88A, a type I isotherm is observed only for the polymer prepared at 400°C, whereas other frameworks synthesized at higher temperatures exhibit a type IV isotherm with an associated H_2 type hysteresis, indicating the formation of additional mesopores. Pore size distribution (Fig. 2.88B) shows a marked increase of mesopores with sizes of about 5.5 nm. S_{BET}, V_{total}, and V_{meso} increased with increasing synthesized temperature, from 930 m^2/g, 0.47, and 0.16 cm^3/g for the 400°C-carbonized one to 2530 m^2/g, 2.26, and 2.06 cm^3/g for the 700°C-carbonized one, respectively. The relative content of mesopores increased markedly with increasing temperature, from 34% for 400°C- to 91% for the 700°C-carbonized ones. Mesoporous carbons thus prepared were activated using KOH at 750°C to obtain microporous carbons, which had high performances for gas adsorption, i.e., H_2 storage and CO_2 capture [294]. Terephthalonitrile-derived nitrogen-rich frameworks (TNNs, one of POFs) were synthesized at different temperatures from 400 to 700°C under ionothermal conditions, where $ZnCl_2$ worked as a solvent and also a catalyst [295]. As shown in Fig. 2.89A, the N_2 adsorption–desorption isotherm gradually changes from type I to type IV with increasing

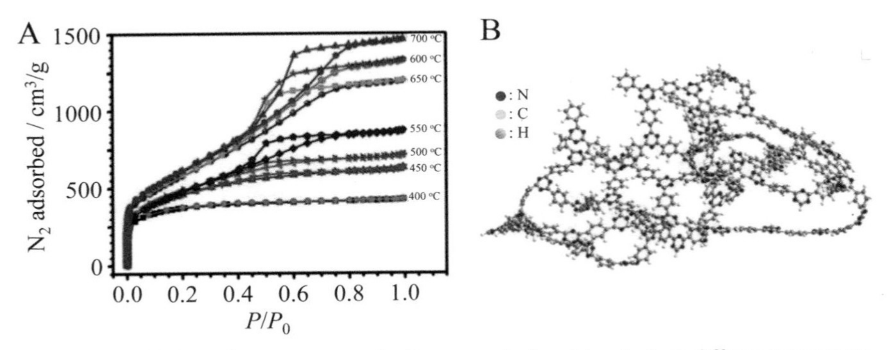

Figure 2.89 N-doped porous organic frameworks heat-treated at different temperatures: (A) N_2 adsorption—desorption curves and (B) illustration of the structure [295].

synthesized temperature, indicative of the formation of mesopores, associated with marked increases in S_{BET} and V_{total} from 1175 m^2/g and 0.66 cm^3/g for 400°C-synthesized to 2237 m^2/g and 2.26 cm^3/g, respectively, for 700°C-synthesized. A structural model of TNNs is illustrated in Fig. 2.89B.

Fibrous mesoporous carbons were prepared from hyperbranched polyphenylene using an anodic aluminum oxide (AAO) membrane (average nanopore diameter of 200 nm and length of 60 μm) at 250°C, followed by carbonization at 800°C [296]. The resultant carbon fibers showed a type IV N_2 isotherm with a pronounced hysteresis, with S_{BET} reaching 1140 m^2/g.

2.2.3.3 Carbon aerogels

Mesoporous carbon aerogels were prepared by the carbonization of organic aerogels of resorcinol-formaldehyde, which were synthesized by aqueous polycondensation with sodium carbonate as the catalyst, followed by supercritical drying [297—299]. In Fig. 2.90A, a TEM image is shown for a carbon aerogel which has been prepared at 1000°C and has a bulk density of 0.43 g/cm^3 [299]. Carbon particles consist of the agglomerates of approximately uniform spheres with sizes of 4—9 nm. Adsorption isotherms for carbon aerogels with different bulk densities are shown in Fig. 2.90B, which belong to type IV and have a clear hysteresis. The analysis of the isotherms using α_s plot demonstrated a marked development of mesopores in the aerogels carbonized up to 1050°C, the aerogel with 0.43 g/cm^3 density having S_{BET} of 577 m^2/g, V_{total} of 1.51 cm^3/g, including V_{meso} of 1.43 cm^3/g. The mesopores in these carbon aerogels are thought to be formed in a three-dimensional network of interconnected minute carbon particles (interparticle pores) and only a small amount of micropores is

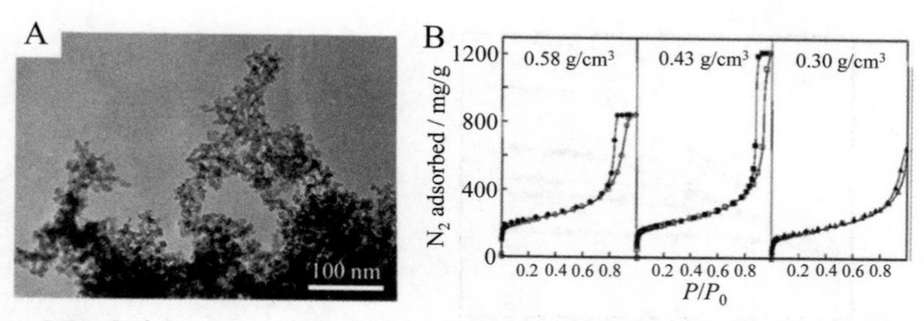

Figure 2.90 Carbon aerogels prepared at 1050°C: (A) TEM image of an aerogel and (B) N_2 adsorption–desorption isotherms of those with different bulk densities [299].

formed in primary carbon particles. Drastic decreases in S_{BET} and micropore volume measured by using CO_2 adsorption were reported on the carbon aerogels derived from RF aerogels and carbonized above 1050°C [300], of which the structural characteristics were observed to be glass-like carbon by Raman spectroscopy [298].

The pore structure of carbon aerogels is governed by that of precursor organic aerogels, in the case of RF resin, which is controlled by the mole ratios of resorcinol to formaldehyde (R/F), to water (R/W) and to basic catalyst of sodium carbonate (R/C). Aqueous gels of RFs synthesized with different R/C ratios of 25–800, R/W ratios of 0.06–0.25 g/cm^3, and R/F ratios of 0.25–1.0, are dried under supercritical conditions with CO_2 and then carbonized at 950°C [301]. In Table 2.18, the surface area and pore volume are shown for the pristine RF aerogels and the carbon aerogels derived from the corresponding RF aerogels. The RF aerogels with various R/C and constant R/W ratio of 0.11 g/cm^3 and R/F of 0.5 show maximum S_{BET} and V_{meso} at around R/C of 100 and a decrease with increasing R/C up to 800. All aerogels, both pristine organic aerogels and carbon aerogels, are mesoporous, depending greatly on the pristine aerogels; a large R/C ratio of 500 is not advantageous and large R/F ratio of 0.67 are not desirable for producing mesopores. By carbonization, V_{meso} decreases, while S_{BET}, S_{micro}, and V_{micro} become larger than those of pristine RF aerogels, and as a consequence the ratio V_{meso}/V_{micro} becomes smaller than the pristine, although mesopores are still principal in the carbon aerogels. Pore-size distributions in the mesopore region are shown for both the pristine RF aerogels and the resultant carbon aerogels for different R/C ratios in Fig. 2.91, revealing that the R/C ratio has to be large to achieve a sharp pore size distribution in the carbon aerogels. The effect of the R/W ratio was also studied by N_2 adsorption–desorption isotherms

Table 2.18 Surface area and pore volume of RF and carbon aerogels [301].

Condition			RF aerogel					Carbon aerogel				
R/C	R/W	R/F	S_{BET}	V_{meso}	S_{micro}	V_{micro}	V_{meso}/V_{micro}	S_{BET}	V_{meso}	S_{micro}	V_{micro}	V_{meso}/V_{micro}
25	0.11	0.50	834	3.07	87.9	0.05	61	1270	1.88	668	0.24	7.8
200	0.11	0.50	537	1.03	127	0.06	17	1810	1.78	1370	0.51	3.5
200	0.11	0.34	355	1.95	207	0.01	195	1920	1.72	1310	0.49	3.5
200	0.11	0.67	568	1.12	171	0.08	14	1740	1.46	1290	0.46	3.2
200	0.25	0.50	612	2.99	103	0.04	75	1040	2.41	734	0.25	9.6
500	0.11	0.50	377	0.40	115	0.05	8	1270	0.74	1190	0.41	1.8

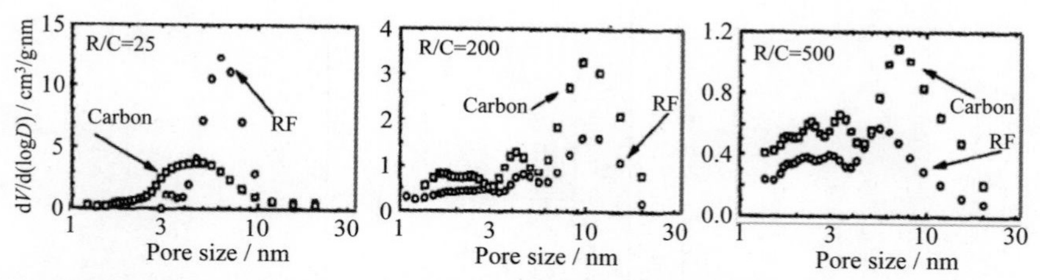

Figure 2.91 RF-derived carbon aerogels prepared using different R/C ratios at 950°C in comparison with pristine RF aerogels [301].

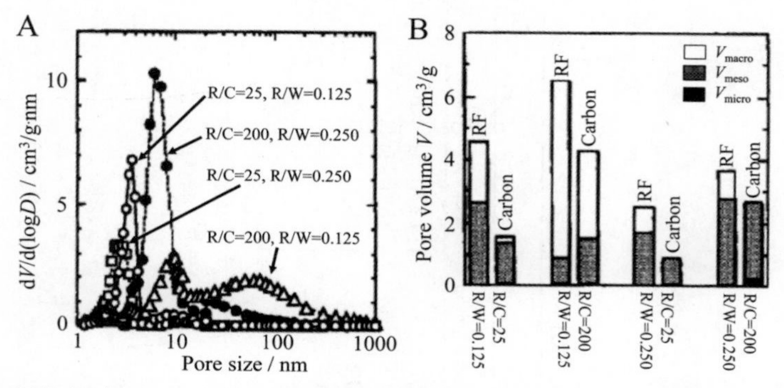

Figure 2.92 Carbon aerogels prepared with different R/C and R/W ratios at R/F = 0.5: (A) pore size distributions of carbon aerogels and (B) pore constructions in comparison with the pristine RF aerogels [303].

and Hg porosimetry [302,303]. A low R/W ratio of 0.125 g/cm^3 results in a broad size distribution ranging from about 5 to 400 nm for the carbon gels after carbonization at 1000°C, as shown in Fig. 2.92A. As shown, in the pore construction in carbon aerogels by comparing with RF aerogels as histograms in Fig. 2.92B, V_{macro} decreases by carbonization probably due to shrinkage and as a consequence V_{meso} becomes much larger than V_{macro} in carbon aerogels, except the case of R/W of 0.125 g/cm^3.

RF aerogels synthesized by curing and drying of an alcohol solution of R and F with hexamethylenetetramine (H) were carbonized at 900°C in N_2 flow [304]. The resultant carbon aerogels are composed of small particles, of which the sizes depend strongly on the molar ratio of R/H, as shown in the SEM images of the aerogels synthesized using different R/H ratios in Fig. 2.93. The R/H ratio also influences V_{meso} as shown in Fig. 2.93, the aerogel with R/H of 25 giving large particles and very small V_{meso} but that with R/H of 45 resulting in much smaller particles and much larger V_{meso} while keeping the same V_{micro}.

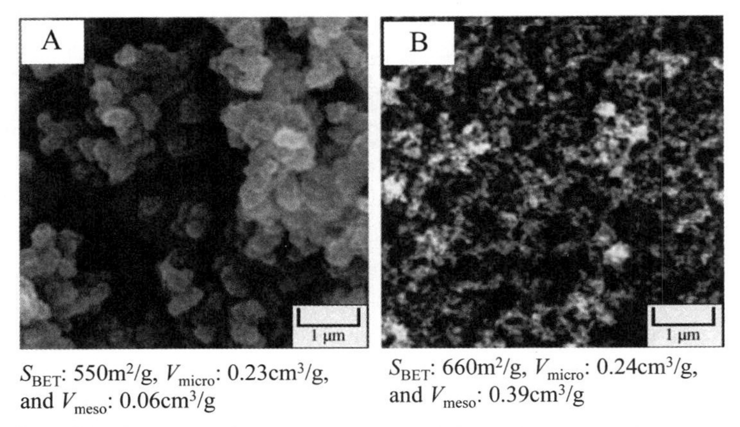

Figure 2.93 SEM images and pore parameters of the carbon aerogels prepared from resorcinol (R)/furfural aerogels with different contents of hexamethylenetetramine (H): (A) R/H (by mole) of 25 and (B) that of 45 [305].

Instead of supercritical drying of aerogels, different drying methods were applied, including freeze drying, hot air drying, and vacuum drying [305]. RF cryogels were also mesoporous with V_{meso} of more than 0.58 cm^3/g, whereas their S_{BET} and V_{meso} were lower than those of the aerogels. On the carbon cryogels, the conditions for the preparation of RF gels seemed not to have a marked effect on the pore structure of carbon aerogels, as shown in Fig. 2.94A. Cryogel prepared by the conditions of R/C = 200, R/W = 0.25 g/cm^3, and R/F = 0.5 exhibited S_{BET} of 908 m^2/g and V_{meso} of 1.39 cm^3/g. In Fig. 2.94B, pore size distributions of the RF gels

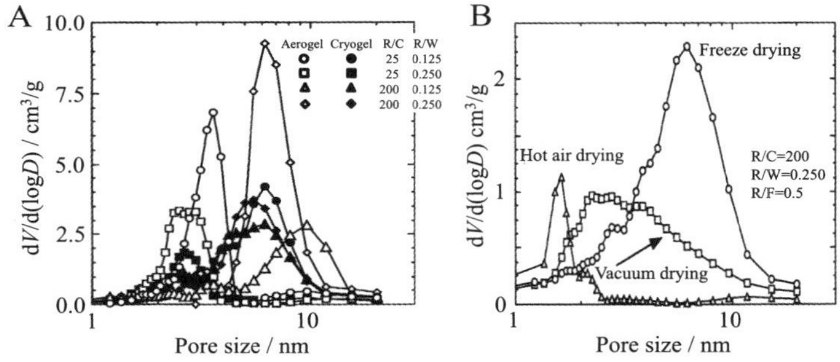

Figure 2.94 Pore size distributions of the cryogels and aerogels prepared from the RF gels with different R/C and R/W at R/F of 0.5: (A) comparison between carbon cryogels and carbon aerogels prepared from RF gel (R/C = 200, R/W = 0.25 g/cm^3, and R/F = 0.5) and (B) pore size distributions of the RF gels dried by different methods [305].

prepared by different drying methods are shown, revealing that the drying method has a strong effect on pore structure of the RF gels, and suggesting a strong effect on the pore structures in the resultant carbon gels. The freeze-drying conditions on RF gels and the resultant carbon gels are studied in detail in order to control mesoporosity [306–308]. RF gels are prepared from its hydrogels by either freeze drying, microwave drying, or hot air drying, and converted to carbon gels by heating at 1000°C [309]. RF gels were prepared with different R/W ratios, constant R/C of 200, and using different drying methods. Freeze drying was the most effective method for the fabrication of mesoporous carbon gels. The effect of the drying process on pore structure in the resultant carbons was discussed using RF resin in water containing sodium carbonate by comparing supercritical drying, freeze-drying, and evaporative drying [310].

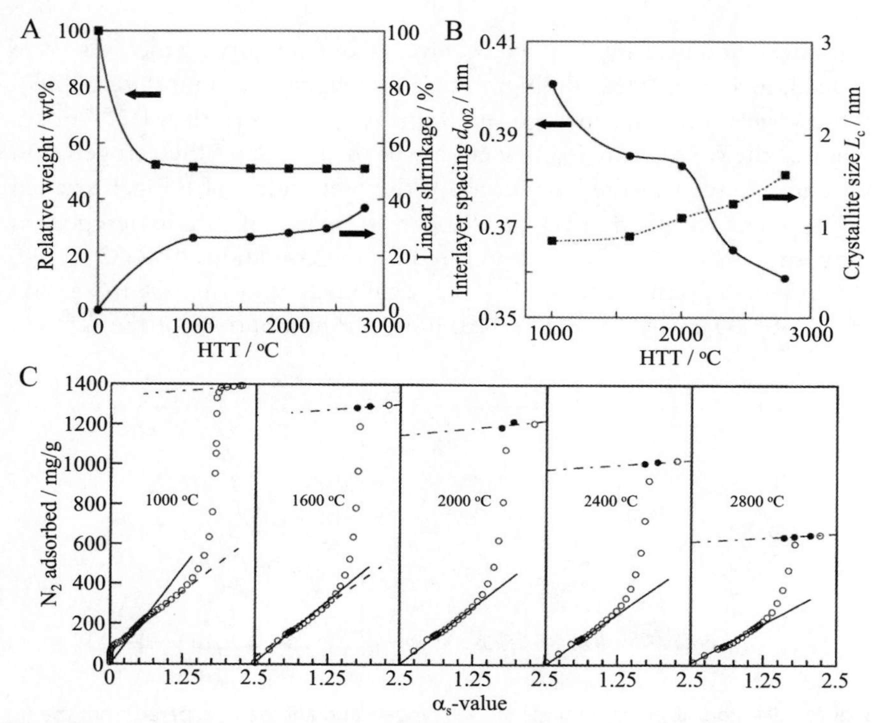

Figure 2.95 RF-derived carbon aerogels heat-treated at high temperatures (using HTT): (A) HTT dependences of weight loss and shrinkage, (B) those of interlayer spacing d_{002} and crystallite size L_c, and (C) α_s plots for the carbon gels heat-treated at different temperatures [311].

Table 2.19 Effect of HTT on pore structure parameters of carbon aerogels [311].

HTT (°C)	S_{total} (m^2/g)	V_{total} (cm^3/g)	S_{micro} (m^2/g)	V_{micro} (cm^3/g)	S_{meso} (m^2/g)	V_{meso} (cm^3/g)	S_{ext} (m^2/g)
1000	850	1.63	416	0.12	361	1.51	73
1600	493	1.54	1.5	0.06	333	1.48	55
2000	456	1.44	0	0	396	1.44	60
2400	425	1.11	0	0	383	1.11	42
2800	325	0.76	0	0	285	0.76	40

Changes in pore structure parameters of RF-derived carbon aerogels with heat treatment at high temperatures have been reported [311]. The changes in structural parameters, i.e., weight change, linear shrinkage, interlayer spacing d_{002}, and crystallite size L_c, with increasing HTT are shown in Fig. 2.95A and B. The weight loss with HTT occurred mainly below 1000°C, while the changes in linear shrinkage and d_{002} occurred in two steps, below 1000°C and then above 2000°C. L_c increases above 2000°C. In Fig. 2.95C, the α_s plots obtained from N_2 adsorption–desorption isotherms are shown for the aerogels heat-treated at different temperatures. Micropores disappear above 2000°C, of which the volume is evaluated from the intercept of the broken line in the medium α_s region. Total surface area and volume, which were calculated from the slope of the straight line passing through the origin (solid line) and the intercept of the chained line in the high α_s region, decrease with increasing HTT, which are mainly due to the decrease in micropores. However, S_{ext} calculated from the slope of the chained line, which is mainly due to mesopores, decreases slightly with increasing HTT. Pore structure parameters calculated on the aerogels with different HTTs are shown in Table 2.19. The

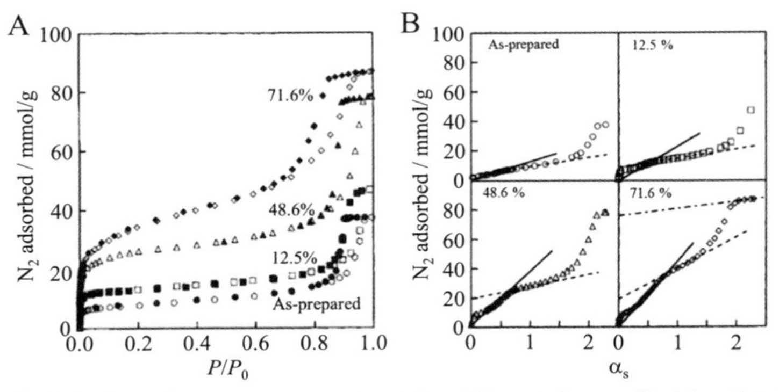

Figure 2.96 Activated carbon aerogels with different burn-off: (A) adsorption–desorption isotherms of N_2 at 77K and (B) α_s plots [312].

Table 2.20 Pore structure parameters for activated carbon aerogels [312].

Activation time (h)	Burn-off (mass%)	S_{total} (m^2/g)	V_{total} (mL/g)	S_{micro} (m^2/g)	V_{micro} (mL/g)	S_{meso} (m^2/g)	V_{meso} (mL/g)
0	As-prepared	713	1.301	366	0.115	347	1.186
1	12.5	1314	1.625	960	0.296	354	1.329
5	48.6	2260	2.714	1750	0.676	510	2.038
7	71.6	2600	2.634	1390	0.656	940	1.978

heat treatment of the carbon aerogel above 2000°C gives a carbon containing only mesopores.

Mesoporous carbon aerogels were activated at 800°C in CO_2 flow for different periods of 1—7 h to increase microporosity [312]. Adsorption—desorption isotherms after activation are characterized by the uptake at low P/P_0 (less than 0.1) and the hysteresis above P/P_0 of 0.7 (Fig. 2.96A). Both characteristics become more noticeable with increasing burn-off, i.e., increasing activation degree, suggesting the development of micropores together with mesopores by the activation. As shown by the α_s plots in Fig. 2.96B, the upward deviation from the solid straight line passing though the origin below α_s of 0.5 becomes marked, indicating the development of micropores, and the linear region of the broken line in the range of α_s of 0.7—1.5 becomes narrower, and simultaneously the slope of the line increases with increasing burn-off, indicating the development of mesopores. The numerical results are listed in Table 2.20, together with the activation time and burn-off. As the activation progresses, S_{total} and V_{total} increase up to 2600 m^2/g and 2.6 cm^3/g, respectively, mainly due to the increases in S_{micro} and V_{micro}, accompanied by the tendencies to saturate. However, S_{meso} and V_{meso} increase more markedly with intense activation at burn-off of more than 40%, suggesting a partial transformation of micropores to mesopores. The detailed studies on adsorption of N_2 at 77K and of water vapor at 303K on activated carbon aerogels, whose surface functional groups are nil, show clearly that the amount of adsorbed water corresponds mainly to V_{micro}, not to V_{meso} [313].

Ce/Zr-co-loaded carbon aerogels were prepared by mixing $Ce(NO_3)_3$ and $ZrO(NO_3)_2$ with RF in aqueous solution, followed by gelation at either pH 7 or 3, drying with supercritical CO_2, and carbonization at 1050°C [314,315]. The resultant carbon aerogels through the gelation at pH 7 and 3 gave different particle sizes, the former consisting of loosely bound particles with about 20 nm diameter, while the latter was of spherical particles with diameter around 3 μm [314]. The N_2

adsorption–desorption isotherms were also different, with the former exhibiting an adsorption isotherm of type IV with hysteresis H3 and the latter exhibiting type I, i.e., the former having micropores with some mesopores (S_{total} of 800 m^2/g with S_{ext} of 280 m^2/g) while the latter only had micropores (S_{total} of 500 m^2/g with S_{ext} of 11 m^2/g).

2.2.3.4 Ionic liquids

Ionic liquids (ILs) are defined as semiorganic salts that exist in a liquid state below 100°C and can have strong Coulombic interactions between the constituent ions, giving intrinsic null vapor pressure, which seems to be favorable for carbonization to form cross-linking of monomeric ILs and also carbon precursors dissolved into IL solvents. Some ILs have been used as carbon precursors. In Fig. 2.97A, TG curves of four ILs (molecular structure of cations and anions with their notations are shown in Table 2.21) are shown [316], revealing that the IL [BMIm][Tf_2N] decomposes quickly and leaves negligible residues, while ILs composed from [BCNIm] cation give certain amounts of porous carbons and the thermal stability of these ILs depends greatly on the nature of the anion. The IL [BCNIm][Cl] gave a very high carbonization yield of 53% at 800°C with a carbon yield of 46.0% (i.e., 46% carbon atoms in the precursor were recovered as solid carbon), while it was essentially nonporous (S_{BET} of 15 m^2/g). Carbonization yields for the ILs having the same cation coupled with the fluorinated anions [Tf_2N]$^-$ and [beti]$^-$ were the same, as well as carbon yields, and gave high S_{BET}'s near 650 m^2/g. N_2 adsorption–desorption isotherms of the carbons prepared from four ILs at 800°C in N_2 are shown in Fig. 2.97B. The anions of these ILs have a profound influence on the pore structure in the resultant carbons; simply replacing [Tf_2N]$^-$ with Cl^- for a fixed cation [BCNIm]$^+$ results in complete loss of porosity, [BCNIm][Tf_2N] and [BCNIm][beti]

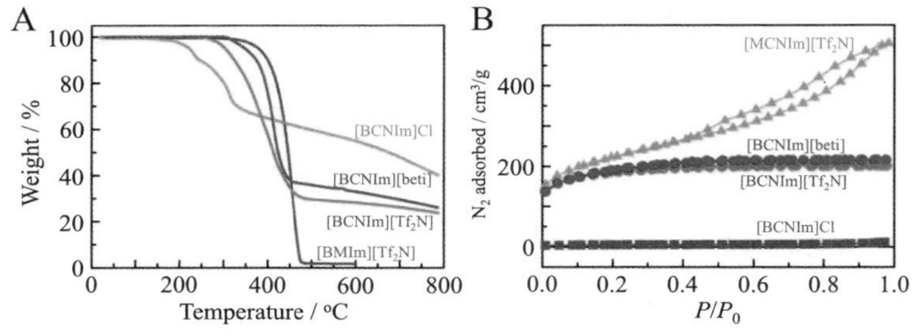

Figure 2.97 Carbonization of various ionic liquids: (A) TG curves and (B) N_2 adsorption–desorption isotherms of the carbons obtained at 800°C [316].

Table 2.21 Cations and anions used to compose ionic liquids.

Cation		
1-Cyanomethyl-3-methylimidazolium	$[MCNIm]^+$	
1,3-Bis(cyanomethyl)-imidazolium	$[BCNIm]^+$	
1-Butyl-3-methylimidazolium	$[BMIm]^+$	
1-Ethyl-3-methylimidazolium	$[EMIm]^+$	
1-Hexyl-3-methylimidazolium	$[C6MIm]^+$	
1-Methyl-3-nonylimidazolium	$[C9MIm]^+$	
1-Decyl-3-methylimidazolium	$[C10MIm]^+$	
1-Vinyl-3-ethylimidazolium	$[VEIm]^+$	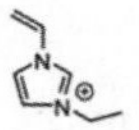
1-Vinyl-3-cyanomethylimidazolium	$[VCMIm]^+$	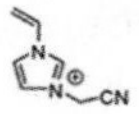
1-Vinyl-3-cyanopropylimidazolium	$[VCPIm]^+$	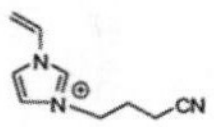
3-Methyl-N-butyl-pyridinium	$[3MBP]^+$	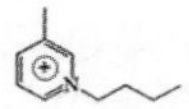
1—Butyl-3-methyl-pyridinium	$[BMP]^+$	
Anion		
Bis(trifluoromethylsulfonyl) imide	$[Tf_2N]^-$	$N(CF_3SO_2)_2^-$
	$[beri]^-$	$N(C_2F_6SO_2)_2^-$
	Cl^-	Cl^-
	Br^-	Br^-
	$C(CN)_3^-$	$C(CN)_3^-$
Dicyanamide	$[dca]^-$	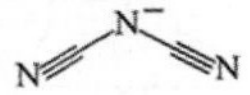
Tetracyanoborate	$[tcb]^-$	

giving S_{BET} of about 650 m^2/g while [BCNIm][Cl] was about 5 m^2/g. The cation also impacts on the pore structure of the resultant carbon, as shown by [MCNIm][Tf_2N], which shows a typical type IV isotherm with H_2 hysteresis indicating the presence of mesopores (S_{BET} of 780 m^2/g), whereas $[BCNIm][Tf_2N]$ is strictly microporous. The ILs composed of the imidazolium cations with the alkyl groups (having different lengths) coupled with $[C(CN)_3]^-$ were carbonized at 800°C [317]. S_{BET} of the carbons derived from $[C_9MIm][C(CN)_3]$ and $[C_{10}MIm][C(CN)_3]$ was lower than that from $[C_6MIm][C(CN)_3]$, which might be caused by a partial collapse of pore structure during carbonization, but greater than that from $[BMIm][C(CN)_3]$. All of these ILs have melting points below room temperature and decomposed in the range of 209.7–329.0°C, except $[BCNIm][C(CN)_3]$ which is a pale-brown powder at room temperature.

Four ILs, [VEIm][Br], [VEIm][dca], [VCMIm][Br], and [VCPIm][Br], were carbonized at 1000°C in N_2 with or without mixing $FeCl_2.4H_2O$ (12 wt.% to IL) [318]. The resultant carbon from $FeCl_2$-mixed ILs after washing with HCl to remove Fe gave type IV N_2 adsorption–desorption isotherm with H_2 hysteresis loop was shown to be mesoporous and had a developed lamellar structure, and as a consequence had relatively low electrical resistance, being even lower than the graphite reference.

ILs were employed as carbon precursors for the synthesized porous carbons, particularly N-doped ones, by using templates, SBA-15 and SiO_2 nanospheres (Ludox) [319,320], and eutectic salts [232]. ILs composed of $[EMIm]^+$ and $[3MBP]^+$ with $[dca]^-$ gave N-doped carbons with relatively high crystallinity and low electrical resistivity [320]. As shown in the XRD profiles in Fig. 2.98A and B, both ILs give relatively sharp 002 diffraction line at about 26.4°C even after carbonization at 500°C and no pronounced change with higher temperatures. The 10 band at around 43° becomes detectable with increasing temperature while its profile is very broad even after 1000°C-treatment, suggesting turbostratic stacking of carbon layers with small lateral sizes. As shown in Fig. 2.98C, the resistance of the carbons derived from these ILs decreases drastically with the increase in

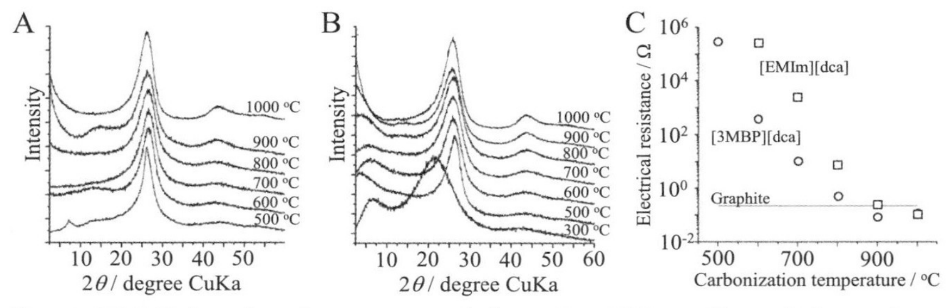

Figure 2.98 N-doped carbons prepared from ILs; XRD profiles of the carbons (A) derived from [EMIm][dca] and (B) derived from [3MBP][dca] at different temperatures, and (C) the resistance change with carbonization temperature [320].

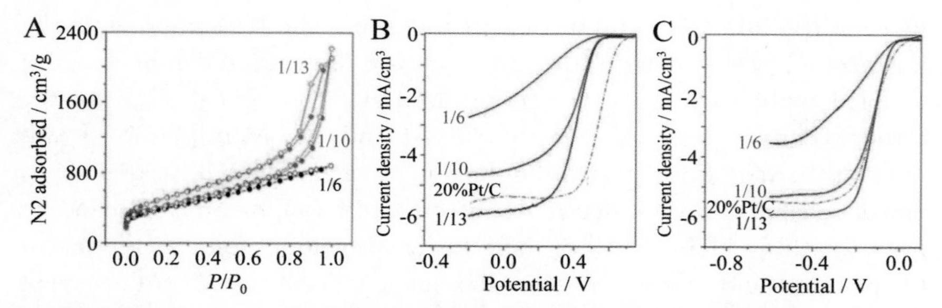

Figure 2.99 Mesoporous N-doped carbon prepared from [EMIm][dca] using eutectic salt template with different IL/salt ratios: (A) N_2 adsorption–desorption isotherms and polarization curves (B) in O_2-saturated 0.05 M H_2SO_4, and (C) in O_2-saturated 0.1 M KOH in comparison with commercial 20 wt.% Pt/C [321].

carbonization temperature, the carbon derived from [3MBP][dca] showing a faster decrease than that derived from [EMIm][dca], and becoming comparable to synthetic graphite above 900°C. From the IL [3MBP][dca] as carbon precursor, N-doped carbons with ordered mesopores and homogeneous spherical mesopores were synthesized using SBA-15 and SiO_2 nanospheres as templates, respectively, and N-doped carbon nanotubes using AAO membrane [319,320]. N-doped porous carbons were synthesized using [BMP][dca] and [EMIm][dca] as carbon precursors using eutectic salt templates at 1000°C, and also N/B-co-doped carbons from [EMIm][tcb] at 1400°C [232]. A couple of [EMIm][dca] as carbon precursor with NaCl/$ZnCl_2$ eutectic salt mixture as template delivered mesoporous N-doped carbons, which exhibit high ORR activities [321]. The mixing ratio of IL/salt (precursor/template ratio) governed the pore structure and ORR activity of the resultant carbons, as shown in Fig. 2.99. N_2 isotherms in Fig. 2.99A suggest that the interparticle pores increase with increasing salt amount (decreasing IL/salt ratio), while the intraparticle pores (mainly micropores) are not affected. The N content and S_{BET} were measured as 4.5 wt.% and 1410 m^2/g for the carbon prepared with a ratio of 1/6, 3.7 wt.% and 1770 m^2/g for that with 1/10, and 3.8 wt.% and 1550 m^2/g for that with 1/13, respectively, where the configuration of N atoms doped was mainly pyridinic and graphitic. As shown in Fig. 2.99B and C, the carbon prepared with a ratio of 1/13 delivered ORR activity comparable with commercial 20 wt.% Pt/C catalysis in acidic and alkaline solutions, respectively, although onset potential E_{onset} was a little lower. Mesoporous N-doped carbons were synthesized from [EMIm][dca] with additional nitrogen precursors and SiO_2 nanosphere template by carbonization at 1000°C, followed by washing out of SiO_2 [322]. The resultant carbons exhibited high ORR activity in alkaline solution.

2.2.3.5 Others

Mesoporous carbons were synthesized from poly(styrene sulfonic acid-*co*-maleic acid) salts containing Na, Fe, Co, or Ni [323]. Na-salt was converted to transition metal-salt by cation exchange using transition metal nitrate at 200°C for 4 h. The salts were carbonized at 800°C for 40 min in N_2 flow, followed by washing with distilled water. The resultant carbons contained metal particles, which were dissolved out by washing with 18% HCl for 24 h. The carbon derived from the pristine Na-salt delivered S_{BET} of 1348 m^2/g and V_{total} of 0.89 cm^3/g including V_{micro} of 0.47 cm^3/g, whereas the carbons derived from transition-metal-salts delivered slightly lower S_{BET} (700–1000 m^2/g) and similar V_{total} (0.58–0.89 cm^3/g) including lower V_{micro} (0.20–0.30 cm^3/g).

2.3 Macroporous carbons

2.3.1 Carbonization with blowing

Blowing procedures of carbon precursor polymers to prepared carbon foams were divided into three: under pressure, using some chemical additives (blowing agents), and self-blowing without pressure. The first procedure consists of the saturation by decomposition gases from the precursor itself in a closed vessel, followed by a rapid reduction of pressure, which has been applied mainly on pitches. The second procedure has been applied on different polymers, including pitches, with different blowing

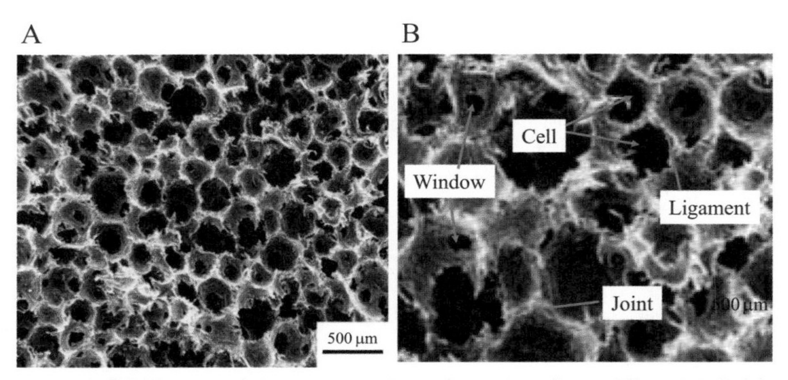

Figure 2.100 SEM image of the cross-section of a carbon foam (A) and definitions of the terms to characterize the pore structure in the carbon foam (B).

chemicals. In the third procedure, the precursor blows by itself due to the decomposition accompanied by evolving a relatively large amount of gas on the pyrolysis step before carbonization. A carbon foam prepared by carbonization with blowing is shown in Fig. 2.100A by the SEM image of its cross-section. In Fig. 2.100B, the terminology customarily employed to characterize carbon foams is illustrated. Two different pores are usually seen in an SEM image: one is a macropore surrounded by a carbon wall, which is called a "cell", and the other is a hole formed in the carbon wall to connect neighboring cells and is called a "window", the walls of cells consist of two parts, "ligament" and "joint", and both parts often contain micropores, mesopores, and even macropores smaller than cells in some cases. The carbon foams were reviewed by focusing on their preparation and applications [324].

For these macroporous carbons, various expressions have been used, such as carbon foams, carbon sponges, three-dimensional carbon frameworks (3D carbon frameworks), even carbon aerogel, etc. In this book, the term "carbon foam(s)" is employed in most cases.

2.3.1.1 Pyrolysis under pressure

The process had been developed for the production of polymer foams, such as polystyrene and polyurethane, and called the "microcellular foaming technique" [325–327]. Carbon foams were prepared from pitches by blowing in a pressure chamber, followed by carbonization. During the controlled heating of the precursors under pressure in an inert atmosphere, the evolving volatiles from the light fractions and the thermally decomposed fractions of the precursors serve as foaming agents to create cells in the highly viscous precursor material.

Commercial mesophase pitches A (softening point T_s of 237°C) and B (T_s of 355°C) were heated in an autoclave up to a temperature slightly above its T_s at a constant rate and kept for a relatively long time [328,329]. During this heating, the pressure was controlled either by built-up pressure due to volatile matters from the precursor pitch or by adding an inert gas up to a few MPa. After being kept under pressure at a certain temperature for a certain time, the product was cooled to room temperature and then the pressure was released quickly. The product was carbonized at around 1000°C in Ar and then graphitized at 2800°C. SEM images of the cross-sections of the resultant carbon foams prepared from pitch A and B under two different pressures are shown in Fig. 2.101 [328]. Bulk density and cell size depended greatly on the precursor pitch. The foam derived from pitch A shown in Fig. 2.101A had a bulk density of 0.22 g/cm^3 and average

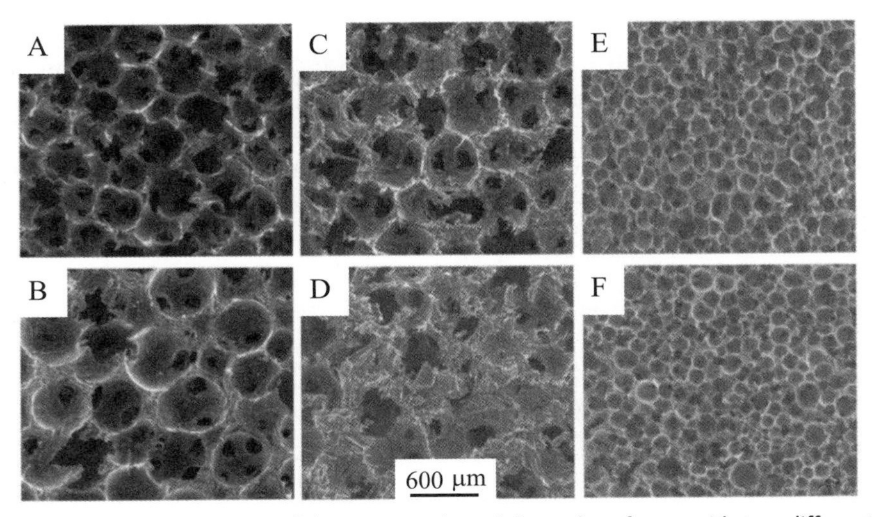

Figure 2.101 SEM images of the cross-section of the carbon foams with two different bulk densities: (A, B) foams from pitch A and (E, F) those from pitch B. (C and D) after graphitization of (A) and (B), respectively [328].

cell size of around 280 μm, while that from pitch B shown in Fig. 2.101E had 0.33 g/cm^3 and around 50 μm, respectively. The bulk density of the foam increased with increasing pressure. By long-time heating above T_s under pressure, the stabilization of the foam before carbonization to maintain foam morphology was not needed. The foam morphology is kept even after graphitization at 2800°C, as shown in Fig. 2.101C and D. The heating rate (foaming rate) also affected bulk density after carbonization at 1000°C [329]. A slow foaming rate as 3.5°C/min on pitch A under

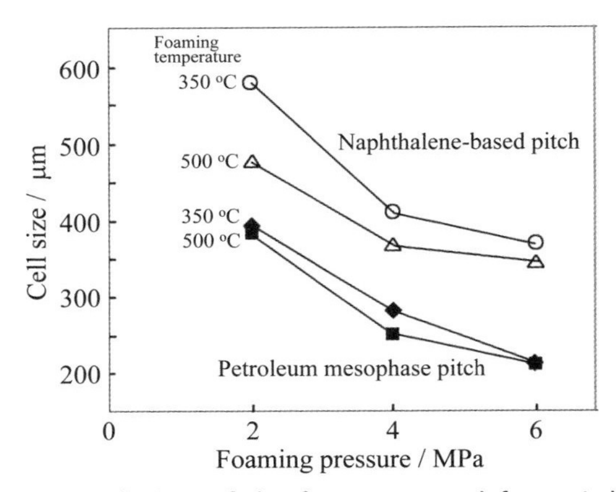

Figure 2.102 Average cell sizes of the foams prepared from pitches by different foaming temperatures and pressures [330].

6.9 MPa pressure resulted in a bulk density of 0.474 g/cm^3, whereas a fast rate such as 10°C/min resulted in a bulk density of 0.434 g/cm^3. From a petroleum mesophase pitch (T_s of 291°C) and a naphthalene-based pitch (T_s of 283°C), carbon foams with homogeneous cell sizes were obtained by foaming at 350 and 500°C under 2, 4, and 6 MPa pressure and following carbonization at 1050°C [330]. The average cell size depends on the precursor pitch and foaming pressure, as shown in Fig. 2.102. The carbon derived from a coal tar pitch (T_s of 94°C), however, had an inhomogeneous structure, the upper part containing spherical closed cells with thick cell walls, whereas the lower part was imperforated pitch block. A bituminous coal with T_s of 414°C was used for the fabrication of carbon foams at 450–550°C under 3–7 MPa [331,332]. The resultant carbon foams carbonized at 1100°C consisted of interconnected cells with sizes of 100–200 μm and had a bulk density of 0.42–0.52 g/cm^3 and bending failure strength of 6–8 MPa. Carbon foams were prepared from a naphthalene-based mesophase pitch (T_s of 275°C) using toluene as the supercritical agent at a temperature of 320°C under pressure of 12 MPa, followed by stabilization at 280°C in air flow and then carbonization at 860°C in N_2 [333]. The resultant carbon foams have a bulk density of about 0.4 g/cm^3, containing spherical cells with sizes of 20–200 μm.

Various pitches and those pretreated by solvents, such as tetrahydrofurane (THF), quinoline, toluene, and *N*-methyl-2-pyrrolidone (NMP), were used as the precursors for carbon foam fabrication [334,335]. Extraction of a heavy fraction in a coal tar pitch by quinoline (quinoline-insoluble fraction, QI) resulted in decreases in bulk density, porosity, and compressive strength of the resultant carbon foam; QI-free

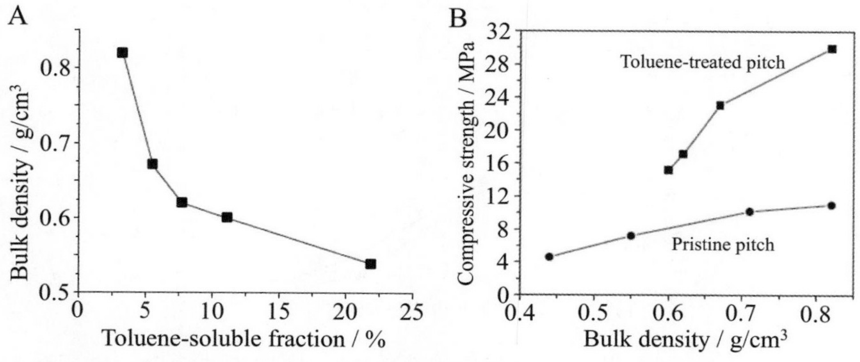

Figure 2.103 Toluene treatment of a synthetic pitch: (A) bulk density versus toluene-soluble fraction and (B) compressive strength versus bulk density [335].

pitch delivered 0.56 g/cm^3, 72%, and 8.0 MPa, whereas the pristine pitch (containing QI) gave 0.67 g/cm^3, 64.8%, and 18.2 MPa, respectively [334]. The content of toluene-soluble fraction (**TS**) in a naphthalene-based synthetic pitch (T_s of 283°C) influenced the bulk density of the resultant foams (foaming at 250°C under 3 MPa and carbonized at 1300°C) and consequently compressive strength, as shown in Fig. 2.103A and B, respectively [335]. Thermal pre-treatment of the precursor pitch at 420°C in N_2 flow was shown to be effective to improve the properties of the resultant carbon foams [336], during the pre-treatment with an agitation of 70 rpm being applied for removal of volatile fractions. Pretreatment in H_2SO_4 and HNO_3 at 120°C was applied on the precursor coal-derived pitch, followed by heating at 350°C after washing [337]. The foams prepared from these modified pitches at 580°C under 1 MPa were carbonized at 1000°C in N_2 and then heat-treated at 2000°C in Ar. The resultant carbon foams prepared from pretreated pitches in HNO_3 delivered a bulk density of 0.59 g/cm^3, porosity of 68.1%, and compressive strength of 17.4 MPa. To create micropores in the matrix carbon, carbon foams prepared by using these pretreated pitches were subjected to steam activation at 800°C [338]. Carbon foams prepared from HNO_3-pretreated pitches gave marked increases in S_{BET} and V_{micro}, from 22 to 933 m^2/g and from 0.011 to 0.389 cm^3/g, respectively, by steam activation.

Effects of foaming conditions, temperature, pressure, and pressure release time, on bulk density, compressive strength, and total porosity measured by mercury porosimetry were studied using a mesophase pitch

Table 2.22 Effects of foaming conditions on the properties of the resultant foams derived from a mesophase pitch [339].

Foaming conditions			Bulk density (g/cm^3)	Compressive strength (MPa)	Total porosity (%)
Temperature (°C)	Pressure (MPa)	Release time (s)			
280	6.8	5	0.380	1.47	45
283	6.8	5	0.390	1.59	45
293	6.8	5	0.410	2.10	45
300	3.8	5	0.500	1.87	65
300	5.8	5	0.540	2.52	69
300	6.8	5	0.560	3.31	80
300	7.8	5	0.580	3.52	86
300	6.8	80	0.510	2.92	70
300	6.8	190	0.370	2.59	58
300	6.8	600	0.240	2.16	37

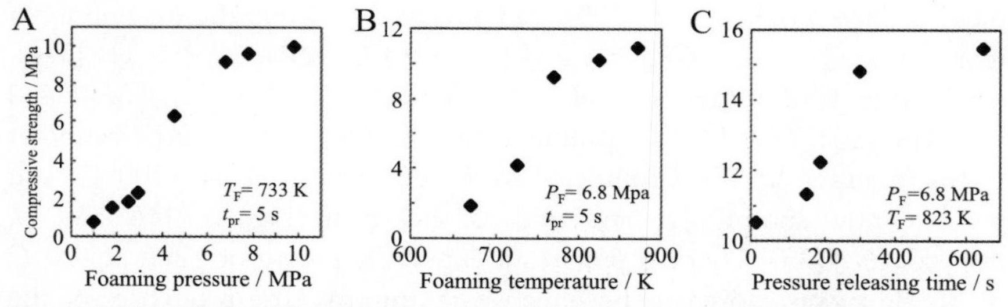

Figure 2.104 Effects of foaming pressure (P_F), foaming temperature (T_F), and pressure releasing time (t_{pr}) on the compressive strength of the carbon foams prepared from an asphaltite pitch; changes with (A) pressure, (B) temperature, and (C) pressure releasing time [340].

with T_s of 275–295°C [339]. The pitch foams prepared at different foaming conditions were stabilized at 390°C in air flow and then carbonized at 1050°C in N_2. The results are summarized in Table 2.22. Using a high foaming temperature, cells become more uniform and more spherical, and bulk density, compressive strength, and porosity become higher. The foam prepared at 300°C delivers a bulk density of 0.560 g/cm^3, compressive strength of 3.31 MPa, and total porosity of 80%. An increase in foaming pressure from 3.8 to 7.8 MPa also causes increases in bulk density and compressive strength, from 0.500 to 0.580 g/cm^3 and from 1.87 to 3.52 MPa, respectively. In contrast, the increase in releasing time from 5 to 600 s causes decreases in both density and strength, from 0.560 to 0.240 g/cm^3 and from 3.31 to 2.16 MPa, respectively.

The effects of foaming pressure, foaming temperature, and pressure releasing time were studied on an asphaltite pitch [340]. Carbon foams were prepared at temperatures of 400–600°C (673–873K) under pressures of 1–9.8 MPa with pressure releasing times of 5–650 s, followed by carbonization at 1050°C in N_2. In Fig. 2.104, the effects of these foaming conditions are shown on the compressive strength of the resultant carbon foams. The compressive strength of the foam increases substantially with increasing foaming pressure up to 6.8 MPa and then levels off (Fig. 2.104A). It increases sharply with increasing temperature, reaching 10 times at around 750K and then tends to level off (Fig. 2.104B), and also increases sharply with increasing pressure releasing time from 100 to 300 s and then tends to level off (Fig. 2.104C). For the bulk density of the foams, similar dependences on foaming pressure, temperature, and pressure releasing time were observed. The asphaltite-derived carbon foam has higher bulk density and compressive strength than those derived from coal

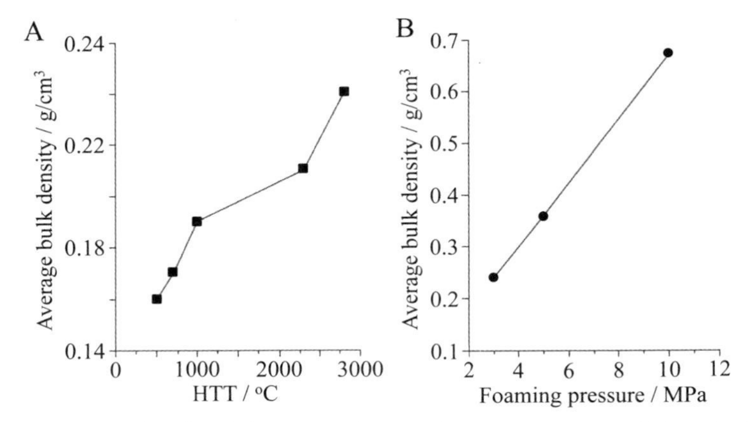

Figure 2.105 Bulk density of the foams derived from a mesophase pitch under 3 MPa pressure: (A) change with HTT and (B) change with foaming pressure on the foams after heat treatment at 2800°C [341].

tar pitch, mesophase pitch, and naphthalene-derived synthetic mesophase pitch.

A mesophase pitch (T_s of 285°C) was foamed under different pressures at 500°C, followed by carbonization at 700°C in N_2, and graphitization at 2300–2800°C in Ar [341]. With increasing foaming pressure from 3 to 10 MPa, the cell sizes decreased from 600–700 to 300–400 μm and the cross-sections of the cells in the foam prepared under 3 MPa were elliptic. The bulk density of the foam increased with increasing HTT and foaming pressure, as shown in Fig. 2.105A and B, respectively. The compressive strength and modulus of the resultant carbon foams depend on the foaming pressure, as well as the bulk density, and also on HTT, the foam carbonized at 1000°C exhibiting high strength and modulus of 1.02 and 36.8 MPa, respectively. By heat treatment at 2800°C, the foam prepared under high pressure (10 MPa) exhibited a high strength of 1.26 MPa and high modulus of 110 MPa.

In order to improve the mechanical properties of carbon foams, different additives into the precursor pitches were used. Graphite particles (<30 mm size) were mixed into a mesophase pitch, followed by blowing under 8 MPa, carbonization at 1200°C and graphitization at 3000°C, to prepare graphite foam [342]. With the addition of 5 wt.% graphite, the compressive strength was improved from 3.7 to 12.5 MPa and thermal conductivity from 70.2 to 107.4 W/m•K. Carbon foams were prepared by adding mesocarbon microbeads (MCMBs) (particle size of about 1 μm) into a mesophase pitch (T_s of 283°C) and foaming at 250°C under 3 MPa,

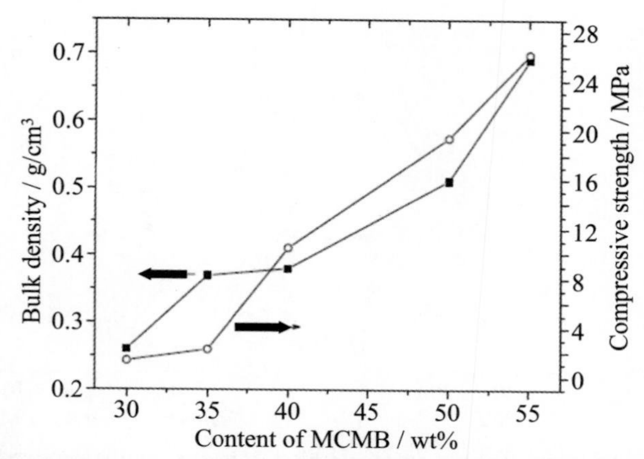

Figure 2.106 Changes in bulk density and compressive strength of carbon foam with MCMB content [343].

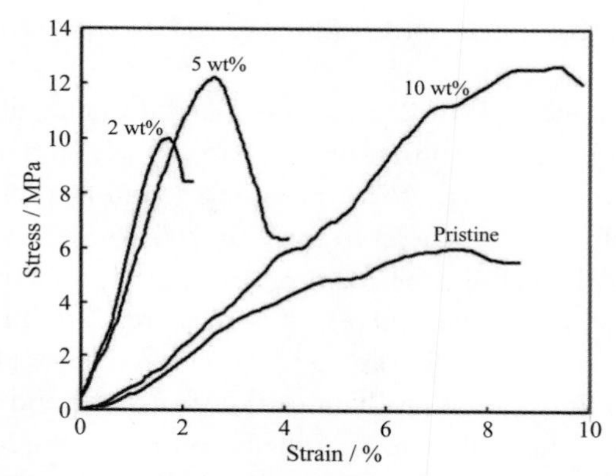

Figure 2.107 Compressive stress–strain curves for clay/carbon composite foams [344].

followed by stabilization at 460°C and carbonization at 1300°C [343]. The bulk density and compressive strength of the resultant carbon foams increased markedly, as shown in Fig. 2.106, whereas thermal conductivity decreased with increasing MCMB content. A clay (Na-montmorillonite) was also used as an additive to a coal-tar-based mesophase pitch to prepare carbon foams at 500°C under 8 MPa [344]. By adding a small amount of clay (<5 wt.%), ultimate strain decreases and breaking strength increases markedly, as shown by the stress–strain curves in Fig. 2.107. The porosity and thermal conductivity of the resultant foams decreased with increasing

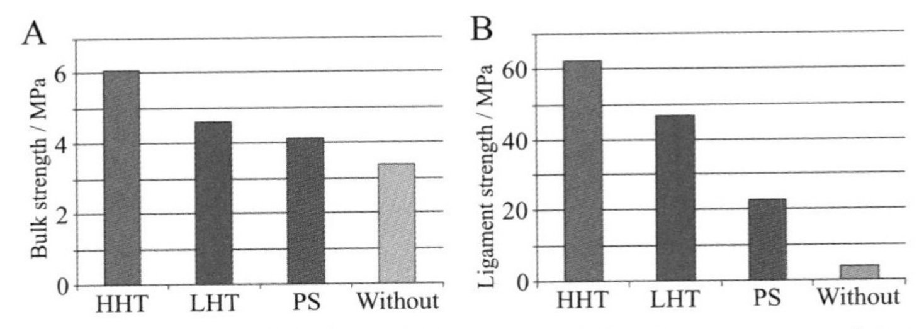

Figure 2.108 Strengths of bulk and ligament of the foams prepared by mixing different carbon nanofibers (CNFs) [HTT, low-temperature-treated (LHT) and as-carbonized (PS)] and without mixing CNFs: (A) bulk strength and (B) ligament strength. The ligament strength is an average of 20 specimens [347].

clay content, from 75% and 2 W/m•K for the foam prepared from pristine pitch to 45% and 0.25 W/m•K for that from the pitch containing 10 wt.% clay. The addition of 3 wt.% pitch fluoride, which has a composition of CH_xF_y ($0 < y < 1.8$), to a mesophase pitch markedly enhanced the thermal conductivity of the resultant carbon foam heat-treated at 3000°C, with the specific thermal conductivity increasing from 82 to 155 W/m•K with the addition of only 1.5 wt.% [345].

The addition of 4.6 vol.% vapor-grown carbon fibers (fiber diameter of 78 nm), which were graphitized at 3000°C and chopped to 10 μm in length, was effective in improving the crush strength of the foams, from 2.13 to 3.5 MPa, while the thermal conductivity decreased from about 68 to 58 W/m•K passing through a maximum of about 78 W/m•K at about 1 vol% fiber content [346]. Carbon foams were prepared at 275°C under 1000 psi pressure, followed by stabilization and carbonization at 1000°C, from the mixtures of a synthetic mesophase pitch with 0.5—5 wt.% carbon nanofibers (CNFs) with different heat treatment temperatures, i.e., as-carbonized (PS), treated at low temperature (LHT). and high temperature (HHT) [347]. The compressive strengths were measured on the bulk of the foam and ligaments sampled from the foams. In Fig. 2.108A and B, strengths are compared for the CNFs used together with the foam without using CNFs on the bulk specimens and also on the ligaments (20 specimens), respectively. Reinforcement of the foam is achieved by mixing CNFs, depending on the CNFs, as shown in Fig. 2.108A, with high-temperature-treated CNFs (HHT) being the most effective. The reinforcement effect is demonstrated more markedly on the ligaments (Fig. 2.108B), suggesting that reinforcement of bulk strength is mainly due to that of the ligands of the foams. The same results were obtained from the measurements of stiffness of the foams.

Anisotropy in pore morphology was observed in the carbon foams prepared from a mesophase pitch under 3–10 MPa, followed by carbonization at 700–1000°C and graphitization at 2300–2800°C: pores extended parallel to the direction of gravity and, as a consequence, compressive strength and modulus were higher [341]. The detailed nanotexture of carbon layer alignment in the ligament and junction was analyzed by a polarized-light optical microscope [348,349].

The formation mechanism of carbon foams from mesophase pitch was discussed in relation to the foaming conditions [350]. With increasing temperature, small bubbles are formed thanks to the release of decomposition gases from the pitch and then their coalescence occurs with neighboring bubbles to form cells in the resultant foam, for which the driving force is thought to be the surface tension of the molten pitch. The shear stress originated from bubble growth makes pitch molecules orient their aromatic planes preferentially along the bubble wall. This oriented nanotexture is preserved in the ligaments and joints, because of the high viscosity of the partially decomposed pitch. Therefore, the viscosity and surface tension of molten pitch are key factors in controlling not only the cell size of the resultant foam but also the nanotexture in ligaments and junctions of the foams, which influences greatly the mechanical properties and also graphitizability of the foam. Theoretical discussions on the mechanism of carbon foams have been presented [351,352].

Carbon foams were prepared from novolac-type PF resin by blowing its ethanol solution with different concentrations at 170°C under pressure of 4 MPa, followed by carbonization at 800°C [353]. The PF concentration had strong influences on bulk density and properties of the resultant carbon

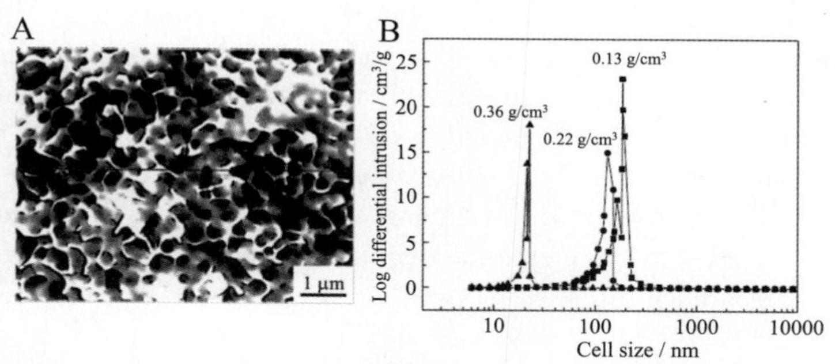

Figure 2.109 Carbon foams prepared from phenol-formaldehyde resin with different concentrations in ethanol solution: (A) SEM image of carbon foam prepared from 0.13 g/cm^3 concentration solution and (B) cell size distributions of the foams prepared by different PF concentrations [353].

foams; **with** the change in concentration from 0.13 to 0.36 g/cm^3 resulted in increases in the bulk density from 0.24 to 0.73 g/cm^3, in compressive strength from 13.1 to 98.3 MPa, and in thermal conductivity from 0.06 to 0.24 W/m•K. Cell sizes of the resultant carbon foam are very homogeneous, as shown in Fig. 2.109A, and their distributions are very sharp and depend greatly on the PF concentration, as shown in Fig. 2.109B. The average cell size is about 180, 120, and 20 nm for the foam prepared from the PF solution with 0.13, 0.22, and 0.36 g/cm^3 concentrations, respectively. The bulk density and compressive strength of the foams depend also on the starting PF concentration, the 0.13 g/cm^3 solution resulting in 0.24 g/cm^3 and 13.1 MPa, and the 0.36 g/cm^3 solution giving 0.73 g/cm^3 and 98.3 MPa, respectively.

2.3.1.2 Addition of blowing agents

Carbon foams were prepared from a commercial mesophase pitch using barium salt of 5-phenyltetrazote as the blowing agent [354]. The resultant carbon foams consisted of large cells, more than a few hundred μm, and much larger than those prepared by blowing under pressure (about 10 μm), and could be graphitized, d_{002} reaching 0.336 nm after 2400°C treatment. Carbon foams were prepared from coal liquefaction residues (T_s of 193°C) mixed with pentane (blowing agent) by foaming under 3—6 MPa at 230°C, followed by air stabilization at 350°C and carbonization at 700°C [355]. On the surfaces of cell walls of the carbon foams thus prepared, carbon nanofibers (CNFs) were deposited by CVD of C_2H_4 in H_2 at 800°C using metallic impurities in the precursor residues as catalysts. SEM images of the resultant foam are shown in Fig. 2.110. Carbon foam consists of interconnected open cells with a size of 300—600 μm and the surfaces of the cell walls are covered by CNFs with diameters of 50—100 nm and lengths of several tens of micrometers. CNF deposition on the cell walls of carbon foams, which were derived from a bituminous coal (T_s of 391°C) at 450°C

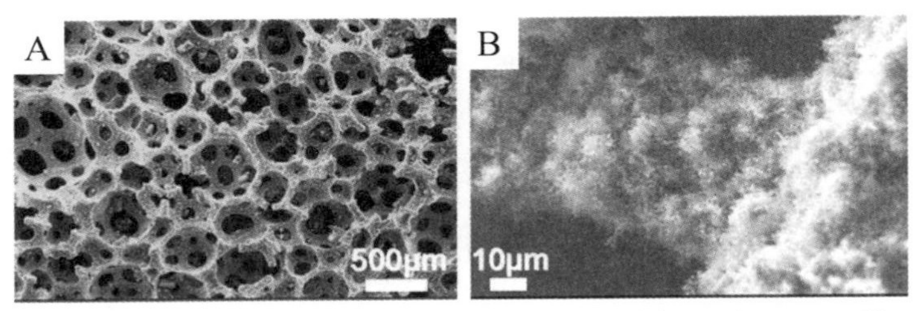

Figure 2.110 SEM images of the carbon foam deposited by carbon nanofibers [355].

under pressure, followed by carbonization at 1100°C and heat treatment at 2200°C in Ar, was also performed by CVD of C_2H_4 using Ni catalyst at 700°C [356].

Carbon foam membranes were prepared through coaxial electrohydrodynamic atomization (EHDA) using polyacrylonitrile (PAN) and borneol ($C_{10}H_{18}O$) as carbon precursors and a blowing agent, respectively [357]. An acetone solution of borneol with a small amount of methylene blue was used as the core solution and a dimethylacetamide (DMAc) solution of PAN as the sheath solution for coaxial EHDA under 15 kV at 22°C with a relative humidity of 61% to obtain PAN/borneol composite membrane on Al foil. The PAN membrane was converted to foam by heating with blowing, followed by oxidative stabilization and carbonization to obtain carbon foam. The resultant carbon foams consisted of spherical cells with an average diameter of 2.34 μm.

Carbon foam was prepared from poly(arylacetylene) (PAA), which had a high carbonization yield, via chemical blowing [358,359]. PAA prepolymer mixed with pentane (blowing agent), Tween 80 (bubble stabilizer), and sulfuric acid (catalyst) were heated slowly up to 100°C and kept for 40 h to blow pentane, and then heated up to 350°C step by step to complete polymerization, followed by carbonization at 1000°C [358]. The resultant carbon foam had a bulk density of 0.6 g/cm^3 and compressive strength of 25.8 MPa, slightly higher than the mesophase pitch-based carbon foam reinforced by the addition of MCMB. The ligament thickness of carbon foams depended on the concentration of sulfuric acid added; changing from 10 μm to more than 100 μm by tuning the concentration of sulfuric acid from 70 to 90 wt.%. To achieve homogeneous foaming, the mixing ratios of pentane and Tween 80 had to be controlled in shallow ranges, the optimal ratio of pentane and Tween 80 to PAA prepolymer was reported to be 0.3 [359].

Phenolic resin precursor containing 3—15 wt.% aluminosilicate (particle size of 100—200 μm) was converted to foam at 80°C using sodium bicarbonate as the blowing agent and then carbonized at 800°C [360]. The bulk density and compressive strength of the foam increased with increasing aluminosilicate content; a change from 0 to 15 mass% resulted in increases in bulk density from 0.261 to 0.316 g/cm^3 and compressive strength from 3.78 to 4.71 MPa. The weight loss of the foams by oxidation at 400°C in air decreased with increasing aluminosilicate content, from 82 wt.% loss to 49 wt.% loss by adding 11 wt.% aluminosilicate.

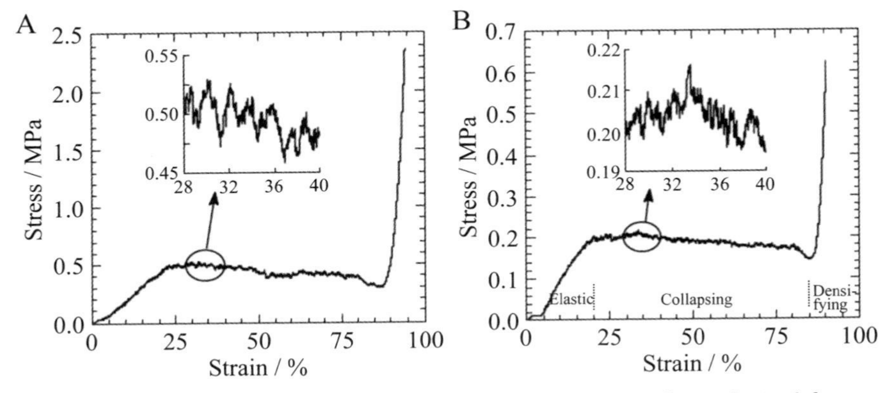

Figure 2.111 Compressive stress–strain curves of the carbon foam derived from a tannin extract: (A) along the *z*-direction and (B) x/y-direction of the foam [364].

Polyflavonoid tannins, which originated from vegetables and were composed of 75%–85% of polyflavonoids (reactive phenolic nuclei), were converted to rigid foams by mixing with formaldehyde and carbonized to carbon foams [361–363]. Mimosa tannin extract, which contained 84% phenolic materials, mixed with formaldehyde (cross-linking agent), furfuryl alcohol (strengthening agent), diethylether (blowing agent), and toluene-4-sulfonic acid (catalyst) in water, was changed to foam with a bulk density of 0.05–0.08 g/cm^3 at 40°C, which could be converted to carbon foam at 900°C [364,516]. The resultant carbon foam was composed from glass-like carbon, of which the bulk density was 0.067 g/cm^3 and porosity was 96.4% [364]. Since the foam was grown vertically in the container, the cells were slightly elongated along the vertical direction. The cell sizes were relatively uniform and slightly anisotropic, average size along the vertical axis of the foam (*z*-direction) of 250 μm, while that along the perpendicular direction (x/y-direction) was 135 μm. Compressive behaviors are shown along two directions on the carbon foam in Fig. 2.111. The stress–strain curve observed is composed of three regions, elastic, collapsing, and densifying regions; elastic deformation occurs up to about 20% strain, collapsing deformation due to successive cell wall fracture is prolonged typically from 20% to 80% strain, and densifying deformation beyond 80% strain is associated with a sharp stress increase. A Young's modulus of 2.26 MPa, densification strain of 90%, and yield strength of 0.52 MPa were obtained along the *z*-direction, and 1.18 MPa, 87.5%, and 0.21 MPa, respectively, along the x/-direction. The amount of foaming agent (diethyl ether) governs the bulk density of the green foam (tannin foam) and as a

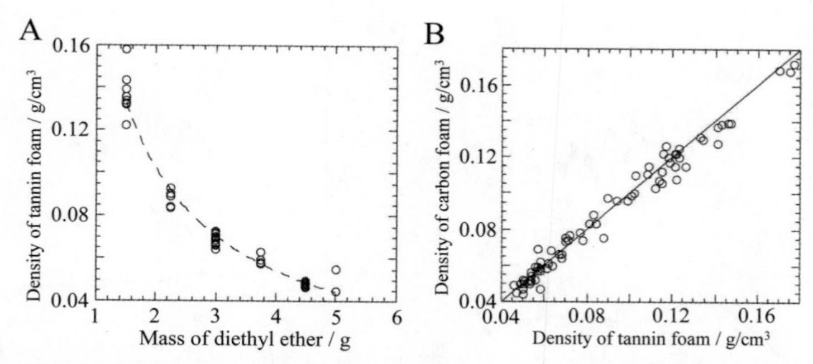

Figure 2.112 Bulk densities of the tannin foams and carbon foams: (A) bulk density of tannin foam versus mass of diethyl ether and (B) bulk density of carbon foam versus that of tannin foam [365].

consequence the bulk density of the final carbon foam prepared at 900°C in N_2, as shown in Fig. 2.112 [365]. The bulk density of the green foam depended also on the content of furfuryl alcohol: a maximum appeared at around 8 g for 30 g tannin, which was explained as a result of two competitive effects, hinder of foaming due to fast toughening of the tannin matrix and highly exothermic polymerization of furfuryl alcohol which accelerated the evolution of volatiles from the tannin matrix. The addition of a small amount of isocyanate into the starting tannin solution allowed considerable changes in the pore structure of the resultant carbon foams [366], an addition of 14 wt.% 4,4′-methylene diphenyl diisocyanate resulted in a marked increase in cell size from 250 to 500 μm.

The mechanical properties of tannin-derived carbon foams were studied in detail as a function of bulk density of the foam [367–369]. The definitions of elastic modulus E, compressive strength σ_{pl}, and densification strain ε_d of foams are shown in Fig. 2.113A, and the stress–strain curves

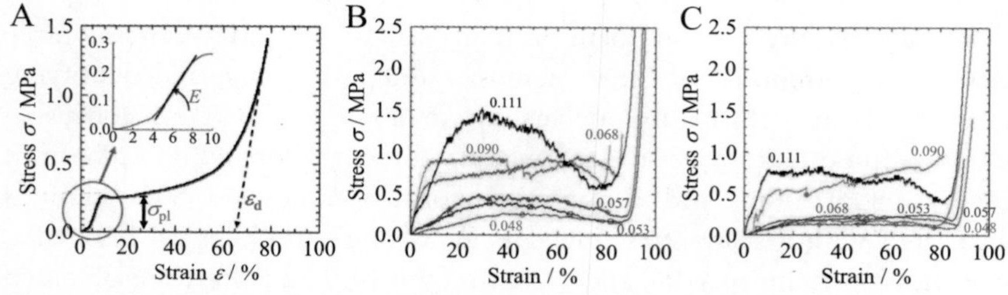

Figure 2.113 Compressive stress–strain behaviors of the tannin-derived carbon foams: (A) definitions of elastic modulus E, compressive strength σ_{pl}, and densification strain ε_d, (B) and (C) stress–strain curves observed along the z-direction and x–y-direction, respectively, for the foams with different relative densities [367].

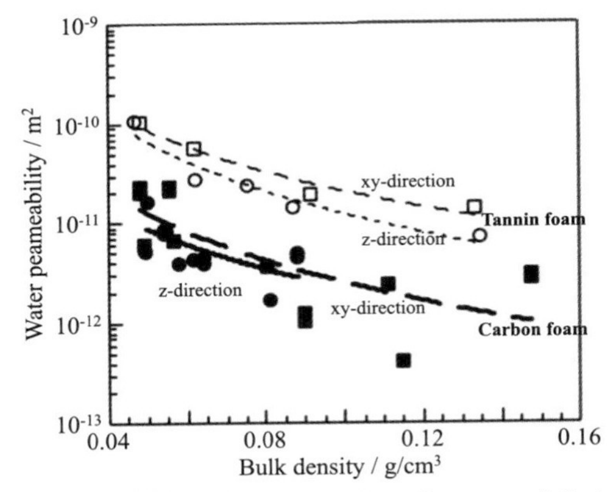

Figure 2.114 Water permeability of tannin-derived foams and their carbonized foams [368].

observed on the tannin-derived carbon foams with different bulk densities are shown in Fig. 2.113B and C [367]. Mechanical properties determined from stress—strain curves were mostly on the basis of the "relative density" of the foams, which was calculated as the ratio of bulk density against helium density measured on the powder. E measured along the z-direction was slightly larger than that measured along the x—y-direction and increased with increasing relative density, in other words, with increasing bulk density, a linear relation being observed between log E and relative density. σ_{pl} measured along the z-direction was also higher than σ_{pl} along the x/y-direction, both increasing with increasing relative density. ε_d of the carbon foams was around 0.8%—0.9% with a slight dependence on relative density.

On a 17-mm cube of the tannin-derived carbon foams, permeability for water was measured [368]. In Fig. 2.114, water permeability is plotted against bulk density of two kinds of foams, original tannin foams and their carbon foams carbonized at 900°C. The carbon foams present 10 times lower permeabilities than those of the tannin foams. This is thought to be due to the shrinkage that occurred during pyrolysis and carbonization, which make cells smaller and possibly windows are closed, and so only part of the porosity is available for fluid flow.

To create micropores and mesopores, tannin-derived carbon foams were prepared by thickening the ligaments via repeated impregnation of furfuryl alcohol (FA) and polymerization in HCl aqueous solution, followed by carbonization at 550°C [369]. The resultant thickened carbon foams were activated by steam at 800°C. With increasing burn-off due to

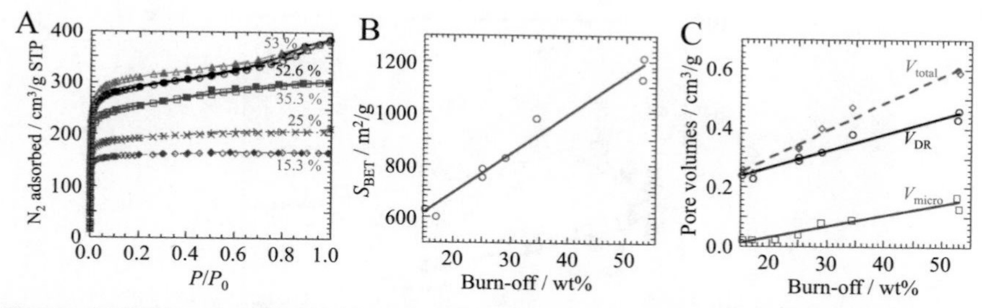

Figure 2.115 Steam-activated tannin-derived foams coated by FA-derived carbon: (A) N_2 adsorption–desorption isotherms for the foams with different burn-off values, (B) S_{BET} versus burn-off, and (C) pore volumes V_{DR}, V_{total}, and V_{micro} versus burn-off [369].

steam activation, the N_2 adsorption–desorption isotherm changes, as shown in Fig. 2.115A, and pore parameters, S_{BET}, V_{total}, and V_{micro}, increase almost linearly with increasing burn-off, as shown in Fig. 2.115B and C. With increasing burn-off, micropore sizes also increased, about 0.8 nm after 53 wt.% burn-off. With this coverage by FA-derived carbon, the compressive strength of the foams was not increased while anisotropy was considerably decreased, whereas further activation reduced the compressive strength markedly.

Preparation of carbon foams from sucrose was also performed using aluminum nitrate $Al(NO_3)_3$ as the blowing agent [370]. By using $Al(NO_3)_3$, foaming and setting times could be shortened; by increasing the $Al(NO_3)_3$ concentration from 0.5 to 4 wt.%, the foaming time decreased from 8 to 1 h and the setting time from 15 to 7 h at a foaming temperature of 150°C.

2.3.1.3 Self-blowing

Foaming of pitches was also performed in a restricted space. Petroleum-derived mesophase pitch (T_s of 305°C) powders were restricted in a cylindrical mold of 30 mm diameter and 30 mm height with a heavy lid [371]. By changing the amount of charged pitch of 4 and 5 g, different free spaces (15.0 and 12.5 mm in height, respectively) in the mold were achieved. Foaming was performed by keeping the temperature of the mold at 460°C for 0.5 h. The obtained pitch foams were stabilized at 350°C in air flow and then carbonized at 1000°C in N_2. The resultant carbon foams had bulk densities of 0.323 and 0.380 g/cm^3, respectively. The SEM images of cross-sections of the foams are shown in Fig. 2.116. Since there was a clearance between the sidewalls and the lid of the mold (about 0.2 mm), the mold lid only offered size restriction and gas could pass in and out through the clearance.

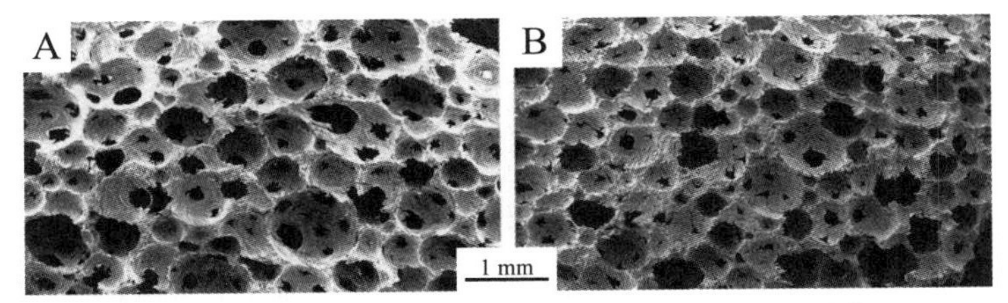

Figure 2.116 SEM images of the cross-section of the foams derived from a mesophase pitch in a restricted space: (A) with the initial free-space of 15.0 mm height and (B) with 12.5 mm height (changing the pitch of 4 and 5 g, respectively, in the mold) [371].

Some precursors can blow by themselves due to the evolution of decomposition gases during pyrolysis where the carbonaceous products maintain suitable viscosity. Acidic aqueous solution of sucrose was concentrated and turned to a dark viscous resin by heating at 110°C on a hot plate, which could foam itself by staying at 110°C [372]. The green foam after complete foaming and drying was heated at temperatures of 300–950°C in Ar. The appearance and SEM images of the cross-sections of the carbon foams are shown in Fig. 2.117. Carbonization of the green foam occurred in two steps, at 150–340°C accompanying a large gas evolution, and at 340–430°C. The resultant carbon foams exhibited high trapping efficiency for ^{137}Cs. The formation of dark viscous resin with a viscosity of 637–670 MPa•s occurred slowly by holding sucrose at a temperature of 120°C [373]. On the surface of the resin kept steady at 120°C, a skin was observed to be formed, which was thought to restrict further escape of volatiles and to result in slow foaming. The resin took nearly 8 h for foaming and 40 h for setting into the solid green foam. While the carbonization temperature increased from 600 to 1200°C, marked increases in shrinkage and bulk density of the green foams were observed, volume shrinkage from 31% to 46% and bulk density from 0.115 to 0.142 g/cm^3. Neither cracks in the ligands and junctions of the foams nor detectable

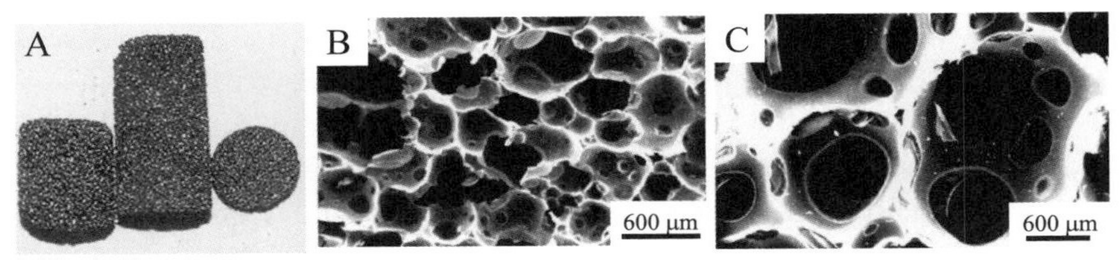

Figure 2.117 Carbon foams prepared from sucrose: (A) appearance of the blocks prepared at 950°C, and SEM images of the cross-section of the foam prepared (B) at 600°C and (C) at 950°C [372].

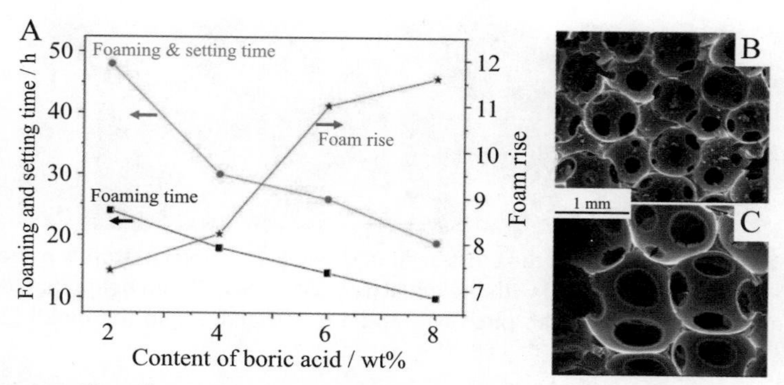

Figure 2.118 The effects of boric acid addition on the foaming process: (A) changes of foaming and setting time and foam rise at 120°C and SEM images of the cross-sections of the foams (B) with 2 wt.% and (C) with 8 wt.% boric acid [376].

deformation in cellular texture were observed during heat treatment up to 1400°C. An abrupt increase in the electrical conductivity of the foams was observed at the carbonization temperature between 800 and 1000°C, from 0.02 to 0.17 S/cm. Carbon foams were prepared from glucose by mixing with NH_4Cl and heating up to 1350°C for 3 h with a constant heating rate of 4°C/min, of which cell walls were said to consist of a single or few layers of graphene [374]. Activated carbon (AC)-dispersed carbon foams were prepared by using sucrose as the precursor for the foam matrix via the self-blowing technique [375]. The carbon foam with a mass ratio of AC powder to sucrose of 1/10 exhibited the highest density of 0.22 g/cm^3, the highest compressive strength of 3.4 MPa and the lowest cell size of about 0.5 mm. To improve the oxidation resistance, carbon foams were prepared at 120°C from sucrose aqueous solution containing 2–8 wt.% boric acid [376]. The addition of boric acid influenced greatly the foaming conditions and also the size of cells, as shown in Fig. 2.118. With increasing boric acid content, the foaming and setting time are shortened, shortening of foaming time being marked, and foam rise (the ratio of the final height of the foam to the initial height of the resin) increasing (Fig. 2.118A). The cell sizes become larger with increasing boric acid content, as shown on the foams prepared using 2 and 8 wt.% boric acid in Fig. 2.118B and C, respectively, and average cell size was determined as 0.67 and 1.17 mm, respectively.

Carbon foams were also prepared by direct pyrolysis of cyanate ester resin at 900°C under ambient pressure after curing of bisphenol A cyanate ester at 220°C [377]. As shown in Fig. 2.119A and B, the shape of the resin block is retained even after carbonization at 900°C. Self-blowing occurred

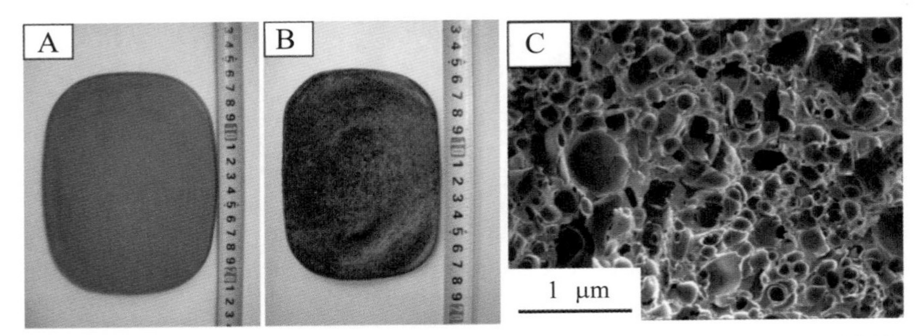

Figure 2.119 Carbon foam prepared from cyanate ester resin; appearance of (A) the precursor resin, (B) after carbonization at 900°C, and (C) SEM image of the cross-section of the carbon foam [377].

mainly in the temperature range of 400–600°C. The resultant foam consists of cells with an average size of about 200 μm, as shown in Fig. 2.119C, and delivers a compressive strength of 3.15 MPa with a bulk density of 0.26 g/cm^3.

2.3.2 Template-assisted carbonization

In 1987, carbon foams consisting of glass-like carbon were prepared using sintered NaCl as a template [378]. Commercially available NaCl with an average particle size of 17 μm was molded into a bar with a size of $20.0 \times 2.5 \times 1.0\ cm^3$ and about 35% porosity, and then sintered at 710°C. Phenol polymer was impregnated into the sintered NaCl bar through its THF solution and then carbonized at 700°C. Template NaCl was removed by careful washing with water and HNO_3, and then the carbon bar was freeze-dried. The carbon foam thus prepared contained macropores with a size of about 8 μm and had a bulk density of 0.035–0.075 g/cm^3.

Reticulated vitreous carbons (RVCs) with glass-like carbon characteristics (carbon foams with honeycomb-type open cell structure) were prepared by carbonization of polyurethane (PU) foam after the impregnation of either phenol resin, FA, or epoxy resin [379–382], as shown in Fig. 2.120A. The cell size and fraction of open cells of the PU foam were controlled by adding a clay (montmorillonite) of 1–4 pphp (parts per 100 parts of matrix) in a polyol with a foaming agent for foaming in a mold of stainless steel [380]. The PU foams were impregnated by phenolic resin and converted to carbon foams by heat treatment at a high temperature of 600–3000°C in inert atmosphere. PU foams with controlled cell sizes using different amounts of clays were impregnated by FA and carbonized at

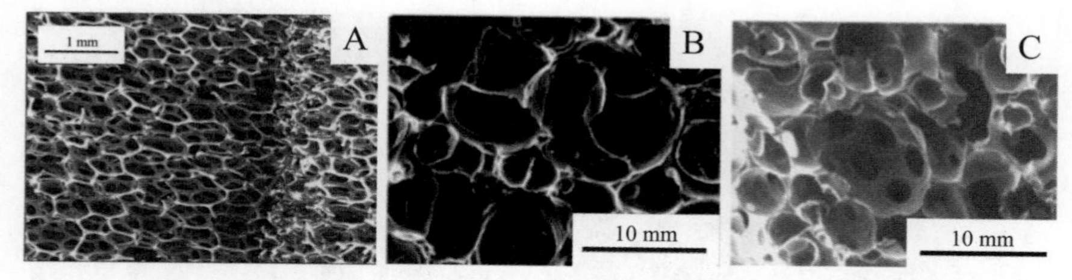

Figure 2.120 SEM images of reticulated vitreous carbon (RVC): (A) phenolic resin impregnated, and furfuryl alcohol impregnated using polyurethane foam containing (B) 2 pphp clay and (C) 4 pphp clay. The arrow in (A) points to the vertical wall formed during mechanical cutting [380,382].

900°C [382]. SEM images of the cross-sections of the RVCs are shown in Fig. 2.120B and C. By changing the clay content to the precursor polyol from 0 to 4 pphp, the bulk density of the resultant RVCs changed from 0.49 to 0.87 g/cm^3. These RVCs have a wide range of applications in electrochemical devices, for example, as optically transparent electrodes [381]. Low-cost RVCs were produced using household cleaning pad waste as a template and sucrose as the carbon precursor [383]. From commercial melamine foam, RVCs were obtained by simple carbonization at 600−1200°C in an inert atmosphere [384]. The carbonization yield from melamine foam was about 15 wt.% at 800°C and about 7 wt.% at 1200°C. By carbonization, the melamine foam shrank markedly while maintaining foam morphology and certain resilience even after 1200°C carbonization. The resultant carbon foams contained a relatively large amount of nitrogen, about 7 wt.% N in the foam heated at 1000°C.

Carbon foams were prepared using either PU or melamine foam as template with impregnation of poly(amic acid) composed of pyromellitic dianhydride (PMDA) and 4,4-diaminodiphenyl ether (DDE), followed by imidization and carbonization [385]. PU, as well as melamine, is pyrolyzed at a much lower temperature than carbonization of PI, as shown by the comparison with TG curves in Fig. 2.121A and gave only a small amount of carbon residues after carbonization. Since the PI (PMDA/DDE) used has a higher glass transition temperature than its pyrolysis temperature (above 500°C) and maintains a solid state, the honeycomb-type open cell structure of the template PU foam is preserved in the resultant carbonized PI, i.e., carbon foam, as shown in the SEM image of the cross-section in Fig. 2.121B. For this process, PU consisting of ester bonding had to be selected, because the other type of PU consisting of ether bonding is dissolved into the solvent for poly(amic acid), i.e., DMAc. The carbon foams

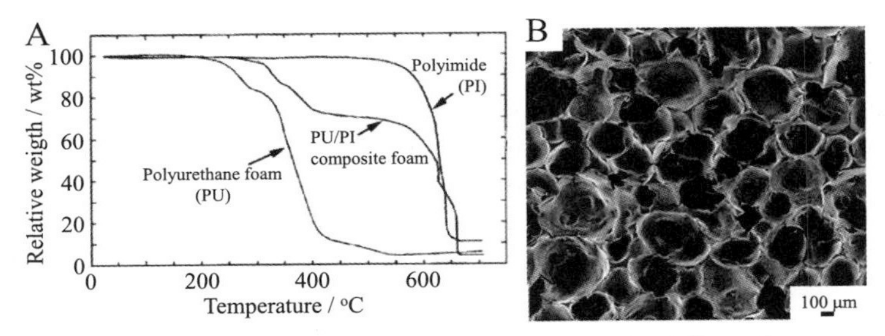

Figure 2.121 Carbon foam derived from polyurethane (PU) foam impregnated by polyimide (PI): (A) TG curves of PI-impregnated PU foam and each component polymers, and (B) SEM image of the cross-section of the carbon foam prepared [385].

with a bulk density of 0.35–0.47 g/cm^3 were synthesized using the ester-type PU foam with a bulk density of 0.17 g/cm^3, and the carbon foam with 0.29 g/cm^3 from PU foam with 0.06 g/cm^3. Because of the presence of macropores, air activation was efficiently performed to increase the micropores in the carbon foams, which resulted in the acceleration of water vapor adsorption. The carbon foams synthesized from PI (PMDA/DDE) were graphitizable by keeping the foam texture, the interlayer spacing d_{002} reducing to 0.3367 nm after 3000°C treatment. By using fluorinated PI, 6FDA/TFMB, as a carbon precursor with melamine foam template, microporous carbon foams were prepared by a single-step carbonization at 900°C, which delivered S_{total} of 1340 m^2/g and V_{micro} of 0.44 cm^3/g. Their microporous structure was enhanced by air activation at 400°C, reaching S_{total} of 1749 m^2/g and V_{micro} of 0.63 cm^3/g [386].

Resol-type PF resin was impregnated into ether-type PU foams via its ethanol solution together with block copolymer F127 [387]. After frequent squeezing and rotating in flowing air to remove air bubbles inside the foam for uniform coating of the PF resin on the cell wall of the PU foam, followed by complete evaporation of ethanol at 28°C, the impregnated foams were carbonized at temperatures of 350–900°C in N_2 to obtain carbon foams. The resultant carbon foams exhibited macropores (cells) of 100–450 μm and ordered mesoporous channels with diameters of 3.8–7.5 nm in the ligaments and joints of the foam. PU foams were impregnated into a water/ethanol mixed solution of resorcinol, formaldehyde, and block copolymer F127 with a small amount of HCl and then treated at 50°C for 2 days in an autoclave, followed by carbonization at 800°C in N_2 to prepare carbon foams [388]. The carbonized foams were loaded by iron nitrate in ethanol solution and then heat-treated at 1000°C in N_2 flow. The resultant Fe-loaded foams exhibited higher electrical conductivity of 11 S/cm and thermal conductivity of 0.33 W/m•K than the foams prepared without iron nitrate (0.35 S/cm and 0.05 W/m•K, respectively). They also exhibited higher S_{BET} of 445 m^2/g and V_{meso} of 0.25 cm^3/g than the foams prepared without the addition of the surfactant F127 (130 m^2/g and 0.075 cm^3/g, respectively). Mesoporous carbon/silica composite foams were synthesized by coating a mixture of resol, TEOS, and block copolymer F127 onto PU foam to obtain an interconnected open cell structure [389].

PU foams were impregnated into water slurry dispersed small particles (<45 μm) of a petroleum pitch (a high T_s) with a surfactant and then heat-treated at 450°C in N_2 [390]. The pitch-loaded PU foams thus prepared were stabilized at 350°C in air flow, followed by carbonization at 1000°C and graphitization at 2800°C in Ar to be converted to carbon foams. As shown in Fig. 2.122A, the pore structure of the template PU foam is preserved well, with uniform cells with large windows in the walls and the ligaments of dense carbon matrix (Fig. 2.122B), where the initial pitch particulates are recognized even after graphitization.

The effects of the PU template and the concentration of precursor slurries were studied using a mesophase pitch which had T_s of 232°C, quinoline insolubles of 30.2%, and toluene insolubles of 64.5% and mesophase content of 85%–90% [391]. PU foams with different cell numbers of 30, 40, and 50 PPI (pores per inch) were impregnated by the aqueous slurry of different contents of the mesophase pitch with particle sizes of <30 μm containing a small amount of PVA, followed by pressing to obtain uniform distribution of slurry and to remove excess slurry and then dried at 100°C. The impregnated foams were stabilized in air at 300°C, followed by carbonization at 1400°C in N_2. In Fig. 2.123A and B, the bulk

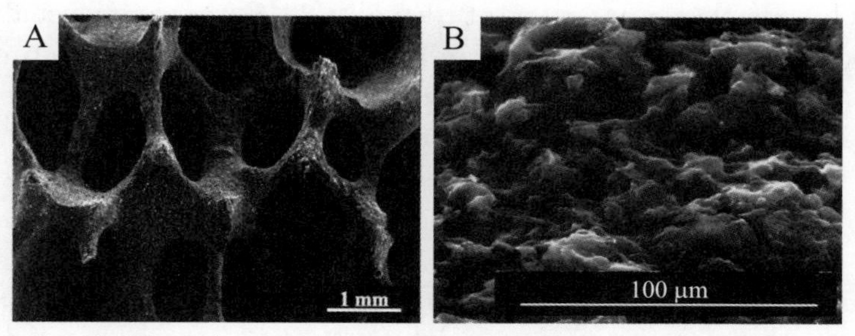

Figure 2.122 Pitch-based carbon foam prepared using PU template: (A) SEM image of the cross-section and (B) enlarged image of the ligament [390].

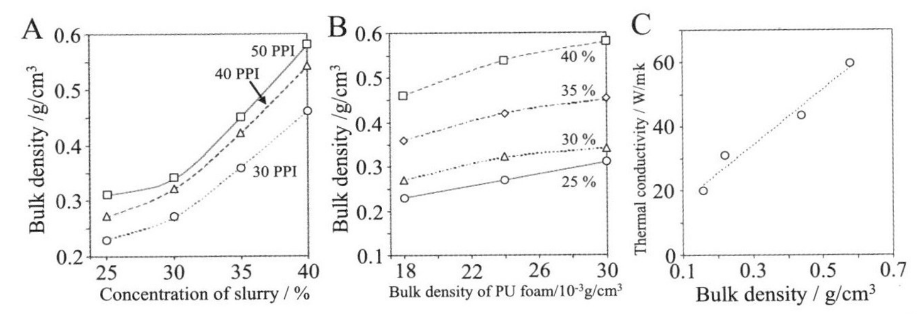

Figure 2.123 Pitch-based carbon foams prepared using PU foam template: (A) bulk density of carbon foam versus concentration of pitch slurry as a function of cell number of unit length (PPI: pores per inch), (B) bulk density of carbon foam versus bulk density of PU foam as a function of concentration of the slurry, and (C) thermal conductivity versus bulk density for the carbon foam [391].

density of the resultant carbon foams is plotted against the concentration of the slurry when the PU foams with different cell numbers are used and against the bulk density of the template foam when the slurry with different concentrations of the pitch are used, respectively. The effect of the template PU foam characterized by cell number and bulk density is clearly demonstrated; the PU foam with high cell number, i.e., high bulk density, is desired to obtain a high bulk density of the carbon foams. However, the effect of the slurry concentration is more pronounced, the higher slurry concentration giving much higher bulk density of the carbon foam. High bulk density of carbon foam is important to achieve high thermal conductivity of the resultant foam, as shown in Fig. 2.123C.

RF-derived carbon microhoneycombs (microcellular foams) (Fig. 2.124A) were prepared from the mixtures of resorcinol (R),

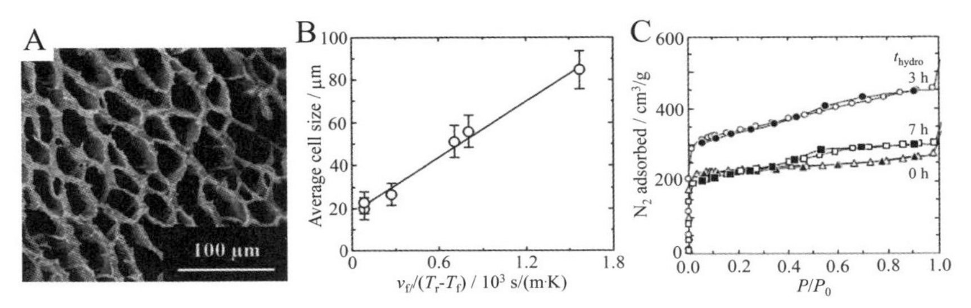

Figure 2.124 Microhoneycombs prepared from RF resin by ice templating: (A) SEM image of the cross-section, (B) change in cell size with icing condition parameter $v_f/(T_r - T_f)$ (see the text), and (C) N_2 adsorption–desorption isotherms of the honeycombs prepared via hydrothermal treatment at 90°C for different times (t_{hydro}) [392,393].

formaldehyde (F), and catalyst (C) in water (W) [392,393]. The ratios R/W, R/F, and R/C were 0.2, 0.5, and 100, respectively, by curing and following gelation for 24 h in a glass mold, followed by unidirectionally freezing at a constant temperature T_f with a constant immersion rate v_f in a polypropylene tube (ice-templating) and freeze-drying at −10°C. The microcellular foams of RF cryogel thus prepared were carbonized at 1000°C in N_2 flow. The cell size of the honeycombs depends strongly on the ice templating conditions, such as v_f and T_f, with average cell size being able to be linearly related to the parameter $v_f/(T_r - T_f)$ (T_r is room temperature), as shown in Fig. 2.124B. By insertion of hydrothermal treatment at 90°C between ice-templating and following freeze-drying, mesopores were created in the RF matrix, as shown in the N_2 adsorption−desorption isotherms in Fig. 2.124C, S_{BET} and V_{meso} increasing from 930 m^2/g and 0.14 cm^3/g to 1280 m^2/g and 0.36 cm^3/g, respectively, by hydrothermal treatment for 3 h.

Carbon foams were prepared from the mixtures of kraft lignin (L), resorcinol (R), formaldehyde (F) with sodium hydroxide catalyst, and poly(methyl methacrylate) (PMMA, average particle size of 6.4 μm) spheres as sacrificial template by aging at 80°C, eliminating PMMA template at 400°C, and finally carbonized at 800°C [394]. In the precursor mixture, the ratio of (L + R)/F was kept constant at 1/2 and L content in (L + R) was changed from 0 to 80 wt.%, and the content of PMMA was controlled to be 100, 200, and 300 wt.% to the sum of resin precursors, i.e., (L + R + F). The bulk density and porosity of the resultant carbon foams do not depend apparently on the content of lignin and slightly depend on PMMA content, as shown in Fig. 2.125A. However, the mechanical properties and electrical conductivity depend strongly on both lignin and PMMA contents, as

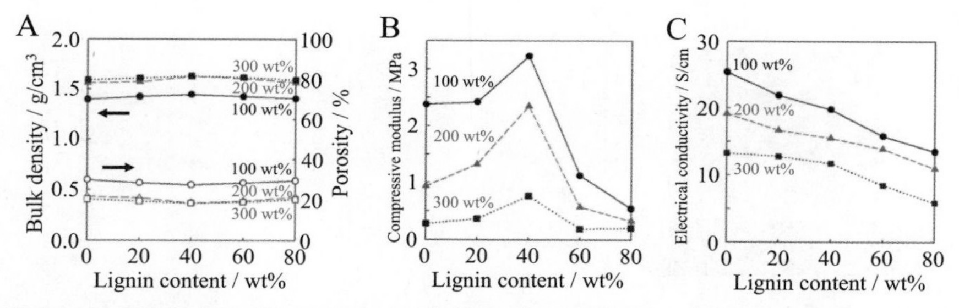

Figure 2.125 Carbon foams prepared from the mixtures of lignin and resorcinol-formaldehyde with different lignin contents using PMMA template at 800°C; the lignin-content dependences of (A) bulk density and porosity, (B) compressive modulus, and (C) electrical conductivity as a function of PMMA-contents [394].

shown in Fig. 2.125B and C, respectively. The compressive modulus and strength increase with increasing lignin content up to 40 wt.% but decrease above 40 wt.%, although the bulk density does not change, and the electrical conductivity decreases gradually with increasing lignin content. With increasing PMMA content, however, the compressive modukus and strength as well as conductivity decreased. The experimental results suggest that natural lignin can replace resorcinol, even improving the mechanical properties. PMMA honeycomb, which was prepared by standing PMMA emulsion for 7 days at room temperature, was used as template to prepare carbon honeycomb [395]. PMMA honeycombs were impregnated by R and F, followed by polymerization at 90°C and carbonization at 850°C in N_2, as shown in the cross-section of as-carbonized honeycomb in Fig. 2.126A. Onto the carbon honeycombs, nanoparticles of Ni_3Co alloy were loaded by impregnation into aqueous solution of nickel acetate and cobalt acetate (3/1 by molar ratio) with dicyandiamide (DCD) (a mass ratio of 10/1 to metal salts), controlling the total metal (Ni + Co) mass percentage of 20%, 30%, and 40% in the solution, and then heating at 850°C in N_2. The alloy nanoparticles are homogeneously dispersed on the honeycomb, as shown in Fig. 2.126B. The resultant composites exhibited high ORR activity.

N-doped carbon foams were synthesized from the mixture of dopamine hydrochloride with colloidal silica by polymerization at room temperature and then carbonization at 800°C in N_2 [396]. The resultant carbon had S_{BET} of 1056 m^2/g and V_{total} of 2.56 cm^3/g with about 8 wt.% N at a mainly graphitic configuration.

Carbon nanofiber (CNF) gels were prepared from glucose using tellurium (Te) nanowires as template [397]. Hydrothermal treatment of the

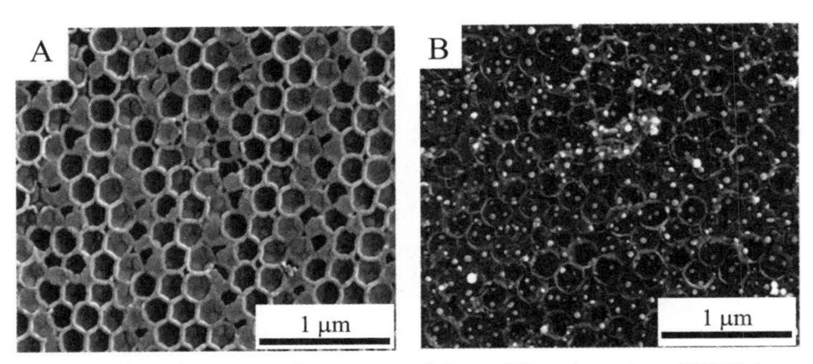

Figure 2.126 Carbon honeycombs prepared from RF resin using PMMA honeycomb template: (A) as-carbonized and (B) after loading of Ni_3Co nanoparticles [395].

mixture of glucose with Te nanowires in solution at 180°C for 12–48 h results in a mechanically robust monolithic gel-like product, which was washed and chemical etching used to remove the template, followed by freeze-drying to obtain CNF aerogel. The resultant CNF aerogel had high flexibility: a compression strain as high as 80% could be applied with almost complete recovery of its original volume after releasing the compression.

Graphite foams were prepared from the mixtures of natural graphite powder (average particle size of 87 μm) and sucrose by adding either $NaHCO_3$ (foaming agent) or NaCl (template) [398]. The mixtures were molded to 35 mm diameter discs and then carbonized at 600°C for 2 h, followed by dissolving $NaHCO_3$ and NaCl with water. Sucrose worked as the binder between graphite flakes and as a foaming agent. The bulk density, apparent porosity, and thermal diffusivity are listed with the preparation conditions of the precursor mixtures in Table 2.23. The addition of NaCl is effective for reducing the thermal diffusivity of the foams: mixture T5 delivered the highest thermal diffusivity of 7.1 mm^2/s (the highest thermal conductivity of 3.7 W/m•K) probably due to prohibition of contact between graphite particles by macropores formed by the NaCl template. The thermal diffusivity of the foam decreases with increasing sucrose content, which is caused by the increase in the content of carbon binder formed from sucrose. The compressive strength of the foam depended greatly on the sucrose content. The foam derived from mixture T2 delivered the highest strength of about 1.2 MPa, and other mixtures having the same sucrose content (F1 and T1) gave similar strength, while the foams derived from the mixtures T5–T7 gave very low strength (about 0.4 MPa).

2.3.3 Precursor design

2.3.3.1 Polymer blend

Polymer blend membranes of PI with PU were prepared by mixing PI precursor PAA (PMDA/ODA) with PU prepolymer (ester-type polyol) in different ratios in NMP, followed by casting on a glass plate and imidizing at 200°C [399–401]. The PU/PI membranes [poly(urethane-imide) membranes] with different PU contents (0–70 wt.%) were preliminarily heat-treated by sandwiching between glass plates at 300 and 400°C to obtain porous PI membranes, in which thermally less stable PU domains decomposed [399]. The porous PI membranes thus prepared were carbonized at 900°C in N_2 flow [400,401]. TG curves of the membranes with 30 wt.% PU pretreated at 200–400°C (Fig. 2.127A) demonstrate

Table 2.23 Preparation conditions of the precursor mixture and the properties of the resultant graphite foams [398].

Code	Graphite (g)	Sucrose (g)	Distilled water (g)	$NaHCO_3$ (g)	NaCl (g)	Bulk density (g/cm^3)	Porosity (%)	Diffusivity (mm^2/s)
F1	5.0	6.0	3.0	0.2	—	0.500	69.39	6.0
F2		6.0	3.0	0.0	—	0.471	65.09	6.8
T1	5.0	6.0	0.0	—	7.5	0.522	61.42	1.1
T2		6.0	3.0	—	7.5	0.498	65.14	1.6
T7		1.25	1.25	—	7.5	0.674	41.83	5.6
T6		0.625	0.625	—	7.5	0.795	42.10	6.7
T5		0.025	0.025	—	7.5	0.728	42.60	7.1

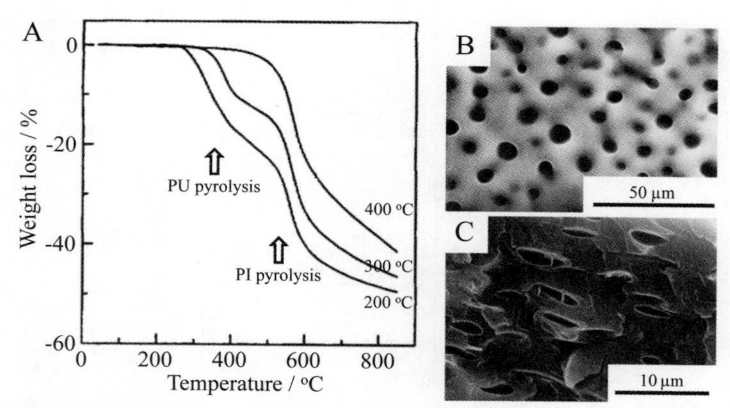

Figure 2.127 Porous carbon membranes prepared from PU/PI membranes: (A) TG curves for the PU/PI membranes with 30 wt.% PU and pretreated at 200–400°C, and SEM images of (B) the surface and (C) the cross-section of the membrane with 50% PU heat-treated at 400°C [400,401].

clearly that the pre-heat-treatment at 400°C is effective in removing PU by decomposition and leaving pores in the PI matrix. The formation of macropores is clearly observed in the PU/PI membranes pre-treated at 400°C, as shown by the SEM images of the surface and cross-section of the membranes in Fig. 2.127B and C, respectively. In Fig. 2.128A and B, size distributions of macropores on SEM images of the surface of PI membranes after pre-treatment at 400°C and carbon membranes are shown as a function of PU content in the starting PU/PI membranes, respectively. In PI membranes (Fig. 2.128A), pore size distribution becomes broader with increasing PU content, accompanied by a shift of the peak size to larger side, average size of 0.6 mm for 10% PU and 10 μm for 70% PU, which is

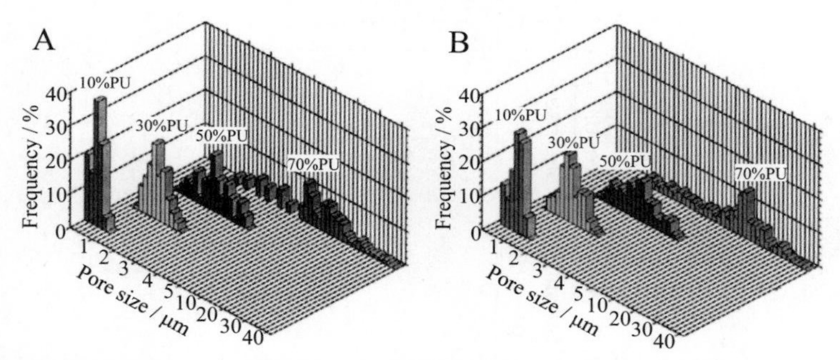

Figure 2.128 Size distributions of macropores on the membrane surface as a function of PU content: (A) PI membranes and (B) carbon membranes prepared at 900°C [401].

reasonably supposed due to the interconnection of PU domains. In the carbonized membranes at 900°C (Fig. 2.128B), the size of macropores and their distribution are almost the same as for PI membranes. The experimental results suggested that the size and volume of macropores in carbon membranes are possible to control by changing the PU content. The carbon membranes synthesized from PU/PI membranes contained micropores, S_{micro} determined by α_s plots of N_2 adsorption isotherms increasing from 260 to 350 m^2/g by increasing the PU content above 30% and to 690 m^2/g by increasing the carbonization temperature to 800°C.

Novolac-type PF resins mixed with 3–11 wt.% hexamine were partially cured at 150°C, followed by grinding to less than 40 μm particles and then mixed with additional hexamine to reach total hexamine of 11 wt.% [402]. The doughs thus prepared were extruded to form green tubes after adding a small amount of methyl cellulose as additives and then cured at 120°C, followed by carbonization at 800°C in Ar. In partially cured doughs, macropores are formed, depending greatly on the amount of hexamine added in the first step, as shown in Fig. 2.129, the average pore size increasing from 0.10 to 0.56 μm with increasing hexamine content in the first step from 3 to 11 wt.%. The polymer blends of anionic RF sols prepared using different catalysts with cationic polystyrene spheres having diameters of 35–37 nm were pyrolyzed and carbonized at 800°C in Ar to produce mesocellular carbon foams [403].

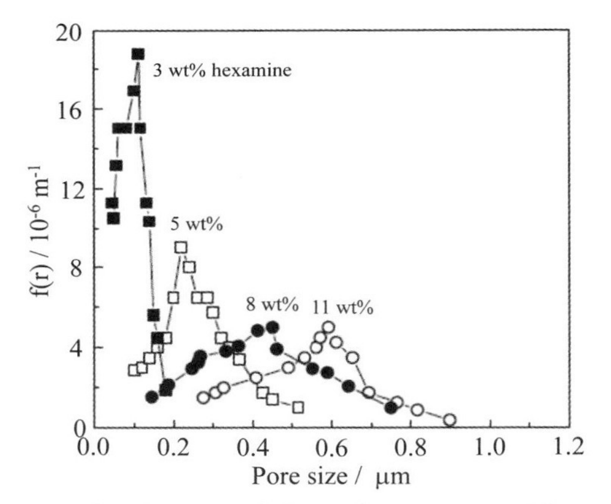

Figure 2.129 Pore size distributions of the carbons prepared from partially cured PF resins with different hexamine contents (1–11 wt.%) [402].

2.3.3.2 Exfoliation of graphite oxides

Exfoliated graphite (EG) is an important industrial raw material for the production of flexible graphite sheets, which are widely used as gaskets, seals, and packings [404–407]. EG is usually prepared by rapid heating of residue compounds of natural graphite flakes with sulfuric acid to about 1000°C [406]. It consists of fragile worm-like particles, which are formed by exfoliating preferentially along the normal to the basal plane of graphite flakes, as shown in Fig. 2.130A and B. Worm-like particles contain macropores which can be approximated to be elliptic, as shown in the SEM image of its cross-section in Fig. 2.130C.

EG was fabricated from natural graphite flakes through two sequential steps: (1) oxidation in concentrated inorganic acid solution and (2) thermal decomposition by rapid heating to high temperature. For the first step, various oxidants have been proposed to accelerate the intercalation of HSO_4^- and SO_4^{2-} into interlayer spaces of graphite, and to promote further oxidation of graphite layers: $NaNO_3$ and $KMnO_4$ with concentrated H_2SO_4 [408], HNO_3 and $KClO_3$ with concentrated H_2SO_4 [409], $KClO_3$ with fuming HNO_3 [410], and HNO_3 with H_2SO_4 [411]. In this process, graphite changes to so-called graphite oxide through an intermediate stage of graphite intercalation compound with either sulfate or nitrate ions. In the intermediate compound, the layer structure of graphite is still maintained and various oxygen-containing functional groups, such as −COOH, −OH, etc., are attached to the carbon layer by covalent bonding in the final product of this process, these products having been called graphite oxide by referring to graphite fluoride in which carbon atoms bind with fluorine atoms covalently. For the second step, the products of the first step were heated quickly to a high temperature of around 1000°C to be exfoliated and reduced by evolution of SO_x, NO_x, and other oxygen-containing gases, the graphitic structure being partly recovered. Since the main purpose of the preparation of these exfoliated worm-like particles was

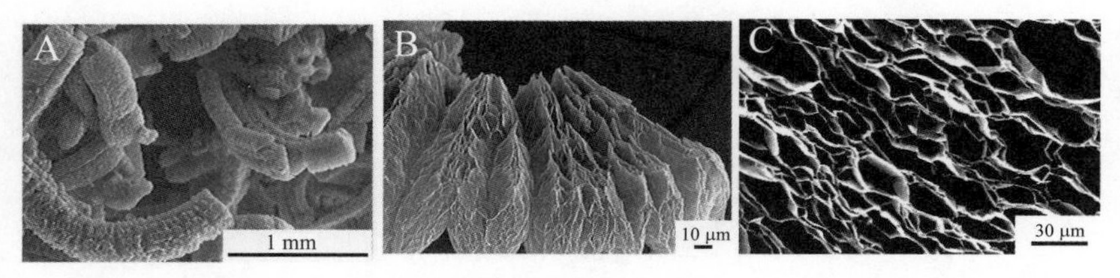

Figure 2.130 SEM images of worm-like particle of EG: (A) aggregated particles, (B) a particle, and (C) cross-section [406].

industrial production of flexible graphite sheets through compression and rolling, which is the only way to obtain large-area graphite sheets, the characteristics of natural graphite flakes have to be retained in the final graphite sheets and exhausting gases have to be as low as possible because some of them are harmful. Therefore, the final products of the first step, i.e., graphite oxide, are often kept in air to decompose the structure of intercalation compounds to residue compounds to make the intercalates in the product as small as possible but sufficient for exfoliation of the graphite layer structure.

Hammers method [408] is frequently employed with some modifications to synthesize a kind of graphene flakes from natural graphite by aiming to have thin flakes, hopefully in single layers. Therefore, the oxidation of natural graphite in acidic solution is performed for a long time with extensive stirring and only the oxidation products in the supernatant are recovered to be sent to the next process, i.e., either thermal or chemical reduction, to obtain graphene flakes. The oxidation products in the supernatant are composed of less than a few carbon layers, which may be called "graphene oxide," and the products after their reduction are called "reduced graphene oxide" (rGO). Graphene oxide has to be differentiated from graphite oxide which retains the memory of the precursor graphite, as explained above. As a consequence, rGO has to be differentiated from graphene synthesized by CVD and peeling of graphite, because the former contains many structural defects and some functional groups in the graphene layer, but the latter retains much higher crystallinity.

In Fig. 2.131A and B, exfoliation volume and weight loss are plotted against exfoliation temperature, respectively, on two runs of experiments

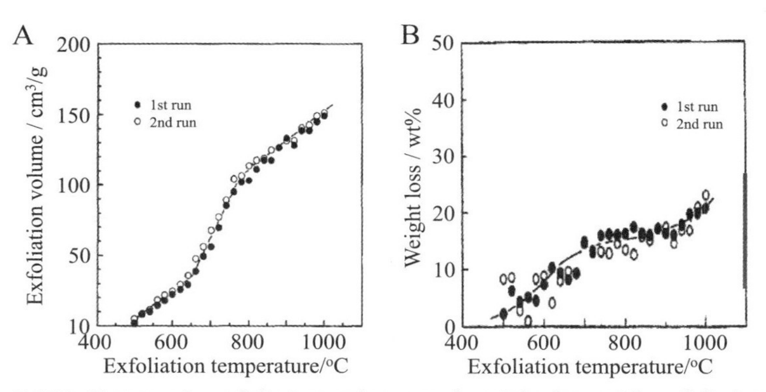

Figure 2.131 Changes in exfoliation volume and weight loss with exfoliation temperature [412].

on the same residue compound [412]. The exfoliation volume increases with increasing exfoliation temperature, as shown in Fig. 2.131A. Two kinks are clearly observed in the dependence of exfoliation volume on temperature, at around 650 and 800°C. Below 650°C and above 800°C, the increasing rate of volume is smaller than that in an intermediate temperature range, i.e., 650–800°C. As shown in Fig. 2.131B, the same kinks at around 650 and 800°C were also observed on weight loss, below 650°C and above 800°C rapid increase in weight loss, while in the intermediate temperature range only slight increase in weight loss. Exfoliation volume is a parameter for a lump of EG, including macropores inside the worm-like particles (intraparticle pores) and those among the particles (interparticle pores), although the pores detected by N_2 adsorption–desorption isotherms (micropores and mesopores) are in negligibly small amounts in most EG lumps.

As shown in Fig. 2.130C, the worm-like particles contain a large amount of intraparticle macropores, of which the cross-sections may be approximated by ellipses. The cross-sectional area, the length along the major axis (mostly perpendicular to the worm length), and that along the minor axis (mostly parallel to the worm length) of intraparticle pores were measured using image processing of SEM images [413,414]. In Fig. 2.132, the distributions of cross-sectional area, and lengths of the major axis and the minor axis of elliptic pores in worm-like particles (intraparticle pores) are shown on the graphite exfoliated at 1000°C for 60 s, as an example, and the averaged values of pore parameters for EG prepared at 600, 800, and 1000°C are listed in Table 2.24, together with the number of intraparticle pores measured. Pore structures in the worm-like particles of 600- and 800°C EGs are difficult to differentiate from averaged values of pore

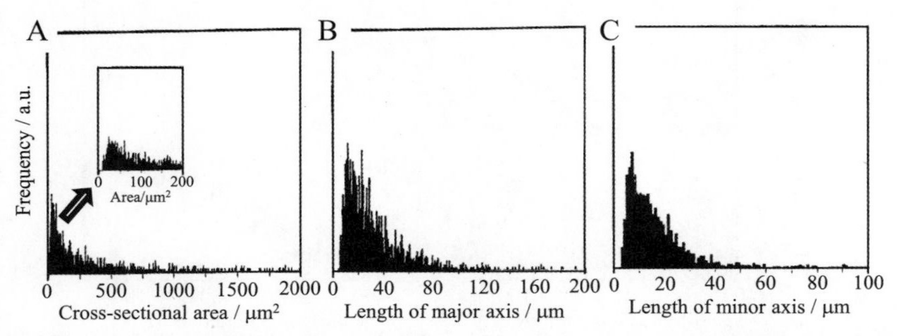

Figure 2.132 Distributions of intraparticle pores of a graphite exfoliated at 1000°C: (A) cross-sectional area, (B) length of the major axis, and (C) length of minor axis [413].

Table 2.24 Parameters for pores inside the worm-like particles (intraparticle pores) of EGs prepared at 600, 800, and 1000°C [412].

Exfoliation temperature		600°C	800°C	1000°C
Exfoliation volume (cm^3/g)		25	114	152
Averaged pore parameters	Cross-sectional area (μm^2)	193	217	321
	Major axis (μm)	24.4	26.0	31.2
	Minor axis (μm)	8.8	9.7	11.2
	Aspect ratio	0.412	0.424	0.412
	Fractal dimension	1.09	1.10	1.09
Number of pores measured		2583	2161	2059

parameters, whereas their exfoliation volume increases markedly from 25 to 100 cm^3/g. On the other hand, EG prepared at 1000°C has much larger averaged pore parameters than those prepared at low temperatures (Table 2.24). Since the EG prepared at 1000°C had slightly broadened distributions of pore parameters than those prepared at low temperatures, these differences in averaged values of pore parameters were supposed to be due to the increase in frequency of large pores.

Development of the intraparticle pores with exfoliation of graphite oxides proceeds in three steps. In the first step, below 650°C, the exfoliation volume increases with temperature through exfoliation of each graphite flakes to worm-like particles, in which ellipsoidal pores are developed. In the second step, above 650°C, the main process may be the introduction of complicated entanglement of worm-like particles and result in the increase in exfoliation volume, because growth of pores inside the particles is not as marked but the exfoliation volume increases more rapidly than in the first step. In this temperature range, large open spaces among the particles are expected to grow markedly, which is suggested from quite different exfoliation volumes between 600- and 800°C EG. In the third step, above 800°C, the development of pores inside the worm-like particles is thought to continue with increasing temperature, because the increase in exfoliation temperature from 800 to 1000°C causes mainly widening of pores inside the particles, but only a relatively small increase in exfoliation volume.

The content of intercalates remaining in the residue compound has a marked influence on exfoliation behavior. The residue compounds are prepared by electrochemical intercalation of sulfuric acid [413,415]. In Fig. 2.133A, the volatiles content in the starting residue compounds and

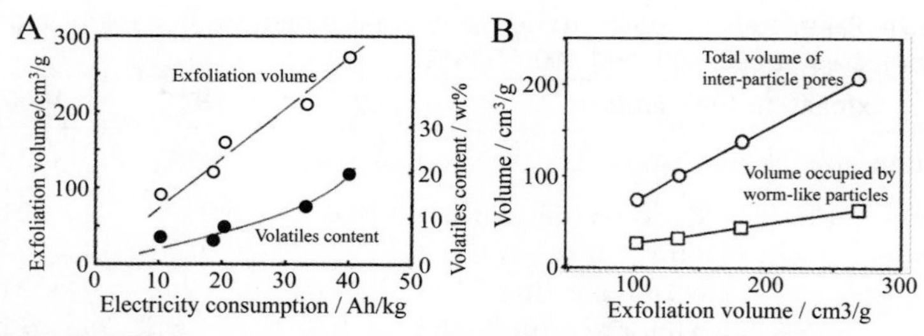

Figure 2.133 Exfoliation of residue compounds prepared by electrolysis: (A) changes in volatiles content of the residue compounds and exfoliated volume of the resultant EG with electrical consumption for intercalation, and (B) changes in total volume of interparticle pores and volume occupied by worm-like particles with exfoliation volume [413].

exfoliation volume of the resultant graphite exfoliated at 1000°C are plotted against electricity consumed during electrolysis to prepare intercalation compounds. With increasing electricity consumption, the volatiles content, i.e., the content of intercalates, and exfoliation volume increase almost proportionally. The analyses on the intraparticle pores using image analysis showed that the pore parameters depended only slightly on electricity consumption and volatiles content in the residue compounds, even though the exfoliation volume of the lumps of EG changed with increasing electricity consumption. Large spaces among the particles (interparticle pores) were measured by impregnating paraffin into a lump of EG, solidifying, and then slicing into thin sections to apply image analysis [416]. Averaged parameters for the interparticle pores are tabulated in Table 2.25 [417]. After 60 s of heating at 1000°C, interparticle pores become notably larger than those after 20 s heating. The total volume of the interparticle pores and the volume occupied by worm-like particles are plotted against the exfoliation volume in Fig. 2.133B. Both the volume of interparticle pores and the volume occupied by particles increase linearly with increasing exfoliation volume, while the latter is in a narrow range of 73%–77%, in other words, the interparticle pores are predominant in exfoliation volume.

Table 2.25 Average values of parameters of interparticle pores for EG prepared at 1000°C for the residence time of 20 and 60 s [417].

Residence time at 1000°C		20 s	60 s
Average pore parameters	Area (mm^2)	0.070	0.104
	Periphery length (mm)	0.647	0.847
	Aspect ratio	0.76	0.76
	Fractal dimension	1.72	1.74
Number of pores used		9678	8194

Intercalation compounds of graphite bisulfate with first to fourth stage structures were subjected to thermal exfoliation at 600, 800, and 1000°C and the crystalline structure of the resultant EGs was studied [418]. The crystallite size along c-axis L_c measured on EG from 006 diffraction line decreased from the pristine graphite, depending on stage number, although no pronounced difference in intraparticle pore structure was reported. The value of $L_c(006)$ decreased from 59 nm for the pristine graphite down to 12–15 nm by intercalation and exfoliation. The intercalation and exfoliation processes resulted in less damage to the crystallinity of the matrix graphite with increasing stage number of intercalation compounds.

To perform the exfoliation of graphite at low temperatures, different couples of intercalating reagents with oxidation agents were proposed, $FeCl_3$ with H_2O_2 [419], CrO_3, and H_2O_2 [420], concentrated H_2SO_4 with ammonium persulfate $(NH_4)_2S_2O_8$ [421], $KMnO_4$ with $HClO_4/NH_4NO_3$ [422], and concentrated H_2SO_4 with H_2O_2 [423]. Into concentrated H_2SO_4 solution dispersing natural graphite flakes and keeping at 5°C, H_2O_2 was added dropwise, kept at 5°C for 0.5 h and then transferred to a 30°C atmosphere [423]. The graphite flakes in the solution started to exfoliate gradually at 30°C, taking about 4 h for exfoliation to be completed. In Fig. 2.134A–C, the changes in the final exfoliation volume with molar ratio of H_2SO_4/H_2O_2, amount of the concentrated H_2SO_4, and ambient temperature during exfoliation are shown for 0.5 g natural graphite (lateral size of less than 300 μm and carbon content >99%). Based on these results, the optimum condition for the graphite was decided as a molar ratio of 7, H_2SO_4 of more than 7 mL, and ambient temperature higher than 30°C. In order to avoid the decomposition of H_2O_2, the solution of the mixture had to keep a low temperature (<5°C) and the

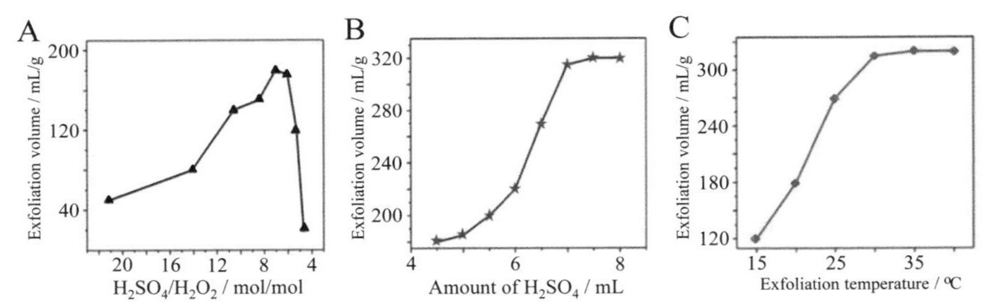

Figure 2.134 Changes in exfoliation volume with (A) molar ratio of H_2SO_4/H_2O_2 at 30°C, (B) amount of H_2SO_4 (H_2SO_4/H_2O_2 of 7/1) at 30°C, and (C) exfoliation temperature (H_2SO_4/H_2O_2 of 7/1 and amount of H_2SO_4 of 7 mL) [423].

exfoliation was recommended to be performed at a temperature higher than 30°C for completing the exfoliation. When the coupling of H_2SO_4 with $(NH_4)_2S_2O_8$ was employed [421], the interaction and exfoliation could be performed in one step, while the exfoliation seemed to be not completed. When $FeCl_3$ with H_2O_2 was selected [419], the $FeCl_3$-graphite intercalation compounds had to be synthesized by a conventional two-zone vapor transport technique (at 380°C for 24 h) before being subjected to the reaction with H_2O_2 at room temperature to exfoliate the host graphite. In the case of the coupling of CrO_3 with H_2O_2 [420], the CrO_3-graphite intercalation compounds had to be synthesized and the exfoliation was performed using 30% H_2O_2. Both intercalation reaction and following exfoliation could proceed at room temperature, while sonication in the NMP solution of the resultant EG was needed to remove impurities in the EGs. In the case of $KMnO_4$ with $HClO_4/NH_4NO_3$ [422], graphite flakes were added into the mixed solution of $KMnO_4$ with NH_4NO_3 and $HClO_4$ at 30°C and then subjected to exfoliation at 400°C, with a complete exfoliation being obtained.

Graphite residue compounds are successfully exfoliated under microwave irradiation [424,425]. The microwave exfoliation can be performed in a short time and no heating furnace is needed, but care has to be taken in applying this technique because it might result in inhomogeneity, with some flakes exfoliating to worm-like particles but some flakes having no exfoliation. The exfoliated particles are easily separated from unexfoliated particles by a simple sieving method [425].

EG was found to have very high sorption capacity of spilled heavy oils, reaching 80 g/g [426–429], and also high capacity of biological proteins and body fluids [430,431] into its macropores. The heavy oils sorbed into EG were easily recovered by filtration for reuse.

Sulfur-free EG was prepared from natural graphite by a two-step chemical oxidation process, the first chemical intercalation in $KMnO_4$ and $HClO_4$ and the second intercalation in KH_2PO_4, followed by exfoliation at 950°C [432]. The resultant EG had a high sorption capacity for oils.

2.3.4 Graphene foams

2.3.4.1 Assemblage of reduced graphene oxide

2.3.4.1.1 Hydrothermal treatment

Foams of reduced graphene oxide (rGO) were prepared by hydrothermal reduction of graphene oxides (GOs) as suspension in aqueous solution prepared from natural graphite by oxidation in acid solutions, such as H_2SO_4/HNO_3 and H_2SO_4 with $KMnO_4$. To obtain thin flakes of GO, the precursor graphite has to be oxidized almost completely with strong shearing by stirring to separate the flakes to thin them, and often just supernatants are used to select thinner flakes consisting of single- to few-layered graphene flakes. The reduction of GO to rGO changes the particles from hydrophilic to hydrophobic, resulting in the assembly of these flexible nanoflakes into three-dimensional porous networks (rGO foams) under hydrothermal condition, which is usually stabilized by freeze-drying (lyophilization). Unlike carbon foams from polymeric precursors, the fabrication of rGO foams can eliminate the energy-consuming carbonization process.

From homogeneous GO aqueous dispersion after removing aggregated GO by centrifugation, rGO foams were prepared by treatment under hydrothermal conditions at 180°C, followed by freeze-drying [433]. The properties of the foams depended strongly on the GO concentration of the starting suspension and the hydrothermal reaction time. An SEM image of the foam and the effect of reaction time on the appearance of cylindrical foams and their properties are shown in Fig. 2.135, for a GO concentration of 2 mg/cm^3. With increasing reaction time, the height of the as-prepared foam cylinder decreases clearly within the initial 6 h and subsequently changes a little, water content just after hydrothermal treatment decreases slightly, and compressive strength and modulus increase markedly. Electrical conductivity increased also with increasing reaction time, from 0.23 mS/cm for a 1 h reaction to 4.9 mS/cm for a 12 h treatment. By using natural graphite flakes with lateral sizes of 500–600 and 40–60 μm, rGO foams were prepared by hydrothermal treatment of the GO suspensions

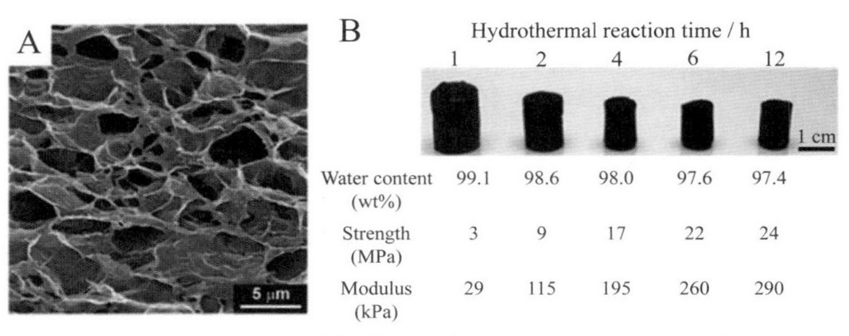

Figure 2.135 rGO foams prepared by hydrothermal treatment: (A) SEM image of the cross-section and (B) appearance and properties of the rGO foams prepared from 2 mg/mL GO suspension for different treatment times [433].

with the assistance of thiourea at 180°C [434]. The resultant foams of self-assembled rGO flakes were dipped into distilled water to remove the residual thiourea and then freeze-dried to remove water. The thiourea added played different roles, making GO flakes separately by its decomposition gases, reducing GO to rGO, and giving new functional groups, such as $-NH_2$ and $-SO_3H$. The foams showed very high processability, structural stability, and high mechanical properties. They could be cut into different shapes and those with a size of 22.9 mm diameter and 11.9 mm height supported a weight of 500 g without any deformation. The maximum compression stress that a foam is able to sustain before degradation reaches 140 kPa with a strain of about 82%, as shown in Fig. 2.136A, and good cyclic performance for compressive loading and unloading of 14 kPa weight at a maximum strain of about 25%, as shown in Fig. 2.136B. Ammonia treatment of GO hydrogels was effective in strengthening the resultant rGO foams [435]. GO hydrogels obtained via hydrothermal treatment at 180°C were immersed into ammonia solution (14 vol%) at 25, 60, and 90°C, followed by freeze-drying. The resultant rGO foams via ammonia treatment at 90°C delivered a bulk density of about 16 mg/cm^3, compressive strength of about 0.7 kPa, and modulus of about 4.8 kPa with improved resilience for compression.

N-doped rGO foams with an ultra-low density of 2.1 mg/cm^3 were prepared from a low-concentration GO aqueous suspension (0.35–0.4 mg/cm^3) mixed with 5 vol% pyrrole by hydrothermal treatment at 180°C, followed by freeze-drying and annealing at 1050°C in an Ar

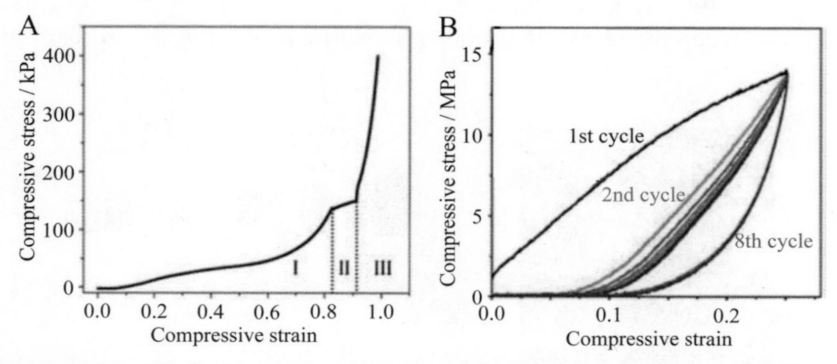

Figure 2.136 rGO foams prepared under hydrothermal treatment with thiourea assistance at 180°C: (A) compressive stress–strain curve for a foam with a size of 23.6 mm diameter and 12.7 mm thickness and (B) cyclic performances at a maximum strain of 25% in air [434].

atmosphere [436]. Despite the low density, the resultant foam had a high conductivity of 1.2×10^3 S/m and relatively high S_{BET} of 280 m^2/g. The foams with an ultra-low density of 0.45 mg/cm^3 were prepared by freeze-drying an aqueous suspension of large-sized GO with CNTs, followed by chemical reduction with hydrazine vapor [437]. The resultant CNT/rGO composite foams showed nearly complete recovery after 50%–82% compression and compressive stress–strain curves consisting of distinct two stages, a linear-elastic regime at strain of <29% and a nonlinear regime at 29%–50% strain. Cylindrical GO hydrogels, which were prepared from GO suspension with a concentration of 2 mg/cm^3 by hydrothermal treatment, were subjected to vacuum drying at room temperature and then annealed at 800°C, the products having a high density of 1.6 g/cm^3 and being microporous, while the same GO hydrogels could give low-density rGO foams by being subjecting to freeze-drying [438]. rGO blocks having relatively high bulk densities of 1.4–1.6 g/cm^3 were obtained by controlling the pH value at 10.1 from the GO suspensions via hydrothermal treatment with NH_3 and NaOH [439]. The resultant rGO blocks exhibited compressive strength ranging between 318–401 MPa and electrical conductivity of 7.1–8.2 S/cm.

rGO foams were prepared from mixtures of GO nanoflakes, consisting mainly of single-layered flakes using the supernatant, with glucose and noble-metal salts, such as $PdCl_2$, $HAuCl_4$, $RhCl_3$, etc., by hydrothermal treatment at 120°C, followed by freeze-drying [440]. Metal nanoparticles were dispersed homogeneously in the foam and worked to inhibit the stacking of rGO layers and the collapsing of foam frameworks, and glucose was necessary for reducing metal salts and also for reinforcing the mechanical strength of the self-assembled framework. The mixture of GO with poly(vinyl alcohol) (PVA) was converted to self-assembled cylindrical foam under hydrothermal conditions at 150°C, and then to rGO foam by heating at 800°C for the reduction of GO and carbonization of PVA [441]. The resultant rGO foams consisted of a wide range of pores, micropores, mesopores, and two kinds of macropores (cells), including spheroidal pores. The foams exhibited high performances in adsorption of dye pollutants and oils [441,442]. Composite foams of rGO with carbon were prepared from a mixture of GO aqueous suspension (1 mg/cm^3 concentration, pH of 3) with ethanol solution of resol (50 wt.%) by hydrothermal treatment at 180°C, followed by freeze-drying and carbonization at 900°C [443]. The resultant composite foam had a bulk density of 7.9 mg/cm^3, S_{BET} of 1019 m^2/g, and electrical conductivity of 13.1 S/m.

2.3.4.1.2 Freeze-drying

GO foams were prepared from an aqueous suspension of GO (about 4 mg/cm^3) by freezing at −18°C for 2 days and then freeze-drying at a sublimation temperature of −53°C under a pressure of less than 10 Pa for 3 days, followed by reducing and annealing at 800°C to obtain rGO foams [444]. For comparison, the precursor GO was heated at 800°C in N_2 flow, without the freezing process. SEM and TEM images are compared of rGO prepared with and without the freezing process in Fig. 2.137, revealing that freezing of GO aqueous suspension is effective in exfoliation of the starting graphite flakes and to obtain thin flakes. rGO obtained via the freezing process gave a foam which exhibited high performance as the electrodes in capacitive deionization devices.

Reducing agents mixed into GO suspensions had an influence on foaming via freeze-drying [445]. GO suspensions at concentrations of 0.1−2.0 mg/cm^3 after adding various reducing agents of 27−54 mmol/L were heated at 95°C without stirring to give hydrogels which were converted to aerogels by freeze-drying after dialysis in deionized water. In Table 2.26, the bulk density, atomic ratio C/O by elemental analysis, and electrical conductivity of the rGO foams are listed for each reducing agent. The reduction using HI aqueous solution results in high electrical conductivity of the foam, as 110 S/m, which may be due to the relatively high

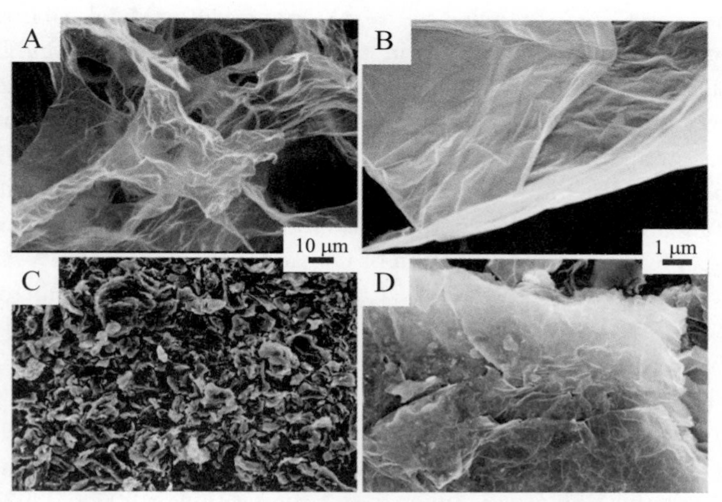

Figure 2.137 SEM images (A and C) and TEM images (B and D) of rGO foam prepared via freezing (A and B) and of rGO prepared without the freezing process (C and D) [444].

Table 2.26 The effect of reducing agents on the rGO foams [445].

Reducing agent	Vitamin C	Na_2S	HI[a]	$NaHSO_3$	$NaHSO_3$ + 400°C[b]
Bulk density (g/cm^3)	0.021	0.015	0.037	0.018	0.015
C/O atomic ratio	6.8	5.8	8.5	7.0	10.1
Conductivity (S/m)	23	6.8	110	17	87

[a]HI in 80 vol.% acetic acid aqueous solution.
[b]heated at 400°C for 5 h in a vacuum.

bulk density. The heat treatment of the $NaHSO_3$ reduced foam at 400°C is effective for removing oxygen-containing groups. L-ascorbic acid was used as a reducing agent for GO suspensions with 4.0 mg/cm^3 concentration to obtain rGO hydrogel and then the hydrogel was converted to aerogel (foam) by either freeze-drying or supercritical CO_2 drying [446]. L-ascorbic acid does not evolve gaseous products during reduction of GO and so it is possible to obtain a uniform hydrogel. Freeze-drying delivered rGO foams with bulk densities of 12–38 mg/cm^3, while supercritical CO_2 drying gave rGO foams with bulk densities of 23–96 mg/cm^3. The utilization of a mixture of hypophosphorous acid-iodine as a reducing agent for GO suspensions at 90°C resulted in rGO foams with a high electrical conductivity of 500 S/m and high C/O atomic ratio of 14.7 [447].

2.3.4.1.3 Templating

rGO foams with long straight cells packed into a honeycomb lattice were prepared using ice as a template [448]. GO aqueous suspension containing ascorbic acid (GO/ascorbic acid of 1/2 by weight) was frozen at dry ice temperature for 0.5 h, thawed at room temperature, and then heated in a boiling water bath for 8 h to reduce GO, followed by sequential dialysis in water, freeze-drying, and thermal annealing at 200°C in air to obtain rGO foams. The starting GO flakes had lateral sizes of 0.2–10 μm and their concentration in the suspension was 0.5–7.0 mg/cm^3. SEM images of the cross-sections of the resultant rGO foam in two orthogonal directions are shown with an enlarged view in Fig. 2.138A–C. Their bulk densities could be controlled in the range of 0.5–7.0 mg/cm^3. The rGO foam with a bulk density of 5.10 mg/cm^3 exhibited a compressive strength of 18 kPa and a high resilience at 80% strain. rGO foams with cellular texture were prepared by emulsifying GO suspension with toluene and then unidirectional

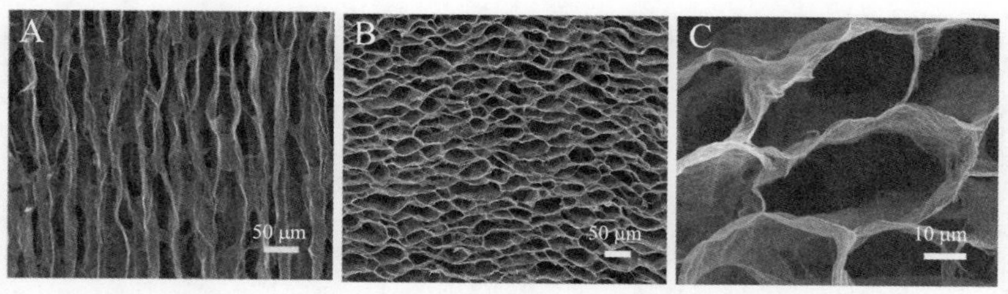

Figure 2.138 SEM images of the rGO foam prepared via ice templating: (A) side view, (B) top view, and (C) enlarged top view [448].

freezing in Teflon cylinders, followed by freeze-drying [449]. The resultant GO cellular foams were obtained as discs of 18–20 mm in diameter and 9–15 mm height and with densities of 0.003–0.015 g/cm^3, which were subjected to thermal reduction at 300–1000°C in H_2(10%)/Ar and heat treatment at 1000–2400°C in a vacuum.

rGO foams were prepared by vacuum filtering of a mixture of rGO and 10 wt.% polystyrene colloidal particles (about 2 μm size) in their water dispersion, which was composed of the interconnected spherical cells with uniform size of around 1 μm and wall thickness of 10 nm [450]. rGO foams were prepared from a mixture of the aqueous dispersion of GO with polystyrene microspheres with sizes of 280 nm as sacrificial templates by vacuum filtration on a millipore filter, followed by annealing at 550°C and then carbonization at 900°C in N_2 [451]. During these heat-treatment processes, GO was reduced to rGO and polystyrene spheres were pyrolyzed simultaneously by leaving macropores (cells) composed of the wall of thin rGO flakes, as shown in the SEM and TEM images in Fig. 2.139. The resultant rGO foams demonstrated high capacitive deionization performances.

Polystyrene-grafted GO flakes (PS-GO) dispersed in benzene were drop-casted onto SiO_2 substrate and exposed to a stream of humid air, where spontaneous condensation and close packing of aqueous droplets at the organic solution surface occurred due to the endothermic evaporation of benzene and accelerated self-assemblage of GO flakes around the aqueous droplets [452]. The resultant PS-GO films were converted to rGO foam films by evaporation of benzene and thermal reduction. Cell size could be controlled by changing the concentration of PS-GO in the precursor solution and the chain length of grafted PS at the GO flakes, as shown in Fig. 2.140. When a PS with M_w of 17,600 g/mol was used, the resultant carbon foam prepared at the PS-GO concentration of 3 mg/cm^3

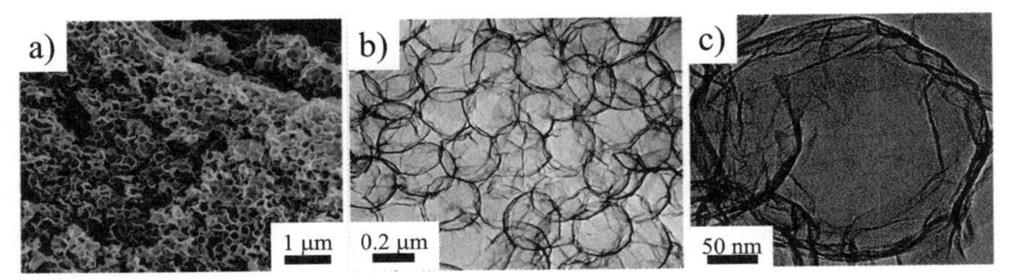

Figure 2.139 rGO foam prepared using polystyrene template: (A) SEM, (B) low-magnification TEM, and (C) high-magnification TEM images [451].

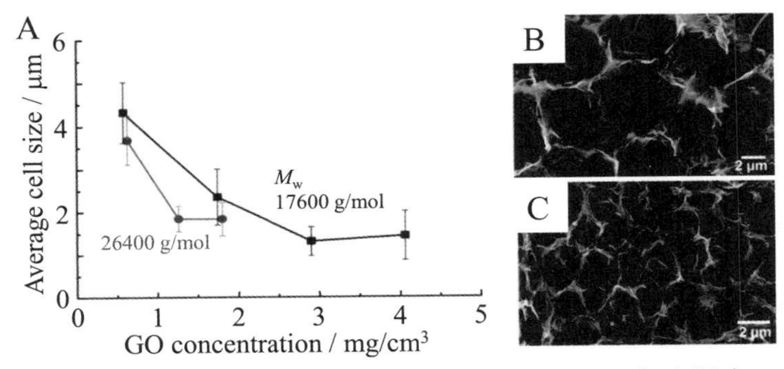

Figure 2.140 rGO foam films prepared from the polystyrene-grafted GO by aqueous droplet templating: (A) changes in average cell size with GO concentration using polystyrene with different M_w, and (B) and (C) SEM images of the cross-sections of the rGO foams prepared using PS with M_w of 17,600 g/mol with concentrations of 1 and 5 mg/cm^3 [452].

exhibited hexagonal-shaped cells with an average size of 3.7 μm. With a PS having larger M_w of 26,400 g/mol, however, cell sizes of the carbon foam became smaller and had irregular shapes.

Polycrystalline α-Fe_2O_3 pseudo-cubic microparticles were used as the template to prepare rGO foams [453]. α-Fe_2O_3 particles prepared from NaOH aqueous solution of $FeCl_3$ were mixed with GO, where self-assemblage occurred via the electrostatic interaction between negatively charged GO flakes and positively charged α-Fe_2O_3 particles in an aqueous solution. The GO-wrapped Fe_2O_3 thus prepared were converted to crumpled rGO foam by reducing GO with HI and etching Fe_2O_3 with HCl. The resultant rGO foams showed a N_2 adsorption–desorption isotherm of type IV with S_{BET} of 884 m^2/g and V_{total} of 2.91 cm^3/g.

2.3.4.1.4 Solvent evaporation

rGO foams were prepared from aqueous suspensions of GO nanoflakes in the concentration of 2–5 mg/cm^3 by vacuum evaporation at 40°C, followed by heating up to 800°C in H_2/Ar flow, although evaporation at 80°C led to the formation of GO films [454]. Obtaining either GO foam or film was thought to result from the balance between van der Waals force to form networks of GO flakes and evaporation force of the solvent (water) at different temperatures. At low temperature such as 40°C, the former force is superior to the latter force and as a consequence a three-dimensional network of GO flakes, i.e., GO foam, is formed.

Microwave irradiation of rGO foams was effective for enhancing their compressibility [455]. A basic and weak reducing agent, ethylenediamine (EDA), was added into the GO aqueous suspension to functionalize GO flakes and initiate their assembly into a three-dimensional network (foaming). The GO hydrogels thus prepared were converted to aerogels via freeze-drying and then subjected to microwave irradiation for 1 min in Ar to make rGO aerogels (rGO foams). As-prepared GO showed C/O (by mole) of 1.9, which increased to 2.3 after the EDA-mediated assembly process (after freeze-drying), demonstrating a slight reduction in GO flakes. After microwave irradiation, however, the C/O ratio of the foams increased up to 23.0, indicating the elimination of most of the functional groups. The bulk density of the as-prepared foams can be controlled from 8.3 to 10.9 mg/cm^3 by changing the EDA/GO ratio from 0.6 to 2.4, while subsequent microwave irradiation makes the bulk density increase, ranging from 3.0 to 5.4 mg/cm^3, as shown in Fig. 2.141A, and the porosity can reach as high as 99.7%–99.8%. The foams after irradiation had high compressibility, and could be squeezed into a pellet under pressure but

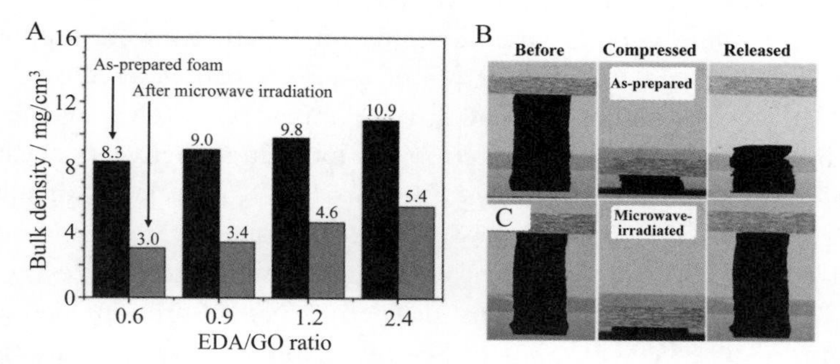

Figure 2.141 Effect of microwave irradiation on rGO foams prepared using ethylenediamine (EDA) as a reducing agent: (A) bulk density versus EDA concentration, (B) and (C) compression behaviors [455].

could recover almost completely by releasing the pressure (enhanced resilience), while the foams before irradiation recovered only partially, as shown in Fig. 2.141B.

2.3.4.1.5 Cross-linking

Cross-linking the individual graphene sheets was effective for obtaining foams with a low density, high surface area, relatively high electrical conductivity, and high mechanical properties. An aqueous suspension of GO was mixed with resorcinol (R) and formaldehyde (F) with sodium carbonate as a catalyst (C) to achieve sol–gel polymerization including GO flakes to form GO aerogel, which could be converted to rGO foam consisting of covalent cross-linking between rGO layers through simultaneous reduction of GO and carbonization of RF resin at 1050°C [456]. The resultant rGO foams had a bulk density of 0.01 g/cm^3, bulk electrical conductivity of about 100 S/m, and high pore structure parameters (S_{BET} of 584 m^2/g and V_{total} of 2.96 cm^3/g). Their pore structure depends strongly on the content of RF: their IV-type N_2 isotherms suggest the formation of mesopores, and H3-type hysteresis loop observed particularly on the carbon obtained by 4 wt.% RF addition shows the presence of macropores, as shown in Fig. 2.142A [457]. Pore size distribution curves (Fig. 2.142B) show that the formation of the pores had a wide range of sizes from mesopores to macropores, the sizes of mesopores depending strongly on the content of RF. Fig. 2.142C shows the stress–strain curves measured by using flat-punch nanoindentation, demonstrating marked hysteresis with loading–unloading cycles [458]. The foam had a Young's modulus of 51 MPa and failure stress of 10.4 MPa. The foam was extraordinarily stiff and exhibited supercompressive behavior with indentation strain up to more than 100% and a complete recovery for low strains.

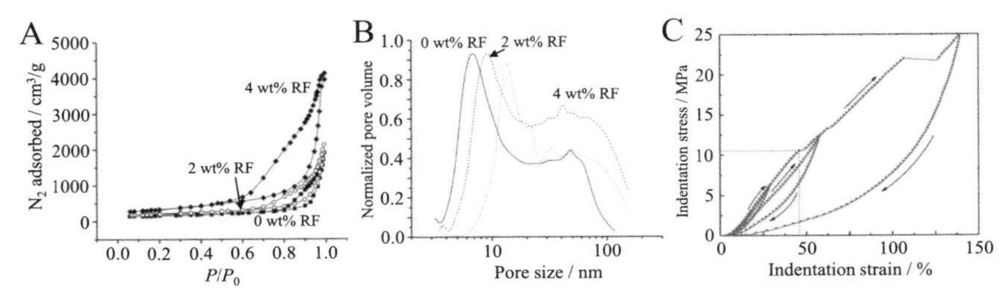

Figure 2.142 rGO foams prepared by covalent cross-linking using different resorcinol-formaldehyde additions: (A) N_2 adsorption–desorption isotherms, (B) pore size distributions, and (C) stress–strain curves with loading–unloading cycles on the foam with bulk density of 0.1 g/cm^3 [457,458].

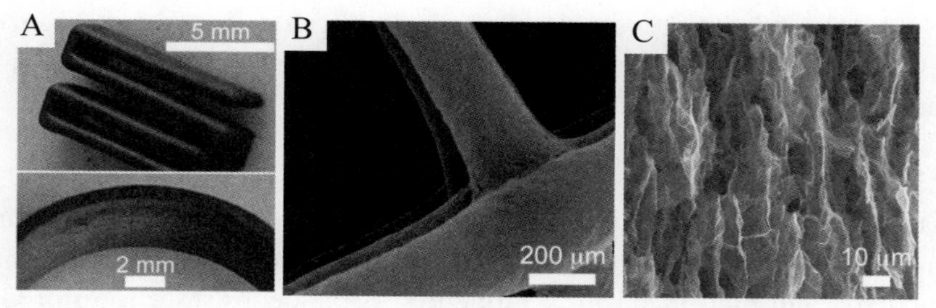

Figure 2.143 3D-printed rGO: (A) photographs of a filament pile and ring, (B) SEM image of the woodpile structure, and (C) SEM image of the fractured surface [460].

2.3.4.1.6 3D-printing

rGO-foams with 3D periodic structure, as well as rGO nanowires, were fabricated by direct-ink writing [459] via a 3D printing technique, in which the preparation of an extrudable GO-based composite ink was key.

GO suspension mixed with a branched copolymer surfactant (BCS) was subjected to printing after lowering the pH with glucono-δ-lactone (GδL), where GO content to total organics (BCS+DδL) was fixed at 1.3 [460]. After printing, thermal treatments at 900−1000°C in ***q*** a 10% H_2/Ar atmosphere was used to reduce GO and to decompose BCS and GδL. The reduction was accompanied by a 76%−81% weight loss, which depended on the GO concentration in the inks (ranging from 1.7−3 wt.% GO), and by macroscopic shrinkage of 4%−7%, while their shape, structural integrity, and inner texture were retained, as shown in some example photographs and SEM images in Fig. 2.143. The electrical conductivity of the rGO for the filament pile (Fig. 2.143A) with density of 0.006 g/cm^3 reached 0.4 S/cm. rGO ink for 3D printing was prepared by mixing GO (lateral size of 150−400 nm) gel with R−F aqueous solution (cross-linking agent) include fumed silica (viscosifier) [461]. 3D lattices, such as microlattices, honeycomb, etc., formed by a 3D printer were freeze-dried after gelation, carbonized, and then etching of silica. The resultant microlattices had high S_{BET} (700−1100 m^2/g), large mesopore volumes (2−4 cm^3/g) with macropores, and high electrical conductivity (90−280 S/m), as well as remarkable flexibility.

An aqueous suspension of GO with a relatively high concentration of 40 mg/cm^3 was used for spinning by adding 4.2 wt.% rGO nanoparticles, 12.5 wt.% fumed SiO_2, and 11 wt.% RF with sodium carbonate catalyst [462]. rGO nanoparticles worked to improve the electrical conductivity of the final rGO products, hydrophilic fumed SiO_2 served as a viscosifier to meet the rheology requirements of reliable flow through a fine nozzle under shear force and shape retention after deposition, and RF was added to form a homogeneous, highly viscous, and thixotropic ink. The composite ink was extruded through a micronozzle to patterned 3D structures, followed by gelation, freeze-drying, carbonization, and etching of the silica with HF. The rGO products prepared from GO without rGO nanoparticles exhibited S_{BET} of 739 m^2/g and resistance of 61.1 Ω/sq. and that from GO with rGO nanoparticles exhibited slightly lower S_{BET} of 418 m^2/g and much lower resistance of 0.96 Ω/sq. An aqueous suspension of GO flakes with an average lateral size of 1, 3, and 5 μm and thickness of about 0.9 nm (concentration of 1 g/L) was successfully spun to GO nanowires through a 1.3-μm nozzle, which were converted to rGO nanowires by either reducing at 400°C in a vacuum or reducing with hydrazine at 120°C [463]. The diameter of the GO nanowires depended greatly on the pulling rate from the nozzle, irrespective of the lateral size of GO flakes.

2.3.4.2 Assemblage of graphene nanoflakes

Graphene foams were synthesized by CVD of CH_4 at 1000°C for 5 min under ambient pressure using Ni foam (about 320 g/m^2 in areal density and about 1.2 mm in thickness) as a substrate and also template [464]. After CVD, a thin layer of PMMA was deposited on the surface of the graphene layer on the Ni surface to prevent collapsing during etching of Ni substrate by a hot HCl solution. After the PMMA was carefully removed by a hot acetone, a graphene monolith of three-dimensional network (graphene foam) was obtained. The resultant graphene foams are free-standing, light weight, and consist of an interconnected flexible network, as shown in Fig. 2.144A and B. The ligaments and joints were composed of few-layered graphenes, of which ripples and wrinkles are preserved even after removing the Ni templates, as shown SEM image in Fig. 2.144C. The average

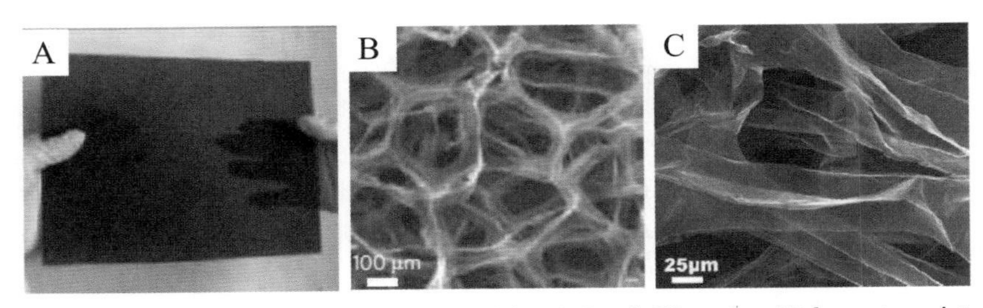

Figure 2.144 Graphene foam synthesized by CVD of CH_4 using Ni-foam template: (A) photograph of a 170 × 220 mm^2 free-standing foam, (B) and (C) SEM images [464].

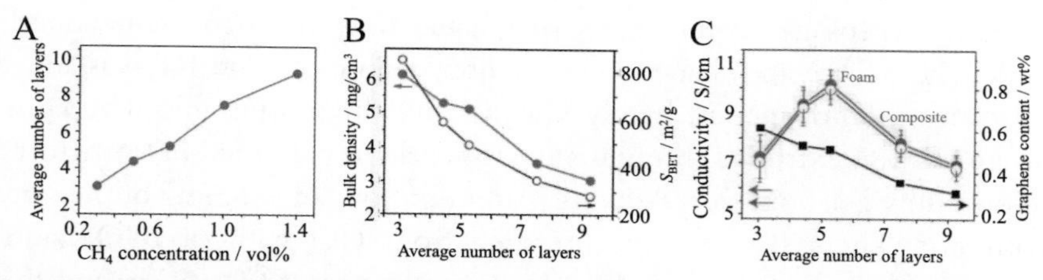

Figure 2.145 Graphene foam synthesized by CVD of CH_4 using Ni-foam template: (A) average number of layers versus CH_4 concentration, (B) bulk density and S_{BET} versus average number of layers, and (C) electrical conductivity of graphene foam and its composite with PDMS, and graphene content versus average number of layers [464].

number of graphene layers determined by high-resolution TEM and Raman spectrum increases with increasing CH_4 concentration in the precursor gas, as shown in Fig. 2.145A, with the CH_4 concentration of 0.7 vol% resulting in the deposition of few-layered graphene. The bulk density and S_{BET} of the foams depend greatly on the average number of graphene layers, as shown in Fig. 2.145B the foam consisting of the thinnest thickness of deposited graphene (three layers) with a bulk density of about 6 mg/cm^3 and a relatively high S_{BET} of about 850 m^2/g. The composite of the graphene foams with poly(dimethyl siloxane) (PDMS) was fabricated by infiltrating with PDMS prepolymer, followed by degassing and curing at 80°C. The composite foam delivers a very high electrical conductivity of about 10 S/cm (Fig. 2.145C), which is the same as the pristine graphene foam. In contrast, the mechanical properties were improved by compositing with PDMS even with a low loading of the graphene foam, such as 0.5 wt.% (0.22 vol%). The electrical resistance of the composite foam showed a small increase with increasing bend radius up to 2.5 mm and could perfectly recover after straightening, associated with a small increase (only about 2.7%) after 10,000 cycles of bending to 2.5 mm radius.

Graphene foams were also prepared by CVD of ethanol at 1000°C in Ar flow using Ni foam as template, deposited graphene consisting of a few layers and having high crystallinity with no D-band and a distinct G′-band [465]. Onto the graphene foams, $Ni(OH)_2$ was electrochemically deposited in the electrolyte of 0.2 mol/L $Ni(NO_3)_2 \cdot 6H_2O$ and 0.2 mol/L hexamethylenetetramine ($C_6H_{12}N_4$), followed by annealing at 300°C in Ar to convert to NiO/graphene foam composites. An SEM image and Raman spectrum of NiO/graphene composite foam are shown in Fig. 2.146A and B, respectively. The foam morphology and high crystallinity of the pristine graphene remain even after NiO deposition. The resultant NiO/graphene

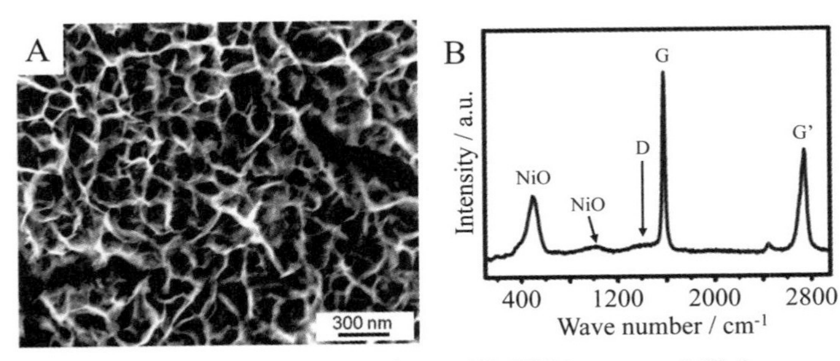

Figure 2.146 NiO/graphene composite foam: (A) SEM image and (B) Raman spectrum [465].

composite foams exhibited a high capacitance (about 816 F/g at a scan rate of 5 mV/s) and a stable cycling performance as electrodes of a supercapacitor with 3 M KOH aqueous electrolyte.

Graphene nanoflakes were deposited by microwave plasma CVD of CH_4 on the surface of rGO foams, which were prepared from GO hydrogel by hydrothermal treatment at 120°C with EDA, followed by thermal annealing at 1000°C in Ar, and had a microcellular texture [466]. The resultant graphene foam is shown in Fig. 2.147. They had a bulk density of 48.2 mg/cm^3, in which graphene nanoflakes are deposited homogeneously by making their basal planes vertical to the cell walls of the template rGO foam. Raman spectrum measured on the resultant foam gave intensity ratios I_D/I_G of 0.81, probably due to the template rGO foam, and $I_{G'}/I_G$ of 1.23, suggesting that graphene flakes deposited by CVD are very thin, consisting of a small number of layers. The foams exhibited high electrical conductivity, as high as 1000 S/m, and high elasticity, Young's modulus of 89 kPa, and related stress of 97 kPa in the elastic region of small strain less than 12%.

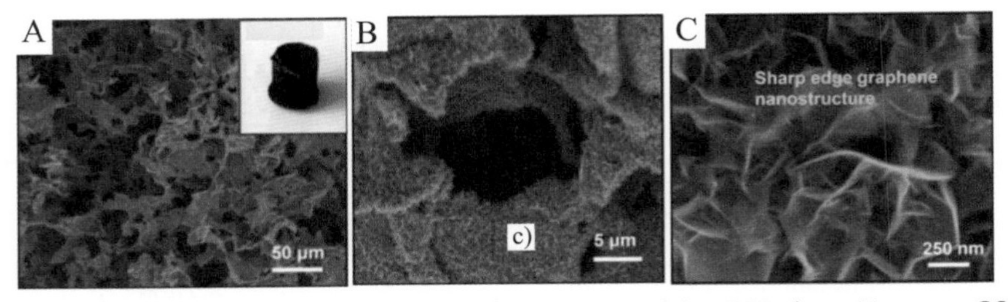

Figure 2.147 SEM images of graphene foam prepared by CVD deposition on rGO foam [466].

Ultra-lightweight graphene aerogels were synthesized by CVD of toluene vapor at 760°C on ZnO template [467]. Various morphologies of graphene aerogels were achieved by changing the morphology of the ZnO template, including foam morphology. Template ZnO was excluded from the graphene deposits by heating in H_2 flow, where ZnO was reduced to metallic Zn and evaporated out of the graphene. Similar ultra-flyweight rGO aerogels were obtained from aqueous suspensions of large-sized GO flakes with CNTs via freeze-drying, followed by chemical reduction of GO with hydrazine vapor [437]. The bulk density of the resultant aerogels was less than 0.16 mg/cm^3, which is lighter than air at ambient conditions. Ultralight carbon aerogels were also prepared from the mixture of CNF-GO/glucose-kaolin composites by carbonizing at 700°C [468].

Graphene foams were reviewed mainly on those derived from rGO flakes [469,470].

2.3.5 Assemblage of carbon nanotubes

Films composed of vertically aligned MWCNTs exhibited super-compressible foam-like behavior, providing a highly resilient open-cell foam system [471]. Individual CNTs act as strong nanoscale ligaments and the inter-CNT space as interconnected open cells. As shown by the schematic illustration and SEM images in Fig. 2.148, CNTs are buckled and folded by compression, with buckling starting from their bottom, and they recover to almost their original lengths upon being released, like springs. Through repeated compressions, these CNT ligaments can be squeezed to less than 15% of their original lengths.

MWCNTs with diameters of 80−100 nm were successfully dispersed in an aqueous solution of sodium carboxymethylcellulose (CMC) (anionic surfactant), as shown in the phase diagram of a CNT/cellulosic salt system in Fig. 2.149A [472]. CNT/carbon foams were prepared from the

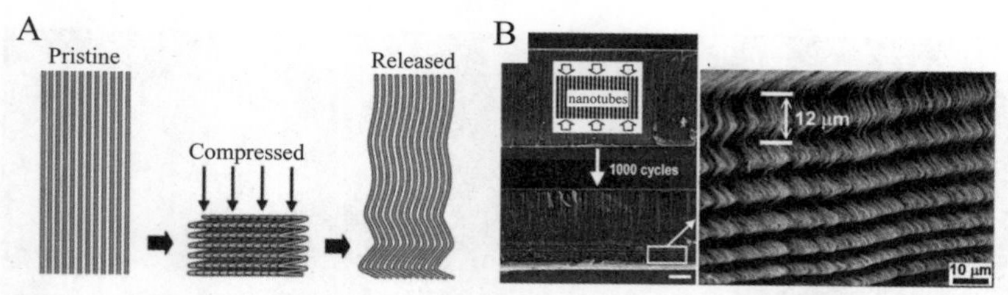

Figure 2.148 Foam-like carbon nanotube films: (A) schematic illustration of the change in carbon nanotube film by compression and releasing and (B) SEM images of buckled carbon nanotubes under compression [471].

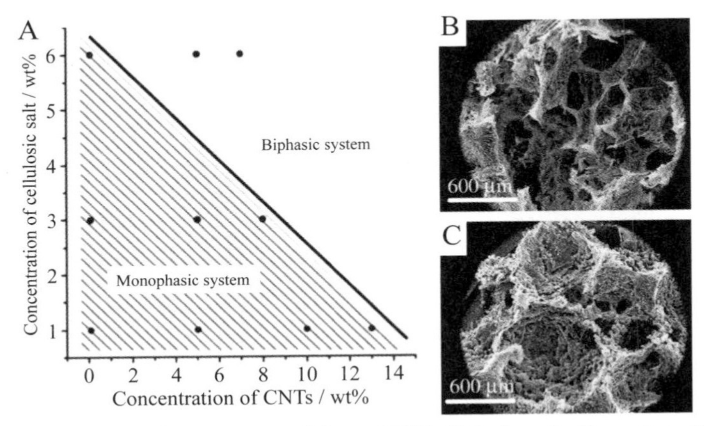

Figure 2.149 MWCNT-foam prepared from CNT/cellulosic salt dispersion: (A) phase diagram of the MXCNT/CMC system, (B) SEM image of CNT/cellulosic foam, and (C) that of CNT/carbon foam [473].

MWCNT/CMC aqueous dispersion using foaming agents (sodium dodecyl sulfate), with freeze-drying at −60°C and stabilized at 400°C, followed by carbonization at 1200°C in N_2 flow [473], as shown in the SEM images of CNT/CMC and CNT/carbon composite foams in Fig. 2.149B and C, respectively. The resultant foams composed of large cells with mesopores have homogeneous sizes at around 40 nm.

N-doped carbon nanotube sponges were obtained at 1020°C by feeding the precursor solutions through two independent sprayers [474]. One sprayer feeds a solution made of benzylamine, ferrocene, and thiophene and another sprayer has a solution made of ethanol-acetone, ferrocene, and thiophene. The produced sponges were mixtures of various morphology of CNTs and carbon particles depending on the position of the reactor tube and also on precursor solutions and were produced with three different acetone concentrations (acetone/ethanol = 0, 1, and 3). The increase in the acetone concentration resulted in an increase in CNT diameter, and the formation of Fe_3C nanoparticles on the surface of CNTs. The sponges contained about 3 wt.% nitrogen.

2.3.6 Other processes

Carbon foams were prepared from olive stone ground to a uniform size of 3.5 mm by heating to 500°C under 1.0 MPa pressure of N_2 in the presence of either water or NaCl aqueous solution [475]. In most cases examined, the resultant carbonized blocks were composed of the internal part with the

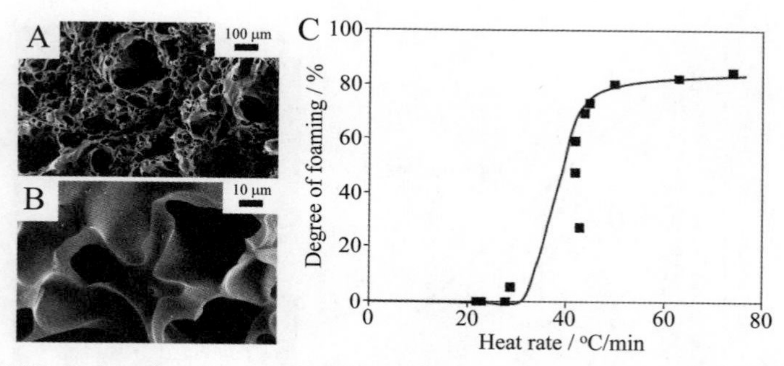

Figure 2.150 Carbon foams prepared from olive stone: (A) and (B) SEM images of the cross-section, and (C) degree of foaming versus heating rate [475].

aspect of foam and the external part with a granular aspect, with the two parts being easily separated. The degree of foaming is determined as the weight percentage of the foam part. The foams have a bulk density of 0.2–0.3 g/cm^3 and cell sizes are in a wide range, as shown in the SEM images of their cross-section in Fig. 2.150A and B. Heating as a fast rate of more than 40°C/min is essential to obtain the product with 80% yield of carbon foams, as shown by the dependence of degree of foaming on heating rate in Fig. 2.150C. Carbon foams were prepared from a sponge-like natural product, pomelo peel, after drying at 80°C in a vacuum, by direct carbonization at 900°C with a heating rate of 5°C/min. The resultant carbon foams are similar to so-called RVC derived from PU foam (Fig. 2.120A) and showed a high porosity of up to 97% and a low electrical resistivity of 5.29 Ω•m [476].

Foams of thermosetting resins were able to be converted to carbon foams by a simple carbonization process, most of which were composed of nongraphitizing carbon matrix. A residue with foam morphology was prepared from an aqueous solution of resorcinol (water phase), formaldehyde, and liquid paraffin (oil phase) with composite surfactants (Span 80 and Tween 80) by emulsifying and polymerizing by adding NaOH, followed by washing with water to remove NaOH and Tween80, and then with ether to dissolve out liquid paraffin and Span80, before finally being converted to carbon foam at 1000°C in N_2 (oil-in-water emulsion method) [477]. The resultant carbon foams were composed of cells with a very narrow size range of about 2–3 μm and had a bulk density of 0.25 g/cm^3 with S_{BET} of about 700 m^2/g. Resorcinol/formaldehyde precursor emulsified using sodium dodecylbenzenesulfonate surfactant with silicone oil in aqueous solution was aged at 85°C and carbonized at 800°C after removing

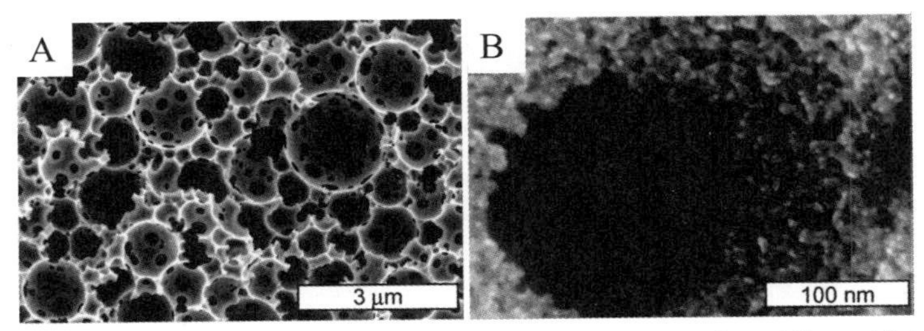

Figure 2.151 SEM images of carbon foam prepared from resolcinol/formaldehyde by emulsifying with silicon oil and carbonized at 800°C [478].

silicon oil by chloroform to obtain carbon foam [478]. The foam is composed of macropores (cells) with mesoporous walls, as shown in Fig. 2.151A and B, mesopores and macropores were independently controlled by changing the concentration of precursor and silicon oil, respectively, in the emulsion. Cross-linking of resorcinol/formaldehyde aerogel with isocyanate-derived cross-linkers resulted in the formation of carbon foam similar to RVC after carbonization at 800°C [479]. The resultant carbon foams were composed of glass-like carbon and had a bulk density of 0.22–0.25 g/cm^3 and porosity of 86%–88%.

Carbon foams were also prepared by carbonization of the mixture of either hollow phenolic spheres or hollow carbon spheres with furfuryl alcohol (FA) under compression in a mold at 900°C [480] or with phenolic resin at 900–1000°C [481,482,484]. Carbon foams prepared from phenolic resin-derived carbon spheres were composed of cells with about 30 μm diameter and cell walls of 1–5 μm thickness, having a density of 0.3 g/cm^3 and about 80% porosity [482]. Their compressive test was performed at different temperatures up to 3100°C. Porous films of aromatic polyimides BPDA/PDA and BPDA/ODA with about 50% porosity were converted to carbon foams by simple heat treatment at 900°C [485]. The resultant carbon foams can be graphitized, interlayer spacing d_{002} becoming 0.3362 nm by heat treatment at 3000°C, which is slightly higher than that obtained from Kapton film. Carbon monolith consisting of square channels with 630 μm size was prepared by extruding phenolic resin, followed by carbonization and activation, as shown in Fig. 2.152 [486]. The wall of square channels consists of granules and contains micropores, the foam having S_{BET} of 806 m^2/g and V_{micro} of 0.371 cm^3/g.

Carbon foam RVC with ultra-low density was prepared by pulsed laser deposition [487]. The beam of a Nd:YAG laser was directed on a pyrolytic graphite target with a 45 degrees angle of incidence and the ablated carbon

Figure 2.152 Carbon monolith with square channels prepared from phenolic resin [486].

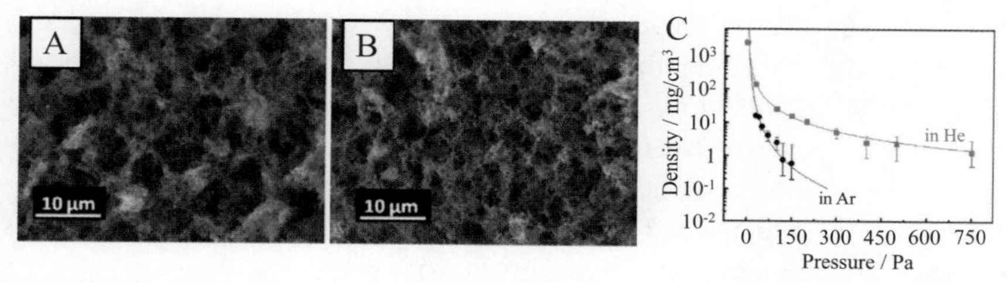

Figure 2.153 Carbon foam layer prepared by pulse laser deposition: (A) and (B) SEM images (top view) of the foams deposited with 0.03 and 3 mg/s gas flow, respectively, and (C) mean density versus gas pressure [487].

species expanded onto a Si(100) substrate at room temperature. Cell size in the foam depends greatly on the gas flow rate, as shown in the SEM images in Fig. 2.153A and B. The mean density of carbon foam is tunable in the range of 1—1000 mg/cm^3, depending greatly on the gas, either Ar and He, as shown in Fig. 2.153C. The thickness of the foams is also tunable in the range of 5—80 μm.

2.4 Hierarchically porous carbons

Most porous carbons contain micropores, mesopores, and macropores, irrespective to their amounts, in other words, these carbons have hierarchical pore structures. In previous sections in this chapter, syntheses conditions of porous carbon are reviewed on the basis of what kind of pores is created in the carbon and what kind of pores predominate in the carbon, micropores, mesopores, or macropores. Through the development of the techniques for the control of pore sizes, particularly template-assisted carbonization and precursor design techniques, some researchers are trying to control the sizes and amounts of each pore independently, in other words, to synthesize carbons with hierarchical pore structures, which could be described as hierarchically porous carbons.

Here, the results reporting certain successes in independent control of macropores, mesopores, and micropores are reviewed. These hierarchical controls of the pores in carbons have often been achieved by applying more than two techniques for pore structure control from activation, template-assisted carbonization, and precursor design technique on a carbon precursor, for example, colloidal silica (hard template) for macropore control while blending a block copolymer for mesopore control and finally activated by CO_2 for micropore control. Reviews focused on hierarchically porous carbons have been published [488,489].

2.4.1 Carbonization with dual assistances

Hierarchically porous carbons have been synthesized from the mixtures of phenol-formaldehyde (PF) as a carbon precursor with three kinds of templates, polystyrene latex (PS) for macropore formation, a block copolymer Pluronic F127 for ordered mesopores, and TEOS for additional mesopores through solution mixing by carbonization at 900°C [490]. Four combinations of the additives for pore control are shown in Table 2.27, together with the pore parameters obtained after carbonization. N_2 adsorption–desorption isotherms, and pore size distributions on the resultant porous carbons are shown in Fig. 2.154A and B, respectively. Carbon D has a hierarchical pore structure, macropores owing to PS, mesopores with sharp size distributions at around 4 and 6 nm owing to the templates of SiO_2 derived from TEOS and F127, respectively, in addition to micropores formed intrinsically in matrix carbon derived from PF. Mesopore formation is also recognized by the increases in S_{BET} and V_{total}, which are associated with a slight increase in micropores (S_{micro} and V_{micro}).

Table 2.27 Effect of additives, polystyrene latex (PS), block copolymer (Pluronic F127), and TEOS, on pore parameters of PF-derived carbons [490].

Code	Additives	S_{BET} (m^2/g)	S_{micro} (m^2/g)	V_{total} (cm^3/g)	V_{micro} (cm^3/g)
A	PS	155	94	0.42	0.04
B	PS + F127	655	355	0.50	0.16
C	F127 + TEOS	1078	505	1.46	0.19
D	PS + F127 + TEOS	1112	654	1.18	0.29

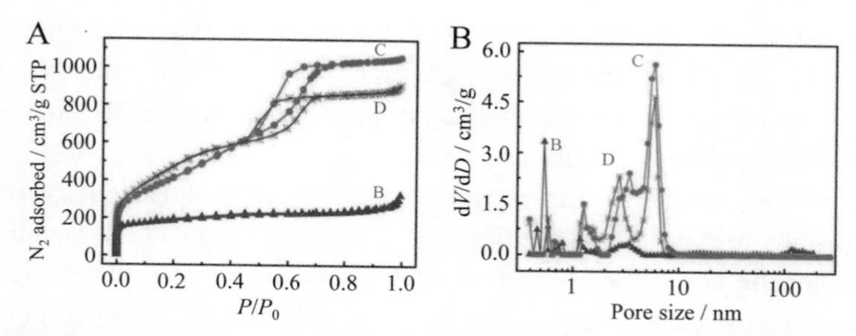

Figure 2.154 Porous carbons prepared from phenol-formaldehyde using four kinds of templates by carbonizing at 900°C: (A) N_2 adsorption–desorption isotherms and (B) pore size distributions [490].

Template PS resulted in macropores with sizes of 110–140 nm. Hierarchically porous carbon D delivered an excellent energy-power capability, and an energy density of 4–10 Wh/kg over the power density range of 1–14 kW/kg.

Hierarchically porous carbons were prepared from PF prepolymer mixed with TEOS, F127, and PMMA colloidal particles (416 nm) [491]. The ethanol solution of PF prepolymer, TEOS, and F127 was infiltrated into PMMA solid particles at room temperature and then the solvent ethanol was removed under vacuum. The composites thus prepared were heated at 100°C for polymerization and then carbonized at 400°C for 3 h and then at 900°C for another 2 h, followed by removing SiO_2 formed from TEOS by 10 wt.% HF. The resultant porous carbons are composed of ordered macropores (about 359 nm diameter), which are interconnected through windows (about 127 nm diameter), and mesopores with sizes of 5.4 and 7.3 nm in macropore walls, as shown in Fig. 2.155. The carbons

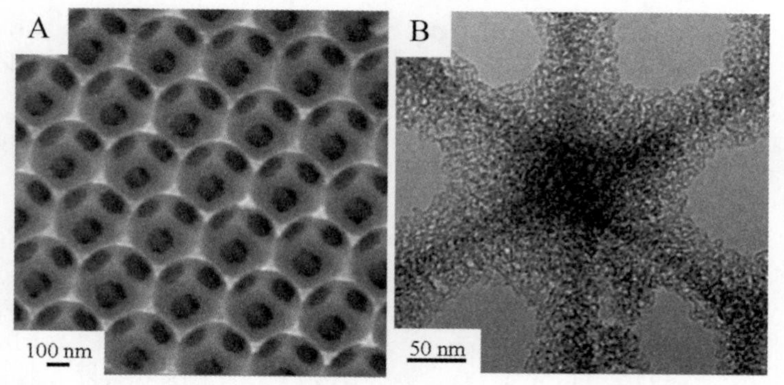

Figure 2.155 Hierarchically porous carbons prepared from phenol resin using TEOS, F127, and PMMA spheres: (A) SEM image and (B) TEM image [491].

delivered S_{BET} of 1900 m^2/g and V_{total} of 1.25 cm^3/g consisting of 68 vol.% mesopores and 28 vol.% micropores. By changing the concentration of F127, the morphology of the mesopores could be changed; bimodal with larger nanocubes and smaller nanospheres by high F127 concentration and only nanospheres with bimodal size distribution by adding organic cosolvent (1,3,5-trimethylbenzene TMB). Carbons with ordered macropores, of which walls contained ordered mesopores, were synthesized via a dual-templating technique using PMMA spheres and triblock copolymer as templates (silica-free synthesis route) [492]. A solution of PF prepolymer and F127 was added into sedimented colloidal spheres of PMMA, followed by evaporation of the solvent, cross-linking, and heating at 450°C to obtain phenol resin monoliths. Carbon monoliths were obtained by heating up to 900°C. The sizes of macropores formed by PMMA template and that of mesopores by F127 template were 342—404 and around 3 nm, respectively. Mesopores were formed in either face-centered cubic or two-dimensional hexagonal symmetry.

N-doped hierarchically porous carbon was prepared by carbonization of melamine foam at 800°C for 1 h in Ar [493]. Melamine foams were fabricated from a melamine resin, which was prepared by mixing para-formaldehyde, melamine, and block copolymer P123 with NaOH in aqueous solution refluxing at 85°C, by adding octylphenolpolyoxyethylene ether (OP-10), pentane, and formic acid, followed by vigorous stirring and solidifying at 80°C for 4 h. For comparison, a macroporous carbon (carbon foam) with limited micropores was prepared with the same procedures but without P123, and a microporous carbon was also prepared without adding OP-10 or pentane. The resultant hierarchically porous carbon consists of macropores with various sizes, as shown in the SEM images in Fig. 2.156A and B, with a large amount of micropores and a small amount of

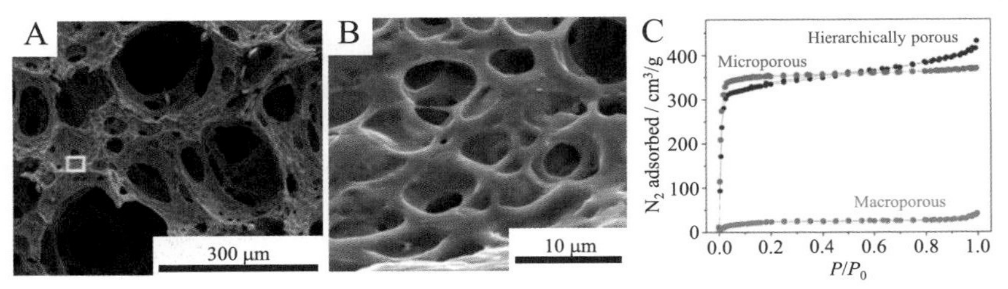

Figure 2.156 Hierarchically porous carbon prepared from melamine foam: (A) and (B) SEM images of the cross-section, and (C) N_2 adsorption isotherm in comparison with those of the microporous and macroporous carbons [493].

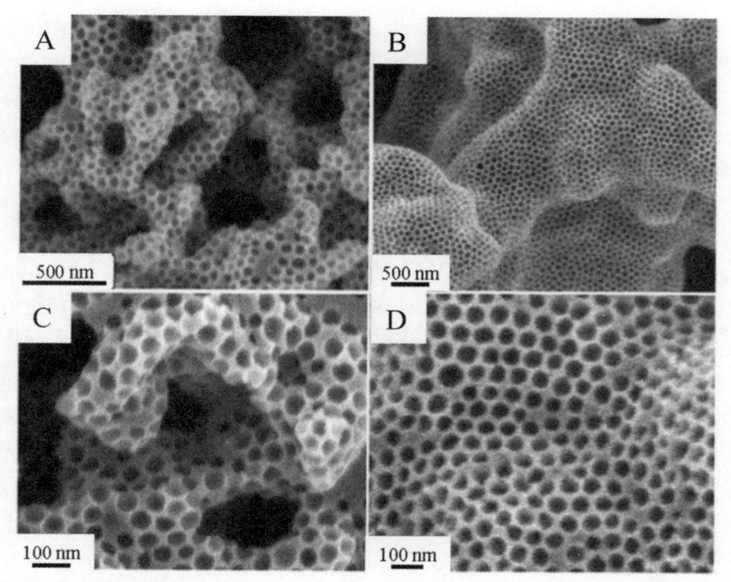

Figure 2.157 SEM images of the coral-like porous carbons: (A) and (C) hydrothermal carbonization at 130°C followed by calcination at 550°C and (B) and (D) hydrothermal carbonization at 180°C followed by 550°C calcination [494].

mesopores, as suggested from the N_2 isotherm in Fig. 2.156C. The hierarchical carbons had a bulk density of about 0.050 g/cm^3 (porosity of 97.2%) with S_{BET} of 1175 m^2/g and can accommodate a compressive stress of 0.163 MPa without collapse.

Coral-like porous carbons, which had trimodal pore structures, as shown in the SEM images in Fig. 2.157, were synthesized from D-fructose by dual templating using block copolymer F127 and monodisperse nanoparticles of polystyrene (PS) latex through hydrothermal carbonization [494]. The resultant porous carbons had a high surface area and large pore volume, consisting of highly layered 3D continuous macropores (2–5 μm), spherical large mesopores (50–60 nm), and micropores (<2 nm) in pore walls. The mixtures of D-fructose, F127, and PS latex prepared in aqueous solution were subjected to hydrothermal carbonization in an autoclave at either 130 or 180°C, and the resultant carbons were calcined at either 550 or 900°C to remove polymer templates. N_2 adsorption–desorption isotherms of the resultant porous carbons and pore size distributions measured by mercury porosimeter are shown in Fig. 2.158. For the carbon synthesized by hydrothermal carbonization at 180°C and calcination at 550°C (refer to Fig. 2.158A and C), pore parameters determined from N_2

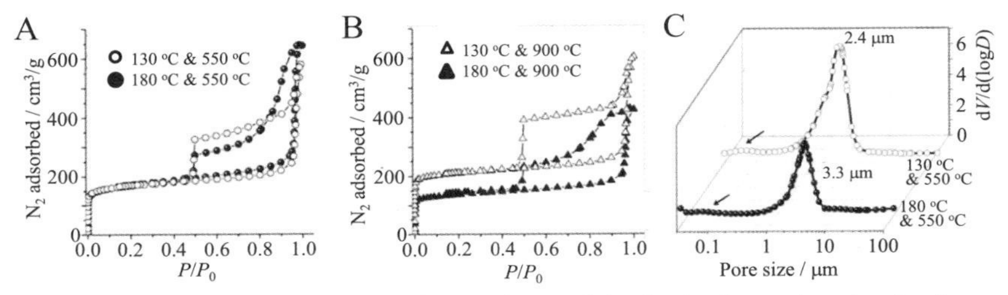

Figure 2.158 Coral-like porous carbons obtained by hydrothermal carbonization at 130 and 180°C followed by calcination at 550 and 900°C: (A) and (B) N_2 adsorption–desorption isotherms of the carbon obtained under different conditions, and (C) macropore size distribution [494].

isotherms were S_{BET} of 634 m^2/g, V_{total} of 1.00 cm^3/g including V_{micro} of 0.23 cm^3/g and V_{ext} of 0.77 cm^3/g, and macropore size of around 3.3 μm. With high-temperature treatments of up to 3000°C, these porous carbons were partially graphitized by keeping the original pore structure [495]. The resultant carbons exhibited high performance as the anode materials for lithium-ion batteries, particularly at high charge–discharge rates.

The hierarchically porous carbons were synthesized from resol-type PF mixing with block copolymer F127 and colloidal SiO_2 microspheres with 240, 320, and 450 nm diameters by carbonization at 800°C in N_2 followed by dissolution of SiO_2 [496]. The resultant carbons were composed of macropores ordered in face-centered cubic symmetry with tunable sizes of 230–430 nm by templating SiO_2 microspheres and interconnected windows with a size of 30–65 nm, together with ordered channels with mesopore-size diameters (11–12.5 nm) and micropores in the matrix carbon derived from PF. Resol-type PF mixed with the hollow microspheres of polystyrene, of which hollow space was loaded by oxalic acid in advance, was carbonized at 800°C for 2 h in N_2 [497]. Upon heating to carbonization, macropores with about 450 nm diameters were formed by templating of the microspheres (500 nm) at around 450°C, and then pyrolysis/carbonization of the matrix PF at higher temperatures, resulting in the formation of micropores in the matrix, associated with collapsing and deformation of the macropores. The carbon obtained at 800°C possessed a type-I N_2 adsorption–desorption isotherm with H4 hysteresis as a result of the coexistence of several different types of pores.

Starting from a mixture of phenol, TEOS, and colloidal silica with F127, porous carbons consisting of macropores of 20 or 50 nm caused by colloidal silica, mesopores of about 12 nm by F127, and micropores of about 2 nm size by silica formed from TEOS were prepared [498].

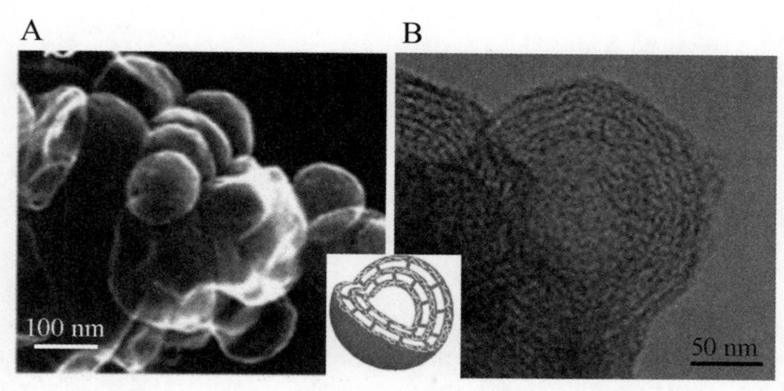

Figure 2.159 Onion-like carbon vesicles: (A) SEM and (B) TEM images with a schematic illustration of the structure [499].

Activation with CO_2 at 850°C resulted in marked increases in micropores and consequently in S_{BET} to 2800 m^2/g and V_{total} to 6.0 cm^3/g. Onion-like carbon-silica composite vesicles were prepared through aqueous emulsion of a low-molecular-weight resol-type phenol resin as a carbon precursor, TEOS as a silica source, F127 as a template, and 1,3,5-trimethylbenzene as an organic cosolvent [499]. These vesicles could be converted into onion-like mesoporous carbon vesicles with bimodal pores of 4–23 and 66–82 nm sizes, as shown by their SEM and TEM images together with a schematic illustration in Fig. 2.159. Carbon foams were synthesized from the mixture of chitosan, SiO_2 spheres (chitosan/SiO_2 of 1/1 by weight), and glutaraldehyde by gelation, freeze-drying, and carbonization at 800°C in N_2, followed by washing out of SiO_2 with 2 M NaOH [500]. The carbon foams were activated using KOH at 800°C, resulting in carbon foams with a hierarchical pore structure, having S_{BET} of 2906 m^2/g and V_{micro} of 0.930 cm^3/g, of which the nonmicropore (>2 nm) volume reached 3.4 cm^3/g.

Hierarchically porous carbons were synthesized by one-step carbonization of a mixture of phenol resin and $Ni(OH)_2$ in ethanol via heat treatment at 200°C in ammonia atmosphere and then at 600°C, followed by dissolving out of Ni compound by 3 M HCl [234]. The resultant carbon had the hierarchical pore structure composed of micropores with sizes of <1 nm and 1–2 nm, mesopores with 5–50 nm, and macropores of 60–100 nm, as shown in the N_2 adsorption–desorption isotherm and pore sized distribution in Fig. 2.160A and B, respectively. This carbon delivered high performances as the electrodes of symmetric supercapacitors with aqueous and nonaqueous electrolytes, energy density of 5.3 Wh/kg, and power density of 25 kW/kg in 6 M KOH.

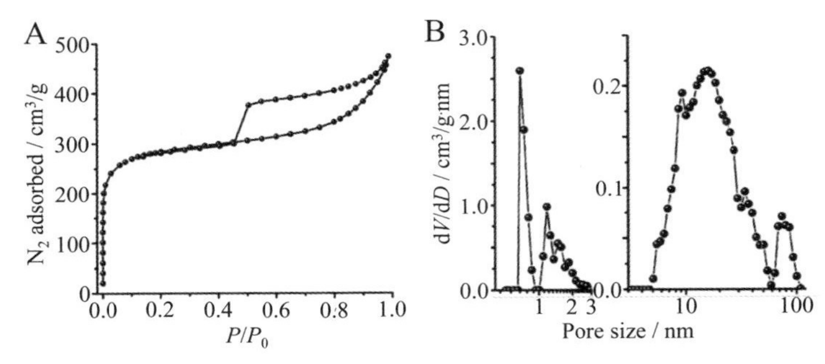

Figure 2.160 Hierarchically porous carbons prepared from phenol resin using $Ni(OH)_2$ template: (A) N_2 adsorption–desorption isotherm and (B) pore size distributions [234].

Hierarchically porous carbons were prepared by impregnation of ethanol solutions of furfuryl alcohol (FA) with different concentrations of 5–30 vol% into a silica template, followed by carbonization at 850°C in N_2 flow [501]. The template silica having hierarchical pore structures was synthesized from TEOS using block copolymer P123 under hydrothermal conditions at 120°C, followed by calcination at 530°C [502]. The pore structure of the resultant porous carbons depended greatly on the concentration of FA solution used [501]. When diluted solutions as 5 and 7.5 vol.% FA solution were used, the resultant carbons were macro-/mesoporous and delivered S_{meso} of 1354 and 1229 m^2/g and V_{meso} of 1.49 and 1.37 cm^3/g with V_{macro} of 48.6 and 32.9 cm^3/g, respectively, with no micropores being detected. When 30 vol% FA solution was used, however, micropores were formed together with meso-/macropores, S_{micro} of 260 m^2/g, V_{micro} of 0.12 cm^3/g, together with S_{meso} of 469 m^2/g, V_{meso} of 0.55 cm^3/g, and V_{macro} of 10.3 cm^3/g. These porous carbons showed excellent performance in sorption of spilled oils and organics with high sorption capacity, rate, and stability. Mesoporous silica SBA-15 was impregnated by FA with oxalic acid and then polymerized at 80°C for 24 h [503]. The composite PFA/SBA-15 was mixed with a powder of NaCl (about 0.2 μm size) and then molded to a cylinder under 1000 kg/cm^2, followed by curing at 150°C and carbonization at 800°C in Ar. The carbon monoliths were obtained after removing NaCl with water and silica by HF or NaOH solution. The carbon obtained using only SBA-15 (no NaCl addition) delivered S_{BET} of 1270 m^2/g and V_{total} of 1.102 cm^3/g with a pore size peak at 3.7 nm, of which V_{macro} measured by a mercury porosimeter was 1.50 cm^3/g with a size of 0.34 μm. By the addition of NaCl template with NaCl/SBA-15 weight ratio of 1.0, the resultant carbon contained more macropores, V_{macro} of 2.02 cm^3/g with a larger size of 0.55 mm without marked changes in nanopore structure (S_{BET} of 1155 m^2/g and V_{total} of 1.072 cm^3/g).

Flower-like carbons with hierarchical pore structure were synthesized by mixing flower-like ZnO particles with pitch in THF solution, followed by carbonization at 500°C and by activation with KOH [504].

2.4.2 Carbonization process design

rGO foams prepared by hydrothermal treatment of GO suspension and freeze-drying were immersed into an aqueous solution containing cetyltrimethylammonium bromide (CTAB), ethanol, and NaOH, followed by addition of TEOS and heating at 800°C in Ar to result in SiO_2/rGO composite foams, where mesoporous SiO_2 was deposited on the surface of rGO flakes [505]. Into this SiO_2/rGO composite foam, an ethanol solution of sucrose was repeatedly impregnated, freeze-dried, and then carbonized at 700°C in Ar to result in the composite foams of rGO flakes coated by mesoporous carbon. The N_2 adsorption−desorption isotherm and pore size distribution are shown in Fig. 2.161A and B, respectively. The resultant composite foams contain a large amount of mesopores with 2.0−3.5 nm sizes and macropores due to rGO foam. It delivered a high capacity of 168 F/g at the current density of 2 A/g and excellent cycle stability up to 5000 cycles. From the SiO_2/rGO composite foams, various composite rGO foams, Co_3O_4/rGO and RuO_2/rGO, could be prepared.

Hierarchically porous carbons were prepared from the mixture of chitosan with GO by freeze-drying and then carbonization at 700°C in N_2 [506]. In Fig. 2.162A−C, N_2 adsorption−desorption isotherms at 77K,

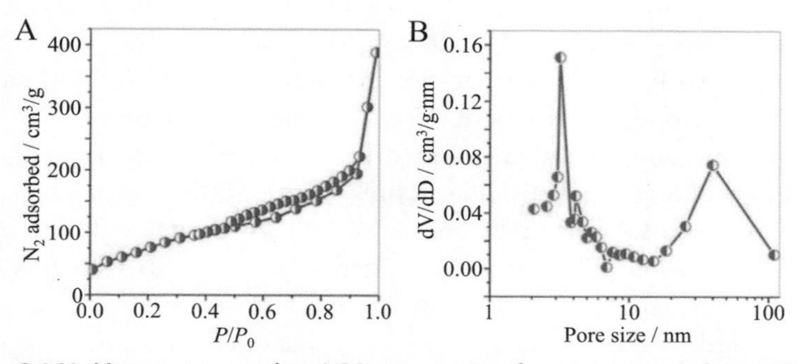

Figure 2.161 Mesoporous carbon/rGO composite foam prepared from SiO_2/rGO composite foam by sucrose infiltration and carbonization at 700°C: (A) N_2 adsorption−desorption isotherm and (B) pore size distribution [505].

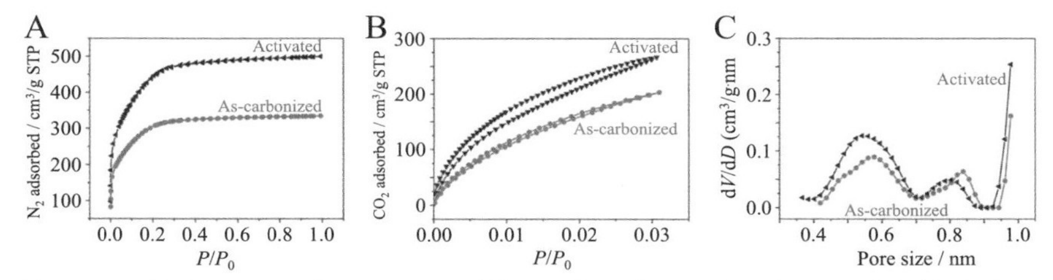

Figure 2.162 Porous carbons prepared from the mixture of chitosan with GO before and after KOH activation: (A) N_2 adsorption–desorption isotherms, (B) CO_2 adsorption–desorption isotherms, and (C) pore size distribution determined from CO_2 isotherms [506].

CO_2 adsorption–desorption isotherms at 273K, and pore size distribution from CO_2 isotherms by DFT method, respectively, for the as-carbonized carbon and the carbon after KOH activation at 700°C, are shown. TEM observations demonstrated that rGO flakes derived from GO were constructed in three-dimensional frameworks (foams) and coated by chitosan-derived porous carbons. By KOH activation, S_{BET} of the resultant porous carbons increased from 1008 to 1511 m^2/g, with an increase in the content of pores at around 2 nm size, while the pore structure consisting of macro-, meso-, and micropores was retained. The chitosan-derived carbons contained 3.97 at% N atoms after carbonization, while they had 5.73 at% N after activation mainly with pyridinic and graphitic configurations.

N-doped porous carbons were fabricated from the mixture of dicyandiamide, $FeCl_2$, and SiO_2 spheres (about 7 nm size) with GO by carbonization at 900°C in N_2, followed by washing with HF and HCl [507]. The resultant porous carbons were composed of three-dimensional macroporous frameworks (foams), of which N_2 adsorption–desorption isotherms are type IV and the pore size distribution curves exhibited a feature of multipeaks in the range of 0–60 nm, confirming the formation of the hierarchical porous structure. The resultant porous carbons delivered ORR activity close to that of commercial catalyst Pt/C.

Hierarchically porous carbons were prepared by carbonization of RF gels mixed with a cationic polyelectrolyte, poly(diallyldimethylammonium chloride), PDADMAC (P) [508]. Into 10 wt.% methanol/water solution of resorcinol (R) and formaldehyde (F) (37 wt% R) with R/water ratio of 0.5 g/cm^3, sodium carbonate (a basic catalyst) and P with molar ratio P/R of 3.5, 7, and 14 were added. The solution was polymerized at 70°C to gel in a closed system. The gel was carbonized at 800°C in Ar, resulting in hierarchically porous carbon. Their N_2 adsorption–desorption isotherms

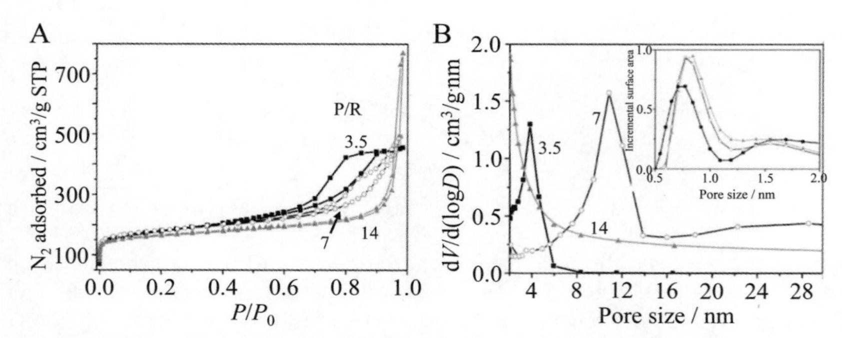

Figure 2.163 The carbons derived from RF mixed with a cationic electrolyte (P) in different P/R ratios: (A) N_2 adsorption–desorption isotherms and (B) pore size distributions [508].

and pore size distributions calculated by BJH method are shown in Fig. 2.163A and B, respectively. Since RF primary particles are negatively charged in basic media due to the phenolic groups, cationic electrolyte P interact electrostatically with RF particles to form self-assembled RF clusters, resulting in larger clusters of RF with increasing amount of P, and consequently the formation of large mesopores and/or macropores, as shown in Fig. 2.163. The N_2 isotherms for the carbons prepared from P/R of 14 suggest the presence of large mesopores and/or macropores. The carbon prepared using a P/R ratio of 7 delivered V_{total} of 0.750 cm^3/g with V_{meso} of 0.559 cm^3/g at around 12 nm size and V_{micro} of 0.180 cm^3/g at around 0.8 nm.

Hierarchically porous carbons were prepared from a slurry mixed polypyrrole (PPy) microsheets with KOH after carbonization at 700°C [509]. PPy microsheets were prepared by a modified oxidative template assembly route using cetrimonium bromide (CTAB) $(C_{16}H_{33})N(CH_3)_3Br$, and ammonium persulfate (APS). PPy microsheets were dispersed in KOH solution at 80°C to obtain jelly-like slurry, followed by carbonization at 700°C in N_2. The N_2 adsorption–desorption isotherm of the resultant porous carbon is shown in Fig. 2.164, with an SEM image of its cross-section. The carbon was composed of macropores as seen in the SEM image (inset), with mesopores centered at 2.73 nm and continuously distributed in a range of 5–50 nm, and micropores centered at 0.68 and 1.26 nm, which delivered S_{BET} of 2870 m^2/g and V_{total} of 2.19 cm^3/g. The carbon also had a high N content of 7.7 wt.% and excellent electrical conductivity of 5.6 S/cm.

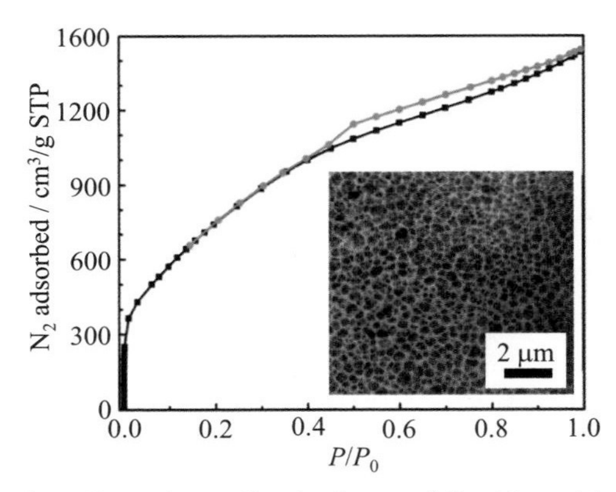

Figure 2.164 N_2 adsorption–desorption isotherm of the hierarchically porous carbon prepared from polypyrrole microsheets at 700°C, with an SEM image of its cross-section (inset) [509].

2.4.3 Inheritance of precursor texture

Carbon foams were prepared from hydrogels of chitosan with a small amount of glutaraldehyde (gelation agent) by carbonization at 800°C in N_2 [510]. By adding K_2CO_3 to the hydrogel, a hierarchical pore structure was obtained by simple carbonization, followed by washing with 0.1 M HCl. The resultant carbon foams are composed of carbon sheets of which 3D aggregation forms macropores with uniform diameters of about 1 μm, as shown by the SEM images in Fig. 2.165A and B. The carbon sheets are as thin as about 90 nm in the carbon foam prepared by adding K_2CO_3, whereas it is much thicker at about 500 nm in the carbon foam without adding K_2CO_3, and as a consequence the former has much smaller bulk density than the latter. This difference was thought to be caused by activation by K_2CO_3. The N_2 adsorption–desorption isotherm shown in Fig. 2.165C for the activated foam suggests the coexistence of micropores with mesopores, the isotherm providing S_{BET} of 1013 m^2/g, V_{total} of 0.576 cm^3/g including V_{micro} of 0.461 cm^3/g and V_{meso} of 0.154 cm^3/g. The nonactivated foam was nonporous, as shown in Fig. 2.165C.

Lignin-derived hierarchically porous carbons were prepared by drying the mixed aqueous solution of lignin and KOH (KOH/lignin of 5/2 by weight) and carbonization at 700°C in N_2 [511]. KOH crystallized in lignin during evaporation of water and acted both as a template and activating agent during carbonization. The resultant carbon was composed of macropores with sizes of approximately 100 nm and nanopores characterized by S_{BET} of 907 m^2/g, V_{total} of 0.515 cm^3/g including V_{micro} of 0.408 cm^3/g.

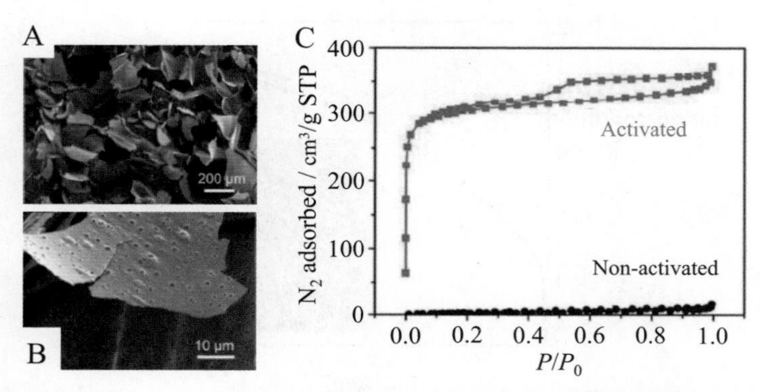

Figure 2.165 Chitosan-derived carbon foams: (A) and (B) SEM images of activated foam and (C) N_2 adsorption–desorption isotherms for activated and nonactivated foams [510].

Hierarchically porous carbons were prepared from sugarcane bagasse coated by PF resin mixed with F127 by carbonization at 600 and 1000°C [512]. In the resultant carbons, macropores were inherited from the sugarcane bagasse scaffold, mesoporous channels from F127, and micropores from PF carbon precursor. The carbon prepared at 600°C delivered V_{total} of 0.283 cm^3/g including V_{micro} of 0.163 cm^3/g (58%).

Porous carbons were prepared from mushroom (shiitake) mixed with H_3PO_4 (HP) by carbonization at 500°C in N_2 flow, followed by washing with water and ethanol [513]. For comparison, the resultant carbon was activated using KOH with KOH/carbon ratio of 3/1 by weight at 800°C for 2 h in N_2 flow, and the mushroom was carbonized directly at 800°C without H_3PO_4 and then activated with KOH at 800°C. The elemental composition and pore structure parameters are shown for these carbons together with their preparation conditions in Table 2.28. The mushroom-derived carbons were confirmed to be composed of macropore ranges from tens to a few hundred nanometers with a pore wall thickness of 8–20 nm by SEM and TEM observations. The coexistence of HP during carbonization of the mushroom was effective in developing nanopores, particularly mesopores, by retaining the macroporous texture, in other words, it was possible to develop hierarchical pore structures. The carbon prepared from mushroom without HP was not porous and converted to microporous carbon by KOH activation at 800°C, with no formation of mesopores. By additional activation using KOH of the carbon obtained by HP activation at 500°C, further development of micropores was possible. The resultant hierarchically porous carbons delivered high performances as electrode materials for supercapacitors with aqueous and nonaqueous electrolytes.

Table 2.28 Porous carbons prepared from shiitake mushroom [513].

Preparation conditions		Elemental composition (at%)					
Carbonization	**Further treatment**	**C**	**O**	**P**	S_{BET} (m^2/g)	V_{total} (cm^3/g)	V_{micro} (cm^3/g)
500°C with H_3PO_4	—	90.1	7.2	2.7	1341	2.02	0.28
	KOH-activated at 800°C	91.1	8.9	—	2988	1.76	0.71
	800°C-treated	92.2	6.0	1.8	1315	2.02	0.28
800°C without H_3PO_4	KOH-activated at 800°C	—	—	—	1081	0.55	—

Powdered wheat husks (WHs) mixed with Teflon in Teflon/WH weight ratio of 1.7—2.5 were carbonized at 900°C for 2—9 h in N_2 flow [514]. The pore structure in the carbons depended on the Teflon/WH weight ratio and carbonization time. The carbon obtained from the mixture with a ratio of 2.5 gave the N_2 adsorption—desorption isotherm of type I with H4 hysteresis, S_{BET} reaching 1059 m^2/g with S_{micro} and S_{ext} (owing to meso-/macropores) of 773 and 287 m^2/g, respectively.

The hierarchically porous carbons were prepared from shrimp shells by mixing with KOH and heat-treated at 700°C, in which $CaCO_3$ (one of the pyrolysis products from shells) worked as a template for macropores and mesopores [515]. The preparation procedures are shown schematically in Fig. 2.166A; the heat treatment of the mixture of shrimp shells with KOH

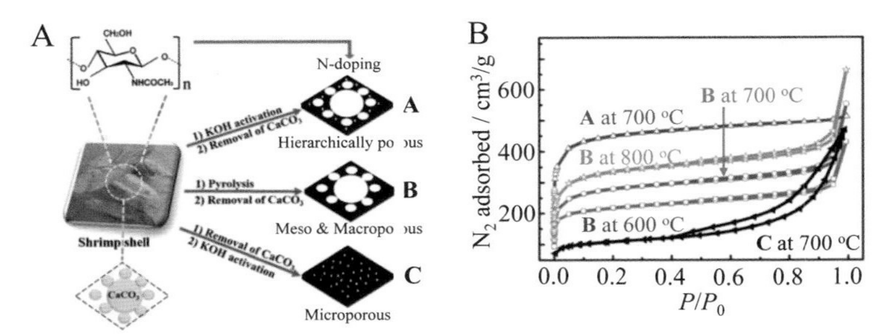

Figure 2.166 Porous carbons from shrimp shells: (A) scheme of preparation and (B) N_2 adsorption—desorption isotherms for the carbons prepared at different procedures and temperatures [515].

at high temperatures and removal of $CaCO_3$ by acetic acid results in a hierarchically porous carbon (denoted as **A**). The heat treatment of shrimp shells without KOH at high temperature and following removal of $CaCO_3$ results in a carbon containing macropores and mesopores (**B**), and the heat treatment of shrimp shells after complete removal of $CaCO_3$ and then mixed with KOH resulted in a microporous carbon (**C**). The N_2 adsorption—desorption isotherms of these carbons are compared in Fig. 2.166B. The hierarchical carbon obtained at 700°C had S_{BET} of 2032 m^2/g with S_{micro} and S_{meso} of 1735 and 297 m^2/g, respectively.

References

[1] Mahajan OP, Yarzab R, Walker Jr PL. Unification of coal-char gasification reaction mechanisms. Fuel 1978;57:643—6.
[2] Ramachandran PA, Doraiswamy LK. Modeling of noncatalytic gas-solid reactions. AIChE J 1982;28:881—900.
[3] Raghunathan K, Yang RYK. Unification of coal gasification data and its applications. Ind Eng Chem Res 1989;28:518—23.
[4] Inagaki M, Suwa T. Nano-structure control in carbon spheres by air oxidation - derivation of master curves-. Mol Cryst Liq Cryst 2002;386:197—203.
[5] Kim MI, Yun CH, Kim YJ, Park CR, Inagaki M. Changes in pore properties of phenol formaldehyde- based carbon with carbonization and oxidation conditions. Carbon 2002;40:2003—12.
[6] Nishikawa T, Inagaki M. Air oxidation of carbon spheres. I. Master curves of pore parameters. Adsorpt Sci Tech 2005;23:827—37.
[7] Inagaki M, Nishikawa T, Oshida K, Fukuyama K, Hatakeyama Y, Nishikawa K. Air oxidation of carbon spheres. II. Micropore development. Adsorpt Sci Tech 2006;24:55—64.
[8] Culverad RV, Heath NS. Saran charcoals Part 1, -activation and adsorption studies. Trans Faraday Soc 1955;51:1569—75.
[9] Bekyarova E, Kankeo K, Kasuya D, Murata K, Yudasaka M, Iijima S. Oxidation and porosity evaluation of budlike single-wall carbon nanohorn aggregates. Langmuir 2002;18:4138—41.
[10] Bekyarova E, Kaneko K, Yudasaka M, Murata K, Kasuya D, Iijima S. Micropore development and structure rearrangement of single-wall carbon nanohorn assemblies by compression. Adv Mater 2002;14:973—5.
[11] Bekyarova E, Murata K, Yudasaka M, Kasuya D, Iijima S, Tanaka H, Kahoh H, Kaneko K. Single- wall nanostructured carbon for methane storage. J Phys Chem B 2003;107:4681—4.
[12] Yoshizawa N, Maruyama K, Yamada Y, Zielinska-Blajet M. XRD evaluation of CO_2 activation process of coal- and coconut shell-based carbons. Fuel 2000;79:1461—6.
[13] Marsh H, Rand B. The process of activation of carbons by gasification with CO_2 — I. Gasification of pure polyfurfuryl alcohol carbon. Carbon 1971;9:47—61.
[14] Marsh H, Rand B. The process of activation of carbons by gasification with CO_2 — II. The role of catalytic impurities. Carbon 1971;9:63—77.
[15] Holmes J, Emmett PH. Alteration of the size and distribution of pores in charcoals. J Phys Colloid Chem 1947;57:1276—307.

[16] Lyubchik SB, Benoit R, Beguin F. Influence of chemical modification of anthracite on the porosity of the resulting activated carbons. Carbon 2002;40:1287–94.
[17] Hesas RH, Daud WMAW, Sahu J, Arami-Niya A. The effects of a microwave heating method on the production of activated carbon from agricultural waste: a review. J Anal Appl Pyrol 2013;100:1–11.
[18] Guo J, Lua AC. Preparation of activated carbons from oil-palm-stone chars by microwave-induced carbon dioxide activation. Carbon 2000;38:1985–93.
[19] Surez-Garcıa F, Martınez-Alonso A, Tascon J, Ruffine L, Furdin G, Mareche JF, Celzard A. Characterization of porous texture in composite adsorbents based on exfoliated graphite and polyfurfuryl alcohol. Fuel Process Tech 2002;77–78:401–7.
[20] Alcaniz-Monge J, Cazorla-Amoros D, Linares-Solano A, Yoshida S, Oya A. Effect of the activating gas on tensile strength and pore structure of pitch-based carbon fibers. Carbon 1994;32:1277–83.
[21] Marsh H, Yan DS, O'Grady TM, Wennerberg A. Formation of active carbons from cokes using potassium hydroxide. Carbon 1984;22:603–11.
[22] Hirst EA, Taylor A, Mokaya R. A simple flash carbonization route for conversion of biomass to porous carbons with high CO_2 storage capacity. J Mater Chem A 2018;6:12393–403.
[23] Deng H, Li G, Yang H, Tang J, Tang J. Preparation of activated carbons from cotton stalk by microwave-assisted KOH and K_2CO_3 activation. Chem Eng J 2010;163:373–81.
[24] Deng H, Yang L, Tao GH, Dai JL. Preparation and characterization of activated carbon from cotton stalk by microwave assisted chemical activation - application in methylene blue adsorption from aqueous solution. J Hazard Mater 2009;166:1514–21.
[25] Ji YB, Li TH, Zhu L, Wang XX, Lin Q. Preparation of activated carbons by microwave heating KOH activation. Appl Surf Sci 2007;254:506–12.
[26] Sevilla M, Al-Jumialy ASM, Fuertes AB, Mokaya R. Optimization of the pore structure of biomass- based carbons in relation to their use for CO_2 capture under low- and high- pressure regimes. ACS Appl Mater Interf 2018;10:1623–33.
[27] Bayamba-Ochir N, Shim WG, Balathanigaimani MS, Moon H. Highly porous activated carbon prepared from carbon rich Mongolian anthracite by direct NaOH activation. Appl Surf Sci 2016;379:331–7.
[28] Guan T, Zhao J, Zhang G, Zhang D, Han B, Tang N, Wang J, Li K. Insight into controllability and predictability of pore structures in pitch-based activated carbons. Microp Mesop Mater 2018;271:118–27.
[29] Teng H, Wang S-C. Preparation of porous carbons from phenol–formaldehyde resins with chemical and physical activation. Carbon 2000;38(6):817–24.
[30] Wang SW, Tristan F, Minami D, Fujimori T, Cruz-Silva R, Terrones M, Takeuchi K, Teshima K, Rodrıguez-Reinoso F, Endo M, Kaneko K. Activation routes for high surface area graphene monoliths from graphene oxide colloids. Carbon 2014;76:220–31.
[31] Almansa C, Molina-Sabio M, Rodrıguez-Reinoso F. Adsorption of methane into $ZnCl_2$-activated carbon derived discs. Microp Mesop Mater 2004;76:185–91.
[32] Molina-Sabio M, Almansa C, Rodrıguez-Reinoso F. Phosphoric acid activated carbon discs for methane adsorption. Carbon 2003;41:2113–9.
[33] Suzuki M. Activated carbon fiber: Fundamentals and applications. Carbon 1994;32:577–86.
[34] Kaneko K, Suzuki S, Kakei K. Evaluation of micropore width of activated carbon fibers by multi- stage micropore filling analysis. TANSO 1989;140:288–94.

[35] Kaneko K, Sato N, Suzuki T, Fujiwara Y, Nishikawa K, Jaroniec M. Surface fractal dimension of microporous carbon fibres by nitrogen adsorption. J Chem Soc Faraday Trans 1991;87:179–84.

[36] Kaneko K, Katori T, Shimizu K, Shindo N, Maeda T. Changes in the molecular adsorption properties of pitch-based activated carbon fibres by air oxidation. J Chem Soc Faraday Trans 1992;88:1305–9.

[37] Kakei K, Ozeki S, Suzuki T, Kaneko K. Multi-stage micropore filling mechanism of nitrogen on microporous and micrographitic carbon. J Chem Soc Faraday Trans 1990;86:371–6.

[38] Suzuki T, Kaneko K. The dynamic structural change of high surface area microcrystals of graphite with N_2 adsorption. J Colloid Interface Sci 1990;138:590–2.

[39] Jaroniec M, Gilpin RK, Kaneko K, Choma J. Evaluation of energetic heterogeneity and microporosity of activated carbon fibers on the basis of gas adsorption isotherms. Langmuir 1991;7:2719–22.

[40] Shiratori N, Lee KJ, Miyawaki J, Hong SH, Mochida I, An B, Yokogawa K, Jang J, Yoon S-H. Pore structure analysis of activated carbon fiber by microdomain-based model. Langmuir 2009;25(13):7631–7.

[41] Kuwabara H, Suzuki T, Kaneko K. Ultramicropores in microporous carbon fibres evidenced by helium adsorption at 4.2 K. J Chem Soc Faraday Trans 1991;87:1915–6.

[42] El-Merraoui M, Aoshima M, Kaneko K. Micropore size distribution of activated carbon fiber using the density functional theory and other methods. Langmuir 2000;16:4300–4.

[43] Martınez-Alonso A, Jamond M, Montes-Moran M, Tascon JMD. Microporous texture of activated carbon fibres prepared from aramid fiber pulp. Microporous Mater 1997;11:303–11.

[44] Blanco Lopez MC, Martınez-Alonso A, Tascon JMD. Microporous texture of activated carbon fibres prepared from Nomex aramid fibres. Microp Mesop Mater 2000;34:171–9.

[45] Villar-Rodil SM, Denoyel R, Rouquerol J, Martınez-Alonso A, Tascon JMD. Porous texture evolution in Nomex-derived activated carbon fibers. J Colloid Interface Sci 2002;252:169–76.

[46] Suarez-Garcıa F, Paredes JI, Martınez-Alonso A, Tascon JMD. Preparation and porous texture characteristics of fibrous ultrahigh surface area carbons. J Mater Chem 2002;12:3213–9.

[47] Suarez-Garcia F, Martinez-Alonso A, Tascon JMD. Activated carbon fibers from Nomex by chemical activation with phosphoric acid. Carbon 2004;42:1419–26.

[48] Zhang S-J, Yu H-Q, Feng H-M. PVA-based activated carbon fibers with lotus root-like axially porous structure. Carbon 2006;44:2059–68.

[49] Qiao WM, Yoon SH, Korai Y, Mochida I, Inoue S, Sakurai T, Shimohara T. Preparation of activated carbon fibers from polyvinyl chloride. Carbon 2004;42:1327–31.

[50] Phan NH, Rio S, Faur C, Le Coq L, Le Cloirec P, Nguyen TH. Production of fibrous activated carbons from natural cellulose (jute, coconut) fibers for water treatment applications. Carbon 2006;44:2569–77.

[51] Kim BH, Bui NN, Yang KS, Cruz MED, Ferraris JP. Electrochemical properties of activated polyacrylonitrile/pitch carbon fibers produced using electrospinning. Bull Korean Chem Soc 2009;30:1967–72.

[52] Kyotani T. Control of pore structure in carbon. Carbon 2000;38:269–86.

[53] Kyotani T, Tomita A. Preparation of novel porous carbons using various zeolites as templates. J Jpn Petrol Inst 2002;45:261–70.

[54] Kyotani T, Ma Z, Tomita A. Template synthesis of novel porous carbons using various types of zeolites. Carbon 2003;41:1451–9.
[55] Nishihara H, Itoi H, Kogure T, Hou P-X, Touhara H, Okino F, Kyotani T. Investigation of the ion storage/transfer behavior in an electrical doublelayer capacitor by using ordered microporous carbons as model materials. Chem Eur J 2009;15:5355–63.
[56] Ma ZX, Kyotani T, Tomita A. Preparation of a high surface area microporous carbon having the structural regularity of Y zeolite. Chem Commun 2000:2365–6.
[57] Ma ZX, Kyotani T, Liu Z, Terasaki O, Tomita A. Very high surface area microporous carbon with a three-dimensional nano-array structure: synthesis and its molecular structure. Chem Mater 2001;13:4413–5.
[58] Ma Z, Kyotani T, Tomita A. Synthesis methods for preparing microporous carbons with a structural regularity of zeolite Y. Carbon 2002;40:2367–74.
[59] Matsuoka K, Yamagishi Y, Yamazaki T, Setoyama N, Tomita A, Kyotani T. Extremely high microporosity and sharp pore size distribution of a large surface area carbon prepared in the nanochannels of zeolite Y. Carbon 2005;43:876–9.
[60] Hou P-X, Yamazaki T, Orikasa H, Kyotani T. An easy method for the synthesis of ordered microporous carbons by the template technique. Carbon 2005;43:2624–7.
[61] Hou P-X, Orikasa H, Yamazaki T, Matsuoka K, Tomita A, Setoyama N, Fukushima Y, Kyotani T. Synthesis of nitrogen-containing microporous carbon with a highly ordered structure and effect of nitrogen doping on H_2O adsorption. Chem Mater 2005;17:5187–93.
[62] Hou P-X, Orikasa H, Itoi H, Nishihara H, Kyotani T. Densification of ordered microporous carbons and controlling their micropore size by hot-pressing. Carbon 2007;45:2011–6.
[63] Itoi H, Nishihara H, Kyotani T. Effect of heteroatoms in ordered microporous carbons on their electrochemical capacitance. Langmuir 2016;32:11997–2004.
[64] Enzel P, Bein T. Poly(acrylonitri1e) chains in zeolite channels: polymerization and pyrolysis. Chem Mater 1992;4:819–24.
[65] Johnson SA, Brigham ES, Ollivier PJ, Mallouk TE. Effect of micropore topology on the structure and properties of zeolite polymer replicas. Chem Mater 1997;9:2448–58.
[66] Meyers CJ, Shah SD, Patel SC, Sneeringer RM, Bessel CA, Dollahon NR, Leising RA, Takeuchi ESJ. Templated synthesis of carbon materials from zeolites (Y, beta, and ZSM-5) and a montmorillonite clay (K10): physical and electrochemical characterization. J Phys Chem B 2001;105:2143–52.
[67] Kyotani T, Nagai T, Inoue S, Tomita A. Formation of new type of porous carbon by carbonization in zeolite nanochannels. Chem Mater 1997;9:609–15.
[68] Rodriguez-Mirasol J, Cordero T, Radovic LR, Rodriguez JJ. Structural and textural properties of pyrolytic carbon formed within a microporous zeolite template. Chem Mater 1998;10:550–8.
[69] Nishiyama H, Yang Q-H, Hou P-X, Unno M, Yamauchi S, Saito R, Paredes JI, Marinez-Alonso A, Tascon JMD, Sato Y, Terauchi M, Kyotani T. A possible buckybowl-like structure of zeolite templated carbon. Carbon 2009;47:1220–30.
[70] Kopelevich Y, da Silva RR, Torres JHS, Penicaud A, Kyotani T. Local ferromagnetism in microporous carbon with the structural regularity of zeolite Y. Phys Rev B 2003;68:092408.
[71] Takai K, Suzuki T, Enoki T, Nishihara H, Kyotani T. Fabrication and characterization of magnetic nanoporous zeolite templated carbon. J Phys Chem Solids 2010;71:565–8.
[72] Nishihara H, Imai K, Itoi H, Nomura K, Takai K, Kyotani T. Formation mechanism of zeolite templated carbons. TANSO 2017;280:169–74.

[73] Su F, Zhao XS, Lv L, Zhou Z. Synthesis and characterization of microporous carbons templated by ammonium-form zeolite Y. Carbon 2004;42:2821−31.
[74] Garsuch A, Klepel O, Sattler RR, Berger C, Glaeser R, Weitkamp J. Synthesis of a carbon replica of zeolite Y with large crystallite size. Carbon 2006;44:593−6.
[75] Gaslain FOM, Parmentier J, Valtchev VP, Patarin J. First zeolite carbon replica with a well resolved X-ray diffraction pattern. Chem Commun 2006;2006:991−3.
[76] Yang Z, Xia Y, Sun X, Mokaya R. Preparation and hydrogen storage properties of zeolite-templated carbon materials nanocast via chemical vapor deposition: effect of the zeolite template and nitrogen doping. J Phys Chem B 2006;110:18424−31.
[77] Bandosz TJ, Jagiello J, Putyera K, Schwarz JA. Pore structure of carbon-mineral nanocomposites and derived carbons obtained by template carbonization. Chem Mater 1996;8:2023−9.
[78] Santos C, Andrade M, Vieira AL, Martins A, Pires J, Freire C, Carvalho AP. Templated synthesis of carbon materials mediated by porous clay heterostructures. Carbon 2010;48:4049−56.
[79] Han B-H, Zhou W, Sayant A. Direct preparation of nanoporous carbon by nanocasting. J Am Chem Soc 2003;125:3444−5.
[80] Kim YK, Park HB, Lee YM. Carbon molecular sieve membranes derived from thermally labile polymer containing blend polymers and their gas separation properties. J Membr Sci 2004;243:9−17.
[81] Kim YK, Park HB, Lee YM. Gas separation properties of carbon molecular sieve membranes derived from polyimide/polyvinylpyrrolidone blends: effect of the molecular weight of polyvinylpyrrolidone. J Membr Sci 2005;251:159−67.
[82] Inagaki M, Ohta N, Hishiyama Y. Aromatic polyimides as carbon precursors. Carbon 2013;61:1−21.
[83] Le TH, Yang Y, Huang Z, Kang F. Preparation of microporous carbon nanofibers from polyimide by using polyvinyl pyrrolidone as template and their capacitive performance. J Power Sources 2015;278:683−92.
[84] Le TH, Yang Y, Yu L, Gao TJ, Huang ZH, Kang F. Polyimide-based porous hollow carbon nanofibers for supercapacitor electrode. J Appl Polym Sci 2016;133:43397.
[85] Coutinho EB, Salim VMM, Borges CP. Preparation of carbon hollow fiber membranes by pyrolysis of polyetherimide. Carbon 2003;41:1707−14.
[86] Salleh WNW, Ismail AF. Carbon hollow fiber membranes derived from PEI/PVP for gas separation. Sep Purif Technol 2011;80:541−8.
[87] Lee HJ, Suda H, Haraya K, Moon SH. Gas permeation properties of carbon molecular sieving membranes derived from the polymer blend of polyphenylene oxide (PPO)/polyvinylpyrrolidone (PVP). J Membr Sci 2007;296:139−46.
[88] Itta AK, Tseng HH, Wey MY. Fabrication and characterization of PPO/PVP blend carbon molecular sieve membranes for H_2/N_2 and H_2/CH_4 separation. J Membr Sci 2011;372:387−95.
[89] Sun DD, Qin GT, Lü M, Wei W, Wang N, Jiang L. Preparation of mesoporous polyacrylonitrile and carbon fibers by electrospinning and supercritical drying. Carbon 2013;63:585−8.
[90] Zhang J, Zhong XF, Chen HB, Gao Y, Li HM. Synthesis and electrochemical performance of porous carbon by carbonizing PF/PMMA interpenetrating polymer networks. Electrochim Acta 2014;148:203−10.
[91] Yamazaki M, Kayama M, Ikeda K, Alii T, Ichihara S. Nanostructured carbonaceous material with continuous pores obtained from reaction-induced phase separation of miscible polymer blends. Carbon 2004;42:1641−9.
[92] Inomata K, Otake Y. Activation-free preparation of porous carbon by carbonizing phenolic resin containing pore-forming substance. Energy Porcedia 2012;14:626−31.

[93] Rao MM, Li WS, Cairns EJ. Porous carbon-sulfur composite cathode for lithium/sulfur cells. Electrochem Commun 2012;17:1–5.
[94] Zhang XY, Hu HQ, Zhu YD, Zhu SW. Carbon molecular sieve membranes derived from phenol formaldehyde novolac resin blended with poly (ethylene glycol). J Membr Sci 2007;289:86–91.
[95] Yang DS, Chaudhari S, Rajesh KP, Yu JS. Preparation of nitrogen-doped porous carbon nanofibers and the effect of porosity, electrical conductivity, and nitrogen content on their oxygen reduction performance. ChemCatChem 2014;6(5):1236–44.
[96] Ozaki J, Endo N, Ohizumi W, Igarashi K, Nakahara M, Oya A. Novel preparation method for the production of mesoporous carbon fiber from a polymer blend. Carbon 1997;35:1031–3.
[97] Wang MX, Huang ZH, Kang F, Liang KM. Porous carbon nanofibers with narrow pore size distribution from electrospun phenolic resins. Mater Lett 2011;65:1875–7.
[98] Wang L, Huang ZH, Yue M, Li M, Wang M, Kang F. Preparation of flexible phenolic resin-based porous carbon fabrics by electrospinning. Chem Eng J 2013;218:232–7.
[99] Kim BH, Yang YK, Kim YA, Kim YJ, An B, Oshida K. Solvent-induced porosity control of carbon nanofiber webs for supercapacitor. J Power Sources 2011;196:10496–501.
[100] Grzyb B, Machnikowski J, Weber JV, Koch A, Heintz O. Mechanism of co-pyrolysis of coal-tar pitch with polyacrylonitrile. J Anal Appl Pyrol 2003;67:77–93.
[101] Grzyb B, Machnikowski J, Weber JV. Mechanism of co-pyrolysis of coal-tar pitch with polyvinylpyridine. J Anal Appl Pyrol 2004;72:121–30.
[102] Machnikowski J, Grzyb B, Machnikowska H, Weber JV. Surface chemistry of porous carbons from N-polymers and their blends with pitch. Microp Mesop Mater 2005;82:113–20.
[103] Inagaki M, Takeichi T, Hishiyama Y, Oberlin A. High quality graphite films produced from aromatic polyimides. In: Chemistry and physics of carbon, vol. 26. Marcel Dekker; 1999. p. 245–333.
[104] Ohta N, Nishi Y, Morishita T, Tojo T, InagakiM. Carbonization of aromatic polyimides and pore development in carbon films. TANSO 2008;233:174–80.
[105] Ohta N, Nishi Y, Morishita T, Tojo T, Inagaki M. Preparation of microporous carbon films from fluorinated aromatic polyimides. Carbon 2008;46:1350–7.
[106] Ohta N, Nishi Y, Morishita T, Tojo T, Inagaki M. Water vapor adsorption of microporous carbon films prepared from fluorinated aromatic polyimides. Ads Sci Technol 2008;26:373–82.
[107] Xu Z, Zhuang X, Yang C, Cao J, Yao Z, Tang Y, Jiang J, Wu D, Feng X. Nitrogen-doped porous carbon superstructures derived from hierarchical assembly of polyimide nanosheets. Adv Mater 2016;28(10):1981–7.
[108] Yeo H, Jung J, Song HJ, Choi Y-M, Wee J-H, You N-H, Joh H-I, Yang C-M, Goh M. Preparation and formation mechanism of porous carbon cryogel. Microp Mesop Mater 2017;245:138–46.
[109] Jansta J, Dousek FP. Electrochemical corrosion of polytetrafluorethylene contacting lithium amalgam. Electrochim Acta 1973;18:673–4.
[110] Jansta J, Dousek FP, Patzelova V. Low temperature electrochemical preparation of carbon with a high surface area from polytetrafluoroethylene. Carbon 1975;13:377–80.
[111] Dousek FP, Jansta J. Reactivity of polymeric carbon chains reduced from poly (tetrafluoroethylene). Carbon 1980;18:13–20.

[112] Shiraishi S, Kurihara, Tsuboi H, Oya A, Soneda Y, Yamada Y. Electric double layer capacitance of highly porous carbon derived from lithium metal and polytetrafluoroethylene. Electrochem Solid State Lett 2001;4:A5.
[113] Yamashita J, Shioya M, Kikutani T, Hashimoto T. Activated carbon fibers and films derived from poly(vinylidene fluoride). Carbon 2001;39:207–14.
[114] Yang Y, Centrone A, Chen L, Simeon F, Hatton TA, Gregory CR. Highly porous electrospun polyvinylidene fluoride (PVDF)-based carbon fiber. Carbon 2011;49:3395–403.
[115] Ben T, Li YQ, Zhu LK, Zhang DL, Cao DP, Xiang ZH, Yao XD, Qiu SL. Selective adsorption of carbon dioxide by carbonized porous aromatic framework (PAF). Energy Environ Sci 2012;5:8370–6.
[116] Tian Z, Huang J, Zhang X, Shao G, He Q, Cao S, Yuan S. Ultra-microporous N-doped carbon from polycondensed framework precursor for CO_2 adsorption. Microp Mesop Mater 2018;257:19–26.
[117] Pachfule P, Dhavale VM, Kdambeth S, Kurungot S, Banerjee R. Porous-organic-framework- templated nitrogen-rich porous carbon as a more proficient electrocatalyst than Pt/C for the electrochemical reduction of oxygen. Chem Eur J 2013;19:974–80.
[118] Zhang YM, Li BY, Williams K, Gao WY, Ma SM. A new microporous carbon material synthesized via thermolysis of a porous aromatic framework embedded with an extra carbon source for low- pressure CO_2 uptake. Chem Commun 2013;49:10269–71.
[119] Liu X, Zhou L, Zhao Y, Bian L, Feng X, Pu Q. Hollow, spherical nitrogen-rich porous carbon shells obtained from a porous organic framework for the supercapacitor. ACS Appl Mater Interf 2013;5(20):10280–7.
[120] Kong F, Fan X, Zhang X, Wang L, Kong A, Shan Y. Soft-confinement conversion of Co-Salen- organic-frameworks to uniform cobalt nanoparticles embedded within porous carbons as robust trifunctional electrocatalysts. Carbon 2019;149:471–82.
[121] Kou Y, Xu YH, Guo ZQ, Jiang DL. Supercapacitive energy storage and electric power supply using an aza-fused π-conjugated microporous framework. Angew Chem Int Ed 2011;50:8753–7.
[122] Feng X, Liang Y, Zhi L, Thomas A, Wu D, Lieberwirth I, Kolb U, Mullen K. Synthesis of microporous carbon nanofibers and nanotubes from conjugated polymer network and evaluation in electrochemical capacitor. Adv Funct Mater 2009;19:2125–9.
[123] Gogotsi YG, Jeon JD, McNallan MJ. Carbon coatings on silicon carbide by reaction with chlorine- containing gases. J Mater Chem 1997;7:1841–8.
[124] Ersoy D, McNallan MJ, GogotsiYG. Carbon coatings produced by high temperature chlorination of silicon carbide ceramics. Mater Res Innov 2001;5:55–62.
[125] Kyutt RN, Smorgonskaya ÉA, Danishevski AM, Gordeev SK, Grechinskaya AV. Structural study of nanoporous carbon produced from polycrystalline carbide materials: small-angle x-ray scattering. Phys Solid State 1999;41:1359–63.
[126] Gordeev SK, Kukushkin SA, Osipov AV, Pavlov YV. Self-organization in the formation of a nanoporous carbon material. Phys Solid State 2000;42:2314–7.
[127] Gogotsi Y, Nikitini A, Ye H, Zhou W, Fischer J, Yi B, Folley HC, Barsoum M. Nanoporous carbide- derived carbon with tunable pore size. Nat Mater 2003;2:591–4.
[128] Chmiola J, Yushin G, Dash R, Gogotsi Y. Effect of pore size and surface area of carbide derived carbons on specific capacitance. J Power Sources 2006;158:765–72.
[129] Jaenes A, Permann L, Arulepp M, Lust E. Electrochemical characteristics of nanoporous carbide- derived carbon materials in non-aqueous electrolyte solutions. Electrochem Commun 2004;6:313–8.

[130] Laudisio G, Dash RK, Singer JP, Yushin G, Gogotsi Y, Fischer JE. Carbide-derived carbons: a comparative study of porosity based on small-angle scattering and adsorption isotherms. Langmuir 2006;22:8945−50.
[131] Chmiola J, Yushin G, Dash RK, Hoffman EN, Fischer JE, Barsoum MW, Gogotsi Y. Double-layer capacitance of carbide derived carbons in sulfuric acid. Electrochem Solid State Lett 2005;8:A357−60.
[132] Dash RK, Yushin G, Gogotsi Y. Synthesis, structure and porosity analysis of microporous and mesoporous carbon derived from zirconium carbide. Microp Mesop Mater 2005;86:50−7.
[133] Dash RK, Nikitin A, Gogotsi Y. Microporous carbon derived from boron carbide. Microp Mesop Mater 2004;72:203−8.
[134] Kravchik AE, Kukushkina JA, Sokolov VV, Tereshchenko GF. Structure of nanoporous carbon produced from boron carbide. Carbon 2006;44:3263−8.
[135] Dimovski S, Nikitin A, Ye H, Gogotsi Y. Synthesis of graphite by chlorination of iron carbide at moderate temperatures. J Mater Chem 2004;14:238−43.
[136] Leis J, Perkson A, Arulepp M, Nigu P, Svensson G. Catalytic effects of metals of the iron subgroup on the chlorination of titanium carbide to form nanostructural carbon. Carbon 2002;40:1559−64.
[137] Rose M, Kockrick E, Senkovska I, Kaskel S. High surface area carbide-derived carbon fibers produced by electrospinning of polycarbosilane precursors. Carbon 2010;48:403−7.
[138] Tamai H, Kakii T, Hirota Y, Kumamoto T, Yasuda H. Synthesis of extremely large mesoporous activated carbon and its unique adsorption for giant molecules. Chem Mater 1996;8:454−62.
[139] Oya A, Yoshida S, Alcaniz-Monge J, Linares-Solano A. Formation of mesopores in phenolic resin- derived carbon fiber by catalytic activation using cobalt. Carbon 1995;33:1085−90.
[140] Oya A, Yoshida S, Alcaniz-Monge J, Linares-Solano A. Preparation and properties of an antibacterial activated carbon fiber containing mesopores. Carbon 1996;34:53−7.
[141] Tamai H, Kojima S, Ikeuchi M, Mondori J, Kanata T, Yasuda H. Preparation of mesoporous activated carbon fibers and their adsorption properties. TANSO 1996;175:243−8 [in Japanese].
[142] Tamai H, Ikeuchi M, Kojima S, Yasuda H. Extremely large mesoporous carbon fibers synthesized by the addition of rare earth metal complexes and their unique adsorption behaviors. Adv Mater 1997;9:55−8.
[143] Yoshizawa N, Yamada Y, Furuta T, Shiraishi M, Kojima S, Tamai H, Yasuda H. Coal-based activated carbons prepared with organometallics and their mesoporous structure. Energy Fuels 1997;11:327−30.
[144] Freeman JJ, Gimblett FGR, Roberts RA, Sing KSW. Studies of activated charcoal cloth I. Modification of adsorptive properties by impregnation with boron-containing compounds. Carbon 1987;25:559−63.
[145] Freeman JJ, Gimblett FGR, Roberts RA, Sing KSW. Studies of activated charcoal cloth III. Mesopore development induced by phosphate impregnants. Carbon 1988;26:7−11.
[146] Freeman JJ, Gimblett FGR. Studies of activated charcoal cloth. II. Influence of boron-containing impregnants on the rate of activation in carbon dioxide gas. Carbon 1987;25:565−8.
[147] Myglovets M, Poddubnaya OI, Sevastyanova O, Lindstrom ME, Gawdzik B, Sobiesiak M, Tsyba MM, Sapsay VI, Klymchuk DO, Puziy AM. Preparation of carbon adsorbents from lignosulfonate by phosphoric acid activation for the adsorption of metal ions. Carbon 2014;80:771−83.

[148] Suarez-Garcia F, Martınez-Alonso A, Toscon JMD. Porous texture of activated carbons prepared by phosphoric acid activation of apple pulp. Carbon 2001;39:1103–16.
[149] Romero-Anaya AJ, Lillo-Rodenas MA, Salinas-Martınez de Lecea C, Linares-Solano A. Hydrothermal and conventional H_3PO_4 activation of two natural bio-fibers. Carbon 2012;50:3158–69.
[150] Kikuchi K, Yamashita R, Sakuragawa S, Saeki T, Oikawa K, Kume T. Pore structure and chemical composition of activated carbon derived from composted spent coffee grounds. TANSO 2017;278:118–22.
[151] Huang L, Sun Y, Wang W, Yue Q, Yang T. Comparative study on characterization of activated carbons prepared by microwave and conventional heating methods and application in removal of oxytetracycline (OTC). Chem Eng J 2011;171:1446–53.
[152] Huang L, Sun Y, Yang T, Li L. Adsorption behavior of Ni (II) on lotus stalks derived active carbon by phosphoric acid activation. Desalination 2011;268:12–9.
[153] Li D, Wang Y, Zhou J, Wang J, Liu X, Tian Y, Zhang Z, Qiao Y, Wei L, Li J, Wen L. Drying before microwave-assisted H_3PO_4 activation to produce highly mesoporous activated carbons. Mater Lett 2018;230:61–3.
[154] Ryoo R, Joo SH, Kruk M, Jaroniec M. Ordered mesoporous carbons. Adv Mater 2001;13:677–81.
[155] Lee J, Han S, Hyeon T. Synthesis of new nanoporous carbon materials using nanostructured silica materials as templates. J Mater Chem 2004;14:478–86.
[156] Lu A-H, Schueth F, Nanocasting. A versatile strategy for creating nanostructured porous materials. Adv Mater 2006;18:1793–805.
[157] Ryoo R, Joo SH, Jun S. Synthesis of highly ordered carbon molecular sieves via template-mediated structural transformation. J Phys Chem B 1999;103:7743–6.
[158] Lee J, Yoon S, Hyeon T, Oh SM, Kim KB. Synthesis of a new mesoporous carbon and its application to electrochemical double-layer capacitors. Chem Commun 1999;1999:2177–8.
[159] Yoon SB, Kim JY, Yu J-S. Synthesis of highly ordered nanoporous carbon molecular sieves from silylated MCM-48 using divinylbenzene as precursor. Chem Commun 2001;2001:559–60.
[160] Yang H, Shi Q, Liu X, Xie S, Jiang D, Zhang F, Yu C, Tu B, Zhao D. Synthesis of ordered mesoporous carbon monoliths with bicontinuous cubic pore structure of *Ia3d* symmetry. Chem Commun 2002;2002:2842–3.
[161] Kaneda M, Tsubakiyama T, Carisson A, Sakamoto Y, Ohsuna T, Terasaki O, Joo SH, Ryoo R. Structural study of mesoporous MCM-48 and carbon networks synthesized in the spaces of MCM-48 by electron crystallography. J Phys Chem B 2002;106:1256–66.
[162] Joo SH, Jun S, Ryoo R. Synthesis of ordered mesoporous carbon molecular sieves CMK-1. Microp Mesop Mater 2001;44–45:153–8.
[163] Kruk M, Jaroniec M, Ryoo R, Joo SH. Characterization of ordered mesoporous carbons synthesized using MCM-48 silicas as templates. J Phys Chem B 2000;104:7960–8.
[164] Tian B, Che S, Liu Z, Liu X, Fan W, Tatsumi T, Terasaki O, Zhao D. Novel approaches to synthesize self-supported ultrathin carbon nanowirearrays templated by MCM-41. Chem Commun 2003;2003:2726–7.
[165] Jun S, Joo SH, Ryoo R, Kruk M, Jaroniec M, Liu Z, Ohsuna T, Terasaki O. Synthesis of new, nanoporous carbon with hexagonally ordered mesostructured. J Am Chem Soc 2000;122:10712–3.
[166] Shin HJ, Ryoo R, Kruk M, Jaroniec M. Modification of SBA-15 pore connectivity by high- temperature calcination investigated by carbon inverse replication. Chem Commun 2001;2001:349–50.

[167] Lee JS, Joo SH, Ryoo R. Synthesis of mesoporous silicas of controlled pore wall thickness and their replication to ordered nanoporous carbons with various pore diameters. J Am Chem Soc 2002;124:1156–7.
[168] Yu C, Fan J, Tian B, Zhao D, Stucky GD. High-yield synthesis of periodic mesoporous silica rods and their replication to mesoporous carbon rods. Adv Mater 2002;14:1742–5.
[169] Lu A-H, Li W-C, Schmidt W, Schueth F. Fabrication of hierarchically structured carbon monoliths via self-binding and salt templating. Microp Mesop Mater 2006;95:188–93.
[170] Xia Y, Mokaya R. Synthesis of ordered mesoporous carbon and nitrogen-doped carbon materials with graphitic pore walls via a simple chemical vapor deposition method. Adv Mater 2004;16:1553–8.
[171] Fuertes AB. Synthesis of ordered nanoporous carbons of tunable mesopore size by templating a SBA- 15 silica materials. Microp Mesop Mater 2004;67:273–81.
[172] Kim S-S, Pinnavaia TJ. A low cost route to hexagonal mesostructured carbon molecular sieves. Chem Commun 2001;2001:2418–9.
[173] Alvarez S, Fuertes AB. Template synthesis of mesoporous carbons with tailorable pore size and porosity. Carbon 2004;42:433–6.
[174] Fuertes AB, Nevskaia DM. Control of mesoporous structure of carbons synthesized using a mesostructured silica as template. Microp Mesop Mater 2003;62:177–90.
[175] Lee HI, Kim JH, You DJ, Lee JE, Lim JM, Ahn W-S, Pak C, Joo SH, Chang H, Seung D. Rational synthesis pathway for ordered mesoporous carbon with controllable 30- to 100-angstrom pores. Adv Mater 2008;20:757–62.
[176] Lee J, Yoon S, Oh SM, Shin C-H, Hyeon T. Development of a new mesoporous carbon using an HMS aluminosilicate template. Adv Mater 2000;12:359–62.
[177] Lee HI, Kim JH, Joo SH, Chang H, Seung D, Joo O-S, Suh DJ, Ahn W-S, Pak C, Kim JM. Ultrafast production of ordered mesoporous carbons via microwave irradiation. Carbon 2007;45:2851–4.
[178] Kim TW, Park IS, Ryoo R. A Synthetic route to ordered mesoporous carbon materials with graphitic pore wall. Angew Chem Int Ed 2003;42:4375–9.
[179] Vix-Guterl C, Saadallah S, Vidal L, Reda M, Parmentier J, Patarin J. Template synthesis of a new type of ordered carbon structure from pitch. J Mater Chem 2003;13:2535–9.
[180] Kruk M, Dufour B, Celer EB, Kowalewski T, Jaroniec M, Matyjaszewski K. Synthesis of mesoporous carbons using ordered and disordered mesoporous silica templates and polyacrylonitrile as carbon precursor. J Phys Chem B 2005;109:9216–25.
[181] Fuertes AB, Alvarez S. Graphitic mesoporous carbons synthesised through mesostructured silica templates. Carbon 2004;42:3049–55.
[182] Fuertes AB. Synthesis of ordered nanoporous carbons of tunable mesopore size by templating SBA- 15 silica materials. Microp Mesop Mater 2004;67:272–81.
[183] Seo Y, Kim K, Jung Y, Ryoo R. Synthesis of mesoporous carbons using silica templates impregnated with mineral acids. Microp Mesop Mater 2015;207:156–62.
[184] Xia K, Gao Q, Jiang J, Hu J. Hierarchical porous carbons with controlled micropores and mesopores for supercapacitor electrode materials. Carbon 2008;46(13):1718–26.
[185] Kim ND, Kim W, Joo JB, Oh S, Kim P, Kim Y, Yi J. Electrochemical capacitor performance of N- doped mesoporous carbons prepared by ammoxidation. J Power Sources 2008;180(1):671–5.
[186] Banham D, Feng F, Burt J, Alsrayheen E, Birss V. Bimodal, templated mesoporous carbons for capacitor applications. Carbon 2010;48:1056–63.
[187] Oda Y, Fukuyama K, Nishikawa K, Namba S, Yoshitake H, Tatsumi T. Mesocellular foam carbons: aggregates of hollow carbon spheres with open and closed wall structures. Chem Mater 2004;16:3860–6.

[188] Yang X, He P, Xia Y. Preparation of mesocellular carbon foam and its application for lithium/oxygen battery. Electrochem Commun 2009;11:1127–30.
[189] Zakhidov AA, Baughman RH, Iqbal Z, Cui C, Khayrullin I, Dantes SO, Marti J, Ralchenko G. Carbon structures with three-dimensional periodicity at optical wavelengths. Science 1998;282:897–901.
[190] Sevilla M, Fuertes AB. Catalytic graphitization of templated mesoporous carbons. Carbon 2006;44:468–74.
[191] Taguchi A, Smatt JH, Linden M. Carbon monoliths processing a hierarchical fully interconnected porosity. Adv Mater 2003;15:1209–11.
[192] Han S, Lee KT, Oh M, Hyeon T. The effect of silica template structure on the pore structure of mesoporous carbons. Carbon 2003;41:1049–56.
[193] Li Z, Jaroniec M. Colloidal imprinting: a novel approach to the synthesis of mesoporous carbons. J Am Chem Soc 2001;123:9208–9.
[194] Li Z, Jaroniec M, Lee Y-J, Radovic LR. High surface area graphitized carbon with uniform mesopores synthesised by a colloidal imprinting method. Chem Commun 2002:1346–7.
[195] Han S, Hyeon T. Novel silica-sol mediated synthesis of high surface area porous carbons. Carbon 1999;37:1645–7.
[196] Han S, Hyeon T. Simple silica-particle template synthesis of mesoporous carbons. Chem Commun 1999;1999:1955–6.
[197] Kawashima D, Aihara T, Kobayashi Y, Kyotani T, Tomita A. Preparation of mesoporous carbon from organic polymer/silica nanocomposite. Chem Mater 2000;12:3397–401.
[198] Pang J, Li X, Wang D, Wu Z, John VT, Yang Z, Lu Y. Silica-templated continuous mesoporous carbon films by a spin-coating technique. Adv Mater 2004;16:884–6.
[199] Gierszal KP, Jaroniec M. Carbons with extremely large volume of uniform mesopores synthesized by carbonization of phenolic resin film formed on colloidal silica template. J Am Chem Soc 2006;128(31):10026–7.
[200] Gierszal KP, Jaroniec M, Liang C, Dai S. Electron microscopy and nitrogen adsorption studies of film-type carbon replicas with large pore volume synthesized by using colloidal silica and SBA-15 as templates. Carbon 2007;45:2171–7.
[201] Woo S-W, Dokko K, Nakano H, Kanamura K. Preparation of three dimensionally ordered macroporous carbon with mesoporous walls for electric double-layer capacitors. J Mater Chem 2008;18:1674–80.
[202] Lee J, Sohn K, Hyeon T. Fabrication of novel mesocellular carbon foams with uniform ultralarge mesopores. J Am Chem Soc 2001;123:5146–7.
[203] Zhang HW, Noonan O, Huang XD, Yang YN, Xu C, Zhou L, Yu CZ. Surfactant-free assembly of mesoporous carbon hollow spheres with large tunable pore sizes. ACS Nano 2016;10:4579–86.
[204] Krawiec P, Kockrick E, Borchardt L, Geiger D, Corma A, Kaskel S. Ordered mesoporous carbide derived carbons: novel materials for catalysis and adsorption. J Phys Chem C 2009;113(18):7755–61.
[205] Inagaki M, Kobayashi S, Kojin F, Tanaka N, Morishita T, Tryba B. Pore structure of carbons coated on ceramic particles. Carbon 2004;42:3153–8.
[206] Morishita T, Tsumura T, Toyoda M, Przepiórski J, Morawski AW, Konno H, Inagaki M. A review of the control of pore structure in MgO-templated nanoporous carbons. Carbon 2010;48:2690–707.
[207] Morishita T, Wang L, Tsumura T, Toyoda M, Konno H, Inagaki M. Pore structure and application of MgO-templated carbons. TANSO 2010;242:60–8 [in Japanese].
[208] Inagaki M, Toyoda M, Soneda Y, Tsujimura S, Morishita T. Templated mesoporous carbons: synthesis and applications. Carbon 2016;107:448–73.
[209] http://www.toyotanso.com/Products/new_developed_products/cnovel.html.

[210] Morishita T, Ishihara K, Kato M, Tsumura T, Inagaki M. Mesoporous carbons prepared from mixtures of magnesium citrate with poly(vinyl alcohol). TANSO 2007;226:19–24.
[211] Morishita T, Ishihara K, Kato M, Tsumura T, Inagaki M. Preparation of a carbon with a 2 nm pore size and of a carbon with a bi-modal pore size distribution. Carbon 2007;45:209–11.
[212] Morishita T, Suzuki T, Nishikara T, Tsumura T, Inagaki M. Preparation of porous carbons by carbonization of the mixtures of thermoplastic precursors with MgO. TANSO 2005;219:226–31 [in Japanese].
[213] Funabashi H, Takeuchi S, Tsujimura S. Hierarchical meso/macro-porous carbon fabricated from dual MgO templates for direct electron transfer enzymatic electrodes. Sci Rep 2017;7:45147.
[214] Morishita T, Suzuki T, Nishikawa T, Tsumura T, Inagaki M. Preparation of mesoporous carbons by carbonization of the mixtures of poly(vinyl alcohol) with magnesium salts. TANSO 2006;223:220–6 [in Japanese].
[215] Inagaki M, Kato M, Morishita T, Morita K, Mizuuchi K. Direct preparation of mesoporous carbon from a coal tar pitch. Carbon 2007;45:1121–4.
[216] Wang X, Chen M, Matsumura K, Toyoda M, Wang C. MgO-templated mesoporous carbons using pitch-based thermosetting carbon precursor. RSC Adv 2016;6:100546–53.
[217] Liu W, Wang CY, Wag JZ, Jia F, Zheng JM, Chen MM. Preparation of mesoporous MgO-templated carbons from phenolic resin and their applications for electric double-layer capacitors. Chinese Sci Bull 2013;58:992–7.
[218] Przepiórski J, Karolczyk J, Takeda K, Tsumura T, Toyoda M, Morawski AW. Porous carbon obtained by carbonization of PET mixed with basic magnesium carbonate: pore structure and pore creation mechanism. Ind Eng Chem Res 2009;48:7110–6.
[219] Karolczyk J, Janus M, Przepiorski J. Adsorption of humic acid on mesoporous carbons prepared from poly(ethylene terephthalate) templated with magnesium compounds. Pol J Chem Technol 2012;14:95–9.
[220] Konno H, Onishi H, Yoshizawa N, Azumi K. MgO-templated nitrogen-containing carbons derived from different organic compounds for capacitor electrodes. J Power Sources 2010;195:667–73.
[221] Zhang WF, Huang Z-H, Cao GP, Kang F, Yang Y. A novel mesoporous carbon with straight tunnel- like pore structure for high rate electrochemical capacitors. J Power Sources 2012;204:230–5.
[222] Zhang WF, Huang Z-H, Zhou C, Cao G, Kang F, Yang Y. Porous carbon for electrochemical capacitors prepared from a resorcinol/formaldehyde-based organic aquagel with nano-sized particles. J Mater Chem 2012;22:7158–63.
[223] Zhang WF, Huang Z-H, Cao GP, Kang F, Yang Y. Coal tar pitch-based porous carbon by one dimensional nano-sized MgO template. J Phys Chem Solids 2012;73:1428–31.
[224] Chen XY, Song H, Zhang ZJ, He YY. A rational template (MgO) carbonization method for producing highly porous carbon for supercapacitor application. Electrochim Acta 2014;117:55–61.
[225] Ma J, Liu J, Song J, Tang T. Pressurized carbonization of mixed plastics into porous carbon sheets on magnesium oxide. RSC Adv 2018;8:2469–76.
[226] Jung M-J, Im JS, Jeong E, Jin H, Lee Y-S. Hydrogen adsorption of PAN-based porous carbon nanofibers using MgO as the substrate. Carbon Lett 2009;10:217–20.
[227] Kim Y, Cho S, Lee S, Lee YS. Fabrication and characterization of porous non-woven carbon based highly sensitive gas sensors derived by magnesium oxide. Carbon Lett 2012;13:254–9.

[228] Le TH, Tian H, Cheng J, Huang Z-H, Kang F, Yang Y. High performance lithium-ion capacitors based on scalable surface carved multi hierarchical construction electrospun carbon fibers. Carbon 2018;138:325–36.
[229] Li C, Zhang X, Wang K, Sun X, Liu G, Li J, Tian H, Li J, Ma Y. Scalable self-propagating high- temperature synthesis of graphene for supercapacitors with superior power density and cyclic stability. Adv Mater 2017;29:1604690.
[230] Zhou J, Yuan X, Xing W, Si W, Zhuo S. Mesoporous carbons derived from citrates for use in electrochemical capacitors. New Carbon Mater 2010;25:370–5.
[231] Fechler N, Wohlgemuth SA, Fellinger TP, Jaker P, Antonietti M. Salt and sugar: direct synthesis of high surface area carbon materials at low temperatures via hydrothermal carbonization of glucose under hypersaline conditions. J Mater Chem A 2013;1:9418–21.
[232] Fechler N, Fellinger TP, Antonietti M. "Salt templating": a simple and sustainable pathway toward highly porous functional carbons from ionic liquids. Adv Mater 2013;25:75–9.
[233] Fellinger TP, Thomas A, Yuan J, Antonietti M. 25th anniversary article: "cooking carbon with salt": carbon materials and carbonaceous frameworks from ionic liquids and poly(ionic liquid)s. Adv Mater 2013;25:5838–55.
[234] Wang DW, Li F, Liu M, Lu GQ, Cheng HM. 3D aperiodic hierarchical porous graphitic carbon material for high-rate electrochemical capacitive energy storage. Angew Chem Int Ed 2008;47(2):373–6.
[235] Filho CA, Zarbin AJG. Hollow porous carbon microspheres obtained by the pyrolysis of TiO_2/poly(furfuryl alcohol) composite precursors. Carbon 2006;44:2869–76.
[236] Zhou J, Yuan X, Xing W, Si W, Zhuo S. Capacitive performance of mesoporous carbons derived from the citrates in ionic liquid. Carbon 2010;48:2765–72.
[237] Shao J, Ma F, Wu G, Dai C, Geng W, Song S, Wan J. In-situ MgO ($CaCO_3$) templating coupled with KOH activation strategy for high yield preparation of various porous carbons as supercapacitor electrode materials. Chem Eng J 2017;321:301–13.
[238] Liu G-W, Chen T-Y, Chung C-H, Lin H-P, Hsu C-H. Hierarchical micro/mesoporous carbons synthesized with a ZnO template and petroleum pitch via a solvent-free process for a high- performance supercapacitor. ACS Omega 2017;2:2106–13.
[239] Ferrero GA, Fuertes AB, Sevilla M, Titirici MM. Efficient metal-free N-doped mesoporous carbon catalysts for ORR by a template-free approach. Carbon 2016;106:179–87.
[240] Wang Z, Xiong YC, Guan SY. A simple $CaCO_3$-assisted template carbonization method for producing nitrogen doped porous carbons as electrode materials for supercapacitors. Electrochim Acta 2016;188:757–66.
[241] Cui Y, Liu W, Lyu Y, Zhang Y, Wang H, Liu Y, Li D. All-carbon lithium capacitor based on salt crystal- templated, N-doped porous carbon electrodes with superior energy storage. J Mater Chem A 2018;6(37):18276–85.
[242] Yang J, Zou LD. Recycle of calcium waste into mesoporous carbons as sustainable electrode materials for capacitive deionization. Microp Mesop Mater 2014;183:91–8.
[243] Nishihara H, Simura T, Kobayashi S, Nomura K, Berenguer R, Ito M, Uchimura M, Iden H, Arihara K, Ohma A, Hayasaka Y, Kyotani T. Oxidation-resistant and elastic mesoporous carbon with single- layer graphene walls. Adv Funct Mater 2016;26:6418–27.
[244] Yeletsky PM, Yakovlev VA, Melgunov MS, Parmon VN. Synthesis of mesoporous carbons by leaching out natural silica templates of rice husk. Microp Mesop Mater 2009;121:34–40.

[245] Ma C, Cao E, Li J, Fan Q, Wu L, Song Y, Shi J. Synthesis of mesoporous ribbon-shaped graphitic carbon nanofibers with superior performance as efficient supercapacitor electrodes. Electrochim Acta 2018;292:364–73.
[246] Nguyen-Thanh D, Bandosz TJ. Metal-loaded carbonaceous adsorbents templated from porous clay heterostructures. Microp Mesop Mater 2006;92:47–55.
[247] Lu Y, Wang L, Preuß K, Qiao Mo, Titirici M-M, Varcoe J, Cai Q. Halloysite-derived nitrogen doped carbon electrocatalysts for anion exchange membrane fuel cells. J Power Sources 2017;372:82–90.
[248] Inagaki M, Orikasa H, Morishita T. Morphology and pore control in carbon materials via templating. RSC Adv 2011;1:1620–40.
[249] Yang Y, Le TH, Kang F, Inagaki M. Polymer blend techniques for designing carbon materials. Carbon 2017;111:546–68.
[250] Tanaka S, Nishiyama N, Egashira Y, Ueyama K. Synthesis of ordered mesoporous carbons with channel structure from an organic–organic nanocomposite. Chem Commun 2005:2125–7.
[251] Simanjuntak FH, Jin J, Nishiyama N, Egashira Y, Ueyama K. Ordered mesoporous carbon films prepared from 1,5-dihydroxynaphthalene/triblock copolymer composites. Carbon 2009;47:2531–3.
[252] Wang X, Liang C, Dai S. Facile synthesis of ordered mesoporous carbons with high thermal stability by self-assembly of resorcinol-formaldehyde and block copolymers under highly acidic conditions. Langmuir 2008;24:7500–5.
[253] Meng Y, Gu D, Zhang F, Shi Y, Yang H, Li Z, Yu C, Tu B, Zhao D. Ordered mesoporous polymers and homologous carbon frameworks: amphiphilic surfactant templating and direct transformation. Angew Chem Int Ed 2005;44:7053–9.
[254] Tanaka S, Katayama Y, Tate MP, Hillhouse HW, Miyake Y. Fabrication of continuous mesoporous carbon films with face-centered orthorhombic symmetry through a soft templating pathway. J Mater Chem 2007;17:3639–45.
[255] Liang C, Dai S. Dual phase separation for synthesis of bimodal meso-/microporous carbon monoliths. Chem Mater 2009;21:2115–24.
[256] Jin J, Nishiyama N, Egashira Y, Ueyama K. Vapor phase synthesis of ultrathin carbon films with a mesoporous monolayer by a soft-templating method. Chem Commun 2009:1371–3.
[257] Liu L, Wang FY, Shao GS, Yuan ZY. A low-temperature autoclaving route to synthesize monolithic carbon materials with an ordered mesostructure. Carbon 2010;48:2089–99.
[258] Liu D, Lei JH, Guo LP, Deng KJ. Simple hydrothermal synthesis of ordered mesoporous carbons from resorcinol and hexamine. Carbon 2011;49:2113–9.
[259] Liang CD, Dai S. Synthesis of mesoporous carbon materials via enhanced hydrogen-bonding interaction. J Am Chem Soc 2006;128:5316–7.
[260] Zhang FQ, Meng Y, Gu D, Yan Y, Yu CZ, Tu B, Zhao DY. A facile aqueous route to synthesize highly ordered mesoporous polymers and carbon frameworks with *Ia3hd* bicontinuous cubic structure. J Am Chem Soc 2005;127:13508–9.
[261] Zhang F, Meng Y, Gu D, Yan Y, Chen Z, Tu B, Zhao D. An aqueous cooperative assembly route to synthesize ordered mesoporous carbons with controlled structures and morphology. Chem Mater 2006;18:5279–88.
[262] Carriazo D, Pico F, Gutierrez MC, Rubio F, Rojo JM, del Monte F. Block-copolymer assisted synthesis of hierarchical carbon monoliths suitable as supercapacitor electrodes. J Mater Chem 2010;20:773–80.
[263] Lin H-P, Chang-Chien C-Y, Tang C-Y, L C-Y. Synthesis of p6mm hexagonal mesoporous carbons and silicas using Pluronic F127–PF resin polymer blends. Microp Mesop Mater 2006;93:344–8.

[264] Kowalewski T, Tsarevsky NV, Matyjaszewski K. Nanostructured carbon arrays from block copolymers of polyacrylonitrile. J Am Chem Soc 2002;124:10632–3.
[265] Meng Y, Gu D, Zhang F, Shi Y, Cheng L, Feng D, Wu Z, Chen Z, Wan Y, Stein A, Zhao D. A family of highly ordered mesoporous polymer resin and carbon structures from organic-organic self- assembly. Chem Mater 2006;18:4447–64.
[266] Jin J, Nishiyama N, Egashira Y, Ueyama K. Pore structure and pore size controls of ordered mesoporous carbons prepared from resorcinol/formaldehyde/triblock polymers. Microp Mesop Mater 2009;118:218–23.
[267] Song L, Feng D, Fredin NJ, Yager KG, Jones RL, Wu Q, Zhao D, Vogt BD. Challenges in fabrication of mesoporous carbon films with ordered cylindrical pores via phenolic oligomer self-assembly with triblock copolymers. ACS Nano 2010;4:189–98.
[268] Tanaka S, Nakatani N, Doi A, Miyake Y. Preparation of ordered mesoporous carbon membranes by a soft-templating method. Carbon 2011;49:3184–7.
[269] Yu T, Deng Y, Wang L, Liu R, Zhang L, Tu B, Zhao D. Ordered mesoporous nanocrystalline titanium- carbide/carbon composites from in situ carbothermal reduction. Adv Mater 2007;19:2301–6.
[270] Liang C, Hong K, Guiochon GA, Mays JW, Dai S. Synthesis of a large-scale highly ordered porous carbon film by self-assembly of block copolymers. Angew Chem Int Ed 2004;43:785–9.
[271] Kosonen H, Valkama S, Nykänen A, Toivanen M, Brinke G, Ruokolainen J, Ikkala O. Functional porous structures based on the pyrolysis of cured templates of block copolymer and phenolic resin. Adv Mater 2006;18:201–5.
[272] Gorka J, Zawislak A, Choma J, Jaroniec M. KOH activation of mesoporous carbons obtained by soft- templating. Carbon 2008;46:1159–61.
[273] Jin J, Tanaka S, Egashira Y, Nishiyama N. KOH activation of ordered mesoporous carbons prepared by a soft-templating method and their enhanced electrochemical properties. Carbon 2010;48:1985–9.
[274] Liu R, Shi Y, Wan Y, Meng Y, Zhang F, Gu D, Chen Z, Tu B, Zhao D. Triconstituent co-assembly to ordered mesostructured polymer-silica and carbon-silica nanocomposites and large-pore mesoporous carbons with high surface areas. J Am Chem Soc 2006;128:11652–62.
[275] Hatori H, Kobayashi T, Hanzawa Y, Yamada Y, Iimura Y, Kimura T, Shiraishi M. Mesoporous carbon membranes from polyimide blended with poly(ethylene glycol). J Appl Polym Sci 2001;79:836–41.
[276] Kim C, Jeong Y, Ngoc B, Yang KS, Kojima M, Kim YA, Endo M, Lee J-W. Synthesis and characterization of porous carbon nanofibers with hollow cores through the thermal treatment of electrospun copolymeric nanofiber webs. Small 2007;3(1):91–5.
[277] Guo T, Zhu J, Chen X, Song M, Zhang B. Novel monolithic mesoporous foamed carbons prepared using micro-colloidal particles as templates. J Non-Cryst Solids 2007;353:2893–9.
[278] Yang XJ, Teng DH, Liu BX, Yu YH, Yang XP. Nanosized anatase titanium dioxide loaded porous carbon nanofiber webs as anode materials for lithium-ion batteries. Electrochem Commun 2011;13:1098–101.
[279] Ozaki J, Ohizumi W, Endo N, Oya A, Yoshida S, Iizuka T, Roman-Martinez MC, Linares-Solano A. Preparation of platinum loaded carbon fiber by using a polymer blend. Carbon 1997;35:1676–7.
[280] Rasines G, Lavela P, Macıas C, Zafra MC, Tirado JL, Parra JB, Ania C. N-doped monolithic carbon aerogel electrodes with optimized features for the electrosorption of ions. Carbon 2015;83:262–74.

[281] Sun J-K, Xu Q. Functional materials derived from open framework templates/precursors: synthesis and applications. Energy Environ Sci 2014;7:2071–100.
[282] Liu B, Shioyama H, Akita T, Xu Q. Metal-organic framework as a template for porous carbon synthesis. J Am Chem Soc 2008;130:5390–1.
[283] Liu B, Shioyama H, Jiang H, Zhang X, Xu Q. Metal–organic framework (MOF) as a template for syntheses of nanoporous carbons as electrode materials for supercapacitor. Carbon 2010;48:456–63.
[284] Yuan D, Chen J, Tan S, Xia N, Liu Y. Worm-like mesoporous carbon synthesized from metal-organic coordination polymers for supercapacitors. Electrochem Commun 2009;11:1191–4.
[285] Hu J, Wang H, Gao Q, Guo H. Porous carbons prepared by using metal–organic framework as the precursor for supercapacitors. Carbon 2010;48:3599–606.
[286] Deng H-G, Jin S-L, Zhan L, Wang Y, Lu B, Qiao W, Ling L. Synthesis of porous carbons derived from metal-organic coordination polymers and their adsorption performance for carbon dioxide. New Carbon Mater 2012;27:194–9.
[287] Shi X, Chen Y, Lai Y, Zhang K, Li J, Zhang Z. Metal organic frameworks templated sulfur-doped mesoporous carbons as anode materials for advanced sodium ion batteries. Carbon 2017;123:250–8.
[288] Zou G, Jia X, Huang Z, Li S, Liao H, Hou H, Huang L, Ji X. Cube-shaped porous carbon derived from MOF-5 as advanced material for sodium-ion batteries. Electrochim Acta 2016;196:413–21.
[289] Jiang H-L, Liu B, Lan Y-Q, Kuratani K, Akita T, Shioyama H, Zong F, Xu Q. From metal-organic framework to nanoporous carbon: toward a very high surface area and hydrogen uptake. J Am Chem Soc 2011;133:11854–7.
[290] Zhong S, Zhan C, Cao D. Zeolitic imidazolate framework-derived nitrogen doped porous carbons as high performance supercapacitor electrode materials. Carbon 2015;85:51–9.
[291] Zhang P, Sun F, Xiang Z, Shen Z, Yun J, Cao D. ZIF-derived in situ nitrogen-doped porous carbons as efficient metal-free electrocatalysts for oxygen reduction reaction. Energy Environ Sci 2014;7:442–50.
[292] Zhang X, Zhu G, Wang M, Li J, Lu T, Pan L. Covalent-organic-frameworks derived N-doped porous carbon materials as anode for superior long-life cycling lithium and sodium ion batteries. Carbon 2017;116:686–94.
[293] Kuhn P, Forget A, Su DS, Thomas A, Antonietti M. From microporous regular frameworks to mesoporous materials with ultrahigh surface area: dynamic reorganization of porous polymer networks. J Am Chem Soc 2008;130:13333–7.
[294] Yang ZX, Xia YD, Zhu YQ. Preparation and gases storage capacities of N-doped porous activated carbon materials derived from mesoporous polymer. Mater Chem Phys 2013;141:318–23.
[295] Hao L, Luo B, Li XL, Jin MH, Fang Y, Tang ZH, Jia YY, Liang MH, Thomas A, Yang JH, Zhi LJ. Terephthalonitrile-derived nitrogen-rich networks for high performance supercapacitors. Energy Environ Sci 2012;5:9747–51.
[296] Liang YY, Feng XL, Zhi LJ, Kolb U, Mullen K. A simple approach towards one-dimensional mesoporous carbon with superior electrochemical capacitive activity. Chem Commun 2009:809–11.
[297] Pekara RW. Organic aerogels from the polycondensation of resorcinol with formaldehyde. J Mater Sci 1989;24:3221–7.
[298] Pekara RW, Alviso CT, Kong FM, Hulsey SS. Aerogels derived from multifunctional organic monomers. J Non-Cryst Solids 1992;145:90–8.
[299] Hanzawa Y, Kaneko K, Yoshizawa N, Pekala RW. The pore structure determination of carbon aerogels. Adsorption 1998;4:187–96.

[300] Reichenauer G, Emmerling A, Fricke J, Pekala RW. Microporosity in carbon aerogels. J Non-Cryst Solids 1998;225:210–4.
[301] Tamon H, Ishizaka H, Mikami M, Okazaki M. Porous structure of organic and carbon aerogels synthesized by sol-gel polycondensation of resorcinol with formaldehyde. Carbon 1997;35:791–6.
[302] Tamon H, Ishizaka H, Araki T, Okazaki M. Control of mesoporous structure of organic and carbon aerogels. Carbon 1998;36:1257–62.
[303] Tamon H, Ishizuka H. Porous characterization of carbon aerogels. Carbon 1998;36:1397–409.
[304] Wu D, Fu R, Zhang S, Dresselhaus MS, Dresselhaus G. Preparation of low density carbon aerogels by ambient pressure drying. Carbon 2004;42:2033–9.
[305] Tamon H, Ishizaki H, Yamamoto T, Suzuki T. Preparation of mesoporous carbon by freeze drying. Carbon 1999;37:2049–55.
[306] Tamon H, Ishizaka H, Yamamoto T, Suzuki T. Influence of freeze-drying conditions on the mesoporosity of organic gels as carbon precursors. Carbon 2000;38:1099–105.
[307] Yamamoto Y, Nishimura T, Suzuki T, Tamon H. Control of mesoporosity of carbon gels prepared by sol-gel polycondensation and freeze drying. J Non-Cryst Solids 2001;288:46–55.
[308] Yamamoto T, Nishimura T, Suzuki T, Tamon H. Effect of drying conditions on mesoporosity of carbon precursors prepared by sol–gel polycondensation and freeze drying. Carbon 2001;39:2374–6.
[309] Yamamoto T, Nishimura T, Suzuki T, Tamon H. Effect of drying method on mesoporosity of resorcinol-formaldehyde drygel and carbon gel. Drying Tech 2001;19:1319–33.
[310] Job N, Thery A, Pirard R, Marien J, Kocon L, Rouzaud JN, Beguin F, Pirard J-P. Carbon aerogels, cryogels and xerogels: influence of the drying method on the textural properties of porous carbon materials. Carbon 2005;43:2481–94.
[311] Hanzawa Y, Hatori H, Yoshizawa N, Yamada Y. Structural changes in carbon aerogels with high temperature treatment. Carbon 2002;40:575–81.
[312] Hanzawa Y, Kaneko K, Pekala RW, Dresselhaus MS. Activated carbon aerogels. Langmuir 1996;12:6167–9.
[313] Kaneko K, Hanzawa Y, Iiyama T, Kanda T, Suzuki T. Cluster-mediated water adsorption on carbon nanopores. Adsorption 1999;5:7–13.
[314] Bekyarova E, Kaneko K. Structure and physical properties of tailor-made Ce,Zr-doped carbon aerogels. Adv Mater 2000;12:1625–8.
[315] Bekyarova E, Kaneko K. Adsorption of supercritical N_2 and O_2 on pore-controlled carbon aerogels. J Colloid Interface Sci 2001;238:357–61.
[316] Lee JS, Wang XQ, Luo HM, Baker GA, Dai S. Facile ionothermal synthesis of microporous and mesoporous carbons from task specific ionic liquids. J Am Chem Soc 2009;131:4596–7.
[317] Lee JS, Wang X, Luo H, Dai S. Fluidic carbon precursors for formation of functional carbon under ambient pressure based on ionic liquids. Adv Mater 2010;22:1004–7.
[318] Yuan JY, Giordano C, Antonietti M. Ionic liquid monomers and polymers as precursors of highly conductive, mesoporous, graphitic carbon nanostructures. Chem Mater 2010;22:5003–12.
[319] Paraknowitsch JP, Thomas A, Antonietti M. A detailed view on the polycondensation of ionic liquid monomers towards nitrogen doped carbon materials. J Mater Chem 2010;20:6746–58.
[320] Paraknowitsch JP, Zhang J, Su D, Thomas A, Antonietti M. Ionic liquids as precursors for nitrogen- doped graphitic carbon. Adv Mater 2010;22:87–92.

[321] Elumeeva K, Fechler N, Fellinger TP, Antonietti M. Metal-free ionic liquid-derived electrocatalyst for high-performance oxygen reduction in acidic and alkaline electrolytes. Mater Horiz 2014;1:588−94.
[322] Yang W, Fellinger TP, Antonietti M. Efficient metal-free oxygen reduction in alkaline medium on high-surface-area mesoporous nitrogen-doped carbons made from ionic liquids and nucleobases. J Am Chem Soc 2011;133:206−9.
[323] Hines D, Bagreev A, Bandosz TJ. Surface properties of porous carbon obtained from polystyrene sulfonic acid-based organic salts. Langmuir 2004;20(8):3388−97.
[324] Inagaki M, Qiu J, Guo Q. Carbon foam: preparation and application. Carbon 2015;87:128−52.
[325] Young JR, Suh NM. Processing of microcellular polyester composites. Polym Compos 1985;6(3):175−80.
[326] Colton JS, Suh NP. The nucleation of microcellular thermoplastic foam with additives: Part II: experimental results and discussion. Polym Eng Sci 1987;27(7):493−8.
[327] Ramesh NS, Rasmussen DH, Campbell GA. Numerical and experimental studies of bubble growth during the microcellular foaming process. Polym Eng Sci 1991;31(23):1657−65.
[328] Klett J, Hardy R, Romine E, Walls C, Burchell T. High-thermal conductivity, mesophase-pitch- derived carbon foams: effect of precursor on structure and properties. Carbon 2000;38:953−73.
[329] Klett J, McMillan AD, Gallego NC, Burchell TD, Walls CA. Effects of heat treatment conditions on the thermal properties of mesophase pitch-derived graphite foams. Carbon 2004;42:1849−52.
[330] Min Z, Cao M, Zhang S, Wang X, Wang Y. Effect of precursor on the pore structure of carbon foams. New Carbon Mater 2007;22:75−9.
[331] Calvo M, Garcia R, Arenillas A, Suarez I, Moinelo SR. Carbon foams from coals. A preliminary study. Fuel 2005;84:2184−9.
[332] Calvo M, García R, Moinelo SR. Carbon foams from different coals. Energy Fuel 2008;22:3376−83.
[333] Li J, Wang C, Zhan L, Qiao W-M, Liang XY, Ling L-C. Carbon foams prepared by supercritical foaming method. Carbon 2009;47:1204−6.
[334] Chen C, Kennel EB, Stiller AH, Stansberry PG, Zondlo JW. Carbon foam derived from various precursors. Carbon 2006;44:1535−43.
[335] Li S, Guo Q, Song Y, Liu Z, Shi J, Liu L, Yan X. Carbon foams with high compressive strength derived from mesophase pitch treated by toluene extraction. Carbon 2007;45(14):2843−5.
[336] Wang M, Wang C-Y, Li T-Q, Hu Z-J. Preparation of mesophase-pitch-based carbon foams at low pressures. Carbon 2008;46:84−91.
[337] Tsyntsarski B, Petrova B, Budinova T, Petrov N, Krzesinska M, Pusz S, Majewska J, Tzvetkov P. Carbon foam derived from pitchers modified with mineral acids by a low pressure foaming process. Carbon 2010;48:3523−30.
[338] Tsyntsarski B, Petrova B, Budinova T, Petrov N, Velasco L, Parra J, Ania C. Porosity development during steam activation of carbon foams from chemically modified pitch. Microp Mesop Mater 2012;154:56−60.
[339] Eksilioglu A, Gencay N, Yardim MF, Ekinci E. Mesophase AR pitch derived carbon foam: effect of temperature, pressure and pressure release time. J Mater Sci 2006;41(10):2743−8.
[340] Baran D, Yardim MF, Atakül H, Ekinci E. Synthesis of carbon foam with high compressive strength from an asphaltene pitch. New Carbon Mater 2013;28:127−33.
[341] Ge M, Shen ZM, Chi WD, Liu H. Anisotropy of mesophase pitch-derived carbon foams. Carbon 2007;45:141−5.

[342] Zhu J, Wang X, Guo L, Wang YI, Wang YA, Yu M, Lau K. A graphite foam reinforced by graphite particles. Carbon 2007;45:2547–50.
[343] Li SZ, Song YZ, Song Y, Shi JL, Liu L, Wei XH, Guo QG. Carbon foams with high compressive strength derived from mixtures of mesocarbon microbeads and mesophase pitch. Carbon 2007;45:2092–7.
[344] Wang XY, Zhong JM, Wang YM, Yu MF. A study of the properties of carbon foams reinforced by clay. Carbon 2006;44:1560–4.
[345] Li S, Guo Q, Song Y, Shi J, Liu L. Effects of pitch fluoride on the thermal conductivity of carbon foam derived from mesophase pitch. Carbon 2010;48:1312–20.
[346] Fawcett W, Shetty DK. Effects of carbon nanofibers on cell morphology, thermal conductivity and crush strength of carbon foam. Carbon 2010;48:68–80.
[347] Beechem T, Lafdi K. Novel high strength graphitic foams. Carbon 2006;44:1548–59.
[348] Wang MX, Wang CY, Li YL, Zhang C. The use of optical microscopy to detect the bubble shape of carbon foams. Carbon 2007;45:687–9.
[349] Fethollahi B, Zimmer J. Microstructure of mesophase-based carbon foam. Carbon 2007;45:3057–9.
[350] Li S, Tian Y, Zhong Y, Yan X, Song Y, Guo Q, Shi J, Liu L. Formation mechanism of carbon foams derived from mesophase pitch. Carbon 2011;49:618–24.
[351] Rosebrock G, Elgafy A, Beechem T, Lafdi K. Study of the growth and motion of graphitic bubbles. Carbon 2005;43:3075–87.
[352] Beechem T, Lafdi K, Elgafy A. Bubble growth mechanism in carbon foams. Carbon 2005;43:1055–64.
[353] Lei S, Guo Q, Shi J, Liu L. Preparation of phenolic-based carbon foam with controllable pore structure and high compressive strength. Carbon 2010;48:2644–6.
[354] Mehta R, Anderson DP, Hager JW. Graphitic open-celled carbon foams: processing and characterization. Carbon 2003;41(11):2174–6.
[355] Xiao N, Zhou Y, Qiu J, Wang Z. Preparation of carbon nanofibers/carbon foam monolithic composite from coal liquefaction residue. Fuel 2010;89(5):1169–71.
[356] Calvo M, Arenillas A, García R, Moinelo SR. Growth of carbon nanofilaments on coal foams. Fuel 2009;88(1):46–53.
[357] Yang J, Yang G, Yu D, Wang X, Zhao B, Zhang L, Du P, Zhang X. Carbon foams from polyacrylonitrile-borneol films prepared using coaxial electrohydrodynamic atomization. Carbon 2013;53:231–6.
[358] Liu M, Gan L, Zhao F, Fan X, Xu H, Wu F, Xu Z, Hao Z, Chen L. Carbon foams with high compressive strength derived from polyarylacetylene resin. Carbon 2007;45:3055–7.
[359] Zhang S, Liu M, Gan L, Wu F, Xu Z, Hao Z, Chen L. Synthesis of carbon foams with a high compressive strength from arylacetylene. New Carbon Mater 2010;25:9–14.
[360] Wu X, Liu Y, Fang M, Mei L, Luo B. Preparation and characterization of carbon foams derived from alum inosilicate and phenolic resin. Carbon 2011;49:1782–6.
[361] Meikleham N, Pizzi A. Acid- and alkali-catalyzed tannin-based rigid foams. J Appl Polym Sci 1994;53:1547–56.
[362] Pizzi A, Tondi G, Pasch H, Celzard A. Matrix-assisted laser desorption/ionization time-of-flight structure determination of complex thermoset networks: polyflavonoid tannin-furanic rigid foams. J Appl Polym Sci 2008;110:1451–6.
[363] Tondi G, Pizzi A, Pasch H, Celzard A. Structure degradation, conservation and rearrangement in the carbonisation of polyflavonoid tannin/furanic rigid foams - a MALDI-TOF investigation. Polym Degrad Stab 2008;93(5):968–75.
[364] Tondi G, Fierro V, Pizzi A, Celzard A. Tannin-based carbon foam. Carbon 2009;47:1480–92. Erratum: Carbon 2009; 47: 2761.

[365] Zhao W, Pizzi A, Fierro V, Du G, Celzard A. Effect of composition and processing parameters on the characteristics of tannin-based rigid foams. Part I: cell structure. Mater Chem Phys 2010;122(1):175–82.
[366] Li X, Basso MC, Braghiroli FL, Fierro V, Pizzi A, Celzard A. Tailoring the structure of cellular vitreous carbon foams. Carbon 2012;50(5):2026–36.
[367] Celzard A, Zhao W, Pizzi A, Fierro V. Mechanical properties of tannin-based rigid foams undergoing compression. Mater Sci Eng A 2010;527(16–17):4438–46.
[368] Zhao W, Fierro V, Pizzi A, Du G, Celzard A. Effect of composition and processing parameters on the characteristics of tannin-based rigid foams. Part II: physical properties. Mater Chem Phys 2010;123(1):210–7.
[369] Zhao W, Fierro V, Pizzi A, Celzard A. Bimodal cellular activated carbons derived from tannins. J Mater Sci 2010;45(21):5778–85.
[370] Narasimman R, Prabhakaran K. Preparation of low density carbon foams by foaming molten sucrose using an aluminium nitrate blowing agent. Carbon 2012;50(5):1999–2009.
[371] Li TQ, Wang CY, An BX, Wang H. Preparation of graphitic carbon foam using size-restriction method under atmospheric pressure. Carbon 2005;43:2030–2.
[372] Jana P, Ganesan V. Synthesis, characterization and radionuclide (^{137}Cs) trapping properties of a carbon foam. Carbon 2009;47(13):3001–9.
[373] Prabhakaran K, Singh P, Gokhale N, Sharma S. Processing of sucrose to low density carbon foams. J Mater Sci 2007;42(11):3894–900.
[374] Wang XB, Zhang YJ, Zhi CY, Wang X, Tang DM, Xu YB, Weng QH, Jiang XF, Mitome M, Golberg D, Bando Y. Three-dimensional strutted graphene grown by substrate-free sugar blowing for high- power-density supercapacitors. Nat Commun 2013;4:2905.
[375] Narasimman R, Prabhakaran K. Preparation of carbon foams by thermo-foaming of activated carbon powder dispersions in an aqueous sucrose resin. Carbon 2012;50:5583–93.
[376] Narasimman R, Prabhakaran K. Preparation of carbon foams with enhanced oxidation resistance by foaming molten sucrose using a boric acid blowing agent. Carbon 2013;55:305–12.
[377] Lin Q, Luo B, Qu L, Fang C, Chen Z. Direct preparation of carbon foam by pyrolysis of cyanate ester resin at ambient pressure. J Anal Appl Pyrolysis 2013;104:714–7.
[378] Pekala R, Hopper RW. Low-density microcellular carbon foams. J Mater Sci 1987;22:1840–4.
[379] Wang J. Reticulated vitreous carbon - a new versatile electrode material. Electrochim Acta 1981;26:1721–6.
[380] Chakovskoi AG, Hunt CE, Forceberg G, Nillson T, Persson P. Reticulated vitreous carbon field emission cathodes for light source applications. J Vac Sci Tech B 2003;21:571–5.
[381] Friedrich JM, Ponee-de-Leon C, Reade GW, Walsh FC. Reticulated vitreous carbon as an electrode material. J Electroanal Chem 2004;561:203–17.
[382] Harikrishnan G, Patro TU, Khakhar DV. Reticulated vitreous carbon from polyurethane foam–clay composites. Carbon 2007;45:531–5.
[383] Jana P, Fierro V, Celzard A. Ultralow cost reticulated carbon foams from household cleaning pad wastes. Carbon 2013;62:517–20.
[384] Kodama M, Yamashita J, Soneda Y, Hatori H, Kamegawa K. Preparation and electrochemical characteristics of N-enriched carbon foam. Carbon 2007;45:1105–7.
[385] Inagaki M, Morishita T, Kuno A, Kito T, Hirano M, Suwa T, Kusakawa K. Carbon foams prepared from polyimide using urethane foam template. Carbon 2004;42:497–502.

[386] Ohta N, Nishi Y, Morishita T, Ieko Y, Ito A, Inagaki M. Preparation of microporous carbon foams for adsorption/desorption of water vapor in ambient air. New Carbon Mater 2008;23:216–20.
[387] Xue C, Tu B, Zhao D. Facile fabrication of hierarchically porous carbonaceous monoliths with ordered mesostructured via organic–organic self-assembly. Nano Res 2009;2:242–53.
[388] Karthik M, Faik A, Doppiu S, Roddatis V, D'Aguanno B. A simple approach for fabrication of interconnected graphitized macroporous carbon foam with uniform mesopore walls by using hydrothermal method. Carbon 2015;87:434–43.
[389] Xue C, Tu B, Zhao D. Evaporation-induced coating and self-assembly of ordered mesoporous carbon- silica composite monoliths with macroporous architecture on polyurethane foams. Adv Funct Mater 2008;18:3914–21.
[390] Chen Y, Chen B, Shi X, Xu H, Hu Y, Yuan Y, Shen N-B. Preparation of pitch-based carbon foam using polyurethane foam template. Carbon 2007;45:2132–4.
[391] Yadav A, Kumar R, Bhatia G, Verma GL. Development of mesophase pitch derived high thermal conductivity graphite foam using a template method. Carbon 2011;49(11):3622–30.
[392] Nishihara H, Mukai SR, Tamon H. Preparation of resorcinol-formaldehyde carbon cryogel microhoneycombs. Carbon 2004;42:899–901.
[393] Mukai SR, Nishihara H, Yoshida T, Taniguchi K, Tamon H. Morphology of resorcinol-formaldehyde gels obtained through ice-templating. Carbon 2005;43:1563–5.
[394] Seo J, Park H, Shin K, Baeck SH, Rhym Y, Shim SE. Lignin-derived macroporous carbon foams prepared by using poly(methyl methacrylate) particles as the template. Carbon 2014;76:357–67.
[395] Li G, Yi Q, Yang X, Chen Y, Zhou X, Xie G. Ni-Co-N doped honeycomb carbon nano-composites as cathodic catalysts of membrane-less direct alcohol fuel cell. Carbon 2018;140:557–68.
[396] Tang J, Wang T, Salunkhe RR, Alshehri SM, Malgras V, Yamauchi Y. Three-dimensional nitrogen- doped hierarchical porous carbon as an electrode for high-performance supercapacitors. Chem Eur J 2015;21(48):17293–8.
[397] Liang HW, Guan QF, Chen LF, Zhu Z, Zhang WJ, Yu SH. Macroscopic-scale template synthesis of robust carbonaceous nanofiber hydrogels and aerogels and their applications. Angew Chem Int Ed 2012;51(21):5101–5.
[398] Duan S, Wu X, Zeng K, Tao T, Huang Z, Fang M, Liu Y, Min X. Simple routes from natural graphite to graphite foams: preparation, structure and properties. Carbon 2020;159:527–41.
[399] Takeichi T, Yamazaki Y, Ito A, Zuo M. Preparation and properties of porous polyimide films prepared by the pyrolysis of poly(urethane-imide) films. J Photopolym Sci Technol 1999;12:203–8.
[400] Takeichi T, Yamazaki Y, Fukui T, Matsumoto A, Inagaki M. Preparation and characterization of porous carbonized films by pyrolysis of poly(urethane-imide) films. TANSO 2000;195:388–94 [in Japanese].
[401] Takeichi T, Yamazaki Y, Zuo M, Ito A, Matsumoto A, Inagaki M. Preparation of porous carbon films by the pyrolysis of poly(urethane-imide) films and their pore characteristics. Carbon 2001;39:257–65.
[402] Wei W, Hu HQ, Qin GT, You LB, Chen GH. Pore structure control of phenol-formaldehyde based carbon microfiltration membranes. Carbon 2004;42:679–81.
[403] Lukens WW, Stucky GD. Synthesis of mesoporous carbon foams templated by organic colloids. Chem Mater 2002;14:1665–70.
[404] Chung DDL. Exfoliation of graphite. J Mater Sci 1987;22:4190–8.

[405] Furdin G. Exfoliation process and elaboration of new carbonaceous materials. Fuel 1998;77:479−85.
[406] Inagaki M, Toyoda M, Kang F. Exfoliation of graphite via intercalation compounds. Chem Phys Carbon 2004;29:1−69.
[407] Chung DDL. A review of exfoliated graphite. J Mater Sci 2016;51:554−68.
[408] Hummers W, Offeman R. Preparation of graphitic oxide. J Am Chem Soc 1958;80:1339.
[409] Staudenmaier L. Verfahren zur Darstellung der Graphitsäur. Ber Dtsch Chem Ges 1898;31:1481−7.
[410] Brodie BC. Surle poids atomique du graphite. Ann Chim Phys 1860;59:466−72.
[411] Hofmann U, Konig E. Untersuchungen über Graphitoxyd. Z Anorg Allg Chem 1937;234:311−36.
[412] Inagaki M, Saji N, Zheng Y-P, Kang F, Toyoda M. Pore development during exfoliation of natural graphite. TANSO 2004;215:258−64.
[413] Inagaki M, Toyoda M, Kang F, Zheng Y, Shen W. Pore structure of exfoliated graphite. New Carbon Mater 2003;18:241−9.
[414] Inagaki M, Tashiro R, Washino Y, Toyoda M. Exfoliation process of graphite via intercalation compounds with sulfuric acid. J Phys Chem Solids 2004;65:133−7.
[415] Kang F, Zheng Y-P, Wang H-N, Nishi Y, Inagaki M. Effect of preparation conditions on the characteristics of exfoliated graphite. Carbon 2002;40:1575−81.
[416] Zheng YP, Wang HN, Kang FY, Wang L-N, Inagaki M. Sorption capacity of exfoliated graphite for oil-sorption in and among worm-like particles. Carbon 2004;42:2603−7.
[417] Inagaki M, Tashiro R, Toyoda M, Zheng YP, Kang FY. Pore structure of exfoliated graphite prepared from residue compounds with sulfuric acid. J Ceram Soc Japan 2004;112:S1513−6.
[418] Ivanov AV, Maksimova NV, Kamaev AO, Malakho AP, Avdeev VV. Influence of intercalation and exfoliation conditions on macrostructure and microstructure of exfoliated graphite. Mater Lett 2018;228:403−6.
[419] Geng XM, Guo YF, Li DF, Li WW, Zhu C, Wei XF, Chen ML, Gao S, Qiu SQ, Gong YP, Wu LQ, Long MS, Sun MT, Pan GB, Liu LW. Interlayer catalytic exfoliation realizing scalable production of large-size pristine few-layer graphene. Sci Rep 2013;3:6.
[420] Lin S, Dong L, Zhang JJ, Lu HB. Room-temperature intercalation and similar to ~1000-fold chemical expansion for scalable preparation of high-quality graphene. Chem. Mat 2016;28:2138−46.
[421] Liu T, Zhang R, Zhang X, Liu K, Liu Y, Yan P. One-step room-temperature preparation of expanded graphite. Carbon 2017;119:544−7.
[422] Peng T, Liu B, Gao X, Luo L, Sun H. Preparation, quantitative surface analysis, intercalation characteristics and industrial implications of low temperature expandable graphite. Appl Surf Sci 2018;444:800−10.
[423] Hou S, He S, Zhu T, Li J, Ma L, Du H, Shen W, Kang F, Huang Z-H. Environment-friendly preparation of exfoliated graphite and functional graphite sheets. J Materiomics 2021;7:136−45.
[424] Inagaki M, Kobayashi N, Tryba B. Pore structure inside the particles of exfoliated graphite prepared by microwave orradiation. TANSO 2004;215:249−51.
[425] Tryba B, Morawski AW, Inagaki M. Preparation of exfoliated graphite by microwave irradiation. Carbon 2005;43:2397−429.
[426] Toyoda M, Aizawa J, Inagaki M. Sorption and recovery of heavy oil by using exfoliated graphite. Desalination 1998;115:199.
[427] Toyoda M, Inagaki M. Heavy oil sorption using exfoliated graphite: new application of exfoliated graphite to protect heavy oil pollution. Carbon 2000;38:199−210.

[428] Inagaki M, Toyoda M, Iwashita N, Nishi Y, Konno H. Exfoliated graphite for spilled heavy oil recovery. Carbon Sci 2001;2:1–8.
[429] Inagaki M, Toyoda M, Iwashita N, Nishi Y, Konno H, Fujita J, Kihara T. Sorption, recovery and recycle of spilled heavy oils using carbon materials. TANSO 2002;201:16–25.
[430] Shen WE, Wen SZ, Cao NZ, Zheng L, Zhou W, Liu Y, Gu J. Expanded graphite A new kind of biomedical material. Carbon 1999;37:356–8.
[431] Kang F, Zheng Y, Zhao H, Wang H, Wang L, Shen W, Inagaki M. Sorption of heavy oils and biomedical liquids into exfoliated graphite – research in China. New Carbon Mater 2003;18:161–73.
[432] He J, Song L, Yang H, Ren X, Xing L. Preparation of sulfur-free exfoliated graphite by a two-step intercalation process and its application for adsorption of oils. J Chem 2017:5824976.
[433] Xu YX, Sheng KX, Li C, Shi GQ. Self-assembled graphene hydrogel via a one-step hydrothermal process. ACS Nano 2010;4:4324–30.
[434] Zhao J, Ren W, Cheng HM. Graphene sponge for efficient and repeatable adsorption and desorption of water contaminations. J Mater Chem 2012;22:20197–202.
[435] Han Z, Tang ZH, Li P, Yang GZ, Zheng QB, Yang JH. Ammonia solution strengthened three- dimensional macro-porous graphene aerogel. Nanoscale 2013;5:5462–7.
[436] Zhao Y, Hu CG, Hu Y, Cheng HH, Shi GQ, Qu LT. A versatile, ultralight, nitrogen-doped graphene framework. Angew Chem Int Ed 2012;51:11371–5.
[437] Sun HY, Xu Z, Gao C. Multifunctional, ultra-flyweight, synergistically assembled carbon aerogels. Adv Mater 2013;25:2554–60.
[438] Tao Y, Xie X, Lv W, Tang D, Kong D, Huang Z, Nishihara H, Ishii T, Li B, Golberg D, Kang F, Kyotani T, Yang Q-H. Towards ultrahigh volumetric capacitance: graphene derived highly dense but porous carbons for supercapacitors. Sci Rep 2013;3:2975.
[439] Bi H, Yin K, Xie X, Zhou Y, Wan N, Xu F, Banhart F, Sun L, Ruoff RS. Low temperature casting of graphene with high compressive strength. Adv Mater 2012;24:5124–9.
[440] Tang ZH, Shen SL, Zhuang J, Wang X. Noble-metal-promoted three-dimensional macroassembly of single-layered graphene oxide. Angew Chem Int Ed 2010;49:4603–7.
[441] Tao Y, Kong DB, Zhang C, Lv W, Wang MX, Li BH, Huang ZH, Kang FY, Yang QH. Monolithic carbons with spheroidal and hierarchical pores produced by the linkage of functionalized graphene sheets. Carbon 2014;69:169–77.
[442] Lv W, Zhang C, Li Z, Yang Q-H. Self-assembled 3D graphene monolith from solution. J Phys Chem Lett 2015;6:658–68.
[443] Qian Y, Ismail IM, Stein A. Ultralight, high-surface-area, multifunctional graphene-based aerogels from self-assembly of graphene oxide and resol. Carbon 2014;68:221–31.
[444] Xu X, Pan L, Liu Y, Lu T, Sun Z, Chua DHC. Facile synthesis of novel graphene sponge for high performance capacitive deionization. Sci Rep 2015;5:8458.
[445] Yan LF, Chen WF. In situ self-assembly of mild chemical reduction graphene for three-dimensional architectures. Nanoscale 2011;3:3132–7.
[446] Zhang XT, Sui ZY, Xu B, Yue SF, Luo YJ, Zhan WC, Liu B. Mechanically strong and highly conductive graphene aerogel and its use as electrodes for electrochemical power sources. J Mater Chem 2011;21:6494–7.
[447] Pham HD, Pham VH, Cuong TV, Nguyen-Phan TD, Chung JS, Shin EW, Kim S. Synthesis of the chemically converted graphene xerogel with superior electrical conductivity. Chem Commun 2011;47:9672–4.

[448] Qiu L, Liu JZ, Chang SLY, Wu YZ, Li D. Biomimetic superelastic graphene-based cellular monoliths. Nat Commun 2012;3:1241.
[449] Barg S, Perez FM, Ni N, do V, Pereira P, Maher RC, Garcia-Tuñon E, Eslava S, Agnoli S, Mattevi C, Saiz E. Mesoscale assembly of chemically modified graphene into complex cellular networks. Nat Commun 2014;5:4328.
[450] Choi BG, Yang M, Hong WH, Choi JW, Huh YS. 3D macroporous graphene frameworks for supercapacitors with high energy and power densities. ACS Nano 2012;6:4020−8.
[451] Wang H, Zhang D, Yan T, Wen X, Zhang J, Shi L, Zhong Q. Three-dimensional macroporous graphene architectures as high performance electrodes for capacitive deionization. J Mater Chem A 2013;1:11778−89.
[452] Lee SH, Kim HW, Hwang JO, Lee WJ, Kwon J, Bielawski CW, Ruoff RS, Kim SO. Three- dimensional self-assembly of graphene oxide platelets into mechanically flexible macroporous carbon films. Angew Chem Int Ed 2010;49:10084−8.
[453] Kim E, Kim H, Park BJ, Han YH, Park JH, Cho J, Lee S-S, Son JG. Etching-assisted crumpled graphene wrapped spiky iron oxide particles for high-performance Li-ion hybrid supercapacitor. Small 2018;14:1704209.
[454] Liu F, Seo TS. A controllable self-assembly method for large-scale synthesis of graphene sponges and free-standing graphene films. Adv Funct Mater 2010;20:1930−6.
[455] Hu H, Zhao Z, Wan W, Gogotsi Y, Qiu J. Ultralight and highly compressible graphene aerogels. Adv Mater 2013;25:2219−23.
[456] Worsley MA, Pauzauskie PJ, Olson TY, Biener J, Satcher Jr JH, Baumann TF. Synthesis of graphene aerogel with high electrical conductivity. J Am Chem Soc 2010;132:14067−9.
[457] Worsley MA, Olson TY, Lee JR, Willey TM, Nielsen MH, Roberts SK, Pauzauskie PJ, Biener J, Satcher Jr JH, Baumann TF. High surface area, sp^2-cross-linked three-dimensional graphene monoliths. J Phys Chem Lett 2011;2:921−5.
[458] Worsley MA, Kucheyev SO, Mason HE, Merrill MD, Mayer BP, Lewicki J, Valdez CA, Suss ME, Stadermann M, Pauzauskie PJ, Satcher JH, Biener J, Baumann TF. Mechanically robust 3D graphene macroassembly with high surface area. Chem Commun 2012;48:8428−30.
[459] Lewis JA. Direct ink writing of 3D functional materials. Adv Funct Mater 2006;16:2193−204.
[460] García-Tuñon E, Barg S, Franco J, Bell R, Eslava S, D'Elia E, Maher RC, Guitian F, Saiz E. Printing in three dimensions with graphene. Adv Mater 2015;27:1688−93.
[461] Zhu C, Han TYJ, Duoss EB, Golobic AM, Kuntz JD, Spadaccini CM, Worsley MA. Highly compressible 3D periodic graphene aerogel microlattices. Nat Commun 2015;6:6962.
[462] Zhu C, Liu T, Qian F, Han TY, Duoss EB, Kuntz JD, Spadaccini CM, Worsley MA, Li Y. Supercapacitors based on three-dimensional hierarchical graphene aerogels with periodic macropores. Nano Lett 2016;16(6):3448−56.
[463] Kim JH, Chang WS, Kim D, Yang JR, Han JT, Lee GW, Kim JT, Seol SK. 3D printing of reduced graphene oxide nanowires. Adv Mater 2015;27:157−61.
[464] Chen Z, Ren W, Gao L, Liu B, Pei S, Cheng HM. Three-dimensional flexible and conductive interconnected graphene networks grown by chemical vapour deposition. Nat Mater 2011;10:424−8.
[465] Cao XH, Shi YM, Shi WH, Lu G, Huang X, Yan QY, Zhang QC, Zhang H. Preparation of novel 3D graphene networks for supercapacitor applications. Small 2011;7:3163−8.
[466] Zhang Q, Wang Y, Zhang B, Zhao K, He P, Huang B. 3D superelastic graphene aerogel-nanosheet hybrid hierarchical nanostructures as high-performance supercapacitor electrodes. Carbon 2018;127:449−58.

[467] Mecklenburg M, Schuchardt A, Mishra YK, Kaps S, Adelung R, Lotnyk A, Kienle L, Schult K, Aerographite. Ultra lowightweight, flexible nanowall, carbon microtube material with outstanding mechanical performance. Adv Mater 2012;24:3486−90.
[468] Long S, Feng Y, He F, He S, Hong H, Yang X, Zheng L, Liu J, Gan L, Long M. An ultralight, supercompressible, superhydrophobic and multifunctional carbon aerogel with a specially designed structure. Carbon 2020;158:137−45.
[469] Li C, Shi GQ. Three-dimensional graphene architectures. Nanoscale 2012;4:5549−63.
[470] Wu Y, Zhu J, Huang L. A review of three-dimensional graphene-based materials: synthesis and applications to energy conversion/storage and environment. Carbon 2019;143:610−40.
[471] Cao A, Dickell PL, Sawer WG, Ghamesi-Nejhad MN, Ajayan PM. Super-compressible foamlike carbon nanotube films. Science 2005;310:1307−10.
[472] Minami N, Kim Y, Miyashita K, Kazaoui S, Nalini B. Cellulose derivatives as excellent dispersants for single-wall carbon nanotubes as demonstrated by adsorption and photoluminescence spectroscopy. Appl Phys Lett 2006;88:093123.
[473] Leroy CM, Carn F, Backov R, Trinquecoste M, Delhaes P. Multiwalled-carbon-nanotube-based carbon foams. Carbon 2007;45:2307−20.
[474] Cortes-Lopez AJ, Munoz-Sandova E, Lopez-Urí F. Efficient carbon nanotube sponges production boosted by acetone in CVD-synthesis. Carbon 2018;135:145−56.
[475] Rios RVRA, Martinez-Escandell M, Molina-Sabio M, Rodriguez-Reinoso F. Carbon foam prepared by pyrolysis of olive stones under steam. Carbon 2006;44:1448−54.
[476] Chen S, Liu Q, He G, Zhou Y, Hanif M, Peng W, Wang S, Hou H. Reticulated carbon foam derived from a sponge-like natural product as a high performance anode in microbial fuel cells. J Mater Chem 2012;22:18609−13.
[477] Liu M, Gan L, Zhao F, Xu H, Fan X, Tian C, Wang X, Xu Z, Hao Z, Chen L. Carbon foams prepared by an oil-in-water emulsion method. Carbon 2007;45:2710−2.
[478] Gross AF, Nowak AP. Hierarchical carbon foams with independently tunable mesopore and macropore size distributions. Langmuir 2010;26:11378−83.
[479] Mulik S, Sotiriou-Leventis C, Leventis N. Macroporous electrically conductive carbon networks by pyrolysis of isocyanate-cross-liked resolcinol-formaldehyde aerogels. Chem Mater 2008;20:6985−97.
[480] Benton ST, Schmitt CR. Preparation of syntactic carbon foams. Carbon 1972;12:185−90.
[481] Nicholson J, Thomas CR. Syntactic carbon foams. Carbon 1973;11:65−6.
[482] Bruneton E, Tallaron C, Gras-Naulin N, Cosculluela A. Evolution of the structure and mechanical behaviour of a carbon foam at very high temperatures. Carbon 2002;40:1919−27.
[484] Zhang L, Ma J. Processing and characterization of synthetic carbon foams containing hollow carbon microspheres. Carbon 2009;47:1451−6.
[485] Kaburagi Y, Aoki H, Yoshida A. Highly oriented porous graphite film prepared from porous aromatic polyimide film. TANSO 2012;253:95−9.
[486] Crittenden B, Patton A, Jouin C, Perera S, Tennison S, Echevarris JAB. Carbon monoliths: a comparison with granular materials. Adsorption 2005;11:537−41.
[487] Zani A, Dellasega D, Russo V, Passoni M. Ultra-low density carbon foams produced by pulsed laser deposition. Carbon 2013;56:358−65.
[488] Fu R, Li Z, Liang Y, Li F, Xu F, Wu D. Hierarchical porous carbons: design, preparation, and performance in energy storage. New Carbon Mater 2011;26:171−9.

[489] Dutta S, Bhaumik A, Wu KC-W. Hierarchically porous carbon derived from polymers and biomass: effect of interconnected pores on energy applications. Energy Environ Sci 2014;7:3574–92.
[490] Chou T-C, Huang C-H, Doong R-A, Hu C-C. Architectural design of hierarchically ordered porous carbons for high-rate electrochemical capacitors. J Mater Chem A 2013;1(8):2886–95.
[491] Wang Z, Stein A. Morphology control of carbon, silica, and carbon/silica nanocomposites: from 3D ordered macro-/mesoporous monoliths to shaped mesoporous particles. Chem Mater 2008;20:1029–40.
[492] Wang Z, Kiesel ER, Stein A. Silica-free syntheses of hierarchically ordered macroporous polymer and carbon monoliths with controllable mesoporosity. J Mater Chem 2008;18:2194–200.
[493] Chen J, Xu J, Zhou S, Zhao N, Wong C-P. Nitrogen-doped hierarchically porous carbon foam: a free- standing electrode and mechanical support for high-performance supercapacitors. Nano Energy 2016;25:193–202.
[494] Kubo S, White RJ, Tauer K, Titirici MM. Flexible coral-like carbon nanoarchitectures via a dual block copolymer-latex templating approach. Chem Mater 2013;25:4781–90.
[495] Maruyama J, Maruyama S, Fukuhara T, Chashiro K, Uyama H. Ordered mesoporous structure by graphitized carbon nanowall assembly. Carbon 2018;126:452–5.
[496] Deng Y, Liu C, Yu T, Liu F, Zhang F, Wan Y, Zhang L, Wang C, Tu B, Webley PA, Wang H, Zhao D. Facile synthesis of hierarchically porous carbons from dual colloidal crystal/block copolymer template approach. Chem Mater 2007;19:3271–7.
[497] Yang M, Wang G. Synthesis of hierarchical porous carbon particles by hollow polymer microsphere template. Colloid Surf A 2009;345:121–6.
[498] Gorka J, Jaroniec M. Hierarchically porous phenolic resin-based carbons obtained by block copolymer-colloidal silica templating and post-synthesis activation with carbon dioxide and water vapor. Carbon 2011;49:154–60.
[499] Gu D, Bongard H, Deng Y, Feng D, Wu Z, Feng Y, Mao J, Tu B, Schueth F, Zhao D. An aqueous emulsion route to synthesize mesoporous carbon vesicles and their nanocomposites. Adv Mater 2010;22:833–7.
[500] Zhang F, Liu T, Li M, Yu M, Luo Y, Tong Y, Li Y. Multiscale pore network boosts capacitance of carbon electrodes for ultrafast charging. Nano Lett 2017;17(5):3097–104.
[501] Tao G, Zhang L, Hua Z, Chen Y, Guo L, Zhang J, Shu Z, Gao J, Chen H, Wu W, Liu Z, Shi J. Highly efficient adsorbents based on hierarchically macro/mesoporous carbon monoliths with strong hydrophobicity. Carbon 2014;66:547–59.
[502] Le Hua Z, Gao JH, Bu WB, Zhang LX, Chen HR, Shi JL. Hierarchically macro/mesoporous silica monoliths constructed with interconnecting micrometer-sized unit rods. J Sol Gel Sci Technol 2009;50(1):22–7.
[503] Lu A-H, Li W-C, Schmidt W, Schüth F. Fabrication of hierarchically structured carbon monoliths via self-binding and salt templating. Microp Mesop Mater 2006;95:187–92.
[504] Wang Q, Yan J, Wang Y, Wei T, Zhang M, Jing X, Fan Z. Three-dimensional flower-like and hierarchical porous carbon materials as high-rate performance electrodes for supercapacitors. Carbon 2014;67:119–27.
[505] Wu Z, Sun Y, Tan Y, Yang S, Feng X, Muellen K. Three-dimensional graphene-based macro- and mesoporous frameworks for high-performance electrochemical capacitive energy storage. J Am Chem Soc 2012;134(48):19532–5.

[506] Sun G, Li B, Ran J, Shen X, Tong H. Three-dimensional hierarchical porous carbon/graphene composites derived from graphene oxide-chitosan hydrogels for high performance supercapacitors. Electrochim Acta 2015;171:13–22.
[507] Zhou X, Bai Z, Wu M, Qiao J, Chen Z. 3-Dimensional porous N-doped graphene foam as a non- precious catalyst for the oxygen reduction reaction. J Mater Chem A 2015;3:3343–50.
[508] Balach J, Tamnorini L, Sapag K, Acevedo DF, Barbero CA. Facile preparation of hierarchical porous carbons with tailored pore size obtained using a cationic polyelectrolyte as a soft template. Colloid Surf A 2012;415:343–8.
[509] Qie L, Chen W, Xu H, Xiong X, Jiang Y, Zou F, Hu X, Xin Y, Zhang Z, Huang Y. Synthesis of functionalized 3D hierarchical porous carbon for high-performance supercapacitors. Energy Environ Sci 2013;6:2497–504.
[510] Zhang F, Liu T, Hou G, Kou T, Yue L, Guan R, Li Y. Hierarchically porous carbon foams for electric double layer capacitors. Nano Res 2016;9(10):2875–88.
[511] Zhang W, Lin H, Lin Z, Yin J, Lu H, Liu D, Zhao M. 3D hierarchical porous carbon for supercapacitors prepared from lignin through a facile template free method. ChemSusChem 2015;8(12):2114–22.
[512] Huang C-H, Doong R-A. Sugarcane bagasse as the scaffold for mass production of hierarchically porous carbon monoliths by surface self-assembly. Microp Mesop Mater 2012;147:47–52.
[513] Cheng P, Gao S, Zang P, Yang X, Bai Y, Xu H, Liu Z, Lei Z. Hierarchically porous carbon by activation of shiitake mushroom for capacitive energy storage. Carbon 2015;93:315–24.
[514] Zhang W, Chen L, Xu L, Dong H, Hu H, Xiao Y, Zheng M, Liu Y, Liang Y. Advanced nanonetwork- structured carbon materials for high-performance formaldehyde capture. J Colloid Interface Sci 2019;537:562–8.
[515] Gao F, Qu JY, Geng C, Shao GH, Wu MB. Self-templating synthesis of nitrogen-decorated hierarchical porous carbon from shrimp shell for supercapacitors. J Mater Chem A 2016;4:7445–52.
[516] Tondi G, Pizzi A, Masson E, Celzard A. Analysis of gases emitted during carbonization degradation of polyflavonoid tannin/furanic rigid foams. Polym Degrad Stab 2008;93(8):1539–43.

CHAPTER 3

Porous carbons for energy storage and conversion

Various electrochemical devices have been developed, and their positions are often shown on a plot of energy density against power density (Ragone plot), as shown in Fig. 3.1. Various batteries, such as Pb//acid, Ni//Cd, and Ni//hydrogen batteries have greatly contributed to the marked development of science and technology, and the development of lithium-ion rechargeable batteries (LIBs) has accelerated the development of modern science and technology. Electric double-layer capacitors (EDLCs), represented by supercapacitors using carbon materials as electrodes, are now used as useful devices for supporting the development of various electronic devices. Recently, energy storage devices based on the hybridization of capacitor-type and battery-type functionalities have attracted attention, because energy density comparable to LIBs and power density comparable to EDLCs are expected, as shown in Fig. 3.1. Carbon materials, including porous carbons, graphite and graphene, have played important roles in these electrochemical devices, not only as electrochemically active materials for intercalation/deintercalation reaction and formation of electric double layers, but also as scaffolds for electrochemically active materials by suppressing their agglomeration and/or buffering their large volume change during charge/discharge processes. These electronic devices contributed greatly to making various electronic devices smaller and lighter, which improves the quality of our lives.

The principal mechanisms for energy storage in three devices, rechargeable batteries, supercapacitors, and hybrid cells, are illustrated by using $LiPF_6$ electrolyte in Fig. 3.2. In rechargeable batteries, the principal electrochemical reaction is either intercalation/deintercalation (insertion/deinsertion) or redox reaction (Faradaic reaction) at the electrodes. For the former mechanism, graphite is often employed in the negative electrode with Li^+-containing electrolyte and these are called Li-ion rechargeable batteries (often Li-ion batteries, LIBs). For the latter mechanism, electrochemically active materials, such as silicon and sulfur, are held on a carbon scaffold, either by

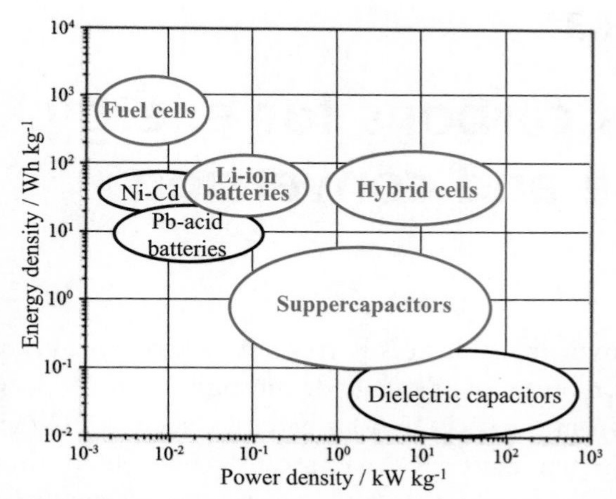

Figure 3.1 Comparison of energy and power densities of storage devices for electrical energy.

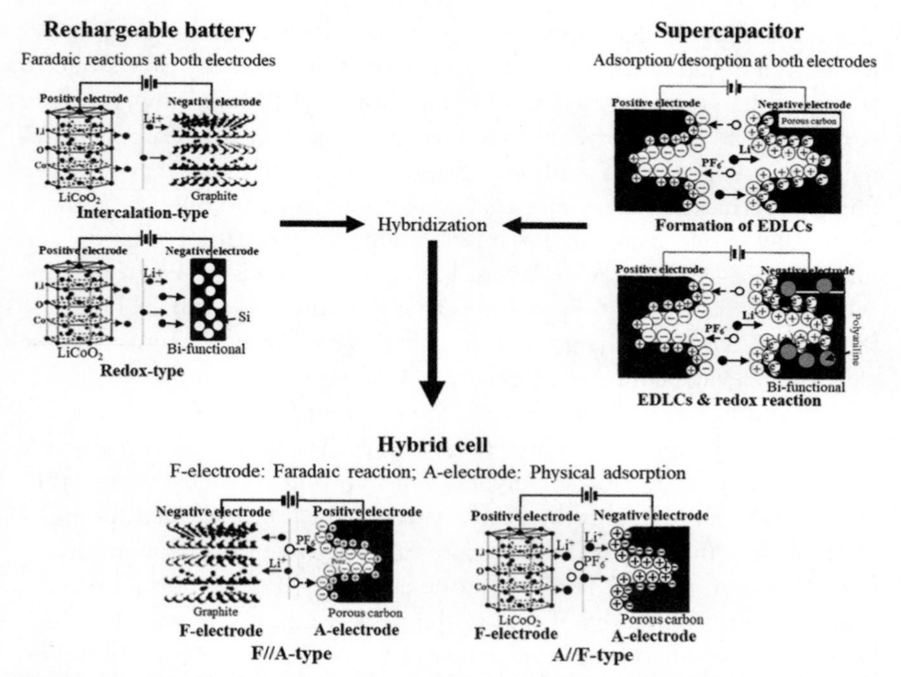

Figure 3.2 Classification of energy-storage electrochemical devices using $LiPF_6$ electrolyte.

deposition on the carbon substrate or confinement into the pores of the carbon matrix, consequently these electrodes often have the functions of battery and supercapacitor, i.e., they are "bifunctional." In the case of sulfur, the term "Li—S batteries" has often been used. In supercapacitors, the principal reaction is physical adsorption/desorption via formation of electric double layers (EDLs) on the surface of the carbon electrode. In supercapacitors, active materials are also held in the pores of electrode carbon to introduce a redox interaction due to the confined active materials (called pseudocapacitance) and as a consequence the cell capacitance is enhanced. In hybrid cells, the principal reaction in one electrode is the Faradaic reaction (e.g., intercalation/deintercalation) with charge transfer between ions of the electrolyte and active material in the electrode, and in another electrode there is physical adsorption/desorption of electrolyte ions to form electric double layers on the surface of the electrode material. The Faradaic reaction can occur either on a negative or positive electrode, and so the hybrid cells are classified into two types, as illustrated in Fig. 3.2. The Faradaic reaction occurs at the negative electrode and physical adsorption/desorption at the positive electrode (F//A-type), and adsorption/desorption at the negative electrode and a Faradaic reaction at the positive electrode (A//F-type). These hybrid cells are expected to achieve much greater energy densities than EDLC without sacrificing the advantage of higher power density over batteries. In addition, these hybrid cells are expected to have cycle and rate performances comparable to those of EDLCs. The hybrid cells are also called "BatCaps" because of the hybridization of battery and capacitor functionalities. The hybrid cells with Li ions as one of the electrolyte ions often have been called lithium-ion capacitors (LICs), in contrast to LIBs. Various reviews on these energy-storage devices have been published from different viewpoints.

In fuel cells where thermal energy can be converted to electrical energy, carbon materials are also important materials. In addition to these devices, rechargeable batteries, supercapacitors, their hybrid cells, and fuel cells, carbon materials are an important material for hydrogen and methane storage, which are ultimate energy sources for the near future. The storage of thermal energy is also an important technique in relation to energy problems.

3.1 Rechargeable batteries

Rechargeable batteries are classified into two types on the bases of Faradaic reactions at the negative electrode, either intercalation/deintercalation (insertion/extraction) or redox reaction, as illustrated in Fig. 3.2, denoted as intercalation type and redox type, respectively. In the intercalation-type rechargeable batteries, lithium ions have mostly been employed, while sodium ions are now attracting attention because of the natural resources of lithium. In this section on rechargeable batteries, a division of lithium-ion batteries and sodium-ion batteries is employed, and the former section is further divided into intercalation type and redox type on the bases of electrochemical reactions at the electrodes. The so-called Li—S batteries in which carbon materials have an important role as the scaffold for S are described separately.

3.1.1 Intercalation-type lithium-ion batteries

3.1.1.1 Graphitized carbons

Discharge curves are shown on coke and mesocarbon microbeads (MCMBs), which have been heat-treated at high temperatures, at the negative electrode in Fig. 3.3A and B, respectively [1,2], revealing that the performance of carbon materials in lithium-ion batteries greatly depends on the crystalline structure of carbon, in other words, on the development of the graphite structure. For well-graphitized carbon with high crystallinity, such as 3000°C-treated coke in Fig. 3.3A, a rather flat discharge plateau is observed at a potential of nearly 0 volt, with the discharge capacity determined at the potential up to 2 V being close to the theoretical value of 372 mAh/g for graphite. A similar change in discharge curves on MCMBs is observed with different heat treatment temperatures (HTTs), as shown in Fig. 3.3B. With increasing HTT above 2000°C, the plateau in the discharging process becomes clear and longer, i.e., there is an increase in the discharge capacity.

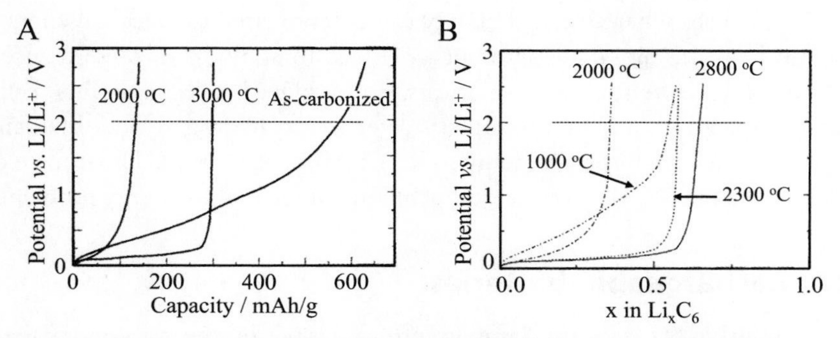

Figure 3.3 Discharge curves for graphitizing carbons heat-treated at different temperatures: (A) coke and (B) mesocarbon microbeads [1,2].

In Fig. 3.4A, the experimental data for the lithium storage capacity of various carbons, which were measured using three-electrode cells with a lithium metal counter electrode, are collected as a function of the HTT for carbonization and graphitization [2]. Three regions are selected to discuss the lithium storage capacities of carbon materials, and representative first cycles of charge–discharge curves for each region are shown in Fig. 3.4B. Region 1 corresponds to highly crystalline graphite synthesized from graphitizing carbons by heat treatment at high temperatures above 2500°C (graphitized carbons), of which the structure is represented by flaky natural graphite. The graphitized carbons in this region deliver capacity close to the theoretical value of 372 mAh/g with negligibly small irreversible capacity. The mechanism for Li storage is well understood by intercalation of Li^+ into the interlayer spaces of a graphite structure. However, non-graphitizing carbons do not have high capacity, even after heat treatment at high temperatures, because of poor crystallinity development in these carbons. Region 2 exhibits high capacities, which are much higher than the theoretical capacity, by the carbons heat-treated at low temperatures of around 600–800°C (ungraphitized carbons), which are the carbons just after pyrolysis of organic carbon precursors, such as pitches, and have amorphous structure. The mechanism of Li storage is thought to be mainly due to storage in nanopores formed in these highly disorganized structures, and a large portion of Li stored in these nanopores remains even after the discharge process, which is observed as a large irreversible capacity after

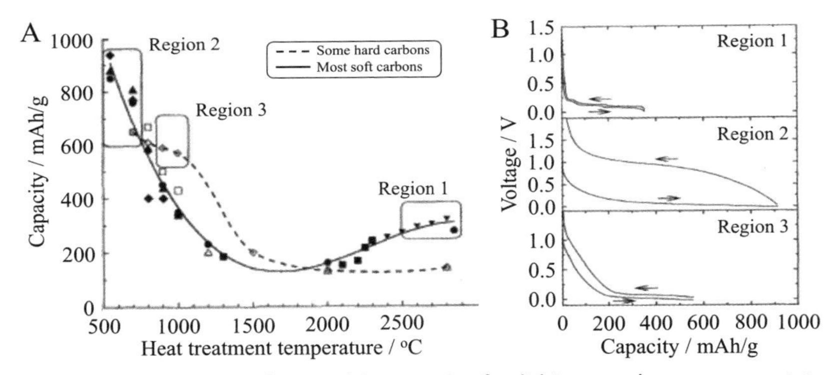

Figure 3.4 Dependence of reversible capacity for lithium on heat treatment temperature on various carbon materials with representative charge–discharge cycles for regions 1, 2, and 3. Open symbols are hard carbons and closed ones are soft carbons [2].

the first cycle. Region 3 corresponds to some nongraphitizing carbons heat-treated at around 1000°C, which give higher capacity than graphitizing carbons heat-treated at the same temperature, which is explained by the adsorption of Li on the surfaces of single carbon layers thought to be formed in nongraphitizing carbons [2]. In Fig. 3.5A, the change in reversible capacity with crystallinity of the electrode carbon is schematically shown, and is approximated as a curve giving a minimum. The position of the capacity minimum is dependent on the parameters used for evaluation of the crystallinity of carbon, at around 1500°C by HTT, as shown in Fig. 3.4A, at around 10 nm by crystallite size *Lc*(002) and at around 0.1−0.2 by graphitization degree P_1. This scheme can be explained by two mechanisms for the storage of Li in the carbon at the negative electrode, intercalation/deintercalation of Li^+ into the interlayer spaces of carbon and adsorption/desorption of Li^+ into the nanopores of carbon, as shown by the curves in Fig. 3.5A. The former mechanism becomes marked with increasing crystallinity because of the increase in the accommodation places (interlayer spaces) for Li^+. The latter mechanism contributes much more at a low crystallinity region. Most carbons, particularly carbons with random nanotexture and poor crystallinity (represented by nongraphitizing carbons), contain a large amount of nanopores, which can adsorb Li^+ during the charging process although they are rapidly diminished with improving crystallinity by heat treatment.

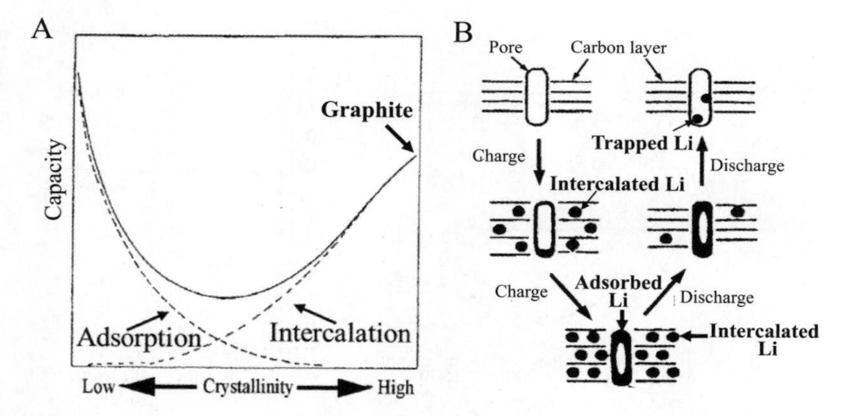

Figure 3.5 Schematic illustrations of charge−discharge performance of carbon at the negative electrode: (A) capacity change with the crystallinity of the carbon as a result of the competition between two Li-storage mechanisms, adsorption and intercalation, and (B) illustration of the two mechanisms during charge and discharge processes.

Some adsorbed Li may not be able to be recovered from nanopores by discharging, i.e., trapped Li, which causes irreversible capacity. The two mechanisms are illustrated in Fig. 3.5B.

Carbon coating of graphite materials for the negative electrode, including natural graphite, synthetic graphite, carbon fiber clothes, MCMBs, and nongraphitizing carbons, was carried out by either chemical vapor deposition (CVD) under different conditions or heat treatment of their mixtures with an organic precursor at a high temperature. Carbon deposition in a range of 8.6–17.6 wt.% was successfully performed at 950–1000°C on natural graphite powder under fluidizing in a flow of carrier gas containing toluene vapor [3–6], ethylene [7], propane [8], and methane [9]. The battery performance of natural graphite with spherical morphology prepared by impact milling (Fig. 3.6A) is slightly improved by carbon coating at 1000°C, as shown in Fig. 3.6B [6]. The carbon deposited by the CVD technique had a relatively high density of 1.86 g/cm^3 and Raman intensity ratio I_D/I_G of 0.77, in comparison with I_D/I_G of 0.07 for the pristine graphite, suggesting the deposition of disordered carbon [7].

Heating of the powder mixture of synthetic graphite with either poly(vinyl alcohol) (PVA) or poly(vinyl chloride) (PVC) at 900°C resulted in carbon-coated graphite powders [10–12]. The morphology of the particles did not change appreciably before and after carbon coating, except that the edges of the particles became round and the size increased slightly, with no marked coagulation of particles being observed. The carbon coating proved to be disordered and porous from the measurements of Raman spectrum, immersion density, and N_2 adsorption/desorption isotherm [10]. Its thickness, which was calculated from the difference in

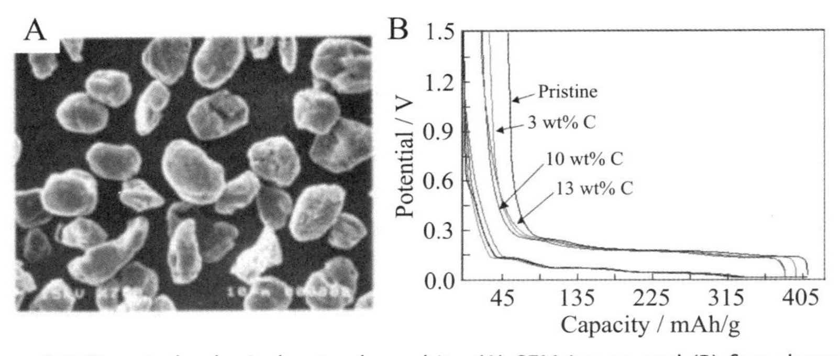

Figure 3.6 C-coated spherical natural graphite: (A) SEM image and (B) first charge–discharge curves for different amounts of C coating [6].

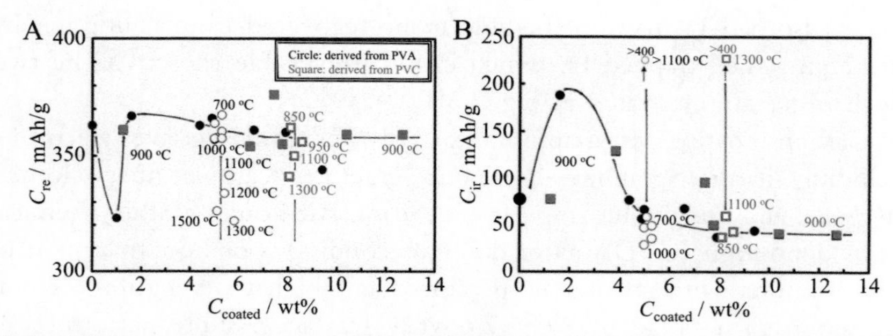

Figure 3.7 C-coated synthetic graphite using PVA and PVC precursors: (A) reversible capacity C_{re} and (B) irreversible capacity C_{ir} for the first charge–discharge cycles against C_{coated} graphite [10].

particle size distributions before and after the carbon coating, was shown to be proportional to the amount of coated carbon C_{coated} and so to be controlled by changing the mixing ratio of PVA. In Fig. 3.7A and B, discharge and irreversible capacities for the first cycle in 1 M $LiClO_4$/(EC + PC) electrolyte are plotted against C_{coated} as functions of the precursor of coated carbon, either PVA or PVC, and of carbonization temperature [10]. Since PVC gives a higher carbon yield than PVA, the relation between capacity and C_{coated} for PVC shifts to the higher side of C_{coated} than that for PVA. Carbon coating of a small amount (<5 wt.%) at 900°C tends to increase the irreversible capacity, although the discharge capacity either decreases or does not change appreciably. The carbonization at a temperature higher than 1100°C gives very high irreversible capacity and low discharge capacity. Therefore, a carbon coating of more than 5 wt.% on natural graphite has to be done at the carbonization temperature of 700–1000°C in order to obtain a low irreversible capacity by keeping the discharge capacity at more than 350 mAh/g.

Coral-like porous carbons synthesized from the mixtures of fructose, polystyrene latex (PSL) with a particle size of about 63 μm, and triblock copolymer F127 by hydrothermal treatment at 180°C and following carbonization at 900°C have a hierarchical pores structure [13]. These coral-like carbons were found to be graphitized partially by heat treatment at high temperatures by retaining their ordered pore structures, the 3000°C-treated carbon delivering a high Li storage capacity [14]. The charge–discharge profiles at different rates for the 3000°C-treated coral-like carbon are shown in Fig. 3.8 by comparing them with natural graphite. At a slow rate such as C/10, its electrochemical Li^+ intercalation

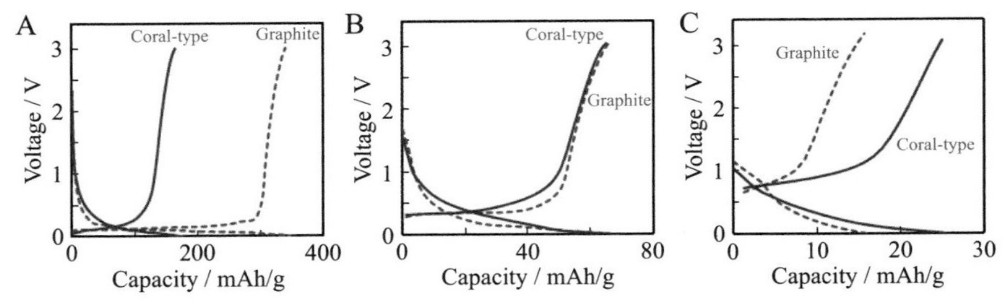

Figure 3.8 Charge–discharge profiles at the fifth cycle for coral-like porous carbons heat-treated at 3000°C at different rates, in comparison with graphite: charge–discharge rates of (A) C/10, (B) 1C, and (C) 5C [14].

and deintercalation reversibly occur, although its capacity was lower and the irreversible capacity was higher than for graphite, as shown in Fig. 3.8A. The capacity as well as the irreversible capacity were improved by oxidation in air at 750°C, probably due to partial removal of the amorphous carbons. Its capacity was almost equivalent to graphite at 1C (Fig. 3.8B). At a high rate as 5C, however, it becomes higher than graphite, as shown in Fig. 3.8C, probably because of the hierarchical pore structure and partial graphitic structure in the carbons.

Reduced graphene oxide flakes (rGO) were prepared by dispersing GO in deionized water with hydrazine at 100°C and by recovering as supernatant, which delivered a reversible capacity of 650 mAh/g in the first cycle in 1 M $LiPF_6$/(EC + DMC) electrolyte at a current density of 1C within the range of 0.02–3.0 V [15]. Doping of N and B into carbon was shown to be effective for improving LIB performance as a negative electrode using rGO [16]. N-doped rGO (3.06 at.% N) prepared in a NH_3/Ar flow at 600°C and B-doped one (0.88 at.% B) in BCl_3/Ar at 800°C gave high reversible capacity in LIB at a low current density of 50 mA/g, almost constant capacity of 870, and 1220 mAh/g during cycling after the fifth cycle up to the 30th cycle, as shown in Fig. 3.9A and B, respectively. No plateau was observed on charge/discharge curves within the potential windows employed (0.01–3.0 V). These materials for negative electrodes could achieve relatively high capacities at a high current density of 25 A/g, 199 and 235 mAh/g for N- and B-doped rGOs, respectively. N-doped rGO prepared by thermal exfoliation and reduction of GO at 1050°C, followed by annealing in NH_3, exhibited unusual cycling behavior [17]; the discharge capacity decreased in the first 17 cycles and then tended to increase up to the 501st cycle, reaching a maximum capacity of 684 mAh/g.

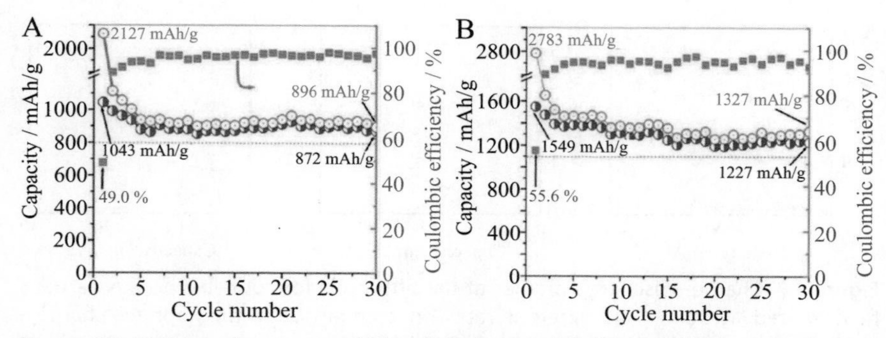

Figure 3.9 Cycle performances of N- and B-doped rGO, (A) and (B) respectively, at a current density of 50 mA/g in a voltage range of 0.01–3.0 V with 1 M $LiPF_6$/(EC + DEC) electrolyte [16].

Recently, various batteries have been required to be lightweight and compact, and as a consequence, porous carbons, such as activated carbons (ACs), are not employed in the negative electrodes of LIBs because of their low bulk density. However, some attempts have been made to use porous carbons, including rGO foams, as the negative electrode for LIBs, in order to meet their improved performances. N-doped porous carbons were tested as the negative electrode of LIBs, as well as sodium-ion batteries (SIBs), by using 1 M $LiPF_6$/(EC + PC) and $NaClO_4$/(EC + PC) electrolytes, respectively [17]. N-doped porous carbons were prepared by carbonization of a covalent organic framework (COF) synthesized from hexamethylene-tetramine, phloroglucinol, and *p*-phenylenediamine at 500, 600, and 700°C in N_2, delivering high performances in both LIBs and SIBs. Cycle performances of the LIBs are shown in Fig. 3.10A, where the

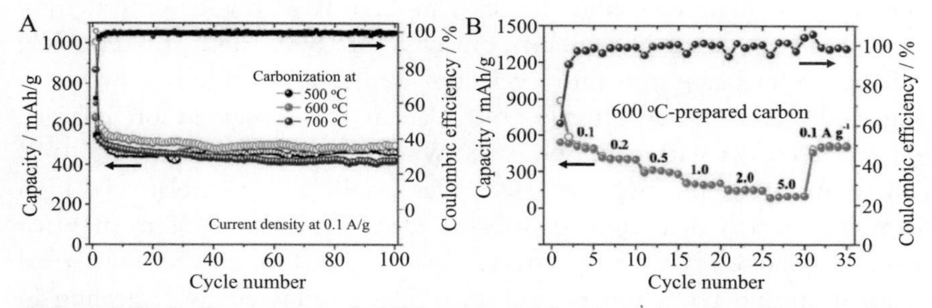

Figure 3.10 Performances of LIBs composed of N-doped porous carbons prepared from a covalent organic framework (COF): (A) cycle performances for the carbon carbonized at different temperatures, and (B) rate performances for the 600°C-carbonized carbon [18].

600°C-carbonized carbon with S_{BET} of 482 m^2/g delivers a relatively high capacity of 488 mAh/g at a current density of 0.1 A/g, with the rate performance as shown in Fig. 3.10B. This porous carbon exhibited superior long-life cycling stability, even after 5000 cycles at a high current density of 5 A/g. N- and B-doped soot particles were also tested as the LIB negative electrode [19]. The mesoporous carbons prepared from Mg-citrate at 900°C were heat-treated at different temperatures up to 3000°C and the LIB performances as the negative electrode studied [20]. With increasing HTT, micropores and mesopores were collapsed, resulting in a low S_{BET}, particularly above 2000°C. The carbons heated at 2700 and 3000°C had thick graphitized carbon layers and retained a mesopore structure showing charge—discharge curves with three plateaus at approximately 0.2, 1.0, and 1.5 V versus Li/Li^+. The 3000°C-treated carbon exhibited the best performances, with low irreversible capacity, high Columbic efficiency, and high rate capability.

Microwave irradiation assisted the graphitization of a xerogel, which was prepared from RF resin by carbonization at 700°C, and consequently improved the LIB performances [21]. Irradiation with 750—1500 W power resulted in a partial graphitization of the xerogel particles giving a composite 002 diffraction profile.

3.1.1.2 Nongraphitized carbons

Carbon with poor crystallinity, i.e., as-carbonized coke in region 2 of Fig. 3.4A, shows a gradual potential change, no plateau is detected and its discharge capacity determined at 2 V is much higher than for the 3000°C-treated coke, and even higher than the theoretical value, suggesting that another mechanism for energy storage is additionally working in these carbons. In addition, these low-temperature-treated cokes showed certain hysteresis between charge and discharge curves, whereas the cokes heat-treated at high temperatures gave only a small difference between the charge and discharge profiles [22]. Nongraphitizing carbons, which have low crystallinity, gave much more marked crystallinity dependences of capacity and hysteresis in charge—discharge curves. An example on the nongraphitizing carbon derived from polyacene [23] is shown in Fig. 3.11, revealing a gradual change of potential in both charge and discharge processes, a marked hysteresis at the first cycle giving a large irreversible capacity of 230 mAh/g, and a high discharge capacity (reversible capacity) of 580 mAh/g, which was higher than the theoretical one for graphite.

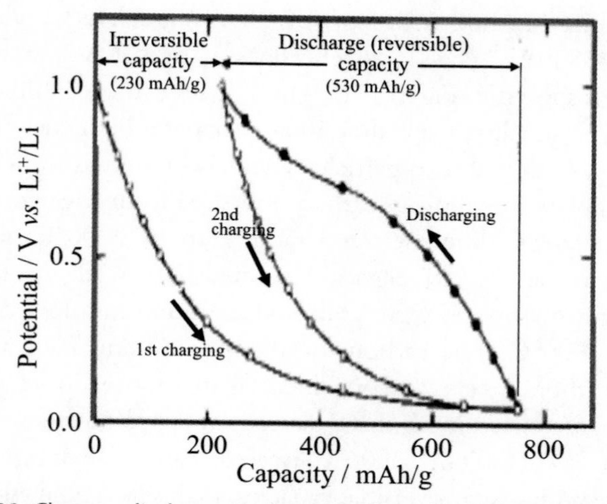

Figure 3.11 Charge–discharge cycles on carbon derived from polyacene [23].

The carbons heat-treated at 1800–2400°C show almost minimum capacities at low current densities, lower than the 1C rate, as shown in Fig. 3.12A, while they are found to have higher capacities and cyclability at high rates, as compared with the discharge profiles at different rates for coke heat-treated at 2200°C and natural graphite in Fig. 3.12B [24]. At the 0.3C (1.125 mA/g) rate, natural graphite gives 370 mAh/g, which is nearly equal to the theoretical value, but the 2200°C-treated coke exhibited only 300 mAh/g. At high rates, higher than 3C (11.25 mA/g), however, the coke delivers higher capacities than natural graphite, as

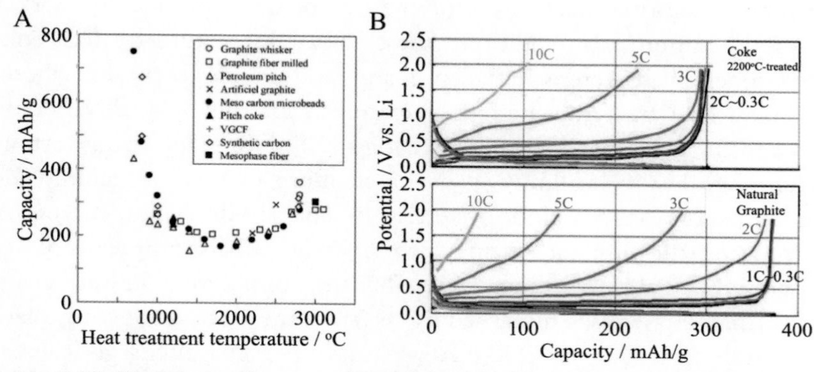

Figure 3.12 (A) Relation between HTT and reversible capacity for various carbon materials, and (B) discharge curves at different rates for coke heat-treated at 2200°C and natural graphite in 1 M $LiPF_6$/(EC + MEC) electrolyte [24].

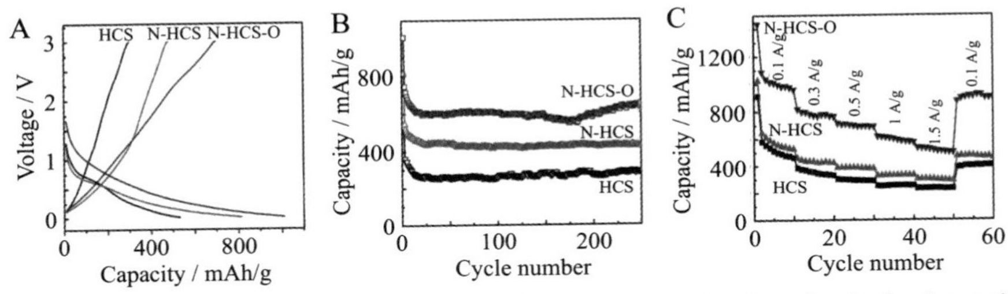

Figure 3.13 Hollow carbon spheres doped by N and oxidized under hydrothermal condition, HCS, N-HCS. and N-HCS-O: (A) first cycle charge–discharge curves, (B) cycle performances at 0.5 A/g rate, and (C) rate performances [25].

shown in Fig. 3.12B. The detailed studies using cyclic voltammetry and X-ray diffraction profile change of the coke suggested that the first and second stage intercalation compounds are directly formed in the coke particles without forming intermediate higher stage compounds.

Hollow carbon spheres (HCSs) with diameters of about 300 nm were synthesized using SiO_2 nanospheres template by CVD of pitch with or without CH_3CN (nitrogen source) at 900°C [25]. The carbon spheres obtained without CH_3CN were composed of an amorphous carbon shell with a thickness of about 30 nm (HCS) and the spheres obtained with CH_3CN were composed of N-doped graphitic carbon with a thickness of about 20 nm (N-HCS). The N-HCS spheres were hydrothermally treated in HNO_3 at 160°C (N-HCS-O). First cycle charge–discharge curves, cycle performances, and rate performances for HCS, N-HCS, and N-HCS-O are shown in Fig. 3.13A–C, respectively. N-HCS-O exhibits the best battery performances among these three, giving the initial reversible capacity of 741 mAh/g with an irreversible capacity of 268 mAh/g at a current density of 0.5 A/g in the voltage range of 0.01–3.00 V with a capacity retention of 616 mAh/g after 250 cycles. The high performances of N-HCS-O were attributed to the existence of graphitic carbon on the surface of hollow spheres, in addition to N-doping and O-containing functional groups.

3.1.1.3 MXenes

Layered ternary carbides and/or nitrides $M_{n+1}AX_n$ (M: an early transition metal, A: an A group element, X: C and/or N) attracted attention for their relatively high electrical conductivities, in which M_3X_2 layers composed of edge-sharing MX_6 octahedra are separated by A [26,27]. From these $M_{n+1}AX_n$ compounds, layer-structured transition metal carbides and/or nitrides were easily prepared by extraction of A-element, which consisted of a large family having a general formula $M_{n+1}X_n$ and which were called MXenes because they were easily intercalated and exfoliated to thin flakes, just like graphene from graphite. Ti_3C_2, for example, was obtained by the extraction of Al from Ti_3AlC_2 by either HF treatment [28] or HCl solution of LiF [29]. On their surfaces, some functional groups, such as OH, F, etc. remained and so it was expressed as $Ti_3C_2T_x$ (or f-Ti_3C_2). From various $M_{n+1}AX_n$ compounds, such as Ti_2AlC, Ta_4AlC_3, $(Ti_{0.5}Nb_{0.5})_2AlC$, $(V_{0.5}Cr_{0.5})_3AlC_2$, and Ti_3AlCN, the corresponding MXenes were prepared using HF [30]. The resultant layered $Ti_3C_2T_x$ particles could be intercalated by hydrazine monohydrate (HM), N,N-dimethylformamide (DMF), and dimethyl sulfoxide (DMSO) at 80°C; the DMSO-intercalated one was possible to be exfoliated by sonication in deionized water [29]. $Ti_3C_2T_x$ flakes were also shown to be electrochemically intercalated by a variety of cations, including Li^+, Na^+, K^+, NH_4^+, Mg^{2+}, and Al^{3+}, offering high volumetric capacitance over 300 F/cm^3 [31]. In the KOH electrolyte, few-layered $Ti_3C_2T_x$ flakes delivered about 340 F/cm^3 with a high retention at 1 A/g rate after 10,000 cycles [31] and a rolled $Ti_3C_2T_x$ sheet exhibited 900 F/cm^3 (245 F/g) at 2 mV/s rate [29]. Other MXenes, Nb_2C and V_2C, exhibited reversible capacities of 170 and 260 mAh/g at 1C, and 110 and 125 mAh/g at 10C, respectively [32]. A theoretical discussion on the storage of Li, Na, K, and Ca on Ti_3C_2 has been reported [33]. Storage capacities were calculated to be 447.8, 351.8, 191.8, and 319.8 mAh/g, respectively.

To avoid stacking of MXene layers in the film, mixing of CNTs was effective to enhance capacitive performances. $Ti_3C_2T_x$/SWCNT sheet yielding around 345 F/cm^3 at 5 A/g rate with high retention even after 10,000 cycles in 1 M $MgSO_4$ aqueous solution [34], and the deposition of Sn^{4+} nanoparticles (6−7 nm sizes) on Ti_3C_2 flakes resulted in an increase in the capacity to 1375 mAh/cm^3 (635 mAh/g) at 216.5 mA/cm^3 (100 mA/g) rate in 1 M $LiPF_6$/(EC + DMC) electrolyte [35].

3.1.2 Redox-type lithium-ion batteries

The composites of carbon materials, porous carbon and graphene (including rGO), with organic molecules, metals, and their oxides, which are electrochemically active through redox reactions with Li^+, were applied to the electrode of LIBs. Some improved performances of the batteries were reported. Most active materials are either loaded on the surface of carbon materials or confined in carbon materials, where carbon materials have important roles to keep the particle of the active material small and to buffer the large volume change of the active material during lithiation/delithiation, as well as to work as a current collector. In this section, active materials are reviewed by focusing on their battery performances owing to the composite formation with carbon materials, mostly porous carbons.

3.1.2.1 Lithium compounds

A porous carbon coating of active materials has been successfully applied for different electrode materials of LIBs. Although spinel-type $LiCoO_2$ and $LiMn_2O_4$ have been used for positive electrodes of LIBs, other lithium compounds also have attracted attention as positive electrode material, such as $LiFePO_4$, $LiMnPO_4$, $Li_3V_2(PO_4)_3$, Li_2MnSiO_4, and Li_2FeSiO_4, with increasing research activity on energy-storage devices because of the strong demand from electric vehicle applications. Their carbon coating was demonstrated to be effective for improving their performances in LIBs.

3.1.2.1.1 For the positive electrode

The advantages of $LiFePO_4$ (LFP) having an olivine-type structure are its flat voltage profile, low material cost, abundant material supply, and better environmental compatibility, compared to others [36–40]. However, its disadvantages are also mentioned as low theoretical capacity of 170 mAh/g and low density of 3.60 g/cm^3, which are much lower than conventional $LiCoO_2$ (274 mAh/g and 5.05 g/cm^3, respectively), in addition to poor electrical conductivity and low ionic diffusivity. The C coating of LFP particles was reported to be effective (1) to reduce Fe^{3+} to Fe^{2+}, (2) to keep the particles small, (3) to increase the electrical conductivity of the sheet formed as electrode, and (4) to enhance the ionic diffusivity.

C-coated LFP was prepared by heat-treating the mixture of CH_3COOLi, $(CH_3COO)_2Fe$, and $NH_4H_2PO_4$ in stoichiometric ratio with resorcinol-formaldehyde gel at 700°C in N_2 [41]. The resultant composite (about 15 wt.% C coated) exhibited a flat plateau and capacity of 160 mAh/g, close to the theoretical capacity, at C/5 rate (1C = 0.17 A/g) with 99.9% retention after 100 cycles, as shown in Fig. 3.14A. At a much faster rate than 5C, the capacity increased slightly during the initial cycles to reach a maximum of 120 mAh/g and decreased gradually to 92% retention after 800 cycles, as shown in Fig. 3.14B. Micrometer-sized spheres of C-coated LFP were prepared from a stoichiometric mixture of LiOH,

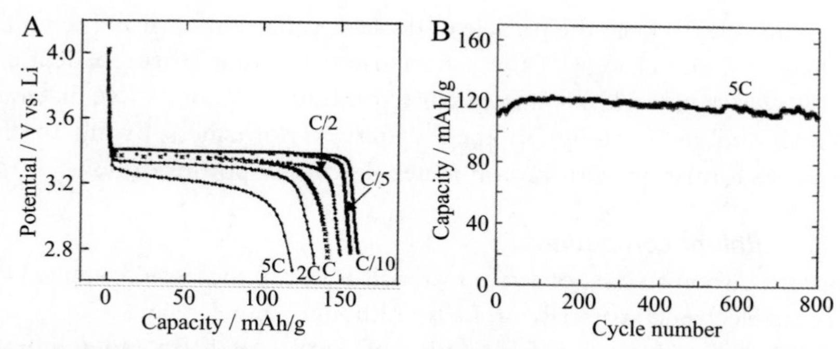

Figure 3.14 C-coated LFP: (A) discharge curves at different rates and (B) cycle performance at 5C rate [41].

$NH_4H_2PO_4$, and FeC_2O_4˙ in an aqueous solution either by spray drying at 240°C or by drying in a vacuum at 130°C, followed by carbonization at 650°C in Ar [42]. The C-coated LFP prepared by spray-drying delivered a higher capacity of 110 mAh/g at 10C (1.7 A/g) rate than that prepared by vacuum-drying. Natural ores of triphylite ($LiFePO_4$) from different mining localities exhibited discharge capacities of about 85 mAh/g and their C coating was effective for improving the cycle performance [43].

C coating of LFP particles using graphite, carbon black, and acetylene black (5 wt.%) at 700°C in a 5 vol% H_2/Ar atmosphere after mechanochemical activation in a planetary mill was shown to be effective in improving the electrical conductivity to 10^{-2}–10^{-4} S/cm and consequently improving the cyclic performance, where the graphite coating was the most effective [44]. C coating was performed by mixing LFP powder synthesized with pyromellitic acid and ferrocene, followed by heat treatment at 600°C [45]. The resultant C-coated LFP exhibited an improved rate performance. This performance improvement was explained by the catalytic graphitization of coated carbon due to ferrocene, although the Raman spectrum observed seemed not to support this explanation. C-coated LFP composite prepared from the mixture of LFP precursor with sucrose by the calcination at 550°C after the mechanochemical activation in a planetary mill exhibited better rate and cyclic performances than that prepared without mechanochemical activation [46]. The addition of sucrose during mechanochemical activation was also effective for preparing uniformly C-coated LFP of micrometer-sized particles [47,48]. The synthesis of LFP powders was performed by solid-state reaction at 700°C of the stoichiometric mixture of Li_2CO_3,

FeC_2O_4, and $NH_4H_2PO_4$ (LFP precursors) without the addition of sucrose, with sucrose, and with sucrose followed by mechanochemical activation using a high-energy ball mill [49]. As shown by the first charge–discharge curves in Fig. 3.15, C coating coupled with mechanochemical activation is very effective for increasing capacity, giving a capacity of 174 mAh/g at 0.1C rate and 117 mAh/g at 20C. Too low temperature and insufficient time for the heat treatment resulted in the formation of carbon residues containing hydrogen, and as a consequence poor battery performance of the resultant C-coated LFP.

Hydrothermal treatment of the mixture of LiOH, $FeSO_4$, and H_3PO_4 with glucose at 180°C, followed by annealing at 750°C in N_2, resulted in C-coated LFP (in situ carbon coating) [50,51], which gave a better cycle performance than that prepared by heat treatment at 750°C of the mixture of LFP with glucose (mixed coating). In situ carbon coating is a simple one-step process and makes the carbon coating for small primary particles of LFP possible, with consequent effective improvement of battery performances. C-coated LFPs with various carbon contents were prepared by an in situ coating process using Li_2CO_3, FeC_2O_4, $(NH_4)_2HPO_4$, and sugar at 800°C, and their capacities in 1 M $LiPF_6$/(EC + DEC) were measured as a function of the amount of carbon coated [52]. The resultant composites exhibited a marked increase in capacity with increasing carbon content up to 1 wt.% and tended to be saturated above 2 wt.%, but the tap

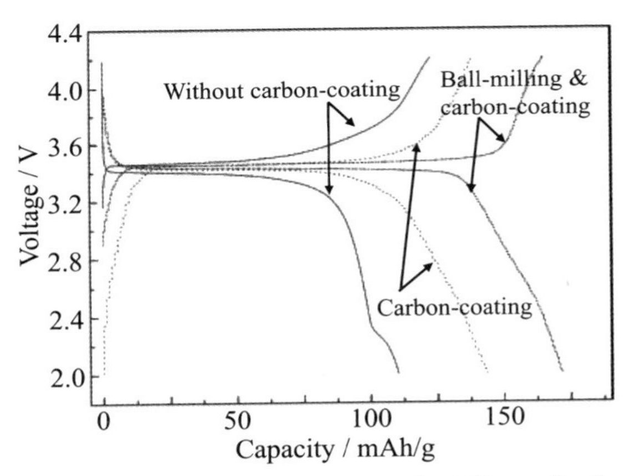

Figure 3.15 Charge–discharge curves at 0.1C rate for LFPs synthesized by solid-state reaction at 700°C of LFP precursor mixture without sucrose, with sucrose, and with sucrose and high-energy ball-milling [49].

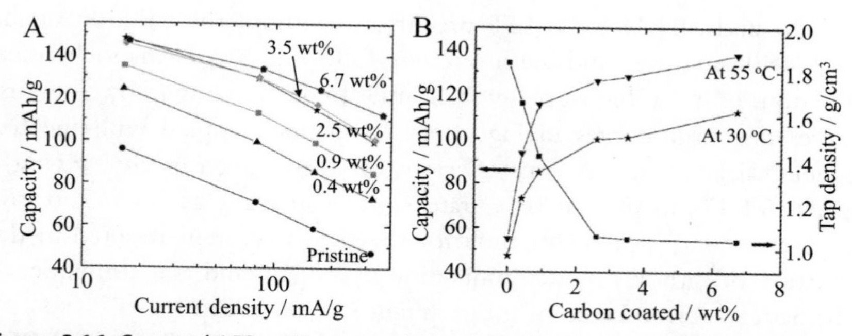

Figure 3.16 C-coated LFPs: (A) rate performances for the composites with different carbon contents and (B) dependences of capacities with 2C rate at 30–55°C and tap density on carbon content [52].

density decreased markedly with increasing carbon content, as shown in Fig. 3.16A and B. Sintered pellet of C-coated LFP with diameter of about 10 mm and thickness of about 2 mm was prepared by compressing the precursor mixtures under 20 MPa, followed by sintering at 800°C [53]. Its electrical conductivity increased from about 5×10^{-8} to over 10^{-1} S/cm with increasing carbon content from 0 to 31 wt.%. The thickness of the coated carbon was also an important factor to improve battery performances, being demonstrated by the deposition of additional carbon via a mixed coating process using malonic acid onto carbon-coated LFP particles prepared by an in situ coating process using polystyrene [54]. The thickness of the coated carbon increased from 2–6 to 4–8 nm with increasing carbon content from 1.25 to 2.28 wt.%, so that the discharge capacity at the seventh cycle increased from 137 to 151 mAh/g, whereas the increase in carbon content to 2.54 wt.% and thickness to 10–25 nm resulted in a capacity decrease to 143 mAh/g.

C-coated LFPs were synthesized through the sol–gel method using ethylene glycol (carbon source and solvent) with FeC_2O_4, $LiNO_3$, and $NH_4H_2PO_4$ with heating at 700°C in 5 vol.% H_2/N_2, which contained various amounts of carbon and conductive FeP [55]. The composite containing 1.5 wt.% C and 1.2 wt.% FeP had discharge capacities of 155 and 111 mAh/g at the rates of 0.1 and 1C, respectively. The effect of carbon source was studied through heat treatment at 650°C for the LFP precursor prepared by the sol–gel method by mixing with acetylene black, sucrose, and glucose [56]. The discharge voltage and discharge capacity depended strongly on the carbon source used, with glucose being the most efficient.

Spray pyrolysis of an aerosol precursor resulted in spherical LFP particles with diameters of 50–100 nm and coated by carbon layers with thicknesses of 2–4 nm, which could give a relatively high tap density of 1.2–1.6 g/cm^3, probably due to the spherical morphology of the particles [57]. A cell constructed from this C-coated LFP at the positive electrode and C-coated $L_4Ti_5O_{12}$, which was prepared by the same procedure, at the negative electrode exhibited a high rate performance in a wide range of discharging rates from 0.045 mA/cm^2 (C/25) to 5.5 mA/cm^2 (5C).

C-coated LFP was prepared through LFP precipitation in molten stearic acid and following calcination at 600–800°C in Ar atmosphere [58], where stearic acid served as the chelating agent and carbon precursor. Several aromatic anhydrides were used as the carbon precursor for C coating of amorphous LFP at 750°C [59]. Benezene-1,2,4,5-tetracarboxylic acid could give the best performance on nano-sized LFP. Polyethylene oxide, polybutadiene, polystyrene, and block copolymer (styrene-butadiene-styrene) were also tested as carbon precursor with the heat treatment at 600°C [60]. Polystyrene was the most effective for improvement in the rate performance in 1 M $LiPF_6$/(EC + DMC).

$LiMnPO_4$ has the same olivine-type structure and electrochemical activity as LFP [36]. However, it has attracted little attention because of its poor cyclic performance, although it has a theoretical capacity of 170 mAh/g and high redox potential of 4.0 V versus Li^+/Li. To improve its electrochemical performance, the preparation of $Li(Fe_xMn_{1-x})PO_4$ [61,62] and of composite with carbon nanotubes [63] have been reported.

Hollow microspheres of C-coated $LiCoPO_4$ were prepared by spray pyrolysis, of which the inside and outside surfaces were coated by carbon [64]. C-coated $LiCoPO_4$ hollow microspheres having about 70 nm diameters delivered stable charge–discharge performances at least up to 2C rate.

Through the sol–gel process of ethanol solution of Li_2CO_3, $FeCl_2$, H_3PO_4, and NH_4VO_3 with citric acid as a carbon precursor, a single phase of C-coated $LiFe_{1-x}V_xPO_4$ solid solution was obtained in the range of $0 \leqq x \leqq 0.07$ [65]. With increasing V content up to x = 0.07, electrical conductivity and apparent lithium ion diffusion coefficient increased and, as the consequence, the discharge capacity in 1 M $LiPF_6$/(EC + DEC) electrolyte increased at a high discharge rate of 10C.

Electrochemical performances at low temperatures were measured in the electrolyte 1 M $LiPF_6$/(EC + DMC) on C-coated $Li_3V_2(PO_4)_3$ (LVP)

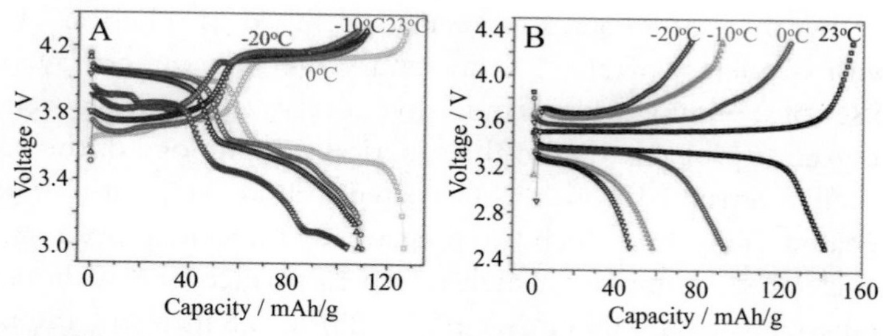

Figure 3.17 Charge–discharge curves of the first cycle of (A) C-coated LVP and (B) C-coated LFP at different temperatures with 0.3C rate [66].

by comparing with C-coated LFP [66]. LVP exhibits stable charge–discharge curves at low temperatures, whereas LFP shows a marked temperature dependence, as shown in Fig. 3.17. At −20°C, C-coated LVP exhibits a stable discharge capacity of 104 mAh/g, but C-coated LFP only 45 mAh/g, while at 23°C the former has 127 mAh/g and the latter 142 mAh/g.

3.1.2.1.2 For the negative electrode

Crystalline $Li_4Ti_5O_{12}$ (LTO) can work as a negative electrode of LIBs by intercalating Li ions to form $Li_{4+x}Ti_5O_{12}$ ($0 < x < 3$) accompanied by a flat voltage range, theoretical capacity of 175 mAh/g, and no structure change during charge–discharge cycles, but its poor electrical conductivity (less than 10^{-13} S/cm) has to be overcome. Carbon coating was applied to improve the conductivity of an LTO electrode by different processes. Coating by CVD of toluene vapor at 800°C improved the conductivity to 2.05 S/cm and as a consequence improved the rate performance [67]. Carbon-coated spherical powders were prepared through spray drying of a slurry of the mixture of LiOH and TiO_2, followed by calcination at 900°C and then at 750°C after mixing with a pitch, which realized a high density of 1.15 g/cm^3 [68]. As shown in Fig. 3.18A and B, powder with 3.25 wt.% C delivers an improved rate and cycle performances, respectively, 170 mAh/g at a current density of 175 mA/g (1C rate) with a capacity retention of about 98% after 100 cycles and 81.7 mAh/g at 17.5 A/g (100C rate). Spherical LTO particles were coated with N-doped porous carbon via heat treatment at 600°C after mixing with an ionic liquid, 1-ethyl-3-methylimidazolium dicyanamide (EMIm-dca) [69]. They exhibited much better rate and cycle performances than those of the pristine LTO and carbon-coated LTO without N-doping. Carbon coating

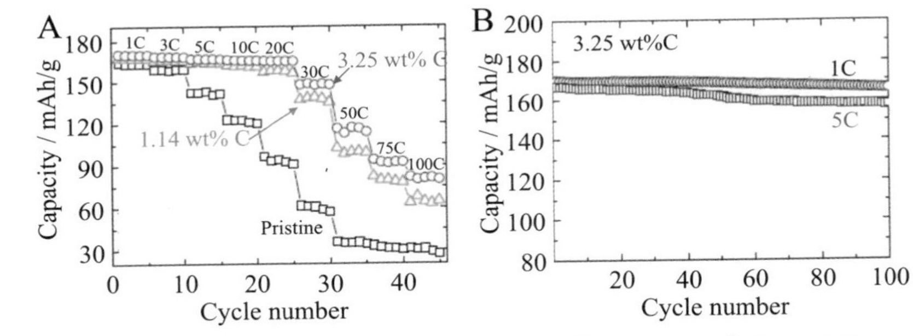

Figure 3.18 C-coated LTO spheres: (A) rate performances of the pristine and C-coated LTOs and (B) cycle performances of a 3.25 wt.% C-coated one at two different rates [68].

on LTO particles, which were synthesized by hydrothermal treatment of aqueous solution of LiOH and $TiCl_4$ at 170°C followed by calcination at 800°C, was performed by mixing in an aqueous solution of amphiphilic carbonaceous materials at 80°C and then carbonizing at 800°C in N_2 [70]. The resultant C-coated LTO gave a capacity of 137 mAh/g at 20C rate with a capacity retention of 91% after 100 cycles.

Carbon coating was also carried out during the synthesis of LTO through solid-state reaction (in situ carbon coating) on a mixture of lithium polyacrylate (lithium and carbon sources) and TiO_2 (titanium source) at 800°C [71], giving discharge capacities of 148 and 116 mAh/g at 0.1 and 0.5 A/g rates with capacity retentions of about 94% and 88%, respectively, after 50 cycles. C-coated LTO, which was prepared from a gel mixture of lithium acetate, tetrabutyl titanate [$Ti(OC_4H_9)_4$], and citric acid at 800°C, delivered improved performances of about 86% capacity retention after 50 cycles at 0.8 A/g rate [72]. Carbon-coated LTOs were synthesized in two steps; carbon-coated TiO_2 was prepared from the mixture of TiO_2 and sucrose by carbonization at 500°C after high-energy ball milling, and then it was heat-treated at 750°C by mixing with Li_2CO_3 [73]. They delivered a capacity of 160 mAh/g at 0.5C, which was almost the same as the pristine LTO, while having a better rate performance. From a gel mixture of lithium acetate, tetrabutyl titanate, and rGO flakes prepared by the reduction at 1050°C, rGO-coated LTO powders were prepared by calcination at 800°C [74]. This composite gave a reversible capacity of 174 mAh/g with Coulombic efficiency of 99% at 0.2C rate in a three-electrode cell with 1 M $LiPF_6$/(EC + DEC) electrolyte, which was slightly higher than the 165 mAh/g and 94% efficiency for the pristine LTO.

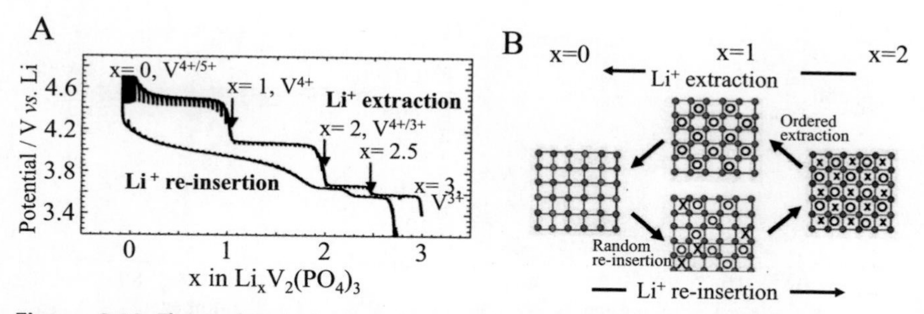

Figure 3.19 Electrochemical extraction and reinsertion process of $Li_{3-x}V_2(PO_4)_3$: (A) potential changes with x-value and (B) illustration of Li sites during the process [75,76].

From monoclinic α-$Li_3V_2(PO_4)_3$ (LVP), Li ions can be extracted by electrochemical oxidation to $V_2(PO_4)_3$ through intermediate crystalline phases $Li_{3-x}V_2(PO_4)_3$ with $x = 1$, 2, 2.5, and 3, and reinserted by reduction to the composition $Li_3V_2(PO_4)_3$ [75,76]. The process of electrochemical extraction and reinsertion shows hysteresis, as shown in Fig. 3.19A; the formation of each intermediate phase was clearly recognized as a potential step where two phases coexisted on the extraction process, but the reinsertion up to the phase with $x = 1.0$ becomes obscure due to the random insertion of Li^+, as illustrated in Fig. 3.19B [76].

3.1.2.2 Metallic silicon

Metallic Si is known to have a high capacity for Li storage; at room temperature the theoretical capacity of Si is 3579 Ah/kg (8344 Ah/L) due to the formation of $Li_{15}Si_4$ phase [77], this being substantially larger than that of graphite (371 Ah/kg, 830 Ah/L). However, its large volume change during lithiation/delithiation and relatively low electrical conductivity have to be overcome for practical applications as the negative electrode of LIBs. Recently, carbon coating of Si particles was reported to be effective for obtaining high reversible capacity and high cyclability [78—80].

Carbon coating on nano-sized Si (thickness of the carbon layer of ca. 10 nm) by a pressure-pulsed CVD improves the performances. The carbon layer is observed to be deformed and taken into the wrinkles on the particles of the matrix Si by scanning TEM with EDS mapping [81], as shown in the TEM images and illustrations in Fig. 3.20. A similar structural change was also observed on ball-milled Si [78]. As shown in Fig. 3.21A, the initial reversible capacity reached 3 Ah/g in 1 M $LiPF_6$/(EC + DEC) electrolyte. By restricting the discharge capacity to up to 1.5 Ah/g, cycling

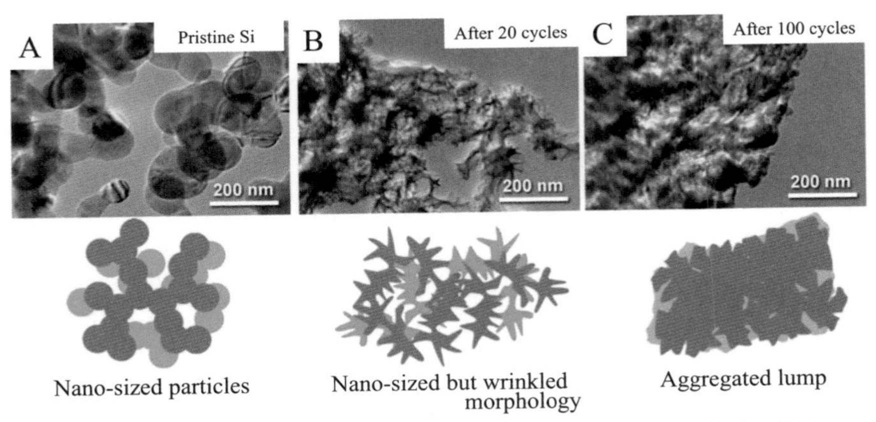

Figure 3.20 TEM images and morphology illustrations of nano-sized Si after lithiation/delithiation cycling [78].

with different current densities of 0.2−5.0 A/g is possible on the carbon-coated Si for at least up to 100 cycles, whereas without a carbon coating marked fading is observed, as shown in Fig. 3.21B. Flaky Si nanoparticles (thickness of ca. 16 nm and lateral size of 0.2−1 μm) were produced from Si sawdust by bead-milling in isopropyl alcohol. Since the Si sawdust contained 4 wt.% graphite, which came from graphite substrate used for the cutting process of Si ingots, the Si particles after ball-milling were coated with thin carbon layers. They could be transformed to wrinkled

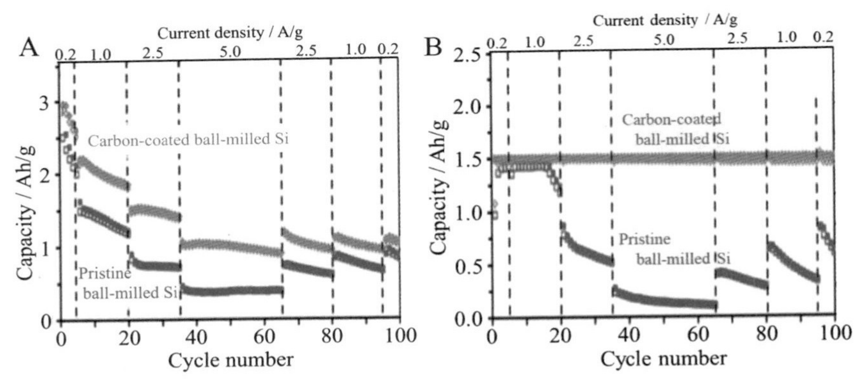

Figure 3.21 Capacity changes with cycling with different current densities on pristine ball-milled Si and a carbon-coated one: (A) without and (B) with a capacity restriction to 1.5 Ah/g [78].

morphology by lithiation/delithiation cycling and demonstrated better performances as a negative electrode of LIBs [79]. The carbon layer coated on Si particles is thought to work as a buffer between neighboring Si nanoparticles even with marked morphology change from spherical or flaky to wrinkled, in other words, to confine Si nanoparticles by coated carbon. Carbon coating of Si nanoparticles (<50 nm size) was carried out by mixing with different amounts of an ionic liquid, 1-ethyl-3-methylimidazolium dicyanamide (EMIM-DCA), and following carbonization at 600°C in Ar [82]. Si nanoparticles coated with 34 wt.% N-doped carbon exhibited a reversible capacity of 1145 mAh/g over 100 cycles with 71% retention at 1 A/g rate.

Thin Si flakes were synthesized by CVD of silane (SiH_4) gas on cubic crystals of NaCl at 550°C, followed by dissolution of NaCl, the thickness and lateral size of which were controlled to be 50 nm and 1 μm, respectively [80]. Carbon coating of the Si flakes were performed by heating in acetylene gas at 900°C, resulting in a carbon content of 7 wt.% and thickness of about 10 nm. Both the pristine Si flakes and the C-coated ones showed relatively small irreversible capacity, much less than commercial Si nanoparticles, and initial reversible capacity of 2943 and 2255 mAh/g, respectively, in the cell coupled with $LiCoO_2$ positive electrode in 1.3 M $LiPF_6$/(EC + DEC) electrolyte. As shown in Fig. 3.22, the C-coated Si flakes deliver a steady rate performance of around 2000 mAh/g up to 10C rate and their full cell shows a steady cycle performance at 0.5C rate. In contrast, the reversible capacity of the pristine Si flakes decayed drastically above 10C rate, and even at low rates the capacity decay occurred gradually

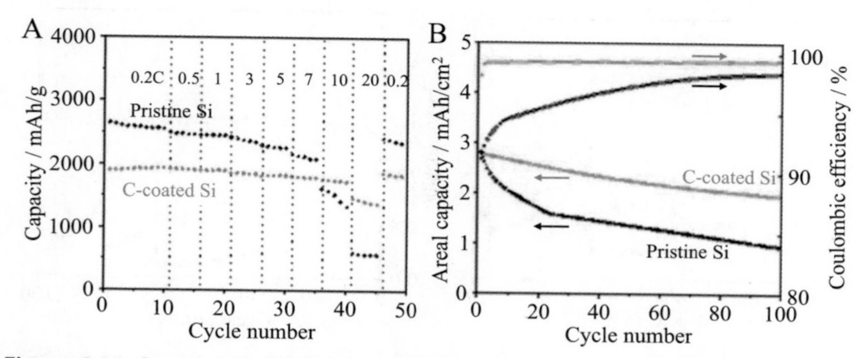

Figure 3.22 C-coated Si flakes in comparison with pristine Si flakes: (A) reversible capabilities at different C-rates and (B) cycle performance of the cell coupled with $LiCoO_2$ positive electrode at 0.5C rate [80].

with cycling. Carbon layer coated on Si flakes works not only to improve the electrical conductivity in the negative electrode but also to buffer the large volume expansion of Si by lithiation. The structural deformation of the carbon-coated Si flakes was monitored by in situ TEM observation under a bias of −3 and 3 V [80].

B-doped Si/SiO_2 composites were prepared from a mixture of ball-milled SiO_2 and B_2O_3 powder (20:1 by mole) at 950°C in Ar, where B-doped crystalline Si nanoparticles of about 15 nm were dispersed in amorphous SiO_2 particles (particles of about 3 μm) [83]. Carbon coating of the composite by CVD of acetylene at 700°C results in high cycle performance of the hybrid cell coupled with mesoporous carbon spheres as the positive electrode with 1 M $LiPF_6$/(EC + DEC + DMC), which delivered a maximum energy density of 128 Wh/kg and maximum power density of 9.7 kW/kg with a potential window of 2.0−4.5 V. Amorphous SiO_2 matrix in the composites was supposed to work as a buffer for a large volume change of Si during lithiation/delithiation. Si-included carbon composites were prepared from wasted glass (Si source), Mg (reducing reagent), and wasted polyvinyl butyral (carbon source) by carbonization at 500°C, of which the negative electrode performances in 1 M $LiPF_6$/(EC + PC) electrolyte were examined [84]. The reversible capacity in the first charge/discharge cycle reached about 3.3 Ah/g with an irreversible capacity of about 0.5 Ah/g at 210 mA/g rate, but it decayed rapidly with cycling, down to less than 1 Ah/g after 50 cycles. The composite of Si/rGO was prepared from a colloidal mixture of waste Si sludge containing SiC with GO by aerosol spray pyrolysis with thermal reduction of GO [85]. The composite prepared from a mixture with 0.1 wt.% GO delivered a capacity of about 1.6 A/g in the initial charge−discharge cycles by retaining the capacity of 1.2 A/g after 50 cycles and that with 0.4 wt.% GO gave much better cyclability although the capacity was reduced to about 0.8 A/g.

Carbon-coated Si nanowires were synthesized from the mixture of bis(bis(trimethylsilyl)amino)tin [Sn(HMDS)$_2$] and monophenylsilane (MPS) with different mole ratios (Si/Sn of 16−64) in anhydrous toluene solution at 490°C and 10.3 MPa, followed by the carbonization at 900°C under N_2/H_2 gas flow [86]. Sn(HMDS)$_2$ decomposed to Sn nanoparticles, which worked as a seed for the growth of polyphenylsilane-coated Si nanowires by the decomposition of a part of MPS, and polyphenylsilane was converted to carbon shells during heat treatment at 900°C. The carbon-coated Si nanowire exhibited high capacities of over 2 Ah/g, with nearly 96%

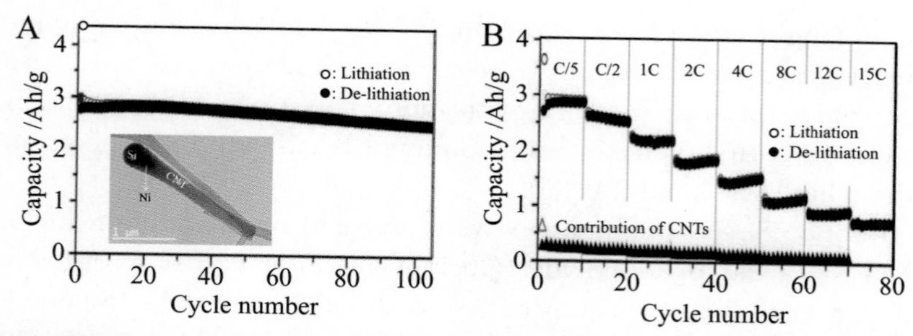

Figure 3.23 Core–shell structure of Si/CNT composites: (A) cycle performance at C/5 rate with TEM image of the composite (inset) and (B) rate capability [87].

Coulombic efficiency when cycled at a slow rate of 0.1C for 100 cycles and a stable capacity of about 1.3 Ah/g with greater than 98% Coulombic efficiency when cycled at 1C.

Metallic Si was deposited from the top of vertically aligned CNT arrays, where Si deposited along the length of CNTs with thickness gradient, as shown in the TEM image of a single CNT as an inset in Fig. 3.23A [87]. The resultant composites demonstrate good cyclability and rate capability, as shown in Fig. 3.23A and B, respectively, together with high capacity of close to 3 Ah/g at C/5 rate and 718 mAh/g even at 15C (63 A/g). This high performance was thought to be mainly caused by the advantages in strain accommodation and electrolyte access due to large interwire spacing.

3.1.2.3 Metallic germanium

The theoretical capacity of metallic Ge is 1625 Ah/kg (8643 Ah/L) due to the formation of the alloy $Li_{22}Ge_5$ and it was experimentally shown that the final product of electrochemical lithiation was $Li_{15}Ge_4$ corresponding to 1385 mAh/g (7366 Ah/L) using an evaporated Ge film [88]. Amorphous Ge nanoparticles (sizes from 100 nm to submicrometers) were electrochemically lithiated to crystalline $Li_{15}Ge_4$ through an amorphous Li_xGe alloy, associated with a high total volume expansion (about 260%). These Ge particles, however, were reported through in situ TEM observation not to be cracked during electrochemical lithiation and delithiation cycles, in other words, Ge nanoparticles are sufficiently tough for this type of cycling [89], in contrast to fragile degradation of Si particles described in the previous section. Ge particles with different morphologies were synthesized to improve the battery performances; amorphous Ge nanoparticles with sizes of about 10 nm stabilized by butyl groups [90], Ge nanowires synthesized by different processes [91–93], mesoporous Ge prepared by mechanochemical reaction of GeO_2 and Mg powders followed by HCl etching [94], Ge hollow spheres (220 nm diameter) and their 3D assembly by SiO_2 sphere templating [95], Ge nanocrystals by the gas-phase photolysis of tetramethyl germanium [96], etc. Commercial Ge powder after milling and after being C-coated using PVA at 900°C gave almost the same Li storage behaviors in 1 M $LiPF_6$/(EC + DEC), exhibiting two distinct voltage plateaus corresponding to the formation of Li_9Ge_4 and Li_7Ge_2 and forming a mixture of $Li_{15}Ge_4$ and $Li_{22}Ge_5$ phases as the final product [97]. C-coated Ge showed good cyclability and retained fairly large capacities of 579 mAh/g after 50 cycles.

Composite formations of Ge nanoparticles with various carbon materials are reported to result in the enhancement of battery performances of Ge. Microporous and mesoporous carbons containing Ge nanoparticles were synthesized by hydrothermal treatment of the mixture of GeO_2 and glucose with NaOH at 180°C, followed by heating in H_2/Ar flow at 680°C [98]. The resultant carbons were spherical with an average diameter of 57 nm and contained Ge particles of about 16.6 wt.%, which exhibited stable cycle performance with a capacity of about 600 mAh/g after the second charge–discharge cycle at 0.1C rate in 1 M $LiPF_6$/(EC + DMC + DEC). The composites of Ge nanoparticles with carbon were prepared by casting the mixture of *n*-butyl-capped Ge nanoparticles (10 nm diameter) with poly(styrene-*b*-isoprene) (PS-PI) block copolymer onto a stainless steel substrate, followed by curing at 65°C and carbonization at 800°C in Ar/H_2 flow [99]. The composite delivered a discharge capacity of about 1600 mAh/g up to 50 cycles at 1C rate (1C = 1.6 A/g). GeO_2-encapsulated mesoporous carbon (mesoporous GeO_2/C composite) was prepared from the mixture of resol, tetramethoxygermane (TMOG), and block copolymer PEO-*b*-PS (template) in THF solution with concentrated HCl, followed by carbonization at 600°C in Ar [100]. The resultant GeO_2/C composite was further heat-treated at 600°C in H_2/Ar flow for partial reduction of GeO_2 to metallic Ge (Ge/GeO_2/C composite). This mesoporous Ge/GeO_2/C composite exhibits the most stable battery performance among the three mesoporous composites prepared, GeO_2, GeO_2/C, and Ge/GeO_2/C, as shown in Fig. 3.24A, and delivers a high reversible capacity of 1.6 Ah/g in 1.3 M $LiPF_6$/(EC + EMC) electrolyte at 0.1 A/g rate. Its charge–discharge profiles in Fig. 3.24B suggest a relatively small irreversible capacity at the first cycle and high Coulombic efficiency in each cycle. Its rate capability was also the highest, with a reversible capacity of

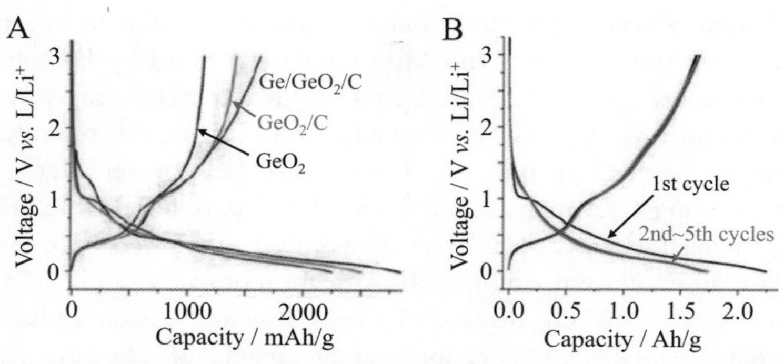

Figure 3.24 Charge–discharge performances of the mesoporous Ge/GeO_2/carbon composite: (A) comparison with the mesoporous GeO_2/C composite and GeO_2 and (B) charge–discharge cycle profiles for Ge/GeO_2/C composite [100].

428 mAh/g at 8 A/g rate and 1.2 Ah/g when the rate was changed from 8 to 1 A/g. C-coated Ge (58 wt.% C) was prepared by carbonization of tetraallylgermane at 700°C in vacuum, which exhibited a highly stable and reversible capacity of about 900 mAh/g in 1 M $LiPF_6$/(EC + DMC) electrolyte [101]. C-coated Ge particles were fabricated from GeO_2 powders with different sizes, <100 and 700 nm, by CVD of C_2H_2, followed by reduction of GeO_2 to Ge at 620°C in H_2 flow [102]. The C-coated Ge powder prepared from small-sized GeO_2 was composed from the large clusters of small C-coated Ge particle aggregates and that from large-sized GeO_2 was composed from large aggregates of small C-coated Ge particles, but did not result in clustering, as shown in the inset in Fig. 3.25 by coding "Clustered" and "Nonclustered." The latter shows

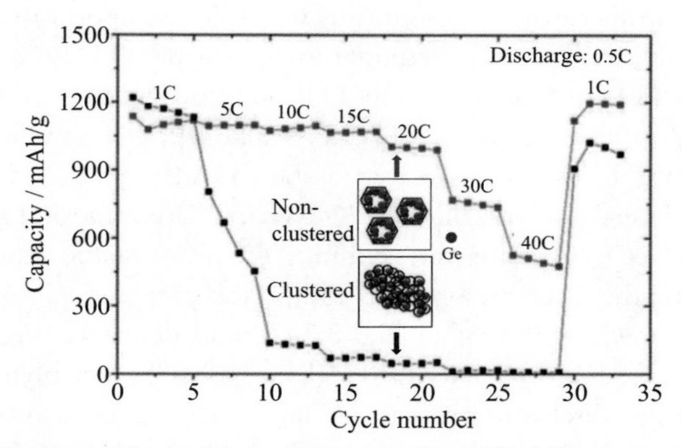

Figure 3.25 Rate capabilities of C-coated Ge powders with different textures, clustered and nonclustered [102].

better rate capability, particularly in high rates above 5C, than the former, although both deliver almost the same cycle performance at rates lower than 1C, as shown in Fig. 3.25. At 0.5C rate, the reversible capacity stabilized at about 900 mAh/g, even after 120 cycles. C coating of Ge nanowires was also reported to be effective for cycle performance improvement [103].

Composites of Ge nanoparticles with rGOs were prepared to study battery performances by constructing an LIB cell using it at the negative electrode [104–109]. Ge particles were deposited via its vapor on rGO flakes prepared by thermal exfoliation and reduction of GO flakes at 1050°C [107]. The reversible capacity of the composite at 400 mA/g rate was 795 mAh/g and it retained 675 mAh/g after 400 cycles. Ge-loaded rGO flakes were fabricated by reducing the mixture of GeO_2 and GO flakes with $NaBH_4$ at 60°C, which delivered stable cycling performance, more stable than that of the pure Ge, with a capacity of about 832 mAh/g after 50 cycles [108]. C-coated Ge-loaded rGO flakes were prepared by heating the mixture of GeI_2 and oleylamine-stabilized GO at 275°C, followed by carbonization at 500°C in H_2/Ar flow after flocculation by adding ethanol and toluene [109]. The battery performances were studied by constructing a full cell of this composite with the $LiCoO_2$-positive electrode in 1 M $LiPF_6$/(EC + DEC) electrolyte. Ge particles in the composite are crystallized in cubes with an average size of 70 nm, of which the content was 79 wt.%. As shown in Fig. 3.26A, the composite exhibits steady charge–discharge profiles, discharge capacity of 988 mAh/g, with a large irreversible capacity as 1300 mAh/g at the first cycle. After the second cycle, however, reversible capacity increases slightly with cycling, 1055 and

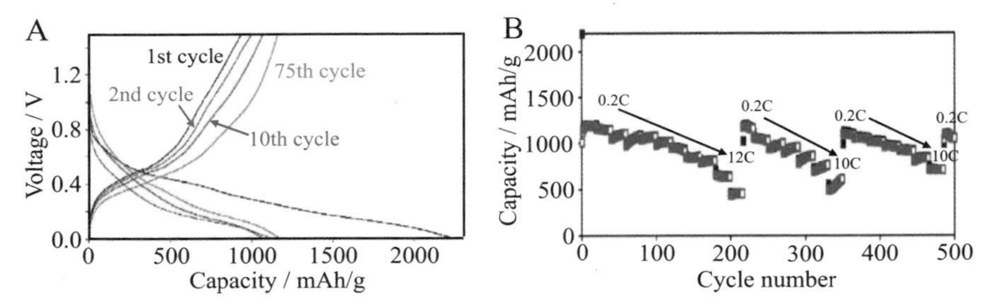

Figure 3.26 C-coated Ge-loaded rGO flakes: (A) charge–discharge profiles in 1 M $LiPF_6$/(EC/DMC) electrolyte at 0.2C rate and (B) rate capability by changing the rate from 0.2 to 12C repeatedly [109].

1166 mAh/g for the second and 75th cycles, respectively, with a high Coulombic efficiency of more than 99% (i.e., small irreversible capacity). The reversible capacity decreases with increasing charge–discharge rate, from 1182 mAh/g at 0.2C to 453 mAh/g at 12C, while the capacity is rapidly recovered to 1050 mAh/g by returning the rate to 0.2C, as shown in Fig. 3.26B.

Ge-encapsulated carbon nanofibers were prepared by electrospinning of tetramethoxy-germane (TMOG) and polyacrylonitrile (PAN), followed by stabilization at 280°C in air and heat treatment at 650°C in Ar/H_2 for the carbonization of PAN and reduction of GeO_2 formed from TMOG [110], which delivered a reversible capacity of about 1.42 Ah/g after 100 cycles at 0.15C. C-coated Ge nanoparticles synthesized by CVD of phenyl-trimethylgermane at 800°C were loaded onto single-walled carbon nanotube (SWCNT) paper through sonication in N-cyclohexylpyrrolidinone [111]. The full cell composed of the resultant composite fixed on Ti foil and $LiFePO_4$ in 1.2 M $LiPF_6$/(EC + EMC) electrolyte delivered a stable reversible capacity of 750 mAh/g at 100 mA/g rate. Loading of Ge nanoparticles onto SWCNTs was also performed by vacuum filtration [112].

Deposition of Ge onto vertically grown CNT arrays was employed to buffer large volume change of Ge particles by flexibility of the CNT array [113–115]. Onto vertically grown multiwalled carbon nanotube (MWCNT) arrays, amorphous Ge was deposited using a low-pressure CVD and crystalline Ge by radio frequency sputtering. The former delivered a capacity of 1096 mAh/g at 162 mA/g rate after 100 cycles, while the latter had a capacity of 730 mAh/g [114]. Amorphous Ge was deposited onto a CNT array, which was vertically grown on a few-layered graphene sheet by electron beam under a high vacuum of 3×10^{-6} Torr [115]. The Ge content in the composites was controlled at 39, 52, and 61 wt.%. The composite with 52 wt.% Ge delivered a reversible capacity of 1.5 Ah/g after 10 cycles at 1 A/g rate in 1 M $LiPF_6$/(EC + DEC). The capacity decreases to 0.8 Ah/g with an increasing rate up to 40 A/g, while it is recovered to 1.3 Ah/g by reducing the rate at 1 A/g, as shown in Fig. 3.27A. Charge–discharge profiles at different current densities in Fig. 3.27B reveal relatively small irreversible capacities, even at the first cycle, and flat voltage plateaus with three steps at 0.44, 0.36, and 0.20 V.

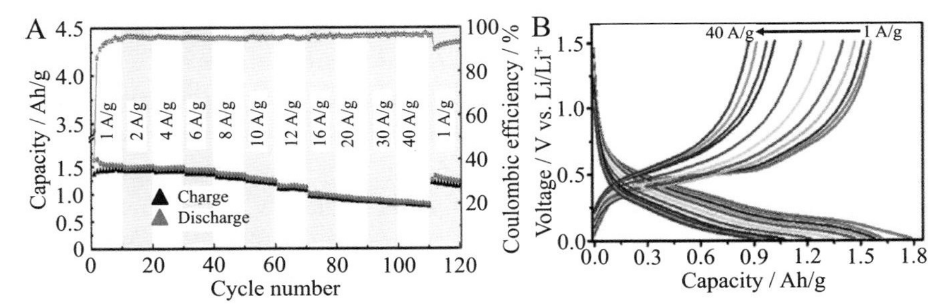

Figure 3.27 Ge-deposited CNT arrays: (A) rate capability and (B) charge–discharge profiles at different current densities [115].

3.1.2.4 Tin oxide and metallic tin

Tin oxide SnO_2 is known to have a high theoretical capacity (1494 mAh/g) and its reactions with Li^+ are expressed by the following two equations:

$$SnO_2 + 4Li^+ + 4e^- \quad Sn + 2Li_2O$$

$$Sn + xLi^+ + xe^- \quad Li_xSn \ (x \leq 4.4)$$

The first reaction can possibly occur reversibly for SnO_2 nanoparticles, although it is irreversible for bulk SnO_2, and the second reaction is perfectly reversible. In addition, Sn-based compounds have numerous advantages, including low cost, abundance, environmental benignity, etc. However, they have serious disadvantages when used as negative electrodes in LIBs, such as a large volume change (>300%) during Li-alloying and dealloying, marked aggregation of their particles during alloying, etc.

SnO_2 nanowires synthesized on Si substrate by the deposition of a vapor evaporated from the mixture of SnO and Sn at 900°C showed a large irreversible capacity of 1134 mAh/g in the first cycle, which was caused by the reaction of the formation of metallic Sn and Li_2O [116]. These SnO_2 nanowires delivered a discharge capacity that decreased gradually with cycling at 100 mA/g rate, from about 900 mAh/g at the first cycle to about 300 mAh/g after 50 cycles, whereas the bulk SnO_2 showed a marked decrease from about 570 mAh/g at the first cycle to less than 100 mAh/g after 50 cycles. Flower-like SnO_2 particles with approximately 6 mm sizes were synthesized from $SnCl_4$ via a solvent-induced and surfactant-assisted self-assembly technique at ambient temperature, followed by annealing at

500°C [117]. These flower-like SnO_2 particles delivered an initial discharge capacity of 1782 mAh/g at C/5 rate, but the reversible capacity decreased gradually to about 200 mAh/g after 20 cycles. Carbon coating of these particles by heating with glucose at 750°C could improve cycle performance slightly, to about 300 mAh/g after 20 cycles. Loading of SnO_2 nanoparticles on the surfaces of cross-stacked CNT yarns, which were spun from a CNT array after treatment by HNO_3 and polyvinyl pyrrolidone (PVP), was performed under hydrothermal conditions with $SnCl_4$ sol at 100°C, followed by annealing at 450°C in Ar [118]. The resultant composite delivered an initial capacity of 851 mAh/g. Through further cycling, the capacity steadily increased to a maximum of 1004 mAh/g at the 48th cycle.

C-coated SnO_2 hollow spheres (240−250 mm diameter), as illustrated in Fig. 3.28A, were prepared using silica nanospheres (template) and glucose at 500°C [119]. The resultant composite exhibited a good cycle performance for up to 200 cycles with relatively small irreversible capacity at the first cycle and stable reversible capacity around 500 mAh/g, as shown in Fig. 3.28.

The composites of N-doped rGO with SnO_2 nanocrystals were prepared through the reduction of GO by hydrazine in an aqueous dispersion with SnO_2, where SnO_2 nanocrystals were encapsulated by rGO flakes with thicknesses of about 5 nm [120]. The resultant composite exhibited high battery performances in 1 M $LiPF_6$/(EC + DMC) electrolyte. After the initial few cycles, the reversible capacity increased gradually with cycling and reached 1346 mAh/g after 500 cycles with a high Coulombic efficiency of more than 97%. Even at a very high current density of 20 A/g, the reversible capacity of 417 mAh/g was obtained. SnO_2 nanoflowers were deposited

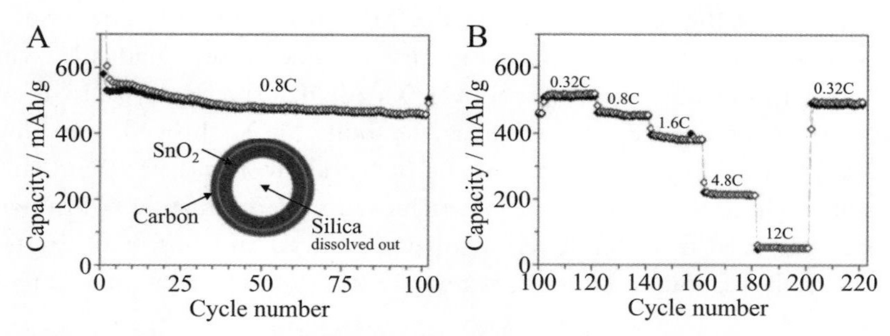

Figure 3.28 C-coated SnO_2 hollow spheres: (A) cycle performance at 0.8C with illustration of sphere construction and (B) rate capability [119].

under hydrothermal conditions at 180°C in $SnCl_2$ aqueous solution with sodium citrate onto N-doped CNFs, which were prepared from PAN by electrospinning [121]. The resultant composites exhibited slightly better performance in $LiPF_6$/(EC + DMC) than the SnO_2 nanoflowers and much better than the pristine CNFs. In 1 M $NaClO_4$/(EC + DMC) electrolyte (SIB), they exhibited slightly higher capacity than the pristine N-doped CNFs and much higher than SnO_2 nanoflowers. The composite of SnO_2 with a foam of porous rGO flakes could obtain a high SnO_2 loading of 12 mg/cm^2 and delivered a high areal capacity of up to 14.5 mAh/cm^2 at 0.2 mA/cm^2 rate and stable areal capacity of 9.5 mAh/cm^2 at 2.4 mA/cm^2 rate [122].

SnO_2 nanoparticles were deposited in the pores of mesoporous carbon CMK-5 by impregnation of $SnCl_4$ via its aqueous solution, followed by hydrothermal treatment at 90°C and heat treatment up to 550°C in air [123]. The resultant SnO_2@CMK-5 composite exhibited an improved cycle stability and a high reversible capacity, a very high and stable reversible capacity of about 978 mAh/g in the initial 10 cycles and 1039 mAh/g after 100 cycles at a current density of 200 mA/g and voltage of 0.005–3 V.

Metallic Sn nanoparticles were encapsulated into carbon hollow spheres from tributylphenyltin (TBPT, Sn source)/resorcinol-formaldehyde (carbon source) core–shell colloids stabilized by a surfactant (cetyltrimethylammonium bromide, CTAB) by carbonization at 700°C in Ar [124]. The size of carbon hollow spheres can be controlled by changing the ratio resorcinol/CTAB, at about 1 μm for the ratio of 4.1 and about 0.5 μm for 2.4 and the size of metallic Sn particles depended on the TBPT/CTAB ratio, being smaller with a decreasing ratio. Sn-encapsulated carbon hollow spheres with sizes of 1 mm and Sn content of 24 wt.% delivered a discharge capacity of 452 mAh/g in the first cycle in the voltage range of 0.0–2.0 V at a current density of 50 mA/g. Fine particles of metallic Sn could be dispersed in the pores of MgO-template mesoporous carbons [125]. MgO (particle size of about 100 nm) and SnO_2 (particle size of about 200 nm) were mixed with PVA in powder and carbonized at 900°C, followed by dissolution of MgO by diluted HCl, fine Sn particles being kept in the pores, as shown in Fig. 3.29A and B. On heating the mixture, SnO_2 was reduced to form molten Sn, which was supposed to adhere to the surface of the solid MgO particles and to be hindered from agglomeration during the carbonization of PVA. After the dissolution of MgO particles, fine particles of metallic Sn were kept in the pores. This carbon-coated Sn was

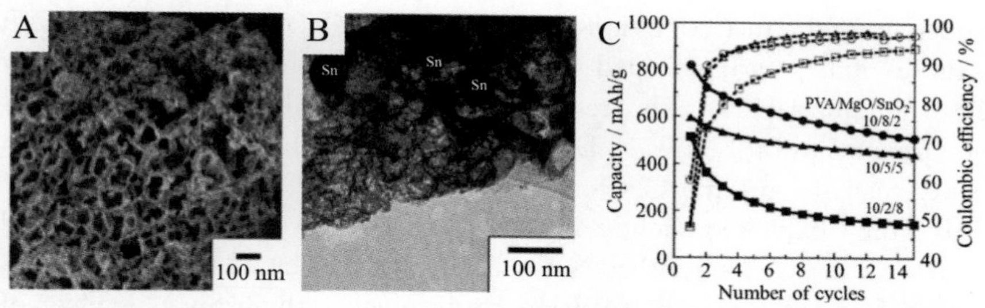

Figure 3.29 C-coated Sn: (A) SEM image, (B) TEM image, and (C) changes in capacity and Coulombic efficiency for different mixing ratios of PVA/MgO/SnO_2 by weight [125].

successfully applied to the negative electrode of LIB. The space neighboring Sn particle was thought to absorb a marked volume expansion due to lithiation and carbon shell surrounded Sn particles hindered the agglomeration of Sn particles during lithiation/delithiation cycles. The experimental results in Fig. 3.29C reveal that the mixing of the smaller amount of SnO_2 (PVA/MgO/SnO_2 of 10/8/2 by weight) results in the higher capacity of LIB, which may be due to the formation of smaller sizes of Sn and more efficient lithiation of Sn particles. Optimizations of the mixing ratio of MgO/SnO_2 and of the size of template MgO particles are required. From the mixtures of SnO_2 with PVA without MgO, large spherical particles of Sn metal were obtained separately from the carbon. Loading of SnO_2 nanoparticles was performed by heating a mixture of $SnCl_2$ with porous carbon at 320°C for 24 h [126]. The porous carbon was prepared from PF resin by colloidal silica template. Loading below 69 wt.% SnO_2 resulted in an improvement in capacities to 734–855 mAh/g and high capacity retentions of 70%–86% after 30 cycles. By coupling of this SnO_2-loaded porous carbon with a $LiNi_{1/3}Co_{1/3}Mn_{1/3}O_2$-based positive electrode, an all-solid-state LIB was proposed [127].

The composite of Sn nanoparticles with N-doped rGO was fabricated from a mixture of SnO_2 nanocrystals with GO by the reduction of GO with hydrazine vapor at 120°C and by the thermal reduction of SnO_2 at 500°C in H_2/Ar atmosphere [128]. The Sn content was determined to be about 2 at.% by XPS and the battery performances of these composites were studied in 1 M $LiPF_6$/(EC + DMC). The broad peak at 0.48 V and sharp peaks at 0.63, 0.72, and 0.79 V on the charge process of the CV curves (Fig. 3.30A) correspond to the phase transitions from Li_xSn alloy to Sn and on the discharging process the peaks at 0.60, 0.50, and 0.30 V corresponding to reversed phase transitions are observed, the corresponding

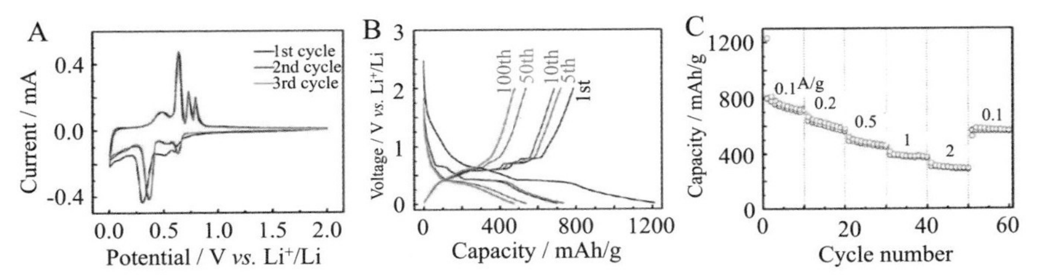

Figure 3.30 Sn nanoparticles encapsulated by N-doped rGO flakes: (A) CV curves at 0.1 mV/s rate, (B) charge–discharge profiles at 0.1 A/g rate, and (C) rate capability [128].

voltage steps being observed in charge–discharge profiles (Fig. 3.30B). The composite exhibits a high reversible capacity of 788 mAh/g with a relatively large irreversible capacity (about 800 mAh/g) in the first cycle at 0.1 A/g rate and its reversible capacity decreases gradually with the cycling to 481 mAh/g after 100 cycles. With increasing current density, the reversible capacity decreases from 628 mAh/g at 0.2 A/g rate to 307 mAh/g at 2 A/g rate, as shown in Fig. 3.30C.

3.1.2.5 Titanium oxides

Titanium oxide (TiO_2) has different polymorphs, among them the phases rutile, anatase, brookite, and TiO_2–B (monoclinic) having been reported to have a capacity for electrochemical lithium storage. These polymorphs consist of TiO_6 octahedra by either sharing their edges or corners. The electrochemical reaction of TiO_2 with Li^+ is expressed as:

$$xLi^+ + TiO_2 + xe^- \leftrightarrow Li_xTiO_2.$$

In the lithiated product Li_xTiO_2, not only the insertion of Li^+ in the crystal lattice, mostly at tetrahedral sites of closely packed oxygen atoms, but also the partial change in Ti valences from Ti^{4+} to Ti^{3+} compensate for interstitially inserted Li^+.

A monoclinic phase of TiO_2 (called TiO_2–B), which has a lower density than other crystalline phases, was synthesized from layered titanates $A_2Ti_nO_{2n+1}$ (A = Na, K, Cs and $3 \leqq n \leqq 6$) by hydrolysis in acidic solution, followed by heat treatment at 500°C [129]. It was applied to energy-storage devices as electrode material [130–133]. TiO_2–B was also synthesized as nanowires by a hydrothermal reaction between NaOH and TiO_2 at 170°C, followed by annealing at 400°C [130], which was able to store Li electrochemically up to the composition of $Li_{0.82}TiO_2$ (corresponding to the capacity of 275 mAh/g).

Rutile-type TiO_2 nano-sized rods with diameters of about 10 nm and lengths of about 200 nm, which were prepared by hydrolysis of $TiCl_4$ in HCl solution, could be inserted by Li ions up to $Li_{0.85}TiO_2$, the content of Li slightly increasing by repetition of insertion [134], as shown in Fig. 3.31A. On the other hand, a bulk crystal could accept only 0.03Li and nano-sized commercial TiO_2 powder 0.23Li. The cycle performance is compared on these three TiO_2 with different particle sizes in Fig. 3.31B, revealing a much higher capacity for the nano-sized rods. These results suggest a marked size effect of TiO_2 particles for Li storage. Mesoporous TiO_2 microspheres revealed stable capacitive performance in 1 M $LiPF_6$/(EC + DEC + DMC), which were synthesized by solvothermal treatment of tetrabutyltitanate in acetic acid at 150°C, followed by heat treatment at different temperatures of 400–1000°C in air [135]. TiO_2 microspheres heat-treated at 400°C, which was a mixture of TiO_2–B and anatase, gave the highest performance in the cells coupled with AC. Its charge–discharge curves and cycle performance are shown in Fig. 3.32A and B, respectively. With increasing heat-treatment temperature, the relative content of the anatase phase increased up to 800°C, and above 900°C the rutile phase became predominant.

Composite spheres were prepared from mesoporous TiO_2 nanoparticles (with average diameters of 350 nm) mixed with glucose by heat treatment under hydrothermal conditions at 180°C, followed by annealing at 400°C in Ar [136]. Making a composite of TiO_2 with porous carbon is very effective for improving the rate capability and cycle performance in 1 M $LiPF_6$/(EC + DEC + DEC), particularly at very high rates of 100C, as shown in Fig. 3.33. A mesoporous composite of TiO_2 with carbon was

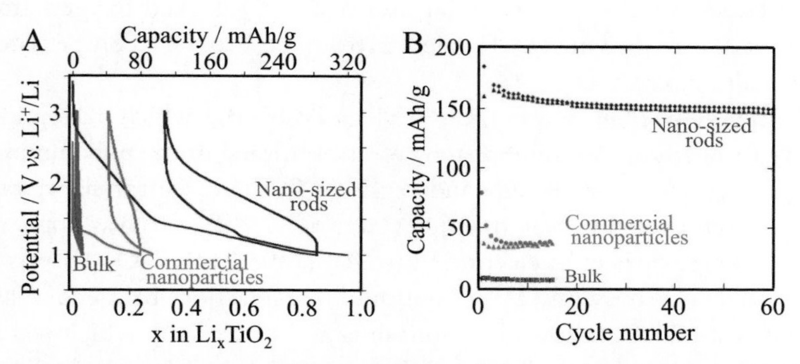

Figure 3.31 LIB performances of three rutile-type TiO_2 particles: (A) charge–discharge curves and (B) cycle performances [134].

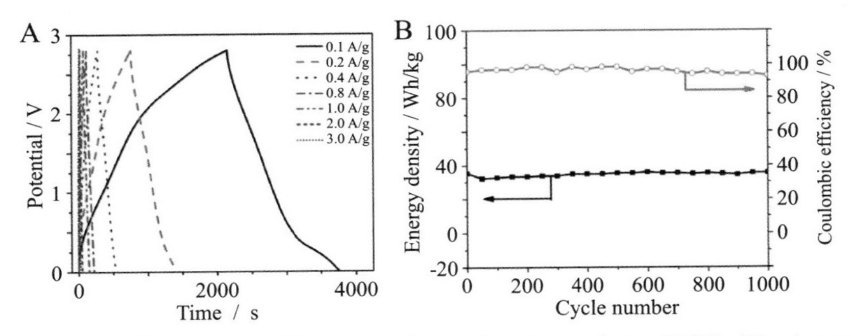

Figure 3.32 Mesoporous TiO_2 microspheres heat-treated at 400°C: (A) charge–discharge curves at different rates and (B) cycle performance at 1.0 A/g rate in 1 M $LiPF_6$/(EC + DEC + DMC) electrolyte [135].

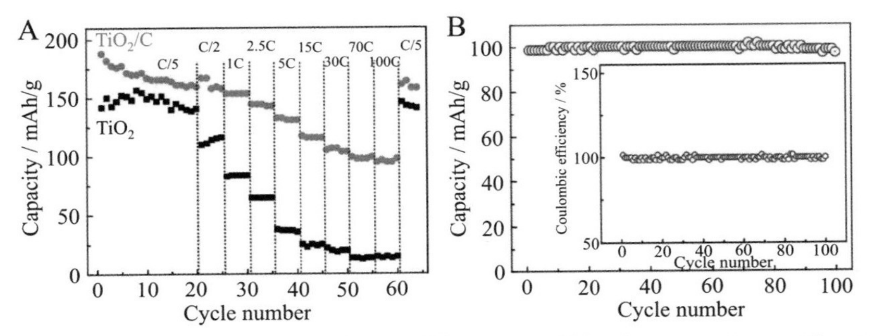

Figure 3.33 TiO_2/C composite spheres: (A) rate capability in comparison with neat TiO_2 spheres and (B) cycle performance at 100C rate [136].

prepared from the mixture of resol and $TiCl_4$ with triblock copolymer F127 with different carbonization temperatures from 500 to 700°C [137]. The composite containing 70 wt.% TiO_2 and carbonized at 600°C delivered a reversible capacity of 197 mAh/g at 50 mA/g rate and could maintain the reversible capacity as high as 109 mAh/g even at a 1 A/g rate. Porous carbon fiber webs loaded by anatase-type TiO_2 nanoparticles (about 13 nm size) were prepared by electrospinning of the polymer blends of PAN copolymer fibril, PMMA, and $TiO(OAc)_2$, followed by stabilization and carbonization at 600°C [138], which could deliver a high rate performance, and high capacity retention of about 200 mAh/g at a current density as high as 800 mA/g.

The flakes of rGO were used to hold anatase-type TiO_2 nanoparticles to construct the negative electrode in LIBs. The composites of rutile- and anatase-type TiO_2 with rGO flakes were fabricated by precipitation from an aqueous solution of $TiCl_4$ with rGO flakes, which were prepared by thermal exfoliation and reduction followed by functionalization using sodium dodecyl sulfate surfactant in aqueous solution [139]. The composites with rutile- and anatase-type TiO_2 exhibited stable cycle performances at capacities of about 170 and 160 mAh/g, respectively, at 1C rate.

Composites of TiO_2 nanotubes with carbon nanofibers (CNFs) were fabricated through hydrothermal treatment of a CNF web containing TiO_2 nanoparticles in 10 M NaOH solution at 150°C, followed by immersing in 0.1 M HNO_3 and then calcination at 420°C in Ar [140]. The composites consisted of TiO_2 nanotubes grown on CNFs and constructed a three-dimensional network, as shown in the schematic illustration as an inset in Fig. 3.34B. The starting CNF web containing TiO_2 was prepared by electrospinning of an ethanol/acetic-acid mixed solution of PVP and titanium tetraisopropoxide ($Ti(OiPr)_4$, followed by calcination at 900°C in an H_2/Ar atmosphere. The resultant composite exhibited stable cycle performance, particularly at a high rate such as 30C up to 1000 cycles, as shown in Fig. 3.34A and B.

3.1.2.6 Transition metal oxides

Transition metal oxides have electrochemical activity for Li ion storage through their redox reactions with Li ions. To make their activity effective in LIB cells, sufficient electrical conductivity and enough surface area for the redox reaction have to be kept in the negative electrode. Various composites of the oxides with carbon materials have been proposed.

C-coated α-Fe_2O_3 nanorods were prepared by carbonization of the mixture of citric acid with Fe_2O_3 nanorods [141]. The nanorods were synthesized by hydrothermal treatment of an aqueous solution of $FeCl_3$

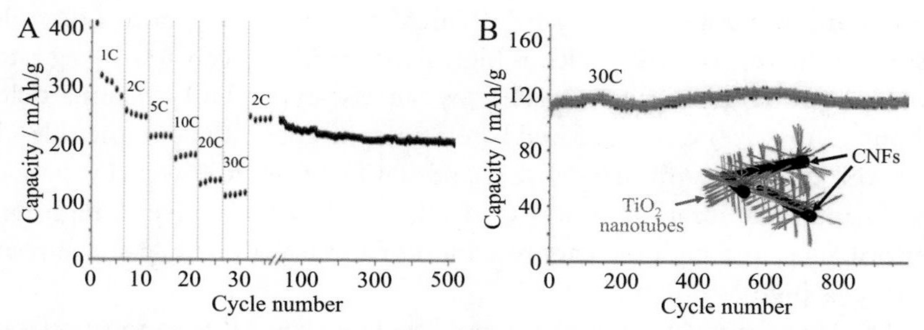

Figure 3.34 Composites of TiO_2 nanotubes with carbon nanofibers (CNFs): (A) rate capability and (B) cycle performance at 30C rate with an illustration of the composite (inset) [140].

with urea at 120°C and had diameters of 30–50 nm and lengths of a few hundred nm. The resultant composites had a coated carbon layer thickness of 2–5 nm and delivered an initial lithium storage capacity of 1120 mAh/g and a reversible capacity of 394 mAh/g after 100 cycles at 0.1C rate in 1 M $LiPF_6$/(EC + DMC) electrolyte. C coating was performed on crystalline powder of α-Fe_2O_3 (particle size of 15–40 nm) by mixing with malic acid ($C_4H_6O_5$), followed by carbonization at 300°C [142]. Cycle performance and voltage profile are shown on the C-coated α-Fe_2O_3 with a carbon content of 10.5 wt.%, in comparison with neat α-Fe_2O_3 powder in Fig. 3.35A and B, respectively. The composite shows excellent long-term cycling properties, with capacity increasing gradually with cycling to reach 2112 mAh/g after 100 cycles at C/2 rate. The reversible capacity at the 10th cycle of the composite is 1576 and 584 mAh/g at C/2 and 20C rates, respectively, while that of the neat α-Fe_2O_3 is 1223 and 245 mAh/g, respectively. C-coated α-Fe_2O_3 nanoparticles were prepared by pyrolysis of ferrocene vapor at 1050°C in Ar flow, followed by oxidation at different temperatures of 430–700°C [143]. The composite oxidized at 430°C contained about 10 wt.% C and that at 700°C 0.04 wt.% C; the latter steadily delivered a reversible capacity of 820 mAh/g in the potential range of 0.08–3.0 V with Coulombic efficiency of 99% for up to 50 cycles.

Fe_3O_4-encapsulated carbon composites were fabricated by carbonization of the mixture of $Fe(NO_3)_3$ and NaCl with glucose at 750°C in Ar, followed by annealing at 250°C in air and then dissolving NaCl, in which Fe_3O_4 nanoparticles (about 18 nm size) were coated with carbon and homogeneously embedded on carbon nanosheets [144]. Their LIB performances were investigated in 1 M $LiPF_6$/(EC + DMC + DEC)

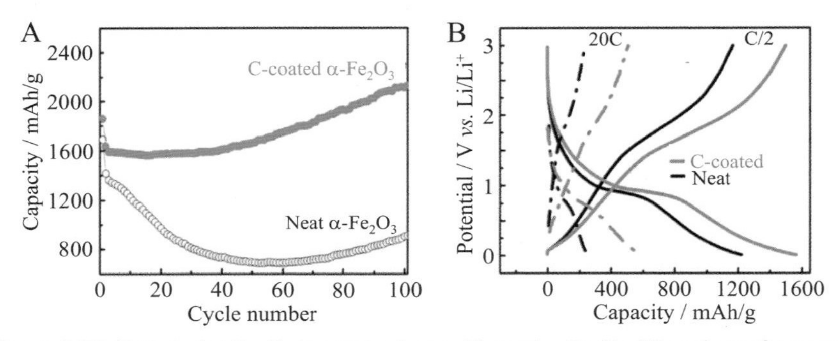

Figure 3.35 C-coated α-Fe_2O_3 in comparison with neat α-Fe_2O_3: (A) cycle performance at C/2 and (B) voltage profiles at the 10th cycle at C/2 and 20C rates [142].

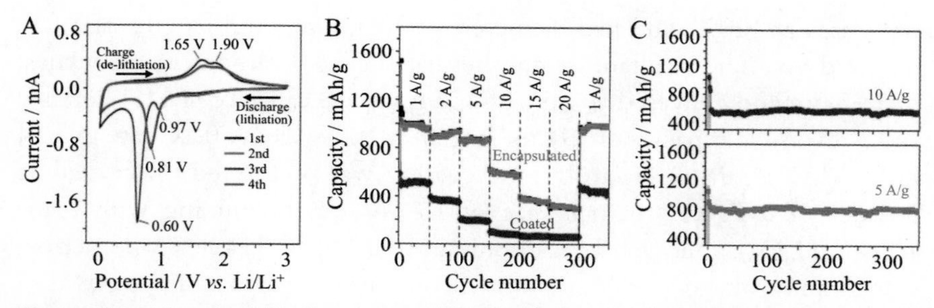

Figure 3.36 C-coated Fe_3O_4 nanoparticles loaded on carbon sheets (CFC): (A) CV curves, (B) rate performance in comparison with C-coated Fe_3O_4, and (C) cycle performances [144].

electrolyte by comparing with C-coated Fe_3O_4. As shown in Fig. 3.36A, CVs after the second cycle show two well-defined peaks at 0.97 and 0.60 V during discharging and 1.65 and 1.90 V during charging, which correspond to the two-step reaction between Li^+ and Fe_3O_4. The composite of C-coated Fe_3O_4 nanoparticles loaded on the carbon sheets (denoted as CFC) delivers a capacity of 963 mAh/g at 1 A/g rate and 570 mAh/g at 10 A/g after 200 cycles, which is a much higher capacity than the C-coated Fe_3O_4 (no-loading on the carbon sheets), as shown in Fig. 3.36B. Even at high rates of 15 and 20 A/g, the CFC could deliver 349 and 297 mAh/g, respectively. Cycle performances on this CFC are shown in Fig. 3.36C, where the first three cycles are performed at 1 A/g and later cycles up to 353 cycles are done at 5 and 10 A/g rates, revealing steady cycling with reversible capacity of 792 and 556 mAh/g, respectively, even after the 350th cycle.

A composite of Fe_3O_4 with rGO (Fe_3O_4/rGO composite) was prepared from the mixture of rGO flakes with spindle-shaped FeOOH by heat treatment at 600°C in Ar, in which Fe_3O_4 nanoparticles were encapsulated between rGO flakes and firmly attached to the flakes [145]. The cycle performance and rate capability of the composite in 1 M $LiPF_6$/(EC + DMC + EMC) electrolyte are compared with Fe_2O_3 powder prepared using the same procedures without rGO and commercial Fe_3O_4 powder in Fig. 3.37, revealing a marked improvement in battery performances by encapsulation between rGO flakes.

Composites of V_2O_5 with MWCNTs were fabricated using different processes, by coating of the surfaces of MWCNTs through the CVD of $VO(OC_3H_7)_3$ vapor [146], by heat treatment of MWCNTs mixed with hexadecylamine ($C_{16}H_{33}NH_2$) at 400°C in air, by hydrothermal treatment

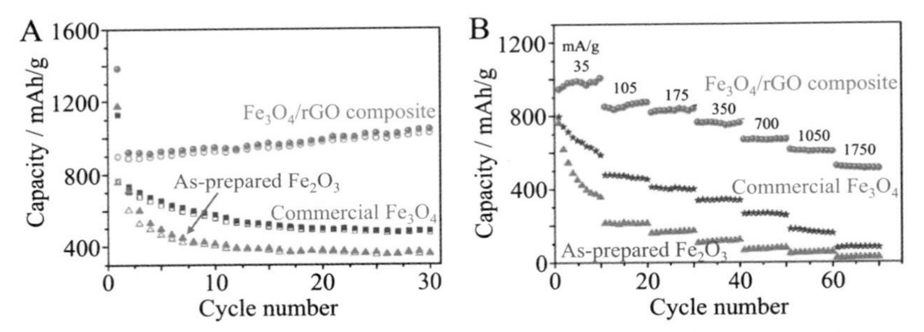

Figure 3.37 Fe_3O_4/rGO composites: (A) cycle performance at the current density of 35 mA/g and (B) rate capability in comparison with commercially available Fe_3O_4 and Fe_2O_3 prepared by the same procedure but without rGO [145].

of the mixture of acid-treated MWCNTs, V_2O_5 dissolved into H_2O_2 solution and hexadecylamine [147], and by hydrothermal treatment of the mixture of acid-treated CNTs and vanadium oxytriisopropoxide dispersed in isopropyl alcohol with surfactant at 190°C [148]. The V_2O_5/CNT composites demonstrated improved battery performances. As shown in the example in Fig. 3.38, the composite delivers much higher reversible capacity, particularly at high rates such as 20C and 30C [148]. Loading of V_2O_5 nanosheets onto rGO flakes [149] and onto graphene nanoribbons derived from unzipped CNTs [150] was also reported to be effective for the improvement of battery performance. An N-doped rGO foam loaded with 40 wt.% V_2O_5 was prepared through hydrothermal treatment of an aqueous dispersion of V_2O_5 and GO with pyrrole at 180°C, followed by heat treatment in Ar flow at 500°C to reduce GO to rGO and then in air at

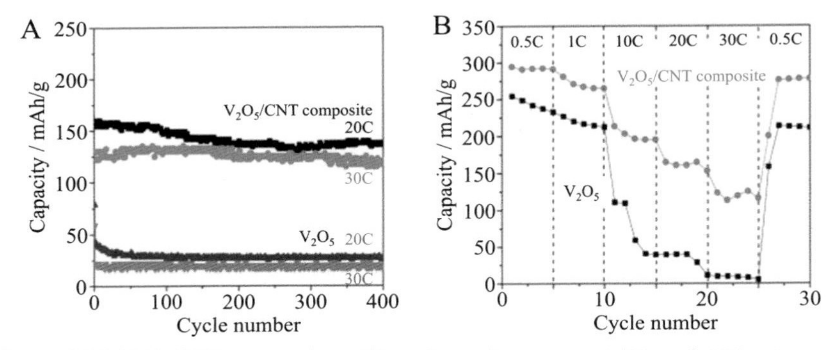

Figure 3.38 V_2O_5/CNT composites: (A) cycle performance at 20 and 30C rates and (B) rate capability in comparison with V_2O_5 microflowers [148].

300°C to obtain orthorhombic V_2O_5 [151]. This V_2O_5-loaded N-doped rGO foam exhibited a reversible capacity of 183 mAh/g at a current density of 500 mA/g after 100 cycles with a capacity retention of 82%.

MnO_2 has relatively high theoretical capacity of 1230 mAh/g in LIBs. Crystalline MnO_2 nanoparticles were deposited on hollow carbon spheres through the reduction of MnO_4^- using carbons on the sphere surface [152]. The hollow carbon spheres were synthesized using PMMA emulsion as a template and resorcinol-formaldehyde as a carbon precursor at 800°C, which had homogeneous diameters of about 220 nm. The composite with 47 wt.% MnO_2 delivered a stable reversible capacity of about 690 mAh/g during 100 cycles at 0.1 A/g rate and 420 mAh/g at 1 A/g. Crystalline MnO_2 leaves were deposited onto onion-like carbon particles [153], which could deliver a stable reversible capacities of 404 and 102 mAh/g at rates of 200 and 2000 mA/g, respectively. MnO_2 nanoflakes were assembled on graphene prepared on an Ni foam by CVD of CH_4 at 1000°C [154]. The resultant composite, of which SEM and TEM images are shown in Fig. 3.39A as insets, delivered a high capacity of about 1200 mAh/g at a current density of 500 mA/g after 300 cycles and a high-rate capability, with capacity higher than 500 mAh/g at 5 A/g rate, as shown in Fig. 3.39A and B, respectively. Carbon nanohorns loaded by MnO_2 flakes (42 wt.%) delivered a reversible capacity of 565 mAh/g at 450 mA/g rate after 60 cycles in 1 M $LiPF_6$/(EC + EMC + DMC) electrolyte [155].

Coaxial fibrous composites of MnO_2 and CNT were fabricated using AAO membrane (nanopore diameter of about 200 nm and length of about 50 μm) as the template, first depositing MnO_2 by vacuum infiltration and then carbon by CVD [156]. The resultant composites delivered a reversible

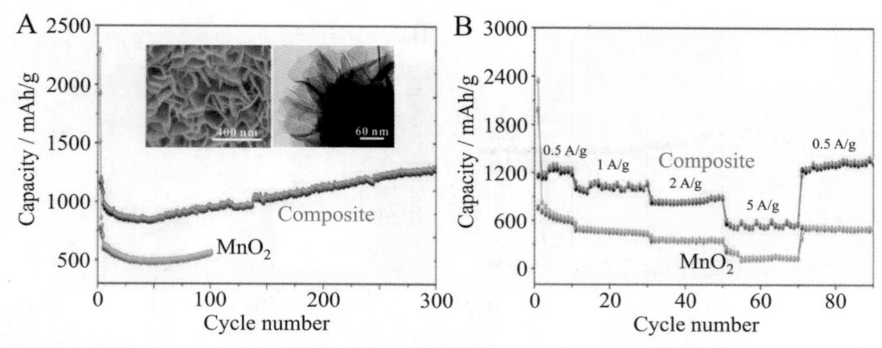

Figure 3.39 A composite of MnO_2 nanoflakes loaded on graphene foam: (A) cycle performance at 0.5 A/g rate with SEM and TEM images (insets) and (B) rate capability in comparison with MnO_2 [154].

capacity of about 500 mAh/g at 50 mA/g rate after 15 cycles with a large irreversible capacity at the first cycle. Deposition of flaky MnO_2 onto acid-treated MWCNTs was performed by hydrothermal treatment of CNT dispersion in $KMnO_4$ aqueous solution at 150°C [157]. The composite exhibited no capacity decay in the first 20 cycles and a gradual decrease to 620 mAh/g after 50 cycles at 200 mA/g rate, although MnO_2 without CNT showed marked fading by cycling up to 15th.

3.1.2.7 Other compounds

Self-assembled nanosheets of carbon-coated highly crystalline Cu_3P were prepared by heating of phosphorus-containing resin with copper foam at 900 and 1000°C by recovering from the surface of Cu foam, which had a high storage capability of lithium ions [158]. For comparison, Cu_3P was synthesized by heating red phosphorus with Cu powder at 900°C. Cycle performances at 37 mA/g rate with a voltage window of 0.01–3.0 V in 1 M $LiPF_6$/(EC + DMC + EMC) electrolyte are shown in Fig. 3.40A, revealing that C coating is effective for improving the capacity and cycle performance greatly. The 900°C-prepared composite delivers a discharge capacity of 689 mAh/g in the first cycle and 546 mAh/g after 50 cycles (78% retention) at a 37 mA/g rate, which is much better than Cu_3P without C coating. In Fig. 3.40B, its CV curves are shown, exhibiting clear reduction peaks at 0.85 V (peak B) and 0.74 V (peak C) corresponding to the formation of $LiCu_2P$ and Li_2CuP, respectively. The broad peak below 0.25 V appears to originate from the reduction of Li_2CuP to Li_3P and Li^+ adsorption into the coated carbon.

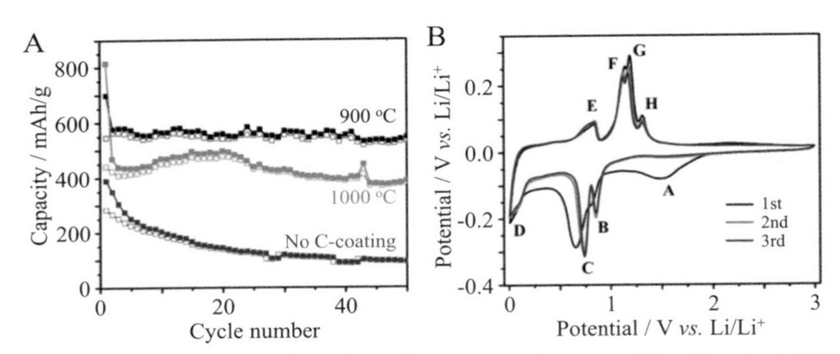

Figure 3.40 C-coated Cu_3P: (A) cycle performances of C-coated Cu_3P prepared at 900 and 1000°C in comparison with that of Cu_3P without C coating at a current density of 37 mA/g, and (B) CV curves of the 900°C-prepared composite at the first to third cycles [158].

A composite of Bi/rGO was prepared through electrodeposition at 25°C onto a carbon cloth in a solution of GO and $Bi(NO_3)_3$, in which Bi was deposited on the rGO scaffold by applying −0.9 V for 30 min [159]. The composite loaded with 40 wt.% Bi delivered an areal capacity of 3.51 mAh/cm^2 and a high-capacity retention of 83% after 50,000 cycles at 100 mV/s rate.

3.1.3 Lithium-sulfur and lithium-oxygen batteries

3.1.3.1 Lithium-sulfur batteries

Lithium-sulfur (Li—S) rechargeable batteries have attracted attention mainly because of their high capability for energy storage, of which the theoretically predicted energy density is 2.57 kWh/kg (2.20 kWh/L). The electrochemical reaction is expressed as

$$2Li + S = Li_2S,$$

and nonaqueous electrolyte using 1,2-dimethoxyethane (DME) and 1,3-dioxolane (DOL) often is employed. Sulfur in the negative electrode has to be held often in the pores of carbon materials, being confined in porous carbons. The performance of Li—S batteries was reviewed by comparing them with LIB and Li—O_2 batteries [160—162]. For their practical application, however, there are many issues to be solved, such as short lifespan, low efficiency and safety, insulating nature of sulfur, etc. Various carbon materials, including rGO, have been used for the negative electrode materials by making a composite with S to improve the electrical conductivity of sulfur, to immobilize sulfur at the negative electrode, and to retain the polysulfide diffusion within the pores of carbon [163].

Li—S cells were constructed using a mixture of elemental sulfur with poly(ethylene oxide) (PEO) and acetonitrile at the negative electrode and Li foil at the positive electrode with polymer electrolyte composed of PEO and $LiBF_4$ [164]. As shown in Fig. 3.41, cell performance was strongly influenced by the preparation process of the polymer electrolyte; the slurry of polymer PEO with $LiBF_4$ in acetonitrile is mixed by stirring for 24 h (stirring), ball-milling for 12 h (ball-milling), and ball-milling after adding nanoparticles of Al_2O_3 (ball-milling + Al_2O_3). Stepwise changes in cell voltage during discharge are clearly recognized, particularly for the electrolyte prepared by ball-milling with 10 wt.% Al_2O_3 additives and explained by the reactions between Li^+ and solid S, as expressed in Fig. 3.41. The initial discharge capacity of the cell using this electrolyte

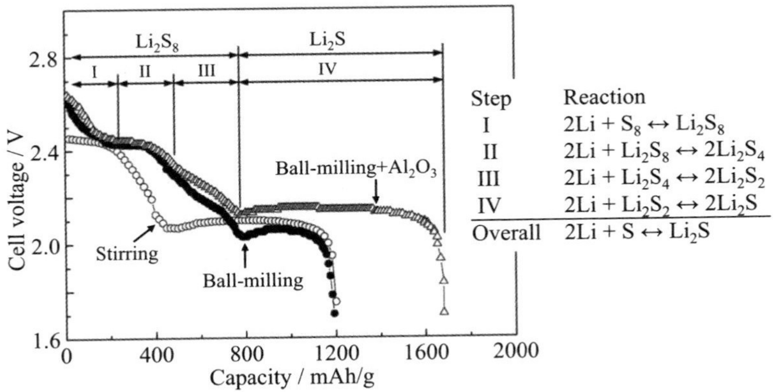

Figure 3.41 First discharge profiles of Li—S cells at a current density of 0.07 mA/cm^2 at 80°C with $(PEO)_6LiBF_4$ electrolyte prepared with three different processes [164].

(ball-milling with Al_2O_3 addition) was 1670 mAh/g, approximately the same as the theoretical value and better than those of the electrolytes prepared by a single processing, stirring and ball-milling.

Composites of sulfur with activated carbon (S/AC) were prepared by impregnating commercial ACs derived from coal, coconut shell, and sawdust into dimethyl sulfoxide (DMSO) solution of sulfur at 90°C, followed by washing with ethanol [165]. The Li—S battery using coal-derived AC as a negative electrode, which has S_{BET} of 854 m^2/g and V_{total} of 0.256 cm^3/g and can contain about 30 wt.% S, giving a slightly better rate performance than that using coconut shell-derived AC with S_{BET} of 1134 m^2/g and V_{total} of 0.379 cm^3/g and can contain about 31 wt.% S, as shown in Fig. 3.42A, of which the charge—discharge curves are shown in Fig. 3.42B. Mesoporous carbon CMK-3 after impregnating S at 155°C, in which the sulfur content reached about 70 wt.%, was used as the negative electrode with 1.2 M $LiPF_6$/ethyl-methyl-sulfone electrolyte [166]. The cell delivered an initial capacity of 1000 mAh/g at 168 mA/g rate (C/10 rate) at room temperature and retained about 800 mAh/g capacity after 20 cycles, as shown in Fig. 3.43. CMK-3 carbon was functionalized with a carboxylic group by oxidation in concentrated HNO_3 solution before S impregnation and then treated in an aqueous solution of polyethylene glycol (PEG) to trap the highly polar polysulfide species on the carbon surface. The resultant negative electrode gives a higher reversible capacity (around 1100 mAh/g), as shown in Fig. 3.43. By holding S in porous carbon CMK-3, the S content in the electrolyte solution was markedly reduced and was further reduced by PEG modification of CMK-3/S composite.

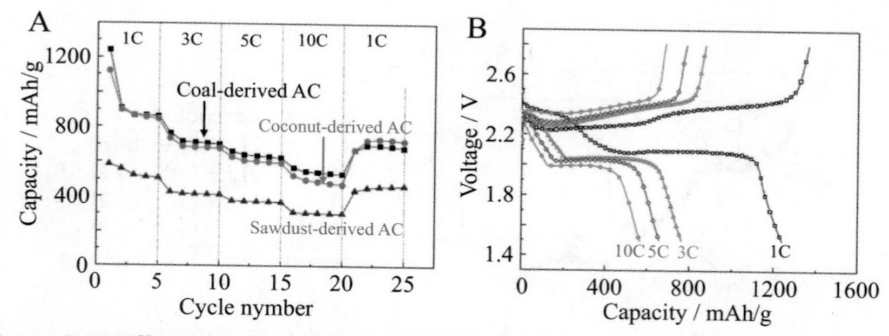

Figure 3.42 Effect of activated carbon (AC) for the preparation of the negative electrode of Li–S batteries: (A) rate performances of the cells with the negative electrodes composed from different ACs and (B) charge–discharge curves with different rates for the cell using the coal-derived AC at the negative electrode [165].

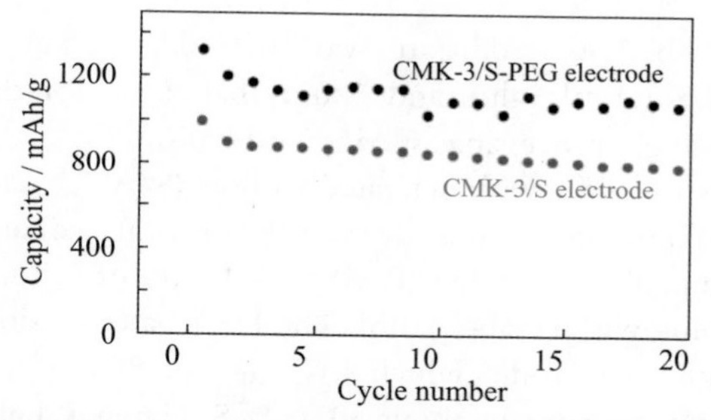

Figure 3.43 Cycle performance of Li–S cells using CMK-3/S and its PEG-modification (CMK-3/S-PEG) at the negative electrode [166].

Mesoporous carbon prepared from a mixture of sodium silicate and sucrose at 850°C was used to hold elemental sulfur at the negative electrode of an Li–S battery with different electrolyte solutions, 1 M lithium bis(trifluoromethanesulfonyl)imidate (LiTFSI) in poly(ethylene glycol) dimethyl ether (PEGDME) [167]. A carbon nanorod array was synthesized from glucose using ZnO array as the template under hydrothermal conditions at 180°C, followed by heating at 800°C in Ar, and employed to hold S at the negative electrode of an Li–S cell with 0.1 M $LiNO_3$/(DOL + DME) electrolyte, which could deliver a discharge capacity of 1050 mAh/g at 0.2C and 892 mAh/g at 2.0C [168].

A composite was prepared by heating a mixture of commercial MWCNTs, which had diameters of 10–60 nm and 0.1–5 μm length, with sublimed sulfur powder in the mass ratio of 1/5 and studied as the negative electrode of an Li–S battery with an Li foil positive electrode in

$LiPF_6$/(EC + DMC + EMC) electrolyte [169]. The pristine MWCNT web had a large amount of pores at 3–80 nm, as shown in Fig. 3.44A, which was reasonably supposed to occur by entanglement of CNTs and had S_{BET} of 238 m^2/g, which was markedly reduced by impregnation of S, suggesting that impregnated S was held in these pores, not inside the CNTs. The resultant composite delivered much higher capacity and better cyclability than a conventional S electrode without MWCNTs, as shown in Fig. 3.44B; the capacity at the first cycle reaches 700 mAh/g and reduces gradually with cycling down to about 500 mAh/g at 60th cycle.

Composite webs prepared by sulfurization (vulcanization) at 400°C in S powder (S/PAN/CNT composite) exhibited high cycle and rate performances, as shown in Fig. 3.45A and B, respectively [170]. The S/PAN/CNT composite webs were fabricated by electrospinning of DMF solution

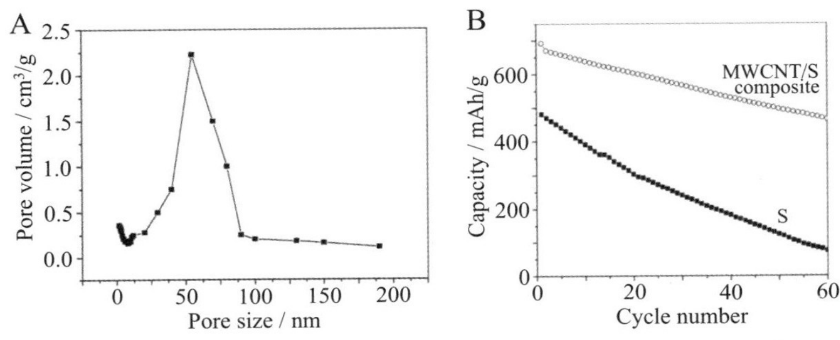

Figure 3.44 MWCNT/S composite electrode: (A) pore-size distribution on the pristine MWCNTs web and (B) cycle performance of the Li–S cell with the negative electrode of the composite at a current density of 2 mA/cm^2 [169].

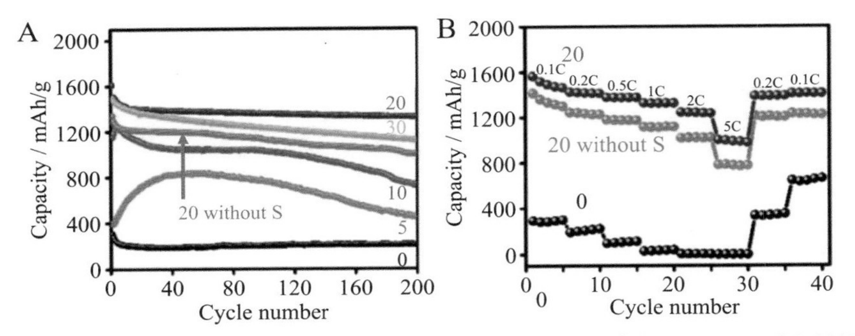

Figure 3.45 S/CNT composites prepared by electrospinning of the mixtures of S, PAN, and CNTs with different CNT contents: (A) cycle and rate performances. (B) The performances of the web prepared from the mixture of PAN, 20 wt% CNTs without S are shown for comparison [170].

of the mixture of sulfur, PAN, and CNT (diameter of 7–11 nm and length of about 1 μm) with a constant S/PAN ratio of 3/7 and different CNT contents to PAN of 5–30 wt.%, which were vulcanized at 400°C by embedding in sulfur powder and then assembled in a cell with 1 M $LiPF_6$/(EC + DMC + DEC) electrolyte. The composite web containing 20% CNT exhibited the highest discharge capacity of 1314 mAh/g for over 250 cycles at 0.5C rate, much higher than the web prepared from the mixture of S and PAN without CNTs. The composite web spun from a DMF solution with PAN, 20 wt.% CNTs without S exhibits slightly inferior Li–S battery performances.

A composite of elemental sulfur with porous rGO flakes with S content of 17.6 wt.% was prepared by heating the mixture up to 300°C, where rGO flakes were assumed to be completely coated by S [171]. This composite at the negative electrode in Li–S cell delivered higher capacity and better cycle performance than a pure S electrode, initial capacity of 1600 mAh/g and around 600 mAh/g after 40 cycles at a current density of 50 mA/g. rGO foams, which were prepared by Co^{2+}-assisted hydrothermal reduction of GO at 120°C could hold 73 wt.% S and deliver a first discharge capacity of 1260 mAh/g [172]. The capacity decreases with cycling, while became stable at about 700 mAh/g after 40 cycles. The composites of S with rGO for the electrode of Li–S batteries were fabricated through two routes: (1) mixing polysulfide with GO in NaOH solution, followed by reduction of GO with $NaBH_4$ at 80°C, and (2) mixing rGO reduced by using $NaBH_4$ in NaOH solution at 80°C with $Na_2S_2O_5$ solution [173]. The latter composite delivered an initial capacity of 819 mAh/g at C/2 rate (836 mA/g), retaining 541 mAh/g after 10 cycles and 504 mAh/g after 100 cycles. The deposition of TiC onto the graphene layer was shown to block effectively the shuttle of Li-polysulfides and to improve the sulfur utilization and cycling performance of Li–S cells [174]. Deposition was achieved using (Ar + 5% H_2)/$TiCl_4$ gas mixture at 1200°C by changing the deposition time in a range of 1.5–4 h to control the amount of TiC in 18–40 wt.%.

N-doped carbons were found to be able to anchor Li-polysulfide complexes, which resulted in suppression of their shuttling between the negative and positive electrodes, and in improvement of the cycling performance of Li–S batteries. N-doped microporous carbon spheres were prepared from polypyrrole (PPY) spheres by carbonization/activation at 900°C for 2 h in N_2 flow after mixing with $ZnCl_2$ with $ZnCl_2$/PPY of 4/1 by weight, followed by washing with HCl and water [175]. PPY spheres were synthesized by mixing pyrrole and $(NH_4)_2S_2O_8$ in aqueous solution at 0–5°C for 24 h. The resultant carbon spheres had diameters of about 200 nm and S_{BET} of 1958 m^2/g with V_{total} of 1.15 cm^3/g and could hold 53 wt.% S. It could give excellent cycling stability to Li–S cell and retain a reversible capacity of 1002 mAh/g after 200 cycles. N/S-co-doped microporous carbon spheres prepared from a mixture of D-glucose and L-cysteine by hydrothermal treatment, followed by KOH activation could provide marked affinity with Li-polysulfides for their confinement in the carbon electrode [176].

N-doped rGO flakes were prepared by heating GO at 400°C in a flow of NH_3, and its performance in an Li–S battery was compared with undoped pristine rGO prepared in Ar gas flow [177]. Much better rate and cycle performances were obtained for the N-doped rGO than the pristine rGO, as shown in Fig. 3.46. The N-doped rGO negative electrode can deliver a capacity of 330 mAh/g even at a very high current density of 6 A/g. The effective role of doped nitrogen in anchoring the soluble Li-polysulfide complexes was theoretically confirmed by DFT calculation with van der Waals correction [178]; polysulfide molecules, such as S_8, bind more strongly with pyridinic N substituted in graphene layer than with solvent molecules. Reduction and in situ N-doping process of GO was found to affect the battery performances of the resultant N-doped rGOs [179]. rGO flakes reduced by ethylenediamine and by urea had almost the same total N contents of 12.5 and 11.5 at.%, respectively, while about 33% N atoms had the pyridinic configuration in the former and almost no pyridinic N in the latter, and as a consequence the former delivered better electrochemical performances. The former gave a discharge capacity of

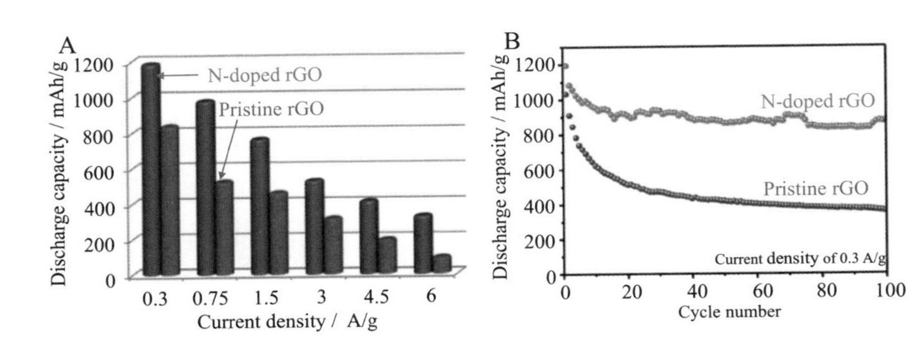

Figure 3.46 Li–S battery performances of the N-doped rGO in comparison with the pristine rGO: (A) rate and (B) cycle performances [177].

1357 mAh/g at a current density of 0.1C and long cycle stability with a capacity of 579 mAh/g at 1C up to 500 cycles. N-doped rGO flakes were prepared from a mixture of GO flakes with poly(diallyldimethylammonium chloride) (PDDA) by heating in Ar either at 700 or 850°C [180]. The rGO prepared at 700°C contained 75% pyrrolic-N and 25% pyridinic-N and that at 850°C contained 30% pyrrolic-N, 19% pyridinic-N, and 51% graphitic-N. When they were tested as the negative electrode in an Li–S battery at the rate of 0.2C, the capacity of the former was initially about 1100 mAh/g, increasing to 1300 mAh/g after five cycles, and then decreasing gradually to about 1000 mAh/g, whereas for the latter initial capacities of 814 mAh/g and 650 mAh/g after 100 cycles were obtained.

Carbon foams (carbon sponges) have the advantage of higher sulfur content than porous carbons containing micro- and mesopores. N-doped rGO-foams, which were prepared by hydrothermal treatment of GO dispersion in ammonia solution at 200°C and contained 10.1 at.% N (33.9% pyridinic-, 37.9% pyrrolic-, and 28.2% graphitic-N), were converted to the composites with S via a solution route using $Na_2S_2O_3$ and HCl, of which the S content reached 87.6 wt.% [181]. The resultant composites exhibited a discharge capacity of 792 mAh/g after 145 cycles at 600 mA/g rate, and 671 mAh/g after 200 cycles at 1.5 A/g rate. Codoping of N and S (N/S codoping) into rGO foams was carried out using a thiourea solution with GO dispersion under hydrothermal conditions at 180°C, in which 5.4 at.% N and 3.9 at.% S were doped [182]. N-doped rGO with 5.1 at.% N and S-doped rGO with 0.6 at.% S were also prepared using urea and Na_2S, respectively. These rGO foams were immersed into LiS_6 catholyte solution to load S with 4.6 mg/cm^2. The electrochemical performances of the composite of N/S-codoped rGO foams were determined and compared with those of N-doped, S-doped, and pristine rGO foams. As shown in Fig. 3.47, the composite of S/N-codoped rGO foam with LiS_6 delivered much better performances, a high capacity of 1200 mAh/g at 0.2C rate and 430 mAh/g at 2C rate, as well as high cycling stability up to 500 cycles by maintaining a higher capacity.

Wrapping of sulfur nanoparticles by N-doped rGO nanoflakes was also effective in improving battery performances [183], and delivering a high discharge capacity at high rate and long cycle life; at 0.2C about 1167 mAh/g and at 5C about 606 mAh/g with capacity retention (0.028% per cycle) even after more than 2000 cycles.

N-doped porous carbon loaded on flaky rGO substrate was prepared from the aqueous suspension of GO with glucose and pyrrole by heating

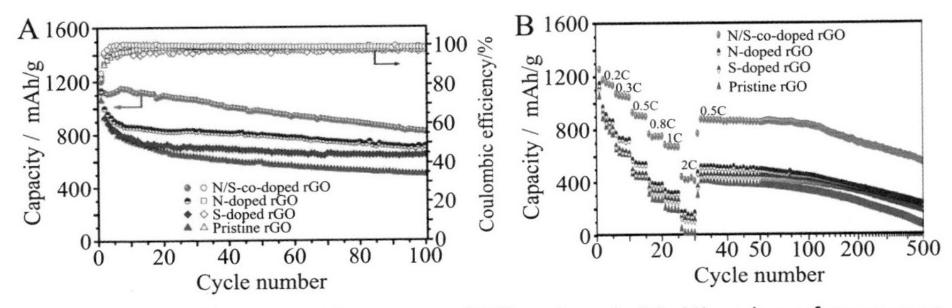

Figure 3.47 Li—S battery performances of N/S-codoped rGO: (A) cycle performance at 0.2C rate and (B) rate performance, in comparison with related rGOs [182].

under hydrothermal conditions at 180°C, followed by KOH activation at 900°C [184]. The resultant composites had high S_{BET} of 2680 m^2/g and V_{total} of 1.82 cm^3/g, and a high sulfur loading of about 60—70 wt.%. The composites after S-loading delivered improved electrochemical performance of an Li—S battery. Composites of rGO flakes loaded with N/P-codoped porous carbon were prepared from a mixture of goat hair (a precursor for porous carbon) with GO in an H_3PO_4 solution by carbonization at 600°C, where the porous carbon derived from goat hair contained 9.17 at.% N and 3.24 at.% P [185]. The composite containing 3 wt.% rGO delivered a reversible capacity of 503 mAh/g after 500 cycles at 1.0C rate with a low capacity decay with cycling.

The composite of N-doped aligned CNTs with graphene flakes was fabricated by CVD of the mixture of C_2H_4 and NH_3 onto exfoliated vermiculite-loaded FeMo catalyst particles at 750°C, followed by graphene deposition via CVD of CH_4 at 950°C, which were employed for the negative electrode of Li—S batteries with 0.5 M $LiNO_3$ solution in tetraethylene glycol dimethyl ether (TEGDME) and 1 M LiTFSI/(DOL + DME) electrolytes after impregnation of S [186]. Charge—discharge curves at the first and 50th cycles and rate performance for the composite are shown by comparing with aligned CNTs without N-doping and graphene deposition in Fig. 3.48A and B, respectively. The composite exhibited much better battery performances, delivering a high initial reversible capacity of 1152 mAh/g at 1C rate, and retaining about 880 mAh/g after 80 cycles.

To avoid serious shuttling of polysulfides, the coating of an ultrathin blocking layer onto the surface of the negative electrode was proposed [187]. A mixture of a carbon black with poly(3,4-ethylenedioxythiophene):poly(styrene sulfonate) (PEDOT:PSS) was deposited by an electrostatic spray technique onto the electrode of the

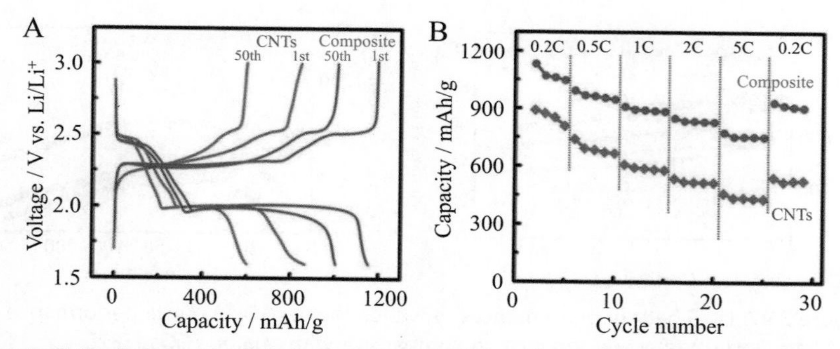

Figure 3.48 Composite of N-doped aligned CNTs with graphene in comparison with aligned CNTs: (A) charge–discharge curves with 1C rate at the first and 50th cycles, and (B) rate performances [186].

composite of rGO flakes covered by N-doped porous carbon. The coated electrode delivers more stable capacity with charge–discharge cycles than the uncoated one, as shown in Fig. 3.49A and B, respectively. At the PEDOT:PSS-coated electrode, the capacity Q_H due to the conversion of sulfur to higher order lithium polysulfides (Li_2S_n, $4 \leq n \leq 8$) at around 2.30 V is almost constant with cycling, and the capacity Q_L due to the further reduction to Li_2S_2/Li_2S decreases only slightly with cycling, although both Q_H and Q_L decrease with cycling at the uncoated electrode. The PEDOT:PSS layer coated with a thickness of 40 μm was effective.

Rice husks were carbonized at 700°C, resulting in the carbon containing 37 wt.% SiO_2, of which the content could be controlled by washing with HF [188]. The resultant carbon (RC) containing 15 wt.% SiO_2 was

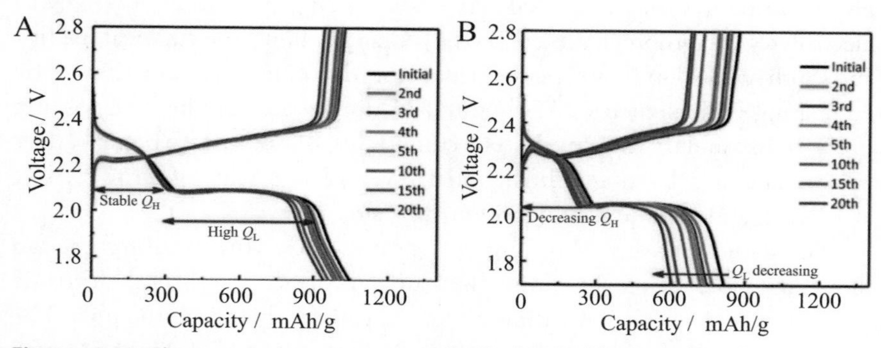

Figure 3.49 Charge–discharge curves of (A) PEDOT:PSS-coated rGO composite negative electrode and (B) uncoated rGO composite electrode [187].

loaded by S at 155°C after KOH activation because of the presence of sulfur-philic SiO_2, while it could also be loaded with Li after the modification by F-containing surface groups. Li–S cells were assembled using Li-loaded RC (Li/RC) at the positive electrode and sulfur-loaded RC (S/RC) at the negative electrode with 1 M LiTFSI/(DOL + DME) containing 0.1 M $LiNO_3$ (Li/RC//S/RC cell) and compared with the cell composed of Li-foil and S/RC (Li-foil//S/RC cell). Typical voltage profiles at two rates and rate performances of these two Li–S cells are shown in Fig. 3.50A–C. Li/RC composite affords better battery performances than Li-foil, with the Li/RC//S/RC cell delivering an initial capacity of about 1200 mAh/g and retaining about 800 mAh/g after 300 cycles at 0.5C rate.

3.1.3.2 *Lithium-oxygen batteries*

Lithium-oxygen batteries ($Li{-}O_2$ batteries) or lithium-air batteries have attracted attention mainly because of their high theoretical energy density of approximately 13,000 Wh/kg. The fundamental reaction in $Li{-}O_2$ batteries is expressed by

$$2Li + O_2 \leftrightarrows Li_2O_2.$$

The formation and dissolution of lithium peroxide, Li_2O_2, on the surface of the electrode (air electrode) materials upon repeating discharge and charge cycles of the cell were observed by in situ TEM analysis using $LiCF_3SO_3$ in tetraethylene glycol dimethyl ether ($LiCF_3SO_3$-TEGDME) electrolyte [189].

The essential prerequisite to the batteries is to design a porous air electrode (negative electrode) with highly open spaces for fast oxygen

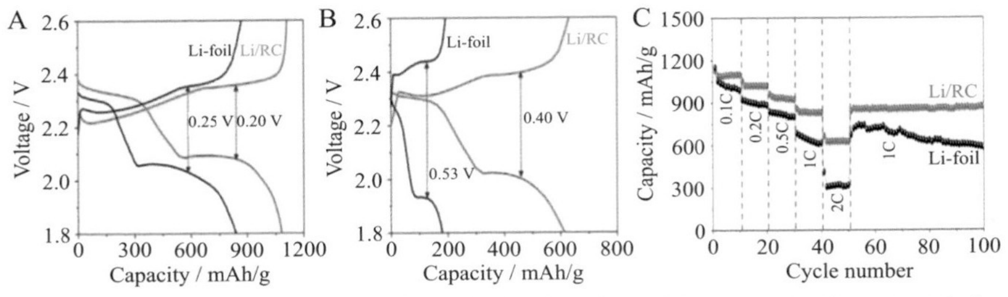

Figure 3.50 Li–S cells using rice-husk-derived carbon (RC) as electrode materials by loading Li (Li/RC) and S (S/RC), Li-foil//S/RC and Li/RC//S/RC; voltage profiles at (A) 0.1C and (B) 2C, and (C) rate performances [188].

diffusion, substantial nanopores to provide triple-phase regions (liquid–solid–gas) for oxygen reduction, and sufficient pore volume to accommodate the discharge products (Li_2O_2). In addition, a binder-free electrode is in great demand to avoid the deposition of insulating reaction products due to oxidation of organic binders. Various porous carbon materials have been applied as electrodes in batteries. The materials for the air electrode and electrolytes for $Li-O_2$ batteries and their battery performances have been reviewed and discussed elsewhere [190–193].

$Li-O_2$ batteries were assembled using Li foil at the positive electrode and carbon black (acetylene black) at the negative electrode with $LiPF_6$/(PAN + EC + PC) electrolyte, which could deliver 1410 mAh/g at a current density of 0.1 mA/cm^2 and 600 mAh/g at 2.0 mA/cm^2 rate in O_2 flow [194]. Mesocellular carbons synthesized from sucrose by carbonization using mesocellular silica template at 900°C were used as the electrode material with a binder PTFE for $Li-O_2$ cell with 1 M $LiClO_4$/PC electrolyte [195]. The cell could deliver a capacity of 2500 mAh/g at 0.1 mA/cm^2 rate, which is higher than that for the cell composed from carbon black (Super P). Carbon aerogels were prepared by the pyrolysis of RF hydrogels, which were prepared by a constant R/F molar ratio of 0.50 and different R/C molar ratio (C: catalyst) of 100–600, at 600–1000°C, followed by CO_2 activation at 800–1100°C [196]. The electrode of the $Li-O_2$ cell was constructed from the paste of the carbon aerogel, MnO_2, a binder, and PC by spreading it into a 200 μm thick film. The discharge capacity of the carbon aerogels in 1 M $LiPF_6$/PC electrolyte was governed by the pore structure of the carbon evaluated by S_{BET} and burn-off during CO_2 activation, as shown by the discharge curves at the first cycle in Fig. 3.51.

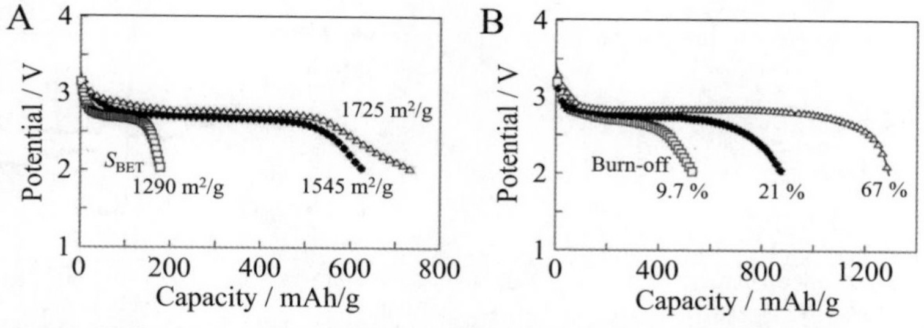

Figure 3.51 Carbon aerogels derived from RF hydrogels: first discharge curves of the carbons with (A) different S_{BET}s and (B) different burn-off [196].

Enhancement of the performance of Li—O_2 cells by N-doping using NH_3 at 1000°C was experimentally shown using commercially available carbon black (Ketjenblack KB) and activated carbon (Calgon activated CA), and their blend (KB/CA of 40/60 by weight) [197]. The N-doped carbon blend had S_{BET} of 1385 m^2/g with V_{total} of 1.8 cm^3/g and N content of 5.2 at%. The composites were prepared on the Ni foam by pasting the PTFE mixture of this N-doped carbon blend with 5 and 10 wt.% LAGP [$Li_{1.5}Al_{0.5}Ge_{1.5}(PO_4)_3$] and used for solid-state Li—O_2 batteries as the air electrode [198]. The composite containing 10 wt.% LAGP, of which S_{BET} was 974 m^2/g and V_{total} was 1.09 cm^3/g, delivered a much higher discharge capacity than the composite with 5 wt.% LAGP and the pristine N-doped carbon blends, as shown in Fig. 3.52. This significant enhancement in discharge capacity was supposed to be attributed to higher electrocatalytic activity and fast lithium ion conduction ability of LAGP in the air electrode.

Ultra-lightweight N-doped carbon foams with a bulk density of 0.002 g/cm^3 were prepared by carbonization of ultra-lightweight polyaniline foams at different temperatures of 500—1000°C in N_2, from which the electrodes for the electrochemical cells were fabricated by adhering to stainless steel mesh using 5% Nafion at 80°C under 10 MPa compression (binder-free electrode) [199]. By using this foam as the air electrode, an Li—O_2 cell with Li foil electrode and a ceramic separator was constructed using the electrolyte, 1 M $LiClO_4$/(EC + DEC) and 1 M KOH aqueous solution. The first discharge capacity of the cell depends on the carbonization temperature of the carbon foam, as shown in Fig. 3.53A, as high

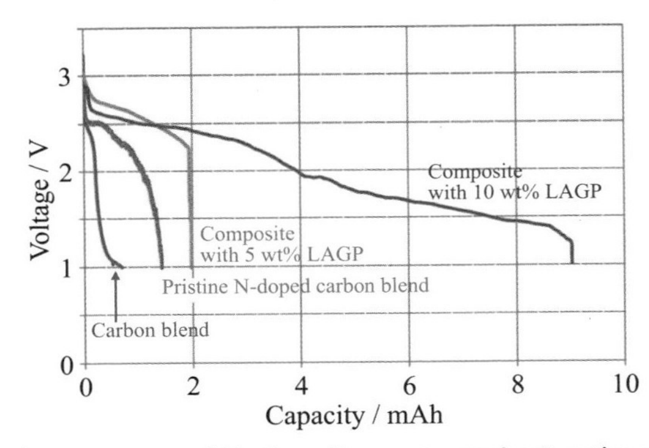

Figure 3.52 Discharge curves of Li—O_2 cells constructed using the composites of N-doped carbon black with LAGP at a discharge current of 0.2 mA at 75°C [198].

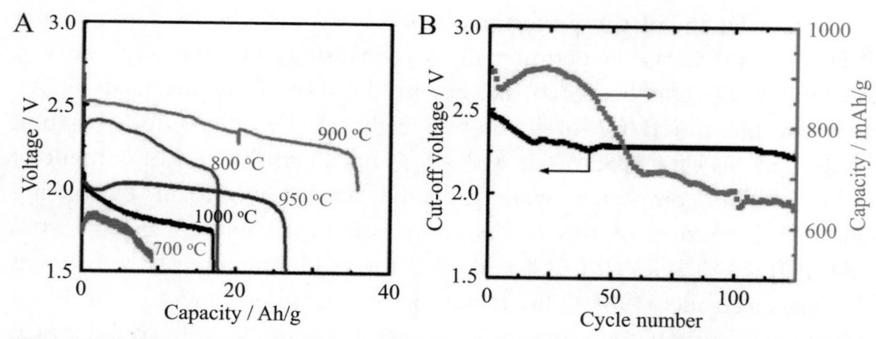

Figure 3.53 Ultra-lightweight N-doped carbon foams: (A) discharge voltage profiles of the first cycle at 1 mA/cm^2 rate for the foams prepared at different carbonization temperatures, and (B) cut-off voltage and capacity for the foam prepared at 900°C [199].

as 35.8 Ah/g at a current density of 1 mA/cm^2 for the foam carbonized at 900°C. The cycling lifetime reaches more than 250 h after 125 cycles, as shown in Fig. 3.53B. The three-dimensional pore structure of the carbon foam was able to achieve efficient diffusion of both O_2 gas and electrolyte ions in the cell.

rGO fabricated by thermal exfoliation and reduction of GO at 1050°C delivered high discharge capacity in Li–O_2 cell with 0.1 M $LiPF_6$/TEGDME electrolyte: 8530, 5333, and 3090 mAh/g at current densities of 75, 150, and 300 mA/g, respectively [200]. N-doping of this rGO by annealing in NH_3 enhanced the discharge capacity more: 11,660, 6640, and 3960 mAh/g at 75, 150, and 300 mA/g rates, respectively. Commercial rGO powders with different oxygen contents, C/O ratios of 14 and 100, were used as the air electrode by using 25 wt.% PTFE binder (thickness of 0.35 mm) with LiTFSI/Tri-EGDME electrolyte [201]. The cell using the rGO with C/O of 14 delivered a discharge capacity of 15,000 mAh/g with a plateau at around 2.7 V at 0.1 mA/cm^2 rate in oxygen pressure of 2 atm, whereas that using the rGO with C/O of 100 delivered less than 8000 mAh/g. SEM and TEM observations on electrode materials after discharging showed the deposition of Li_2O_2 particles, about 10 nm-sized on the rGO with C/O of 14, while much larger particles on the rGO with C/O of 100.

N-doped CNTs were synthesized by floating catalyst CVD of imidazole at 850°C with ferrocene as a catalyst precursor [202]. N-doped CNTs had characteristic bamboo-like texture with diameters of 50–60 nm, of which the N content was measured as about 10 at.% by XPS, while the nondoped

CNTs (pristine CNTs) had diameters of 40–50 nm without a bamboo-like texture. $Li-O_2$ cells with Li foil electrode in 1 M $LiPF_6$/(PC + EC) electrolyte were constructed using the air electrode prepared by casting a mixture of either the pristine or N-doped CNTs with 10 wt.% binder PVDF onto a separator. The first cycle discharge capacities were calculated from discharge profiles to be 590 and 866 mAh/g for the pristine and N-doped CNTs, respectively. The discharge capacities for the N-doped CNTs decreased to 264 and 133 mAh/g at the second and third cycles, respectively, while those for the pristine CNTs dropped more markedly to less than 45 mAh/g.

Galvanostatic discharge–charge profiles in 1 M $LiPF_6$/TEGDME are compared for the $Li-O_2$ cells using orthogonally woven MWCNTs, carbon black (Ketjenblack), and commercial CNT powder at the air electrode [203]. The profiles up to 20th cycle are presented in Fig. 3.54A–C, respectively. The cell using woven MWCNT sheet delivers a discharge capacity of about 2500 mAh/g with comparable charge capacity. The discharge–charge cycles could be continued with high Coulombic efficiency up to 20 cycles, while it exceeded over 100% after the 10th cycle, which suggested electrolyte (TEGDME) decomposition, and as a consequence the capacity decreased abruptly after 20 cycles. On the other hand, Ketjenblack and the CNT powder exhibit reasonably high discharge capacities at the first cycle, but rapid capacity decays after only a few cycles, which was attributed to the accumulation of discharge products and their incomplete decomposition during cycles. The woven MWCNTs exhibited high cyclability by limiting the discharge capacity at 1000 mAh/g even at a high current density of 1 and 2 A/g. Free-standing foam-like films of CNTs were fabricated by mixing acid-modified CNT suspension with PS colloids

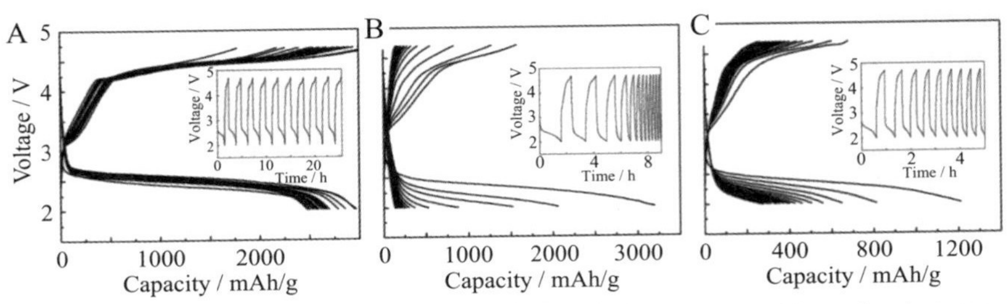

Figure 3.54 Discharge–charge profiles of the $Li-O_2$ cells for 20 cycles between 2.0–4.7 V at a current density of 2 A/g: (A) woven MWCNTs, (B) Ketjenblack, and (C) CNT powder [203].

in deionized water, where negatively charged CNTs and PS colloids ensured a highly stable suspension, followed by vacuum filtration and annealing at 500°C in N_2 [204]. The resultant films had a very low bulk density of 0.42 g/cm^3, corresponding to a high porosity of 81%, and delivered a high capacity of 4683 mAh/g at the current density of 50 mA/g.

N-doped MWCNTs were synthesized onto Ni foam substrate by catalyst floating CVD of ethylene with different amounts of melamine, which were directly used as the air electrode for the Li–O_2 cells [205]. The discharge–charge curves for the CNTs with different N contents are compared by limiting the discharge capacity to about a half of the full discharge capacity in Fig. 3.55A, and the cycle performance of the cell used CNTs containing 3.92 at.% N is shown by the discharge–charge curves in Fig. 3.55B. The discharge capacity increases with increasing N content up to 3.92 at%, which can maintain the discharge capacity up to the 26th cycle. The discharge capacity of the pristine CNTs began to decrease at the 14th cycle. Effectiveness of N-doping into MWCNTs on Li–O_2 cell performance was confirmed in two electrolyte solutions, 1 M $LiPf_6$/(PC + EC) and LiTFSI/(DOL + DME) [206].

Electrodes fabricated by carbon nanofibers (CNFs) grown by CVD of C_2H_4 at 700°C on AAO substrates coated with thin layers of Ta and Fe were used as the air electrode of Li–O_2 cells with 0.1 M $LiClO_4$/DME electrolyte [207]. CNFs with diameters of about 30 nm consisting of cup-staked microtexture along the fiber axis were grown perpendicular to the substrate with slight entanglement near the substrate, resulting in a carpet-like film. The Li–O_2 cells delivered a discharge capacity of about

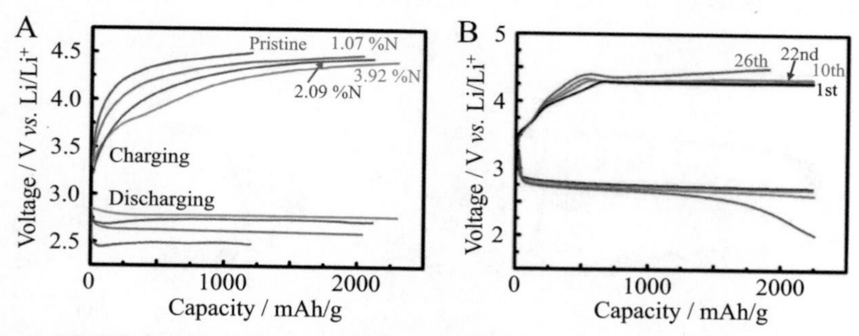

Figure 3.55 N-doped CNTs: (A) discharge–charge curves at the first cycle for the CNTs with different N contents and (B) cycling performance for CNTs containing 3.92% N at 0.3 mA/cm^2 rate by limiting the discharge capacity to about a half of full discharge [205].

4720 mAh/g at low current densities less than 70 mA/g, but the capacity decreased markedly with increasing current density to less than 1000 mAh/g at 1000 mA/g rate. By using the CNF electrode, the deposition of Li_2O_2 particles on the sidewalls of the fibers and their morphological changes during discharging and charging (deposition and disappearance) were visually confirmed. Li_2O_2 particles appeared as small spheres with diameters of less than 100 nm by discharging up to 350 mAh/g, grown to a toroidal shape with an average particle size of 400 nm by further discharging up to 1880 mA/g, after full discharging the discrete particles merged into large and dense particles (about 1 μm size).

Loading of Pd nanoparticles on the electrode carbon was shown to be effective in improving the performance of $Li-O_2$ batteries using a commercial carbon black (Super P) [208]. Pd loading was performed by atomic layer deposition (ALD) technique of palladium hexafluoro-acetylacetonate at 200°C. To control the amount of Pd deposited, the ALD process was cycled 1, 3, and 10 times (with each cycle, approximately 1 wt.% Pd was deposited), the particle sizes of Pd increasing from 2.6 to 8 nm after 1 cycle to 10 cycles, respectively. As shown in the voltage profiles in 1 M $LiCF_3SO_3$/TEGDME electrolyte at 100 mA/g rate in Fig. 3.56, the carbon black after three cycles of Pd deposition gave the highest discharge capacity of about 6500 mAh/g, in comparison with the pristine carbon black (about 1500 mAh/g) and that after 10 cycles deposition (about 1700 mAh/g). The discharge capacity decreased dramatically with increasing cycle numbers, decreasing to about 3000 mAh/g at the second cycle and about 500 mAh/g at the seventh cycle.

Co_3O_4 nanoparticles were loaded onto commercial rGO flakes (lateral size of 0.5–2 μm, thickness of about 0.8 nm, and oxygen content of approximately 7 wt.%) through hydrothermal treatment in an aqueous

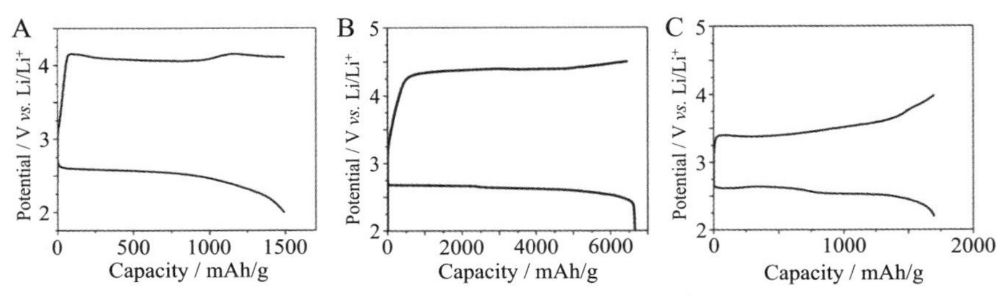

Figure 3.56 The first cycle discharge–charge profiles of carbon blacks with different Pd-loading in 1 M $LiCF_3SO_3$/TEGDME electrolyte at 100 mA/g: (A) the pristine carbon black, (B) after three cycles, and (C) after 10 cycles of Pd deposition [208].

solution of $Co(NO_3)_2$ at 100°C, followed by annealing at 150°C [209]. The composite Co_3O_4/rGO was cast onto a glassy carbon plate by dispersing in ethanol with 0.05 wt.% Nafion to be used as the air electrode of an Li–O_2 cell with O_2-saturated 0.1 M KOH electrolyte. This cell could be reversibly charged and discharged up to 50 cycles at a current density of 160 mA/g. Co_3O_4 nanofibers were prepared by electrospinning of DMF solution of cobalt acetate with a small amount of PVP, followed by calcination at 600°C in air, and used to fabricate the composites with rGO and graphene flake (GF) (Co_3O_4/rGO and Co_3O_4/GF, respectively) through their suspensions [210]. GFs were noncovalently functionalized by 1-pyrenebutyric acid (PBA) in advance to achieve uniform loading of Co_3O_4 nanofibers, and as a consequence to prevent the restacking of graphene layers. As shown by the discharge–charge curves at the first cycle in LiTFSI/TEGDME electrolyte in Fig. 3.57A, the composite Co_3O_4/GF delivers a high capacity of 10,500 mAh/g at a current density of 200 mA/g, compared to Co_3O_4 nanoparticles (2000 mAh/g), Co_3O_4 nanofibers (10,000 mAh/g), and Co_3O_4/rGO composite (5000 mAh/g), and exhibits high reversibility. In addition, Co_3O_4/GF composite has high cyclability, as compared with cycle performances of others up to 80th cycle and its discharge–charge profiles up to the fifth cycle under the limitation of the discharge capacity to 1000 mAh/g are shown in Fig. 3.61B and C, respectively. High discharge capacity and cyclability of the Co_3O_4/GF composite was supposed to be achieved by improved catalytic activity of Co_3O_4 nanofibers, associated with facile electron transport via interconnected GFs, and fast O_2 diffusion through porous Co_3O_4 nanofiber networks. Loading of α-MnO_2 nanorods onto a carbon black (Super P) was

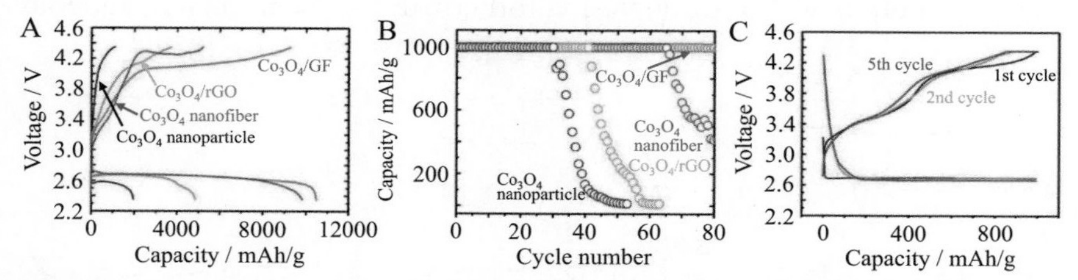

Figure 3.57 Composite of Co_3O_4 nanofibers with graphene flakes (Co_3O_4/GF): (A) the first cycle discharge–charge curve at 200 mA/g rate, (B) cycle performance under the limitation of discharge capacity of 1000 mAh/g in a voltage window between 4.35 and 2.0 V at 200 mA/g rate in comparison with Co_3O_4 nanoparticles, Co_3O_4 nanofibers, and the composite Co_3O_4/rGO, and (C) discharge–charge profiles at the first, second, and fifth cycles under the limitation discharge capacity of 1000 mAh/g [210].

carried out via $KMnO_4$ solution at room temperature [211]. The composite containing 10 wt.% MnO_2 delivered the first cycle discharge capacity of about 1400 mAh/g at 100 mA/g rate in 1 M $LiCF_3SO_3$/TEGDME electrolyte, which exhibited reversible cycling up to the 50th cycle under the capacity limitation of 500 mAh/g. Loading of TiN nanoparticles onto carbon black (Vulcan XC-72) was also reported to be effective in improving the performance of $Li-O_2$ batteries [212].

Hydrophobic ionic liquids were shown to be effective with a high discharge capacity and a long cycle life by using mesoporous carbon (S_{BET} of 1329 m^2/g and the same V_{total} and V_{meso} of 2.01 cm^3/g) [213]. They delivered a high first cycle discharge capacity of 2433 mAh/g in air at a rate of 0.1 mA/cm^2 in 1 M $LiClO_4$/(EC + PC). They could also deliver a much higher capacity, as high as 5360 mAh/g, in hydrophobic ionic liquid electrolyte EMITFSI, probably because EMITFSI prevented vaporization of the electrolyte and hydrolysis of the Li electrode.

3.1.4 Sodium-ion batteries

Lithium-ion rechargeable batteries (LIBs) have commonly been used in various devices mainly due to their long lifetime and high energy density. However, there is increasing concern about the cost and limitation of lithium reserves, and sodium-ion rechargeable batteries (SIB) have been proposed as an alternative to LIB, because of the low cost and high abundance (it is the fourth most abundant element on Earth). A brief comparison between lithium and sodium is shown in Table 3.1. Many reviews have been published on SIBs by focusing on electrode materials relating to battery performance [214–228].

The operating principles of the SIBs are very similar to those with LIB, with the "rocking-chair" mechanism of Na^+ ions between positive and

Table 3.1 Characteristics of lithium and sodium.

Electrolyte ion	Capacity (Ah/g)	Voltage (V vs. SHE)	Atomic weight (g/mol)	Ionic radius (nm)	Melting point (°C)	Coordination preference
Li	3.86	−3.04	6.9	0.069	180.5	Octahedral and tetrahedral
Na	1.16	−2.70	23	0.098	97.7	Octahedral and prismatic

negative electrodes. In the negative electrode of SIBs, reversible extraction/insertion of Na^+ ions at a high redox potential, usually >2 V versus Na^+/Na, are required with low volume change. Therefore, a wide group of materials, including layered oxides, polyanionic phosphates, and pyrophosphates has been proposed. The layered metal oxides Na_xMO_2 (M is one or more transition metals with multiple oxidation states) have been applied, such as Na_xCoO_2, Na_xFeO_2, $Na_xFe_{1/2}Co_{1/2}O_2$, $Na_x[Ni_{1/3}Fe_{1/3}Mn_{1/3}]O_2$, etc., in which MO_6 octahedra form layers by sharing their edges and Na is preferentially located in either trigonal prismatic or octahedral positions in interlayer spaces. In the negative electrode of SIBs, the safety issues resulting from the growth of dendrites are more serious than LIBs and so it is necessary and critical to find a suitable sodium host material. So far, various materials have been proposed, including carbonaceous materials, sodium alloys, metal oxides/sulfides, phosphorus, and phosphides. Since the principal purposes of this book are focused on porous carbons, the application of carbon materials, even nonporous graphite, to the negative electrode of SIBs is explained and discussed in this section.

Graphite has been widely employed on the negative electrode of commercial LIBs, while it cannot be used in SIBs because of its low acceptability for Na^+. However, it was reported that exfoliated graphite can accept a certain amount of Na^+ [229]. Exfoliated graphite was prepared by thermal exfoliation and simultaneous reduction at 600°C for either 1 or 5 h (EG-1h and EG-5h), respectively, of graphite oxide (GO) synthesized via modified Hammers method. On charge and discharge cycles, cell voltage changes gradually, there is no plateau, and EG-1h delivers a large reversible capacity of 284 mAh/g at a current density of 20 mA/g after 30 cycles, as shown in Fig. 3.58A and B. Reversible capacity is retained as 184 mAh/g at 100 mA/g rate with 74% retention after 2000 cycles, as shown in Fig. 3.58C. The experimental facts that exfoliated graphite EG-1h still contains a large amount of functional groups, such as C—OH, C—O—C, and C=O, and its charge—discharge curves are sloped, and there is no plateau, may suggest that the principal mechanism of Na^+ storage is interaction of Na^+ ions with functional groups in the EG, not intercalation/deintercalation. Electrode performances of mesocarbon microbeads (MCMBs) heat-treated at 700°C were studied in SIB with 1 M $NaClO_4$/(EC + DMC) electrolyte [230]. The MCMBs delivered an initial capacity of 232 mAh/g at a current density of 25 mA/g and retained a reversible capacity of 161 mAh/g after 60 cycles. The charge—discharge curves did not show a plateau at a voltage range of 0—1 V.

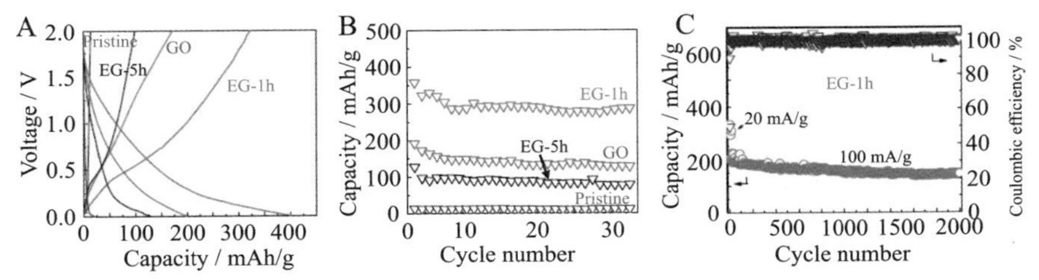

Figure 3.58 SIBs with exfoliated graphite heated for 1 h (EG-1h) in comparison with pristine graphite, graphite oxide (GO), and EG heated for 5 h (EG-5h): (A) charge–discharge curves at the second cycle, (B) cycle performances at 20 mA/g rate, and (C) long time cycle performance of EG-1h [229].

Different nongraphitized carbons, including soft and hard carbons, were studied as the negative electrode of SIBs. Hard carbons were used as the negative electrode of SIBs [18,231–236]. Hard carbons prepared by carbonization at 1300°C from phloroglucinol and glyoxylic acid mixed using either water, ethanol, or their mixture exhibited high performance in SIBs [235]. The charge–discharge curves are shown in Fig. 3.59A for the hard carbons, which have no appreciable effect on the solvent used for the preparation of the carbon precursor and are characterized by the presence of the plateau at a voltage lower than 0.2 V. As shown in Fig. 3.59B, reversible capacities of the hard carbons are stabilized at about 250–270 mAh/g with Coulombic efficiency of almost 100% at a current density of 37.2 mA/g (C/10), while they showed relatively high irreversible capacities in the initial few cycles. The same plateaus were reported on the hard carbon prepared from resol-type PF resin by carbonization at 900°C, followed by

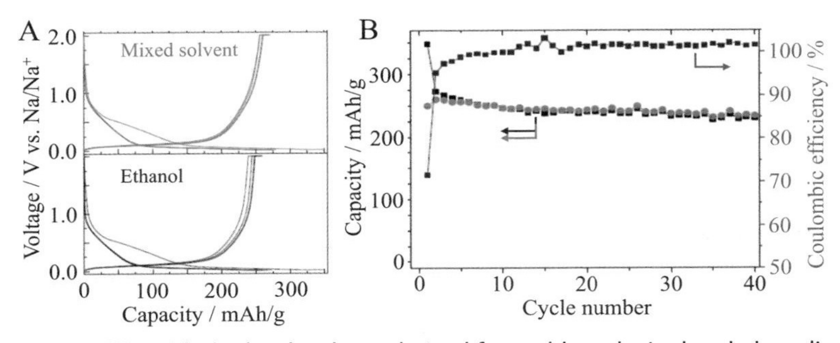

Figure 3.59 SIBs with the hard carbons derived from phloroglucinol and glyoxylic acid by carbonization at 1300°C: (A) charge–discharge curves of the carbons derived from the resins prepared by using mixed solvent of ethanol with water and ethanol and (B) cycle performance at C/10 rate for the carbon from the resin using ethanol [235].

heat treatment at different temperatures in the rage of 1100−1500°C [236]. The capacities measured on the cell were discussed by separating to the capacity due to the plateau region and that due to the sloping region.

N-doping for the porous carbons was effective in improving the SIB performances [18]. The carbons were prepared by heating covalent organic frameworks (COFs) at 500, 600, and 700°C in N_2 flow, which contained 11.5, 9.7, and 8.38 at.% N, respectively. A SIB cell composed of 600°C-carbonized N-doped carbons exhibited good cyclic and rate performances, as shown in Fig. 3.60A and B, respectively, with 240 mAh/g at the current density of 50 mA/g and 127 mAh/g at 2.5 A/g rate, and a capacity retention of 88.8 mAh/g after the 5000 cycles.

Battery performances of microporous carbons prepared by carbonization of ZIF-8 at 930°C in N_2 (ZIF-C) were compared with a mesoporous carbon CMK-3 (from sucrose using SBA-15 at 930°C in N_2) in 1 M $NaClO_4$/PC electrolyte [237]. S_{BET} of ZIF-C and CMK-3 was 1251 and 1056 m^2/g, respectively, and the average pore size was about 0.5 and 4.0 nm, respectively. As shown in Fig. 3.61A, ZIF-C delivered much higher reversible capacity than CMK-3, and the initial capacities of ZIF-C and CMK-3 were 164 and 95 mAh/g, respectively. The rate and cycle performances are comparable, as shown in Fig. 3.61B and C, with ZIF-C exhibiting stable cycling by retaining the capacity of 136 mAh/g after 50 cycles at 50 mA/g rate.

The carbon nanofibers (CNFs) prepared from cellulose nanofibers stabilized at 240°C in air by carbonization at 1000°C in Ar exhibited high electrochemical properties as an electrode of SIBs with 1 M $NaClO_4$/(EC + PC) [238]. They delivered a high reversible capacity of 255 mAh/g at a current density of 40 mA/g with good rate stability of

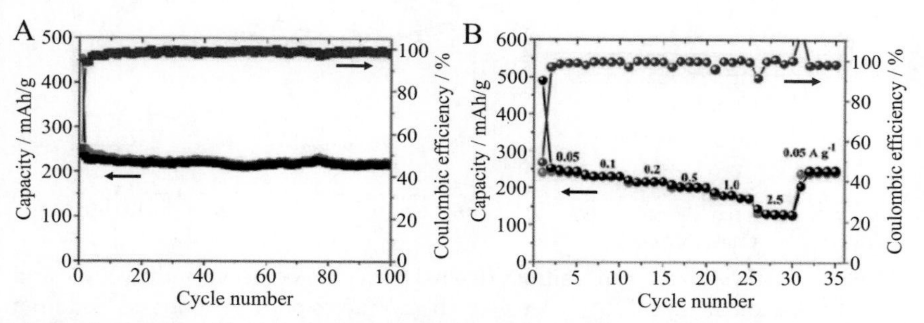

Figure 3.60 SIBs with N-doped porous carbons derived from covalent organic frameworks at 600°C: (A) cycle performance at 50 mA/g rate and (B) rate capability [18].

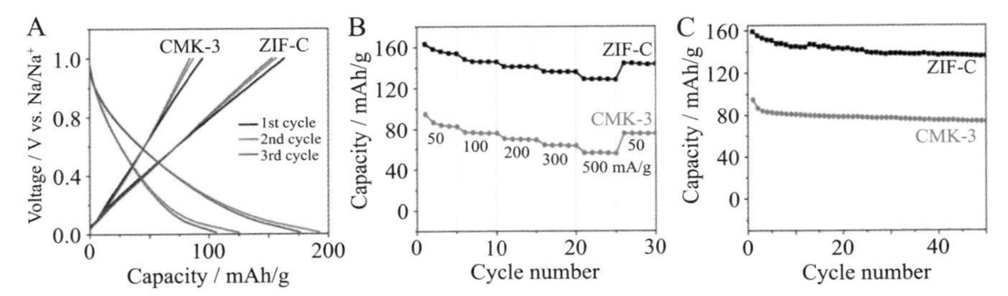

Figure 3.61 Microporous carbon ZIF-C and mesoporous carbon CMK-3: (A) initial charge—discharge profiles, (B) rate capabilities, and (C) cycle performances at 50 mA/g rate [237].

85 mAh/g at 2 A/g and cyclability of 176 mAh/g at 200 mA/g over 600 cycles. The carbon prepared by the pyrolysis of hollow polyaniline nanowires at 1150°C, which maintained the morphology of hollow wires with 20—40 nm inner diameter and interlayer spacing of about 0.37 nm, delivered discharge curves composed of a plateau at 0—0.2 V and a gradual increase in voltage in the range of 0.2—1 V [239]. The plateau part of the discharge curve at a low voltage was thought to be due to the intercalation of Na^+ ions mainly into the interlayer space of the crystallites and the sloped part above 0.2 V due to the Faradaic interaction of Na^+ with defect sites and/or functional groups on the surface of the carbon. The charge—discharge cycle gave a reversible capacity of about 251 mAh/g with almost the same irreversible capacity at the first cycle. After 400 cycles at a current density of 50 mA/g, the capacity retention of about 82%, i.e., reversible capacity of 206 mAh/g, was obtained. N-doped CNFs prepared by carbonization of polypyrrole nanofibers at 650°C exhibited a reversible capacity of 243 mAh/g after 100 cycles at 50 mA/g rate in 1 M $NaClO_4$/PC electrolyte [240]. Activation of these CNFs by KOH was not effective in improving battery performances. SnO_2 nanoflowers were deposited onto N-doped CNFs prepared from PAN by electrospinning, with SnO_2 deposition being performed under hydrothermal conditions at 180°C in $SnCl_2$ aqueous solution with sodium citrate [121]. The resultant composites exhibited slightly higher capacity in 1 M $NaClO_4$/(EC + DMC) electrolyte than the pristine N-doped CNFs, while they had much higher capacity than SnO_2 nanoflowers. In contrast, they exhibited slightly better performances in $LiPF_6$/(EC + DMC) than SnO_2 nanoflowers and much better than the pristine CNFs. Hollow carbon spheres were prepared from RF resin by hydrothermal treatment at 100°C with TEOS and either with or without ethylenediamine (EDA),

followed by carbonization at 750°C [241]. The resultant hollow carbon spheres had diameters of 200–400 nm with thin walls of about 18 nm thickness. They delivered a reversible capacity of 334 mAh/g after 100 cycles at 50 mA/g rate, which was much higher than for the spheres prepared under the same procedure without EDA (nondoped carbon) and commercial AC, and a high cycle performance even at 500 mA/g for up to 1200 cycles.

Carbon derived from various biomasses was also tested as the negative electrode of SIBs [242–248]. The carbons were prepared from banana peel by carbonizing in an Ar flow at 800–1400°C, followed by washing with 20% KOH at 70°C and 2 M HCl at 60°C [242]. A part of the resultant carbons was activated by mixing with KOH at 300°C. The banana-peel-derived carbons delivered reversible capacity of 300–350 mAh/g at 50 mA/g rate in 1 M $NaClO_4$/(EC + DEC). The carbons were prepared from sucrose by hydrothermal heating at 190°C, followed by carbonization at 1000, 1300, and 1600°C [243]. The resultant carbons demonstrated discharge curves consisting of plateau and sloped parts clearly, in which the carbon obtained at 1600°C gives the highest plateau capacity of 220 mAh/g and high cycle performance with a capacity retention of 93% after 100 cycles with 0.1C (30 mA/g) rate, as shown in Fig. 3.62. With increasing carbonization temperature, the capacity calculated from the plateau part increases, while that from the sloped part decreases. A hard carbon prepared from corn cobs at 1000, 1300, and 1600°C showed discharge curves composed of plateau and sloped parts [244]. The carbon prepared at 1600°C delivered the highest plateau capacity of 225 mAh/g and the smallest slope capacity of 61 mAh/g at 0.1C rate, the former increasing but

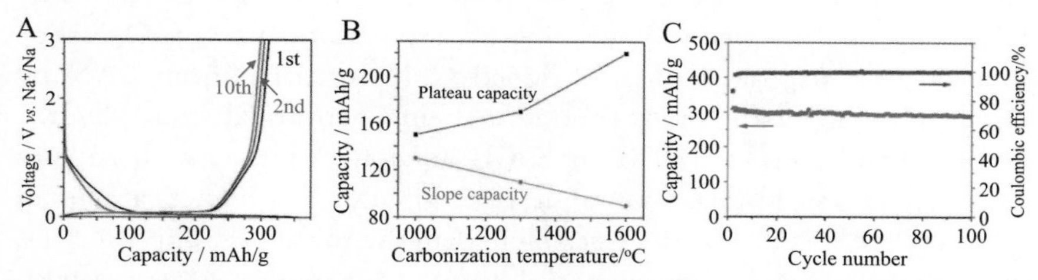

Figure 3.62 SIB performances at 0.1C rate for carbon derived from sucrose: (A) charge–discharge profiles, (B) dependences of capacities calculated from the plateau and sloped parts of discharge curves, and (C) cycle performance of the carbon carbonized at 1600°C [243].

the latter decreasing with increasing carbonization temperature. Carbon fiber webs were prepared by electrospinning of fibers from a mixture of fulvic acid with 70 wt.% PAN, followed by stabilization at 280°C in air and carbonization at a temperature of 800–1500°C [245]. The carbon fibers prepared at 1300°C delivered a reversible capacity of 248 mAh/g after 100 cycles at a current density of 100 mA/g (capacity retention of 91%). The carbons were prepared from leonardite humic acid by carbonization at different temperatures of 900–1600°C [246], all of which exhibited steady performance, as shown in the charge–discharge curves at 0.1C and cycle performance at 0.4C rate in Fig. 3.63A and C, respectively. These humic-acid-derived carbons demonstrate discharge curves composed clearly of plateau and following sloped parts, and so capacity is calculated from the two parts separately. The carbon prepared at 1500°C exhibits the highest capacity of 345 mAh/g with the highest contribution of the plateau part, as shown in Fig. 3.63B. The carbon derived from lychee seeds, which was carbonized at 500°C and contained large macropores due to the cell structure of the precursor, delivered a reversible capacity of 200 mAh/g at the second cycle and 146 mAh/g at the 100th cycle with a current density of 200 mA/g [247]. N-doped micro/mesoporous carbons were obtained from jackfruit rags separated from jackfruits by carbonization at 800°C for 2 h in Ar [248]. The resultant carbon had S_{BET} of 957 m^2/g and V_{total} of 0.887 cm^3/g, consisting of almost the same volumes of micropores and mesopores, and contained 2.60 wt.% N. It delivered a stable reversible capacity of 122 mAh/g at 1 A/g rate with a high capacity retention of 99.1% after 2000 cycles.

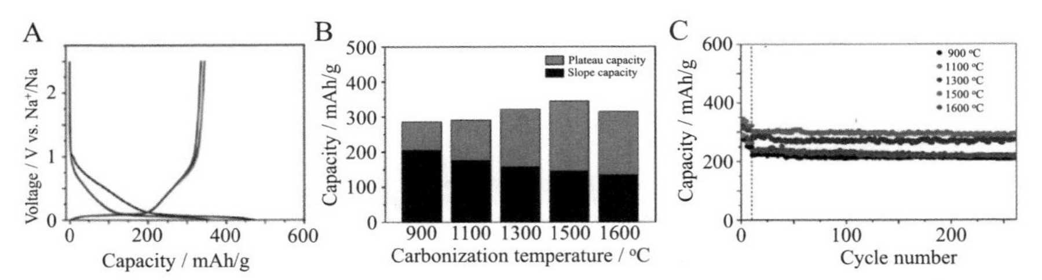

Figure 3.63 SIB performances of the carbons prepared from humic acid at different temperatures: (A) charge–discharge curves for one to three cycles with 0.1C rate for the carbon prepared at 1500°C, (B) reversible capacity measured from the plateau and slope parts of the discharge curve with 0.1C rate, and (C) cycle performances with 0.4C rate [246].

The sodium ion storage capacity of anatase-type TiO_2 was improved by loading its mesocages onto rGO flakes by microwave-assisted solvothermal treatment [249]. In Fig. 3.64A, the rate performances of the composites with different rGO contents are shown, revealing that the discharge capacities decrease continuously from a low rate of 0.2C with increasing rate; the composite with 10 wt.% rGO delivered a reversible capacity of 268 mAh/g at 0.2C (67 mA/g) to 104 mAh/g at 20C (6.7 A/g). The advantage of the composites using rGO becomes evident at high rate regions with long-term cyclability: at 10C rate the composite with 10 wt.% rGO maintains about 126 mAh/g capacity even after 18,000 cycles without noticeable decay, whereas at 5C the composite is degraded after around 8000 cycles, although it delivers a slightly higher capacity of about 150 mAh/g, as shown in Fig. 3.64B.

Necklace-like hollow N-doped carbons shown in the TEM image in Fig. 3.65A were prepared by carbonization of the webs electrospun from mixtures of ZIF-67/PAN at 700°C in H_2/Ar, which were further heated by mixing with $SbCl_2$ under hydrothermal conditions at 60°C, followed by annealing at 300°C [250]. The resultant composites contain 45.7 wt.% Sb nanoparticles (about 6.8 nm in size), as shown in Fig. 3.65B, and exhibit an initial discharge capacity of 1552 mAh/g with a Coulombic efficiency of 55.7%, as shown in the charge—discharge curves at different cycles in Fig. 3.65C. The reversible capacity of the composite decreased rapidly in the initial few cycles and stabilized at about 579 mAh/g after 100 cycles at 0.1 A/g rate and 401 mAh/g after 6000 cycles at 1 A/g rate. Na storage into the composite was thought to proceed in the multistep transformation from Sb to Na_3Sb.

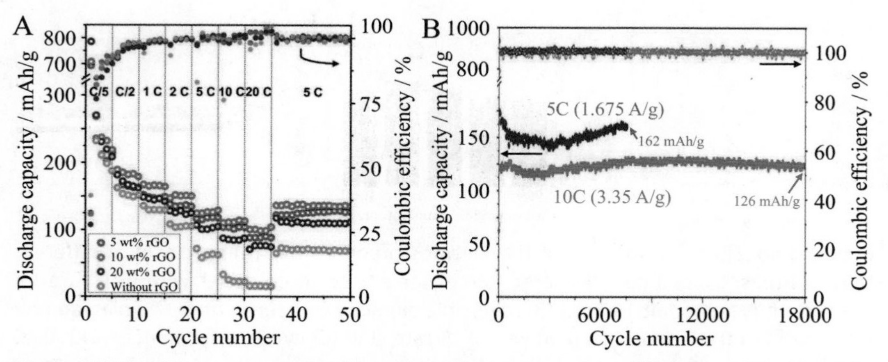

Figure 3.64 TiO_2-mesocage-loaded rGO composite with different rGO contents: (A) rate performances and (B) long-term cycling performance of the composite containing 10 wt.% rGO [249].

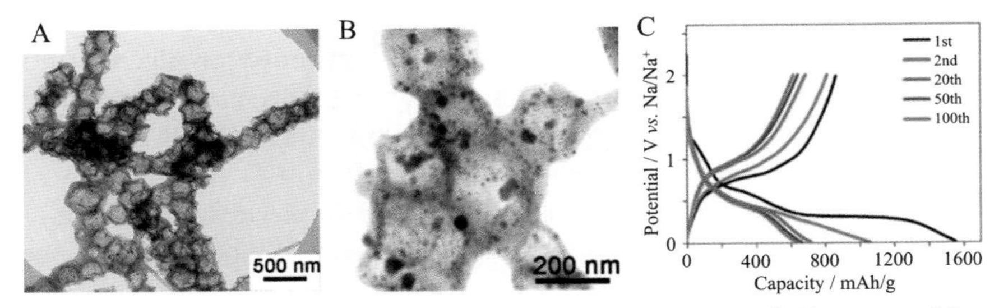

Figure 3.65 The composite of N-doped carbon nano-necklaces with Sb nanoparticles: (A) and (B) TEM images and (C) charge–discharge curves for the composite in 1 M $NaClO_4$/(EC + DMC + EMC) at 20 mA/g rate [250].

By the selection of the solvents for sodium-containing electrolyte, Na^+ ions were able to intercalate into graphite to work as the negative electrode of SIBs, where Na^+ was cointercalated with solvent molecules. Ether-based solvents, tetra-EGDME (TEGDME), di-EGDME, and DME with different electrolytes, $NaPF_6$, $NaClO_4$, and $NaCF_3SO_3$, were employed to prepare the electrolyte solutions to construct a half cell [251]. As shown by the charge–discharge curves with different electrolytes and solvents in Fig. 3.66A and B, Na^+ could be cointercalated with these ether-based solvents, regardless of the chain length of the solvent (DME to tetra-EGDME in Fig. 3.66A), although there was a large irreversible capacity at the first cycle. The graphite in these electrolyte solutions delivered a

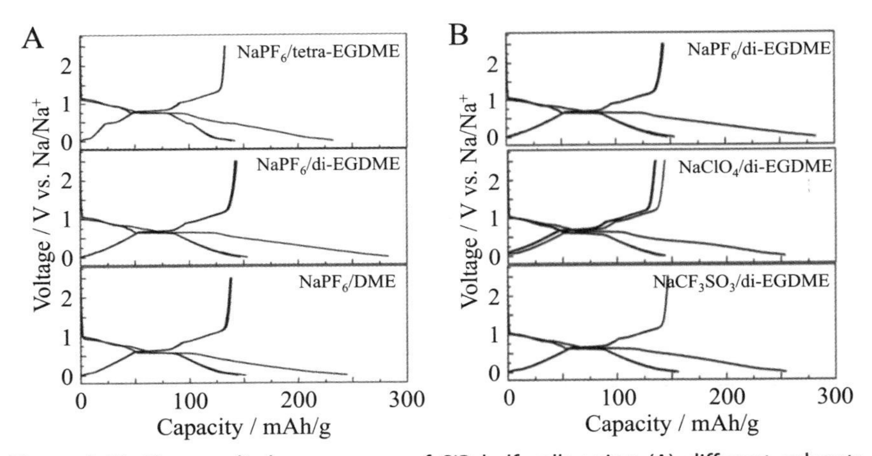

Figure 3.66 Charge–discharge curves of SIB half cells using (A) different solvents, tetra-EGDME, di-EGDME, and DME with $NaPF_6$, and (B) different electrolytes, $NaPF_6$, $NaClO_4$ and $NaCF_3SO_3$ with di-EGDME [251].

reversible capacity of about 150 mAh/g with a cycle stability for 300 cycles, even at 60°C. The rate capability, however, depended on the solvent; di-EGDME showed the most promising rate capability among the three systems, as shown for the electrolyte solutions using the di-EGDME solvent in Fig. 3.66B. A full SIB cell graphite//$Na_{1.5}VPO_{4.8}F_{0.7}$ (weight ratio of 1.5/1.0) was constructed using $NaPF_6$/di-EGDME electrolyte, delivering an energy density of about 120 Wh/kg with capacity retention of about 70% after 250 cycles.

Co-intercalation of alkali metals and solvent molecules into graphite was used as the electrode reaction at the negative electrode in LIBs and SIBs [252]. In Fig. 3.67A and B, charge–discharge characteristics are shown on the electrolytes based on diglyme with lithium triflate ($LiCF_3SO_3$, LiOTf) or sodium triflate (NaOTf), respectively. Both Li^+ and Na^+ can be intercalated into graphite at the negative electrode and gave a reversible capacity of 80–120 mAh/g during cycling up to the 50th cycle. However, only Li^+ ions can intercalate into graphite in $LiPF_6$/(EC + DMC) electrolyte but not intercalation of Na^+ ions (Fig. 3.67C). The occurrence of the intercalation was confirmed by a marked shift of the 002 diffraction peak for graphite toward the low diffraction angle side, showing an increase in interlayer spacing between graphite layers.

3.1.5 Potassium-ion batteries

Potassium-ion batteries (PIBs) have attracted attention recently mainly due to their abundant resources and low cost, in which carbons are often selected as electrode materials and their performances are discussed in comparison with LIBs and SIBs. Potassium rechargeable batteries, including K–S and K–O_2 batteries, have been reviewed [253].

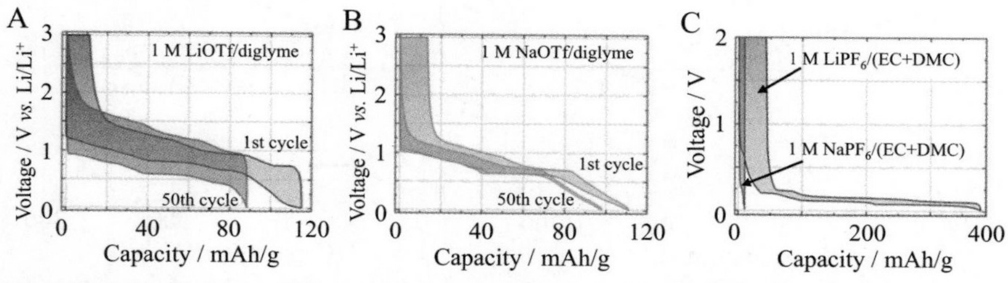

Figure 3.67 Charge–discharge characteristics of lithium//graphite and sodium// graphite cells at the current density of 37.2 mA/g in different electrolytes shown in each figure [252].

Electrochemical intercalation of K into a synthetic graphite was performed by a half cell with 0.8 M KPF_6/(EC + DEC) electrolyte [254]. The reversible (depotassiation) capacity was obtained as 273 mAh/g, which is very close to the theoretical capacity of 279 mAh/g for the formation of the first-stage intercalation compound KC_8. Graphite could exhibit a high capacity at low current rates, but its capacity decreased dramatically at high rates, with reversible capacities of 263, 234, 172, and 80 mAh/g at C/10, C/5, C/2, and C/1, respectively (1C = 280 mA/g). Ex situ XRD studies confirmed the sequential formation of KC_{36}, KC_{24}, and KC_8 upon potassiation and the opposite phase sequence to recover graphite during depotassiation. On a high-purity and highly crystalline flaky graphite (KS4), electrochemical potassiation and depotassiation behavior were studied in three electrolytes, 1 M KPF_6 in (EC + PC), (EC + DEC), and (EC + DMC) mixed solvents (1/1 by volume) [255]. As shown in Fig. 3.68A and B, the K-storage capacity was measured to be 246 mAh/g and 89% capacity retention of the initial capacity after 200 cycles in a potential range of 0−1.5 V at 20 mA/g rate using a half cell with KPF_6/(EC + PC) electrolyte. In the KPF_6/(EC + DMC) electrolyte, the capacity depressed markedly after 70 cycles, as shown in Fig. 3.68B, which was caused by the gradual decomposition of the solvent DMC.

Soft carbons prepared from 3,4,9,10-perylenetetracarboxylic dianhydride by carbonization at 700, 900, and 1100°C for 5 h in Ar were applied to the negative electrode in a half cell with 1.0 M KPF_6/(EC + DEC) electrolyte [256]. As shown by the charge−discharge curves in Fig. 3.69A, they show only shallow plateaus at around 0.8 V and the carbon prepared at

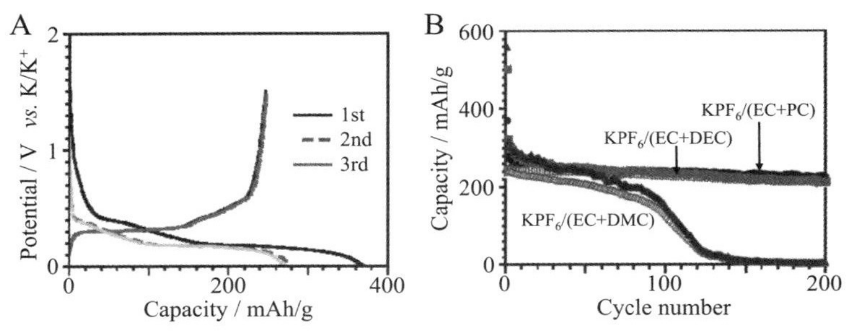

Figure 3.68 Highly crystalline graphite: (A) charge−discharge curves up to the third cycle in KPF_6/(EC + PC) electrolyte and (B) cycle performances with the electrolytes of KPF_6 in three different solvents [255].

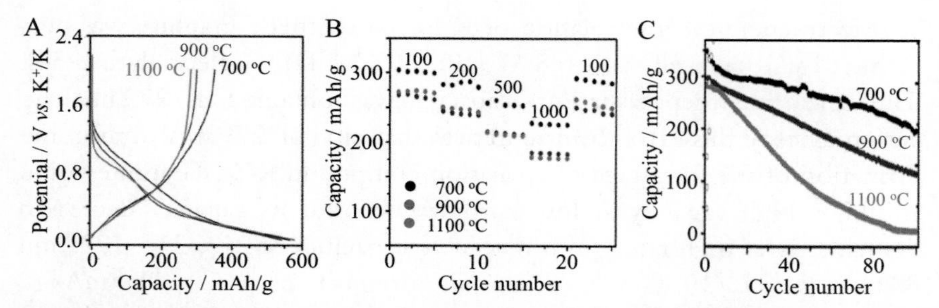

Figure 3.69 PIB performances of a soft carbon carbonized at 700, 900, and 1100°C: (A) charge–discharge curves at 20 mA/g rate, (B) rate capability in the current density range of 100–1000 mA/g, and (C) cycle performances at 20 mA/g [256].

700°C gave the highest reversible capacity (depotassiation capacity) of 350 mAh/g at 20 mA/g rate in the potential range of 2 V with Coulombic efficiency of 62% for the first cycle. It exhibits also the highest performances in the rate capability and cyclability, as shown in Fig. 3.69B and C, respectively, and even at 1000 mA/g rate the capacity retention was 75% of that at 100 mA/g rate and 25% retention after 100 charge–discharge cycles with 20 mA/g rate. These carbons could store Na ions also, but the storage capacity was inferior to K storage; the reversible Na-ion storage capacity of the carbon prepared at 700°C was measured as 216 mAh/g with an initial Coulombic efficiency of 60%.

Various hard carbons were tested for the negative electrode of PIBs, PF resin [236], sucrose [257,258], glucose [259], a seafood waste (chitin) [260], a MOF [261], sugar, and 3,4,9,10-perylenetetracarboxylic dianhydride (hard–soft carbons composite) [262], polymer microspheres [263], and a mixture of sodium citrate with urea [264]. Hard carbons prepared from PF resin by carbonization at 1100, 1300, and 1500°C delivered high K^+ storage capacity [236]. Reversible capacity increased from 295 to 336 mAh/g upon increasing the HTT. Hard carbon microspheres were prepared from sucrose by hydrothermal treatment at 195°C for 5 h, followed by carbonization at 1100°C for 5 h in Ar [257]. These carbon microspheres delivered a reversible capacity of 262 mAh/g at C/10 rate and 136 mAh/g at 5C rate in 0.8 M KPF_6/(EC + DEC) electrolyte. This PIB could retain the capacity of 216 mAh/g after 100 cycles at C/10 rate, with 83% capacity retention. The SIB composed of these carbon microspheres exhibited a capacity of 322 mAh/g at C/10 rate but dropped to 73 mAh/g at 5C rate. Chitin microspheres were carbonized at 800°C for 2 h in Ar to obtain N-doped carbon microspheres, which exhibited a high rate capability and cycle performance [260]. The PIB composed of these carbon microspheres delivered a reversible capacity of about 250 mAh/g at 0.12C rate and 154 mAh/g at a high rate of 72C (20 A/g), as shown in Fig. 3.70A. The rate performance of this PIB is compared with that of SIB composed of the same carbon, and the PIB possesses higher capacity, particularly very high C-rates above 18C (5 A/g), as shown in Fig. 3.70B. The PIB also has a high cycle performance (Fig. 3.70C), and the capacity is retained, even increasing slightly, after 4000 cycles at 1.8C rate. N-doped porous carbon was prepared from a mixture of sodium citrate and urea at 900°C for 1 h [264]. It exhibited high K-storage capacity and cycling stability: a reversible capacity of 420 mAh/g at 50 mA/g and 185 mAh/g at an extremely high rate of 10.0 A/g with a high capacity retention of 343 mAh/g after 500 cycles at 100 mA/g rate.

On a hard carbon prepared from sucrose, its performances in LIB, Na-SIB, and PIB were compared [258]. The hard carbons were prepared from sucrose by carbonization at 700—1300°C for 1 h in N_2 or Ar and the carbons obtained at 1000 or 1200°C were further heat-treated at 1500 and 1600°C for 1 h and 1800 and 2000°C for 0.5 h in N_2. The three electrode cells were assembled by 1 M $LiPF_6/(EC + DMC)$, 1 M $NaPF_6/PC$, and $KN(SO_2F)_2/(EC + DEC)$ electrolytes with respective metal foil counter and reference electrodes and their performances were compared with each other. In Fig. 3.71A—C, charge—discharge curves are shown for LIB, SIB, and PIB, respectively, composed of the sucrose-derived hard carbons heat-treated at high temperatures (1300—2000°C). In Fig. 3.72A—C, discharge capacities at the second cycle are shown as a function of HTT applied for the electrode carbons by separating to the potential ranges for the plateau and sloping regions. In the LIB, the carbons with HTTs

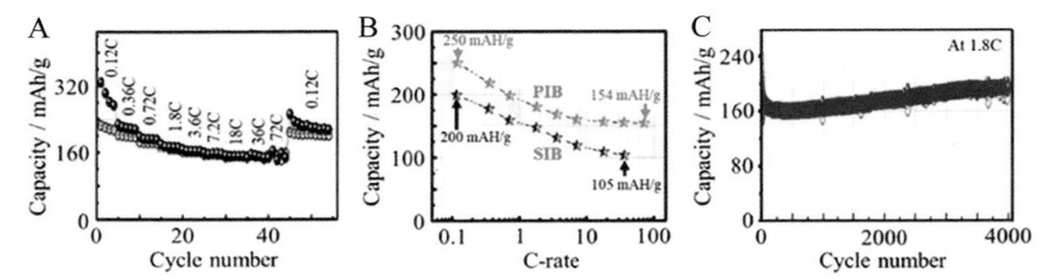

Figure 3.70 The performances of PIB composed of the chitin-derived carbon microspheres: (A) rate capability, (B) rate performance in comparison with SIB composed of the same carbon, and (C) cycle performance at 1.8C rate [260].

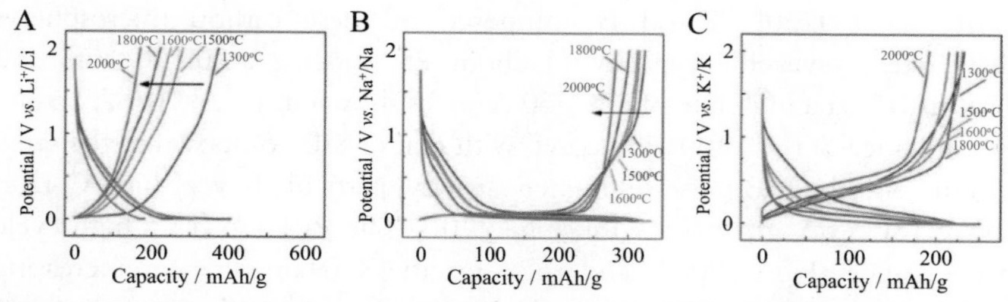

Figure 3.71 Charge–discharge curves for the sucrose-derived hard carbons heat-treated at 1300–2000°C: (A) LIB, (B) SIB, and (C) PIB [258].

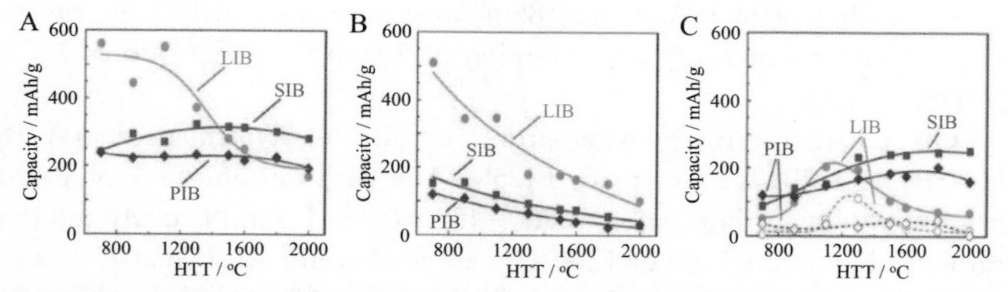

Figure 3.72 Reversible capacities at the second cycle for LIB, SIB, and PIB constructed from the sucrose-derived hard carbons with different HTTs: (A) total capacities in the potential range of 0.002–2.0 V, (B) capacities in 0.25–2.0 V (sloping region), and (C) capacities in 0.002–0.25 V for LIB and SIB, and in 0.002–0.7 V for PIB (plateau region) with those (*dashed lines*) in 0.002–0.05 V for LIB and SIB and in 0.002–0.20 V for PIB [258].

less than 1100°C delivered a reversible capacity of around 500 mAh/g (Fig. 3.72A) and its dominant part is due to the sloping region (Fig. 3.72B), which is associated with the interaction of Li ions with edge sites of carbon layers and functional groups in the electrode carbon and as a consequence a large irreversible capacity at initial cycles (not shown). On the carbons obtained at 1100 and 1300°C, a long flat potential region appears at about 0.05 V during discharging and corresponding capacity in the plateau region reaches a maximum (Fig. 3.72C). In SIB, both a sloping region above 0.25 V and a plateau region at 0.1 V are observed on all the carbons obtained at 700–2000°C (Fig. 3.71B) and the capacity for the sloping region decreases gradually and that for the plateau region increases gradually (Fig. 3.72B and C, respectively), and consequently the total capacity changes only slightly with HTT (Fig. 3.72A). The total capacity for SIB became larger than that of LIB when the carbons with HTT higher than 1500°C were used as the negative electrode. In PIB, only sloping profiles are observed without a plateau region, although plateau-like features are seen at the potential below 0.1 V on the charging process (potassiation process) and a linear region with a small slope is observed for the carbons with HTT above 1600°C (Fig. 3.71C). For the sucrose-derived hard carbon with different HTTs from 700 to 2000°C, the total reversible capacity of LIBs decreases from 561 to 169 mAh/g with increasing HTTs, whereas SIBs and PIBs deliver almost constant capacities with maximum values of 324 and 238 mAh/g, respectively.

Carbon nanofibers (CNFs) and nanotubes (CNTs) were also applied to the negative electrode of PIBs, PAN-based CNF webs [265,266], CNFs derived from a MOF [267], N-doped CNFs derived from polypyrrole nanofibers [268], and cup-stacked CNTs [269]. CNF webs were prepared from a DMF solution of PAN by electrospinning and following stabilization at 350°C in air and carbonization at 800°C for 1 h in N_2 [265]. The CNFs were rich in N (10.0 at.%) and afforded a free-standing flexible electrode. They delivered the reversible capacity of 230 mAh/g at C/10 rate and 110 mAh/g at 10C rate. They also show a stable cycling performance, reversible capacity being retained as 170 mAh/g even after 1900 cycles at 1C rate. CNF webs were prepared from a DMF solution of PAN and PMMA (4/3 by weight) by electrospinning, followed by stabilization at 280°C for 5 h in air and carbonization at 1000°C for 2 h in N_2 [266]. The resultant CNF webs contained 4.71 at% N and were rich in narrow micropores, as confirmed by CO_2 adsorption at 0°C. These CNF webs delivered a reversible capacity of 270 mAh/g and a stable cyclability, with no capacity decay over 80 cycles at 0.02 A/g rate. Highly N-doped CNFs were prepared by carbonization of polypyrrole nanofibers at 650, 950, and 1100°C for 2 h in N_2 [268]. PIB cells composed of these N-doped CNFs carbonized at 650°C, which contained the highest N content of 13.8 at%, delivered the highest reversible capacity of 368 mAh/g at the first cycle and 248 mAh/g after 100 cycles at 25 mA/g rate. The capacities were discussed by separating the surface-dominant K^+ storage (in other words, capacitive contribution) from the observed cell capacity. On these N-doped CNFs, the capacitive contribution for K^+ storage was predominant, reaching 67%–90%. The cup-stacked CNTs synthesized by CVD were applied to the negative electrode [269]. The slope of the depotassiation curves changed at around 0.6 V and a large irreversible capacity was observed for the initial few cycles, as shown in Fig. 3.73A and B. The reversible capacity was calculated to be 324 mAh/g. These CNTs can retain a high capacity of 236 mAh/g after 100 cycles at 20 mA/g rate, as shown in Fig. 3.73B, and decrease to 75 mAh/g at 1000 mA/g rate, as shown in Fig. 3.73C.

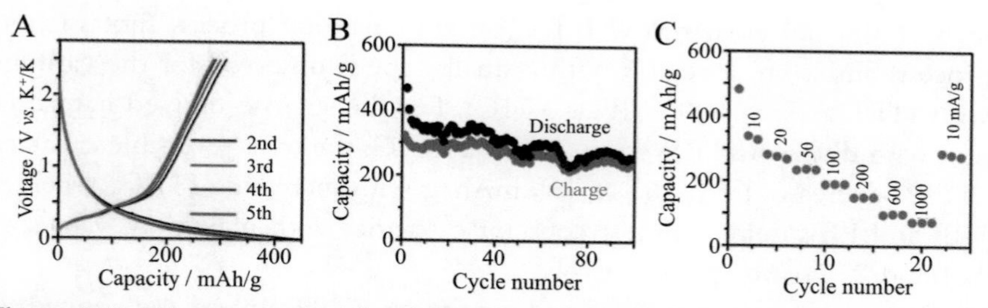

Figure 3.73 PIB composed of the N-doped cup-stacked CNTs: (A) charge–discharge curves for the second to fifth cycles at 10 mA/g rate, (B) cycle performance at 20 mA/g rate, and (C) rate capability [269].

N-doped carbon nanotubes (N-doped CNTs) were synthesized by carbonization of a Co-containing MOF (ZIF-67) at 1100°C for 3.5 h, followed by washing with 0.5 M H_2SO_4 to remove Co components, which were subjected to study electrochemical storage of K ions [267]. ZIF-67 nanoparticles were synthesized by the reaction of 2-methylimidazole with $Co(NO_3)_2$ in CH_3OH/C_2H_5OH solution at room temperature. The resultant N-doped CNTs delivered a reversible capacity of 297 mAh/g and the capacity was stabilized at 293 mAh/g after 10 cycles at 50 mA/g rate, whereas they had a large irreversible capacity, reaching 900 mAh/g at the first cycle. They exhibited an improved rate capability by a high-capacity retention of 102 mAh/g at a high current density of 2000 mA/g and a good cyclability without evident capacity loss over 500 cycles at 2000 mA/grate.

Few-layered graphene (FLG) was grown on Ni foam substrate by CVD of C_2H_2 at 650°C for 5 min and N-doped FLG by CVD of CH_3CN vapor at 750°C for 5 min [270]. Their PIB performances were measured by assembling a half cell with 0.8 M KPF_6/(EC + DEC) electrolyte. Charge–discharge curves at 50 and 100 mA/g rates are shown for FLG and N-doped FLG in Fig. 3.74A and B, respectively, revealing that N-doping of 2.2 at.% markedly enhances the capacity for K storage. N-doped FLG delivered a capacity of 350 mAh/g, which is much higher than the theoretical capacity of 278 mAh/g. Cycle performances at 100 mA/g rate in Fig. 3.74C indicate a stable, higher storage capacity for N-doped FLG compared to FLG. N-doped FLG initially possesses a capacity near 270 mAh/g and after 100 cycles remains above 210 mAh/g, while FLG exhibits an initial capacity near 190 mAh/g and after 100 cycles only 150 mAh/g. P/O-codoping was shown to be effective to enhance PIB performances using rGO [271]. The doping was carried out by mixing

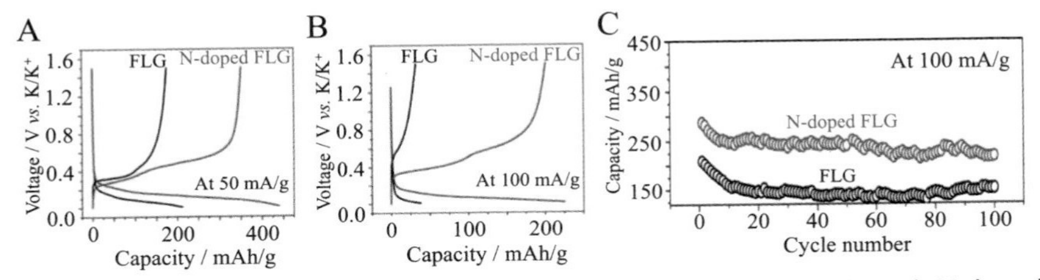

Figure 3.74 PIB performances of the few-layered graphene (FLG) and N-doped few-layered graphene (N-doped FLG); charge–discharge curves at (A) 50 mA/g and (B) 100 mA/g rate, and (C) cycle performances at 100 mA/g rate [270].

GO with triphenylphosphine in water/ethanol solution, followed by heating at 900°C for 1 h. At the first cycle, the codoped rGO delivered a depotassiation capacity of 556 mAh/g with a potassiation capacity of 2054 mAh/g, showing a large irreversible capacity. By cycling at 50 mA/g rate, however, the depotassiation capacity was stabilized above 15 cycles and retained at about 470 mAh/g for up to 50 cycles. Stabilization of capacity was confirmed by long time cycling to 600 cycles at 500, 1000, and even 2000 mA/g rates, stabilized capacity being 310, 222, and 165 mAh/g, respectively. A porous carbon monolith containing a high pyridinic N was synthesized by carbonization of GO nanoflakes wrapped by melamine-formaldehyde resin at 900°C for 1 h in Ar [272]. The resultant carbon contained 18.9 at.% N, of which 53% was at the pyridinic site (i.e., 10.1 at.% pyridinic N) and had S_{BET} of 443 m^2/g. It delivered a high initial depotassiation capacity of 487 mAh/g at a current density of 20 mA/g in 0.8 M KPF_6/(EC + DEC) electrolyte. In addition, it exhibited high-rate capability and cyclability; reversible capacity was 388 mAh/g at 50 mA/g, while at 5000 mA/g rate the capacity could retain 178 mAh/g, and the initial reversible capacity of 325 mAh/g at 500 mA/g rate decreased gradually but stabilized at about 200 mAh/g after 180 cycles.

N-doped porous CNF webs were used as scaffold for loading MoP nanoparticles, which could be alloyed with potassium [273]. The N-doped CNFs were prepared from PAN mixed with ZIF-8 nanoparticles in DMF solution by electrospinning, followed by stabilization at 300°C in air and carbonization at 900°C for 5 h in N_2. The CNF web was immersed into a saturated ammonium molybdate for 24 h and then mixed with sodium hypophosphite (NaH_2PO_2) with CNF/NaH_2PO_2 at a ratio of 1:5 by weight, and then annealed at 600°C for 5 h in an Ar/H_2 atmosphere to obtain self-standing MoP-loaded N/P-codoped CNF web. The resultant composite web delivered a high capacity of 320 mAh/g at 100 mA/g rate with a superior rate capability maintaining 220 mAh/g at 2 A/g and a capacity retention of more than 90% even after 200 cycles at 100 mA/g.

3.2 Supercapacitors

In supercapacitors, the principal mechanism for energy storage is physical adsorption of electrolyte ions onto the surface of electrodes to form EDLs, and as a consequence their capacitances depend strongly on the surface area of the electrode material and so they have often been called EDLCs. Carbon materials can have a wide range of surface areas from a few hundreds to more than 3000 m^2/g and their pore structure, i.e., pore sizes, volume, and even morphology, can be controlled by various techniques, as explained in the last Chapter 2 of this book. This is the main reason for the application of porous carbon materials to the electrode of supercapacitors, in addition to their light weight and high electric conductivity. However, the capacitances observed on various supercapacitors depend on different factors, electrolyte solutions, charge/discharge rate, and temperature of measurement, in addition to the pore structure of carbon materials in the electrodes. In this section, supercapacitors are explained by classifying them on the bases of porous carbons for the electrodes, activated carbons and activated carbon fibers, template-assisted porous carbons, and precursor-designed porous carbons. Often the porous carbons were modified with mechanical treatment, such as grinding and agglomeration, with chemical modification, such as oxidation, foreign atom doping, and modification by functional groups, and with composite formation, such as composites with other materials, including other carbon materials, and with other compounds. In this section, the experimental results related to supercapacitors are collected.

In this section, mutual comparison of the values of capacitance among papers was avoided because the calculations of capacitance values in many papers were not carried out correctly, although this has been often pointed out by different authors. The capacitance measured using two-electrode cells has to be calculated on the basis of the total mass of the electrochemically active materials represented by porous carbons, in other words, the masses of two electrodes, irrespective of whether they are symmetric or asymmetric cells, whereas it has been calculated on the basis of the mass of a single electrode in some papers (regrettably not a small number of them).

Some papers described how the calculation was done but some did not. Therefore, there are high risks of misunderstanding the experimental results when comparing the figures reported in various papers.

3.2.1 Activated carbons and activated carbon fibers

The capacitances observed on various carbon materials, C_{obs}, were attempted to be correlated with the surface area, such as S_{BET}, but it was not successful in obtaining the common relations for various carbon materials, although some relations were reported on limited carbon materials, such as activated carbon fibers. To analyze the effect of pore structure on the capacitance of supercapacitors, the following relation has been proposed [274] by differentiating surface areas due to micropores and other pores, measured as the microporous surface area, S_{micro} and external surface area, S_{ext}, respectively, using either t- or α_s-plots;

$$C_{obs} = C_{micro} \times S_{micro} + C_{ext} \times S_{ext},$$

where C_{micro} and C_{ext} are the contributions (in the unit of F/m^2) from micropores and other larger pores, mainly mesopores, respectively. In Fig. 3.75A and B, as examples, the plots of C_{obs}/S_{ext} against S_{micro}/S_{ext} for various activated carbons in nonaqueous electrolyte (1 M $TEMABF_4$/PC) and aqueous electrolyte (1 M H_2SO_4), respectively, at two current densities [275], demonstrate well-defined linear relations. Several papers [274–282] have provided the parameters needed for this analysis, i.e., C_{obs}, S_{micro}, and S_{ext}, however the conditions for the measurement of capacitance and the evaluation method of surface areas are different.

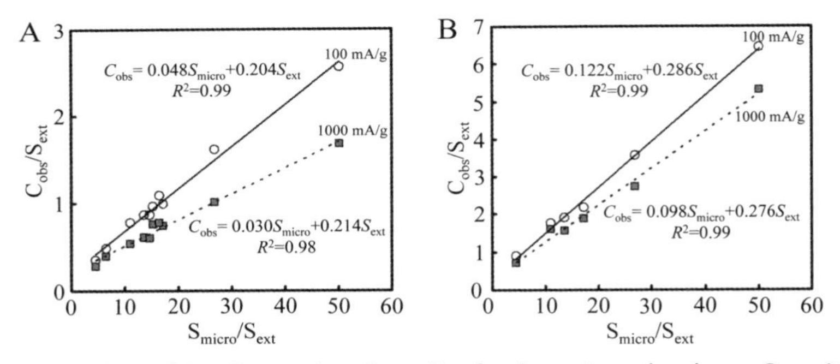

Figure 3.75 Plots of C_{obs}/S_{ext} against S_{micro}/S_{ext} for the activated carbons. Capacitance C_{obs} was measured in (A) nonaqueous electrolyte (1 M $TEMABF_4$/PC) and (B) aqueous electrolyte (1 M H_2SO_4) [275].

Therefore, the figures of C_{micro} and C_{ext} derived from these data were scattered, while it can be concluded that C_{ext}/C_{micro} was about 2 in aqueous electrolytes and about 4–7 in nonaqueous electrolytes [283]. This conclusion suggests that the capacitance contribution of mesopores is much larger than that of micropores, particularly in nonaqueous electrolytes. Changes in C_{micro} and C_{ext} with current density are shown for 1 M $TEMABF_4$/PC and 1 M H_2SO_4 electrolytes in Fig. 3.76A and B, respectively. C_{micro} is almost one order of magnitude larger in H_2SO_4 (aqueous electrolyte) than in $TEMABF_4$/PC (nonaqueous electrolyte), although C_{ext} is almost the same in these two electrolytes, probably because of a large difference in the sizes of cations H^+ and $TEMA^+$. The same analysis was applied to a series of the carbons derived from resorcinol-formaldehyde (RF) at 950°C [284]. The results illustrated that a steric or energetic limitation works on the accessibility of electrolyte ions into micropores, which is resolved with increasing voltage applied, while the storage of ions in larger pores, of which the surface area is expressed by S_{ext}, has no such restriction and so is accessible in all voltage windows.

Activated carbons contain certain amounts of functional groups, mostly oxygen- and nitrogen-containing functional groups, at the structural defects, including edges of carbon layers. These functional groups are expected to have an influence on the capacitive behavior in supercapacitors, most of them giving additional capacitance due to redox interaction with electrolyte ions (pseudocapacitance) at the surface of the electrode carbon. Commercially available ACF clothes with different surface areas (S_{BET} of 700–2000 m^2/g) were subjected to O_2/Ar plasma with different O_2 concentrations (0–20 vol.%) to control the concentration of functional groups on the surface of the fibers, which was evaluated by Boehm's

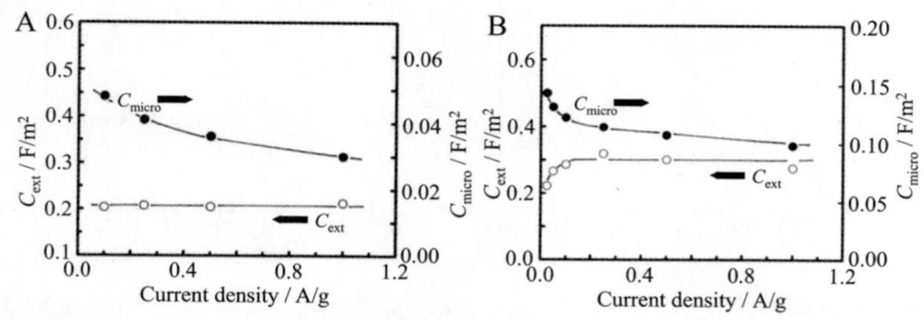

Figure 3.76 Changes to the capacitance contributions of microporous and external surfaces, C_{micro} and C_{ext}, of various carbons with current density: (A) in 1 M $TEMABF_4$/PC and (B) in 1 M H_2SO_4 electrolytes [282].

titration method [285]. As shown in the cyclic voltammogram (CV) curves measured in 0.5 M H_2SO_4 electrolyte in Fig. 3.77A, plasma irradiation with a low O_2 concentration reduced the capacitance, while that with a high O_2 concentration made the capacitance increase in the whole potential range swept, with no clear peak being observed. The capacitance cannot be related to the total functional groups concentration, as shown in Fig. 3.77B, nor to S_{BET}, but it shows much better correlation with the concentration of quinone groups, as shown in Fig. 3.77C. Oxidation in 0.1 M HNO_3 with and without a dc voltage of 3 V and also a reduction in H_2 flow were applied on ACFs to study the effect of functional groups on the capacitance in 1 M H_2SO_4 and 0.5 M $LiClO_4$/PC electrolytes [286]. In aqueous electrolyte, the capacitance depended on the concentration of oxygen-containing functional groups, showing poor dependences on the pore structure parameters. In nonaqueous electrolyte, in contrast, pore structure seemed to be a more dominant factor than functional groups. An AC derived from wood was modified by either urea or melamine at 950°C to introduce nitrogen-containing functional groups [287]. In 1 M H_2SO_4, the urea-modified AC delivered a capacitance of 224 F/g at 1 A/g rate, while the pristine AC gave only 40 F/g, although at a low rate of 500 mA/g both gave almost the same capacitance (260 and 253 F/g, respectively), suggesting a marked effect of urea and melamine treatment on the rate performance. The gravimetric capacitance was discussed in relation to the number of basic functional groups and to the content of quaternary (graphitic) and pyridinic nitrogen species on the surface. Surface modification of coconut-shell-derived carbon was performed by HNO_3 treatment and heat treatment in melamine and urea at 950°C to investigate the effect

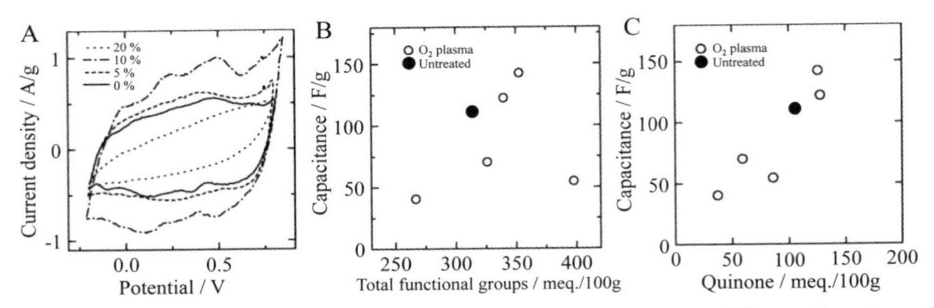

Figure 3.77 ACF clothes treated by O_2-plasma: (A) CV curves at 5 mV/s as a function of oxygen concentrations in plasma, (B) capacitance in 0.5 M H_2SO_4 electrolyte as a function of total functional groups concentration and (C) as a function of quinone group concentration [285].

of N- and O-containing functional groups on the carbon surfaces on capacitive performances in 1 M H_2SO_4 aqueous electrolyte [288]. It was concluded that the most important functional groups affecting energy-storage performance were pyrrolic and pyridinic nitrogen along with quinone oxygen. The capacitances measured at the current densities of 0.05, 0.1, 0.5, and 1 A/g on six carbons modified could be linearly related to the sum of functional groups containing pyrrolic and pyridinic nitrogen and quinone oxygen with R^2 of more than 0.99.

On 15 porous carbons, including commercial ACs and ACFs derived from different carbon precursors and synthetic porous carbons prepared by the templating method, gravimetric capacitances were measured in 1 M $LiClO_4$/PC electrolyte [289]. The carbons were classified into two types on the basis of the pore morphologies, either slit-shape or worm-like, with the latter class including micro-/mesoporous carbons. The optimum pore structure for high EDL capacitance was concluded to be the carbon having worm-like pores. For the carbons having worm-like pores, the new parameter R_{ssa} was proposed to be introduced, which is the ratio of the surface area with pore width larger than 1.5 nm ($S_{>1.5\ nm}$) to the surface area with pore width of 0.55−1.5 nm ($S_{0.55-1.5\ nm}$); the former $S_{>1.5\ nm}$ corresponding to the space for solvated ions and the latter $S_{0.55-1.5\ nm}$ to the space for nonsolvated ions. The ideal structure for porous carbon with high capacitance was concluded to be a hierarchical porous structure with R_{ssa} of around 0.5 and worm-like shape pores.

The addition of only 2 wt.% rGO as a binder to form an electrode sheet from coffee-derived ACs was found to be very effective in enhancing the capacitive performance in 6 M KOH electrolyte [290]. AC was prepared from coffee grounds mixed with KOH by carbonization and simultaneous activation at 700°C in Ar. To test its capacitive performance, the AC powder was mixed with GO suspension and then coated onto a current collector (carbon paper), followed by calcination at 900°C to reduce GO to rGO. As shown in Fig. 3.78A, the electrode formed using rGO binder has much higher capacitance than that formed in the conventional way (using 10 wt.% PVDF binder), demonstrating almost rectangular-shaped CV curves. The capacitance for the former electrode increased almost linearly with increasing S_{BET} of the AC; the capacitance at 0.5 A/g rate reached 512 F/g for the AC with S_{BET} of 2620 m^2/g (volumetric capacitance of 230 F/cm^3), while that for the latter electrode increases slightly with increasing S_{BET} of the AC, tending to saturate at around 130 F/g above 1500 m^2/g. The former electrode maintains a high capacitance of

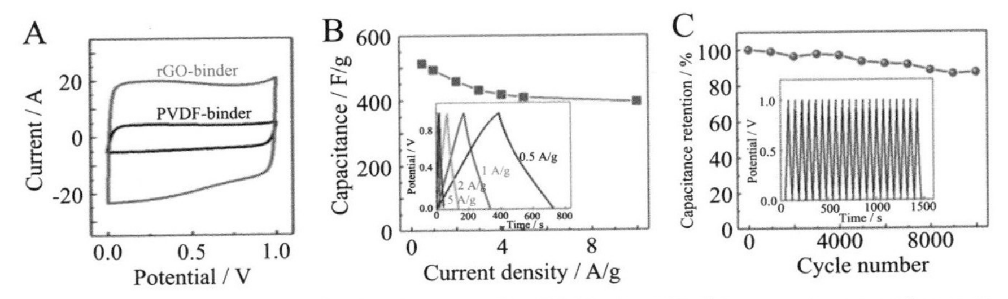

Figure 3.78 Coffee-derived AC using 2 wt.% rGO binder: (A) CV curve at a scanning rate of 10 mV/s in comparison with the AC using 10 wt.% PVDF binder, (B) rate performance determined from galvanostatic charge–discharge curves (inset), and (C) cycle performance at 5 A/g rate with galvanostatic charge–discharge curves (inset) [290].

394 F/g (177 F/cm^3) at the current density of 10 A/g (Fig. 3.78B) and exhibits excellent cyclability, about 87% retention after 10,000 cycles (Fig. 3.78C). Mixing of DWCNTs into AC was reported to improve the rate performance of the EDLCs in 1.5 M NEt_4BF_4/AN electrolyte, although the capacitance decreased slightly [291].

Microporous carbons were fabricated by carbonization of the mixture of a coal tar and an ionic liquid ($BMIMBF_4$) in equal weights with different amounts of KOH at 800°C in Ar [292]. As shown by the charge–discharge curves and rate performances in Fig. 3.79A–C, the carbon prepared using a KOH/tar ratio of 4, which has S_{BET} of 1593 m^2/g with V_{total} of 0.85 cm^3/g including V_{micro} of 0.63 cm^3/g, exhibits the highest capacitance of 314 F/g at 0.05 A/g, and 195 F/g at 100 A/g with a large IR drop. From the mixtures of a coal tar with melamine and KOH, where the weight ratios of melamine/tar and KOH/tar were 1/2 and 2/1, respectively, microporous carbons composed of lamellar particles were obtained by carbonization at 800–900°C in Ar and N_2 [293]. The carbons prepared at 800°C in N_2 had S_{BET} of 1517 m^2/g with S_{micro} of 1477 m^2/g and exhibited a capacitance of 274 F/g at 0.05 A/g and 191 F/g at 20 A/g in 6 M KOH electrolyte. The formation of lamellar carbon particles was explained by the confinement of polycyclic aromatic hydrocarbon molecules from coal tar in the layered spaces of graphitic C_3N_4 nanosheets, which were formed by the polymerization of melamine at the beginning of carbonization (around 550°C).

ACF webs were prepared from a PAN/pitch mixture by electrospinning, followed by stabilization at 300°C and carbonization/activation in a steam/N_2 flow at temperatures of 700–900°C [294]. The webs activated at 900°C exhibited high S_{BET} of 1725 m^2/g with V_{total} of 1.11 cm^3/g,

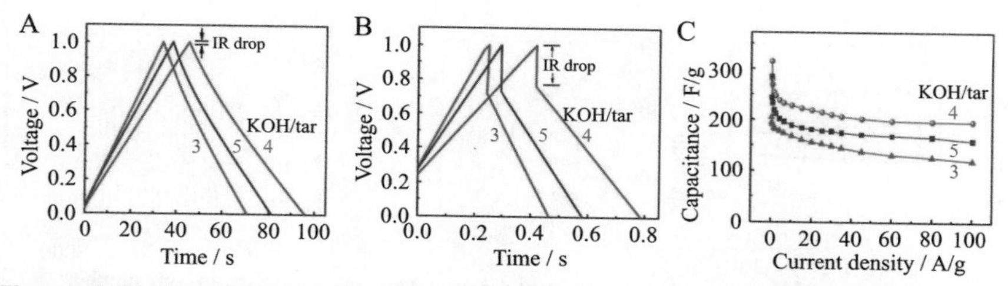

Figure 3.79 Capacitive performance of microporous carbons fabricated from the mixture of coal tar and an ionic liquid: (A) charge–discharge curves at 1.2 A/g rate and (B) those at 100 A/g, and (C) rate performances [292].

consisting of almost equal V_{micro} and V_{meso}, and as a consequence delivered a capacitance of 143 F/g at 10 mV/s rate and about 120 F/g at 500 mV/s rate (15% reduction).

Hollow carbon spheres were synthesized by hydrothermal carbonization of alginic acid prepared from brown algae at 200°C, which were activated by KOH at 700–900°C in N_2 [295]. The resultant carbons composed of hollow spheres had outer diameters of 100–200 nm with walls of 30–50 nm thickness, as shown in the TEM image in Fig. 3.80A, and exhibited type-IV N_2 adsorption–desorption isotherms with H4 type hysteresis. The pore structure parameters of the carbons depended greatly on the activation temperature, 900°C-activated carbon had the highest values as S_{BET} of 2421 m^2/g and V_{total} of 1.61 cm^3/g including V_{micro} of 0.87 cm^3/g, of which capacitance was 314 F/g at 1 A/g rate in 6 M KOH electrolyte. A slight deviation from the linear relation in galvanostatic charge–discharge curves, as shown in Fig. 3.80B, suggests some contribution of pseudocapacitance. As shown in Fig. 3.80C,

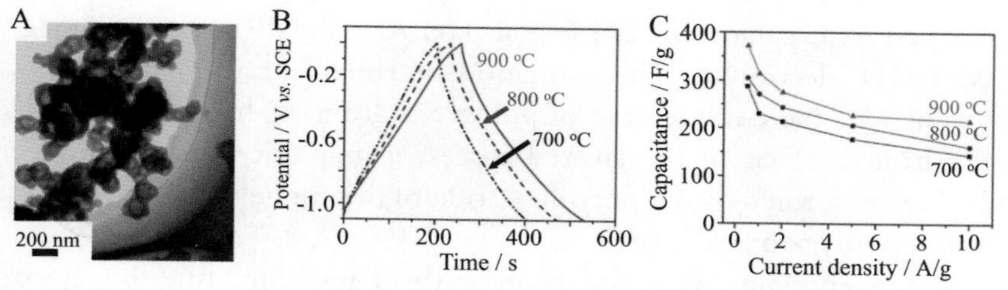

Figure 3.80 Porous carbon hollow spheres derived from alginic acid and activated by KOH at different temperatures: (A) TEM image, (B) charge–discharge curves at 1 A/g rate, and (C) rate performances in 6 M KOH electrolyte [295].

900°C-activated carbon exhibits better rate performance than the 700°C- and 800°C-activated ones, with capacitance retention of 58% as current density increases from 0.5 to 10 A/g, probably owing to the high percentage of mesopores (V_{meso} of 0.74 cm^3/g) in the 900°C-activated carbon.

N-doped porous carbons were prepared from chitosan (N-containing biomass) mixed with $ZnCl_2$ in a weight ratio $ZnCl_2$/chitosan of 10/1 by carbonization at temperatures between 400 and 700°C [296]. In comparison, the mixture of chitosan with $ZnCl_2$ in a 4/1 weight ratio was also carbonized at 600°C. Their capacitive performances are shown in Fig. 3.81A–C. The carbon prepared at 600°C with $ZnCl_2$/chitosan of 10/1 exhibited the highest S_{BET}, V_{total}, and V_{meso} of 1582 m^2/g, 1.23 cm^3/g, and 0.86 cm^3/g, respectively, and as a consequence the highest capacitance in 6 M KOH aqueous electrolyte, 252 and 145 F/g at 0.5 and 50 A/g rate, respectively. It demonstrates excellent cyclability up to 10,000 cycles with a high current density of 8 A/g (Fig. 3.81C). N-doped porous carbons were prepared from gelatin via NaOH-activation at 600°C, which delivered a high capacitance of up to 385 F/g in 6 M KOH with excellent rate capability (235 F/g even after 200 cycles at 50 A/g) and cycle durability [297].

N-doping by NO was applied on a steam-activated carbons (PF-based) at 800°C in He gas containing 2000 ppm NO, aiming to widen the potential window and expecting an enhancement in the energy density of the supercapacitors [298,299]. The effect of N-doping on capacitive performances was compared to the effect of high-temperature annealing.

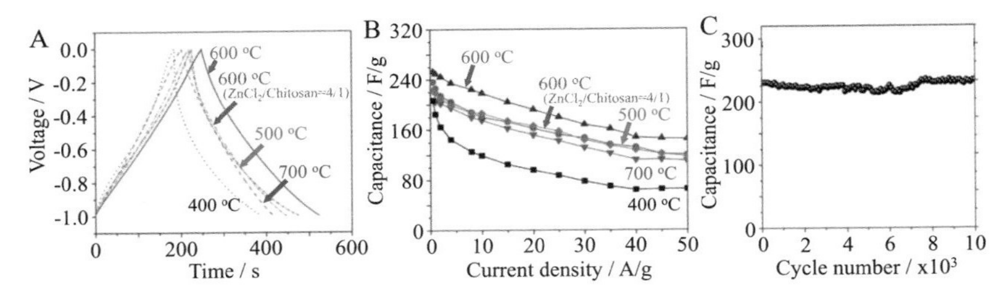

Figure 3.81 Capacitive performances of N-doped porous carbons derived from chitosan with $ZnCl_2$/chitosan of 10/1: (A) charge–discharge curves and (B) rate performances of the carbon prepared at different temperatures in comparison with the carbon prepared at 600°C with $ZnCl_2$/chitosan of 4/1, and (C) cycle performance at 8 A/g rate for carbon prepared at 600°C [296].

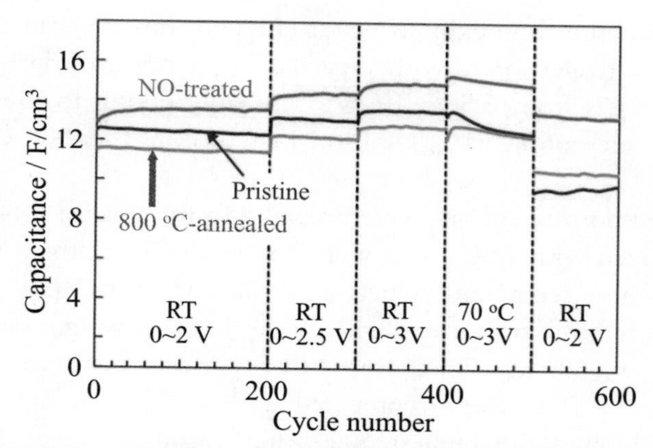

Figure 3.82 Discharge capacitance at 80 mA/g rate with different potential windows for the pristine AC, AC treated by NO at 800°C, and AC annealed at 800°C (without NO) [299].

As shown in Fig. 3.82, discharging capacitance increased after NO treatment (N-doping) even at a widened potential window (0—3 V) more markedly than pristine AC [298]. The enhanced capacitance of the NO-treated AC is maintained, and even slightly increased, by increasing the ambient temperature from room temperature to 70°C, whereas the capacitance of the pristine AC is markedly depressed at 70°C. After cycling associated with changing the potential window and temperature, the NO-treated AC recovered the initial capacitance but the pristine and 800°C-annealed ACs give depressed capacitances.

N/P-codoped microporous carbons were prepared from a mixture of phosphoric acid (H_3PO_4) with polyaniline by carbonization at 800°C, containing 2.7 at.% pyrrolic-N and 1.4 at.% P. The resultant carbon exhibited a high capacitance retention of about 90% at a high current density of 2 A/g in 6 M KOH aqueous electrolyte [300].

3.2.2 Templated carbons

3.2.2.1 Zeolite-templated carbons

Zeolite-templated carbons (ZTCs) were prepared by CVI of acetylene at 600°C using different zeolites as templates, followed by heat treatment at either 850 or 900°C [301]. The resultant carbons were highly microporous, giving very high S_{BET} of 2700—3000 m^2/g and high V_{total} of 1.3—1.6 cm^3/g, with high fraction of V_{micro} as about 70%—93%, as shown in the N_2 adsorption—desorption isotherms in Fig. 3.83A and listed in

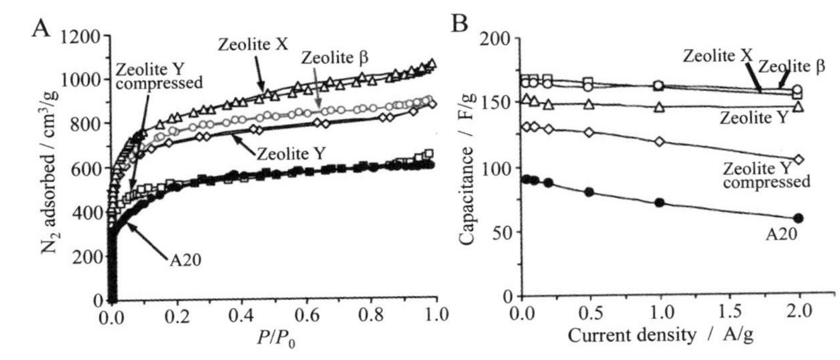

Figure 3.83 Ordered microporous carbons prepared using different zeolites as templates: (A) N_2 adsorption–desorption isotherms and (B) rate performances in 1 M $TEABF_4$/PC, in comparison with commercial AC (A20) [301].

values in Table 3.2. Sizes of micropores are homogeneous and ordered, with averaged pore diameter of about 0.93 nm. ZTCs can deliver high capacitances, as shown by the gravimetric capacitances at 50 mA/g rate in Table 3.2, and excellent rate performance up to 2 A/g rate, as shown in Fig. 3.83B. The capacitance retention calculated by the ratio of the capacitance with 2 A/g to that with 50 mA/g is more than 90%. High gravimetric capacitance of ZTCs is attributed to high S_{BET} with a certain contribution of pseudocapacitance caused by surface functional groups containing oxygen (Table 3.2). ZTC synthesized using zeolite Y with hot-pressing at 300°C (Zeolite Y compressed) exhibits slightly lower pore parameters and lower capacitance. High rate performance is reasonably supposed to be due to an interconnection between ordered micropores. ZTCs were synthesized by CVI of either ethylene at 550–750°C or acetonitrile vapors at 700–900°C [302]. ZTCs prepared from acetonitrile exhibited more developed microporous structures than those from ethylene, S_{BET} of 1122–1987 m^2/g with V_{total} of 0.81–1.08 cm^3/g for the former, while the latter had 853–1032 m^2/g with 0.46–0.74 cm^3/g, respectively, and as a consequence the highest capacitance of 146 F/g was obtained for the ZTC with the largest S_{BET}.

N-doped ZTC was synthesized using Na–Y zeolite template and two-step carbonization, in the first step acrylonitrile (AN) impregnation via its aqueous solution and carbonized at 700°C and in the second step CVI of propylene at 700 °C, followed by annealing at 900°C in N_2 flow [303]. The resultant N-doped ZTC contained a relatively large amount of N predominantly at graphitic and pyridinic configurations and had S_{BET} of

Table 3.2 Pore structure parameters, capacitive performances, and electrical resistance of ZTCs prepared using different zeolites in comparison with commercial AC (A20) [301].

	S_{BET} (m^2/g)	V_{total}[b] (cm^3/g)	V_{micro}[c] (cm^3/g)	d_{micro}[d] (nm)	O- content[e] (wt.%)	Capacitance[f] (F/g)	Resistance[g] (Ω)
Zeolite Y	2670	1.33	1.13	0.87	6.0	153	0.67
Zeolite X	3040	1.59	1.10	0.98	5.5	168	1.30
Zeolite Y compressed	1950	0.97	0.78	0.79	3.9	131	0.31
Zeolite β	2750	1.36	1.26	0.95	3.3	164	0.53
A20[a]	1650	0.91	0.62	1.11	3.4	90	1.31

[a]Commercial activated carbon.
[b]Calculated from N_2 adsorbed at $P/P_0 = 0.96$.
[c]Calculated by DR equation.
[d]Average micropore size obtained from α_s plot.
[e]Determined by elemental analysis as a difference.
[f]1 M $TEABF_4$/PC electrolyte with 50 mA/g.
[g]Electrical resistance of the electrode sheet.

1680 m^2/g with V_{total} of 0.86 cm^3/g, including V_{micro} of 0.51 cm^3/g. In Fig. 3.84A, the CV curves in 1 M H_2SO_4 electrolyte for the resultant N-doped ZTC are compared with those for ZTC synthesized using acetylene (no N-doping) and a commercial AC with similar pore structures, revealing that ZTCs show a wide reversible hump, indicating a large pseudocapacitance due to surface functional groups, although AC demonstrates quite a rectangular shape. N-doped ZTC shows a more marked hump, which contributes to an increase in the overall capacitance of the cell, 340 F/g, in comparison to 240 F/g for ZTC and 120 F/g for the AC. As shown in Fig. 3.84B, both N-doped ZTC and AC could be cycled with the maximum voltage of 0.6 V at a current density of 100 mA/g up to 10^4 cycles. Under severe conditions of a maximum voltage of 1 V and a current density of 100 mA/g, AC could not be cycled after about 40 cycles. For the N-doped ZTC, however, the cyclability even under severe conditions of a maximum voltage of 1.2 V at 200 mA/g rate remains parallel to the case of the maximum voltage of 0.6 V at 100 mA/g rate, although the capacitance is slightly lower, as shown in Fig. 3.84B.

Two kinds of ZTC were synthesized using the same Zeolite Y as template by carbonization of FA-impregnated zeolite at 700°C followed by CVI of propylene at 700°C, and by CVI of acetylene at 600°C, followed by annealing at 900°C [304,305]. The former has a small number of carbon atoms per spherical micropore (N_C of 38.9) and is coded as ZTC-S, and the latter has a large number of N_C of 46.2 and is coded as ZTC-L, with these two ZTCs having almost the same micropore structure, as shown in Fig. 3.85A, with the micropores being ordered and interconnected with each other. Capacitive performances with very high current densities of up to 20 A/g in nonaqueous electrolyte (1 M Et_4NBF_4/PC) of these

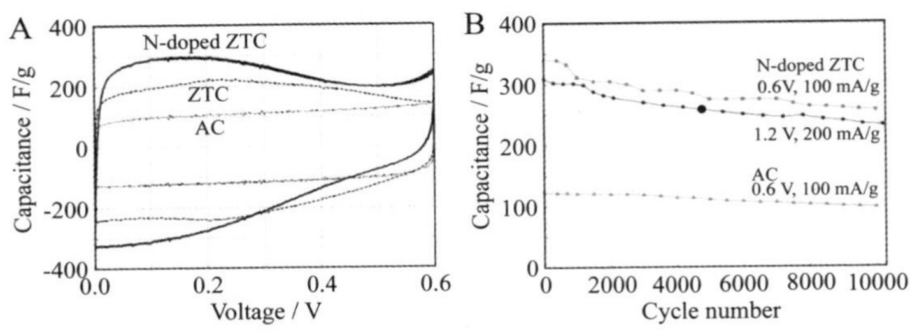

Figure 3.84 N-doped ZTC in 1 M H_2SO_4 electrolyte: (A) CV curves at 2 mV/s in comparison with ZTC and AC and (B) cycle performances with different current densities, in comparison with AC [303].

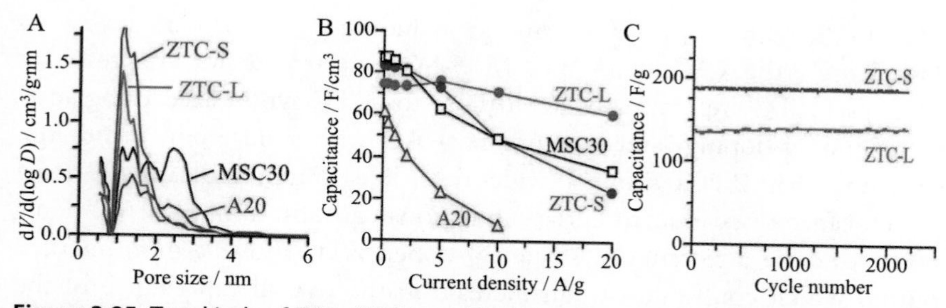

Figure 3.85 Two kinds of ZTCs (ZTC-L and ZTC-S) in comparison with two commercial ACS (MSC30 and A20): (A) pore size distributions, (B) rate performances, and (C) cycle performances at 1 A/g rate in 1 M Et_4NBF_4/PC electrolyte at 25°C [304].

two ZTCs were compared with two commercial ACs (MSC30 and A20, of which the pore size distributions are shown in Fig. 3.85A). At very low rates, the ZTCs and ACs show larger volumetric capacitances, as shown in Fig. 3.85B. At higher rates, A20 cannot retain the capacitance, while MSC30 retains 40% of its initial capacitance even at an ultrahigh rate of 20 A/g, which may be attributed to the presence of small mesopores (around 3 nm size) (refer to Fig. 3.85A). In contrast, ZTCs can well retain the capacitance at the high rates, even though they do not contain small mesopores, which may be because of their 3D and mutually connected micropore arrangement. At a relatively low rate of 1 A/g, ZTCs have excellent cycle performance up to 2500 cycles, as shown in Fig. 3.85C.

3.2.2.2 MgO-templated carbons

Mesoporous carbons were synthesized through MgO template technique using different MgO precursors, such as MgO, $Mg(OH)_2$, Mg-acetate, and Mg-citrate, with different carbon precursors, such as PVC, PVA, and pitches, and subjected to investigations into supercapacitor applications [306–311]. The MgO-templated carbons are rich in mesopores, depending greatly on the precursors of MgO and of carbon, as well as the mixing method; the carbon derived from Mg-acetate with PVA in a 7/3 ratio by powder mixing gave S_{meso} of 715 m^2/g and S_{meso}/S_{total} of 0.71 and that by solution mixing gave 1624 m^2/g and 0.85, respectively, while the carbon derived from Mg-citrate with PVA in a 7/3 ratio, even by powder mixing, gave S_{meso} of 1190 m^2/g and S_{meso}/S_{total} of 0.77 [306]. As a consequence of this characteristic of pore structure, the MgO-templated carbons could deliver high capacitances and high cycle performances in H_2SO_4 aqueous electrolyte, as shown in Fig. 3.86A and B. These high capacitance and high cycle performances were also confirmed in nonaqueous electrolyte, with

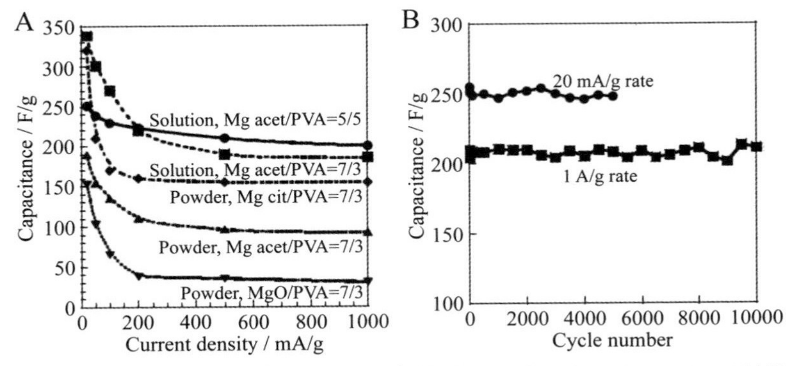

Figure 3.86 Capacitive performances of MgO-templated carbons in 1 M H_2SO_4 aqueous electrolyte: (A) rate performances for the carbons prepared from mixtures of Mg-acetate, Mg-citrate, and MgO with PVA by solution and powder mixing, and (B) cycle performances of the carbon prepared from the mixture of Mg-acetate with PVA in a 7/3 ratio by solution mixing [306].

the rate performance being much better in 1 M $TENBF_4$ [307]. Mesoporous carbons synthesized from the mixture of Mg-acetate and -citrate with a pitch at 950°C were studied by measuring the capacitive performance in 30% KOH electrolyte [309]. By using needle-like $Mg(OH)_2$ as the MgO precursor, the effect of the morphology of mesopores on the performances of supercapacitors was studied [311].

N-doped mesoporous carbons were synthesized from the mixtures of N-containing carbon precursors, PVP (12.6 wt.% N), polyacrylamide (PAA, 19.7 wt.% N) and trimethylolmelamine (TMM, 38.4 wt.% N), with Mg acetate, by carbonization at 800–1000°C in N_2 flow, with the carbon precursor/MgO ratio being controlled at 3/7 or 5/5 by weight [312]. Rate performances in capacitors with 1 M H_2SO_4 electrolyte are shown in Fig. 3.87A, revealing that most of the synthesized carbons exhibit similar changes in gravimetric capacitance C_M with increasing scanning rate, and that TMM-derived carbons do not have high capacitance in comparison with PVP- and PAA-derived carbons, whereas the former had a higher N content than the latter two. As shown in Fig. 3.87B, C_M cannot be related to N content but C_A (capacitance normalized by S_{BET}, C_M/S_{BET}) shows linear relations against N content, which seem to depend on the mixing ratio.

Mesoporous carbon was synthesized by sealing a mixture of MgO and Mg powders (MgO/Mg of 8 by weight) in CO_2, where the reaction of Mg with CO_2 ($Mg + CO_2 = 2MgO + C$) could be self-propagating by an

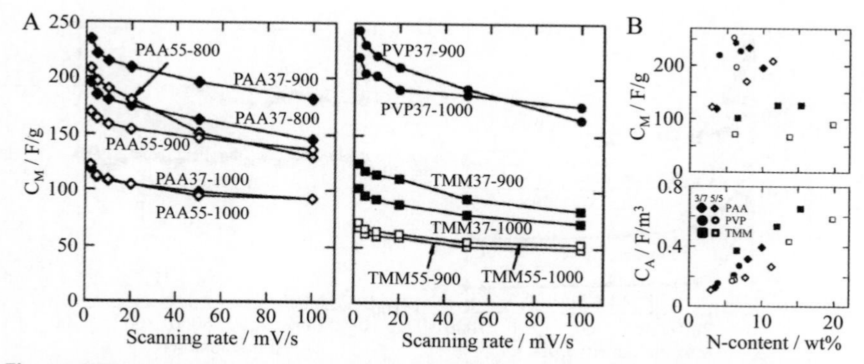

Figure 3.87 Capacitive performance of N-doped mesoporous carbon prepared from different carbon precursors using MgO template at different temperatures: (A) rate performances of gravimetric capacitance C_M and (B) relations of C_M and C_A (C_M/S_{BET}) with N content [312].

initial thermal stimulus to deposit C onto MgO particles (self-propagating high-temperature synthesis, SHS) [313]. The resultant carbon after dissolution of MgO showed a type-IV N_2 adsorption−desorption isotherm with H3 hysteresis, of which S_{BET} was measured as 709 m^2/g. It delivered high capacitance in ionic liquid electrolytes as 244 F/g at 2 A/g and 113 F/g at 500 A/g.

By coupling MgO-templated mesoporous carbons with different S_{BET} with microporous carbons (commercial ACs, S_{BET} of 1906 m^2/g), asymmetrical supercapacitor cells were constructed with 1 M $TEMABF_4$/PC electrolyte, of which two electrodes composed of different carbons, and their capacitive performance, were investigated [282,314,315]. In Fig. 3.88A and B, capacitances measured at the current densities of 0.1 and

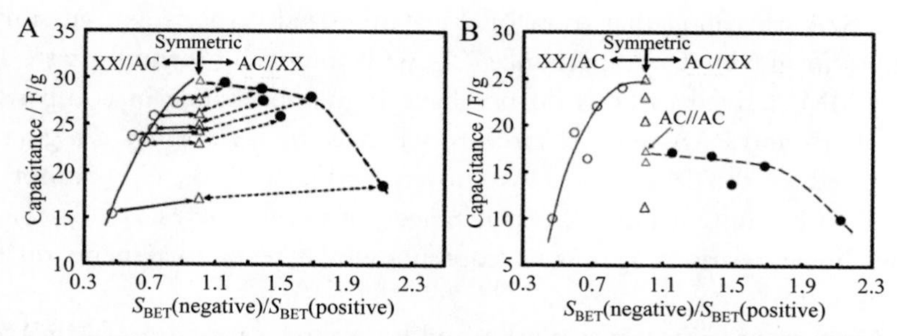

Figure 3.88 Changes in capacitance with the ratio S_{BET}(negative)/S_{BET}(positive) of asymmetric and symmetric electrodes assembles in 1 M $TEMABF_4$/PC electrolyte: at current densities of (A) 0.1 and (B) 1 A/g [282].

1 A/g, respectively, are plotted against the ratio in S_{BET}'s of negative to positive electrodes [S_{BET}(negative)/S_{BET}(positive)]. The asymmetric cells are expressed by AC//XX and XX//AC, together with symmetric cells (XX//XX and AC//AC) [282]. In the figures, symmetric cells are located at a S_{BET} ratio of 1.0, of which capacitances depend on the S_{BET} of the carbon materials used. Asymmetric cells used the AC in the positive electrode and other carbon XX in the negative electrode give almost the same capacitance as that observed on the symmetric cell of XX, as shown by the full lines in Fig. 3.88. On the other hand, asymmetric cells with AC in the negative electrode and other carbon XX in the positive electrodes deliver higher capacitances than symmetrical cells of XXs at a low current density of 0.1 A/g, as shown by the broken lines in Fig. 3.88A. At a high current density of 1 A/g, however, the capacitances of asymmetric cells are the same or even lower than those of the symmetric cell of AC, as shown in Fig. 3.88B. On the asymmetric AC//XX cell side, the performance rating calculated as the ratio of capacitance at 1 A/g to that at 0.1 A/g, $C_1/C_{0.1}$, was almost constant, which was the same as that for symmetric AC//AC cell, while on the asymmetric XX//AC cell side $C_1/C_{0.1}$ depends on the carbon used at the negative electrode. Almost the same results on the relations between capacitance and S_{BET} ratio for various combinations of microporous and mesoporous carbons were obtained [312,314]. These experimental results suggest that the carbon in the negative electrode governs predominantly the performance of supercapacitors with 1 M $TEMABF_4$/PC nonaqueous electrolyte.

The carbon synthesized from Ba-citrate by carbonization at 700°C was rich in large mesopores (about 13 nm size) with negligibly small V_{micro} demonstrating a slightly better rate performance in 30% KOH and an ionic liquid ($EMImBF_4$) electrolyte than the carbon derived from Mg-citrate composed of small mesopores with micropores [309].

Porous carbons were synthesized from fluorene via MgO (about 50 nm diameter) templating and simultaneous activation by KOH at 900°C [316]. As shown by the N_2 adsorption–desorption isotherms in Fig. 3.89A, the mixing ratio of KOH/fluorene by weight governs the development of pore structure in the resultant carbons. The resultant carbons composed of thin wrinkled sheets (graphene-like) show net-like morphology, which was supposed to be inherited from the assemblage of fluorene in the restraint spaces between MgO nanoparticles. The carbon synthesized using a KOH/ fluorene ratio of 3 exhibited S_{BET} of 2550 m^2/g with V_{total} of 1.57 cm^3/g including V_{micro} of 1.06 cm^3/g, and high capacitance of 313 F/g at 0.05 A/g

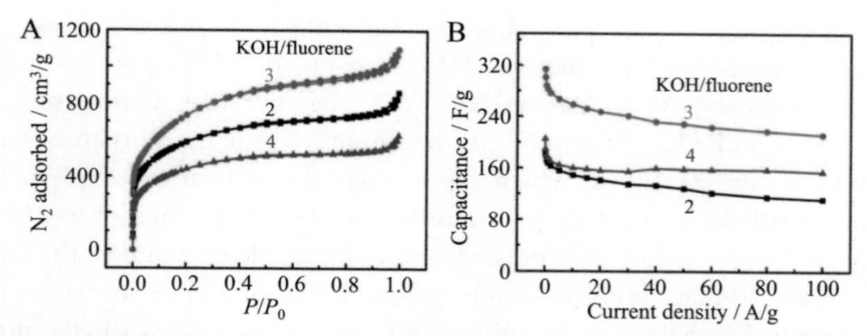

Figure 3.89 Fluorene-derived mesoporous carbons using MgO template and simultaneous KOH activation with different KOH/fluorene weight ratios: (A) N_2 adsorption–desorption isotherms and (B) cycle performance of the carbons activated [316].

and 211 F/g at 100 A/g, as shown by the rate performance of the carbons in Fig. 3.89B. A similar carbon morphology was obtained from petroleum pitch with MgO and KOH [317]. Capacitive performance was studied on the carbons prepared from coal tar pitch by coupling with simultaneous KOH activation using the templates of MgO [318,319], $Mg(OH)_2$ [320], as well as ZnO [321] and Fe_2O_3 [322].

The porous carbons were fabricated from a mixture of rice husk-derived carbon (RHC) and coal tar with MgO nanoparticles and KOH, where the weight ratios of RHC/tar, MgO/RHC, and KOH/tar were 1/1, 8/1, and 4/1–8/1, by carbonization and simultaneous activation at 950°C [323]. The resultant carbons were the composites of two kinds of carbons, RHC- and tar-derived carbons, and highly mesoporous, S_{BET} of 1833–3120 m^2/g with V_{total} of 2.17–3.00 cm^3/g including V_{micro} of 0.41–0.44 cm^3/g. As shown in Fig. 3.90, they delivered a high capacitance

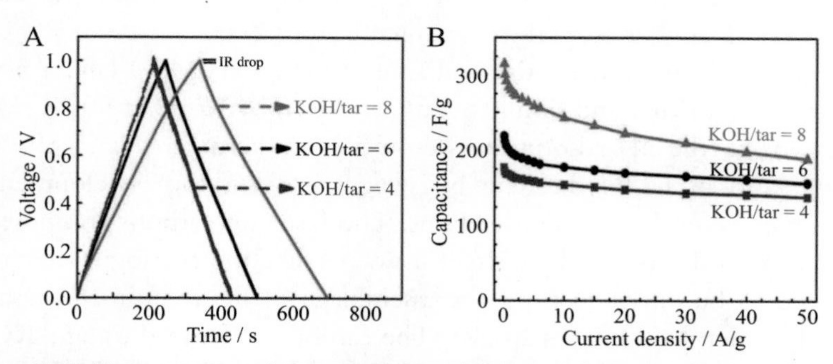

Figure 3.90 Mesoporous carbon composites fabricated from rice husk and coal tar with MgO templating and simultaneous KOH activation: (A) charge–discharge curves at 0.2 A/g rate and (B) rate performances [323].

of 315 F/g at 0.1 A/g rate and 189 F/g at 50 A/g with high cycling stability (95.8% retention after 10,000 cycles at 5 A/g rate) for the carbon prepared using KOH/tar ratio of 8/1.

3.2.2.3 Silica-templated carbons

Ordered mesoporous carbons (OMCs) were synthesized mostly from thermosetting resins using mesoporous silicas as templates and investigated as the electrodes of supercapacitors [324–336]. OMCs synthesized from different carbon precursors, propylene gas, sucrose, and pitch using different mesoporous silicas, MCM-48, SBA-15, and MSU-1, had their capacitive performances studied in an aqueous electrolyte (1 M H_2SO_4) and nonaqueous electrolyte (1.0 or 1.4 M $TEABF_4$/AN) [331]. The capacitances are correlated with V_{micro} measured by CO_2 [$V_{micro}(CO_2)$], as shown in Fig. 3.91A, suggesting the important role of micropores evaluated by $V_{micro}(CO_2)$ in the carbon electrode for the formation of EDLs. The carbons were synthesized from sucrose using silica MCM-48 (the carbon denoted as OMC-1) using SBA-15 (OMC-2) and using aluminated SBA-15 (OMC-3) [334]. The carbon OMC-1 contained homogeneous size mesopores which connected three-dimensionally with each other, OMC-2 contained a two-dimensional array of mesopores between ordered carbon nanorods and OMC-3 contained carbon nanopipes rigidly interconnected. The dependences of capacitance in 30 wt.% KOH aqueous electrolyte are shown in Fig. 3.91B on these OMCs by comparing with commercial AC (Maxsorb). OMCs have much better rate performances, particularly OMC-2 and OMC-3, than AC, although S_{BET} and V_{total} for the OMCs are much smaller (1700 m^2/g and 1.39 cm^3/g for

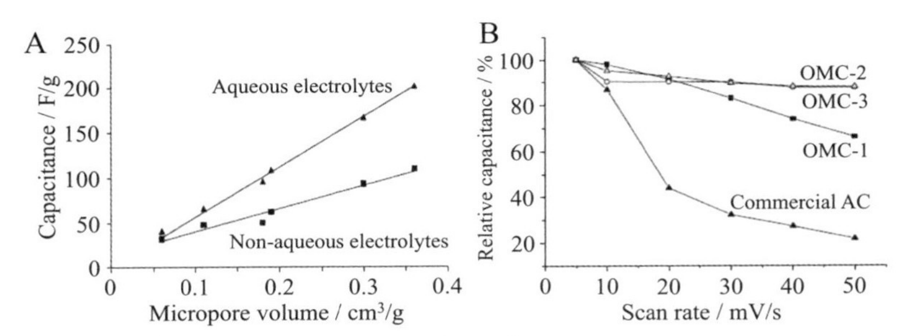

Figure 3.91 Capacitive performances of the ordered mesoporous carbons in aqueous and nonaqueous electrolytes: (A) dependences of capacitance on $V_{micro}(CO_2)$ and (B) rate performances in comparison with a commercial AC [331,334].

OMC-3, respectively) than the AC (3300 m^2/g and 1.75 cm^3/g, respectively). These experimental results suggest that the mesopores, particularly ordered mesopores, accelerate the diffusion of electrolyte ions in the carbon electrode. The carbons with different sizes of mesopores and V_{meso} of 1.1–2.0 cm^3/g were synthesized using a series of SBA-type silica templates from FA at 1000°C, which exhibited better rate performances than an AC (PX-21) [332].

N-doping of mesoporous carbons synthesized from sucrose with SBA-15 template was effective in improving capacitive performances [337]. N-doping was performed by ammoxidation at 300°C under a flow of the mixed gas of NH_3 and air (1/1 by volume) either before or after dissolution of the template SBA-15 by HF from the 850°C-carbonized sucrose/SBA-15 composite. The former resulted in carbon with 8 wt.% N, almost equally distributed in graphitic, pyrrolic, and pyridinic configurations, and the latter condition gave 1.6 wt.% N, mainly in a pyrrolic configuration. As compared with the pristine mesoporous carbon and a commercial microporous AC (S_{BET} of 1117 m^2/g) by CV curves in 1 M H_2SO_4 electrolyte in Fig. 3.92, N-doped mesoporous carbons deliver better capacitive performance, with 8 wt.% N-doped carbon giving a capacitance of 182 F/g and a 1.6 wt.% doped one giving 136 F/g, in comparison to 119 and 95 F/g for the pristine mesoporous carbon and a microporous AC.

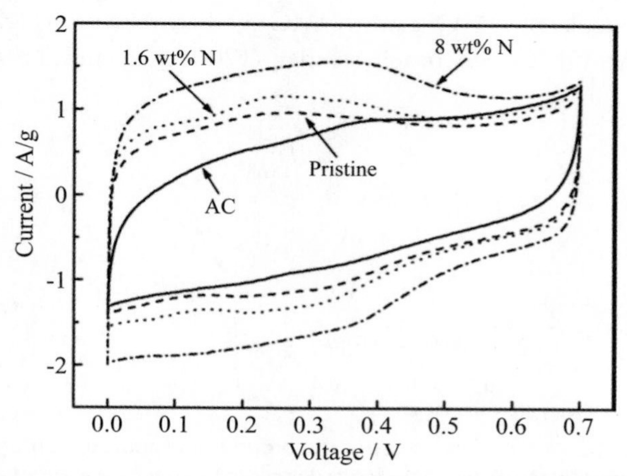

Figure 3.92 CV curves at 10 mV/s in 1 M H_2SO_4 electrolyte for N-doped mesoporous carbons, in comparison with the pristine mesoporous carbon and a commercial AC [337].

B-doping also altered the capacitive behavior of the carbon electrode [338,339]. B-doped mesoporous carbons were prepared by coimpregnation of sucrose and boric acid in mesopores of SBA-15 silica template and following carbonization at 900°C [338]. The carbon prepared from a mixture with a low boric acid content (0.075 weight ratio to sucrose) exhibited an improvement in capacitance measured in 1 M H_2SO_4 electrolyte. Carbons with boron contents of 1–3.4 wt.% and S_{BET} of 850–1360 m^2/g were prepared from glucose–borate complexes under hydrothermal conditions, which gave a capacitance of about 230 F/g in 1 M H_2SO_4 and about 100 F/g in 1 M Na_2SO_4 at 2 mV/s rate [339]. Boron was found to be mainly in the form of C–B–O bonding and to give a broad peak corresponding to pseudocapacitance in CV.

OMCs were synthesized from sucrose at 900°C using hexagonal mesoporous silicas (HMSs), which were prepared from TEOS using a surfactant $C_nH_{2n+1}NH_2$ with different n values of 8, 10, 12, and 16, and composed of a wormhole structure, being coded as OMC-8, -10, -12, and -16 [340]. By comparing the CV curves of these OMCs with a commercial carbon black (VC-72R) in Fig. 3.93A, four OMCs are forming EDLs associated with pseudocapacitive peaks at around 0.6 V, which is thought to be due to a redox reaction on the functional groups attached to carbon particles. Capacitances of OMC-8, -10, -12, and -16 are calculated as 130, 170, 175, and 260 F/g, respectively, at a scanning rate of 10 mV/s in a voltage range of 0.05–1.10 V, which are related to S_{BET} by a line passing through the origin, as shown in Fig. 3.93B.

3.2.2.4 Alumina-templated carbons

Carbon foams composed of mesopore-sized cells (mesosponges) were synthesized by CVD of CH_4 at 900°C using alumina Al_2O_3 nanospheres as

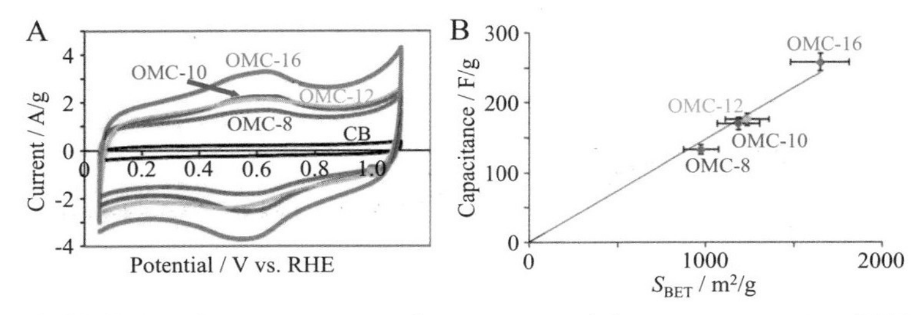

Figure 3.93 Ordered mesoporous carbons prepared from sucrose using HMS templates and the surfactant $C_nH_{2n+1}NH_2$ with different n-values (n = 8–16): (A) CV curves in 0.5 M H_2SO_4 at 10 mV/s and (B) relation between capacitance and S_{BET} [340].

template, followed by heat treatment at 1800°C [341]. Its capacitive performances are compared with a commercial AC (YP−50F), as shown in Fig. 3.94. The mesosponge heat-treated at 1800°C had a relatively high S_{BET} of 1940 m^2/g with very high V_{total} of 2.79 cm^3/g including a small V_{micro} of 0.47 cm^3/g. The wall of this mesosponge was considered to be composed of graphene sheets, supposedly up to 74% being single-layer graphene, by comparing the observed S_{BET} to theoretical S_{BET}'s of single- and double-layered graphene sheets. Mesosponge sheets were synthesized using a sheet (40 × 90 × 0.2 mm) of Al_2O_3 nanoparticles (diameter of 5−7 nm) as a template through CVD of CH_4 and heat-treated at 1800°C, which had S_{BET} of 1800 m^2/g with V_{total} of 2.6 cm^3/g [342]. From the resultant sheets, a symmetric supercapacitor cell was constructed with 1.5 M $TEMABF_4$/PC electrolyte at 25°C and capacitive performances were measured by applying a high voltage of up to 4.4 V by comparing with a commercial AC, SWCNTs, rGO, and the powder of this mesosponge. In Fig. 3.95A and B, Ragone plots and cycle performances are

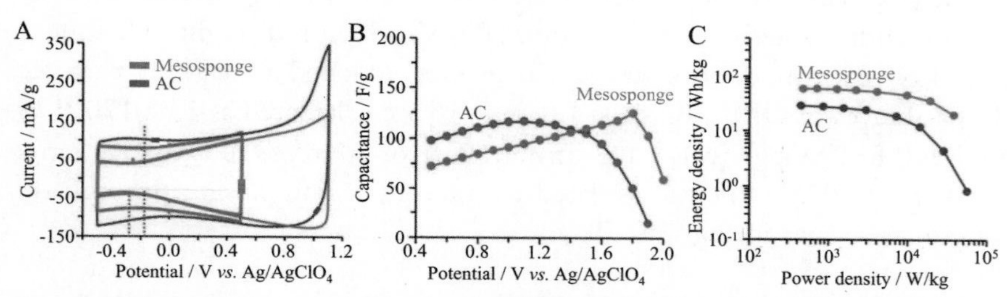

Figure 3.94 Capacitive performances of Al_2O_3-templated carbon mesosponge in comparison with an AC: (A) CV curves, (B) capacitance vs. upper-limit potential, and (C) Ragone plots [341].

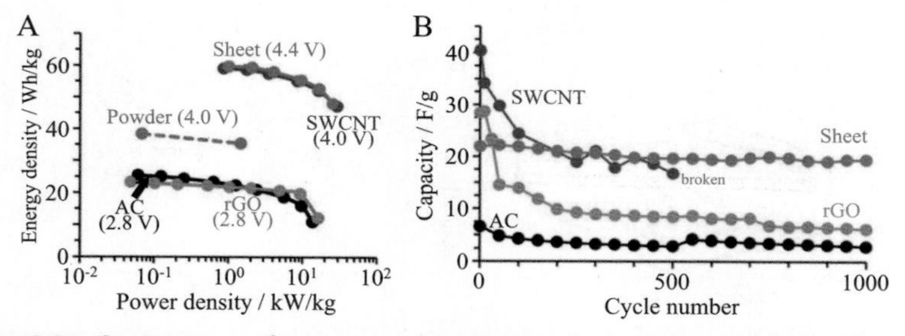

Figure 3.95 Capacitive performances of symmetrical supercapacitor cells of a mesosponge sheet in comparison with its powdered form and SWCNTs, rGO, and AC: (A) Ragone plots and (B) cycle performances measured at 0−4.4 V and 25°C [342].

shown. The mesosponge sheet, by applying 4.4 V, delivers high energy densities, which are much higher than the powder of the mesosponge, much superior to rGO and an AC, and almost comparable with SWCNTs with 4.0 V applied. However, the electrode of SWCNTs is broken after 500 cycles at 0—4.4 V although the mesosponge sheet remains unbroken even after 1000 cycles. The symmetric supercapacitors constructed from this mesosponge demonstrated excellent stability under 4.4 V at 25°C and 3.5 V at 60°C conditions.

AAO films, which had straight channels with pore sizes of about 24 nm and lengths of about 90 μm, were ground into small particles with sizes of 25—50 μm and then coated by carbon through CVD of propylene and acetonitrile at 800°C to prepare the AAO particles coated by the nondoped and N-doped carbons [343]. Coating by B-doped carbon was performed by deposition of B-containing carbon through the second CVD of the mixed gas of benzene and BCl_3 at 725°C on the AAO film powder coated by carbon via the CVD of propylene. Each of the C-coated AAO film powders was annealed at 900°C in an inert atmosphere. Capacitive performances of these C-coated AAO film particles are compared with each other in aqueous and nonaqueous electrolytes, 1 M H_2SO_4 and 1 M Et_4NBF_4/PC, as shown by the CV curves in Fig. 3.96A and B, respectively. Both N- and B-doping of carbon make the capacitance of the carbon-coated AAOs increase, more markedly in H_2SO_4, which was concluded to be mainly due to the pseudocapacitance.

Fibrous mesoporous carbons were prepared by synthesizing alkyl-substituted hyperbranched polyphenylene in channels of AAO membrane

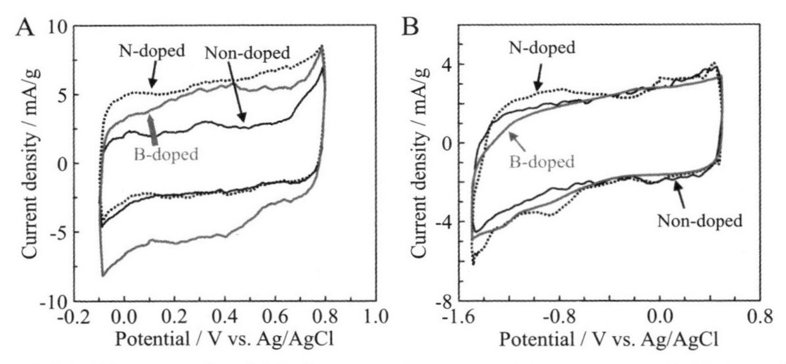

Figure 3.96 CV curves for AAO film powders coated by nondoped, N-doped, and B-doped carbons: (A) in 1 M H_2SO_4 aqueous electrolyte and (B) 1 M Et_4NBF_4/PC at scan rate of 1 mV/s [343].

at 250°C and then carbonizing at 600–800°C, followed by dissolution of AAO in a 3 M NaOH [344]. The resultant fibrous carbon prepared at 800°C had S_{BET} of 1140 m^2/g with large-sized mesopores (10–20 nm size) and delivered a capacitance of 304 F/g at 5 mV/s rate.

3.2.2.5 $CaCO_3$-templated carbons

N-doped porous carbons were synthesized from the mixture of ethylenediamine (EDA) and carbon tetrachloride (CTC) using $CaCO_3$ nanoparticles (about 50 nm size) at 700°C in N_2 [345]. The carbon prepared using a mixing ratio ($CaCO_3$/(EDA + CTC) of 0.3, which had S_{BET} of 1276 m^2/g with V_{total} of 0.75 cm^3/g and N content of 13.5 wt.%, delivered a capacitance of 226 F/g at 0.1 A/g rate and 132 F/g at 30 A/g. The capacitive performance of N-doped porous carbons prepared from melamine-formaldehyde as a carbon precursor with Ca-acetates as $CaCO_3$ template precursor was compared to those prepared from the same carbon precursor with $MgCl_2$ as MgO template precursor. The former dlivered a capacitance of 311 F/g and the latter of 329 F/g at 1 A/g rate after KOH activation [346]. The carbons were prepared from anthracene oil and $CaCO_3$ with KOH by carbonization and simultaneous activation at the temperature of 850–1000°C in Ar flow [347]. The resultant mesoporous carbons were called crumpled carbon nanonets possessing high S_{BET} up to 1822 m^2/g with V_{total} up to 4.11 cm^3/g. As shown by the charge–discharge curves in an ionic liquid ($BMIMPF_6$) electrolyte in Fig. 3.97A, the carbons exhibit linear and symmetrical profiles with only small IR drops (less than 0.02 V) and their capacitance depends greatly on the preparation

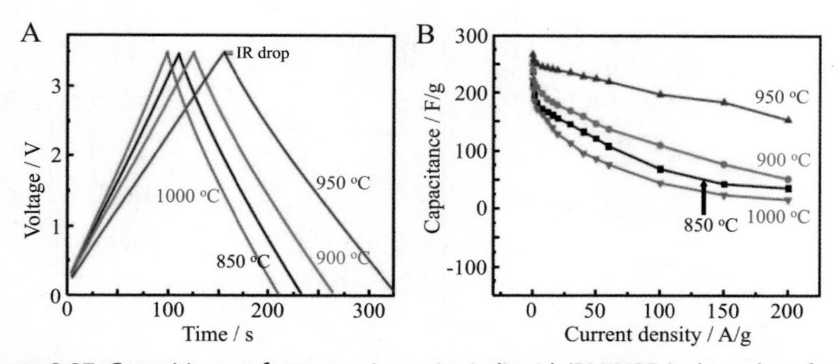

Figure 3.97 Capacitive performance in an ionic liquid ($BMIMPF_6$) electrolyte for the $CaCO_3$-templated carbons prepared from anthracene oil at different temperatures: (A) galvanostatic charge–discharge curves at the current density of 1.2 A/g and (B) rate performances [347].

temperature; the 950°C-prepared one gave the highest capacitance of 266 F/g at 0.05 A/g. Capacitance decreases with increasing rate, quickly at low rates and then gradually above 10 A/g, as shown in Fig. 3.99B: decreasing to197 F/g at 10 A/g and then gradually to about 150 F/g at 200 A/g for the 950°C-prepared one. The hierarchically porous carbons prepared from shrimp shells by mixing with KOH and heat-treated at 700°C, in which $CaCO_3$ (one of the pyrolysis products from shells) worked as the template for macropores and mesopores [348]. The resultant carbon having S_{BET} of 2032 m^2/g with S_{micro} of 1735 m^2/g and S_{meso} of 297 m^2/g exhibited capacitances of 239 and 201 F/g at 0.05 and 1.0 A/g, respectively, in 6 M KOH electrolyte.

Wrinkled porous carbon nanosheets were fabricated from methylnaphthalene oil using $CaCO_3$ template and simultaneous KOH activation at 900°C [349]. The resultant carbons were mesoporous and delivered capacitances of 286 F/g at 0.05 A/g and 201 F/g at 20 A/g.

3.2.3 Precursor-designed carbon

3.2.3.1 Polymer-blended carbons

Mesoporous carbons prepared from the mixture of resorcinol-formaldehyde (RF) and block copolymer P127 by carbonization at 800°C were applied to the electrodes of a supercapacitor with aqueous and nonaqueous electrolytes, and gave a higher capacitance than commercial AC [350]. As-prepared carbon had S_{BET} of 694 m^2/g with V_{total} of 0.66 cm^3/g including V_{meso} of 0.48 cm^3/g and that after KOH activation had 1685 m^2/g with 1.25 cm^3/g including 0.75 cm^3/g, respectively, demonstrating a marked increase in V_{micro} from 0.16 to 0.47 cm^3/g. N_2 adsorption—desorption isotherms are shown in Fig. 3.98A. The pore structure of the as-prepared carbon was less developed than that of the AC, but was markedly developed by KOH activation, even more markedly than AC after KOH activation. As shown in Fig. 3.98B and C, KOH activation is effective in improving the capacitance in both aqueous and nonaqueous electrolytes, although the improvement in rate performance is not marked. Two mesoporous carbons, as-prepared and activated, could deliver much higher capacitance than AC used as a reference.

Mesoporous carbons were prepared from the mixture of sulfonated pitch (SP) with block copolymer F127 and KOH by carbonization/activation at 800°C in N_2 [351]. The carbon prepared with KOH/SP of 5 (by weight) was microporous, S_{BET} of 2095 m^2/g with V_{total} of 1.07 cm^3/g including V_{meso} of 0.31 cm^3/g, while that with KOH/SP of 6 was rich in mesopores,

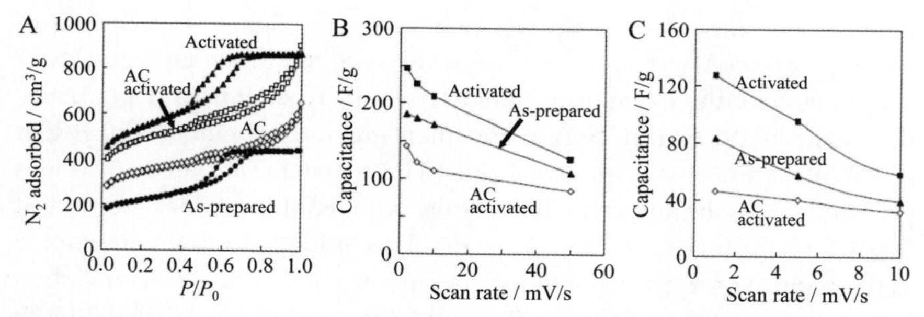

Figure 3.98 Mesoporous carbons prepared from RF and P127 before and after KOH activation in comparison with AC before and after KOH activation: (A) N_2 adsorption—desorption curves, (B) rate performances in 1 M H_2SO_4, and (C) those in 1 M Et_4NBF_4/PC [350].

S_{BET} reaching 3006 m^2/g with V_{total} of 1.61 cm^3/g including V_{meso} of 0.85 cm^3/g, as compared their N_2 adsorption—desorption isotherms in Fig. 3.99A. The latter carbon can apply high voltage, up to 3.5 V, and demonstrates steady charge—discharge curves with linear relations in $EMIMBF_4$ electrolyte, as shown in Fig. 3.99B, and delivers high capacitance, 157.8 F/g (57.4 F/cm^3) at 20 A/g, and high cycle performance, 86.6% capacitance retention after 10,000 charge—discharge cycles at 2.5 A/g, higher than the former, as shown in Fig. 3.99C. These high capacitive performances of the latter were thought to be due to the highly developed pore structure (appropriate balance between mesopores and micropores) and high electric conductivity of 135 S/m. Its energy density could retain up to 67.1 Wh/kg at a high power density of 17.5 kW/kg. Mesoporous carbon spheres were prepared from resol-type PF resin with F127 under hydrothermal conditions at 100°C, followed by carbonization at 700°C [352].

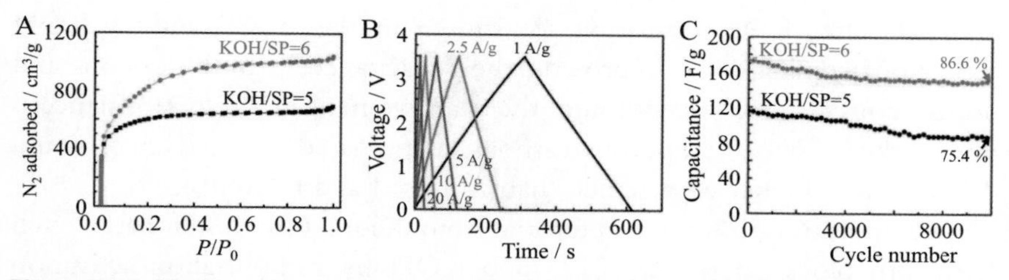

Figure 3.99 Mesoporous carbons synthesized from sulfonated pitch (SP) with F127 and KOH (KOH/SP of 5 and 6): (A) N_2 adsorption—desorption isotherms, (B) galvanostatic charge—discharge curves at 0—3.5 V and different current densities for the carbon prepared using KOH/SP = 6, and (C) cycle performances at 2.5 A/g [351].

Resol-F127 micelle formation time before hydrothermal treatment was found to influence the capacitive performance in 6 M KOH electrolyte, and micelle formation over 22 h resulted in a much better rate performance than that during 17 h, although capacitance at the current density below 20 mA/cm^2 was slightly lower.

Porous CNFs were prepared from a blend of polybenzimidazole (PBI) with a polyimide (Matrimid5218) (3/1 by weight) in DMAc/NMP solution by electrospinning, followed by stabilization at 400°C, carbonization at 950°C in N_2, and activation by steam at 800°C [353]. The resultant CNFs exhibited a capacitance of 126 F/g, and maximum energy and power densities of 49 Wh/kg and 6.5 kW/kg, respectively, in an ionic liquid (EMITFSI) electrolyte.

Porous hollow CNFs were manufactured from DMF solutions of styrene-acrylonitrile copolymer (SAN) as core and polyacrylic acid (PAA) with PVP (pore-forming agent) as shell by a coaxial electrospinning technique, followed by imidization of PAA at 300°C and carbonization at 900°C in N_2 [354]. The flow rate of the core solution was controlled at 0−0.6 cm^3/L. Marked improvement in capacitive performance by preparing the CNFs with a porous carbon shell was confirmed by assembling a symmetric supercapacitor with 6 M KOH electrolyte. The CNFs synthesized using a flow rate of 0.4 cm^3/L during electrospinning delivered a capacitance of 221 F/g at 0.2 A/g and high cyclability with 95% capacitance retention after 5000 cycles. MnO_2 nanocrystals confined into microporous CNFs delivered high capacitance in 1 M Na_2SO_4 aqueous electrolyte, giving a capacitance of 1282 F/g at 0.2 A/g rate [355]. Microporous CNFs (S_{BET} of 645 m^2/g with V_{total} of 0.303 cm^3/g including V_{micro} of 0.273 cm^3/g) were synthesized from the mixture of polyimide with PVP via electrospinning, followed by carbonization at 900°C. In order to confine Mn_2O nanocrystals preferentially in the nanopores, the porous CNFs were immersed in a diluted aqueous solution of $KMnO_4$ (less than 0.024 g/L). A higher concentration of $KMnO_4$ solution resulted in MnO_2 deposition mostly on the external surface of the CNFs. The MnO_2 crystals confined into the nanopores were thought to have different crystal habits from those deposited on the external surface and delivered much higher capacity than the latter. A symmetric supercapacitor composed of these MnO_2-confined CNFs could deliver an energy density of 36 Wh/kg at a power density of 0.039 kW/kg and maintain 7.5 Wh/kg at 10.3 kW/kg. Electrospun composite fibers of a polyimide (PMDA/ODA) mixed with PVP were immersed into

$Mg(NO_3)_2$ aqueous solution after plasma treatment, followed by carbonization at 900°C using different heating rates, 3 and 10°C/min, and dissolution of MgO with HCl to make the surface of the carbon fibers mesoporous, by keeping the core of the fibers microporous [356]. As shown in Fig. 3.100, the carbon fibers prepared using a fast heating rate of 10°C/min exhibit much improved capacitive performances; a straight charge—discharge curve without noticeable IR drop, high capacitance retention from 140 F/g at 0.1 A/g rate to 66 F/g at 50 A/g, and high cyclability with 91% retention after 3000 cycles at 1 A/g rate.

3.2.3.2 Carbide-derived carbons

Heat treatment of different metal carbides in Cl_2 flow resulted in microporous carbons (carbide-derived carbons, CDCs), which were used as the electrodes of supercapacitors [357—363]. Micropore sizes are governed by the starting carbides and chlorination temperature, where metal atoms were excluded by vaporization as metal chlorides, as shown by the chlorination temperature dependences of S_{BET} and the capacitance in 1 M H_2SO_4 aqueous electrolyte on Ti_2AlC and B_4C in Fig. 3.101A and B,

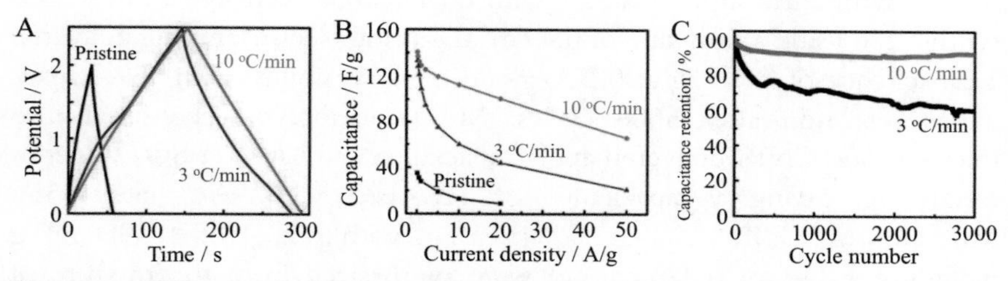

Figure 3.100 Polyimide-derived carbon fibers with MgO treatment: (A) charge—discharge curves at 1 A/g rate, (B) rate performances, and (C) cycle performances [356].

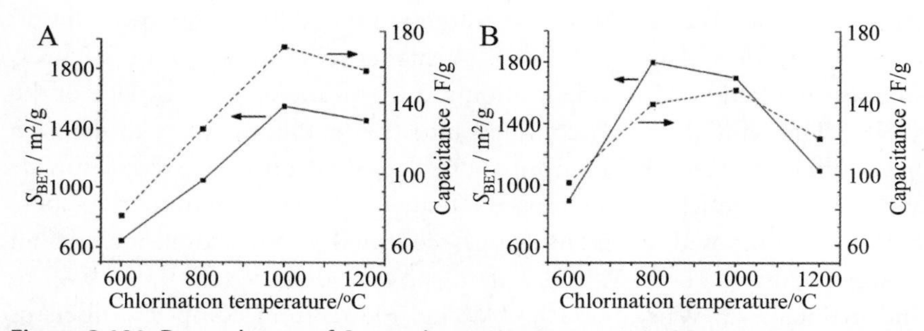

Figure 3.101 Dependences of S_{BET} and capacitance in 1 M H_2SO_4 aqueous electrolyte on chlorination temperature for (A) Ti_2AlC- and (B) B_4C-derived carbons [359].

respectively [359]. In Ti_2AlC-derived carbons, S_{BET} reaches a maximum of about 1550 m^2/g and correspondingly capacitance also reaches a maximum of about 175 F/g. In B_4C-derived carbons, S_{BET} shows a maximum of around 1700 m^2/g at about 800–1000°C and capacitance shows a maximum of about 140 F/g at the same chlorination temperature range. Changes in capacitance and surface areas, S_{micro} and S_{meso}, with chlorination temperature were studied for TiC- and ZrC-derived CDCs, revealing that capacitance changed in parallel with S_{micro} in both CDCs [360].

CDCs were prepared from the mixture of TiC (average particle size of 1.3–3 μm) with 10 wt.% TiO_2 (particle size of about 1 μm) by chlorination at 700, 800, and 900°C, followed by treatment with hydrogen at 800°C to dechlorinate thoroughly as well as to remove oxygen-containing functional groups from the surface of the resultant carbon (TiC/TiO_2-derived CDCs) [364]. The carbons were characterized by narrow pore size distributions at around 0.8 nm and a noticeably smaller amount of pores below 0.7 nm compared to the carbon derived from pure TiC without TiO_2 (TiC-derived CDCs). The capacitive performances of these two CDCs were investigated in 1.2 M $TEMABF_4$/AN electrolyte by comparing with a commercial AC prepared from coconut shell. CV curves for these carbon electrodes were nearly rectangular, although coconut-derived AC and TiC-derived CDCs possessed lower capacitance. The TiC-derived CDCs contained a significant fraction of micropores less than 0.7 nm size, which were not accessible for the electrolyte cations ($TEMA^+$). As shown in the Ragone plots in Fig. 3.102A, the volumetric energy densities at moderately low power density are fairly similar for all CDCs, although AC shows lower energy

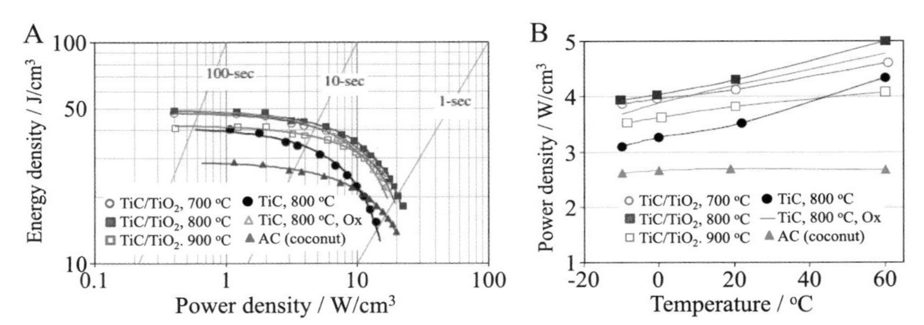

Figure 3.102 Capacitive performances of $TiCTiO_2$-derived CDCs at different temperatures: (A) Ragone plots and (B) temperature dependences of power density at discharge time, of 10 s, in comparison with TiC-derived CDCs and a coconut-derived AC [364].

density due to its low bulk density. TiC/TiO_2-derived CDCs prepared at 800 and 700°C deliver the highest power and energy density, and demonstrate a significant improvement in the power density, in comparison with TiC-derived CDCs and that additionally oxidized (TiC-derived CDCs oxidized). In Fig. 3.102B, the temperature dependences of powder density are shown by selecting a discharge time at 10 s, revealing the effectiveness of TiO_2-assisted chlorination to prepared CDCs for a supercapacitor electrode.

Supercapacitor characteristics for CDCs prepared from TiC, α-SiC, Mo_2C, Al_4C_3 and B_4C were studied in various nonaqueous electrolyte solutions, 1 M various tetraalkylammonium tetrafluoroborate, as $(C_2H_5)_3CH_3NBF_4$, $C_2H_5(CH_3)_3NBF_4$, $(C_2H_5)_3C_3H_7NBF_4$, and $(C_2H_5)_3C_4H_9NBF_4$, with various solvents, acetonitrile (AN), γ-butyrolactone, and propylene carbonate (PC) [357,358,365,366].

N/S-codoped CDCs prepared using Cr_3C_2 with thiourea under hydrothermal conditions delivered excellent cycle performance in 6 M KOH electrolyte at room temperature [367]. The CDCs with 5.10 at.% N and 0.56 at.% S delivered the highest capacity of 270 F/g, though the charge—discharge curves suggested the presence of pseudocapacitance, and a high capacitance retention of 95% at 0.5 A/g after 6000 cycles.

3.2.3.3 MOF- and COF-derived carbons

Porous carbons were prepared by carbonization of MOF-5 after impregnation of FA, of which the pore structure depended greatly on a carbonization temperature from 530 to 1000°C [368]. As shown CV curves in 1.0 M H_2SO_4 for the carbons prepared at 530, 800, and 1000°C in Fig. 3.103A—C, S_{BET}, V_{total}, and V_{meso} change with increasing carbonization temperature through the minimums at 800°C. The carbon prepared

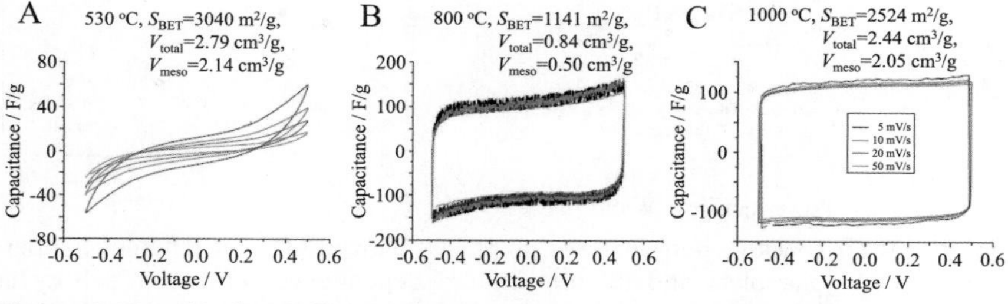

Figure 3.103 CVs for MOF-templated carbon prepared from furfuryl alcohol at different temperatures [368].

at 530°C shows very high S_{BET} and V_{meso}, while V_{micro} calculated as the balance between V_{total} and V_{meso} is almost negligible and as a consequence gives a capacitance of only 12 F/g and poor rate performance. The carbon prepared at 1000°C gives a relatively high capacitance of 120 F/g with almost ideal rectangular CV curves and good rate performances, suggesting the importance of a balanced existence of micropores and mesopores. The carbons synthesized from zeolite-type MOF (ZIF-8) with FA at 1000°C delivered S_{BET} of 3405 m^2/g and capacitance of about 200 F/g in 1 M H_2SO_4 [369].

Mesoporous carbons were synthesized from MOF-5 with or without additional carbon precursors (phenol resin and a mixture of CCl_4 with EDA) by carbonization at 900°C in N_2 flow [370]. The resultant carbons were activated with KOH (KOH/carbon at a ratio of 5/1 by weight) at 700°C. In Fig. 3.104A and B, galvanostatic charge–discharge curves at the current density of 0.25 A/g in aqueous and nonaqueous electrolytes, respectively, are shown for the resultant carbons. The addition of the carbon precursors depressed the development of pore structure in the resulting carbons, particularly in the case of the addition of CCl_4 with EDA; S_{BET} decreased from 1812 m^2/g for the carbon without additional carbon precursor to 384 m^2/g, and as a consequence the capacitances measured in aqueous and nonaqueous electrolytes are very low, at 72 and 4 F/g, respectively. On the other hand, activation worked markedly on this carbon (with CCl_4 and EDA added), S_{BET} increased to 2222 m^2/g and the capacitances at 0.25 A/g rate became the highest in the carbons prepared, at 274 F/g in aqueous electrolyte and 168 F/g in nonaqueous electrolyte. The activation on the carbon prepared from MOF-5 without any additives reduced S_{BET} to 1673 m^2/g, while made the capacitances grow to 222 F/g

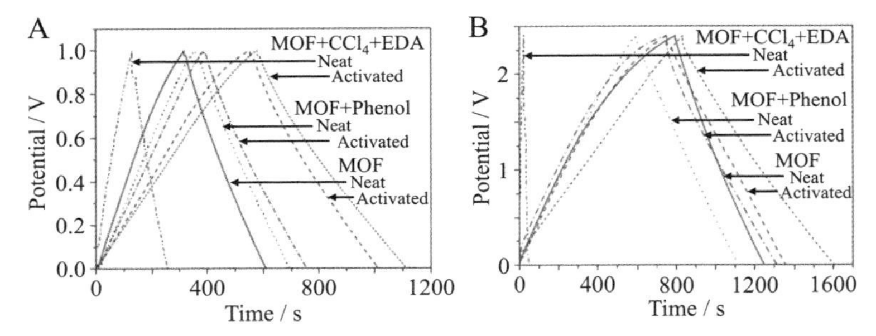

Figure 3.104 Galvanostatic charge–discharge curves of MOF-derived carbons without and with KOH activation: (A) in 6 M KOH and (B) 1.5 M Et_4NBF_4/AN [370].

in aqueous electrolyte and 126 F/g in nonaqueous electrolyte. Mesoporous carbon synthesized from glycerol with MOF-5 showed a high capacitance of 344 F/g in 6 M KOH electrolyte at a current density of 0.05 A/g, although the capacitance observed included a large amount of pseudocapacitance [371].

N-doped porous carbons were synthesized from ZIF-8 (nitrogen-containing zeolitic imidazolate framework) with additional carbon precursors (melamine, urea, xylitol, and sucrose) by carbonization at 950°C [372]. The carbon synthesized from the mixture of ZIF-8 and sucrose gave the highest capacitance, 286 F/g at 0.1 A/g and 208 F/g at 2 A/g in 6 M KOH electrolyte, probably owing to the largest N content with graphitic, pyridinic, and pyrrolic configurations and proper micro-/mesoporous structure.

COF synthesized from the mixture of 1,3,5-triformylbenzene with *p*-phenylenediamine was carbonized at 700, 800, and 900°C in N_2. Resulting in hollow N-doped porous carbon shells [373]. The carbon prepared at 800°C contained 4.73% N in pyridinic and graphitic configurations and exhibited the highest capacitance of 230 F/g at a current density of 0.5 A/g and 190 F/g at 3 A/g with a high capacitance retention of 98% after 1500 cycles.

π-Conjugated microporous polymers (CMPs) ionothermally synthesized by a condensation reaction of 1,2,4,5-benzenetetramine with triquinoyl hydrate at 300, 350, 400, 450, and 500°C (aza-CMPs) delivered very high capacitances in 1 M H_2SO_4 [374]. Elemental analysis gave a composition of 68.4% C, 4.25% H, and 22.40% N for the 350°C-synthesized one and 67.04% C, 4.64% H, and 24.04% N for the 450°C-synthesized one. The capacitance of these aza-CMPs depends greatly on the current density, as shown in Fig. 3.105, and reaches 946 F/g at 0.1 A/g, 439 F/g at 1 A/g, and 273 F/g at 10 A/g for the 450°C-synthesized one. They showed also excellent cycle performance, with no decline in capacitance after the first 2000 cycles at 1 A/g and following 3000 cycles at 5 A/g. Terephthalonitrile-derived nitrogen-rich networks (TNNs) were synthesized by polymerization at 550—700°C under ionothermal conditions using $ZnCl_2$ as the solvent and catalyst [375]. Capacitance measured by the three-electrode cells with 1 M H_2SO_4 electrolyte depends on the synthesis temperature, with 298 and 243 F/g for 550°C- and 700°C-synthesized TNNs at 0.2 A/g rate. Symmetrical supercapacitors were constructed using TNNs synthesized at different temperatures and their capacitive performances were investigated.

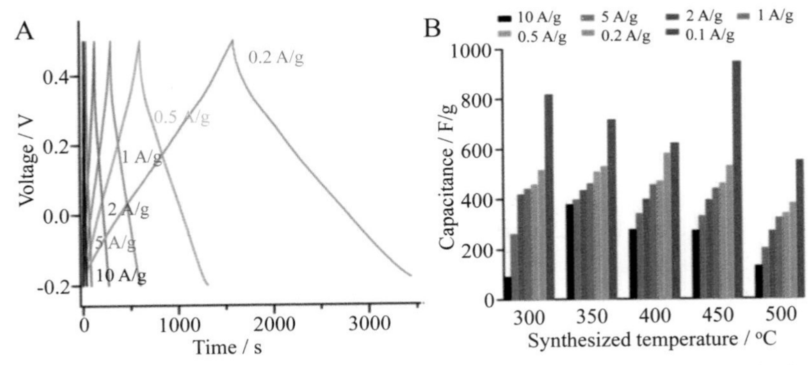

Figure 3.105 CMPs synthesized at different temperatures: (A) charge—discharge curves for the 350°C-synthesized CMP and (B) capacitance observed at different current densities [374].

3.2.3.4 Carbon foams

N-doped carbon foam prepared from the mixture of glucose and urea exhibited a high capacitance of 305 F/g at 0.2 A/g rate in 6 M NaOH and a high capacitance retention of 89% with 5 A/g rate after 10,000 cycles [376]. N-doped carbon foams prepared from ferric citrate and NH_4Cl delivered capacitances of 242 F/g at 1 A/g rate and 155 F/g at 30 A/g in 6 M KOH and high capacitance retention of 92% at 1 A/g rate after 5000 cycles [377].

Carbon foams were prepared from commercial melamine foams by carbonization at 800—1200°C (reticulated vitreous carbon, RVC) [378]. A 800°C-carbonized carbon foam, which contained about 20 wt.% N, delivered capacitances of 241 F/g at 0.2 A/g rate and 222 F/g at 1 A/g in 1 M H_2SO_4 electrolyte.

N-doped carbon foams were prepared from a mixture of g-C_3N_4 with glucose by heating under hydrothermal conditions at different temperatures (120, 140, and 160°C) for 10 h, followed by annealing at 900°C for 1 h in N_2 [379]. The capacitive performances of the foams were determined by construction of a solid-state symmetric cell with PVA-KOH gel electrolyte. The carbon foam obtained at 120°C had S_{BET} of 954 m^2/g and contained 11% N, and delivered a capacitance of 90 F/g at 5 mV/s rate. It possessed excellent cyclability, 95% capacity retention after 5000 cycles and 90% after 10,000 cycles at 1 A/g rate.

Carbon foam was prepared by mixing octylphenol polyoxyethylene ether (OP-10) with pentane (blowing reagent) into an NaOH solution of paraformaldehyde, melamine, and block copolymer P123 and then carbonizing by heating up to 800°C, which had a hierarchical pore structure, macropores formed by blowing of pentane, mesopores due to P123, and micropores due to matrix resin (hierarchical carbon foam, H-carbon foam) [380]. Its capacitive performances were compared with carbons prepared by the same process but without P123 (carbon foam) and without OP-10 and pentane (mesoporous carbon monolith). H-carbon foam exhibits high capacitance in 3 M KOH aqueous electrolyte, as shown in Fig. 3.106A, 222 F/g at 2 mV/s rate and 136 F/g at 50 mV/s. Galvanostatic charge–discharge curves for the symmetrical supercapacitor composed of H-carbon foam show a triangle shape and small voltage drops (RI drops), as shown in Fig. 3.106B, giving a capacitance of 238 F/g at 0.5 A/g rate. The H-carbon foam worked successfully as the support for pseudocapacitive materials, such as $NiCo_2S_4$ and Fe_2O_3. N-doped carbon foams with hierarchical pore structure were also synthesized from the mixtures of chitosan and glutaraldehyde with K_2CO_3 at 800°C in N_2, which could deliver a capacitance of 247 F/g at 0.5 A/g rate, 197 F/g at 10 A/g, and 166 F/g at 100 A/g [381].

3.2.3.5 Graphene and reduced graphene oxide

Graphene yarns were prepared by picking up graphene film grown by CVD of CH_4 from its ethanol surface using tweezers and self-shrinking into fiber with evaporation of ethanol [382]. MnO_2-loading onto this graphene yarns was performed by emerging into $KMnO_4$ ethanol solution, followed by heating at 80°C. The graphene yarns with 20–50 μm diameters delivered a capacitance of 1.4 mF/cm^2 in 1 M Na_2SO_4 electrolyte at 10 mV/s rate and the MnO_2-loading onto the yarns demonstrated marked

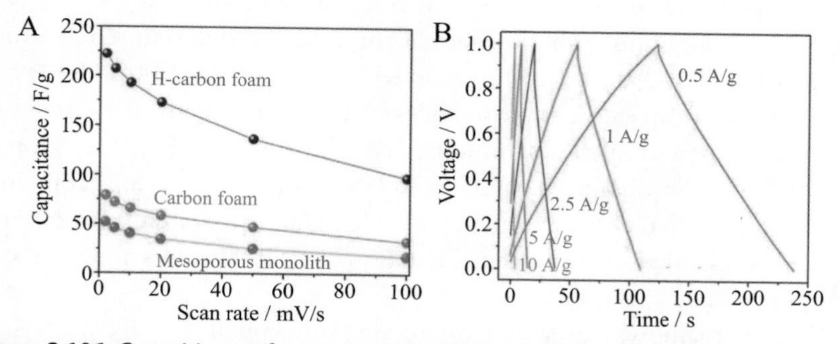

Figure 3.106 Capacitive performances in 3 M KOH electrolyte for hierarchically porous carbon (H-carbon foam), in comparison with a carbon foam without mesopores and mesoporous carbon monolith: (A) rate performances for three carbons prepared and (B) galvanostatic charge–discharge curves for H-carbon foam at different rates [380].

enhancements in the capacitance up to 12.4 mF/cm^2 and in cycle stability. Symmetric supercapacitor wires delivered 443 $\mu F/cm^2$ (527 mF/cm^3) at a scan rate of 50 mV/s with high cyclability as 105% retention after 10,000 cycles [383]. The rGO yarns composed of rGO core and porous rGO sheath (Fig. 3.107A) were prepared by electrolysis of the core rGO fiber in 3 mg/mL GO aqueous suspension containing 0.1 M $LiClO_4$ at a potential of −1.2 V [384]. A symmetric supercapacitor wire was assembled by interwinding these two rGO fibers after solidifying with PVA/H_2SO_4 gel electrolyte, as shown in Fig. 3.107B, which delivered a capacitance of 1.2−1.7 mF/cm^2 (25−40 F/g) during 500 straight−bending−straight cycles, as shown in Fig. 3.107C. This supercapacitor is highly compressible and stretchable and so can be woven into a textile for wearable electronics.

Capacitive performance was compared on three rGO flakes, N-doped, thermally reduced, and chemically-reduced rGOs, in 0.5 M H_2SO_4 aqueous and 1 M $TEABF_4$/PC nonaqueous electrolytes [385]. N-doping was performed by rapid heating of the mixture of GO with melamine up to 900°C under vacuum, with thermal reduction of GO by rapid heating to 900°C without melamine and chemical reduction of GO using ethylene glycol at 90°C. As shown by the CV curves for the N-doped rGO in Fig. 3.108A, the contribution of pseudocapacitance to the observed capacitance is not negligible. The contribution of pseudocapacitance was estimated to be about 16%, 16%, and 30% in H_2SO_4 and about 26%, 23%, and 56% in $TEABF_4$ for the N-doped rGO, thermally reduced rGO, and chemically reduced rGO, respectively. In other words, N-doped and thermally reduced rGOs had almost the same contribution of pseudocapacitance, while the pseudocapacitance contribution was particularly large for the chemically reduced rGO. N-doped rGO delivers the highest capacitance of about 234 F/g at 5 mV/s rate and capacitance retention of

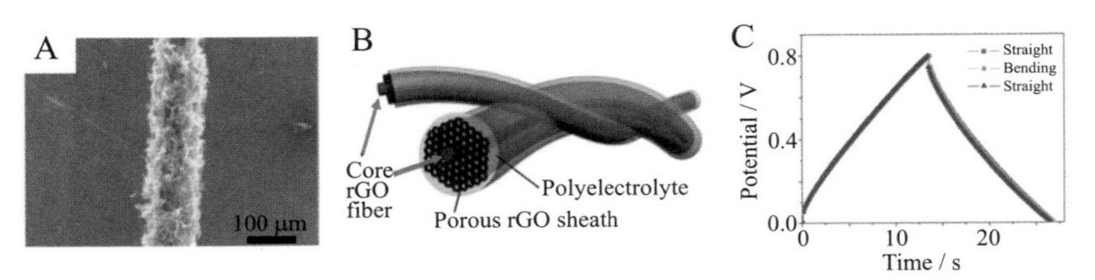

Figure 3.107 A symmetric supercapacitor wire composed of core−sheath-structured rGO fiber: (A) SEM image, (B) scheme of symmetric supercapacitor wire, and (C) charge−discharge curves for a straight−bending−straight cycle [384].

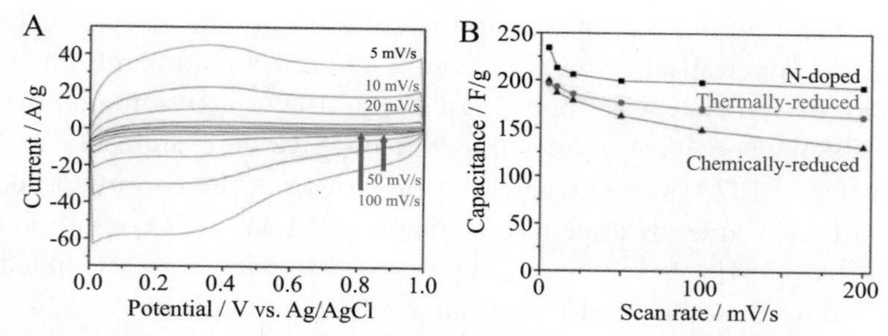

Figure 3.108 rGO flakes: (A) CV curves at different scanning rates for N-doped rGO and (B) rate performances of N-doped rGO, thermally reduced rGO, and chemically reduced rGO in 0.5 M H_2SO_4 electrolyte [385].

about 82% at 200 mV/s in 0.5 M H_2SO_4, as shown in Fig. 3.108B. High capacitance of the N-doped rGO was thought to be mainly due to the large electrolyte-accessible surface area (S_{BET} of 677 m^2/g), crumpled structure, and nitrogen-doping. In 1 M $TEABF_4$/PC, N-doped rGO exhibited also the highest capacitance of 188 F/g at 5 mV/s with about 71% retention at 200 mV/s.

Foams of rGO were prepared by hydrothermal treatment of GO with or without H_2O_2 at 180°C, followed by further reduction with 1 M sodium ascorbate aqueous solution at 100°C (denoted as HGF or GF, respectively) [386]. The rGO foams after immersing in the electrolyte were compressed onto Pt or Al foils under 150 MPa pressure to form about 14 mm thick film. Both rGO foams adhered firmly to the metal foil and could be directly used as electrodes to assemble a supercapacitor cell without any other additives. The capacitive performances of these rGO foams, HGF and GF, in 6 M KOH aqueous electrolyte and in an ionic liquid nonaqueous electrolyte ($EMIMBF_4$/AN) are shown in Fig. 3.109.

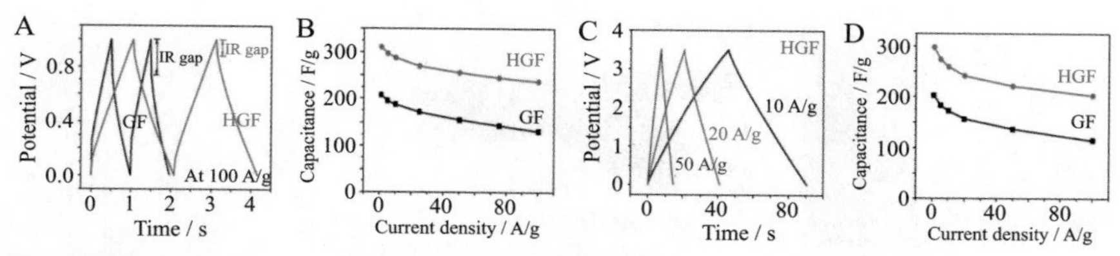

Figure 3.109 Capacitive performances of rGO foam with and without H_2O_2 treatment (HGF and GF, respectively): (A) charge–discharge curves, (B) and (D) rate performances and (C) charge–discharge curves for HGF; (A) and (B) in 6 M KOH and (C) and (D) in $EMIMBF_4$/AN electrolytes [386].

HGF delivers higher capacitance in aqueous and nonaqueous electrolytes (Fig. 3.109B and D, respectively) than GF. H_2O_2 treatment under hydrothermal conditions is effective in enhancing capacitance in both electrolytes and reducing the IR drop in aqueous electrolyte (Fig. 3.109A and C). HGF demonstrated high cyclic stability, about 95% capacitance retention after more than 20,000 cycles at 25 A/g rate in aqueous electrolyte, and about 91% retention after 10,000 cycles at 20 A/g in nonaqueous electrolyte.

Chemical modification of the surface of rGO flakes was reported to be effective in improving the capacitive performance, by keeping electrical conductivity as high as about 2×10^2 S/m (close to that of pristine graphite) [387]. rGO flakes were modified by an ionic liquid poly(1-vinyl-3-ethylimidazolium) bis(trifluoromethylsulfonyl)amide, which delivered improved capacitive performance with no marked pseudocapacitance peaks in CV curves and a slight deviation from the linear voltage changes [388]. The foams of chemically modified rGO (CM foams) were fabricated as films with thicknesses of about 55 μm from the suspension mixture of chemically reduced rGO flakes and polystyrene (PS) spheres (2 μm diameter) in deionized water with controlled pH at 6, where rGO flakes charged negatively and PS particles positively, followed by dissolution of PS with toluene [389]. The loading of MnO_2 nanoparticles on the CM-foam films was performed by dipping into a solution of 0.1 M $NaMnO_4$/0.1 M Na_2SO_4 at a neutral pH at room temperature. Chemical modification of rGO flakes and MnO_2 loading on rGO flakes are effective in improving the capacitive performance in 1 M Na_2SO_4 electrolyte, as shown by the CV curves and rate performances in Fig. 3.110A and B, respecively, revealing

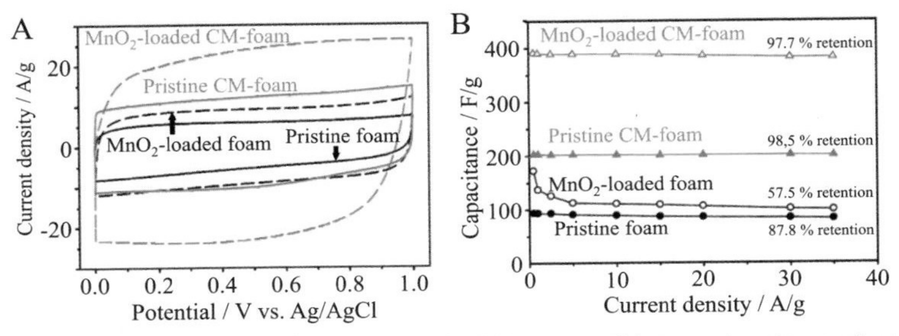

Figure 3.110 Capacitive performances of rGO foams fabricated using polystyrene colloids with chemical surface modification (CM foam) and with MnO_2 loading (MnO_2-loaded foam): (A) CV curves at a scan rate of 50 mV/s and (B) rate performance [389].

the effectiveness of chemical modification and MnO_2 loading for capacitive performances. An asymmetric cell was constructed using CM foam as the negative electrode and MnO_2-loaded CM foam as the positive electrode with 1 M Na_2SO_4 electrolyte. The cell delivered an energy density of 44 Wh/kg at a power density of 11.2 kW/kg and a power density of 25 kW/kg at an energy density of 39.1 Wh/kg with an operating potential of 2.0 V, accompanying a high cyclability as 95% retention after 1000 cycles at 1 A/g rate. A similar asymmetric capacitor was also reported by constructing from rGO and MnO_2-nanowires/rGO composite [390]. Self-assemblage of rGO flakes and CNTs by chemical modifications of the surfaces of both components was successfully applied to fabricate three-dimensional networks of N-doped rGO with CNTs [391] and composite films of rGO with CNTs [392] for supercapacitor electrodes.

N-doping of rGO by nitrogen plasma (500 W) after the reduction of GO by hydrogen plasma worked effectively to increase the capacitance in 6 M KOH aqueous and 1 M $TEABF_4$ nonaqueous electrolytes [393]. N-doped rGOs containing 1.86 at.% N showed high capacitance of 282 F/g at 1 A/g rate and even 165 F/g at 33 A/g, which is much higher than the pristine rGO. N-doped graphene-like carbon flakes prepared from quinoline and melamine through a mica-templating process delivered relatively high capacitances of about 200 F/g at 20 mA/g rate in 1 M H_2SO_4 electrolyte, although these flakes had S_{BET} of less than 250 m^2/g [394,395]. N/B-codoped rGO foams were synthesized by hydrothermal treatment of the mixture of GO flakes with NH_3BF_3 at 180°C, which delivered a capacitance of 239 F/g at 1 mV/s rate and 132 F/g at 100 mV/s in 1 M H_2SO_4 electrolyte, which are higher than N-, B-, and nondoped rGO foams prepared using the same procedure [396]. All-solid-state symmetric supercapacitors constructed using this N/B-codoped rGO foam exhibited a capacitance of 62 F/g at 1 mV/s with an energy density of 8.7 Wh/kg.

GO suspension was mixed with bacteria (*E. coli*) with $FeCl_3$, followed by freeze-drying and then calcination at 700°C in N_2, to obtain rGO with bacteria-promoted hierarchical texture [397]. The resultant rGO was applied to the electrode material for a supercapacitor with 1 M H_2SO_4 aqueous electrolyte. It delivered a capacitance of 327 F/g at 1 A/g rate and 160 F/g at 5 A/g. By cycling with 1 A/g rate, the capacitance increased slightly after several cycles, perhaps due to full exposure of electrochemically active materials to the electrolyte, and then it was stabilized at 340 F/g after 1000 cycles, suggesting that the pseudocapacitance due to foreign atoms in the rGO was very stable. Its energy density reached 45 Wh/kg at a power density of 0.5 kW/kg and still remained at 22 Wh/kg and 2.5 kW/kg.

rGO foams prepared by 3D printing technique were used as the electrodes of a supercapacitor in 3 M KOH electrolyte [398]. The foam was prepared from an ink of a mixture of 3.3 wt.% GO, 4.2 wt.% graphene nanoparticles, and 12.5 wt.% SiO_2 with RF resin, followed by gelation, supercritical drying, carbonization, and etching of SiO_2 with HF. It delivered a capacitance of 72 F/g at 0.5 A/g rate.

Onto large-sized rGO flakes prepared under hydrothermal condition at 120°C, few-layered graphene nanosheets were grown by microwave plasma CVD of CH_4 with 500 W power (graphene/rGO composites) [399]. The deposited graphene sheets are grown vertical to the rGO flakes with sharp edges. The compressive strength of rGO aerogel was markedly improved by graphene deposition, with more than 700% improvement, and so the resultant composites had recoverable compressive strains of up to 87% under a stress of up to 96.9 kPa and high electric conductivity of about 1000 S/m. The composites delivered relatively high capacitance of 245 F/g at 1 A/g rate and 201 F/g at 20 A/g in 1 M H_2SO_4 electrolyte. The symmetric all-solid-state supercapacitors constructed from this composite as electrodes delivered an areal capacitance of 0.42 F/cm^2 and an excellent long-term cyclic stability over 5000 cycles with a low internal resistance. The deposition of graphene-like carbon onto mesoporous graphitic carbon nitride (g-C_3N_4) powder was performed by plasma-enhanced CVD of CH_4 at 500°C, followed by heat treatment at 750−950°C [400]. The resultant composite exhibited a high conductivity of 693 S/m, large S_{BET} of 1277 m^2/g with high V_{total} as 4.35 cm^3/g, abundant mesopores, and high N content of 8.75 wt.%. It delivered a capacitance of 261 F/g at 1 A/g rate and 189 F/g at 100 A/g with cycle performance of 97% capacitance retention at 10 A/g after 20,000 cycles in 6 M KOH electrolyte.

3.2.3.6 Carbon nanotubes and nanofibers

The electrochemical behaviors of SWCNTs were studied on their films cast through their dispersion in acetone on Pt foil and compared with fullerene C_{60} [401]. CV curves in 0.1 M TBAPF$_6$/AN are shown on SWCNTs and C_{60} in Fig. 3.111A and B, respectively. SWCNTs exhibit capacitive behavior without any Faradaic reaction peak, in contrast C_{60} demonstrates well-defined peaks due to the reduction of C_{60} to C_{60}^{-} and C_{60}^{2-} with

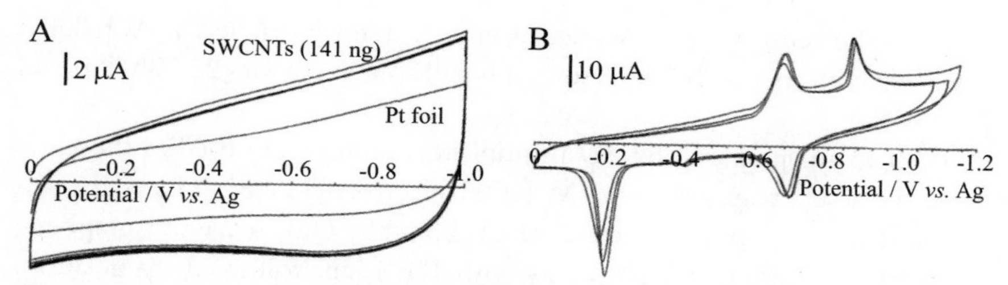

Figure 3.111 CV curves in 0.1 M $TBAPF_6$/AN at 50 mV/s: (A) SWCNTs in comparison with the Pt soil (substrate) and (B) fullerenes C_{60} [401].

corresponding oxidation peaks. As shown in Fig. 3.111A, the capacitance increases substantially with the addition of the SWCNT film, the effective capacitance of SWCNTs being calculated to be 283 F/g. Capacitive behaviors of SWCNT films were also measured in 3 M NaCl aqueous electrolyte and 0.01 M $AgClO_4$/0.1 M tetrabutylammonium perchlorate/AN nonaqueous electrolyte [402]. On three different CNT films, SWCNTs, as-prepared MWCNTs, and 450°C-treated MWCNTs, electrochemical behaviors were compared in 0.10 M KCl solution containing 5.0 mM $K_3Fe(CN)_6$ electrolyte [403]. CV curves consisted of the capacitive current due to the surfaces of CNTs and the Faradaic current due to the redox reaction of $K_3Fe(CN)_6/K_4Fe(CN)_6$ presented in the solution as a benchmark. SWCNTs delivered the highest capacitance, probably due to their high packing in the film, and as-prepared MWCNTs delivered the lowest capacitance, probably because the surface of the as-produced MWCNTs was blocked by impurities, mainly consisting of amorphous carbon, which can be removed by prolonged heat treatment at 450°C.

CNTs were grown on the surface of graphene foam by catalytic CVD using NiCo catalyst at 750°C in a flow of C_2H_4/H_2/Ar mixed gas (CNT/foam composite) [404]. Graphene foam was fabricated by CVD of CH_4 at 1000°C on Ni foam substrate, followed by dissolution of Ni substrate with the $FeCl_3$/HCl mixed solution to obtain free-standing graphene foam. Onto the composite film, MnO_2 nanosheets were deposited by hydrothermal treatment of the composite film in $KMnO_4$ solution (MnO_2-loaded composite). Polypyrrole (PPy) was also loaded onto the composite via chemical polymerization of pyrrole at 0–5°C (PPy-loaded composite). Asymmetric capacitors were assembled using MnO_2-loaded composite as the positive electrode and PPy-loaded composite as the negative electrode with 0.5 M Na_2SO_4 electrolyte. It delivered high energy and power densities of 22.8 Wh/kg at 0.86 kW/kg and 6.2 Wh/kg at 2.7 kW/kg, with an output voltage of 1.6 V.

Wires and webs of CNFs were prepared by electrospinning of an NMP solution of a coal char with 5% PANMA [poly(acrylonitrile)-co-methyl acrylate] and 5% PVP onto a drum rotating at different speeds passing through a parallel wire collector, followed by stabilization at 300°C for 3 h in air and carbonization at 1000°C for 1 h in N_2 [405]. For comparison, the as-spun fibers before oxidative stabilization were hot-stretched up to 1.5 times their length at 135°C. The capacitive performances of these CNF webs and wires were measured in 6 M KOH aqueous electrolyte using a three-electrode cell. The charge—discharge curves and rate performances for the five CNFs in webs and wires are shown in Fig. 3.112A and B, respectively. The change in rotation speed of the drum at 200, 800, and 1600 rpm, which correspond to a linear velocity of 1.1, 2.1, and 4.2 m/s, respectively, does not affect the capacitances or rate dependence. In contrast, hot-stretching gives a noticeable influence on the capacitor, but no effect on the rate performance. The hot-stretched wire delivered an areal capacitance of 1457 mF/cm^2 at 4 mA/cm^2 rate and gravimetric capacitance of 366 F/g at 1 A/g rate, which is higher than those for the nonstretched wire (1310 mF/cm^2 and 329 F/g, respectively).

Webs of bamboo-like CNFs were synthesized from a DMF solution of PAN and TEOS by electrospinning, followed by carbonization at 1200°C in H_2/Ar (5/95 by volume) and etching SiO_2 in HF solution [406]. Symmetric supercapacitors with a gel electrolyte were constructed by sandwiching a cellulose separator between two identical web electrodes,

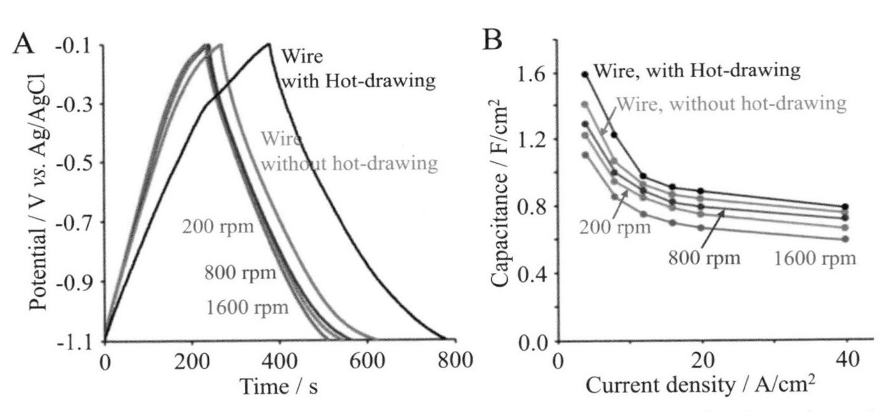

Figure 3.112 Capacitive performances in 6 M KOH aqueous solution for the webs and wires electrospun from a coal char using different rotation speeds of the collecting drum and with or without hot-stretching at 135°C, followed by stabilization at 300°C and carbonization at 1000°C [405].

which served as the active material and current collector. Because of the superior mechanical durability, the webs did not require any flexible substrates. The supercapacitors delivered a volumetric capacitance of 2.1 F/cm^3 at 33 mA/cm^3 rate and superior mechanical durability and electrochemical stability even after 10,000 charge–discharge cycles, as well as mechanical stability for in situ bending and twisting operations.

N-doped tubular mesoporous carbons were prepared via an electrostatic assembly approach using MnO_2 rod as the template, RF as the carbon precursor, TEOS as the pore-forming agents, EDA as the base catalyst and N precursor [407]. All materials were mixed in water and then heat-treated at 800°C for 4 h in N_2, followed by dissolution of MnO_2 rods and SiO_2 with concentrated HCl and HF, respectively. The resultant tubular carbons had outer diameters of about 66 nm and inner diameters of 30–35 nm with lengths of about 0.6 μm, of which the pore structure parameters were S_{BET} of 833 m^2/g with V_{total} of 1.50 cm^3/g including V_{micro} of 0.12 cm^3/g and the N-content was measured as 1.66 wt.% by XPS. This carbon delivered a capacitance of 307 F/g at 1 A/g in 6 M KOH aqueous electrolyte with an excellent rate performance (78.2% capacitance retention at 20 A/g) and a high cyclability (95.1% capacitance retention after 10,000 cycles). The symmetric cell with 1 M Na_2SO_4 electrolyte constructed from this carbon exhibited a capacitance of 201 F/g at 1 A/g and an energy density of 72.4 Wh/kg at a power density of 3.2 kW/kg.

Commercial carbon fiber clothes oxidized in acidic solution were doped by N using the aqueous solution of NH_3 and hydrazine (N_2H_4) hydrate (concentrations of 0.17 and 1.7 wt.%, respectively) under hydrothermal conditions and their supercapacitor performances were studied [408]. In Fig. 3.113, the charge–discharge curves and rate dependences of areal

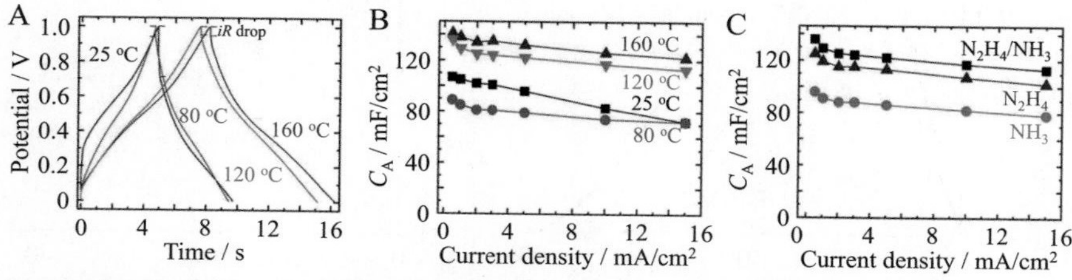

Figure 3.113 N-doped carbon fiber clothes: (A) charge–discharge curves at current density of 15 mA/cm^2 and (B) rate dependences of areal capacitance C_A for the clothes doped by N using an N_2H_4/NH_3 mixture at different hydrothermal temperatures (80–160°C), and (C) rate dependences of C_A for the clothes doped by N using different dopants at 120°C [408].

capacitance C_A are shown for the clothes prepared at different temperatures using different N-dopants. The capacitive performances depended greatly on the doping temperature and N-dopants; N-doping at 120–160°C using a mixture of N_2H_4 and NH_3 gave high capacitance and good rate performance.

3.2.3.7 Other precursors

Protic salt of 5-carboxybenzene-1,3-diamine sulfate [DABA][2HSO$_4$] was synthesized by adding a DMF solution of 5-carboxybenzene-1,3-diamine (DABA) dropwise into H_2SO_4, which was carbonized at temperatures of 700–1000°C in N_2 flow to obtain N/S-co-doped carbon [409]. The carbon obtained at 900°C (NSC-900) contained large amounts of N and S of 5.80 and 0.44 at.%, respectively, and had a hierarchical pore structure, S_{BET} of 1543 m^2/g with V_{total} of 1.11 cm^3/g including V_{micro} of 0.79 cm^3/g. The protic salt [DABA][2HSO$_4$] served as the carbon source and heteroatom dopants as well as the pore-forming agent during carbonization. The NSC-900 exhibited a capacitance of 282 F/g at 1 A/g rate and 202 F/g at 20 A/g rate in 6 M KOH electrolyte, and long-term cycling life with 96% capacitance retention after 10,000 cycles at 2 A/g rate. An asymmetric supercapacitor composed of NSC-900 at the negative electrode and MnO_2-loaded NSC-900 at the positive electrode with 1 M Na_2SO_4 electrolyte delivered the energy density of 39 Wh/kg at a power density of 901 W/kg with a wide voltage widow of 1.8 V.

The microelectrode for the supercapacitor was fabricated by laser-induced carbonization of a polyimide film (Kapton), as shown in the scheme of the fabrication procedure in Fig. 3.114A [410]. The laser with a wavelength of 522 nm and beam spot of about 3 μm diameter was irradiated with a pulse duration of about 500 fs and a repetition rate of 1 MHz under ambient conditions of 15–20°C and about 25% relative humidity. A carbonized part in polyimide film is shown by the SEM images in Fig. 3.114B and C. An all-solid-state microsupercapacitor was constructed using the microelectrode sheets (1.5 × 1.5 mm^2) with PVA-H_3PO_4 gel electrolyte. The capacitor demonstrated the characteristics of a supercapacitor, capacitance of about 800 μF/cm^2 at 10 mV/s rate and about 150 μF/cm^2 at 1 V/s, associated with reasonable mechanical bendability. Fig. 3.114D shows the effect of laser scanning conditions on the capacitive performance of the microsupercapacitor.

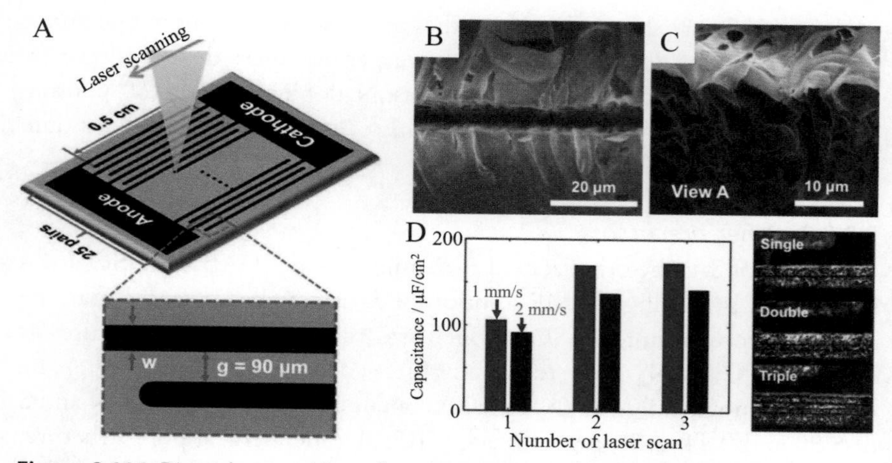

Figure 3.114 Direct laser writing of porous carbon in polyimide: (A) scheme of the microelectrode fabrication by laser carbonization of polyimide film, (B) and (C) SEM images of top and side views of the carbonized part, respectively, and (D) capacitance at a scan rate of 1 V/s for the microsupercapacitor fabricated by single-, double-, and triple-laser scanning at two different speeds (1 and 2 mm/s) with optical images [410].

N/S-codoped succulent-like carbons were synthesized by carbonization of the mixture of potassium citrate tribasic monohydrate ($K_3C_6H_5O_7.H_2O$, PCT) with thiourea (TU) (weight ratios PCT/TU of 3/4, 4/4, and 5/4) at 800°C in Ar, coded by PT-3, -4, and -5, respectively [411]. N/S-codoping and forming of three-dimensional succulent-like carbon particles, as shown in Fig. 3.115A, could be achieved by supramolecular polymerization due to noncovalent self-assembly of potassium citrate and N/S-rich small organic molecules (thiourea). The resultant carbon PT-4 contained 8.5 at.% N and 2.4 at.% S with 8.6 at.% O. Its galvanostatic charge—discharge curves at different current densities measured by three-electrode cells at the potential

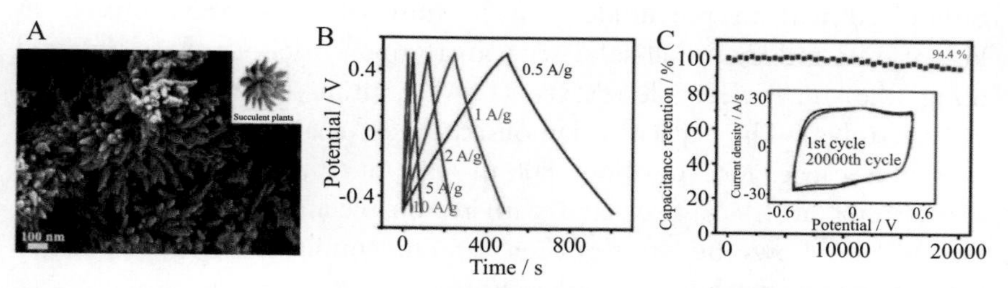

Figure 3.115 Succulent-like carbons PT-4: (A) SEM image, (B) charge—discharge curves at different current densities, and (C) cycle performance at 100 mV/s with CV curves (inset) [411].

range of −0.5 to 0.5 V in 1 M H_2SO_4 electrolyte are shown in Fig. 3.115B, suggesting a slight contribution of pseudocapacitance and exhibiting a capacitance of about 260 F/g at 0.5 A/g rate and 207 F/g at 10 A/g. The CV curves shown as an inset in Fig. 3.115C also suggest the presence of some pseudocapacitance. From a separation of the observed capacitance into electric double-layer capacitance and pseudocapacitance by assuming that the former is independent from the scan rate but the latter is dependent on it, the contribution of pseudocapacitance was estimated to be about 37% of the observed capacitance. The carbon PT-4 exhibited high cyclability, the capacitance retention of 94% after 20,000 cycles at 100 mV/s rate, as shown in Fig. 3.115C. By using the carbon PT-4 in the electrodes, a flexible symmetric supercapacitor device was constructed with PVA/H_2SO_4 gel electrolyte and stainless-steel foils as substrate and current collector, which delivered a capacitance of 73 F/g at a current density of 0.5 A/g with a slight IR drop and 54 F/g at 10 A/g.

3.2.4 Mechanical modification of carbon texture

Mesophase-pitch-based carbon fibers after heat treatment at 3000°C were exfoliated at 1000°C after oxidation and intercalation in concentrated HNO_3, and studied for performance as a supercapacitor electrode [412–414]. The resultant exfoliated carbon fiber delivered a huge capacitance of more than 500 F/g in 18 M H_2SO_4 aqueous electrolyte [412,413]. Its S_{BET} and the capacitances in 1 and 18 M H_2SO_4 electrolytes are compared to a commercial AC and ACF in Table 3.3. Although the exfoliated carbon fibers have much smaller S_{BET} than ACF and AC, they delivered a remarkably high capacitance of 555 F/g. After exfoliation, the carbon fibers exhibited a clear hysteresis in adsorption–desorption

Table 3.3 Capacitances measured by charge–discharge measurements and S_{BET} [412].

	S_{BET} (m^2/g)	Capacitance in 1 M H_2SO_4	Capacitance in 18 M H_2SO_4
Exfoliated carbon fibers (mesophase-pitch-based)	295	117	555
Activated carbon fibers (phenol-based)	1950	143	186
Activated carbon (coconuts-derived)	1484	80	146

isotherm, suggesting the formation of mesopores [414]. Even in high concentration H_2SO_4, no redox peaks due to intercalation–deintercalation of ions were observed on the CV curves [413].

Partially exfoliated carbon fiber cloth was successfully applied to the electrodes of a supercapacitor [415]. The fiber surface of commercial carbon fiber cloth was converted to graphite oxide via modified Hammers method and then reduced by hydrazine at 80°C, followed by annealing at 1000°C in ammonia flow. The fiber surface is shown by SEM and high-resolution TEM in Fig. 3.116A and B, respectively. The resultant surface-exfoliated fibers could retain the capacitance with fast charging–discharging; about 70 mF/cm^2 at 1000 mV/s. A symmetric supercapacitor was assembled by using these surface-exfoliated carbon fiber cloths as electrodes with PVA/H_2SO_4 polymer gel as the solid electrolyte. This solid-state supercapacitor delivers high capacitive performance, as shown in Fig. 3.116C. The capacitance is retained at 81% with increasing current density from 1 to 20 mA/cm^2. The charge–discharge curve at 10 mA/cm^2 shows a linear and symmetrical profile without an obvious IR drop, as shown in the inset in Fig. 3.116C. The assembled supercapacitor was flexible, and no obvious change in the quasi-rectangular CV curves and areal capacitance values were observed by bending up to 180 degrees or by wrapping.

Milling of natural graphite was shown to be effective for increasing the surface area and as a consequence improving capacitive performance. Milling of high-purity natural graphite (lateral size of about 100 μm) was

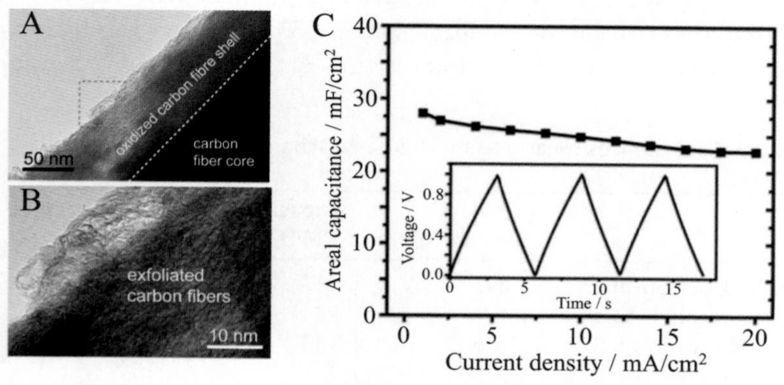

Figure 3.116 Surface-exfoliated carbon fiber cloth: (A) SEM image and (B) high-resolution TEM image of the surface of the carbon fiber, and (C) capacitive performance in PVA/H_2SO_4 polymer gel electrolyte with charge–discharge profile (inset) [415].

performed in air using a zirconia vessel with zirconia balls (1 mm in diameter) for 1–750 min at 500–900 rpm [416]. S_{BET} of the graphite reached its maximum at a rotation speed of 700 rpm for a milling time of 150 min. The capacitances of graphite measured in 1 M $TEABF_4$/PC electrolyte changed with milling time at 700 rpm rotation, giving a maximum at about 100–300 min, as shown in Fig. 3.117A, where these dependences of capacitance on milling time were the same as that of S_{BET} on milling time. By milling for more than 100 min, interlayer spacing d_{002} increased gradually with increasing milling time, suggesting a gradual deterioration in crystallinity of the graphite structure. The increases in the contents of O and N in milled graphite were confirmed by XPS, O/C, and N/C increasing from 0.048 and 0.000 for the pristine graphite to 0.102 and 0.015 for the graphite, respectively, after milling for 150 min. Galvanostatic charge–discharge curves were composed of linear triangles, as shown for the graphite after 150 min milling in Fig. 3.117B. The rate performance of volumetric capacitance for the graphite after 150 min milling is much better than that for a commercial AC (YP50F), as shown in Fig. 3.117C, whereas this milled graphite gives gravimetric capacitances comparable with AC. Ball-milled graphite was also tested in aqueous electrolyte (6 M KOH), demonstrating higher capacitance than the reference AC [417]. Milling was performed using a stainless steel jar and balls (10 mm diameter) with a rotation speed of 400 rpm in air which was necessary to be interrupted every 30 min and rested for 30 min. S_{BET} of the graphite decreased with the increase in milling time after passing through its maximum of 580 m^2/g at about 20 h milling, while gravimetric capacitance increased gradually, even

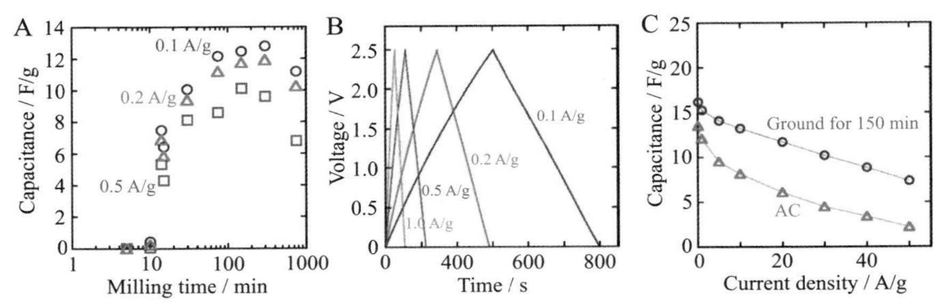

Figure 3.117 Capacitive performances of ball-milled natural graphite: (A) change in gravimetric capacitances in 1 M $TEABF_4$/PC at different current densities with milling time, (B) and (C) charge–discharge curves at different current densities and change in volumetric capacitance with current density, respectively, for the graphite milled for 150 min at 700 rpm in comparison with a commercial AC [416].

after passing S_{BET} maximum, the graphite after 70 h milling delivering a capacitance of 207 F/g with S_{BET} of 313 m^2/g. In comparison with the graphite after 70 h milling, the graphite after 30 h milling retained almost constant capacitance (about 170 F/g) at current densities of 1−10 A/g and high capacitance retention, more than 90% retention after 5000 cycles at 1 A/g rate. High-energy ball milling of natural graphite flakes in a zirconia vessel with zirconia balls in air resulted in an amorphous structure containing a high oxygen content of 11.5 at.% [418]. A symmetric supercapacitor composed of this milled graphite delivered a capacitance of 81 F/g in 1 M Na_2SO_4 electrolyte with a cell voltage of 1.6 V. High-energy ball milling was also applied to high-purity natural graphite for 3 and 6 h, and milled graphite powders were subjected to a study of the capacitive performance in 1 M H_2SO_4 [419]. Milling for 3 h resulted in a capacitance of 44 F/g and prolonged milling to 6 h gave 55 F/g at 0.1 A/g rate, although the neat graphite gave only 0.46 F/g.

3.2.5 Bifunctional electrodes

Different electrochemically active materials via redox interaction with electrolyte ions have been tried to be kept in nano-sized pores of porous carbons to enhance the capacitance and/or improve the rate and cycle performances of supercapacitors. The holding of active materials in nanopores is often effective in confining the materials by inhibiting their coagulation to a large particle, by buffering a large volume change, and by keeping the structure of the material active during charge−discharge cycles. On the other hand, the porous carbon used for the holding active material can still adsorb and desorb the electrolyte ions into its unoccupied nanopore surfaces to form electric double layers. Therefore, the porous carbon electrode holding additional active material is bifunctional, i.e., capacitor-type functionality using pore surfaces of the carbon and battery-type functionality using active material held into some pores of the carbon.

Various organic active materials were tried to be confined into nanopores of different porous carbons to enhance the capacitive performances [420−423]. Polypyrene (PPY) and polyaniline (PANI) were confined into nanopores in a commercial AC (S_{BET} of 3160 m^2/g, V_{micro} of 0.99 cm^3/g, and V_{meso} of 0.60 cm^3/g) by adsorption of the vapors of pyrene (PY) and aniline (ANI) at 150 and 25°C, respectively, followed by electrochemical polymerization in 1 M H_2SO_4 [422]. The maximum adsorption amounts of PY and ANI were 60.2 and 62.4 wt.%, respectively, for the AC.

The capacitive performances are shown on the ACs with different amounts of PPY and PANI in Fig. 3.118. Anodic and cathodic peaks at 0.3–0.4 V and 0.4–0.6 V due to the redox reaction of PPY and PANI confined in the AC, respectively, were clearly observed in CVs (Fig. 3.118A and D, respectively). By the confinement of 30–50 wt.% PPY and PANI, gravimetric capacitances of the cells are markedly enhanced, as shown in Fig. 3.118B and E, maximum capacitance reaches 545 and 419 F/g, respectively, accompanying excellent rate performances with capacitance retentions of more than 80% at 10 A/g rate, in comparison with the pristine AC having a maximum capacitance of 348 F/g and retention of 53% even at 0.1 A/g rate. The maximum volumetric capacitance for the PPY-confined AC reached 314 F/cm^3 and for the PANI-confined AC 299 F/cm^3. Although most increments in observed capacitance were reasonably supposed to be due to the redox reactions of PPY and PANI, the rate performances of the cells are also improved, as shown in Fig. 3.118C and F, respectively, which was explained by a high utilization of minute particles of PPY and PANI in the nanopores. The particles of PPY and PANI confined in the nanopores of AC are reasonably supposed to be restricted by their swelling and shrinkage during the charge–discharge processes by the pore walls. For the ACs confined by maximum amount of PPY and PANI, their capacitance as well as rate and cycle

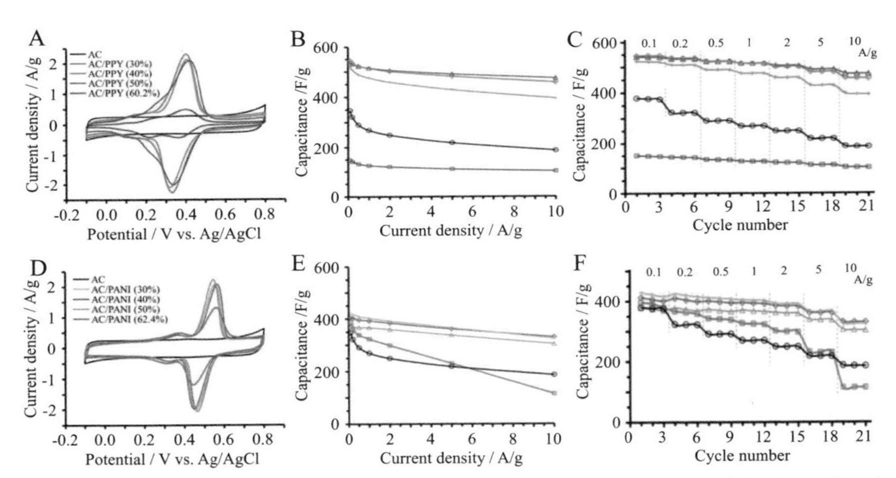

Figure 3.118 Capacitive performances of PPY-confined AC (A–C) and PANI-confined AC (D–F): (A) and (D) CV curves with 1 mV/s rate, (B) and (E) rate performances of gravimetric capacitance determined by charge–discharge measurement, and (C) and (F) rate capabilities [422].

performances are far inferior to the pristine AC, because almost all nanopores are filled with either PYY or PANI, which is illustrated by a marked reduction in the capacitive contribution of AC matrix and by a marked decrease in the utilization of PPY and PANI particles, as clearly seen in CVs in Fig. 3.118A and D.

The composites with PANI were proposed using porous carbons [424–426], in which the relative position of PANI to carbon matrix was different from the above-mentioned cases (confinement in nanopores in carbon matrix). PANI was deposited on the surface of carbon hollow spheres prepared by carbonization of hollow spheres synthesized from the copolymer of styrene, methylacrylic acid, and divinylbenzene [424]. Fibrous (rod-like) PANI particles were formed by polymerization on the surfaces of mesoporous carbon spheres, which were synthesized using $CaCO_3$ template [425]. The resultant composites delivered high capacitances in a current density of 0.5–5 A/g, which is much higher than pure PANI and the matrix carbon spheres, as compared in the charge–discharge curves in Fig. 3.119A. The charge–discharge profiles for the PANI/carbon composite are characteristic of a bifunctional electrode, with a linear change in potential making a triangular shape overlapped with potential plateaus at around 0.2 V; the former is due to the formation of EDLs on the carbon surfaces and the latter to the redox reaction of PANI on the carbon surfaces with electrolyte ions. No improvements in cycle or rate performances were obtained, as shown in Fig. 3.119B. Porous carbon spheres prepared by carbonization of polypyrrole monodispersed nanospheres at 750°C, followed by KOH activation (S_{BET} of 2818 m^2/g and V_{total} of 2.06 cm^3/g) were impregnated by ANI monomers through ethanol solution, followed

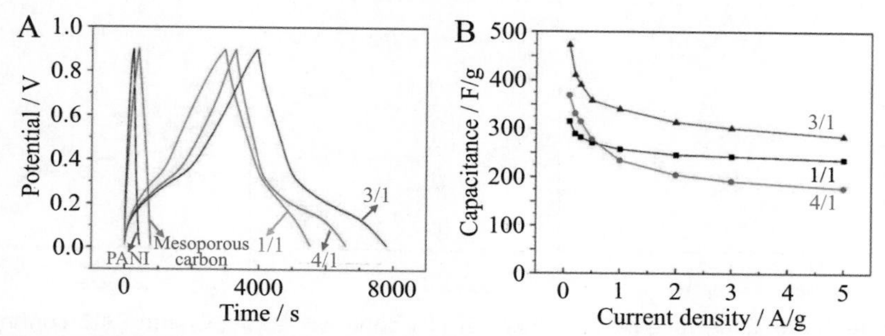

Figure 3.119 PANI-nanofiber-deposited mesoporous carbon with different weight ratios of PANI/carbon: (A) charge–discharge curves, in comparison with PANI nanofiber and the pristine mesoporous carbon, and (B) rate performances [425].

by polymerization of ANI to obtain PANI/carbon composite [426]. In the resultant composite, PANI nanorods were grown on the surface of carbon spheres, of which the average diameter was around 20–30 nm. Its capacitance in 1 M H_2SO_4 aqueous electrolyte was improved to 589 F/g at 0.2 A/g rate with capacitance retention of about 70% after 200 cycles at 5 A/g rate. In contrast, pristine carbon spheres delivered a lower capacity (320 F/g) at 0.2 A/g, while there was almost 100% capacitance retention even after 1000 cycles at 5 A/g rate.

2,5-Dichloro-1,4-benzoquinone (DCBQ) was confined to the pores of mesoporous Ketjenblack (KB, S_{BET} of 1340 m^2/g, V_{micro} of 0.48 cm^3/g, and V_{meso} of 1.24 cm^3/g) at a slightly higher temperature than its sublimation temperature (92°C) [421]. The weight content of DCBQ in KB was tuned to 5, 10, 20, 40, and 60.1 wt.%. The confinement of DCBQs in both micropores and mesopores of the KB was confirmed from the changes in N_2 adsorption–desorption curves; V_{micro} and V_{meso} were reduced with increasing DCBQ content to 0.17 and 0.71 cm^3/g, respectively, by 40 wt.% DCBQ confinement and the adsorption saturation (60.1 wt.% DCBQ) gave V_{micro} of 0.03 cm^3/g and V_{meso} of 0.15 cm^3/g. DCBQ-confined KBs exhibit clearly the presence of a redox reaction due to DCBQ by a plateau in galvanostatic charge–discharge curves in 1 M H_2SO_4 aqueous electrolyte, as shown in Fig. 3.120A. In Fig. 3.120B, gravimetric capacitance is plotted against current density, revealing that the excellent rate performance of the pristine KB is retained even in DCBQ-confined KBs and capacitance increases with increasing content of DCBQ up to 40 wt.%. Gravimetric capacitance becomes the highest with the addition of 40 wt.% DCBQ, but the KB confined 60.1 wt.% DCBQ

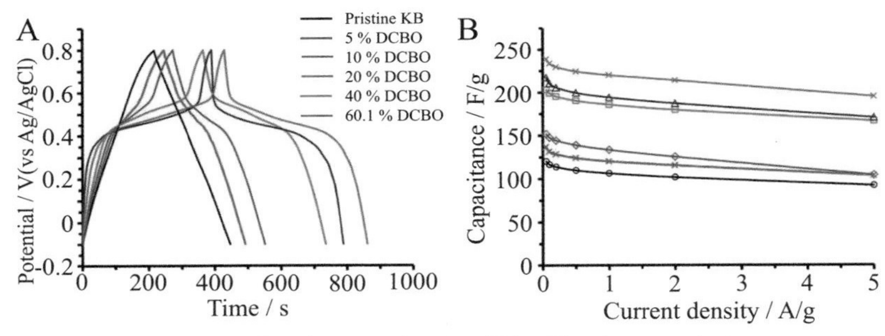

Figure 3.120 DCBQ-confined Ketjenblack (KB): (A) charge–discharge curves in 1 M H_2SO_4 electrolyte with a current density of 50 mA/g, (B) rate performances of gravimetric capacitance [421].

gave the highest volumetric capacitance. The resulting capacitor was characterized by high capacitance, particularly high volumetric capacitances that were 4.7 times higher than those of the pristine KB, and a high rate capability of up to 5 A/g, along with an excellent cycle lifetime of up to 10,000 cycles. It has to be pointed out that the confinement processes employed for these electrochemically active organics are simple solvent-free processes and the resultant composites are composed of metal-free electrode materials.

One of the derivatives of 2,2,6,6-tetramethylpiperidine-N-oxyl (TEMPO), i.e., 4-hydroxy-TEMPO benzoate (HTB), was confined to the nanopores of an AC through its vapor at 150°C, and the capacitive performance of the resultant HTB-confined AC with different HTB contents (20–50 wt.%) was studied in 1 M H_2SO_4 electrolyte [423]. A marked reduction in pore structure parameters with increasing HTB content suggested the confinement of HTB molecules into nanopores of AC, high S_{BET}, V_{micro}, and V_{meso} for the pristine AC (3160 m^2/g, 0.99 cm^3/g, and 0.60 cm^3/g, respectively) being reduced to 250 m^2/g, 0.07 cm^3/g, and 0.05 cm^3/g, respectively, by adsorption of HTB up to 50 wt.%. Anodic and cathodic peaks for the redox reaction of HTB were clearly observed at around 0.7 V. The AC-confined 30 wt.% HTB exhibited high energy and power densities, 39.5 Wh/kg at 16.5 W/kg and 14.1 Wh/kg at 8.2 kW/kg, which are much higher than the pristine AC, 18.0 Wh/kg at 16.5 W/kg and 1.1 Wh/kg at 8.2 kW/kg. The confinement of HTB molecules into nanopores of AC was effective in preventing dissolution of HTB in an aqueous electrolyte and improving markedly the contact of HTB with conductive carbon wall, in addition to fine dispersion of minute aggregates of HTB molecules, which was thought to allow the fast redox reactions of HTB with electrolyte ions.

An organometallocene, ruthenocene $RuCp_2$ ($Cp = \eta^5$-C_5H_5), was adsorbed into the micropores of a commercial AC at 100°C in a glass ampoule and then electrochemically oxidized in an aqueous H_2SO_4 electrolyte ($RuCp_2$/AC composites) [427]. The composites with different $RuCp_2$ contents exhibited a reversible and rapid redox reaction without dimerization or disproportionation reactions in H_2SO_4 aqueous electrolyte. The charge–discharge curves of the composites at a current density of 1 A/g are shown in Fig. 3.121A, revealing that the redox reaction of $RuCp_2$ observed by a potential plateau at about 0.3 V increases the cell capacity, of which the contribution increases with increasing $RuCp_2$ content. The enhancement in the gravimetric capacity is markedly

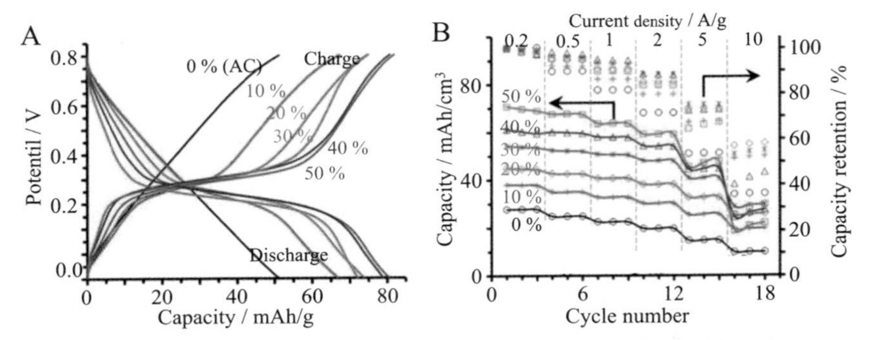

Figure 3.121 Electrochemical performance of the composite of AC with ruthenocene $RuCp_2$ in different contents: (A) charge–discharge curves at the current density of 1 A/g and (B) dependences of volumetric capacity and capacity retention on current density [427].

observed by loading only a small $RuCp_2$ content of 10 wt.% (Fig. 3.121A), while the volumetric capacity increases gradually with increasing $RuCp_2$ content, as shown by the rate capabilities of volumetric capacitance in Fig. 3.121B. The composites had high cycle performances with about 70% capacity retention after 10,000 cycles at 1 A/g rate. These results were due to a large contact area of $RuCp_2$ molecules with the nanopore walls of conductive carbon which enabled rapid charge transfer at the contact interface. The composite $RuCp_2$/AC can be used as a high-power-density positive electrode for proton-based batteries.

MnO_2/CNT composites were fabricated by impregnating $KMnO_4$ into a CNT array with its aqueous solution, in which MnO_2 nanosheets were firmly deposited onto the surface of air-oxidized CNTs at 120°C, where the content of MnO_2 was controlled at 50 wt.% [428]. The resultant composite was obtained as a self-standing film and so it could be used as the electrode of a supercapacitor without any additional binder or current collector. The composite demonstrates bifunctional Li storage in 1 M $LiPF_6$/(EC + DEC) electrolyte, as shown by the CV curves at various scan rates of 0.2–20 mV in Fig. 3.122A. By analyzing the current response in CV as a function of scan rate, the contribution of capacitive function was estimated to be about 66% at a scan rate of 20 mV/s, as shown in Fig. 3.122B. The bifunctional performance of the composite is also seen on its discharge curves, as shown in Fig. 3.122C.

Loading of MnO_2 nanoparticles onto rGO was carried out by electrodeposition of MnO_2 onto the rGO surfaces modified with polyacrylic

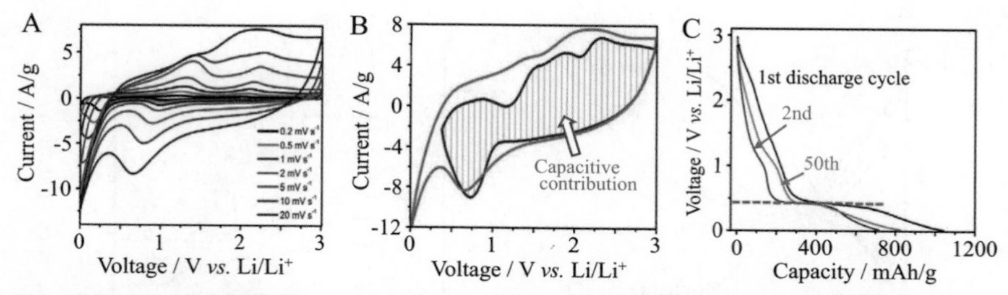

Figure 3.122 MnO_2/CNT composites: (A) CV curves at different scan rates, (B) CV curve at 20 mV/s separated into the contributions of two functions, capacitor- and battery-types, and (C) discharge curves [428].

acid (PAA), resulting in MnO_2/PAA-rGO composite foams with different MnO_2 contents [429]. The composites exhibited stable charge—discharge behavior in 1 M Na_2SO_4 electrolyte with a voltage window of −0.2 to 0.8 V, showing bifunctional character. A symmetric supercapacitor composed of the composite with 26 wt.% MnO_2 delivered the highest capacitance of 123 F/g at 0.5 A/g rate with a capacitance retention of about 86% after 5000 cycles.

The composite foams of MoO_3/rGO prepared by hydrothermal heating of the mixture of MoO_3 with GO at 200°C exhibited bifunctional behavior in 1 M $LiPF_6$/(EC + DEC) electrolyte [430]. The capacitance observed on the composites using a three-electrode cell was separated into the capacitive contribution due to formation of EDLs and Faradaic contribution due to redox reaction of MoO_3 with functional groups in rGO flakes. Fig. 3.123A shows a CV curve at 0.5 mV/s rate separated into

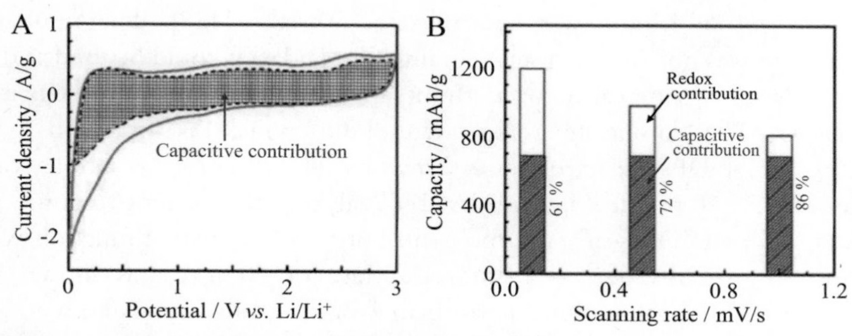

Figure 3.123 MoO_3/rGO composites: (A) CV curve at 0.5 mV/s separated into the contributions of two functions and (B) capacitive contributions in the overall capacity at different scan rates [430].

two contributions, with the capacitive contribution reaching 72% in total capacitance measured. The relative contribution of capacitive function increases with increasing scan rate, as shown in Fig. 3.123B.

3.3 Hybrid cells

Energy-storage devices based on the hybridization of capacitor-type with battery-type functionalities are attracting increased attention, in which the energy density is expected to be comparable to rechargeable batteries based on Faradaic reactions, such as intercalation–deintercalation and reduction–oxidation (redox reaction), and power density to be comparable with supercapacitors based on physical adsorption–desorption to form EDLs. In this section, these energy-storage devices (hybrid cells) are classified into two types, Faraday reaction at the negative electrode (as a consequence, adsorption–desorption at the positive electrode) and Faraday reaction at the positive electrode (as a consequence, adsorption–desorption at the negative electrode), as shown schematically in the relation to rechargeable batteries and supercapacitors in Fig. 3.124. The electrode working mainly with Faraday reaction is expressed as the F-electrode and that working mainly with adsorption–desorption is expressed as the

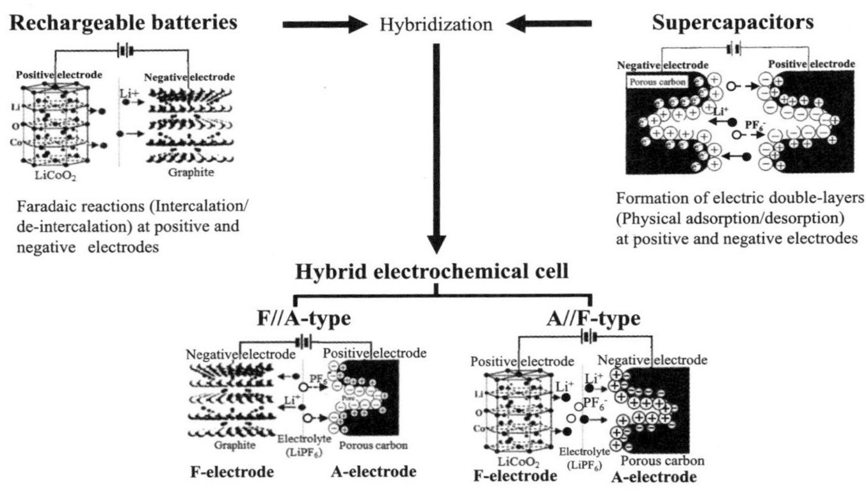

Figure 3.124 Classification of hybrid cells and relations to rechargeable batteries and supercapacitors.

A-electrode, and so these two types may be abbreviated to F//A-type and A//F-type, respectively. Each type of hybrid cell is explained by focusing on the electrochemically active materials at the F-electrode.

3.3.1 Faraday reaction at negative electrode (F//A-type)

3.3.1.1 Carbon materials

For electrochemically active materials in the A-electrode in hybrid cells, porous carbons, mostly activated carbons (ACs), were used because of the principal mechanism being adsorption–desorption of electrolyte ions to build mainly electric double layers (EDLs) on their surfaces. With the use of carbon materials as active materials in the F-electrode, electrochemical activation of F-electrode carbons is required to achieve steady charge–discharge performance of the hybrid cells. When lithium compounds, such as $LiPF_6$, are employed as the electrolyte, doping of lithium in the F-electrode carbons has to be carried out in advance.

3.3.1.1.1 Asymmetric combination of carbons in two electrodes

A hybrid cell is constructed from graphite at the F-electrode (negative electrode), where the intercalation–deintercalation of Li^+ ions was the principal reaction, with an activated carbon (AC) at the A-electrode (positive electrode), where adsorption–desorption of PF_6^- ions was the principal reaction, in a 1 M $LiPF_6$/(EC + DMC) electrolyte [431,432]. The hybrid cell delivered a maximum energy density of 145.8 Wh/kg in a potential range of 1.5–5 V and a maximum power density of above 10 kW/kg. The galvanostatic charge–discharge profiles of the hybrid cell composed of spheroidal particles of graphite (F-electrode) and a commercially available AC with S_{BET} of 1400 m^2/g (A-electrode) with the mass ratio m_F/m_A of 1/5 are shown in Fig. 3.125A [431]. The potential at the F-electrode is kept at a constant due to the intercalation–deintercalation, where the operating conditions are optimized to allow full intercalation of graphite. Since the graphite at the F-electrode has a capacity of about 380 mAh/g and the AC at the A-electrode has a capacity of about 75 mAh/g, the optimization between two electrodes is important; the energy and power densities relative to the respective maximum values of the hybrid cell are plotted against the mass ratio of active materials in the F-electrode to that in the A-electrode m_F/m_A in Fig. 3.125B. Since the principal function was the physical adsorption of anions to form EDLs at the surface of AC, the capacity of the A-electrode depended strongly on the current density for charge/discharge, revealing that high S_{BET} was not

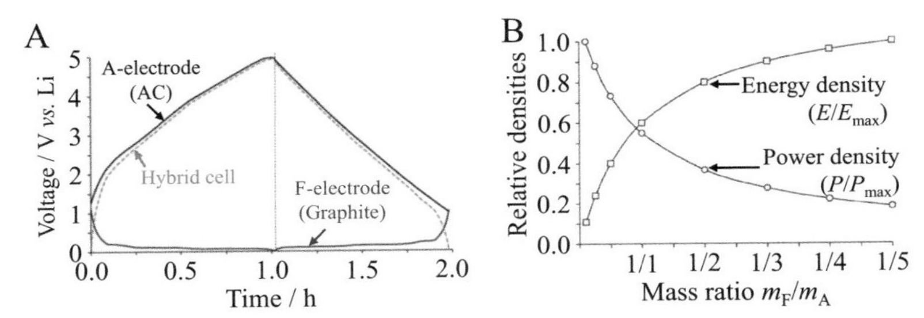

Figure 3.125 Hybrid cell graphite//AC ($m_F/m_A = 1/5$): (A) galvanostatic charge–discharge profiles for individual electrodes and the hybrid cell at 1C rate in 1 M $LiPF_6$/(EC + DMC) electrolyte and (B) changes in relative power and energy densities with the mass ratio of the F- to A-electrode, m_F/m_A [431].

always advantageous for the A-electrode as presenting important capacitance fading. The dependences of capacity of graphite at the F-electrode, which was prepared by coating of graphite with carboxymethyl cellulose (CMC) onto a 30 μm copper foil at thicknesses of 24–100 μm, on the rate of intercalation and deintercalation of Li^+ were studied using half-cells with various commercial synthetic graphite materials [433]. As shown in Fig. 3.126A, deintercalation of the Li^+ intercalated by a slow rate of 0.1C is possible with high rates, although the thickness of the graphite film of the F-electrode must be kept at less than 24 μm. On the other hand, intercalation of Li^+ into graphite depends greatly on the current density, as shown in Fig. 3.126B, suggesting that the rate performance of the hybrid cell is

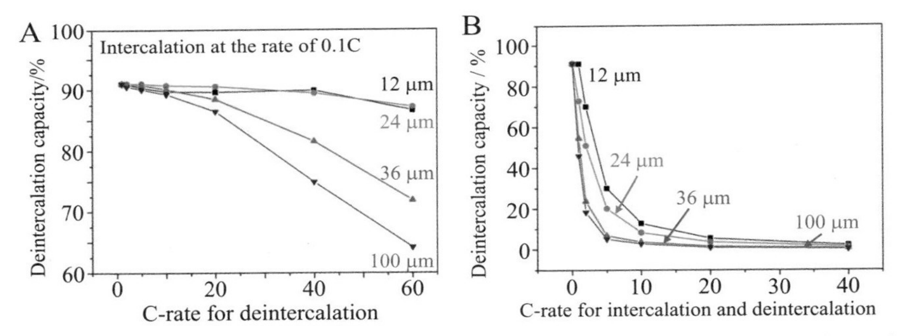

Figure 3.126 Dependence of deintercalation capacity of graphite F-electrode with different thicknesses on the rate of Li intercalation and deintercalation: (A) deintercalation capacity with a constant intercalation with 0.1C rate and deintercalation with various C-rates, and (B) both intercalation and deintercalation at the same C-rate [433].

governed by the current density for the intercalation. The effects of ball milling of graphite were also studied for use at the F-electrode [434]. Mesoporous carbons were prepared from teak wood sawdust by subjecting them to hydrothermal treatment at 200°C after mixing with H_2O_2, followed by activation with $ZnCl_2$ at 105°C [435]. It had S_{BET} of 2108 m^2/g with V_{total} of 2.2 cm^3/g including V_{meso} of 2.03 cm^3/g. From this carbon, two hybrid cells were constructed by coupling with electrochemically prelithiated graphite (C_6Li) and LTO with 1 M $LiPF_6$/(EC + DMC) electrolyte. The hybrid cell coupled with C_6Li delivered a maximum energy density of 111 Wh/kg, which was higher than 53 Wh/kg for the hybrid cell coupled with LTO and much higher than 37 Wh/kg for the symmetric supercapacitor of the mesoporous carbon.

A graphitized soft carbon, which was prepared from a petroleum coke and had d_{002} of 0.3446 nm, was coupled with an AC (S_{BET} of 1400 m^2/g) [436]. This hybrid cell delivered higher capacity and better cycle performance than that using prelithiated synthetic graphite (d_{002} of 0.336 nm), as shown in Fig. 3.127A and B. Polyimide-derived carbon microspheres with diameters of 3–5 μm carbonized at 700, 800, and 900°C, which contained 5.3, 3.95, and 3.5 %N mainly at pyridinic and pyrrolic configurations, were used at the F-electrode of the hybrid cell coupled with an AC (S_{BET} of 1600 m^2/g) in 1 M $LiPF_6$/(EC + DMC) electrolyte [437]. The hybrid cell constructed using 900°C-carbonized microspheres delivered a capacity of 328 mAh/g at 0.1 A/g rate and an energy density of 28.5 Wh/kg at the power density of 0.35 kW/kg, with the capacity retention of about 97% after 5000 cycles with 0.5 A/g rate, as shown in Fig. 3.128, its cyclability being much better than 700°C- and 800°C-carbonized microspheres.

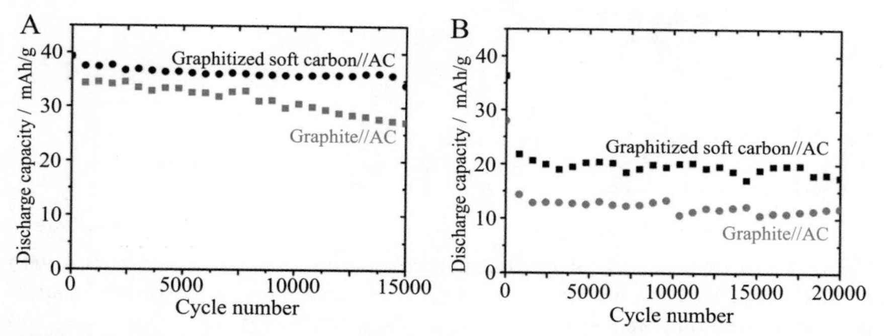

Figure 3.127 Hybrid cells of soft carbon//AC and synthetic graphite//AC; cycle performances with (A) 5C (0.74 A/g) and (B) 15C (2.2 A/g) rates [436].

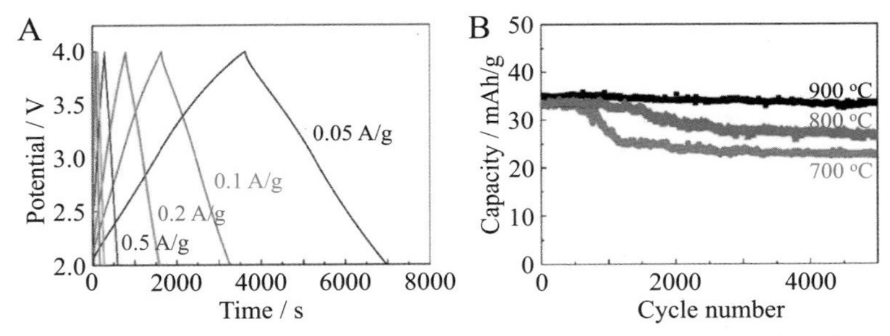

Figure 3.128 Hybrid cells of N-doped carbon microspheres coupled with an AC: (A) charge–discharge curves with different current densities on the cell constructed using the 900°C-prepared carbon microspheres and (B) cycle performances at 0.5 A/g for the hybrid cells using microspheres carbonized at different temperatures [437].

In order to have steady charge–discharge cycles of the hybrid cells using carbon materials, such as graphite, at the F-electrode, prelithiation was recommended. Prelithiation was achieved by external short-circuiting with the Li metal and graphite electrodes in a three-electrode cell of Li//graphite//AC [432], by repeating the charge–self-discharge cycles in a 2 M LiTFSI/(EC + DMC) electrolyte [438], and also by mixing sacrificial lithium ion source, such as $Li_{0.65}Ni_{1.35}O_2$, Li_5ReO_6, etc., into AC of the A-electrode, Li ions being irreversibly extracted from these sources and intercalated into the material of F-electrode during the first cycle [439,440].

A porous composite (S_{BET} of 2675 m^2/g) prepared by carbonization of a mixture of GO with sucrose at 500°C and following KOH activation at 800°C was used as the A-electrode for the construction of a hybrid cell with prelithiated graphite [441]. It delivered an energy density of 233 Wh/kg at a power density of 0.45 kW/kg and 143.8 Wh/kg at 15.7 kW/kg in 1 M $LiPF_6$/(EC + DEC + DMC). The reducing process of GO to rGO influenced the electrode performance at the A-electrode [442]. The reduction of GO flakes prepared from synthetic graphite was performed in a water dispersion by adding either urea or hydrazine at 98°C. Urea-reduced rGO had a high N content of 8.10 at.% (atomic ratio C/N of 9.80). It gave higher capacity than hydrazine-reduced rGO, and both rGOs giving higher capacity and better rate performance than AC, as shown in Fig. 3.129A and B. The hybrid cell composed of lithiated graphite//urea-reduced rGO delivered a maximum energy density of 106 Wh/kg and a maximum power density of 4.2 kW/kg with almost 100% retention up to 1000 cycles at a current density of 0.14 A/g in the

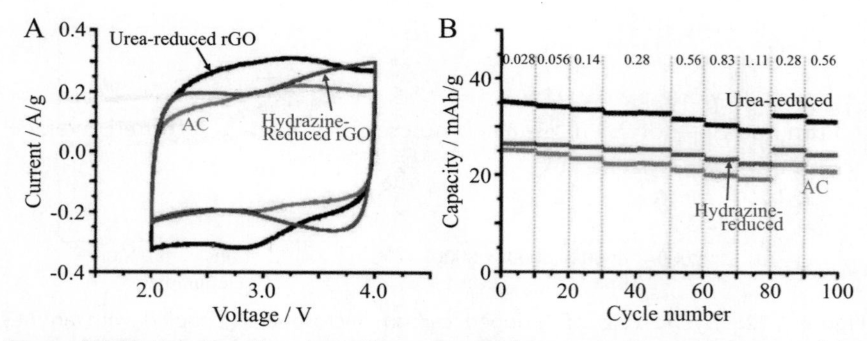

Figure 3.129 Hybrid cells using rGOs prepared by a reduction with urea and hydrazine and AC at the A-electrode coupled with the F-electrode of prelithiated graphite with 1 M $LiPF_6$/(EC + DEC): (A) CV curves at a scan rate of 5 mV/s and (B) rate performances [442].

potential range of 2.2—3.8 V. A symmetric supercapacitor constructed from urea-reduced rGO delivered capacitance of 105 F/g, which was higher than that from hydrazine-reduced rGO (82 F/g).

A hybrid cell rGO film//rGO foam with m_F/m_A of 1/3.5 in 1 M $LiPF_6$/(EC + DEC + DMC) electrolyte exhibited high performances, with a high operating voltage (4.2 V), high energy density of 148 Wh/kg at power density of 0.141 kW/kg, and 72 Wh/kg at 7.8 kW/kg and a long cycle life [443]. The rGO films were prepared by vacuum filtration of aqueous dispersion of GO flakes, followed by shining with a simple digital camera photoflash for the reduction, and the rGO foams were prepared from a mixture of GO flakes with phenol-formaldehyde (PF/GO weight ratio of 16) by heat treatment under hydrothermal conditions at 180°C, followed by KOH activation at 900°C, giving a high S_{BET} of 3523 m^2/g. The rGO film had a high conductivity of about 320 S/m and exhibited a potential window of 0—3 V versus Li/Li^+. LIB composed of this rGO film and Li exhibited no potential plateau in the charge—discharge process, giving a reversible capacity of more than 660 mAh/g at 0.372 A/g rate. The symmetric supercapacitor composed of the rGO foams with $LiPF_6$ electrolyte delivered a capacitance of 160 F/g at 0.1 A/g and about 130 F/g at 5 A/g, corresponding to an energy density of 41 and 27.7 Wh/kg, respectively.

A hybrid cell assembled by a KOH-activated soft carbon at the F-electrode and a hard carbon at the A-electrode delivered high energy and power densities [444]. The soft carbon was prepared by carbonization of a commercial pitch (AR pitch) at 800°C, followed by activation by mixing with KOH in a KOH/coke weight ratio of 4 at 800°C, with an average particle size of 11.6 μm and S_{BET} of 46 m^2/g. The hard carbon was a commercial nongraphitizable carbon, with an average particle size of 9.3 μm and S_{BET} of 2.9 m^2/g. A commercial AC (steam-activated, average particle size of 2.9 μm and S_{BET} of 1500 m^2/g) was also employed for comparison. In Fig. 3.130A, a charge–discharge curve of a symmetric supercapacitor composed of the soft carbon is shown by comparing it with that of AC. It demonstrates that the lithiation process is clearly observed on the soft carbon in the first charging process (called "electrochemical activation"), but not on AC, and the soft carbon delivers a higher capacitance even after the second cycle than the AC. Ragone plots for the hybrid cells using this KOH-activated soft carbon at the F-electrode coupled with the A-electrode of either the AC or a hard carbon are shown together with a symmetrical supercapacitor of AC//AC for comparison in Fig. 3.130B. The hybrid cells clearly show higher energy and power densities than the symmetric supercapacitor AC//AC. The energy and power densities of the hybrid cell of soft carbon//hard carbon are about 8.5 times and 4.6 times higher than those of the AC//AC, respectively, and even higher than the hybrid cell of soft carbon//AC. The effects of electrolytes of 1 M nonaqueous electrolyte solutions of various lithium salts [$LiPF_6$, $LiBF_4$, $LiClO_4$, $LiCF_3SO_3$, $Li(CF_3SO_2)_2N$, $LiB(C_2O_4)_2$ and $Li(CF_3SO_2)_3C$] with

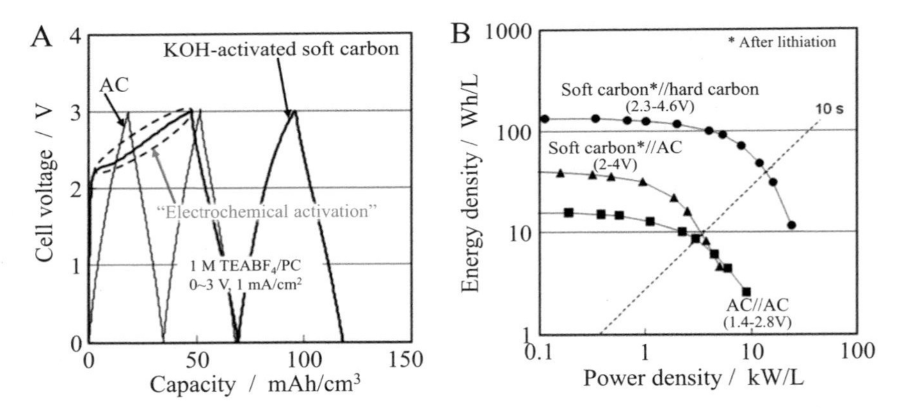

Figure 3.130 (A) Charge–discharge curves for symmetric supercapacitors of KOH-activated soft carbon and an AC in 1 M $TEABF_4$/PC electrolyte and (B) Ragone plots for two hybrid cells using soft carbon at the F-electrode and a hard carbon and an AC at the A-electrode, in comparison with the symmetric supercapacitor composed of AC [444].

various solvents (EC, PC, BC, DMC, EMC, DEC and GBL) were also studied. The KOH-activated soft carbon showed high performances in $LiPF_6$/(EC + DMC) electrolyte, delivering high capacitance, good potential stability, and low internal resistance.

The hybrid cell was constructed using a petroleum coke (S_{BET} of about 4 m^2/g) at the F-electrode (negative electrode) with a commercial AC (Norit, S_{BET} of 1400 m^2/g) at the A-electrode (positive electrode) [445]. Its cycling performance was steady and displayed a capacity of 37 mAh/g up to 15,000 cycles with a slight decrease (about 5%). The cell voltage profiles together with the individual electrode potentials at the 100th and 15,000th cycles are shown in Fig. 3.131A, demonstrating how the charge−discharge process of the cell is occurring steadily. In Fig. 3.131B, a Ragone plot for this hybrid cell is compared to the symmetric supercapacitor and lithium-ion battery (LIB) composed from the same coke, revealing that the hybrid cell can deliver significantly higher power density compared to the LIB, and significantly higher energy density than the supercapacitor in the range of high current densities (0.74−8.9 A/g). Upon charge−discharge with 30C rate (4.5 A/g), the hybrid cell exhibited a stable cycle performance up to 50,000 cycles.

N-doped porous carbons worked as the A-electrode material could give improved performance of hybrid cells. Composites of N-doped mesoporous carbon nanospheres with rGO flakes (lateral size of about 70 μm) (N-GMCSs) were prepared from a mixture of phenol, formaldehyde, and triblock copolymer F123 with GO in an NMP solution by treating under hydrothermal conditions at 100−150°C, followed by carbonization at

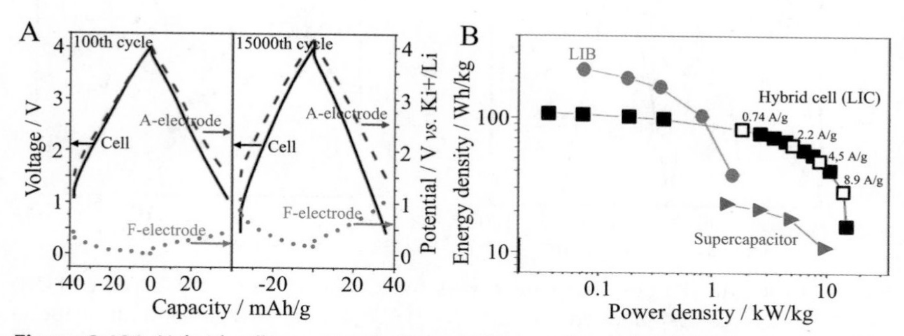

Figure 3.131 Hybrid cell composed of a petroleum coke at the F-electrode with AC at the A-electrode: (A) voltage profiles at the 100th and 15,000th cycles at a current density of 0.74 A/g and (B) a Ragone plot comparing the LIB and the supercapacitor composed of the same coke [445].

700°C, KOH activation at 700°C, and then heat treatment at 800°C in an NH_3 atmosphere [446]. The composite was used at the A-electrode of the hybrid cell coupled with prelithiated spheroidal graphite at the F-electrode with 1 M $LiPF_6$/(EC + DEC) electrolyte (N-GMCS//graphite), which delivered a maximum energy density of 80 Wh/kg (66.7 Wh/L) at a power density of 0.152 kW/kg and a maximum power density of 352 kW/kg (292 kW/L) with a capacitance retention of 93% after 4000 cycles. In Fig. 3.132A, the rate performance of this hybrid cell is compared to the hybrid cells of GMCS//graphite and AC//graphite, exhibiting the effect of N-doping on the capacitance retention. Even at an ultrahigh current density of 80 A/g, N-doped carbon composite at the A-electrode gives much better retention than nondoped composite and AC. The Ragone plot for this hybrid cell is compared with the symmetric cell N-GMCS//N-GMCS and even a hybrid cell graphite//$LiFePO_4$ in Fig. 3.132B. The hybrid cell graphite//N-GMCS delivers much higher energy and power densities than the symmetric cell N-GMCS//N-GMCS. N-doped carbon foam coated by rGO prepared from melamine foam delivered an energy density of about 400 Wh/kg by coupling with AC A-electrode in $LiPF_6$/(EC + DEC) [447].

A hybrid cell was constructed using an amorphous carbon at the F-electrode (positive electrode) and an N-doped porous carbon at the A-electrode (positive electrode) [448]. The carbon at the F-electrode was prepared from glucose by hydrothermal treatment at 180°C and following carbonization at 800°C in N_2, and the carbon at the A-electrode was prepared by simple carbonization of the mixture of glucose with urea at 950°C in N_2. The cell with a m_F/m_A ratio of 2/3 worked at the voltage

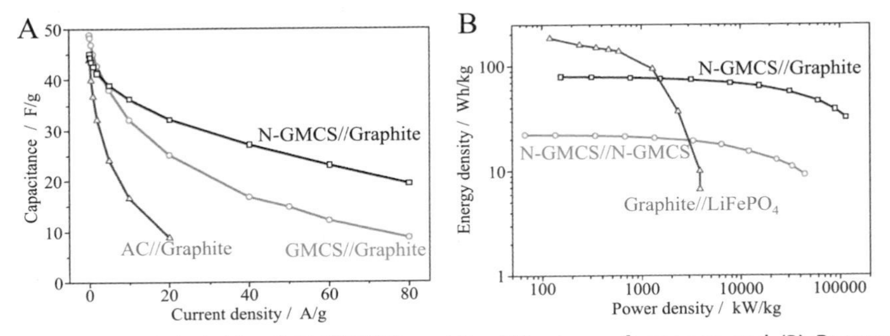

Figure 3.132 A hybrid cell N-GMCS//graphite: (A) rate performance and (B) Ragone plot in comparison with other hybrid cells [446].

range of 0.5−4 V by exhibiting 65 F/g at the current density of 0.1 A/g and 39 F/g at 10 A/g. It delivered an energy density of 133 Wh/kg at a power density of 0.21 kW/kg with capacitance retention of 82% after 5000 cycles at a current density of 5 A g.

3.3.1.1.2 Symmetric combinations of carbons in two electrodes

A symmetric AC//AC cell could be operated as a hybrid cell by using 1 M Li_2SO_4 aqueous electrolyte containing 0.5 M KI [449]. KOH-activated AC having S_{BET} of 2180 m^2/g and V_{micro} of 0.812 cm^3/g was selected. In Fig. 3.133A, CVs for individual electrodes are shown, demonstrating reversible peaks at around 0.5 V due to the redox reactions (Faradaic reaction) associated with the $2I^-/I_2$ system at the A-electrode with the potential window widened to 1.6 V. CVs in (Li_2SO_4 + KI) electrolyte with a potential window of 1.6 V are shown in Fig. 3.133B by comparing with those in Li_2SO_4 electrolyte (without KI). With the window up to 1.6 V, the current reaches a maximum at around 0.9 V and then decreases slightly. This increase in capacitance in (Li_2SO_4 + KI) electrolyte is understood mainly due to the evolution observed by the CV of the positive electrode in Fig. 3.133A. With the rate of 0.2 A/g, the cell using (Li_2SO_4 + KI) electrolyte displayed a high capacitance of 75 F/g, in which the energy density reached 26 Wh/kg, close to 30.9 Wh/kg for a symmetric AC//AC supercapacitor using Et_4NBF_4/AN nonaqueous electrolyte. A similar hybrid cell was constructed using an AC//AC cell with 2 M $MnSO_4$ + 0.5 M KI aqueous electrolyte [450], in which the capacitance was twice as high as the cell using $MnSO_4$ aqueous electrolyte (without KI)

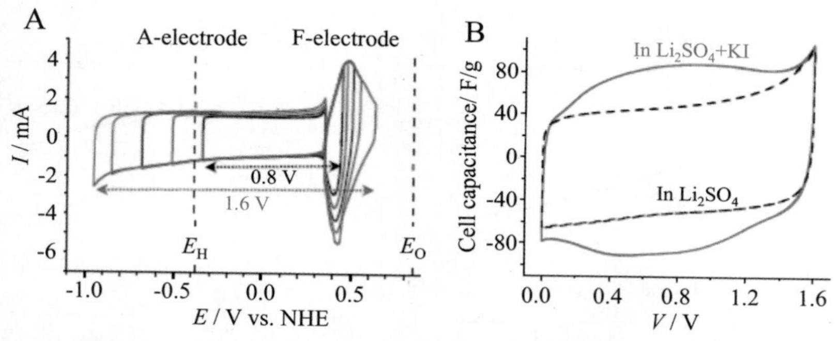

Figure 3.133 CVs on an AC//AC cell with (Li_2SO_4+KI) electrolyte: (A) CVs of individual electrodes with potential windows from 0.8 to 1.6 V, and (B) CVs of hybrid cells with 1.6 V potential window in (Li_2SO_4 + KI) and Li_2SO_4 electrolytes [449].

and the cycle performance with 0.5 A/g rate up to 1.5 V was excellent, with 92% retention after 10,000 cycles. In the AC used in the positive electrode after 10 h at 1.5 V, iodine was confirmed to be confined in its micropores as I^{3-} and I^{5-} by Raman spectroscopy, while no iodine was detected in the AC at the negative electrode.

Symmetric hybrid cells were assembled using N-doped mesoporous carbons, the carbon at the F-electrode being prelithiated [451]. The N-doped mesoporous carbons were synthesized from a mixture of methyl cellulose, urea, $NaHCO_3$, and KOH by carbonization at 800°C, where $NaHCO_3$ worked as the template for mesopore formation, and had S_{BET} of 1748 m^2/g with V_{total} of 1.47 cm^3/g. The performance of the hybrid cell depended greatly on the m_A/m_F as shown by the charge–discharge curves and Ragone plots in Fig. 3.134A and B, respectively. The cells exhibited stable charge–discharge performance at a current density of 5 A/g with a voltage window of 0–4 V, even with the 100 A/g rate IR drop being negligible, and high cyclability, 84.6% capacity retention after 5000 cycles at 10 A/g rate. The cell with m_A/m_F of 1/2 delivers an energy density of 116 Wh/kg at a power density of 0.167 kW/kg and it can deliver a power density of 66 kW/kg by keeping the energy density at 70 Wh/kg.

N-doped porous carbons were synthesized from the aerosol droplets of the mixture of a copolymer of melamine, phenol, and formaldehyde with colloidal silica prepared at 450°C by carbonization at 900°C in N_2, followed by HF washing, which had S_{BET} of 1560 m^2/g, V_{total} of 2.56 cm^3/g, and N content of 14.5 at.% [452]. Symmetric hybrid cells were assembled using this carbon at both electrodes with 1 M $LiPF_6$/AN electrolyte, in

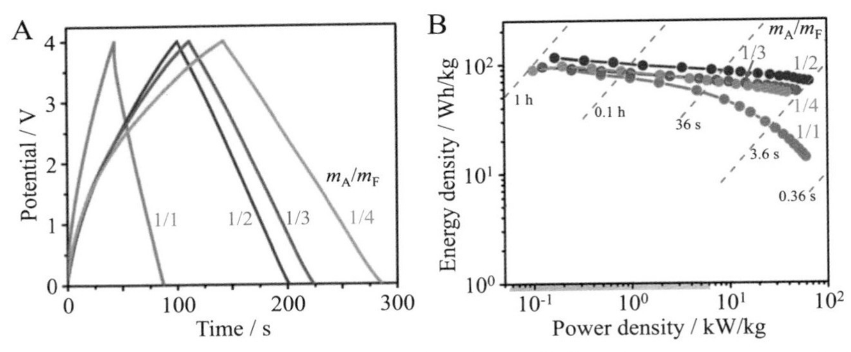

Figure 3.134 Symmetric hybrid cells based on N-doped mesoporous carbon: (A) charge–discharge curves at 5 A/g rate and (B) Ragone plots for the cells with different m_A/m_F ratios [451].

which Li^+ intercalation–deintercalation occurred at the F-electrode (negative electrode) and PF_6^- adsorption–desorption at the A-electrode (positive electrode). The cell delivered an energy density of 208 Wh/kg at a power density of 0.225 kW/kg and 115 Wh/kg at 22.5 kW/kg with 86% capacity retention after 10,000 cycles at 2 A/g rate. Symmetric hybrid cells were also constructed from mesoporous carbons derived from sheep bone powder at 1100°C with 1.0 M $LiPF_6$/(EC + DEC + DMC) electrolyte [453]. The cell delivered an energy density of 106 Wh/kg at a power density of 88.8 kW/kg, with cyclability as capacitance retention of 88.3% after 8000 cycles at 2 A/g rate. N-doped porous carbons were synthesized from deacetylated chitosan by mixing with Mg-acetate and different amounts of K-acetate, followed by carbonization at 800°C in Ar flow [454]. Their N_2 adsorption–desorption isotherms display a combination of type-I and type-IV isotherms with an H3 hysteresis and had S_{BET} of 1223–2350 m^2/g. This carbon was used to fabricate a hybrid cell as the A-electrode (positive electrode) by coupling with the F-electrode (negative electrode) of mesoporous graphene synthesized by CVD of CO_2. The graphene at the F-electrode had to be prelithiated in advance and the electrolyte selected was 1 M $LiPF_6$/(EC + DEC + DMC). The cell exhibited a maximum energy density of 146 Wh/kg at a power density of 650 W/kg and maintained 103 Wh/kg at 52 kW/kg with a capacitance retention of 91% after 40,000 cycles at 4 A/g rate.

N/B-codoped CNF webs were synthesized from a commercial bacterial cellulose (BC) membrane after immersion into $NH_4HB_4O_7$ aqueous solutions with different concentrations (0–0.15 M) by carbonization at 1000°C in Ar/H_2 [455]. Symmetric hybrid cells were constructed using these N/B-codoped webs at both electrodes with $LiPF_6$/(EC + DEC) electrolyte. The hybrid cell exhibits steady charge–discharge cycles with a capacitance of 78.2 F/g at a current density of 0.1 A/g and 36.8 F/g at 10 A/g with a capacitance retention of 81% after 5000 cycles at 2 A/g rate. It delivers an energy density of 220 Wh/kg at a power density of 0.225 kW/kg and 104 Wh/kg at of 22.5 kW/kg with a wide voltage window of 0–4.5 V.

Hybrid cells were fabricated by coupling rGO foam at the A-electrode with rGO film at the F-electrode [443]. The rGO foam was prepared by hydrothermal treatment of the mixture of GO with PF resin (PF/GO weight ratio of 16) at 180°C and following carbonization with simultaneous KOH activation at 900°C, and the rGO film by shining the GO film (15–20 μm thick) with a camera photoflash. Charge–discharge curves at different current densities are shown for the hybrid cell with m_A/m_F of 3.5 in Fig. 3.135B, with a representative CV curve in Fig. 3.135A, showing no clear peak due to redox reaction in CV and slight deviation of the linear voltage profile in charge–discharge curves, which is thought to be due to approximately linear voltage profiles without a plateau of the rGO film used. As shown in Fig. 3.135C, energy and power densities depend greatly on the m_A/m_F and the cell with optimized ratio of 3.5 delivers a maximum energy density of 128 Wh/kg. Hybrid cells of N-doped rGO foam coupled with commercial AC delivered a maximum energy density of 39 Wh/kg at a powder density of 0.15 kW/kg and 11 Wh/kg at 1.2 kW/kg [456]. Hybrid cells were assembled using as-prepared rGO flakes at the A-electrode (negative electrode) and the armored rGO flakes, which was electrochemically coated by lithium difluoro(oxalate)borate, at the F-electrode (positive electrode) with 1 M $LiPF_6$/(EC + DMC) electrolyte [457]. The cell delivered a maximum energy density of 160 Wh/kg and high capacity retention of about 90% after 1000 cycles with a voltage window of 0–4.3 V.

Hybrid cells were constructed from the composite of commercial graphene nanoribbons with MWCNTs grown on their surfaces by CVD, placing the as-prepared composite at the A-electrode (negative electrode) and the prelithiated composite at the F-electrode (positive electrode) with 1 M $LiPF_6$/(EC + DMC + DEC) electrolyte [458]. The cell delivered a maximum energy density of 121 Wh/kg at a power density of about 0.1 kW/kg and about 29 Wh/kg at about 20.5 kW/kg with a voltage window of 0.01–4.3 V and capacity retention of 89% after more than 10,000 cycles at 1 A/g rate. A composite MWCNT/rGO where two

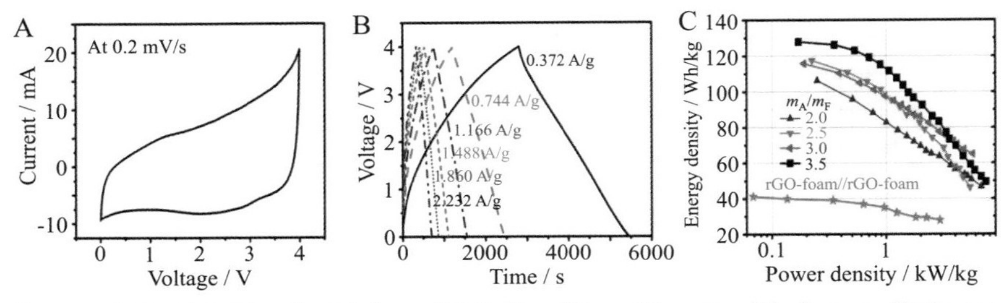

Figure 3.135 Hybrid cell rGO foam//rGO film: (A) a CV curve, (B) charge–discharge curves at different current densities for the cell with m_A/m_F of 3.5, and (C) Ragone plots for the hybrid cell with different m_A/m_F ratios, in comparison with a symmetric supercapacitor of rGO foam [443].

components were chemically bonded through surface modification of MWCNTs by liquid oxidation and rGO flakes was doped by N using *N*—hydroxyl succinimide ester [459]. The symmetric hybrid cell based on this composite was compared with the asymmetric hybrid cell coupled with N-doped rGO without MWCNTs. Two hybrid cells worked with the potential window of 1.5—4.5 V and high cyclability with about 93% capacitance retention after 1000 cycles at 200 mV/s rate. As shown in Fig. 3.136, the hybrid cell based on MWCNT/rGO composite exhibits a slightly higher capacitance than that on rGO, delivering an energy density of 124.9 Wh/kg at a power density of 1.26 kW/kg. The composites of SWCNTs and single-layer rGO were fabricated by mixing commercially available SWCNTs with yellow-colored dispersion of GO in deionized water, followed by reduction with hydrazine monohydrate [460]. The symmetric hybrid cell based on this composite after the lithiation of the composite at the F-electrode (negative electrode) delivered an energy density of 222 Wh/kg at a power density of 0.41 kW/kg under a voltage widow of 0—4.1 V.

A quasisymmetric hybrid cell was constructed using polyimide-derived carbon microspheres with 1 M $LiPF_6$/(EC + DMC) electrolyte, as-carbonized microspheres at the F-electrode (negative electrode), and the same carbon microspheres after KOH activation at the A-electrode [461]. The hybrid cell delivered an energy density of 95 Wh/kg at a power density of 0.30 kW/kg and 48 Wh/kg at 15 kW/kg with a voltage window of 2—4 V.

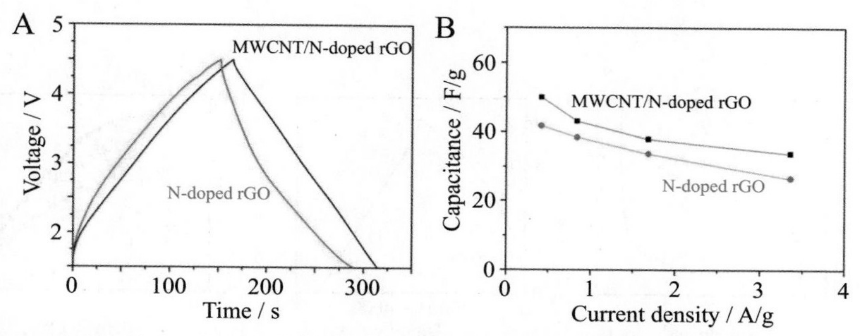

Figure 3.136 Two symmetric hybrid cells constructed from MWCNT/N-doped rGO composite and N-doped rGO: (A) charge—discharge curves at 0.84 A/g rate and (B) rate performances [460].

3.3.1.2 Lithium compounds

Various lithium compounds, $Li_4Ti_5O_{12}$ (LTO), $LiCoO_2$ (LCO), and $LiMn_2O_4$ (LMO), etc., have been used at the F-electrode of hybrid cells coupled with ACs at the A-electrode. The performance of the hybrid cell (LTO//AC) was compared with those of the symmetric supercapacitor (AC//AC) and different LIBs, a commercial one, LTO//LCO, and LTO//LMO [462]. Ragone plots and cycle performances of these cells are compared in Fig. 3.137A and B, respectively, revealing that the hybrid cell (LTO//AC) has a lower energy density but much better cyclability than LIBs. However, it has higher energy density but poorer cyclability than the supercapacitor (AC//AC).

A hybrid cell composed of LTO at the F-electrode with mesoporous ACs at the A-electrode delivered an energy density of about 69 Wh/kg in 1 M $LiPF_6$/(EC + DMC) electrolyte with a high cyclability up to 2000 cycles with 0.1 A/g rate [463]. The mesoporous carbon was prepared from coconut shells by hydrothermal treatment after mixing with $ZnCl_2$ at 275°C, followed by carbonization in N_2 flow and activation in CO_2 flow at 800°C, which had S_{BET} of 1650 m^2/g with V_{total} of 1.29 cm^3/g including V_{meso} of 0.77 cm^3/g. A microporous carbon prepared from the same coconut shells without hydrothermal treatment gave somewhat poor pore development, of which the hybrid cell coupled with LTO at the F-electrode delivered slightly lower capacity. A hybrid cell composed of LTO and an AC prepared from corncob via KOH activation delivered an energy density of 79.6 Wh/kg at a power density of 0.2 kW/kg and 32.7 Wh/kg at 4 kW/kg in 1 M $LiPF_6$/(EC + DEC + DMC) electrolyte [464]. A hybrid cell of LTO coupled with activated mesoporous

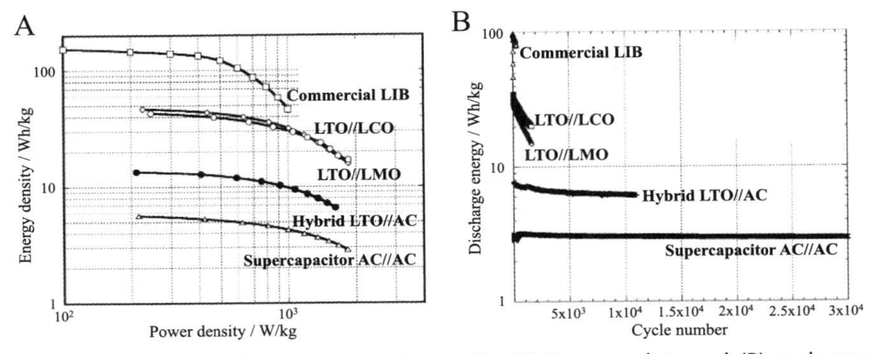

Figure 3.137 Comparison among various cells: (A) Ragone plots and (B) cycle performances [462].

carbon (m_F/m_A of 1/2.3) with 1 M $LiPF_6$/(EC + DEC) delivered a capacitance of 166 F/g at a current density of 1 A/g with a capacitance retention of about 97% after 6000 cycles and an energy density of 63 Wh/kg [465]. The mesoporous carbon was prepared from 1,2,3,4-benzene tetracarboxylic acid by using colloidal SiO_2 as the template at 800°C, followed by KOH activation. An LTO//AC hybrid cell with ionic-liquid-based electrolytes [1 M Li-TFSI/(PMPyrr-TFSI) and 0.8 M Na-TFSI/(PMPyrr-TFSI)] delivered a maximum energy density of 98 and 90 Wh/kg and a maximum power density of 1.9 and 261.8 kW/kg, respectively, with a maximum cell voltage of 4 V. However, the same LTO//AC cell with 1 M $LiClO_4$/ACN electrolyte delivered 30 Wh/kg and 14.8 kW/kg with a voltage of 2.5 V [466]. The hybrid cells using IL-based electrolytes exhibited stable behavior at 80°C for more than 3000 cycles at a high rate of 2.5 A/g rate.

Composites of lithium compounds, LTO, LMO, etc., with AC [464,467–472], CNFs [473], and rGO [474,475] were used at the F-electrode of the hybrid cells with AC at the A-electrode to improve energy storage performances. The composite electrode materials, for example LMO/AC, are bifunctional, battery-type function- due to LMO and capacitor-type function due to AC. The potential profile observed on the electrode of the composite consisting of the segments of LMO and AC with equal total amounts is the same as that on the composite of powder mixture of LMO and AC [469], as shown in Fig. 3.138A. The charging process can be divided into three parts, as shown in Fig. 3.138B and C: (i) a capacitor-behavior as a straight potential increase up to 3.9 V (the current goes through mainly AC), (ii) a battery-behavior in the potential range of 3.9–4.2 V (the current is transported mainly by LMO), and (iii) a capacitor-behavior above 4.2 V where the current is transported by AC again because of the full lithiation of LMO. The discharge profile is just a reverse image. A hybrid cell with LMO nanorod (diameter of about 360 nm and length of 2.3 μm) at the F-electrode and AC at the A-electrode with m_F/m_A of 1/2 exhibited steady charge–discharge cycles with 98% capacity retention after 1000 cycles at 2C (33.16 mA/g) rate [476]. The cell delivered an energy density of 33 Wh/kg at a power density of 16.6 W/kg and 13 Wh/kg at 395 W/kg.

Coating of Li-compounds by porous carbon was very effective in improving the performances of the hybrid cells [477,478]. C-coating was simultaneously performed during LTO synthesis from tetrabutyl titanate $Ti(OC_4H_9)_4$ by adding hexadecyltrimethyl-ammonium bromide

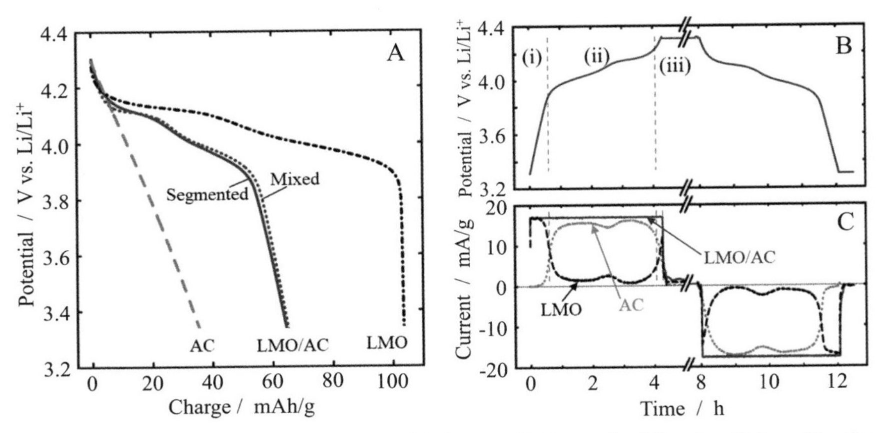

Figure 3.138 Contributions of LMO and AC in the F-electrode: (A) potential profiles for the electrodes of AC, LMO, their mixture, and their segmented composite during discharging at 17.5 mA/g rate, and (B) potential profile for the electrode of the segmented configuration, and (C) current profiles for the electrode components [469].

under hydrothermal conditions at 180°C, followed by calcination at 700°C in Ar [478]. A hybrid cell was constructed using this C-coated LTO at the F-electrode and rGO foam, which was prepared by hydrothermal treatment of GO flakes at 150°C, at the A-electrode with 1 M $LiPF_6$/(EC + DMC) electrolyte. The capacity of the hybrid cell is optimized at 2 by changing the mass ratio of m_A/m_F, as shown in Fig. 3.139A, which agrees with the mass ratio 2.08 calculated on the bases of capacities at each electrode material. Charge–discharge curves of the hybrid cell, C-coated LTO//rGO foam, suggest the coexistence of two functions, EDL formation corresponding to the linear part of the curves at the voltage

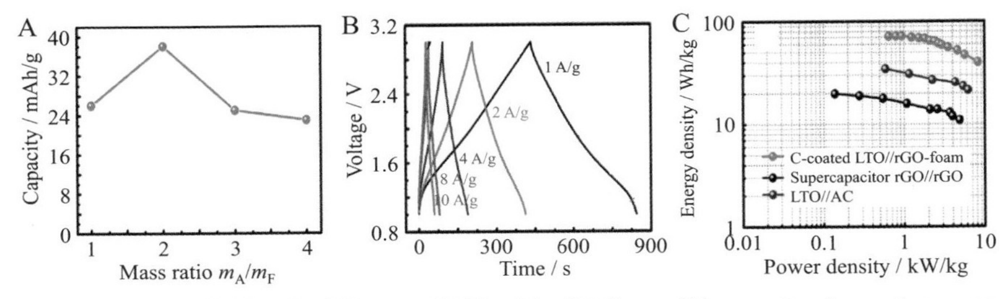

Figure 3.139 Hybrid cell of C-coated LTO with rGO-foam: (A) capacity dependence on m_A/m_F, (B) charge–discharge curves at different current densities, and (C) Ragone plot in comparison with a supercapacitor composed of rGO foam and a hybrid cell LTO//AC [478].

range of 1.6—3 V and Li-ion storage corresponding to a small plateau at around 1.55 V, as shown in Fig. 3.139B. The Ragone plots in Fig. 3.139C clearly show that the hybrid cell (C-coated LTO//rGO foam) exhibits significant enhancement in both energy and power densities compared with the supercapacitor (rGO foam//rGO foam) and the hybrid cell using LTO without C-coating (LTO//AC). C-coated LTO was fabricated by calcination of the mixture of mesoporous TiO_2 with a stoichiometric amount of Li_2CO_3 and 20 wt.% pitch at 900°C in Ar, which was composed of aggregation of 10—50 nm sized primary particles of C-coated LTO and of which the carbon content was 5.2 wt.% [477]. Charge—discharge curves of the C-coated LTO in a half-cell displayed clearly a flat plateau at 1.5 V, corresponding to the Li^+ intercalation—deintercalation process and a capacity of 163 mAh/g at 1C (0.17 A/g) rate, close to theoretical capacity, 170 mAh/g. The hydric cell composed of this C-coated LTO at the F-electrode (negative electrode) was coupled with an AC at the A-electrode, which delivered a maximum energy density of 35.5 Wh/kg at 0.17 A/g rate, and stable capacity retention of 95% at 1.7 A/g rate after 1000 cycles. In Fig. 3.140A and B, gravimetric and volumetric energy and power densities of the hybrid cell (C-coated LTO//AC) are compared with an LIB composed of LTO and a supercapacitor AC//AC, revealing markedly high volumetric capacity of the hybrid cell due to a high tap density of the C-coated LTO (1.31 g/cm^3). LMO microspheres (diameters of 4—8 μm) wrapped by rGO flakes (rGO-wrapped LMO-S) were used as the F-electrode to construct hybrid cells by coupling with AC in $LiPF_6$/(EC + DEC + DMC) electrolyte [479].

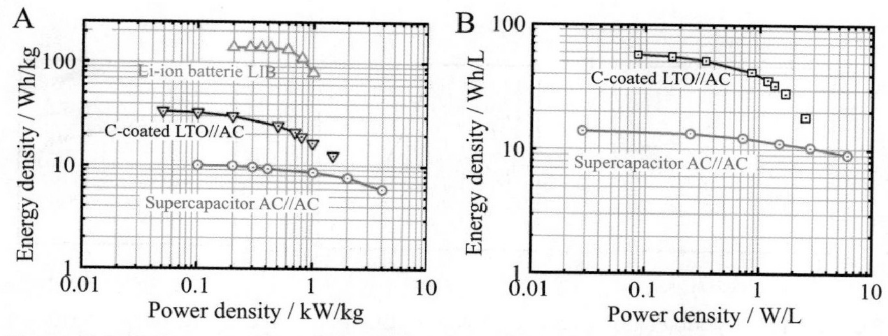

Figure 3.140 Ragone plots of the hybrid cells of C-coated LTO//AC, in comparison with a LIB using LTO and a supercapacitor AC//AC: (A) gravimetric and (B) volumetric energy and power densities [477].

In Fig. 3.141A and B, CV and Ragone plots of this hybrid cell are compared with the hybrid cells composed of commercial LMO and LMO microspheres without rGO wrapping (LMO-S) with AC (LMO//AC and LMO-S//AC, respectively), revealing the effectiveness of rGO wrapping for improving the performances of the cell.

$LiFePO_4$ (LFP) mixed with 80 wt.% AC (S_{BET} of 1602 m^2/g) was used at the F-electrode to construct a hybrid cell with a commercial hard carbon at the A-electrode (m_A/m_F of 3/1) with 1 M $LiPF_6$/(EC + DMC) electrolyte [480]. As shown in Fig. 3.142A, the characteristic shape of the charge—discharge curves due to the coexistence of two different functions, i.e., capacitive- and battery-functions, is kept even at very high rates of 53C. Capacity measured on the hybrid cell was separated to the contributions from two functions, as shown by plotting capacities against C-rate in Fig. 3.142B. Capacity due to battery-function reaches 135 mAh/g at

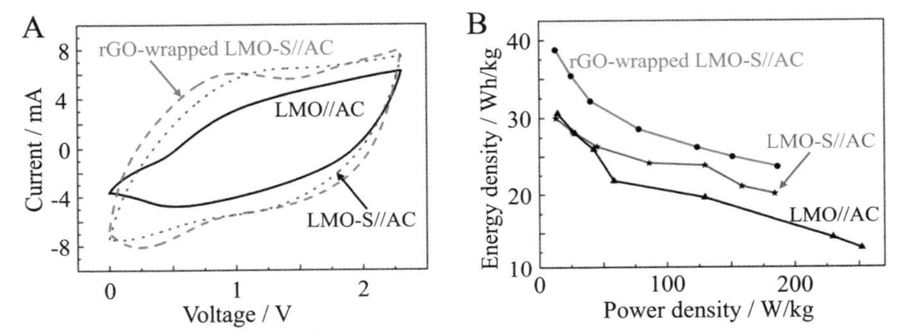

Figure 3.141 Three hybrid cells composed of a commercial LMO, LMO microspheres (LMO-S), and rGO-wrapped LMO-S coupled with AC: (A) CV curves and (B) Ragone plots [479].

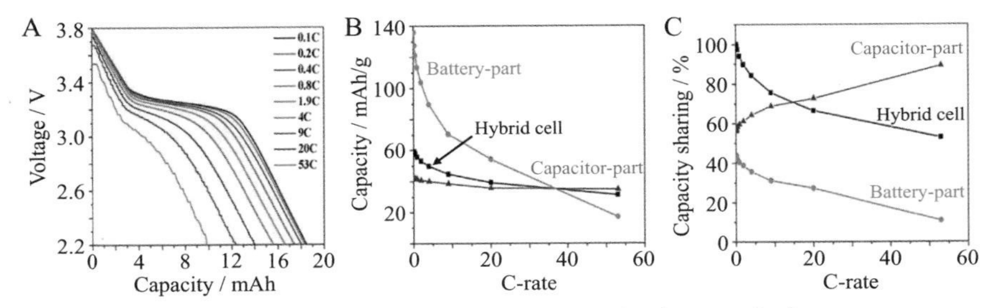

Figure 3.142 Hybrid cell LFP/AC//hard carbon: (A) charge—discharge curves at different C-rates, (B) rate performance with separation into capacitor and battery parts, and (C) dependences of relative fractions of two parts on C-rate [480].

0.1C rate, but decreases rapidly with increasing C-rate to 17 mAh/g. On the other hand, capacity due to capacitor-function is 44 mAh/g at 0.1C and decreases gradually to 34 mAh/g at 53C rate. As a consequence, relative contribution of battery-function is rapidly decreased and that of capacitor-function increases gradually with increasing C-rate, as shown in Fig. 3.142C. The hybrid cell exhibited high cyclability, 94% capacity retention after 1000 cycles at 1C rate, and 92% retention after 100,000 cycles at 60C rate.

$LiNi_{0.5}Mn_{1.5}O_4$ nanofibers were prepared by electrospinning of the methanol solution of the mixture of anhydrous lithium acetate, nickel nitrate hexahydrate, and anhydrous manganese acetate with PVP, followed by calcination at 600°C [481]. By coupling with AC prepared from coconut shell with S_{BET} of about 2141 m^2/g, a hybrid cell was constructed using the $LiNi_{0.5}Mn_{1.5}O_4$ nanofibers at the F-electrode. It delivered a maximum energy density of about 19 Wh/kg with a voltage window of 1.5–3.2 V and high cyclability, and capacity retention of about 81% after 3000 cycles at 1 A/g rate.

Since α-$Li_3V_2(PO_4)_3$ (LVP) can serve as both a negative electrode (V^{3+}/V^{2+}) and positive electrode (V^{3+}/V^{5+}), two hybrid cells were constructed with 1 M $LiPF_6$/(EC + DEC), using C-coated LVP as the negative electrode and positive electrode by coupling with an AC as the counter electrodes [482]. In the former cell (C-coated LVP//AC), the principal reactions during charge–discharge are the insertion/deinsertion of Li^+ at LVP as the negative electrode and the adsorption/desorption of PF_6^- at AC as the positive electrode. In the latter (AC//C-coated LVP), however, the deinsertion–reinsertion of Li^+ at LVP as the positive electrode and the adsorption–desorption of Li^+ at AC as the negative electrode. Charge–discharge curves of two hybrid cells are compared in Fig. 3.143A and B,

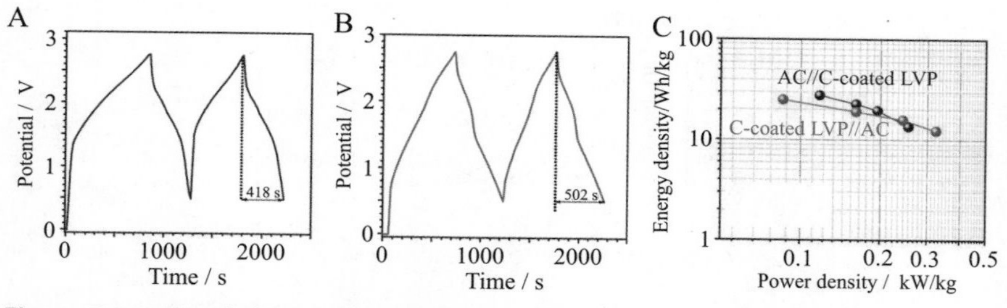

Figure 3.143 Two hybrid cells using C-coated LVP: (A) the first and second charge–discharge curves for AC//C-coated LVP (positive electrode), (B) those for C-coated LVP (negative electrode)//AC, and (C) Ragone plots for two hybrid cells [482].

with the prelithiation process being observed in the first charging on both cells. Even though the principal electrochemical reaction is different between the two cell configurations, energy and power densities are not very different, as shown in Fig. 3.143C. C-coating on Li_2MnSiO_4 was also effective in obtaining improved performance of a hybrid cell coupled with AC [483].

3.3.1.3 Metals

N-doped mesoporous carbons containing metallic Sn nanoparticles in their pores were fabricated from the mixture of the carbons with $SnCl_2$ via an ethanol solution and following heat treatment up to 700°C in Ar [484]. TEM images of the Sn-embedded mesoporous carbon are shown in Fig. 3.144A and B. A hybrid cell was constructed using this Sn-embedded carbon at the F-electrode and microporous AC at the A-electrode with 1 M $LiPF_6$/(EC + DEC) electrolyte. The mesoporous carbon was derived from pomelo peel by carbonization at 700°C and subsequent KOH activation at 900°C, possessing S_{BET} of 2167 m^2/g with V_{total} of 0.98 cm^3/g. The voltage profiles of the hybrid cell at different current densities in Fig. 3.144C reveal a small deviation from the linear slope, suggesting the coexistence of two functions in the F-electrode. It delivered high-performance, high-energy densities of 195.7 and 84.6 Wh/kg at power densities of 0.731and 24.4 kW/kg, respectively, with 70% capacity retention after 5000 cycles at 2 A/g rate.

B-doped Si/SiO_2 composites, where B-doped crystalline Si nanoparticles (the size of about 15 nm) were dispersed in amorphous SiO_2 particles (particle sizes of about 3 μm) were used after carbon coating at the F-electrode to construct hybrid cells coupled with mesoporous carbon

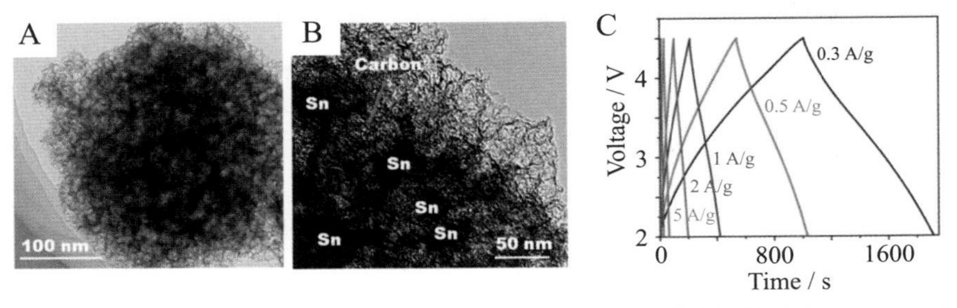

Figure 3.144 Sn-embedded mesoporous carbon: (A) and (B) TEM images, and (C) charge–discharge curves of the hybrid cell coupled with AC [484].

spheres at the A-electrode with 1 M $LiPF_6$/(EC + DEC + DMC) electrolyte [83]. Carbon coating of the composite was carried out by CVD of acetylene at 700°C. The hybrid cell delivered an energy density of 128 Wh/kg at a power density of 1.23 kW/kg and even 89 Wh/kg at 9.7 kW/kg with a potential window of 2.0–4.5 V. It exhibited a high cycle performance, about 93% capacity retention after 100 cycles at 0.6 A/g rate. Amorphous SiO_2 matrix in the composites was thought to work as a buffer for a large volume change of metallic Si during lithiation and delithiation.

3.3.1.4 Metal oxides

Metal oxides, MnO_2, Fe_3O_4, TiO_2, V_2O_5, $MnFe_2O_4$, etc. are electrochemically active through redox reaction of metals with electrolyte ions, but their nanoparticles were strongly demanded for delivering their activity efficiently. These metal oxide nanoparticles were proposed to be loaded onto the surface of porous carbons or confined into nanopores in the carbons to achieve successfully high performances for the hybrid cells.

3.3.1.4.1 Iron oxides

A hybrid cell was composed of Fe_3O_4-loaded rGO nanocomposite at the F-electrode with rGO foam at the A-electrode with m_A/m_F of 4.5/1 in 1 M $LiPF_6$/(EC + DEC + DMC) electrolyte [485]. It works in the potential range of 1.0–4.0 V and gives an energy density of 204 Wh/kg at a power density of 0.055 kW/kg and 122 Wh/kg at about 1 kW/kg, being high in comparison with the symmetric supercapacitor of rGO foams, as shown in Fig. 3.145A. The capacity retention of the hybrid cell is about 70% after

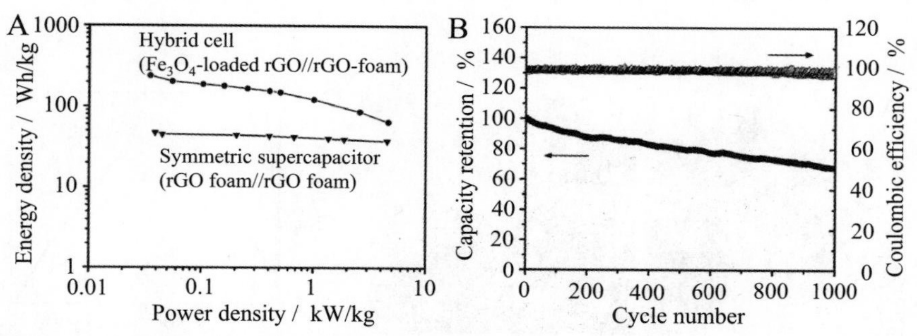

Figure 3.145 Hybrid cell of Fe_3O_4-loaded rGO//rGO-foam: (A) Ragone plot in comparison with the symmetric supercapacitor of rGO foam and (B) cycle performance of the hybrid cell at 2 A/g rate [485].

1000 cycles at 2 A/g rate with a Coulombic efficiency of about 100% during cycles, as shown in Fig. 3.145B. The web of the electrospun PAN-based carbon fibers, of which the pore structure was controlled using an Fe nanoparticle template and by KOH activation to have S_{BET} of more than 2000 m^2/g, was used as the A-electrode and Fe_3O_4 at the F-electrode [486]. The resultant hybrid cell exhibited an energy density of 124.6 Wh/kg at a power density of 0.094 kW/kg and 104 Wh/kg at 4.69 kW/kg. A hybrid cell was constructed from F_3O_4-loaded gelatin-derived carbon at the F-electrode and a microporous carbon derived from egg-white at the A-electrode, which delivered an energy density of 125 Wh/kg at a power density of 17.0 kW/kg with a capacity retention of 88% after 2000 cycles at 5 A/g rate [487].

C-coated α-Fe_2O_3 nanoparticles (sizes of 50–150 nm) were used at the F-electrode together with an AC (S_{BET} of 2190 m^2/g) at the A-electrode in m_A/m_F of 3.8/1.1 to construct a hybrid cell with 1 M $LiPF_6$/PC electrolyte [488]. Carbon-coated α-Fe_2O_3 was prepared by the pyrolysis of ferrocene vapor at 1050°C in Ar flow, followed by oxidation in an O_2/N_2 flow at 430–700°C, which was successfully used as electrode material for LIB [143]. The hybrid cell could be operated in a potential range of 0–3.4 V, but cycling stability was limited to about 2500 cycles because of the degradation of the F-electrode [488]. A hybrid cell of the F-electrode of C-coated Fe_2O_3 pyrolyzed at 800°C and the A-electrode of MgO-templated porous carbon after heating at 800°C in a NH_3 flow delivered an energy density of 65 Wh/kg at a power density of 0.37 kW/kg and 31 Wh/kg at 9.2 kW/kg [489]. rGO-wrapped α-Fe_2O_3 was prepared in GO aqueous suspension with polyethylenimine-coated Fe_2O_3 through the electrostatic interaction between negatively charged GO flakes and positively charged α-Fe_2O_3 microparticles in an aqueous solution, followed by HI reduction [490]. By dissolving out Fe_2O_3 particles from rGO-wrapped Fe_2O_3 by HCl, crumpled rGO foam was obtained. A hybrid cell was assembled by placing rGO-wrapped α-Fe_2O_3 at the F-electrode and crumpled rGO foam at the A-electrode with 1 M $LiPF_6$/(EC + DEC + DMC) electrolyte. It displays charge–discharge curves consisting of a triangular shape with negligible IR drop in the potential range of 1–4 V and a rate performance with the capacitance retention of about 66%, capacitance of 238 F/g at a current density of 0.2 A/g decreasing to about 160 F/g at a very high current of 50 A/g, as shown in Fig. 3.146A and B. It shows good cycle stability with a capacity retention of 87% after 2000 cycles at 1 A/g rate, as shown in Fig. 3.146C.

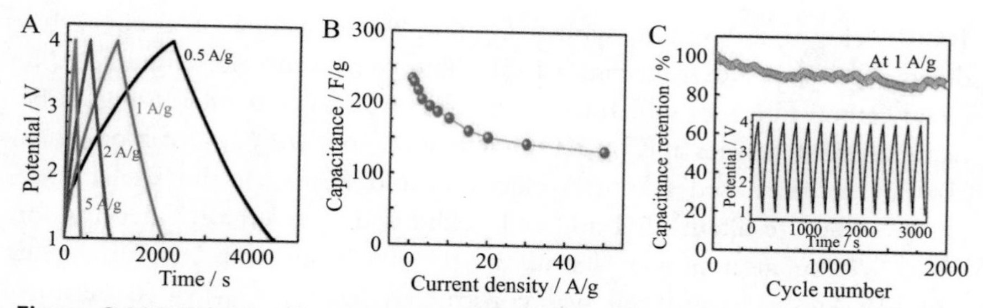

Figure 3.146 Hybrid cell rGO-wrapped α-Fe_2O_3//rGO foam: (A) charge–discharge curves at different current densities, (B) rate performance, and (C) cycle performance at 1 A/g rate with the potential profile (inset) [490].

It delivered an energy density of 121 Wh/kg at a power density of 0.2 kW/kg and 60.1 Wh/kg at 18.0 kW/kg. Nanoparticles of β-FeOOH worked also as an F-electrode by hybridizing with the A-electrode of an AC in 1 M $LiPF_6$/(EC + DMC) electrolyte, delivering an energy density of 45 Wh/kg in the potential range of 0–2.8 V with a capacity retention of about 96% after 800 cycles with 4C rate [491].

3.3.1.4.2 Manganese oxides

Three cells were constructed by the composite of MnO_2 with MWCNT (M/M composite) and MWCNT with 1 M $LiSO_4$/(EC + DEC) electrolyte, i.e., symmetric cells of M/M//M/M and MWCNT//MWCNT, and an asymmetric hybrid cell composed of M/M at the F-electrode and MWCNTs at the A-electrode [492]. The asymmetric hybrid cell M/M//MWCNT delivered a higher capacitance than two symmetric cells of MWCNTs and M/M composites in the range of discharge current density employed (1–10 mA/cm^2) with similar cycle performance, as shown in Fig. 3.147A and B. A hybrid cell composed of MnO_2 at the F-electrode and

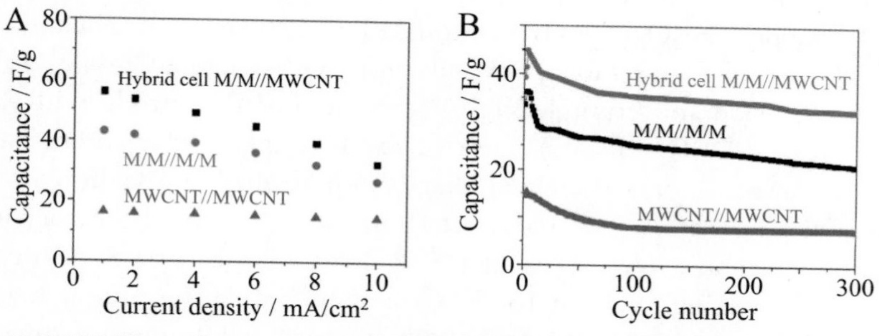

Figure 3.147 The cells composed of MnO_2/MWCNT composite (M/M) and MWCNT, i.e., hybrid cell of M/M composite//MWCNT and two symmetric cells of M/M//M/M and MWCNT//MWCNT: (A) rate and (B) cycle performances [492].

an AC (S_{BET} of 2800 m^2/g) at the A-electrode with 0.1 M K_2SO_4 aqueous electrolyte delivered an energy density of 17.3 Wh/kg at a power density of 0.605 kW/kg in the potential rang of 0–2.2 V [493].

Polyester textile fibers were coated by graphene flakes by sonicating with graphite powder in $NaClO_4$ aqueous solution, on which MnO_2 nanoparticles were electrochemically deposited. The resultant composites were applied to a hybrid cell at the F-electrode by coupling with SWCNTs at the A-electrode in 0.5 M Na_2SO_4 aqueous electrolyte [494]. The polyester textile used was intertwined fabric fibers, and the graphene-coated flakes had lateral sizes of about 0.3–1 μm and thickness of about 5 nm, and MnO_2 nanoparticles deposited were 300–800 nm. The capacitance of the graphene/MnO_2 composite textile was measured using a half-cell with different scan rates up to 100 mV/s which is much higher than graphene-coated textile without MnO_2. A hybrid cell composed of this composite textile with SWCNTs showed a maximum energy density of 12.5 Wh/kg and a maximum power density of 110 kW/kg with cycle performance of about 95% retention over 5000 cycles.

MnO_2-nanowire/rGO composite was fabricated by mixing colloidal exfoliated MnO_2 with GO flakes (less than three layers) in solution, which possessed mesoporous structure with a broad pore size distribution of 2–50 nm [390]. CV curves are measured on the rGO and the composite MnO_2-nanowire/rGO using three-electrode cells with 1 M Na_2SO_4 aqueous electrolyte, as shown in Fig. 3.148A, revealing a typical EDL formation behavior with a voltage window of −1.0 to 0.4 V for rGO and a quite different profile for MnO_2/rGO composite with a voltage window of 0.0–1.0 V which is due to the coexistence of pseudocapacitance of MnO_2

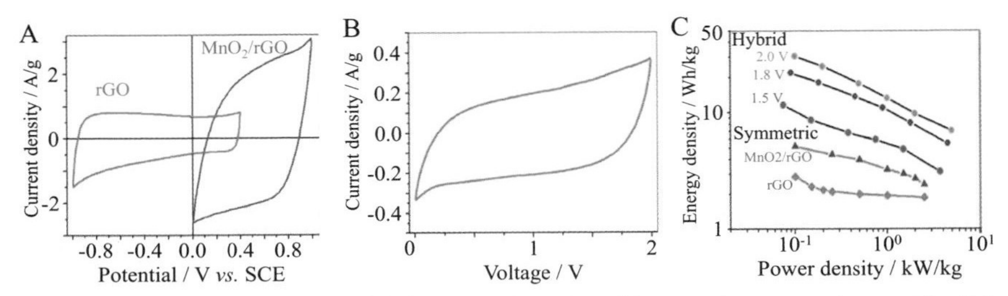

Figure 3.148 MnO_2-nanowire/rGO composites: (A) CV curve in comparison with rGO in a three-electrode cell in a 1 M Na_2SO_4 solution at a scanning rate of 10 mV/s, (B) CV curve of an asymmetric cell of the composite//rGO, and (C) Ragone plots for the asymmetric cell with different voltage ranges, in comparison with two symmetric cells of the composite and rGO [390].

and EDL formation on rGO flakes. Hybrid cells were assembled using the composite at the F-electrode and rGO at the A-electrode with m_F/m_A of 7/13. A representative CV curve on the hybrid cell is shown in Fig. 3.148B, possessing a wide voltage window of 0–2 V. The charge–discharge curves in this voltage range were nearly linear and stable, retaining a capacitance of 24.5 F/g at 0.5 A/g rate, even after 100 cycles. In Fig. 3.148C, Ragone plots for the hybrid cell measured with different voltage windows, up to 2.0, 1.8, and 1.5 V, at 0.1–5 A/g rate are compared with symmetric supercapacitors of the MnO_2-nanowire/rGO composite and rGO. The hybrid cell delivers a maximum energy density of 30.4 Wh/kg with a cell voltage of 2.0 V, which is much higher than two symmetric cells using the same electrode materials. Hybrid cells were constructed from MnO_2 powder at the F-electrode and AC (S_{BET} of 2800 m^2/g) at the A-electrode with 0.1 M K_2SO_4 aqueous electrolyte, which delivered a capacitance of 31 F/g with a voltage window of 0–2.2 V [493]. A hybrid cell composed of MnO_2/graphene composite and ACFs with 1 M Na_2SO_4 aqueous electrolyte delivered a capacitance of 113.5 F/g at 1 mV/s rate and a high capacitance retention of 97.3% after 1000 cycles [495]. Its energy density reaches 51 Wh/kg at a power density of 0.102 kW/kg, and still remains at 8.2 Wh/kg at a high power density of 16.5 kW/kg.

The composite MnO_2-nanowire/SWCNT was used to construct hybrid cells by coupling with In_2O_3 nanowire/SWCNT composite that exhibited a capacitance of 184 F/g, maximum energy density of 25.5 Wh/kg, and maximum power density of 50.3 kW/kg [496]. MnO_2/SWCNT composites were fabricated by injection of the solution containing SWCNTs, MnO_2, and sodium cholate into a rotating ethanol coagulation bath through a nozzle to produce gel-like fibers, followed by keeping them in water to eliminate residual sodium cholate and drying at 60°C [497]. CV curves measured by a three-electrode cell with 1 M LiCl electrolyte are shown for the composite with different MnO_2 contents from 0 to 75 wt.% in Fig. 3.149A, revealing that the curves have a quasi-rectangular shape up to an MnO_2 content of 67 wt.%. Galvanostatic charge–discharge curves were nearly symmetrical and the volumetric capacitance increased with increasing MnO_2 content, from 39 F/cm^3 for the pristine SWCNT fibers (without MnO_2) to 428 F/cm^3 for the composite of SWCNTs with 67 wt.% MnO_2. Solid-state hybrid cells constructed from MnO_2/SWCNT//SWCNTs using PVA-LiCl gel electrolyte exhibited a volumetric capacitance of 74.8 F/cm^3, cycling stability of 94% capacitance retention over 5000 cycles at 5 A/cm^3, and energy and power densities of 10.4 mWh/cm^3 and

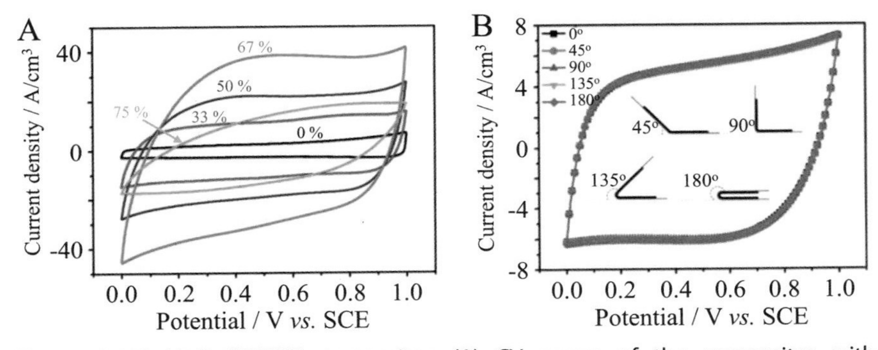

Figure 3.149 MnO_2/SWCNT composites: (A) CV curves of the composites with different MnO_2 contents in 1 M LiCl at a scan rate of 100 mV/s, and (B) CV curves of the solid-state hybrid cell measured for different bending angles [497].

5 W/cm^3, respectively, in addition to their flexible and wearable characteristics. No changes in CV curves are observed on the cell at different bending angles, as shown in Fig. 3.149B.

MnO-loaded carbon nanosheets (73 wt.% MnO with about 42 nm particle size) were prepared by reducing MnO_2 nanoparticles loaded on carbon nanosheets in an H_2 flow [498]. Carbon nanosheets were obtained from agricultural wastes (hemp bast fibers) by carbonization under hydrothermal conditions at 180°C, followed by KOH activation at 700–800°C [499]. A hybrid cell was constructed from these MnO-loaded carbon nanosheets at the F-electrode and the pristine carbon nanosheets at the A-electrode, which delivered an energy density of 184 Wh/kg at a power density of 0.083 kW/kg and 90 Wh/kg at 15 kW/kg with 76% capacity retention after 5000 cycles [498].

C-coated $MnFe_2O_4$ nanocubes, which were synthesized from a mixture of $MnCl_2$ and $FeCl_3$ with sodium oleate ($C_{18}H_{33}NaO_2$) via their aqueous solution by calcining at 600°C in N_2. Were used as the F-electrode by coupling with an N-doped porous carbon prepared from glucose with urea at 950°C at the A-electrode [500]. C-coated $MnFe_2O_4$ was confirmed using a half-cell to give a high discharge capacity of 1237 mAh/g with a small irreversible capacity (about 99 mAh/g). This capacity observed on C-coated $MnFe_2O_4$ was higher than the theoretical capacity (917 mAh/g), probably due to a contribution from coated carbon. A hybrid cell of this C-coated $MnFe_2O_4$ with N-doped porous carbon can operate in a wide range of potential (about 4 V) and deliver much higher capacitance, in comparison with hybrid cells of this F-electrode coupled with either an rGO or a commercial AC, as shown in Fig. 3.150A and B. The hybrid cell

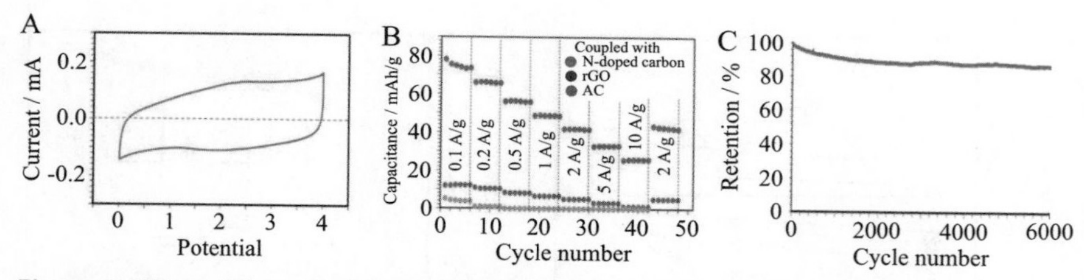

Figure 3.150 Hybrid cell of C-coated $MnFe_2O_4$//N-doped porous carbon: (A) CV curve at scan rate at 1 mV/s, (B) rate performance in comparison with the same F-electrode coupled with other A-electrodes, and (C) cycle performance at a current density of 2 A/g [500].

has also high cyclability with capacity retention of 86.5% after 6000 cycles at 2 A/g rate, as shown in Fig. 3.150C. The 4 V assemblage of these hybrid cells delivered high energy density of 157 Wh/kg at a power density of 0.2 kW/kg.

$K_{0.27}MnO_2 \cdot 0.6H_2O$, which was synthesized by heat treating the mixture of K_2CO_3 and MnO_2 in a molar ratio of 1/2 at 550°C for 8 h, was found to show the reversible electrochemical intercalation/deintercalation of K^+ in K_2SO_4 aqueous electrolyte [501]. The hybrid cell of $K_{0.27}MnO_2 \cdot 0.6H_2O$ at the F-electrode (positive electrode) and an AC (S_{BET} of 2800 m^2/g) at the A-electrode (negative electrode) with 0.5 M K_2SO_4 aqueous electrolyte delivered an energy density of 25.3 Wh/kg at a power density of 0.14 kW/kg and 17.6 Wh/kg at 2 kW/kg with a capacitance retention of more than 98% after 10,000 cycles at 25C rate between 0 and 1.8 V.

3.3.1.4.3 Vanadium oxides

The intertwined composite of MWCNTs with V_2O_5 nanowires was prepared from HCl solution of NH_4VO_3 with the surfactant P123 and MWCNTs under hydrothermal conditions at 120°C [502]. MWCNTs had diameters of around 2–30 nm and lengths up to the micrometer scale, which were functionalized by attaching carboxylic groups on their surfaces, V_2O_5 nanowires had diameters of 20–50 nm with lengths up to tens of micrometers. Electrochemical performances of the hybrid cell composed of this composite V_2O_5/MWCNT (MWCNT content of 33 wt.%) with silica-templated mesoporous carbon loaded by MnO_2 nanoparticles (MnO_2/carbon composite) with 1 M Na_2SO_4 aqueous electrolyte were studied. The hybrid cells have much higher capacitance than the symmetric cells constructed from V_2O_5/MWCNT composite and MWCNTs, but it demonstrates a strong dependence of capacitance on current density, as

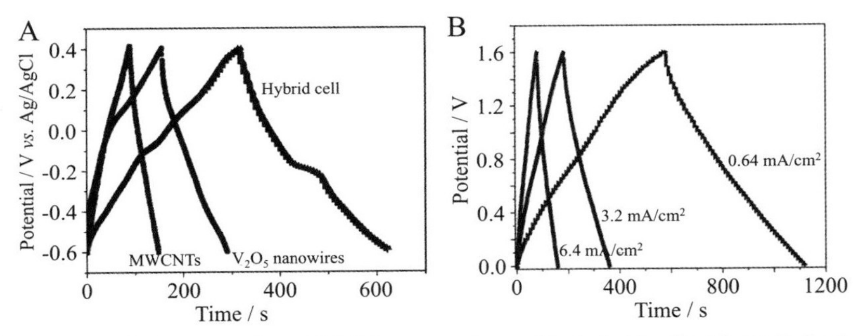

Figure 3.151 V_2O_5 nanowires/MWCNT composite and its hybrid cell with MnO_2/carbon composite: (A) charge–discharge curves of the hybrid cell in comparison with symmetric cells of the V_2O_5 nanowires and MWCNTs, and (B) those of the hybrid cell with different current densities [502].

shown in Fig. 3.151A and B, respectively. For the hybrid cell, relatively high energy densities of 16 and 5.5 Wh/kg at a power density of 0.075 and 3.75 kW/kg, respectively, were obtained. A hybrid cell using this V_2O_5/MWCNT composite coupled with an AC (S_{BET} of 1900 m^2/g) was also studied in 1 M $LiClO_4$/PC electrolyte, which gave energy densities of 40 and 6.9 Wh/kg at power densities of 0.21 and 6.3 kW/kg, respectively [503].

A hybrid cell was composed of amorphous V_2O_5, which was prepared by quenching the melt of crystalline V_2O_5 powder at 800°C into water, with an AC (V_2O_5/AC weight ratio of 1) in aqueous electrolytes of sulfates and nitrates of alkali metals, Li, Na, and K [504]. The electrolyte had some effects on the capacitive performances, as well as current density, potential range, and optimum mass ratio m_A/m_F. Capacitance values of the hybrid cells were larger in nitrate solutions than in the corresponding sulfate solutions, for example, it was 32.5 F/g in 2 M $NaNO_3$ while it was 29.3 F/g in Na_2SO_4 at 200 mA/g rate, probably due to the smaller size and lower quantity of electric charge of NO_3^- than SO_4^{2-}.

3.3.1.4.4 Molybdenum oxides

MoO_2 films with different crystallinities were synthesized from MoO_3 powder by solvothermal treatment at different temperatures (160, 180, and 200°C) in a water/ethylene glycol mixture [505]. As shown in Fig. 3.152A, the CV curve of the LIB half-cell shows two cathodic peaks at 1.50 and 1.23 V which can be attributed to stepwise insertion of Li^+ into the monoclinic MoO_2 crystal and the peaks at 1.44 and 1.74 V in the anodic

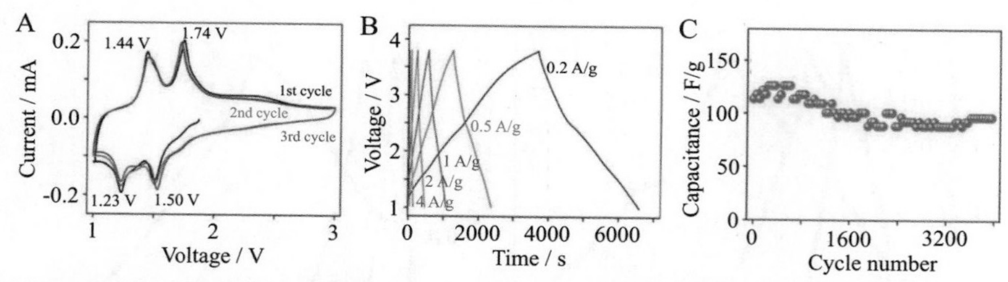

Figure 3.152 MoO_2 film electrode: (A) CV curve at 0.2 mV/s measured on MpO_2 by a half-cell, (B) charge–discharge curves at different current densities, and (C) cycle performance for the hybrid cell of MoO_2//AC with 1 M $LiPF_6$/(EC + DMC) at 5 A/g rate [505].

sweep can be attributed to reversible extraction of Li^+. A hybrid cell of MoO_2 at the F-electrode with an AC (S_{BET} of 1800 m^2/g) at the A-electrode achieves steady charge–discharge performances, as shown in Fig. 3.152B and C. It delivered an energy density of 150 Wh/kg at a power density of 0.2 kW/kg and 30 Wh/kg at about 7 kW/kg.

The composite foams of MoO_3 with rGO were prepared by hydrothermal treatment of the mixture of MoO_3 nanobelts with GO at 200°C in different ratios [430]. The composite foams of the PANI nanotube with rGO were prepared by the same process at 180°C. The hybrid cells composed of these two composites delivered an energy density of 128 Wh/kg at a power density of 0.182 kW/kg and 44 Wh/kg at 13.5 kW/kg, with an operating voltage range of 0.0–3.8 V and a long-term cycle life (about 90% capacity retention after 3000 cycles at 1 A/g rate). On the composite MoO_3/rGO containing about 11 wt.% rGO, CV curves were attempted to be divided into the capacitive-type part and battery-type part. The fraction of the latter decreased from about 39% to 14% with increasing current density from 0.1 to 1.0 A/g.

Hybrid cells were constructed from MoO_3/rGO composite at the F-electrode and MnO_2/rGO composite at the A-electrode with H_3PO_4/PVA gel electrolyte [506]. Conformation of a flexible hybrid cell from fiber-shaped composites is shown in Fig. 3.153A. CV curves measured in 1 M H_2SO_4 are shown for each fibrous composite in Fig. 3.153B, revealing that the stable voltage windows of the MoO_3/rGO and MnO_2/rGO electrodes are −0.8 to 0 and 0 to 0.8 V (vs. Hg/Hg_2SO_4), respectively. The hybrid cell composed of these two composites exhibits a stable operating window up to 1.6 V, with the capacitance increasing from 27 to 43 F/cm^3 with increasing voltage window from

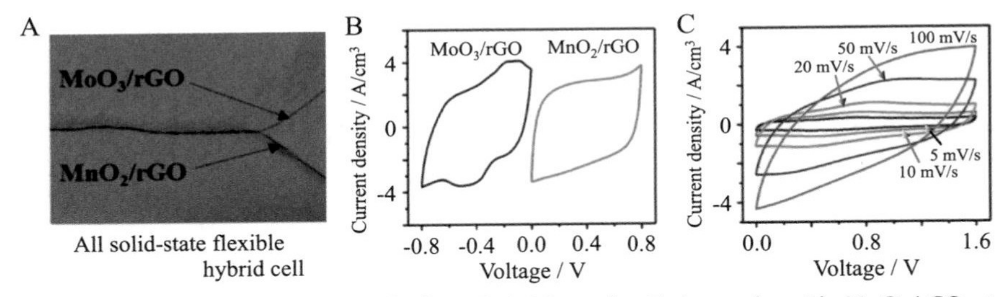

Figure 3.153 Fibrous hybrid cell of MoO_3/rGO at the F-electrode with MnO_2/rGO at the A-electrode with H_3PO_4/PVA gel electrolyte: (A) photo of the cell, (B) CV curves for each composite measured by three-electrode cells in 1 M H_2SO_4, and (C) CV curves of the hybrid cell at different scan rates [506].

0.8 to 1.6 V. The CV curves of the hybrid cell with the voltage window of 0.0−1.6 V at different scan rates (Fig. 3.153C) show typical Faradaic pseudocapacitive shape at a low scan rate, suggesting fast redox reactions of the metal oxides. The hybrid cell delivered a volumetric energy density of 18.2 mWh/cm^3 at a power density of 76.4 mW/cm^3.

3.3.1.4.5 Titanium oxides

Anatase-type TiO_2 nanoparticles were deposited onto the surface of GO flakes in an aqueous solution of $TiCl_4$ with a following reduction with hydrazine and ammonia [507]. Its hybrid cell coupled with AC delivered an energy density of 41 Wh/kg at a power density of 0.8 kW/kg and 0.9 Wh/kg at 8 kW/kg in $LiPF_6$/(EC + DMC) electrolyte. Ultrafine TiO_2 nanoparticles embedded in mesoporous carbon nanofibers, which were prepared by electrospinning from the mixture of tetrabutyltitanate with PVP and TEOS as template precursors, delivered an improved performance in its hybrid cell coupled with an AC [508]. Anatase-type TiO_2 mesocages deposited on rGO were prepared by microwave-assisted solvothermal treatment of the mixture of $TiOSO_4$ with commercial rGO flakes (O content of 7−7.5 at.%) at 110°C [249]. Its hybrid cell coupled with AC at the A-electrode in 1 M $NaClO_4$/(EC + PC) delivered an energy density of 64 Wh/kg at a power density of 0.056 kW/kg and 26 Wh/kg at 1.36 kW/kg with more than 90% retention after 10,000 cycles at 10C. Sodium titanate ($Na_2Ti_3O_7$) nanotubes, which were synthesized by hydrothermal treatment of commercial TiO_2 nanoparticles (P25, particle size of about 30 nm) with 10 M NaOH solution at 150°C, were used at the F-electrode in a hybrid cell with AC at the A-electrode in 1.5 M $NaClO_4$/(PC + DMC) [509]. The cell gave a maximum capacity of about

70 mAh/g at the optimized m_A/m_F above 4. It delivered an energy density of 34 Wh/kg at a power density of 0.889 kW/kg with a capacity retention of 80% after 1000 cycles.

$H_2Ti_6O_{13}$ nanowires, which were synthesized from a 15 M KOH solution of dispersed TiO_2 in an autoclave at 170°C and were 5–7 nm in diameter with 50–150 nm in length, were coupled with an ordered mesoporous carbon CMK-3 (A-electrode) in 1 M $LiPF_6$/(EC + DEC) [510]. As shown in Fig. 3.154A, CMK-3 gives a linear potential change at the range of 3.0–4.5 V due to adsorption of PF_6^- and $H_2Ti_6O_{13}$ nanowires give a sharp potential decrease at the range of 3.0–2.0 V and then a gradual decrease down to 1.0 V due to insertion of Li^+ into $H_2Ti_6O_{13}$ nanowire host at the F-electrode. The hybrid cell composed from these two electrode materials can give a high energy density of 90 Wh/kg and a maximum power density of 11 kW/kg, much higher than those achieved by the symmetric supercapacitor of CMK-3//CMK-3, as shown in Fig. 3.154B. H_2-treated TiO_2 nanoparticles wrapped by polypyrrole and SWCNTs delivered improved performance of the hybrid cell coupled with an AC [511].

Nanowire arrays of titanate and TiO_2 were synthesized on Ti-foil in 1 M NaOH aqueous solution under hydrothermal conditions at 220°C, followed by annealing at 150, 300, and 450°C in air, to obtain self-supported nanowire arrays of $H_2Ti_8O_{17}$ (HTO), TiO_2–B and a mixture of anatase-type TiO_2 and TiO_2–B (TiO_2-(A + B)), respectively [512]. Hybrid cells were constructed from these nanowire arrays at the F-electrode

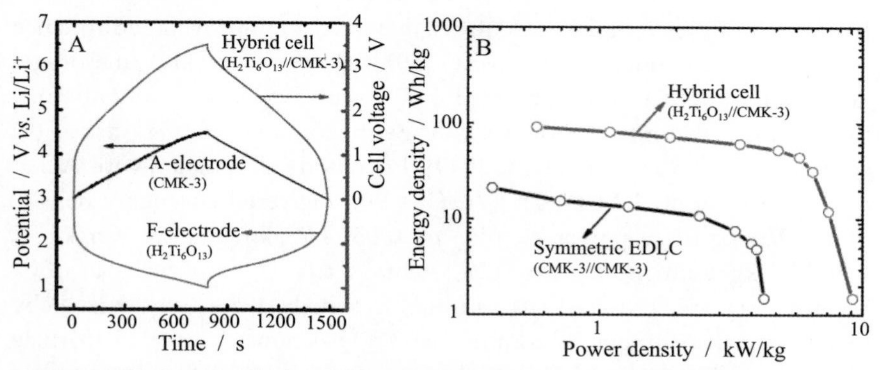

Figure 3.154 Hybrid cell of $H_2Ti_6O_{13}$ and CMK-3: (A) charge–discharge curves of the individual electrode ($H_2Ti_6O_{13}$ and CMK-3), along with the voltage profile of the hybrid cell at a current density of 0.2 A/g, and (B) Ragone plots for the hybrid cell and the symmetric supercapacitor of CMK-3 [510].

and an AC at the A-electrode with 1 M $LiPF_6$/(EC + DMC) electrolyte. The hybrid cell HTO//AC exhibits much higher capacity than other related hybrid cells, TiO_2−B//AC and TiO_2-(A + B)//AC, as shown by the rate and cycle performances in Fig. 3.155. The cell HTO//AC achieved a relatively high energy density of 93.8 Wh/kg with a capacity retention of 78.8% after 3000 cycles at a high current density of 5.0 A/g with a potential window of 0.0−3.0 V. C-coated $H_2Ti_{12}O_{25}$ worked as the F-electrode in a hybrid cell with AC in 1.5 M $LiBF_4$/(EC + DMC) [513]. $H_2Ti_{12}O_{25}$ powder prepared from $Na_2Ti_3O_7$ via Na^+/H^+ ion exchange reaction using a 1 M HCl solution at 60°C was coated with carbon through calcination of the mixture with β-cyclodextrin at 800°C. The hybrid cell using 2.5 wt.% C-coated $H_2Ti_{12}O_{25}$ gave an energy density of 38.8 Wh/kg at a power density of 0.182 kW/kg and 5.7 Wh/kg at 5.4 kW/kg with stable cycle performance.

A hybrid cell was fabricated from TiO_2−B nanowires at the F-electrode by coupling with MWCNTs at the A-electrode, the former being synthesized via hydrothermal treatment of anatase-type TiO_2 in NaOH solution at 180°C and the latter by catalytic CVD of ethylene at 550°C [514]. The hybrid cell had better capacitive performances than the symmetric supercapacitor of MWCNT//MWCNT; higher energy and power densities with comparable cycle performance. TiO_2−B nanotubes were also used in a hybrid cell by coupling with MWCNTs [515]. A hybrid cell of TiO_2−B powder and an AC (S_{BET} of 2300 m^2/g, V_{total} of 1.32 cm^3/g including V_{meso} of 0.72 cm^3/g) with 1 M $LiPF_6$/(EC+2DEC) exhibited an energy density of 45−80 Wh/kg with a power density of 0.24−0.42 kW/kg in the voltage range of 2.75−3.5 V [516]. A hybrid cell of TiO_2−B nanorods and an AC (m_A/m_F of 5.9/1) delivered the maximum

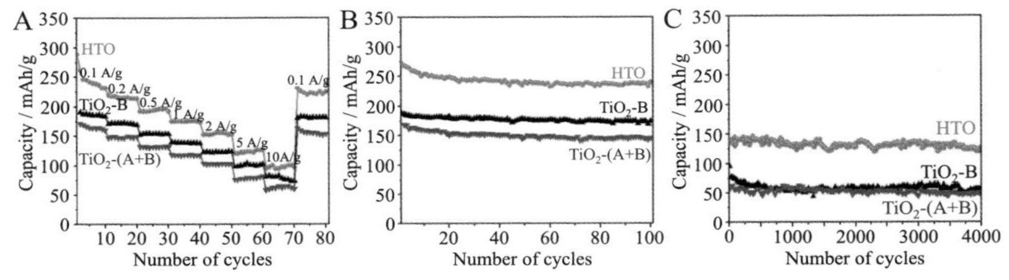

Figure 3.155 Hybrid cells of either HTO, TiO_2−B, or TiO_2-(A + B) at the F-electrode and an AC at the A-electrode with 1 M $LiPF_6$/(EC + DMC) electrolyte: (A) rate performances, and (B) cycle performances at 0.2 A/g rate and (C) those at 5 A/g [512].

energy and power densities of around 23 Wh/kg and 2.8 kW/kg, respectively [517]. Li^+ insertion behavior of mesoporous TiO_2–B, which was synthesized using a block copolymer (P123), was compared with that of nonporous TiO_2–B. The former exhibited a higher overall capacity although both showed pseudocapacitive behavior [518].

3.3.1.4.6 Other oxides

Mesoporous Nb_2O_5/C composites were prepared from the mixture of niobium ethoxide and niobium chloride with a block copolymer (PEO-*b*-PS) by carbonization at 700°C in N_2 [519]; the wall of mesopores of Nb_2O_5 was coated by carbon formed from the block copolymer (Fig. 3.156A). The hybrid cells composed from this composite at the F-electrode and an AC at the A-electrode with 1 M $LiPF_6$/(EC + DMC) exhibited an energy density of 15 Wh/kg at a power density of 18.5 kW/kg with a potential window of 1.0–3.5 V and a marked rate dependence, as shown in Fig. 3.156B and C. With 1 A/g rate, capacity retention was about 90% after 1000 cycles. The composite of Nb_2O_5-loaded rGO showed high electrochemical performance in 1 M $LiPF_6$/(EC + DMC) and its hybrid cell coupled with an AC at the A-electrode delivered maximum energy and power densities of 29 Wh/kg and 2.9 kW/kg, respectively [520]. A composite of carbon cloth coated by Nb_2O_5 nanorods was fabricated through hydrothermal treatment at 200°C and following annealing at 450–800°C [521]. Hybrid cells composed of this composite at the F-electrode and an AC at the A-electrode with m_A/m_F of 3.5/1 delivered an energy density of 65 Wh/kg at a power density of 5.4 kW/kg. They were flexible, maintaining their charging–discharging features without the capacity decay during bending from 180 to 30 degrees. The composite of Nb_2O_5 nanowires covered with ultrathin graphene sheets by CVD at 500°C showed a fast pseudocapacitive reaction process and its hybrid cell

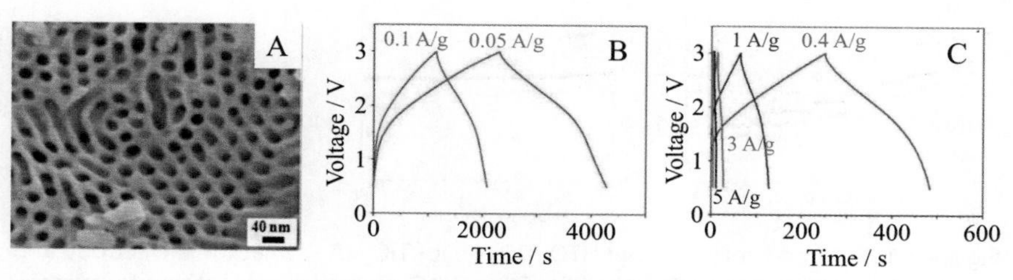

Figure 3.156 Hybrid cell of Nb_2O_5/C composite with AC: (A) SEM image of the composite, (B) and (C) charge–discharge curves with different rates [519].

coupled with an AC (m_A/m_F of 4/1) in 1 M $NaClO_4$/(EC + DMC) electrolyte presented an energy density of 113 Wh/kg at a power density of 0.080 kW/kg and 62 Wh/kg at 5.33 kW/kg [522].

Composite fibers prepared from a DMF solution of polyacrylonitrile (PAN), niobium pentachloride ($NbCl_5$), and tetrabutyltitanate [$Ti(OC_4H_9)_4$] by electrospinning, followed by heat treatment at 1000°C in air (TNO/C composite fibers) [523]. A hybrid cell of these composite fibers at the F-electrode with electrospun PAN-based carbon fibers at the A-electrode with 1 M $LiPF_6$/(EC + DMC) was constructed. It exhibited an energy density of 110 Wh/kg at a power density of 0.1 kW/kg and 20 Wh/kg even at 5.46 kW/kg, in the potential window of 1.0–3.0 V.

TiP_2O_7 crystal could accommodate Li ions electrochemically, as many lithium compounds, Fe_3O_4, V_2O_5, etc. do [524]. Galvanostatic charge–discharge curves of a half-cell of its nanoparticles are shown at a constant current density of 15 mA/g between 2 and 3.4 V at room temperature in Fig. 3.157A, revealing steady charge–discharge cycles up to the 100th cycle with a capacity of 60 mAh/g. A hybrid cell was constructed using this TiP_2O_7 at the F-electrode with an AC (S_{BET} of 880 m^2/g) at the A-electrode in 1 M $LiPF_6$/(EC + DMC). Its CV curves at different scan rates are shown in Fig. 3.157B. It delivered a discharge capacitance of 29 F/g, stable in the range of 100–500 cycles, and a Ragone plot showed maximum energy and power densities of 13 Wh/kg and 0.37 kW/kg, respectively.

3.3.1.5 Other materials

A hybrid cell was constructed using N-doped mesoporous carbon encapsulating metal carbide Co_3ZnC nanoparticles at the F-electrode and a

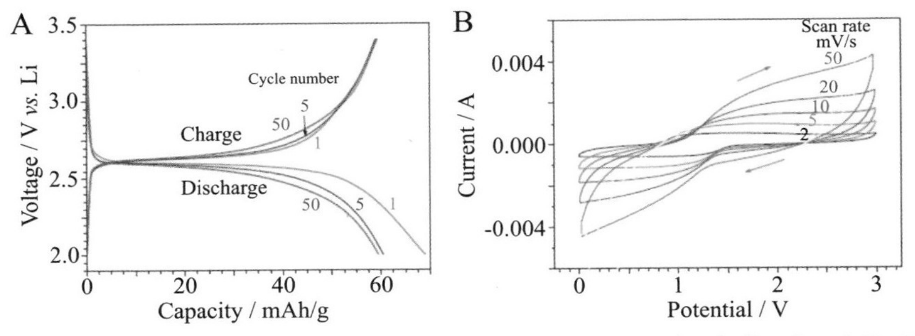

Figure 3.157 TiP_2O_7 electrode: (A) charge–discharge curves of its half-cell and (B) CV curves of its hybrid cell with an AC [524].

microporous carbon at the A-electrode, which exhibited a maximum energy density of 141 Wh/kg and maximum power density of 10.3 kW/kg with a wide operating voltage range of 1.0–4.5 V [525].

Co_9S_8-loaded rGO composite was prepared by hydrothermal treatment of the mixture of GO, cobalt nitrate, urea, sodium hydrophosphate, and sodium disulfide at 160°C, which was used as the F-electrode to construct a hybrid cell with rGO at the A-electrode [526]. The composite exhibits bifunctional electrochemical behavior, as shown by the charge–discharge curves measured using its half-cell in Fig. 3.158A. Fig. 3.158B shows CV curve for the rGO at the A-electrode, showing capacitive behavior, and that for the composite at the F-electrode, bifunctional behavior. CV curves of the hybrid cell at different scan rates are shown in Fig. 3.158C, revealing that the cell works at the voltage window of 0.0–1.5 V and its capacity depends greatly on the scan rate. The hybrid cell delivered an energy density of 37 Wh/kg at a power density of 0.17 kW/kg and 15 Wh/kg at 12 kW/kg.

MoS_2-loaded rGO was prepared by the solvothermal process of the mixture of GO with ammonium tetrathiomolybdate in DMF solution at 190°C, followed by vacuum freeze-drying and annealing at 800°C in Ar [527]. LIB and SIB using MoS_2-loaded rGO could deliver reversible capacities of 1210 and 585 mAh/g, respectively. A hybrid cell was assembled using this MoS_2-loaded rGO at the F-electrode with N-doped rGO foam at the A-electrode with either 1 M $LiPF_6$ or $NaClO_4$ in EC + DMC solution. At the A-electrode of the hybrid cell, mainly an adsorption reaction of PF_6^- or ClO_4^- occurs with a linear increase in potential from 2 to 4.5 V, and at the F-electrode intercalation reaction of Li^+ or Na^+ occurs, resulting in a smooth potential plateau from 0.5 to near 0 V, as shown schematically in Fig. 3.159A. The hybrid device, therefore, exhibited nearly

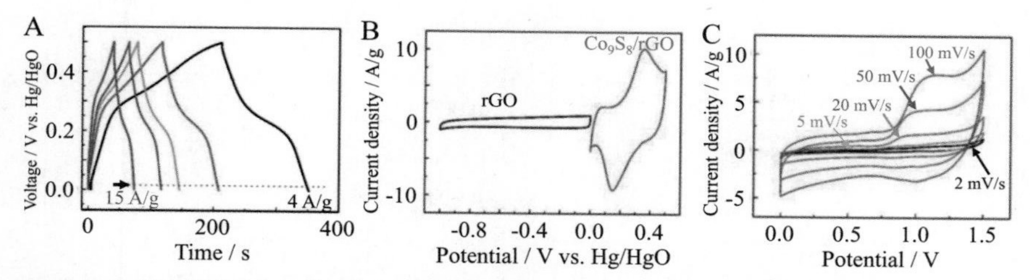

Figure 3.158 Hybrid cell of Co_9S_8-loaded rGO composite with rGO: (A) charge–discharge curves for the composite (half-cell), (B) CV curves for the rGO at the A-electrode and for the composite at the F-electrode, and (C) CV curves of the hybrid cell at different scan rates [526].

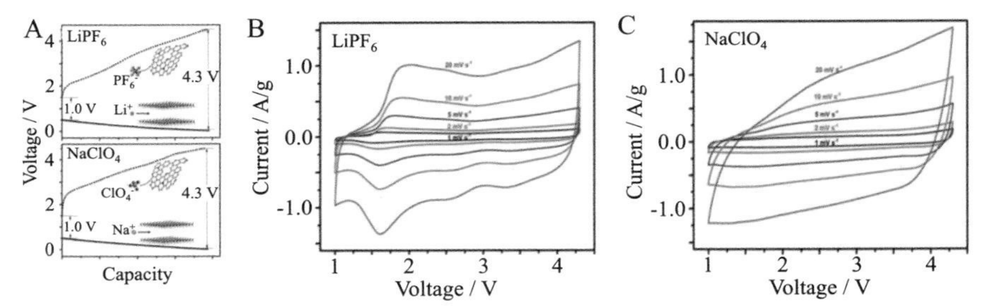

Figure 3.159 Hybrid cells of MoS_2-loaded rGO with N-doped rGO foam with $LiPF_6$/(EC + DMC) and $NaClO_4$/(EC + DMC): (A) charge–discharge curves with schema of interaction of cation and anion and CV curves with different scan rates in (B) $LiPF_6$ and (C) $NaClO_4$ [527].

linear charge–discharge curves over a voltage window of 1–4.3 V. Fig. 3.159B and C show the CV curves at various scan rates in $LiPF_6$ and $NaClO_4$, respectively. At low scan rates, the curves show rectangular shapes in both electrolytes. With increasing scan rate, however, the CV curve in $LiPF_6$ gradually deviates from the ideal rectangular shape, but the CV curve in $NaCiO_4$, by contrast, is not seriously distorted. Both hybrid cells showed high energy densities of over 100 Wh/kg, as well as high power densities of over 50 kW/kg. Composites of MoS_2, carbon fiber (CF), and carbon (C) were fabricated [528]. MoS_2/CF composite was prepared by electrospinning of the DMF solution of MoS_2 powder and PAN, followed by stabilization at 220°C in air and subsequent carbonization at 700°C in Ar. This MoS_2/CF composite was immersed into an aqueous solution of $Na_2MoO_4\cdot$ with thiourea (CH_4N_2S) and hydrothermally treated at 200°C to prepare MoS_2/CF@MoS_2 composites. Finally, the composite MoS_2/CF@MoS_2 was coated with carbon through heat treatment with oleic acid at 700°C (MoS_2/CF@MoS_2@C composite). These composites were used as the A-electrode of the hybrid cell coupled with graphite at the F-electrode in 1 M $NaPF_6$/(EC + DMC) electrolyte. The hybrid cell composed of MoS_2/CF@MoS_2@C could deliver an initial discharge capacity of 112.3 mAh/g at 0.2 A/g rate and retained a reversible capacity of 90.5 mAh/g at 0.5 A/g after 500 cycles.

3.3.2 Faraday reaction at the positive electrode (A//F-type)

Hybrid cells were constructed using various carbon materials with different graphitization degrees and different particle morphologies, including graphite, at the F-electrode (positive electrode), where intercalation/deintercalation of anions occurred, and an AC (S_{BET} of 1050 m^2/g) at the

A-electrode (negative electrode), where the adsorption/desorption of cations occurred, with 1.5 M $TEMAPF_6$/PC and 1.5 M $TEMABF_4$/PC electrolytes [529,530]. Typical charge–discharge curves for the initial three cycles are shown for the hybrid cell AC//graphite in Fig. 3.160A and those for the symmetric supercapacitor (AC//AC) in Fig. 3.160B. In contrast to the latter cell (supercapacitor) which shows sloping straight charge and discharge lines, the curves for charging and discharging of the former cell (hybrid cell) are composed of at least two lines with different slopes. The line in the range of 2.0–3.5 V is confirmed by in situ XRD measurement to be due to intercalation of anions (PF_6^-) into graphite, as shown in Fig. 3.160C, and the potential range up to 2.0 V is due to adsorption of anions, which contributes to increased energy and power densities for the hybrid cell. Some of the carbon materials used were confirmed to occur the intercalation of anions clearly by XRD, as Fig. 3.160C, but some were not, despite their interlayer spacing d_{002} of about 0.336 nm. In these hybrid cells, the intercalation of anions was the principal reaction at the F-electrode [531], as shown by the

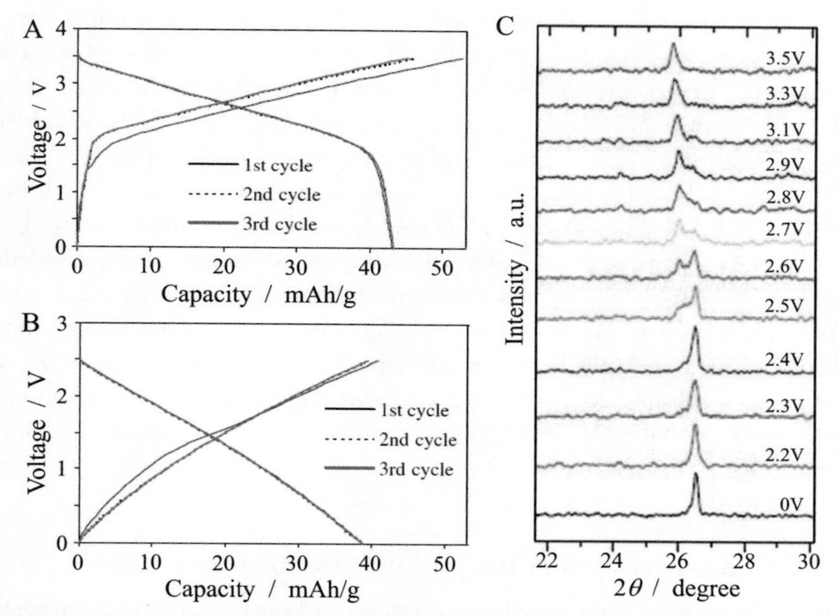

Figure 3.160 Hybrid cell AC//graphite with 1.5 M $TEMAPF_6$/PC electrolyte: (A) typical charge–discharge curves of the initial three cycles for the hybrid cell graphite//AC, (B) that for the symmetric cell AC//AC, and (C) change in XRD pattern for graphite in the F-electrode with cell voltage [529].

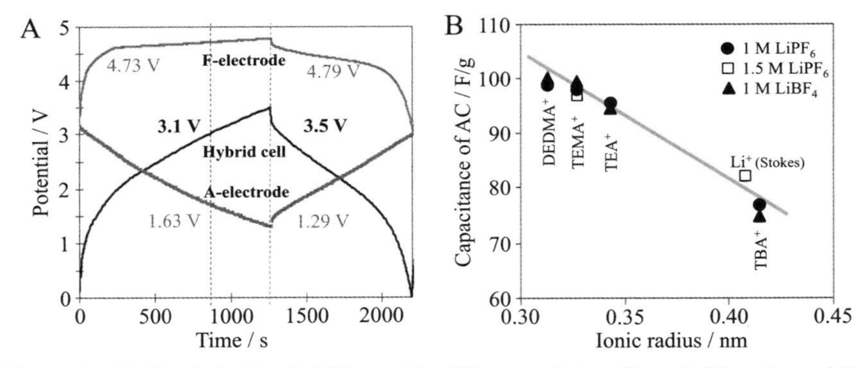

Figure 3.161 The hybrid cell AC//graphite: (A) potential profiles of AC and graphite electrodes at the first cycle in 1.5 M $LiPF_6$/PC electrolyte and (B) relation between the capacitance of AC electrode (A-electrode) and the ionic radii of organic cations in PC [531].

potential profiles at each electrode and the profile for the hybrid cell in Fig. 3.161A. Therefore, the size of cation, which form EDLs on the surface of AC at the A-electrode, was found to govern the cell performance by using quaternary alkyl ammonium cations, such as $TEMA^+$ (triethylmethyl ammonium cation), TEA^+ (tetraethyl ammonium cation), and TBA^+ (tetrabutyl ammonium cation) together with Li^+. As shown by the relation between the capacitance of AC and ionic radii of these cations in Fig. 3.161B, the cation radius is the smaller, the capacitance for AC at the A-electrode (negative electrode) is the larger, and as a consequence the capacity of the hybrid cell is the higher.

The intercalation behaviors of anions, such as BF_4^- (tetrafluoroborate anion), $DFOB^-$ [difluoro(oxalato)borate anion], and BOB^- [bis(oxalato) borate anion], at the graphite F-electrode were also investigated by in situ XRD measurements [532]. The intercalate anion was the larger, the cell voltage needed to be the higher, and as a consequence the smaller discharge capacity was delivered by the hybrid cell.

3.4 Fuel cells

A fuel cell is an electrochemical energy conversion device which can convert chemical energy directly into electric power with much higher efficiency and lower greenhouse gas emissions than conventional techniques based on fossil fuel combustion. It is considered to be one of the power sources for next-generation vehicles and portable electronics.

Polymer electrolyte membrane fuel cells (PEMFCs) and direct methanol fuel cells (DMFCs) are particularly attractive because of their lower operation temperature and longer life span. The key component of fuel cells is a Pt-containing catalyst anchored to a porous conductive material that acts as the catalyst for hydrogen oxidation reaction (HOR) or alcohol oxidation reaction (AOR) at the anode and oxygen reduction reaction (ORR) at the cathode. The performance of fuel cells is often evaluated by the activity and performance for ORR at the cathode. ORR occurred via either a four-electron ($4e^-$) path to directly produce H_2O (in acidic medium) and OH^- (in alkaline medium) or a two-step two-electron ($2e^-$) path with the intermediate species H_2O_2 (in acidic medium) and HO_2^- (in alkaline medium), as shown in Table 3.4, depending on the catalysts. The composite of Pt nanoparticles dispersed on a carbon black (Pt/C) has been used as a reference with effective $4e^-$ ORR catalyst. For practical applications of fuel cells, many proposals have been presented to replace Pt partly or completely and to enhance the ORR activity by modifying the carbon supports, mainly because of the high cost and weak durability for contamination of Pt-based catalysts, as explained in many reviews [533–541]. Recently, however, carbons doped by heteroatoms (e.g., N, P, S) are considered to be one of the promising alternatives to Pt-based catalysts for ORR (metal-free catalysts), because of their low cost, excellent antipoison capability, and durability, whereas they are still discussed by comparing them with the commercially available Pt/C catalysts. Metal-free catalysts were reviewed by focusing on N-doped carbons, including graphene relatives [542–545]. Microbial fuel cells have attracted attention as promising sustainable and green energy sources to convert chemical energy in organic wastes into electricity, although many problems remain to be overcome, such as low bacterial loading capacity, low extracellular electron transfer efficiency between the bacteria and the anode material, etc. For these microbial cells, metal-free carbon catalysts are one of the important components.

Table 3.4 ORR pathways in acidic and alkaline media.

Pathway	In acidic medium	In alkaline medium
Four-electron ($4e^-$) path	$O_2 + 4H^+ + 4e^- \rightarrow 2H_2O$	$O_2 + 2H_2O + 4e^- \rightarrow 4OH^-$
Two-electron ($2e^-$) path	$O_2 + 2H^+ + 2e^- \rightarrow H_2O_2$ $H_2O_2 + 2H^+ + 2e^- \rightarrow 2H_2O$	$O_2 + H_2O + 2e^- \rightarrow HO_2^- + OH^-$ $H_2O + HO_2^- + 2e^- \rightarrow 3OH^-$

In this section, porous carbons working not only as supports of active metal particles but also as active materials in fuel cells are explained by dividing them into three subsections, noble metal-based catalysts, metal-free catalysts (i.e., carbon materials), and transition metal-based catalysts. In metal-based catalysts (noble metal-based and transition metals-based), the structure and textures of carbon supports have a strong influence on the ORR activities of the metals, particularly N-doping into carbon. The development of synthesis techniques for porous carbons (i.e., template-assisted carbonization and polymer blending techniques, associated with N-doping techniques, in addition to activation) has promoted the research into the carbon supports for ORR active materials to achieve higher performances of fuel cells.

3.4.1 Noble metal-based catalysts

Optimization of the platinum efficiency at the cathode is one of key factors for practical applications of fuel cells by selecting the particle size of Pt and its alloys with other noble metals, and by controlling the support (substrate) for the particles. A carbon black, Vulcan XC-72, has been used as a substrate for Pt and Pt alloys, which have been commercialized as the electrode of fuel cells and used as the reference to evaluate the activity for ORR. The commercial Pt-loaded Vulcan series (often expressed as Pt/C) is quoted as having Pt particle sizes of 2.0 nm at 10 wt.% loading, 3.2 nm at 30 wt.%, and 8.8 nm at 60 wt.%. The alloy formation and dispersion as nanoparticles of noble metals of Pt, Pd, Ru etc., are also important processes for catalyst synthesis. Various processes have been proposed in addition to conventional vapor deposition and precipitation processes, such as radiolysis [546], precipitation in ethylene glycol [547], etc. Syntheses of nanoparticles of Pt and Pt–Ru alloy have been reviewed by focusing on loading onto carbon black, Vulcan XC-72, for applications in fuel cells [548].

3.4.1.1 Templated porous carbons

Pt nanoparticles were loaded onto zeolite-templated carbon (ZTC) with an ordered microporous carbon by the reduction of chloroplatinic acid (H_2PtCl_6) with $NaBH_4$ in aqueous suspension of ZTC (Pt/ZTC) [549]. The substrate ZTC with a core–shell texture was prepared by carbonization of furfuryl alcohol (FA) impregnated into zeolite template at 900°C, followed by CVD of benzene at 900°C, which had S_{BET} of 1722 m^2/g and V_{total} of 1.11 cm^3/g including V_{micro} of 0.76 cm^3/g. Pt nanoparticles were homogeneously dispersed on the ZTC, of which the average particle size was about 4.7 nm, a little larger than 3.8 nm on the commercial Pt/C, as shown in Fig. 3.162A and B. In Fig. 3.162C, the cyclic voltammogram for the Pt-loaded ZTC (Pt/ZTC) is compared with that for a commercial Pt/C. Both catalysts give peaks at about 0.63 V in the forward scan and at about 0.47 V in the reverse scan, which can be attributed to the methanol electro-oxidation on the Pt surface and to the reactivation of oxidized Pt. The Pt nanoparticles in ZTC substrate are possibly more chemically active than those in Pt/C, and the ordered microporous network with a high surface area of ZTC is reasonably supposed to facilitate the transport of methanol and reaction product CO_2 gas, as compared to that of carbon black (Vulcan XC-72) of Pt/C. Loading of Pt nanoparticles was performed by calcination of the Na form of zeolite X; Na^+ in the zeolite was exchanged by $(NH_3)_4Pt^{2+}$ using the tetramine nitrate salt $(NH_3)_4Pt(NO_3)_2$ in aqueous solution and then it was calcined in air at 350°C, resulting in the decomposition of Pt complex to partially oxidized Pt nanoparticles [550,551]. The Pt-loaded zeolite obtained thus was impregnated with ethanolic solution of FA and then heated at 400°C in Ar flow for the polymerization of FA, which was densified by CVD of propylene at 800°C, followed by the carbonization of polymerized FA in the zeolite and then cooled down under 10% H_2/N_2 to reduce Pt to its metallic state. Pt/ZTC catalysts were obtained after removing the zeolite template by concentric HF solution from the carbonized products. The resultant Pt/ZTC

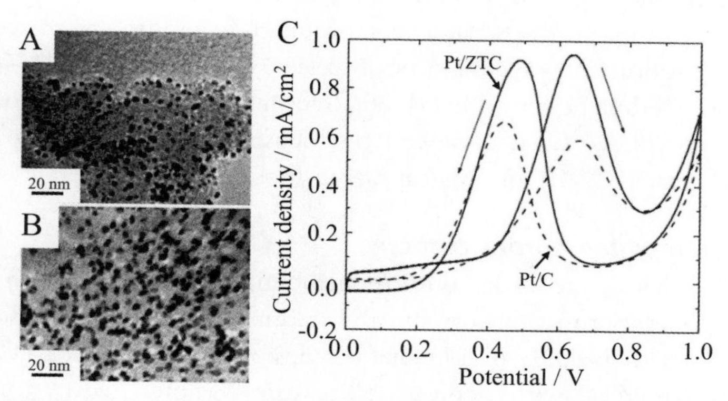

Figure 3.162 Pt-loaded zeolite-derived carbon (Pt/ZTC) and a commercially available Pt/C: TEM images of (A) commercial Pt/C and (B) Pt/ZTC, and (C) cyclic voltammograms with a scan rate of 20 mV/s at room temperature in 1 M CH_3OH+0.5 M H_2SO_4 electrolytes [551].

containing less than 17 wt.% Pt, of which sizes were about 1.5 nm, delivered an ORR activity almost comparable to the commercial 20 wt.%Pt/C (E-TEK) with 2.5 nm-sized Pt.

Pt loading into ordered mesoporous carbon prepared using SBA-15 template was performed by impregnation of H_2PtCl_6 through its acetone solution, followed by heating at 300°C for 2 h in H_2 flow to pyrolyze Pt salt and then outgassing at the same temperature for another 2 h to desorb H_2 from the resultant Pt particles [552]. The sizes of Pt particles could be controlled to below 3 nm and to disperse homogeneously without any functionalization of the carbon surfaces. The ORR activities for the carbons with 20–50 wt.% Pt loading are much higher than those of the commercial Pt/C, the maximum activity being obtained on the carbon with 30 wt.% Pt loading.

Pt nanoparticles were loaded on mesoporous silica SBA-15 by impregnation into $Pt(NH_3)_4(NO_3)_2$ aqueous solution and calcination at 400°C. The Pt-loaded SBA-15 thuse prepared was immersed into sucrose aqueous solution containing H_2SO_4, and then heated at 100 and 160°C to polymerize sucrose in the pores of SAB-15, followed by carbonization at 900°C in N_2 [553]. From the composite C/Pt/SBA-15 thus prepared, Pt-loaded ordered mesoporous carbons (Pt/CMK-3) were recovered by washing with 1.0 M NaOH in 50% ethanol/water solution at 100°C. Pt–Ru nanoparticles were loaded by immersion of CNK-3 into an ethylene glycol solution with their dispersion (Pt–Ru/CMK-3). The amount of Pt nanoparticles loaded on CMK-3 was controlled at 8.7 and 10.4 wt.% by changing the concentration of $Pt(NH_3)_4(NO_3)_2$ aqueous solution, which also affected the sizes of Pt, at about 40 and 4 nm, respectively. The size and dispersion state of metal particles in Pt/CMK-3 and Pt–Ru/CMK-3 are compared with commercial catalysts Pt/C by TEM images in Fig. 3.163A–C, and size distributions of two kinds of Pt/CMK-3 and Pt–Ru/CMK-3 are shown in Fig. 3.163D and E. Pt/CMK-3 with 10.4 wt.% Pt provides a similar content and size of Pt particles, but better electrocatalytic performance for ORRs to commercial Pt/C, probably because of the higher accessible surface area supplied by uniform mesopores of carbon substrate. In Pt–Ru/CMK-3, small particles of the alloy with a Pt/Ru ratio of 3/1 were homogeneously dispersed on CMK-3 substrate with a content of 15.3 wt.%, whereas the electrocatalytic performance was slightly inferior to a commercial Pt–Ru/C catalyst.

Bimodal porous carbons, which were synthesized from the mixture of divinylbenzene containing a small amount of azobisisobutyronitrile with

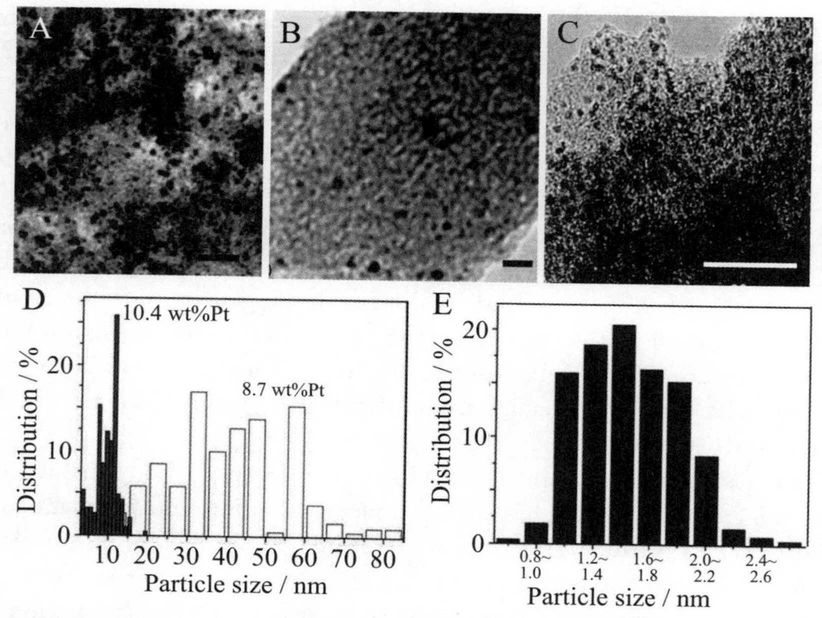

Figure 3.163 Nanoparticles dispersed on carbon (CMK-3): TEM images of (A) commercial Pt/C, (B) Pt/CMK-3 with 10.4 wt.%Pt, (C) Pt—Ru/CMK-3, and particle size distributions of (D) two Pt/CMK-3 and (E) Pt-Ru/CMK-3. Each scale bar in TEM images indicates 20 nm [553].

monodispersed polystyrene spheres and silica nanoparticles, were used as the substrates for Pt—Ru alloys [554]. The bimodal porous carbons were composed of ordered macropores with uniform diameters of about 330 nm together with mesopores with sizes of less than 10 nm in the carbon walls. The deposition of the alloy particles on the bimodal porous carbon was carried out in the mixture of H_2PtCl_6 and $RuCl_3$ in equal molar ratio with $NaBH_4$ at room temperature. Polarization curves at 30 and 70°C for the Pt—Ru/bimodal carbon are compared with those for two commercial Pt—Ru/C catalysts (Cabot and E-TEK) in Fig. 3.164. The methanol-oxidation activity of the Pt—Ru/bimodal carbons could deliver a much higher cell voltage and power density than two references: at 30°C about 0.65 V and 65 mW/cm^2, respectively, for Pt—Ru/bimodal carbon. High activity is supposed to be partly due to the higher surface area and the larger pore volume of the substrate carbon, which allow for a greater degree of active metals dispersion, and partly due to the three-dimensionally interconnected bimodal pore structure with macropores and mesopores.

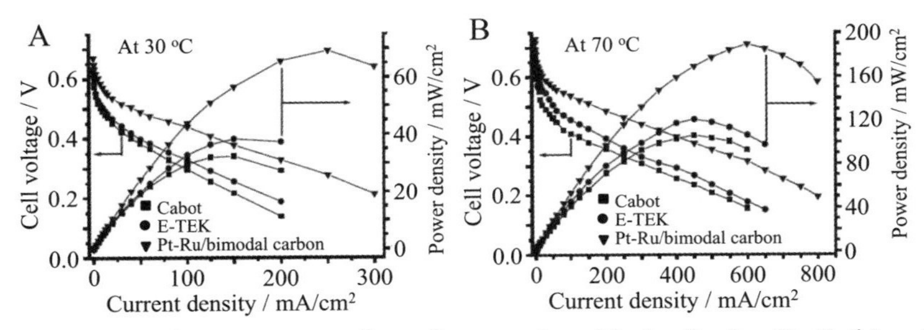

Figure 3.164 Polarization curves for a direct methanol fuel cell using Pt–Ru/bimodal carbon catalyst in comparison with two Pt–Ru/C references (Cabot and E-TEK): (A) at 30 and (B) 70°C [554].

3.4.1.2 Carbon nanotubes and nanofibers

Various types of CNFs (platelet, ribbon, and herring-bone types) were used as the substrates for Pt nanoparticles [555]. Pt loading was performed by immersing CNFs into an ethanol solution of $(NH_3)_2Pt(NO_2)_2$ at 110°C, followed by calcination at 250°C and reduction at 300°C. The catalysts thus prepared were cooled to room temperature in He and, prior to removal from the reactor, they were passivated in 2% air/He to produce a protective oxide layer over the metal particle surface to prevent bulk oxidation of Pt. The Pt particles loaded on CNFs were relatively thin and supposed to have an interaction with the edge sites of CNFs, whereas Pt particles loaded on the carbon black (Vulcan) were denser and thought to have a weak interaction with the surfaces of carbon black particles. In Fig. 3.165, the methanol oxidation current was plotted against the content

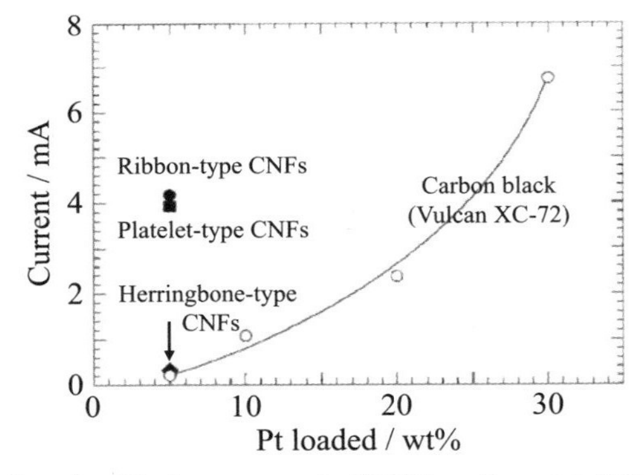

Figure 3.165 Methanol oxidation current in (0.5 M methanol + 0.5 M H_2SO_4) solution at 40°C versus Pt loaded on various types of carbon nanofibers (CNFs) and carbon black [555].

of Pt loaded on the catalysts, revealing the importance of the types of substrate CNFs (i.e., nanotexture in the CNFs) on the activity of the Pt. A 5 wt.% Pt loading on the platelet- and ribbon-type CNFs corresponds to approximately 25 wt.% Pt loading on the carbon black. Loading of Pt–Ru alloy particles on herringbone-type CNFs was done by repeated deposition of the precursor (η-C_2H_4) (Cl)Pt(μ-Cl)$_2$Ru(Cl) at 250°C and heating at 650 C in N_2 [556]. The resultant composite had a total metal content of 42 wt.% with a Pt/Ru atomic ratio of about 1/1 and average particle sizes of about 7 nm. The anode performance of Pt–Ru/CNFs was enhanced by about 50% relative to that of unsupported Pt–Ru colloids.

3.4.1.3 Reduced graphene oxides

Reduced graphene oxides (rGOs) were shown to be an effective supporting material for Pt nanoparticles by the reduction of H_2PtCl_6 with $NaBH_4$ in a suspension of graphene oxide (GO) flakes [557]. Loading of Pt nanoparticles was also carried out by reduction of the mixture of K_2PtCl_4 with either GO, carbon black (Vulcan XC-72, CB), or graphite powder in water with ascorbic acid at 85°C [558]. Cyclic voltammograms for the resultant composites, Pt/rGO, Pt/CB, and Pt/graphite, are compared in Fig. 3.166. The forward peak is due to the oxidation of methanol and the backward peak is attributed to the removal of the incompletely oxidized carbonaceous species generated in the forward scan. For methanol oxidation, the Pt/rGO composite electrode has a low onset potential E_{onset} of

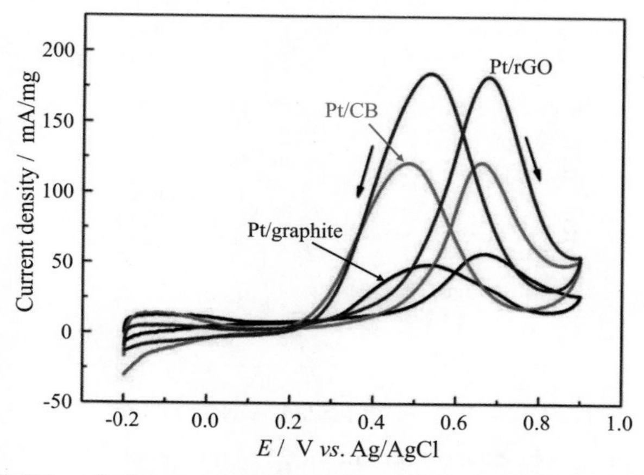

Figure 3.166 CVs of three composite electrodes in N_2-saturated solution of 0.5 M H_2SO_4 containing 0.5 M methanol [558].

about 0.15 V, which is lower than those for the other two composites, and a high peak current density of 182 mA/mg, which is much higher than those for the other two, revealing that methanol was more easily oxidized on the surface of Pt/rGO. Long-term durability of Pt/rGO composite was confirmed by chronoamperometry in 0.5 M H_2SO_4 and 0.5 M methanol for 1200 s at a fixed potential of 0.6 V. This high performance of Pt/rGO composite for oxidation of methanol was thought to be due to the small size and high dispersion of Pt nanoclusters, in addition to the high specific surface area and excellent electrical conductivity of rGO.

Pt nanoparticles were deposited on GO flakes using either ethylene glycol, ascorbic acid, or $NaBH_4$ as the reducing agents in an aqueous solution of H_2PtCl_6 [559]. When ethylene glycol was used, the mixed solution of GO, H_2PtCl_6, and ethylene glycol was refluxed at 150°C. In Fig. 3.167A–C, TEM images of the resultant Pt/rGO catalysts prepared by using three reducing agents are shown in comparison with the commercial Pt/C catalyst, with the Pt particle size distribution histogram with average particle size (*D*) (insets). The deposition-reduction process using ethylene glycol reflux is the most effective to deposit Pt nanoparticles with small sizes of around 2 nm homogeneously on rGO flakes, whereas ascorbic acid reduction results in the deposition of large-sized Pt and $NaBH_4$ in the aggregation of Pt nanoparticles. The effectiveness of the process (reduction via ethylene glycol reflux) was confirmed on another substrate, carbon black Vulcan XC-72 (Fig. 3.167D). Thanks to homogeneously distributed small Pt nanoparticles, the Pt/rGO with ethylene glycol reflux delivers high performance of fuel cells, as compared in the CVs in Fig. 3.168, which is much better than that of commercial Pt/C. The Pt/rGO gave the highest current density among the catalysts prepared from different substrates of graphene nanoplatelet, Vulcan XC-72, and thermally reduced GO. Pt/rGO composites were prepared by mixing either GO or $NaBH_4$-reduced GO (rGO) with H_2PtCl_6 in an ethylene glycol–water solution, with the one prepared from rGO exhibiting high activity [560]. N-doped rGO was also reported to be the supporting material effective for the enhancement of catalytic activity of Pt nanoparticles loaded by deposition and reduction of H_2PtCl_6 at 130°C [561]. The N-doped rGO was prepared from HCl solution of aniline monomers with suspended GO flakes by stabilization at 250°C in air and following reduction at 900°C in N_2 flow.

CNTs were successfully grown on rGO flakes by catalytic CVD of acetylene at 500°C using $Fe(NO_3)_3$ [562]. This CNT-rGO composite, in which CNTs acted as a spacer to prevent the aggregation of rGO layers

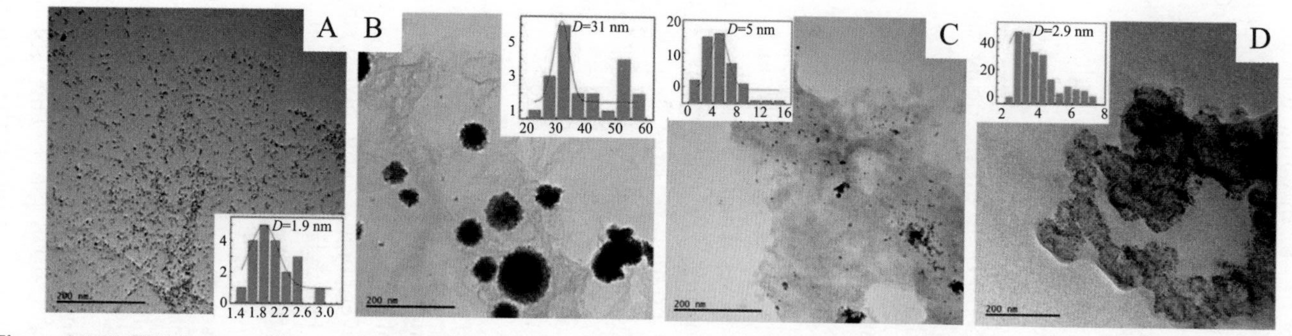

Figure 3.167 TEM images of Pt particles on rGO flakes: (A) by ethylene glycol reflux, (B) by ascorbic acid reduction, (C) by $NaBH_4$ reduction, and (D) on carbon black by ethylene glycol reflux. On the histograms (insets) in each image, the ordinate is frequency and the abscissa is particle size, with the averaged particle size *D* [559].

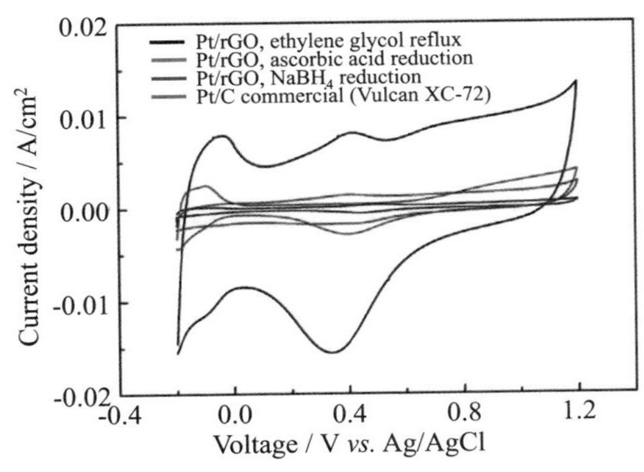

Figure 3.168 CVs of Pt/rGO synthesized by various deposition–reduction methods, in comparison with commercial Pt/C [559].

(foam-like texture) and worked as a support for Pt nanoparticles with diameters of 2–5 nm. The ORR activity of the resultant composite could be superior to the commercial Pt/C catalyst. This composite had a high electrical conductivity (144.4 S/cm) and a high electrochemically active surface area (77.4 m^2/g), and consequently gave a maximum power density of 32.0 mW/cm^2.

Since Pt is a very expensive metal, great efforts have been devoted to discovering alternative catalysts. Recent works on catalytic activities of various metals supported by different carbon materials for fuel cells have been reviewed [538]. Pd nanoparticles were indicated as a promising alternative to Pt and were found to be active when they were placed on rGO flakes, and were more active on a carbon black [563]. Pd nanoparticles were loaded on rGO prepared by hydrazine reduction and a carbon black (Vulcan XC-72) (CB) by ethylene glycol reflux process at 100°C using Na_2PdCl_4. Loading of Pt nanoparticles was carried out using a similar procedure with H_2PtCl_6 as the precursor. Pd were deposited as nanorods with sizes of $4 \times (4–10)$ nm^2, while the sizes of Pt nanoparticles were 2.9–3.1 nm. Both Pd and Pt particles were well dispersed on the rGO flakes and CB. In Fig. 3.169A and B, polarization curves are shown for Pd-based and Pt-based composites, respectively. The Pd/rGO composite electrode exhibits a much higher maximum power density than Pd/CB, at 2.2 times higher. On the contrary, Pt/rGO gives a 2.4 times lower maximum power density than Pt/CB. Pd nanoparticles were loaded on rGO by the reduction of GO and K_2PdCl_4 using ascorbic acid in the presence of PVP at 85°C (Pd/PVP-rGO composite) [564]. The composite

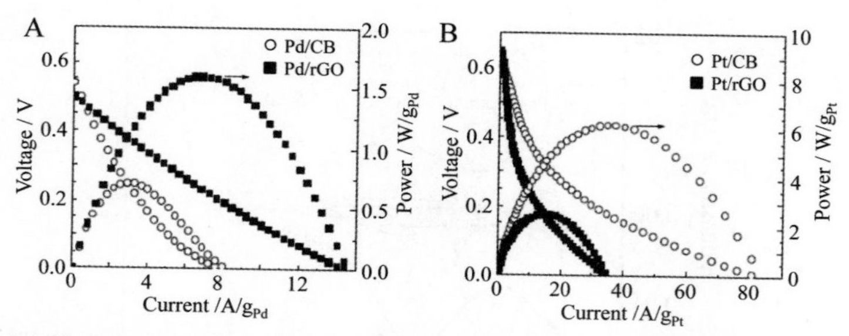

Figure 3.169 Polarization curves in 1 M methanol solution at 300K for (A) Pd-based catalysts (Pd/rGO and Pd/CB) and (B) Pt-based catalysts (Pt/rGO and Pt/CB) [563].

exhibits high electrocatalytic activity and stability toward electro-oxidation of alcohols (methanol and ethanol), as shown by comparing it with commercially available Pd/CB catalyst in Fig. 3.170. The alcohol oxidation peak can be clearly observed in the CVs at about −0.1 V for both Pd/PVP-rGO and Pd/CB electrodes, which was assigned to the oxidation of freshly adsorbed alcohols. The current density of methanol oxidation peak for the former is about 1.08 mA/cm^2, which is much larger than that for the latter (0.563 mA/cm^2). The CVs for ethanol exhibit a similar difference in two electrodes. Pd/PVP-rGO was experimentally proved to have sufficient long-term stability and multicycle repeatability for electro-oxidation of methanol and ethanol. These results demonstrate that the Pd/PVP-rGO has a superior electrocatalytic activity over Pd/CB. High dispersion of Pd nanoparticles on rGO was accomplished by simultaneous reduction of H_2PdCl_4 and GO using the photocatalyst SnP, which exhibited higher electrocatalytic activity than the commercial Pd/C catalyst for methanol electro-oxidation in alkaline media [565].

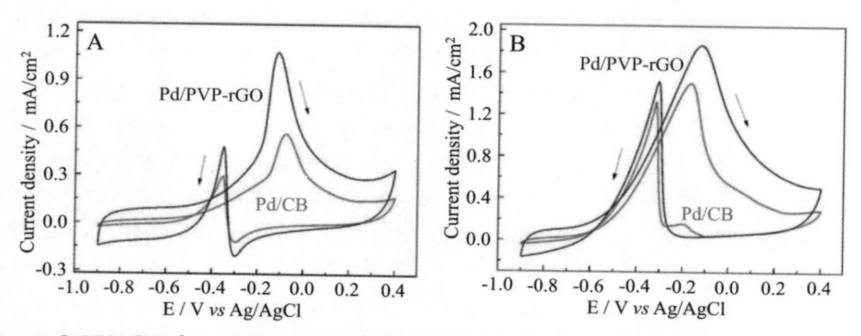

Figure 3.170 CVs for Pd/PVP-rGO and Pd/CB electrodes: in N_2-saturated 0.5 M NaOH solution containing (A) 0.5 M methanol and (B) 0.5 M ethanol [564].

Bimetallic alloy nanoparticles Pd—Ag exhibited a high catalytic activity for methanol oxidation on rGO, which was higher than Pd on rGO and also much higher than on GO [566]. The composites of the alloy nanoparticles loaded on rGO were prepared by mixing GO, $AgNO_3$, and K_2PdCl_4 in an aqueous solution, where GO worked as the reducing agent to form metals, followed by heating at 200°C to reduce GO to rGO. In Fig. 3.171A and B, CVs for the electrodes composing of Pd—Ag alloys were measured in 1 M KOH containing 1 M methanol and ethanol, respectively. The oxidation peaks in the forward scans due to the oxidation of freshly chemisorbed alcohols are observed at around −0.1 V. rGO is effective support for Pd and Pd—Ag alloys, and more effective than GO. For methanol oxidation (Fig. 3.171A), Pd—Ag(1:1)/rGO shows the highest peak current density of 630 mA/cm^2 and the lowest peak potential at −0.12 V, indicating that the methanol is the most easily oxidized. For ethanol oxidation (Fig. 3.171B), the composite [Pd—Ag(1:1)/rGO]$_{SB}$, which was prepared by the reduction using $NaBH_4$ (SB) for comparison, has a peak current density inferior to Pd—Ag(1:1)/rGO, suggesting the procedure for the reduction of metal is also an important factor to attain high activity.

N-doping into carbon was experimentally demonstrated to cause an improvement in the Pt nanoparticle dispersion and a significant increases in catalytic activity and durability for ORR by using a model catalytic system with N-doped HOPG [567]. N-doped HOPG substrate was also shown to enhance the durability of the Pt nanoparticle cluster for H_2-D_2 exchange reaction [568].

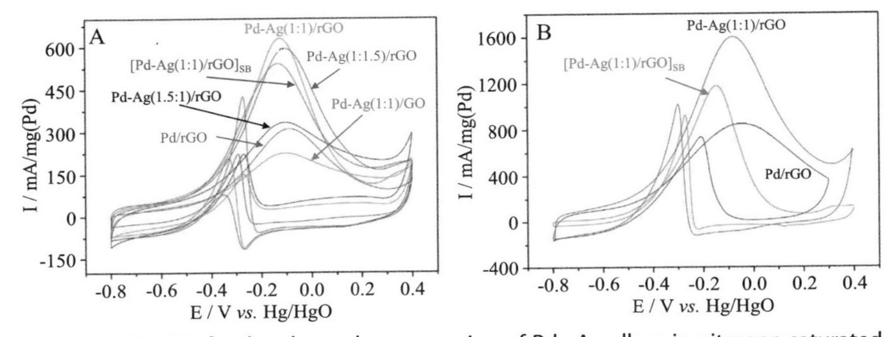

Figure 3.171 CVs for the electrodes composing of Pd—Ag alloys in nitrogen saturated solution of 1 M KOH containing (A) 1 M methanol and (B) 1 M ethanol [566].

3.4.2 Metal-free catalysts

3.4.2.1 Templated porous carbons

N-doped mesoporous carbons were synthesized by carbonizing *o*-phenylenediamine (*o*PD) mixed with colloidal SiO_2 (12 nm diameter) at 900°C in Ar flow, followed by washing with 2 M NaOH and activation in NH_3 at 900°C [569]. Carbon precursor *o*PD with a high N/C atomic ratio of 0.34 resulted in the formation of mesoporous carbon with a high N content of 9.50 at.% and S_{BET} of 685 m^2/g with S_{micro} of 324 m^2/g and S_{meso} of 350 m^2/g (meso-P*o*PD). By NH_3 activation, S_{BET} increased to 1280 m^2/g, associated with increases in S_{micro} and S_{meso} (515 and 751 m^2/g, respectively) (micro/meso-P*o*PD). In comparison, the carbons were prepared without templating and NH_3 activation (non-P*o*PD, S_{BET} of 40 m^2/g) and without templating but with NH_3-activation (micro-P*o*PD, S_{BET} of 277 m^2/g and S_{micro} of 251 m^2/g). The micro/meso-P*o*PD catalyst delivers a high ORR activity in alkaline electrolyte, comparable to the commercial 20 wt.%Pt/C catalyst, and better cycle performance, as shown in Fig. 3.172A and B, respectively. Templating using SiO_2 colloids and NH_3 activation are effective in developing pore structure and, as a consequence, enhancing ORR activity. The micro/meso-P*o*PD also outperformed the Pt-based catalyst as an air electrode of a zinc-air battery. N-doped carbons with hierarchical pore structure were synthesized from the mixtures of melamine, resorcinol, and hexamethylenetetramine with colloidal SiO_2 by gelation at 120°C, followed by carbonization at 600°C and then 900°C in N_2 flow [570]. The resultant N-doped carbon exhibited better ORR activity in O_2-saturated 0.1 M KOH than the carbon produced without melamine, suggesting the importance of N species doped in carbon matrix.

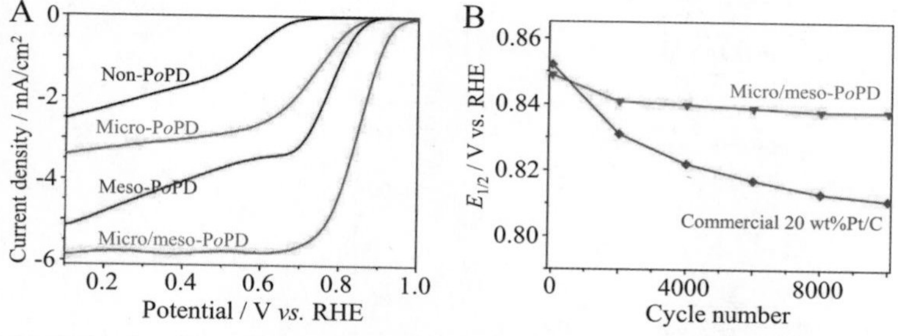

Figure 3.172 Mesoporous carbons derived from *o*-phenylenediamine with templating and NH_3 activation: (A) polarization curves and (B) cycle performances evaluated by half-wave potential $E_{1/2}$ in O_2-saturated 0.1 M KOH [569].

P-doped ordered mesoporous carbons were synthesized using SBA-15 as template by infiltration of mixtures of triphenylphosphine as the P and C source and phenol as the C source, followed by carbonization at 900°C in Ar and removal of the SiO_2 by HF [571]. A small amount of P-doping gave excellent ORR activity via a four-electron pathway in an alkaline medium, associated with greatly enhanced stability and alcohol tolerance, in comparison to those of a Pt/C.

N-doped porous carbons were synthesized by the carbonization of an ionic liquid (IL), 1-ethyl-3-methylimidazolium dicyanamide (EMIM-dca) mixed with a salt melt, $NaCl/ZnCl_2$ (NZ), at 1000°C in N_2 (salt templating), followed by washing with water [572]. The carbons obtained were microporous, having S_{BET} of 1400–1800 m^2/g with V_{total} of 1.2–2.9 cm^3/g, and relatively high N content of 3.7–4.5 wt.%, depending on the weight ratio of NZ to IL. In Fig. 3.173A and B, linear sweep voltammetry (LSV) polarization curves in O_2-saturated 0.05 M H_2SO_4 and 0.1 M KOH electrolytes, respectively, are presented for the catalysts prepared with different NZ/IL ratios ($NaCl/ZnCl_2$ ratios of 6, 10, and 13), together with those measured on the commercial 20 wt.% Pt/C catalyst. N-doped microporous carbons using NZ/IL ratios of 10 and 13 have almost the same ORR activities as the commercial 20 wt.% Pt/C in alkaline electrolytes. These IL-derived N-doped carbons revealed also comparably high stability in both H_2SO_4 and KOH. From the catalytic activities at different rotation speeds, an electron transfer number n was

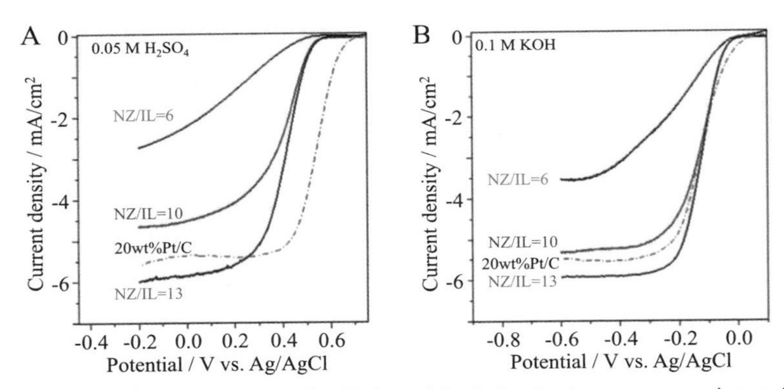

Figure 3.173 Polarization curves for N-doped IL-derived microporous carbons: (A) in O_2-saturated 0.05 m H_2SO_4 and (B) in O_2-saturated 0.1 KOH electrolytes. The measurements were performed at a scan rate of 10 mV/s and rotation speed of 1600 rpm [572].

calculated to be around 4. N-doped hierarchically porous carbons were synthesized from lignin, which was nitrogen-functionalized in advance, by carbonization in a eutectic $KCl/ZnCl_2$ melt at 850°C [573]. The resultant porous carbons exhibited S_{BET} of about 1600 m^2/g and V_{total} of 0.744 cm^3/g including V_{micro} of 0.595 cm^3/g, together with macropores, and contained 5–6 wt.% N at graphitic and pyridinic sites. They delivered a strongly enhanced ORR activity with electron-transfer number n of 3.6–4.0 in O_2-saturated 0.1 M KOH. Nucleobases (adenine, guanine, cytosine, thymine, and uracil), which were isolated from biomass using bacteria, were dissolved into an IL (EMIM-deca) together with SiO_2 nanoparticles and carbonized at 1000°C, to obtain mesoporous N-doped carbons with high S_{BET} up to 1500 m^2/g and high N content of 12 wt.% [574]. The resultant carbons exhibited a low E_{onset} for ORR in alkaline medium and a high methanol tolerance.

N-doped mesoporous carbons were fabricated from zinc and calcium citrates by carbonization at 800°C in N_2, followed by washing with HCl and heating at 800°C, together with melamine (melamine/carbon of 4/1) (ZnC and CaC, respectively) [575]. In O_2-saturated KOH solution, the ORR activity of CaC was much better than that of ZnC, although the latter was comparable to the commercial 20 wt.%Pt/C, and both CaC and ZnC exhibited much better long-term stability than Pt/C. In O_2-saturated 0.5 M H_2SO_4, both CaC and ZnC were inferior to Pt/C, although their cyclabilities were much better.

Halloysite $[Al_2Si_2O_5(OH)_4 \cdot 2H_2O]$ was used for the synthesis of N-doped porous carbons as the template from mixtures of glucose with urea (1/1 by weight) by carbonization at 1000°C in N_2 [576]. The mixture of furfural with urea was also used as carbon and nitrogen precursors using the template halloysite acidified using H_2SO_4. The carbon derived from glucose gave S_{BET} of 118 m^2/g and V_{total} of 0.22 cm^3/g and that from furfural exhibited 824 m^2/g and 1.23 cm^3/g, respectively, while their ORR activities were very similar with each other, while a slightly inferior to the commercial Pt/C.

3.4.2.2 Precursor-designed porous carbons

N-doped carbon nanoflakes were synthesized by carbonization of the mixture of 1,8-diaminonaphthalene with $FeCl_3$ at 800–1000°C in Ar flow [577]. S_{BET} was measured as 1042, 1584, and 1560 m^2/g for the carbon prepared at 800, 900, and 1000°C, respectively, the 900°C-prepared one having mesopores centered at around 3 nm size. As shown by the

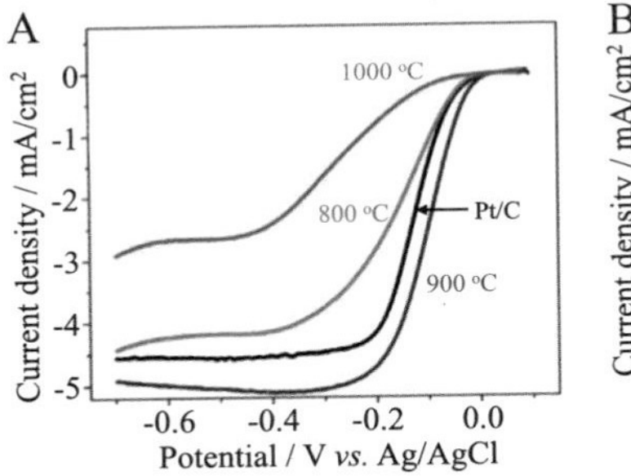

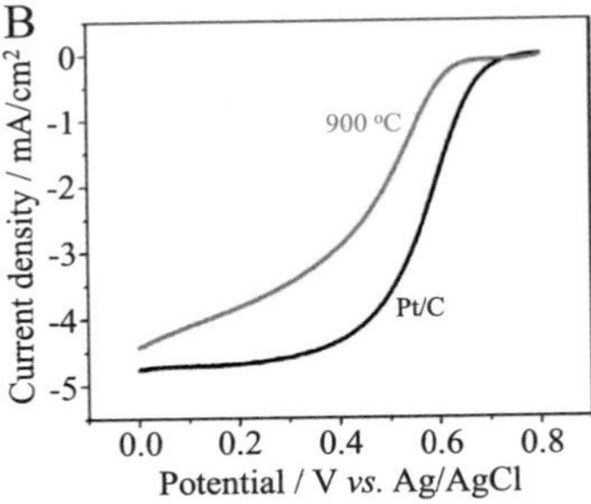

Figure 3.174 Polarization curves of N-doped carbons prepared from the mixture of 1,8-diaminonaphthalene with $FeCl_3$ at different temperatures: (A) in O_2-saturated 0.1 M KOH and (B) 0.1 M $HClO_4$ electrolytes, in comparison with commercial Pt/C catalyst [577].

polarization curves in Fig. 3.174, the 900°C-prepared carbon demonstrated better ORR activity in 0.1 M KOH electrolyte than the commercial Pt/C catalyst, while it was inferior in 0.1 M $HClO_4$ electrolyte.

N-doped porous carbons were synthesized from a MOF, ZIF-7 ($C_{14}H_{16}N_4O_3Zn$), by mixing with additional carbon precursor, glucose, through either solution or powder mixing and carbonizing at 950°C in Ar [578]. Mixing of glucose to ZIF-7 is effective in increasing micropores, particularly solution mixing, although ZIF-7 itself also gives microporous carbon and glucose itself gives a nonporous solid, as shown in Fig. 3.175A. The carbon obtained by solution mixing of ZIF-7 with glucose exhibited V_{total} of 0.56 cm^3/g including V_{micro} of 0.37 cm^3/g, although ZIF-7 itself gave 0.29 and 0.22 cm^3/g, respectively. As shown by the polarization

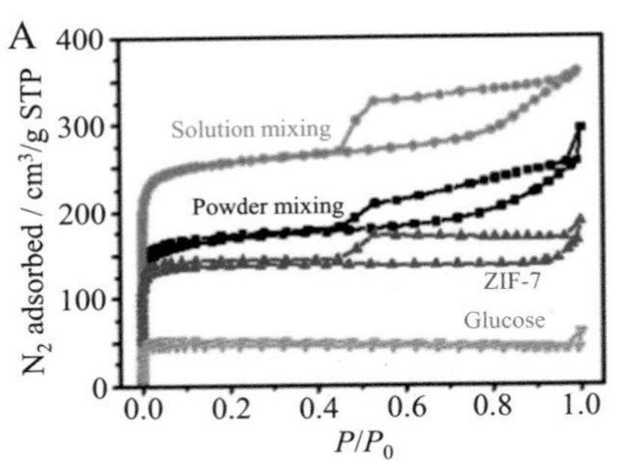

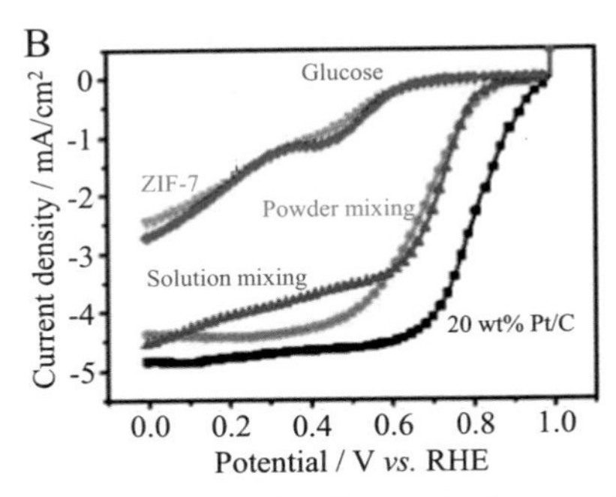

Figure 3.175 Porous carbons prepared from the mixture of ZIF-7 and glucose via solution and powder mixing: (A) N_2 adsorption–desorption isotherms and (B) polarization curves in O_2-saturated 0.1 M KOH [578].

curves in Fig. 3.175B, the carbon prepared by solution mixing delivers a limiting current density of −4.6 mA/cm^2 and electron transfer number *n* of 3.68, comparable to commercial 20 wt.%Pt/C (−4.9 mA/cm^2 and 3.97, respectively). The carbons exhibited excellent ability to avoid methanol perturbation.

Porous carbons were synthesized by CVD of CO_2 at 680°C on the powder of Mg metal and the powder mixture of Mg and Cu metals (1:1 by mole), followed by washing with HCl and NH_3 at room temperature to remove MgO and Cu [579]. The resultant carbons have high graphitic crystallinity, particularly the carbon deposited on Mg/Cu alloy (C−Mg/Cu), as shown by the Raman spectrum in Fig. 3.176A, and are mesoporous, S_{BET} of C−Mg being 829 m^2/g and C−Mg/Cu being 706 m^2/g, as shown in Fig. 3.176B. The carbon C−Mg/Cu, having high crystallinity, delivered higher power density in a microbial fuel cell than that with C−Mg, as shown in Fig. 3.176C.

Ultra-lightweight N-doped carbon foams with a bulk density of 0.002 g/cm^3 were prepared by carbonization of ultra-lightweight polyaniline foams at different temperatures of 500−1000°C in N_2 [199]. The carbonization temperature governed its activity for ORR; the activity of the foam carbonized at 900°C in O_2-saturated 1 M KOH aqueous solution was comparable with that of commercial Pt/C catalyst, its onset potential was close to Pt/C, and its limiting current density was much higher than that of Pt/C. Reticulated vitreous carbon foams (RVC) prepared by direct carbonization of the sponge-like natural product pomelo peel at 900°C were tested as an anode in microbial fuel cells [580]. The resultant RVCs gave a current density of over 4.0 mA/cm^2, which is much higher than the commercial RVC and graphite felt.

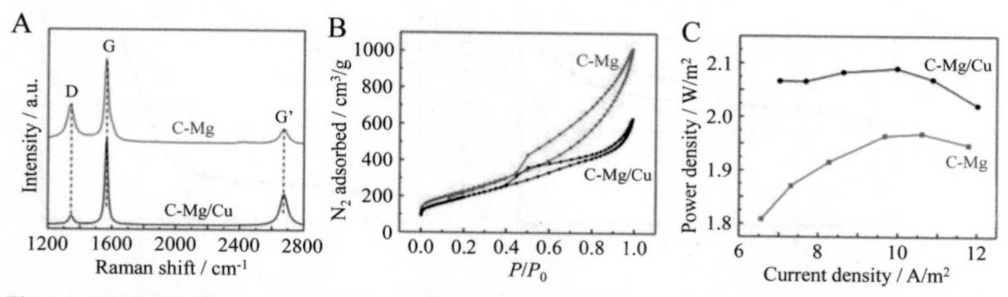

Figure 3.176 Porous carbon by CVD of CO_2 on Mg and Mg/Cu alloy (C−Mg and C-Mg/Cu, respectively): (A) Raman spectra, (B) N_2 adsorption-desorption isotherms, and (C) power density versus current density of microbial fuel cells [579].

3.4.2.3 Carbon nanotubes and nanofibers

N-doped CNTs synthesized as their vertically grown arrays by catalytic CVD of ferrocene with NH_3 demonstrated ORR activity much better than commercial 20 wt.%Pt/C (E-TEK) [581]. N content in the resultant CNTs was measured as 4—6 at.% by XPS. As shown by comparing with nondoped CNTs and commercially available 20 wt.%Pt/C catalyst in Fig. 3.177, N-doping is not only effective in improving activity but also in donating insensitivity to CO poisoning.

N-doped MWCNTs were prepared by heating the pristine MWCNTs mixed with nitrogen-containing precursors, i.e., 1H-benzotriazole (BTA), 1H-1,2,3-triazolo[4,5-b]pyridine] (TAPy), and 5-(2-pyridyl)-1H-tetrazole (PyTZ), in weight ratio of N-precursor/MWCNTs of 2/1, at 700°C in Ar [582]. The resultant N-doped MWCNTs delivered ORR activities comparable to commercial Pt/C in O_2-saturated 0.1 M KOH, associated with better stability and n in a narrow range from 3.35 to 3.60.

N/S-codoped CNTs were prepared from the mixtures of carbonaceous tubes with cysteine ($C_3H_7O_2NS$) in different ratios (cysteine/CNT of 1/2, 1/1, and 2/1 by weight) by heating at different temperatures of 700—1000°C in 5 vol%H_2/Ar flow [583]. The carbonaceous tubes were obtained by hydrothermal treatment of the water/ethanol solution (2/1 by volume) of $MnCl_2$ with KNO_3 at 180°C, which had diameters of 200—600 nm and the wall was highly porous with abundant functional groups. The MnO_x nanorods were formed during hytrothermal treatment and worked as the template for the formation of CNTs. The codoping of N and S was efficient in improving the ORR performance in acidic media.

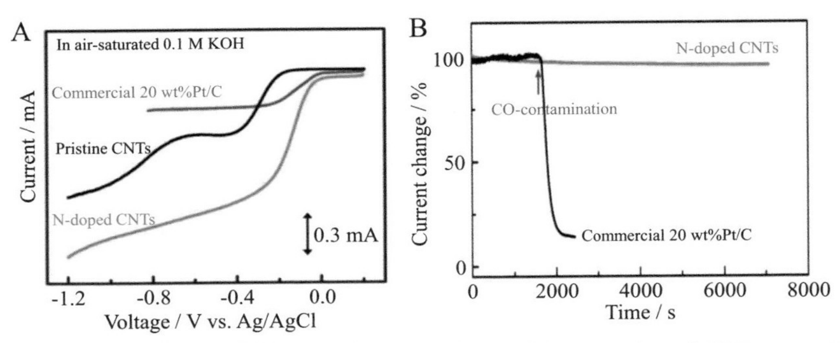

Figure 3.177 N-doped CNT array in comparison with a nondoped CNT array and commercial 20 wt.% Pt/C catalyst: (A) polarization curves and (B) effect of CO contamination [581].

The CNTs prepared at 900°C, which have N and S contents of 2.75 and 0.18 at%, delivered the highest activity in O_2-saturated 0.5 M H_2SO_4, with E_{onset} at 851 mV and high stability. These N/S-codoped catalysts also exhibited improved performance in hydrogen evolution.

3.4.2.4 Graphene and reduced graphene oxides

N-doped graphene films were reported to have high ORR activity, even higher than commercial Pt/C. N-doped graphene was synthesized by CVD from the mixture of NH_3, CH_4, H_2, and Ar ($NH_3/CH_4/H_2$/Ar of 10/50/65/200) at 1000°C, which was composed of two to eight graphene layers and contained about 4 at.% N with pyridinic and pyrrolic configurations [584]. N-doped graphene electrode exhibits a one-step, four-electron pathway for the ORR. The steady-state catalytic current density for the N-doped graphene is much higher than that of the commercial Pt/C catalyst, as shown in Fig. 3.178A. Chronoamperometric responses by the introduction of 2 wt.% methanol are shown for the N-doped graphene and the Pt/C electrodes; the current changes about 40% for the Pt/C, but for the N-doped graphene no change in current is detected, as shown in Fig. 3.178B. Fig. 3.178C demonstrates that the Pt/C electrode is rapidly poisoned by the introduction of 10 vol.% CO in air, but the N-doped graphene is insensitive to CO contamination. These experimental results reveal that the N-doped graphene acts as a metal-free electrode with a much better electrocatalytic activity and long-term operation stability than Pt for ORR via a four-electron pathway in alkaline electrolyte. N-doped single-layer graphene films were synthesized by CVD in a flow of C_2H_4 and H_2 at 0.6 kPa with a rate of 30 sccm and 10 sccm, respectively, by introducing NH_3 diluted in He (10 vol% of NH_3)

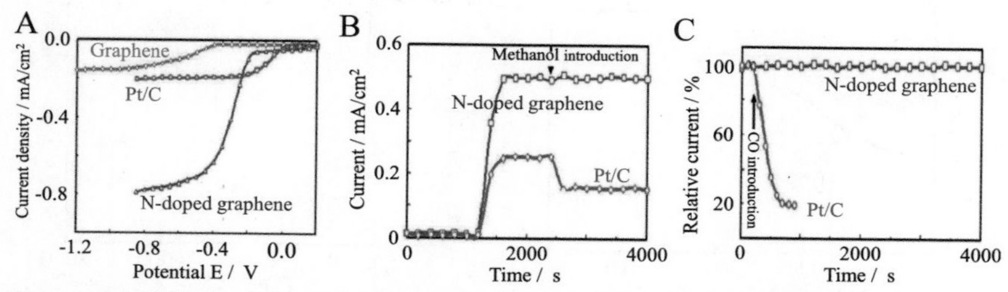

Figure 3.178 N-doped graphene electrode for ORR in air-saturated 0.1 M KOH, in comparison with graphene without N-doping and commercial Pt/C electrodes: (A) polarization curves, (B) chronoamperometric responses at −0.4 V with the addition of 2 wt.% methanol, and (C) those with the introduction of 10 vol.% CO [584].

with a flow rate of 3–12 sccm [585]. The N content of the resultant graphene films depends on the flow rate of NH_3, at a flow rate of 6 sccm a maximum N/C ratio of 0.16, and most N atoms are pyridinic. The ORR activity of the resultant graphene films, however, was lower than for the commercial Pt/C even after N doping, although the pyridinic-N was generally considered to be responsible for the ORR activity. N-doped graphene deposited on Ni mesh was prepared by CVD in a mixture of C_2H_2/H_2 with 2/45 at 1000°C, and applied to the cathode in a microbial fuel cell, which delivered a maximum power density of 147 mW/m^2, which was 32% higher than that of the Pt/C [586].

N-doped rGO flakes were prepared by heating rGO in NH_3 at 973K [587]. Three rGO flakes with different N contents were obtained by changing reducing reagents for GO, either $NaBH_4$ or hydrazine: N-doped rGO-A containing 1.7 at.% N (pyridinic-N of 0.72 at.%), rGO-B 2.4 at.% (1.9 at.%), and rGO-C 8,1 at.% (6.3 at.%). ORR activity increases with increasing N content, as shown in Fig. 3.179A. The current density at different potentials, 0.5, 0.6, and 0.7 V, was related closely to the content of pyridinic-N, as shown in Fig. 3.179B. No clear relations of current density with total N content and with the content of graphitic-N were obtained. These experimental results suggested that the pyridinic-N in graphene is responsible for high ORR activity, which is higher than that of Pt. A mechanism for ORR activity of N-doped carbon materials was proposed by considering the active sites of carbon atoms with Lewis basicity next to pyridinic-N.

N-doped rGO foam was prepared from the mixture of GO with either dicyanamide, melamine, or urea (nitrogen precursor), $FeCl_2$ (catalyst) and the template SiO_2 spheres (diameters of 7 nm) by heating at 900 C,

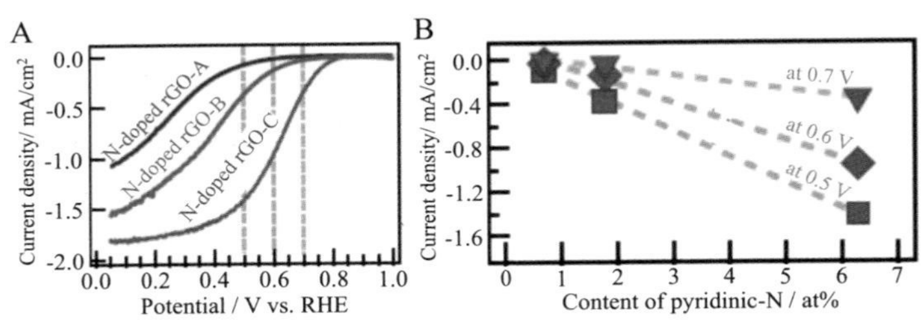

Figure 3.179 ORR performance of N-doped rGOs: (A) polarization curves in O_2-saturated solutions and (B) current density against the content of pyridinic-N at different potentials [587].

followed by dissolving out of Fe and SiO_2 [588,589]. The resultant foams exhibited high ORR activity in both alkaline and acidic media. In Fig. 3.180, voltammograms and chronoamperometric responses of the foams are compared with the commercial 20 wt.%Pt/C in alkaline and acidic media. In KOH (Fig. 3.180A), the E_{onset} for the N-doped carbon foam is 1.03 V, which is almost the same as that of Pt/C (1.04 V), but the diffusion-limiting current of the former is much higher than that of the latter, about 9 mA/cm^2 for the former and about 5.7 mA/cm^2 for the latter. In $HClO_4$ (Fig. 3.180B), the E_{onset} and limiting current are 0.81 V and 9.9 mA/cm^2 for the N-doped foam, respectively, whereas they are 1.0 V and 5.4 mA/cm^2 for the Pt/C, respectively. As shown in Fig. 3.180C and D, the N-doped foam shows almost perfect tolerance for the introduction of 3.0 M methanol, although typical inverse methanol oxidation current can be observed on Pt/C in both KOH and $HClO_4$ media. An accelerated durability test showed that the rGO foam was superior to Pt/C.

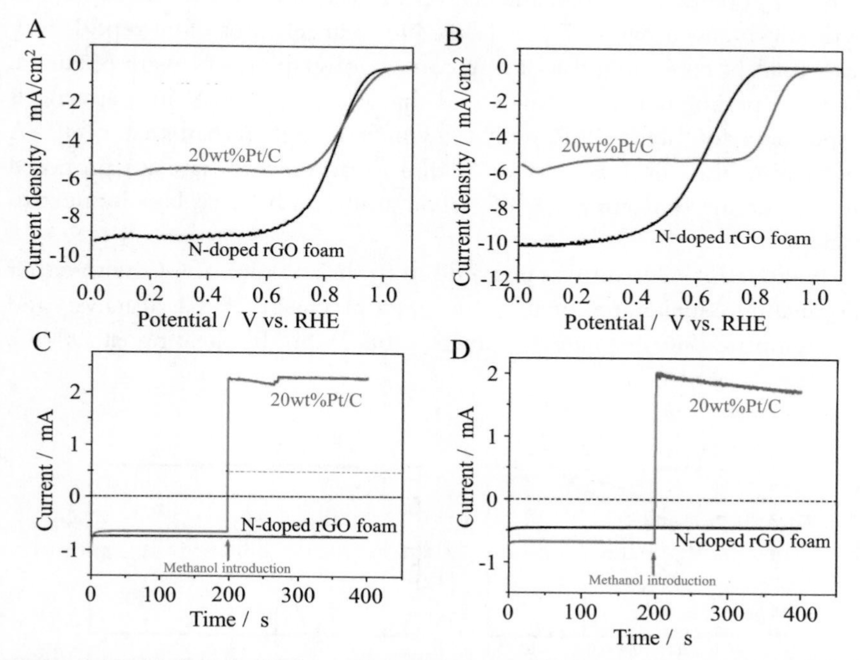

Figure 3.180 N-doped rGO foams in comparison with the commercial 20 wt.%Pt/C: (A) and (B) polarization curves, and (C) and (D) chronoamperometric responses for methanol introduction; (A) and (C) in O_2-saturated 0.1 M KOH, and (B) and (D) in O_2-saturated 0.1 M $HClO_4$ [589].

These experimental results reveal that N-doped rGO foam shows good ORR activity, high methanol tolerance, and excellent long-term stability in both alkaline and acidic media, which are better than commercial 20 wt.%Pt/C catalyst. N-doped rGO foam fabricated from GO using 3-aminopropyltriethoxysilane as a template precursor exhibited perfect methanol tolerance and better ORR performance than the commercial 20 wt.%Pt/C catalyst [590]. N-doped ultralight rGO foams were fabricated by hydrothermal treatment of the aqueous suspension of GO with 5 vol.% pyrrole at 180°C, followed by freeze-drying and annealing at 1050°C [591]. ORR activity of an N-doped rGO was enhanced by implantation of nitrogen active sites of mesoporous g-C_3N_4 [592].

Graphene is expected to provide some opportunities for improving cell performances. rGO foams covered by hydrophilic conducting polymer PANI gave improved performance of microbial fuel cells [593]. Crumpled rGO flakes were prepared by nebulizing GO dispersions to aerosol droplets through an ultrasonic atomizer, followed by rapid heating at 400°C under microwave irradiation [594]. Microbial fuel cell using this crumpled rGO exhibited high short-circuit current density and maximum power density, which were much higher than for the flat rGO flakes and AC. Porous rGO electrodes were prepared by simultaneous electrodeposition and electroreduction of GO on glassy carbon rotating disk electrodes and functionalized with aryl diazonium derivatives to immobilize the enzymes [595]. The electrodes thus fabricated gave high capacitive area and high current density. N-doped graphene flakes having the N/C ratio of 0.125 with three N configurations (graphitic, pyridinic, and pyrrolic) were synthesized through detonation reaction of cyanuric chloride and trinitrophenol at 320°C with a momentary pressure of 60 MPa [596]. The resultant flakes delivered a maximum power density of 1350 mW/m^2 as a cathode of microbial fuel cell.

Graphene foams were synthesized by CVD of ethanol with Ni foam as template, on which PANI was deposited under acidic conditions with ammonium persulfate as the catalyst [593]. A microbial fuel cell constructed from this PANI-coated graphene foam as anode delivered a maximum power density of about 768 mW/m^2, whereas the cell constructed from carbon cloth gave about 158 mW/m^2. PANI/rGO composites were prepared by electrochemical reduction of GO on carbon cloth, followed by immersion into PANI solution [597]. Microbial fuel cells constructed from this composite as the anode and carbon felt as the cathode yielded a maximum powder density of 1390 mW/m^2. Similar composites gave a maximum power density of 884 mW/m^2, which is much higher than for PANI-coated carbon cloth [598]. N-doped rGO flakes prepared by carbonization of the mixture of GO with ammonium persulfate at 850°C, followed by KOH activation at 850°C, delivered a maximum power density of 1159 mW/m^2 [599]. Polypyrrole/GO composites prepared by electropolymerization of pyrrole were used to construct microbial fuel cells after loading on carbon felt, which delivered a maximum power density of 1326 mW/m^2 [600].

N-doped rGO/CNT composite, which was prepared by heat treatment of the mixture of CNTs and GO at 800°C in NH_3, delivered ORR activity slightly better than the commercial 20 wt.%Pt/C catalyst in alkaline electrolyte; its onset potential E_{onset} is as high as 1.08 V, which is nearly 80 mV higher than that of commercial Pt/C, as shown in Fig. 3.181A [601]. It gave much better activity than N-doped CNT and rGO in acidic electrolyte ($HClO_4$), although it was not so different in alkaline electrolyte (KOH), as shown in Fig. 3.181C and B, respectively. rGO/MWCNT composites were prepared from the mixture of GO with hydrophilic MWCNTs coated by poly(diallyldimethylammonium chloride) (PDAC), where homogeneous dispersion was achieved via electrostatic interactions between the two oppositely charged constituent components, followed by converting to rGO using the reducing agent hydrazine [602]. The composites exhibited enhanced ORR activity.

3.4.2.5 Enhancement mechanism of N-doping

The enhancement of ORR activity on N-doped carbons was discussed on the bases of the experimental results on model catalysts composed from HOPGs doped by N via different doping processes [587]. The current densities at different potentials depend on the content of pyridinic N in the

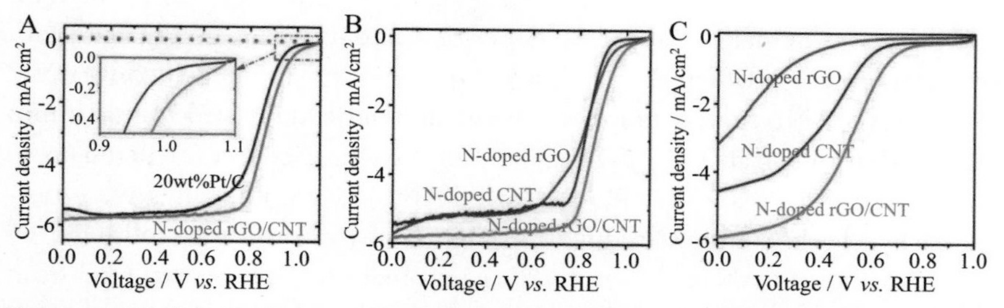

Figure 3.181 Polarization curves of N-doped rGO/CNT in comparison with N-doped rGO and CNT: (A) and (B) in O_2-saturated 0.1 M KOH, and (C) in 0.1 M $HClO_4$ [601].

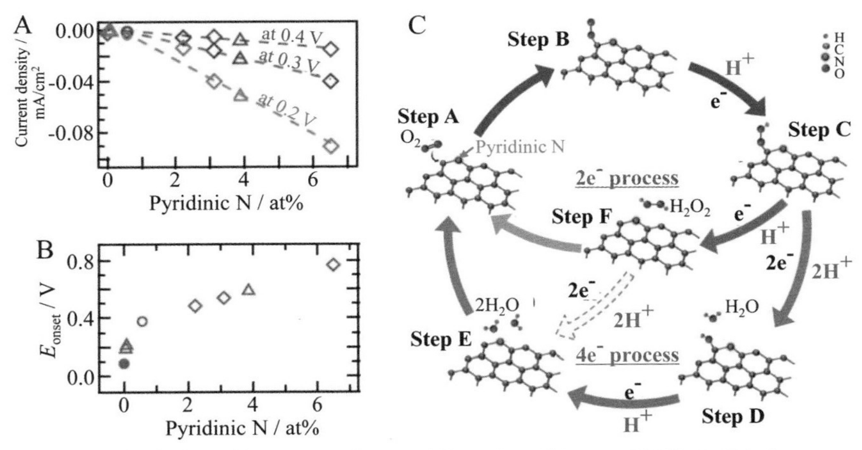

Figure 3.182 ORR performance of a model catalyst (N-doped HOPG): (A) changes of current density at different potentials and (B) those of onset potential E_{onset} with the content of pyridinic N, and (C) schematic illustration of pathways for ORR on N-doped carbon [587].

model catalyst prepared from HOPG, as shown in Fig. 3.182A. The same dependences are obtained on N-doped rGO (refer to Fig. 3.179B). The onset potential E_{onset} also depends heavily on the content of pyridinic N, independent of the preparation method (Fig. 3.182B). From these experimental results on the model catalysts using HOPG and on rGO, a possible mechanism for the ORR on N-doping of carbon was proposed, as illustrated in Fig. 3.182C. Since the Lewis base site is created by pyridinic N, an O_2 molecule is first adsorbed at the carbon atom next to the pyridinic N (step B), followed by protonation of the adsorbed O_2 (step C). From step C, two pathways are possible: the two-electron pathway via step F and the four-electron pathway via steps D and E. In either pathway, the C atom next to pyridinic N with Lewis basicity plays an important role as the active sites where an oxygen molecule is adsorbed as the initial step of the ORR.

3.4.2.6 Other dopants

B/N- and P/N-codoped carbons were prepared from the mixtures of dicyandiamide (DCDA) with boric acid and phosphoric acid and small amounts of $FeCl_3$ and $CoCl_2$ by pyrolysis at 900°C in Ar flow, followed by dissolving out Fe and Co particles in the carbon with aqua regia [603]. Surface dopant concentrations relative to C, N/C, B/C, and P/C in at.%, were 3.6 in N-doped, 3.4 and 3.8 in B/N-doped, <0.6 and 5.5 in P/N-doped, and 4.3, <0.6, and 9.0 in B/P/N-doped carbons (measured by XPS). Although the catalysts prepared demonstrated similar E_{onset} close to 0.83 V, the performance of the ORRs was improved by doping of B and P in addition to N; B/P/N-doped catalyst shows the highest activity, as shown in Fig. 3.183. The performance improvement was associated with an increase in the *n*-value to 4. The B/P/N-doped catalyst demonstrated much higher stability for 10 h in 1 M $HClO_4$ with continuous oxygen bubbling. Additional P-doping into N-doped carbon up to P/C of 2.8 was performed [604].

Doping of N and codoping of S, P, B with N was performed by mixing of a carbon black (Ketjenblack, KB), with IL precursors in the weight ratio KB/IL of 40/55 and by heating at 900°C in an inert atmosphere [605]. IL precursors used were EMIM-dca for N-doping, THIA-dca for S/N-doping, (EMIM-dca + TEPBr) for P/N-doping, and EMIM-tcb for B/N-doping. The same KB/IL mixtures after adding 5−27 wt.% $FeCl_3$ as a mediator were also carbonized at 900°C, followed by washing with H_2SO_4 to leach out residual metal and then annealing at 900°C. In Fig. 3.184A and B, polarization curves on the doped carbons were shown by comparing with the pristine KB. N-doping of KB improved ORR activity in alkaline electrolyte, and addition of $FeCl_3$ into KB + IL mixtures followed by carbonization enhances the activity improvement due to N-doping. The addition of 27 wt.% $FeCl_3$ can outperform commercial Pt/C, as shown in Fig. 3.184A. The effect of codoping of S, P, and B with N using 27 wt.% $FeCl_3$ is demonstrated in Fig. 3.184B, revealing that S/N-codoping markedly enhances ORR activity by shifting onset potential to 0.93 V, higher than 0.86 V for N-doping and commercial Pt/C. The catalyst codoped with S/N by using

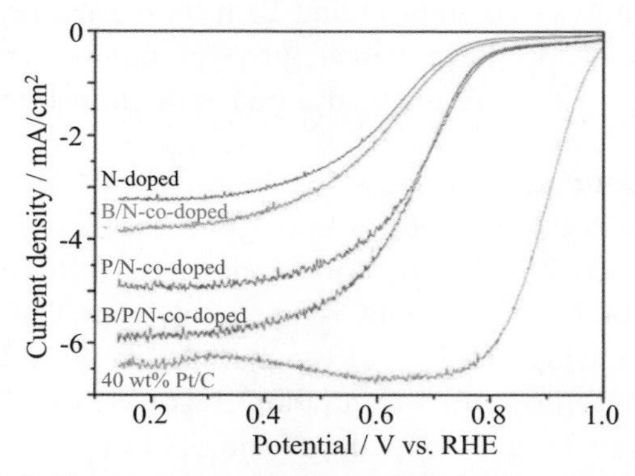

Figure 3.183 Polarization curves of N/B/P-codoped carbons in 1 M $HClO_4$ [603].

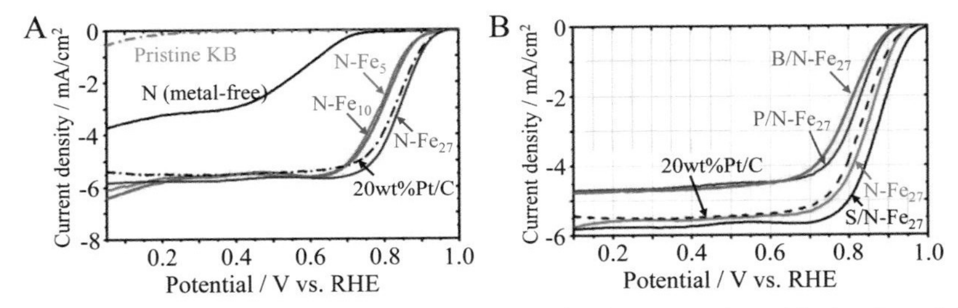

Figure 3.184 Polarization curves in O_2-saturated KOH on carbon blacks doped by (A) N using $FeCl_3$ and (B) N and S (S/N), N and P (P/N), and N and B (B/N) using $FeCl_3$ in comparison with commercial Pt/C and the pristine Ketjenblack (KB) [605].

27 wt.% $FeCl_3$ contained 1.51 wt.% N, 0.65 wt.% S, and 1.54 wt.% Fe, in comparison with that doped N using $FeCl_3$ contained 1.24 wt.% N and 1.44 wt.% Fe. Another carbon black (BP-2000) was also used for N-doping matrix by mixing with different amounts of melamine and $FeCl_3$ and heat-treating at 900°C in Ar flow [606]. Codoping of N and F was achieved via a two-step process on BP-2000 and acetylene black; heating of the mixtures of a carbon black with melamine at 900°C in Ar flow for N-doping and then heating of the mixture of the N-doped carbon black with ammonium fluoride (NH_4F) by step-wise heating at 400 and 900°C in Ar flow [607]. N/F-doped carbon blacks gave enhanced ORR activity: E_{onset} of about −0.15 V and four-electron process.

N/B-codoped rGO was synthesized from GO by two-step doping: first, N was doped by heating in 20% NH_3/Ar flow at either 500 or 900°C, and then B was introduced by pyrolysis of H_3BO_3 in the presence of N-doped rGO at 900°C [608]. Heating at the low temperature in the first step resulted in a lower reduction degree of GO with more defects, and as a consequence in a higher concentration of N. After heating in the second step, it was necessary to wash with hot water to remove the residue B_2O_3. The direct heating of GO mixed with H_3BO_3 in 20% NH_3/Ar at 900°C resulted in the formation of *h*-BN/rGO composite. Fig. 3.185A reveals that N/B-codoped rGO has E_{onset} of 0.06 V, which is closer to that of Pt/C than N- and B-doped rGO, suggesting a synergistic interaction between the two dopants, N and B, and the four-electrons dominated ORR pathway by *n*-value of 3.97, comparable to Pt/C ($n = 3.98$). N/B-doped rGO exhibited excellent methanol tolerance, as shown in Fig. 3.185B.

Halogen-doping into carbon materials was also reported to improve ORR performances in order to fabricate metal-free catalyst. In the resultant catalysts, halogen atoms, such as a cluster of I_3^-, were reasonably supposed to

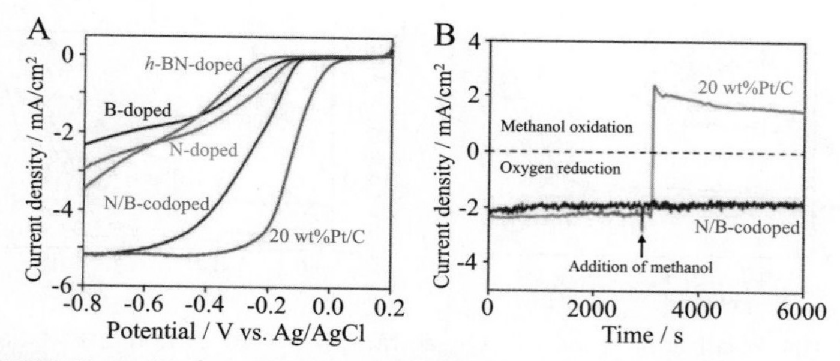

Figure 3.185 N- and B-doped rGO: (A) polarization curves and (B) chronoamperometric responses in O_2-saturated 0.1 M KOH by the introduction of methanol [608].

make a bond with the active site at the edge of carbon layer, which was formed by ball-milling (therefore the term "doping" might not be appropriate). I-doping into ACs were performed by mechanochemical ball-milling of the mixture of AC and iodine (AC/I of 1/4 by weight) in Ar, followed by washing with 1 M HCl to remove remained iodine and metallic impurities, and then by freeze-drying at −120°C [609]. I-doping in the AC does not have a marked influence on its pore structure (Fig. 3.186A) while it increases current density much more than the pristine AC and even more than commercial Pt/C, as shown in Fig. 3.186B. Both the pristine and I-doped ACs exhibit excellent stability for methanol addition (Fig. 3.186C), as well as against CO poisoning. Halogen-doped graphite nanoflakes were prepared by mechanochemical ball-milling of graphite flakes in the presence of Cl_2, Br_2, and I_2, and their ORR activities have been investigated [610]. In Fig. 3.187A, CV of I-doped graphite flakes in O_2-saturated 0.1 M KOH shows an obvious oxygen reduction peak while no peak is seen in the N_2-saturated solution. As shown in Fig. 3.187B, I-doping is the most effective in enhancing ORR activity in

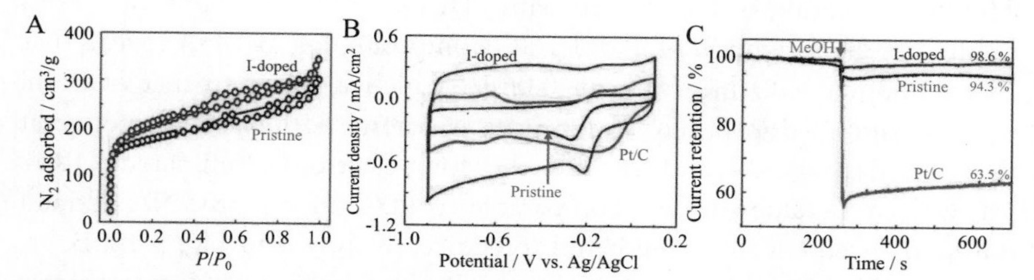

Figure 3.186 I-doped AC and pristine AC: (A) N_2 adsorption–desorption isotherms, (B) CVs in O_2 saturated 0.1 M KOH, and (C) chronoamperometric response at −0.3 V with the addition of 3 M methanol [609].

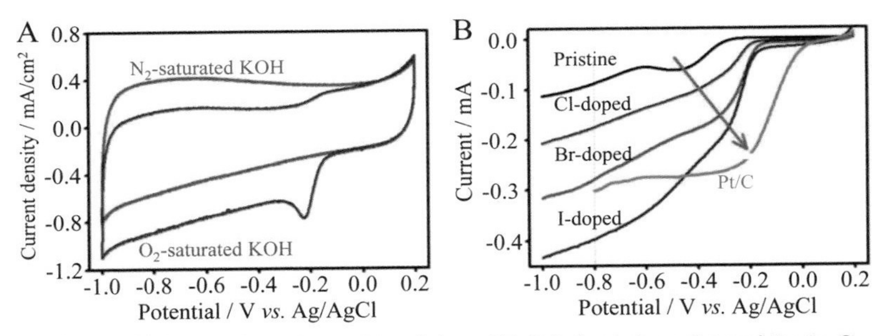

Figure 3.187 Halogen-doped graphite flakes: (A) CVs for I-doped graphite in O_2- and N_2-saturated 0.1 M KOH and (B) polarization curves for halogen-doped graphite flakes in O_2-saturated 0.1 M KOH in comparison with the pristine graphite and commercial Pt/C [610].

comparison with the doping of other halogens. All halogen-doped graphite flakes delivered high tolerance to methanol crossover/CO poisoning and longer term stability.

I-doping onto rGO was carried out by annealing the mixture of GO and I_2 at 500–1100°C in Ar [611]. The contents and bonding configurations of iodine were adjusted by varying either the mixing ratios of GO and iodine or the annealing temperature. I-doped rGO prepared at 900°C exhibited the most positive E_{onset} and the highest current density. By increasing the annealing temperature from 500 to 900°C, the I content decreased from 1.21 to 1.05 wt.%, associated with the transformation from I_5^- to I_3^-, suggesting that I_3^- contributed to enhancement of ORR activity more than I_5^-. With the increase in annealing temperature from 900 to 1100°C, the proportion of I_3^- changed slightly, while the overall iodine content clearly decreased, causing the lower activity.

3.4.3 Transition metal-based catalysts

Marked enhancement of the effect of N-doping on ORR activity was reported by transition metal loading onto N-doped porous carbons.

m-Phenylenediamine monomers (*m*PDA) were polymerized on the surfaces of GO flakes using ammonium persulfate as the oxidant, followed by impregnation of $FeCl_3$ and heat treatment for carbonization of PDA and reduction of GO to rGO at 900°C in N_2 [612]. The resultant composites were composed of rGO flakes covered by PDA-derived porous carbons codoped by N- and S. In Fig. 3.188, polarization curves are compared for three composites, Fe-loaded PDA-derived carbon/rGO, Fe-loaded PDA-derived carbon (without rGO), and PDA-derived carbon/rGO

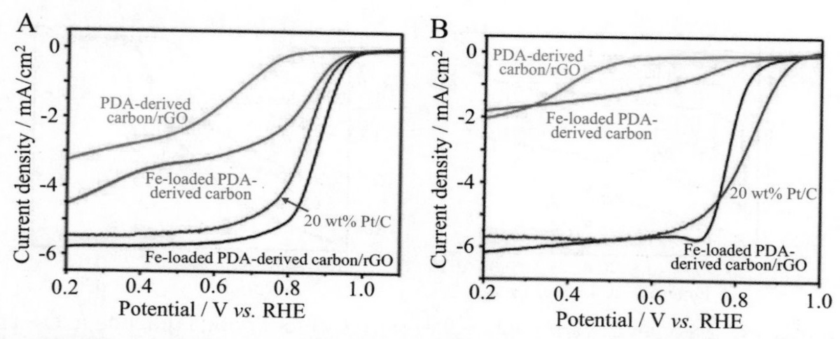

Figure 3.188 Polarization curves for four catalysts, (A) in O_2-saturated 0.1 M KOH and (B) in O_2-saturated 0.1 M $HClO_4$ electrolytes [612].

(without Fe-loading) to a commercial 20 wt.% Pt/C. Loading of both Fe and N/S-codoped PDA carbon onto rGO flakes are very efficient in improving ORR activity not only in alkaline but also acidic electrolytes.

N-doped porous carbon spheres were prepared by carbonization of polydopamine spheres at 800°C, followed by activation with KOH at 700°C [613]. Dopamine hydrochloride, after dissolving in water, was injected into ammonia solution mixed with ethanol to form polydopamine (POD) spheres, which could be carbonized by keeping spherical morphology at 800°C. The diameter of POD spheres was easily tuned in the range of 120−780 nm by adjusting the amount of ammonia aqueous solution. The resultant carbon spheres deliver much higher ORR activity than a glassy carbon substrate using activity measurement, and the porous carbon spheres were prepared using the same procedures from phenol-formaldehyde (PF), as shown in Fig. 3.189A. As shown in Fig. 3.189B,

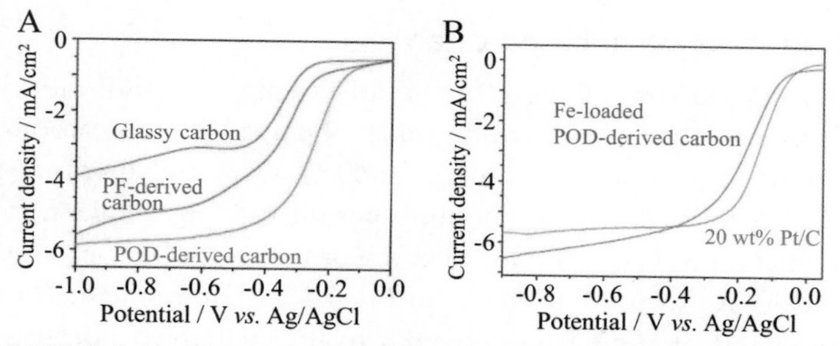

Figure 3.189 Polarization curves in O_2-saturated 0.1 M KOH for polydopamine (POD)-derived carbon: (A) comparison with phenol-formaldehyde (PF)-derived carbon and glassy carbon substrate and (B) comparison of POD-derived carbon after Fe-loading to a commercial Pt/C [613].

the ORR activity of the POD-derived carbon spheres could be improved by loading Fe, which was achieved by carbonization of POD spheres mixed with Fe-acetate at 800°C. The loaded Fe species existed as Fe_3C and metallic Fe in N-doped POD-derived carbon spheres. A marked enhancement of ORR activity by Fe-loading was not observed on PF-derived porous carbon spheres, suggesting the importance of the interaction between Fe and doped N. Enhancement of ORR activity in alkaline electrolyte by Fe-loading was also observed on N-doped rGO [614].

N-doped mesoporous carbons were prepared from porphyrinic metal-organic frameworks (MOF-545 or PCN-222) containing different amounts and different states of Fe by carbonization at 800°C [615]. The precursors PCN-222 with different contents of Fe (Fe_x-PCN-222) were synthesized by mixing Fe-TCPP [TCPP = tetrakis(4-carboxyphenyl) porphyrin] or H_2-TCPP with $ZrOCl_2$ and a small amount of CF_3COOH. The precursors Fe_0-, Fe_{20}-, and Fe_{40}-TCPP were carbonized at 800°C, followed by washing with 20 wt.% HF to exclude ZrO_2, and resulted in the mesoporous carbon with no Fe-loading, that loaded by Fe as single atoms (atomic Fe-loading), and that loaded by Fe as nano-sized aggregates (aggregated Fe-loading), respectively. Polarization curves in alkaline and acidic electrolytes are compared to a commercial Pt/C in Fig. 3.190. The N-doped mesoporous carbon loaded by single-atom Fe gave $E_{1/2}$ of 0.891 V, which was higher than those with no Fe loading and loaded by nanoparticles (0.795 and 0.889 V, respectively) and even the Pt/C (0.848 V) in O_2-saturated 0.1 KOH electrolyte. In acidic electrolyte (0.1 M $HClO_4$), Fe-loading and also the state of loaded Fe govern the ORR

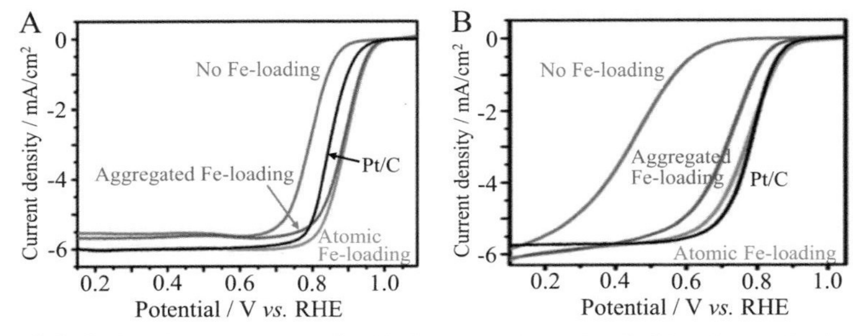

Figure 3.190 Polarization curves for N-doped carbons loaded by Fe as single atoms and aggregated nanoparticles, in comparison with no Fe loaded and a commercial Pt/C: (A) in O_2-saturated 0.1 M KOH and (B) 0.1 M $HClO_4$ [615].

activity more markedly, with the carbon-loaded Fe as a single-atomic state giving $E_{1/2}$ of 0.776 V, which is much higher than for the other two mesoporous carbons and is even comparable to Pt/C.

N-doped carbon nanotube/nanoparticle composites loaded by Fe (N–Fe–CNT/CNP) were prepared from mixtures of cyanamide, iron acetate, and carbon black (Black Pearls 2000) by carbonizing at 950°C in N_2 [616]. The resultant composites composed of N-doped CNTs, which had diameters of 20–30 nm, lengths of about 10 μm, and bamboo-like texture (characteristic for N-doped CNTs), and N-doped carbon nanoparticles, these two components being homogeneously distributed. Metallic Fe nanoparticles were encapsulated by carbon either inside of the CNTs or nanoparticles. ORR polarization curves in O_2-saturated 0.1 M NaOH are shown in Fig. 3.191A in comparison with a commercial 20 wt.% Pt/C catalyst (Pt-loading of 60 μg/cm^2), revealing that the composite at a Fe loading of 0.2 mg/cm^2 gives $E_{1/2}$ of 0.87 V, which is lower than that for the Pt/C of 0.91 V, while E_{onset} is about 40 mV higher than for Pt/C (inset figure). The composite has excellent cycle durability in the potential range of 0.6–1.0 V at a scan rate of 50 mV/s, as shown in Fig. 3.191B. N-doped CNTs with bamboo-like nanotexture could encapsulate Fe nanoparticles inside, which exhibits high ORR activity with E_{onset} of 1.10 V and $E_{1/2}$ of 0.93 V, as well as low H_2O_2 yield (<1%) in O_2-saturated 0.1 M NaOH [617]. They were synthesized from a mixture of a carbon black (KJ600) with three N-precursors (benzoguanamine, cynuric acid, and melamine) and $FeCl_3$ by stepwise heating at 700, 800, and 900°C with intermittent washing with H_2SO_4 to remove Fe_3C and FeS.

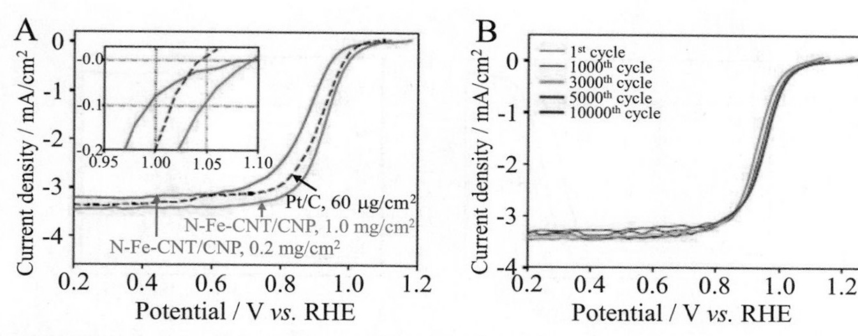

Figure 3.191 Polarization curves of N–Fe–CNT/CNP composites: (A) with different loading of Fe, in comparison with commercial Pt/C and (B) cycle performance of the composite with Fe-loading of 0.2 mg/cm^2 [616].

MnO nanoparticles coated by N-doped mesoporous carbon were synthesized from the mixture of $Mn(NO_3)_2$ and aniline with $KMnO_4$ in aqueous solution by treating under hydrothermal conditions at 180°C, followed by carbonization at 900°C in Ar flow and finally etching in 0.1 M H_2SO_4 [618]. MnO nanoparticles are encapsulated into carbon walls of mesopores, although a small number of large MnO nanoparticles were observed, as shown by the TEM in Fig. 3.192A and B. As shown by the cyclic voltammograms in Fig. 3.192C, the MnO nanoparticles show a cathodic broad peak at 0.56 V, and the N-doped mesoporous carbon shows a cathodic peak at 0.75 V in O_2-saturated 1 M KOH electrolyte. In contrast, the MnO encapsulated into mesoporous carbon (the composite) delivers a much more positive cathodic peak at 0.82 V with a higher cathodic current, suggesting the synergetic ORR catalytic activity of two components, MnO nanoparticles and N-doped mesoporous carbon, in the composite. The cathodic peak for the composite is higher than that for a mechanical mixture of the two components, probably due to better coupling between them in the composite. The electron transfer number n for the composite was calculated to be 3.84 at potentials from 0.45 to 0.65 V, suggesting the dominant four-electron oxygen reduction process, whereas n values for the two components were 2.56–2.65. The composite also exhibited high stability and methanol tolerance, which were superior

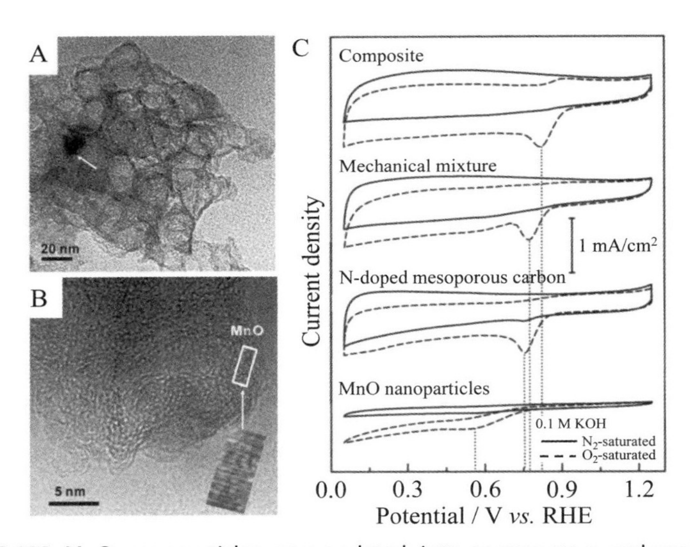

Figure 3.192 MnO nanoparticles encapsulated into mesoporous carbon: (A) and (B) TEM images and (C) CVs in N_2- and O_2-saturated 0.1 M KOH electrolyte [618].

to Pt/C. MnO-encapsulated mesoporous carbon nanofibers were prepared by electrospinning of the DMF solution of manganese acetylacetonate, polyacrylonitrile (PAN), and PVP, followed by stabilization at 280°C and carbonization at 800°C in Ar [619]. Pore structure control by adding PVP and loading of MnO onto N-doped carbon nanofibers was shown to be effective in improving the ORR activities of N-doped carbon nanofibers.

N-doped porous carbons were synthesized from the powder mixtures of ethylenediaminetetraacetic acid (EDTA), melamine, KOH, and various amounts of $Co(NO_3)_2 \cdot 6H_2O$ at 700°C [620]. The carbon synthesized using 0.45 g $Co(NO_3)_2 \cdot 6H_2O$ per 4 g EDTA (Co = 45) had a N/C weight ratio of 0.115 and S_{BET} of 1485 m^2/g, while that without Co compound (Co = 0) had 0.184 and 1678 m^2/g, respectively. In Fig. 3.193A, CVs in N_2- and O_2-saturated 1 M KOH electrolytes for the N-doped carbons with and without Co are compared with the commercial 20 wt.%Pt/C. In O_2-saturated electrolyte, Co-loaded carbon shows a similar curve to Pt/C, and a more enhanced ORR peak at the potential of 0.79 V than the carbon without Co-loading, in contrast to the curves in N_2-saturated electrolyte. The polarization curves in Fig. 3.193B demonstrate the maximum enhancement in ORR activity by Co-loading up to the carbon Co = 45, which is comparable with Pt/C. The electron transfer number *n* increases also with increasing Co content up to Co = 45, which is almost 4 in the potential range of 0.15–0.65 V, as shown in Fig. 3.193C. Enhancement of ORR activity by N-doping and Co-loading was demonstrated on hollow carbon spheres synthesized from the mixture of 2,4-dihydroxybenzoic acid (DA) and hexamethylenetetramine (HMT) using two surfactants, P123 and sodium oleate, under hydrothermal conditions at 160°C, followed by carbonization at 650°C in an NH_3 atmosphere and then by heating at above 750°C after mixing with $CoCl_2$ [621]. Co-loaded N-doped porous carbons were synthesized from the

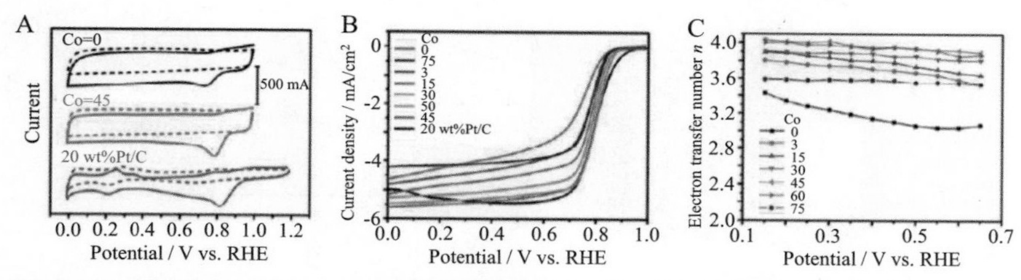

Figure 3.193 N-doped porous carbons loaded by Co: (A) CVs in O_2- and N_2-saturated KOH (*solid and dotted lines*, respectively), (B) polarization curves in O_2-saturated electrolyte, and (C) electron transfer number *n* in the potential range of 0.15–0.65 V [620].

mixtures of 1,3,5-tris(40-hydroxy-50-formylphenyl) benzene (Salen-based covalent organic framework) with 1,2-diaminobenzene and cobalt (II) acetate via DMF solution by carbonization at 700, 800, and 900°C in N_2 [622]. The carbon obtained at 800°C exhibited a significantly higher ORR activity in O_2-saturated 0.1 M KOH than 700- and 900°C-treated ones and much higher than the carbon prepared without Co. Co_3Ni-loaded N-doped honeycomb carbons were prepared from the mixture of nickel acetate and cobalt acetate (in 3/1 mole ratio) and dicyandiamide with honeycomb carbon by calcination at 850°C in N_2 [623]. The composite loaded by 40 wt.% metal alloy exhibited a similar polarization curve with a commercial Pt/C catalyst in O_2-saturated 1 M NaOH.

3.5 Hydrogen storage

Hydrogen storage is a key technological barrier to the development and widespread use of fuel cell power technologies in transportation, stationary, and portable applications. The US Department of Energy (DOE) listed the following materials for hydrogen storage: adsorbents (e.g., MOF-5), liquid organics (e.g., BN-methylcyclopentane), interstitial hydrides (e.g., $LaNi_5H_6$), complex hydrides (e.g., $NaAlH_4$), and chemical hydrogen (e.g., NH_3BH_3), in addition to physical techniques, such as compression and liquefaction [624]. Microporous carbons have been considered as one of the candidates in "adsorbents" of which the adsorption capacity has been targeted as 9.0 wt.%. In a recent report by the DOE, however, carbon-based adsorbents for hydrogen are not mentioned and the target for hydrogen storage for automobile applications was defined on the bases of the total mass/volume of the whole storage system, i.e., the media for storage including stored hydrogen and the system components, such as 0.045 kg H_2 per kg system for the year 2020. These target values, such as 0.045 kg/kg (i.e., 4.5 wt.%) and 0.03 kg/L, are very severe figures for carbon materials if taking the heavy weight of the storage system into consideration. However, different carbon materials, not only activated carbons (ACs), including activated carbon fibers (ACFs), but also carbon nanofibers and carbon nanotubes, including those doped by alkali metals, have been studied for hydrogen storage, whereas most of the results were not high in adsorption capacity, with very poor reproducibility in capacity values. The porous carbons produced by new techniques, such as template-assisted carbonization and polymer-blend methods, were often used to test the hydrogen adsorption behaviors. It was experimentally shown using an

AC that, to reach a capacity of 0.041 kg/L of the storage vessel, it needs to compress the gas up to a pressure as high as 75 MPa at 298K, whereas it is up to 15 MPa by compression at 77K [625]. Therefore, investigations into hydrogen storage have often been performed at 77K to evaluate the storage performance of carbon materials.

3.5.1 Physical adsorption storage

3.5.1.1 Porous carbons

Hydrogen adsorption isotherms were carefully measured on various carbon materials, ACs, ACFs, single-wall carbon nanohorns (SWCNHs), SWCNTs, and CNFs, and even their adsorption capacities were not high. The data on hydrogen adsorption capacities on various ACs are summarized as a function of V_{micro} in Fig. 3.194A, suggesting that H_2 adsorption is mainly governed by micropores, but the procedure to determine V_{micro}, either using N_2 adsorption isotherms at 77K or CO_2 isotherms at 273K [$V_{micro}(N_2)$ or $V_{micro}(CO_2)$, respectively] must be selected [626]. The plots of the maximum capacities, which are either measured directly as a plateau of the adsorption isotherms or as extrapolated values of the isotherm using an approximated equation, against $V_{micro}(CO_2)$ give much better linearity of the relation, as shown in Fig. 3.194B, revealing the importance of ultramicropores (<0.7 nm in diameter) for hydrogen adsorption. The importance of micropores, particularly narrow micropores evaluated by CO_2 adsorption at 273K, for hydrogen adsorption was demonstrated by detailed studies into various carbon materials [627,628]. An AC (Maxsorb) showed the best hydrogen storage capacity of 0.67 wt.%

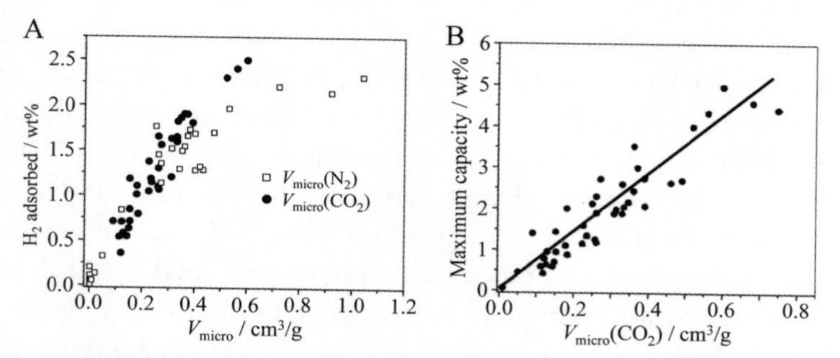

Figure 3.194 Relation between hydrogen adsorption capacity and micropore volume on various carbons: (A) H_2 adsorbed at 77K under 0.1 MPa against micropore volume V_{micro} measured by N_2-adsorption at 77K [$V_{micro}(N_2)$] and by CO_2 at 273K [$V_{micro}(CO_2)$] and (B) H_2 adsorption capacity against $V_{micro}(CO_2)$ [626].

at room temperature, while it gave 5.7 wt.% at 77K under a low pressure of 3 MPa. Hydrogen adsorption at 77K under pressures of up to 0.1 MPa was studied on ACs derived from different precursors and different pretreatments [629]. H_2 storage capacity was well related to $V_{micro}(CO_2)$ and functional groups on ACs had a small detrimental effect on hydrogen adsorption. A linear relation between H_2 uptake at 77K and the volume of narrow micropores (<1.3 nm) was confirmed using ACs, porous carbons, and SWCNTs [630]. The highest sorption capacity of 4.5 wt.% at 77K was observed on the AC having S_{BET} of 2560 m^2/g and the volume of narrow micropores of 0.75 cm^3/g.

Monoliths of anthracite-derived ACs, which were prepared by compression with a polymeric binder, followed by carbonization of the binder at 750°C, gave a volumetric hydrogen capacity of 29.7 g/L at 77K and 4 MPa [631]. H_2-adsorption capacities at 77K under pressures of up to 0.1 MPa and at 303K up to 3.5 MPa were measured on the phenolic-resin-derived ACFs heat treated at 1000–2300°C (S_{BET} 980–2250 m^2/g), two SWCNTs (as-received and activated in 4 M HNO_3 followed by 700°C annealing in H_2 flow) and two kinds of zeolites (H–Y-type and H-ZSM-5), the results being discussed in relation to V_{micro} [632]. H_2-uptakes at 77K and 0.1 MPa and at 303K and 3.1 MPa measured on four ACFs could be placed on a curve against V_{micro}, which is common to two zeolites. However, the pristine and activated SWCNTs exhibited much higher H_2-uptakes than ACFs although their V_{micro}'s were much smaller; activated SWCNT having V_{micro} of 0.25 cm^3/g delivered H_2-uptake at 303K and 3.1 MPa of about 2.5 wt.%, while ACF having a similar V_{micro} of 0.38 cm^3/g delivered H_2-uptake of about 1.2 wt.%. A commercially available AC (S_{BET} = 1585 m^2/g and V_{micro} = 0.59 cm^3/g) was subjected to additional activation using CO_2 and KOH, and the results showed that KOH activation was more effective, increasing V_{micro} to 1.09 cm^3/g and H_2-uptake to 7.08 wt.% at 77K under 2 MPa [633]. H_2-uptakes were compared among two ACs, i.e., a commercial AC (Maxsorb 3000) and the AC made in a laboratory from anthracite by KOH activation, and two kinds of MOFs, i.e., MOF-5 and IRMOF-1 [634]. Gravimetric and volumetric isotherms of H_2 adsorption are compared in Fig. 3.195A and B, respectively. Although these three adsorbents had similar H_2-uptake on a gravimetric basis, two ACs have higher H_2-uptake than MOF-5 on a volumetric basis.

The amount of H_2 adsorbed at different temperatures into MWCNTs prepared by catalytic CVD of acetylene was compared with an AC having

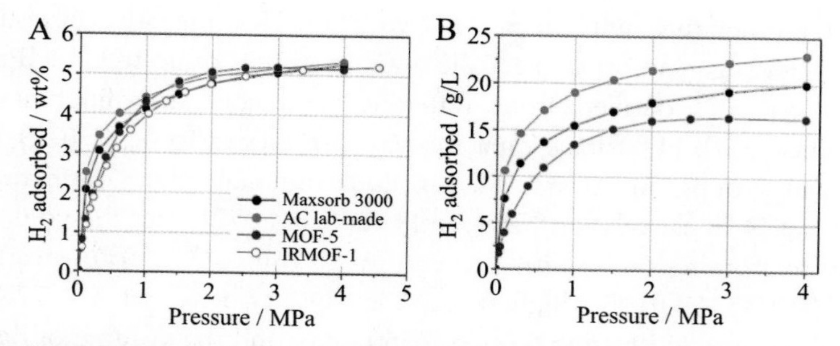

Figure 3.195 H_2 adsorption isotherms at 77K for two ACs (Maxsorb 3000 and a laboratory-made one) and two MOFs (MOF-5 and IRMOF): (A) gravimetric and (B) volumetric scales [634].

S_{BET} of 3000 m^2/g (AX-21) [635]. The AC has a much higher adsorption capacity than MWCNTs, but has much stronger dependence of capacity on temperature, as shown in Fig. 3.196. A double-walled carbon nanotube (DWCNT) was shown to adsorb relatively large amounts of H_2 at 77K under pressure of less than 0.1 MPa, which is higher than SWCNTs, even though DWCNTs have much lower S_{BET} and V_{micro} [636]. Hydrogen storage in CNTs was reported to be less than 1.7 wt.% under the pressure up to 12 MPa at room temperature [637].

Monoliths were prepared from CNFs produced by the polymer blend technique from a novolac-type phenolic resin, followed by activation using CO_2, KOH, and NaOH [638]. Hydrogen storage at 77K under 0.11 MPa

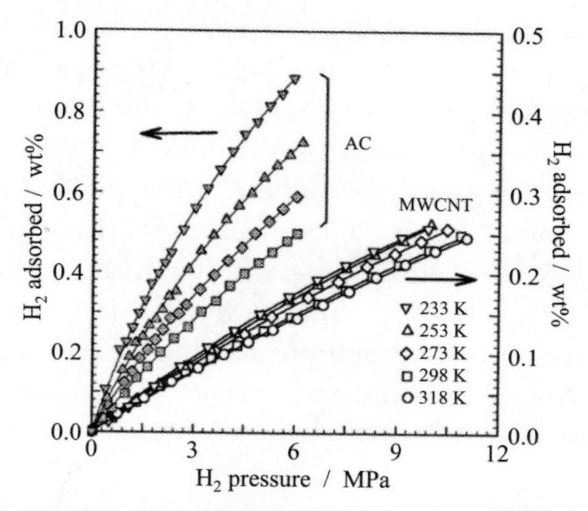

Figure 3.196 Amount of H_2 adsorbed into an AC and an MWCNT at different temperatures [635].

is plotted against S_{BET} and $V_{micro}(CO_2)$ by comparing with commercial ACFs in Fig. 3.197A and B, respectively, revealing that H_2 adsorption is related to $V_{micro}(CO_2)$ and there was no marked difference between activated CNFs and ACFs. Activated CNFs had sufficient compressibility to achieve a high packing density. Electrospun PAN-based carbon nanofibers were also studied after activation with NaOH and K_2CO_3 at 750°C [639]. SWCNT oxidized in HNO_3 (activated SWCNT) gave higher S_{BET} and V_{micro} (710 m^2/g and 0.25 cm^3/g, respectively) and its H_2-uptake at a given V_{micro} was higher than that of the zeolite and ACFs, suggesting that acid treatment of SWCNTs may increase the sites with a high interaction potential for hydrogen adsorption [632].

Hydrogen adsorption behaviors of zeolite-templated microporous carbons (ZTCs) with different S_{BET} of 1610–3800 m^2/g and $V_{micro}(N_2)$ of 0.6–1.58 cm^3/g were studied at temperatures of 30–150°C under H_2 pressures of 1–34 MPa by comparison with commercial ACs [S_{BET} of 1700–2680 m^2/g and $V_{micro}(N_2)$ of 0.74–1.20 cm^3/g] [640]. ZTCs were prepared by FA impregnation, followed by carbonization at 700°C, and then densification by CVD of either propylene or acetonitrile at 700–850°C. H_2 adsorption isotherms on ZTCs prepared under different conditions are shown in Fig. 3.198A, suggesting the necessity of the optimization of preparation conditions. In Fig. 3.198B, the H_2 adsorption capacity is plotted against surface area calculated from N_2 isotherm by the subtracting pore effect (SPE) method (S_{SPE}) under different H_2 pressures up to 10 MPa, revealing marked pressure dependence and no difference between ZTCs and ACs. However, the ZTC prepared via propylene CVD at 700°C for 2 h with the following annealing at 900°C exhibited

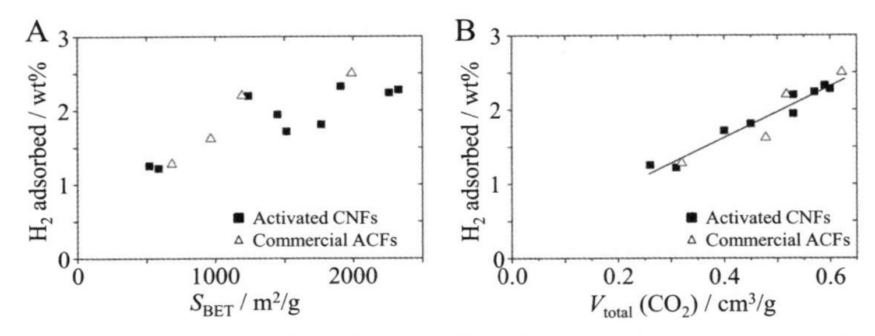

Figure 3.197 H_2-uptake at 77K under 0.11 MPa of activated CNFs in comparison with commercial ACFs as a function of (A) S_{BET} calculated from the adsorption isotherms of N_2 at 77K and (B) $V_{micro}(CO_2)$ calculated by DR equation from CO_2 adsorption isotherms at 298K [638].

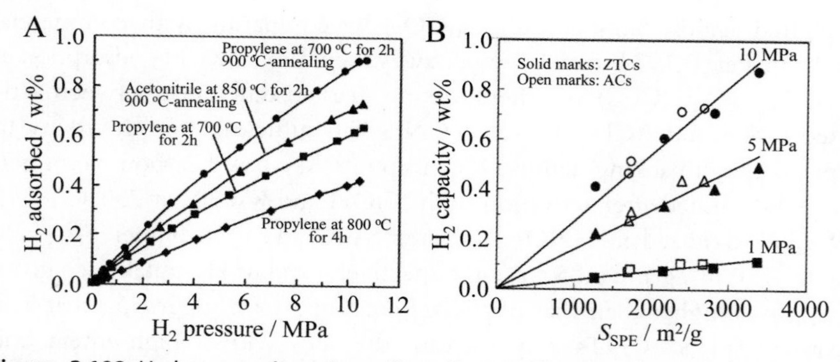

Figure 3.198 Hydrogen adsorption of zeolite-templated microporous carbons in comparison with commercial ACs: (A) adsorption isotherms at 30°C and (B) adsorption capacity at 30°C under different H_2 pressures versus surface area of the carbons calculated by SPE method (S_{SPE}) [640].

H_2-uptake as high as 2.2 wt.% at 30°C under high pressure of 34 MPa, much higher than for commercial ACs, which is mainly due to a marked contribution of the uniform micropores with a diameter of 1.2 nm. ZTCs prepared by CVD of either acetonitrile or ethylene on zeolite 13X and Y delivered H_2-uptakes of 2.0–4.5 wt.% at 77K under 2 MPa [641]. A ZTC synthesized by CVD of acetonitrile at 800°C, which contained 6 wt.% N and S_{BET} of 1825 m^2/g, exhibited the highest H_2-uptake of 4.5 wt.%. A ZTC prepared using acetonitrile at 850°C, followed by annealing at 850°C in N_2 flow, was reported to give an H_2-uptake of 6.9 wt.% at 77K under 2 MPa [642]. It was also used after KOH activation, reporting the increases in V_{micro} from 0.74 to 1.01 cm^3/g and in hydrogen uptake from 4.81 to 6.30 cm^3/g at 77K under 2 MPa [643].

Carbide-derived microporous carbons (CDCs) have also been studied on hydrogen storage. CDCs prepared from different carbides by chlorination gave H_2 storage up to 3.0 wt.% at 77K under 0.1 MPa pressure after annealing in H_2 for removing trapped Cl_2 [644]. Fig. 3.199A shows that S_{BET} of the CDCs from various metal carbides depends strongly on the precursor carbides and their chlorination temperatures. However, H_2-uptake of the CDCs was not uniquely related to S_{BET}; the capacity of the CDC derived from B_4C at 800°C (giving the highest S_{BET}) was 1.91 wt.%, while that derived from TiC at the same temperature was 2.55 wt.%. In Fig. 3.199B, dependences of S_{micro}, S_{meso}, and H_2-uptake are plotted against chlorination temperature, revealing clearly that micropores (<2 nm size) govern H_2-uptake. CDCs demonstrated higher H_2-uptake compared to MOF-5, SWCNTs, and MWCNTs. On the CDC prepared from B_4C

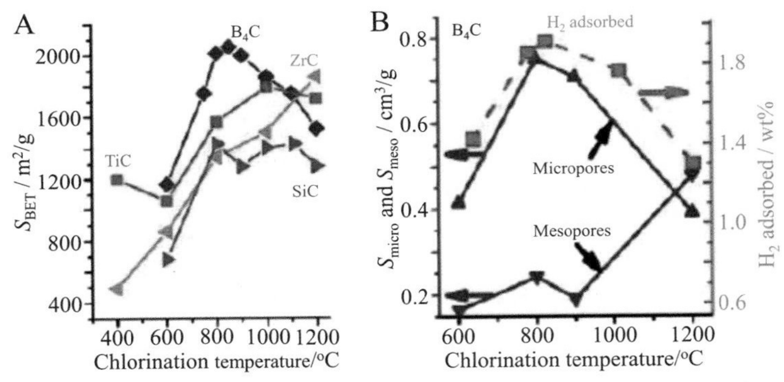

Figure 3.199 Carbide-derived microporous carbons: (A) changes in S_{BET} for various carbides and (B) S_{micro}, S_{meso}, and H_2 adsorbed on B_4C-derived carbon with chlorination temperature [644].

at 1000°C, H_2 capacity was reported to be 1.06 wt.% at 77K under 0.1 MPa [363]. Microporous CNFs prepared by electrospinning of THF solution of polycarbosilane, followed by chlorination to decompose SiC, gave a high S_{BET} of 3116 m^2/g with H_2 storage of 3.86 wt.% at 77K under 1.7 MPa [645]. CDC derived from TiC by chlorination at 800°C exhibited an H_2 uptake of 2.8 wt.% at 77K [646]. By using CDCs derived from various metal carbides at different chlorination temperatures of 600–1200°C, pores with sizes of less than 1 nm were experimentally demonstrated to be efficient for H_2 adsorption at 77K under ambient pressure, while pores above 1 nm did not contribute much [647]. Annealing of as-prepared CDCs in H_2 was effective in increasing the H_2 uptake. H_2 uptakes at 77K and 10 MPa were reported to be 4.2 and 6.2 wt.% for the CDC derived from SiC and Al_4C_3, respectively [648].

MOFs were reported to have a gravimetric adsorption of 4.33–5.05 wt.% and volumetric capacity of 33.2–37.8 g/L at 77K and around 2 MPa [649]. MOF-5 [$Zn_4O(O_2CC_6H_4CO_2)_3$] gave highly porous carbon by carbonization at 1000°C, with S_{BET} of 2872 m^2/g, V_{total} of 2.06 cm^3/g, and H_2 uptake of 2.6 wt.% at 77K under 0.1 MPa [650]. Different MOFs (IRMOF-1, -3, and -8) were carbonized at 900°C in N_2 [651]. The resultant carbons are rich in ultramicropores and as a consequence exhibit relatively high H_2 uptake, as summarized in Table 3.5, together with those of the precursor MOFs. The resultant carbons have higher H_2 uptake than the precursor MOFs. Porous carbon with S_{BET} of 3405 m^2/g and V_{total} of 2.58 cm^3/g was prepared from the mixture of a zeolite-type MOF (ZIF-8) (as a template and carbon precursor) and FA

Table 3.5 Pore structure parameters and H_2 uptake of MOFs and carbons derived from them at 900°C [651].

	MOF					Carbon derived from MOF				
	S_{BET} (m^2/g)	V_{total} (cm^3/g)	V_{micro} (cm^3/g)	V_{ultra} (cm^3/g)	H_2 uptake[a] (wt.%)	S_{BET} (m^2/g)	V_{total} (cm^3/g)	V_{micro} (cm^3/g)	V_{ultra} (cm^3/g)	H_2 uptake[a] (wt.%)
IRMOF-1	3447	1.45	1.40	0.17	1.3	3174	4.06	1.01	0.63	3.25
IRMOF-3	2351	0.90	0.90	0.18	1.4	1678	2.01	0.66	0.49	2.1
IRMOF-8	1735	0.69	0.69	0.40	1.9	1978	1.92	0.78	0.54	2.41

[a]At 77K under 0.1 MPa pressure.

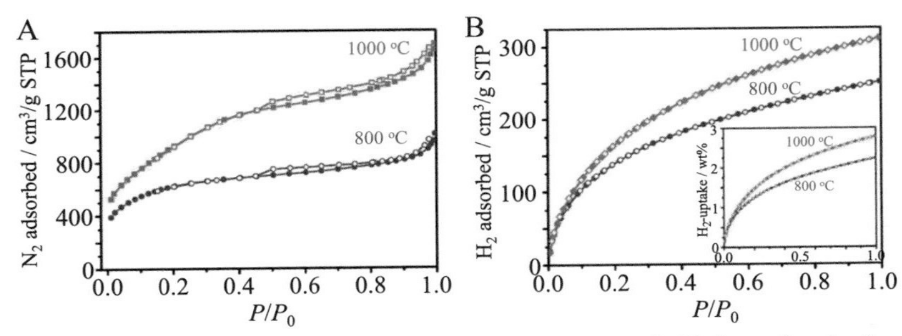

Figure 3.200 Porous carbons from the mixture of ZIF-8 with FA by carbonization at 800 and 1000°C: (A) N_2 adsorption–desorption isotherms at 77K and (B) H_2 adsorption isotherms at 77K [369].

(additional carbon precursor) by carbonization [369]. Isotherms of N_2 adsorption–desorption at 77K and H_2 adsorption at 77K are shown for the resultant carbons in Fig. 3.200. The carbons obtained at 800 and 1000°C exhibited S_{BET} of 2160 and 3405 m^2/g and V_{total} of 1.50 and 2.58 cm^3/g, and as a consequence delivered H_2 uptake of 2.23 and 2.77 wt.% at 77K and 0.1 MPa, respectively. Mesoporous polymer framework (polytriazine) prepared from 1,4-dicyanobenzene with $ZnCl_2$ was carbonized at 400°C and activated by mixing with KOH to obtain N-doped micro-/mesoporous carbons [652]. The pore structure in the resultant carbons depends greatly on the KOH/C weight ratio in the range of 3/1–7/1 and activation temperature from 650 to 850°C, S_{BET} in 1611–2406 m^2/g including S_{micro} of 1264–2203 m^2/g and V_{total} of 0.88–1.38 cm^3/g including V_{micro} of 0.57–1.08 cm^3/g. H_2 uptake increased from 3.04 to 4.77 wt.% proportionally with increasing S_{BET}.

Hydrogen adsorption of mesoporous carbons prepared via mesoporous silica templates (MCM-48, SBA-15, etc.) was also studied, in which H_2 diffusion is facilitated by the interconnected mesopores but a large content of micropores is demanded to enhance the H_2 uptake. CO_2 activation at 950°C for 2, 4, and 6 h was applied on the ordered mesoporous carbon CMK-3 prepared from sucrose using SBA-15 template at 900°C [653]. The pore structure parameters and H_2 uptake at 77K under 0.1 MPa are listed in Table 3.6. V_{micro} is divided into the micropores with size less than 1 nm ($V_{<1\ nm}$) and those with size of 1–2 nm ($V_{1\sim2\ nm}$). After activation for 6 h, S_{BET} increases to 2749 m^2/g and V_{micro} increases markedly to 0.96 cm^3/g, and consequently H_2 uptake to 2.24 wt.% at 77K and 0.1 MPa. Between H_2 uptake and $V_{<1\ nm}$, a strictly linear relation was obtained. H_2 uptake was measured at 77K under 1 MPa on ordered

Table 3.6 Pore structure parameters and H_2 uptake for CO_2-activated CMK-3 [653].

	S_{BET} (m^2/g)	V_{total} (cm^3/g)	V_{meso} (cm^3/g)	V_{micro} (cm^3/g)	$V_{1\sim2\ nm}$ (cm^3/g)	$V_{<1\ nm}$ (cm^3/g)	H_2 uptake (wt.%)
Pristine (CMK-3)	984	1.09	0.72	0.37	0.22	0.15	0.98
950°C, 2 h	1814	1.76	1.07	0.69	0.41	0.28	1.84
950°C, 4 h	2110	2.01	1.22	0.79	0.49	0.30	1.95
950°C 6 h	2749	2.09	1.13	0.96	0.62	0.34	2.24

mesoporous carbons synthesized from a petroleum pitch, sucrose, and propylene was well related to their $V_{micro}(CO_2)$ [654]. The carbon synthesized by CVD of propylene into SBA-15 gave the lowest H_2 uptake of about 1.4 wt.% and the one by repeated impregnation of sucrose into MCM-48 gave the highest uptake of about 3.5 wt.%. H_2 uptake of nitrogen-doped mesoporous carbons prepared from ethylenediamine and CCl_4 using SBA-15 increased by KOH activation at 750°C [655]. When H_2 gas pressure was 2 MPa, H_2 uptake at 77K increased up to 6.84 wt.% after KOH activation from 1.27 wt.% for the pristine carbon, while under a low gas pressure of 0.1 MPa H_2 uptake at 77K increased up to 2.34 wt.% after KOH activation from 0.53 wt.% for the pristine carbon. It was also reported, however, that the treatment of CMK-3 in NH_3 flow for 5 h at 400, 500, and 700°C made no marked change in H_2 uptake at around 1.2–1.3 wt.% at 77K under 0.1 MPa, although it was effective in increasing the N content in the carbon, with increasing temperature N content increased up to 3.9 wt.% [656]. An ordered mesoporous carbon derived from SiC using SBA-15 and chlorination exhibited a high S_{BET} of 2819 m^2/g and V_{total} of 2.26 cm^3/g including V_{micro} of 0.51 cm^3/g, of which H_2 uptake was 2.54 wt.% at 77K and 0.1 MPa [657].

Milling of high-purity natural graphite in a hydrogen atmosphere resulted in high uptake of hydrogen, although S_{BET} decreased to a negligibly small amount after passing a maximum (about 400 m^2/g) at around 2–4 h milling [658]. The H_2 content increases gradually with increasing milling time, accompanying some changes in 002 X-ray diffraction profile, shifting to lower angle side, broadening and weakening, and reaches about 7.4 wt.% (corresponding to the composition of $CH_{0.95}$), as shown in Fig. 3.201A. Desorption of H_2 from the milled graphite occurs in two steps

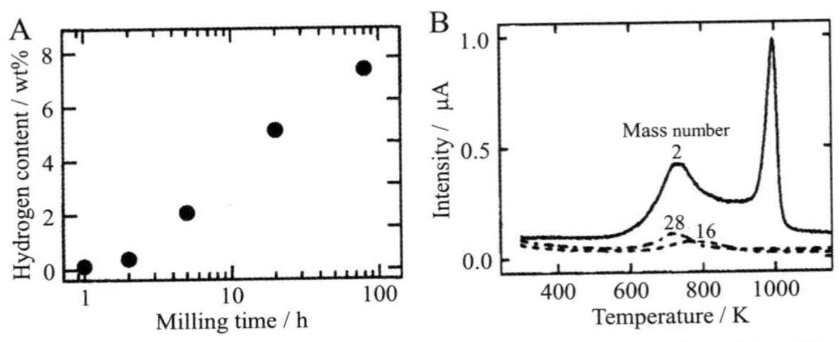

Figure 3.201 Natural graphite milled in H_2: (A) change in H_2 uptake with milling time and (B) thermal desorption mass spectrograms of 80-h milled graphite for different mass numbers [659].

at around 700 and 1000°C, as shown by the thermal desorption spectrograms on 80 h-milled graphite in Fig. 3.201B [659]. At the first desorption step, about 80% of adsorbed hydrogen (corresponding to about 6 wt.% hydrogen) was released as H_2 molecules, associated with small amounts of hydrocarbons (mass numbers of 16 and 28). The remaining hydrogen (about 20 wt.%) was sharply released at around 1000K, which was supposed to be covalently bonded to the carbon atom at the edge of the carbon layer during milling in a hydrogen atmosphere. On SWCNTs, graphite and diamond after sonication in 5 M HNO_3, H_2 storage reached 1.5 wt.% at room temperature [660]. Desorption of hydrogen was found to occur in two steps, sharply at around 340K and gradually above 350K to about 650K with a peak at around 500K.

3.5.1.2 Metal-loaded porous carbons

Loading of metals, mostly Ni, was applied on various porous carbons, some papers reporting an enhancement in H_2 uptake but some even a depression in uptake. Ni loading onto an AC was performed by impregnation of either nickel acetate [$Ni(CH_3COO)_2$] or nitrate [$Ni(NO_3)_2$], and then heat-treated in H_2 flow at a high temperature (350−600°C) before measurement of the H_2 adsorption isotherms [661]. The effect of Ni-loading on the AC was not clearly observed in the loading range of 1−10 wt.%, although the pore parameters became lower with an increase in Ni content. The reduction of Ni-acetate in AC was performed using hydrazine (N_2H_4) at 80°C, instead of H_2 flow at high temperatures [662]. For 1 wt.% Ni-loaded AC, hydrazine reduction resulted in a slight increase in H_2 uptake under 2 MPa to 0.51 wt.%, and to 0.31 wt.% by H_2-reduction. Ni-loading onto commercial ACFs was performed by a vacuum impregnation process using acetone solution of $Ni(NO_3)_2$, followed by reduction at 100°C in H_2 [663]. Ni-loading was effective to improve H_2 uptake at 298K, more markedly than surface modification in diluted F_2 flow, the uptake reaching about 1.3 wt.% under 10 MPa, as shown by the isotherms in Fig. 3.202. The effects of Ni-loading and N-doping of mesoporous carbons on H_2 adsorption were studied using ordered mesoporous carbons synthesized using sucrose and SBA-15 at 900°C (CMK-3), in comparison with mesoporous graphitic carbon nitride (g-C_3N_4) synthesized using CCl_4 and ethylenediamine in mesopore channels of SBA-15 [664]. Ni-loading was performed by impregnation of $Ni(NO_3)_2$ into CMK-3 and following calcination in He and reduction in H_2 flow at 500°C, and N-doping by heat treatment of CMK-3 in NH_3 flow at 700°C. Pore parameters and H_2 uptakes at 77 and 298K under 3 MPa pressure are summarized in Table 3.7. Ni-loading on the pristine CMK-3 is not effective in enhancing the H_2 uptake at 77K, but effective in increasing the uptake at 298K. On the other hand, N-doping increases the uptake slightly at 77K and markedly at 298K. Mesoporous carbons were prepared from sucrose using the synthesized mesoporous zeolite ZSM-5 as template at 900°C, on which Ni was loaded by vacuum impregnation of $NiSO_4$ and following calcination at 450°C in N_2 [665]. Ni-loading onto the resultant mesoporous carbons made pore structure parameters and also H_2 uptake at ambient temperature lower; the carbon loaded by 15 wt.% Ni exhibited S_{BET} of 737 m^2/g, V_{total} of 0.544 cm^3/g, and H_2 uptake of 0.633 wt.%, although the pristine carbon (no Ni-loading) gave 910 m^2/g, 0.713 cm^3/g, and 2.124 wt.%, respectively.

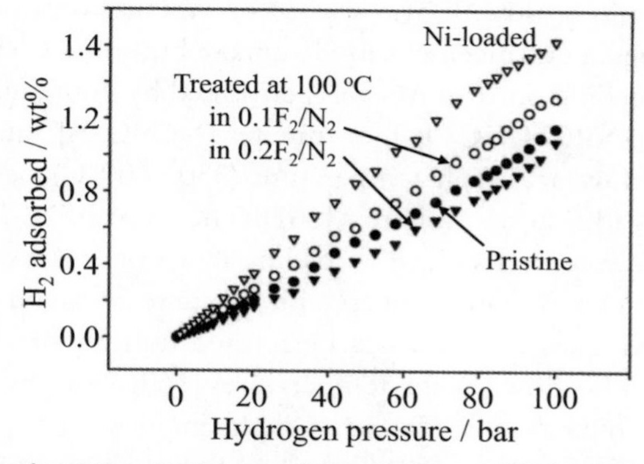

Figure 3.202 H_2 absorption isotherms of the pristine and Ni-loaded activated carbon fibers at 303K, in comparison with functionalized fibers in diluted F_2/N_2 flows [663].

Table 3.7 Pore structure parameters and H_2 uptake of mesoporous carbons and *g*-C_3N_4 with N-doping and Ni-loading [664].

			Pore structure parameter			H_2 uptake (wt.%)	
	Ni-loading (wt.%)	N-doping (wt.%)	S_{BET} (m^2/g)	V_{meso} (cm^3/g)	$V_{micro}(CO_2)$ (cm^3/g)	77K, 3 MPa	298K 3 MPa
CMK-3	—	—	825	1.14	0.22	2.3	0.12
	5.6	—	682	0.85	0.24	2.1	0.20
	—	3.9	942	1.06	0.22	2.8	0.22
	5.5	3.9	984	1.18	0.20	2.3	0.26
g-C_3N_4	—	19.6	286	0.45	0.12	0.9	0.04
	13.8	—	376	0.65	0.11	0.9	0.04

Recovering (desorption) of H_2 stored in Ni-loaded mesoporous carbons requires heating at high temperatures because of chemisorption of hydrogen atoms on the carbon surface as spilt-over species. Carbon aerogels were prepared from RF at 800°C in N_2, of which the averaged mesopore size was controlled to be 16 and 45 nm (coded as CA-A and CA-B, respectively). Ni-loading onto these CAs was achieved either by conventional impregnation of $Ni(NO_3)_2$ via its aqueous solution or electrostatic interaction (strong electrostatic adsorption, SEA) between porous carbon and Ni^{2+} after KOH activation, and then surface modification (oxidation) by HNO_3 of the substrate carbon [666]. The conventional impregnation method could load a larger amount of Ni than the newly developed SEA method, but neither had a serious effect on pore structure parameters. Hydrogen adsorption isotherms were measured at 298K up to 20 MPa and at 77K up to 4 MPa after activation of the adsorbents at 400°C in H_2 flow for 2 h. H_2 desorption from these adsorbents was evaluated by temperature programmed desorption (TPD) by heating the sample up to 800°C after adsorbing H_2 at 400°C for 2 h. TPD profiles observed on Ni-loaded CAs are shown by separating the temperature region below 400°C (Fig. 3.203A and B) and above 350°C (Fig. 3.203C). On the CAs loaded with Ni via impregnation (Fig. 3.203A and C), the TPD profiles show two different regions of H_2 evolution: below 250°C and above 350°C. The first region is considered to correspond to hydrogen adsorbed on Ni particles, which do not interact with the substrate carbon, and the second region is assigned to hydrogen atoms adsorbed more strongly, probably due to the spill-over effect resulting from the Ni–carbon interaction. On the CAs

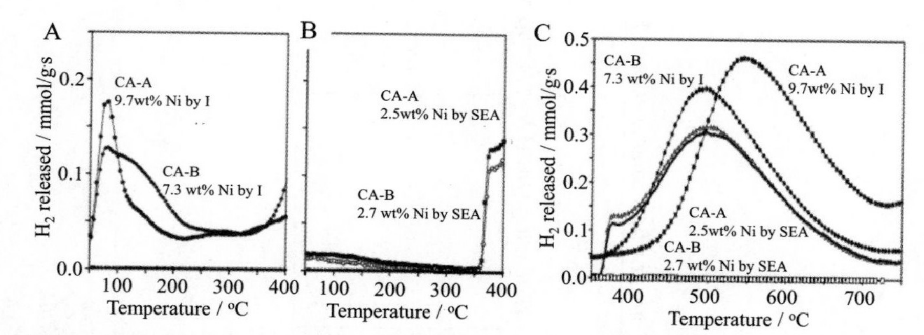

Figure 3.203 Temperature-programmed desorption profiles of H_2 desorption for Ni-loaded mesoporous carbons (CA-A and CA-B): (A) in a temperature range of 50–400°C for the Ni-loaded by impregnation (I) and (B) for the Ni-loaded by SEA, and (C) in a temperature range of 350–750°C [666].

loaded with Ni via SEA (Fig. 3.203B and C), however, the release of H_2 occurs at high temperatures (above 370°C), with no release at low temperature, suggesting a strong interaction of almost all Ni particles with the surface of the substrate carbon. The peaks of TPD did not appear on all the CAs without Ni-loading. Even at high-temperature desorption, the CAs loaded with Ni by impregnation released large amounts of H_2, larger than those by SEA, demonstrating that some Ni particles loaded by impregnation have a strong interaction with the substrate carbon but some do not, as pointed out before [667], while all Ni particles loaded using the SEA method have a strong interaction. The desorption of H_2 was also studied by the TPD technique on Ni-loaded carbon spheres prepared from FA via impregnation and physical mixing [668].

Ni-loading was applied on MWCNTs to study H_2 adsorption and desorption [669]. MWCNTs were synthesized by microwave plasma-enhanced CVD of CH_4 with Co catalyst at 700°C, on which Ni nanoparticles were loaded via $Ni(NO_3)_2$ acetone solution, followed by reduction using H_2 gas flow. H_2 adsorption was performed at 300K under 4 MPa, and desorption was probed by gas chromatography. The H_2 releasing (desorption) profiles are shown in Fig. 3.204, revealing that H_2 release from Ni-loaded MWCNTs occurs under high temperatures, which are higher than those for pristine (as-grown) MWCNTs, and continued up to 550K. From the H_2 evolution from MWCNTs with Ni-loading of 3, 6, 13, and 40 wt.%, H_2 storage was calculated to be 1.2, 2.8, 1.8, and 0.4 wt.%, respectively, compared with an H_2 uptake of 0.9 wt.% for the pristine MWCNTs without Ni-loading.

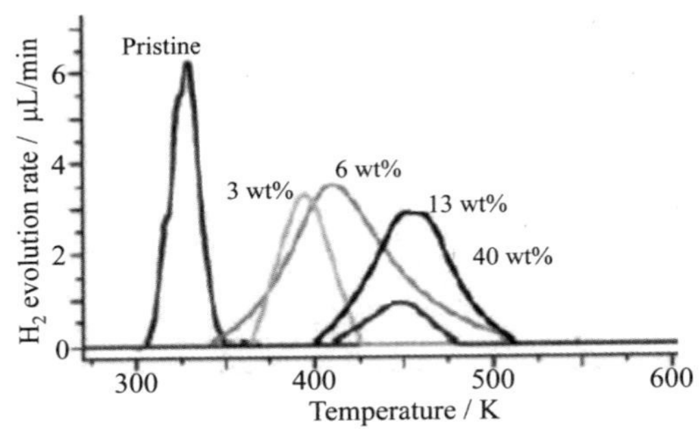

Figure 3.204 H_2-releasing profiles of Ni-loaded MWCNTs with different amounts of Ni [669].

Loading of Pt was performed on an AC (AX-21) by mixing with H_2PtCl_6 in acetone, followed by the heat treatment at 300°C in H_2 flow [670]. H_2 adsorption isotherms were measured at 298K up to the pressure of 10 MPa. By 0.6 wt.% Pt-loading, the pore structure parameters changed only slightly, from S_{BET} of 2880 m^2/g and V_{total} of 1.27 cm^3/g for the pristine AC to 2518 m^2/g and 1.22 cm^3/g for 5.6 wt% Pt-loaded one. H_2 uptake at 298K under 10 MPa, however, almost doubled, from 0.6 wt.% for the pristine to about 1.2 wt.% for 5.6 wt.% Pt-loaded. The H_2 adsorption–desorption isotherm for Pt-loaded AC showed only a slight hysteresis. By evacuation to a pressure of 1 Pa for 12 h at 298K, complete desorption occurred, and the second adsorption isotherm was the same as the first one.

V-loaded carbon fibers were prepared by electrospinning of DMF solution of PAN mixed with V_2O_5, followed by stabilization at 250°C in air and carbonization at 1050°C in N_2, where V_2O_5 was reduced to metallic V during carbonization [671]. The pristine carbon fibers and V-loaded ones were activated using KOH at 750°C. KOH activation was effective in developing pores even on V-loaded carbon fibers, so that S_{BET} increased drastically from 43 m^2/g before activation to 2780 m^2/g after activation. V contents before and after KOH activation were 24.5 and 47.8 wt.%, respectively, and H_2 uptakes at 298K under 10 MPa were 1.78 and 2.41 wt.%.

In practice, most adsorbents are packed into a container and then H_2 gas is introduced under pressure, where adsorption of H_2 into the adsorbent occurs but at the same time as H_2 gas filling the voids between adsorbent particles occurs too [631,666,672]. Therefore, the total H_2 stored in the adsorbent container is higher than the H_2 adsorption capacity measured on the adsorbent itself. An example is presented in Fig. 3.205, where an AC with S_{BET} of 2374 m^2/g and $V_{micro}(CO_2)$ of 0.63 cm^3/g is packed with a density of 0.61 g/cm^3 into a 1-L container and compressed by H_2 gas up to 4 MPa at a temperature of 77K [631]. With increasing pressure applied, the amount of H_2 filled in the voids between particles increases linearly (compressed), but the amount of H_2 adsorbed physically into AC increased rapidly at the low-pressure region and tends to saturate at high pressure (adsorbed). The total storage capacity of this container is measured to be 39.3 g H_2 at 4 MPa and 77K, which is roughly three times the amount stored in the same container just by compression (without packing AC), revealing the importance of the adsorbent, but contribution of the voids between adsorbent particles is not negligible.

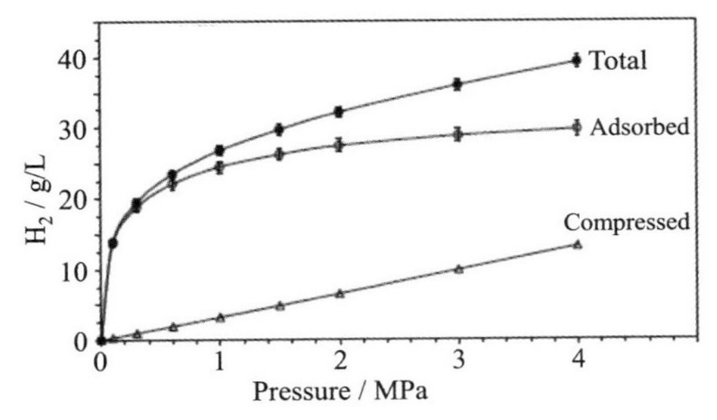

Figure 3.205 Total hydrogen storage capacity of AC monolith packed into a 1-L container at 77K [631].

Theoretical calculations for hydrogen storage into carbons were also reported, but their results were scattered over a wide range [627,673–678].

3.5.1.3 *Graphitic carbon nitrides*

The hydrogen adsorption capacity of *g*-C_3N_4 was experimentally investigated only in a few researches, although several theoretical calculations have predicted relatively high capacity [679–684]. It has to be pointed out that the material "*g*-C_3N_4" mentioned in most of the literature is not well characterized in chemical composition, crystallinity, pore structure, etc. [685].

N-enriched carbons were synthesized from a mixture of cyanuric chloride and melamine by stepwise condensation reactions at 0, 25, and 120°C under atmospheric pressure [686]. For the final products, XRD consisted of 002, 10, 004, and 11 diffraction peaks, clear TEM lattice fringes and spotty electron diffraction pattern were observed. The products had an N/C ratio of 1.12 (1.33 for C_3N_4) by elemental analysis and H_2 storage capacity of 0.34 wt.% at 298K under a pressure of 10 MPa, which is much higher than MWCNTs. At room temperature under a low pressure of 0.1 MPa, however, the capacity for N-enriched carbon was much lower than for MWCNTs. Carbon nitride nanobelts prepared by microwave plasma-assisted CVD, of which the N content was up to 10 at.%, were reported to have an H_2 uptake of 8 wt.% at 300°C under ambient pressure [687].

Relatively high H_2 storage capacities were reported mostly on metal-decorated *g*-C_3N_4. Pd loading was performed by mixing $PdCl_2$ with melamine-derived *g*-C_3N_4 in ethylene glycol by refluxing at 125°C, by which about 20 wt.% crystalline Pd nanoparticles having about 10 nm sizes were loaded [688]. The resultant Pd-loaded *g*-C_3N_4 exhibited H_2 storage capacities of 3.8 and 3.4 wt.% at 0 and 25°C, respectively, under 4 MPa pressure, whereas the pristine g-C_3N_4 had capacities of 2.6 and 2.2 wt.%, respectively, as shown by the isotherms in Fig. 3.206. About 90% H_2 adsorbed could be recovered by reducing the pressure after the saturation of adsorption at room temperature.

3.5.2 Electrochemical storage

Hydrogen storage during water electrolysis was also proposed using various carbon materials (electrochemical H_2 storage).

Electrochemical H_2 storage was studied on viscose-based carbon cloths in alkaline and acidic media, 3 M KOH and H_2SO_4, at room temperature [689]. The carbon cloths were activated by CO_2 at 1000°C to fabricate the microporous carbon with S_{BET} of 1390 m^2/g, V_{total} of 0.522 cm^3/g, and $V_{micro}(CO_2)$ of 0.291 cm^3/g. As shown by the galvanostatic charge–discharge curves in Fig. 3.207, H_2 storage proceeds spontaneously during

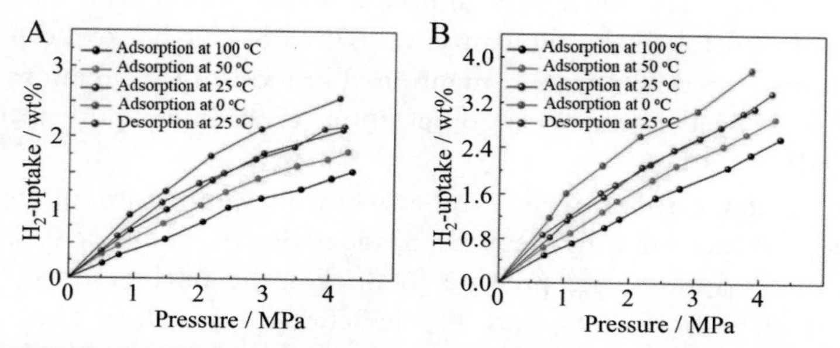

Figure 3.206 H_2 adsorption isotherms at different temperatures for (A) pristine g-C_3N_4 and (B) Pd-loaded g-C_3N_4 [688].

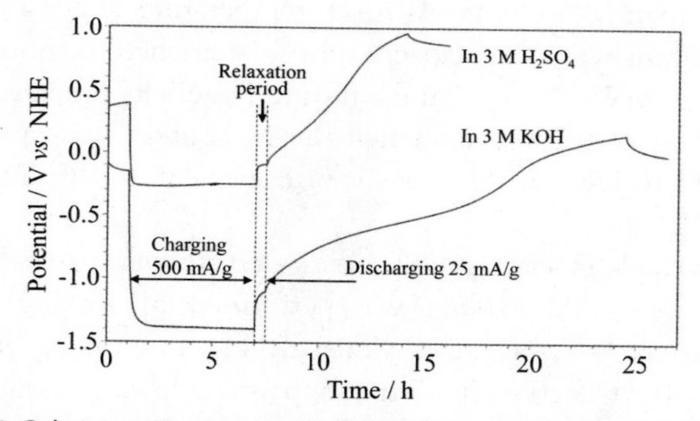

Figure 3.207 Galvanostatic charge–discharge curves of hydrogen in an AC [689].

the decomposition of the electrolyte solution using a high current density of 500 mA/g but it was kept for a long time (6 h) to ensure full saturation. H_2 adsorption is more favored in an alkaline medium than an acidic one. After allowing 30 min for relaxation, H_2 desorption (hydrogen oxidation) is performed by a current density 25 mA/g. In an alkaline medium, discharging occurs in two steps. With increasing charge capacity, discharge capacity (reversible capacity) increases with decreasing efficiency (i.e., discharge capacity/charge capacity) and tends to saturate, with the maximum reversible capacity being approximately 500 mAh/g (1.85 wt.% H_2). This electrochemical H_2 uptake of the cloth was much higher than the uptake measured at 273K under 7 MPa H_2 pressure [690,691]. Electrochemical H_2 storage was measured on a wide range of ACs, in a range of S_{BET} of 631−1450 m^2/g, in 30 wt.% KOH [692].

Electrochemical H_2 storage was studied using ordered mesoporous carbons synthesized from different carbon precursors, propylene (Pr), sucrose (S) and pitch (P), using mesoporous silica templates, MCM-48 (48) and SBA-15 (15) [331]. The electrochemical measurements were performed in a three-electrode cell of working, counter, and reference electrodes of porous carbon, Ni, and Hg/HgO, respectively, with 6 M KOH aqueous electrolyte. For H_2 storage experiments, five porous carbons were synthesized from combinations with different couples of carbon precursor and template, i.e., Pr/15, S/48, S/15, P/48, and P/15, and the charging process was carried out with a great excess of charge, 500 mA/g, to achieve full saturation by H_2. Pore structure parameters and H_2 uptake are listed for five mesoporous carbons in Table 3.8, where V_{micro}'s measured using N_2 and CO_2 are shown for information on the supermicropore volume, and electrochemically determined H_2 uptakes are

Table 3.8 Pore structure parameters and H_2 uptake of ordered mesoporous carbons [331].

Carbon precursor/ template	S_{BET} (m^2/g)	$V_{micro}(CO_2)$ (cm^3/g)	$V_{micro}(N_2)$ (cm^3/g)	H_2-uptake (mAh/g)	H_2 uptake[a] (wt.%)
Pr/15	713	0.11	0.21	100	0.37
S/48	2000	0.36	0.58	388	1.44
S/15	1470	0.30	0.45	311	1.15
P/48	1300	0.18	0.29	246	0.91
P/15	923	0.19	0.25	195	0.72

[a]Calculated by assuming 270 mAh/g = 1 wt.%.

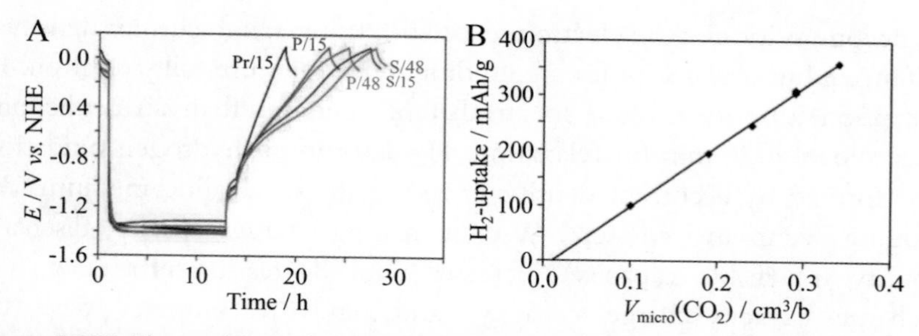

Figure 3.208 Electrochemical H_2 storage of ordered mesoporous carbons prepared from different combinations of carbon precursor/template (see text): (A) charge–discharge curves in 6 M KOH and (B) H_2-uptake versus $V_{micro}(CO_2)$ [331].

converted in the unit of wt.% by assuming that a 270 mAh/g capacity increase corresponds to 1 wt.% H_2. Charge—discharge curves in alkaline electrolyte are shown in Fig. 3.208A for these porous carbons, revealing strong dependence of the electrochemical storage of H_2 on the electrode (adsorbent) carbons. Fig. 3.208B confirms the fact that H_2 storage in carbon materials is governed by narrow micropores in the carbon.

Electrochemical storage of H_2 formed by electrolysis of water in an electrolyte solution was investigated using a three-electrode cell with a commercial ACF cloth as a working electrode in 6 M KOH by focusing on the releasing behavior of H_2 stored electrochemically [693]. An AC cloth having S_{BET} of 1390 m^2/g was selected because it was self-standing and able to avoid side effects due to additives, such as binders. In Fig. 3.209A,

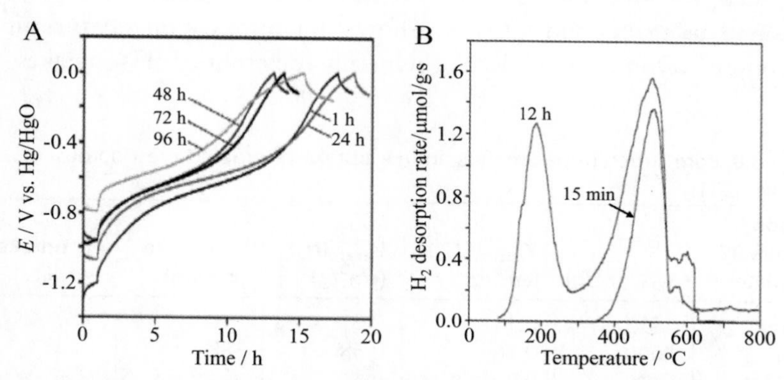

Figure 3.209 Electrochemical storage of H_2 formed by electrolysis of water in an electrolyte solution by ACF cloth: (A) discharge curves in 6 M KOH with +25 mA/g of the cloths which were held in fresh 6 M KOH for different periods after charging with −500 mA/g, (B) TPD profile of H_2 evolution after charging for 15 min and 12 h [693].

discharge curves with +25 mA/g in fresh 6 M KOH on the cloths, which were kept in another 6 M KOH for different periods after being fully charging with −500 mA/g, are compared. This result demonstrates that the reversible storage is stable, even after 4 days out of the cell, suggesting the formation of stable hydrogen bonding in the pores of carbon. Fig. 3.209B shows TPD curves obtained for the cloths charged with −500 mA/g over 15 min and 12 h, electrochemically charged cloths being immediately transferred in the TPD chamber without washing. The desorption peak at around 400°C is supposed to be due to the hydrogen produced by the reaction of KOH with carbon at high temperature and so must not be included in the electrochemically formed hydrogen. After 12 h charging, a new peak appears at around 200°C which is reasonably assumed to be due to H_2 stored electrochemically, and no H_2 evolution is detected below 100°C. These experimental results suggest a weak chemical interaction between stored hydrogen and adsorbent carbon, which may be possible because of the formation of nascent hydrogen during water decomposition.

3.6 Storage of methane and methane hydrate

Methane (CH_4), the major component of natural gas, has a high H/C ratio and provides a greater energy density per unit mass and less carbon dioxide emissions during combustion than gasoline does. These properties led to methane being nominated as a premium fuel for a low-carbon society. Methane storage is based either on adsorption into porous materials or physical compression and/or liquefaction. The latter techniques are undesirable from safety and energy-saving points of view, and the former exhibits certain limitations to fulfill the new target figure of 263 v/v (0.5 g/g) for vehicle applications at 25°C as proposed by US Department of Energy (DOE) [694]. Various microporous materials, such as zeolites, MOFs, and other compounds containing micropores in their lattices, were tested for methane adsorption and storage. On some of the MOFs, the methane storage capacity was reported to be 230 v/v at 290K and 3.5 MPa [695] and 270 v/v at 6.5 MPa [696]. On various porous carbons, many research works have been carried out on the adsorption of methane under different conditions, and their results have been reviewed in the literature [697−700]. The theoretical calculation of storage capacity of CH_4 for porous carbons was predicted to be 200−270 v/v.

3.6.1 Methane storage

The adsorption performance of various adsorbents was reviewed in 1998 [697]. In this review, the published data of adsorption capacity of CH_4 on various microporous adsorbents were discussed mainly on the basis of the surface area of adsorbents, even though the presence of micropores was emphasized to be an important factor for CH_4 adsorption. Adsorption capacity per mass of the adsorbent can be approximated by a linear relation to the surface area, as shown in Fig. 3.210A. By expressing the capacity per volume of the adsorbent, however, monoliths of ACs and ACFs give relatively high values, which are higher than granular ACs and mesoporous silica MCM-41 (Fig. 3.210B). In another review published in 2002 [698], the linear dependence of gravimetric capacity for CH_4 on micropore volume was presented for various activated carbon materials.

ACs were prepared from poly(vinylidene chloride) (PVDC) and coconut shell powders by carbonization at 700°C in N_2 after conforming into pellets (1.9 cmΦ × 0.6 cm), followed by activation using steam at 800°C and CO_2 at 850°C [701]. CH_4 adsorption was measured on the resultant AC pellets at 298K up to a pressure of 3.4 MP. Gravimetric CH_4 uptake (adsorption per unit mass of carbon) at 3.4 MPa increases with increasing burn-off at the activation process because the density of the carbon decreases as burn-off proceeds, as shown in Fig. 3.211A. However, volumetric uptake does not depend on burn-off for both ACs, and is almost constant, as shown in Fig. 3.211B. In other words, it is not an advantage that CH_4 adsorption increases burn-off.

Three commercially available ACs (AC-1—AC-3) and three laboratory-made ACs were compared with MOF $Cu_2(C_{12}H_4O_8)$ on CH_4 storage [702]. The laboratory-made ACs are prepared from olive stones by KOH

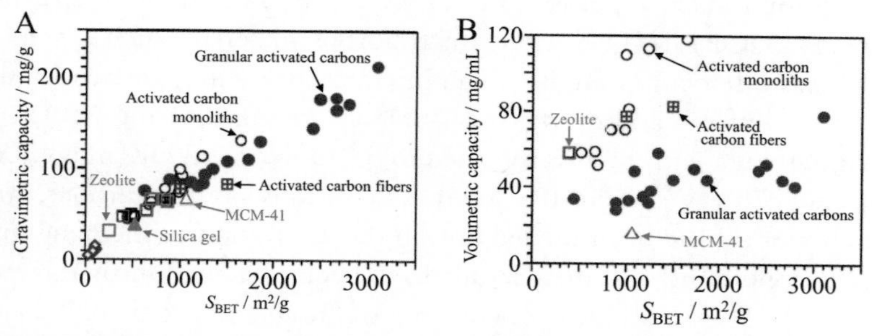

Figure 3.210 Adsorption capacity of various adsorbents for CH_4 at 298K under 3.5 MPa as a function of S_{BET}: (A) gravimetric capacity and (B) volumetric capacity versus S_{BET} [697].

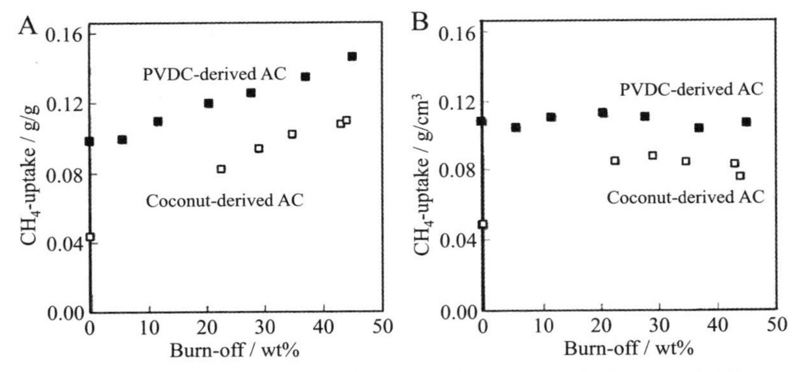

Figure 3.211 CH_4 uptake of PVDC-derived and coconut shell-derived ACs against burn-off during the activation process: (A) gravimetric and (B) volumetric CH_4 uptakes [701].

activation (KOH/C of 5/1) at 800°C (Lab-A), and from mesophase pitch by KOH activation (KOH/C of 8/1 and 6/1) at 800°C (Lab-B and Lab-C, respectively). Laboratory-made ACs (Lab-A, -B, and -C) had high S_{BET} (3290–3551 m^2/g) with high $V_{micro}(N_2)$ (1.00–1.11 cm^3/g) and so had gravimetric CH_4 uptakes comparable to an MOF, and even higher. However, volumetric uptakes of these adsorbents are lower than for the MOF, as shown in Fig. 3.212, mainly because of their low packing densities (0.4–0.6 g/cm^3). It has to be pointed out that some of them, Lab-B and -C, show H_2 uptakes close to the new DOE target.

Isotropic-pitch-based carbon fibers were activated with either CO_2 or steam at 890°C to prepare ACFs with different pore structures, on which

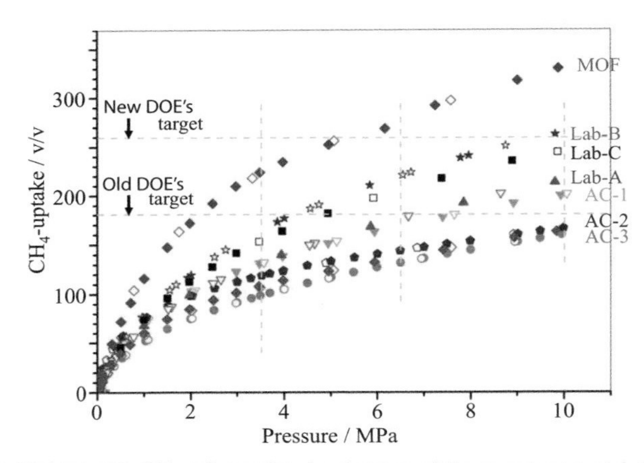

Figure 3.212 Volumetric CH_4 adsorption isotherms of three commercial ACs and three laboratory-made ACs in comparison with MOF [702].

CH_4 uptake was gravimetrically measured at 298K under pressures up to 4 MPa [703]. CH_4 uptake at 4 MPa is well related to the total micropore volume $V_{total-micro}$, including the pristine carbon fibers, as shown in Fig. 3.213A, where $V_{total-micro}$ is defined as $V_{micro}(CO_2)$ if $V_{micro}(N_2)$ is lower than $V_{micro}(CO_2)$ and $V_{micro}(N_2)$ if $V_{micro}(N_2)$ is higher than $V_{micro}(CO_2)$. The plot of CH_4 uptake against $V_{total-micro}$ gave much better linearity than the plots against $V_{micro}(N_2)$ and $V_{micro}(CO_2)$. For practice, the amount of CH_4 released from the CH_4-stored AC is more important. In Fig. 3.213B, the released CH_4 is shown as a function of burn-off during activation of the pristine carbon fibers. These results demonstrate that CH_4 release increases with increasing burn-off, in accordance with the increase in CH_4 uptake and that ACFs activated with CO_2 tend to release more CH_4 than those activated with steam. The mixture of chopped isotropic-pitch-based carbon fibers with phenol resin was hot-pressed to densities of 0.7–0.92 g/cm^3, carbonized at 650°C, and activated at 800 and 900°C with CO_2 to prepare ACF pellets with about 23 mm diameter, on which CH_4 storage was studied at ambient temperature and 3.5 MPa pressure [704]. The CH_4 storage capacity of 150 v/v was obtained by the carbon after 57.8 wt.% burn-off, which had S_{BET} of 2451 m^2/g and $V_{micro}(N_2)$ of 0.84 cm^3/g. CH_4 uptake was measured at 258–298K under pressure of up to 1.8 MPa on three ACFs with different S_{BET} values [705]. The ACF with the highest S_{BET} of 1511 m^2/g gave the highest CH_4 uptakes of 9.83 wt.% at 298K and 13.75 wt.% at 258K under 1.8 MPa.

The adsorption capacity of CH_4 at 3 MPa was related to the volumes of micropores, $V_{micro}(N_2)$ measured by N_2 adsorption at 77K and $V_{micro}(CO_2)$ measured by CO_2 adsorption at 273K [706]. By taking into account

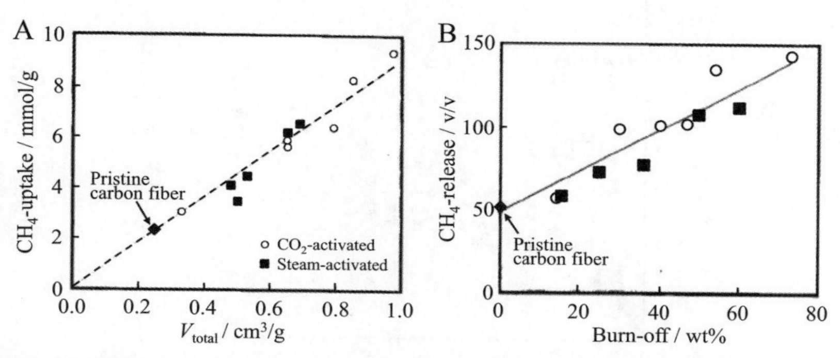

Figure 3.213 CH_4 uptake and release of isotropic-pitch-based carbon fibers activated by CO_2 and steam: (A) CH_4 uptake against $V_{total-micro}$ (refer to the text) and (B) CH_4 release at 298K against burn-off during activation [703].

that micropores are divided into two, one corresponding to narrow micropores (up to twice the molecular size of CH_4), which can be evaluated as $V_{micro}(CO_2)$, and another corresponding to wide micropores (up to five CH_4 molecules), which may be evaluated as $\{V_{micro}(N_2)-V_{micro}(CO_2)\}$, the mass of adsorbed CH_4 (m) is expressed as:

$$m = d_n V_{micro}(CO_2) + d_w\{V_{micro}(N_2) - V_{micro}(CO_2)\}$$

where d_n and d_w are the density of CH_4 in narrow and wide micropores, respectively. Based on this relation, d_n and d_w at different pressures up to 3 MPa were calculated from the experimental data on 35 ACs with different pore structures from different carbon precursors by different activation procedures. The relation obtained from the experimental data at 3 MPa is shown, as an example, in Fig. 3.214A, and the dependences of d_n and d_w on pressure are shown by comparing with the density of compressed CH_4 in Fig. 3.214B. The density in narrow micropores d_n increases quickly up to 1 MPa and tends to saturate at around 0.21 g/cm^3, whereas that in wide micropores d_w increases continuously up to 3 MPa, with both d_n and d_w being much higher than the density of compressed CH_4 and much lower than the density of liquefied CH_4 at 112°C, i.e., 0.42 g/cm^3. A simulation predicted the increase in the density of the adsorbed CH_4 with increasing pressure, reaching 0.27 g/cm^3 at 3 MPa (corresponding to two layers of molecules) [707], this predicted density being somewhat larger than d_n and much larger than d_w determined assuming the above relation, 0.21 and 0.09 g/cm^3, respectively.

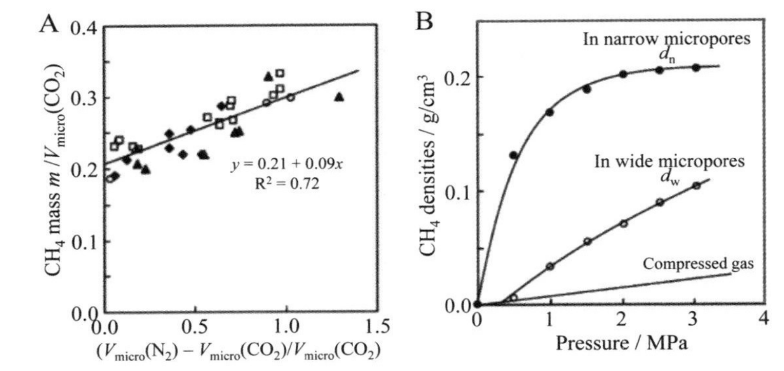

Figure 3.214 An example for the determination of the densities of CH_4 adsorbed in micropores at 3 MPa and 25°C (A) and changes in CH_4 densities in narrow and wide micropores with pressure at 25°C (B) [706].

Single-walled carbon nanohorns (SWCNHs) synthesized by CO_2 laser ablation of carbon at room temperature without a metal catalyst delivered a high CH_4 storage capacity of 160 v/v at 30°C under 3.5 MPa [708]. SWCNHs were oxidized in either H_2O_2 or O_2 after dispersion in ethanol, followed by repeated compression and crushing, in order to obtain a sufficiently high micropore volume. After nine cycles of compression under 50 MPa and crushing, a thin disk with a bulk density of 0.97 g/cm^3 was obtained, which had S_{BET} of 1097 m^2/g and $V_{micro}(N_2)$ of 0.55 cm^3/g, although as-grown SWCNHs had 1030 m^2/g and 0.50 cm^3/g, respectively. Repetitive compression and crushing caused significant distortion in the structure, such as increasing of the Raman D-band and broadening of the G-band, accompanying pore structure development and, as a consequence, enhanced CH_4 uptake.

CH_4 adsorption of carbide-derived porous carbons was also studied. CH_4 uptake of TiC-derived carbon, which was chlorinated at 800°C followed by annealing at 600°C in NH_3, increased with decreasing temperature from room temperature and saturated at −148°C; the uptake of 13.8 wt.% at room temperature under 6 MPa and 26 wt.% at −148°C were obtained [709]. The ordered mesoporous carbide-derived carbons were synthesized by chlorination of ordered mesoporous SiC which were prepared by infiltration of polycarbosilane into ordered mesoporous silica SBA-15, followed by pyrolysis at 1000°C and HF treatment [710,711]. The resultant carbide-derived carbons were microporous with ordered mesopores, S_{BET} of 2415−2914 m^2/g, V_{total} of 1.61−1.91 cm^3/g with V_{micro} of 0.47−0.59 cm^3/g, and gave a high CH_4 adsorption of 19.1 wt.% at 25°C under 10 MPa. Further treatment of SiC-derived carbon in a flow of hydrogen at 600°C was effective in removing any remaining metal particles and enhancing CH_4 adsorption.

MgO-loaded ACFs, which were prepared by immersing pitch-based ACFs into a saturated solution of $Mg(NO_3)_2$ with pH 10 at 30°C followed by heating at 300°C, delivered an enhanced saturated adsorption capacity by Langmuir approximation, 118 mg/g for MgO-loaded ACF and 79 mg/g for the pristine ACF, associated with an increase in V_{micro} from 0.45 to 0.49 cm^3/g [712].

3.6.2 Methane hydrate storage

A noticeable enhancement of CH_4 adsorption by the preadsorbed water in micropores was found by using steam-activated pitch-based ACFs with different S_{BET} (900–1800 m^2/g) and V_{total} (0.336–0.946 cm^3/g) and KOH-activated pitch-based carbon with a high surface area, S_{BET} of 2290 m^2/g and V_{total} of 1.33 cm^3/g [713]. Although CH_4 adsorption on the microporous carbons at 30°C was less than 9.4 mg/g at 101 kPa, the presence of the preadsorbed water noticeably enhanced CH_4 adsorption at 30°C even under subatmospheric pressure. The increment in CH_4 adsorption depends on pressure and reaches saturation after 20–50 h through a maximum at 1–2 h, as shown in Fig. 3.215A. It also depends on the fractional filling of micropores by the preadsorbed water, ϕ_w, as shown in Fig. 3.215B on the ACF with S_{BET} of 1800 m^2/g. The CH_4 adsorption capacity of the ACF increased linearly with increasing ϕ_w until 0.35, suggesting the formation of the stable methane–water clathrate (methane hydrate).

Water molecules confined into nanopores of AC could adsorb methane under pressures above 3.5 MPa at 2°C to form methane hydrate [714]. In nature, crystalline methane hydrate is found with three types of structures, type-I, -II, and -H in regions experiencing high pressure (more than 6 MPa) and relatively low temperature (slightly below room temperature). The type-I hydrate, most widely distributed in nature, has a cubic structure including 46 water molecules and at most eight methane molecules, expressed as $CH_4 \cdot 575H_2O$ [$(CH_4)_8 \cdot 46H_2O$]. The AC used was prepared from mesophase-pitch at 450°C, followed by the activation using KOH (KOH/C of 6/1 by weight) at 800°C, which had S_{BET} of 3670 m^2/g,

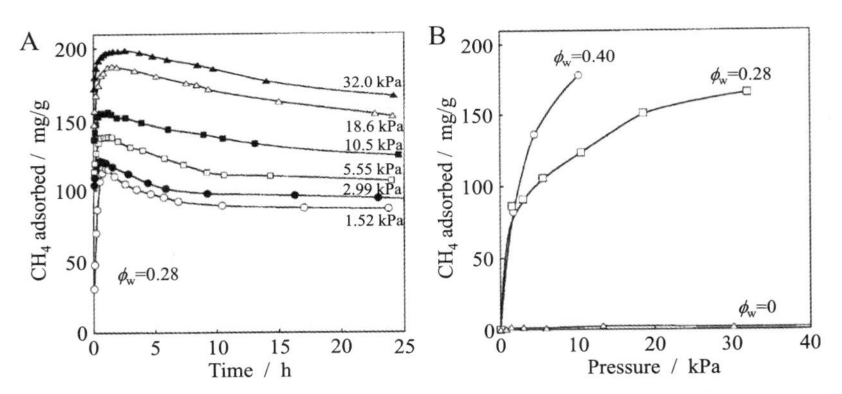

Figure 3.215 Enhancement of CH_4 adsorption at 30°C by preadsorbed water on the ACF with S_{BET} of 1800 m^2/g: (A) dependence on adsorption time under different pressures at constant ϕ_w (fractional filling of micropores by preadsorbed water) and (B) that on CH_4 pressure with different ϕ_w's [713].

V_{micro} of 1.20 cm^3/g, and V_{meso} of 1.24 cm^3/g. In Fig. 3.216A, CH_4 adsorption–desorption isotherms at 2°C are shown for the ACs with different preadsorbed water contents (R_w in g/g). The CH_4 adsorption capacity depended greatly on R_w, and was much larger than the dry AC ($R_w = 0$). At low CH_4 pressure, the wet ACs show lower adsorption capacity than dry ones, particularly for those with high R_w, probably due to pore blocking by preadsorbed water. Above 3 MPa pressure, however, the adsorption capacity of CH_4 shows sigmoidal increases. The formation of a methane hydrate structure was proved by inelastic neutron scattering and synchrotron X-ray powder diffraction. This condition of methane hydrate formation (3.5 MPa and 2°C) in nanopores of the AC is much milder and hydration proceeds much faster (finishing within minutes) than its occurrence in nature. The ACs with high water contents, R_w of 2.9 and 4.1 (oversaturated), give two-step adsorption isotherms with pronounced hysteresis (Fig. 3.216B), suggesting that adsorption–desorption mechanisms for the CH_4/wet-AC systems are different in the dry ones, and the methane hydrate formed in carbon nanopores is stable. In Fig. 3.217, X-ray powder diffraction patterns on the AC oversaturated with D_2O (R_w of 4.1) are shown. The AC adsorbed CH_4 under 3 MPa at 2 and −3°C gives rise to the characteristic diffraction peaks of the hexagonal ice crystal (denoted by "ice" in Fig. 3.217), while that under 5 MPa at 2 and 3°C shows clearly the peaks of the type-I methane hydrate (denoted by H in Fig. 3.217), revealing that the threshold pressure for methane hydrate formation in the pores of the AC is around 3 MPa, agreeing with the results of adsorption–desorption isotherm measurements (Fig. 3.216B).

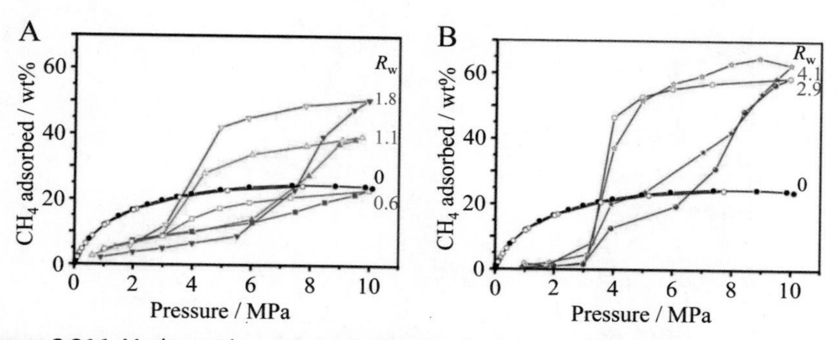

Figure 3.216 Methane absorption–desorption isotherms of the activated carbon with different preabsorbed water content R_w at 2°C [714].

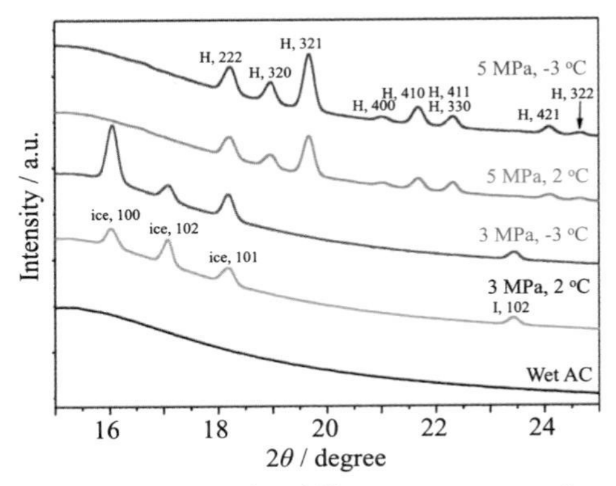

Figure 3.217 Synchrotron X-ray powder diffraction patterns of wet-deuterated AC at 25°C (in the absence of methane) and wet-deuterated carbon at 3 and 5 MPa of methane at 2 and −3°C; "ice" indicates an ice crystal formed by adsorbed water and "H" indicates a crystal of methane hydrate [714].

The effect of preadsorbed water on the adsorption and desorption of CH_4 was studied on different ACs under various conditions. The adsorption isotherm depended greatly on the water content R_w of AC [715]. In Fig. 3.218A, isotherms observed at 2°C on an AC prepared from coconut shells are shown. The amount of saturated CH_4 uptake increases with increasing R_w to 1.4 but too much water (i.e., 3.0) disturbs CH_4 adsorption. It is characteristic for the CH_4 adsorption into AC containing preadsorbed water to show a marked inflection point. This inflection was reasonably

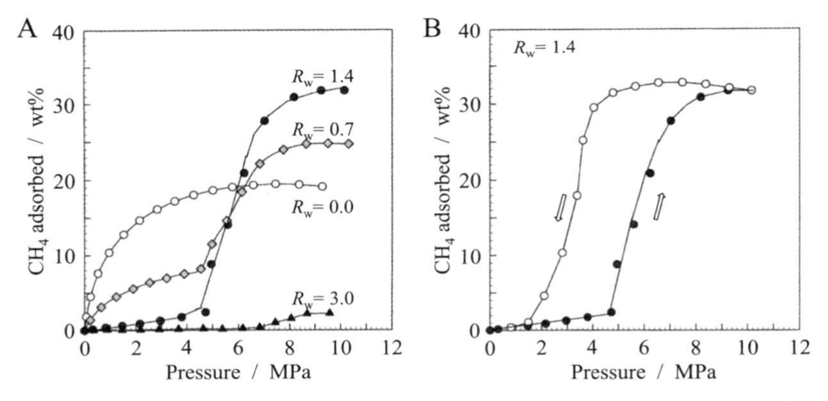

Figure 3.218 Adsorption isotherms of CH_4 into an AC at 2°C: (A) adsorption isotherms with different water contents R_w, and (B) an example of the marked adsorption–desorption hysteresis observed [715].

supposed to occur due to the formation of methane hydrate in the micropores, because this inflection pressure around 4 MPa at 2°C is a little higher than the formation pressure of the hydrate. In Fig. 3.218B a marked hysteresis in adsorption-desorption isotherms is shown on an AC with R_w of 1.4. On a series of ACs prepared from corncob particulates by KOH activation and then controlled pore volume in a range of 1.29−1.77 cm^3/g, pronounced inflections in CH_4 adsorption isotherms at a little less than 5 MPa were observed and hystereses in the desorption isotherms were also observed on ACs preadsorbed water in a R_w range of 1.5−3.75 [716]. Gravimetric CH_4 storage capacity showed a maximum of about 60 wt.% (based on dry AC) at R_w of around 3 and experimental volumetric capacity of 204 v/v was obtained at 2°C and 9 MPa.

CH_4 adsorption isotherms for four ACs were compared under different pressures up to 8 MPa [717]. Three ACs were microporous and derived from coconut shells with different degrees of steam activation (coconut-A, -B, and -C) and the fourth AC was mesoporous and derived from pinewood by phosphate activation (pinewood-A). CH_4 adsorption was performed at 2°C on the dry ACs after being outgassed at 200°C, as well as on the ACs after being wetted by water up to an R_w of around 1. Pore structure parameters, packing densities, and CH_4 storage capacities under 8 MPa are summarized for four ACs in Table 3.9. Adsorption isotherms based on volumetric uptake are compared in Fig. 3.219, revealing that the isotherms for wet ACs are very similar to each other, although the pore

Table 3.9 Pore structure parameters, packing density and CH_4 storage capacity of dry and wet ACs [726].

				Packing density (g/cm^3)		Storage capacity under 8 MPa[a]	
ACs	S_{BET} (m^2/g)	V_{micro} (cm^3/g)	V_{meso} (cm^3/g)	**Dry**	**Wet**	**Dry**	**Wet**
Coconut-A	1000	0.405	0.049	0.4	1.27	107 (12.8)	177 (10.5)
Coconut-B	1587	0.570	0.088	0.34	0.98	158 (21)	216 (16.5)
Coconut-C	2031	0.814	0.141	0.29	0.85	121 (23.1)	223 (21.9)
Pinewood-A	1967	0.65	0.72	0.16	0.54	164 (40)	227 (35.7)

[a]Unit in v/v and the figures in the parentheses in mol/kg.

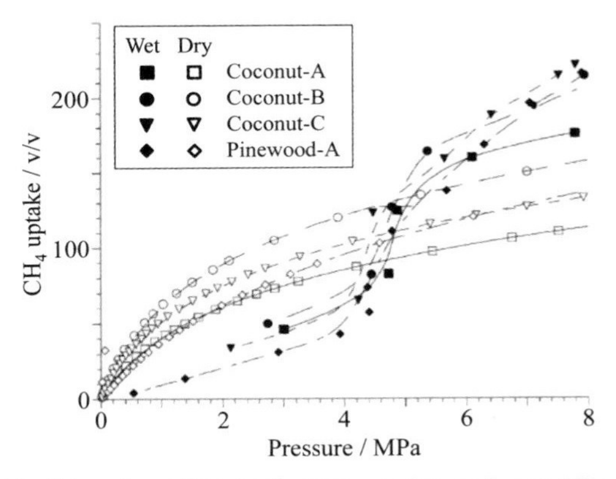

Figure 3.219 CH_4 adsorption isotherms on dry and wet ACs at 2°C [717].

structure is quite different and the packing densities of wet ACs are widely different, from 0.54 g/cm^3 for pinewood-A to 1.27 g/cm^3 for coconut-A. The CH_4 storage capacity of wet ACs under 8 MPa were measured as 177–227 v/v. The threshold pressure for the formation of methane hydrate in the pores of carbon at 2°C is clearly demonstrated to be 5 MPa pressure of CH_4. The investigation into the kinetics of methane hydrate formation predicted that the threshold pressure of its formation depended greatly on temperature; threshold pressure of about 3 MPa at 1.2°C, around 6 MPa at 8°C, and around 8 MPa at 11°C [718].

Releasing of CH_4 from wet carbon was studied by using an AC derived from coconut shell with CO_2-activation [719]. Pore structure of the AC was characterized as S_{BET} of 2585 m^2/g and $V_{micro}(CO_2)$ of 1.13 cm^3/g. The maximum capacity of water adsorption of the AC was evaluated by R_W as 1.4 and CH_4 adsorption and desorption was measured volumetrically on water-saturated AC particles (the sizes smaller than 0.154 mm) packed into a steel container. As shown in Fig. 3.220A, CH_4 released from the CH_4-adsorbed wet AC increases with increasing pressure during adsorption up to 152 v/v at 8 MPa. Below 5 MPa, the amount released from the wet AC is lower than that from the container without AC (empty container), showing CH_4 adsorbed into wet AC is lower than CH_4 adsorbed into the container, in other words, there is no advantage to the use of AC for CH_4 storage. The storage capacity in AC was closely related to the packing density of wet carbon in the container, increasing slightly with increasing packing density, but decreasing drastically after passing a maximum at about 0.6 g/cm^3 density, as shown in Fig. 3.220B. In Fig. 3.220C, the apparent molar ratio of water/CH_4 adsorbed into the

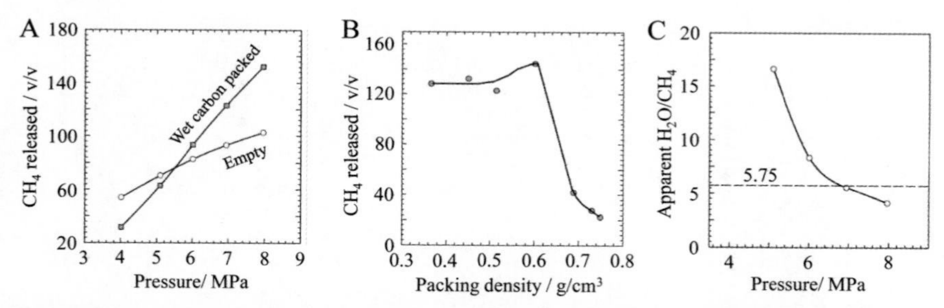

Figure 3.220 CH_4 release from coconut-shell-derived ACs: (A) CH_4 release from wet carbon packed and empty containers under different pressures, (B) dependence of CH_4 release on the packing density at 2°C under 7 MPa for the wet carbon, and (C) dependence of the apparent water/CH_4 ratio on pressure [719].

container is plotted against CH_4 pressure during adsorption. By taking into account that type-I methane hydrate has a molar ratio of 5.75, there is excess water in comparison to the amount of CH_4 adsorbed in the system (AC and container) when the CH_4 adsorption is performed under a pressure of less than 6 MPa. In the AC-adsorbed CH_4 under pressure greater than 7 MPa, in contrast, water is needed slightly more to complete hydration of adsorbed CH_4.

3.7 Thermal energy storage

A phase change in materials has been proposed to be used for thermal energy storage as a latent heat associated with the phase change (latent heat thermal storage, LHTS). The use of a liquid—solid phase change is more practical, because its volume change is much smaller than those by gas—liquid and gas—solid phase changes. The materials used for this thermal energy storage are called phase change materials (PCMs), which are generally divided into two groups, organic and inorganic compounds. At low temperatures below 100°C, organic PCMs, such as various paraffin waxes, have several advantages, including their abilities of congruent melting, self-nucleation, and noncorrosive behavior. Paraffin waxes consisting of mixtures of normal alkanes could provide a wide range of melting points T_m's. For LHTS at high temperatures, however, inorganic salts, such as $NaNO_3$ and KNO_3, are currently employed, but their low thermal conductivity is a more serious problem than for LHTS at low temperatures. The LHTS technique has attracted attention in relation to energy saving, particularly techniques at high temperature for solar energy storage recently, and many reviews have been published from various viewpoints, focusing on PCMs and their applications [720,721], on enhancement of thermal conductivity of PCMs [722], and on applications of PCM/carbon-foam composites in building parts, such as walls, roofs, and floors [723–725].

Porous carbons, particularly macroporous carbons (carbon foams), work well as a thermal enhancer and a container of organic PCMs, particularly at low temperatures, mainly due to their high thermal conductivity, low density, low thermal expansion coefficient, and chemical inertness, although different materials have been proposed as containers, such as porous metal, porous silica, expanded perlite, etc. The advantage of using graphite foam as a container for organic PCMs is demonstrated in Fig. 3.221, where the thermal conductivity κ of paraffin (about 0.24 W/m.K) is possible to be offset and to make heat exchange faster, with quick thermal energy storage and release [726].

Carbon foams, particularly graphite foams, such as the foams composed of exfoliated graphite and graphene, have the advantages of high thermal conductivity and the possibility of keeping PCMs in their macropores, which have led to excellent shape stability of LHTS units by holding molten PCMs [727]. Different carbon materials, such as natural graphite, carbon nanofibers (CNFs), carbon nanotubes (CNTs), etc., also have been proposed as additives to PCMs, in other words, by mixing with PCMs to improve the thermal conductivity, whereas the resultant mixtures have to be kept in a container. Although some of the carbon additives for PCM, such as graphite platelets, are not porous, they are included in this section to compare the LHTS performances with graphite foams and activated carbons (ACs).

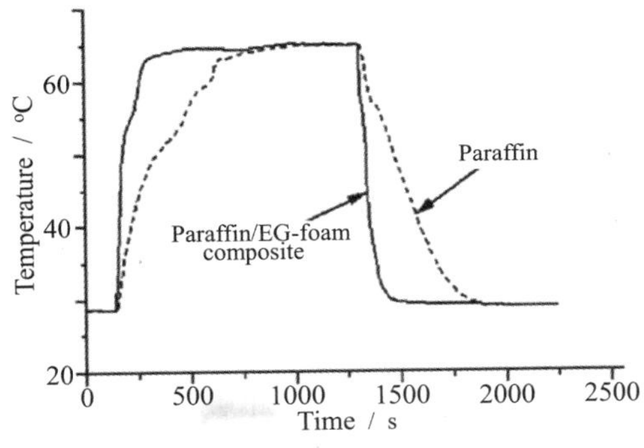

Figure 3.221 Heat storage and release curves for a paraffin/graphite-foam composite in comparison with paraffin itself [726].

In addition, adsorption heat pumps are briefly explained, because they have attracted attention as an environmentally friendly system for effective usage of relatively low-level heat sources such as solar energy, geothermal energy, and industrial waste thermal energy.

3.7.1 Thermal energy storage

3.7.1.1 Graphite and exfoliated graphite

3.7.1.1.1 Impregnation of PCMs

The graphite foams, which were prepared by compression of commercially available exfoliated graphite (EG) into a cubic mold to obtain various bulk densities, were impregnated by soaking into molten paraffin waxes either with a melting point T_m of 73–80°C, −9°C, or 18.1°C to prepare paraffin/EG-foam composites [728]. The amount of PCM impregnated depends principally on the bulk density of the EG-foam, as shown in Fig. 3.222A, revealing that the amount of paraffin impregnated decreases almost linearly with increasing bulk density of matrix EG-foam, but almost whole pores in the EG-foams are possible to be impregnated. The weight ratios of paraffin to EG are 65–95 wt.% and the volume ratios 60–90 vol.%. The thermal conductivity κ of the composites parallel and perpendicular to the compression axis during compression process of the matrix EG-foams, κ_{para} and κ_{perp}, are plotted against bulk density of the composites in Fig. 3.222B, showing that the overall thermal conductivity of the composites is very similar to that of the matrix EG-foams and as a consequence a marked

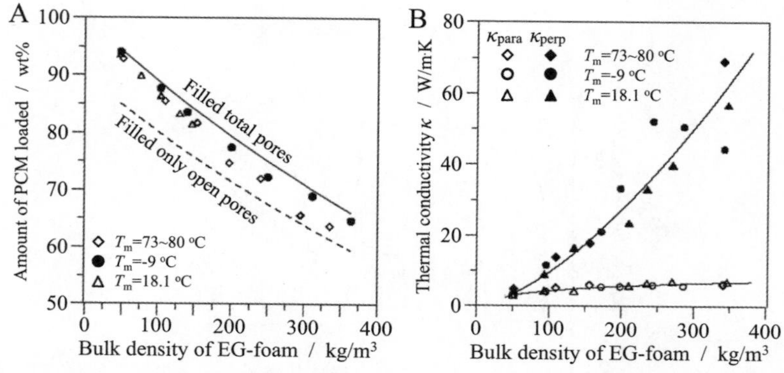

Figure 3.222 The composites of EG-foam with paraffins having different melting points T_m: (A) dependences of loading of PCM and (B) changes in κ of the composite along parallel and perpendicular directions to compression force, κ_{para} and κ_{perp}, with bulk density of the matrix EG-foam [728].

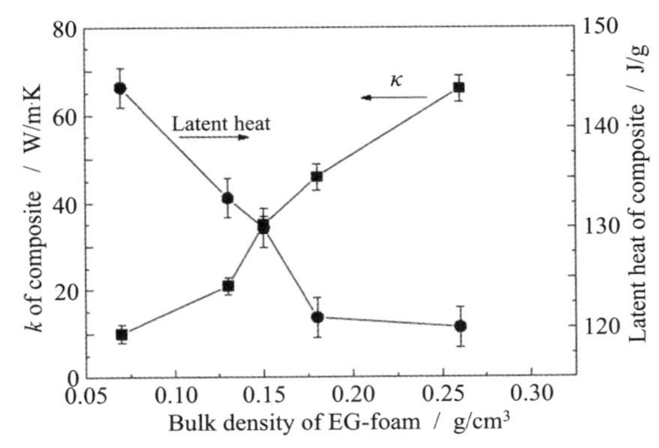

Figure 3.223 Dependences of thermal conductivity κ and latent heat (energy stored) of the paraffin/EG-foam composites on bulk density of the matrix EG-foam [729].

anisotropy in κ mainly due to the preferred orientation of graphite layers in the foams. In Fig. 3.223, κ and latent heat of the paraffin/EG-foam are plotted against bulk density of the matrix EG-foam [729]. The EG-foams with bulk density of 0.07–0.26 g/cm^3 were fabricated by compressing an exfoliated graphite into a cubic mold and the paraffin having T_m of 323K and latent heat for melting of 167 J/g was impregnated at 343K. With increasing bulk density of the matrix EG-foam, κ of the composite increases markedly, to 28–180 times that of paraffin itself (0.35 W/m.K), but the latent heat of the composite, i.e., energy stored in the composite, decreases because of the decrease in the amount of paraffin loaded. The composites of EG-foam with paraffin (melting point T_m of 48–50°C) showed a faster heat exchange with air than paraffin itself, as demonstrated in Fig. 3.221, due to the higher thermal conductivity of the foam, which had a large thermal storage capacity and did not experience liquid leakage during its solid–liquid phase change [726].

By using EG-foams prepared at 900°C, the optimal amount of paraffin (*n*-docosane) for the fabrication of a form-stable composite was investigated using composites containing 2–10 wt.% EG with sizes of 105 × 32 × 22 mm^3 [730]. The composite with 10 wt.% EG, i.e., containing 90 wt.% paraffin, was considered to be the most promising, with its form-stability allowing no leakage of melted paraffin during the solid–liquid phase change due to capillary and surface tension forces of EG. In addition, it has a high κ of 0.83 W/m.K, almost the same T_m of 40.2°C, and a satisfying latent heat storage capacity of 178 J/g, in comparison with 0.22 W/m.K, 41.6°C, and 194.6 J/g for pure paraffin, respectively. Using EG-foams prepared at 700°C and a paraffin with T_m of 55.8°C, the content of 7 wt.% EG in the composite was shown to give proper performance for LHTS from the heat storage/retrieval rates of the system [731]. Graphite oxide flakes prepared using nitric acid were exfoliated at 900°C and the resultant EG (S-free EG-foam) was molded at room temperature and then impregnated by a paraffin (T_m of about 61°C) under 0.02 MPa at 100°C to obtain paraffin/EG-foam composites [732].

The composite with a bulk density of 0.789 g/cm^3, κ of 16.6 W/m.K, T_m of 55°C, and latent heat of 185 J/g was applied to a lithium-ion battery (LIB) [733]. EG-foams were prepared by compaction of an EG into the plates with 7 × 7 × 1 cm^3 size and different bulk densities, into which a molten paraffin was impregnated at 80°C to form PCM/EG-foam composites. In Fig. 3.224A, the temperature profiles during discharging with a 2.1C rate at 30°C are shown for the pack of LIBs with and without PCM/EG-foam composite. A rapid increase in cell temperature is observed on the pack without the composite due to the exothermic reactions in the cell. In the pack with the composite, however, temperature increases much slower, with a shallow plateau, demonstrating effective heat dissipation through the composite, particularly by melting of PCM (paraffin) when the temperature of the pack exceeded its T_m (around 55°C). The temperature of the pack increases steeply after complete melting of the paraffin in the composite. In Fig. 3.224B, cycle performances of LIB at C/1 discharge rate and C/3 charge rate are also compared for the LIB packs with and without the composite. The discharge capacity decreases rapidly with cycling, about 0.20 Wh per cycle, on the pack without the composite, while the capacity fading rate for the pack with the composite is much lower, at only about

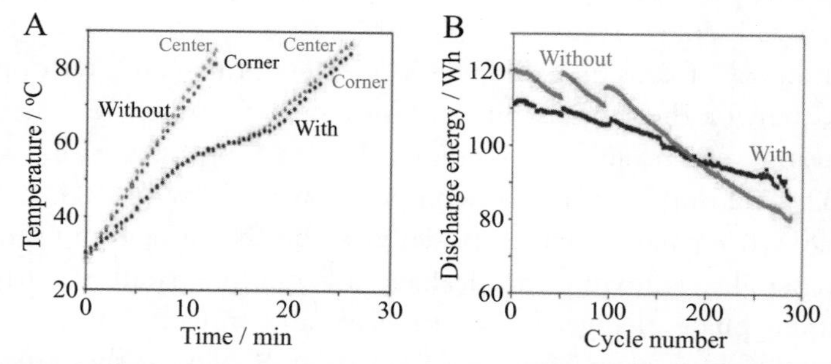

Figure 3.224 Performance of a pack of lithium-ion batteries with or without PCM/EG-foam composite: (A) temperature profiles of the pack during 2.1C discharging rate at 30°C and (B) cycle performances at C/1 discharge rate and C/3 charge rate [733].

0.09 Wh per cycle. For thermal management of LIBs, the thermo-mechanical properties of paraffin/EG-foam composites have been studied also [734]. The content of paraffin in the composites was shown to be important; with increasing content, the tensile and compression strengths of the cells increase, while burst strength under a gas pressure up to 1.2 MPa weakens at room temperature.

Graphitized carbon foams (graphite foams) were prepared by carbonization of the mixture of mesocarbon microbeads (MCMBs; about 1 μm diameter) with a mesophase pitch under pressures of up to 3 MPa at 1300°C, followed by graphitization at 2600°C [735]. Macropore sizes in the resultant carbon foams depend on the content of MCMB, decreasing from about 1 mm for 30 wt.% MCMBs to about 15 μm for 55 wt.% MCMBs. Paraffin wax with T_m of 58 °C and latent heat of 143.7 J/g was impregnated into four kinds of the foams at 80°C [736]. The thermo-physical properties are summarized in Table 3.10 for the pristine foams and their composite impregnated by paraffin. The latent heat of the composites increases with increasing paraffin content in the composite. Thermal diffusivities of the matrix foams and their composites are much higher in comparison with that of paraffin (0.13 mm^2/s), suggesting a predominant contribution of the matrix foams to the thermal diffusivity of the composites, whereas the diffusivity of the composite could not be related to that of the matrix carbon foam. It was concluded that the pore size and thickness of ligaments of the carbon foam played an important role in heat transferring in the composite. Infiltration of a graphite foam by a pitch (softening point of 200°C) at 300°C was shown to be effective in enhancing κ of the foam, as a consequence, enhancing κ of the composite with a paraffin [737].

The composite prepared by impregnation of Wood's alloy (50Bi/27Pb/13Sn/10Cd) (T_m of 71°C) into graphite foam (bulk density of 0.35 g/cm^3) resulted in a high κ of 193 W/m.K, which is higher than the 58.9 W/m.K for Wood's alloy itself, and a small thermal expansion coefficient of 7.8 ppm/K, which is smaller than 24.8 ppm/K for Wood's alloy, but a latent heat of 29.2 J/g which was almost the same as Wood's alloy itself [738].

Composites of graphite foams with either $NaNO_3$ (T_m of 309°C and latent heat of 177 J/g) or a binary mixture of hydroxides (T_m of 314°C and a high latent heat of 530 J/g) were prepared by vacuum infiltration at 350–400°C [739]. Three commercial graphite foams prepared from mesophase pitch by graphitization were used and two carbon foams prepared from a coal were also used for comparison. The compressive modulus

Table 3.10 Thermophysical properties of the pristine foams and their composites with paraffin [736].

	Pristine graphite foam				Composite with paraffin			
	Pore size (μm)	**Bulk density (g/cm^3)**	**Porosity (%)**	**Thermal diffusivity (mm^2/s)**	**Paraffin content (wt.%)**	**Density (g/cm^3)**	**Thermal diffusivity (mm^2/s)**	**Latent heat (J/g)**
A	600	0.20	90.5	142	76	1.02	36.0	109.7
B	500	0.35	83.3	231	65	1.08	74.3	92.8
C	400	0.48	77.1	95	74	1.13	24.9	104.9
D	300	0.57	72.9	90	47	1.19	65.8	65.3

and strength of graphite and carbon foams became higher with infiltration of PCM. The compressive modulus of the graphite foam was 0.9 and 0.46 GPa parallel and perpendicular to the exfoliation direction during the foaming process, respectively, that increased to 1.61 and 0.82 GPa after $NaNO_3$ infiltration. The changes in mechanical properties with the infiltration PCMs were discussed taking into account debonding at the interface between the PCM and carbon wall of the matrix foam during PCM crystallization and formation of the crack network in the PCM. The EG-foam was compared with the foams of copper and copper—steel alloys as the containers for $NaNO_3$ [740]. Heat transfer can be significantly enhanced using both foams, thereby reducing the heat-charging and heat-discharging period (thermal energy storing and releasing period) markedly. It was concluded that the overall performance of metal foams was superior to that of EG-foam.

3.7.1.1.2 Mixing into PCMs

EG nanoplatelets with lateral size of about 15 μm and thickness of less than 10 nm were mixed with a paraffin (T_m of 48—50°C) [741]. Mixing of EG nanoplatelets was effective in improving the electrical and thermal conductivities of the composites. As shown in Fig. 3.225A, the electrical resistivity of the composites decreases abruptly from around 10^9 Ωcm for 1 wt.% addition to 10^4 Ωcm for 2 wt.%, and then decreases gradually, revealing that the percolation threshold locates at very low EG content, between 1 and 2 wt.%. The thermal conductivity κ of the composites increased linearly from 0.26 W/m · K for pure paraffin to 0.8 W/m · K for 7 wt.% EG addition. The DSC profile of the composite consists of two peaks, corresponding to solid—solid phase transition in paraffin at around 35°C and solid—liquid phase change in paraffin at around 55°C, as shown in the inset in Fig. 3.225B. The composites showed very similar DSC profiles to the neat paraffin. Latent heat of the composite at these phase transition and change are almost the same as for neat paraffin, as shown in Fig. 3.225B, probably because the EG content is very small, at less than 7 wt.%. Effectiveness of EG nanoplatelet addition was checked using nanoplatelets of about 1 μm size, instead of those with 15 μm size [742]. Improvements in electrical and thermal conductivities of the composites using 1 μm-sized EG occurred at a little higher EG content, around 2 wt.%, whereas there was no difference in melting/crystallization behavior, i.e., temperature and latent heat. EG exfoliated by microwave with 800 W power showed a maximum adsorption of 92 wt.% of a paraffin with T_m of

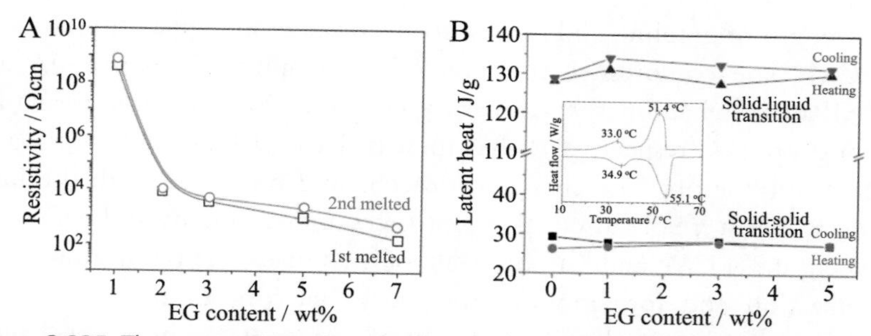

Figure 3.225 The composites of mixed EG nanoplatelets with paraffin: (A) electrical resistivity vs. EG content and (B) latent heats at phase transition and change versus EG content. The inset is the DSC profile of the composite with 3 wt.% EG [741].

52–54°C in its melt at 80°C, of which the solid–liquid phase change occurred at about 52.5°C, which is very close to 52.2°C for the neat paraffin, with a latent heat of 170.3 J/g, that is close to 173.1 J/g calculated by the paraffin content in the composite [743]. The thermal energy storage duration (charging duration) for the composite was shortened to about one-third of that of the neat paraffin. Nanoplatelets of EG was mixed with a mixed PCM, coconut oil with *n*-hexadecane, in different proportions by vacuum impregnation [744]. By compositing with EG nanoplatelets, the latent heat capacity of the PCM increased to 89, 104, and 125 J/g with an EG content of 70, 50, and 30 wt.%, respectively, which is higher than 82 J/g for coconut oil and 96 J/g for *n*-hexadecane. The κ for the composite PCM was more than 280% higher than for coconut oil and *n*-hexadecane.

The preferred orientation of EG platelets in PCM governs the thermal conductivity of the composites [745]. EG nanoplatelets with thickness of 30–100 nm were fabricated by microwave exfoliation of the intercalation compounds of natural graphite. The mixtures of a paraffin with 0.1–5 wt.% EG nanoplatelets were compressed to a film with size of $1.0 \times 1.0 \times 0.1\ cm^3$ under a pressure of $100\ kg/cm^2$ (random composites). The mixture was rolled to a thin sheet with a thickness of 0.3–0.5 mm, followed by stacking sheet by sheet to form a film with thickness of 1.2–1.3 cm and then cutting perpendicular to the rolling plane into a film with the size of $1.0 \times 1.0 \times 0.1\ cm^3$ (oriented composites). In the film of the oriented composite, EG nanoplatelets were preferentially oriented by making their basal plane perpendicular to the film surface. The thermal conductivity κ of the composite increased linearly with increasing content of nanoplatelets up to 3 wt.% and tended to saturate. The random composites gave much higher κ along the parallel direction to the film surface than the oriented composites, by 5.0 wt.% mixing of nanoplatelets the former and the latter delivering κ of 1.68 and 4.47 W/m · K, respectively. The melting point and the solid–liquid phase transition temperature of the paraffin matrix were not affected significantly by mixing the nanoplatelets, at approximately 53 and 60°C, respectively, while the latent heat of the composites decreased with increasing nanoplatelets content.

Stearic acid (SA, T_m of 54°C with latent heat of 182 J/g) was mixed with EG (exfoliated by microwave irradiation and had a lateral size of 300 μm) by stirring in a beaker [746]. The composites exhibited the same phase transition characteristics as the neat SA and their latent heats were approximately the same as the values calculated by the weight content of SA. The thermal diffusivity of the composites was markedly improved by EG addition, 4.88 mm^2/s for the composite with an SA/EG weight ratio of 1/1 while it was 0.11 mm^2/s for neat SA.

Eutectic salt $KNO_3/NaNO_3$ (1/1 by mol), which had a T_m of 223°C with a latent heat of 106 J/g, was mixed with EG and then compressed either uniaxially or isostatically [747,748]. The eutectic salt employed has certain advantages, such as negligible undercooling, chemical stability, no phase segregation, low corrosion, hygroscopicity, etc. During uniaxial compression under 5–10 MPa, worm-like particles of EG were collapsed and mixed with salt particles to form alternate stacking of graphite domains and salt domains; in the graphite domains basal planes of graphite particles were highly oriented perpendicular to the compression direction. On the other hand, the composites prepared by isostatic compression look like a random distribution of salt spheres (0.5–3 mm diameter) within a continuous and homogeneous matrix of graphite, as shown in the SEM image on the cross-section in Fig. 3.226A, and as a consequence they are almost isotropic. Thermal conductivity was improved in both composites prepared by uniaxial and isostatic cold compressions, the addition of 15–20 wt.% EG leading to κ of the composites close to 20 W/m · K (20 times greater than κ of the neat salt) [747]. These two kinds of composites showed quite different behaviors on melting and freezing. On the composites prepared by uniaxial compression, the addition of EG even by 20 wt.% did not modify the phase transition behavior (temperature of melting and latent heat). The composite prepared by isostatic compression, however, gives a very broad and shifted profile. particularly at the first heating cycle, as shown in Fig. 3.226B, suggesting that the melting of the

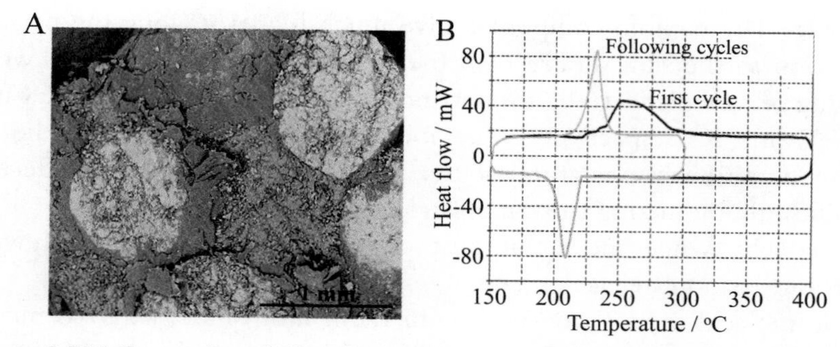

Figure 3.226 Composites of EG with $KNO_3/NaNO_3$: (A) SEM image of the cross-section of the composite prepared by isostatic compression and (B) DSC profiles of the composite prepared by isostatic compression with 20 wt.% EG with heat charging and discharging cycles [748].

salt occurs gradually, place by place, and the energy delivered by the composite is lower and slower at the first crystallization [748]. Powders of KNO_3 and $NaNO_3$ were uniformly mixed in different weight ratios and then EG prepared by heating graphite oxide was added to the nitrate mixtures with 5, 10, 20, and 50 wt.% to test as PCMs [749]. On packed PCM powders, apparent thermal conductivity was measured by the transient hot-wire method. Latent heat per mass of composites (LHM) measured by DSC was shown as functions of $NaNO_3$ content in the nitrate mixture and EG content in Fig. 3.227. The addition of 5 wt.% EG reduces

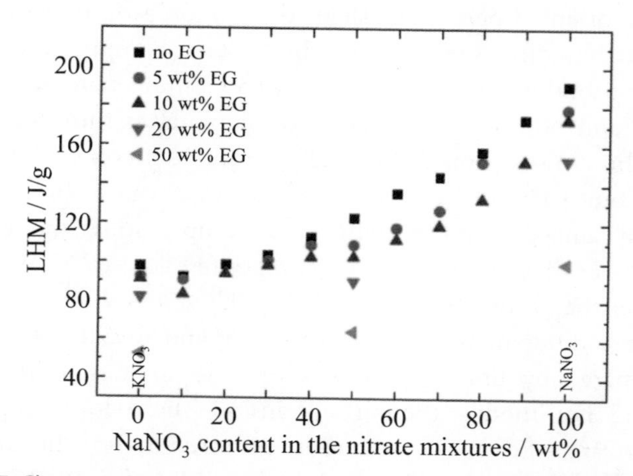

Figure 3.227 Change of latent heat per mass of the composite (LHM) with $NaNO_3$ content as a function of EG mixed [749].

LHM of the composites slightly (by about 2%–13%) from those of the nitrate mixtures without EG, whereas greater addition of EG decreases LHM more markedly, e.g., mixing of 10 wt.% EG to KNO_3 decreases LHM from 97.9 to 90.9 J/g and more EG addition reduces LHM more.

Various carbon materials were proposed as additives for improving thermal and electrical conductivities of PCMs and compared their effectiveness. The efficiency of EG, which was prepared by exfoliation of graphite oxide at 1100°C, was compared to reduced graphene oxide (rGO), which were prepared by reduction of graphene oxide at 1100°C, by mixing with a paraffin (T_m of 61.6°C) [750]. The dependences of resistivity and κ at 25°C (paraffin is solid) and at 80°C (paraffin is melted) on the content of EG and rGO are shown in Fig. 3.228A and B. rGO is more effective to reduce resistivity at 25°C, from 10^8 Ωcm for neat paraffin to 1 Ωcm for 5 wt.% addition of rGO, although the same 5 wt.% addition of EG reduced resistivity to 10^5 Ωcm (Fig. 3.228A). On the other hand, EG is more effective to increase κ than rGO, even under melting state of paraffin (80°C) (Fig. 3.228B). The improvement of κ depends on mixing methods of additives into paraffin, although the advantage of EG is kept, as shown in Fig. 3.228C, where three mixing methods are employed: (1) additive-dispersed toluene solution to toluene solution of paraffin, (2) additives directly to toluene solution of paraffin, and (3) mechanical mixing of the additive to solid paraffin by using roll milling. The addition of commercial graphene nanoflakes with lateral size of 5–10 μm and thickness of 4–20 nm (up to 30 layers stacked) was reported to improve κ of the composite with *n*-eicosane [751].

The addition of graphite nanoplatelets with lateral size of 35 nm was shown to improve κ markedly, from 0.126 W/m·K for the neat paraffin to 0.936 W/m·K (7.4 times higher) for the composite with 10 wt.% graphite

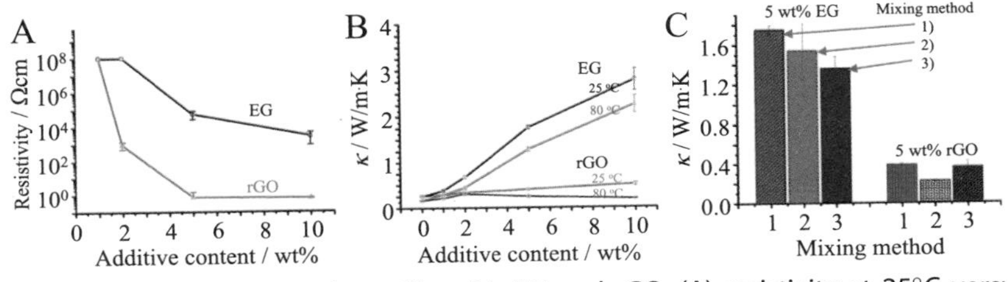

Figure 3.228 Composites of paraffin with EG and rGO: (A) resistivity at 25°C versus. additive content, (B) thermal conductivity κ versus. additive content, and (C) κ of the composites prepared by different mixing methods (refer to the text) [750].

addition, accompanied by a slight reduction in latent heat, from 209 to 182 J/g, and shift of phase change temperature from 30.9 to 29.7°C [752]. Graphite nanoplatelets (lateral size of 510 μm and thickness of 420 nm) were compared with short MWCNTs (diameter of 815 nm and length of 0.52 μm), long MWCNTs (diameter of 3050 nm and length of 513 μm), and CNFs (diameter of 200 nm and length of 1030 μm) by making composites with a paraffin (T_m of around 59°C) [753]. The relative enhancement of κ by the addition of graphite is the highest among the additives selected, as shown in Fig. 3.229.

The composites of inorganic salt PCMs with various graphites were studied by applying them in temperature and power ranges for concentrated solar light [754]. Inorganic salts of nitrates ($NaNO_3$ and KNO_3), hydroxides (NaOH and KOH), and chlorides ($ZnCl_2$, NaCl and KCl) are used as PCMs composed of single salts and binary eutectics, such as $KNO_3/NaNO_3$ (1/1) and $ZnCl_2$/KCl (0.319/0.681). Different kinds of graphite, natural graphite flakes with the lateral size of 400 μm, synthetic graphites with different particle sizes (6, 15, 44, 75, and 150 μm), compacted EG under 3 kg/cm^3, and powdered EG by grinding to the particle sizes of 50 and 500 μm, were used as additives for enhancement of κ of the composites. The composites were prepared either by mechanical dispersion within the molten PCMs at temperatures above their T_m's or by mechanical mixing graphite and PCM powders. The performance of thermal cycles for the composite of $KNO_3/NaNO_3$ salt with 5 wt.% synthetic graphite having 75 μm lateral size is shown in Fig. 3.230A, revealing that tens of thermal cycles are needed to reach a stable behavior.

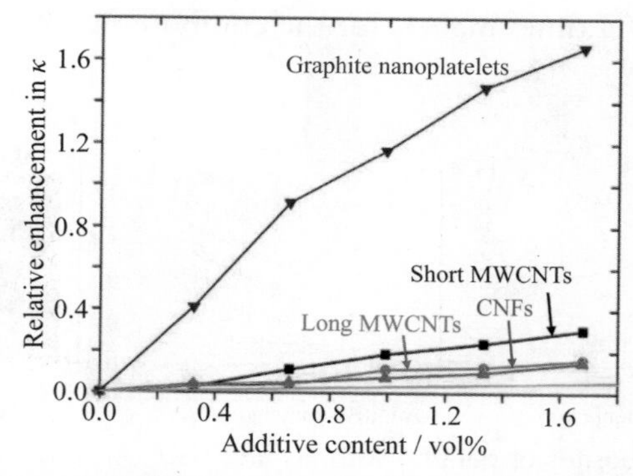

Figure 3.229 Enhancement of thermal conductivity of a paraffin by different additives [753].

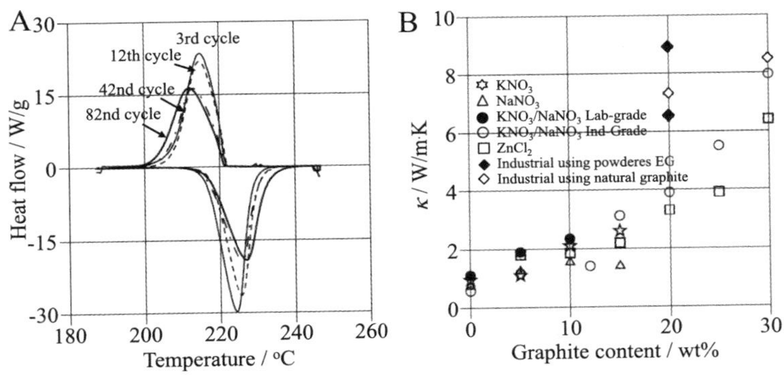

Figure 3.230 Graphite/PCM composites: (A) heating–cooling cycle performance of the composite of $KNO_3/NaNO_3$ with 5 wt.% synthetic graphite with 75 μm particle size, and (B) the dependences of κ of the composites of various PCMs with synthetic graphite, in comparison with κ of the composites fabricated in the industry using powdered EG of 500 μm particles and natural graphite of 400 μm particles [754].

In Fig. 3.230B, the dependences of κ on the content of the synthetic graphite of 75 μm particles are shown on the composites of various PCMs including the composites fabricated in the industry. κ increases with increasing synthetic graphite content, irrespective of the PCMs, but the industrial-grade composites using different graphites, powdered EG of 500 μm particles and natural graphite flakes of 400 μm particles, deliver higher κ than the composites using ground EG, probably because the larger particle size is favorable to create the better conductive network within the PCM.

3.7.1.2 Graphene and reduced graphene oxides

3.7.1.2.1 Impregnation of PCMs

The composites were fabricated by vacuum impregnation of melted octadecanoic acid (OA, T_m of 56°C) into rGO-foam at 80°C [755]. rGO-foams were prepared by hydrothermal treatment of GO flakes at 160°C, followed by freeze-drying and annealing at 1000°C in Ar, and had a bulk density of 0.227 g/cm^3. The composite with 15 wt.% rGO-foam delivered a high heat storage capacity of 182 J/g, very close to the neat OA (186 J/g), and high κ of 2.64 W/m.K. By impregnation of a paraffin (docosane) into rGO-foam (bulk density of 0.003 g/cm^3) at 80°C, simultaneous enhancement of κ and latent heat were resulted [756]. For the composite, the latent heat increases slightly to 263 J/g and κ increases markedly to 0.59 W/m·K, in contrast to 256 J/g and 0.26 W/m · K, respectively, for the neat docosane, whereas the melting and freezing points of the

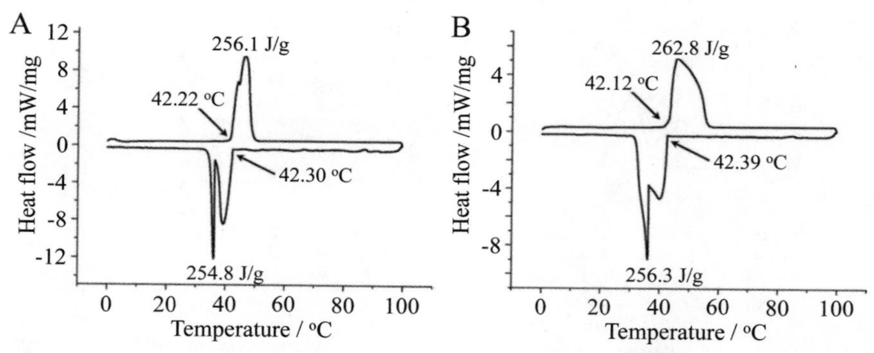

Figure 3.231 DSC curves: (A) neat docosane and (B) docosane/rGO-foam composite [756].

composite are hardly affected by the use of rGO-foam, as shown in Fig. 3.231. The composite of ultra-flyweight rGO aerogels with bulk density of less than 0.001 g/cm^3 with paraffin exhibited 7.8% and 28.7% higher latent heat for melting and freezing, respectively, than neat paraffin [757].

Lightweight and flexible rGO/cellulose aerogels with 3D networks were successfully used as a container of polyethylene glycol (PEG) [758]. The rGO flakes had a lateral size of 48 μm and were prepared by thermal exfoliation of GO at 1050°C, followed by annealing at 2200°C in Ar. The rGO/cellulose aerogel was fabricated by mixing various amounts of rGO with microcrystalline cellulose together with epichlorohydrin (cross-link agent) in an aqueous solution of NaOH and urea at −12°C, followed by freeze-drying at −55°C. PEG was impregnated into these aerogels at 80°C under vacuum. The composite with 5.3 wt.% rGO showed a T_m of 62.8°C with a latent heat of 158 J/g, close to the neat PEG (62.5°C and 169 J/g, respectively). The κ of the composite was 0.96 W/m · K at 30°C, although κ of the cellulose without rGO was 0.24 W/m · K. rGO aerogel mixed with a small amount of GO was also employed for PEG impregnation, which had a high κ, enough shape stability and improved energy storage density; the composite containing only 0.45 wt.% GO with 1.79 wt.% rGO delivered κ of 1.43 W/m · K and T_m at 41.8°C with latent heat (energy stored) of about 185.6 J/g, while the neat PEG had 0.31 W/m · K and 39.5°C with 179.5 J/g, respectively [759]. These composites could work also as a light-to-thermal energy converter.

3.7.1.2.2 Mixing into PCMs

Graphene and rGO were experimentally shown to be very effective additives for enhancement of κ in PCMs. The addition of rGO nanoflakes was shown to drastically improve the κ of the 1-octadecanol (stearyl alcohol, T_m of 66°C and κ of 0.38 W/m · K) with a small reduction in the heat storage capacity [760]. By the addition of 4 wt.% rGO, κ is enhanced to about 0.9 W/m · K but the storage capacity is reduced to about 210 J/g (140% increase in conductivity but 15% reduction in storage capacity). The composites of a paraffin with three kinds of rGO nanoflakes were fabricated [761], a single layer with an average lateral size of 550 nm (rGO-A), rGO with thickness of about 1 nm (about three layers stacked) and about 10 μm lateral size (rGO-B), and rGO with about 8 nm thickness (20–30 layers stacked) and about 550 nm lateral size (rGO-C). The paraffin used was mainly composed of *n*-alkanes with C_{34}–C_{35} and a T_m of about 70°C. In Fig. 3.232A and B, κ and specific heat of the composites were shown as a function of temperature, respectively, revealing marked increases in κ from 0.25 W/m · K for the pristine paraffin to 45 W/m · K for the composite with 20 wt.% rGO-B and rGO-C, and a little larger specific heat by the addition of rGO nanoflakes. These composites were shown to be effective in reducing the temperature rise due to charge–discharge cycles in Li-ion batteries (LIBs). Temperature rises during 10 cycles of charge–discharge with a current of 16 and 4 A are compared for six 4-V LIBs packed in neat paraffin, paraffin mixed with 1 wt.% rGO-A, 20 wt.% rGO-B, and 20 wt.% rGO-C, together with a package without any paraffin in Fig. 3.233. By using the neat paraffin, the temperature of the package of six LIBs is kept low, and the effect is enhanced by mixing rGO into paraffin, probably due to a combination of the latent heat storage by the paraffin together

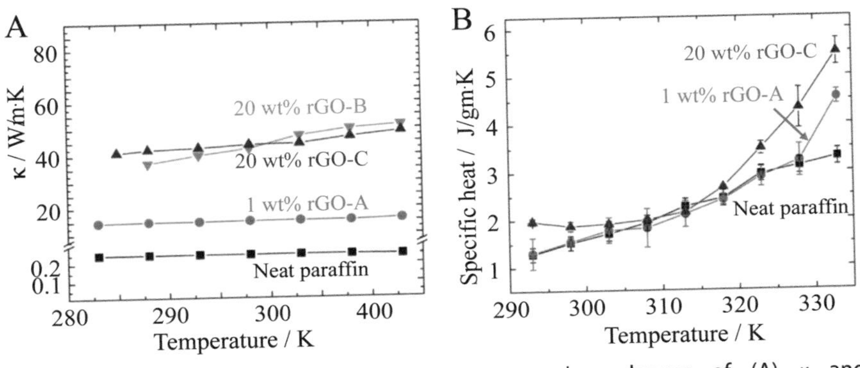

Figure 3.232 rGO/paraffin mixtures; temperature dependences of (A) κ and (B) specific heat of the mixtures [761].

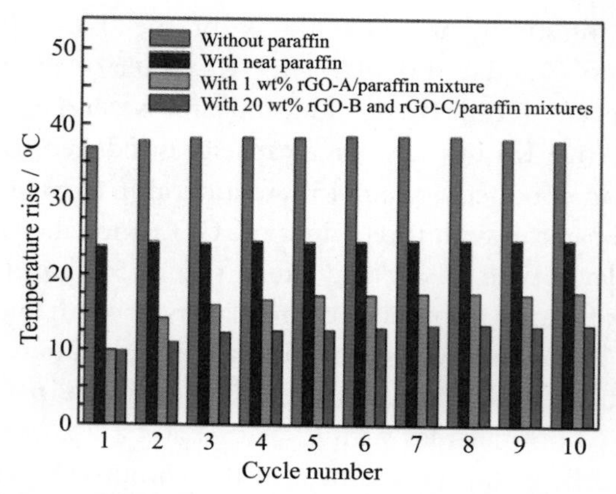

Figure 3.233 Temperature rise of the inside of the LIB packages with and without paraffin composites during charge—discharge cycles [761].

with the improved heat conduction to the outside of the battery package by the mixtures. The κ-value of lauric acid was increased linearly with an increasing fraction of commercial multilayered graphene, from 0.215 W/m · K for the neat lauric acid to 0.489 W/m · K by the addition of only 1 vol.% graphene [762].

The efficiency of the addition of either graphene or EG for κ of the composites and phase change (melting/crystallization) of PCM in the composites was investigated on a eutectic mixture of palmitic acid and stearic acid (62/38 by weight) as PCM [763]. EG was prepared via microwave heating and graphene was a commercial one, of which particles had thicknesses of 4—20 nm (composed of less than 20 layers per stack) and lateral size of 5—10 μm. Either graphene or EG was mixed with melted eutectic acid using PVP as the dispersion stabilizer. As shown in Fig. 3.234A, EG is more effective for enhancing κ of the composites than graphene, probably because the porous worm-like structure of EG is beneficial in forming effective heat conduction pathways. No apparent difference is observed on DSC profiles for the phase change in the composites (Fig. 3.234B) and the neat eutectic acid, while latent heats at this phase change (melting and crystallization) decreased with the increase in the content of graphene and EG. Decreases in the latent heat are larger than the value calculated from the decreased amount of eutectic acid in the composite, which may be attributed to the carbon materials inhibiting crystallization of the acid.

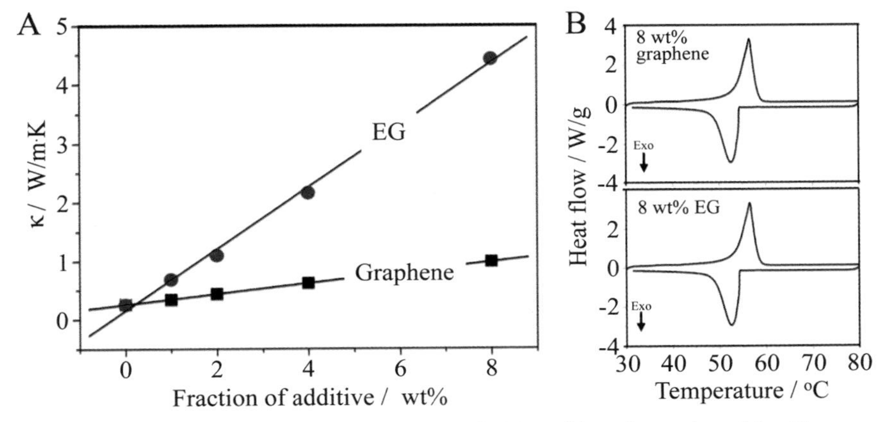

Figure 3.234 The composites of eutectic palmitic acid and stearic acid with commercial graphene and EG: (A) changes in κ with graphene and EG contents, and (B) DSC profiles of the composites with 8 wt.% graphene and EG [763].

Effectiveness of the enhancement of κ by the addition of commercial graphene was experimentally demonstrated using PCMs, stearic acid (SA, $C_{18}H_{36}O_2$) [764], and a eutectic salt (62 mol% Li_2CO_3 + 38 mol% K_2CO_3) [765]. Only 0.01 vol.% additives makes κ of the SA increased, as shown in Fig. 3.235A [764]. The addition of 0.1 vol.% into SA makes the difference among carbon additives clear, with graphene being more effective than other additives. The addition of PVP together with graphene results in more marked improvement in κ, as shown in Fig. 3.235B. Graphene particles are easily dispersed in SA, easier than with other additives, and PVP may assist graphene dispersion more and, as a consequence, the κ of the SA becomes higher. The effects of the addition of MWCNTs,

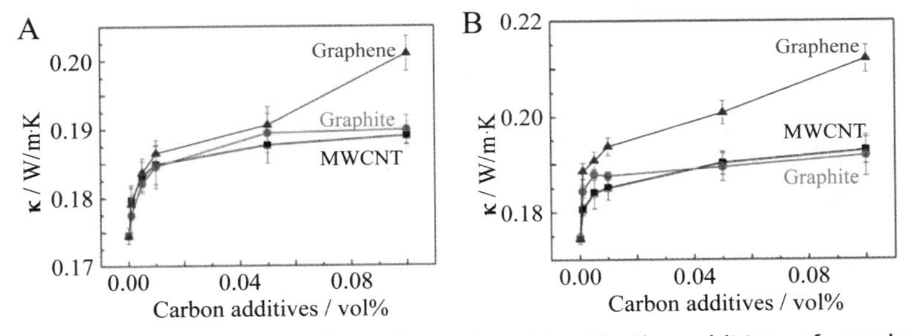

Figure 3.235 Enhancement of κ of stearic acid with the addition of graphene, graphite, and MWCNTs; (A) without PVP and (B) with PVP [764].

SWCNTs, C_{60}, and graphene on energy stored and κ of the PCM (the eutectic salt) were investigated as a function of the fraction of carbon additives up to 2.5 wt.% [765]. Specific heat increased with increasing fraction of carbon additives up to 1.5 wt.% both in the solid (240–280°C) and liquid region (560–580°C); graphene showed a 16.8% enhancement in the solid region. κ of the PCM increased also with the addition of SWCNTs, MWCNTs, and graphene by increasing their fraction up to 1.5 wt.%, about 50% enhancement by the addition of SWCNTs and MWCNTs, and about 20% enhancement by graphene. However, the addition of C_{60} decreased κ of the PCM by increasing its fraction.

Porous composites were prepared by the carbonization of absorbent cotton in H_2/Ar atmosphere at 1200°C after forming into disks under 10 MPa pressure, followed by CVD of CH_4 at the same temperature to deposit graphene flakes on the fiber surface [766]. The carbon fibers derived from cotton are hollow, with diameters of 1–10 μm and wall thicknesses of several hundred nanometers, as shown in Fig. 3.236A. The graphene nanoflakes are deposited almost vertically and uniformly on the inside and

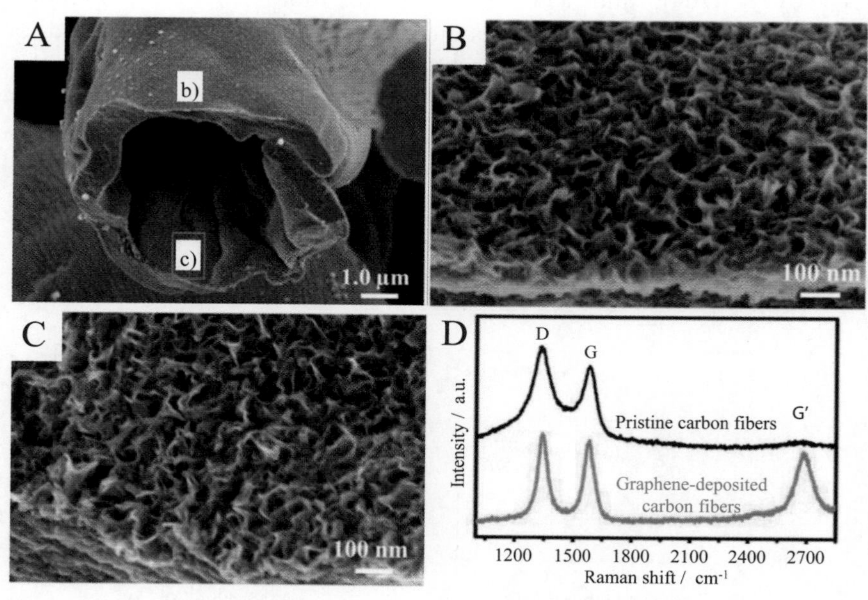

Figure 3.236 Composite of cotton-derived carbon fiber with graphene: SEM images of (A) cotton-derived carbon fiber, (B) and (C) graphene flakes deposited on the outside and inside surfaces of the fiber, respectively, and (D) Raman spectra of carbon fiber before and after graphene deposition [766].

outside surfaces of the fibers, as shown in Fig. 3.236B and C, respectively, of which the content in the composite disks is about 9.2 wt.%. The graphene flakes deposited are supposed to be few-layered from the Raman spectrum shown in Fig. 3.236D, where $I_{G'}/I_G$ is 0.82. Into these composite disks and also the pristine carbon fiber disks (without graphene deposition), melted SA was impregnated at 80°C to fabricate the composites, and their thermal properties were investigated. Graphene deposition was very effective in improving the performance of the disks as PCMs; κ and thermal diffusivity of the composite disk was 0.69 W/m · K and 0.448 mm^2/s, although κ for the pristine carbon fiber disk and the neat SA were 0.28 and 0.16 W/m · K, and thermal diffusivity 0.185 and 0.086 mm^2/s, respectively. The heat storage capacity of the composite disk was about 174 J/g, which is very close to the value of the neat SA (about 186 J/g) although the density of the composite (0.85 g/cm^3) is lower than that of SA (0.94 g/cm^3). The composite disk impregnated by SA showed good thermal reliability, even after 500 melting—freezing cycles.

3.7.1.3 *Other carbons*

3.7.1.3.1 Impregnation of PCMs

Commercial ACs, which were derived from coconut shells at 1000°C and had S_{BET} of 1200 m^2/g and V_{total} of 0.79 cm^3/g, were also used as a container for PEG with different average molecular weights of 1500—10,000 [767]. Impregnation of 30—90 wt.% PEG into the AC was performed through its ethanol solution, followed by evaporation of ethanol. The resultant composites with PEG content less than 70 wt.% showed shape stability and no leakage of melted PEG even at 80°C (above the melting point of PEG). The dependences of melting and crystallization temperatures on the content of PEG and on the average molecular weight of the PEG are shown in Fig. 3.237A and B, respectively, revealing marked supercooling of PEG. Temperatures of melting and crystallization depend slightly on the PEG content in the composites but mostly on the molecular weight of PEG, although the degree of supercooling is almost the same in all composites. Enthalpies for the phase change of PEG in the composites were quite different on melting and crystallization, 81.3 J/g on melting and 72.8 J/g on crystallization for the PEG with a molecular weight of 1500 in the composite. The S_{BET} and V_{total} decreased markedly by impregnation of PEG and became negligibly small by impregnation of more than 50 wt.%. These experimental results suggest that PEG molecules are largely confined to the micropores and mesopores of the AC.

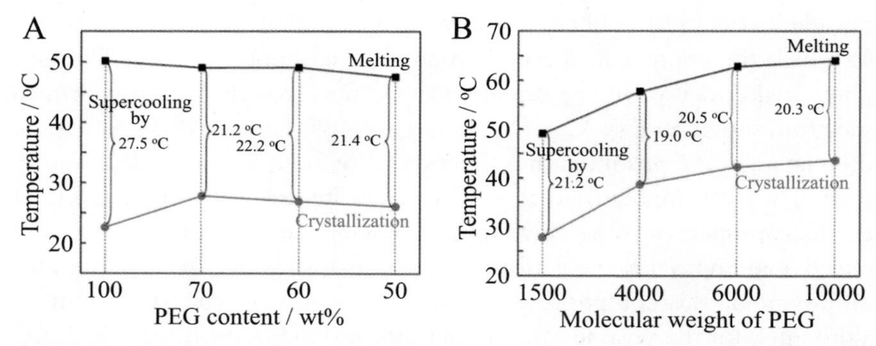

Figure 3.237 Melting and crystallization temperatures of the composites of an AC with PEG: (A) effects of PEG content with average molecular weight of 1500 and (B) effects of average molecular weight of PEG for the composite with a PEG content of 70 wt.% [767].

Composites of carbon with paraffin were fabricated using four commercial carbon foams with different thermal conductivities and porosities, of which the thermal characteristics of a thermal protection system were studied by experiments and numerical calculation [768]. The disks of carbon foam with diameters of 51 mm and thicknesses of 13 mm were fully impregnated in the molten paraffin wax and then packed into a Teflon container. On this PCM unit, the temperature distributions in the composites during heat charging and discharging were measured by thermocouples set in the composite. The results demonstrated that the high thermal conductivity of the carbon foam increases the heat exchange rate of the composite. A thermal energy storage unit was constructed by alternative stacks of PCM (a paraffin) plate and the composite plate composed of paraffin-impregnated graphite with a size of $38.5 \times 14.5 \times 2.5$ cm^3 [769]. By using a graphite matrix embedded by PCM, the charging–discharging time was noticeably shortened and the power consumption for the dissipation of heat from the unit was also reduced. The composite prepared from carbon foam with 97% porosity gave five times larger energy storage output power than neat PCM [770,771]. Vacuum impregnation of melted PCM was shown to be effective using a carbon foam with porosity of about 85.0% and a paraffin (T_m of 60.0–62.0°C) [772].

3.7.1.3.2 Mixing into PCMs

The addition of CNFs less than 10 wt.% was found to improve the thermal performance of a paraffin, *n*-tricosane (T_m of 56°C with latent heat of 220 J/g), by examining the effect of the size and aspect ratio of the CNFs [773]. CNFs were synthesized by catalytic CVD of hydrocarbons and of CO on an iron catalyst, which had a diameter of 2–100 nm with length of up to 100 μm and herringbone nanotexture, which had been found to enhance effectively the thermal conductivity of the composites [774]. CNFs tend to be settled at the bottom part of composites during thermal cycling, while this was shown to be avoided by adding a high-density polyethylene (HDPE) as a stabilizing polymer [775]. With the addition of 10 wt.% HDPE into the melted paraffin before immersion of CNFs, settling of CNFs in PCM could be avoided. The effects of the addition of CNFs and CNTs with diameters of about 30 nm and lengths of about 50 μm into soy wax were studied [776]. The thermal conductivities of the composites at room temperature increase with an increase in the CNF and CNT contents, but CNFs were more effective than CNTs, from 0.324 W/m·K for the neat soy wax to 0.467 and 0.403 W/m K with the addition of 10 wt.% CNFs and CNTs, respectively, probably because of the better dispersion of CNFs in the matrix soy wax. DSC profiles of the composites of CNFs and CNTs are shown in Fig. 3.238A and B, respectively. The neat soy wax shows one main phase change peak in the endothermic process (heating process) and two phase change peaks in the exothermic process (freezing process), these characteristics being kept in the composites with no significant difference being observed. The effects of the addition of chopped carbon fibers (CFs) with diameters of 6 μm and lengths of 5 mm into SA (stearic acid, T_m of 67–70°C) was compared with EG with a lateral size of 35–75 μm prepared by thermal exfoliation and reduction at 900°C [777]. κ of the composite increased linearly by increasing

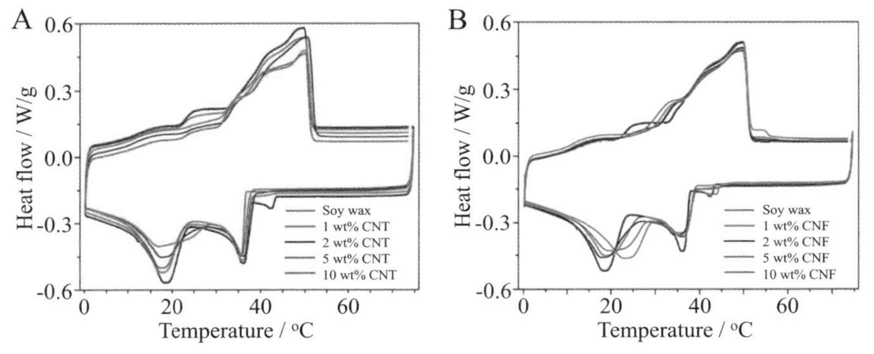

Figure 3.238 DSC profiles of the composites of CNFs and CNTs with soy wax: (A) CNT addition and (B) CNF addition [776].

the content of EGs and CFs. The enhancement of κ by the addition of EG was slightly higher than that by CF addition, increasing by 279% and 217% by the addition of 10 wt.% EGs and CFs, respectively. This was thought to be due to the higher dispersibility and better compatibility of EG with SA. The latent heat (energy stored) decreases by 8% and 7%, from 198.8 J/g for neat SA to 183.1 and 184.6 J/g with the addition of 10 wt.% EGs and CFs, respectively, while the melting time for the composites was shortened, from 95 min for neat SA to 75–78 min by adding only 2 wt.% EGs and CFs, as a consequence of the enhanced κ.

3.7.2 Adsorption heat pumps

Adsorption heat pumps (AHPs) have been actively studied to use effectively low-temperature waste heat (thermal energy) at around 100°C or lower. Various adsorbates, such as water, ethanol, methanol, and CO_2 have been studied by coupling with various adsorbents, such as silica gel, zeolite, and activated carbons (ACs). In Fig. 3.239, an ice maker using solar energy is schematically shown, using a commercial AC as adsorbent and either methanol or ethanol as adsorbate (refrigerant) [778]. Using solar radiation energy, the bed of the AC (adsorbent) is warmed up to the desorption temperature and begins to desorb the refrigerant. The desorbed refrigerant vapor is condensed into liquid in a condenser and flows directly into the evaporator. After complete desorption of the adsorbent is reached, the adsorbent adsorbs the refrigerant from the evaporator. During this adsorption process, the water was cooled down to freezing point due to the refrigerant evaporation and ice is obtained in the water tank placed inside a

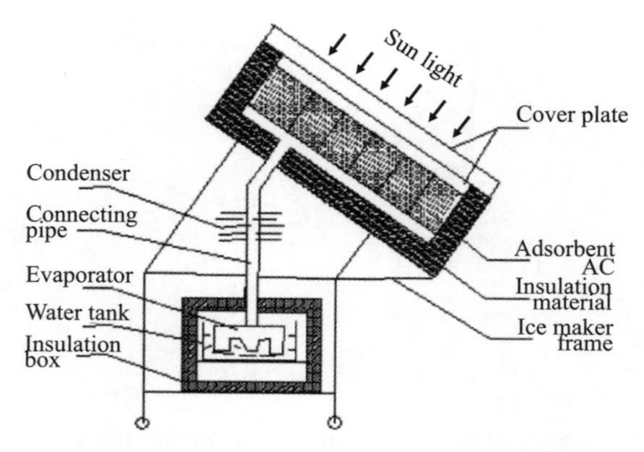

Figure 3.239 A scheme of a system of a solar ice maker [778].

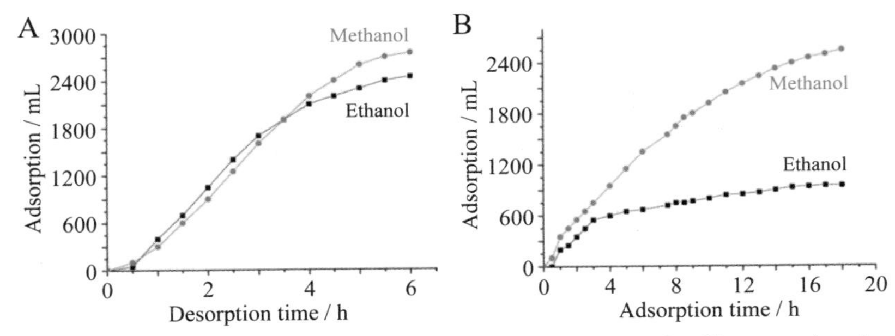

Figure 3.240 Changes in desorbed and adsorbed amounts of refrigerants (methanol and ethanol) for an AC in a solar ice maker: (A) desorption and (B) adsorption in the first cycle at a solar energy of 19.44 MJ [778].

thermally insulated container. This desorption–adsorption process is cycled. The adsorbate (refrigerant) governs the cycle efficiency. In Fig. 3.240A and B, the desorbed mass and adsorbed mass of the refrigerant (methanol or ethanol) are plotted against desorption and adsorption time, respectively, when 19.44 MJ of radiation energy is supplied to the adsorbent bed. With 19.44 MJ of radiation energy, the AC desorbs 2750 mL of methanol and 2450 mL of ethanol during 6 h in the first cycle, whereas it adsorbs 2550 mL of methanol while only 950 mL of ethanol during 18 h. With a lower energy input of 14.58 MJ, methanol desorption and adsorption were 2150 and 2050 mL, respectively, during 4.5 h, while ethanol desorption and adsorption were only 800 and 400 mL, respectively, over 18 h. Using ethanol as a refrigerant, ice could not be obtained using this solar ice maker. A pair of an AC (Maxsorb III) –ethanol could achieve a specific cooling effect of about 420 kJ/kg at an evaporator temperature of 7°C along with a heat source of temperature 80°C [779]. The result suggested that the AC–ethanol pair could be used for solar cooling applications.

Functional groups of carbon materials had some effect on the adsorption of adsorbates. An AC (Maxsorb III) was modified by treatment in a flow of H_2/Ar mixed gas (20 vol%H_2) at 600°C for 24 h and the H_2-treated AC was further treated by mixing with KOH (KOH/AC of 2/1 by weight) at 600°C for 1 h in N_2 [780]. The former (H_2-treated AC) had a lower oxygen content of 1.75 wt.%, while the latter (KOH-treated AC) had a higher oxygen content of 10.46 wt.% in comparison to the pristine AC (oxygen content of 4.35 wt.%). These 3 ACs, the pristine, H_2-treated, and KOH-treated, exhibited similar pore structure parameters, S_{BET} of

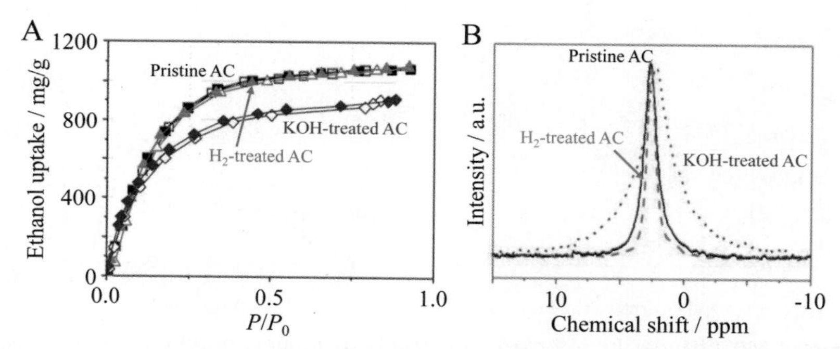

Figure 3.241 The pristine, H_2-treated, and KOH-treated ACs: (A) ethanol adsorption–desorption isotherms at 30°C and (B) ^{2}H solid-state NMR spectra after ethanol adsorption [780].

3045, 3029, and 2992 m^2/g and V_{micro} of 1.70, 1.73, and 1.65 cm^3/g. The ethanol adsorption–desorption isotherms for these three ACs are shown in Fig. 3.241A, revealing a significant decrease in the ethanol adsorption on the KOH-treated AC with a high oxygen content. In Fig. 3.241B, ^{2}H solid-state NMR spectra are shown on these three ACs after ethanol-adsorption. A marked broadening of the spectrum is observed on the KOH-treated AC on which spinning side bands are also detected (not shown). The experimental results suggested that ethanol molecules in micropores of ACs strongly interact with the oxygen-containing surface functional groups of carbon, blocking their adsorption into micropores. The desorption–adsorption behavior was studied on these three ACs by the measurements using evaporator temperatures ranging from −14 to 77°C and adsorption temperatures between 20 and 80°C [781]. The KOH-treated AC has a lower adsorption capacity for ethanol, but faster adsorption kinetics, in comparison with the pristine and H_2-treated ACs. Desorption–adsorption performances of ethanol were also studied on the phenol-resin-derived microporous carbon spheres [782]. The carbon spheres were carbonized at 600°C for 1 h in N_2, followed by activation at 900°C for 1 h in N_2 with KOH (KOH/carbon of 4/1 and 6/1 by weight). The former (prepared using KOH/carbon of 4/1) had a high V_{total} of 1.90 cm^3/g including V_{micro} of 1.85 cm^3/g and high ethanol uptake of 1.43 g/g. The application of biomass-derived carbons resulted in high performance for an adsorption heat pump using ethanol [783,784]. The microporous carbons were prepared from waste palm trunk and mangrove by carbonization at 600°C for 1 h in N_2, followed by KOH activation at

900°C for 1 h in N_2 with a mixing ratio of KOH/carbon of 6/1 by weight. The carbon derived from a palm trunk delivered V_{total} of 2.51 cm^3/g including V_{micro} of 2.41 cm^3/g and ethanol uptake of 1.9 g/g.

Adsorption heat pump systems were reviewed by comparing them with the vapor compression and absorption heat pumps and by discussing the adsorbent–adsorbate pairs [785]. By focusing on activated carbon fibers (ACFs) as the adsorbent, the performance of the pump systems using water, ammonia, CO_2, acetone, methanol, ethanol, and gasoline vapor as the refrigerant has been reviewed [786].

References

[1] Tatsumi K, Iwashita N, Sakaebe H, Shioyama H, Higuchi S, Mabuchi A, Fujimoto H. The influence of the graphitic structure on the electrochemical characteristics for the anode of secondary lithium batteries. J Electrochem Soc 1995;142:716–20.
[2] Dahn JR, Zheng T, Liu Y, Xue JS. Mechanisms for lithium insertion in carbonaceous materials. Science 1995;270:590–3.
[3] Yoshio M, Wang H, Fukuda K, Hara Y, Adachi Y. Effect of carbon coating on electrochemical performance of treated natural graphite as lithium-ion battery anode material. J Electrochem Soc 2000;147:1245–50.
[4] Wang H, Yoshio M. Carbon-coated natural graphite prepared by thermal vapor decomposition process, a candidate anode material for lithium-ion battery. J Power Sources 2001;93:123–9.
[5] Yoshio M, Wang H, Fukuda K. Spherical carbon-coated natural graphite as a lithium-ion battery- anode material. Angew Chem Int Ed 2003;42:4203–6.
[6] Yoshio M, Wang H, Fukuda K, Umeno T, Abe T, Ogumi Z. Improvement of natural graphite as a lithium-ion battery anode material, from raw flake to carbon-coated sphere. J Mater Chem 2004;14:1754–8.
[7] Natarajan C, Fujimoto H, Tokumitsu K, Mabuchi A, Kasuh T. Reduction of the irreversible capacity of a graphite anode by the CVD process. Carbon 2001;39:1409–13.
[8] Han Y-S, Lee J-Y. Improvement on the electrochemical characteristics of graphite anode by coating of the pyrolytic carbon using tumbling chemical vapor deposition. Electrochim Acta 2003;48:1073–9.
[9] Ding Y-S, Li W-N, Iaconetti S, Shen X-F, DiCarlo J, Galasso FS, Suib SL. Characteristics of graphite anode modified by CVD carbon coating. Surf Coat Tech 2006;200:3041–5.
[10] Nozaki H, Nagaoka K, Hoshi K, Ohta N, Inagaki M. Carbon-coated graphite for anode of lithium ion rechargeable batteries: carbon coating conditions and precursors. J Power Sources 2009;194:486–93.
[11] Ohta N, Nagaoka K, Hoshi K, Bitoh S, Inagaki M. Carbon-coated graphite for anode of lithium ion rechargeable batteries: graphite substrates for carbon coating. J Power Sources 2009;194:985–90.
[12] Hoshi K, Ohta N, Nagaoka K, Bitoh S, Yamanaka A, Nozaki H, Okuni T, Inagaki M. Production and advantages of carbon-coated graphite for the anode of lithium ion rechargeable batteries. TANSO 2009;240:213–20.

[13] Kubo S, White RJ, Tauer K, Titirici MM. Flexible coral-like carbon nano-architectures via a dual block copolymer-latex templating approach. Chem Mater 2013;25:4781−90.
[14] Maruyama J, Maruyama S, Fukuhara T, Chashiro K, Uyama H. Ordered mesoporous structure by graphitized carbon nanowall assembly. Carbon 2018;126:452−5.
[15] Wang GX, Shen XP, Yao J, Park J. Graphene nanosheets for enhanced lithium storage in lithium ion batteries. Carbon 2009;47:2049−53.
[16] Wu Z-S, Ren W, Xu L, Li F, Cheng H-M. Doped graphene sheets as anode materials with super high rate and large capacity for lithium ion batteries. ACS Nano 2011;7:5463−71.
[17] Li X, Geng D, Zhang Y, Meng X, Li R, Sun X. Superior cycle stability of nitrogen-doped graphene nanosheets as anodes for lithium ion batteries. Electrochem Commun 2011;13:822−5.
[18] Zhang X, Zhu G, Wang M, Li J, Lu T, Pan L. Covalent-organic-frameworks derived N-doped porous carbon materials as anode for superior long-life cycling lithium and sodium ion batteries. Carbon 2017;116:686−94.
[19] Nakajima T, Koh M, Katsube T. Structure, chemical bonding and electrochemical behavior of heteroatom-substituted carbons prepared by arc discharge and chemical vapor deposition. Solid State Sci 2000;2:17−29.
[20] Tsumura T, Arikawa A, Kinumoto T, Arai Y, Morishita T, Orikasa H, Inagaki M, Toyoda M. Structure of heat-treated mesoporous carbon and its electrochemical lithium intercalation behavior. Mater Chem Phys 2014;147:1175−82.
[21] Canal-Rodríguez M, Arenillas A, Menendez JA, Beneroso D, Rey-Raap N. Carbon xerogels graphitized by microwave heating as anode materials in lithium-ion batteries. Carbon 2018;137:384−94.
[22] Mabuchi A, Tokumitsu K, Fujimoto H, Kasuh T. Charge-discharge characteristics of the mesocarbon miocrobeads heat-treated at different temperatures. J Electrochem Soc 1995;142:1041−6.
[23] Yata S, Kinoshita H, Komori M, Ando N, Kashiwarnura T, Harada T. Structure and properties of deeply Li-doped polyacenic semiconductor materials beyond C6Li stage. Synth Met 1994;62:153−8.
[24] Fujimoto H. Development of efficient carbon anode material for a high-power and long-life lithium ion battery. J Power Sources 2010;195:5019−24.
[25] Ma Q, Wang L, Xia W, Jia D, Zhao Z. Nitrogen-doped hollow amorphous carbon spheres@graphitic shells derived from pitch: new structure leads to robust lithium storage. Chem Eur J 2016;22(7):2339−44.
[26] Barsoum MW. The MN+1AXN phases: a new class of solids; Thermodynamically stable nanolaminates. Prog Solid State Chem 2000;28:201−81.
[27] Wang XH, Zhou YC. Layered machinable and electrically conductive Ti_2AlC and Ti_3AlC_2 ceramics: a review. J Mater Sci Technol 2010;26:385−416.
[28] Naguib M, Kurtoglu M, Presser V, Lu J, Niu J, Heon M, Hultman L, Gogotsi Y, Barsoum MW. Two- dimensional nanocrystals produced by exfoliation of Ti_3AlC_2. Adv Mater 2011;23:4248−53.
[29] Ghidiu M, Lukatskaya MR, Zhao MQ, Gogotsi Y, Barsoum MW. Conductive two-dimensional titanium carbide ‘clay’ with high volumetric capacitance. Nature 2014;516:78−81.
[30] Naguib M, Mashtalir O, Carle J, Presser V, Lu J, Hultman L, Gogotsi Y, Barsoum MW. Two- dimensional transition metal carbides. ACS Nano 2012;6:1322−31.

[31] Lukatskaya MR, Mashtalir O, Ren CE, Dall'Agnese Y, Rozier P, Taberna PL, Naguib M, Simon P, Barsoum MW, Gogotsi Y. Cation intercalation and high volumetric capacitance of two-dimensional titanium carbide. Science 2013;341:1502–5.
[32] Naguib M, Halim J, Lu J, Cook KM, Hultman L, Gogotsi Y, Barsoum MW. New two-dimensional niobium and vanadium carbides as promising materials for Li-ion batteries. J Am Chem Soc 2013;135:15966–9.
[33] Er D, Li J, Naguib M, Gogotsi Y, Shenoy VB. Ti3C2 MXene as a high capacity electrode material for metal (Li, Na, K, Ca) ion batteries. ACS Appl Mater Interfaces 2014;6:11173–9.
[34] Zhao MQ, Ren CE, Ling Z, Lukatskaya MR, Zhang C, Van Aken KL, Barsoum MW, Gogotsi Y. Flexible MXene/carbon nanotube composite paper with high volumetric capacitance. Adv Mater 2015;27:339–45.
[35] Luo J, Tao X, Zhang J, Xia Y, Huang H, Zhang L, Gan Y, Liang C, Zhang W. Sn^{4+} ion decorated highly conductive Ti_3C_2 MXene: promising lithium-ion anodes with enhanced volumetric capacity and cyclic performance. ACS Nano 2016;10:2491–9.
[36] Padhi AK, Najundaswamy KS, Goodenough JB. Phospho-olivines as positive-electrode materials for rechargeable lithium batteries. J Electrochem Soc 1997;144:1188–94.
[37] Ohzuku T, Brodd RJ. An overview of positive-electrode materials for advanced lithium-ion batteries. J Power Sources 2007;174:449–56.
[38] Jugovic D, Uskokovic D. A review of recent developments in the synthesis procedures of lithium iron phosphate powders. J Power Sources 2009;190:538–44.
[39] Scrosati B, Garche J. Lithium batteries: status, prospects and future. J Power Sources 2010;195:2419–30.
[40] Zhang W-J. Structure and performance of $LiFePO_4$ cathode materials: a review. J Power Sources 2011;196:2962–70.
[41] Huang H, Yin SC, Nazar LF. Approaching theoretical capacity of $LiFePO_4$ at room temperature at high rates. Electrochem Solid State Lett 2001;4:A170–2.
[42] Huang B, Zheng XD, Jia DM, Lu M. Design and synthesis of high-rate micron-sized, spherical $LiFePO_4$/C composites containing clusters of nano/microspheres. Electrochim Acta 2010;55:1227–31.
[43] Ravet N, Chouinard Y, Magnam JF, Besner S, Gauthier M, Armand M. Electroactivity of natural and synthetic triphylite. J Power Sources 2001;97:503–7.
[44] Shin HC, Cho WI, Jang H. Electrochemical properties of carbon-coated $LiFePO_4$ cathode using graphite, carbon black, and acetylene black. Electrochim Acta 2006;52:1472–6.
[45] Doeff MM, Wilcox JD, Kostecki R, Lau G. Optimization of carbon coatings on $LiFePO_4$. J Power Sources 2006;163:180–4.
[46] Franger S, Le Cras F, Bourbon C, Rouault H. $LiFePO_4$ synthesis routes for enhanced electrochemical performance. Electrochem Solid State Lett 2002;5:A231–2.
[47] Kim JK, Cheruvally G, Ahn JH, Hwang GC, Choi JB. Electrochemical properties of carbon-coated $LiFePO_4$ synthesized by a modified mechanical activation process. J Phys Chem Solids 2008;69:2371–7.
[48] Kim JK, Cheruvally G, Ahn JH, Ahn HJ. Electrochemical properties of $LiFePO_4$/C composite cathode material: carbon costing by the precursor method and direct addition. J Phys Chem Solids 2008;69:1257–60.
[49] Wang K, Cai R, Yuan T, Yu X, Ran R, Shao Z. Process investigation, electrochemical characterization and optimization of $LiFePO_4$/C composite from mechanical activation using sucrose as carbon source. Electrochim Acta 2009;54:2861–8.

[50] Liang G, Wang L, Ou X, Zhao X, Xu S. Lithium iron phosphate with high-rate capability synthesized through hydrothermal reaction in glucose solution. J Power Sources 2008;184:538–43.
[51] Murugan AV, Muraliganth T, Manthiram A. One-pot microwave-hydrothermal synthesis and characterization of carbon-coated $LiMPO_4$ (M=Mn, Fe, and Co) cathodes. J Electrochem Soc 2009;156:A79–83.
[52] Chen Z, Dahn JR. Reducing carbon in $LiFePO_4$/C composite electrodes to maximize specific energy, volumetric energy, and tap density. J Electrochem Soc 2002;149:A1184–9.
[53] Bewlay SI, Konstantinov K, Wang GX, Dou SX, Liu HK. Conductivity improvements to spray- produced $LiFePO_4$ by addition of a carbon source. Mater Lett 2004;58:1788–91.
[54] Cho YD, Fey GTK, Kao HM. The effect of carbon coating thickness on the capacity of $LiFePO_4$/C composite cathodes. J Power Sources 2009;189:256–63.
[55] Lin Y, Gao MX, Zhu D, Liu YF, Pan HG. Effects of carbon coating and iron phosphides on the electrochemical properties of $LiFePO_4$/C. J Power Sources 2008;184:444–8.
[56] Chen ZY, Zhu HL, Ji S, Fakir R, Linkov V. Influence of carbon sources on electrochemical performances of $LiFePO_4$/C composites. Solid State Ionics 2008;179:1810–5.
[57] Jaiswal A, Horne CR, Cheng O, Zhang W, Kong W, Wang L, Chern T, Doeff MM. Nanoscale $LiFePO_4$ and $Li_4Ti_5O_{12}$ for high rate Li-ion batteries. J Electrochem Soc 2009;156:A1041–6.
[58] Jugovic D, Mitric M, Kuzmanovic M, Cvjeticanin N, Skapin S, Cekic B, Ivanovski V, Uskokovic D. Preparation of $LiFePO_4$/C composites by co-precipitation in molten stearic acid. J Power Sources 2011;196:4613–8.
[59] Ong C-W, Lin Y-K, Chen J-S. Effect of various organic precursors on the performance of $LiFePO_4$/C composite cathode by coprecipitation method. J Electrochem Soc 2007;154:A527–33.
[60] Nien YH, Carey JR, Chen JS. Physical and electrochemical properties of $LiFePO_4$/C composite cathode prepared from various polymer-containing precursors. J Power Sources 2009;193:822–7.
[61] Yamada A, Hosoya M, Chung SC, Kudo Y, Hinokuma K, Liu KY, Nishi Y. Olivine-type cathodes. Achievements and problems. J Power Sources 2003;119–121:232–8.
[62] Hong J, Wang F, Wang X, Graetz J. LiFexMn1-xPO_4: a cathode for lithium-ion batteries. J Power Sources 2011;196:3659–63.
[63] Kaymaksiz S, Kaskhedikar N, Sato N, Roth S, Dettlaff-Weglikowska U. Evaluation of synthesis routes for pure and composite $LiMnPO_4$ with carbonaceous additives for Li-ion battery. Electrochem Soc Trans 2010;25:187–97.
[64] Liu J, Conry TE, Song X, Yang L, Doeff MM, Richardson TJ. Spherical nanoporous $LiCoPO_4$/C composites as high performance cathode materials for rechargeable lithium-ion batteries. J Mater Chem 2011;21:9984–7.
[65] Ma J, Li B, Du H, Xu C, Kang F. The effect of vanadium on physicochemical and electrochemical performances of $LiFePO_4$ cathode for lithium battery. J Electrochem Soc 2011;158:A26–32.
[66] Rui XH, Jin Y, Feng XY, Zhang LC, Chen CH. A comparative study on the low-temperature performance of $LiFePO_4$/C and $Li_3V_2(PO_4)_3$/C cathodes for lithium-ion batteries. J Power Sources 2011;196:2109–14.
[67] Cheng L, Li XL, Liu HJ, Xiong H-M, Zhang P-W, Xia Y-Y. Carbon-coated $Li_4Ti_5O_{12}$ as a high rate electrode material for Li-ion intercalation. J Electrochem Soc 2007;154:A692–7.

[68] Jung HG, Kim J, Scrosati B, Sun Y-K. Micron-sized, carbon-coated $Li_4Ti_5O_{12}$ as high power anode material for advanced lithium batteries. J Power Sources 2011;196:7763−6.
[69] Zhao L, Hu YS, Li H, Wang Z, Chen L. Porous $Li_4Ti_5O_{12}$ coated with N-doped carbon from ionic liquids for Li-ion batteries. Adv Mater 2011;23:1385−8.
[70] Guo X, Wang C, Chen M, Wang J, Zheng J. Carbon coating of $Li_4Ti_5O_{12}$ using amphiphilic carbonaceous material for improvement of lithium-ion battery performance. J Power Sources 2012;214:107−12.
[71] Hu X, Lin Z, Yang K, Deng Z, Suo J. Influence factors on electrochemical properties of $Li_4Ti_5O_{12}$/C anode material pyrolyzed from lithium polyacrylate. J Alloys Comps 2010;506:160−6.
[72] Wang J, Liu X-M, Yang H, Shen X-D. Characterization and electrochemical properties of carbon- coated $Li_4Ti_5O_{12}$ prepared by a citric acid sol−gel method. J Alloys Compd 2011;509:712−8.
[73] Yuan T, Cai R, Shao ZP. Different effect of the atmospheres on the phase formation and performance of $Li_4Ti_5O_{12}$ prepared from ball-milling-assisted solid-phase reaction with pristine and carbon- precoated TiO_2 as starting materials. J Phys Chem C 2011;115:4943−52.
[74] Xiang H, Tian B, Lian P, Li Z, Wang H. Sol−gel synthesis and electrochemical performance of $Li_4Ti_5O_{12}$/graphene composite anode for lithium-ion batteries. J Alloys Compd 2011;509:7205−9.
[75] Yin SC, Grondey H, Strobel P, Huang H, Nazar LF. Charge ordering in lithium vanadium phosphates: electrode materials for lithium-ion batteries. J Am Chem Soc 2003;125:326−7.
[76] Yin SC, Grondey H, Strobel P, Anne M, Nazar LF. Electrochemical property: structure relationships in monoclinic $Li_{3-y}V_2(PO_4)_3$. J Am Chem Soc 2003;125:10402−11.
[77] Obrovac MN, Christensen L, Le DB, Dahn JR. Alloy design for lithium-ion battery anodes. J Electrochem Soc 2007;154:A849−55.
[78] Kasukabe T, Nishihara H, Iwamura S, Kyotani T. Remarkable performance improvement of inexpensive ball-milled Si nanoparticles by carbon-coating for Li-ion batteries. J Power Sources 2016;319:99−103.
[79] Kasukabe T, Nishihara H, Kimura K, Matsumoto T, Kobayash H, Okai M, Kyotani T. Beads-milling of waste Si sawdust into high-performance nanoflakes for lithium-ion batteries. Sci Rep 2017;7:42734.
[80] Ryu J, Chen T, Bok T, Song G, Ma J, Hwang C, Luo L, Song H-K, Cho J, Wang C, Zhang S, Park S. Mechanical mismatch-driven rippling in carbon-coated silicon sheets for stress-resilient battery anodes. Nat Commun 2018;9:2924.
[81] Iwamura S, Nishihara H, Kyotani T. Fast and reversible lithium storage in a wrinkled structure formed from Si nanoparticles during lithiation/delithiation cycling. J Power Sources 2013;222:400−9.
[82] Sasidharachari K, Na B-K, Woo S-G, Yoon S, Cho KY. Facile conductive surface modification of Si nanoparticle with nitrogen-doped carbon layers for lithium-ion batteries. J Solid State Electrochem 2016;20(10):2873−8.
[83] Yi R, Chen S, Song J, Gordin ML, Manivannan A, Hai D. High-performance hybrid supercapacitor enabled by a high-rate Si-based anode. Adv Funct Mater 2014;24:7433−9.
[84] Choi M, Kim J-C, Kim D-W. Waste windshield-derived silicon/carbon nanocomposites as high- performance lithium-ion battery anodes. Sci Rep 2018;8:960.
[85] Kim SK, Kim H, Chang H, Cho B-G, Huang J, Yoo H, Kim H, Jan HD. One-step formation of silicon- graphene composites from silicon sludge waste and graphene oxide via aerosol process for lithium ion batteries. Sci Rep 2016;6:33688.

[86] Bogart TD, Oka D, Lu X, Gu M, Wang C, Korgel BA. Lithium ion battery performance of silicon nanowires with carbon skin. ACS Nano 2014;8:915–22.
[87] Fan Y, Zhang Q, Xiao Q, Wang X, Huang K. High performance lithium ion battery anodes based on carbon nanotube-silicon core-shell nanowires with controlled morphology. Carbon 2013;59:264–9.
[88] Baggetto L, Notten PHL. Lithium-ion (de)insertion reaction of germanium thin-film electrodes: an electrochemical and in situ XRD study. J Electrochem Soc 2009;156:A169–75.
[89] Liang W, Yang H, Fan F, Liu Y, Liu XH, Huang JY, Zhu T, Zhang S. Tough germanium nanoparticles under electrochemical cycling. ACS Nano 2013;7(4):3427–33.
[90] Lee H, Kim MG, Choi CH, Sun YK, Yoon CS, Cho J. Surface stabilized amorphous germanium nanoparticles for lithium storage material. J Phys Chem B 2005;109(44):20719–23.
[91] Chan CK, Zhang XF, Cui Y. High capacity Li ion battery anodes using Ge nanowires. Nano Lett 2008;8(1):307–9.
[92] Yuan FW, Yang HJ, Tuan HY. Alkanethiol-passivated Ge nanowires as high-performance anode materials for lithium ion batteries: the role of chemical surface functionalization. ACS Nano 2012;6(11):9932–42.
[93] Chockla AM, Klavetter KC, Mullins CB, Korgel BA. Solution grown germanium nanowire anodes for lithium-ion batteries. ACS Appl Mater Interfaces 2012;4(9):4658–64.
[94] Yang LC, Gao QS, Li L, Tang Y, Wu YP. Mesoporous germanium as anode material of high capacity and good cycling prepared by a mechanochemical reaction. Electrochem Commun 2010;12(3):418–21.
[95] Park MH, Kim K, Kim J, Cho J. Flexible dimensional control of high-capacity Li-ion-battery anodes: from 0D hollow to 3D porous germanium nanoparticle assemblies. Adv Mater 2010;22(3):415–8.
[96] Kim CH, Im HS, Cho YJ, Jung CS, Jang DM, Myung Y, Kim HS, Back SH, Lim YR, Lee C-W, Park J. High-yield gas-phase laser photolysis synthesis of germanium nanocrystals for high-performance photodetectors and lithium ion batteries. J Phys Chem C 2012;116(50):26190–6.
[97] Yoon S, Park CM, Sohn HJ. Electrochemical characterizations of germanium and carbon-coated germanium composite anode for lithium-ion batteries. Electrochem Solid State Lett 2008;11(4):A42–5.
[98] Xiao Y, Cao M, Ren L, Hu C. Hierarchically porous germanium-modified carbon materials with enhanced lithium storage performance. Nanoscale 2012;4(23):7469–74.
[99] Jo G, Choi I, Ahn H, Park MJ. Binder-free Ge nanoparticles carbon hybrids for anode materials of advanced lithium batteries with high capacity and rate capability. Chem Commun 2012;48(33):3987–9.
[100] Hwang J, Jo C, Kim MG, Chun J, Lim E, Kim S, Jeong S, Kim Y, Lee J. Mesoporous Ge/GeO_2/carbon lithium-ion battery anodes with high capacity and high reversibility. ACS Nano 2015;9:5299–309.
[101] Cui G, Gu L, Zhi L, Kaskhedikar N, van Aken PA, Müllen K, Maier J. A germanium-carbon nanocomposite material for lithium batteries. Adv Mater 2008;20:3079–83.
[102] Seng KH, Park MH, Guo ZP, Liu HK, Cho J. Self-assembled germanium/carbon nanostructures as high-power anode material for the lithium-ion battery. Angew Chem Int Ed 2012;51(23):5657–61.

[103] Seo MH, Park M, Lee KT, Kim K, Kim J, Cho J. High performance Ge nanowire anode sheathed with carbon for lithium rechargeable batteries. Energy Environ Sci 2011;4(2):425−8.
[104] Chockla AM, Panthani MG, Holmberg VC, Hessel CM, Reid DK, Bogart TD, Harris JT, Mullins CB, Korgel BA. Electrochemical lithiation of graphene supported silicon and germanium for rechargeable batteries. J Phys Chem C 2012;116(22):11917−23.
[105] Cheng J, Du J. Facile synthesis of germanium−graphene nanocomposites and their application as anode materials for lithium ion batteries. CrystEngComm 2012;14(2):397−400.
[106] Xue DJ, Xin S, Yan Y, Jiang K-C, Yin YX, Guo YG, Wan L-J. Improving the electrode performance of Ge through Ge@C core-shell nanoparticles and graphene networks. J Am Chem Soc 2012;134(5):2512−5.
[107] Ren J-G, Wu Q-H, Tang H, Hong G, Zhang W, Lee S-T. Germanium-graphene composite anode for high-energy lithium batteries with long cycle life. J Mater Chem A 2013;1:1821−6.
[108] Zhong C, Wang J-Z, Gao X-W, Wexler D, Liu H-K. In situ one-step synthesis of a 3D nanostructured germanium-graphene composite and its application in lithium-ion batteries. J Mater Chem A 2013;1:10798−804.
[109] Yuan F-W, Tuan H-Y. Scalable solution-grown high-germanium-nanoparticle-loading graphene nanocomposites as high-performance lithium-ion battery electrodes: an example of a graphene-based platform toward practical full-cell applications. Chem Mater 2014;26:2172−9.
[110] Li W, Yang Z, Cheng J, Zhong X, Gu L, Yu Y. Germanium nanoparticles encapsulated in flexible carbon nanofibers as self-supported electrodes for high performance lithium-ion batteries. Nanoscale 2014;6:4532−7.
[111] DiLeo RA, Frisco S, Ganter MJ, Rogers RE, Raffaelle RP, Landi BJ. Hybrid germanium nanoparticle- single-wall carbon nanotube free-standing anodes for lithium ion batteries. J Phys Chem C 2011;115:22609−14.
[112] Wang J, Wang J-Z, Sun Z-Q, Gao X-W, Zhong C, Chou S-L, Liu H-K. A germanium/single-walled carbon nanotube composite paper as a freestanding anode for lithium-ion batteries. J Mater Chem A 2014;2:4613−8.
[113] Wang X, Susantyoko RA, Fan Y, Sun L, Xiao Q, Zhang Q. Vertically aligned CNT-supported thick Ge films as high-performance 3D anodes for lithium ion batteries. Small 2014;10:2826−9. 2742.
[114] Susantyoko RA, Wang X, Sun L, Pey KL, Fitzgerald E, Zhang Q. Germanium coated vertically- aligned multiwall carbon nanotubes as lithium-ion battery anodes. Carbon 2014;77:551−9.
[115] Gao C, Kim ND, Salvatierra RV, Lee S-K, Li L, Li Y, Sha J, Lopez Silva GA, Fei H, Xie E, Tour JM. Germanium on seamless graphene carbon nanotube hybrids for lithium ion anodes. Carbon 2017;123:433−9.
[116] Park M-S, Wang G-X, Kang Y-M, Wexler D, Dou S-X, Liu H-K. Preparation and electrochemical properties of SnO_2 nanowires for application in lithium-ion batteries. Angew Chem Int Ed 2007;46:750−3.
[117] Jiang L-Y, Wu X-L, Guo Y-G, Wan L-J. SnO_2-based hierarchical nano-microstructures: facile synthesis and their applications in gas sensors and lithium-ion batteries. J Phys Chem C 2009;113:14213−9.
[118] Zhang H-X, Feng C, Zhai Y-C, Jiang K-L, Li Q-Q, Fan S-S. Cross-stacked carbon nanotube sheets uniformly loaded with SnO_2 nanoparticles: a novel binder-free and high-capacity anode material for lithium-ion batteries. Adv Mater 2009;21:2299−304.

[119] Lou XW, Li CM, Archer LA. Designed synthesis of coaxial SnO_2@carbon hollow nanospheres for highly reversible lithium storage. Adv Mater 2009;21:2536–9.

[120] Zhou X, Wan L-J, Guo Y-G. Binding SnO_2 nanocrystals in nitrogen-doped graphene sheets as anode materials for lithium-ion batteries. Adv Mater 2013;25:2152–7.

[121] Liang J, Yuan C, Li H, Fan K, Wei Z, Sun H, Ma J. Growth of SnO_2 nanoflowers on N-doped carbon nanofibers as anode for Li- and Na-ion batteries. Nano-Micro Lett 2018;10:21.

[122] Liang J, Sun H, Zhao Z, Wang Y, Feng Z, Zhu J, Guo L, Huang Y, Duan X. Ultrahigh areal capacity realized in three-dimensional holey graphene/SnO_2 composite anodes. iScience 2019;19:728–36.

[123] Han F, Li W-C, Li M-R, Lu A-H. Fabrication of superior-performance SnO_2@C composites for lithium-ion anodes using tubular mesoporous carbon with thin carbon walls and high pore volume. J Mater Chem 2012;22:9645–51.

[124] Lee KT, Jung YS, Oh SM. Synthesis of tin-encapsulated spherical hollow carbon for anode material in lithium secondary batteries. J Am Chem Soc 2003;125:5652–3.

[125] Morishita T, Hirabayashi T, Okuni T, Ota N, Inagaki M. Preparation of carbon-coated Sn powders and their loading onto graphite flakes for lithium ion secondary battery. J Power Sources 2006;160:638–44.

[126] Oro S, Urita K. Nanospace control of SnO_2 nanocrystallites-embedded nanoporous carbon for reversible electrochemical charge-discharge reactions. J Phys Chem C 2016;120:25717–24.

[127] Notohara H, Urita K, Yamamura H, Moriguchi I. High capacity and stable all-solid state Li ion battery using SnO_2-embedded nanoporous carbon. Sci Rep 2018;8:8747.

[128] Zhou X, Bao J, Dai Z, Guo Y-G. Tin nanoparticles impregnated in nitrogen-doped graphene for lithium-ion battery anodes. J Phys Chem C 2013;117:25367–73.

[129] Marchand R, Brohan L, Tournoux M. TiO_2(B) a new form of titanium dioxide and the potassium octatitanate $K_2Ti_8O_{17}$. Mater Res Bull 1980;15:1129–33.

[130] Armstrong AR, Armstrong G, Canales J, Bruce PG. TiO_2-B nanowires. Angew Chem Int Ed 2004;43:2286–8.

[131] Armstrong AR, Armstrong G, Canales J, García R, Bruce PG. Lithium-ion intercalation into TiO_2-B nanowires. Adv Mater 2005;17:862–5.

[132] Armstrong G, Armstrong AR, Canales J, Bruce P. Nanotubes with the TiO_2-B structure. Chem Commun 2005:2454–6.

[133] Kobayashi M, Petrykin VV, Kakihana M. One-step synthesis of TiO_2(B) nanoparticles from a water- soluble titanium complex. Chem Mater 2007;19:5373–6.

[134] Baudrin E, Cassaignon S, Koesch M, Jolivet J-P, Dupont L, Tarascon JM. Structural evolution during the reaction of Li with nano-sized rutile type TiO_2 at room temperature. Electrochem Commun 2007;9:337–42.

[135] Cai Y, Zhao B, Wang J, Shao Z. Non-aqueous hybrid supercapacitors fabricated with mesoporous TiO_2 microspheres and activated carbon electrodes with superior performance. J Power Sources 2014;253:80–9.

[136] Cao FF, Wu XL, Xin S, Guo YG, Wan LJ. Facile synthesis of mesoporous TiO_2-C nanosphere as an improved anode material for superior high rate 1.5 V rechargeable Li ion batteries containing $LiFePO_4$-C cathode. J Phys Chem C 2010;114:10308–13.

[137] Ishii Y, Kanamori Y, Kawashita T, Mukhopadhyay I, Kawasaki S. Mesoporous carbon-titania nanocomposites for high-power Li-ion battery anode material. J Phys Chem Solids 2010;71:511–4.

[138] Yang XJ, Teng DH, Liu BX, Yu YH, Yang XP. Nanosized anatase titanium dioxide loaded porous carbon nanofiber webs as anode materials for lithium-ion batteries. Electrochem Commun 2011;13:1098–101.

[139] Wang DH, Choi DW, Li J, Yang ZG, Nie ZM, Kou R, Hu DH, Wang CM, Saraf LV, Zhang JG, Aksay IA, Liu J. Self-assembled TiO_2-graphene hybrid nanostructures for enhanced Li-ion insertion. ACS Nano 2009;3:907−14.
[140] Zhao B, Jiang S, Su C, Cai R, Ran R, Tadé MO, Shao Z. A 3D porous architecture composed of TiO_2 nanotubes connected with a carbon nanofiber matrix for fast energy storage. J Mater Chem A 2013;1:12310−20.
[141] Liu H, Wang G, Wang J, Wexler D. Magnetite/carbon core-shell nanorods as anode materials for lithium-ion batteries. Electrochem Commun 2008;10:1879−82.
[142] Hassan MF, Rahman MM, Guo ZP, Chen ZX, Liu HK. Solvent-assisted molten salt process: a new route to synthesis α-Fe_2O_3/C nanocomposite and its electrochemical performance in lithium-ion batteries. Electrochim Acta 2010;55:5006−13.
[143] Brandt A, Balducci A. Ferrocene as precursor for carbon-coated α-Fe_2O_3 nanoparticles for rechargeable lithium batteries. J Power Sources 2013;230:44−9.
[144] He C, Wu S, Zhao N, Shi C, Liu E, Li J. Carbon-encapsulated Fe_3O_4 nanoparticles as a high-rate lithium ion battery anode material. ACS Nano 2013;7:4459−69.
[145] Zhou GM, Wang DW, Li F, Zhang LL, Li N, Wu ZS, Wen L, Lu GQ, Cheng H-M. Graphene-wrapped Fe_3O_4 anode material with improved reversible capacity and cyclic stability for lithium ion batteries. Chem Mater 2010;22:5306−13.
[146] Chen X, Zhu H, Chen Y-C, Shang Y, Cao A, Hu L, Rubloff GW. MWCNT/V_2O_5 core/shell sponge for high areal capacity and power density Li-ion cathodes. ACS Nano 2012;6(9):7948−59.
[147] Zhou X, Wu G, Wu J, Yang H, Wang J, Gao G, Cai R, Yan Q. Multiwalled carbon nanotubeseV_2O_5 integrated composite with nanosized architecture as a cathode material for high performance lithium ion batteries. J Mater Chem A 2013;1(48):15459−68.
[148] Yu R, Zhang C, Meng Q, Chen Z, Liu H, Guo Z. Facile synthesis of hierarchical networks composed of highly interconnected V_2O_5 nanosheets assembled on carbon nanotubes and their superior lithium storage properties. ACS Appl Mater Interfaces 2013;5(23):12394−9.
[149] Cheng J, Wang B, Xin HL, Yang G, Cai H, Nie F, Huang H. Self-assembled V_2O_5 nanosheets/reduced graphene oxide hierarchical nanocomposite as a high-performance cathode material for lithium ion batteries. J Mater Chem A 2013;1:10814−20.
[150] Yang Y, Li L, Fei H, Peng Z, Ruan G, Tour JM. Graphene nanoribbon/V_2O_5 cathodes in lithium-ion batteries. ACS Appl Mater Interfaces 2014;6:9590−4.
[151] Gao X-T, Liu Y-T, Zhu X-D, Yan D-J, Wang C, Feng Y-J, Sun K-N. V_2O_5 nanoparticles confined in three-dimensionally organized, porous nitrogen-doped graphene frameworks: flexible and free- standing cathodes for high performance lithium storage. Carbon 2018;140:218−26.
[152] Zang J, Chen JJ, Zhang CL, Qian H, Zheng MS, Dong QF. The synthesis of a core−shell MnO_2/3D- ordered hollow carbon sphere composite and its superior electrochemical capability for lithium ion batteries. J Mater Chem A 2014;2:6343−7.
[153] Wang Y, Han ZJ, Yu SF, Song RR, Song HH, Ostrikov K, Yang HY. Core-leaf onion-like carbon/MnO_2 hybrid nano-urchins for rechargeable lithium-ion batteries. Carbon 2013;64:230−6.
[154] Deng J, Chen L, Sun Y, Ma M, Fu L. Interconnected MnO_2 nanoflakes assembled on graphene foam as a binder-free and long-cycle life lithium battery anode. Carbon 2015;92:177−84.
[155] Lai H, Li JX, Chen ZG, Huang ZG. Carbon nanohorns as a high performance carrier for MnO_2 anode in lithium-ion batteries. ACS Appl Mater Interfaces 2012;4:2325−8.
[156] Reddy ALM, Shaijumon MM, Gowda SR, Ajayan PM. Coaxial MnO_2/carbon nanotube array electrodes for high performance lithium batteries. Nano Lett 2009;9:1002−6.

[157] Xia H, Lai MO, Lu L. Nanoflaky MnO_2/carbon nanotube nanocomposites as anode materials for lithium ion batteries. J Mater Chem 2010;20:6896–902.
[158] Zhu J, Wu Q, Key J, Wu M, Shen PK. Self-assembled superstructure of carbon-wrapped, single- crystalline Cu3P porous nanosheets: one-step synthesis and enhanced Li-ion battery anode performance. Energy Storage Mater 2018;15:75–81.
[159] Wang M, Xie S, Tang C, Fang X, Liao M, Wang L, Zhao Y, Wen Y, Ye L, Wang B, Peng H. In situ intercalation of bismuth into 3D reduced graphene oxide scaffolds for high capacity and long cycle- life energy storage. Small 2019:1905903.
[160] Ji X, Nazar LF. Advances in Li-S batteries. J Mater Chem 2010;20:9821–6.
[161] Bruce PG, Freunberger SA, Hardwick LJ, Tarascon JM. Li–O_2 and Li–S batteries with high energy storage. Nat Mater 2012;11:19–29.
[162] Yin Y-X, Xin S, Guo Y-G, Wan L-J. Lithium-sulfur batteries: electrochemistry, materials, and prospects. Angew Chem Int Ed 2013;52:13186–200.
[163] Li Z, Huang Y, Yuan L, Hao Z, Huang Y. Status and prospects in sulfur–carbon composites as cathode materials for rechargeable lithium–sulfur batteries. Carbon 2015;92:41–63.
[164] Jeong SS, Lim YT, Choi YJ, Cho GB, Kim KW, Ahn HJ, Cho KK. Electrochemical properties of lithium sulfur cells using PEO polymer electrolytes prepared under three different mixing conditions. J Power Sources 2007;174:745–50.
[165] Park J-W, Kim I, Kim K-W, Nam T-H, Cho K-K, Ahn J-H, Ryu H-S, Ahn H-J. Effect of commercial activated carbons in sulfur cathodes on the electrochemical properties of lithium/sulfur batteries. Mater Res Bull 2016;82:109–14.
[166] Ji X, Lee KT, Naza LF. A highly ordered nanostructured carbon–Sulphur cathode for lithium–sulphur batteries. Nat Mater 2009;8:500–6.
[167] Wang J, Chew SY, Zhao ZW, Ashraf S, Wexler D, Chen J, Ng SH, Chou SL, Liu HK. Sulfur- mesoporous carbon composites in conjunction with a novel ionic liquid electrolyte for lithium rechargeable batteries. Carbon 2008;46:229–35.
[168] Li S, Xia X, Wang X, Tu J. Free-standing sulfur cathodes composited with carbon nanorods arrays for Li-S batteries application. Mater Res Bull 2016;83:474–80.
[169] Zheng W, Liu YW, Hu XG, Zhang CF. Novel nanosized adsorbing sulfur composite cathode materials for the advanced secondary lithium batteries. Electrochim Acta 2006;51(7):1330–5.
[170] Razzaq AA, Yao Y, Shah R, Qi P, Miao L, Chen M, Zhao X, Peng Y, Deng Z. High-performance lithium sulfur batteries enabled by a synergy between sulfur and carbon nanotubes. Energy Storage Mater 2019;16:194–202.
[171] Wang JZ, Lu L, Choucair M, Stride JA, Xu X, Liu H-K. Sulfur–graphene composite for rechargeable lithium batteries. J Power Sources 2011;196:7030–4.
[172] Xu C, Wu Y, Zhao X, Wang X, Du G, Zhang J, Tu J. Sulfur/three-dimensional graphene composite for high performance lithium-sulfur batteries. J Power Sources 2015;275:22–5.
[173] Moo JGS, Omar A, Jaumann T, Oswald S, Balach J, Maletti S, Giebeler L. One-pot synthesis of graphene-sulfur composites for Li-S batteries: influence of sulfur precursors. J Carbon Res 2018;4:2.
[174] Zhou T, Zhao Y, Zhoud G, Lv W, Sun P, Kang F, Li B, Yang Q-H. An in-plane heterostructure of graphene and titanium carbide for efficient polysulfide confinement. Nano Energy 2017;39:291–6.
[175] Niu S, Zhou G, Lv W, Shi H, Luo C, He Y, Li B, Yang Q-H, Kang F. Sulfur confined in nitrogen doped microporous carbon used in a carbonate-based electrolyte for long-life, safe lithium-sulfur batteries. Carbon 2016;109:1–6.
[176] Niu S, Lv W, Zhou G, He Y, Li B, Yang Q-H, Kang F. N and S co-doped porous carbon spheres prepared using L-cysteine as a dual functional agent for high-performance lithium–sulfur batteries. Chem Commun 2015;51:17720–3.

[177] Li L, Zhou G, Yin L, Koratkar N, Li F, Cheng H-M. Stabilizing sulfur cathodes using nitrogen-doped graphene as a chemical immobilizer for Li-S batteries. Carbon 2016;108:120–6.
[178] Yin L-C, Liang J, Zhou G-M, Li F, Saito R, Cheng H-M. Understanding the interactions between lithium polysulfides and N-doped graphene using density functional theory calculations. Nano Energy 2016;25:203–10.
[179] Wang XW, Zhang ZA, Qu YH, Lai YQ, Li J. Nitrogen-doped graphene/sulfur composite as cathode material for high capacity lithium sulfur batteries. J Power Sources 2014;256:361–8.
[180] Han K, Shen JM, Hao SQ, Ye HQ, Wolverton C, Kung MC, Kung HH. Free-standing nitrogen-doped graphene paper as electrodes for high-performance lithium/dissolved polysulfide batteries. ChemSusChem 2014;7:2545–53.
[181] Wang C, Su K, Wan W, Guo H, Zhou HH, Chen JT, Zhang X, Huang Y. High sulfur loading composite wrapped by 3D nitrogen-doped graphene as a cathode material for lithium-sulfur batteries. J Mater Chem A 2014;2:5018–23.
[182] Zhou G, Paek E, Hwang GS, Manthiram A. Long-life Li/polysulphide batteries with high sulphur loading enabled by lightweight three-dimensional nitrogen/sulphur-codoped graphene sponge. Nat Commun 2015;6:7760.
[183] Qiu YC, Li WF, Zhao W, Li GZ, Hou Y, Liu MN, Zhou L, Ye F, Li H, Wei Z, Yang S, Duan W, Ye Y, Guo J, Zhang Y. High-rate, ultralong cycle-life lithium/sulfur batteries enabled by nitrogen-doped graphene. Nano Lett 2014;14:4821–7.
[184] Niu S, Lv W, Zhang C, Li F, Tang L, He Y, Li B, Yang Q-H, Kang F. A carbon sandwich electrode with graphene filling coated by N-doped porous carbon layers for lithium–sulfur batteries. J Mater Chem A 2015;3:29218–24.
[185] Ren J, Xia L, Zhou Y, Zheng Q, Liao J, Lin D. A reduced graphene oxide/nitrogen, phosphorus doped porous carbon hybrid framework as sulfur host for high performance lithium-sulfur batteries. Carbon 2018;140:30–40.
[186] Tang C, Zhang Q, Zhao M-Q, Huang J-Q, Cheng X-B, Tian G-L, Peng H-J, Wei F. Nitrogen-doped aligned carbon nanotube/graphene sandwiches: facile catalytic growth on bifunctional natural catalysts and their applications as scaffolds for high-rate lithium-sulfur batteries. Adv Mater 2014;26:6100–5.
[187] Niu S, Lv W, Zhoud G, Shi H, Qin X, Zheng C, Zhou T, Luo C, Deng Y, Li B, Kang F, Yang Q-H. Electrostatic-spraying an ultrathin, multifunctional and compact coating onto a cathode for a long-life and high-rate lithium-sulfur battery. Nano Energy 2016;30:138–45.
[188] Jin C, Sheng O, Zhang W, Luo J, Yuan H, Yang T, Huang H, Gan Y, Xia Y, Liang C, Zhang J, Tao X. Sustainable, inexpensive, naturally multi-functionalized biomass carbon for both Li metal anode and sulfur cathode. Energy Storage Mater 2018;15:218–25.
[189] Jung H-G, Kim H-S, Park J-B, Oh I-H, Hassoun J, Yoon CS, Scrosati B, Sun Y-K. A transmission electron microscopy study of the electrochemical process of lithium-oxygen cells. Nano Lett 2012;12:4333–5.
[190] Christensen A, Albertus P, Sanchez-Carrera RS, Lohmann T, Kozinsky B, Liedtke R, Ahmed J, Kojic A. A critical review of Li/air batteries. J Electrochem Soc 2012;159:R1–30.
[191] Luntz C, McCloskey BD. Nonaqueous Li-air batteries: a status report. Chem Rev 2014;114:11721–50.
[192] Lu J, Li L, Park JB, Sun YK, Wu F. Aprotic and aqueous Li-O_2 batteries. Chem Rev 2014;114:5611–40.
[193] Kim H, Lim H, Kim J, Kang K. Graphene for advanced Li/S and Li/air batteries. J Mater Chem A 2014;2:33–47.

[194] Abraham KM, Jiang Z. A polymer electrolyte-based rechargeable lithium/oxygen battery. J Electrochem Soc 1996;143:1–5.
[195] Yang X, He P, Xia Y. Preparation of mesocellular carbon foam and its application for lithium/oxygen battery. Electrochem Commun 2009;11:1127–30.
[196] Mirzaeian M, Hall P. Preparation of controlled porosity carbon aerogels for energy storage in rechargeable lithium oxygen batteries. Electrochim Acta 2009;54:7444–51.
[197] Kichambare P, Kumar J, Rodrigues S, Kumar B. Electrochemical performance of highly mesoporous nitrogen doped carbon cathode in lithium-oxygen batteries. J Power Sources 2011;196:3310–6.
[198] Kichambare P, Rodrigues S, Kumar J. Mesoporous nitrogen-doped carbon-glass ceramic cathodes for solid-state lithium-oxygen batteries. ACS Appl Mater Interfaces 2012;4:49–52.
[199] Nong J, Xie P, Zhu AS, Rong MZ, Zhang MQ. Highly conductive doped carbon framework as binder- free cathode for hybrid Li-O_2 battery. Carbon 2019;142:177–89.
[200] Li Y, Wang J, Li X, Geng D, Banis M, Li R, Sun X. Nitrogen-doped graphene nanosheets as cathode materials with excellent electrocatalytic activity for high capacity lithium-oxygen batteries. Electrochem Commun 2012;18:12–5.
[201] Xia J, Mei D, Li X, Xu W, Wang D, Graff GL, Bennett WD, Nie Z, Saraf LV, Aksay IA, Liu J, Zhang J-G. Hierarchically porous graphene as a lithium-air battery electrode. Nano Lett 2011;11:5071–8.
[202] Li Y, Wang J, Li X, Liu J, Geng D, Yang J, Li R, Sun X. Nitrogen-doped carbon nanotubes as cathode for lithium-air batteries. Electrochem Commun 2011;13:668–72.
[203] Lim HD, Park KY, Song H, Jang EY, Gwon H, Kim J, Kim YH, Lima MD, Robles RO, Lepro X, Baughman RH, Kang K. Enhanced power and rechargeability of a Li-O_2 battery based on a hierarchical-fibril CNT electrode. Adv Mater 2013;25:1348–52.
[204] Liu S, Wang Z, Yu C, Zhao Z, Fan X, Ling Z, Qiu J. Free-standing, hierarchically porous carbon nanotube film as a binder-free electrode for high-energy Li–O_2 batteries. J Mater Chem A 2013;1:12033–7.
[205] Mi R, Li S, Liu X, Liu L, Li Y, Mei J, Chen Y, Liu H, Wang H, Yana H, Lau W-M. Electrochemical performance of binder-free carbon nanotubes with different nitrogen amounts grown on the nickel foam as cathodes in Li-O_2 batteries. J Mater Chem A 2014;2(44):18746–53.
[206] Mi R, Liu H, Wang H, Wong K, Mei J, Chen Y, Liu W, Yan H. Effects of nitrogen-doped carbon nanotubes on the discharge performance of Li-air batteries. Carbon 2014;67:744–52.
[207] Mitchell R, Gallant B, Thompson C, Shao-Horn Y. All-carbon-nanofiber electrodes for high-energy rechargeable Li-O_2 batteries. Energy Environ Sci 2011;4:2952–8.
[208] Lei Y, Lu J, Luo X, Wu T, Du P, Zhang X, Ren Y, Miller JT, Sun Y-K, Elam JW. Synthesis of porous carbon supported palladium nanoparticle catalysts by atomic layer deposition: application for rechargeable lithium-O_2 battery. Nano Lett 2013;13:4182–9.
[209] Sun C, Li F, Ma C, Wang Y, Ren Y, Yang W, Ma Z, Li J, Chen Y, Kim Y, Chen L. Graphene-Co_3O_4 nanocomposite as an efficient bifunctional catalyst for lithium-air batteries. J Mater Chem A 2014;2:7188–96.
[210] Ryu W-H, Yoon T-H, Song SH, Jeon S, Park Y-J, Kim I-D. Bifunctional composite catalysts using Co_3O_4 nanofibers immobilized on nonoxidized graphene nanoflakes for high capacity and long-cycle Li-O_2 batteries. Nano Lett 2013;13:4190–7.

[211] Qin Y, Lu J, Du P, Chen Z, Ren Y, Wu T, Miller JT, Wen J, Miller DJ, Zhang Z, Amine K. In situ fabrication of porous-carbon-supported α-MnO_2 nanorods at room temperature: application for rechargeable Li-O_2 batteries. Energy Environ Sci 2013;6:519−31.
[212] Li F, Ohnishi R, Yamada Y, Kubota J, Domen K, Yamada A, Zhou H. Carbon supported TiN nanoparticles: an efficient bifunctional catalyst for non-aqueous Li−O_2 batteries. Chem Commun 2013;49:1175−7.
[213] Kuboki T, Okuyama T, Ohsaki T, Takami N. Lithium-air batteries using hydrophobic room temperature ionic liquid electrolyte. J Power Sources 2005;146:766−9.
[214] Palomares V, Serras P, Villaluenga I, Hueso KB, Carretero-Gonzalez J, Rojo T. Na-ion batteries, recent advances and present challenges to become low cost energy storage systems. Energy Environ Sci 2012;5:5884−901.
[215] Slater MD, Kim D, Lee E, Johnson CS. Sodium-ion batteries. Adv Funct Mater 2013;23:947−58.
[216] Palomares V, Casas-Cabanas M, Castillo-Martinez E, Han MH, Rojo T. Update on Na-based battery materials. A growing research path. Energy Environ Sci 2013;6:2312−37.
[217] Pan H, Hu Y-S, Chen L. Room-temperature stationary sodium-ion batteries for large-scale electric energy storage. Energy Environ Sci 2013;6:2338−60.
[218] Hong SY, Park KY, Choi A, Choi NS, Lee KT. Charge carriers in rechargeable batteries: Na ions vs. Li ions. Energy Environ Sci 2013;6:2067−81.
[219] Kubota K, Yabuuchi N, Yoshida H, Dahbi M, Komaba S. Layered oxides as positive electrode materials for Na-ion batteries. MRS Bull 2014;39:416−22.
[220] Dahbi M, Yabuuchi N, Kubota K, Tokiwa K, Komaba S. Phys Chem Chem Phys 2014;16:15007−28.
[221] Yabuuchi N, Kubota K, Dahbi M, Komaba S. Research development on sodium-ion batteries. Chem Rev 2014;114:11636−82.
[222] Kim Y, Ha K-H, Oh SM, Lee KT. High-capacity anode materials for sodium-ion batteries. Chem Eur J 2014;20:11980−92.
[223] Han MH, Gonzalo E, Singh G, Rojo T. Na-ion batteries, recent advances and present challenges to become low cost energy storage systems. Energy Environ Sci 2015;8:81−102.
[224] Ponrouch A, Monti D, Boschin A, Steen B, Johansson P, Palacin MR. Non-aqueous electrolytes for sodium-ion batteries. J Mater Chem A 2015;3:22−42.
[225] Wang LP, Yu LH, Wang X, Srinivasan M, Xu ZJ. Recent developments in electrode materials for sodium-ion batteries. J Mater Chem A 2015;3:9353−78.
[226] Kubota K, Komaba S. Review - practical issues and future perspective for Na-ion batteries. J Electrochem Soc 2015;162:A2538−50.
[227] Kang H, Liu Y, Cao K, Zhao Y, Jiao L, Wang Y, Yuan H. Update on anode materials for Na-ion batteries. J Mater Chem A 2015;3:17899−913.
[228] Lin Q, Zhang J, Lv W, Ma J, He Y, Kang F, Yang Q-H. A functionalized carbon surface for high- performance sodium-ion storage. Small 2019:1902603.
[229] Wen Y, He K, Zhu YJ, Han FD, Xu YH, Matsuda I, Ishii Y, Cumings J, Wang C. Expanded graphite as superior anode for sodium-ion batteries. Nat Commun 2014;5:4033.
[230] Song LJ, Liu SS, Yu BJ, Wang CY, Li MW. Anode performance of mesocarbon microbeads for sodium-ion batteries. Carbon 2015;95:972−7.
[231] Yuan Z, Si L, Zhu X. Three-dimensional hard carbon matrix for sodium-ion battery anode with superior-rate performance and ultralong cycle life. J Mater Chem A 2015;3:23403−11.

[232] Wang H, Shi Z, Jin J, Chong C, Wang C. Properties and sodium insertion behavior of phenolic resin- based hard carbon microspheres obtained by a hydrothermal method. J Electroanal Chem 2015;755:87−91.
[233] Hasegawa G, Kanamori K, Kannari N, Ozaki J, Nakanishi K, Abe T. Studies on electrochemical sodium storage into hard carbons with binder-free monolithic electrodes. J Power Sources 2016;318:41−8.
[234] Ghimbeu CM, Gorka J, Simone V, Simonin L, Martinet S, Vix-Guterl C. Insights on the Na- ion storage mechanism in hard carbon: discrimination between the porosity, surface functional groups and defects. Nano Energy 2018;44:327−35.
[235] Beda A, Taberna P-L, Simon P, Ghimbeu CM. Hard carbons derived from green phenolic resins for Na-ion batteries. Carbon 2018;139:248−57.
[236] Kamiyama A, Kubota K, Nakano T, Fujimura S, Shiraishi S, Tsukada H, Komaba S. High-capacity hard carbon synthesized from microporous phenolic resin for sodium-ion and potassium-ion battery. ACS Appl Energy Mater 2020;3:135−40.
[237] Qu Q, Yun J, Wan Z, Zheng H, Gao T, Shen M, Shao J, Zheng H. MOF-derived microporous carbon as a better choice for Na-ion batteries than mesoporous CMK-3. RSC Adv 2014;4(110):64692−7.
[238] Luo W, Schardt J, Bommier C, Wang B, Razink J, Simonsen J, Ji X. Carbon nanofibers derived from cellulose nanofibers as a long-life anode material for rechargeable sodium-ion batteries. J Mater Chem A 2013;1:10662−6.
[239] Cao YL, Xiao LF, Sushko ML, Wang W, Schwenzer B, Xiao J, Nie Z, Saraf LV, Yang Z, Liu J. Sodium ion insertion in hollow carbon nanowires for battery applications. Nano Lett 2012;12:3783−7.
[240] Fu L, Tang K, Song K, van Aken PA, Maier J. Nitrogen doped porous carbon fibres as anode materials for sodium ion batteries with excellent rate performance. Nanoscale 2014;6:1384−9.
[241] Qu Y, Zhang Z, Du K, Chen W, Lai Y, Liu Y, Li J. Synthesis of nitrogen-containing hollow carbon microspheres by a modified template method as anodes for advanced sodium-ion batteries. Carbon 2016;105:103−12.
[242] Lotfabad EM, Ding J, Cui K, Kohandehghan A, Kalisvaart WP, Hazelton M, Mitlin D. High-Density sodium and lithium ion battery anodes from banana peels. ACS Nano 2014;8:7115−29.
[243] Li YM, Xu SY, Wu XY, Yu JZ, Wang YS, Hu YS, Li H, Chen L, Huang X. Amorphous monodispersed hard carbon microspherules derived from biomass as a high performance negative electrode material for sodium-ion batteries. J Mater Chem A 2015;3:71−7.
[244] Lin P, Li YM, Hu YS, Li H, Chen LQ, Huang XJ. A waste biomass derived hard carbon as a high performance anode material for sodium-ion batteries. J Mater Chem A 2016;4:13046−52.
[245] Zhao PY, Zhang J, Li Q, Wang CY. Electrochemical performance of fulvic acid-based electrospun hard carbon nanofibers as promising anodes for sodium-ion batteries. J Power Sources 2016;334:170−8.
[246] Zhu Y, Chen M, Li Q, Yuan C, Wang C. High-yield humic acid-based hard carbons as promising anode materials for sodium-ion batteries. Carbon 2017;123:727−34.
[247] Raj KA, Panda MR, Dutta DP, Mitra S. Bio-derived mesoporous disordered carbon: an excellent anode in sodium-ion battery and full-cell lab prototype. Carbon 2019;143:402−12.
[248] Zhao B, Ding Y, Wen Z. From jackfruit rags to hierarchical porous N-doped carbon: a high-performance anode material for sodium-ion batteries. Trans Tianjin Univ 2019;25:429−36.

[249] Le Z, Liu F, Nie P, Li X, Liu X, Bian Z, Chen G, Wu HB, Lu Y. Pseudocapacitive sodium storage in mesoporous single-crystal-like TiO_2-graphene nanocomposite enables high-performance sodium-ion capacitors. ACS Nano 2017;11:2952−60.
[250] Jing WT, Zhang Y, Gu Y, Zhu YF, Yang CC, Jiang Q. N-doped carbon nano-necklaces with encapsulated Sb as a sodium-ion battery anode. Matter 2019;1:720−33.
[251] Kim H, Hong J, Park Y-U, Kim J, Hwang I, Kang K. Sodium storage behavior in natural graphite using ether-based electrolyte systems. Adv Funct Mater 2015;25:534−41.
[252] Jache B, Adelhelm P. Use of graphite as a highly reversible electrode with superior cycle life for sodium-ion batteries by making use of co-intercalation phenomena. Angew Chem Int Ed 2014;53:10169−73.
[253] Eftekhari A, Jian Z, Ji X. Potassium secondary batteries. ACS Appl Mater Interfaces 2017;9:4404−19.
[254] Jian Z, Luo W, Ji X. Carbon electrodes for K-ion batteries. J Am Chem Soc 2015;137:11566−9.
[255] Zhao J, Zou X, Zhu Y, Xu Y, Wang C. Electrochemical intercalation of potassium into graphite. Adv Funct Mater 2016;26:8103−10.
[256] Li Z, Shin W, Chen Y, Neuefeind JC, Greaney PA, Ji X. Low temperature pyrolyzed soft carbon as high capacity K-ion anode. ACS Appl Energy Mater 2019;2:4053−8.
[257] Jian Z, Xing Z, Bommier C, Li Z, Ji X. Hard carbon microspheres: potassium-ion anode versus sodium-ion anode. Adv Energy Mater 2016;6(3):1501874.
[258] Kubota K, Shimadzu S, Yabuuchi N, Tominaka S, Shiraishi S, Abreu-Sepulveda M, Manivannan A, Gotoh K, Fukunishi M, Dahbi M, Komaba S. Structural analysis of sucrose-derived hard carbon and correlation with the electrochemical properties for lithium, sodium, and potassium insertion. Chem Mater 2020;32:2961−77.
[259] Liu L, Chen Y, Xie Y, Tao P, Li Q, Yan C. Understanding of the ultrastable K-ion storage of carbonaceous anode. Adv Funct Mater 2018;28(29):1801989.
[260] Chen C, Wang Z, Zhang B, Miao L, Cai J, Peng L, Huang Y, Jiang J, Huang Y, Zhang L, Xie J. Nitrogen-rich hard carbon as a highly durable anode for high-power potassium-ion batteries. Energy Storage Mater 2017;8:161−8.
[261] Yang J, Ju Z, Jiang Y, Xing Z, Xi B, Feng J, Xiong S. Enhanced capacity and rate capability of nitrogen/oxygen dual-doped hard carbon in capacitive potassium-ion storage. Adv Mater 2018;30:1700104.
[262] Jian Z, Hwang S, Li Z, Hernandez AS, Wang X, Xing Z, Su D, Ji X. Hard-soft composite carbon as a long-cycling and high-rate anode for potassium-ion batteries. Adv Funct Mater 2017;27(26):1700324.
[263] Chen M, Wang W, Liang X, Gong S, Liu J, Wang Q, Guo S, Yang H. Sulfur/oxygen co-doped porous hard carbon microspheres for high-performance potassium-ion batteries. Adv Energy Mater 2018;8(19):1800171.
[264] Li D, Ren X, Ai Q, Sun Q, Zhu L, Liu Y, Liang Z, Peng R, Si P, Lou J, Feng J, Ci L. Facile fabrication of nitrogen-doped porous carbon as superior anode material for potassium-ion batteries. Adv Energy Mater 2018;8(34):1802386.
[265] Adams RA, Syu JM, Zhao Y, Lo CT, Varma A, Pol VG. Binder-free N- and O-rich carbon nanofiber anodes for long cycle life K-ion batteries. ACS Appl Mater Interfaces 2017;9:17872−81.
[266] Zhao X, Xiong P, Meng J, Liang Y, Wang J, Xu Y. High rate and long cycle life porous carbon nanofiber paper anodes for potassium-ion batteries. J Mater Chem A 2017;5:19237−44.

[267] Xiong P, Zhao X, Xu Y. Nitrogen-doped carbon nanotubes derived from metal-organic frameworks for potassium-ion battery anodes. ChemSusChem 2018;11:202–8.
[268] Xu Y, Zhang C, Zhou M, Fu Q, Zhao C, Wu M, Lei Y. Highly nitrogen doped carbon nanofibers with superior rate capability and cyclability for potassium ion batteries. Nat Commun 2018;9:1720.
[269] Zhao X, Tang Y, Ni C, Wang J, Star A, Xu Y. Free-standing nitrogen-doped cup-stacked carbon nanotube mats for potassium-ion battery anodes. ACS Appl Energy Mater 2018;1:1703–7.
[270] Share K, Cohn AP, Carter R, Rogers B, Pint CL. Role of nitrogen-doped graphene for improved high- capacity potassium ion battery anodes. ACS Nano 2016;10:9738.
[271] Ma G, Huang K, Ma J-S, Ju Z, Xing Z, Zhuang Q-C. Phosphorus and oxygen dual-doped graphene as superior anode material for room-temperature potassium-ion batteries. J Mater Chem A 2017;5:7854–61.
[272] Xie Y, Chen Y, Liu L, Tao P, Fan M, Xu N, Shen X, Yan C. Ultra-high pyridinic N-doped porous carbon monolith enabling high-capacity K-ion battery anodes for both half-cell and full-cell applications. Adv Mater 2017;29:1702268.
[273] Yi Z, Liu Y, Li Y, Zhou L, Wang Z, Zhang J, Cheng H, Lu Z. Flexible membrane consisting of MoP ultrafine nanoparticles highly distributed inside N and P co-doped carbon nanofibers as high- performance anode for potassium-ion batteries. Small 2019:1905301.
[274] Shi H. Activated carbons and double layer capacitance. Electrochim Acta 1996;41:1633–9.
[275] Wang L, Toyoda M, Inagaki M. Dependence of electric double layer capacitance of activated carbons on the types of pores and their surface areas. New Carbon Mater 2008;23:111–5.
[276] Gryglewicz G, Machnikowski J, Lorenc-Grabowska E, Lota G, Frackowiak E. Effect of pore size distribution of coal-based activated carbons on double layer capacitance. Electrochim Acta 2005;50:1197–206.
[277] Wen Y, Cao G, Cheng J, Yang Y. Correlation of capacitance with the pore structure for nanoporous glassy carbon electrodes. J Electrochem Soc 2005;152:A1770–5.
[278] Barbieri O, Hahn M, Koetz R. Capacitance limits of high surface area activated carbons for double layer capacitors. Carbon 2005;43:1303–10.
[279] Wang L, Fujita M, Inagaki M. Relationship between pore surface areas and electric double layer capacitance in non-aqueous electrolytes for air-oxidized carbon spheres. Electrochim Acta 2006;51:4096–102.
[280] Ito E, Mozia S, Okuda M, Nakano T, Toyoda M, Inagaki M. Nanoporous carbons from cypress II. Application to electric double layer capacitors. New Carbon Mater 2007;22:321–6.
[281] Jaenes A, Kurig H, Lust E. Characterisation of activated nanoporous carbon for supercapacitor electrode materials. Carbon 2007;45:1226–33.
[282] Wang L, Inagaki M, Toyoda M. Contributions of micropores and mesopores in electrode carbon to electric double layer capacitance. TANSO 2009;240:230–8.
[283] Inagaki M, Kang F, Toyoda M, Konno H. Advanced materials science and engineering of carbon. Elsevier and Tsinghua University Press; 2013.
[284] Lederer I, Baizer C, Reichenauer G. Contributions of storage sites located in micro- and meso/macropores to the capacitance of carbonaceous double layer capacitor electrodes. Electrochim Acta 2018;281:753–60.
[285] Okajima K, Ohta K, Sudoh M. Capacitance behavior of activated carbon fibers with oxygen-plasma treatment. Electrochim Acta 2005;50:2227–31.
[286] Oda H, Yamashita A, Minoura S, Okamoto M, Morimoto T. Modification of the oxygen-containing functional group on activated carbon fiber in electrodes of an electric double-layer capacitor. J Power Sources 2006;158:1510–6.

[287] Seredych M, Hulicova-Jurcakova D, Lu GQ, Bandosz TJ. Surface functional groups of carbons and the effects of their chemical character, density and accessibility to ions on electrochemical performance. Carbon 2008;46:1475–88.
[288] Hulicova-Jurcakova D, Seredych M, Lu GQ, Bandosz TJ. Combined effect of nitrogen- and oxygen- containing functional groups of microporous activated carbon on its electrochemical performance in supercapacitors. Adv Funct Mater 2009;19:438–47.
[289] Urita K, Urita C, Fujita K, Horio K, Yoshida M, Moriguchi I. The ideal porous structure of EDLC carbon electrodes with extremely high capacitance. Nanoscale 2017;9:15643–9.
[290] Choi J-H, Lee C, Cho S, Moon GD, Kim B, Chang H, Jang HD. High capacitance and energy density supercapacitor based on biomass-derived activated carbons with reduced graphene oxide binder. Carbon 2018;132:16–24.
[291] Portet C, Taberna PL, Simon P, Flahaut E, Laberty-Robert C. High power density electrodes for carbon supercapacitor applications. Electrochim Acta 2005;50:4174–81.
[292] Xie X, He X, Zhang H, Wei F, Xiao N, Qiu J. Interconnected sheet-like porous carbons from coal tar by a confined soft template strategy for supercapacitors. Chem Eng J 2018;350:49–56.
[293] Xie X, He X, Shao X, Dong S, Xiao N, Qiu J. Synthesis of layered microporous carbons from coal tar by directing, space-confinement and self-sacrificed template strategy for supercapacitors. Electrochim Acta 2017;246:634–42.
[294] Kim BH, Bui NN, Yang KS, Cruz MED, Ferraris JP. Electrochemical properties of activated polyacrylonitrile/pitch carbon fibers produced using electrospinning. Bull Korean Chem Soc 2009;30:1967–72.
[295] Fan Y, Liu PF, Huang ZY, Jiang TW, Yao KL, Han R. Porous hollow carbon spheres for electrode material of supercapacitors and support material of dendritic Pt electrocatalyst. J Power Sources 2015;280:30–8.
[296] Deng X, Zhao B, Zhu L, Shao Z. Molten salt synthesis of nitrogen-doped carbon with hierarchical pore structures for use as high-performance electrodes in supercapacitors. Carbon 2015;93:48–58.
[297] Xu B, Hou SS, Cao GP, Wu F, Yang YS. Sustainable nitrogen-doped porous carbon with high surface areas prepared from gelatin for supercapacitors. J Mater Chem 2012;22:19088–93.
[298] Shiraishi S. Heat-treatment and nitrogen-doping of activated carbons for high voltage operation of electric double layer capacitor. Key Eng Mater 2012;497:80–6.
[299] Shiraishi S. Highly-durable carbon electrode for electrochemical capacitors. Bol Grupo Espanol Carbon 2013;28:18–24.
[300] Wang C, Sun L, Zhou Y, Wan P, Zhang X, Qiu J. P/N co-doped microporous carbons from H_3PO_4- doped polyaniline by in situ activation for supercapacitors. Carbon 2013;59:537–46.
[301] Nishihara H, Itoi H, Kogure T, Hou P-X, Touhara H, Okino F, Kyotani T. Investigation of the ion storage/transfer behavior in an electrical double layer capacitor by using ordered microporous carbons as model materials. Chem Eur J 2009;15:5355–63.
[302] Portet C, Yang Z, Gogotsi KY, Mokaya R, Yushin G. Electrical double-layer capacitance of zeolite- templated carbon in organic electrolyte. J Electrochem Soc 2009;156:A1–6.
[303] Ania CO, Khomenko V, Raymundo-Pinero E, Parra JB, Beguin F. The large electrochemical capacitance of microporous doped carbon obtained by using a zeolite template. Adv Funct Mater 2007;17:1828–36.

[304] Itoi H, Nishihara H, Kogure T, Kyotani T. Three-dimensionally arrayed and mutually connected 1.2- nm nanopores for high-performance electric double layer capacitor. J Am Chem Soc 2011;133:1165–7.
[305] Itoi H, Nishihara H, Kyotani T. Effect of heteroatoms in ordered microporous carbons on their electrochemical capacitance. Langmuir 2016;32:11997–2004.
[306] Morishita T, Soneda Y, Tsumura T, Inagaki M. Preparation of porous carbons from thermoplastic precursors and their performance for electric double layer capacitors. Carbon 2006;44:2360–7.
[307] Fernandez JA, Morishita T, Toyoda M, Inagaki M, Stoeckli F, Centeno TA. Performance of mesoporous carbons derived from poly(vinyl alcohol) in electrochemical capacitors. J Power Sources 2008;175:675–9.
[308] Wang Y, Wang C. Templated mesoporous carbons and their performance for electric double layer capacitors. New Carbon Mater 2010;25:376–81.
[309] Zhou J, Yuan X, Xing W, Si W, Zhuo S. Mesoporous carbons derived from citrates for use in electrochemical capacitors. New Carbon Mater 2010;25:370–5.
[310] Zhang WF, Huang Z-H, Cao GP, Kang F, Yang Y. A novel mesoporous carbon with straight tunnel- like pore structure for high rate electrochemical capacitors. J Power Sources 2012;204:230–5.
[311] Zhang WF, Huang Z-H, Zhou CJ, Cao G, Kang F, Yang Y. Porous carbon for electrochemical capacitors prepared from a resorcinol/formaldehyde-based organic aquagel with nano-sized particles. J Mater Chem 2012;22:7158–63.
[312] Konno H, Onishi H, Yoshizawa N, Azumi K. MgO-templated nitrogen-containing carbons derived from different organic compounds for capacitor electrodes. J Power Sources 2010;195:667–73.
[313] Li C, Zhang X, Wang K, Sun X, Liu G, Li J, Tian H, Li J, Ma Y. Scalable self-propagating high- temperature synthesis of graphene for supercapacitors with superior power density and cyclic stability. Adv Mater 2017;29:1604690.
[314] Wang L, Morishita T, Toyoda M, Inagaki M. Asymmetric electric double layer capacitors using carbon electrodes with different pore size distributions. Electrochim Acta 2007;53:882–6.
[315] Wang L, Toyoda M, Inagaki M. Performance of asymmetric electric double layer capacitors - predominant contribution of the negative electrode. Adsorpt Sci Tech 2008;26:491–5.
[316] He X, Xie X, Wang J, Ma X, Xie Y, Gu J, Xiao N, Qiu J. From fluorene molecules to ultrathin carbon nanonets with an enhanced charge transfer capability for supercapacitors. Nanoscale 2019;11:6610–9.
[317] He X, Zhang N, Shao X, Wu M, Yu M, Qiu J. A layered-template-nanospace-confinement strategy for production of corrugated graphene nanosheets from petroleum pitch for supercapacitors. Chem Eng J 2016;297:121–7.
[318] He X, Li R, Qiu J, Xie K, Ling P, Ma Y, Zhang X, Zheng M. Synthesis of mesoporous carbons for supercapacitors from coal tar pitch by coupling microwave-assisted KOH activation with a MgO template. Carbon 2012;50:4911–21.
[319] He X, Zhang H, Zhang H, Li X, Xiao N, Qiu J. Direct synthesis of 3D hollow porous graphene balls from coal tar pitch for high performance supercapacitors. J Mater Chem A 2014;2:19633–40.
[320] He X, Ma H, Wang J, Xie Y, Xiao N, Qiu J. J Porous carbon nanosheets from coal tar for high- performance supercapacitors. J Power Sources 2017;357:41–6.
[321] He X, Li X, Ma H, Han J, Zhang H, Yu C, Xiao N, Qiu J. ZnO template strategy for the synthesis of 3D interconnected graphene nanocapsules from coal tar pitch as supercapacitor electrode materials. J Power Sources 2017;340:183–91.

[322] He X, Zhao N, Qiu J, Xiao N, Yu M, Yu C, Zhang X, Zheng M. Synthesis of hierarchical porous carbons for supercapacitors from coal tar pitch with nano-Fe_2O_3 as template and activation agent coupled with KOH activation. J Mater Chem A 2013;1:9440–8.
[323] Dong S, He X, Zhang H, Xie X, Yu M, Yu C, Xiao N, Qiu J. Surface modification of biomass-derived hard carbon by grafting porous carbon nanosheets for high-performance supercapacitors. J Mater Chem A 2018;6:15954–60.
[324] Lee J, Yoon S, Hyeon T, Oh SM, Kim KB. Synthesis of a new mesoporous carbon and its application to electrochemical double-layer capacitors. Chem Commun 1999:2177–8.
[325] Lee J, Yoon S, Oh SM, Shin C-H, Hyeon T. Development of a new mesoporous carbon using an HMS aluminosilicate template. Adv Mater 2000;12:359–62.
[326] Zhou H, Zhu S, Hibino M, Honma I. Electrochemical capacitance of self-ordered mesoporous carbon. J Power Sources 2003;122:219–23.
[327] Fuertes AB, Pico F, Rojo JM. Influence of pore structure on electric double-layer capacitance of template mesoporous carbons. J Power Sources 2004;133:329–36.
[328] Fuertes AB, Lota G, Centeno TA, Frackowiak E. Templated mesoporous carbons for supercapacitor application. Electrochim Acta 2005;50:2799–805.
[329] Jurewicz K, Vix-Guterl C, Frackowiak E, Saadallaha S, Reda M, Parmentier J, Patarin J, Beguin F. Capacitance properties of ordered porous carbon materials prepared by a templating procedure. J Phys Chem Solids 2004;65:287–93.
[330] Centeno TA, Sevilla M, Fuertes AB, Stoeckli F. On the electrical double-layer capacitance of mesoporous templated carbons. Carbon 2005;43:3012–5.
[331] Vix-Guterl C, Frackowiak E, Jurewicz K, Friebe M, Parmentier J, Beguin F. Electrochemical energy storage in ordered porous carbon materials. Carbon 2005;43:1293.
[332] Alvarez S, Blanco-Lopez MC, Miranda-Ordieres AJ, Fuertes AB, Centeno TA. Electrochemical capacitor performance of mesoporous carbons obtained by templating technique. Carbon 2005;43(4):866–70.
[333] Frackowiak E, Lota G, Machnikowski J, Vix-Guterl C, Beguin F. Optimisation of supercapacitors using carbons with controlled nanotexture and nitrogen content. Electrochim Acta 2006;51:2209–14.
[334] Xing W, Qiao SZ, Ding RG, Li F, Lu GQ, Yan ZF, Cheng HM. Superior electric double layer capacitors using ordered mesoporous carbons. Carbon 2006;44:216–24.
[335] Sevilla M, Alvarez S, Centeno TA, Fuertes AB, Stoeckli F. Performance of templated mesoporous carbons in supercapacitors. Electrochim Acta 2007;52:3207–15.
[336] Xia K, Gao Q, Jiang J, Hu J. Hierarchical porous carbons with controlled micropores and mesopores for supercapacitor electrode materials. Carbon 2008;46:1718–26.
[337] Kim ND, Kim W, Joo JB, Oh S, Kim P, Kim Y, Yi J. Electrochemical capacitor performance of N- doped mesoporous carbons prepared by ammoxidation. J Power Sources 2008;180(1):671–5.
[338] Wang D-W, Li F, Chen Z-G, Lu GQ, Cheng H-M. Synthesis and electrochemical property of boron- doped mesoporous carbon in supercapacitor. Chem Mater 2008;20:7195–200.
[339] Ito T, Ushiro M, Fushimi K, Azumi K, Konno H. Preparation of boron-containing carbons from glucose-borate complexes and their capacitive performance. TANSO 2009;239:156–61 [in Japanese)].
[340] Banham D, Feng F, Burt J, Alsrayheen E, Birss V. Bimodal, templated mesoporous carbons for capacitor applications. Carbon 2010;48:1056–63.

[341] Nishihara H, Simura T, Kobayashi S, Nomura K, Berenguer R, Ito M, Uchimura M, Iden H, Arihara K, Ohma A, Hayasaka Y, Kyotani T. Oxidation-resistant and elastic mesoporous carbon with single- layer graphene walls. Adv Funct Mater 2016;26:6418–27.

[342] Nomura K, Nishihara H, Kobayashi N, Asada T, Kyotani T. 4.4 V supercapacitors based on super- stable mesoporous carbon sheet made of edge-free graphene walls. Energy Environ Sci 2019;12:1542–9.

[343] Kwon T, Nishihara H, Itoi H, Yang Q-H, Kyotani T. Enhancement mechanism of electrochemical capacitance in nitrogen-/boron-doped carbons with uniform straight nanochannels. Langmuir 2009;25:11961–6.

[344] Liang Y, Feng X, Zhi L, Kolb U, Mullen K. A simple approach towards one-dimensional mesoporous carbon with superior electrochemical capacitive activity. Chem Commun 2009:809–11.

[345] Wang Z, Xiong YC, Guan SY. A simple $CaCO_3$-assisted template carbonization method for producing nitrogen doped porous carbons as electrode materials for supercapacitors. Electrochim Acta 2016;188:757–66.

[346] Shao JQ, Ma FW, Wu G, Dai CC, Geng WD, Song SJ, Wan JF. In-situ MgO ($CaCO_3$) templating coupled with KOH activation strategy for high yield preparation of various porous carbons as supercapacitor electrode materials. Chem Eng J 2017;321:301–13.

[347] Wei F, He X, Zhang H, Liu Z, Xiao N, Qiu J. Crumpled carbon nanonets derived from anthracene oil for high energy density supercapacitor. J Power Sources 2019;428:8–12.

[348] Gao F, Qu JY, Geng C, Shao GH, Wu MB. Self-templating synthesis of nitrogen-decorated hierarchical porous carbon from shrimp shell for supercapacitors. J Mater Chem A 2016;4:7445–52.

[349] Zhang H, He X, Gu J, Xie Y, Shui H, Zhang X, Xiao N, Qiu J. Wrinkled porous carbon nanosheets from methylnaphthalene oil for high performance supercapacitors. Fuel Process Technol 2018;175:10–6.

[350] Jin J, Tanaka S, Egashira Y, Nishiyama N. KOH activation of ordered mesoporous carbons prepared by a soft-templating method and their enhanced electrochemical properties. Carbon 2010;48:1985–9.

[351] Chang P, Matsumura K, Zhang J, Qi J, Wang C, Kinumoto T, Tsumura T, Chen M, Toyoda M. 2D porous carbon nanosheets constructed of few-layer graphene sheets by "medium-up" strategy for ultrahigh power-output EDLCs. J Mater Chem A 2018;6:10331–9.

[352] Yu X, Wang JG, Huang ZH, Shen W, Kang F. Ordered mesoporous carbon nanospheres as electrode materials for high-performance supercapacitors. Electrochem Commun 2013;36:66–70.

[353] Jung KH, Ferraris JP. Preparation and electrochemical properties of carbon nanofibers derived from polybenzimidazole/polyimide precursor blends. Carbon 2012;50:5309–15.

[354] Le TH, Yang Y, Yu L, Gao TJ, Huang ZH, Kang F. Polyimide-based porous hollow carbon nanofibers for supercapacitor electrode. J App Polym Sci 2016;133:43397.

[355] Le TH, Yang Y, Yu L, Huang Z-H, Kang F. In-situ growth of MnO2 crystals under nanopore-constraint in carbon nanofibers and their electrochemical performance. Sci Rep 2016;6:37368.

[356] Le TH, Tian H, Cheng J, Huang Z-H, Kang F, Yang Y. High performance lithium-ion capacitors based on scalable surface carved multi hierarchical construction electrospun carbon fibers. Carbon 2018;138:325–36.

[357] Lust E, Janes A, Arulepp M. Influence of solvent nature on the electrochemical parameters of electrical double layer capacitors. J Electroanal Chem 2004;562:33–42.

[358] Jaenes A, Permann L, Arulepp M, Lust E. Electrochemical characteristics of nanoporous carbide- derived carbon materials in non-aqueous electrolyte solutions. Electrochem Commun 2004;5:313–8.
[359] Chmiola J, Yushin G, Dash RK, Hoffman EN, Fischer JE, Barsoum MW, Gogotsi Y. Double-layer capacitance of carbide derived carbons in sulfuric acid. Electrochem Solid State Lett 2005;8:A357–60.
[360] Chmiola J, Yushin G, Dash R, Gogotsi Y. Effect of pore size and surface area of carbide derived carbons on specific capacitance. J Power Sources 2006;158:765–72.
[361] Chmiola J, Yushin G, Gogotsi Y, Portet C, Simon P, Taberna PL. Anomalous increase in carbon capacitance at pore sizes less than 1 nanometer. Science 2006;313:1760–3.
[362] Lin R, Taberna PI, Chmiola J, Guay D, Gogotsi Y, Simon P. Microelectrode study of pore size, ion size, and solvent effects on the charge/discharge behavior of microporous carbons for electrical double-layer capacitors. J Electrochem Soc 2009;156:A7–12.
[363] Wang H, Gao Q. Synthesis, characterization and energy-related applications of carbide-derived carbons obtained by the chlorination of boron carbide. Carbon 2009;47:820–8.
[364] Leis J, Arulepp M, Kuura A, Latt M, Lust E. Electrical double-layer characteristics of novel carbide- derived carbon materials. Carbon 2006;44:2122–9.
[365] Jaenes A, Permann L, Nigu P, Lust E. Influence of solvent nature on the electrochemical characteristics of nanoporous carbon/1 M $(C_2H_5)_3CH_3NBF_4$ electrolyte solution interface. Surf Sci 2004;560:145–57.
[366] Jaenes A, Lust E. Electrochemical characteristics of nanoporous carbide-derived carbon materials in various nonaqueous electrolyte solutions. J Electrochem Soc 2006;153:A113–6.
[367] Liu J, Wang X, Lu Q, Yu R, Chen M, Cai S, Wang X. Synthesis of nitrogen and sulfur co-doped carbon derived from chromium carbide for the high performance supercapacitor. J Electrochem Soc 2016;163:A2991–8.
[368] Liu B, Shioyama H, Jiang H, Zhang X, Xu Q. Metal–organic framework (MOF) as a template for syntheses of nanoporous carbons as electrode materials for supercapacitor. Carbon 2010;48:456–63.
[369] Jiang H-L, Liu B, Lan Y-Q, Kuratani K, Akita T, Shioyama H, Zong F, Xu Q. From metal-organic framework to nanoporous carbon: toward a very high surface area and hydrogen uptake. J Am Chem Soc 2011;133:11854–7.
[370] Hu J, Wang H, Gao Q, Guo H. Porous carbons prepared by using metal-organic framework as the precursor for supercapacitors. Carbon 2010;48:3599–606.
[371] Yuan D, Chen J, Tan S, Xia N, Liu Y. Worm-like mesoporous carbon synthesized from metal-organic coordination polymers for supercapacitors. Electrochem Commun 2009;11:1191–4.
[372] Zhong S, Zhan C, Cao D. Zeolitic imidazolate framework-derived nitrogen doped porous carbons as high performance supercapacitor electrode materials. Carbon 2015;85:51–9.
[373] Liu X, Zhou L, Zhao Y, Bian L, Feng X, Pu Q. Hollow, spherical nitrogen-rich porous carbon shells obtained from a porous organic framework for the supercapacitor. ACS Appl Mater Interfaces 2013;5:10280–7.
[374] Kou Y, Xu YH, Guo ZQ, Jiang DL. Supercapacitive energy storage and electric power supply using an aza-fused π-conjugated microporous framework. Angew Chem Int Ed 2011;50:8753–7.
[375] Hao L, Luo B, Li XL, Jin MH, Fang Y, Tang ZH, Jia YY, Liang MH, Thomas A, Yang JH, Zhi LJ. Terephthalonitrile-derived nitrogen-rich networks for high performance supercapacitors. Energy Environ Sci 2012;5:9747–51.

[376] Tian W, Zhang H, Sun H, Tade MO, Wang S. Template-free synthesis of N-doped carbon with pillared- layered pores as bifunctional materials for supercapacitor and environmental applications. Carbon 2017;118:98–105.
[377] Zhu J, Xu D, Wang C, Qian W, Guo J, Yan F. Ferric citrate-derived N-doped hierarchical porous carbons for oxygen reduction reaction and electrochemical supercapacitors. Carbon 2017;115:1–10.
[378] Kodama M, Yamashita J, Soneda Y, Hatori H, Kamegawa K. Preparation and electrochemical characteristics of N-enriched carbon foam. Carbon 2007;45:1105–7.
[379] Qu Y, Tang Z, Duan L, Li X, Zhang X, Lüz W. Synthesis of three dimensional porous carbon materials using g-C_3N_4 as template for supercapacitors. J Electrochem Soc 2019;166:A3564–9.
[380] Chen J, Xu J, Zhou S, Zhao N, Wong C-P. Nitrogen-doped hierarchically porous carbon foam: a free- standing electrode and mechanical support for high-performance supercapacitors. Nano Energy 2016;25:193–202.
[381] Zhang F, Liu T, Hou G, Kou T, Yue L, Guan R, Li Y. Hierarchically porous carbon foams for electric double layer capacitors. Nano Res 2016;9(10):2875–88.
[382] Li X, Zhao T, Wang K, Yang Y, Wei J, Kang F, Wu D, Zhu H. Directly drawing self-assembled, porous, and monolithic graphene fiber from chemical vapor deposition grown graphene film and its electrochemical properties. Langmuir 2011;27:12164–71.
[383] Yu J, Wang M, Xu P, Cho S-H, Suhr J, Gong K, Meng L, Huang Y, Byun J-H, Oh Y, Yan Y, Chou T-W. Ultrahigh-rate wire-shaped supercapacitor based on graphene fiber. Carbon 2017;119:332–8.
[384] Meng YN, Zhao Y, Hu CG, Cheng HH, Hu Y, Zhang ZP, Shi G, Qu L. All-graphene core-sheath microfibers for all-solid-state, stretchable fibriform super-capacitors and wearable electronic textiles. Adv Mater 2013;25(16):2326–31.
[385] Li SM, Yang SY, Wang YS, Tsai HP, Tien HW, Hsiao ST, Liao WH, Chang CL, Ma CC, Hu CC. N- doped structures and surface functional groups of reduced graphene oxide and their effect on the electrochemical performance of supercapacitor with organic electrolyte. J Power Sources 2015;278:218–29.
[386] Xu Y, Lin Z, Zhong X, Huang X, Weiss NO, Huang Y, Duan X. Holey graphene frameworks for highly efficient capacitive energy storage. Nat Commun 2014;5:4554.
[387] Stoller MD, Park S, Zhu Y, An J, Ruoff RS. Graphene-based ultracapacitors. Nano Lett 2008;8:3498–502.
[388] Kim TY, Lee HW, Stoller M, Dreyer DR, Bielawski CW, Ruoff RS, Suh KS. High-performance supercapacitors based on poly(ionic liquid)-modified graphene electrodes. ACS Nano 2011;5:436–42.
[389] Choi BG, Yang M, Hong WH, Choi JW, Huh YS. 3D macroporous graphene frameworks for supercapacitors with high energy and power densities. ACS Nano 2012;6:4020–8.
[390] Wu Z-S, Ren W, Wang D-W, Li F, Liu B, Cheng H-M. High-energy MnO_2 nanowire/graphene and graphene asymmetric electrochemical capacitors. ACS Nano 2010;4:5835–42.
[391] You B, Wang LL, Yao L, Yang J. Three dimensional N-doped graphene–CNT networks for supercapacitor. Chem Commun 2013;49:5016–8.
[392] Yu DS, Dai LM. Self-assembled graphene/carbon nanotube hybrid films for supercapacitors. J Phys Chem Lett 2010;1:467–70.
[393] Jeong HM, Lee JW, Shin WH, Choi YJ, Shin HJ, Kang JK, Choi JW. Nitrogen-doped graphene for high-performance ultracapacitors and the importance of nitrogen-doped sites at basal planes. Nano Lett 2011;11:2472–7.

[394] Kodama M, Yamashita J, Soneda Y, Hatori H, Nishimura S, Kamegawa K. Structural characterization and electric double layer capacitance of template carbons. Mater Sci Eng B 2004;108:156−61.
[395] Hulicova D, Yamashita J, Soneda Y, Hatori H, Kodama M. Supercapacitors prepared from melamine- based carbon. Chem Mater 2005;17:1241−7.
[396] Wu Z-S, Winter A, Chen L, Sun Y, Turchanin A, Feng X, Müllen K. Three-dimensional nitrogen and boron co-doped graphene for high-performance all-solid-state supercapacitors. Adv Mater 2012;24:5130−5.
[397] Sun HM, Cao LY, Lu LH. Bacteria promoted hierarchical carbon materials for high-performance supercapacitor. Energy Environ Sci 2012;5:6206−13.
[398] Zhu C, Liu T, Qian F, Han TY, Duoss EB, Kuntz JD, Spadaccini CM, Worsley MA, Li Y. Supercapacitors based on three-dimensional hierarchical graphene aerogels with periodic macropores. Nano Lett 2016;16(6):3448−56.
[399] Zhang Q, Wang Y, Zhang B, Zhao K, He P, Huang B. 3D superelastic graphene aerogel-nanosheet hybrid hierarchical nanostructures as high-performance supercapacitor electrodes. Carbon 2018;127:449−58.
[400] Yang W, Hou L, Xu X, Li Z, Ma X, Yang F, Li Y. Carbon nitride template-directed fabrication of nitrogen-rich porous graphene-like carbon for high performance supercapacitors. Carbon 2018;130:325−32.
[401] Liu C-Y, Bard AJ, Wudl F, Weitz I, Heath JR. Electrochemical characterization of films of single- walled carbon nanotubes and their possible application in supercapacitors. Electrochem Solid State Lett 1999;2(11):577−8.
[402] Barisci JN, Wallace GG, Baughman RH. Electrochemical characterization of single-walled carbon nanotube electrodes. J Electrochem Soc 2000;147(12):4580−3.
[403] Li J, Cassell A, Delzeit L, Han J, Meyyappan M. Novel three-dimensional electrodes: electrochemical properties of carbon nanotube ensembles. J Phys Chem B 2002;106:9299−305.
[404] Liu J, Zhang L, Wu HB, Lin J, Shen Z, Lou XW. High-performance flexible asymmetric supercapacitors based on a new graphene foam/carbon nanotube hybrid film. Energy Environ Sci 2014;7(11):3709−19.
[405] Tan S, Li-Oakey KD. Effect of structural orientation on the performance of supercapacitor electrodes from eletrospun coal-derived carbon nanofibers (CCNFs). J Electrochem Soc 2019;166:A3294−304.
[406] Sun Y, Sills RB, Hu X, Seh ZW, Xiao X, Xu H, Luo W, Jin H, Xin Y, Li T, Zhang Z, Zhou J, Cai W, Huang Y, Cui Y. A bamboo-inspired nanostructure design for flexible, foldable, and twistable energy storage devices. Nano Lett 2015;15(6):3899−906.
[407] Wang W, Liu L, Zong S, Chen A. Nitrogen-doped hollow mesoporous carbon tube for supercapacitors application. J Electrochem Soc 2019;166:A4047−55.
[408] Nakayama M, Komine K, Inohara D. Nitrogen-doped carbon cloth for supercapacitors prepared via a hydrothermal process. J Electrochem Soc 2016;163:A2428−34.
[409] Miao L, Zhu D, Liu M, Duan H, Wang Z, Lv Y, Xiong W, Zhu Q, Li L, Chai X, Dan L. N, S co-doped hierarchical porous carbon rods derived from protic salt: facile synthesis for high energy density supercapacitors. Electrochim Acta 2018;274:378−88.
[410] In JB, Hsia B, Yoo JH, Hyun S, Carraro C, Maboudian R, Grigoropoulos CP. Facile fabrication of flexible all solid-state micro-supercapacitor by direct laser writing of porous carbon in polyimide. Carbon 2015;85:144−51.
[411] Liu C, Yi F, Shu D, Chen W, Zhou X, Zhu Z, Zeng R, Cao A, He C, Li X. In-situ N/S co-doping three- dimensional succulent-like hierarchical carbon assisted by supramolecular polymerization for high- performance supercapacitors. Electrochim Acta 2019;319:410−22.

[412] Soneda Y, Toyoda M, Hashiya K, Yamashita J, Kodama M, Hatori H, Inagaki M. Huge electrochemical capacitance of exfoliated carbon fibers. Carbon 2003;41:2680−3.
[413] Soneda Y, Yamashita J, Kodama M, Hatori H, Toyoda M, Inagaki M. Pseudocapacitance on exfoliated carbon fiber in sulfuric acid electrolyte. Appl Phys A 2006;82:575−8.
[414] Toyoda M, Tani Y, Soneda Y. Exfoliated carbon fibers as an electrode for electric double layer capacitors in a 1 mol/dm^3 H_2SO_4 electrolyte. Carbon 2004;42:2833−7.
[415] Wang G, Wang H, Lu X, Ling Y, Yu M, Zhai T, Tong Y, Li Y. Solid-state supercapacitor based on activated carbon cloths exhibits excellent rate capability. Adv Mater 2014;26(17):2676−82.
[416] Kado Y, Soneda Y, Horii D, Okura K, Suematsu S. Pulverized graphite by ball milling for electric double-layer capacitors. J Electrochem Soc 2019;166(12):A2471−6.
[417] Li H, Wang Y, Wang C, Xia Y. A competitive candidate material for aqueous supercapacitors: high surface-area graphite. J Power Sources 2008;185:1557−62.
[418] Wang Y, Cao J, Zhou Y, Ouyang J, Jia D, Guo L. Ball-milled graphite as an electrode material for high voltage supercapacitor in neutral aqueous electrolyte. J Electrochem Soc 2012;159:A579−83.
[419] Farjami E, Al-Sharab J, Al-Kamal A, Deiner L. Efficient impact milling method to make porous graphitic materials for electric double layer capacitors. J Appl Electrochem 2015;45:385−95.
[420] Itoi H, Hayashi S, Matsufusa H, Ohzawa Y. Electrochemical synthesis of polyaniline in the micropores of activated carbon for high performance electrochemical capacitors. Chem Commun 2017;53:3201−4.
[421] Itoi H, Yasue Y, Suda K, Katoh S, Hasegawa H, Hayashi S, Mitsuoka M, Iwata H, Ohzawa Y. Solvent- free preparation of electrochemical capacitor electrodes using metal-free redox organic compounds. ACS Sus Chem Eng 2017;5:556−62.
[422] Itoi H, Maki S, Ninomiya T, Hasegawa H, Matsufusa H, Hayashi S, Iwata H, Ohzawa Y. Electrochemical polymerization of pyrene and aniline exclusively inside the pores of activated carbon for high-performance asymmetric electrochemical capacitors. Nanoscale 2018;10:9760−72.
[423] Itoi H, Hasegawa H, Iwata H, Ohzawa Y. Non-polymeric hybridization of TEMPO derivatives with activated carbon for high-energy-density aqueous electrochemical capacitor electrodes. Sus Energy Fuel 2018;2:558−65.
[424] Shen K, Ran F, Zhang X, Liu C, Wang N, Niu X, Liu Y, Zhang D, Kong L, Kang L, Chen S. Supercapacitor electrodes based on nano-polyaniline deposited on hollow carbon spheres derived from cross-linked co-polymers. Synth Met 2015;209:369−76.
[425] Liu H, Xu B, Jia M, Zhang M, Cao B, Zhao X, Wang Y. Polyaniline nanofiber/large mesoporous carbon composites as electrode materials for supercapacitors. Appl Surf Sci 2015;332:40−6.
[426] Ning X, Zhong W, Wan L. Ultrahigh specific surface area porous carbon nanospheres and its composite with polyaniline: preparation and application for supercapacitors. RSC Adv 2016;6:25519−24.
[427] Itoi H, Ninomiya T, Hasegawa H, Maki S, Sakakibara A, Suzuki R, Kasai Y, Iwata H, Matsumura D, Ohwada M, Nishihara H, Ohzawa Y. Unusual redox behavior of ruthenocene confined in the micropores of activated carbon. J Phys Chem C 2020. https://doi.org/10.1021/acs.jpcc.0c02965.
[428] Wang D, Wang K, Sun L, Wu H, Wang J, Zhao Y, Yan L, Luo Y, Jiang K, Li Q, Fan S, Li J, Wang J. MnO_2 nanoparticles anchored on carbon nanotubes with hybrid supercapacitor-battery behavior for ultrafast lithium storage. Carbon 2018;139:145−55.

[429] Sun Y, Zeng W, Sun H, Luo S, Chen D, Chan V, Liao K. Inorganic/polymer-graphene hybrid gel as versatile electrochemical platform for electrochemical capacitor and biosensor. Carbon 2018;132:589−97.
[430] Liu WW, Li JD, Feng K, Sy A, Liu YS, Lim L, Lui G, Tjandra R, Rasenthiram L, Go C, Yu A. Advanced Li-ion hybrid supercapacitors based on 3D graphene-foam composites. ACS Appl Mater Interfaces 2016;8(39):25941−53.
[431] Khomenko V, Raymundo-Pinero E, Beguin F. High-energy density graphite/AC capacitor in organic electrolyte. J Power Sources 2008;177:643−51.
[432] Sivakkumar SR, Pandolfo AG. Evaluation of lithium-ion capacitors assembled with pre-lithiated graphite anode and activated carbon cathode. Electrochim Acta 2012;65:280−7.
[433] Sivakkumar SR, Nerkar JY, Pandolfo AG. Rate capability of graphite materials as negative electrodes in lithium-ion capacitors. Electrochim Acta 2010;55:3330−5.
[434] Sivakkumar SR, Milev AS, Pandolfo AG. Effect of ball-milling on the rate and cycle-life performance of graphite as negative electrodes in lithium-ion capacitors. Electrochim Acta 2011;56:9700−6.
[435] Jain A, Jayaraman S, Ulaganathan M, Balasubramanian R, Aravindan V, Srinivasan MP, Madhavi S. Highly mesoporous carbon from teak wood sawdust as prospective electrode for the construction of high energy Li-ion capacitors. Electrochim Acta 2017;228:131−8.
[436] Schroeder M, Menne S, Ségalini J, Saurel D, Casas-Cabanas M, Passerini S, Winter M, Balducci A. Considerations about the influence of the structural and electrochemical properties of carbonaceous materials on the behavior of lithium-ion capacitors. J Power Sources 2014;266:250−8.
[437] Han X, Han P, Yao J, Zhang S, Cao X, Xiong J, Zhang J, Cui G. Nitrogen-doped carbonized polyimide microsphere as a novel anode material for high performance lithium ion capacitors. Electrochim Acta 2016;196:603−10.
[438] Decaux C, Lota G, Raymundo-Piñero E, Frackowiak E, Béguin F. Electrochemical performance of a hybrid lithium-ion capacitor with a graphite anode preloaded from lithium bis(trifluoromethane) sulfonimide-based electrolyte. Electrochim Acta 2012;86:282−6.
[439] Jezowski P, Fic K, Crosnier O, Brousse T, Beguin F. Lithium rhenium(VII) oxide as a novel aterial for graphite pre-lithiation in high performance lithium-ion capacitors. J Mater Chem A 2016;4:12609−15.
[440] Jezowski P, Crosnier O, Deunf E, Poizot P, Béguin F, Brousse T. Safe and recyclable lithium-ion capacitors using sacrificial organic lithium salt. Nat Mater 2018;7:167−73.
[441] Li NW, Du X, Shi J-L, Zhang X, Fan W, Wang J, Zhao S, Liu Y, Xu W, Li M, Guo Y-G, Li C. Graphene@hierarchical meso-/microporous carbon for ultrahigh energy density lithium-ion capacitors. Electrochim Acta 2018;281:459−65.
[442] Lee JH, Shin WH, Ryou MH, Jin JK, Kim J, Choi JW. Functionalized graphene for high performance lithium ion capacitors. ChemSusChem 2012;5:2328−33.
[443] Zhang T, Zhang F, Zhang L, Lu Y, Zhang Y, Yang X, Ma Y, Huang Y. High energy density Li-ion capacitor assembled with all graphene-based electrodes. Carbon 2015;92:106−18.
[444] Aida T, Murayama I, Yamada K, Morita M. High-energy-density hybrid electrochemical capacitor using graphitizable carbon activated with KOH for positive electrode. J Power Sources 2007;166:462−70.
[445] Schroeder M, Winter M, Passerini S, Balducci A. On the cycling stability of lithium-ion capacitors containing soft carbon as anodic material. J Power Sources 2013;238:388−94.

[446] Yu X, Zhan C, Lv R, Bai Y, Lin Y, Huang Z-H, Shen W, Qiu X, Kang F. Ultrahigh-rate and high- density lithium-ion capacitors through hybriding nitrogen-enriched hierarchical porous carbon cathode with prelithiated microcrystalline graphite anode. Nano Energy 2015;15:43−53.

[447] Tjandra R, Liu W, Lim L, Yu A. Melamine based, n-doped carbon/reduced graphene oxide composite foam for Li-ion hybrid supercapacitors. Carbon 2018;129:152−8.

[448] Lee WSV, Huang X, Tan TL, Xue JM. Low Li^+ insertion barrier carbon for high energy efficient lithium-ion capacitor. ACS Appl Mater Interfaces 2018;10:1690−700.

[449] Abbas Q, Babuchowska P, Frackowiak E, Beguin F. Sustainable AC/AC hybrid electrochemical capacitors in aqueous electrolyte approaching the performance of organic systems. J Power Sources 2016;326:652−9.

[450] Przygocki P, Abbas Q, Babuchowska P, Beguin F. Confinement of iodides in carbon porosity to prevent from positive electrode oxidation in high voltage aqueous hybrid electrochemical capacitors. Carbon 2017;125:391−400.

[451] Cui Y, Liu W, Lyu Y, Zhang Y, Wang H, Liu Y, Li D. All-carbon lithium capacitor based on salt crystal- templated, N-doped porous carbon electrodes with superior energy storage. J Mater Chem A 2018;6(37):18276−85.

[452] Sun F, Liu X, Wu HB, Wang L, Gao J, Li H, Lu Y. In situ high-level nitrogen doping into carbon nanospheres and boosting of capacitive charge storage in both anode and cathode for a high-energy 4.5 V full-carbon lithium-ion capacitor. Nano Lett 2018;18(6):3368−76.

[453] Niu J, Shao R, Liu M, Liang J, Zhang Z, Dou M, Huang Y, Wang F. Porous carbon electrodes with battery-capacitive storage features for high performance Li-ion capacitors. Energy Storage Mater 2018;12:145−52.

[454] Li C, Zhang X, Wang K, Sun XZ, Ma YW. High-power and long-life lithium-ion capacitors constructed from N-doped hierarchical carbon nanolayer cathode and mesoporous graphene anode. Carbon 2018;140:237−48.

[455] Xia Q, Yang H, Wang M, Yang M, Guo Q, Wan L, Xia H, Yu Y. High energy and high power lithium- ion capacitors based on boron and nitrogen dual-doped 3D carbon nanofibers as both cathode and anode. Adv Energy Mater 2017:1701336.

[456] Wang X, Wang Z, Zhang X, Peng H, Xin G, Lu C, Zhong Y, Wang G, Zhang Y. Nitrogen-doped defective graphene aerogel as anode for all graphene-based lithium ion capacitor. Chem Select 2017;2:8436−45.

[457] Shan XY, Wang YZ, Wang DW, Li F, Cheng HM. Armoring graphene cathodes for high-rate and long- life lithium ion supercapacitors. Adv Energy Mater 2016;6:1502064.

[458] Salvatierra RV, Zakhidov D, Sha J, Kim ND, Lee S, Raji AO, Zhao N, Tour JM. Graphene carbon nanotube carpets grown using binary catalysts for high-performance lithium-ion capacitors. ACS Nano 2017;11(3):2724−33.

[459] Wang J-A, Li S-M, Wang Y-S, Lan P-Y, Liao W-H, Hsiao S-T, Lin S-C, Lin C-W, Ma C-CM, Hu C-C. Preparation and properties of NrGO-CNT composite for lithium-ion capacitors. J Electrochem Soc 2017;164(14):A3657−65.

[460] Sun Y, Tang J, Qin F, Yuan J, Zhang K, Li J, Zhu D-M, Qine L-C. Hybrid lithium-ion capacitors with asymmetric graphene electrodes. J Mater Chem A 2017;5:13601−9.

[461] Jiang J, Nie P, Ding B, Zhang Y, Xu G, Wu L, Dou H, Zhang X. Highly stable lithium ion capacitor enabled by hierarchical polyimide derived carbon microspheres combined with 3D current collectors. J Mater Chem A 2017;5:23283−91.

[462] Pasquier AD, Plitz I, Menocal S, Amatucci G. A comparative study of Li-ion battery, supercapacitor and nonaqueous asymmetric hybrid devices for automotive applications. J Power Sources 2003;115:171−8.
[463] Jain A, Aravindan V, Jayaraman S, Kumar PS, Balasubramanian R, Ramakrishna S, Madhavi S, Srinivasan MP. Activated carbons derived from coconut shells as high energy density cathode material for Li-ion capacitors. Sci Rep 2013;3:3002.
[464] Li B, Zhang H, Wang D, Lv H, Zhang C. Agricultural waste-derived activated carbon for high performance lithium-ion capacitors. RSC Adv 2017;7:37923−8.
[465] Mhamane D, Aravindan V, Kim MS, Kim HK, Roh KC, Ruan D, Lee SH, Srinivasan M, Kim K-B. Silica-assisted bottom-up synthesis of graphene-like high surface area carbon for highly efficient ultracapacitor and Li-ion hybrid capacitor applications. J Mater Chem A 2016;4(15):5578−91.
[466] Fleischmann S, Widmaier M, Schreiber A, Shim H, Stiemke FM, Schubert TJS, Presser V. High voltage asymmetric hybrid supercapacitors using lithium- and sodium-containing ionic liquids. Energy Storage Mater 2019;16:391−9.
[467] Choi HS, Kim T, Im JH, Park CR. Preparation and electrochemical performance of hyper-networked $Li_4Ti_5O_{12}$/carbon hybrid nanofiber sheets for a battery-supercapacitor hybrid system. Nanotechnol 2011;22:405402.
[468] Choi HS, Im JH, Kim T, Park JH, Park CR. Advanced energy storage device: a hybrid BatCap system consisting of battery—supercapacitor hybrid electrodes based on $Li_4Ti_5O_{12}$-activated carbon hybrid nanotubes. J Mater Chem 2012;22:16986−93.
[469] Cericola D, Ruch PW, Kötz R, Novák P, Wokaun A. Characterization of bi-material electrodes for electrochemical hybrid energy storage devices. Electrochem Commun 2010;12:812−5.
[470] Cericola D, Novak P, Wokaun A, Koetz R. Segmented bi-material electrodes of activated carbon and $LiMn_2O_4$ for electrochemical hybrid storage devices: effect of mass ratio and C-rate on current sharing. Electrochim Acta 2011;56:1288−93.
[471] Cericola D, Novak P, Wokaun A, Koetz R. Mixed bi-material electrodes based on $LiMn_2O_4$ and activated carbon for hybrid electrochemical energy storage devices. Electrochim Acta 2011;56:8403−11.
[472] Cericola D, Novak P, Wokaun A, Keotz R. Hybridization of electrochemical capacitors and rechargeable batteries: an experimental analysis of the different possible approaches utilizing activated carbon, $Li_4Ti_5O_{12}$ and $LiMn_2O_4$. J Power Sources 2011;196:10305−13.
[473] Naoi K, Ishimoto S, Isobe Y, Aoyagi S. High-rate nano-crystalline $Li_4Ti_5O_{12}$ attached on carbon nano- fibers for hybrid supercapacitors. J Power Sources 2010;195:6250−4.
[474] Wang G, Lu C, Zhang X, Wan B, Liu H, Xia M, Gou H, Xin G, Lian J, Zhang Y. Toward ultrafast lithium ion capacitors: a novel atomic layer deposition seeded preparation of $Li_4Ti_5O_{12}$/graphene anode. Nano Energy 2017;36:46−57.
[475] Lu C, Wang X, Zhang X, Peng H, Zhang Y, Wang G, Wang Z, Cao G, Umirov N, Bakenov Z. Effect of graphene nanosheets on electrochemical performance of $Li_4Ti_5O_{12}$ in lithium-ion capacitors. Ceram Int 2017;43:6554−62.
[476] Li J, Guo JQ, Zhang X, Huang Y, Guo L. Asymmetric supercapacitors with high energy and power density fabricated using $LiMn_2O_4$ nano-rods and activated carbon electrodes. Int J Electrochem Sci 2017;12(2):1157−66.
[477] Jung H, Venugopal N, Scrosati B, Sun Y. A high energy and power density hybrid supercapacitor based on an advanced carbon-coated $Li_4Ti_5O_{12}$ electrode. J Power Sources 2013;221:266−71.

[478] Ye L, Liang Q, Lei Y, Yu X, Han C, Shen W, Huang ZH, Kang F, Yang QH. A high performance Li- ion capacitor constructed with $Li_4Ti_5O_{12}$/C hybrid and porous graphene macroform. J Power Sources 2015;282:174−8.

[479] Li J, Zhang X, Peng RF, Huang YJ, Guo L, Qi YC. $LiMn_2O_4$/graphene composites as cathodes with enhanced electrochemical performance for lithium-ion capacitors. RSC Adv 2016;6(60):54866−73.

[480] Shellikeri A, Yturriaga S, Zheng JS, Cao W, Hagen M, Read JA, Jow TR, Zheng JP. Hybrid lithium- ion capacitor with $LiFePO_4$/AC composite cathode - long term cycle life study, rate effect and charge sharing analysis. J Power Sources 2018;392:285−95.

[481] Arun N, Jain A, Aravindan V, Jayaraman S, Ling WC, Srinivasan MP, Madhavi S. Nanostructured spinel $LiNi_{0.5}Mn_{1.5}O_4$ as new insertion anode for advanced Li-ion capacitors with high power capability. Nano Energy 2015;12:69−75.

[482] Satish R, Aravindan V, Ling WC, Madhavi S. Carbon-coated $Li_3V_2(PO_4)_3$ as insertion type electrode for lithium-ion hybrid electrochemical capacitors: an evaluation of anode and cathodic performance. J Power Sources 2015;281:310−7.

[483] Karthikeyan K, Aravindan V, Lee S, Jang I, Lim H, Park G, Yoshio M, Lee Y. Electrochemical performance of carbon-coated lithium manganese silicate for asymmetric hybrid supercapacitors. J Power Sources 2010;195:3761−4.

[484] Sun F, Gao J, Zhu Y, Pi X, Wang L, Liu X, Qin Y. A high performance lithium ion capacitor achieved by the integration of a Sn-C anode and a biomass-derived microporous activated carbon cathode. Sci Rep 2017;7:40990.

[485] Zhang F, Zhang T, Yang X, Zhang L, Leng K, Huang Y, Chen Y. A high-performance supercapacitor- battery hybrid energy storage device based on graphene-enhanced electrode materials with ultrahigh energy density. Energy Environ Sci 2013;6:1623−32.

[486] Shi R, Han C, Xu X, Qin X, Xu L, Li H, Li J, Wong C-P, Li B. Electrospun N-doped hierarchical porous carbon nanofiber with improved degree of graphitization for high-performance lithium ion capacitor. Chem Eur J 2018;24:10460−7.

[487] Shi R, Han C, Li H, Xu L, Zhang T, Li J, Lin Z, Wong C-P, Kang F, Li B. NaCl-templated synthesis of hierarchical porous carbon with extremely large specific surface area and improved graphitization degree for high energy density lithium ion capacitors. J Mater Chem A 2018;6(35):17057−66.

[488] Brandt A, Balducci A. A study about the use of carbon coated iron oxide-based electrodes in lithium- ion capacitors. Electrochim Acta 2013;108:219−25.

[489] Yu X, Deng J, Zhan C, Lv R, Huang Z, Kang F. A high-power lithium-ion hybrid electrochemical capacitor based on citrate-derived electrodes. Electrochim Acta 2017;228:76−81.

[490] Kim E, Kim H, Park BJ, Han YH, Park JH, Cho J, Lee S-S, Son JG. Etching-assisted crumpled graphene wrapped spiky iron oxide particles for high-performance Li-ion hybrid supercapacitor. Small 2018;14:1704209.

[491] Cheng L, Li H-Q, Xia Y-Y. A hybrid nonaqueous electrochemical supercapacitor using nano-sized iron oxyhydroxide and activated carbon. J Solid State Electrochem 2006;10:405−10.

[492] Wang GX, Zhang BL, Yu ZL, Qu MZ. Manganese oxide/MWNTs composite electrodes for supercapacitors. Solid State Ionics 2005;176:1169−74.

[493] Cottineau T, Toupin M, Delahaye T, Brousse T, Belanger D. Nanostructured transition metal oxides for aqueous hybrid electrochemical supercapacitors. Appl Phys A 2006;82:599−606.

[494] Yu GH, Hu LB, Vosgueritchian M, Wang HL, Xie X, McDonough JR, Cui X, Cui Y, Bao ZN. Solution- processed graphene/MnO_2 nanostructured textiles for high-performance electrochemical capacitors. Nano Lett 2011;11:2905−11.

[495] Fan ZJ, Yan J, Wei T, Zhi LJ, Ning GQ, Li TY, Wei F. Asymmetric supercapacitors based on graphene/MnO_2 and activated carbon nanofiber electrodes with high power and energy density. Adv Funct Mater 2011;21:2366–75.

[496] Chen P-C, Shen G, Shi Y, Chen H, Zhou C. Preparation and characterization of flexible asymmetric supercapacitors based on transition-metal-oxide nanowire/single-walled carbon nanotube hybrid thin- film electrodes. ACS Nano 2010;4:4403–11.

[497] Li G-X, Hou P-X, Luan J, Li J-C, Li X, Wang H, Shi C, Liu C, Cheng H-M. A MnO_2 nanosheet/single- wall carbon nanotube hybrid fiber for wearable solid-state supercapacitors. Carbon 2018;140:634–43.

[498] Wang H, Xu Z, Li Z, Cui K, Ding J, Kohandehghan A, Tan X, Zahiri B, Olsen BC, Holt CMB, Mitlin D. Hybrid device employing three-dimensional arrays of MnO in carbon nanosheets bridges battery- supercapacitor divide. Nano Lett 2014;14:1987–94.

[499] Wang H, Xu ZW, Kohandehghan A, Li Z, Cui K, Tan XH, Stephenson TJ, Kingondu CK, Holt CMB, Olsen BC, Tak JK, Harfield D, Anyia AO, Mitlin D. Interconnected carbon nanosheets derived from hemp for ultrafast supercapacitors with high energy. ACS Nano 2013;7:5131–41.

[500] Lee WSV, Peng E, Li M, Huang X, Xue J. Rational design of stable 4V lithium ion capacitor. Nano Energy 2016;27:202–12.

[501] Qua Q, Li L, Tian S, Guo W, Wu Y, Holze R. A cheap asymmetric supercapacitor with high energy at high power: activated carbon//$K_{0.27}MnO_2 \cdot 0.6H_2O$. J Power Sources 2010;195:2789–94.

[502] Chen Z, Qin YC, Weng D, Xiao QF, Peng YT, Wang XL, Li HX, Wei F, Lu YF. Design and synthesis of hierarchical nanowire composites for electrochemical energy storage. Adv Funct Mater 2009;19:3420–6.

[503] Chen Z, Augustyn V, Wen J, Zhang Y, Shen M, Dunn B, Lu Y. High-performance supercapacitors based on intertwined CNT/V_2O_5 nanowire nanocomposites. Adv Mater 2011;23:791–5.

[504] Chen L, Lai Q, Hao Y, Zhao Y, Ji X. Investigations on capacitive properties of the AC/V_2O_5 hybrid supercapacitor in various aqueous electrolytes. J Alloys Compd 2009;467:465–71.

[505] Zhao X, Wang H, Cao J, Cai W, Sui J. Amorphous/crystalline hybrid MoO_2 nanosheets for high- energy lithium-ion capacitors. Chem Commun 2017;53:10723–6.

[506] Ma W, Chen S, Yang S, Chen W, Weng W, Cheng Y, Zhu M. Flexible all-solid-state asymmetric supercapacitor based on transition metal oxide nanorods/reduced graphene oxide hybrid fibers with high energy density. Carbon 2017;113:151–8.

[507] Kim H, Cho M-Y, Kim M-H, Park K-Y, Gwon H, Lee Y, Roh KC, Kang K. A novel high-energy hybrid supercapacitor with an anatase TiO_2-reduced graphene oxide Anode and an activated carbon cathode. Adv Energy Mater 2013;3:1500–6.

[508] Yang C, Lan J, Liu W, Liu Y, Yu Y, Yang X. High-performance Li-ion capacitor based on an activated carbon cathode and well-dispersed ultrafine TiO_2 nanoparticles embedded in mesoporous carbon nanofibers anode. ACS Appl Mater Interfaces 2017;9:18710–9.

[509] Yin J, Qi L, Wang H. Sodium titanate nanotubes as negative electrode materials for sodium-ion capacitors. ACS Appl Mater Interfaces 2012;4:2762–8.

[510] Wang Y, Hong Z, Wei M, Xia Y. Layered $H_2Ti_6O_{13}$-nanowires: a new promising pseudocapacitive material in non-aqueous electrolyte. Adv Funct Mater 2012;22:5185–93.

[511] Tang G, Cao L, Xiao P, Zhang Y, Liu H. A novel high energy hybrid Li-ion capacitor with a three- dimensional hierarchical ternary nanostructure of hydrogen-treated TiO_2 nanoparticles/conductive polymer/carbon nanotubes anode and an activated carbon cathode. J Power Sources 2017;355:1–7.
[512] Que L, Wang Z, Yu F, Gu D. 3D ultralong nanowire arrays with a tailored hydrogen titanate phase as binder-free anodes for Li-ion capacitors. J Mater Chem A 2016;4:8716–23.
[513] Yoon J-R, Baek E, Kim H-K, Pecht M, Lee S-H. Critical dual roles of carbon coating in $H_2Ti_{12}O_{25}$ for cylindrical hybrid supercapacitors. Carbon 2016;101:9–15.
[514] Wang Q, Wen Z, Li J. A hybrid supercapacitor fabricated with a carbon nanotube cathode and a TiO_2- B nanowire anode. Adv Funct Mater 2006;16:2141–6.
[515] Wang G, Liu ZY, Wu JN, Lu Q. Preparation and electrochemical capacitance behavior of TiO_2-B nanotubes for hybrid supercapacitor. Mater Lett 2012;71:120–2.
[516] Brousse T, Marchand R, Taberna P-L, Simon P. TiO_2 (B)/activated carbon non-aqueous hybrid system for energy storage. J Power Sources 2006;158:571–7.
[517] Aravindan V, Shubha N, Ling WC, Madhavi S. Constructing high energy density non-aqueous Li-ion capacitors using monoclinic TiO_2-B nanorods as insertion host. J Mater Chem A 2013;1:6145–51.
[518] Dylla AG, Lee JA, Stevenson KJ. Influence of mesoporosity on lithium-ion storage capacity and rate performance of nanostructured TiO_2(B). Langmuir 2012;28:2897–903.
[519] Lim E, Kim H, Jo C, Chun J, Ku K, Kim S, Lee HI, Nam I-S, Yoon S, Kang K, Lee J. Advanced hybrid supercapacitor based on a mesoporous niobium pentoxide/carbon as high-performance anode. ACS Nano 2014;8:8968.
[520] Wang LP, Yu L, Satish R, Zhu J, Yan Q, Srinivasan M, Xu Z. High-performance hybrid electrochemical capacitor with binder-free Nb_2O_5@graphene. RSC Adv 2014;4:37389–94.
[521] Deng B, Lei T, Zhu W, Xiao L, Liu J. In-plane assembled orthorhombic Nb_2O_5 nanorod films with high-rate Li^+ intercalation for high-performance flexible Li-ion capacitors. Adv Funct Mater 2018;28(1):1704330.
[522] Wang X, Li Q, Zhang L, Hu Z, Yu L, Jiang T, Lu C, Yan C, Sun J, Liu Z. Caging Nb_2O_5 nanowires in PECVD-derived graphene capsules toward bendable sodium-ion hybrid supercapacitors. Adv Mater 2018:1800963.
[523] Wang X, Shen G. Intercalation pseudo-capacitive $TiNb_2O_7$@carbon electrode for high-performance lithium ion hybrid electrochemical supercapacitors with ultrahigh energy density. Nano Energy 2015;15:104–15.
[524] Aravindan V, Reddy MV, Madhavi S, Mhaisalkar SG, Subba Rao GV, Chowdari BVR. Hybrid supercapacitor with nano-TiP_2O_7 as intercalation electrode. J Power Sources 2011;196:8850–4.
[525] Zhu GY, Chen T, Wang L, Ma LB, Hu Y, Chen RP, Wang Y, Wang C, Yan W, Tie Z, Liu J, Jin Z. High energy density hybrid lithium-ion capacitor enabled by Co_3ZnC@N-doped carbon nanopolyhedra anode and microporous carbon cathode. Energy Storage Mater 2018;14:246–52.
[526] Xie B, Yu M, Lu L, Feng H, Yang Y, Chen Y, Cui H, Xiao R, Li J. Pseudocapacitive Co_9S_8/graphene electrode for high-rate hybrid supercapacitors. Carbon 2019;141:134–42.
[527] Zhan C, Liu W, Hu M, Liang Q, Yu X, Shen Y, Lv R, Kang F, Huang Z-H. High-performance sodium- ion hybrid capacitors based on an interlayer expanded MoS_2/rGO composite: surpassing the performance of lithium-ion capacitors in a uniform system. NPG Asia Mater 2018;10:775–87.

[528] Cui C, Wei Z, Xu J, Zhang Y, Liu S, Liu H, Mao M, Wang S, Ma J, Dou S. Three-dimensional carbon frameworks enabling MoS_2 as anode for dual ion batteries with superior sodium storage properties. Energy Storage Mater 2018;15:22–30.
[529] Wang H, Yoshio M, Graphite. A suitable positive electrode material for high-energy electrochemical capacitors. Electrochem Commun 2006;8:1481–6.
[530] Wang H, Yoshio M, Thapa AK, Nakamura H. From symmetric AC/AC to asymmetric AC/graphite, a progress in electrochemical capacitors. J Power Sources 2007;169:375–80.
[531] Wang H, Yoshio M. Effect of cation on the performance of AC/graphite capacitor. Electrochem Commun 2008;10:382–6.
[532] Wang Y, Zheng C, Qi L, Yoshio M, Yoshizuka K, Wang H. Utilization of (oxalate) borate-based organic electrolytes in activated carbon/graphite capacitors. J Power Sources 2011;196:10507–10.
[533] Gasteiger HA, Kocha SS, Sompalli B, Wagner FT. Activity benchmarks and requirements for Pt, Pt- alloy, and non-Pt oxygen reduction catalysts for PEMFCs. Appl Catal, B 2005;56:9–35.
[534] Liu H, Song C, Zhang L, Zhang J, Wang H, Wilkinson DP. A review of anode catalysis in the direct methanol fuel cell. J Power Sources 2006;155:95–110.
[535] Serov A, Kwak C. Review of non-platinum anode catalysts for DMFC and PEMFC application. Appl Catal, B 2009;90:313–20.
[536] Zhang S, Yuan X-Z, Hin JNC, Wang H, Friedrich KA, Schulze M. A review of platinum-based catalyst layer degradation in proton exchange membrane fuel cells. J Power Sources 2009;194:588–600.
[537] Mazumder V, Lee Y, Sun S. Recent development of active nanoparticle catalysts for fuel cell reactions. Adv Funct Mater 2010;20:1224–31.
[538] Huang H, Wang X. Recent progress on carbon-based support materials for electrocatalysts of direct methanol fuel cells. J Mater Chem A 2014;2:6266–91.
[539] Trogadas P, Fuller TF, Strasser P. Carbon as catalyst and support for electrochemical energy conversion. Carbon 2014;75:5–42.
[540] Shahgaldi S, Hamelin J. Improved carbon nanostructures as a novel catalyst support in the cathode side of PEMFC: a critical review. Carbon 2015;94:705–28.
[541] Perivoliotis DK, Tagmatarchis N. Recent advancements in metal-based hybrid electrocatalysts supported on graphene and related 2D materials for the oxygen reduction reaction. Carbon 2017;118:493–510.
[542] Shao Y, Sui J, Yin G, Gao Y. Nitrogen-doped carbon nanostructures and their composites as catalytic materials for proton exchange membrane fuel cell. Appl Catal, B 2008;79:89–99.
[543] Zheng Y, Jiao Y, Ge L, Jaroniec M, Ji JG, Qiao SZ. Nanostructured metal-free electrochemical catalysts for highly efficient oxygen reduction. Small 2012;8(23):3550–66.
[544] Dai L, Xue Y, Qu L, Choi H-J, Baek J-B. Metal-free catalysts for oxygen reduction reaction. Chem Rev 2015;115:4823–92.
[545] Iwan A, Malinowski M, Pasciak G. Polymer fuel cell components modified by graphene: electrodes, electrolytes and bipolar plates. Renew Sust Energy Rev 2015;49:954–67.
[546] Gratiet BL, Remita H, Picq G, Delcourt MO. CO-stabilized supported Pt catalysts for fuel cells: radiolytic synthesis. J Catal 1996;164:36–43.
[547] Wang Y, Ren J, Deng K, Gui L, Tang Y. Preparation of tractable platinum, rhodium, and ruthenium nanoclusters with small particle size in organic media. Chem Mater 2000;12:1622–7.

[548] Chan K-Y, Ding J, Ren J, Cheng S, Tsang KY. Supported mixed metal nanoparticles as electrocatalysts in low temperature fuel cells. J Mater Chem 2004;14:505–16.

[549] Su F, Zeng J, Yu Y, Lv L, Lee JY, Zhao XS. Template synthesis of microporous carbon for direct methanol fuel cell application. Carbon 2005;43:2366–73.

[550] Coker EN, Steen WA, Miller JT, Kropf J, Miller JE. The preparation and characterization of novel Pt/C electrocatalysts with controlled porosity and cluster size. J Mater Chem 2007;17:3330–40.

[551] Coker EN, Steen WA, Miller JT, Kropf AJ, Miller JE. Nanostructured Pt/C electrocatalysts with high platinum dispersions through zeolite-templating. Microp Mesop Mater 2007;101:440–4.

[552] Joo SH, Choi SJ, Oh I, Kwak J, Liu Z, Terasaki O, Ryoo R. Ordered nanoporous arrays of carbon supporting high dispersions of platinum nanoparticles. Nature 2001;412:169–72.

[553] Ding J, Chan K-Y, Ren J, Xiao F-S. Platinum and platinum–ruthenium nanoparticles supported on ordered mesoporous carbon and their electrocatalytic performance for fuel cell reactions. Electrochim Acta 2005;50:3131–41.

[554] Chai GS, Shin IS, Yu J-S. Synthesis of ordered, uniform, macroporous carbons with mesoporous walls templated by aggregates of polystyrene spheres and silica particles for use as catalyst supports in direct methanol fuel cells. Adv Mater 2004;16(22):2057–61.

[555] Bessel CA, Laubernds K, Rodriguez NM, Baker RTK. Graphite nanofibers as an electrode for fuel cell applications. J Phys Chem B 2001;105:1115–8.

[556] Steigerwalt ES, Deluga GA, Cliffel DE, Lukehart CM. A Pt-Ru/graphitic carbon nanofiber nanocomposite exhibiting high relative performance as a direct-methanol fuel cell anode catalyst. J Phys Chem B 2001;105:8097–101.

[557] Seger B, Kamat PV. Electrocatalytically active graphene-platinum nanocomposites. Role of 2-D carbon support in PEM fuel cells. J Phys Chem C 2009;113:7990–5.

[558] Ji K, Chang G, Oyamab M, Shang X, Liu X, He Y. Efficient and clean synthesis of graphene supported platinum nanoclusters and its application in direct methanol fuel cell. Electrochim Acta 2012;85:84–9.

[559] Sanli LI, Bayram V, Yarar B, Ghobadi S, Gursel SA. Development of graphene supported platinum nanoparticles for polymer electrolyte membrane fuel cells: effect of support type and impregnation- reduction methods. Int J Hydrogen Energy 2016;41:3414–27.

[560] Hsieh SH, Hsu MC, Liu WL, Chen WJ. Study of Pt catalyst on graphene and its application to fuel cell. Appl Surf Sci 2013;277:223–30.

[561] Heydari A, Gharibi H. Fabrication of electrocatalyst based on nitrogen doped graphene as highly efficient and durable support for using in polymer electrolyte fuel cell. J Power Sources 2016;325:808–15.

[562] Jhan J-Y, Huang Y-W, Hsu C-H, Teng H, Kuo D, Kuo P-L. Three-dimensional network of graphene grown with carbon nanotubes as carbon support for fuel cells. Energy 2013;53:282–7.

[563] Carrera-Cerritosa R, Baglio V, Aricò AS, Ledesma-García J, Sgroi MF, Pullini D, Pruna AJ, Mataix DB, Fuentes-Ramírez R, Arriaga LG. Improved Pd electro-catalysis for oxygen reduction reaction in direct methanol fuel cell by reduced graphene oxide. Appl Catal, B 2014;144:554–60.

[564] Zhang Y, Shua H, Chang G, Ji K, Oyama M, Liu X, He Y. Facile synthesis of palladium-graphene nanocomposites and their catalysis for electro-oxidation of methanol and ethanol. Electrochim Acta 2013;109:570–6.

[565] Li H, Chang G, Zhang Y, Tian J, Liu S, Luo Y, Asiri AM, Al-Youbi AO, Sun X. Photocatalytic synthesis of highly dispersed Pd nanoparticles on reduced graphene oxide and their application in methanol electro-oxidation. Catal Sci Technol 2012;2:1153−6.
[566] Li L, Chen M, Huang G, Yang N, Zhang L, Wang H, Liu Y, Wang W, Gao J. A green method to prepare Pd-Ag nanoparticles supported on reduced graphene oxide and their electrochemical catalysis of methanol and ethanol oxidation. J Power Sources 2014;263:13−21.
[567] Zhou Y, Pasquarelli R, Holme T, Berry J, Ginley D, O'Hayre R. Improving PEM fuel cell catalyst activity and durability using nitrogen-doped carbon supports: observations from model Pt/Hopg systems. J Mater Chem 2009;19:7830−8.
[568] Kondo T, Suzuki T, Nakamura J. Nitrogen doping of graphite for enhancement of durability of supported platinum clusters. J Phys Chem Lett 2011;2:577−80.
[569] Liang HW, Zhuang X, Bruller S, Feng X, Mullen K. Hierarchically porous carbons with optimized nitrogen doping as highly active electrocatalysts for oxygen reduction. Nat Commun 2014;5:4973.
[570] Hu W, Yoshida N, Hirota Y, Tanaka S, Nishiyama N. Solvothermal co-gelation synthesis of N-doped three-dimensional open macro/mesoporous carbon as efficient electrocatalyst for oxygen reduction reaction. Electrochem Commun 2017;75:9−12.
[571] Yang D-S, Bhattacharjya D, Inamdar S, Park J, Yu J-S. Phosphorus-doped ordered mesoporous carbons with different lengths as efficient metal-free electrocatalysts for oxygen reduction reaction in alkaline media. J Am Chem Soc 2012;134:16127−30.
[572] Elumeeva K, Fechler N, Fellinger TP, Antonietti M. Metal-free ionic liquid-derived electrocatalyst for high-performance oxygen reduction in acidic and alkaline electrolytes. Mater Horiz 2014;1:588−94.
[573] Graglia M, Pampel J, Hantke T, Fellinger T-P, Esposito D. Nitro lignin-derived nitrogen-doped carbon as an efficient and sustainable electrocatalyst for oxygen reduction. ACS Nano 2016;10:4364−71.
[574] Yang W, Fellinger TP, Antonietti M. Efficient metal-free oxygen reduction in alkaline medium on high-surface-area mesoporous nitrogen-doped carbons made from ionic liquids and nucleobases. J Am Chem Soc 2011;133:206−9.
[575] Ferrero GA, Fuertes AB, Sevilla M, Titirici MM. Efficient metal-free N-doped mesoporous carbon catalysts for ORR by a template-free approach. Carbon 2016;106:179−87.
[576] Lu Y, Wang L, Preuß K, Qiao Mo, Titirici M-M, Varcoe J, Cai Q. Halloysite-derived nitrogen doped carbon electrocatalysts for anion exchange membrane fuel cells. J Power Sources 2017;372:82−90.
[577] Ning R, Ge C, Liu Q, Tian J, Asiri AM, Alamry KA, Li CM, Sun X. Hierarchically porous N- doped carbon nanoflakes: large-scale facile synthesis and application as an oxygen reduction reaction electrocatalyst with high activity. Carbon 2014;78:60−9.
[578] Zhang P, Sun F, Xiang Z, Shen Z, Yun J, Cao D. ZIF-derived in situ nitrogen-doped porous carbons as efficient metal-free electrocatalysts for oxygen reduction reaction. Energy Environ Sci 2014;7:442−50.
[579] Xing Z, Gao N, Qi Y, Ji X, Liu H. Influence of enhanced carbon crystallinity of nanoporous graphite on the cathode performance of microbial fuel cells. Carbon 2017;115:271−8.
[580] Chen S, Liu Q, He G, Zhou Y, Hanif M, Peng W, Wang S, Hou H. Reticulated carbon foam derived from a sponge-like natural product as a high performance anode in microbial fuel cells. J Mater Chem 2012;22:18609−13.

[581] Gong K, Du F, Xia Z, Durstock M, Dai L. Nitrogen-doped carbon nanotube arrays with high electrocatalytic activity for oxygen reduction. Science 2009;323:760–4.
[582] Morozan A, Jeqou P, Pinault M, Campidelli S, Jousselme B, Palacin S. Metal-free nitrogen-containing carbon nanotubes prepared from triazole and tetrazole derivatives show high electrocatalytic activity towards the oxygen reduction reaction in alkaline media. ChemSusChem 2012;5:647–51.
[583] Sun T, Wu Q, Jiang Y, Zhang Z, Du L, Yang L, Wang X, Hu Z. Sulfur and nitrogen co-doped carbon tubes as bifunctional metal-free electrocatalysts for oxygen reduction and hydrogen evolution in acidic media. Chem Eur J 2016;22:10326–9.
[584] Qu LT, Liu Y, Baek JB, Dai LM. Nitrogen-doped graphene as efficient metal-free electrocatalyst for oxygen reduction in fuel cells. ACS Nano 2010;4:1321–6.
[585] Luo Z, Lim S, Tian Z, Shang J, Lai L, MacDonald B, Fu C, Shen Z, Yu T, Lin J. Pyridinic N doped graphene: synthesis, electronic structure, and electrocatalytic property. J Mater Chem 2011;21:8038–44.
[586] Wang Q, Zhang X, Lv R, Chen X, Xue B, Liang P, Huang X. Binder-free nitrogen-doped graphene catalyst air cathodes for microbial fuel cells. J Mater Chem A 2016;4:12587–91.
[587] Guo D, Shibuya R, Akiba C, Saji S, Kondo T, Nakamura J. Active sites of nitrogen-doped carbon materials for oxygen reduction reaction clarified using model catalysts. Science 2016;351:361–5.
[588] Zhou X, Bai Z, Wu M, Qiao J, Chen Z. 3-Dimensional porous N-doped graphene foam as a non- precious catalyst for the oxygen reduction reaction. J Mater Chem A 2015;3:3343–50.
[589] Zhou X, Tang S, Yin Y, Sun S, Qiao J. Hierarchical porous N-doped graphene foams with superior oxygen reduction reactivity for polymer electrolyte membrane fuel cells. Appl Energy 2016;175:459–67.
[590] Tang S, Zhou X, Xu N, Bai Z, Qiao J, Zhang J. Template-free synthesis of three-dimensional nanoporous N-doped graphene for high performance fuel cell oxygen reduction reaction in alkaline media. Appl Energy 2016;175:405–13.
[591] Zhao Y, Hu C, Hu Y, Cheng H, Shi G, Qu L. A versatile, ultralight, nitrogen-doped graphene framework. Angew Chem Int Ed 2012;51(45):11371–5.
[592] Feng L, Yang L, Huang Z, Luo J, Li M, Wang D, Chen Y. Enhancing electrocatalytic oxygen reduction on nitrogen-doped graphene by active sites implantation. Sci Rep 2013;3:3306.
[593] Yong Y-C, Dong X-C, Chan-Park MB, Song H, Chen P. Macroporous and monolithic anode based on polyaniline hybridized three-dimensional graphene for high-performance microbial fuel cells. ACS Nano 2012;6:2394–400.
[594] Luo JY, Jang HD, Sun T, Xiao L, He Z, Katsoulidis AP, Kanatzidis MG, Gibson JM, Huang JX. Compression and aggregation-resistant particles of crumpled soft sheets. ACS Nano 2011;5:8943–9.
[595] Bari CD, Goñi-Urtiaga A, Pita M, Shleev S, Toscano MD, Sainz R, De Lacey AL. Fabrication of high surface area graphene electrodes with high performance towards enzymatic oxygen reduction. Electrochim Acta 2016;191:500–9.
[596] Feng L, Chen Y, Chen L. Easy-to-operate and low-temperature synthesis of gram-scale nitrogen- doped graphene and its application as cathode catalyst in microbial fuel cells. ACS Nano 2011;5:9611–8.
[597] Hou J, Liu Z, Zhang P. A new method for fabrication of graphene/polyaniline nanocomplex modified microbial fuel cell anodes. J Power Sources 2013;224:139–44.
[598] Huang L, Li X, Ren Y, Wang X. In-situ modified carbon cloth with polyaniline/graphene as anode to enhance performance of microbial fuel cell. Int J Hydrogen Energy 2016;41:11369–79.

[599] Wen Q, Wang S, Yan J, Cong L, Chen Y, Xi H. Porous nitrogen-doped carbon nanosheet on graphene as metal-free catalyst for oxygen reduction reaction in air-cathode microbial fuel cells. Bioelectrochem 2014;95:23−8.
[600] Lv Z, Chen Y, Wei H, Li F, Hu Y, Wei C, Feng C. One-step electrosynthesis of polypyrrole/graphene oxide composites for microbial fuel cell application. Electrochim Acta 2013;111:366−73.
[601] Shui J, Wang M, Du F, Dai L. N-doped carbon nanomaterials are durable catalysts for oxygen reduction reaction in acidic fuel cells. Sci Adv 2015;1:e1400129.
[602] Lee JS, Jo K, Lee T, Yun T, Cho J, Kim BS. Facile synthesis of hybrid graphene and carbon nanotubes as a metal-free electrocatalyst with active dual interfaces for efficient oxygen reduction reaction. J Mater Chem A 2013;1:9603−7.
[603] Choi CH, Park SH, Woo SI. Binary and ternary doping of nitrogen, boron, and phosphorus into carbon for enhancing electrochemical oxygen reduction activity. ACS Nano 2012;6(8):7084−91.
[604] Choi CH, Park SH, Woo SI. Phosphorus-nitrogen dual doped carbon as an effective catalyst for oxygen reduction reaction in acidic media: effects of the amount of P-doping on the physical and electrochemical properties of carbon. J Mater Chem 2012;22:12107−15.
[605] Ranjbar Sahraie N, Paraknowitsch JP, Göbel C, Thomas A, Strasser P. Noble-metal-free electrocatalysts with enhanced ORR performance by task-specific functionalization of carbon using ionic liquid precursor systems. J Am Chem Soc 2014;136:14486−97.
[606] Liu J, Sun X, Song P, Zhang Y, Xing W, Xu W. High performance oxygen reduction electrocatalysts based on cheap carbon black, nitrogen, and trace iron. Adv Mater 2013;25(47):6879−83.
[607] Sun X, Song P, Zhang Y, Liu C, Xing W. A class of high performance metal-free oxygen reduction electrocatalysts based on cheap carbon blacks. Sci Rep 2013;3:2505.
[608] Zheng Y, Jiao Y, Ge L, Jaroniec M, Qiao SZ. Two-step boron and nitrogen doping in graphene for enhanced synergistic catalysis. Angew Chem Int Ed 2013;52:3110−6.
[609] Jeon I-Y, Kim C, Kim G, Baek J-B. Mechanochemically driven iodination of activated charcoal for metal-free electrocatalyst for fuel cells and hybrid Li-air cells. Carbon 2015;93:465−72.
[610] Jeon I-Y, Choi H-J, Choi M, Seo J-M, Jung S-M, Kim M-J, Zhang S, Zhang L, Xia Z, Dai L, Park N, Baek J-B. Facile, scalable synthesis of edge-halogenated graphene nanoplatelets as efficient metal- free eletrocatalysts for oxygen reduction reaction. Sci Rep 2013;3:1810.
[611] Yao Z, Nie H, Yang Z, Zhou X, Liu Z, Huang S. Catalyst-free synthesis of iodine-doped graphene via a facile thermal annealing process and its use for electrocatalytic oxygen reduction in an alkaline medium. Chem Commun 2012;28:1027−9.
[612] Wang T, Wang J, Wang X, Yang J, Liu J, Xu H. Graphene-templated synthesis of sandwich-like porous carbon nanosheets for efficient oxygen reduction reaction in both alkaline and acidic media. Sci China Mater 2018;61:915−25.
[613] Ai K, Liu Y, Ruan C, Lu L, Lu G. Sp2 C-dominant N-doped carbon sub-micrometer spheres with a tunable size: a versatile platform for highly efficient oxygen-reduction catalysts. Adv Mater 2013;25(7):998−1003.
[614] Parvez K, Yang S, Hernandez Y, Winter A, Turchanin A, Feng X, Muellen K. Nitrogen-doped graphene and its iron-based composite as efficient electrocatalysts for oxygen reduction reaction. ACS Nano 2012;6:9541−50.
[615] Jiao L, Wan G, Zhang R, Zhou H, Yu S-H, Jiang H-L. From metal-organic frameworks to single-atom Fe implanted N-doped porous carbons: efficient oxygen reduction in both alkaline and acidic media. Angew Chem Int Ed 2018;57:8525−9.

[616] Chung HT, Won JT, Zelenay P. Active and stable carbon nanotube/nanoparticle composite electrocatalyst for oxygen reduction. Nat Commun 2013;4:1922.
[617] Rauf M, Chen R, Wang Q, Wang Y-C, Zhou Z-Y. Nitrogen-doped carbon nanotubes with encapsulated Fe nanoparticles as efficient oxygen reduction catalyst for alkaline membrane direct ethanol fuel cells. Carbon 2017;125:605−13.
[618] Tan YM, Xu CF, Chen GX, Fang XL, Zheng NF, Xie QJ. Facile synthesis of manganese-oxide- containing mesoporous nitrogen-doped carbon for efficient oxygen reduction. Adv Funct Mater 2012;22:4584−91.
[619] Shang C, Yang M, Wang Z, Li M, Liu M, Zhu J, Zhu Y, Zhou L, Cheng H, Gu Y, Tang Y, Zhao X, Lu Z. Encapsulated MnO in N-doping carbon nanofibers as efficient ORR electrocatalysts. Sci China Mater 2017;60:937−46.
[620] Liu Z, Zhang G, Lu Z, Jin X, Chang Z, Sun X. One-step scalable preparation of N-doped nanoporous carbon as high- performance electrocatalysts for oxygen reduction reaction. Nano Res 2013;6:293−301.
[621] Xing R, Zhou Y, Ma R, Liu Q, Luo J, Yang M, Wang J. Nitrogen-doped hollow carbon spheres with embedded Co nanoparticles as active non-noble-metal electrocatalysts for the oxygen reduction reaction. J Carbon Res 2018;4:11.
[622] Kong F, Fan X, Zhang X, Wang L, Kong A, Shan Y. Soft-confinement conversion of Co-Salen- organic-frameworks to uniform cobalt nanoparticles embedded within porous carbons as robust trifunctional electrocatalysts. Carbon 2019;149:471−82.
[623] Li G, Yi Q, Yang X, Chen Y, Zhou X, Xie G. Ni-Co-N doped honeycomb carbon nano-composites as cathodic catalysts of membrane-less direct alcohol fuel cell. Carbon 2018;140:557−68.
[624] https://www.energy.gov/eere/fuelcells/hydrogen-storage.
[625] Zhou L, Zhou Y, Sun Y. Enhanced storage of hydrogen at the temperature of liquid nitrogen. Int J Hydrogen Energy 2004;29:319−22.
[626] Thomas KM. Hydrogen adsorption and storage on porous materials. Catal Today 2007;120:389−98.
[627] Texier-Mandoki N, Dentzer J, Piquero T, Saadallah S, David P, Vix-Guterland C. Hydrogen storage in activated carbon materials: role of the nanoporous texture. Carbon 2004;42:2744−7.
[628] Xu W-C, Takahashi K, Matsuo Y, Hattoria Y, Kumagai M, Ishiyama S, Kaneko K, Iijima S. Investigation of hydrogen storage capacity of various carbon materials. Int J Hydrogen Energy 2007;32:2504−12.
[629] Zhao XB, Xiao B, Fletcher AJ, Thomas KM. Hydrogen adsorption on functionalized nanoporous activated carbons. J Phys Chem B 2005;109:8880−8.
[630] Panella B, Hirscher M, Roth S. Hydrogen adsorption in different carbon nanostructures. Carbon 2005;43:2209−14.
[631] Jorda-Beneyto M, Lozano-Castello D, Suarez-Garcia F, Cazorla-Amoros D, Linares-Solano A. Advanced activated carbon monoliths and activated carbons for hydrogen storage. Microp Mesop Mater 2008;112:235−42.
[632] Takagi H, Hatori H, Soneda Y, Yoshizawa N, Yamada Y. Adsorptive hydrogen storage in carbon and porous materials. Mater Sci Eng B 2004;108:143−7.
[633] Wang H, Gao Q, Hu J. High hydrogen storage capacity of porous carbons prepared by using activated carbon. J Am Chem Soc 2009;131:7016−22.
[634] Juan-Juan J, Marco-Lozar JP, Suarez-Garcıa F, Cazorla-Amoros D, Linares-Solano A. A comparison of hydrogen storage in activated carbons and a metal−organic framework (MOF-5). Carbon 2010;48:2906−9.
[635] Zhou L, Zhou Y, Sun Y. A comparative study of hydrogen adsorption on superactivated carbon versus carbon nanotubes. Int J Hydrogen Energy 2004;29:475−9.

[636] Miyamoto J, Hattori Y, Noguchi D, Tanaka H, Ohba T, Utsumi S, Kanoh H, Kim YA, Muramatsu H, Hayashi T, Endo M, Kaneko K. Efficient H_2 adsorption by nanopores of high-purity double-walled carbon nanotubes. J Am Chem Soc 2006;128:12636−7.
[637] Liu C, Chen Y, Wu C-Z, Xu S-T, Cheng H-M. Hydrogen storage in carbon nanotubes revisited. Carbon 2010;48:452−5.
[638] Kunowsky M, Marco-Lozar JP, Oya A, Linares-Solano A. Hydrogen storage in CO_2-activated amorphous nanofibers and their monoliths. Carbon 2012;50:1407−16.
[639] Im JS, Park S, Lee Y. Superior prospect of chemically activated electrospun carbon fibers for hydrogen storage. Mater Res Bull 2009;44:1871−8.
[640] Nishihara H, Hou PX, Li LX, Ito M, Uchiyama M, Kaburagi T, Ikura A, Katamura J, Kawarada T, Mizuuchi K, Kyotani T. High-pressure hydrogen storage in zeolite-templated carbon. J Phys Chem C 2009;113:3189−96.
[641] Yang Z, Xia Y, Sun X, Mokaya R. Preparation and hydrogen storage properties of zeolite-templated carbon materials nanocast via chemical vapor deposition: effect of the zeolite template and nitrogen doping. J Phys Chem B 2006;110:18424−31.
[642] Yang Z, Xia Y, Mokaya R. Enhanced hydrogen storage capacity of high surface area zeolite-like carbon materials. J Am Chem Soc 2007;129:1673−9.
[643] Wang H, Gao Q, Hu J, Chen Z. High performance of nanoporous carbon in cryogenic hydrogen storage and electrochemical capacitance. Carbon 2009;47:2259−68.
[644] Gogotsi Y, Dash RK, Yushin G, Yildirim T, Laudisio G, Fischer JE. Tailoring of nanoscale porosity in carbide-derived carbons for hydrogen storage. J Am Chem Soc 2005;127:16006−7.
[645] Rose M, Kockrick E, Senkovska I, Kaskel S. High surface area carbide-derived carbon fibers produced by electrospinning of polycarbosilane precursors. Carbon 2010;48:403−7.
[646] Dash R, Chmiola J, Yushin G, Gogotsi Y, Laudisio G, Singer J, Fischer J, Kucheyev S. Titanium carbide derived nanoporous carbon for energy-related applications. Carbon 2006;44:2489−97.
[647] Yushin G, Dash R, Jagiello J, Fischer JE, Gogotsi Y. Carbide-derived carbons: effect of pore size on hydrogen uptake and heat of adsorption. Adv Funct Mater 2006;16:2288−91.
[648] Johansson E, Hjorvarsson B, Ekstroem T, Jacob M. Hydrogen in carbon nanostructures. J Alloys Compd 2002;330−332:670−5.
[649] Wang X-S, Ma S, Rauch K, Simmons JM, | Yuan D, Wang X, Yildirim T, Cole WC, López JJ, de Meijere A, Zhou H-C. Metal-organic frameworks based on double-bond-coupled di-isophthalate linkers with high hydrogen and methane uptakes. Chem Mater 2008;20:3145−52.
[650] Liu B, Shioyama H, Akita T, Xu Q. Metal-organic framework as a template for porous carbon synthesis. J Am Chem Soc 2008;130:5390.
[651] Yang SJ, Kim T, Im JH, Kim YS, Lee K, Jung H, Park CR. MOF-derived hierarchically porous carbon with exceptional porosity and hydrogen storage capacity. Chem Mater 2012;24:464−70.
[652] Yang ZX, Xia YD, Zhu YQ. Preparation and gases storage capacities of N-doped porous activated carbon materials derived from mesoporous polymer. Mater Chem Phys 2013;141:318−23.
[653] Xia K, Gao Q, Wu C, Song S, Ruan M. Activation, characterization and hydrogen storage properties of the mesoporous carbon CMK-3. Carbon 2007;45:1989−96.

[654] Gadiou R, Saadallah S, Piquero T, David P, Parmentier J, Vix-Guterl C. The influence of textural properties on the adsorption of hydrogen on ordered nanostructured carbons. Microp Mesop Mater 2005;79:121–8.
[655] Zheng Z, Gao Q, Jiang J. High hydrogen uptake capacity of mesoporous nitrogen-doped carbons activated using potassium hydroxide. Carbon 2010;48:2968–73.
[656] Giraudet S, Zhu Z, Yao X, Lu G. Ordered mesoporous carbons enriched with nitrogen: application to hydrogen storage. J Phys Chem C 2010;114:8639–45.
[657] Krawiec P, Kockrick E, Borchardt L, Geiger D, Corma A, Kaskel S. Ordered mesoporous carbide derived carbons: novel materials for catalysis and adsorption. J Phys Chem C 2009;113(18):7755–61.
[658] Orimo S, Majer G, Fukunaga T, Zuttel A, Schlapbach L, Fujii H. Hydrogen in the mechanically prepared nanostructured graphite. Appl Phys Lett 1999;75:3093.
[659] Orimo S, Matsushima T, Fujii H, Fukunaga T, Majer G. Hydrogen desorption property of mechanically prepared nanostructured graphite. J Appl Phys 2001;90:1545–9.
[660] Hirscher M, Becher M, Haluska M, Dettlaff-Weglikowska U, Quintel A, Duesberg GS, Choi Y-M, Downes P, Hulman M, Roth S, Stepanek I, Ber P. Hydrogen storage in sonicated carbon materials. Appl Phys A 2001;72:129–32.
[661] Zielinski M, Wojcieszak R, Monteverdi S, Mercy M, Bettahar MM. Hydrogen storage on nickel catalysts supported on amorphous activated carbon. Catal Commun 2005;6:777–83.
[662] Zielinski M, Wojcieszak R, Monteverdi S, Mercy M, Bettahar MM. Hydrogen storage in nickel catalysts supported on activated carbon. Int J Hydrogen Energy 2007;32:1024–32.
[663] Lee YS, Kim YH, Hong JS, Suh JK, Cho GJ. The adsorption properties of surface modified activated carbon fibers for hydrogen storages. Catal Today 2007;120:420–5.
[664] Giraudet S, Zhu Z. Hydrogen adsorption in nitrogen enriched ordered mesoporous carbons doped with nickel nanoparticles. Carbon 2011;49:398–405.
[665] Ediati R, Mukminin A, Widiastuti N. Impregnation of nickel on mesoporous ZSM-5 templated carbons as candidate material for hydrogen storage. Indones J Chem 2017;17(1):30–6.
[666] Zubizarreta L, Menendez JA, Job N, Marco-Lozar JP, Pirard JP, Pis JJ, Linares-Solano A, Cazorla- Amoros D, Arenillas A. Ni-doped carbon xerogels for H2 storage. Carbon 2010;48:2722–33.
[667] Silva LMS, Orfo JJM, Figueiredo JL. Formation of two metal phases in the preparation of activated carbon-supported nickel catalysts. Appl Catal A Gen 2001;209:145–54.
[668] Zubizarreta L, Menendez JA, Pis JJ, Arenillas A. Improving hydrogen storage in Ni-doped carbon nanospheres. Int J Hydrogen Energy 2009;34:3070–6.
[669] Kim HS, Lee H, Han KS, Kim J-H, Song M-S, Park M-S, Lee J-Y, Kang J-K. Hydrogen storage in Ni nanoparticle-dispersed multiwalled carbon nanotubes. J Phys Chem B 2005;109:8983–6.
[670] Li YW, Yang RT. Hydrogen storage on platinum nanoparticles doped on superactivated carbon. J Phys Chem C 2007;111:11086–94.
[671] Im JS, Kwon O, Kim YH, Park S-J, Lee Y-S. The effect of embedded vanadium catalyst on activated electrospun CFs for hydrogen storage. Microp Mesop Mater 2008;115:514–21.
[672] Kunowsky M, Marco-Lozar JP, Cazorla-Amoros D, Linares-Solano A. Scale-up activation of carbon fibres for hydrogen storage. Int J Hydrogen Energy 2010;35:2393–402.
[673] Darkrim F, Levesque D. High adsorptive property of opened carbon nanotubes at 77 K. J Phys Chem B 2000;104:6773–6.

[674] Lee SM, Lee TH. Hydrogen storage in single-walled carbon nanotubes. Appl Phys Lett 2000;76:2877–9.
[675] Rzepka M, Lamp P, De la Casa-Lillo MA. Physisorption of hydrogen on microporous carbons and carbon nanotubes. J Phys Chem B 1998;102:10894–8.
[676] de la Casa-Lillo MA, Lamari-Darkrim F, Cazorla-Amoros D, Linares-Solano A. Hydrogen storage in activated carbons and activated carbon fibers. J Phys Chem B 2002;106:10930–4.
[677] Lee SM, Park KS, Choi YC, Park YS, Bok JM, Bae DJ, Nahm KS, Choi YG, Yu SC, Kim N, Frauenheim T, Lee YH. Hydrogen adsorption and storage in carbon nanotubes. Synth Met 2000;113:209–16.
[678] Kayanuma M, Nagashima U, Nishihara H, Kyotani T, Ogawa H. Adsorption and diffusion of atomic hydrogen on a curved surface of microporous carbon: a theoretical study. Chem Phys Lett 2010;495:251–5.
[679] Zhang Y, Sun H, Chen C. New template for metal decoration and hydrogen adsorption on graphene- like C3N4. Phys Lett A 2009;373:2778–81.
[680] Wang YS, Li M, Wang F, Sun Q, Jia Y. Li and Na Co-decorated carbon nitride nanotubes as promising new hydrogen storage media. Phys Lett A 2012;376:631–6.
[681] Wu M, Wang Q, Sun Q, Jena P. Functionalized graphitic carbon nitride for efficient energy storage. J Phys Chem C 2013;117:6055–9.
[682] Zhu G, Lu K, Sun Q, Kawazoe Y, Jena P. Lithium-doped triazine-based graphitic C_3N_4 sheet for hydrogen storage at ambient temperature. Comput Mater Sci 2014;81:275–9.
[683] Wei J, Huang C, Wu H, Kan E. High-capacity hydrogen storage in Li-adsorbed g-C_3N_4. Mater Chem Phys 2016;180:440–4.
[684] Zhang W, Zhan Z, Zhang F, Yang W. Ti-decorated graphitic-C_3N_4 monolayer: a promising material for hydrogen storage. Appl Surf Sci 2016;386:247–54.
[685] Inagaki M, Tsumura T, Kinumoto T, Toyoda M. Graphitic carbon nitrides (g-C_3N_4) with comparative discussion to carbon materials. Carbon 2019;141:580–607.
[686] Yang SJ, Cho JH, Oh GH, Nahm KS, Park CR. Easy synthesis of highly nitrogen-enriched graphitic carbon with a high hydrogen storage capacity at room temperature. Carbon 2009;47:1585–91.
[687] Bai XD, Zhong DY, Zhang GY, Ma XC, Liu S, Wang EG, Chen Y, Shaw DT. Hydrogen storage in carbon nitride nanobells. Appl Phys Lett 2001;79:1552–4.
[688] Nair AAS, Sundara R, Anitha N. Hydrogen storage performance of palladium nanoparticles decorated graphitic carbon nitride. Int J Hydrogen Energy 2015;40:3259–67.
[689] Jurewicz K, Frackowiak E, Beguin F. Towards the mechanism of electrochemical hydrogen storage in nanostructured carbon materials. Appl Phys A 2004;78:981–7.
[690] Jurewicz K, Frackowiak E, Beguin F. Enhancement of reversible hydrogen capacity into activated carbon through water electrolysis. Electrochem Solid State Lett 2001;4:A27–9.
[691] Jurewicz K, Frackowiak E, Beguin F. Electrochemical storage of hydrogen in activated carbons. Fuel Process Technol 2002;77–78:415–21.
[692] Qu DY. Mechanism for electrochemical hydrogen insertion in carbonaceous materials. J Power Sources 2008;179:310–6.
[693] Beguin F, Friebe M, Jurewicz K, Vix-Guterl C, Dentzer J, Frackowiak E. State of hydrogen electrochemically stored using nanoporous carbons as negative electrode materials in an aqueous medium. Carbon 2006;44:2392–8.
[694] http://arpa-e.energy.gov/; Methane Opportunities for Vehicular Energy, Advanced Research Project Agency-Energy, U.S. Department of Energy.

[695] Ma S, Sun D, Simmons JM, Collier CD, Yuan D, Zhou H-C. Metal-organic framework from an anthracene derivative containing nanoscopic cages exhibiting high methane uptake. J Am Chem Soc 2008;130:1012−6.
[696] Peng Y, Krungleviciute V, Eryazici I, Hupp JT, Farha OK, Yildirim T. Methane storage in metal- organic frameworks: current records, surprise findings, and challenges. J Am Chem Soc 2013;135:11887−94.
[697] Menon VC, Komarneni S. Porous adsorbents for vehicular natural gas storage: a review. J Porous Mater 1998;5:43−58.
[698] Lozano-Castello D, Alcaniz-Monge J, de la Casa-Lillo MA, Cazorla-Amorosa D, Linares-Solano A. Advances in the study of methane storage in porous carbonaceous materials. Fuel 2002;81:1777−803.
[699] Zhou L, Liu L, Su W, Sun Y, Zhou Y. Progress in studies of natural gas storage with wet adsorbents. Energy Fuel 2010;24:3789−95.
[700] Makal TA, Li J-R, Lu W, Zhou H-C. Methane storage in advanced porous materials. Chem Soc Rev 2012;41:7761−79.
[701] Quinn DF, MacDonald JA. Natural gas storage. Carbon 1992;30:1097−103.
[702] Casco ME, Martínez-Escandell M, Gadea-Ramos E, Kaneko K, Silvestre-Albero J, Rodríguez- Reinoso F. High-pressure methane storage in porous materials: are carbon materials in the pole position? Chem Mater 2015;27:959−64.
[703] Alcaniz-Monge J, de la Casa-Lilld MA, Cazorla-Amoros D, Lineares-Solano A. Methane storage in activated carbon fibers. Carbon 1997;35:291−7.
[704] Burchell T, Rogers M. Low pressure storage of natural gas for vehicular applications. SAE Tech Pap Ser 2000:01−2205.
[705] Shao X, Wang W, Zhang X. Experimental measurements and computer simulation of methane adsorption on activated carbon fibers. Carbon 2007;45:188−95.
[706] Rodnguez-Reinoso F, Almansa C, Molina-Sabio M. Contribution to the evaluation of density of methane adsorbed on activated carbon. J Phys Chem B 2005;109:20227−31.
[707] Chen XS, McEnaney B, Mays TJ, Alcaniz-Monge J, Cazorla-Amoros D, Linares-Solano A. Theoretical and experimental studies of methane adsorption on microporous carbons. Carbon 1997;35:1251−8.
[708] Bekyarova E, Murata K, Yudasaka M, Kasuya D, Iijima S, Tanaka H, Kahoh H, Kaneko K. Single- wall nanostructured carbon for methane storage. J Phys Chem B 2003;107:4681−4.
[709] Yeon SH, Osswald S, Gogotsi Y, Singer JP, Simmons JM, Fischer JE, Lillo-Ródenasd MA, Linares- Solano Á. Enhanced methane storage of chemically and physically activated carbide-derived carbon. J Power Sources 2009;191:560−7.
[710] Kockrick E, Schrage C, Borchardt L, Klein N, Rose M, Senkovska I, Kaskel S. Ordered mesoporous carbide derived carbons for high pressure gas storage. Carbon 2010;48:1707−17.
[711] Oschatz M, Kockrick E, Rose M, Borchardt L, Klein N, Senkovska I, Freudenberg T, Korenblit Y, Yushin G, Kaskel S. A cubic ordered, mesoporous carbide-derived carbon for gas and energy storage applications. Carbon 2010;48:3987−92.
[712] Kaneko K, Murata K, Shimizu K, Camara S, Suzuki T. Enhancement effect of micropore filling for supercritical methane by MgO dispersion. Langmuir 1993;9:1165−7.
[713] Miyawaki J, Kanda T, Suzuki T, Okui T, Maeda Y, Kaneko K. Macroscopic evidence of enhanced formation of methane nanohydrates in hydrophobic nanospaces. J Phys Chem 1998;102:2187−92.

[714] Casco ME, Silvestre-Albero J, Ramırez-Cuesta AJ, Rey F, Jorda JL, Bansode A, Urakawa A, Peral I, Martınez-Escandell M, Kaneko K, Rodrıguez-Reinoso F. Methane hydrate formation in confined nanospace can surpass nature. Nature Commun 2015;6:6432.
[715] Zhou L, Sun Y, Zhou YP. Enhancement of the methane storage on activated carbon by preadsorbed water. AIChE J 2002;48:2412−6.
[716] Liu J, Zhou Y, Sun Y, Su W, Zhou L. Methane storage in wet carbon of tailored pore sizes. Carbon 2011;49:3731−6.
[717] Perrin A, Celzard A, Mareche JF, Furdin G. Methane storage within dry and wet activated carbons: a comparative study. Energy Fuel 2003;17:1283−91.
[718] Vysniauskas A, Bishnoi PR. A kinetic study of methane hydrate formation. Chem Eng Sci 1983;38:1061−72.
[719] Zhou Y, Wang YX, Chen HH, Zhou L. Methane storage in wet activated carbon: studies on the charging/discharging process. Carbon 2005;43:2007−12.
[720] Zalba B, Marin JM, Cabeza LF, Mehling H. Review on thermal energy storage with phase change materials, heat transfer analysis and applications. Appl Therm Eng 2003;23:251−83.
[721] Farid MM, Khudhair AM, Siddique KR, Al-Hallaj S. A review on phase change energy storage: materials and applications. Energy Convers Manage 2004;45:1597−615.
[722] Fan L, Khodadadi JM. Thermal conductivity enhancement of phase change materials for thermal energy storage: A. review. Renew Sust Energy Rev 2011;15:24−46.
[723] Himran S, Suwono A. Characterization of alkanes and paraffin waxes for application as phase change energy storage medium. Energy Sources 1994;16:117−28.
[724] Kuznik F, David D, Johannes K, Roux J-J. A review on phase change materials integrated in building walls. Renew Sust Energy Rev 2011;15:379−91.
[725] Tyagi VV, Kaushik SC, Tyagi SK, Akiyama T. Development of phase change materials based microencapsulated technology for buildings: a review. Renew Sust Energy Rev 2011;15:1373−91.
[726] Zhang Z, Fang X. Study on paraffin/expanded graphite composite phase change thermal energy storage material. Energy Convers Manage 2006;47:303−10.
[727] Inagaki M, Qiu J, Guo Q. Carbon foam: preparation and application. Carbon 2015;87:128−52.
[728] Py X, Olives R, Mauran S. Paraffin/porous-graphite-matrix composite as a high and constant power thermal storage material. Int J Heat Mass Transfer 2001;44:2727−37.
[729] Zhong Y, Li S, Wei X, Liu Z, Guo Q, Shi J, Liu L. Heat transfer enhancement of paraffin wax using compressed expanded natural graphite for thermal energy storage. Carbon 2010;48:300−4.
[730] Sari A, Karaipekli A. Thermal conductivity and latent heat thermal energy storage characteristics of paraffin/expanded graphite composite as phase change material. Appl Therm Eng 2007;27:1271−7.
[731] Xia L, Zhang P, Wang RZ. Preparation and thermal characterization of expanded graphite/paraffin composite phase change material. Carbon 2010;48:2538−48.
[732] Zhao JG, Guo Y, Feng F, Tong QH, Qv WS, Wang HQ. Microstructure and thermal properties of a paraffin/expanded graphite phase-change composite for thermal storage. Renew Energy 2011;36:1339−42.
[733] Mills A, Farid M, Selman JR, Al-Hallaj S. Thermal conductivity enhancement of phase change materials using a graphite matrix. Appl Therm Eng 2006;26:1652−61.

[734] Alrashdan A, Mayyas AT, Al-Hallaj S. Thermo-mechanical behaviors of the expanded graphite-phase change material matrix used for thermal management of Li-ion battery packs. J Mater Process Tech 2010;210:174−9.

[735] Li S, Song Y, Song Y, Shi J, Liu L, Wei X, Guo Q. Carbon foams with high compressive strength derived from mixtures of mesocarbon microbeads and mesophase pitch. Carbon 2007;45:2092−7.

[736] Zhong Y, Guo Q, Li S, Shi J, Liu L. Heat transfer enhancement of paraffin wax using graphite foam for thermal energy storage. Solar Energy Mater Sol Cells 2010;94:1011−4.

[737] Song J, Guo Q, Zhong Y, Gao X, Feng Z, Fan Z, Shi J, Liu L. Thermophysical properties of high- density graphite foams and their paraffin composites. New Carbon Mater 2012;27:27−34.

[738] Zhong Y, Guo Q, Li S, Shi J, Liu L. Thermal and mechanical properties of graphite foam/wood's alloy composite for thermal energy storage. Carbon 2010;48:1689−92.

[739] Canseco V, Anguy Y, Roa JJ, Palomo E. Structural and mechanical characterization of graphite foam/phase change material composites. Carbon 2014;74:266−81.

[740] Zhao CY, Wu ZG. Heat transfer enhancement of high temperature thermal energy storage using metal foams and expanded graphite. Solar Energy Mater Sol Cells 2011;95:636−43.

[741] Kim S, Drzal LT. High latent heat storage and high thermal conductive phase change materials using exfoliated graphite nanoplatelets. Solar Energy Mater Sol Cells 2009;93:136−42.

[742] Xiang J, Drzal LT. Investigation of exfoliated graphite nanoplatelets (xGnP) in improving thermal conductivity of paraffin wax-based phase change material. Solar Energy Mater Sol Cells 2011;95:1811−8.

[743] Zhang ZG, Zhang N, Peng J, Fang XM, Gao XN, Fang YT. Preparation and thermal energy storage properties of paraffin/expanded graphite composite phase change material. Appl Energy 2012;91:426−31.

[744] Lee H, Jeong SG, Chang SJ, Kang Y, Wi S, Kim S. Thermal performance evaluation of fatty acid ester and paraffin based mixed SSPCMs using exfoliated graphite nanoplatelets (xGnP). Appl Sci 2016;6:106.

[745] Chen Y-J, Nguyen D-D, Shen M-Y, Yip M-C, Tai N-H. Thermal characterizations of the graphite nanosheets reinforced paraffin phase-change composites. Compos Part A 2013;44:40−6.

[746] Fang G, Li H, Chen Z, Liu X. Preparation and characterization of steric acid/expanded graphite composites as thermal energy storage materials. Energy 2010;35:4622−6.

[747] Acem Z, Lopez J, Del Barrio EP. KNO3/NaNO3-graphite materials for thermal energy storage at high temperature: Part I. -Elaboration methods and thermal properties. Appl Therm Eng 2010;30:1580−5.

[748] Lopez J, Acem Z, Del Barrio EP. KNO_3/$NaNO_3$-graphite materials for thermal energy storage at high temperature: Part II. -Phase transition properties. Appl Therm Eng 2010;30:1586−93.

[749] Xiao X, Zhang P, Li M. Thermal characterization of nitrates and nitrates/expanded graphite mixture phase change materials for solar energy storage. Energy Convers Manage 2013;73:86−94.

[750] Shi J-N, Ger M-D, Liu Y-M, Fan Y-C, Wen N-T, Lin C-K, Pu N-W. Improving the thermal conductivity and shape-stabilization of phase change materials using nanographite additives. Carbon 2013;51:365−72.

[751] Fang X, Fan LW, Ding Q, Xiao W, Yao XL, Hou JF, Yu Z-T, Cheng G-H, Hu Y-C, Cen K-F. Increased thermal conductivity of eicosane-based composite phase change materials in the presence of graphene nanoplatelets. Energy Fuel 2013;27:4041−7.

[752] Li M. A nano-graphite/paraffin phase change material with high thermal conductivity. Appl Energy 2013;106:25–30.
[753] Fan LW, Fang X, Wang X, Zeng Y, Xiao YQ, Yu ZT, Xu X, Hua Y-C, Cen K-F. Effects of various carbon nanofillers on the thermal conductivity and energy storage properties of paraffin-based nanocomposite phase change materials. Appl Energy 2013;110:163–72.
[754] Pincemin S, Olives R, Py X, Christ M. Highly conductive composites made of phase change materials and graphite for thermal storage. Sol Energy Mater Sol Cells 2008;92:603–13.
[755] Zhong YJ, Zhou M, Huang FQ, Lin TQ, Wan DY. Effect of graphene aerogel on thermal behavior of phase change materials for thermal management. Sol Energy Mater Sol Cells 2013;113:195–200.
[756] Li JF, Lu W, Zeng YB, Luo ZP. Simultaneous enhancement of latent heat and thermal conductivity of docosane-based phase change material in the presence of spongy graphene. Solar Energy Mater Sol Cells 2014;128:48–51.
[757] Sun H, Xu Z, Gao C. Multifunctional, ultra-flyweight, synergistically assembled carbon aerogels. Adv Mater 2013;25:2554–60.
[758] Yang J, Zhang E, Li X, Zhang Y, Qu J, Yu Z-Z. Cellulose/graphene aerogel supported phase change composites with high thermal conductivity and good shape stability for thermal energy storage. Carbon 2016;98:50–7.
[759] Yang J, Qi G-Q, Liu Y, Bao R-Y, Liu Z-Y, Yang W, Xie B-H, Yang M-B. Hybrid graphene aerogels/phase change material composites: thermal conductivity, shape-stabilization and light-to- thermal energy storage. Carbon 2016;100:693–702.
[760] Yavari F, Fard HR, Pashayi K, Rafiee MA, Zamiri A, Yu ZZ, Ozisik R, Tasciuc TB, Koratkar N. Enhanced thermal conductivity in a nanostructured phase change composite due to low concentration graphene additives. J Phys Chem C 2011;115:8753–8.
[761] Goli P, Legedza S, Dhar A, Salgado R, Renteria J, Balandin AA. Graphene-enhanced hybrid phase change materials for thermal management of Li-ion batteries. J Power Sources 2014;248:37–43.
[762] Harish S, Orejon D, Takata Y, Kohno M. Thermal conductivity enhancement of lauric acid phase change nanocomposite with graphene nanoplatelets. Appl Therm Eng 2015;80:205–11.
[763] Yuan Y, Zhang N, Li T, Cao X, Long W. Thermal performance enhancement of palmitic-stearic acid by adding graphene nanoplatelets and expanded graphite for thermal energy storage: a comparative study. Energy 2016;97:488–97.
[764] Da HC, Lee J, Hong H, Kang YT. Thermal conductivity and heat transfer performance enhancement of phase change materials (PCM) containing carbon additives for heat storage application. Int J Refrig 2014;42:112–20.
[765] Tao YB, Lin CH, He YL. Preparation and thermal properties characterization of carbonate salt/carbon nanomaterial composite phase change material. Energy Convers Manage 2015;97:103–10.
[766] Bi H, Huang H, Xu F, Lin T, Zhang H, Huang F. Carbon microtube/graphene hybrid structures for thermal management applications. J Mater Chem A 2015;3:18706–10.
[767] Feng L, Zheng J, Yang H, Guo Y, Li W, Li X. Preparation and characterization of polyethylene glycol/active carbon composites as shape-stabilized phase change materials. Solar Energy Mater Sol Cells 2011;95:644–50.
[768] Mesalhy O, Lafdi K, Elgafy A. Carbon foam matrices saturated with PCM for thermal protection purposes. Carbon 2006;44:2080–8.

[769] Marin JM, Zalba B, Cabeza LF, Mehling H. Improvement of a thermal energy storage using plates with paraffin-graphite composite. Int J Heat Mass Transfer 2005;48:2561–70.
[770] Lafdi K, Mesalhy O, Shaikh S. The effect of surface energy on the heat transfer enhancement of paraffin wax/carbon foam composites. Carbon 2007;45:2188–94.
[771] Lafdi K, Mesalhy O, Elgafy A. Graphite foams infiltrated with phase change materials as alternative materials for space and terrestrial thermal energy storage applications. Carbon 2008;46:159–68.
[772] Xiao X, Zhang P. Morphologies and thermal characterization of paraffin/carbon foam composite phase change material. Sol Energy Mater Sol Cells 2013;117:451–61.
[773] Sanusi O, Warzoha R, Fleischer AS. Energy storage and solidification of paraffin phase change material embedded with graphite nanofibers. Int J Heat Mass Transfer 2011;54:4429–36.
[774] Weinstein RD, Kopec TC, Fleischer AS, D'Addio E, Bessel CA. The experimental exploration of embedding phase change materials with graphite nanofibers for the thermal management of electronics. J Heat Transfer 2008;130:042401.
[775] Ehid R, Weinstein R, Fleische A. The shape stabilization of paraffin phase change material to reduce graphite nanofiber settling during the phase change process. Energy Convers Manage 2012;57:60–7.
[776] Cui Y, Liu C, Hu S, Yu X. The experimental exploration of carbon nanofiber and carbon nanotube additives on thermal behavior of phase change materials. Solar Energy Mater Sol Cells 2011;95:1208–12.
[777] Karaipekli A, Sarı A, Kaygusuz K. Thermal conductivity improvement of stearic acid using expanded graphite and carbon fiber for energy storage applications. Renew Energy 2007;32:2201–10.
[778] Li M, Haung HB, Wang RZ, Wang LL, Cai WD, Yang WM. Experimental study on adsorbent of activated carbon with refrigerant of methanol and ethanol for solar ice maker. Renew Energy 2004;29:2235–44.
[779] El-Shrakawy II, Saha BB, Koyama S, He J, Ng KC, Yap C. Experimental investigation on activated carbon-ethanol pair for solar powered adsorption cooling application. Int J Refrig 2008;31:1407–13.
[780] Kil H-S, Kim T, Hata K, Ideta K, Ohba T, Kanoh H, Mochida I, Yoon S-H, Miyawaki J. Influence of surface functionalities on ethanol adsorption characteristics in activated carbons for adsorption heat pumps. Appl Therm Eng 2014;72(2):160–5.
[781] El-Sharkawy II UK, Miyazaki T, Saha BB, Koyama S, Kil H-S, Miyawaki J, Yoon S-H. Adsorption characteristics of ethanol onto functional activated carbons with controlled oxygen content. Appl Therm Eng 2014;72(2):211–6.
[782] El-Sharkawy II, Uddin K, Miyazaki T, Saha BB, Koyama S, Kil H-S, Yoon S-H, Miyawaki J. Adsorption of ethanol onto phenol resin based adsorbents for developing next generation cooling systems. Int J Heat Mass Transfer 2015;81:171–8.
[783] Pal A, Kil H-S, Mitra S, Thu K, Saha BB, Yoon S-H, Miyawaki J, Miyazaki T, Koyama S. Ethanol adsorption uptake and kinetics onto waste palm trunk and mangrove based activated carbons. Appl Therm Eng 2017;122:389–97.
[784] Pal A, Thu K, Mitra S, El-Sharkawy II, Saha BB, Kil H-S, Yoon S-H, Miyawaki J. Study on biomass derived activated carbons for adsorptive heat pump application. Int J Heat Mass Transfer 2017;110:7–19.
[785] Demir H, Mobedi M, Ulku S. A review on adsorption heat pump: problems and solutions. Renew Sust Energy Rev 2008;12:2381–403.
[786] Attan D, Alghoul MA, Saha BB, Assadeq J, Sopian K. The role of activated carbon fiber in adsorption cooling cycles. Renew Sust Energy Rev 2011;15:1708–21.

CHAPTER 4

Porous carbons for environment remediation

Porous carbons have high possibility and potentiality to contribute to environment remediation, which is recognized to be an important and urgent problem for life on Earth, as well as addressing energy problems, and energy storage and conversion, as explained in Chapter 3. For remediation of the environment, the coupling of a wide range of materials and various techniques is needed, and in addition, the procedures for remediation are strongly demanded to be low-cost, simple, and easy to scale-up, in addition to being environment friendly. Considering the energy and resources problems faced by the planet, these requirements of low-cost and simplicity have to be met for the production of materials, including raw materials and production processes.

Carbon materials are commonly believed to be able to be produced industrially at low cost. Practically, however, the selection of carbon precursors and the development of production processes for carbon materials with specific functions are not always low cost and so are urgently needed to be reconsidered. The carbon precursors and production processes for functional carbon materials have to be environment-friendly and so the use of biomass and/or wasted materials for the production of carbon materials is required.

In this chapter, therefore, it is attempted to collect as much research data as possible on the syntheses of carbon materials with functions specific for environment remediation by referring to their function data. This chapter consists of five sections, i.e., the removal of pollutants through the adsorption by carbon materials, gas separation by carbon membranes, capacitive deionization of water using carbon electrodes, electromagnetic interference shielding (microwave shielding) by carbon films, and sensing of low-concentration pollutants by carbon-sensing elements. On each application, the data obtained on porous carbons are explained on the basis of the classification employed in the previous chapters, such as activated carbons, templated carbon, etc., and in addition carbon nanotubes and

Table 4.1 Construction and target materials in this chapter.

Section		Target materials
4.1 Adsorption	4.1.1 Inorganic pollutants	SO_2, H_2S, NH_3, etc.
	4.1.2 Organic pollutants	VOCs, phenols, PAHs, dyes, etc.
	4.1.3 Water	Water vapor
	4.1.4 CO_2	CO_2 capture, supports
	4.1.5 Metal ions	Hg, U, Pb, Ni, Cu, Cd, Cr, etc.
	4.1.6 Oils	Heavy oils
4.2 Gas separation		CO_2/N_2, O_2/N_2, CO_2/CH_4, etc.
4.3 Capacitive deionization		NaCl, different metal salts
4.4 Electromagnetic interference shielding		Microwaves
4.5 Sensing	4.5.1Chemical sensors,	NH_3, NO, H_2O, warfare chemicals, etc
	4.5.2 Strain sensors	Tensile, gas pressure, finger motions, etc.
	4.5.3 Biosensors	Biomolecules, glucose, DNA, bacteria, etc.

carbon nanofibers together with graphene and its related materials are also described, because these materials can provide a large amount and different kinds of functional groups and structural defects and also have large surfaces associated with π-electron clouds. Some of the target materials for these application fields of carbon materials are listed in Table 4.1.

4.1 Adsorption

The phenomenon of "adsorption" is commonly observed in most application fields, in many of which adsorption plays important role, even governing the functionalities of materials. Adsorption onto the materials are often explained by two mechanisms: nonelectrostatic and electrostatic. The former is due to filling the adsorbate, either molecules or ions, into pores of the adsorbent, and so fitting the pore size of the adsorbent with the size of the adsorbate is important. The carbon materials have a strong advantage because they can provide a wide range of pore sizes, from macropores to ultra-micropores, and even various pore morphologies, including slit-shaped pores and channels, and their ordered arrangements. The latter mechanism is due to an electrostatic interaction between the surface of the adsorbent and the adsorbate. Most carbon materials retain various functional groups, basic and acidic groups, on their surface, of which the amount increases markedly with the presence of pores, and so various electrostatic interactions occur between the adsorbate and the adsorbent carbon, such as hydrogen bonding, lone-pair electron interaction, $\pi-\pi$ interaction, etc. Therefore, porous carbons contribute to adsorption by providing pores with various sizes and morphologies to accommodate various-sized adsorbates and also provide large amounts of functional groups to facilitate the electrostatic interaction with adsorbates. It has to be pointed out that reduced graphene oxide (rGO) can also keep large amounts of functional groups. Some fibrous carbons, particularly the carbons classified as nanofibers, can provide surfaces for keeping functional groups, and in addition their entanglement can create various-sized pores among the fibers. The presence of a π-electron cloud on the carbon materials also promotes the adsorption of various cations by cation$-\pi$ interaction and that of aromatic organics by $\pi-\pi$ interaction. For effective usage of π-electrons on the hexagonal carbon layers, single-layer graphene is the ultimate adsorbent, with mesoporous and/or macroporous graphene foam being a pathway to realize this situation.

In this section, the carbon materials applied for adsorption are explained by being divided on the bases of adsorbates, either inorganic or organic pollutants, and then on the specific adsorbates, water, CO_2, heavy metals, and oily materials.

4.1.1 Adsorptive removal of inorganic pollutants

For the removal of inorganic pollutants from air and water, adsorption coupled with oxidation of the adsorbates is applied. Here, carbon materials for the removal of SO_2, H_2S and NH_3 in air, and, in addition, for adsorptive removal of arsenic and phosphate ions from water, are the main focus.

4.1.1.1 Sulfur dioxide

Porous carbons are effective for the removal of sulfur dioxide (SO_2) in flue gases, of which the effectiveness is governed by their adsorption capacity of SO_2 and catalytic activity for SO_2 oxidation either to SO_3 or H_2SO_4.

Activated carbons (ACs) were proposed as the adsorbents for SO_2 in flue gas. Many efforts were exerted to achieve high adsorption capacity and rate using PAN-based activated carbon fibers (ACFs) [1,2]. Four commercially available PAN-based ACFs (FE-100, 200, 300, and 400) with a wide range of S_{BET} and V_{micro} were selected and SO_2 adsorption was measured in a flow reactor with a fixed bed of 0.5 g ACFs on the gas containing 1000 ppm SO_2, 5 vol% O_2, 10 vol% H_2O vapor, and balanced N_2 with the different flow rates [1]. The breakthrough curves measured using the ratio of ACF weight to gas flow rate (W/F) of 5×10^{-3} g min/cm^3 are compared on as-received ACFs in Fig. 4.1A, revealing that the FE-100 shows the longest breakthrough time as about 2.5 h, although it has the lowest S_{BET} of 450 m^2/g and V_{micro} of 0.17 cm^3/g. The effect of pre-heat treatment at different temperatures of 150—1000°C in N_2 is shown on one of the ACF (FE-300) using breakthrough curves measured at 100°C using a W/F ratio of 1×10^{-2} g min/cm^3 in Fig. 4.1B. The ACF (FE-300) pre-treated at 800°C delivers the highest breakthrough time of about 15 h, corresponding to the largest SO_2 adsorption capacity of 3.2 mmol/g, which is much higher than the 1.1 mmol/g for the 150°C-treated one. In Fig. 4.1C, the effect of temperature during the breakthrough measurement is shown on FE-300 pretreated at 800°C using a W/F ratio of 5×10^{-3} g min/cm^3. The measurement at 25°C shows the longest breakthrough time as 9.5 h, corresponding to an adsorption capacity of 5.1 mmol/g, while at 200°C the breakthrough time reduced to 4.3 h (4.3 mmol/g). A marked increase in adsorptive removal of SO_2 by ACFs was explained by control of the amount of oxygen-containing functional groups on the pore walls of the adsorbent ACFs. SO_2 adsorbed on the ACF surface could be oxidized, hydrated, and removed from the ACF surface as aqueous H_2SO_4 at an adsorption temperature of 30—100°C [4]. In Fig. 4.2, breakthrough curves of the FE-300 pre-heat-treated at 800°C measured under different conditions of SO_2 gas flow are shown, revealing that adsorptive removal of SO_2 can be continued up to 60 h at a low adsorption

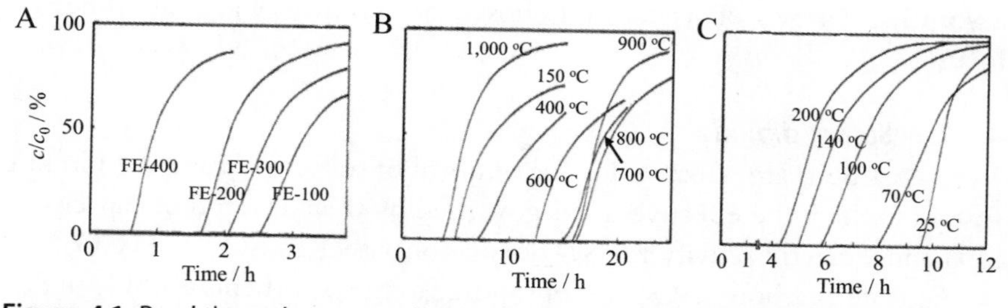

Figure 4.1 Breakthrough curves of PAN-based ACFs for SO_2 removal: (A) four commercial ACFs at 100°C, (B) effect of pretreatment temperature for FE-300 on the SO_2 adsorption at 100°C, and (C) effect of adsorption temperature for FE-300 after pre-treatment at 800°C [1].

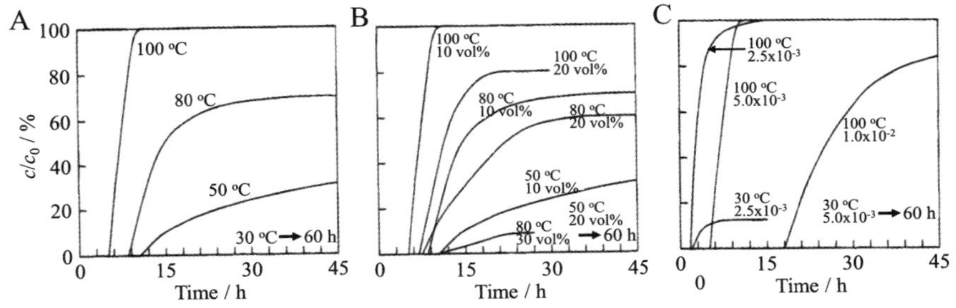

Figure 4.2 Breakthrough curves of the ACF (FE-300) after 800°C-pretreatment: (A) effect of adsorption temperature with 5 vol.% O_2, 10 vol.% H_2O, and W/F of 5.0×10^{-3} g min/cm^3, (B) effect of H_2O content at different temperatures with 5 vol.% O_2 and W/F of 5.0×10^{-3} g min/cm^3, and (C) effect of W/F at different temperatures with 5 vol.% O_2, 10 vol.% H_2O. SO_2 concentration in gas was 1000 ppm [4].

temperature of 30–50°C under the supply of 10–30 vol% H_2O vapor. When the steady removal of SO_2 was achieved, the formation of aqueous H_2SO_4 was observed in the reservoir which flowed out of the ACF bed. The process involves a series of reactions that lead to the formation of H_2SO_4 as the final product. To discuss the roles of surface functional groups for adsorptive removal of SO_2, temperature-programmed desorption (TPD) measurements were performed on FE-100, -200, and -300 after heat treatment at 600 and 800°C [3].

ACFs derived from other precursors, such as pitch, phenol resin, and cellulose, have been also studied [5–9]. Adsorptive removal of SO_2 through oxidation and hydration was performed on the commercial pitch-based ACFs under the control of various factors, including H_2O vapor supply, O_2 content in the gas, gas flow rate, adsorption temperature, and pre-heat-treatment temperature, etc. [5,6,8]. The capacity of the pitch-based, phenol-resin (PF)-based, and coal-based ACFs for adsorptive removal of SO_2 was shown to be enhanced by N-doping, which was carried out by heating the mixture of the pristine ACFs with either urea, NH_3, dicyanodiamine (DCD), or *N,N*-dimethylformamide (DMF) at 300°C in an autoclave, where the N content in ACFs could be controlled in a range of 0.4–4.6 wt.% [9]. In Fig. 4.3A, SO_2 adsorption capacity measured in the presence of excess of O_2 (but no H_2O) is plotted against N content, where the capacity is normalized by S_{BET}, revealing that the capacity increases with increasing N content for three series of ACFs. In Fig. 4.3B, the relative amount of SO_2 adsorbed, which is expressed by $SO_3/(SO_3 + SO_2)$ because the adsorbed SO_2 is reasonably supposed to be oxidized to SO_3, is plotted

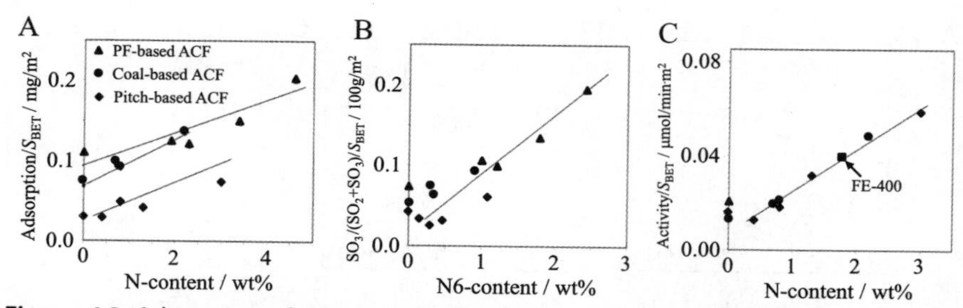

Figure 4.3 Adsorption of SO_2 normalized by S_{BET} on N-doped ACFs: (A) adsorption capacity for SO_2 against N content, (B) oxidation to SO_3 against the N content with pyridinic configuration (N6), and (C) catalytic activity for the oxidation of SO_2 to SO_3 against N content [9].

against the N content located in a pyridinic configuration (N6). The results of three series of ACFs can be fitted to the same linear trend, suggesting an important role of pyridinic N species for the enhancement of SO_2 oxidation to SO_3. Fig. 4.3C presents catalytic activity after being normalized by S_{BET} plotted against N content for the three ACFs, including the result on PAN-based ACF (FE-400), suggesting the common trend to increase the catalytic activity with an increase in N content. On pitch-based ACFs, TPD measurements were applied to evaluate surface functional groups and to discuss the mechanism of SO_2 adsorption and oxidation to SO_3 [10].

The effect of surface modification of ACFs by oxidation and annealing on SO_2 adsorption, oxidation, and conversion to H_2SO_4 was studied using three PF-based ACFs with different burn-off's due to steam/CO_2 activation [11]. For the as-received ACFs, the initial rate of SO_2 adsorption from a model flue gas, containing 5 vol% O_2 and 7 vol% H_2O, depended inversely on the pore size of the ACFs and the saturated amount of SO_2 adsorbed was dependent on both the pore size and pore volume. Oxidation of the ACFs in a nitric/sulfuric acid at room temperature made their adsorption capacity for SO_2 decrease mainly due to a decrease in pore volume and repulsion of the SO_2 from acidic functional groups on the carbon surfaces. By annealing the oxidized ACFs at a temperature of 400−1000°C in N_2, the adsorption capacity increased, which was directly correlated with the amount of CO_2 evolved during annealing of the oxidized ACFs. On cellulose-based ACFs, a correspondence between SO_2 adsorption capacity and CO evolution during calcination at 1000°C was observed [7]. ACFs prepared from a PVC-derived pitch were experimentally shown to have comparable SO_2 adsorption performances to other ACFs [12].

ACs were prepared from a bituminous coal by carbonization at 900°C in N_2, steam activation at 800°C and then treatment in 10 M HNO_3 at 80°C, followed by annealing at different temperatures of 200–925°C to control the amount of oxygen-containing functional groups formed during HNO_3 treatment, on which the process of SO_2 adsorption, oxidation. and hydration was studied [13]. The ACs were also prepared by KOH activation at 800°C in N_2 for comparison. The rate-determining step was pointed out to be the oxidation of SO_2 to SO_3, where three reactions are possible, i.e., the reactions of the adsorbed SO_2 (C–SO_2) with either gas-phase O_2 or with adsorbed oxygen (C–O complex) and the direct reaction of gas-phase SO_2 with C–O complex on the carbon. TPD analyses on the adsorbent carbons suggested the importance of the C–O complex on the surface.

4.1.1.2 Hydrogen sulfide

Adsorptive removal of hydrogen sulfide (H_2S) is also required to keep the environment atmosphere clean, for which ACs are effective adsorbents, as well as for adsorptive removal of SO_2 as described above.

The H_2S adsorption capacity was determined from the breakthrough curves measured on a column of carbon adsorbents by passing a flow of wet air (relative humidity of 80% at 25°C) containing 0.3% (3000 ppm) H_2S with a flow rate of 0.5 L/min at room temperature on two commercial ACs, a peat-based pellet-shaped carbon (RB3) and a coconut shell-based granular carbon (S208c) [14]. H_2S adsorption capacities were 96 and 112 mg/g, respectively, and marked decreases in capacity by preoxidation with 15 M HNO_3 and $(NH_4)_2S_2O_8$ were observed. Adsorption of H_2S at room temperature was studied on commercial ACs, five coal-based ACs with KOH activation and three wood-based ones with H_3PO_4 activation [15]. In the low P/P_0 range (<0.02), the adsorption capacity of ACs for H_2S was independent of the pore structure of ACs, as shown in Fig. 4.4A and B. ACs having lower S_{BET} tended to adsorb more H_2S than those with higher S_{BET}, which was attributed to the presence of smaller micropores and a greater number of functional groups acting as adsorption sites on the surface of AC. KOH activation is more efficient in increasing H_2S adsorption capacity than H_3PO_4 activation, although the latter is more efficient for micropore development, with high S_{BET} and high V_{micro}. For NH_3, the same tendency as H_2S was observed on the same ACs.

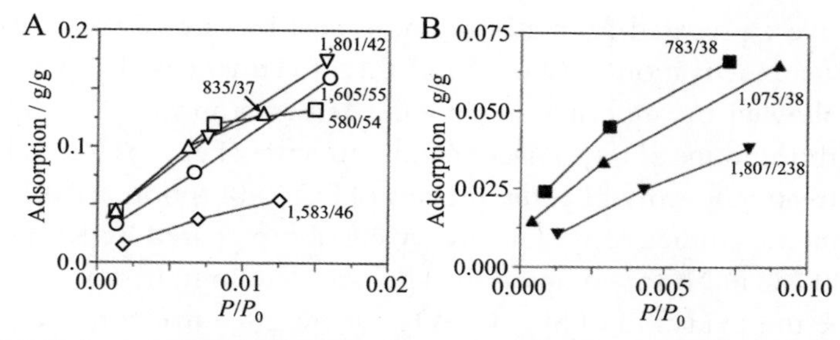

Figure 4.4 H_2S adsorption isotherms of commercial ACs at low P/P_0 range: (A) KOH-activated and (B) H_3PO_4-activated ACs. Each AC is identified by S_{BET}/V_{micro} (in units of m^2/g and cm^3/g, respectively) [15].

Breakthrough curves were measured in a flow of wet air (relative humidity of 80%) containing 3000 ppm H_2S with a 0.5 L/min rate on commercial wood-derived ACs (H_3PO_4-activated) denoted as W1, W2, and W3 [16,17]. Their pore structure parameters, S_{BET} and V_{micro}, are listed together with the surface parameters and H_2S adsorption capacity calculated from the breakthrough curve in Table 4.2, revealing a marked difference in H_2S capacity although the pore and surface parameters are not as different [16]. In Fig. 4.5A–C, DTG curves are compared on three adsorbents (the pristine, exhausted, and leached carbons) for each AC, W1, W2, and W3, respectively. The exhausted carbon adsorbent was obtained after the breakthrough measurement up to an H_2S concentration of 500 ppm and the leached adsorbent was obtained after leaching of the exhausted adsorbents in distilled water for 24 h. DTG profiles of the exhausted adsorbents are quite different on the ACs. The adsorbents W1 and W3 (Fig. 4.5A and C, respectively) exhibit a marked weight loss in the temperature range of 400–700K (about 21% and 15%, respectively), which is thought to be caused by the evaporation of elemental sulfur and the decomposition of adsorbed SO_3 and SO_2 formed by oxidation of H_2S adsorbed. For W2 (Fig. 4.5B), on the contrary, weight loss in this temperature range is negligibly small, at only 1%. By leaching, the exhausted adsorbent gives a marked decrease in weight loss during heating on W1, but a very small change is observed on W3 and almost no change on W2. From the change in DTG profile from the exhausted to leached adsorbents, the weight loss above 800K is thought to be mainly due to vaporization of water-soluble SO_2, which cannot be leached out probably because it is located in the micropores. In the exhausted adsorbents derived from W1 and W2, most SO_2 are leachable, causing differences in weight loss between

Table 4.2 Pore and surface parameters, H_2S adsorption capacity and S-states in the exhausted and leached adsorbents for three commercial ACs. [16,17].

	Pore parameters		Surface parameters				State of S		
	S_{BET} (m^2/g)	V_{micro} (cm^3/g)	pH	Acidic (meq/g)	Basic (meq/g)	H_2S capacity[a]	S^b	$SO_2{}^b$	$SO_2{}^c$
W1	1400	0.561	4.41	1.113	0.125	295	20.6	4.3	1.4
W2	1025	0.359	4.04	1.251	0.100	17	1.6	0.2	0.2
W3	1110	0.410	5.61	1.038	0.325	230	17.9	2.4	0.8

[a]calculated from breakthrough curves,
[b]in the exhausted adsorbents,
[c]in the leached adsorbent.

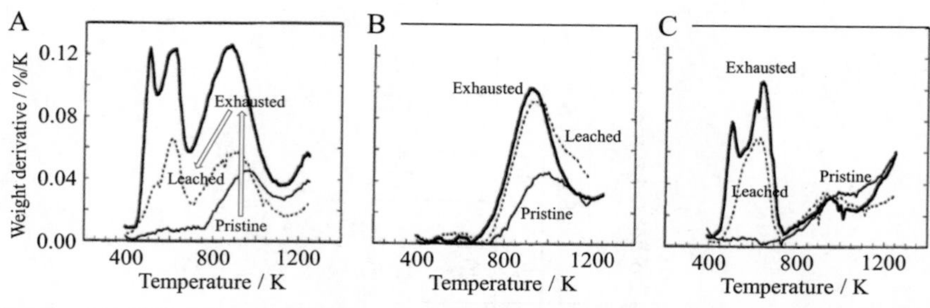

Figure 4.5 DTG curves in N_2 for the pristine, exhausted, and leached adsorbents: (A) adsorbent W1, (B) W2, and (C) W3 [16].

exhausted and leached adsorbents. The H_2S adsorption capacities of these adsorbents (W1, W2, and W3) were measured to be 2.5, 4.5, and 6.5 mg/g, respectively [17]. The experimental results of the adsorption and oxidation of H_2S on various commercial ACs were reviewed [18]. N-doping into W1, which was performed by heating the mixture with urea at 450 and 950°C, was found to enhance the conversion of H_2S to water-soluble species significantly, although no marked effect was observed on adsorption capacity [19]. The enhancement was presumed to be due to high dispersion of basic N-containing functional groups in the carbon adsorbent, where the oxidation of H_2S to sulfur radical resulted in the formation of sulfur oxides and sulfuric acid.

The facilities for sewage sludge treatment and sanitation are important units of city structure and have high responsibility for avoiding the emission of polluting gases, of which H_2S is one. Therefore, there have been great efforts devoted to preparing ACs from sewage sludge which is applicable to the adsorptive removal of H_2S [20–27]. Sewage-sludge-derived fertilizers consisting of granules with 3 mm diameter and about 60% organics were carbonized at different temperatures of 400–900°C in N_2 [21]. The resultant carbons were washed with 18% HCl to eliminate metallic impurities from the precursor sludges. On the sludge-derived carbons after prehumidification in 80% relative humidity, the breakthrough curves were measured at room temperature in wet air (relative humidity 80%) containing 3000 ppm H_2S with a 0.5 L/min flow rate. The breakthrough curves observed are shown in Fig. 4.6A and B. With increasing carbonization temperature, the breakthrough time of the carbon increases significantly. The H_2S adsorption capacity calculated from the breakthrough curve reached 82.6 mg/g for the as-carbonized sludge at 950°C, which adsorbed about 62 mg/g water at the prehumidification stage. After HCl

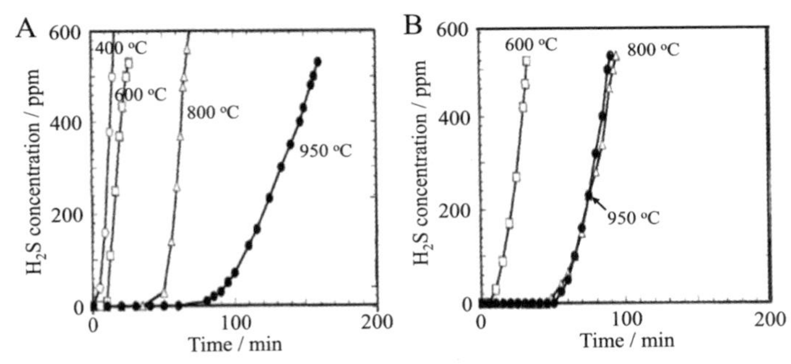

Figure 4.6 H_2S breakthrough curves of the sludge-derived ACs: (A) ACs as-carbonized at different temperatures and (B) ACs washed by HCl after carbonization [21].

washing, the H_2S capacity of the carbon prepared at 950°C was slightly smaller at 57.5 mg/g, and adsorbed 100 mg/g water at prehumidification. These observed capacities were much higher than the AC derived from coconut shells using the same procedure (48.8 mg/g). The carbons prepared from the same fertilizers by carbonization at 600, 800, and 950°C in N_2 after impregnation of a car oil revealed a marked improvement in adsorption capacity [24]. The adsorption capacity for H_2S reached 115 mg/g by carbonization at 950°C after oil impregnation, although there were no significant differences in pore structure parameters.

ACs were prepared from mixtures of wasted paper with dried sludge and KOH in different weight ratios (paper/sludge/KOH) by carbonization at 950°C, of which H_2S breakthrough curves were measured in a flow of wet air (relative humidity of 80% at 25°C) containing 3000 ppm H_2S at a 0.5 L/min rate [28]. As compared with the H_2S breakthrough curves in Fig. 4.7, the carbon obtained at a mixing ratio of 58/38/4 exhibits a high

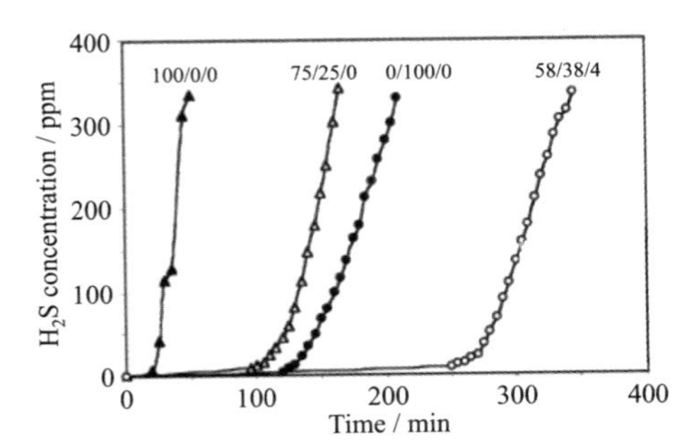

Figure 4.7 H_2S breakthrough curves of ACs prepared from the mixture of wasted paper/dried sludge/KOH (by weight) by carbonization at 950°C [28].

adsorption capacity of 351 mg/g, which is much higher than the ACs prepared from wasted paper and sludge (shown as 100/0/0 and 0/100/0, respectively). The experimental results were discussed on the basis of the composition and arrangement of inorganic impurities, which were evaluated with the ash content reaching up to 70%. The composites of AC with a fibrous silicate, sepiolite, were proposed as adsorbents for H_2S and NH_3, in which sepiolite worked as a binder for AC granules [29].

Carbon fiber composites with molecular sieving function were synthesized from commercially available isotropic pitch-based carbon fibers mixed with a phenolic resin by making water slurry, molding under vacuum, curing at 150°C, and carbonizing at 650°C in an inert gas, followed by activation in moisture-saturated He at a temperature of 800−950°C, in which the burn-off was controlled in a range of 0%−45% [30]. H_2S adsorption capacity calculated from breakthrough curve in H_2S/He mixed gas was 431 mg/g for the composite with 18% burn-off. The composite had also a relatively high CO_2 adsorption capacity of 910 mg/g. Activated carbon fibers (ACFs) loaded by K_2CO_3 nanoparticles were fabricated by centrifugal spinning of a coal tar pitch/KOH mixture and stabilization in air/CO_2 mixed gas at 330°C, where the oxidative stabilization of pitch and the conversion of KOH to K_2CO_3 occurred simultaneously, followed by carbonization and activation at 850°C in CO_2 flow [31]. The resultant ACF possessed S_{BET} of 491 m^2/g and contained about 2 wt.% K as K_2CO_3 nanoparticles at the peripheral region of the fiber. This K_2CO_3-loaded ACF demonstrated a high adsorptive removal efficiency for H_2S. In a flow of humid air (relative humidity of 50% at 25°C) containing 30 ppm H_2S, the removal ratio was kept at 100% even after 800 min, whereas the ACF fabricated without KOH exhibited a removal ratio that decreased down to 30% after 800 min. The adsorbed H_2S was oxidized to elemental S and deposited around the peripheral region of the fiber.

4.1.1.3 Ammonia

Ammonia (NH_3) is widely used as a chemical in industries but its release into atmosphere has to be controlled at less than 25 ppm from both the environmental and human health standpoints. ACs have been used to reduce the release of NH_3 in air, while enhanced adsorption capacity and strength of retention are still demanded. For adsorption of NH_3 into carbons, functional groups and structural defects on the adsorbent surface were demonstrated to play important roles by using bundles of single-walled carbon nanotubes (SWCNTs) [32,33]. Through analyses of NH_3

adsorption isotherms at different temperatures, it was shown that the adsorption of NH_3 by microporous carbons consisted of two mechanisms: interaction with oxygen-containing functional groups and micropore filling [34]. The NH_3 adsorption was experimentally shown to consist of reversible and irreversible components; the former was based on physisorption and the latter on the interaction with functional groups [35].

ACs with lower S_{BET} tended to adsorb more NH_3 than those with higher S_{BET}, the same tendency as for H_2S adsorption described above, which was attributed to the presence of a greater number of functional groups acting as adsorption sites on the surfaces of ACs [15]. The effect of functional groups on the carbon adsorbents was studied by modifying the surface with different processes [36]. A commercial AC was treated with 30 wt.% HCl and 30 wt.% NaOH followed by washing in boiling acetone at 80°C and drying at 85°C. The treatment with HCl slightly enhanced NH_3 adsorption at a low P/P_0 range, while that with NaOH was depressed slightly, although both treatments enhanced CO_2 adsorption. NH_3 adsorption capacity was determined on porous carbons derived from palm shell [37]. The porous carbons were prepared from palm shells with particle sizes of 1−2 mm by heating at 300−700°C for 2 h in N_2 flow after impregnation of H_2SO_4 solutions with different concentrations (5%−40%) (H_2SO_4 activation). For comparison, the palm shells were carbonized at 300−700°C and then activated in CO_2 flow at 500−900°C (CO_2 activation). The sorption capacities for NH_3 at different temperatures (30−100°C) are compared on these two series of ACs in Fig. 4.8A, revealing that the H_2SO_4-activated carbons give higher capacity than

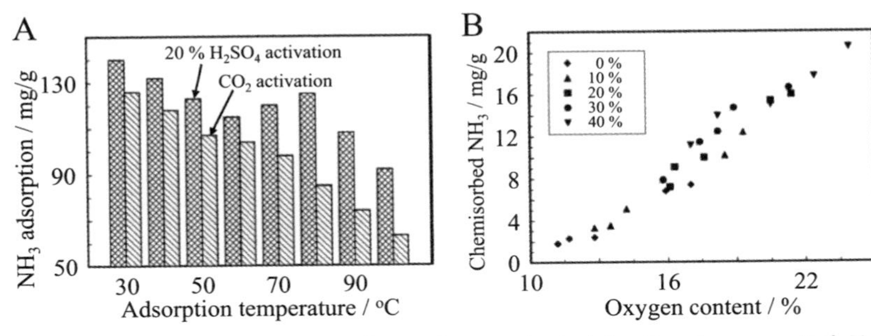

Figure 4.8 NH_3 adsorption into the carbon activated by the impregnated H_2SO_4: (A) change in adsorption capacity with adsorption temperature, in comparison with CO_2-activated carbons, and (B) relation of the NH_3 adsorbed by chemisorption to O content in the carbon adsorbents activated in H_2SO_4 solutions with different concentrations [37].

CO_2-activated ones. The amount of chemisorbed NH_3 is estimated from the amount of NH_3 desorbed at temperatures above 200°C and plotted against oxygen content in the carbon adsorbents activated using different concentrations of H_2SO_4 in Fig. 4.8B, suggesting the contribution of chemisorption by a strong interaction of oxygen-containing functional groups with NH_3.

Two series of porous carbons were prepared from poly(4-styrene sulfonic acid co-maleic acid) sodium salt and poly(sodium 4-styrene sulfonate) by carbonization at 800°C in N_2 (denoted as C-1 and C-2, respectively) [38]. The carbons were subjected to two oxidation treatments by a saturated solution of ammonium persulfate $(NH_4)_2S_2O_8$ in 1 M H_2SO_4, resulting in carbon being denoted as C-1A and C-2A, and by heating at 350°C in air (C-1B and C-2B). These carbons were prehumidified in an air flow with 70% humidity for 2 h (denoted as C-1P, C-1AP, and C-1BP, etc.). On the carbons either without or with prehumidification, adsorptive removal and desorption of NH_3 were studied by dynamic measurements using the air flow containing 1000 ppm NH_3 without or with 70% humidity (dry NH_3 or wet NH_3, respectively). Breakthrough curves were measured by stopping the flow of NH_3-containing dry and wet air when NH_3 concentration reached 100 ppm and then changing immediately to dry air purging to test the desorption for the carbons C-1, C-1A, and C-1B, as shown in Fig. 4.9A–C, respectively. Differences in the behavior of sorbent C-1 by its treatments are observed in the breakthrough time and desorption rate. The longest breakthrough time, i.e., the highest adsorption capacity, is obtained by the carbon preoxidized with $(NH_4)_2S_2O_8$ (C-1A), while the desorption is the slowest. On the carbon C-2, very similar results

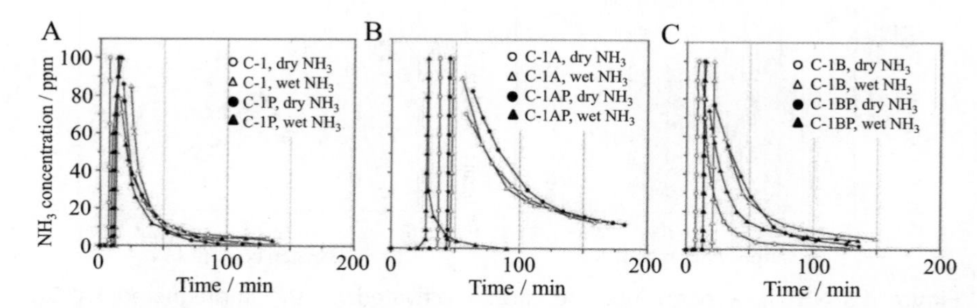

Figure 4.9 Breakthrough curves of the carbon C-1 for dry and wet NH_3: (A) as-prepared, (B) treated in $(NH_4)_2S_2O_8$ and (C) heated at 350°C in air with or without pretreatment in 70% humidity for 2 h (P). The sample notaions are referred to the text [38].

to the adsorption and desorption curves were obtained. On C-1A and C-2A, a marked gap is observed between the adsorption and desorption curves, as shown for C-1A in Fig. 4.9B, which is caused by the fact that the NH_3 concentration is higher than 100 ppm after the breakthrough, indicating a relatively high retention of NH_3 in the adsorbent carbon.

Metal-loading on ACs was shown to be effective in enhancing the NH_3 adsorption [39]. Commercial ACs derived from various precursors were loaded by transition metals, Fe, Co, and Cr, via impregnation into the solutions of their acetates at pH 3 and then calcination at 300°C. In Fig. 4.10A and B, the effect of metal loading on two ACs, coconut-shell-derived (S208) and bituminous-coal-derived (V612) ACs, is shown by breakthrough curves for NH_3. On the former, Cr-loading enhances NH_3 adsorption markedly (Fig. 4.10A) with saturated adsorption capacity increasing to 4.5 mg/g from 0.4 mg/g for the pristine AC. Fe- and Co-loading on this AC result in only a slight change in adsorption capacities. On the latter, modifications by Fe, Co, and Cr have similar improvements in breakthrough time, with Cr giving the longest breakthrough time and the largest capacity of 4.1 mg/g (Fig. 4.10B). On a wood-derived AC (BAX-1500), a marked improvement of adsorption capacity by Fe- and Cr-loading, 19.0 and 14.5 mg/g, respectively, but air oxidation at 300°C gave a more marked improvement reaching 26.7 mg/g, although the pristine AC had 3.9 mg/g capacity. ACs prepared from sewage sludge by carbonization at 700°C and subsequent activation by either KOH or NaOH at 700°C, which contained a relatively high Fe content of 11 mg/g, delivered an NH_3 adsorption capacity of about 12—16 mg/g [40].

Impregnation of metal chlorides, $CuCl_2$, $NiCl_2$, and $ZnCl_2$, through the respective solution onto a commercial AC (BPL) was reported to give marked improvements in NH_3 adsorption. The breakthrough curve for the

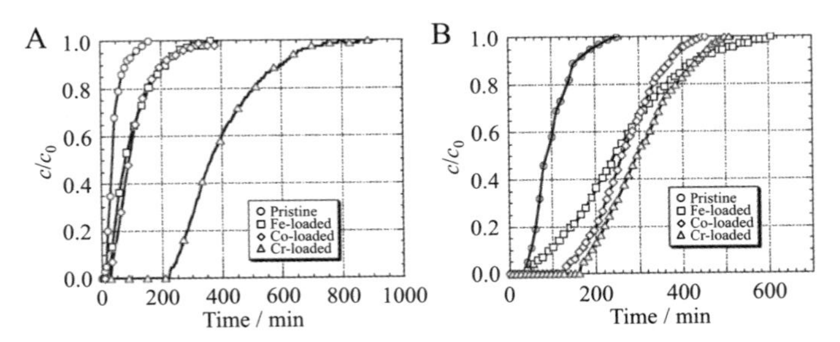

Figure 4.10 Breakthrough curves of metal-loaded ACs for NH_3 adsorption: (A) coconut-shell-derived AC and (B) coal-derived AC [39].

AC bed was measured using a flow of NH_3 gas diluted either in dry (Dry) or wet air (70% humidity) (Wet) with 1000 ppm NH_3 concentration and inlet flow rate of 900 mL/min [41]. The measurements were also done on the ACs with and without prehumidification in 70% humidity (PW and PD, respectively). In Fig. 4.11A, breakthrough curves for the pristine AC with and without prehumidification measured in dry and wet NH_3 flows are compared, revealing the fast interaction of NH_3 with the adsorbent surface. With the introduction of moisture in the air flow to the system, there was a marked increase in breakthrough time, i.e., a marked enhancement in adsorption capacity on the carbons. The loading of metal chlorides onto the ACs also enhanced NH_3 adsorption markedly, with adsorption capacity being 67.7 mg/g for 19.2 wt.% $CuCl_2$-, 58.9 mg/g for16.9 wt.% $NiCl_2$, and 51.7 mg/g for 20.4 wt.% $ZnCl_2$-loaded ACs. Breakthrough curves for the $ZnCl_2$-loaded AC are shown in Fig. 4.11B, showing a marked increase in breakthrough time with the introduction of moisture to the atmospheric gas. The prehumidification slightly enhances the NH_3 adsorption capacity (Fig. 4.11B), while the effect is not as effective as that on the pristine AC (Fig. 4.11A). The desorption process of adsorbed NH_3 in a flow of dry air with a 720 mL/min rate was followed by differential thermal gravimetry (DTG). The results on $ZnCl_2$-loaded ACs after NH_3 adsorption in different conditions are shown in Fig. 4.11C, revealing two-step desorption at 50–130°C and 500–700°C. The first desorption step at low temperature, principal in total desorption, is attributed to the release of weakly adsorbed NH_3 due to either hydrogen bonding, formation of salts with carboxylic groups on the surface, or just van der Waals

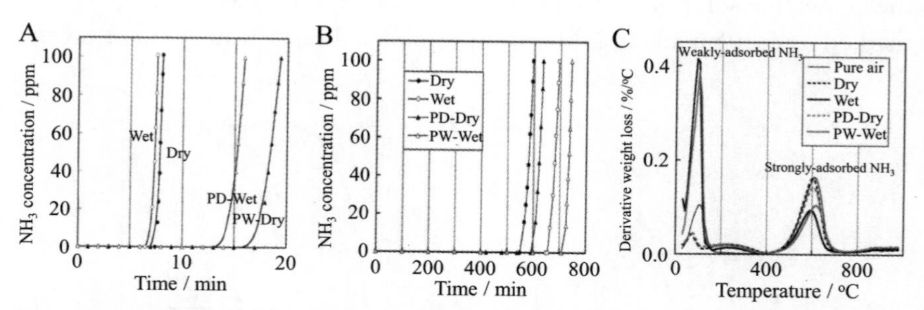

Figure 4.11 NH_3 adsorption of $ZnCl_2$-loaded AC without or with prehumidification (PD or PW, respectively) in dry and wet NH_3 flow (Dry and Wet, respectively): (A) breakthrough curves for the pristine AC, (B) those for $ZnCl_2$-loaded AC, and (C) DTG curves for $ZnCl_2$-loaded AC after NH_3 adsorption [41].

interactions in small pores. The second desorption step at high temperature is attributed to strongly interacted NH_3 due to complex formation with metal ions. Wood-derived ACs (BAX-1500 and -300) were modified by aluminum—zirconium polycations through impregnation into a solution of $Al_{1.2}Zr_{0.3}Cl$ salt, followed by calcination at 300°C, of which the breakthrough curves for NH_3 were measured at room temperature [42]. The adsorption capacity of the modified ACs was reduced in a dry gas stream but increased in a wet gas stream. In a dry gas stream, only Lewis acidic centers associated with Al_2O_3 and ZrO_2, which were formed during the calcination process, were supposed to form weak interactions with NH_3. In a wet gas stream, however, coexisted Brønsted acidic centers provided strong interactions with NH_3. Loading of W_2O_3 and MoO_2 onto BAX-1500 was also effective in increasing NH_3 adsorption capacity in the wet gas stream, where NH_3 was thought to be weakly adsorbed [43].

The adsorption and oxidation of NH_3 on carbon were studied by FTIR spectroscopy using carbon films (about 0.01 mm thick) [44]. Carbon film was prepared by carbonization of cellulose film at 600°C under dynamic vacuum after pyrolysis at 300°C in air. The carbon films prepared were outgassed at 600°C before NH_3 adsorption experiments. The decomposition ratio of NH_3 and formation ratios of oxidation products (N_2, N_2O, and NO) are plotted against reaction temperature in Fig. 4.12A and B, respectively, on the carbon films as-prepared and oxidized at 300°C in 100 kPa O_2 for 1 h. Oxidation of NH_3 on both carbon films starts at around 150°C. At high temperatures, however, the oxidation product on the as-prepared carbon film is predominantly N_2, but those on the oxidized carbon film are N_2 and N_2O, suggesting that the formations of N_2, N_2O, and NO by the oxidation of adsorbed NH_3 are determined by the oxygen-containing functional groups on the adsorbent carbon.

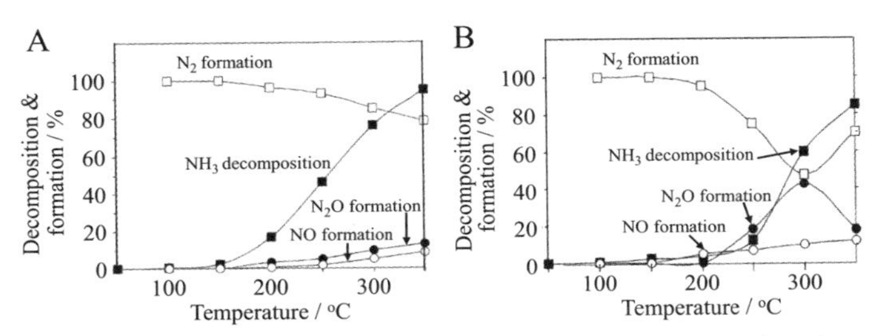

Figure 4.12 NH_3 oxidation on the carbon films: (A) the as-prepared carbon film and (B) the oxidized carbon film [44].

4.1.1.4 Arsenic ions

Arsenic (As), a flagrant metalloid element, has received much attention due to its great toxicity to human health. The World Health Organization (WHO) recommended its content in drinking water being less than 10 μg/L. Literatures reports on the removal of arsenic from water were summarized in 2009 [45]. Since As^{3+} is more toxic and more difficult to remove than As^{5+} although they always coexist in water, most of the works on their removal are proposed to load catalysts for the oxidation of As^{3+} onto the adsorbent to be adsorbed as As^{5+}. In most cases metal oxides worked as catalyst and carbons as adsorbent.

Loading of nanoparticles of $ZrO(OH)_2$ on rGO flakes was found to be effective for simultaneous removal of As^{3+} and As^{5+} from their aqueous solutions [46]. The composites $ZrO(OH)_2$/rGO were prepared by mixing GO suspension with $ZrOCl_2$ in different weight ratios ($ZrO(OH)_2$/GO of 40/1, 100/1, and 150/1 by weight) and diamine hydrate at 90°C under stirring, followed by freeze-drying in a vacuum. Adsorption isotherms of the composites for As^{3+} and As^{5+} were measured at 25.5°C and pH of 7.0 using aqueous solutions of $NaAsO_2$ and Na_2HAsO_4, respectively. The isotherms observed for As^{3+} and As^{5+} are shown in Fig. 4.13A and B, respectively. The composite prepared from a 100/1 mixture has the highest adsorption capacity, of which the maximum adsorption capacities are calculated by Langmuir equation to be 95.2 and 84.9 mg/g, respectively. Leaching of zirconyl ions from the absorbents was negligible and the composites were stable during the arsenic adsorption process. By using this composite (100/1), simultaneous removal of As^{3+} and As^{5+} was possible. The highest total adsorption capacity of 80.1 mg/g (composed of 55.1 and 25.0 mg/g for As^{3+} and As^{5+}, respectively) was obtained from the solution containing equal amounts of two ions. The mechanism for adsorption was

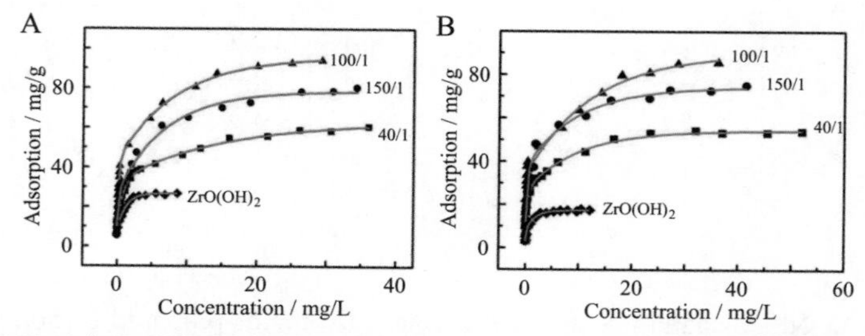

Figure 4.13 Adsorption isotherms of $ZrO(OH)_2$/rGO composites with different mixing ratios: (A) for As^{3+} and (B) for As^{5+} [46].

thought to be due to electrostatic attraction, the negatively charged arsenite AsO_2^- and arsenate $HAsO_4^{2-}$ species being adsorbed on the positive sites of adsorbents in acidic solution. Coloading of $ZrO(OH)_2$ and MnO_2 through hydrothermal coprecipitation at 180°C onto a commercial AC enhanced the sorption of As^{3+} and As^{5+} [47]. The maximum adsorption capacities for As^{3+} and As^{5+} were calculated by Langmuir approximation as 109.3 and 75.4 mg/g at 25°C and 132.3 and 95.6 mg/g at 45°C, respectively.

ZrO_2-embedded carbon nanofibers were prepared from the mixture of polyvinylpyrrolidone (PVP) with $ZrOCl_2$ in DMF solution by electrospinning, followed by carbonization at 600°C in N_2 [48]. Adsorption isotherms for As^{3+} and As^{5+} were measured using aqueous solutions of $NaAsO_2$ and Na_2HAsO_4, respectively, at different temperatures between 20–40°C. ZrO_2 particles on the resultant composite were distributed on the surface of CNFs and had diameters of about 250 nm. Adsorption isotherms for As^{3+} and As^{5+} measured at 20, 30, and 40°C are shown in Fig. 4.14A and B, respectively. The adsorption capacities of the composites for As^{3+} and As^{5+} increase rapidly in the low concentration range and then gradually reach the maximum adsorption capacity which is higher at higher temperatures. The maximum adsorption capacities for As^{3+} and As^{5+} are calculated as 26.8 and 64.5 mg/g at 20°C and 28.6 and 106.6 mg/g at 40°C, respectively. Adsorption capacities depend greatly on the pH of the solution, as shown in Fig. 4.14C, with high capacities being kept in a wide pH range of 3–10.

Coloading of Fe_3O_4 with MnO_2 onto rGO was performed in two steps: first, Fe_3O_4 was loaded onto GO flakes by adding a solution of $FeCl_3$ with $FeSO_4$ and HCl into GO suspension slowly, followed by the addition of ammonia solution quickly, and second, MnO_2 was loaded onto Fe_3O_4-loaded GO by adding $MnSO_4$ with $KMnO_4$, followed by the addition of

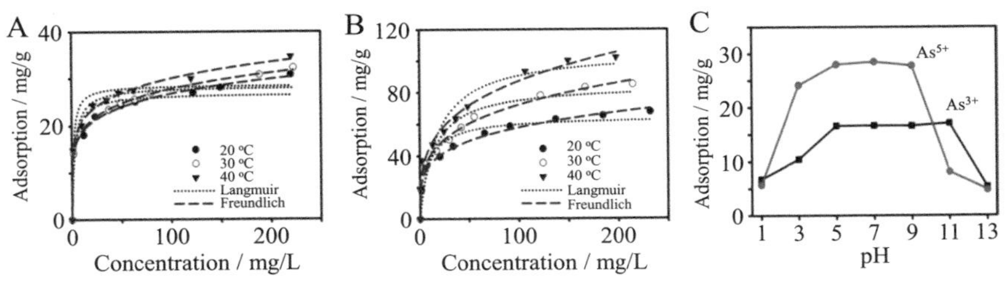

Figure 4.14 Adsorption of As^{3+} and As^{5+} on ZrO_2-embedded CNFs: (A) adsorption isotherms for As^{3+}, (B) those for As^{5+} at pH 6.0, and (C) effect of pH on adsorption capacities [48].

KOH [49]. In $Fe_3O_4-MnO_2$/rGO composites obtained as precipitates, the mass ratios of MnO_2 to rGO were controlled at 3/4, 3/8, and 3/12. The adsorption of As^{3+} and As^{5+} was measured on these composites at 25.5°C and pH 7.0. For As^{3+} and As^{5+}, the adsorption proceeded rapidly within the first 15 min, and increased slowly to reach equilibrium within 40 and 55 min, respectively, meanwhile the adsorption capacity for As^{3+} was higher than that of As^{5+}. The composite prepared using an MnO_2/rGO mass ratio of 3/8 gave the highest capacities for both As^{3+} and As^{5+}, the calculated maximum adsorption capacities being 14.0 and 12.2 mg/g, respectively. Loading of Fe_3O_4 onto rGO was also proposed; the calculated maximum capacity was 10.2 mg/g for As^{3+} and 5.3 mg/g for As^{5+} [50]. The adsorption capacities for both ions decreased slightly with an increase of up to pH 10 and decreased sharply above 10. Loading of the compounds of Fe and Mn oxides, $FeMnO_x$, onto rGO was also performed by mixing $KMnO_4$ and $FeSO_4$ into the rGO suspension, controlling pH in the range of 7.0–8.0 [51]. In the rGO dispersion, a stabilizer, either cetyltrimethylammonium bromide (CTAB), starch, or sodium carboxymethyl cellulose (CMC), was added. The maximum adsorption capacities calculated from adsorption isotherms observed at 25°C and pH 7.0 are summarized in Table 4.3. The loading of the compound $FeMnO_x$ is much more effective in enhancing adsorption capacities for As^{3+} and As^{5+} than the loading of Fe_3O_4 and MnO_2 separately. The addition of a stabilizer is also effective, particularly starch-stabilized composite delivering high adsorption capacities for both ions. A removal mechanism for two ions proposed on a stabilized $FeMnO_x$/rGO composite is shown in Fig. 4.15, where As^{3+} species are oxidized into As^{5+} by MnO_2, associated with the change in valences of Mn, and adsorbed by iron oxides Fe_xO_y. rGO flakes may work to inhibit the aggregation of $FeMnO_x$ particles, as well as to make the adsorption of As ions and the valence change in MnO_2 easier. The effects of the initial pH and coexisted anions, such as SO_4^{2-}, HCO_3^-,

Table 4.3 Maximum adsorption capacities (mg/g) of the composite loaded by $FeMnO_x$ for As^{3+} and As^{5+} which were calculated by Langmuir approximation of the isotherms [51].

	$FeMnO_x$	$FeMnO_x$/rGO	CTAB-$FeMnO_x$/rGO	Starch-$FeMnO_x$/rGO	CMC-$FeMnO_x$/rGO	$Fe_3O_4-MnO_2$/rGO[a]
As^{3+}	39.68	51.55	62.89	78.74	47.85	14.04
As^{5+}	23.26	50.00	41.84	55.56	18.94	12.22

[a]Data for the composite $Fe_3O_4-MnO_2$/rGO (3/8) from Reference [49].

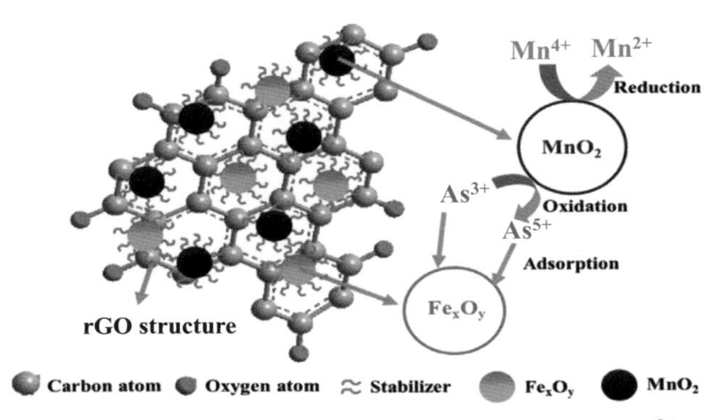

Figure 4.15 Mechanism proposed on the adsorptive removal of As^{3+} and As^{5+} by $FeMnO_x$/rGO composite [51].

and PO_4^{3-}, were also investigated. Over 90% of As^{3+} and As^{5+} could be removed by starch-stabilized $FeMnO_x$/rGO composite during the five consecutive adsorption−desorption cycles.

Nanoparticles of zero-valent Fe, Fe°, were reported to deliver relatively high adsorption capacities for As ions [52], and their loading on different porous carbons was proposed, with a commercial AC [53], silica-templated ordered mesoporous carbons [54], and rGO [55]. Loading of Fe° nanoparticles was performed by mixing a GO suspension with an $FeSO_4$ aqueous solution and then adding an $NaBH_4$ aqueous solution under stirring at 80°C either in N_2 atmosphere or under vacuum [55]. The presence of Fe° together with γ-Fe_2O_3 was confirmed by XPS. The content of Fe° was 66.45%, of which the particles (about 40 nm size) were homogeneously dispersed on the surfaces of rGO flakes. The adsorption isotherms at 25°C and pH 7.0 shown in Fig. 4.16A were well fitted with

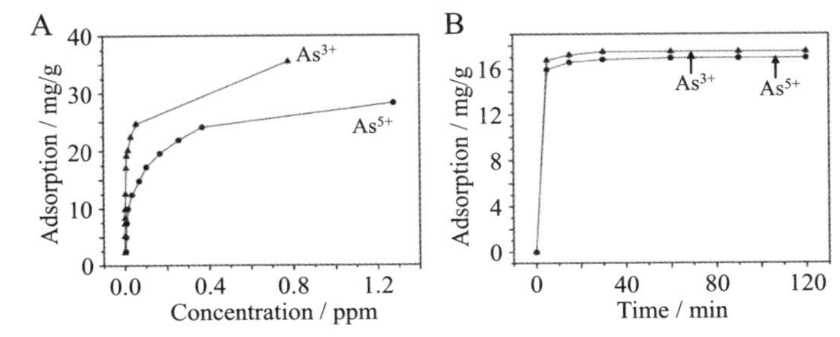

Figure 4.16 Adsorption of As^{3+} and As^{5+} on the rGO loaded by zero-valent Fe nanoparticles at 25°C and pH 7.0: (A) adsorption isotherms and (B) adsorption kinetics [55].

the Langmuir model which gave the maximum adsorption capacity for As^{3+} and As^{5+} of 35.8 and 29.0 mg/g, respectively. As shown in Fig. 4.16B, the adsorption of ions proceeds quickly in the first few minutes and reaches equilibrium within 30–60 min.

4.1.1.5 Phosphate ions

Phosphate is found in river water due to geochemical processes and is an essential nutrient for the growth of microorganisms. The presence of phosphate anions at higher than 2 mM concentrations in water is known to stimulate algal growth. According to WHO, the limit of phosphate in drinking water is 5.0 mg/L. Phosphate in water can be removed by many techniques, including adsorptive removal.

The adsorption of phosphate was studied using rGO [56]. The rGO flakes were obtained by further exfoliation of worm-like particles, which were prepared by thermal exfoliation and reduction of graphite oxide at 900°C in *n*-methyl-2-pyrrolidone (NMP) under ultrasonication and centrifugation. The adsorption of phosphate occurred very rapidly, with the adsorbed amount increasing near equilibrium capacity within the first 10 min and reaching equilibrium in another 10 min. The adsorption capacity increased slightly with an increase in the temperature of the solution, 89.4 mg/g at 30°C and 92.4 mg/g at 70°C. Thermodynamic studies revealed that the adsorption process was spontaneous and endothermic.

4.1.2 Adsorptive removal of organic pollutants

In this section, adsorption for removal of various organic pollutants in air and water by carbon materials is reviewed. The adsorption and removal are classified by the organic pollutants, volatile organic compounds (VOCs), phenol and its derivatives, polycyclic aromatic hydrocarbons (PAHs), humic acid and its halogen derivatives, and dyes. Adsorption of organic molecules from dilute aqueous solutions on carbon materials is reviewed also [57].

4.1.2.1 Volatile organic compounds

VOCs in air may cause serious adverse health effects, and so their removal is important. Adsorption of VOCs by porous materials, particularly porous carbons, is one of most effective methods for removal of VOCs and so the preparation of porous carbons for VOC removal has been studied in depth.

The powders of mesoporous carbons were prepared from Saran (copolymer of vinylidene chloride with vinyl chloride in a molar ratio of

83/17) powder (80–200 μm size) by heat treatment in two steps: preheating at different temperatures of 158–179°C and then carbonization at 760°C in N_2 [58]. The heat-treatment temperature during the first step governed the pore structure of the resultant carbons, and as a consequence it also affected the adsorption behavior for organic vapors. The adsorption isotherms measured at 25°C by TG apparatus are shown in Fig. 4.17A and B on powders preheated at 165 and 179°C. respectively. The powder preheated at 165°C gives isotherms with pronounced and steep hysteresis at P/P_0 of 0.3 for benzene and cyclohexane, as well as *n*-hexane and 2,2-dimethyl-butane (not shown), and slight hysteresis for *i*-octane (Fig. 4.17A). Those preheated at 171°C also resulted in the isotherms with hysteresis. The powder preheated at 179°C (near the melting point of the original polymer), however, did not show hysteresis (Fig. 4.17B). Benzene and cyclohexane were quickly adsorbed and desorbed, except at P/P_0 of 1.0 where it took 60 h to reach the adsorption equilibrium. Adsorption of 2,2-dimethylbutane and *i*-octane was very slow, requiring 100 h at P/P_0 of less than 0.3 and at 1.0, but desorption was much quicker than adsorption. Carbon fibers prepared from Saran fibers (melt-spinning, 110 μm diameter, and about 1 cm length) did not show pronounced hysteresis of their adsorption–desorption isotherms for these organic vapors.

ACs were prepared from mesophase pitch (MP) mixed with KOH in different weight ratios by carbonization at 800°C in N_2 [59]. With increasing KOH/MP ratio from 1/1 to 4/1, $V_{micro}(N_2)$ increased from 0.63 to 0.97 cm^3/g and the narrow micropore volume measured by CO_2 adsorption at 0°C, $V_{micro}(CO_2)$, increased from 0.61 to 0.81 cm^3/g, associated with an increase in V_{meso} from 0.04 to 0.43 cm^3/g. The adsorption of benzene and ethanol at 25°C was measured by dynamic column of the AC with a flow of air containing an adsorbate, and the desorption of the

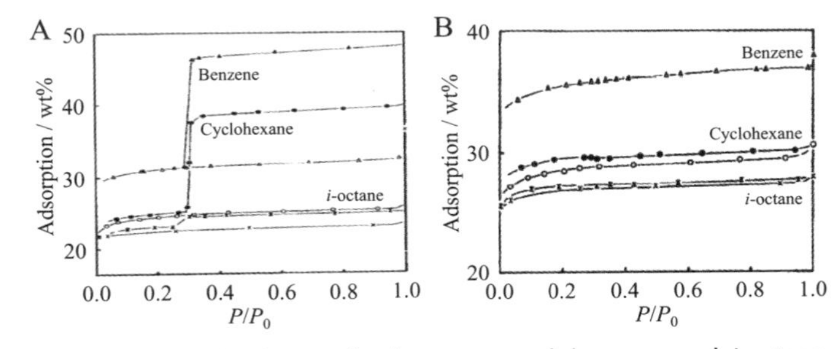

Figure 4.17 Adsorption isotherms for benzene, cyclohexane, and *i*-octane of Saran powder preheated (A) at 165°C and (B) at 179°C [58].

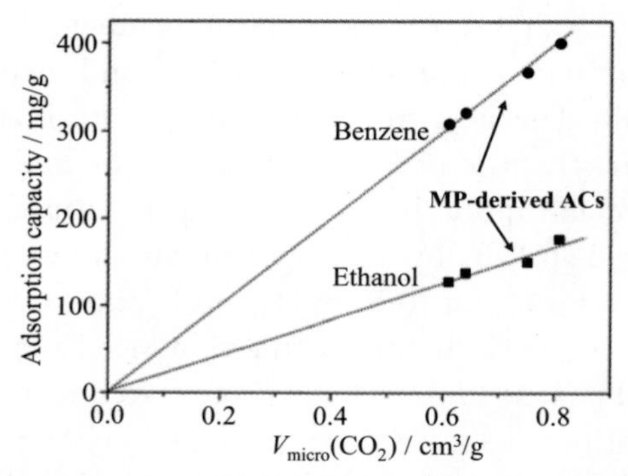

Figure 4.18 Adsorption capacities of ACs derived from mesophase pitch (MP) against $V_{micro}(CO_2)$ [59].

adsorbate was measured by temperature-programmed desorption (TPD) by heating up to 300°C with a flow of clean air. As shown in Fig. 4.18, adsorption capacities for benzene and ethanol are well related to narrow micropore volume $V_{micro}(CO_2)$, including some commercial ACs. Benzene adsorbed into AC could be desorbed by flowing clean air into the column, desorption of 50%–70% benzene at 25°C (weakly adsorbed) and 30%–45% during heating up to 250°C (strongly adsorbed). In the case of ethanol, about 99% was adsorbed weakly and so desorbed quickly at low temperature.

Adsorptive removal of VOCs by commercial activated carbon fiber cloths (ACCs) having S_{BET} of 1000 and 1700 m²/g was studied for toluene and *m*-xylene, focusing on the effects of adsorption temperature, VOC concentration, and gas flow rate on adsorption performance [60]. In Fig. 4.19A, the breakthrough curves at 50°C are compared for toluene and xylene with different concentrations, revealing that the breakthrough times are less than 15 min for both adsorbates, while the adsorption capacity for xylene is larger than that for toluene. In Fig. 4.19B, the breakthrough curves are compared with conventional adsorbents, granular AC, zeolites 5A, and silica gel, to the ACC. The superiority of the ACC to other adsorbents is clearly shown, with a longer breakthrough time and adsorption time (45 and 70 min, respectively) than those of other adsorbents.

ACF monoliths were prepared from a phenol-based carbon fiber (carbonized at 850°C in N_2, 0.1–0.4 mm length) by mixing with phenolic resin powder (3/1 by weight), molding, curing at 180°C, and then carbonizing at 700°C in N_2, followed by activation with steam at 700°C [61].

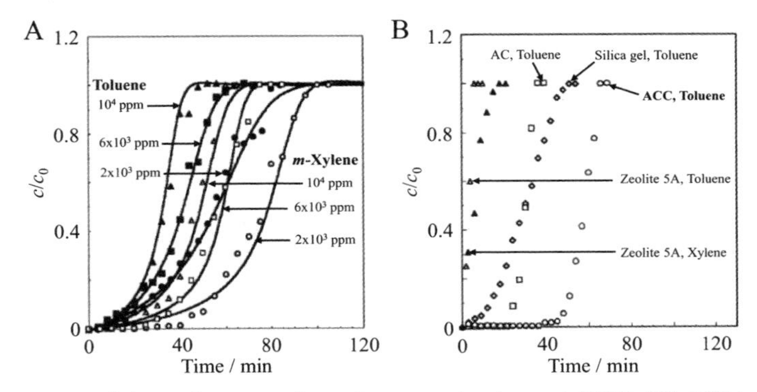

Figure 4.19 Breakthrough curves for toluene and xylene at 50°C: (A) ACC in the adsorbates with different concentrations and (B) comparison with granular AC, zeolite, and silica gel [60].

Burn-off during steam activation was controlled in the range of 0%–40%. In Fig. 4.20A and B, adsorption isotherms for *n*-butane and *n*-hexane of ACF monoliths with different burn-offs are shown in comparison with a commercial AC pellet (RB3, Norit), respectively. At high concentrations of the adsorbates of 10^6 ppm, the adsorption capacity becomes higher with increasing burn-off. At low concentrations below 2×10^3 ppm, the adsorption capacity for *n*-butane does not follow the order of burn-off, the monoliths with burn-off of 10%–20% exhibiting slightly higher capacities than those with higher burn-offs, as shown in the inset of Fig. 4.20A, which might be due to the presence of narrow micropores.

ACs were prepared from the powder mixtures of either natural wheat husk or *Indocalamus* leaf with Teflon by carbonization at 900°C in N_2 flow,

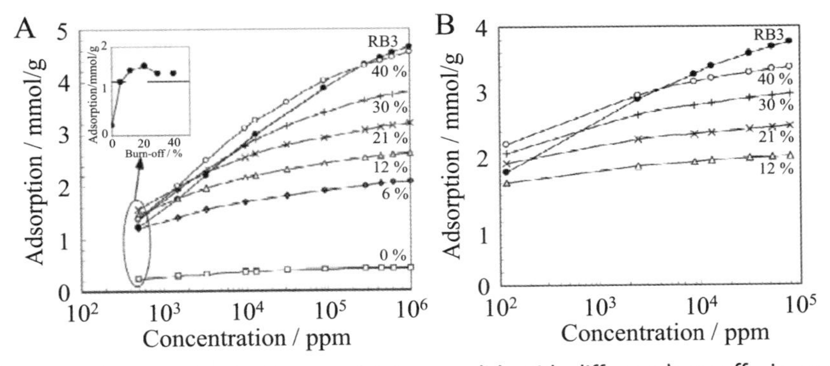

Figure 4.20 Adsorption isotherms of ACF monolith with different burn-offs, in comparison with a commercial AC (RB3): (A) for *n*-butane and (B) *n*-hexane at 30°C [61].

on which adsorption of formaldehyde (HCHO) was investigated [62]. The husk-derived AC had an adsorption capacity of 0.53 mg/g, which had a hierarchical pore structure composed of interconnected micro-, meso-, and macropores (S_{BET}, S_{micro}, and S_{ext} of 1059, 773, and 287 m^2/g, respectively, with V_{total} of 0.90 cm^3/g). An AC was prepared from sewage sludge by carbonization at 450°C followed by $ZnCl_2$ activation at 750°C in N_2 flow, and compared with commercial ACs (wood-, coconut-shell-, and coal-derived ACs) on adsorption capacity for formaldehyde from air with concentrations of 498 and 0.41 mg/m^3 [63]. Surface modification of bone char by impregnating in acetic acid was reported to improve the adsorption of formaldehyde in air flow [64].

Adsorption performances for formaldehyde were compared on commercial ACFs, pitch-based (OG5A and OG15A), rayon-based (KF1500) and PAN-based (FE100, FE200, and FE300) [65]. As shown by the breakthrough curves in Fig. 4.21A and B, PAN-based ACF FE100 exhibits the highest breakthrough time (361 min) and adsorption capacity (0.478 mmol/g), whereas it has the smallest pore parameters; S_{BET} of 378 m^2/g and V_{total} of 0.22 cm^3/g. The adsorption capacities of these ACFs are closely related to the N content in the ACFs measured by XPS, as shown in Fig. 4.21C. However, no relation with oxygen content was obtained. A marked decrease in adsorption ability under humid conditions was observed on PAN-based ACFs (rich in nitrogen), that was thought to be caused by preferential adsorption of water to adsorbents due to water affinity of functional groups containing nitrogen [66]. Pitch-based ACFs (negligible nitrogen), on the contrary, have adsorption capacities under humid conditions that are almost the same as those under dry conditions,

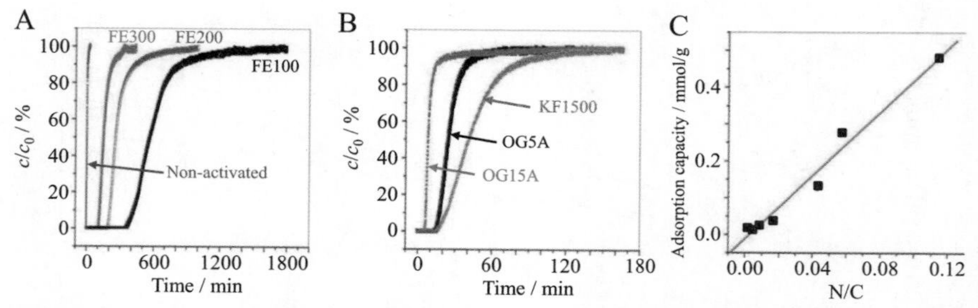

Figure 4.21 Adsorptive removal of formaldehyde by commercial ACFs in a flow of 20 ppm at 30°C: (A) breakthrough curves for the PAN-based ACFs (FE100, 200, and 300) with nonactivated CF, (B) those for pitch-based (OG5A and OG15A) and rayon-based (KF1500) ACFs, and (C) relation between N-content, N/C, and adsorption capacity of ACFs [65].

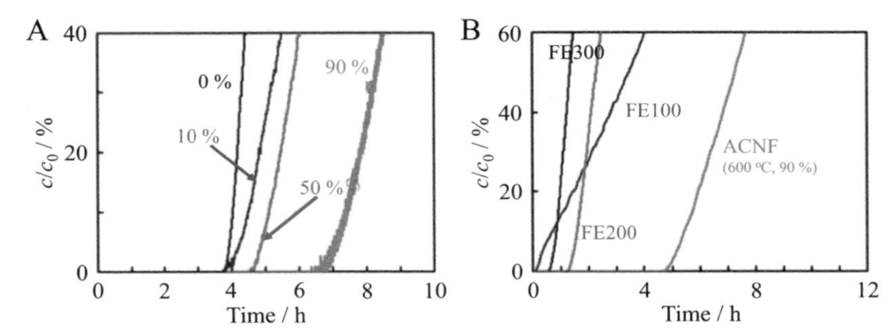

Figure 4.22 Breakthrough curves of the PAN-based carbon nanofibers (ACNFs) for formaldehyde: (A) effect of relative humidity during steam activation on the adsorption in dry 22 ppm formaldehyde and (B) comparison with commercial ACFs in wet 11 ppm formaldehyde [67].

although the time to reach equilibrium is a little longer. Since the porosity of ACFs also influenced the formaldehyde adsorption capacity, the activation under different relative humidities was performed on electrospun PAN nanofibers [67]. As shown in Fig. 4.22A, high humidity conditions of 90% during steam activation at 600°C enhanced formaldehyde adsorption, but humidity of 0%–50% had only a small effect, although the pore structure in the nanofibers was very sensitive to humidity during activation. Extension of the activation time from 1 to 3 h at 600°C did not have an appreciable effect on formaldehyde adsorption. In Fig. 4.22B, the breakthrough curve of the activated carbon nanofibers (ACNFs) for formaldehyde (11 ppm) in wet conditions (relative humidity of 50%) is compared with commercial PAN-based ACFs (FE100, 200, and 300). The breakthrough time for the ACNFs prepared at 600°C in 90% humidity is much longer than for commercial ACFs.

For the adsorption of formaldehyde, on the other hand, a contribution of the hydrophilic oxygen-containing functional groups was reported on ACs derived from coffee residues by activation using different agents [68]. In Fig. 4.23, the relations between adsorption capacity for formaldehyde and its concentration are shown on the ACs prepared at different conditions as indicated, where the carbonization (pyrolysis) and activation are performed at 600°C. The carbon mixed with $ZnCl_2$ followed by a simple pyrolysis has the highest adsorption capacity, which is much higher than a commercial AC (CH-11,000), and those activated by steam and CO_2 have higher capacities than ACs derived from the same precursor without $ZnCl_2$ mixing, though lower than the commercial AC (CH-11,000).

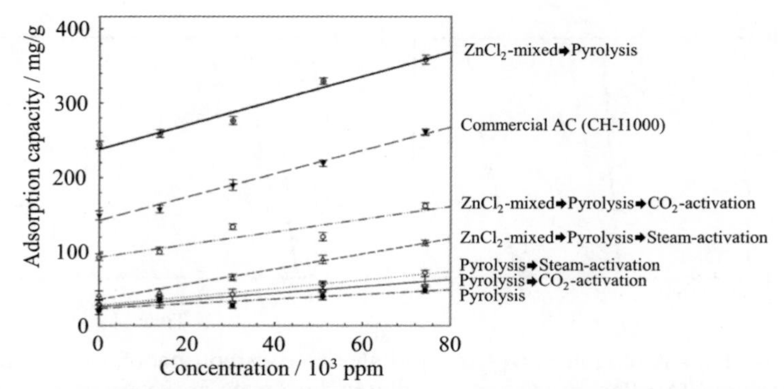

Figure 4.23 Adsorption capacity for formaldehyde versus its concentration for the ACs prepared from coffee residue using different activation reagents [68].

The adsorption capacity of the carbon was found to be significantly influenced by the presence of functional groups of O—H, C=O and C—O, in particular the carbon activated by $ZnCl_2$ which had the largest amount of O—H groups. Formaldehyde adsorption capacity at low concentrations (less than 15 ppm) was shown to be strongly related to the content of basic surface functional groups on three commercial ACs, two granular ACs derived from bituminous coal and one rayon-based ACC [69].

Carbons prepared from bamboo by carbonization at 500—1000°C (particle sizes of 25—125 μm) were studied on their adsorption performance for harmful gases and odorants by focusing the effect of carbonization temperature [70]. Carbonization at 1000°C resulted in the highest adsorption capacity and rate for toluene, formaldehyde, benzene, indole, skatole, and nonenal, while carbonization at 500°C resulted in the highest adsorption for ammonia. Bamboo charcoal prepared at 900°C (S_{BET} of 400 m^2/g) had a high effectiveness for adsorptive removal of nitrate ion in water, and was higher than that for a commercial AC (S_{BET} of 850 m^2/g) [71].

Commercially available viscose rayon-based microporous ACFs with different S_{BET} and V_{micro}, as shown in Table 4.4, were used as adsorbents for the vapors of polar methyl-ethyl-ketone (MEK) and nonpolar benzene [72]. As shown in Fig. 4.24A and B, adsorption capacity increases with increasing S_{BET} of the ACFs, except for MEK at a concentration of less than 100 ppm. The maximum capacities of ACFs were calculated using the Dubinin-Radushkevich equation for MEK and benzene, as listed in Table 4.4. The maximum adsorption capacity of ACF-16 for MEK is greater than that for benzene, but those of ACF-6 and -14 for MEK are

Table 4.4 Characteristics of ACFs and their maximum adsorption capacities for MEK and benzene. [72].

	Pore structure parameters			Maximum adsorption capacity	
	S_{BET} (m^2/g)	V_{micro} (cm^3/g)	Oxygen content C/O	MEK (mg/g)	Benzene (mg/g)
ACF-6	640	0.172	6.7	191.0	213.5
ACF-14	1460	0.320	5.6	534.5	566.2
ACF-16	1680	0.432	4.7	776.0	650.0

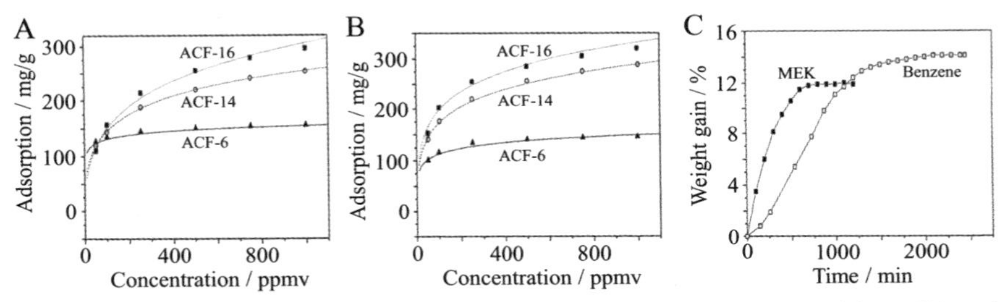

Figure 4.24 Adsorption isotherms of ACFs with different S_{BET} at 25°C: (A) for MEK, and (B) for benzene and (C) weight gains of ACF-14 for 50 ppm MEK and 50 ppm benzene in N_2 [72].

lower than those for benzene, although these two adsorbates have similar boiling points, molecular weights, and saturation vapor pressures, suggesting that polarity of the adsorbate has a significant effect on the adsorption capacity of the adsorbent carbon. Fig. 4.24C shows the weight changes of ACF-14 in N_2 containing about 50 ppm MEK and benzene, showing that a long contact time of up to 2000 min is needed to reach equilibrium, which is a much longer time for nonpolar benzene than polar MEK.

At a low concentration of MEK, ACF-6 has slightly higher adsorption capacity than ACF-14 and -16, although the former has lower S_{BET} than the latter two (Fig. 4.24A and Table 4.4), which is called the cross-over phenomenon. The same cross-over phenomena were reported on phenol-based ACFs with S_{BET} of 900—2420 m^2/g on the adsorption of acetone in air and benzene in N_2 [73]. By using four phenol-based ACFs having S_{BET} of 904, 1420, 1598, and 1978 m^2/g (denoted as ACF-10, -15, -20, and -25, respectively), the cross-over phenomena were observed on the adsorption for normal alkanes, (ethane, propane, butane, and pentane) in N_2 carrier gas [74]. The adsorption isotherms of ACFs for ethane,

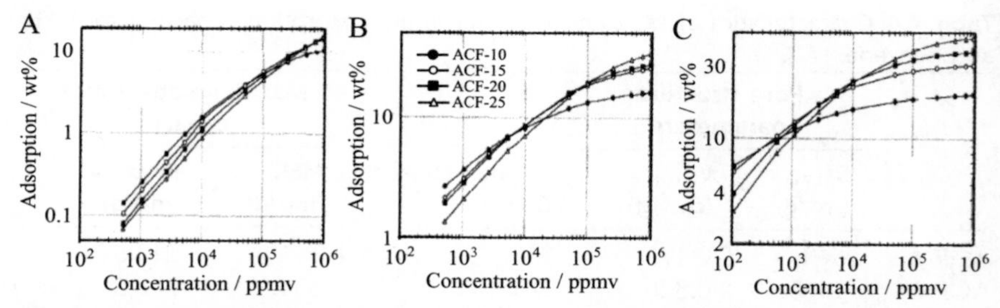

Figure 4.25 Adsorption isotherms of ACFs with different S_{BET} for normal alkanes at 25°C: (A) for ethane, (B) for propane, and (C) for butane [74].

propane, and butane are shown in Fig. 4.25A–C, respectively. The adsorption capacity of each ACF becomes larger with increasing molecular weight of alkanes (with increasing boiling point of alkanes). At high concentrations of alkanes of 10^6 ppmv, the adsorption capacity of the ACFs increases with increasing S_{BET}, while at low concentrations of 10^2 ppmv, in contrast, capacity becomes smaller with increasing S_{BET} of ACFs. The cross-over region seems to shift to a lower concentration with increasing length of alkane molecules. For normal alkane series, the adsorbent with the lower surface area demonstrates the higher adsorption capacity for the alkanes with the lower boiling point at low concentrations. At high concentrations, however, the adsorption capacity of the higher surface area adsorbents is greater for the higher boiling point alkanes due to increasing pore volumes.

4.1.2.2 Phenols

Bisphenol A (BPA) is a common endocrine disrupting chemical (EDC) in wastewater and natural waters. Highly porous carbon was prepared by KOH activation of Gilsonite powder at 400°C in Ar, of which S_{BET}, V_{total}, and V_{micro} reached 3850 m^2/g, 2.366 cm^3/g, and 1.538 cm^3/g, respectively (Gilsonite-derived AC) [75]. Its adsorption performance for BPA in aqueous solution was compared with a commercial AC (S_{BET} of 813 m^2/g, V_{total} and V_{micro} of 0.822 and 0.390 cm^3/g, respectively). Their adsorption isotherms and kinetics are shown in Fig. 4.26A and B, respectively, revealing that adsorption isotherms fit well with the Freundlich equation and adsorption kinetics by a pseudo second-order equation. The maximum adsorption capacity of the Gilsonite-derived AC was 1113 mg/g, which was four times higher than that of commercial AC (271 mg/g) and the adsorption rate constants are slightly different, at 1.66 mg/mg min for the former and 2.63 mg/mg min for the latter.

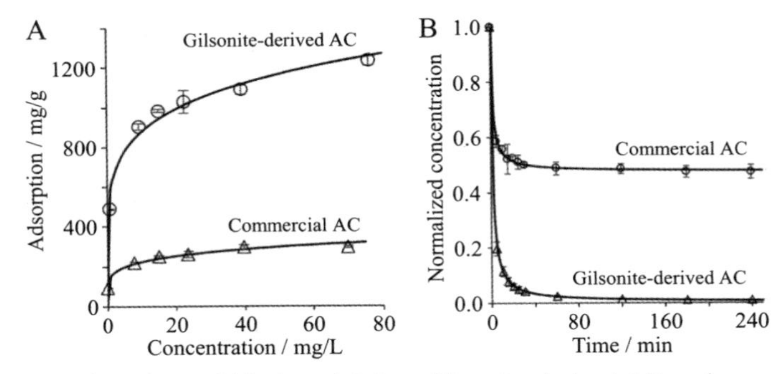

Figure 4.26 Adsorption of bisphenol A into Gilsonite-derived AC and a commercial AC: (A) adsorption isotherms and (B) adsorption kinetics [75].

Adsorption of phenolic compounds, phenol, *p*-cresol, *m*-chlorophenol, *m*-aminophenol, and *p*-nitrophenol, from their aqueous solutions at 25°C on ACs have been studied [76]. The as-received commercial coal-derived AC were demineralized by heat treatment at 850°C in N_2 flow. The as-received and demineralized ACs were re-activated in a steam flow at 840°C for 2.5, 5, and 10 h. Adsorption isotherms for the phenolic compounds were well approximated by the Langmuir equation, as shown on the isotherms for *m*-aminophenol in Fig. 4.27. The maximum adsorption capacity calculated by Langmuir equation for the compounds increased with increasing S_{BET} and V_{total}. The demineralized AC re-activated for 10 h exhibited the highest capacity for each compound, for example, 218 mg/g for *m*-aminophenol and 234 mg/g for *m*-chlorophenol. The affinity of the phenolic compounds toward the AC surface was discussed in

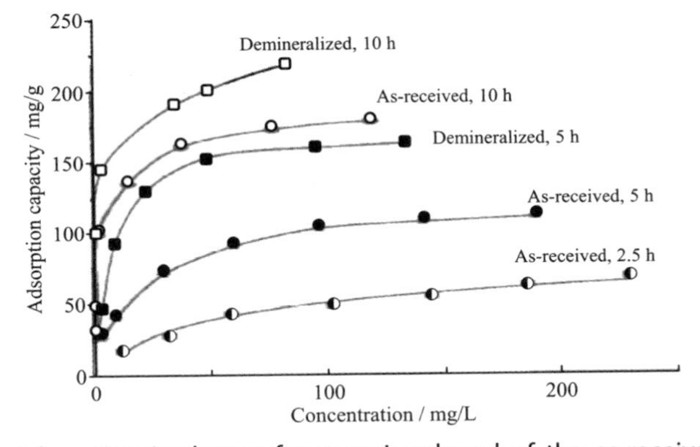

Figure 4.27 Adsorption isotherms for *m*-aminophenol of the as-received and demineralized ACs after reactivation at 840°C for different times in steam [76].

relation to the electron donor–acceptor complexes formed between the basic sites on the carbon and the aromatic ring of the phenol. Adsorptive removal of various hazardous organic pollutants (phenol, *o*-cresol, *p*-nitrophenol, *m*-methoxyphenol, benzoic acid, and salicylic acid) from their aqueous solutions was studied by measuring breakthrough curves using a column of a commercial granular AC (S_{BET} of 970 m^2/g, V_{total} of 0.850 cm^3/g) [77]. Competitive adsorption in bi- and tri-adsorbate systems was also investigated.

Adsorption of phenol, 2-chlorophenol (2-CP), 4-chlorophenol (4-CP), 2,4-dichlorophenol (DCP), 2,4,6-trichlorophenol (TCP), 4-nitrophenol (4-NP), and 2,4-dinitrophenol (DNP) from their aqueous solutions was studied on a pitch-based mesoporous ACF (S_{BET} of 920 m^2/g and V_{micro} of 0.422 cm^3/g) [78]. Compared to phenol, the substituted phenols demonstrate more rapid adsorption at a low concentration range and higher adsorption capacities in the order of TCP > DNP ≈ DCP > 4-NP > 4-CP > 2-CP > phenol, according to the molecular size order (from 0.684 × 0.576 nm^2 for DNP to 0.576 × 0.417 nm^2 for phenol).

ACs were prepared from two lignocellulosic precursors, kenaf (*H. cannabinus*) and rapeseed (*B. napus*), by CO_2 activation at 700°C for different times to control the burn-off between 20 and 80 wt.%, after pyrolysis at 400°C [79]. ACs derived from kenaf and rapeseed activated up to 48 and 38 wt.% burn-off (K-48 and R-38, respectively) were treated in concentrated HNO_3 for 1 h at 80–90°C (K-48-ox and R-38-ox, respectively). Phenol adsorption isotherms of these four ACs are compared in Fig. 4.28A, revealing similar isotherms on two ACs (K-48 and R-38) though precursors and burn-offs are different, and a marked depression of phenol adsorption by HNO_3 oxidation. The maximum adsorption

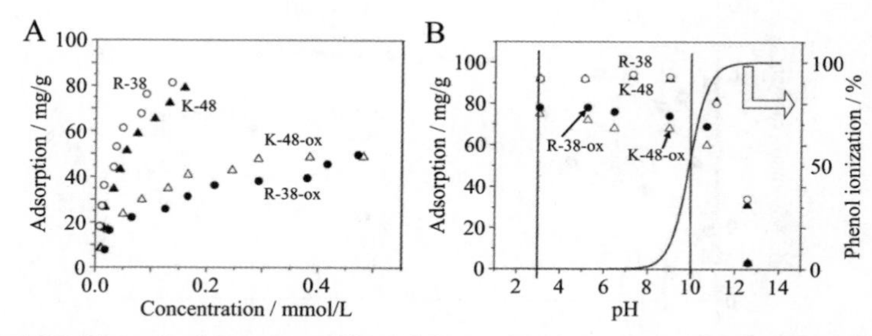

Figure 4.28 Phenol adsorption of kenaf-derived AC (K-48) and rapeseed-derived AC (R-38) and their HNO_3-oxidized ones (K-48-ox and R-38-ox): (A) adsorption isotherms and (B) dependences on pH of the solution [79].

capacities of approximately 80 and 50 mg/g were obtained on the as-received ACs and the HNO_3-oxidized ACs, respectively. The dependences of adsorption capacity on pH can be divided into two ranges, below and above 10, as shown in Fig. 4.28B. In pH less than 10, phenol is predominantly in the molecular form and so the most probable mechanism for the adsorption onto AC was thought to be based on the $\pi-\pi$ interaction between the phenol aromatic ring and the delocalized π electrons in the AC. In pH higher than 10, however, phenol is ionized to phenolate $C_6H_5O^-$ and so the rapid decrease in adsorption capacity is probably due to the electrostatic repulsion.

N-doping of carbon adsorbents was shown to be effective for the adsorption and removal of phenol in water on commercially available ACs with different S_{BET}'s [80]. N-doping of the ACs was performed by heating in ammonia gas flow (50 mL/min) at a temperature of 400–800°C for 2 h. The adsorption capacity for phenol depended greatly on the initial concentration of phenol (150–300 mg/dm^3), but it increased with increasing temperature for N-doping up to 700°C, although S_{BET} and V_{micro} decreased slightly. The phenol adsorption capacity of about 230 mg/g was obtained after 700°C-treatment.

4.1.2.3 Polycyclic aromatic hydrocarbons

Polycyclic aromatic hydrocarbons (PAHs) can be transported over long distances in air and water, and are difficult to biodegrade, mainly because of their chemical persistence and semivolatile nature. Some PAHs are capable of interacting with DNA to promote mutagenic and carcinogenic responses. Most PAHs are hydrophobic, with high boiling and melting points, and possess low water solubility and electrochemical stability. Therefore, they can exist and be accumulated in soils for a long time.

The adsorption of PAHs, naphthalene, fluorene, phenanthrene, pyrene, and fluoranthene, from their aqueous solutions has been studied on ACs prepared by KOH activation at different temperatures, of which S_{BET} ranged between 562–1904 m^2/g and V_{micro} between 0.86–0.91 cm^3/g [81]. Adsorption isotherms of the PAHs on the ACs were well approximated using the Freundlich equation. The maximum capacity for PAHs was larger with lower molecular weight of the adsorbates, as well as a lower molar volume. The adsorption kinetics of some PAHs have been studied on a commercial macroporous AC (S_{BET} of 1000 m^2/g with V_{macro} of 1.5 cm^3/g), which were well simulated by a pseudo-first-order reaction model and characterized by a two-stage process with two different rates [82].

Adsorption and removal of phenanthrene in an aqueous solution was studied on sesame-stalk-derived ACs which were prepared by carbonization at different temperatures (300–700°C), followed by KOH activation [83]. The removal efficiency of phenanthrene approached 100% on the 700°C-carbonized AC, which became slightly lower with the presence of acenaphthene and naphthalene in the solution. Adsorptive removal of phenanthrene in an ethanol aqueous solution was studied on an AC prepared by microwave irradiation of the powder mixture of anthracite with KOH (1/3 by weight) at 693 W for 10 min under vacuum [84].

Vegetable oil has the advantages of being a nontoxic, cost-effective, and biodegradable solvent to extract PAHs from contaminated soils, but the extracted PAHs are required to be removed from the resulting vegetable oil for the reuse of the oil in remediation processes. A commercial AC with a particle size of 0.5–2.0 mm, S_{BET} of 1200 m^2/g, and V_{total} of 0.73 cm^3/g was used for adsorption and removal of PAHs from sunflower oil [85]. Four kinds of PAH-contaminated oils were prepared by mixing 15 PAH molecules, such as acenaphthene (ACN), fluorene (FLU), phenanthrene (PHE), anthracene (ANT), etc., with total PAH contents of 731–2971 mg/L. The effect of the amount of AC used on PAH removal from the oil containing the largest PAH content is shown in Fig. 4.29A, showing that removal efficiency increases with increasing amount of AC. An AC of at least 9 g was required to remove 90% of PAHs in 40 mL of oil. The adsorption isotherm of PAHs is approximated by Langmuir and Freundlich equations, as shown in Fig. 4.29B; the maximum adsorption capacity of the AC is calculated as 86.3 mg/g. The removal efficiencies for individual PAHs passing through the AC column are shown in Fig. 4.29C, revealing efficiencies of higher than 90%, except ACY (acenaphthylene). For total

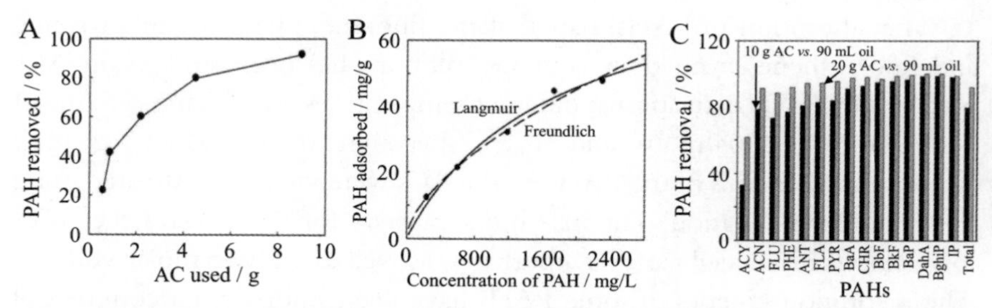

Figure 4.29 Adsorption of PAHs in sunflower oil containing 2971 mg/L PAHs using a commercial AC: (A) effect of the amount of AC used for 40 mL oil, (B) adsorption isotherm for PAHs using 9 g AC in 40 mL oil, and (C) adsorption of individual PAHs and total PAH through AC column [85].

PAHs, the removal efficiency was 79.7% and adsorption capacity was 21.3 mg/g when 10 g AC was used for 90 mL of oil. The adsorptive removal of aromatic compounds from mineral naphthenic oil (containing about 21% total aromatics consisting of one to four benzene rings) using two commercial ACs (S_{BET} of 688 and 1218 m^2/g, and V_{total} of 0.422 and 1.041 cm^3/g, respectively) was studied by batch and column systems [86]. Adsorption and desorption using *n*-hexane as an eluent were steadily cycled at least three times.

Adsorptive removal of PAHs from a hot gas was investigated using binary mixtures of naphthalene and phenanthrene on the ACs with different origins (coke, apricot stone, and lignite) and different pore parameters (S_{BET} of 171−547 m^2/g and V_{micro} of 0.08−0.28 cm^3/g) [87].

4.1.2.4 Humic acid and trihalomethanes

Humic acid was recognized as a precursor of trihalomethanes (THMs) formed during chlorination of water, and these are classified as carcinogenic compounds. The elimination of humic acid in water before chlorination is essential for the water supply.

Mesoporous ACFs were prepared by melt-spinning of the quinoline solution of a pitch (softening point of 85°C) with metallic acetylacetonates, $Y(acac)_3$, $TiO(acac)_2$, and $Al(acac)_3$, stabilization at 360°C in air, and then carbonization and simultaneous activation at 875°C in steam flow [88,89]. On the resultant ACFs, adsorption isotherms for two humic acids denoted as H-1 and H-5 (different suppliers), as well as vitamin B_2, vitamin B_{12}, and α-cyclodextrin, were measured by immersing in adsorbate aqueous solutions at room temperature. The adsorption performances of the resultant ACFs were compared with the commercial pitch-based ACF, A-20, after the same activation process (875°C, 70 min). The pore structure parameters of the resultant ACFs are shown together with the preparation conditions in Table 4.5. ACFs prepared using $Y(acac)_3$ are mesoporous, while those using $TiO(acac)_2$ and $Al(acac)_3$ are microporous. Adsorption isotherms are shown in Fig. 4.30A−D. Mesoporous ACFs (Y-50 and -33) are advantageous for the adsorption of large molecules, humic acid (H-1), vitamin B_{12}, and α-cyclodextrin, as shown in Fig. 4.30A, C, and D, respectively. For vitamin B_{12} adsorption, Y-50 experienced more severe activation conditions and was superior to Y-33 prepared by a shorter activation time, but for α-cyclodextrin two ACFs exhibit comparable adsorption capacities. Small-sized vitamin B_2, as well as a humic acid (H-5) (not shown), are quickly adsorbed into the commercial microporous ACF (A-20).

Table 4.5 Preparation conditions and pore structure parameters of mesoporous ACFs and a commercial ACF (A-20) [88].

Code	Preparation conditions			Pore structure parameters		
	Metallic acetylacetonate	Carbonization and activation	Yield (wt.%)	S_{BET} (m^2/g)	S_{meso} (m^2/g)	Mesopore ratio (%)
Y-50	$Y(acac)_3$	875°C, 50 min	13.3	1370	999	73
Y-33	$Y(acac)_3$	875°C, 33 min	18.7	1198	794	80
Ti	$TiO(acac)_2$	875°C, 40 min	32.7	1280	140	11
Al	$Al(acac)_3$	875°C, 40 min	33.8	1339	24	1.8
A-20	–	875°C, 70 min	17.5	1990	92	4.6

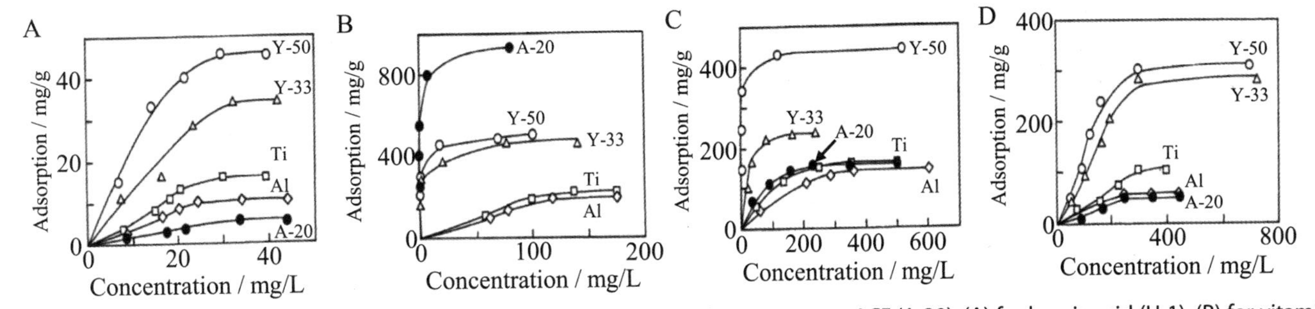

Figure 4.30 Adsorption isotherms of mesoporous ACFs in comparison with microporous ACF (A-20): (A) for humic acid (H-1), (B) for vitamin B_2, (C) for vitamin B_{12}, and (D) for α-cyclodextrin [88].

For adsorptive removal of Leonardite humic acid, mesoporous carbons prepared by MgO templating at 850°C using $MgCO_3$ and $Mg(OH)_2$ as MgO-precursors and PET as carbon-precursor delivered high adsorption capacity and rate [90].

Adsorption of trihalomethanes (THMs; $CHCl_3$, $CHBrCl_2$, and $CHBr_2Cl$) were studied on PF-based and PAN-based ACFs by comparing with a granular AC [91]. The adsorption isotherms were measured by immersing powdered ACFs and granular AC into THM solution for 2 and 5 h, respectively, at a constant temperature. All isotherms were successfully correlated by the Freundlich equation, as shown in $CHCl_3$ at different temperatures. In Fig. 4.31A, the adsorption isotherms of PF-based ACF at different temperatures are shown. THM adsorption capacities of ACFs were comparable with or slightly larger than those of the granular AC. In Fig. 4.31B, the breakthrough curve obtained by flowing a $CHBrC1_2$ solution with a flow rate of 0.381 cm^3/s through a packed column of the PF-based ACFs (packed density of 0.127 g/cm^3) is shown, suggesting that the adsorption equilibrium is quickly established between a THM solution and the ACF.

4.1.2.5 Dyes

Most dyes consist of giant molecules with complex chemical structures and have high solubility in aqueous media, which can have harmful impacts on rivers and lakes by poisoning aquatic animals and plants. Since their adsorptive removal from aqueous solutions requires large pores, mesoporous carbons, mostly templated mesoporous carbons, and carbon foams containing macropores are often employed. Dye adsorption was reviewed by focusing on ACs prepared from biomasses [92].

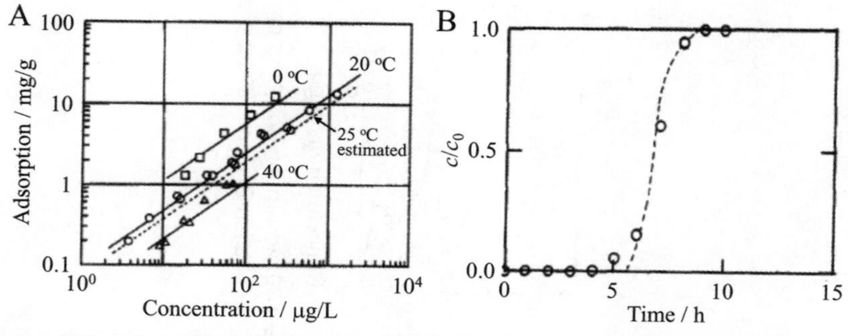

Figure 4.31 Adsorption of the PF-based ACF for trihalomethanes: (A) isotherms for $CHCl_3$ at different temperatures and (B) breakthrough curve for $CHBrCl_2$ (c_0 of 1.18 mg/L) [91].

Adsorption of four anionic dyes, basic red 46, acid brown 283, direct red 89, and direct black 168, was studied in two porous carbons, mesoporous carbons prepared from sewage sludges by carbonization at 700°C after mixing with H_2SO_4 (sludge-derived AC, S_{BET} of 235 m^2/g) and commercial microporous carbon (Chemviron GW, S_{BET} of 1026 m^2/g) [93]. The sludge-derived AC exhibited high adsorption capacities for basic red 46, direct red 89, and direct black 168, which were higher than the commercial microporous carbon, the maximum adsorption capacities calculated by Langmuir equation being 188, 49, and 29 mg/g, respectively. The porous carbons prepared from sewage sludges by carbonization at 625°C in the presence of H_2SO_4 were used as adsorbents after washing with HCl, which were studied as an adsorbent for crystal violet ($C_{25}H_3ClN_3$), indigo carmine ($C_{16}H_8N_2Na_2O_8S_2$), and phenol [94]. The sludge-derived carbons delivered high adsorption capacity and rate for these organics, particularly for crystal violet, with the maximum adsorption capacity of 185—271 mg/g. Preferential adsorption of crystal violet in binary solutions of these adsorbates was observed. The adsorption of methylene blue was studied on the ACs prepared from oil palm fibers by carbonization and activation with KOH at 850°C and following further activation by CO_2 at 850°C for 2 h [95]. The maximum adsorption capacity calculated by Langmuir equation was 228 mg/g at 30°C and 385 mg/g at 50°C.

Ordered mesoporous carbons (CMK-3) were prepared from sucrose with mesoporous SBA-15 template at 880°C, followed by dissolution of silica template by HF [96]. By using three SBA-15 having different mesopore sizes, the carbons with three sizes of mesopores of 3.0, 4.3 and 6.5 nm were synthesized (denoted as CMK-3-3, -4 and -6), on which the adsorption performance of cytochrome C in buffer solutions with different pHs (potassium phosphate buffer with pH 6.5 and sodium bicarbonate buffer with pH 9.6 and 10.5). As shown in Fig. 4.32, the adsorption is significantly affected by the pH of the solution, giving the highest capacity at pH 9.6, near the isoelectric point of cytochrome C (9.8). The maximum adsorption capacities for cytochrome C in the solution with pH 9.6 were calculated by Langmuir approximation as 10.9, 18.5, and 14.5 μmol/g for CMK-3-3, -4, and -6, respectively, of which S_{BET} were 1260, 1250, and 1350 m^2/g and V_{total} were 1.1, 1.3, and 1.6 cm^3/g, respectively.

Porous carbons were fabricated from phenol resin with 40 wt.% diblock copolymer, poly(styrene)-*b*-poly(4-vinylpyridine), by curing at 190°C with hexamethylenetetramine (HMTA) and carbonization at 420°C, in which the adsorption of methylene blue (MB) and rhodamine

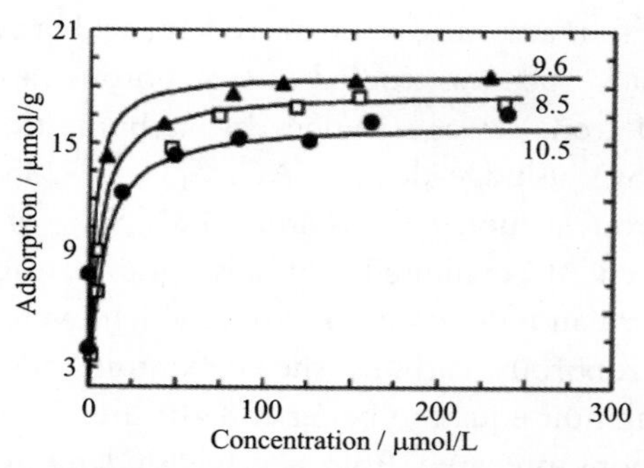

Figure 4.32 Adsorption isotherms of ordered mesoporous carbon CMK-3–4 for cytochrome C in the solutions with different pH [96].

6G was studied [97]. Mesoporous carbons were fabricated from waste polysaccharides (starch and alginic acid) by carbonization at 800°C after gelation of each precursor, of which the adsorption of MB and acid blue 92 in aqueous solutions was studied [98]. The adsorption capacities for MB and acid blue 92 were measured as 52 and 39 mg/g, respectively, for starch-derived carbon and 97 and 108 mg/g, respectively, for alginic-acid-derived carbon, in contrast to 42 and 49 mg/g for commercial microporous AC (Norit). Adsorption of these dyes occurred quickly in the first 15 min. High adsorption capacities were discussed on the bases of hydrogen bonding between oxygen-containing functional groups on the pore surface and dye molecules. Mesoporous carbons prepared from resorcinol-formaldehyde using a colloidal silica (particle size of about 12 nm) as template by carbonization at 850°C exhibited high adsorption capacity for bulky dyes [99].

Exfoliated graphite (EG) was prepared by microwave irradiation of graphite oxide prepared from natural graphite (lateral size of more than 1.25 mm) and subjected to examine the adsorption behavior of different dyes, MB, crystal violet, Congo red, and methyl orange [100]. In Fig. 4.34A, the removal efficiencies of four dyes from their solutions at a concentration of 200 mg/L by the EG are compared; almost 100% of the Congo red was removed from the solution but other dyes could be removed only up to less than 50%, suggesting that the EG has low adsorption capacities for these dyes. As shown in Fig. 4.33B, the adsorption of Congo red proceeds rapidly, with about 84 mg/g in the first 20 min (about 98% removal from the solution), and reaching the saturated capacity of 86 mg/g, whereas the removal of MB occurs gradually and reaches

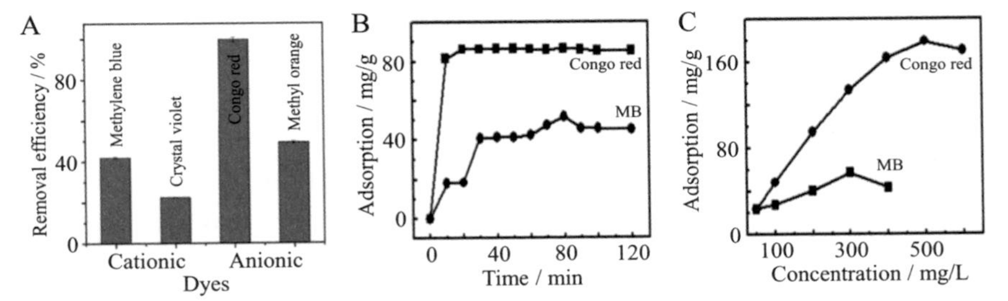

Figure 4.33 Exfoliated graphite for dye adsorption: (A) removal efficiencies for four dyes from their 200 mg/L solutions, (B) effect of contact time in 200 mg/L solutions using 2 g/L EG, and (C) effect of the initial concentration on adsorption capacity [100].

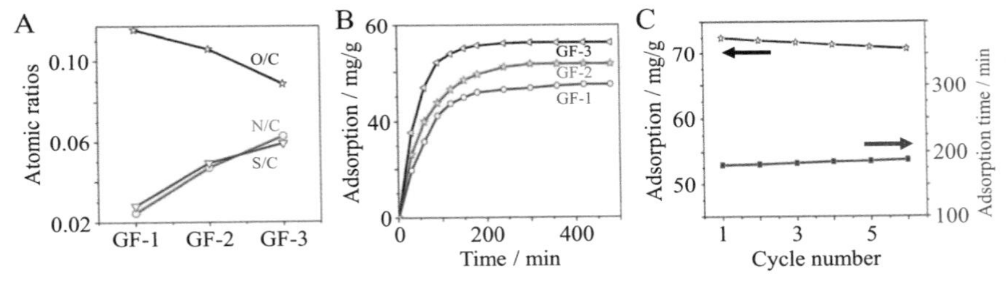

Figure 4.34 rGO foams (GFs) prepared using different amounts of thiourea: (A) ratios of foreign atoms in GFs, (B) adsorption curve of rhodamine B, and (C) cycle performance of GF-3 for rhodamine B with the adsorption and desorption cycle [104].

51 mg/g after 80 min. The adsorption capacity of the EG for Congo red reaches 179 mg/g and that for MB 57 mg/g, as shown in Fig. 4.33C. The recyclability of EG for removal of these dyes was not high mainly due to collapse of the bulky texture, and adsorption capacities decreased to 105 and 27 mg/g for Congo red and MB, respectively, after the first reuse. Slight oxidation of EG by heating at 400°C in air was effective in increasing the adsorption capacity for methyl orange, from 230 mg/g for pristine EG to 300 mg/g for oxidized EG, whereas only a slight increase was observed on MB adsorption [101]. Oxidation of EG in 13% HNO_3 at room temperature resulted in a marked reduction in adsorption capacities for methyl orange and MB. Adsorptive removal of textile dyes (direct and reactive blue) from industrial wastewater was performed using EG nanoplatelets [102]. They were produced via microwave exfoliation at high temperatures and had an average thickness of 6–8 nm with an average lateral size of 25 μm. The removal capability at low pH (about 3.5) and high initial dye concentration (100 mg/L) reached higher than 100 mg/g.

rGO flakes were prepared by reoxidation and reintercalation of commercially available expandable graphite, followed by reduction with hydrazine hydrate at 110°C, which exhibited strong G′-band in Raman spectrum, which was stronger than for the G-band, while their XRD shows the existence of a small amount of graphite [103]. The maximum adsorption capacities calculated from the adsorption isotherms for MB via Langmuir equation was 154 mg/g at 20°C and 204 mg/g at 60°C. rGO foams (GFs) were fabricated from GO (prepared from natural graphite flakes with 500−600 μm lateral size) mixed with thiourea under hydrothermal conditions at 180°C, followed by freeze-drying [104]. By changing the thiourea content in GO dispersion, the resultant GFs have different amounts of foreign atoms (measured by XPS), as shown in Fig. 4.34A, and different adsorption capacities for rhodamine B, as shown in Fig. 4.34B. On the GF-3 (the largest contents of N and S), the adsorption capacities of 184 mg/g for MB and 72.5 mg/g for rhodamine B with a high adsorption rate were obtained, whereas the capacity for methyl orange was 11.5 mg/g. The adsorbed MB and rhodamine B could be easily desorbed in methanol or ethanol. In Fig. 4.34C, cycle performances of adsorption/desorption of rhodamine B are shown for GF-3. The adsorption capacity decreases slightly with cycling, to less than 2.3% after the sixth cycle, associated with a slight increase in the adsorption time.

A composite was fabricated by hydrothermal treatment of the mixture of GO and MWCNTs in ethanol/water solution (1/1 by volume) at 200°C, which delivered a maximum adsorption capacity of 82 mg/g for MB in aqueous solution with the removal efficiency of 97% of the initial MB concentration of 10 mg/L [105].

Loading of magnetic nanoparticles onto carbon adsorbents is reported to make the separation of dyes from their solutions easier, including Ni particles on macro-/mesoporous carbons [106], Fe_3O_4 nanoparticles onto rGO [107], and Fe nanoparticles onto ordered mesoporous carbon [108].

4.1.2.6 Biomolecules

Ordered mesoporous carbide-derived carbon (OM-CDC) prepared from mesoporous SiC was shown to adsorb glucose oxidase from its aqueous solution, with the capacity reaching 800 mg/g, which is higher than ordered mesoporous carbon CMK-3 (600 mg/g) [109]. Templated macro-/mesoporous carbons synthesized from 5 vol.% furfuryl alcohol solution delivered high sorption capacity for bilirubin, with the maximum capacity reaching 885 mg/g [110].

Sorption of macrobiomolecules was studied by using EG [111]. The EG column had a sorption capacity of 4.3 g/g for bovine serum albumin (BSA) from its 5 mg/cm^3 solution, which is superior to a petrolatum gauze. Sorption performances of EG with a bulk density of 5.56 g/L for macrobiomolecules [ovalbumin, serum albumin, bovine serum albumin (BSA), lysine and herring sperm DNA] are summarized [112], with some of the results being shown in Fig. 4.35. The sorption capacity and rate depend greatly on biomolecules (Fig. 4.35A) and sorption capacity is governed by the bulk density of EG. Low density is preferable to obtain high capacity (Fig. 4.35B). The dependence of the sorption capacity of EG is compared with heavy oil, gasoline, and BSA, three sorbates having the same dependency on bulk density of EG and heavy oil being sorbed into EG larger amount (Fig. 4.35C). Loading of $MnFe_2O_4$ onto rGO flakes was efficient in enhancing the removal of glyphosate [N-(phosphonomethyl) glycine, an organophosphorus pesticide] from water [113].

4.1.3 Water vapor adsorption

Although water adsorption is undesirable in some applications, such as VOC removal in air as explained in Section 4.1.2.1, its adsorption is required in applications such as adsorption heat pumps (AHPs) and desiccant humidity conditioners (DHCs) because these processes require a large change in the amount of adsorbed water over the operational pressure range.

Commercially available ACCs produced from phenol resin fibers by carbonization and simultaneous activation with steam and/or CO_2 at 900°C include ACC-15 having S_{BET} of 730 m^2/g with V_{total} of 0.379 cm^3/g, ACC-20 having 1320 m^2/g with 0.694 cm^3/g, and ACC-25 having

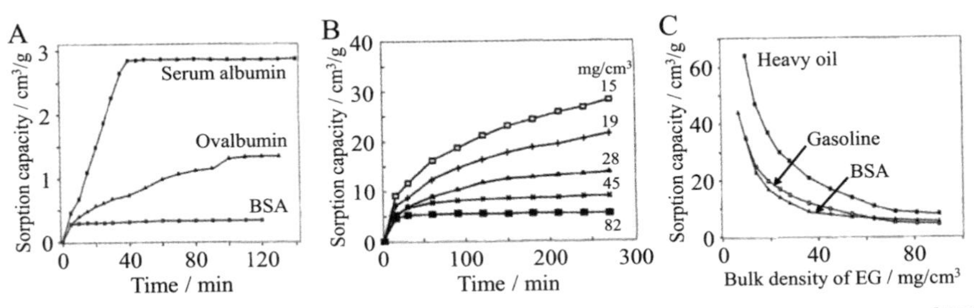

Figure 4.35 Sorption capacity of EG for macrobiomolecules: (A) sorption curves of EG with a bulk density of about 6 mg/cm^3, (B) sorption curves of EG with different bulk densities for BSA, and (C) comparison with heavy oil, gasoline, and BSA [112].

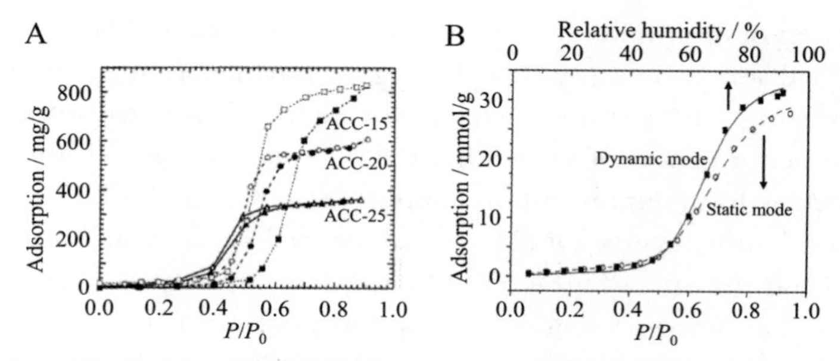

Figure 4.36 Adsorption–desorption isotherms for water vapor: (A) on ACCs with different S_{BET}'s and (B) measured by dynamic and static adsorption modes on a commercial granular AC with S_{BET} of 1493 m^2/g [114,115].

1860 m^2/g with 1.023 cm^3/g, respectively [114]. Adsorption–desorption isotherms of ACCs for water vapor are shown in Fig. 4.36A, presenting a characteristic isotherm for water vapor, a negligibly small amount of water adsorbed at low P/P_0 (<0.3), an abrupt increase in the amount of adsorbed water at P/P_0 of 0.4–0.6, and marked hysteresis at high P/P_0. With increasing S_{BET} of the adsorbent, a rapid increase in adsorption begins at higher P/P_0, the width of the hysteresis loop becomes wider, and the saturated adsorption capacity becomes higher. These characteristics of the adsorption-desorption isotherm for water vapor are quite different from those for organic vapors, such as benzene, etc. Adsorption isotherms were measured by dynamic and static modes on commercial microporous ACs [115]. Isotherms measured with dynamic and static modes on an AC having S_{BET} of 1493 m^2/g are compared in Fig. 4.36B, with both isotherms being similar to each other and revealing characteristically small adsorption at low P/P_0 and an abrupt increase at around 0.5 of P/P_0.

Water adsorption was studied at 25°C on ACFs, which were prepared from commercial pitch-based carbon fibers by activation using either CO_2 (ACF-C) or steam (ACF-S) with different degrees of burn-off [116]. The observed adsorption isotherms are shown in Fig. 4.37A, revealing that the saturated adsorption capacity of the ACFs corresponds to their V_{micro} (shown in parenthesis). In Fig. 4.37B, N_2 adsorption isotherms measured at 77K on the ACFs containing different amounts of water are shown on ACF-C1. The adsorption of even a small amount of water of 4 wt.% had a marked effect on N_2 adsorption because the adsorbed water molecules are preferentially accommodated in micropores. None of the ACFs saturated with water show any N_2 adsorption or CO_2 adsorption, which indicates that all the micropores were filled with water.

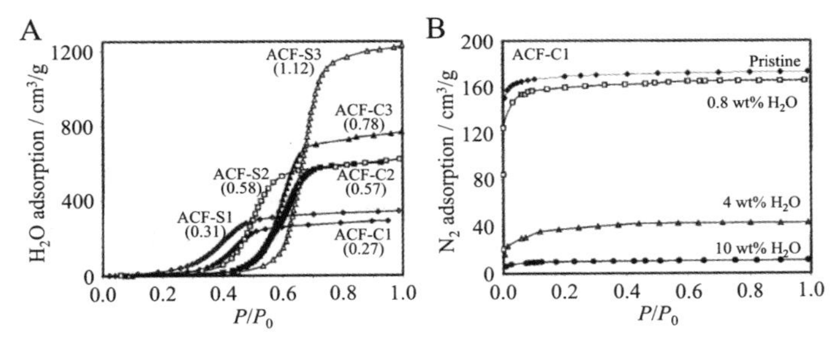

Figure 4.37 Water adsorption of ACFs activated by CO_2 and steam (ACF-Cs and ACF-Ss): (A) adsorption isotherms of different ACFs, V_{micro} in cm^3/g being shown in parentheses for each ACF, and (B) N_2 adsorption isotherms on the ACF-C1 with different water contents [116].

Carbon film, which is prepared by carbonization of commercial polyimide film (Kapton) at 1000°C in Ar flow and has negligibly low S_{BET}, exhibits some adsorption capacity for water vapor, as shown in Fig. 4.38A, although it is much smaller than an ACF with S_{BET} of about 1000 m^2/g [117]. The water adsorption capacity of the carbon film increases by oxidation at 375°C for 1 h in air flow. Microporous carbon films were prepared from the films of polyimides (PIs) containing CF_3 groups in their repeating units by simple carbonization at 600–1000°C in N_2 [118] and the water adsorption behaviors of these microporous carbon films have been studied [119]. The adsorption isotherms of this series of carbon films are shown in Fig. 4.38B, where the adsorption capacity at P/P_0 of 0.95 increases with the increase in F content in the precursor PIs, from 0 to 31.3 wt.% and S_{BET} of carbon films from 418 to 840 m^2/g. In Fig. 4.38C,

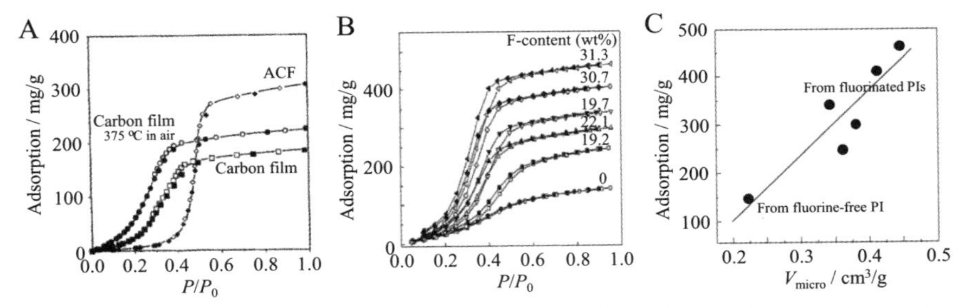

Figure 4.38 Adsorption of water vapor by carbon films derived from various polyimide (PI) films: (A) the adsorption isotherms of the carbon films derived from a polyimide (Kapton) with and without oxidation at 375°C in air, in comparison with a commercial ACF, (B) the isotherms of microporous carbon films derived from fluorinated PI films, and (C) relation between adsorption capacity for water and V_{micro} of the carbon films [117,119].

the adsorption capacity of these carbon films for water is plotted against V_{micro} of the carbon films, suggesting an important contribution of micropores for water adsorption. The carbonization temperature was shown to govern the appearance of the hysteresis loop in the adsorption isotherm for water vapor. Hysteresis in the isotherm appears on the films carbonized at temperature in the range of 1000−1400°C, and the carbon films carbonized at above 1400°C cannot adsorb water vapor because of the negligibly small V_{micro}.

Zeolite-templated carbons (ZTCs) were prepared by two-step carbonization, i.e., impregnation of furfuryl alcohol into zeolite Y, followed by polymerization and carbonization, and then infiltration by CVD of either acetonitrile at 800°C or polypropylene at 700°C [120]. For comparison, the template zeolite was infiltrated by CVD of acetonitrile at 700°C for 1 h, followed by CVD of propylene at 800°C for 0.5 h. The preparation conditions are shown together with the resultant pore structure parameters in Table 4.6. Water adsorption isotherms measured on these ZTCs are shown in Fig. 4.39, revealing sharp increases in the adsorption capacity for water at P/P_0 around 0.5 with a narrow hysteresis loop. The saturated adsorption capacity for water corresponded well to V_{micro} of the adsorbent ZTCs.

Water adsorption into mesoporous carbon was reported to proceed in two steps, associating with a marked enhancement by N-doping into

Table 4.6 CVD conditions and pore structure parameters for zeolite-templated carbons (ZTCs). [120].

	1st step (impregnation)	2nd step (chemical vapor deposition)	Pore structure parameters		
Carbon	Polymer and carbonization temperature	Precursor gas, temperature and time	S_{BET} (m^2/g)	V_{micro} (cm^3/g)	V_{meso} (cm^3/g)
ZTC-A	Furfuryl alcohol, 800°C	Acetonitrile, 800°C, 2 h	3310	1.26	0.33
ZTC-P	Furfuryl alcohol, 800°C	Propylene, 700°C, 1 h	4070	1.78	0.23
ZTC-(P+A)	No impregnation	Propylene, 700°C, 1 h and then acetonitrile, 800°C, 0.5 h	3410	1.35	0.11

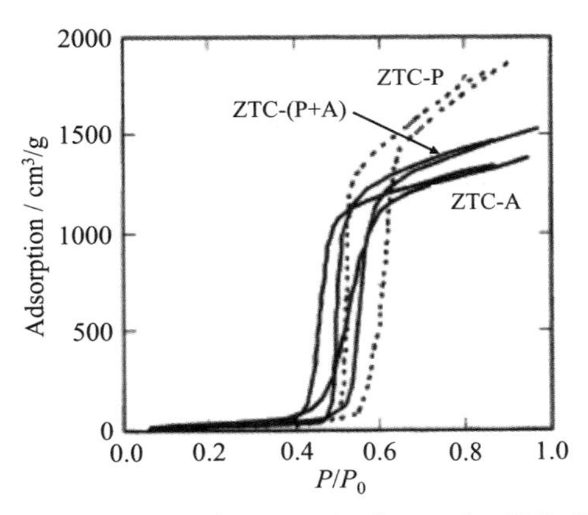

Figure 4.39 Water adsorption isotherms for ZTCs [120].

mesoporous carbon [121,122]. Mesoporous carbon cryogels were prepared by freeze-drying of hydrogels, which were derived from resolcinol (R) and formaldehyde (F) with K_2CO_3 catalyst (C) in R/F molar ratio of 0.50 with different molar ratios of R/W (0.04–0.08) and R/C (50–200), after replacing the water in the hydrogels with *t*-butanol, followed by carbonization at 900°C in N_2 flow. N-doping of the cryogels was performed by carbonization of the hydrogels at 900°C in N_2 flow containing 20% NH_3. In Fig. 4.40, water adsorption isotherms of the RF-derived carbon cryogels are shown comparing the pristine and N-doped gels [122]. The isotherms of both pristine and N-doped carbon gels with R/C = 100 and 200 show two-step isotherms, with the second step occurring at about P/P_0 of

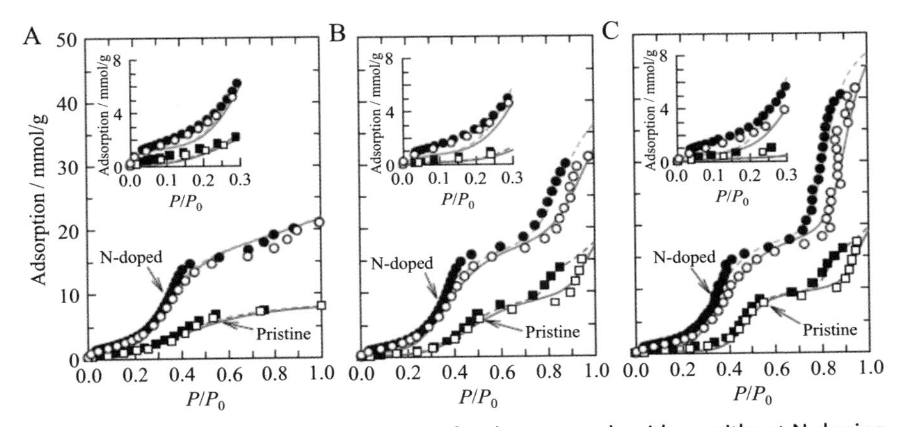

Figure 4.40 Water adsorption isotherms of carbon cryogels with or without N-doping prepared by R/W of 0.04: the gels with R/C of (A) 50, (B) 100, and (C) 200 [122].

0.6—0.8. N-doping was effective in enhancing water adsorption on mesoporous carbon gels prepared with various synthesis conditions (R/W and R/C ratios); the adsorption capacities of the N-doped gels in the medium and high P/P_0 ranges are about twice those for the pristine gels (nondoped), and even three times that in the low P/P_0 range (insets of Fig. 4.40). Higher capacity in a low P/P_0 range was thought to be due to the fact that doped N atoms act as strong adsorption sites for water molecules and, as a consequence, the onset of water adsorption of N-doped carbons shifts to the lower P/P_0. N-doping was reported to be carried out by CVD of pyridine at 800 and 1000°C onto pitch-based carbon fibers, while there was no enhancement in water adsorption, but rather marked depression of capacity was observed, probably because of blocking of the micropores in the carbon fibers with the products of CVD process. Marked decreases in pore structure parameters were also confirmed [123].

Mesoporous and microporous carbons were synthesized by chlorination of mesoporous SiC, which were derived by infiltration of polycarbosilane into ordered mesoporous silica SBA-15, with subsequent pyrolysis at 700 or 1000°C and HF washing [124]. The resultant carbons had ordered mesopores consisting of microporous carbon walls. The water adsorption isotherms of the carbons with different post-treatments (heat treatment in H_2 to eliminate residual Cl and flash heating by IR) are shown by comparing them with SiC-derived microporous carbon (CDC) and ordered mesoporous carbon using SBA-15 template (CMK-3) in Fig. 4.41. Water adsorption isotherms exhibited the characteristic shape on adsorption and marked hysteresis on desorption. Post-treatments do not give

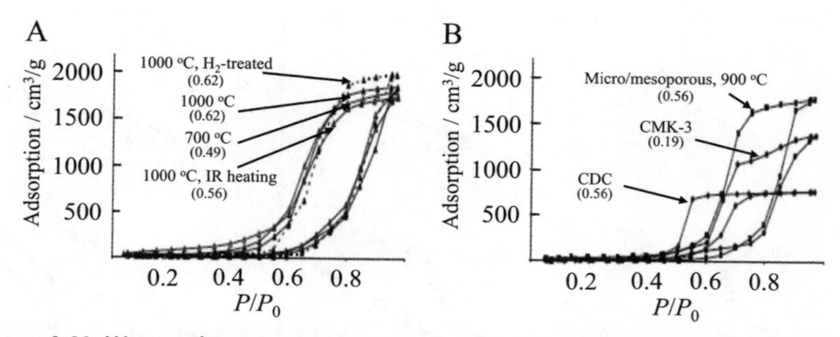

Figure 4.41 Water adsorption isotherms of meso-/microporous carbons: (A) carbon prepared different conditions, and (B) comparison with microporous carbide-derived carbon (CDC) and mesoporous SBA-15-templated carbon (CMK-3). V_{micro} values (cm^3/g) of the carbons are shown in parentheses [124].

appreciable effect on water adsorption–desorption, as shown in Fig. 4.41A. As shown in Fig. 4.41B, the isotherm of CDC without mesopores exhibits raising-up of capacity at lower P/P_0 and a much lower water adsorption capacity than the resultant meso-/microporous carbons, although V_{micro}'s are almost the same. Mesoporous carbon CMK-3 with low V_{micro} gives a larger adsorption capacity than CDC, and similar hysteresis but lower capacity to those of the resultant meso-/microporous carbon.

Commercial ACs derived from wood (W) and peat moss (N) were further oxidized (post-oxidation) either in 73% HNO_3 at 78°C (W1 and N1) or 30% H_2O_2 at 50°C (W2 and N2) to modify the surface functional groups on the AC surface [125]. Water adsorption isotherms are shown in Fig. 4.42, revealing that the post-oxidation depresses the adsorption capacity for water at high P/P_0 range, but enhances those at low P/P_0 range. The post-oxidation also reduced the pore structure parameters, S_{BET} and V_{micro}, as shown in Table 4.6. The water adsorption capacity at high P/P_0 range can be related to V_{micro} of the carbon adsorbents. In contrast, the amounts adsorbed at low P/P_0 range, for example at $P/P_0 = 0.3$ in Fig. 4.42, show correspondence to the density of oxygen-containing functional groups measured by Boehm and potentiometric titration methods, as shown in Table 4.7. These experimental results suggest that water adsorption is due to the interaction of water molecules to these functional groups. On CH_3OH, its adsorption capacity at this low P/P_0 is not related to the existence of functional groups (Table 4.7).

A kinetic study was carried out on the adsorption of water vapor using commercial AC derived from coconut shell by steam activation [126].

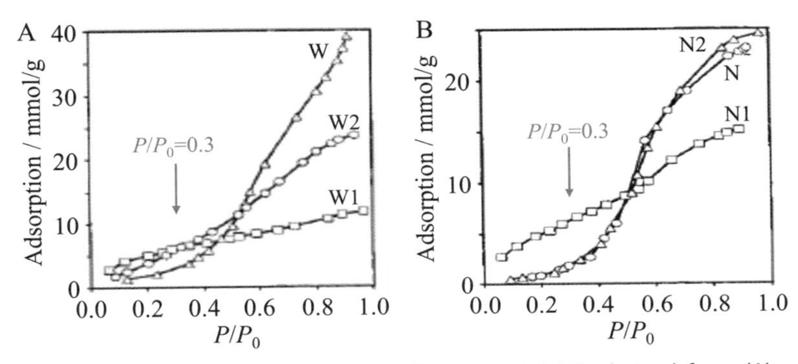

Figure 4.42 Water adsorption isotherms of commercial ACs derived from (A) wood (W) and (B) peat moss (N) at 25°C. ACs were post-oxidized by 73% HNO_3 at 78°C (W1 and N1) and by 30% H_2O_2 at 50°C (W2 and N2) [125].

Table 4.7 Pore structure parameters and adsorbed amount of water and CH_4 at P/P_0 of 0.3 [125].

ACs	Oxidation	Pore parameters		O-containing groups (group/nm^2)		Amount adsorbed (molecule/nm^2)	
		S_{BET} (m^2/g)	Vmicro (cm^3/g)	Boehm method	Potentiometric method	H_2O	CH_3OH
W	As-received	1500	0.642	0.14	0.14	1.31	4.20
W2	H_2O_2	860	0.378	1.05	1.05	4.21	5.89
N	As-received	970	0.490	0.44	0.44	1.08	4.96
N1	HNO_3	625	0.291	2.65	2.65	6.20	6.82
N2	H_2O_2	860	0.426	0.65	0.65	1.32	6.20

The rates of adsorption and desorption varied with the P/P_0 region of the isotherm; the highest rate was observed at the low P/P_0 region and the lowest rate was observed in the region from 0.55 to 0.7. The introduction of nitrogen into the carbon adsorbent by treatment with formamide (1/5 by weight) at 800°C resulted in a marked lowering of the adsorption rate at P/P_0 around 0.4.

Co-adsorption of water vapor with organic vapor was also studied. Water adsorption isotherms of a commercial microporous AC (BPL) using the gases of mixed water vapor with hexane and acetone in different amounts are shown in Fig. 4.43A and B, respectively [127]. The saturated adsorption capacities decrease significantly with increasing hexane and acetone loading, associated with shrinking of the hysteresis loop. For the mixed gas with 0.5 mmol/g hexane, the saturated adsorption capacity for water decreases to about 16 mmol/g from 22 mmol/g for pure water (Fig. 4.43A). By taking into consideration the fact that the adsorption capacity of pure hexane near saturation was approximately 4 mmol/g and by assuming that the same pores are accessible to hexane and water, the pore space available for water adsorption was calculated as 87%. However, the experiment suggests that water fills only 79% of the available pore space. For the gases with 1.0 and 2.0 mmol/g hexane, similarly, water fills only 55% and 21% of the remaining pore space, respectively. For the mixed gases with 1.9 and 3.8 mmol/g acetone, however, approximately 73% and 47% of the pore space is available for water adsorption and practically 90% of

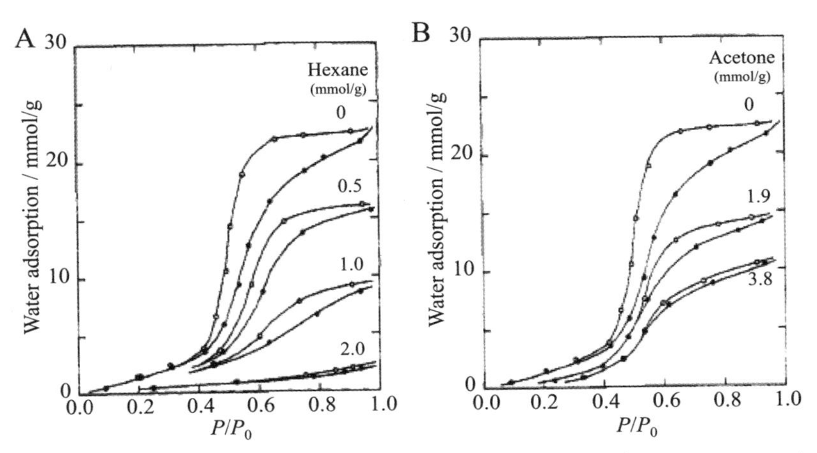

Figure 4.43 Adsorption–desorption isotherms of an AC at 25°C for water vapor mixed with (A) hexane and (B) acetone in different amounts [127].

these available pore spaces are filled with water (Fig. 4.43B). At 100°C, the water filled the available pore space much more effectively even for mixed gases with hexane than at 25°C. Co-adsorption of water with 1,1,2-trichloro-1,2,2-trifluoroethane and dichloromethane was also examined [128].

Carbon foams prepared from polyimides using urethane foam as template demonstrated high cyclability for water adsorption–desorption at room temperature, as shown in Fig. 4.44 [129,130]. Before starting the cycle, the foams were heated at 200°C in dry N_2 to exclude the adsorbed gases. The weight increases in wet N_2 flow due to water adsorption and decreases in dry N_2 flow due to water desorption are highly reproducible. Water uptake of carbon foams was linearly related to V_{micro} of the foam, as obtained microporous carbon films derived from polyimides (refer to Fig. 4.38C).

The mechanism for the adsorption of water vapor into porous carbons has been discussed using theoretical calculations [122,131,132] and by assuming the cluster formation [121,133–135]. Here, water vapor adsorption is briefly discussed phenomenologically on the bases of the experimental results described above. The adsorption isotherm of porous carbons for water vapor may be characterized by four steps: (1) small amount of adsorption at low P/P_0 range (mostly 0.1–0.5), (2) abrupt increase in adsorption at medium P/P_0 range, (3) saturation of adsorbed amount of water at high P/P_0 range, and (4) marked hysteresis between adsorption and desorption. Step (1) is mainly due to the interaction of water

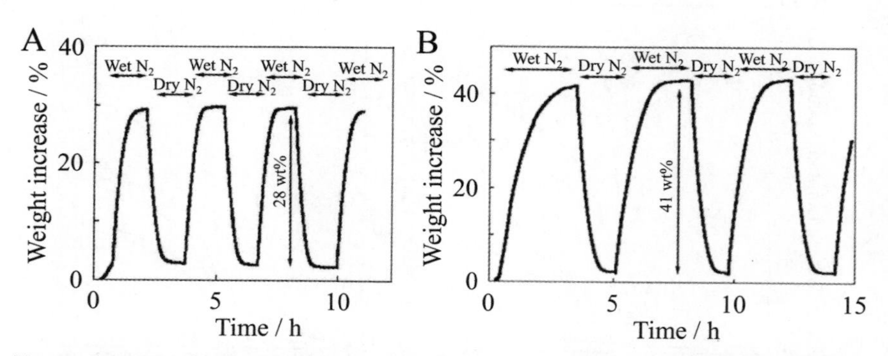

Figure 4.44 Weight changes of the carbon foams from fluorinated PI (6FDA/TFMD) in a flow of dry and wet N_2 gas: (A) carbon foam prepared at 700°C and (B) at 1000°C [130].

vapor with oxygen-containing functional groups on the surface of the carbon adsorbent, as shown in Fig. 4.42 and Table 4.7 [125]. N-doping into the carbon adsorbent depressed the adsorption rate in this step [126]. Step (2) is pore filling of water vapor by forming clusters. The zeolite-templated carbon consisting of ordered micropores with homogeneous sizes demonstrating the isotherm with a sharp increase in adsorption and relatively narrow hysteresis at P/P_0 around 0.5 [120], as shown in Fig. 4.39. Commercial ACs consisting of a broad size distribution of micropores, however, show a sharp increase in capacity at a slightly higher P/P_0, while the hysteresis becomes wider [114], as shown in Fig. 4.36A. Mesoporous carbons present a raising-up of the capacity at much a higher P/P_0 of 0.8 and much wider hysteresis than microporous carbons, as shown by comparing microporous carbon (CDC) and ordered mesoporous carbon (CMK-3) in Fig. 4.41B [124]. When the carbon contains micropores and mesopores in comparable volumes, such as RF-derived carbon cryogels prepared with R/W = 200 and R/C = 0.04, the adsorption isotherm shows two step increases in adsorbed water [122], as shown in Fig. 4.40B and C. In the case of silica-templated SiC-derived carbon which contains micropores derived from the precursor carbide and ordered mesopores templated by mesoporous silica, however, isotherms delivered single and broad hysteresis, as shown in Fig. 4.41A [124]. The saturated adsorption capacity observed in step (3) is mainly determined by the contributions of micropores and mesopores because the contribution of functional groups in step (1) is usually much smaller. The hysteresis in the isotherm during step (4) has been explained by a constriction at the entrance of the pore for releasing water molecules. In zeolite-templated carbon, the constriction is supposed to be homogeneous and weak because of homogeneous sizes and ordered arrangement, which results in narrow hysteresis, as shown in Fig. 4.39. In the cases of carbide-derived carbon, silica-templated mesoporous carbon, and the carbons coexisting with micropores and mesopores shown in Fig. 4.41A and B, the constrictions from the pores are thought to distribute in a wide range, resulting in wide hysteresis.

Water vapor adsorption on porous carbons can have a sharp uptake at medium or high P/P_0 and so carbon adsorbent is possible to provide an ideal system for adsorption heat pumps (AHPs) and desiccant humidity conditioners (DHCs). The application of porous carbons on AHPs was briefly described in the previous chapter (Section 3.7.2) in relation to thermal energy storage.

4.1.4 CO_2 capture

Global warming is now understood to be caused mainly by greenhouse gas emissions, such as CO_2 and CH_4. Particularly, CO_2 is considered to be the main cause of global climate change, because its concentration in the atmosphere has increased significantly, and so the capture and storage of CO_2 (CCS) to reduce CO_2 concentration are urgent and global problems, as well as the reduction of CO_2 emission by substitution of fossil fuels with low or noncarbon fuels, together with improvements in energy efficiency.

The most popular CO_2 capture method is to absorb it by liquid amine and ammonia, but its practical application suffers from many problems such as high regeneration energy consumption, equipment corrosion, and toxicity. As promising alternatives, many porous solids, including zeolites, porous carbons, metal-organic frameworks (MOFs), etc. have been widely investigated and reviewed from various viewpoints [136–145]. Porous carbons were also used to support the adsorbents for CO_2 capture, such as amines. For cost-efficiency, biomass-derived carbons have been proposed for CO_2 capture in many literatures.

CO_2 adsorption is well known to occur by micropore filling and its adsorption capacity is conventionally employed as one of the pore structure parameters, expressed as $V_{micro}(CO_2)$ to evaluate the volume of narrow micropores in carbons. For the characterization of the pore structure of carbon materials, $V_{micro}(CO_2)$ measured at 0°C is commonly employed and its correspondence to $V_{micro}(N_2)$ calculated from the N_2 adsorption isotherm at 77K has been discussed [146–148]. From the viewpoint of CO_2 capture for environment remediation, however, the adsorption temperature and gas pressure are important factors, in addition to the adsorption capacity, for governing the CO_2 capturing process.

4.1.4.1 CO_2 capture by carbons

Microporous carbons prepared from petroleum pitches with different mesophase contents by KOH activation could deliver a CO_2 adsorption capacity comparable with zeolite [149]. Carbon monolith prepared from vacuum residues activated at 800°C, which had S_{BET} of 2450 m^2/g and volume of narrow micropore $V_{micro}(CO_2)$ of 1.42 cm^3/g, could adsorb 380 mg/g CO_2 at 0°C and 185 mg/g at 25°C, although the powder prepared using the same procedure and precursor gave CO_2 adsorption of 236 mg/g at 0°C with S_{BET} of 3100 m^2/g and V_{narrow} of 0.85 cm^3/g. An anthracite with a particle size of 150–250 μm was activated in a flow of steam (65.8%) in a fluidized bed at a temperature between 700 and 890°C

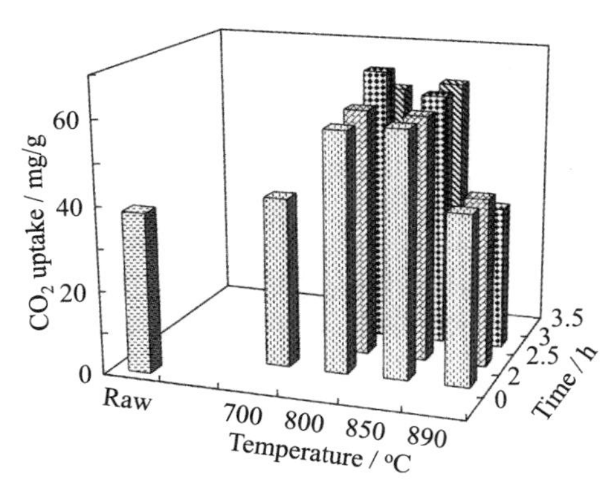

Figure 4.45 CO_2 uptake at 30°C of the anthracites activated by steam at different temperatures and times [150].

for 2–3.5 h, with the CO_2 uptake being measured [150]. As shown in Fig. 4.45, CO_2 uptake at 30°C depends greatly on the activation temperature and time, a maximum of 65.7 mg/g being obtained on the carbon activated at 800°C for 3 h. This activation condition gave S_{BET} of 607 m^2/g, which is much lower than the highest S_{BET} of 1071 m^2/g at 890°C for 2 h. CO_2 uptake depended also on the temperature of the measurement, 58 mg/g at 30°C and 28 mg/g at 75°C for the anthracite activated at 850°C for 2.5 h. A sub-bituminous coal after demineralization was carbonized at 900°C for 1 h in N_2 flow after mixing with K_2CO_3 via its solution (K_2CO_3/coal of 2/100 by weight), followed by activation with CO_2 [151]. The resultant carbon delivered a CO_2 uptake of 4.36 mmol/g at 0°C and 1 bar.

Potassium oxalate monohydrate (PO) mixed with melamine was found to work as an activation agent giving a high S_{BET} of more than 3000 m^2/g and a high V_{micro} of 1 cm^3/g on the carbon derived from glucose [152]. The mixtures of α-D-glucose with PO (PO/glucose of 1.8–3.6 by weight) and with or without melamine (Mel) (Mel/glucose of 0–3 by weight) in powder were carbonized at 800 and 850°C in N_2 flow, followed by washing with diluted HCl. The preparation conditions and pore structure parameters of the resultant carbons are summarized in Table 4.8, together with CO_2 uptake obtained at 25°C and different pressures. By using PO, microporous carbons were obtained, which exhibited high V_{micro}, about 90% of V_{total}, and relatively high CO_2 uptake of about 4.5 mmol/g at 25°C and 1 bar. By using PO with melamine, micro- and mesoporous carbons with extremely high S_{BET} of higher than 3000 m^2/g and high V_{micro} of

Table 4.8 Preparation conditions, pore structure parameters, and CO_2 uptakes at different pressures of the porous carbons prepared from glucose by chemical activation using potassium oxalate (PO) and melamine (Mel) [152].

	Mixing ratio (agent/glucose)		Activation conditions		Pore structure parameters		CO_2 uptake at 25°C (mmol/g)			
Code	PO	Mel	Temperature (°C)	Time (h)	S_{BET} (m²/g)	V_{micro} (cm³/g)	0.15 bar	1 bar	30 bar	50 bar
G800	3.6/1	—	800	1	1270	0.49 (89)[a]	1.4	4.5	—	—
G850			850	1	1330	0.51 (96)	1.2	4.2	—	—
G850-5			850	5	1690	0.67 (93)	1.1	4.5	14.3	15.4
G1.8-2	1.6/1	2/1	800	1	1240	0.47 (68)	—	—	—	—
G2.3-2	2.3/1	2/1	800	1	1520	0.52 (54)	—	2.3	15.8	22.3
G2.7-2	2.7/1	2/1	800	1	3310	1.00 (42)	—	2.1	31.7	45.9
G3.6-2	3.6/1	2/1	800	1	3460	1.00 (37)	—	1.5	32.6	49.1
G3.6-1	3.6/1	1/1	800	1	3470	1.10 (46)	—	2.5	31.9	44.4
G3.6-3	3.6/1	3/1	800	1	870	0.30 (51)	—	—	—	—

[a]Percentage of micropore volume in V_{total}.

1.1 cm^3/g could be obtained, although the percentage of V_{micro} was smaller (about a half of V_{total}), which delivered very high CO_2 uptake under high pressure. The carbon synthesized from the mixture with PO/glucose of 3.6/1 and Mel/glucose of 2/1 delivered high S_{BET} of 3460 m^2/g and V_{micro} of 1.00 cm^3/g, together with high CO_2 uptake of 49.1 mmol/g at 25°C and 50 bar, whereas CO_2 uptake at 1 bar is not as high as 1.5 mmol/g. The results suggest that narrow micropores (<1 nm size) are responsible for CO_2 adsorption under pressures of less than 1 bar, while supermicropores (0.7–2.0 nm size) work as additional sites for storing CO_2 at high pressures of 30–50 bar.

Carbon foams, of which the cell walls and struts contained micropores, were synthesized from sucrose using $Al(NO_3)_3$ (AN) foaming agent and carbonizing at 900°C [153]. As shown in Fig. 4.46A, S_{BET} and V_{total} determined from N_2 adsorption–desorption isotherms increase with increasing AN content, principally due to the development of micropores, revealing that AN not only aids the faster foaming and setting of molten sucrose but also contributes to the formation of micropores in the carbon cell walls and struts of the foam. However, CO_2 uptake of the foams is very similar for the carbons synthesized using different amounts of AN and no marked change in CO_2 adsorption isotherms were detected, as shown in Fig. 4.46B, although the nanopore structures in the cell walls and struts of the foams are quite different, as shown in Fig. 4.46A. Cyclability for CO_2 adsorption–desorption at 25°C is excellent, as shown in Fig. 4.46C, showing the stability of the foam structure. The mixtures of an acid-treated tar pitch and coal in different ratios with KOH were heated in a quartz crucible with a lid at 800°C [154]. The resultant AC contained macropores and had S_{BET} of 725–1670 m^2/g and V_{micro} of 0.32–0.73 cm^3/g, and delivered CO_2 uptake of 7.4 mmol/g at 20°C and 3.5 MPa.

Microporous carbons were prepared from waste celtuce leaves collected from a local vegetable market by air-drying and carbonization at 600°C in Ar, followed by KOH activation at 800°C [155]. The resultant carbon, of which S_{BET} and V_{total} reached 3404 m^2/g and 1.88 cm^3/g, delivered high CO_2 uptakes of 6.04 and 4.36 mmol/g at 0 and 25°C, respectively. Compaction of ACs was experimentally shown to influence CO_2 uptake, adsorption rate, and CO_2/N_2 selectivity using ACs derived from coconut shell and rice husk, volumetric CO_2 uptake, and selectivity at 0°C increasing from 1.85 to 2.65 $mmol/cm^3$ and from 14 to 23, respectively, with compaction up to 887 MPa [156]. Microporous carbons were prepared from a benzoxazine [5-(4-hydroxystyryl)benzene-1,3-diol] mixed

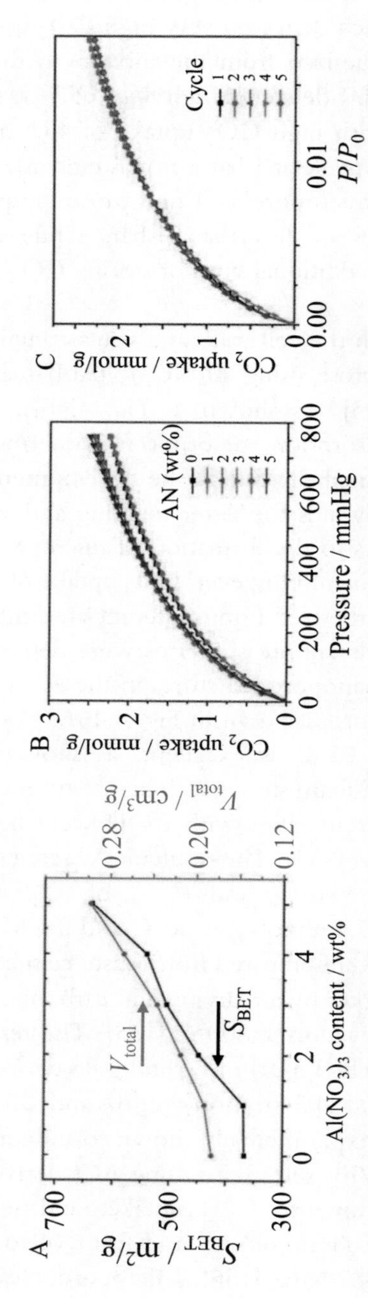

Figure 4.46 Carbon foams synthesized by carbonization of the mixture of sucrose with $Al(NO_3)_3$ (AN) at 900°C: (A) changes in S_{BET} and V_{total} with AN content, (B) CO_2 adsorption isotherms at 25°C for the foams prepared at different AN contents, and (C) cycling performance of CO_2 adsorption at 25°C [153].

with formaldehyde and aniline by curing, polymerization, and carbonization at 500–900°C, followed by activation using KOH. The resultant carbons delivered high CO_2 uptakes and high selectivity from the CO_2/N_2 mixed gas the carbon obtained at 800°C exhibited 8.44 mmol/g and 56 at 0°C, respectively, under 1 bar [157]. An AC prepared from cellulose-rich waste tissue papers mixed with PVA by carbonization at 900°C delivered a CO_2 uptake of 5.14 mmol/g at 0°C [158].

Porous aromatic frameworks (PAFs) are known to have high S_{BET} and high CO_2 uptake at high pressure with high selectivity. One PAF, PAF-1, which was composed of local diamond-like tetrahedral bonding of tetraphenylene methane building units, was found to have an unprecedented high surface area of 7100 m^2/g and high CO_2 uptake of 1.3 g/g at 25°C and 40 bar [159]. For practical applications of CO_2 capture, however, improved CO_2 uptake and selectivity at ambient condition are strongly demanded. To improve CO_2 capture performance, carbonization of PAF-1 was carried out at different temperatures between 350 and 450°C in N_2/O_2 (1.5 vol% O_2) for 6 h [160]. The resultant carbonized PAF-1 was obtained as brown powders, of which S_{BET} and V_{micro} decreased with increasing carbonization temperature from 5300 m^2/g and 1.44 cm^3/g for the pristine to 1191 m^2/g and 0.36 cm^3/g, respectively, for 450°C-carbonized. However, CO_2 uptake at 0°C and 1 bar increased by carbonization, as shown by the adsorption isotherms in Fig. 4.47A, reaching 4.5 mmol/g (165 mg/g) for the carbon obtained at 450°C. This marked increase in CO_2 uptake at 1 bar was thought to be mainly due to the increase in narrow micropores after carbonization, as shown by the cumulative micropore volume distribution in Fig. 4.47B. The volume of micropores with sizes less than 1.2 nm increases with increasing carbonization temperature. The 450°C-carbonized PAF-1 has quite different adsorption capacities for gas species, much higher uptake for CO_2 but very small for N_2 and H_2, as shown in Fig. 4.47C, suggesting high selectivity of CO_2 from N_2 and H_2. Microporous carbon was synthesized from PAF-1 by adding an additional carbon source (FA) and carbonizing up to 900°C, which delivered a CO_2 uptake of 4.1 mmol/g at 22°C and 1 bar [161]. Before mixing with FA, PAF-1 was grafted with sulfonic acid, which served as the catalytic site for the polymerization of FA. Porous carbons synthesized by gelation and carbonization of the mixture of Zn-containing MOF with RF resin exhibited CO_2 uptake of 2.9 mmol/g at 300K and atmospheric pressure, which depended on the condition of gelation of RF resin [162]. A rigid polymer porous framework synthesized via polycondensation reaction of 1,3,5-THB and

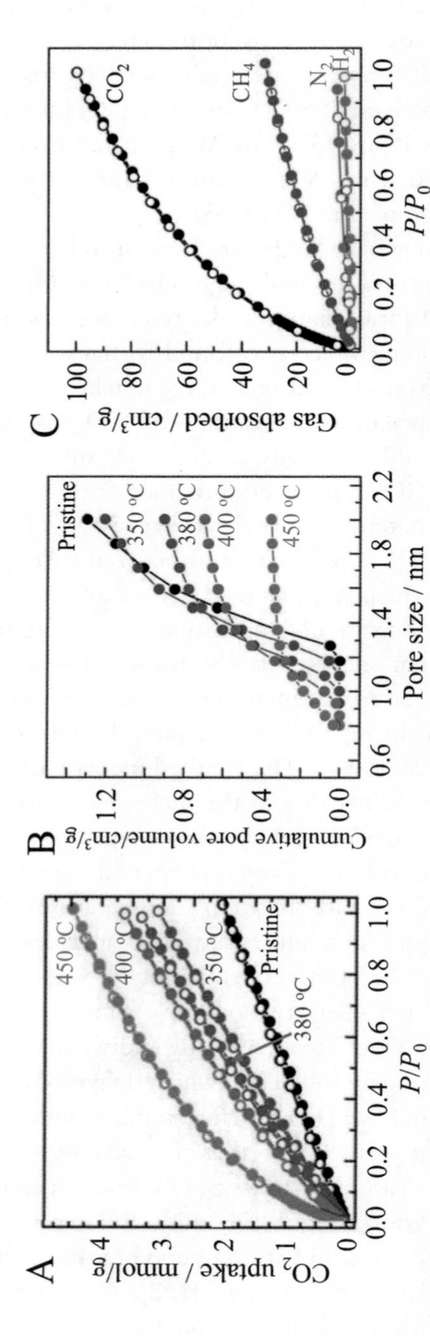

Figure 4.47 Carbonized porous aromatic framework (PAF-1): (A) CO_2 adsorption isotherms and (B) cumulative pore volume distributions in micropore range for PAF-1 carbonized at different temperatures, and (C) adsorption isotherms of 450°C-carbonized PAF-1 for different gases [160].

aniline was converted to N-doped porous carbons by carbonization at 600–800°C for 2 h in N_2 flow after mixing with KOH (KOH/framework of 4/1 by weight) [163]. The carbons obtained at 600 and 700°C delivered CO_2 uptake of 6.0 and 6.3 mmol/g at 0°C under 1 bar and 4.0 and 3.7 mmol/g at 25°C under 1 bar, respectively, which had S_{BET}, S_{micro} and V_{total} of 1129 m^2/g,1065 m^2/g, and 0.47 cm^3/g, respectively, for the former and 2363 m^2/g, 2155 m^2/g, and 1.09 cm^3/g, respectively, for the latter.

Carbide-derived carbons (CDCs) with different surface oxygen contents were prepared from TiC powder by chlorination at 700°C and annealed at 600°C in H_2 [164]. CDCs were oxidized in concentrated HNO_3 at 50 and 80°C for 3.5 h (CDC-50 and -80, respectively), followed by reduction at 800°C in H_2 (CDC-50-HR and -80-HR, respectively). As shown in Fig. 4.48A, HNO_3 treatment at 50°C enhances CO_2 uptake of the CDC and H_2 reduction reduces the uptake, reducing to even smaller than the pristine, which was explained by the increase and decrease in oxygen-containing functional groups through oxidation and reduction. As shown in Fig. 4.48B, CO_2 uptake does not correspond with V_{micro} measured by N_2 adsorption (inset), but the CO_2 uptake per surface area (CO_2 uptake/ S_{BET}) does correspond with the oxygen content of the adsorbent carbon, suggesting a hydrogen bonding interaction between the functional groups on the carbon surface and CO_2 molecules.

N-doping was reported to be effective in enhancing the CO_2 adsorption capacity of carbon materials. The CO_2 adsorption behavior was

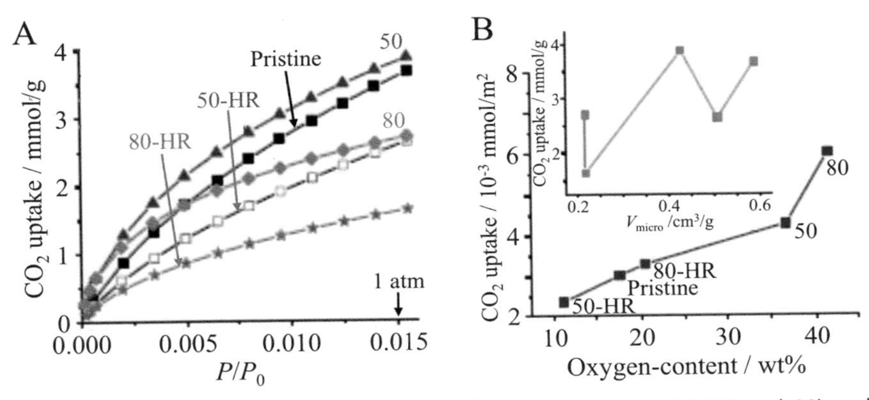

Figure 4.48 TiC-derived carbons oxidized in HNO_3 at 50 and 80°C (50 and 80) and reduced in H_2 (50-HR and 80-HR) with the pristine one: (A) CO_2 adsorption isotherms and (B) relations of CO_2 uptake to oxygen content and V_{micro} (inset) [164].

compared for three microporous ZTCs with and without N-doping [165]. The carbon denoted as Y1 was prepared by polymerization and 700°C carbonization of furfuryl alcohol (FA) in zeolite Y, followed by CVD infiltration of propylene at 900°C, with the carbon Y2 prepared by CVD infiltration of propylene at 800°C, followed by annealing at 900°C, and the carbon E prepared by the same procedure as the carbon Y1 using zeolite EMC-2 as the template instead of zeolite Y. S_{BET} and V_{total} of the carbons Y1, Y2, and E were measured as 3519, 1815, and 3840 m^2/g and 1.8, 1.3, and 1.8 cm^3/g, respectively. N-doping of these three carbons was performed by heating in a flow of acetonitrile-saturated N_2 at 750°C. CO_2 adsorption isotherms are shown on the carbons Y1, Y2, and E without or with N-doping, in Fig. 4.49A and B, respectively, by comparing with a commercially available AC (Maxsorb, S_{BET} of 3311 m^2/g and V_{total} of 1.7 cm^3/g). The CO_2 adsorption capacities of 3.3 and 4.0 mmol/g at 298K and 1 atm were observed on the carbon E and N-doped E, respectively, which has the highest S_{BET} even after N-doping. The CO_2 adsorption and desorption are completely reversible as shown in the N-doped E in Fig. 4.49B. The CO_2 adsorption capacities (at 298K and 1 atm) of the carbons are plotted against their S_{BET} in Fig. 4.49C, revealing linear relations between these two parameters and a marked enhancement in capacity by N-doping. The treatment of commercial ACs in ammonia gas flow (50 mL/min) at high temperatures (200−1000°C) for 2 h was effective in enhancing the CO_2 capture [166]. The CO_2 adsorption capacity of the AC increased to 76 mg/g after N-doping at 400°C in ammonia from the 54 mg/g for the pristine one.

N-doped carbons were prepared from polypyrrole by mixing with KOH and heating at 600−850°C to study the CO_2 adsorption [167]. The carbon prepared by activation at 600°C using KOH/polypyrrole ratio of 2/1 by weight delivered the highest CO_2 adsorption capacity of 169 mg/g at 25°C, although S_{BET} was very inferior to the carbons prepared by other conditions. Porous carbons were synthesized from the mixture of glucose with 5 and 10 wt.% acrylic acid ($C_3H_4O_2$) and 20 wt.% nonionic surfactant Brij 72 by hydrothermal treatment at 180°C (denoted as PC-5 and PC-10, respectively) [168]. Without the surfactant Brij 72, nonporous carbon was obtained (C-10). The carbons PC-5 and PC-10 were doped by N through a reaction with tetraethylenepentamine (TEPA) after the modification of the surface of the carbon using $SOCl_2$ at 70°C, where COOH groups on the surface of the as-prepared carbon were replaced by COCl groups to accelerate the reaction with TEPA. CO_2 adsorption isotherms are

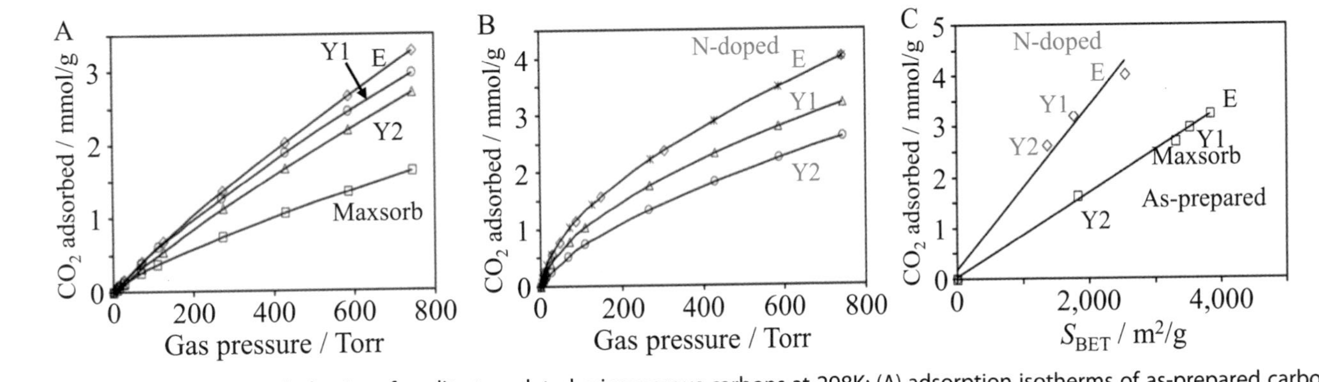

Figure 4.49 CO_2 adsorption behavior of zeolite-templated microporous carbons at 298K: (A) adsorption isotherms of as-prepared carbons and (B) those of N-doped carbons, and (C) relations of CO_2 adsorption capacity to S_{BET} for as-prepared (including Maxsorb) and N-doped carbons. In (B), the desorption isotherm on the N-doped E is also shown by X marks [165].

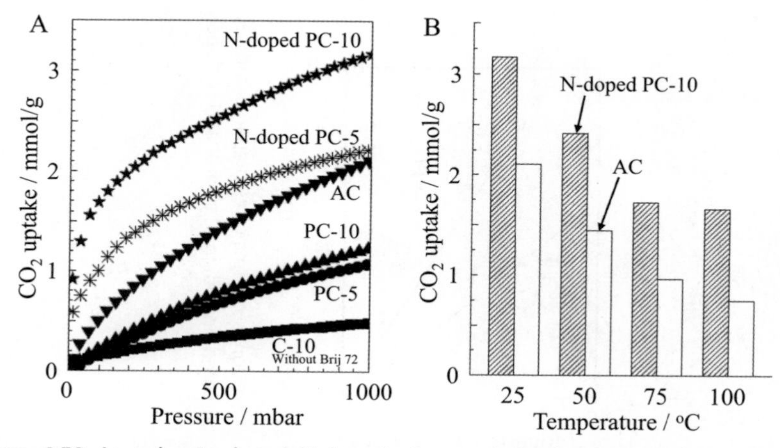

Figure 4.50 As-carbonized and N-doped glucose-derived porous carbons: (A) CO_2 adsorption isotherms at 25°C on as-carbonized carbons (PC-10, PC-5, and C10) and N-doped carbons (N-doped PC-10 and PC-5) and (B) CO_2 adsorption capacity at different temperatures on N-doped PC-10 in comparison with a commercial AC [168].

compared in Fig. 4.50A, showing that N-doping accelerates the CO_2 adsorption and enhances markedly the adsorption capacity. The N-doped PC-10 (S_{BET} of 639 m^2/g and V_{total} of 0.35 cm^3/g) has a much higher CO_2 uptake at the temperature range of 15–100°C and pressure of 1 bar, which is higher than the commercial AC (S_{BET} of 1109 m^2/g and V_{total} of 0.63 cm^3/g) as shown in Fig. 4.50B.

Porous PAN monoliths with a bicontinuous pore structure were carbonized and simultaneously activated in a flow of CO_2/Ar (1/3 v/v) at 800–1000°C [169]. The largest CO_2 uptakes of 5.14 and 11.51 mmol/g at 25 and 0°C under 1 bar were obtained on the carbon activated at 1000°C, which had S_{BET} of 2501 m^2/g and N content of 1.8 wt.%. On the carbon prepared at 900°C with S_{BET} of 1437 m^2/g and N content of 4.4 wt.%, however, CO_2 uptakes of 4.29 and 6.44 mmol/g were obtained at 25 and 0°C under 1 bar, respectively.

N-doped mesoporous carbons were synthesized via a simple heat treatment of the mixtures of chitosan with an activation reagent at high temperatures [170–172]. The mixture of chitosan with K_2CO_3 in weight ratios of 1/1 and 1/2 was carbonized/activated at different temperatures of 600–800°C [170]. With increasing activation temperature from 600 to 800°C, pore structure parameters of the resultant carbons changed from 1180 to 2567 m^2/g on S_{BET} and from 0.47 to 1.25 cm^3/g on V_{micro}, as well as N content from 6.02 to 1.54 wt.%, whereas the CO_2 uptake changed only slightly from 3.57 to 2.78 mmol/g. This change in CO_2 uptake may correspond to a change in relative percentage of micropores, from 96% to 84%. Porous carbons were prepared from the mixture of chitosan with different alkali metal carbonates, Na_2CO_3, K_2CO_3, Rb_2CO_3, and Cs_2CO_3, at 500–700°C in N_2 flow [172]. Pore parameters and N content, as well as CO_2 uptake of the resultant carbons, depended greatly on the alkali metal used and heat treatment temperature. CO_2 uptake at a low pressure of 1 kPa could be almost linearly related to the N content of the carbon. N-doped porous carbons were prepared by carbonization of an ionic liquid, 1,3-bis(cyanomethyl)imidazolium bis((trifluoromethyl)-sulfonyl)amide at 500°C and delivered a high CO_2 adsorption capacity of 13.9 wt.% at 25°C and 1 atm [171]. The resultant carbon exhibited very stable CO_2 adsorption isotherms after treatments in a flow of dry CO_2, air, steam, and N_2 containing 100 ppm SO_2 at 130°C for 24 h, which were severe conditions for carbon materials. N-doped microporous carbons were synthesized from crab shells after the removal of $CaCO_3$ with HCl by activation at 450°C using different KOH/C weight ratios [173]. To achieve high CO_2 uptake, the activation reagent KOH is needed, with a larger weight ratio than 1.0, as shown in Fig. 4.51A. The CO_2 uptake is linearly related to the pore volume of micropores with sizes less than 0.63 nm, as shown in Fig. 4.51B, and adsorption at low temperature is preferable, as shown in Fig. 4.51C. The resultant carbon contained 5.1–8.5 at.% N, most of which had either pyrrolic or pyridinic configurations. Bamboo-derived carbons prepared by KOH activation with a KOH/C mass ratio of 3 at 700°C delivered CO_2 uptake of 7.0 mmol/g at 0°C and 1 bar [174].

N-doped rGO flakes were prepared by activation of polypyrrole-modified rGO flakes mixed with KOH at different temperatures of 400–700°C [175]. The rGO activated at 600°C, which contained 4.8 at% N mainly at pyrrolic configuration and had S_{BET} of 1360 m^2/g with V_{micro} of 0.57 cm^3/g, delivered the highest CO_2 uptake of 4.3 mmol/g at 25°C and 1 bar. It showed a higher selectivity for CO_2 over N_2.

In contrast to the results described above, no marked effect of N-doping on CO_2 capture was reported for the RF-derived carbon spheres [176]. The nondoped and N-doped carbons were synthesized from the spheres using resorcinol and nitrogen-rich copolymer [poly(benzoxazine-*co*-resol)] with formaldehyde by carbonization at 800°C in N_2, followed by activation with CO_2 at around 800°C to obtain different burn-offs ranging between 0 and 47 wt.%. The nondoped and N-doped carbons

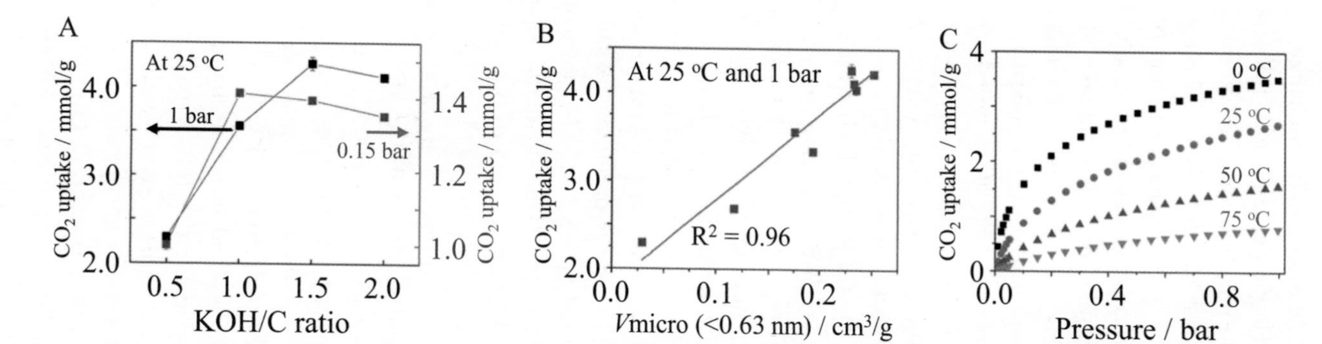

Figure 4.51 CO_2 uptake of N-doped microporous carbons derived from crab shell: (A) CO_2 uptake at 1 and 0.15 bar for carbon activated at 600°C using different KOH/C weight ratios, (B) relation to volume of micropores with sizes less than 0.63 nm, and (C) effect of temperature on adsorption isotherms [173].

(2.53–3.19 wt.% N) had very similar pore structure parameters, such as S_{BET} of 510–1910 m^2/g and $V_{micro}(CO_2)$ at 0°C of 0.26–0.56 cm^3/g depending mainly on the burn-off. The CO_2 adsorption capacity at 0°C increased mostly with increasing burn-off, and tending to saturate above 24% burn-off, while the difference between nondoped and N-doped carbons was very small. N-doped mesoporous carbons synthesized from nitrogen-rich organic frameworks were activated by mixing with KOH in different KOH/carbon weight ratios of 3/1, 5/1, and 7/1 at 650–850°C in Ar [177]. The carbon activated at 650°C with a KOH/C ratio of 5/1 delivered CO_2 uptake of 3.68 mmol/g at 25°C and 1 bar, which contained 1.11 wt.% N and had S_{micro} of 1592 m^2/g with V_{micro} of 0.75 cm^3/g. Meanwhile the pristine mesoporous carbon, which contained much a higher N content (3.37 wt.%) and much smaller S_{micro} and V_{micro} (15 m^2/g and 0.02 cm^3/g, respectively), delivered a CO_2 uptake of 1.95 mmol/g.

Selective adsorption of CO_2 from mixed gases, such as a mixture of CO_2 with N_2 like a post-combustion gas, was studied. For most carbon adsorbents, CO_2 adsorption occurs rapidly, as shown by the breakthrough curves in Fig. 4.52, where commercially available granular AC was introduced into a flow of the mixed gases, 14.8% CO_2 and 85.2% N_2 (Fig. 4.52A) and 10% CO_2, 28% H_2, and 62% He (Fig. 4.52B) at 25°C [178]. With the introduction of the AC into the gas flow, the CO_2 concentration in the effluent decreased quickly to almost zero, with the N_2 concentration increasing simultaneously to 100%, until complete adsorption of CO_2 by the carbon (the breakthrough). Since it is difficult to measure the adsorption amount of each component of a mixed gas experimentally, the ideal adsorption solution theory (IAST) based on the dual-site Langmuir–Freundlich (DSLF) adsorption model was often

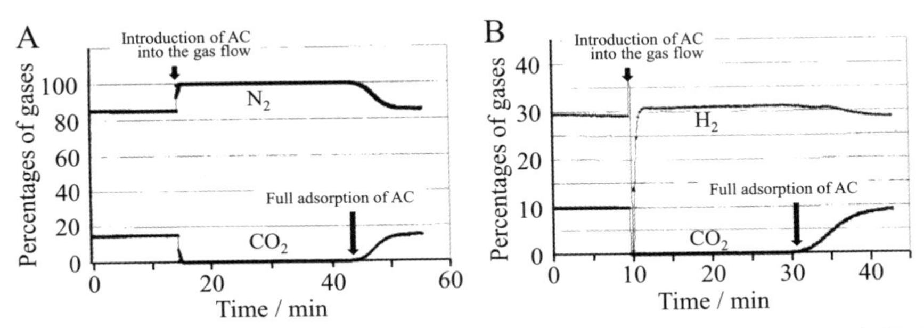

Figure 4.52 CO_2 capture by an AC from mixed gas flow at room temperature: (A) CO_2/N_2 (14.8/85.2) and (B) $CO_2/H_2/He$ (10/28/62) [178].

applied to evaluate the selectivity of the carbon adsorbents. The selectivity $S(I/J)$ is written as the following equation:

$$S(I/J) = (x_I / x_J)/(y_I / y_J),$$

where x_I, x_J and y_I, y_J denote the molar fractions of species I and J in the adsorbed and bulk gases, respectively. Microporous carbons, which were synthesized from the mixture of terephthalaldehyde and melamine by microwave irradiation at 180°C and carbonization at 800°C in N_2 flow and had a CO_2 uptake of 141 mg/g, exhibited a selectivity $S(CO_2/N_2)$ of about 32 at 25°C and 1 bar [179]. Microporous carbons, which were synthesized from the mixture of glucose with 2 and 10 wt.% acrylic acid by hydrothermal carbonization, followed by modification using tris(2-aminoethyl) amine, delivered a CO_2 uptake of 4.1 mmol/g at 25°C and 1 bar with $S(CO_2/N_2)$ of 30.7 at 25°C, 110 at 70°C and 69.4 at −20°C [180].

The selectivity of the 450°C-carbonized PAF-1, which delivered a high CO_2 uptake as shown in Fig. 4.47, was calculated for the mixed gases of CO_2/N_2 (15/85 by volume), CO_2/CH_4 (15/85), and CO_2/H_2 (20/80) at 273K and 1 bar [160]. The selectivity of the 450°C-carbonized PAF-1 is shown as a function of gas pressure in Fig. 4.53. $S(CO_2/N_2)$ at 0°C of carbonized PAF-1 depends on the pressure, increasing from about 50 to 209 with increasing pressure from 0.1 to 1 bar. $S(CO_2/H_2)$ is much larger at 150−309 at 0−1 bar, whereas $S(CO_2/CH_4)$ is much smaller and is almost not pressure dependent at about 7.8−9.8 even at a high pressure range of 0−40 bar.

Irreversible CO_2 adsorption was reported on the carbon spheres, which were prepared from phenol resin spheres by carbonization at 700°C (denoted as APS) and 1000°C (APT) [181]. The CO_2 adsorption−desorption behaviors of these two carbon spheres were studied by

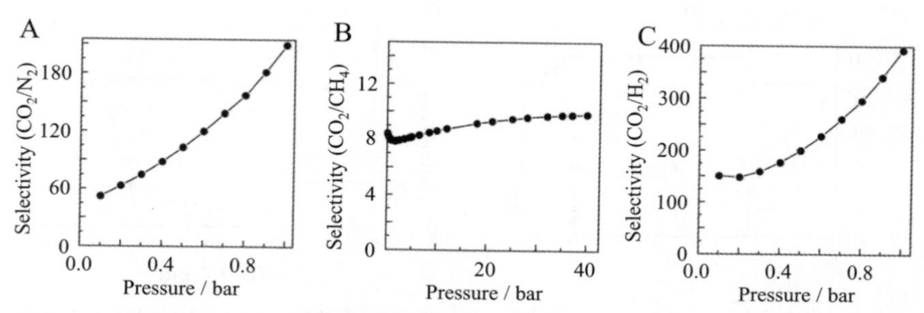

Figure 4.53 CO_2 selectivity of mixed gases at 0°C for the 450°C-carbonized PAF-1 in a mixed gas of (A) CO_2/N_2 (15/85), (B) CO_2/CH_4 (15/85), and (C) CO_2/H_2 (20/80) [160].

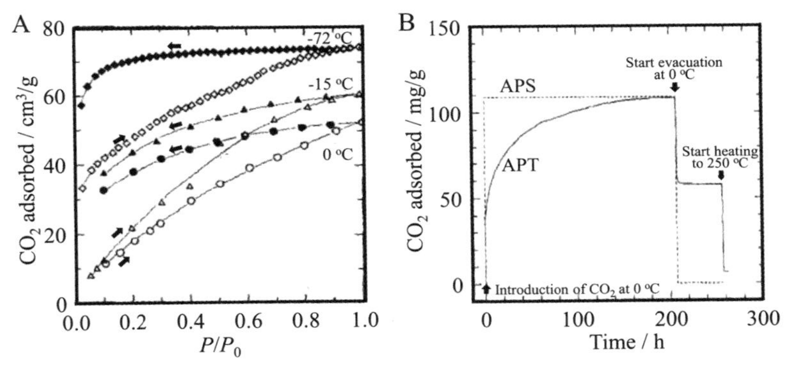

Figure 4.54 CO_2 adsorption of the carbon spheres carbonized at 1000°C (APT) and 700°C (APS): (A) volumetric isotherms on APT at different temperatures, and (B) gravimetric measurements on adsorption and desorption of CO_2 on APT and APS [181].

volumetric and gravimetric measurements, as shown by some results in Fig. 4.54. The carbon APT shows marked hysteresis at temperatures of −72, −15, and 0°C, as shown in Fig. 4.54A. However, the carbon APS exhibited completely reversible isotherms at the same temperature range. Irreversible isotherms were also confirmed on APT by gravimetric measurements on adsorption and desorption at 0°C, and then evacuation at 250°C for 3 h, as shown in comparison with APS in Fig. 4.54B. The adsorption of CO_2 onto APT is very slow, as compared with APS, and desorption by evacuation occurs quickly on both APT and APS. However, the desorption by evacuation at 0°C cannot be completed on APT. Most of the residual CO_2 after evacuation can be removed by heating at 250°C, whereas a small amount of CO_2 seems to be kept in the carbon APT even after heating at 250°C for more than 2 h.

4.1.4.2 Carbon supports of the adsorbents for CO_2

CO_2 capture by adsorption−desorption on polyethyleneimine (PEI) as the adsorbent was proposed by immobilizing PEI in mesoporous silicas, which were described as the "molecular basket" sorbents (MBSs). For example, a basket loaded with 50 wt.% PEI on a mesoporous silica SBA-15 showed a CO_2 capacity as high as 140 mg per g of PEI at a CO_2 partial pressure of 15 kPa. Instead of the expensive mesoporous silica molecular sieves, porous carbons were employed to reduce substantially the MBS preparation cost and to make CO_2 capture more cost-effective.

Five commercial ACs with S_{BET} of 1151–2320 m^2/g and V_{total} of 0.64–2.93 cm^3/g (designated as C1–C5, with no details on the characteristics of the ACs reported) were employed to load PEI by 50 wt.% via wet impregnation and the CO_2 capture performances were studied [182]. An MBS consisting of the porous carbon C4 with 50 wt.% PEI, which gave the highest CO_2 adsorption capacity, was compared with the MBS using mesoporous silica [PEI(50 wt.%)/SBA-15] in Table 4.9. Although the weight-based capacity of PEI(50 wt.%)/C4 is comparable to PEI(50 wt.%)/SBA-15, the former is very advantageous in volume-based capacity and preparation cost. For the former, a slight reduction in the CO_2 adsorption capacity was observed with an increasing adsorption/desorption cycle number, at about an 8% reduction after 10 cycles.

An AC derived from fly ash by steam activation at 850°C, which contained a large amount of ash (56.4 wt.%) and had a low S_{BET} (88 m^2/g), and was used for impregnation of PEI via its methanol solution [183]. At 75°C, adsorption of CO_2 occurs slowly and it is not saturated even after 80 min, whereas desorption proceeds relatively quickly, as shown in the isotherms in Fig. 4.55A. The CO_2 uptake after 80 min increases with increasing amount of PEI loaded; 0.4 wt.% uptake for the pristine AC and 4 wt.% uptake for 60 wt.% PEI-loaded AC. The addition of polyethylene glycol (PEG) enhances CO_2 uptake, as shown on the AC with 60 wt.% PEI and 20 wt.% PEG in Fig. 4.55B. CO_2 uptake of a commercial AC (S_{BET} of 1225 m^2/g and ash content of 0.7 wt.%) decreased with temperature, to about 4 wt.% at 25°C and about 0.6 wt.% at 100°C, whereas the CO_2 uptake of the PEI-loaded AC increased up to 4.5 wt.% at 85°C, which is more than five times that for the pristine AC at the same temperature.

Table 4.9 Comparison of two "molecular basket" sorbents (MBSs). [182].

	CO_2 capacity at 75 °C		Preparation	
MBS	**Mass-based (mg/g)**	**Volume-based (mg/mL)**	**Packing density (g/cm^3)**	**Preparation cost (dollar/kg)**
PEI (50 wt%)/C4	135	47	0.35	ca. 44
PEI (50 wt%)/SBA-15	138	30	0.22	ca. 760

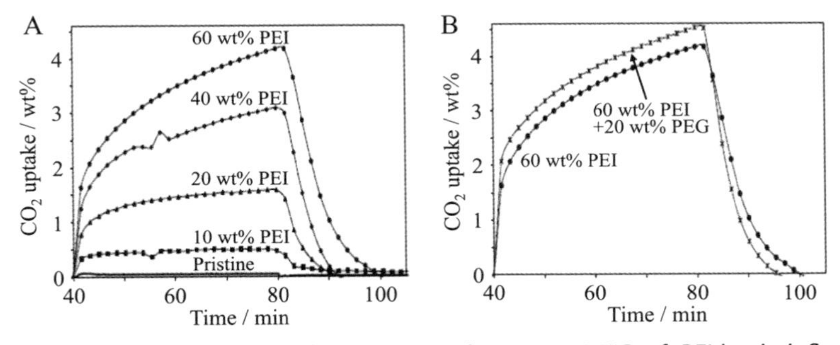

Figure 4.55 CO_2 adsorption–desorption isotherms at 75°C of PEI-loaded fly-ash-derived ACs: (A) effect of PEI loading and (B) effect of PEG addition [183].

Different amines, diethylenetriamine (DETA), pentaethylenehexamine (PEHA), and PEI, were loaded via wet impregnation onto a commercial AC (Norit CGP Super), which was synthesized from wood by chemical activation using H_3PO_4 and had V_{meso} of 0.648 cm^3/g and V_{micro} of 0.204 cm^3/g [184]. As shown in Fig. 4.56A, the impregnation of amines causes a marked reduction in the physical adsorption of CO_2, which is due to the blockage of CO_2 filling the micropores of the carbon. On the amine-impregnated ACs, however, chemisorption of CO_2 with impregnated amine molecules is thought to occur, which is suggested by the appearance of pronounced elbows at very low P/P_0 region and also by hysteresis at low pressure. In Fig. 4.56B, temperature dependences of CO_2 uptake under 50 cm^3/min flow of CO_2 are shown, with all adsorbents showing the highest uptakes at 25°C. The amine impregnation into ACs presents smaller

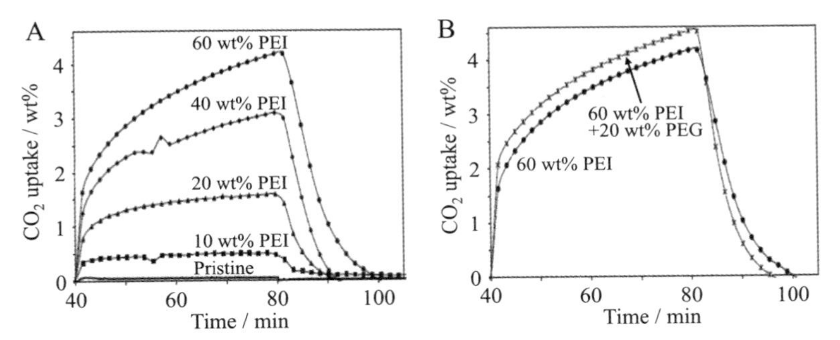

Figure 4.56 CO_2 adsorption of amine-impregnated ACs: (A) adsorption isotherms at 0°C and (B) temperature dependences of CO_2 uptake in 50 cm^3/min flow above 25°C [184].

slopes for the temperature dependence of CO_2 uptake than the pristine AC. The DETA-impregnated AC shows even higher CO_2 uptake above 65°C, which is due to the strong interaction between the acidic gaseous CO_2 and the basic amine groups impregnated in the AC.

Fly-ash-derived carbon purified by washing out the impurities (ashes) in $HCl/HNO_3/HF$ mixed acid was activated with steam at 850°C, onto which monoethanolamine (MEA), diethanolamine (DEA), and methyldiethanolamine (MDEA) were loaded [185]. The impregnation of MDEA, DEA, MEA, and MDEA+MEA resulted in decreases in V_{micro} and V_{meso}, probably due to pore filling, but improved the CO_2 adsorption capacity at 70°C probably due to chemisorption, from 18.5 mg/g for the pristine AC to 30.4, 37.1, 49.8, and 35.8 mg/g after loading of the respective amines.

4.1.5 Metal trapping

Wastewater often contains both metals and organic pollutants. Contamination of water by various metals is regarded as a universal environmental problem, because some are very toxic even at very low concentrations (e.g., mercury, cadmium, and lead) and so their removal is urgent. Although various sorbents have been developed to remove contaminants from wastewater, it would be desirable to develop a cost-effective sorbent that can remove both metals and organics.

The adsorption of metal ions onto the surface of a solid can generally be classified into three processes: physical, ion exchange, and complex/redox adsorptions. Since the accommodation of heavy metal ions in the sorbents due to the latter two processes requires specific functional groups on the surface of adsorbents, porous carbons have certain advantages as the sorbent for heavy metal ions because of the presence of controllable kinds and amounts of functional groups on their surface.

The removal of alkaline metal ions, such as Li, Na, etc., is known as the capacitive deionization (CDI) method and is effective, and so it is explained in a separate section (Section 4.3), including the specific instrumentation required. In this section, therefore, the description is focused on the trapping of heavy metals, mercury, uranium, and other metals.

4.1.5.1 Mercury

Mercury is a pollutant of great concern because of its volatility, persistence, and toxicity, and particularly its neurotoxicity which has an impact on human health and its bioaccumulation in the environment causing serious effects for all life forms. Its emission from coal-fired power plants has attracted increasing attention because of its high toxicity. In flue gas, mercury exists primarily in three states: elemental (Hg^0), oxidized (Hg^{2+}), and particle-bound (Hg^P). Hg^{2+} and Hg^P are easily captured by the current air pollution control devices, whereas Hg^0 is difficult to remove due to its low solubility in water and high volatility. Various methods have been proposed for the elimination of Hg^0, based on adsorption, catalytic oxidation, photochemical removal, plasma removal, wet oxidation, etc. Adsorption processes based on activated carbon (AC) are considered to be among the most effective technologies for Hg^0 removal because of its high removal efficiency. Nevertheless, some disadvantages of AC adsorption processes, mainly high operation costs and adsorbent loss, have been pointed out, and so exploring and developing more cost-effective adsorbents for Hg^0 removal are needed. In order to improve the adsorption efficiency of adsorbents, including ACs and other porous carbons, various modifications have been proposed, which are reviewed [186].

Two commercially available microporous ACs (BPL and WPL) were modified by the following three treatments: treatment at 1200K in N_2, air oxidation at 693K, and treatment in 6N HNO_3 at room temperature [187]. Heat treatment at 1200K in N_2 flow gave no Hg^0 uptake for two ACs, although their S_{BET} and V_{total} did not show marked changes from the as-received ACs. ACs treated in HNO_3 obtained increased oxygen contents, mainly because of marked increases in lactone and carbonyl groups, and exhibited a marked increase in Hg^0 uptake to 1520 mg/g for BPL and 700 mg/g for WPL. In contrast, air oxidation at 693K caused no appreciable change in Hg^0-uptake on BPL, but some decrease for WPL, from 925 μg/g for as-received to 700 μg/g. These experimental results suggested that Hg^0 adsorption depends greatly on the surface characteristics of the carbon, particularly lactone and carbonyl functional groups is thought to be the active sites for adsorption. Ozone treatment at 4.5 wt.% O_3 in dry air dramatically increased the Hg^0 adsorption capacity of mesoporous carbon by up to 134 times when measuring immediately after ozonolysis [188]. The activity for Hg^0 adsorption was easily destroyed by exposure of the ozone-treated carbons either to the atmosphere or water vapor, or by mild heating. The ACs prepared by heating a mixture of waste woods with urea at 800°C exhibited relatively high Hg^0 adsorption capacity, with an average Hg capture efficiency of greater than 96% under different operation conditions using Hg pulse injection in a simulated flue gas containing SO_2 and NO_2 and in a practical flue gas from an operating coal-fired power plant [189]. Their capacities were comparable to that of a commercial Hg^0 adsorbent.

$CuCl_2$-loaded AC was tested in a fixed-bed system and filter-added entrained-flow system, for comparison with a commercial brominated AC, which had been used for Hg^0 removal in power plants [190]. $CuCl_2$-loaded AC could oxidize Hg^0 to Hg^{2+} more markedly than brominated AC, whereas both of the adsorbents have excellent equilibrium adsorption capacities. The former had active sites available for oxidation together with sites for adsorption of the resultant oxidized Hg (Hg^{2+}). $FeCl_3$-loading on AC was also effective in oxidizing Hg^0 to Hg^{2+} and its adsorption removal [191]. As shown by the breakthrough curves in Fig. 4.57A, $FeCl_3$-loaded AC shows the best adsorption behavior among the different loading species, with breakthrough capacity reaching 2.7 mg/g. Loading of $FeCl_3$ is performed via its aqueous solution with different concentrations (Fig. 4.57B), followed by calcination at different temperatures of 105–500°C (Fig. 4.57C). Its loading conditions are optimized for the AC as follows, using $FeCl_3$ aqueous solution with 0.15 mmol/L and following calcination at 300°C. The resultant $FeCl_3$-loaded AC had S_{BET} of 650 m^2/g with V_{total} of 0.34 cm^3/g, which is slightly lower than the pristine AC, suggesting a small degree of pore blockage by $FeCl_3$-loaded AC.

Nonthermal plasma treatments based on dielectric barrier discharge in various atmospheres were employed to improve the Hg^0 trapping capacity of ACs. A commercial AC (particle size of 0.2–0.3 mm, S_{BET} of 163 m^2/g) was treated by nonthermal plasma in air for different times (7, 15, and 30 min), and their adsorption behaviors for Hg^0 were investigated in an N_2 flow containing Hg^0 vapor [192,193]. The removal efficiency of Hg^0 at 30°C was enhanced with increasing plasma treatment time, but the increase in adsorption temperature to 130°C was observed to reduce the final removal efficiency. Treatment at 900°C in N_2 decreased the Hg^0 adsorption efficiency, but the following plasma treatment was effective in enhancing efficiency. Nonthermal plasma treatment of the AC in a flow of N_2 containing 500–3500 ppm Cl_2 was very effective in enhancing the Hg^0 adsorption efficiency, reaching almost 100%, although the plasma treatment in N_2 flow without Cl_2 does not have any effect on Hg^0 adsorption and simple treatment in Cl_2 flow promoted adsorption only a little, as shown in Fig. 4.58A. The removal efficiency depends on the Cl_2 concentration in N_2 flow during plasma treatment, and the adsorption capacity of Hg^0 increases with increasing concentration of Cl_2, reaching about 200 mg/g by 3500 ppm Cl_2 concentration, as shown in Fig. 4.58B and C. SO_2 and H_2O in flue gas inhibited the removal of mercury by the AC, while HCl had a promotional effect. Nonthermal plasma treatment of commercial AC in O_2

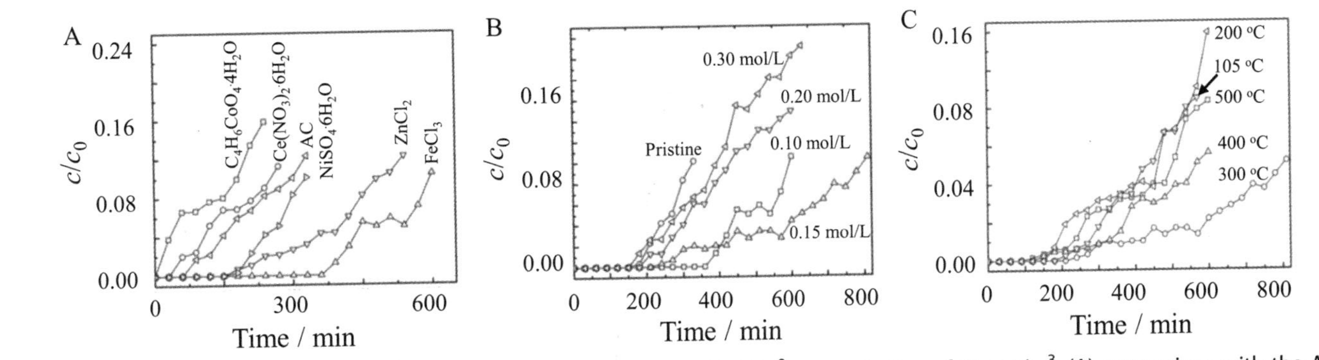

Figure 4.57 Breakthrough curves of Hg^0 on $FeCl_3$-loaded AC with an initial Hg^0 concentration of 30 mg/m^3: (A) comparison with the ACs loaded by various materials at 300°C, (B) effect of the concentration of $FeCl_3$ solution used for loading (calcination temperature of 300°C), and (C) effect of calcination temperature after $FeCl_3$-loading from 0.15 mmol/L solution [191].

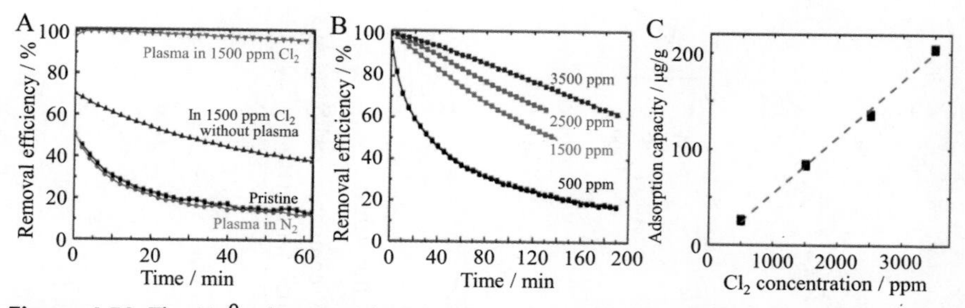

Figure 4.58 The Hg^0 adsorption behavior of ACs treated by nonthermal plasma in N_2 flow containing Cl_2: (A) removal efficiency at 30°C of the ACs treated with different conditions, (B) efficiency at 130°C by the ACs treated by plasma in different Cl_2 concentrations, and (C) dependence of adsorption capacity at 130°C on Cl_2 concentration in plasma treatment [193].

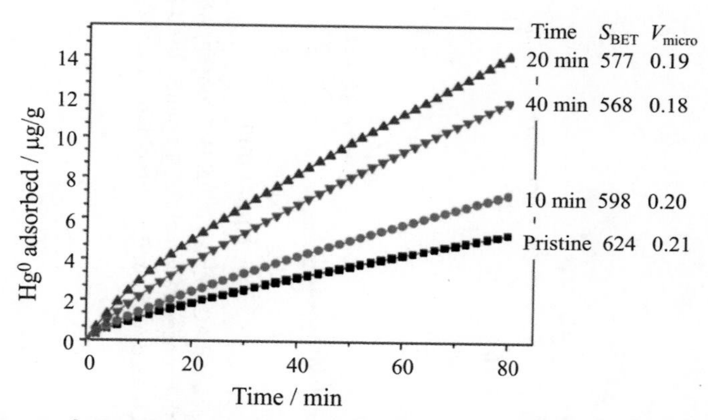

Figure 4.59 Hg^0 adsorption isotherms of ACs treated nonthermal plasma of O_2 flow. The treatment time and the resultant S_{BET} and V_{micro} are shown [194].

flow resulted in an enhancement of Hg^0 adsorption, as shown in Fig. 4.59, together with pore parameters [194]. These experimental results suggested that functional groups of C—Cl, C—O (ester-type), and C=O(carbonyl-type) formed by nonthermal plasma treatments in Cl_2 and O_2 worked active sites for the chemical interaction with Hg^0 (chemisorption). Nonthermal plasma treatments of an AC in O_2, air and HCl flow resulted in a marked increase in Hg^0 adsorption but not that in N_2 flow [195]. Porous carbon spheres, which were synthesized from starch by hydrothermal carbonization at 180°C, annealed at 600°C in N_2 flow, and activated by CO_2 at 1000°C, were modified by nonthermal plasma treatment in a mixed gas of O_2 and SO_2 [196]. The carbon spheres treated for

60 s by the N_2 containing 4% O_2 and 2000 ppm SO_2 delivered much better Hg^0 adsorption than the brominated AC. The equilibrium adsorption capacity calculated by a pseudo-second-order kinetic model reached 13 mg/g, which was much higher than the 1 mg/g for brominated AC.

The adsorptive removal of Hg^{2+} from aqueous solution was also studied using different carbon materials. The ACs prepared from walnut shell by mixing with $ZnCl_2$ ($ZnCl_2$/shell of 1/2 by weight) and carbonization in N_2, of which the maximum adsorption capacity for Hg^{2+} was determined as 151 mg/g at pH 5.0 [197]. On commercial wood-derived AC and rayon-based ACC, Hg^{2+} adsorption was studied through measurement of the breakthrough curves [198]. A composite of rGO with polypyrrole (PPy) prepared by a reduction of the mixture of PPy and 5–20 wt.% GO with hydrazine hydrate at 90°C exhibited a highly selective Hg^{2+} adsorption capacity [199]. On the composites prepared using 15 wt.% GO, Hg^{2+} adsorption capacities are compared to PPY as a function of initial Hg^{2+} concentration and adsorption time in Fig. 4.60A and B, respectively. The composite delivers much higher Hg^{2+} capacity than PPy in whole initial concentrations of Hg^{2+} examined, maximum capacity reaching 980 mg/g, which is much higher than 400 mg/g for PPy. On the composite, adsorption occurs quickly and so is completed within 20 min. The adsorption capacity increased with increasing pH up to pH 5 and with increasing temperature up to 40°C by maintaining the superiority of the composite, while above pH 5 and 40°C the same adsorption capacities were obtained for PPy/rGO and PPy. Palm-shell-derived AC was loaded by an ionic liquid (trioctylmethylammonium thiosalicylate, TOMATS) for adsorptive removal of Hg^{2+} [200]. The maximum adsorption of Hg^{2+} at

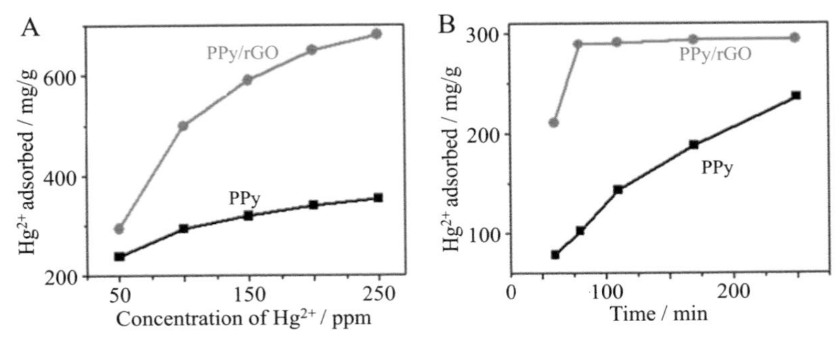

Figure 4.60 Hg^{2+} adsorption of PPy/rGO composites prepared using 15 wt.% GO in comparison with PPY: (A) effect of the initial concentration of Hg^{2+} and (B) that of adsorption time [199].

room temperature was 83 mg/g at pH 8 with a contact time of 3 h and an initial concentration of 10–200 mg/L. Using a carbon aerogel derived from RF resin, the adsorption of Hg^{2+} could be performed more efficiently than other divalent heavy metals (Cd^{2+}, Ni^{2+}, etc.) [201].

4.1.5.2 Uranium

Seawater contains approximately 4.5 billion tons of uranium (U), more than 1000 times the quantity available from terrestrial mining. Extraction of U^{6+} from seawater was performed using an amidoxime-based sorbent on a polyacrylonitrile braid and other sorbents. However, some technological problems remained because of the low overall concentration, the tendency to form a highly stable $[UO_2(CO_3)_3]^{4-}$ species, and the presence of numerous competing ions. Porous solid materials with high surface area, such as mesoporous silicas and carbons, were proposed to be used for the extraction of U^{6+} after appropriate functionalization of their surfaces.

Oxime modification of porous carbons was reported to improve the adsorption capacity for U^{6+}. The ordered mesoporous carbon CMK-5, which was synthesized using SBA-15 after alumination with furfuryl alcohol, exhibited a maximum sorption capacity of 65 mg/g at pH 4.5 after grafting of 4-aminoacetophenone oxime (oxime-CMK-5) [202]. The recovery of uranium was achieved by desorbing from the U^{6+}-adsorbed oxime-CMK-5 with 1.0 M HCl and no significant decrease in adsorption–desorption capability of the oxime-CMK-5 was observed after five cycles.

CMK-3 was synthesized using mesoporous silica SBA-15 as the template and sucrose as the carbon precursor, and then functionalized with carboxyl groups by its simple oxidation using ammonium persulfate (CMK-3-COOH) [203]. The resultant CMK-3-COOH had S_{BET} of 956 m^2/g with V_{total} of 0.96 cm^3/g and contained 1.19 mmol/g COOH groups, although the pristine CMK-3 had 1143 m^2/g, 1.10 cm^3/g, but only 0.20 mmol/g, respectively. The effects of the initial pH of the U^{6+} solution and contact time with the solution are shown in Fig. 4.61A and B, respectively. The functionalized CMK-3 could deliver a U^{6+} adsorption capacity of about 200 mg/g at pH 6, associated with an increase in the hydrophilicity, although the adsorption was thought to occur mainly by physisorption rather than chemisorption. The selective adsorption of U^{6+} from various coexisting cations is markedly improved by CMK-3-COOH, as shown in Fig. 4.61C.

Ordered mesoporous carbons synthesized by the polymer-blending method from resorcinol-formaldehyde using a triblock copolymer F127 were activated using KOH at 850°C and then grafted polyacrylonitrile through sonochemical polymerization at 60–70°C in N_2, followed by amidoximation using hydroxylamine in an H_2O/methanol solution [204]. On the adsorbents thus prepared in different conditions, uranium adsorption was measured using simulated seawater (17 mg of uranyl nitrate $UO_2(NO_3)_2$, 25.6 g of NaCl, and 193 mg of $NaHCO_3$ in 1 L of water) at room temperature. The adsorbent prepared by grafting 20.2% PAN delivered the highest adsorption capacity of 4.62 mg/g. A series of functionalized mesoporous carbons were prepared by covalent grafting with amidoxime, carboxyl, and phosphoryl functional groups [205]. The adsorption capacity for U^{6+} depended greatly on the functional groups grafted on the carbon. The functionalization of mesoporous carbons by phosphoric acid was the most effective, resulting in the highest capacity of 97 mg/g at pH 4 in acidic water and 67 mg/g at pH 8.2 in the seawater simulant, which was calculated by fitting the adsorption isotherms to the Langmuir model.

Trapping of radioactive ^{137}Cs in molten Na was performed using a carbon foam which was prepared from sucrose by heating its aqueous solution at 110°C for 24 h for complete foaming, followed by carbonization at 300–950°C in Ar [206]. The trapping efficiency of ^{137}Cs was 73–77%, which was slightly less than for reticulated vitreous carbon (RVC) which has been used as a trapping material for ^{137}Cs.

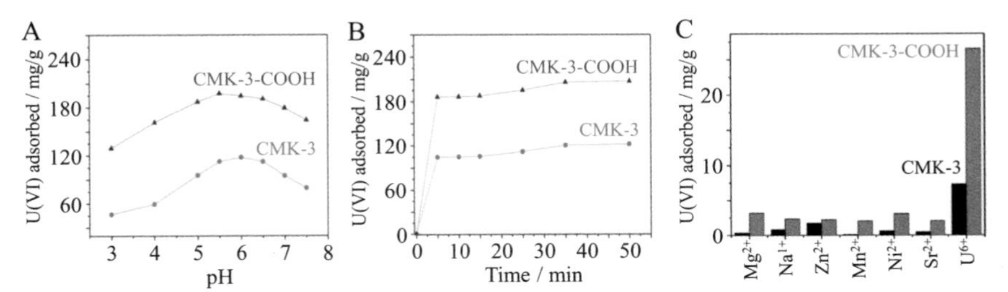

Figure 4.61 U^{6+} adsorption capacity for the mesoporous carbon CMK-3 and that after functionalization (CMK-3-COOH): (A) effect of initial pH of the solution, (B) effect of contact time, and (C) selective adsorption capacity from various coexisting cations [203].

4.1.5.3 Heavy metals

Removal of heavy metals (Pb, Cu, Cr, Cd, Ni, etc.) from industrial wastes using adsorbents is an important process to protect the environment. Since the adsorption of heavy metals to carbon is understood mostly due to the complexation with functional groups on the carbon surface, ACs have been used for this purpose by coupling with various surface modifications. Adsorptive removal of heavy metals, including Hg^{2+}, was reviewed by emphasizing low-cost adsorbents, such as chitosan, waste slurry, lignin, etc. [207].

The adsorptive removal behavior for divalent metal cations in aqueous solutions was studied on a commercial carbon aerogel at room temperature [201]. The effects of the initial metal concentration in the solution, amount of adsorbent, and contact time on the removal percentages of respective metal ions are shown in Fig. 4.62A–C, respectively. The percentage removal decreases with the increase in initial concentration for all heavy metal ions, except Hg^{2+} (Fig. 4.62A). For Ni^{2+}, Cd^{2+}, and Cu^{2+}, the percentage removal decreases gradually to below 80%. On the effects of adsorbent concentration (Fig. 4.62B) and contact time (Fig. 4.62C), Hg^{2+} shows 100% removal by the adsorbent in concentrations above 6 g/L and contact time above 24 h, but other ions need more than 12 g/L adsorbent concentration and at least 72 h contact time to achieve 100% removal.

The adsorption of Pb^{2+} was studied on ACs derived from coconut shell, coal, and petroleum pitch after the introduction of oxygen-containing functional groups by treatments with 7 M HNO_3 and removing these functional groups by heating at high temperatures in high-purity He flow [208]. Heating above 800°C was required to eliminate acidic groups on the

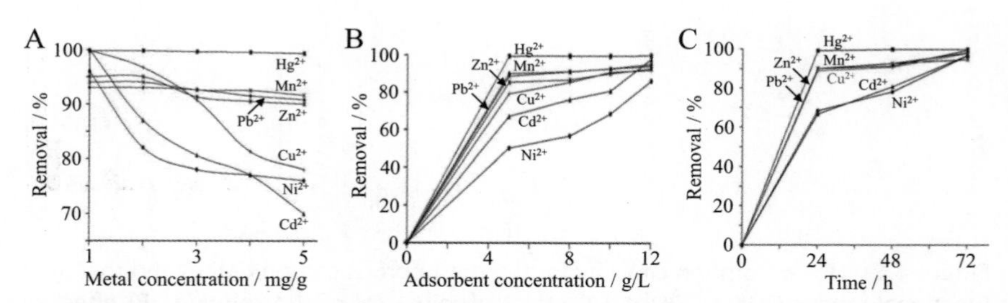

Figure 4.62 Adsorptive removal of different heavy metals by a carbon aerogel from aqueous solution at pH 6: (A) effect of initial metal concentration by 10 mg/L adsorbent for 48 h, (B) effect of adsorbent concentration with initial metal concentration of 3 mg/L and contact time of 48 h, and (C) effect of contact time by 10 mg/L adsorbent and initial metal concentration of 3 mg/L [201].

carbon surface and to expose basic groups for Pb^{2+} adsorption. The experimental results suggested that not only acidic groups, such as carboxylic and lactonic groups, but also basic groups were included for the Pb^{2+} adsorption in aqueous solution. Efficient adsorption and removal of Pb^{2+} ions in water were performed using CNTs formed on the surface of bamboo charcoal, with the maximum adsorption capacity being 47.4 mg/g [209]. CNTs fabricated by catalytic CVD of propane at 750°C delivered the maximum adsorption capacity of 78.74 mg/g after coating with 30 wt.% MnO_2 [210].

Commercial PAN-based carbon cloth (denoted as ACC) and three PF-based carbon felts [denoted as ACF-A (S_{BET} of 1204 m^2/g), ACF-B (S_{BET} of 1014 m^2/g), and ACF-C (S_{BET} of 1542 m^2/g), respectively] were used for the study of Pb^{2+} adsorption [211]. The adsorption isotherms of the carbon adsorbents are compared in Fig. 4.63A, revealing that PAN-based ACC has a slightly smaller adsorption capacity than PF-based ACF-A, but other PF-based ACFs (ACF-B and -C) show very small capacities. Oxidation of these sorbents in 15 vol% HNO_3 at 50–60°C for 2 h increases the Pb^{2+} adsorption capacity, which is more marked on ACC than ACF-A, as shown in Fig. 4.63B. The maximum adsorption capacities before and after oxidation were calculated by Langmuir equation as 36.6 and 52.7 mg/g for ACC and 43.9 and 48.0 mg/g for ACF-A, respectively. The dependences of adsorption capacity on either total acidic groups or carboxylic groups determined by the Boehm method are shown in Fig. 4.63C, suggesting that the adsorption capacity is proportional to the concentrations of total acidic groups, particularly with better proportionality to the concentration of carboxylic groups.

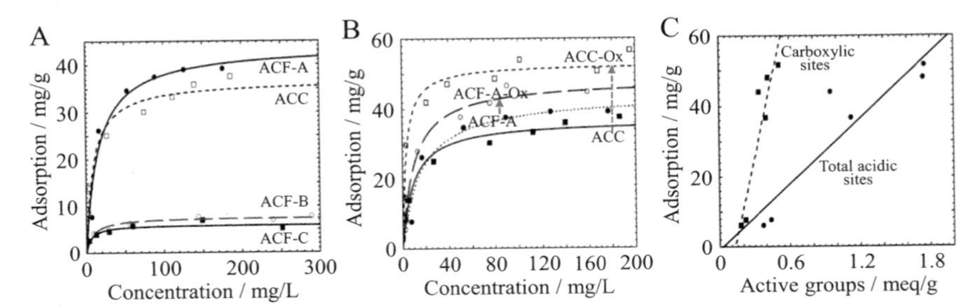

Figure 4.63 Pb^{2+} adsorption by PAN-based ACC and PF-based carbon felts (ACF-A, -B, and -C) at 25°C and pH 4: (A) Pb^{2+} adsorption isotherms, (B) effects of oxidation of ACC and ACF-A on their isotherms, and (C) Pb^{2+} adsorption capacity against total acidic groups and carboxylic groups [211].

Biochar produced from dairy manure (agricultural crop residue) at 200 and 350°C delivered a high adsorption capacity for Pb^{2+} of up to 680 mmol/kg for 200°C-treated biochar [212]. The biochar also adsorbed atrazine ($C_8H_{14}ClN_5$) effectively. When Pb^{2+} and atrazine coexisted (binary solute system), the adsorption capacity of the biochar for each solute was reduced slightly, whereas there was about a 50% reduction on a commercial AC. Porous carbons were prepared by activation of carbonized lotus stalks with phosphoric acid (PPA), trimethyl phosphate (TMP), and tributyl phosphate (TBP) at 450°C (designated as AC-PPA, AC-TMP, and AC-TBP, respectively) [213]. As shown in Fig. 4.64A, the adsorption of Pb^{2+} occurs quickly and is completed after 2 h. The adsorption capacity for Pb^{2+} depends greatly on the activating agents, with PPA giving a capacity of 123 mg/g but TMP and TBP giving 218 and 209 mg/g, respectively. As shown in Fig. 4.64B, adsorption isotherms can be simulated by Freundlich model and maximum capacities are calculated as 139, 241, and 229 mg/g for AC-PPA, AC-TMP, and AC-TBP, respectively. These experimental results suggested that the principal mechanism of Pb^{2+} adsorption involved electrostatic interactions between Pb^{2+} and surface functional groups of the carbons.

The Ni^{2+} adsorption behavior was compared on porous carbons prepared from various carbohydrates, glucose, sucrose and starch, and *Phragmites australis* (PA), by carbonization of the mixture with H_3PO_4 at 450°C (designated as AC-Glu, AC-Suc, AC-Sta, and AC-PA, respectively) [214]. The adsorption of Ni^{2+} occurred rapidly in the first 30 min, then gradually, and reached equilibrium within 3 h for all ACs studied. Saturated Ni^{2+} uptake depends on the initial concentration of the $NiCl_2$ solution, as shown

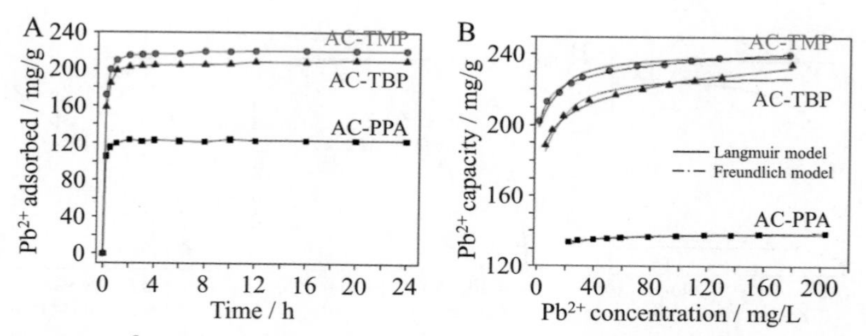

Figure 4.64 Pb^{2+} adsorption of ACs functionalized by PPA, TMP, and TBP: (A) effect of contact time and (B) adsorption isotherms at 22°C after 24 h contact [213].

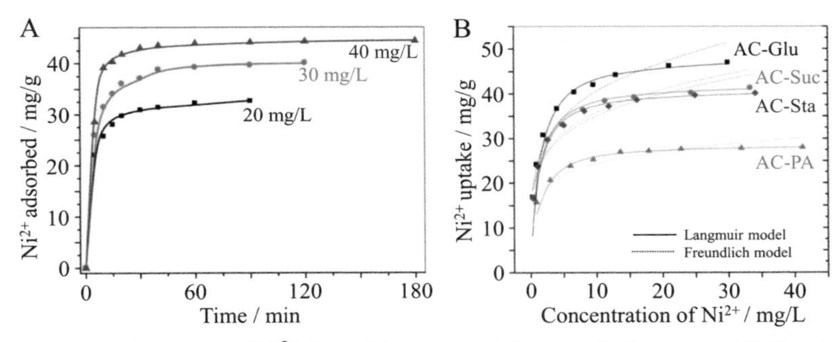

Figure 4.65 Adsorption of Ni^{2+} into ACs prepared from carbohydrates: (A) kinetics on AC derived from glucose (AC-Glu) in $NiCl_2$ solution with different concentrations and (B) isotherms of ACs prepared from various precursors [214].

on AC-Glu in Fig. 4.65A. The adsorption isotherms of ACs are shown in Fig. 4.65B, revealing better fitting by the Langmuir model. The maximum adsorption capacities calculated for AC-Glu, AC-Suc, and AC-Sta (48.5, 42.4, and 41.1 mg/g) were much higher than for AC-PA (28.7 mg/g), although the development in pore structure was much more marked in AC-PA. The high Ni^{2+} adsorption capacities of AC-Glu, AC-Suc, and AC-Sta may be attributed to their larger contents of both acidic and basic groups.

Commercial AC having 100−150 μm particle size and S_{BET} of 1100 m^2/g was modified by tetrabutylammonium (TBA) and sodium diethyl dithiocarbamate (SDDC) via impregnation [215]. As shown in the breakthrough curves in Fig. 4.66A, the SDDC-modified AC has a high adsorption capacity for Cu^{2+} of 38 mg/g, which is four times higher than for the pristine AC. It gave higher adsorption capacities also for Zn^{2+} and Cr^{6+} than the pristine one, reaching 9.9 and 6.84 mg/g, respectively. On the other hand, the TBA-modified AC was effective in removal of CN^- and its sorption capacity was calculated to be 29.2 mg/g, which is much higher than that of the pristine AC (6.6 mg/g). Wastewater from a metal plating industry and containing Cu^{2+} (37 mg/L), Zn^{2+} (27 mg/L), Cr^{6+} (9.5 mg/L), and CN^- (40 mg/L) was passed through the columns of the ACs. The breakthrough curves for metals in this wastewater measured on the columns of SDDC-modified ACs are shown in Fig. 4.66B. The removal capacity of Cu^{2+} (49.3 mg/g) is much higher than that of Cr^{6+} (6.84 mg/g), because the initial concentration of Cu^{2+} was higher than for Cr^{6+}. The functional group from the SDDC modifier was thought to have a relatively stronger affinity for Cu^{2+} than Zn^{2+}.

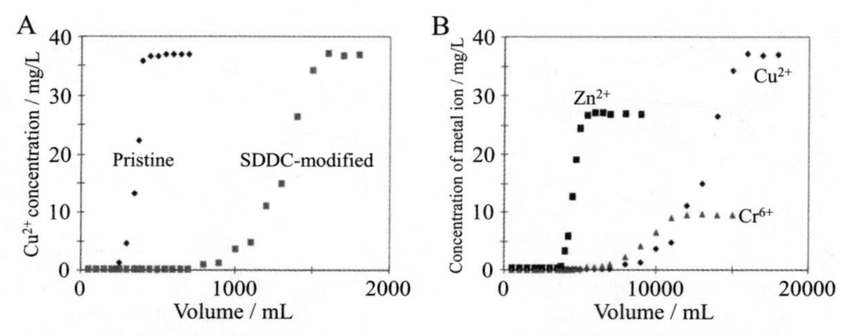

Figure 4.66 Breakthrough curves of the ACs: (A) the pristine and the SDDC-modified ACs for Cu^{2+} in a flow of water containing 37 mg/L Cu^{2+} and (B) the SDDC-modified AC for Cu^{2+}, Zn^{2+}, and Cr^{2+} in a flow of water containing 37 mg/L Cu^{2+}, 27 mg/L Zn^{2+}, and 9.5 mg/L Cr^{6+} [215].

An AC was synthesized from cherry stones (1–2 mm size) by carbonization at 900°C in N_2 and activation with CO_2 at 850°C for 2 h [216]. On this AC, further oxidation treatment was applied to modify the functional groups on the surface, using air at 300°C, air/ozone (1.5 vol.% O_3) mixture at 300°C, 5 N H_2O_2 aqueous solution at room temperature, and 5 N HNO_3 aqueous solution under reflux. Cu^{2+} adsorption isotherms from $Cu(NO_3)_2$ aqueous solution are shown on oxidized ACs in Fig. 4.67A, revealing the marked enhancements of Cu^{2+} uptakes by oxidation, depending greatly on its method. These enhancements of Cu^{2+} uptake were mainly attributed to the fixation of acidic functional groups on the carbon surface. Adsorption isotherms of Co^{2+} from single-solute solution and Co^{2+}/Cu^{2+} binary-solute solution onto the O_3-oxidized AC are shown in Fig. 4.67B. In the binary-solute solution, the uptakes for each

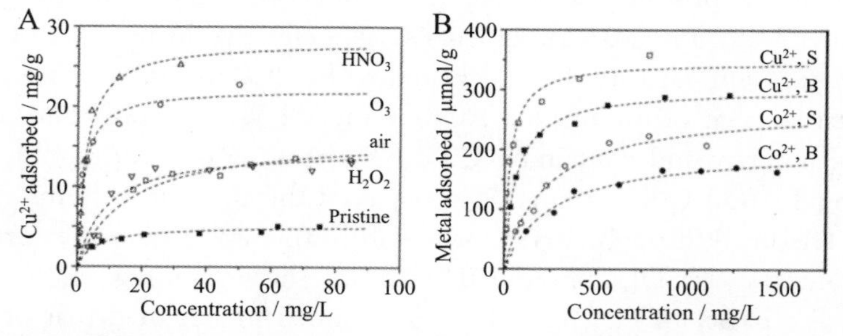

Figure 4.67 Adsorption isotherms of ACs derived from cherry stones: (A) those for Cu^{2+} on the ACs with various oxidation treatments, and (B) those for Cu^{2+} and Co^{2+} in the solutions with single-solute and binary-solute solutions (S and B, respectively) onto the O_3-oxidized AC. The lines are based on the Langmuir model [216].

cation decrease slightly, with the maximum uptakes for Cu^{2+} and Co^{2+} decreasing to 302 and 203 μmol/g for the binary-solute solution from 347 to 273 μmol/g for the single-solute solutions, respectively. Ultrasonic irradiation was found to assist the modification of AC derived from lotus stalks by sodium chlorite ($NaClO_2$) in different concentrations (10%–60%), resulting in an enhancement in Co^{2+} uptake [217].

The effect of functionalization of a commercial AC by carboxyl groups (carboxyl-AC) on the adsorption of Cd^{2+} was studied in comparison with amino- and mercapto-functionalized mesoporous silicas (amino-SiO_2 and mercapto-SiO_2, respectively) [218]. Functionalization was performed by oxidation of the AC in ammonium peroxydisulfate aqueous solution at room temperature and Cd^{2+} adsorption was measured using $Cd(NO_3)_2$ aqueous solution at room temperature. The adsorption capacity increases markedly with functionalization of the AC as well as silicas, by comparing with the isotherms for the pristine AC and silica in Fig. 4.68A. The maximum adsorption capacity for Cd^{2+} was calculated as 0.11 mmol/g for carboxyl-AC and 0.25 and 0.13 mmol/g for amino-SiO_2 and mercapto-SiO_2, respectively. As shown in Fig. 4.68B, the adsorption of Cd^{2+} steeply declined below pH 8; in this pH range Cd^{2+} and $Cd(NO_3)^+$ were the predominant species. The steep declines in Cd^{2+} adsorption below pH around 3 on carboxyl-AC and below pH 4 on mercapto-SiO_2 were attributed to competitive adsorption with protons.

The effect of oxidation treatment on adsorption capacity for Cd^{2+} was studied on carbon nanotubes (CNTs) synthesized by thermal

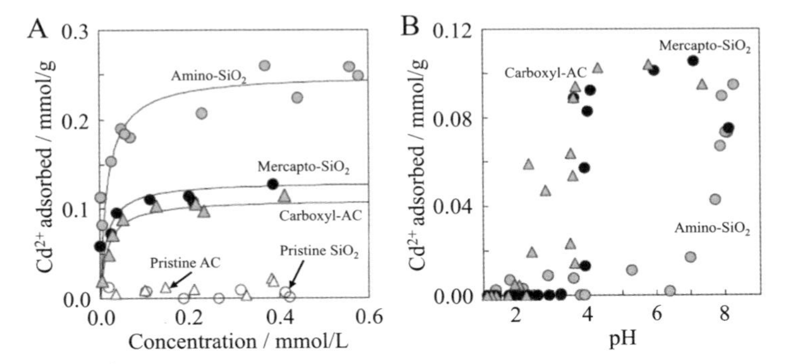

Figure 4.68 Cd^{2+} adsorption of the carboxyl-functionalized AC in comparison with the amino- and mercapto-functionalized silicas: (A) adsorption isotherms at room temperature and (B) pH dependences with the initial Cd^{2+} concentration of 0.22 mmol/g [218].

decomposition of CH_4 in hydrogen flow at 750°C using Ni nanoparticles as catalyst [219]. Oxidation of CNTs was performed by impregnation into $KMnO_4$ and H_2O_2 aqueous solutions at 80°C for 3 h, and by reflexing in HNO_3 solution at 140°C for 1 h. Adsorption isotherms measured on the CNTs of 0.05 g per 100 mL with pH 5.5 are shown in Fig. 4.69A, revealing the enhancement of Cd^{2+} adsorption by oxidative treatments. At Cd^{2+} equilibrium concentration of 4 mg/L, Cd^{2+} adsorption capacity reaches 11.0, 2.6, and 5.1 mg/g for $KMnO_4$-, H_2O_2-, and HNO_3-treated CNTs, respectively, whereas the pristine CNT exhibits only 1.1 mg/g capacity. An exceptionally high capacity of 11.0 mg/g for $KMnO_4$ treatment was thought to be due to Cd^{2+} adsorption by residual MnO_2 particles on the CNTs. In Fig. 4.69B, the effect of pH of the solution on Cd^{2+} adsorption is shown. At alkaline range, high capacity looks to be obtained, but this is mainly due to precipitation as $Cd(OH)_2$. At the acidic region, the adsorption capacities of HNO_3- and $KMnO_4$-treated CNTs are more sensitive to pH changes, reaching 8.0 and 12.6 mg/g at pH 6.0, while the pristine and H_2O_2-treated CNTs exhibit relatively small capacities of 2.0 and 3.3 mg/g. Fig. 4.69C shows the capacity increases with increasing the amount of CNTs used; very slowly for the pristine CNT to 3.5 mg/g, but quickly to 8.4 and 11.8 mg/g for H_2O_2- and HNO_3-treated CNTs, respectively. On $KMnO_4$-treated CNTs, the capacity increases rapidly with increasing CNT dosage and saturates at about 19 mg/g, suggesting the effectiveness of $KMnO_4$ treatment. The adsorption behaviors of Zn^{2+}, Cd^{2+}, and Hg^{2+} from aqueous solutions were studied on an ACC, which was prepared by carbonization and activation of viscose rayon cloth using either CO_2, NH_4Cl, or $ZnCl_2$ at 850°C [220]. With the increasing pH of the solution, the amounts of adsorbed Zn^{2+} and Cd^{2+} increased, whereas the amount of Hg^{2+} remained constant, which involved primarily adsorption of divalent cations (Zn^{2+} and Cd^{2+}) or precipitation of metal

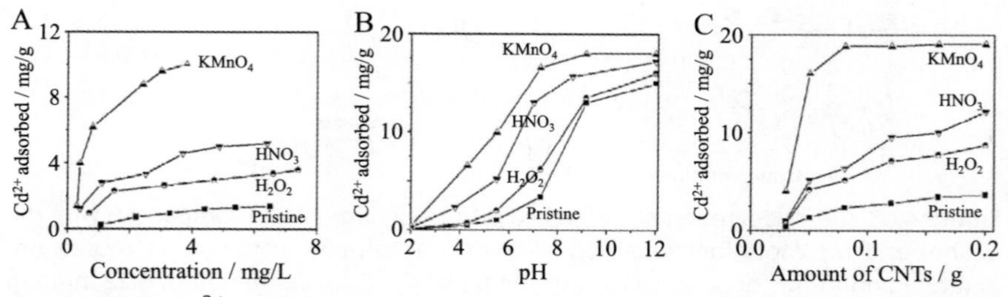

Figure 4.69 Cd^{2+} adsorption by the CNTs oxidized by different oxidants: (A) adsorption isotherms, (B) effect of pH, and (C) effect of the amount of CNTs [219].

hydroxides (Cd^{2+} and Hg^{2+}) on the ACC. The oxidized commercial AC exhibited high performance for the adsorption of Cd^{2+} and Zn^{2+} but poor performance for the aromatic compounds, phenol and nitrobenzene [221].

Adsorption of Cd^{2+} and Ni^{2+} onto a bagasse fly ash was studied to remove these cations from wastewater [222]. As much as 90% removal of Cd^{2+} and Ni^{2+} was possible in about 60 and 80 min, respectively. The maximum adsorption capacities for Cd^{2+} and Ni^{2+} were measured as 1.20 and 1.00 mg/g with an initial concentration of 14 and 12 mg/L at pH 6.0 and 6.5, respectively. The adsorption behaviors of Cd^{2+} and Cu^{2+} in single- and binary-solute solutions were compared on various adsorbents, AC (S_{BET} of 658 m^2/g), kaolin, bentonite, diatomite, compost, cellulose pulp waste, and anaerobic sludge [223]. Bentonite presented the highest adsorption capacities and removal efficiencies for both Cd^{2+} and Cu^{2+}. The maximum adsorption capacities for Cd^{2+} and Cu^{2+} were calculated to be 7.59 and 7.56 mg/g for bentonite and 6.61 and 4.77 mg/g for the AC, respectively.

Chromium has attracted particular attention in the management of industrial wastewater, particularly Cr^{6+}, because of its high toxicity. Adsorptive removal of Cr^{3+} and Cr^{6+} was reviewed by comparing commercial ACs to other low-cost alternatives [224]. Adsorption performances of two commercial granular ACs (AC-1 and AC-2) and two ACFs (ACF-1 and ACF-2) were measured using aqueous solutions of potassium dichromate ($K_2Cr_2O_7$) for the adsorption of Cr^{6+} and chromium chloride ($CrCl_3$) for the adsorption of Cr^{3+} [225]. Carbon adsorbents were preoxidized with HNO_3, ammonium persulfate $(NH_4)_2S_2O_8$, and H_2O_2 in each solution and with gaseous oxygen at 350°C to change the surface oxygen-containing functional groups. The adsorbents were also predegassed at different temperatures between 400 and 950°C to eliminate the surface functional groups gradually. As shown by the adsorption isotherms for Cr^{6+}in Fig. 4.70A, the granular ACs could adsorb larger amounts of Cr^{6+} than the ACFs. The adsorption capacity for Cr^{6+} decreases on oxidation and increases on degassing, as shown in Fig. 4.70B and C, respectively. The Cr^{6+} uptake by each carbon adsorbent was comparatively higher than the Cr^{3+} uptakes, probably because of the smaller size of Cr^{6+} than Cr^{3+}. On Cr^{3+}, however, the adsorbed amounts increased with oxidation and decreased with degassing, in contrast to Cr^{6+}. The experimental results suggested an important contribution of oxygen-containing functional groups on the surface of carbon adsorbents to the adsorptive removal of Cr ions. Surface modification of a carbon adsorbent (coconut-shell-derived) by

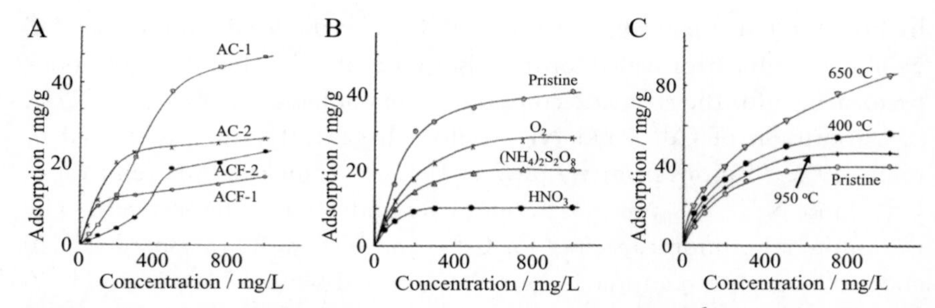

Figure 4.70 Adsorption isotherms of carbon adsorbents for Cr^{6+} (A) comparison of two ACs and two ACFs, (B) effect of preoxidation with different oxidants on AC-1, and (C) effect of predegassing at different temperatures on AC-1 [225].

35 wt.% HCl was shown to improve the adsorption performance for Cr^{6+} in a solution of sodium chromate, Na_2CrO_4 [226]. In contrast, 35 wt.% NaOH treatment of the adsorbent depressed the performance. Ozonation of a commercial AC (bituminous-coal-based) was effective in enhancing the adsorption of Cr^{3+}; the maximum adsorption capacity of the AC increased to 19.2 mg/g through exposure to ozone for 120 min from 7.3 mg/g for the as-received AC, the ozonation being associated with a marked increase in the content of acidic functional groups [227].

Various composite adsorbents were proposed for the adsorptive removal of Cr ions using GO and rGO, rGO with montmorillonite [228], rGO with MoS_2 [229], and GO with chitosan [230]. As shown in Fig. 4.71A, rGO itself has negligibly small adsorption for Cr^{6+}, as well as MoS_2, whereas the MoS_2/rGO composites prepared from the mixtures of ammonium molybdate and GO in different ratios with thioacetamide at 200°C under hydrothermal conditions can have a high adsorption capacity

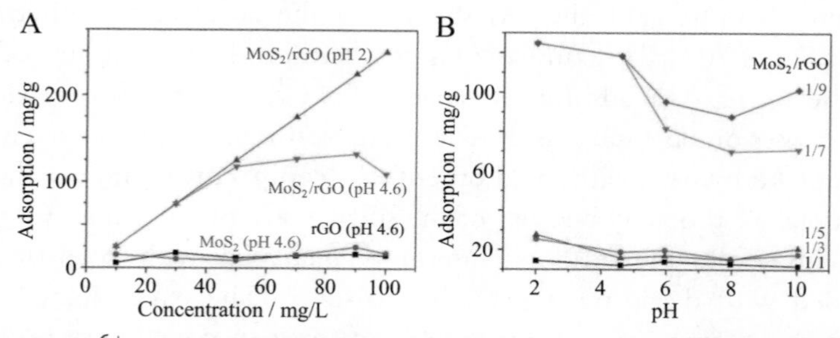

Figure 4.71 Cr^{6+} adsorption of MoS_2/rGO composites: (A) effect of the initial concentration of Cr^{6+} on different adsorbents and (B) the effect of pH on the MoS_2/rGO composites with different mixing ratios [229].

for Cr^{6+} [229]. The adsorption capacity strongly depended on the mixing ratio of MoS_2/rGO and on the pH of the solution. The composite composed of MoS_2/rGO at a ratio of 1/9 exhibits a high adsorption capacity of 125 mg/g at pH 2.0 and 1185 mg/g at pH 4.6, as shown in Fig. 4.71B. The maximum adsorption capacity being calculated by Langmuir equation to be 193 and 269 mg/g, respectively. XPS analysis suggested that Cr ions exist as Cr^{6+} and Cr^{3+} on the surface of the composite. Cr^{6+} adsorption was thought to be due to electrostatic attraction between Cr^{6+} ions and adsorbent, and Cr^{3+} was formed by the reduction of a fraction of Cr^{6+} adsorbed. On the composites of montmorillonite/rGO, a marked pH dependency of Cr^{6+} adsorption was observed, and the maximum adsorption capacity was calculated by Langmuir equation as 12.9 mg/g at 20°C [228]. The composite could be recycled after treatment with 0.1 mol/L NaOH solution.

Trapping of Ag^+ in an aqueous solution was investigated using carbon nanospheres after treatment in NaOH solutions with different concentrations of 0.01−1.0 M at room temperature [231]. Carbon nanospheres were synthesized by carbonization of glucose under hydrothermal conditions at 160°C and had diameters of 400−500 nm. By NaOH treatment, carboxyl and lactone-type functional groups on the sphere surface decreased appreciably. The change in adsorption capacity for Ag^+ with the initial Ag^+ concentration of 7−3200 ppm is shown on carbon nanospheres treated by 0.5 M NaOH, together with corresponding removal efficiency, in Fig. 4.72A. At a low concentration region, the adsorption capacity increases rapidly with increasing concentration. The removal efficiency reaches 99.9% at 7 and 40 ppm, and at concentrations above 200 ppm the capacity

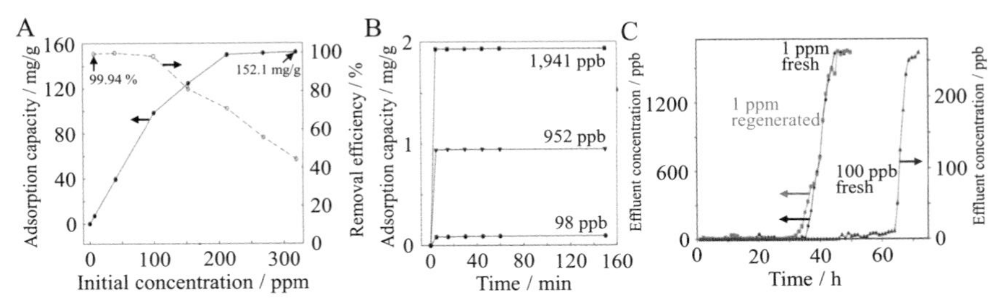

Figure 4.72 Ag^+ adsorption by carbon nanospheres treated by 0.5 M NaOH solution: (A) changes in adsorption capacity and corresponding removal efficiency with the initial concentration of Ag^+, (B) adsorption kinetics with different initial concentrations, and (C) breakthrough curves for a column of carbon nanosphere mixed with silica beads on the initial Ag^+ concentration of 1 ppm and 100 ppb of the flow [231].

saturates to 152 mg/g while the removal efficiency decreases with increasing concentration. The adsorption rate onto carbon nanospheres is very fast at a wide range of concentrations of 98 ppb–202 ppm and adsorption is completed within 6 min, as shown by the kinetic curves on small concentrations in Fig. 4.72B. The main mechanism of the adsorption of Ag^+ was thought to be due to the redox reaction of Ag^+ with functional groups of $-COO^-$ and $-OH$ on the sphere surface. The formation of Ag^0 after the adsorption of Ag^+ from the solution with a high initial concentration of 1080 ppm was confirmed by X-ray diffraction. A continuous mode of Ag^+ removal from an aqueous solution for large-scale application was tested on the carbon nanospheres after 0.5 M NaOH treatment by packing in a fixed bed with colloidal silica. Breakthrough curves for 1 ppm and 100 ppb solutions are shown in Fig. 4.72C. The column can remove almost all Ag^+ with effluent concentrations lower than 10 ppb in the initial 35 h. The adsorbent column was regenerated after saturation and could be reused for adsorption, although its breakthrough time was slightly reduced to 32 h.

4.1.5.4 Competitive adsorption of metal ions

Adsorption of Pb^{2+}, Cd^{2+}, and Hg^{2+} from their aqueous solutions was studied on a carbon aerogel with S_{BET} of 700 m^2/g and V_{meso} of 1.36 cm^3/g, containing mainly carboxyl functional groups (0.784 meq/g) [232]. The maximum adsorption capacities of the carbon for Pb^{2+}, Cd^{2+}, and Hg^{2+} calculated from sorption isotherms of single-solute solution by Langmuir model were 34.7, 15.5, and 35.0 mg/g, respectively. Adsorption isotherms of each cation from binary- and tertiary-solute solutions are shown together with those from a single-solute solution in Fig. 4.73A–C. The adsorption capacity from binary- and tertiary-solutes solutions for each cation is smaller than that from a single-solute solution, for example, maximum capacity for

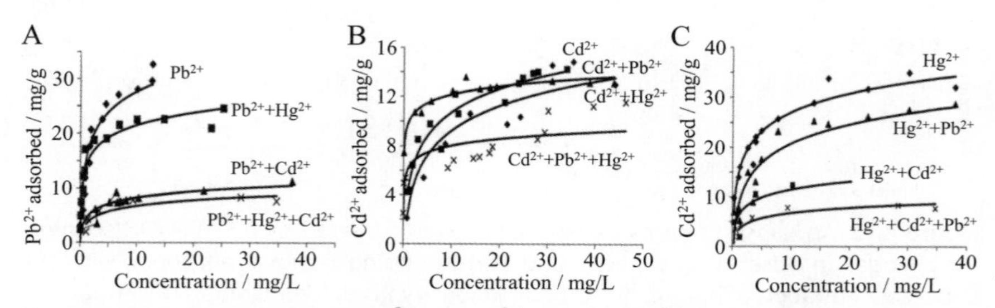

Figure 4.73 Adsorption of (A) Pb^{2+}, (B) Cd^{2+}, and (C) Hg^{2+} from single-, binary-, and tertiary-solute solutions onto a carbon aerogel [232].

Pb^{2+} was calculated as 12.4 and 23.3 mg/g for binary systems with Cd^{2+} and with Hg^{2+}, respectively, and 8.40 mg/g for a tertiary system with Cd^{2+} and Hg^{2+}, whereas it was 34.7 mg/g for a single system (Fig. 4.73A).

The competitive adsorption behaviors of Cu^{2+} with Pb^{2+} were studied on two activated carbon cloths (ACC-1 and ACC-2) and an AC [233]. Both ACC-1 and ACC-2 were prepared from rayon by activation with CO_2 at 1200°C and with steam at 900°C, respectively, and had similar V_{total} (0.665 and 0.506 cm^3/g), while the former was microporous (96.4% micropores) and the latter was mesoporous (68.2% micropores). The AC was derived from coconut shell by activation with steam and CO_2 at 900°C, with V_{total} of 0.471 cm^3/g. On the adsorbents ACC-1, ACC-2, and AC, the adsorption isotherms for Cu^{2+} and Pb^{2+} obtained with the binary-solute solution were compared with the isotherms obtained with the single-solute solutions of each ion, revealing decreases in the adsorption capacity for both ions in binary-solute solution, although the percentages of capacity decreases depended on the metal ion and adsorbent. The adsorption of metal ions (Cu^{2+} and Pb^{2+}) in the binary-solute solution of either Cu^{2+} or $Pb^{2+}$with benzoic acid (C_6H_5COOH, BA) seemed to compete with BA adsorption. At pH 5, BA was thought to be adsorbed onto the carbon adsorbent as $C_6H_5COO^-$, which could interact with metal ions to form a ligand. Therefore, the isotherms of three adsorbents with and without preloading of BA were measured. By preloading BA, the Cu^{2+} adsorption capacity increases markedly on ACC-1, but in contrast it decreases markedly on AC, suggesting different interactions between Cu^{2+} and BA molecules on the surface of the two adsorbents.

The competitive adsorption capacities for Pb^{2+}, Cu^{2+}, and Cd^{2+} were studied using MWCNTs synthesized by CVD of methane using Ni catalyst [234]. The MWCNTs were immersed in HF and concentrated HNO_3 for 24 h to dissolve metal impurities and then refluxed in concentrated HNO_3 at 140°C for 1 h before adsorption measurements. In Fig. 4.74A, adsorption isotherms at room temperature measured on the aqueous solutions of respective metal ions (single-solute systems) with pH 5.0 are compared. The adsorption of Pb^{2+} proceeds significantly at a low concentration and reaches the maximum adsorption capacity of 82 mg/g at a concentration of 16 mg/L, but those of Cu^{2+} and Cd^{2+} proceeded gradually and reach their maximum capacities at lower concentrations. Competitive adsorption of Pb^{2+}, Cu^{2+}, and Cd^{2+} measured using the tertiary-solute solution with equal initial concentrations of each metal ion is shown in Fig. 4.74B. The adsorption capacity for Pb^{2+} increases sharply and attained 27.6 mg/g at an

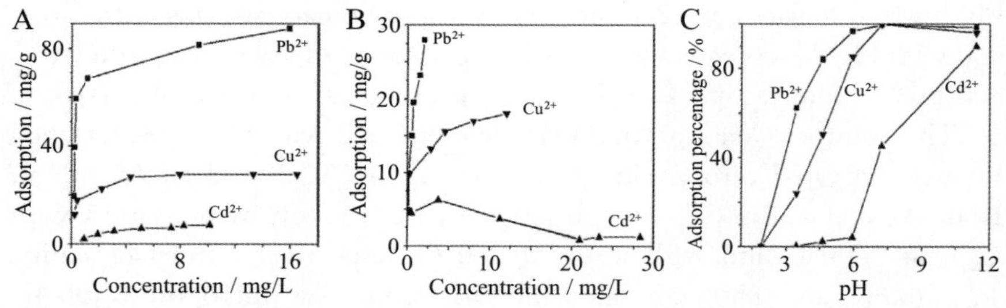

Figure 4.74 Competitive adsorption of Pb^{2+}, Cu^{2+}, and Cd^{2+} on MWCNTs at room temperature: (A) adsorption isotherms measured on single-solute solutions of the respective ions, (B) competitive adsorption isotherms for three ions measured in the tertiary-solute solution with equal concentrations at pH 5.0, and (C) effect of pH on the competitive adsorption at an initial concentration of 30 mg/L [234].

equilibrium concentration of 2.4 mg/L. In comparison, the capacity for Cu^{2+} is 17.6 mg/g at 12.4 mg/L and that for Cd^{2+} is only 7.1 mg/g at 2.9 mg/L. This may be attributed to the fact that the active sites on CNTs were occupied mostly by Pb^{2+} and Cu^{2+}, probably because of the lower affinity of Cd^{2+} than Pb^{2+} and Cu^{2+}. The maximum adsorption capacity of the MWCNTs was calculated by the Langmuir equation for Pb^{2+}, Cu^{2+}, and Cd^{2+}, although the approximation on the Cd^{2+} isotherm was poor, at 97.1, 28.5, and 10.9 mg/g in the single solutions, whereas they were 34.0, 17.0, and 3.3 mg/g in the tertiary-solute solution, respectively. The effect of pH of the solution on the competitive adsorption by the MWCNTs is shown in Fig. 4.74C. The adsorption percentages for three ions are negligible at low pH, and those for Pb^{2+} and Cu^{2+} increase sharply and reach almost 100% above pH 6 although only a small increase is observed for Cd^{2+}.

Pitch-based carbon fibers activated by steam at 900°C (ACFs) were modified by 1 M HNO_3 and 1 M NaOH, of which the adsorption capacities for Ni^{2+} and Cu^{2+} in binary-solute aqueous solution (1 mM $NiCl_2$ and 1 mM $CuCl_2$) were measured with different initial pH values [235]. Treatments of the ACFs with HNO_3 and NaOH gave slight modifications in pore structure, S_{BET} decreasing to 976 and 1226 m^2/g and V_{micro} to 0.39 and 0.49 cm^3/g, respectively, from 1462 m^2/g and 0.59 cm^3/g for the pristine ACF. As shown on the pristine, HNO_3^-, and NaOH-modified ACFs in Fig. 4.75A–C, respectively, the modifications are effective in increasing the adsorption capacities of Ni^{2+} and Cu^{2+} which depend greatly on the pH of the solution, and the higher capacity for Cu^{2+} than

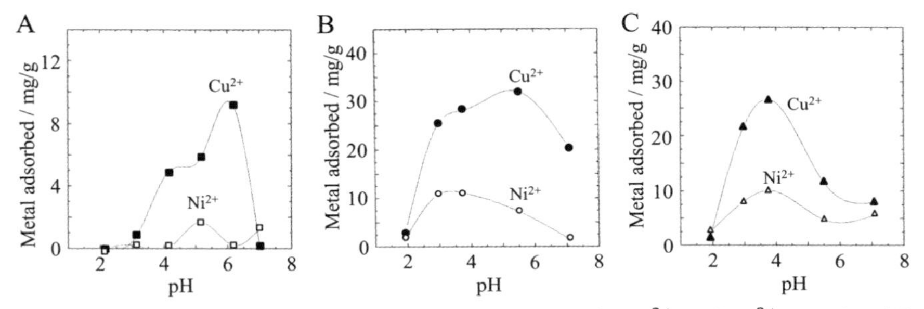

Figure 4.75 Dependences of the adsorption capacities for Ni^{2+} and Cu^{2+} on ACFs: (A) as-received ACF, (B) HNO_3-modified ACF, and (C) NaOH-modified ACF [235].

that for $Ni^{2+}$is kept even after modifications. By HNO_3 modification, the amounts of adsorbed Ni^{2+} and Cu^{2+} increase to about 35% in the low pH region in comparison to the pristine ACF, which is mostly ascribed to the increase in total acidity by the modification. Modification of ACFs increased surface functional groups containing oxygen, such as carboxyl, lactone, and phenol, which caused an increase in total acidity, resulting in an increase in adsorption capacities for Ni^{2+} and Cu^{2+}, whereas there were decreases in S_{BET} and V_{micro}. Particularly, the increase in lactone groups was marked, at 0.603 and 0.516 meq/g by HNO_3 and NaOH modification, respectively, from 0.301 meq/g for the pristine ACF.

4.1.6 Oil sorption and recovery

Oil tanker accidents have serious consequences for the environment, for example, in the accident at Galapagos Islands in January 2001, the amount of spilled oil reached 70×10^4 L and causing anxiety about the ancient ecosystem, and at Mauritius, July 25, 2020, more than 1000 tons was spilt with serious contamination of coral reefs and mangroves. It has to be pointed out, however, that the spilling and loss of oil frequently occur during transportation and storage. Continuous leakage of oil through pipe joints, for example, may produce serious contamination of soil, river water, and sometimes even subterranean water, which has detrimental effects on humans, as well as various plants, fish, and animals. These oil spills result in not only great damage to the environment, but also a great loss of energy resources. A treatment technique used in oil spill accidents on the sea, for example, is containment with large floating barriers (so-called oil fences) followed by skimming with specialized ships that either vacuum the oil off the sea or soak it up with sorbent materials. So far, some porous materials, such as PP, PET, cotton fiber, etc., have been used for the sorption of spilled oil [236]. Their sorption capacity is in the range of 10–30 kg of heavy oil per 1 kg of sorbent, although they sorb water as well as heavy oil.

Exfoliated graphite has been reviewed and was found to sorb heavy oil floating on water preferentially [112,237–240]. In a book written by the present authors (M.I. and F.K.), carbon materials for spilled-oil recovery were reviewed, including some practical tests for their applications [241]. This section, therefore, presents a brief explanation on the sorption of various oils, including heavy oils by carbon materials (exfoliated graphite, carbonized fir fibers, and carbon fibers) with the addition of recent results.

4.1.6.1 Sorption procedure

Exfoliated graphite (EG) can sorb large amounts of heavy oils quickly. As shown in Fig. 4.76A and B, A-grade heavy oil floating on water is completely sorbed into the EG added, and the characteristic brown color of the oil disappears within 1 min after the addition of EG, while the EG is still floating on water after oil sorption. The EG loses its luster with sorbing oil

Figure 4.76 Appearance of sorption and recovery of heavy oil from water using an EG: (A) A-grade heavy oil floating on water, (B) 2 min after the addition of the EG onto the oil that is less than the sorption capacity of EG, (C) after transfer of heavy-oil-sorbed EG onto a white filter paper, (D) the case when the amunt of heavy oil was larger than the adsorption capacity of the EG, (E) dropping A-grade heavy oil onto water-saturated EG, and (F) enlarged view of the water obtained from the EG.

and appears a dark black (Fig. 4.76B). The EG after the sorption of heavy oil can easily be separated from water by conventional filtration. After removing the EG, no contamination appears in the water and also no brown stain is observed after transferring oil-sorbed EG onto white filter paper, as shown in Fig. 4.76C. When the amount of heavy oil is a little more than the sorption capacity of the EG added, the periphery of the EG lump is trimmed by oil. When a large excess of oil is present, the entire EG appears wet and the brown color of oil remains in the water, as shown in Fig. 4.76D.

When A-grade heavy oil was dropped onto one end of a water-saturated EG (water content of about 1.8 g/g), water left from another end of the EG lump, as shown in Fig. 4.76E and F. This shows that heavy oil can replace the water adsorbed into the EG in advance, although complete replacement is not expected because large oil molecules cannot be adsorbed into micropores where water molecules are preferentially adsorbed.

EGs demonstrated high performance for the separation of oil from its water-emulsion, as shown by the setup in Fig. 4.77A [242,243]. Removal of oil from an emulsion was effectively performed by passing the emulsion through an EG column; the oil concentration in the emulsion of 278 mg/L decreased to 1.2 mg/L and that in the emulsion of 66 mg/L became undetectable after passing through the EG column [243].

Sorption of heavy oil into EG was measured by placing EG with different packed densities onto contaminated sands (mixtures of α-alumina

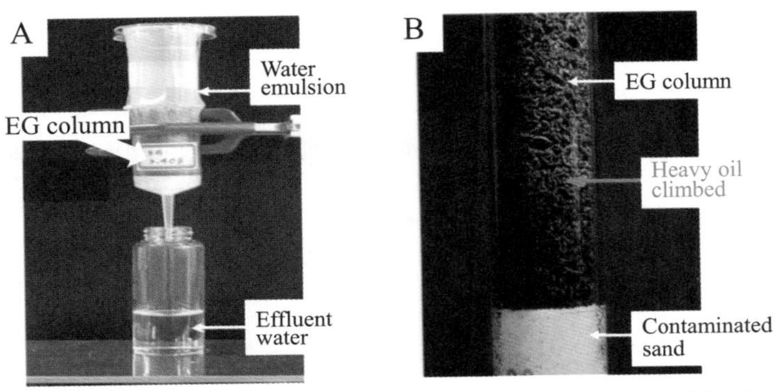

Figure 4.77 Set up for oil sorption: (A) oil separation from a water emulsion by passing through an EG column and (B) pumping up of oil from contaminated sand into the EG column [243,244].

powder having different particle sizes with A-grade heavy oil). The climbing height of heavy oil in the EG was followed with time, as shown in Fig. 4.77B [244]. To achieve the most effective pumping up of the heavy oil, optimization between the particle size of the sand contaminated by heavy oil and the bulk density of EG placed on the sand, in other words, a balance in pore sizes between the sand and the EG, was required. When sand of very small size, e.g., 175 μm, is contaminated by A-grade heavy oil, EG with a bulk density of about 0.035 g/cm^3 gave the highest height, i.e., the highest efficiency of pumping. When the sand particle size was large, for example, 425 μm, the EG with about 0.01 g/cm^3 was suitable for heavy oil pumping.

Oil sorption behaviors were studied using different carbon materials as sorbents, for example exfoliated graphite, carbonized fir fibers, carbon fibers, etc., with different oils in a wide range of viscosities, such as kerosene, grape seed oil, salad oil, motor oil, diesel oil, mineral oil, gasoline, etc.

4.1.6.2 Sorption capacity and sorptivity

The sorption rate of oils into carbon materials was evaluated by applying the so-called wicking method; a carbon sorbent was packed into a glass tube with different packing densities and its mass increase by capillary suction of an oil from the bottom was measured at room temperature as a function of time, as illustrated in Fig. 4.78 [245–247]. Some of the sorption curves for an EG with different packing densities and different oils are shown in Fig. 4.79A and B, respectively. The initial slope depends greatly on the packing density of the EG and viscosity of oils (Fig. 4.79). For A-grade heavy oil (Fig. 4.79A), very rapid suction and saturation are observed for

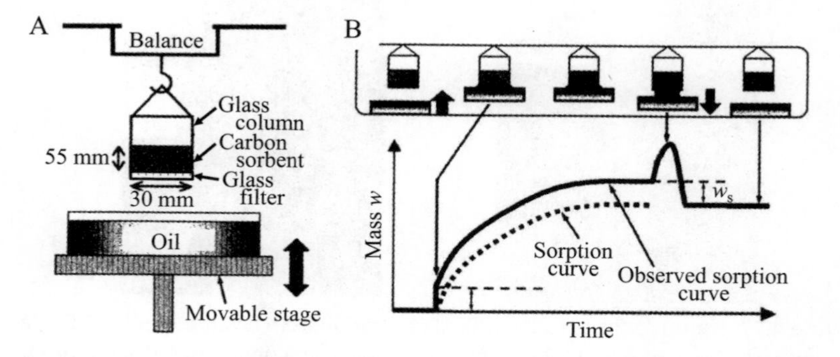

Figure 4.78 Schematic illustration of the measurement of sorption kinetics: (A) set up of the wicking method and (B) measurement of the sorption curve [245].

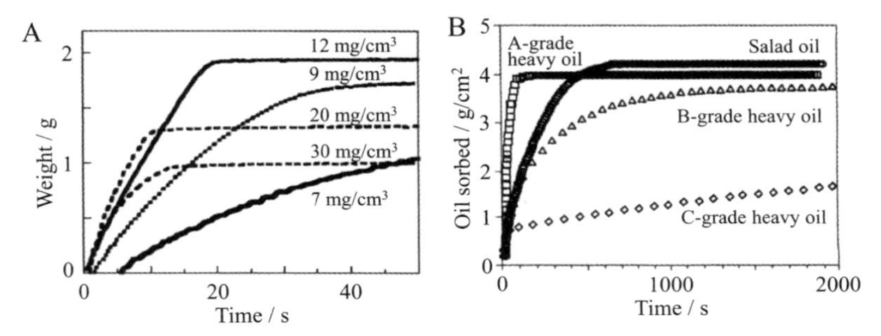

Figure 4.79 Sorption curves: (A) curves observed on the EGs with different packing densities for A-grade heavy oil and (B) curves observed on the EG with packing density of 7 mg/cm^3 for different oils [245,247].

packing densities greater than 12 mg/cm^3 within 20 s, and more gradual suction of the oil for a packing density of 7 mg/cm^3 without reaching saturation even after 50 s. The saturated weight increase shows a maximum at around 12 mg/cm^3, although the initial slope increases with increasing packing density. Very similar dependences on the packing density of an EG were observed for the more viscous C-grade heavy oil, and saturation required much longer time, more than 3000 s. The sorption curve before saturation is well approximated by the equation;

$$m_s = K_s t^{1/2} + B,$$

where m_s is weight increase per cross-sectional area of the sorbent column, t is time, K_s is sorptivity, and B is a constant. Sorptivity K_s and sorption capacity (saturated weight per g of carbon sorbent) were used as the specific parameters for oil sorption. The rate of oil sorption depends greatly on the viscosity of the oil, as compared with the sorption curves for the different grades of heavy oils and a salad oil in Fig. 4.79B.

Sorption capacity of EG depends strongly on the bulk density, decreasing rapidly with increasing bulk density for all oils, as shown on four heavy oils, A-, B-, and C-grade heavy oils and crude oil [248–251]. Sorption capacities for four heavy oils with different viscosities (A-grade: 0.004, B-grade: 0.27, C-grade: 0.35, and crude oil: 0.004 Pa s) are plotted against the bulk density of an EG in Fig. 4.80A. For A-grade heavy oil, sorption capacity reaches up to 80 g/g for the EG with a low bulk density of about 1 mg/cm^3, although it decreases quickly with increasing bulk density, down to less than 10 g/g at the bulk density of 10 mg/cm^3. For more viscous C-grade heavy oil, sorption capacity is smaller than A-grade oil

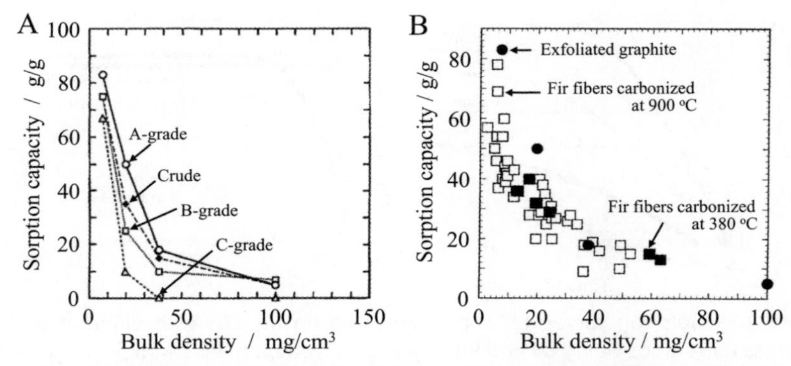

Figure 4.80 Dependences of sorption capacity for A-grade heavy oil on bulk density at room temperature: (A) exfoliated graphite and (B) carbonized fir fibers [251,253].

and cannot be sorbed by EG with a bulk density of more than 40 mg/cm^3. In Fig. 4.80B, dependences of sorption capacity for A-grade heavy oil on a fibrous component extracted from fir trees (fir fibers) carbonized at 380 and 900°C are compared with that on an EG, revealing very similar dependences to EG and no difference due to the carbonization temperature of fir fibers [252]. The carbonized fir fibers with a high bulk density could sorb C-grade oil, with about 15 g/g on the fibers having a bulk density even above 40 mg/cm^3. Carbon fiber felts with a bulk density of 54−77 mg/cm^3, which were prepared from PAN-based and pitch-based carbon fibers, showed sorption capacity of 11−17 g/g for A-grade heavy oil, and the bulk density dependence was almost the same as for EG [253]. Even activated carbon fibers having high S_{BET} did not have high sorption capacity even for low-viscosity A-grade heavy oil. Different granular activated carbons with high S_{BET} (>1000 m^2/g) were also used for heavy oil sorption, but their capacity was also very low, at about 1 g/g or less.

The sorption kinetics for other oils, including kerosene, light oil, various cooking oils, different motor oils, and diesel oil, into a column of EG with a bulk density of about 7 mg/cm^3 was also studied and similar results were obtained [247].

Since the viscosity of oils depends greatly on the temperature, the sorption capacity of the carbon sorbents depends also on temperature during the sorption process. In Fig. 4.81A, the temperature dependences of sorption capacity of the EG with a bulk density of about 6 mg/cm^3 for different heavy oils are shown, revealing marked temperature dependences, particularly for C-grade heavy oil [249]. In Fig. 4.81B, therefore, the

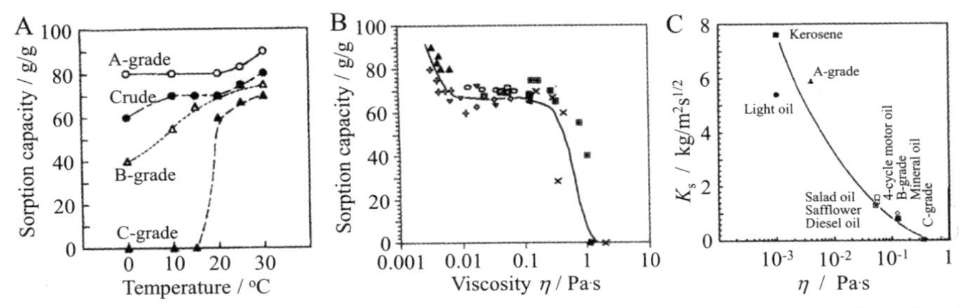

Figure 4.81 The effect of the viscosity of oil on sorption capacity and sorptivity K_s of EG with a bulk density of about 6 mg/cm^3: (A) temperature dependences of sorption capacity for different grades of heavy oils, (B) viscosity dependence on various oils, including heavy oils, and (C) viscosity dependence of K_s at room temperature [247,249].

sorption capacities observed for various oils at different temperatures are plotted against the viscosities of the oils, revealing characteristic dependence of sorption capacity on viscosity of oils, i.e., sharply decreasing with increasing viscosity above 0.2 Pa s and below 0.01 Pa s, giving a plateau in the range of 0.01–0.2 Pa s [247]. The sorption rate of heavy oils onto the EG also showed the strong dependence of the viscosity of the oils. In Fig. 4.81C, sorptivity K_s measured by wicking method on the EG with about 6 mg/cm^3 density is plotted against the viscosity of oils at room temperature, revealing a marked viscosity dependence of K_s.

Two sorption parameters, sorption capacity and sorptivity K_s, are shown as a function of the bulk density of three carbon sorbents, EG, carbonized fir fibers, and carbon fiber felt, in Fig. 4.82A and B, respectively. The EG

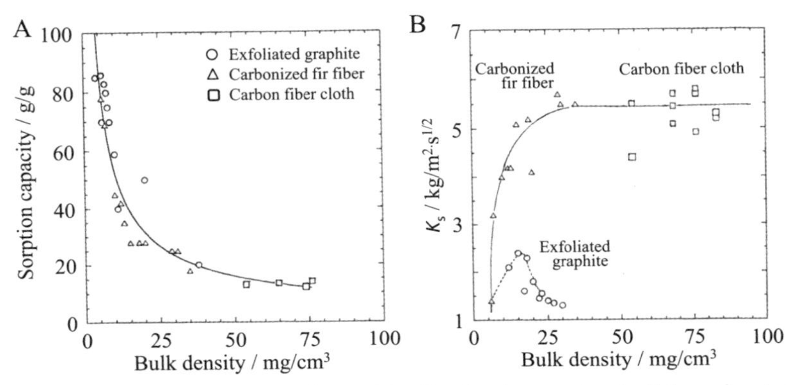

Figure 4.82 Sorption capacity and sorptivity K_s of carbon sorbents, EGs, carbonized fir fibers, and carbon fiber felts, for A-grade heavy oil [241].

and the carbonized fir fibers are in a range of bulk density from 6 to about 30 mg/cm^3 and the carbon fiber cloth covers a range of 50–75 mg/cm^3. Changes in sorption capacity with bulk density for three sorbents can be approximated by a common curve, as shown in Fig. 4.82A. As shown in Fig. 4.82B, the K_s of the carbonized fir fibers drastically increases with increasing bulk density in the region of 8–30 mg/cm^3 and reaches a maximum at around 5.5 kg/m^{2}s$^{1/2}$ when the fibers are densified above 30 mg/cm^3. In the case of carbon fiber felt, K_s is approximately constant at about 5.5 kg/m^{2}s$^{1/2}$, though experimental points are scattered. On the other hand, the EG shows a maximum K_s with about half of that for the others at around 16 mg/cm^3. On the changes of K_s with bulk density was discussed on the bases of effective sorbent porosity, tortuosity of the capillaries, average pore radius of the sorbent, the contact angle of oil and pore wall, etc. [246].

By using a lubricating oil, sorption of oil by the EG was experimentally shown not to be influenced by the salinity of the oil, suggesting that EG was suitable for oil leakage accidents under high-salinity conditions [254].

4.1.6.3 Recycling of heavy oils and carbon sorbents

No marked changes in composition or structure through sorption and desorption (recovery) of heavy oils using EG were detected [255]. No pronounced differences in number-averaged and weight-averaged molecular weights, *M*n and *M*w, between the original and recovered oils were detected, only a slight decrease in *M*w/*M*n for C-grade heavy oil. The fraction of aromatic hydrocarbons F_{arom} tended to decrease a little for the C-grade oil recovered, although there was no detectable difference in the A-grade heavy oil.

The sorption capacity of EG with an initial bulk density of about 6 g/cm^3 is more than 80 g/g and about 50% of the sorbed oil is recovered in the first cycle of sorption/recovery by filtering under suction (under 5–7 kPa pressure) [252]. However, the sorbed and recovered amounts of the oil decrease markedly with increasing number of cycles, as shown in Fig. 4.83A, mainly because of oil retention in the pores of the EG and deterioration of bulky texture due to cycling. A reduction in sorption capacity with cycling was observed more markedly on C-grade heavy oil. Carbonized fir fibers show better cyclability than EG, while the sorption capacity decreases rapidly with cycling, as shown in Fig. 4.83B, because of fragile entanglements of fibrous carbons [252]. In contrast, PAN-based carbon fiber felt can be used repeatedly without marked decreases in the

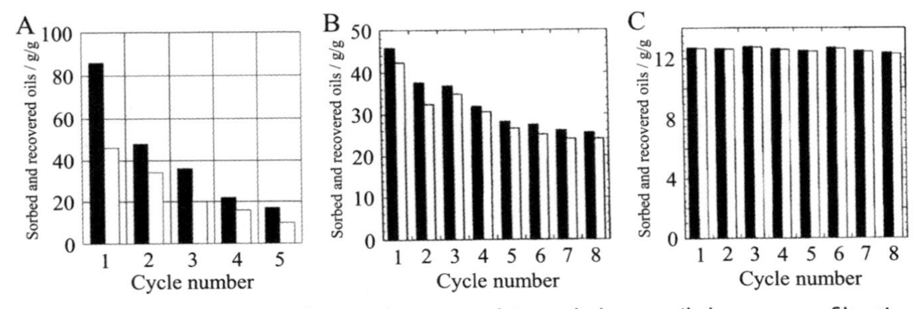

Figure 4.83 Changes in sorbed and recovered A-grade heavy oils by vacuum filtration from (A) EG (initial bulk density of about 6 mg/cm^3), (B) carbonized fir fibers (5.5 mg/ cm^3), and (C) carbon fiber felt (75 mg/cm^3) [237,252,253].

amounts of sorbed and recovered oils, as shown in Fig. 4.83C [253]. For PAN-based carbon fiber felts, the recovery ratio for both A- and C-grade oils by centrifugation at 3800 rpm is almost 100% even after eight cycles. For the felt, the sorbed oil could be repeatedly recovered by washing with *n*-hexane, keeping almost the same sorption capacity, although it is rather low at 12 g/g. The recovery of heavy oil by squeezing and twisting could be also applied for the felts with a small decrease in sorption capacity. Pitch-based carbon fiber felt, on the other hand, could not be squeezed, because of its low mechanical strength.

4.1.6.4 Mechanism of oil sorption

Observation under an optical microscope on an EG showed that, at the beginning of sorption, oil was coming up at the edges of crevices formed on the particle surface to occupy whole crevice-like pores, and then filling large void spaces quickly. Such a complicated pore structure is thought to cause strong holding of the sorbed heavy oil in EG particles. The oleophilic (hydrophobic) nature of the carbon surface is thought to be advantageous for such sorption and occlusion of oil.

Since the carbon sorbents used, i.e., EG, carbonized fir fibers, and carbon fiber felts, were macroporous and fragile, their macropore size distribution had to be evaluated by a specially designed mercury porosimeter, which could be applied on a range of pore diameters of 1−600 μm, as shown in some examples on three sorbents in Fig. 4.84A [256]. The sorption capacity of A-grade heavy oil on the EG is linearly related to total pore volume measured by the specially designed porosimeter, as shown in Fig. 4.84B, proving that the sorption of heavy oil occurs mainly in this range of pore sizes (macropores), most of which are located among the

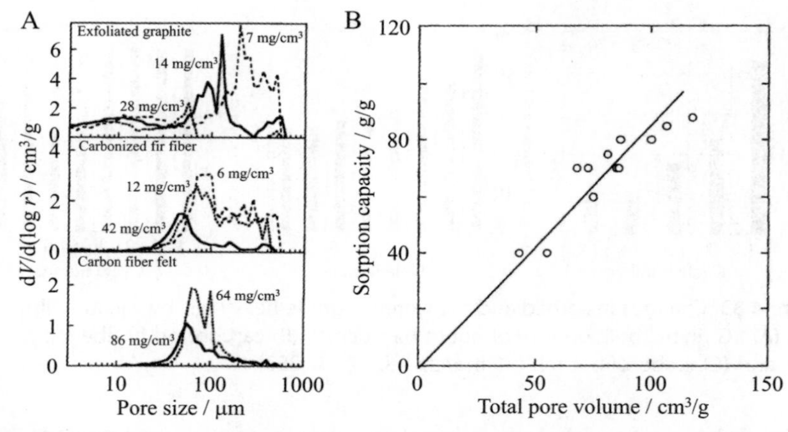

Figure 4.84 Macropore volume and sorption capacity: (A) macropore size distribution measured by a specially designed dilatometer and (B) relation between sorption capacity for A-grade heavy oil and the total pore volume of EG [256].

worm-like particles; in other words, most of the heavy oil is sorbed into the void spaces formed by the entanglement of worm-like particles of EG. Experimental points for the total pore volume above 0.1 m^3/kg tend to deviate from the linear relationship, which suggests that the larger void spaces cannot be filled completely with heavy oil. Large void spaces among worm-like particles were also evaluated by image analysis using thin slices prepared from EG after impregnation with paraffin oil [257]. By comparison with sorption capacity, the volume of these large spaces was found to be responsible for about 70% of the total sorption capacity. About 30% of the sorbed oil is kept at the crevice-like pores on the surface of worm-like particles and the ellipsoidal pores inside the particles.

The oleophilic (hydrophobic) nature of the carbon surface is an important factor for achieving high sorption capacity, in addition to large void spaces among sorbent particles. Fir fibers before carbonization showed a sorption capacity of 17—24 g/g for A-grade oil, but the fibers carbonized at 900°C showed a higher sorption capacity of 80 g/g [252]. This is the case also for rice husks [258,259], which showed sorption capacities of 4.6 and 6.7 g/g for A- and B-grade heavy oil, respectively, and low capacity of less than 1.5 g/g for water after carbonization at 600°C [258]. In the case of natural sorbent, such as milkweed, the presence of wax was pointed out to be important for oil sorption, but the results on fir fibers and rice husks suggested the importance of oleophilicity of the sorbents.

4.1.6.5 Preparation of carbon sorbents

To achieve high sorption capacity for oils, various efforts to modify the preparation procedures of carbon sorbents, including exfoliation of graphite, surface modification of EG, etc., have been developed.

The conditions to prepare EG have been investigated to achieve high sorption capacity for heavy oils [101,260]. Natural graphite was converted electrochemically to intercalation compound in 98% sulfuric acid and then heated to 1000°C for 20–60 s to produce EG. The optimized exfoliation time was determined to be 40 s, resulting in the EG having the lowest bulk density of 4 mg/cm^3 and the highest sorption capacity of 51 g/g for a mineral oil. Heat treatment of a commercial EG at 400°C in air enhanced the sorption capacity for engine oil. However, treatment in 13% HNO_3 at room temperature depressed the sorption capacities for engine oil and dyes (methylene blue and methyl orange) markedly, probably due to the change in surface character by HNO_3 treatment. As shown in Fig. 4.85A, mineral oil in the water emulsions is sorbed quickly into a commercial EG (EG1) with increasing time, oil concentration becoming less than a few mg/L after 60 min stirring [242]. In Fig. 4.85B, four commercial EGs and one of the EGs (EG1) heat-treated at 450°C in air for 10 min are compared on their sorption behaviors for a mineral oil in water-emulsion with commercial AC and vermiculite (one of the sorbents used often), revealing that heat treatment in air enhanced the sorption capacity of EG. Modification of EG surface by cetyltrimethylammonium bromide (CTAB) mixed with KBr enhanced the sorption capacity of EG for various oils slightly [261].

Sulfur-free EG was successfully synthesized using $KMnO_4$, $HClO_4$, and KH_2PO_4, without using H_2SO_4, achieving a high degree of exfoliation and sorption capacities of 123, 77, and 61 g/g for crude oil, diesel oil, and

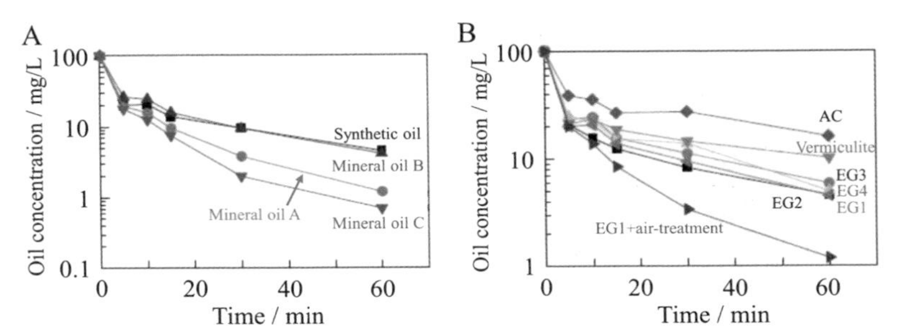

Figure 4.85 Kinetics of oil sorption from its water emulsion on EGs: (A) emulsions with different oils (100 ppm) with EG1 and (B) different EGs in the emulsion with synthetic oil (100 ppm) in comparison with vermiculite [242].

gasoline, respectively [262]. To use sulfur-free EG for heavy oil recovery is useful in avoiding the risk of including S in recovered oils from the adsorbents.

Macro-/mesoporous carbon monoliths with strong hydrophobicity were synthesized from ethanol solutions of 5–30 vol% furfuryl alcohol (FA) by carbonization at 850°C in N_2 using macro-/mesoporous silica as a template, of which sorption capacities for oils were measured by immersing for 20 s [110]. The carbon synthesized from 5 vol% FA solution, which had a bulk density of 17 mg/cm^3 and high hydrophobicity (water contact angle of 140 degrees), exhibited a sorption capacity of about 33 g/g for gasoline and could be reused for more than nine cycles with a small decrease in the capacity. Hollow carbon beads were prepared from the mixture of TEOS with polysulfone and silica powder by pumping the NMP solution of the mixture into water through a needle tip and keeping it in water for the completion of solvent/nonsolvent exchange, followed by carbonization at 550°C in N_2 flow [263]. The resultant hollow carbon beads could sorb various organic liquids, such as motor oil, paraffin oil, gasoline, toluene, and chloroform.

rGO foam was fabricated by hydrothermal treatment of a colloidal dispersion of GO at 180°C, followed by freeze-drying, GO being prepared from commercial expandable graphite by further oxidation through the modified Hammers method [264]. The rGO foam has excellent sorption capacities for various organic liquids, as shown in Fig. 4.86A. The cycle performances of the foam for toluene and dodecane were tested by repeating sorption through immersion and desorption by heating at

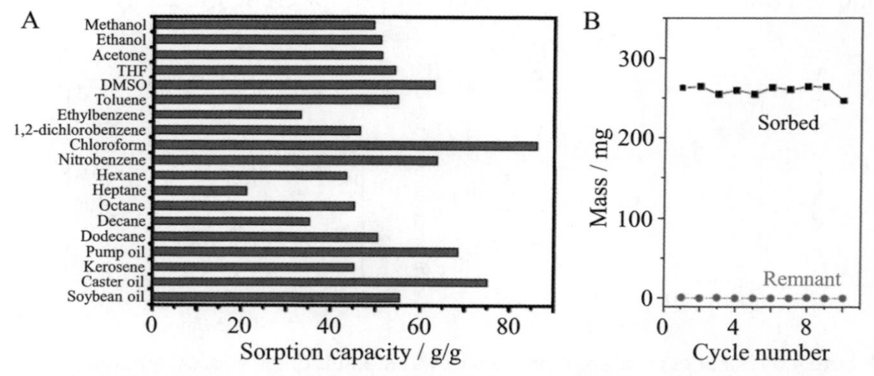

Figure 4.86 Sorption performance of the rGO foam for different chemicals and oils: (A) sorption capacities and (B) sorption/desorption cycles for dodecane [264].

105–200°C, respectively. The residual dodecane in the rGO after desorption is less than 1% in weight, as shown in Fig. 4.86B, revealing the stable recycling performance of the rGO foam. Foams synthesized by hydrothermal treatment of GO mixed with urushiol (1/1 by weight) at 100°C exhibited higher hydrophobicity than the foam without urushiol, the water contact angle being 136 and 105 degrees, respectively, and high sorption capacity for oils of 55.5, 54.5, and 51.0 g/g for pump oil, soybean oil, and lubricating oil, respectively.

CNT foam synthesized by catalytic CVD of 1,2-dichlorobenzene (ODCB) at 860°C was shown to sorb different oils dispersed on the surface of water quickly, using gasoline, pump oil, diesel oil, and vegetable oil [265]. An MWCNT array synthesized by catalytic CVD of C_2H_4 at 750°C was demonstrated to work as a membrane filter for the separation of water from oil, of which the contact angles for diesel oil and water were measured as 0 and 163 degrees, suggesting superoleophilicity of the CNT membrane [266].

A composite of rGO foam with CNTs (CNT/rGO) was prepared by growing CNTs on rGO flakes through CVD of acetylene at 800°C using Fe nanoparticles as catalyst [267]. The resultant composite had a bulk density of about 0.018 cm^3/g, consisting of a hierarchical pore structure and having strongly hydrophobic surface nature. It delivered high sorption capacity for oily liquids of more than 100 g/g. N-doped rGO foams prepared from GO-coated polyurethane foam by heating in an ethanol flame in air showed very high sorption capacity for various organics, as high as 304 g/g for chloroform and about 300 g/g for pump oil [268].

By loading micron-size iron particles on an EG, the recovery of the oil-sorbed EG from the emulsion became easier using a magnet. Loading $CoFe_2O_4$ onto EG was shown to make the recovery easier without marked decreases in sorption capacities for fuel oil, diesel oil, and crude oil [269].

4.2 Gas separation

Carbon membranes are classified according to pore sizes as microfiltration membranes (pore diameter of 100–1000 nm), ultrafiltration membranes (pore diameter of 1–100 nm), and gas separation membranes (pore diameter <1 nm). Molecular sieve carbons (MSCs), including granular and membrane, are important materials for gas separation due to their high selectivity, permeability, and stability in corrosive and high-temperature operations. The MSC membranes for gas separation were reviewed by focusing on the fabrication process [270] and on polyimides as precursors [271]. Porous carbon membranes were reviewed for their syntheses and applications, including gas separation [272–276].

In this section, a brief explanation of the fundamentals of gas separation using MSCs is presented and then the syntheses and characteristics of MSC membranes for gas separation are summarized because the syntheses of the granular MSCs were explained already in Section 2.1. The MSC membranes are classified by their precursors.

4.2.1 Fundamentals

A setup for the measurement of gas-separation performance of MSC membranes is schematically shown in Fig. 4.87. MSC membrane, either the membrane itself or membrane deposited on a substrate, such as porous graphite and alumina, is fixed to a stainless steel pipe using epoxy resin, where a gas permeated through the MSC membrane is transferred by a carrier gas to the detector to determine the permeance and permeability. Using a mixed gas as the permeate gas, permselectivity can be calculated. On the hollow carbon fibers, of which the wall works as an MSC membrane (hollow carbon fiber membrane), a bundle of a few fibers is replaced by the membrane after closing one end of the fibers using epoxy resin, as illustrated in Fig. 4.87.

Gas permeation performance of the membrane is evaluated by either permeability expressed by the unit "barrer" or permeance (permeation rate) expressed by the unit "GPU". One barrer is expressed as 3.35×10^{-16} mol m/(m^2 s Pa) in SI units, which corresponds to gas flux permeated through the membrane per pressure drop across the membrane. One GPU (gas permeance unit) is expressed as 3.35×10^{-10} mol/(m^2 s Pa) and corresponds to the ratio of the permeability to the thickness of the membrane. To evaluate gas separation, permselectivity is usually used,

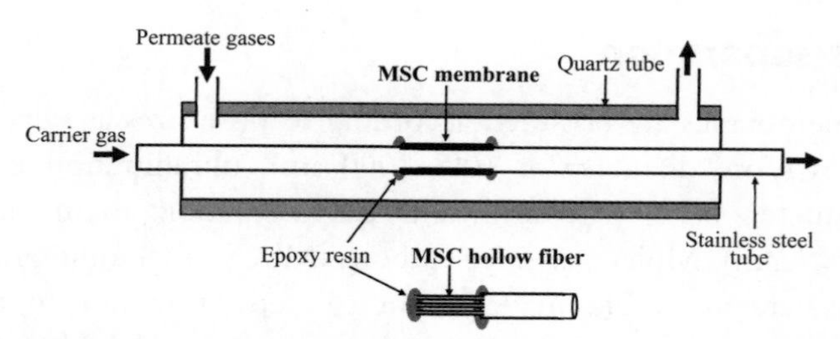

Figure 4.87 Schematic illustration of measurement set up for gas-separation performance of a molecular sieve membrane and fiber.

which is calculated by the ratio of permeability for gas A to that for gas B. Here, permeability of the carbon membrane for gas A is expressed by P_A and its permselectivity for gases A and B is written by P_A/P_B, for example, the separation of CO_2 from air (mixture of CO_2 and N_2) was evaluated by permeability for CO_2 and N_2, i.e., P_{CO_2} and P_{N_2}, and permselectivity P_{CO_2}/P_{N_2}.

The granular MSCs are now applied in gas separation systems. The adsorption rate of gas molecules, such as nitrogen, oxygen, hydrogen, and ethylene, depends greatly on the pore size of MSC, the adsorption rate of a gas becoming slower for MSCs with a smaller pore size. The adsorption temperature also governs the rate of adsorption of a gas because of the activated diffusion of adsorbate molecules in micropores, therefore the higher the temperature the faster the adsorption. Gas separation is performed by controlling (swinging) either the temperature or pressure of the adsorbate gas, temperature swing adsorption (TSA), and pressure swing adsorption (PSA). In Fig. 4.88, the flow diagram for a PSA system is schematically shown for N_2 separation from air, for example. O_2 molecules are adsorbed into the granular MSCs quickly, within 5 min, but N_2 molecules are adsorbed very slowly, at less than 10% of equilibrium adsorption even after 15 min. From the column of MSC granules, therefore, N_2-rich gas comes out on the adsorption process, and O_2-rich gas is obtained on the desorption process. By using more than two columns of MSCs and the repetition of these adsorption/desorption processes, N_2 gas can be isolated

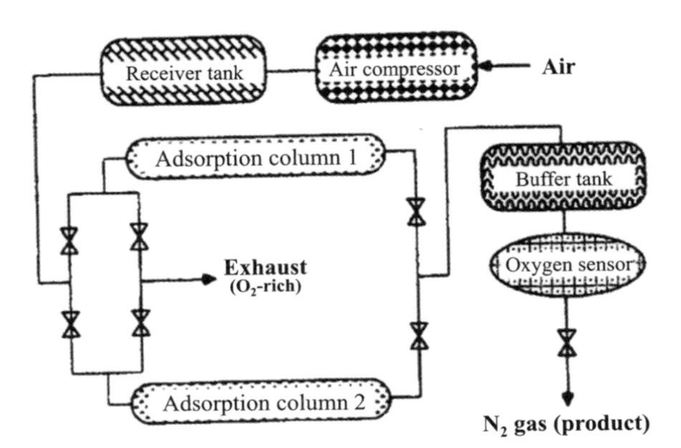

Figure 4.88 Pressure swing adsorption (PSA) method for the separation of N_2 gas from air using MSCs.

from O_2. These swing adsorption methods, mostly PSA, for gas separation have the advantages of low energy cost, room temperature operation, compact equipment, etc., and so are often employed in industries.

4.2.2 Molecular sieve carbon membranes

4.2.2.1 Polyimides

MSC membranes were synthesized by carbonization of a polyimide film of commercially available Kapton (the repeating unit is shown in Fig. 4.89), of which the pore structure and gas adsorption behaviors for He, N_2, O_2, and CO_2 were clarified [277]. Gas permeation and gas separation performance for various mixed gases, such as H_2/N_2, CO_2/N_2, O_2/N_2, CO_2/CH_4, etc., were studied on the carbon membranes derived from polyimides [278–283]. Microporous carbon membranes were prepared from Kapton film with a thickness of 125 μm at different temperatures, of which the micropore volume and gas diffusivity depend greatly on the carbonization temperature, as shown in Fig. 4.90 [282]. Molecules with sizes larger than C_2H_6 (0.40 nm) cannot penetrate the membrane carbonized at 700°C, and only a small amount of CO_2 (0.33 nm) penetrates the membrane

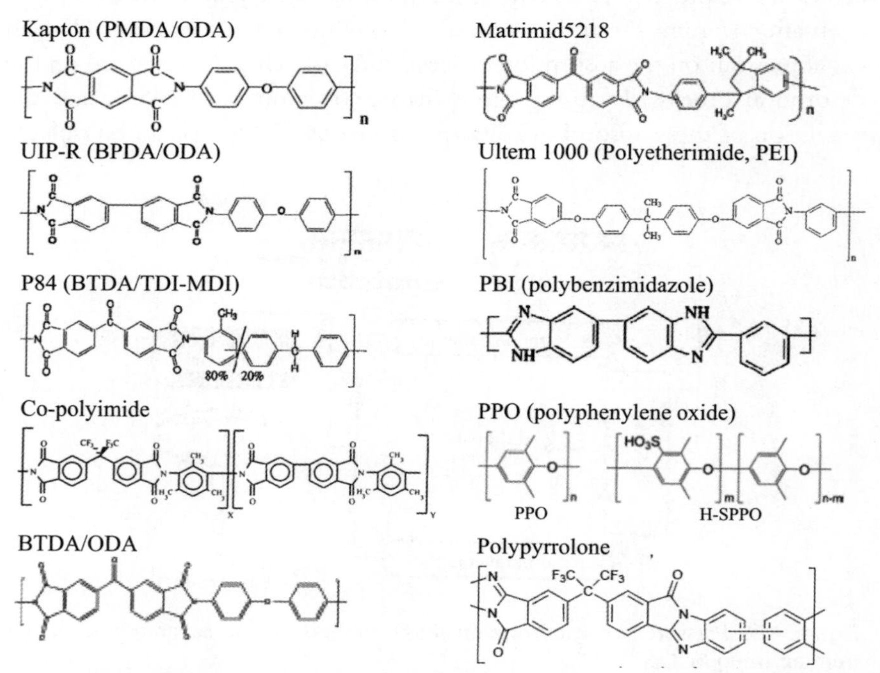

Figure 4.89 Repeating units of polymers used as the precursors for MSC membranes.

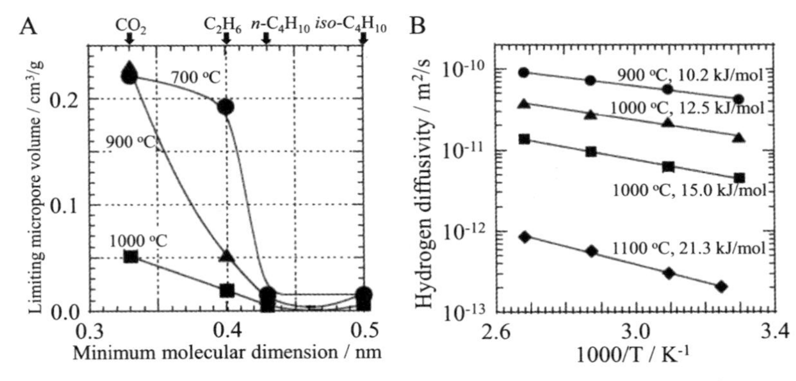

Figure 4.90 Carbon membranes prepared from Kapton films (thickness of 125 μm) at different temperatures: (A) dependences of limiting micropore size in carbon membrane on minimum molecular dimension of the gas and (B) Arrhenius plots of hydrogen diffusivity. The activation energy calculated for surface diffusion is indicated for each membrane [282].

carbonized at 1000°C. The H_2 permeability P_{H_2} and the permselectivity against CO, P_{H_2}/P_{CO}, were very sensitive for the preparation conditions of MSC membrane; a membrane heat-treated at 1000°C gave P_{H_2} of 7.87×10^{-15} mol/m·s·Pa and P_{H_2}/P_{CO} of 1770, but another membrane heat-treated at the same temperature gave 2.40×10^{-15} mol/m.s.Pa and 5900, respectively, suggesting that the heat treatment temperature (HTT) of 1000°C was critical. The selectivity P_{H_2}/P_{CO} of 5900 implies that the membrane can reduce the CO content in H_2 from 1% to 2 ppm. For the membrane heat-treated at 1100°C, P_{H_2} and P_{CO} could not be measured because it was beyond the lower limit of the equipment. In Fig. 4.90B, the dependences of H_2 diffusion coefficient on HTT are shown, together with the activation energy calculated from the slope, revealing that diffusivity decreases markedly with increasing HTT from 900 to 1100°C, associated with a marked increase in activation energy. In addition to the prominent effect of HTT, the heating rate during heat treatment of Kapton membranes (25 μm thick) was reported to influence the permselectivity on different binary gases; with a decreasing heating rate at 950°C from 13.3 to 1.33°C/min, P_{H_2}/P_{N_2}, P_{CO_2}/P_{N_2}, and P_{O_2}/P_{N_2} increased from 1240, 82.7, and 21.6 to 4700, 122, and 36.0, respectively [279]. Permeabilities of the carbon membranes carbonized at 1000°C and slightly oxidized at 400°C in He-containing water vapor for 10 min were measured at 100°C for various organic gases (alkene and alkane) [280]. Permeability is plotted against the

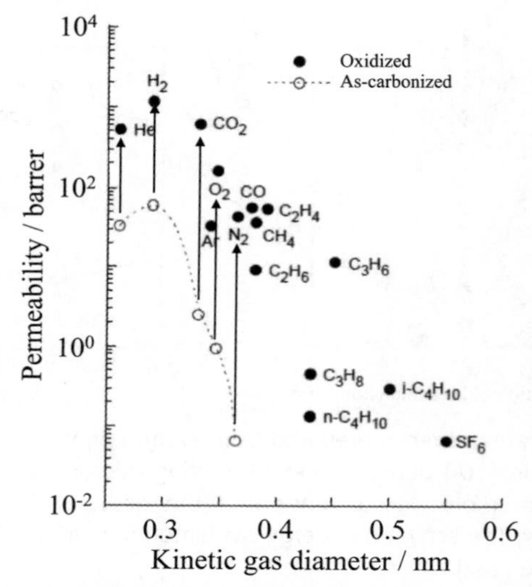

Figure 4.91 Gas permeability of Kapton membranes carbonized at 1000°C (as-carbonized) and slightly oxidized at 400°C (oxidized) for selected gases at 100°C [280].

kinetic diameter of the gas in Fig. 4.91. The carbonized membrane contained only micropores with sizes of less than 0.40 nm, while the slight oxidation made micropore sizes enlarged slightly to about 0.43 nm. Permeabilities of the oxidized membrane are 10–700 times higher than those of the as-carbonized one, although the size enlargement is very small.

Carbon membranes were prepared from a polyamic acid (PAA) of PMDA/ODA (Kapton-type) by spin-coating onto the surface of a macroporous carbon disk, followed by imidization and carbonization, which were confirmed to be suitable for gas separation of O_2/N_2, CO_2/N_2, and CO_2/CH_4 [281]. Carbon membranes on the graphite disk were also prepared from the commercially available polyimide Matrimid, which was soluble in NMP, through the same procedure without imidization, and studied on gas separation. The carbon membrane derived from Kapton-type polyimides was more permeable but less selective than those derived from Matrimid, with permselectivities P_{CO_2}/P_{CH_4} of 16, P_{CO_2}/P_{N_2} of 9, and P_{O_2}/P_{N_2} of 4 for the Kapton-derived membranes, and 33, 15, and 6 for Matrimid-derived ones, respectively. A pretreatment of polyimide membranes (Matrimid 5218 and P84) before carbonization by nonsolvent (methanol, ethanol, 1-propanbol, and 1-butanol) was reported to improve the permselectivity of the membranes after carbonization, with P_{CO_2}/P_{CH_4}

increase from 61 to 169 and from 89 to 139 on Matrimid-derived and P84-derived carbon membranes, respectively, by ethanol pretreatment [284].

Carbon hollow fibers were fabricated by wet-spinning of NMP solution of Matrimid, followed by stabilization at 400°C for 30 min and carbonization at 900°C for 5 min in three different atmospheres, high-purity N_2, N_2 saturated with water vapor, and high-purity CO_2 [285]. The carbon fiber membranes obtained in H_2O and CO_2 atmospheres gave much lower permeances for all gases but significantly higher permselectivities than that obtained in inert atmosphere. The former are more porous and less dense than the latter, and as a consequence they are more permeable and exhibit higher permselectivity. The CO_2 environment is more effective for enhancement of porosity than the H_2O environment, whereas similar permselectivities are obtained by both environments, as shown in Table 4.10. With increasing permeation temperature from 40 to 100°C, the permeance of the membrane tended to increase, as well as permselectivity, on the membrane carbonized in N_2, while it tended to decrease on the membranes carbonized in H_2O and CO_2 environments.

Hollow carbon fiber membranes were synthesized from the hollow fibers of asymmetric polyimides, BPDA (refer to Fig. 4.89) with different aromatic diamines, by carbonization at 600−1000°C, of which the outer and inner diameters were around 300 and 100 μm, respectively, after 1000°C treatment. The resultant carbon hollow fibers had the asymmetric structure of a dense skin layer and porous inside layer, which displayed high gas permeability and permselectivity [286−292]. Carbonization of these polyimide fibers was performed by continuous pulling with a constant rate through a high-temperature region and so the residence time at the carbonization temperature was usually as short as 3.6 min [289]. MSC membranes were prepared by carbonization of hollow fibers (157 μm inner diameter and 231 μm outer diameter) of asymmetric polyimides at 500 and 550°C, of which the permselectivity P_{O_2}/P_{N_2} was in the range of 8.5−14.0 and P_{CO_2}/P_{N_2} of 55−56 [286]. Permselectivity P_{O_2}/P_{N_2} of the membrane was markedly affected by relative humidity in the mixed gases; it was kept at around 10 in the gas with less than 23% relative humidity, while it was reduced to around 8 in the gas with 85% humidity [287]. The reduction in P_{O_2}/P_{N_2} was mainly due to the reduction of P_{O_2} and occurred in the initial period of exposure to the humid gas. The pyrolysis (carbonization) conditions of polyimide hollow fibers influenced greatly the gas

Table 4.10 Pore parameters and permselectivity of hollow carbon fibers fabricated in different atmospheres [285].

Carbonization atmosphere	S_{BET} (m^2/g)	V_{micro}[a] (cm^3/g)	Permselectivity at 40°C				
			P_{H_2}/P_{CH_4}	P_{CO_2}/P_{CH_4}	P_{CO_2}/P_{N_2}	P_{O_2}/P_{N_2}	P_{H_2}/P_{CO_2}
N_2	129	0.091	32.32	0.86	0.7	0.72	37.82
H_2O-saturated N_2	1403	0.542	67.71	20.86	23.6	5.5	3.25
CO_2	1888	0.618	112.18	17.07	21,8	5.5	6.71

[a]Determined by the DR method.

permeability and permselectivity. The pyrolysis in vacuum produced more selective but less permeable membranes than the pyrolysis under inert gas purging. A high purge gas flow rate (i.e., 200 cm^3/min) resulted in membranes with a much higher permeability but lower selectivity, in comparison to a low purge flow rate (20 cm^3/min) [288]. The viscous PAA solution of PMDA/ODA was coated on a polytetrafluoroethylene (PTFE) thread (1 mm diameter) through a nozzle and immersed into the water (coagulant) at 2–40°C and pH of 2–11, and then removed from the PTFE thread after drying, followed by imidization at 400°C and carbonization at 800°C in N_2 to obtain hollow fiber membranes [291]. Permeability and permselectivity of the membrane depended greatly on the condition of gelation of PAA solution (time, temperature, and pH). A high permselectivity P_{CO_2}/P_{CH_4} close to 200 was obtained by gelation for 6 h at 2°C and pH 9.4. A carbon capillary membrane composed of a thin dense top layer and a fine sponge-like matrix was synthesized from PAA of PMDA/ODA by adding the nonsolvent glycerol, followed by imidization at 400°C and carbonization at 950°C [290]. It showed a high permselectivity of 2000 for He/N_2 and about 1000 for H_2/N_2 at 0°C, and even 170 for He/N_2 at 250°C. The hollow fiber membranes prepared from an asymmetric hollow fiber membrane of a polyimide from BPDA and aromatic diamines at temperatures of 600–630°C exhibited high separation performance for olefin/paraffin system at 100°C [292].

On the polyimide membrane of the copolymer shown in Fig. 4.89, the air-separation properties, i.e., permselectivity P_{O_2}/P_{N_2} and permeability P_{O_2}, of the membranes are improved by heat treatment at 535°C, in which the performance reaches the commercially attractive region beyond the upper bound curve from the pristine polyimide film (PI), as shown in Fig. 4.92 [293]. By heat treatment at a higher temperature of 550°C, the permselectivity improves more, though permeability decreases slightly. Heat treatments of Matrimid and P84 up to 800°C (carbonization) were shown to be effective in improving CO_2/CH_4 separation performance (being important for natural gas processing) beyond the upper bound curve to the commercially attractive region [294]. By designing the polyimide structure through hyperbranching various monomers, the improvement of gas separation performance of the carbonized membranes was studied; a high P_{CO_2} of 1085 barrer with P_{CO_2}/P_{CH_4} of 52 was obtained [295].

Carbon membranes from BPDA/ODA-based polyimides (refer to Fig. 4.89) were formed on the surfaces of a porous alumina tube (outer

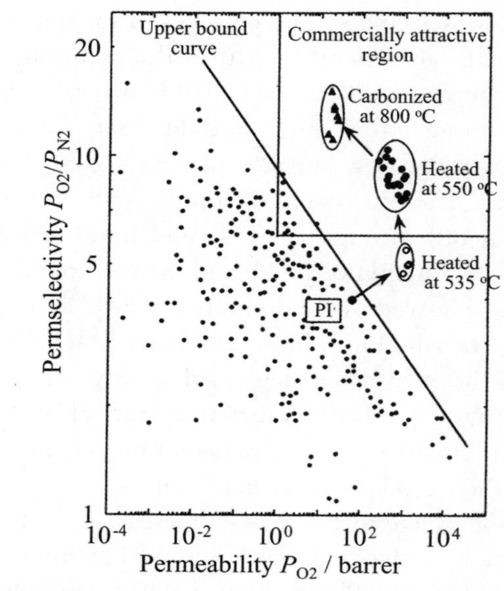

Figure 4.92 Change in O_2/N_2 selectivity of the polyimide film (PI) with heat treatment up to 800°C, in comparison with polymer membranes (dots) with respect to the upper bound curve [293].

diameter 2.4 mm, inner diameter 1.8 mm, void fraction 0.48, and average pore size 140 nm) by dipping into the precursor PAA solution, followed by imidization at 300°C and carbonization at 500–900°C in an inert atmosphere [296–300]. Before carbonization, the coating/imidization cycle was repeated three times. Carbon membranes prepared at 700°C exhibited permselectivities P_{O_2}/P_{N_2} and P_{CO_2}/P_{N_2} of 10 and 47, respectively [299]. These permselectivities were enhanced to 14 and 73, respectively, by carbon deposition onto the inner surface of the membrane by CVD of propylene at 650°C. The membranes exhibited permselectivities of 4–5 for C_2H_4/C_2H_6 system and 25–29 for C_3H_6/C_3H_8 syatem at 100°C [297]. The carbonization temperature of the membranes governed their micropore structure [300]. The membranes prepared at 700°C were oxidized at 300°C by flowing O_2/N_2 mixed gas or pure O_2 for 3 h, which resulted in a significant increase in V_{micro} without an appreciable change in pore size distribution, and as a consequence the permeance increased with no marked change in permselectivity. MSC membranes were formed on a porous alumina disk (47.0 mmΦ × 1.0 mm with mean pore size of 0.14 μm) by spray-coating of NMP solution of 12 wt.% P84, followed by stabilization at

350°C and carbonization at 600−900°C [301]. MSC membranes were fabricated by casting a DMAC solution of 10 wt.% polypyrrolone (refer to Fig. 4.89) onto a glass plate, followed by imidization at 150°C in vacuo, thermal cyclodehydration at 400°C for 10 h, and then carbonization at different temperatures up to 800°C in N_2 [302]. The membrane prepared at 700°C delivered high permselectivity for different gas pairs, such as P_{H_2}/P_{CH_4} of 1200, P_{CO_2}/P_{CH_4} of 180, and P_{O_2}/P_{N_2} of 11.

The polymer membranes fabricated from the blends of various polyimides with polybenzimidazole (PBI, refer to Fig. 4.89) were converted to carbon membranes by heat treatment at 800°C, of which the gas-separation performance was studied as a function of the carbonization conditions [303,304]. The permeabilities and permselectivities of the carbon membranes prepared at 800°C are shown for H_2/CO_2, CO_2/CH_4, and N_2/CH_4 systems in Table 4.11. The carbon membrane derived from PBI/Matrimid (25/75 wt.%) exhibits higher selectivity for all gas systems than the other blends, and higher Matrimid content results in improved permeability and enhanced selectivity, particularly for the separation of N_2 and CO_2 from CH_4. The degree of vacuum during the carbonization process has to be selected at an optimum condition (refer to PBI/P84 blends in Table 4.11). A lower carbonization temperature resulted in membranes with a higher permeability but lower selectivity.

Polymer blends of polyimide with polyvinylpyrrolidone (PVP) were reported to be effective for improvement in permselectivity [305]. The MSC membrane derived from BTDA/ODA containing 10 wt.% PVP and carbonized at 550°C showed a maximum P_{O_2} of 630 barrer and P_{O_2}/P_{N_2} of 10 and that at 700°C gave P_{O_2} of 230 barrer and P_{O_2}/P_{N_2} of 14. The molecular weight of the added PVP influenced gas permeability and permselectivity. The higher the molecular weight resulted in the higher the permeability and the lower the permselectivity [306].

Loading of inorganic particles was shown to enhance the permselectivity of the carbon membrane of polyimides. Nanoparticles of zeolite ZSM-5 (20−50 nm size) were mixed into a PAA of Kapton precursor with different contents (4.76, 9.09, and 16.7 wt.%) in DMAc solution and then cast onto a glass plate, followed by imidization at 350°C and carbonization at temperatures of 600, 700, and 800°C to obtain zeolite-loaded MSC membranes [307]. The permeability of the membranes tended to increase but their permselectivity decreased with the increase in the content of zeolite from 0 to 16.7 wt.%, with P_{O_2}/P_{N_2} decreasing from 14.8 for the

Table 4.11 Gas-separation performances at 35°C of carbon membranes prepared from PBI/polyimide blends at 800°C [303,304].

Preparation conditions			Permeability (barrer)				Permselectivity		
Precursor blend	**Mixing ratio (wt.%)**	**Vacuum (Torr)**	P_{H_2}	P_{N_2}	P_{CH_4}	P_{CO_2}	P_{H_2}/P_{CO_2}	P_{CO_2}/P_{CH_4}	P_{N_2}/P_{CH_4}
PBI/Matrimid	25/75	20	660,2	3.78	0.473	96.5	6.84	203.95	7.99
	50/50		324.0	1.26	0.278	36.6	8.85	131.65	4.53
	75/25		148.4	0.49	0.170	16.1	9.20	94.88	2.88
PBI/Torlon	50/50		970.3	6.24	2.830	279.0	3.47	98.58	2.20
PBI/P84	50/50		355.2	1.83	0.568	60.5	5.87	106.51	3.22
	50/50	10^{-3}	—	5.61	2.266	179.9	—	79.4	2.47
	50/50	10^{-7}	—	4.51	1.801	162.9	—	90.4	2.51
PBI/Kapton	50/50	10^{-3}	—	3.01	1.129	138.1	—	122.3	2.67
PBI/UIP-R	50/50		—	4.94	1.770	189.8	—	107.2	2.79

membrane without zeolite-loading to 8.5 for 16.7 wt.% zeolite-loaded membrane. The highest permeabilities for H_2, CO_2, O_2, and N_2 were obtained on membrane prepared at 700°C. A loading of zeolite KY into a polyimide (Matrimid) enhanced P_{CO_2}/P_{CH_4} to124 from 61 after carbonization at 800°C in a vacuum [308]. Loading of SiO_2 nanoparticles on MSC membranes was performed by adding TEOS into Kapton-type PAA and carbonizing at 600°C, which delivered enhanced selectivity for large molecules, such as C_2H_4/C_2H_6 and C_3H_6/C_3H_8 systems [309].

Polyetherimide (PEI) was also used as the precursor of MSC membranes [310–316]. MSC membranes were formed on the inner surface of porous ceramic tubes (7 mm inner diameter, 10 mm outer diameter, and 4.5 cm long) by dip-coating of the commercial PEI (Ultem 1000, refer to Fig. 4.89) via its 1,2-dichloroethane (DCE) solution, followed by stabilization at 350°C and carbonization at 600°C [311]. The gas-separation performance of the membranes was studied in binary gas mixtures, CO_2/CH_4, H_2/CH_4, and CO_2/H_2, as well as in a ternary mixture of CO_2/CH_4/H_2. Pair selectivity in the ternary mixture is shown in Fig. 4.93A and B as a function of temperature and CH_4 mole fraction, respectively. The permselectivities for CO_2/CH_4 and CO_2/H_2 pairs decrease with increasing temperature, which is the same as their trends in the respective binary mixtures, whereas the behavior of the selectivity for the H_2/CH_4 pair is somewhat different from that observed in the binary gas (Fig. 4.93A), suggesting the effect of the third species, i.e., CO_2 in this case. As shown in Fig. 4.93B, the permselectivity P_{CO_2}/P_{CH_4} as high as 452 and P_{H_2}/P_{CH_4} as low as 20 were obtained for a mixture containing 60% CH_4, 20% CO_2, and 20% H_2. The structural characterization of the MSC membrane formed on ceramic substrate has been studied and discussed [312]. MSC membranes

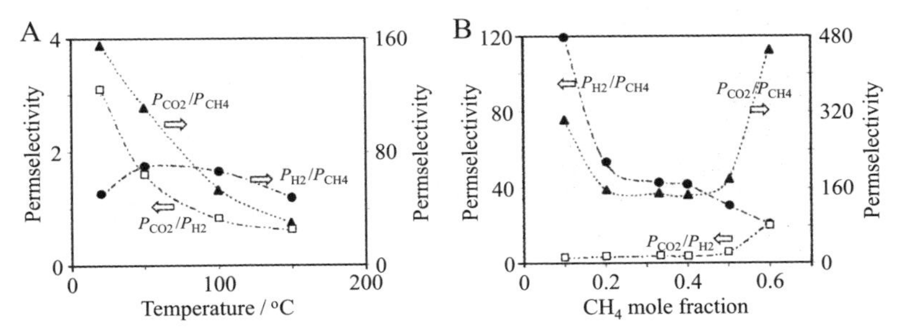

Figure 4.93 Permselectivity of MSC membrane prepared from PEI for a ternary gas mixture of $CO_2/CH_4/H_2$: (A) effect of temperature in the equimolar mixture and (B) effect of gas composition at 20°C [311].

were fabricated from PEI (Ultem) by blending either with PVP or MWCNTs, of which the gas separation performances were studied on a CO_2/N_2 system [314]. PVP blending enhanced permeability but reduced permselectivity, P_{CO_2}/P_{N_2} decreasing to 13.7 from 17.5. In contrast, MWCNT blending increased permeabilities markedly and at the same time increased permselectivity P_{CO_2}/P_{N_2} to 48.8.

MSC hollow fiber membranes were prepared by carbonization of hollow fibers of PEI (Ultem) containing 26 wt.% PVP (PEI/PVP blend) at 650°C with different heating rates in N_2 after being stabilized at 300°C for 30 min in air [315,316]. In Fig. 4.94A and B, SEM images are shown of the surface and cross-section of the fiber, respectively, revealing that on the fiber surface no pores are observed but the fiber wall contains many macropores. The permselectivities P_{CO_2}/P_{CH_4} and P_{CO_2}/P_{N_2} are shown as a function of heating rate from 350 to 650°C in Fig. 4.94C. The membranes prepared from the PEI/PVP blend with different heating rates (1−9°C/min) exhibited much higher permselectivities than those prepared from pure PEI with a heating rate of 3°C/min (shown by indicating PEI on the abscissa), particularly with the rates of 3−5°C/min [314]. Preparation of PEI/PVP hollow fibers by wet-spinning, followed by stabilization and carbonization, was experimentally studied by focusing on the conditions for stabilization [313]. Optimum conditions for obtaining the MSC hollow fiber membrane were determined as follows; stabilization at 400°C for 1 h in He gas containing 5% O_2 and carbonization at 600°C for 1 h in pure He flow with a heating rate of 3°C/min.

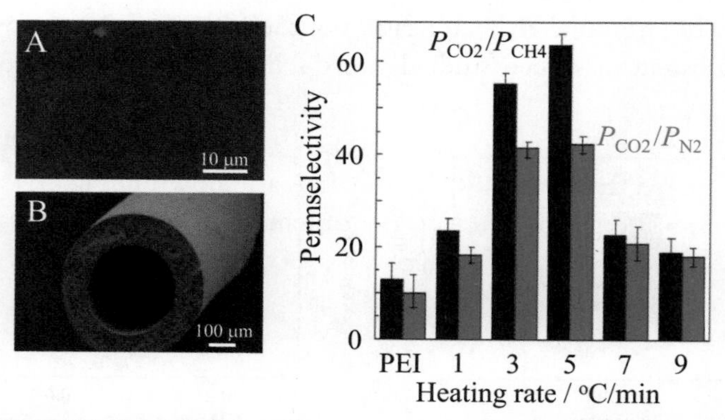

Figure 4.94 MSC hollow fiber membranes prepared from a PEI/PVP blend at 650°C: SEM images of (A) the outer surface and (B) the cross-section, and (C) effect of heating rate on permselectivities of P_{CO_2}/P_{CH_4} and P_{CO_2}/P_{N_2}. PEI on the abscissa indicates the hollow fibers prepared from PEI (without PVP) with a heating rate of 3°C/min [316].

4.2.2.2 Poly(phenylene oxide)

Poly(phenylene oxide) (PPO; refer to Fig. 4.89) was used as the precursor for the preparation of MSC membranes on a porous alumina tube (average pore size of 100 nm) by dip-coating of chloroform solution of 3 wt.% PPO, followed by carbonization in Ar [317–319]. The postoxidation at a temperature of 100–400°C for 0.5–3 h in a flow of air with 200 mL/min rate was reported to result in only a slight increase in permeances with increasing oxidation period at 200°C but in a decrease in permselectivities of CO_2/N_2 and CO_2/CH_4 systems. Blending of 20 wt.% PVP with different molecular weights (10–90K) into PPO was also studied by forming an MSC membrane on a porous alumina tube at 700°C [320]. A marked influence of molecular weight of PVP on permselectivity of the MSC membranes was observed on He/CO_2 and CO_2/CH_4 systems, with a marked increase in P_{He}/P_{CO_2} and a rapid decrease in P_{CO_2}/P_{CH_4} with increasing molecular weight of PVP up to 4×10^4, which was supposed to be due to the decrease in CO_2 permeance.

Marked improvement in permeability and permselectivity was found by sulfonation of PPO (SPPO) [321,322]. Sulfonation of PPO was carried out using chlorosulfonic acid in a chloroform solution under N_2 atmosphere and stored as Na-form (Na-SPPO). A methanol solution of 20 wt.% Na-SPPO was spun into a saturated NaCl aqueous solution (coagulation agent) and then soaked into 1 M HCl to convert to H-SPPO hollow-fiber membrane. H-SPPO membranes were carbonized under vacuum at 650–1000°C for 2 h after stabilization at 280°C in air. H-SPPO could be changed to metal-SPPO (metal: Na^+, Mg^{2+}, Al^{3+}, Ag^+, Cu^{2+}, and Fe^{3+}) by immersing H-SPPO membrane into each metal nitrate solution for 2 h and converted to hollow carbon fiber membranes using the same procedure. The degree of substitution of SPPO was determined as 40%. The permeability and permselectivity for different gases of the carbon membranes prepared from H-SPPO membranes are shown in Fig. 4.95A and B, respectively. The gas permeation performances of most of the carbonized membranes at different temperatures are superior to those of the pristine polymer membranes (SPPOs). The relations between P_{O_2} and P_{O_2}/P_{N_2} on H-SPPO- and Ag-SPPO-derived carbon fiber membranes are plotted in Fig. 4.95C to compare with other data published on Ag-loaded MSC membranes [323,324] and polymeric M-SPPO membrane [325], revealing that sulfonation of PPO membrane can deliver gas separation performance almost comparable to Ag-loaded MSC membranes [323,324] and even

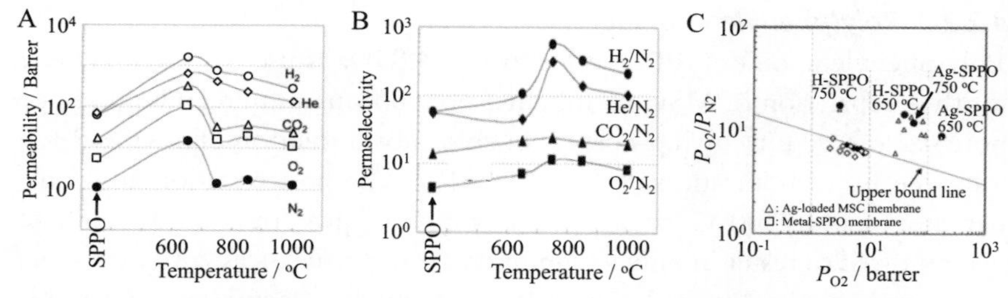

Figure 4.95 Gas permeation performances of the hollow carbon fiber membranes derived from SPPO: (A) carbonization temperature dependences of permeabilities and (B) those of permselectivities, and (C) relation between P_{O_2} and P_{O_2}/P_{N_2} on carbonized H-SPPO and Ag-SPPO membranes [322].

better than that of metal-loading onto polymeric SPPO [325]. The modification of PPO with trimethylsilyl (TMS) substituent improved the permeability markedly, with a slight decrease in permselectivity of the resultant hollow carbon fiber membrane, P_{O_2} and P_{O_2}/P_{N_2} increasing from 54.7 barrer and 11.4 for that derived from the pristine PPO to 125 barrer and 10.0 for the membrane derived from TMS-substituted PPO [326].

4.2.2.3 Other polymers

MSC membranes were prepared from an ethanol solution of novolac-type phenol-formaldehyde (PF) resin with a small amount of hexamethylene tetramine by dip-coating onto an outer surface of the tubular support repeatedly, and then carbonized at 900°C in Ar [327]. The tubular support was fabricated from the same PF resin through curing, pulverizing, and molding after mixing with hexamine and methyl cellulose, the advantage of this support being to have the same shrinkage rate as the coated membrane precursor. The permeance rate and permselectivity of the resultant MSC membrane depended greatly on the concentration of PF in ethanol; the lowest concentration of 30 wt.% resulted in the highest permeation rates of 123 and 120 GPU for N_2 and CO_2, respectively, and the lowest permselectivity of 0.97 for the CO_2/N_2 system. MSC particles prepared from walnut hulls were dispersed in PF resin solution and then dip-coated onto the support tube repeatedly, followed by carbonization at 800°C [328]. The permeation rates for H_2, O_2, and CO_2 increased with the addition of molecular sieve carbon, while those of N_2 and CH_4 decreased. MSC membranes were prepared from the blend of novolac-type PF resin with poly(ethylene glycol) (PEG) by carbonization at 800°C [329]. The MSC membranes derived from PEG/PF blends exhibited permselectivity for O_2/N_2 and H_2/CH_4 in the range of 6–10 and 502–694, respectively, depending on the mixing ratio of PEG and its molecular weight. MSC membranes from novolac-type PF resin were formed on the ceramic tube by dip-coating and carbonization at 700–1000°C under vacuum [330]. The separation of a multicomponent gas mixture made up of hydrocarbons (CH_4, C_2H_6, C_3H_8, n-C_4H_{10}) and N_2 was studied on the membrane prepared at 700°C. The permeation rates of CH_4, C_2H_6, and C_3H_8 were 310, 1033, and 2410 GPU, respectively, in the absence of C_4H_{10}, while the presence of 10% C_4H_{10} in the permeate gases led to competitive adsorption processes, making the permeation rates markedly smaller, 77.5, 312, and 855 GPU, respectively.

Hollow carbon fiber membranes were prepared from an NMP solution of a cellulose acetate with 18 wt.% PVP (10K) by wet-spinning into water at 50°C, followed by deacetylation by immersion in a 0.075 M NaOH/96% ethanol solution at room temperature and carbonization at 650°C for 2 h in CO_2 purge [331]. The resultant hollow carbon fibers have outer diameters of 210–270 μm and wall thicknesses of 25–32 μm. P_{CO_2} and P_{N_2} with P_{CO_2}/P_{N_2} measured at 30°C using single-component gas were 134.5 and 10.2 barrer with 13.2, respectively, while those measured on a mixed gas (10% CO_2 and 90% N_2) were 110.0 and 3.6 barrer with 30.7, respectively.

Hollow carbon fiber membranes were prepared from polyacrylonitrile (PAN) by spinning, thermal stabilization at 250°C for different periods of 10–180 min in either pure oxygen or compressed air [332]. Thermal stabilization was effective in improving the stability and the permeability for O_2 of the membrane, whereas the permselectivity P_{O_2}/P_{N_2} was kept low. PAN was concluded to be not an appropriate precursor for carbon membrane fabrication due to its low gas separation performance.

4.3 Capacitive deionization

Capacitive deionization (CDI) using carbon electrodes has been reported since 1961 using the term "electrochemical parametric pumping" [333–337]. The carbon materials used were commercially available ones, such as graphite, porous carbon, and graphitized carbon black, although their detailed characteristics were not presented. It has to be pointed out, however, that some important experimental results have been reported, such as effective enhancement of electrosorption capacity for Na^+ by the treatments of carbon in a mixed acid and in an oleum [333], instrumentation with the large-sized carbon electrodes in multistages [337], and theoretical expectations on the engineering applicability [334–336].

CDI is a method for removing dissolved salts from brackish water by their adsorption onto electrode surfaces mainly to form electric double-layers (EDLs), and is now one of the potential methods for desalination of drinking water [338–352]. Since the adsorption mechanism of CDI is the same as that of supercapacitors for energy storage, except the electrolyte (brackish water in the case of CDI) is flowing and/or cycling in the cell in most cases, the devices are called flow-through capacitors [339]. After the electrodes become saturated by salt ions, they can easily be regenerated by reversing the potential between the electrodes. The following advantages of CDI have been mentioned: no chemicals are required during the cycles of desalination and regeneration, and the applied voltages are fairly low to avoid the electrolysis of water, which makes this technology nonpolluting, environment-friendly, energy efficient, and cost-effective for water desalination, in comparison with traditional technologies, such as thermal distillation, reverse osmosis, and electrodialysis [338,342]. Thus, the CDI process could play an important role in providing potable water and agricultural water at a low cost without pollution. Desalination technologies for municipal and industrial wastewater were critically reviewed by classifying them into membrane-based, thermal-based, and alternative technologies [347], with CDI being classified into one of several alternative technologies. CDI was reviewed by focusing on the use of activated carbons (ACs) as the electrodes [341].

Since the electrode material is the most important for obtaining a high performance of CDI, which is the same as supercapacitors [353], various carbon electrodes have been tested for CDI, ACs, ACFs, templated porous carbons, carbon aerogels, graphene-related materials, carbon nanotubes (CNTs), and carbon nanofibers (CNFs), including composites with different kinds of carbon materials and metal oxides [351].

Here, carbon materials for the electrodes of the CDI process are explained after a brief explanation of electrochemical cells for CDI. CDI performance of carbon materials is summarized in detail by classifying the carbon materials into porous carbons (ACs, ACFs, templated porous carbons and carbon aerogels), graphene-related materials, and CNTs including CNFs.

4.3.1 Cells for capacitive deionization

The basic concept of CDI is electrochemical adsorption of ions in water, Section 3.2 the cations of a salt, such as Na^+, are adsorbed onto the negative electrode and the anions, such as Cl^-, onto a positive electrode principally by forming EDLs. In Fig. 4.96, three representative cell constructions are schematically shown.

Fig. 4.96A shows the structure of a simple cell, which has been frequently employed for studies into CDI processes. The electrodes are composed from the mixture of a porous carbon (active material), a carbon black (conductive additive), and an organic binder, such as PTFE, as is the case of the electrodes for supercapacitors. A brackish water is made to pass through the gap between two electrodes, where cations are electrochemically adsorbed onto the negative electrode and anions onto the positive electrode, and the deionized water is obtained at the outlet of the cell.

The capacity calculated from the change in the concentration of the salt at the cell outlet is a sum of three adsorption mechanisms: (1) the adsorption due to the formation of EDLs on the surfaces of carbon electrodes under an electric field (electrosorption), (2) adsorption due to reversible Faradaic reaction of salt ions with surface functional groups of carbon electrodes (pseudo-sorption), and (3) physical adsorption of salt ions onto the carbon surface without an electric field (physisorption). The first two occur under an electric field (with polarization) and so depend greatly on the applied voltage. These two are often measured as a sum and are often called

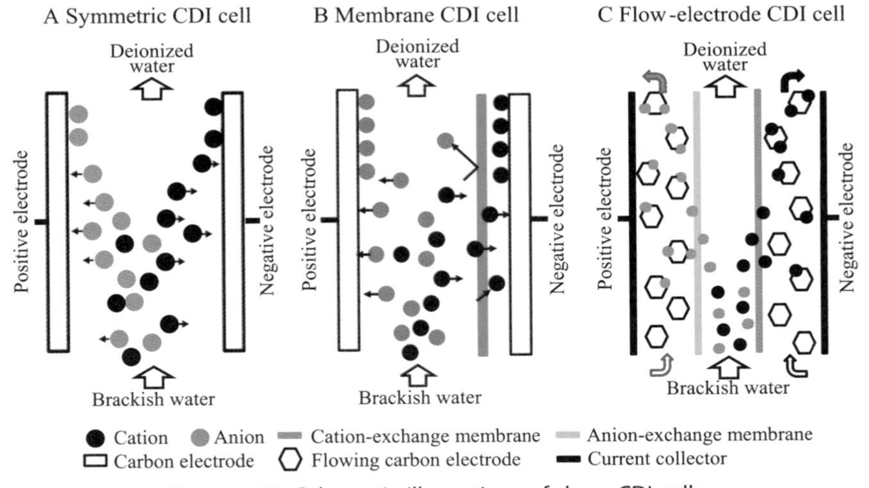

Figure 4.96 Schematic illustrations of three CDI cells.

electrosorption. The third, however, occurs without an electric field (without polarization) and is mainly governed by the salt ions and pore structure of carbon electrodes. In Fig. 4.87A and B, changes in solution conductivity (i.e., concentration of the salt) with time for a carbon aerogel (S_{BET} of 412 m^2/g) are shown for two electrolytes, 100 ppm NaF and 50 ppm $Cu(NO_3)_2$ aqueous solutions [354], the former showing negligibly small physisorption while the latter has marked physisorption. The conductivity of the latter [$Cu(NO_3)_2$ solution] decreases gradually before applying the electric field due to physisorption (it takes time to reach equilibrium) and increases markedly in the beginning of voltage application, due to releasing adsorbed ions having the valence opposite to the polarized electrode, and then decreases due to the electrosorption of ions (Fig. 4.97B). In the case of the former (NaF solution), however, the concentration of the solution decreases only slightly (a short time for equilibrium) and decreases rapidly by the voltage application, and the concentration is recovered rapidly by the voltage turn-off (Fig. 4.97A). The physisorption is predominant in total adsorption measured during the process in the case of $Cu(NO_3)_2$, with the same situation being observed on $NaNO_3$.

A half-cell using ACF cloth (ACC) as a working electrode with Ag/AgCl counter and reference electrodes was also proposed, in which the salt solution was forced to flow through the carbon electrode and gave the fast equilibration of the electrosorption of salt ions [355]. A "flow-through" cell composed of simple cells of ACC (S_{BET} of 1440 m^2/g), as shown in Fig. 4.98, was used for boron removal from water [356]. A simple cell was

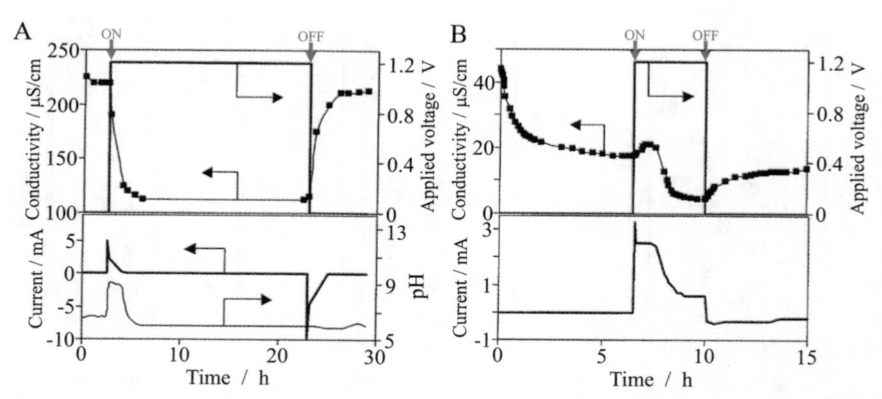

Figure 4.97 Changes in conductivity of the electrolyte solution and current with applying voltage in: (A) NaF and (B) $Cu(NO_3)_2$ solutions [354].

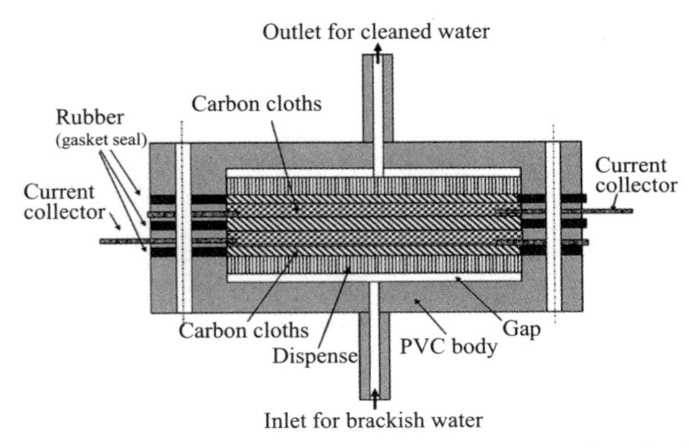

Figure 4.98 Simple CDI cell using "flow-through" electrodes of ACC [356].

constructed from the highly porous carbon electrodes prepared from RF resin using acetic acid catalyst, followed by CO_2 activation at 950°C, where the salt solution flowed through the electrodes, reporting a high CDI performance [357].

Since the size of the cation is different from that of the anion in a salt, electrosorption behavior is quite different between the cation and anion, in other words, it is very rare that one kind of carbon electrode can provide optimum electrosorption conditions for both cation and anion. Therefore, the cells consisting of the electrodes of different active materials, such as carbons with different pore structures and various composites of carbons with metal oxides, have been employed (asymmetric cell), as in supercapacitors for energy storage (refer to Sections 3.2 and 3.3).

By inserting a cation-exchange membrane in front of the negative electrode, as illustrated in Fig. 4.96B salt removal efficiency is markedly improved (membrane CDI, MCDI) because the cations can be selectively passed through the membrane and are subsequently adsorbed by the negative electrode without interference of anions. An anion-exchange membrane in the front of the positive electrode can also improve the electrosorption of anions.

By using the carbon electrode composed of an AC (S_{BET} of 1260 m^2/g) and NaCl aqueous solution (concentration of 200 mg/L) with a flow rate of 20–30 mL/min at the applied voltage of 1.5 V, salt removal efficiency is markedly improved using the cation-exchange membrane, as shown in Fig. 4.99; the efficiency improves from 62.5% for a simple CDI cell (without the membrane) to 83.0% for an MCDI cell [358]. Using anion-exchange and cation-exchange felts at positive and negative electrodes,

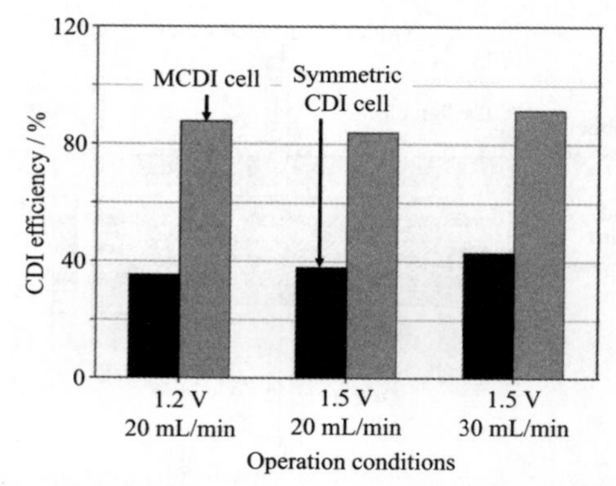

Figure 4.99 NaCl removal efficiencies of a simple CDI cell and MCDI cell with cation-exchange membrane [358].

respectively, the removal efficiency was shown to be improved [359]. As shown in Fig. 4.100, the concentration of NaCl in the effluent decreases with the application of 1.2 V, and is much faster and more marked in an MCDI cell than in a simple CDI cell (without ion-exchange membranes) by using ACFs (S_{BET} of 700 m^2/g) and 1 g/L NaCl solution. The removal efficiency and rate were 60% after 60 min and 440 mg/L h for the MCDI cell, whereas they were 19% and 340 mg/L.h for the simple CDI cell, respectively. Slightly better performance was obtained even by just covering the electrodes with filter papers (Fig. 4.100). By packing the mixture of cation-exchange resin with anion-exchange resin (1/1 by volume) into the S-shaped channel of the gasket, the highest performance was

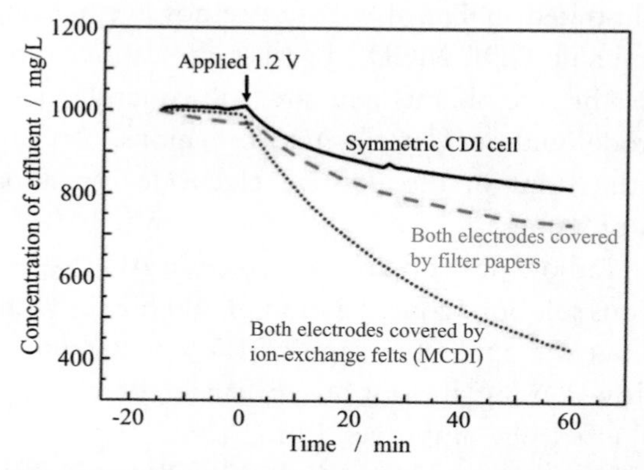

Figure 4.100 Changes in NaCl concentration in the effluent with time for three kinds of CDI cells [359].

obtained, with a removal efficiency of 90% after 60 min and removal rate of 670 mg/L.h. A performance comparison between an MCDI cell and a simple CDI cell was carried out on ACC (S_{BET} of 1197 m^2/g and V_{micro} of 0.35 mL/g) with electrodes using 1 M NaCl solution [360]. In electrosorption capacity, current efficiency, and energy consumption, the MCDI cell was shown to be much superior to the simple CDI cell, revealing the economic advantage of MCDI technology. The efficiency of one MCDI cell using cation- and anion-exchange membranes was shown to be comparable with that of 10 simple CDI cells in series, although the electrosorption rate became a little slower [361].

The process for assembling the anion-exchange resin with carbon electrode was shown to be important for achieving high performance of MCDI cells [362]. The anion-exchange resin was synthesized from a mixture of polyvinyl alcohol (PVA) with glycidyltrimethylammonium chloride (GTMAC) by adding different amounts of 5 wt.% glutaraldehyde (GA) solution (0, 0.4, 0.8, 1.2, and 1.6 mL). The resin film was fixed by either casting via. its solution or mechanical compression of its film onto a carbon electrode made of a mesoporous AC (derived from coconut shell) with a PVDF binder. As shown in Fig. 4.101A and B, mechanical compression is more effective in improving CDI performance, i.e., a higher rate and larger amount of electrosorption than casting, and the CDI performance is much better than that of a conventional simple cell. Bromomethylated poly(2, 6-dimethyl-1, 4-phenylene oxide) (BPPO) was sprayed on carbon cloth followed by sulfonation and amination to form cation exchange and anion exchange layers, respectively [363]. The process to

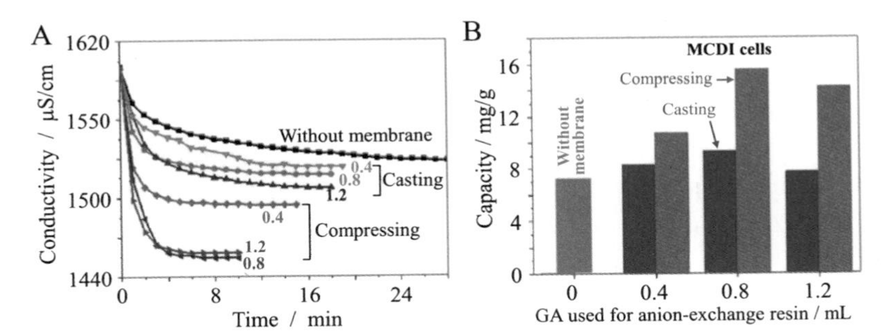

Figure 4.101 Assembling processes of anion-exchange resin with carbon electrode: (A) change in conductivity of the effluent with time and (B) electrosorption capacity as a function of the amount of glutaraldehyde (GA) used for anion-exchange resin [363].

form ion-exchange membranes could reduce the interfacial resistance between the membranes and the carbon electrode (carbon cloth) to enhance salt removal efficiency.

An MCDI cell system with asymmetric active materials, $Na_4Mn_9O_{18}$ negative electrode, and AC positive electrode with anion-exchange membrane, was constructed [364]. This asymmetric MCDI system exhibited an electrosorption capacity of 31.2 mg/g for 1 M NaCl solution with a flow rate of 10 mL/min at the applied voltage of 1.2 V, which is much higher than 13.5 mg/g for the simple CDI cell. For an MCDI pilot-plant set up with 90 cells in series, the performance was experimentally studied as a function of the flow rate and initial salt concentration [365]. The cell was composed from a porous carbon electrode with a size of $10 \times 10\ cm^2$ and the NaCl solution of different initial concentration of 9−90 mM with a flow rate of 1−3.5 L/min were used.

A flow-electrode CDI cell (FCDI cell) has been proposed, as shown in Fig. 4.96C [366,367]. The flow-electrode, which is a 5 wt.% suspension of AC granules in a 0.1 M aqueous solution of NaCl, replaces the fixed carbon electrodes and flows through channels carved on the current collectors. Under a closed cycling of the flow-electrodes, the ions accumulate on the suspended AC granules during desalination. In contrast to the fixed electrodes in other CDI cells (simple CDI and MCDI cells), this system has some advantages, such as being a continuous desalination process and having high removal efficiency from salted water with high concentration, such as seawater, because the flow-electrode has infinite ion adsorption capacity.

A continuous system was constructed, as shown in Fig. 4.102A, which consists of one cell for electrosorption of salt (desalination cell) and another cell for the desorption of salt ions electrosorbed in flowing electrode carbon (regeneration cell) where the salt solution was concentrated [368]. Two slurries of an AC are independently circulating between two cells and the NaCl solution with 1 g/L is supplied to two cells by splitting in different ratios, with one passed through the electrosorption cell giving the desalinated stream and the other through the regeneration cell giving the concentrated stream. In Fig. 4.102B, the dependences of NaCl concentrations in two streams, desalinated and concentrated, on flow rate are shown as a function of the split ratio of the solution at 50/50, 70/30, and 90/10 to the desalination and regeneration cells. In this FCDI cell, the concentration of the desalinated stream increases with increasing flow rate,

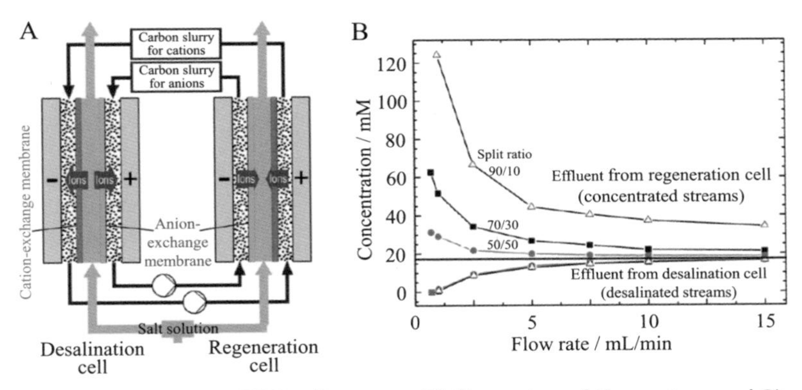

Figure 4.102 Continuous FCDI cell system: (A) illustration of the system and B) concentration changes with flow rate of salt solution (1 g/L) as a function of split ratio between desalination and regeneration cells [368].

as in the simple CDI and MCDI cells, irrespective of the split ratio, but that of the concentrated stream depends greatly on the split ratio, as well as the flow rate.

A proposal to use porous carbon spheres as flow-electrodes has been presented [369], which may make it possible to use the pores in carbon spheres efficiently, which is more efficient than making carbon platelets for the electrodes.

4.3.2 Porous carbons

4.3.2.1 Activated carbons and activated carbon fibers

Commercially available ACs from different origins were studied for the electrosorption of ions of various salts from their aqueous solutions using a simple CDI cell [370]. The amount of Na_2SO_4 electrosorbed during 80 min from its 1 mM aqueous solution with a flow rate of 1.0 mL/min under an applied voltage of 1.0 V depends approximately linearly on S_{BET} of the AC electrode. The effect of cations on the electrosorption capacity of the carbon electrode was studied on a microporous AC (S_{BET} of 2600 m^2/g) in a simple symmetric CDI cell for different salts at an applied voltage of 1.0 V [370–372]. For six commercially available mesoporous ACs, electrosorption capacity was determined using a simple CDI cell with an applied voltage of 1.6 V and NaCl solution with an initial concentration of 50 mg/L [373]. Approximately linear dependences of the capacity on S_{BET} and also on V_{meso} were obtained, whereas the dependence on V_{micro} was poor. On an AC with V_{meso} of 0.44 cm^3/g, the capacity decreased from

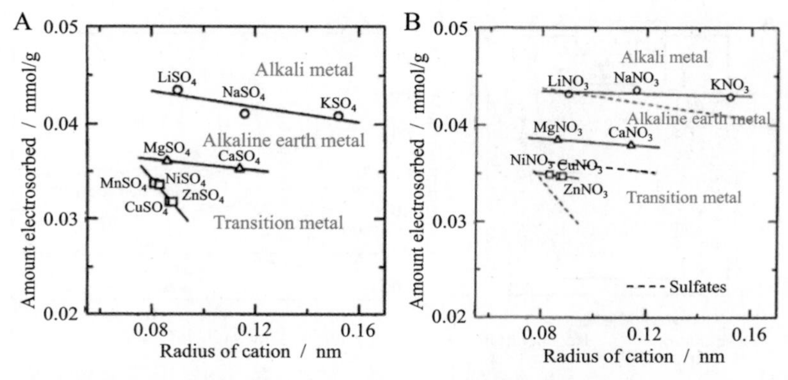

Figure 4.103 Amount of salts electrosorbed over 120 min by an AC in a flow of their aqueous solutions against radius of the cation: (A) for sulfates and (B) for nitrates [370,371].

10.9 to 9.4 mg/g with increasing solution temperature from 16 to 35°C. In Fig. 4.103A and B, the electrosorbed amounts of the salts over 120 min are plotted against ion radius of metal ion (cation) for the aqueous solutions of the salts with an initial concentration of 0.1 mmol/L and 1.0 mL/min flow rate. Electrosorption capacity is higher for monovalent cations (alkali metals) than divalent and trivalent cations (alkaline earth and transition metals). Anions of salts also influence electrosorption capacities, as shown in Fig. 4.103B, the cation radius dependences on sulfates is indicated by broken lines to compare them with those on nitrates. For another AC, electrosorption capacities of metal chlorides were determined as a function of the equilibrium concentration of salt with 10 mL/min flow rate under an applied voltage of 1.0 V in a simple cell [374], revealing that monovalent cations could be electrosorbed into the AC more than divalent cations.

Electrosorptive removal of Cu^{2+} in aqueous solutions was studied using a microporous AC ($S_{BET} = 1124\ m^2/g$) as electrodes [375]. The observed capacity was enhanced to about 25 mg/g by applying a voltage of 0.8 V (electrosorption) from about 5 mg/g at 0.0 V (physisorption). Cu^{2+} was preferentially electrosorbed by the carbon electrode from a mixed solution of $Cu(NO_3)_2$, NaCl, humic acid, and Na_2SiO_3. Ordered mesoporous carbon ($S_{BET} = 1410\ m^2/g$ and $V_{total} = 1.79$ mL/g) showed a high electrosorption capacity for Cu^{2+} of about 56 mg/g, which is much higher than about 16 mg/g for an ACC ($S_{BET} = 1294\ m^2/g$ and $V_{total} = 0.74$ mL/g) [376].

The regeneration of electrode carbon depended also on the salts. The behaviors of electrosorption by applying the voltage and desorption by shutting down the applied voltage during the initial three cycles were compared on sulfates of various metals and for various sodium salts using an AC electrode [372]. A marked difference in the desorption of salts (i.e., regeneration of AC) was observed between alkali and alkaline earth sulfates. The former was recovered more quickly than the latter and the recovery efficiency for the former was 70%–80% but that for the latter 50%–60%. The efficiencies for the heavy metal sulfates ($CuSO_4$ and $ZnSO_4$) were very low at 40%–50%. Among the Na salts studied, Na_2HPO_3 has the highest recovery efficiency, reaching 80% and $NaNO_3$ has the lowest, at about 47%. The electrode of an AC (S_{BET} of 1260 m^2/g) cast as a slurry with PVDF binder onto the substrate using a doctor blade to the thickness of 250 μm exhibited an average NaCl removal efficiency at 1.5 V with a flow rate of 20 mL/min of 77.8% and that with 30 mL/min flow rate of 70% [377]. On this electrode, most ions were desorbed within 1 min by changing the potential to 0 V. The sheet of an AC (S_{BET} of 1124 m^2/g) with 10 wt.% PVDF binder had electrosorption efficiency and capacity of 96.7% and 44.5 mmol/g, respectively, for 2 mM NaCl solution at 1.2 V, together with enough mechanical strength and electrochemical stability [378].

The effect of operational conditions of a cell on CDI performance was studied by measuring the electrosorption capacity and desorption efficiency as functions of flow rate and initial concentration of NaCl solution, using two simple cells in series using an AC (S_{BET} of 800 m^2/g) [379]. As shown in Fig. 4.104A and B, the desorption efficiency of NaCl decreases with increasing flow rate and initial concentration of the solution, and electrosorption capacity of NaCl decreases with increasing flow rate but increases with increasing initial concentration. The desorption efficiency decreased

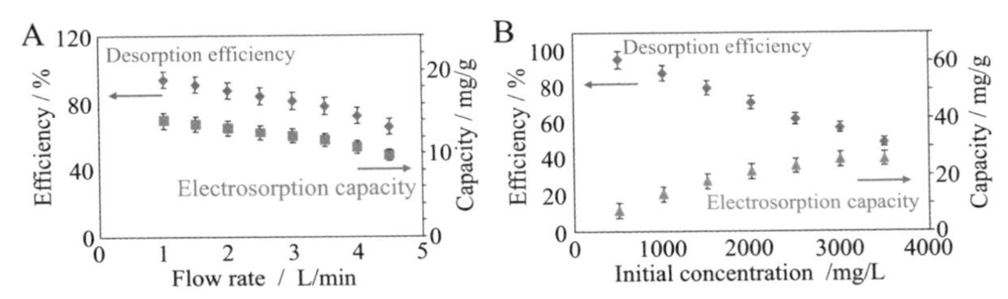

Figure 4.104 Electrosorption capacity and desorption efficiency as a function of (A) flow rate and (B) initial concentration of NaCl solution [379].

with increasing temperature of the solution, from about 90.4% at 20°C to 79.2% at 50°C for the solution of various salts of an initial concentration of 1000 mg/L with a flow rate of 2 L/min. Using a CDI unit consisting of two cells in series, electrosorption kinetics of an AC (S_{BET} of 800 m^2/g) was studied on various cations and anions. The electrosorption kinetics can be approximated by a first-order process, as shown for the NaCl solution in Fig. 4.105A. The rate constant depends greatly on the initial concentration of the solution, 2.07 min^{-1} for 0.5 g/L to 2.53 min^{-1} for 3.5 g/L. The desorption efficiency is different for cations and anions, as shown in Fig. 4.105B with their initial concentration of 1 g/L. The CDI performance of a simple cell composed of a porous carbon was discussed on the bases of the effective surface area for ion storage and charge efficiency of EDLs [380].

A comparison of the electrosorption capacity and energy consumption between two operational conditions, with a constant voltage and constant current, was performed on a simple cell with a commercially available AC and 10 mM NaCl solution with a flow rate of 10 mL/min [381]. Energy consumption and charge efficiency are shown as a function of electrosorption capacity in Fig. 4.106A and B, respectively. The constant current operation consumes approximately 26%–30% less energy than the constant voltage operation, although the charge efficiencies for both operations are the same.

Electrosorption of Na^+ was compared with that of H^+ through the measurements of capacitance by CV for the carbons prepared from pistachio shells by carbonization at 450°C, activation by KOH at 780°C, and then CO_2 activation at 780°C for different times [382]. Capacitances for 1 M $NaNO_3$ and 0.5 M H_2SO_4 are shown as a function of scanning rate

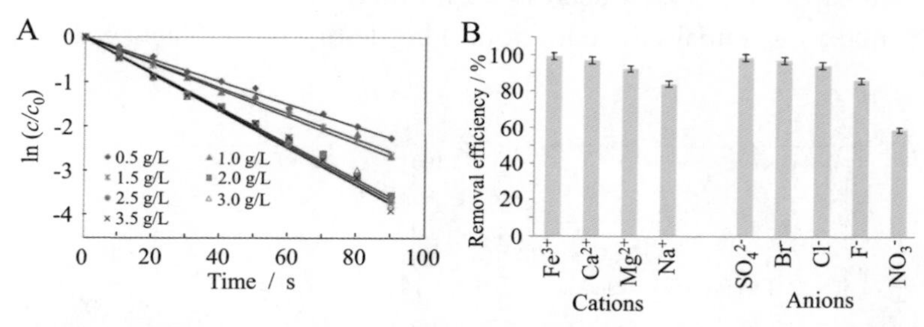

Figure 4.105 Removal kinetics and efficiency with a flow rate of 2 L/min at 24°C: (A) electrosorption kinetics of various initial concentrations of NaCl solution and (B) removal efficiency for various cations and anions with an initial concentration of 1 g/L [379].

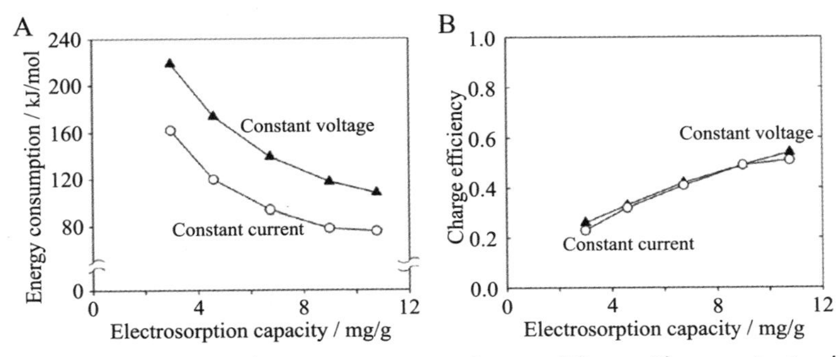

Figure 4.106 Comparison between two operation conditions with a constant voltage and constant current of a simple cell: (A) energy consumption and (B) charge efficiency against electrosorption capacity [381].

during CV measurements in Fig. 4.107A and B, and the pore structure parameters are listed in Table 4.12. With the increase in the time for CO_2 activation, capacitances (i.e., electrosorption capacities) in Na^+ and H^+ are enhanced at a high scan rate, while the optimal activation times are different for two cations, 10 min for Na^+ and 30 min for H^+. The results show that the presence of appropriate V_{meso} in a carbon electrode is important for the electrosorption of Na^+ and H^+, although the optimum conditions are different for two cations. The electrosorption capacity was measured using 0.5–10 mM NaCl solution on the mesoporous carbons prepared from coconut shells after KOH and CO_2 activation at 800°C [383]. The presence of mesopores in the electrode carbon worked not only to facilitate the ion transport but also to improve the accessible surface area, and as a consequence resulted in an improved capacity of capacitive ion storage. The carbon with S_{BET} of 2105 m^2/g with V_{meso}/V_{total} of 71% exhibited an

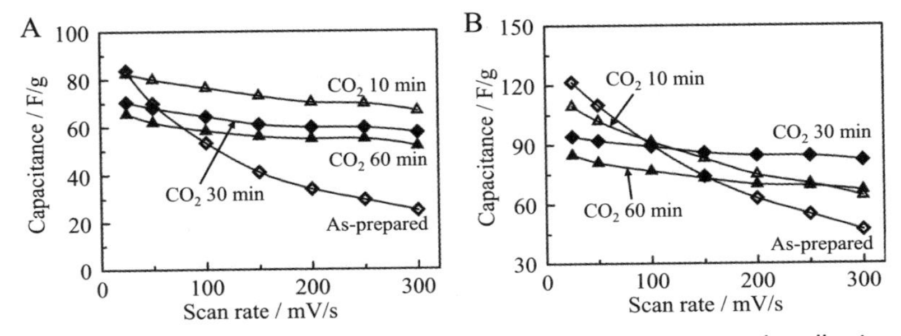

Figure 4.107 Dependences of capacitance on scanning rate on a simple cell using the pistachio-shell-derived carbons activated with CO_2 for different times: (A) in 1 M $NaNO_3$ and (B) in 0.5 M H_2SO_4 [382].

Table 4.12 Pore structure of the pistachio-derived carbons [382].

Activation time	S_{BET} (m^2/g)	S_{micro} (m^2/g)	V_{total} (cm^3/g)	V_{meso} (cm^3/g)
Pristine	1013	933	0.60	0.12
In CO_2, 10 min	1398	1289	0.84	0.15
In CO_2, 30 min	1919	1619	1.16	0.38
In CO_2, 60 min	2145	1742	1.37	0.52

electrosorption capacity of 9.72 mg/g and electrosorption rate constant of 0.060 min^{-1}, in contrast to the carbon having S_{BET} of 2162 m^2/g with V_{meso}/V_{total} of 19% which had the electrosorption capacity of 4.08 mg/g and electrosorption rate constant of 0.029 min^{-1}.

Four MCDI cells based on different combinations of two ACs, the pristine AC and its oxidized one in HNO_3, at the positive and negative electrodes were constructed and their electrosorption performances were compared in a constant voltage operation by cycling between 1.2 and 0 V (short circuit) every 30 min [384]. Although only a small difference was recognized on these two carbons in pore size distribution, electrosorption behaviors were quite different, as shown in Fig. 4.108. The cell composed of the negative electrode of the oxidized AC and the positive electrode of the pristine AC (oxidized//pristine) gave the highest performance for 5 mM NaCl solution (conductivity of ca. 600 μS/cm), but the cell of pristine//oxidized gave the lowest performance. The experimental results suggest that the optimization between two electrodes based on the electrosorption capacities of each electrodes is required to achieve high capacity with steady and long cycles on the CDI cell.

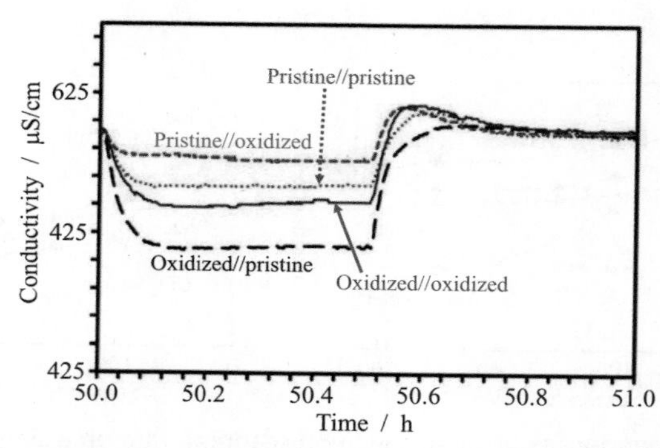

Figure 4.108 Electrosorption and regeneration cycles of four MCDI cells with four combinations of two ACs, pristine and oxidized AC, as electrode materials [384].

Asymmetric CDI cells using ZnO/AC composite and AC electrodes were constructed [385]. An ZnO/AC composite electrode was prepared from a mixture of ZnO nanoparticles, commercial AC and PTFE in a weight ratio of 1/10/0.8 by rolling. The asymmetric cell using ZnO/AC composite electrode as the negative electrode (ZnO/AC//AC) had much better performance for desalination of an NaCl aqueous solution than the symmetric simple cell using AC at both electrodes (AC//AC) and also the asymmetric cell using the composite at the positive electrode (AC//ZnO/AC). The electrosorption capacity for the first cell was 9.4 mg/g and those for the latter two were about 4.5 mg/g. A slight improvement in the electrosorption efficiency of NaCl aqueous solution was observed by the coating of AC particles with TiO_2 nanoparticles via hydrolysis [386,387] and via sol–gel spraying of the titanium tetrabutoxide [$Ti(OC_4H_9)_4$] [388]. For an TiO_2/AC composite prepared from the gel of an AC with titanium tetraisopropoxide ($Ti[OCH(CH_3)_2]_4$), the electrosorption capacity of NaCl was measured using a half cell with 0.5 M NaCl solution as 8.1 mg/g for the composite, while it was 5.4 mg/g for the pristine AC [389]. By employing a carbon felt coated with SiO_2 at the negative electrode and that by γ-Al_2O_3 at the positive electrode, the electrosorption removal of KCl [390] and $CaSO_4$ [391] in aqueous solution was investigated. SiO_2-coating by 1.7 wt.% resulted in a slight decrease in S_{BET} to 1290 m^2/g and Al_2O_3 coating by 0.35 wt.% to 1293 m^2/g from 1630 m^2/g for the pristine carbon felt. The cell exhibited an electrosorption capacity of $CaSO_4$ of 3.4 and 4.4 mg/g at a flow rate of the solution of 45 and 110 mL/min, respectively, after 15 min at the applied voltage of 1.2 V.

By coating an anion-exchange membrane on the positive electrode of an AC, nitrate ions were selectively electrosorbed from the mixed solutions of $NaNO_3$ with NaCl [392]. A commercial anion-exchange membrane was measured to have the selectivity coefficient of 4.37 for nitrate ions when they coexisted with chloride ions in the solution, suggesting that the AC electrode coated by this membrane adsorbed nitrate ions preferentially. On the solutions containing chloride and sulfate ions with nitrate ions, the selective electrosorption of nitrate ions using an anion-exchange membrane was also confirmed [393].

The CDI performance of commercial ACFs with different S_{BET}'s was studied using a simple cell at different voltages [394,395]. The electrosorption capacity and charge efficiency of ACFs are plotted against applied voltage in Fig. 4.109A and B, respectively, suggesting that the electrosorption capacity depends greatly on the charge efficiency, not S_{BET}.

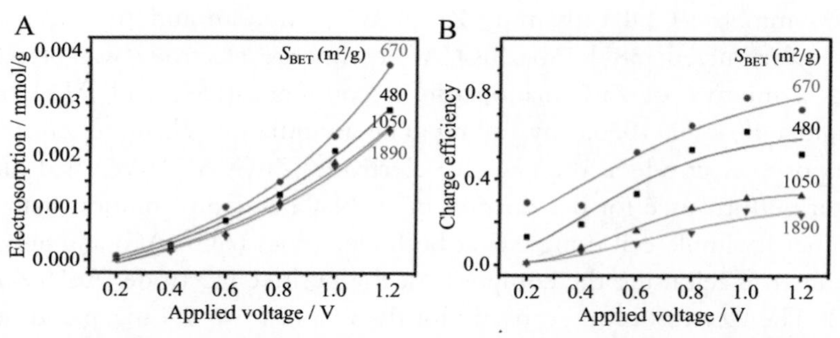

Figure 4.109 ACFs with different S_{BET}'s: (A) amount of electrosorbed NaCl and (B) charge efficiency against applied voltage [394].

Disinfection of water was possible by a simple CDI cell using ACC electrodes with a decrease in viable bacterial cells [396].

ACF webs were prepared from electrospun PAN fiber webs by carbonization and activation in CO_2 at 750–900°C after oxidative stabilization at 280°C and their CDI performances for NaCl were studied using a simple cell [397–400]. The ACF web prepared at 900°C is microporous (Fig. 4.110A), of which S_{BET} and V_{total} are measured as 712 m^2/g and 0.363 cm^3/g, respectively, and shows the highest performance for NaCl solution with a flow rate of 5 mL/min at 25°C (Fig. 4.110B) [397]. The addition of a carbon black into either PAN fiber webs [398] or PAN precursor before electrospinning [400] was effective in improving the CDI performance of the ACF webs for NaCl solution. The addition of rGO in a PAN precursor improved the CDI performance slightly [399].

The effect of functional groups on the CDI performance in NaCl solutions with conductivity of 2 and 6 mS/cm was studied on ACCs after

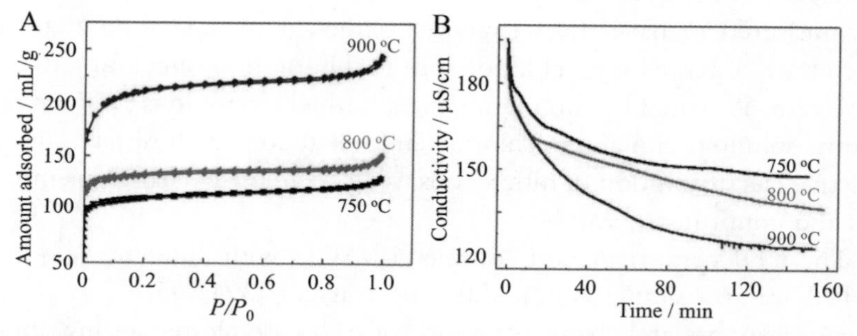

Figure 4.110 ACF webs prepared from electrospun PAN fiber webs carbonized at 700, 800, and 900°C: (A) N_2 adsorption isotherms at 77K and (B) electrosorption behaviors of NaCl at 1.6 V [397].

surface treatment in 1 M KOH and HNO_3 solutions at 90–100°C for 3, 6, and 12 h [401]. XPS measurements showed that KOH treatment was effective in increasing the hydroxyl and carbonyl functional groups, and HNO_3 treatment in increasing the carboxyl, carbonyl and hydroxyl groups. Changes in NaCl concentration with time are shown for the ACC treated in 1 M KOH and HNO_3 for different periods in Fig. 4.111A and B, respectively, revealing that the treatment of the ACC electrode either in KOH or HNO_3 accelerates electrosorption markedly by a short treatment time of 3 h. Surface modification of an ACC in 1 M HNO_3 solution was also reported to be more effective for an improvement in CDI performance than that in 1 M KOH by discussing the contribution of functional groups, especially carbonyl groups, in addition to EDL formation [402]. By increasing the surface acidity through treatment in 10–30 wt.% H_2O_2 solutions, the removal efficiency was shown to have a maximum of around 1.2–1.5 mmol/g on various ACFs [403]. The effect of functional groups of ACFs (S_{BET} of 1400–1560 m^2/g) on electrochemical capacitances for 1 M H_2SO_4 aqueous electrolyte and 0.5 M $LiClO_4$/PC electrolyte was discussed by oxidation of ACFs in 0.1 M HNO_3 with or without DC polarization and in $(NH_4)_2S_2O_8$ aqueous solution with different concentrations, and by a reduction of the H_2 flow at 800°C.

Four carbon materials were compared on electrosorption kinetics of anions by a three-electrode cell using different electrolyte solutions, NaCl, $LiClO_4$, and $MgSO_4$ with different concentrations of 0.03–1 M [404]. Carbon materials used as an electrode were an ACC ($S_{BET} = 1410\ m^2/g$, $V_{micro} = 0.59$ mL/g), a molecular sieve carbon ($S_{BET} = 650\ m^2/g$, $V_{micro} = 0.32$ mL/g), a carbon aerogel ($S_{BET} = 453\ m^2/g$, $V_{micro} = 0.2$ mL/g),

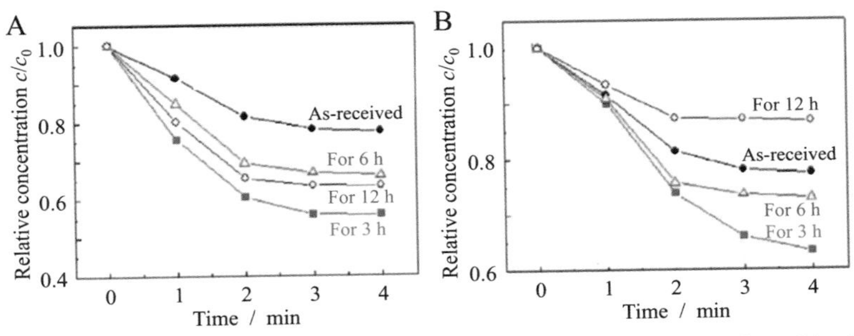

Figure 4.111 Changes in NaCl concentration (the initial conductivity of 2 mS/cm) with time on ACCs: (A) ACCs treated in 1 M KOH and (B) in 1 M HNO_3 for different times [401].

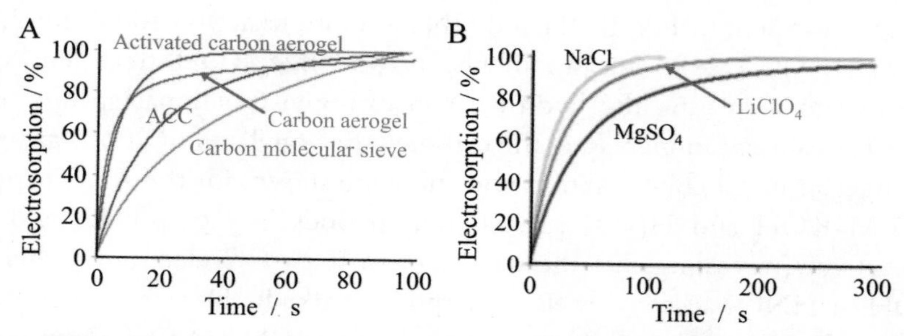

Figure 4.112 Changes in the relative amount of electrosorption evaluated by positive electrode polarization with time: (A) in 0.125 M NaCl solution with the applied voltage of 250 mV for four carbon materials and (B) in different solutions for the ACC [404].

and an AC aerogel ($S_{BET} = 1195\ m^2/g$, $V_{micro} = 0.6\ mL/g$). In Fig. 4.112A, the electrosorption rate is compared in 0.125 M NaCl, revealing that sorption kinetics of anion Cl^- cannot be explained from the pore structure of electrode carbon, because the carbon aerogel with low S_{BET} and V_{micro} electrosorbs more quickly than the ACC with high S_{BET} and V_{micro}. In Fig. 4.112B, the electrosorption rate is compared in three electrolyte solutions for the ACC. The time constant for the electrosorption of the anion into the ACC is a little different for anions, small for Cl^- and large for SO_4^{2-}, but no marked dependences were observed for the other three carbons, although the time constant for an anion was quite different among the four carbon materials.

Selective electrosorption of cations was studied using the mixed solution of 288 mg/L of $CaCl_2$, 220 mg/L of $MgSO_4\ 7H_2O$, and 390 mg/L of $NaHCO_3$ on a commercial ACC [405]. As shown for two different flow rates of the mixed solution in Fig. 4.113A and B, higher selectivity of divalent ions, Ca and Mg, than monovalent ion, Na, was confirmed. After

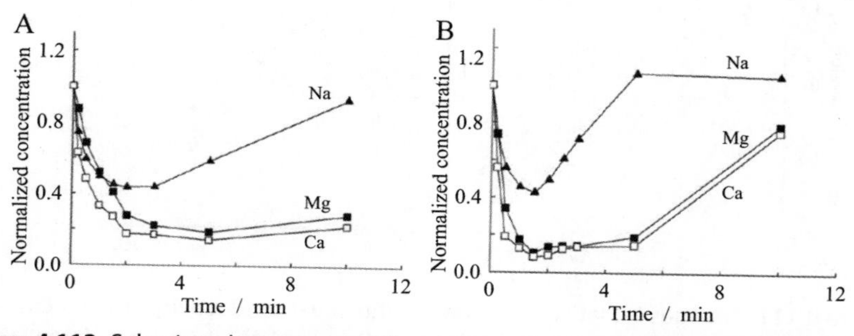

Figure 4.113 Selective electrosorption for cations of an ACC electrode at the applied voltage of 1.5 V: (A) with a flow rate of 16 mL/min and B) 80 mL/min [405].

about 2 min, Na^+ electrosorbed into the ACC is dissolved again into the solution, probably due to the substitution of adsorbed Na^+ by Ca^{2+} and Mg^{2+} in the carbon electrode.

The modification of an ACC (S_{BET} of 1043 m^2/g) by depositing ZnO was reported to be effective for an improvement in CDI performance [406–409]. The ACC was modified by deposition of ZnO nanoparticles (6–8 nm size) by dipping into their colloidal solution and also by growing ZnO nanorods (about 10 μm long and 700 nm diameter) under hydrothermal conditions in a solution of zinc nitrate hexahydrate and hexamethylenetetramine at 90°C [406]. Modification of ACC by ZnO-nanorods was more efficient for the improvement of deionization and regeneration performance in NaCl solution than the pristine ACC and that modified by ZnO-nanoparticles. TiO_2 loading onto an ACC by immersing into a water/ethanol solution of titanium-butoxide was shown to be effective to improve CDI performance [410]. By loading 8.4 wt.% TiO_2, the physisorption capacity decreased while the electrosorption capacity under a voltage of 1.0 V was markedly improved, as shown in Fig. 4.114, although there was almost no change in S_{BET} (from 1980 to 1890 m^2/g).

4.3.2.2 Templated porous carbons

CDI performances of microporous carbons derived from metal carbides (CDCs) were studied using 5 mM NaCl solution (550 μS/cm) flowing through electrodes with a 1 mL/s rate [411,412]. Both CDC-1 and CDC-2, which were primarily composed of micropores with size of

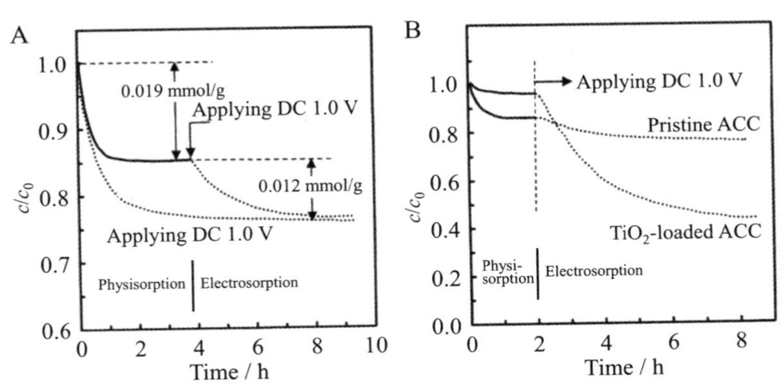

Figure 4.114 Adsorption profiles of NaCl on an ACC: (A) the pristine ACC with and without an electric field and (B) TiO_2-loaded ACC (8.4 wt.% TiO_2) in comparison with the pristine ACC [410].

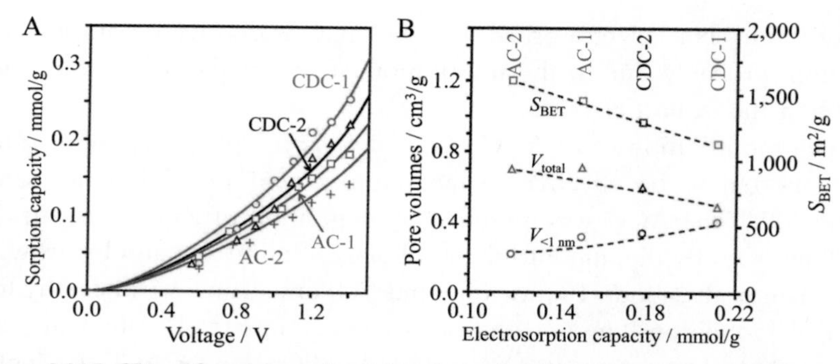

Figure 4.115 CDI performances of carbide-derived carbons (CDCs) in comparison with ACs: (A) electrosorption capacity against applied voltage and (B) S_{BET}, V_{total}, and $V_{<1\ nm}$ of the carbons against their electrosorption capacity at 1.2 V [411].

0.75 nm (the former has relatively more amounts), has a much higher adsorption capacity than commercial ACs (AC-1 and -2), as shown in Fig. 4.115A; CDC-1 has a capacity of 0.21 mmol/g whil AC-2 has about 0.12 mmol/g at a voltage of 1.2 V. In Fig. 4.115B, the pore structure parameters, S_{BET}, V_{total}, and $V_{<1\ nm}$ (volume of the pores having a size of less than 1 nm), of four carbon materials are plotted against their electrosorption capacity at the applied voltage of 1.2 V, revealing that the capacity correlates negatively with S_{BET} and V_{total} (the higher S_{BET} and V_{total} correspond to the lower capacity) while it correlates positively with $V_{<1\ nm}$. The results suggest the relevance of the subnanometer pores (<1 nm size) for electrosorption.

A thin layer of CDC was coated on the graphite rods (current collectors) and used as electrodes of a CDI cell [413]. The CDC electrodes were alternatively dipped into fresh and saline water by applying the voltage to test its performance. By cycling, NaCl was transported from the saline water (NaCl solution) into the fresh water. The amount of transported salt is shown for each cycle at the applied voltages of 1.2 and 1.6 V without and with the cation-exchange membrane at the negative electrode in Fig. 4.116A and B, respectively. Almost the same amount of salt can be transported in each cycle, the higher the cell voltage, the higher the amount of transportable salt (in other words, the higher the adsorption capacity of the carbon electrode), and the transportable salt increases by using the cation-exchange membrane.

Mesoporous carbons prepared from calcium citrate by carbonization at 1000°C, followed by dissolution of Ca with HCl, were applied for CDI of

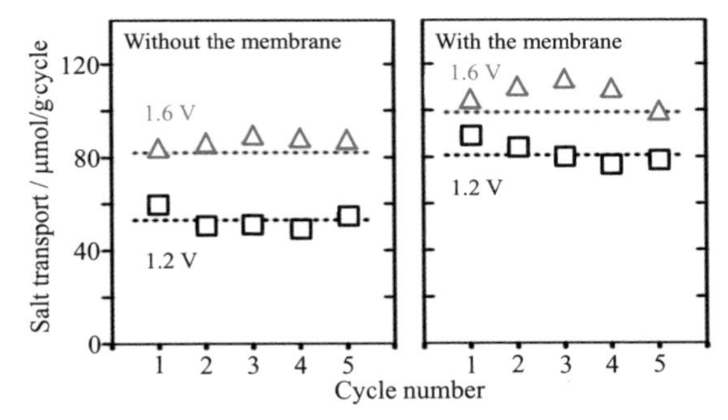

Figure 4.116 The amount of transported salt from the saline water to the fresh water per cycle by the electrode without and with a cation-exchange membrane [413].

NaCl solution [414]. Its electrosorption capacity was about 24 µmol/g and the carbon prepared using recycled Ca-citrate exhibited almost the same electrosorption capacity of 23 µmol/g, and removal efficiency reached 88%–90%.

4.3.2.3 Precursor designed porous carbon

To demonstrate the importance of pore sizes of carbon materials for CDI applications, mesoporous carbons prepared using triblock copolymer were used [415–420]. Ordered mesoporous carbon (OMC) was prepared from the mixture of tetraethyl orthosilicate (TEOS) and triblock copolymer P123 by carbonization at 850°C, followed by the dissolution of silica by HF. Electrosorption measured on the resultant OMCs using a simple CDI cell in 25.6 ppm NaCl solution with an applied voltage of 1.2 V proceeded much faster and gave much higher capacity than an AC. The OMC had lower S_{BET} (844 m^2/g) but higher mesopore fraction (82%) than the AC (968 m^2/g but 41%, respectively) [415]. The addition of Ni salts, $NiSO_4$ and $Ni(NO_3)_2$, into the mixture of TEOS and P123 was effective in increasing V_{meso} and improving electrosorption capacity of Na^+ from 0.1 M NaCl solution [416]. OMCs were also prepared from the mixture of phloroglucinol-glyoxal (PhG) and resorcinol-formaldehyde (RF) with triblock copolymer F127 by carbonization at 850°C [417]. The electrosorption capacity of NaCl measured using a simple cell with the electrodes of graphite sheet coated by the OMC powder was 139.5 mg/g for the PhG-derived OMC and 146.8 mg/g for RF-derived OMC from the NaCl solution with an initial concentration of about 4000 ppm. Phenol-formaldehyde (PF)-derived and phloroglucinol-formaldehyde (PhF)-derived OMCs coated on a graphite plate showed an electrosorption capacity of NaCl of 15.2 and 21 mg/g, respectively, for 35,000 ppm solution, while an aerogel prepared from PhF-derived carbon with KOH activation showed only 5.8 mg/g [418].

OMCs with different symmetries, two-dimensional hexagonal array, three-dimensional cubic array, and three-dimensional bicontinuous array of mesopores were prepared from the mixtures of PF with either triblock copolymer F127 or P123 and NaOH, and their electrosorption performances were studied through CV measurements in the aqueous electrolytes of different metal chlorides [419]. The OMC with a 2-D hexagonal array of mesopores had a higher capacitance (larger electrosorption capacity) than the other two OMCs for monovalent cations, while the OMC with 3-D cubic array showed superior capacitance for bivalent and trivalent cations to the other two OMCs. On the bases of these experimental results, the possible method of cation diffusion was discussed for three types of OMCs.

The mixtures of PF resin with F127 were carbonized at 900°C using sugarcane bagasse as the scaffold, followed by activation using either HNO_3 or CO_2 to attain hierarchical pore structures [420]. In Fig. 4.117A and B, the amount of Ca^{2+} electrosorbed is shown as functions of applied voltage and initial Ca^{2+} concentration, respectively, where a three-electrode cell with a flow rate of the electrolyte of 2.5 mL/min at 25°C is used. The activation in HNO_3 is effective in improving the electrosorption capacity for Ca^{2+}.

The composites of CNTs with OMC were prepared by carbonization of the mixture of PF and triblock copolymer P127 with CNTs (0–20 wt.%) at 600°C [421]. The CV curves in 5 g/L NaCl aqueous solution are shown on the composites with different CNT contents in

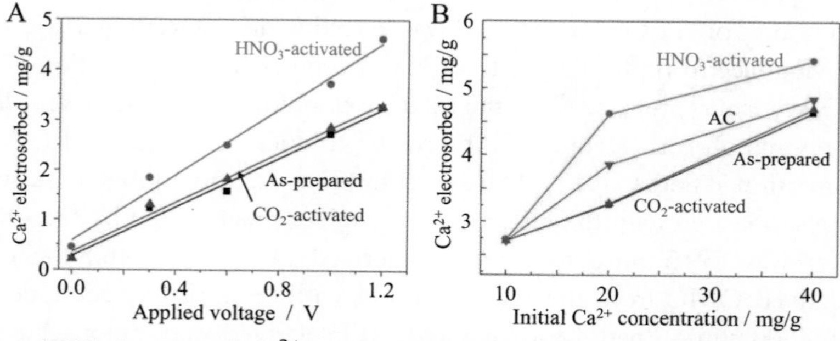

Figure 4.117 Amount of Ca^{2+} electrosorbed from $CaCl_2$ solution by mesoporous carbons: (A) as a function of applied voltage in a 20 mg/L solution, and (B) as a function of the initial Ca^{2+} concentration with a voltage of 1.2 V [420].

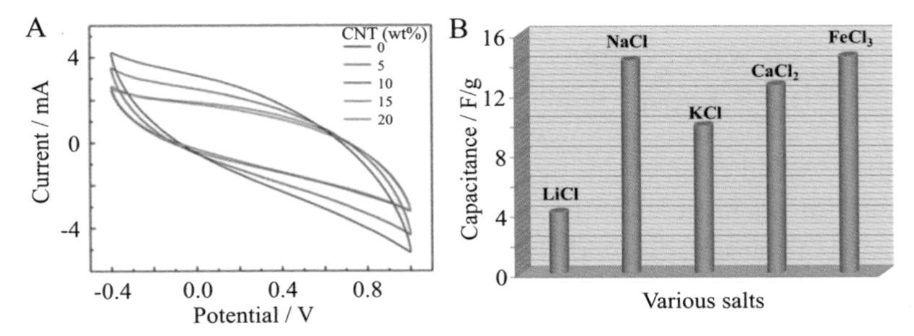

Figure 4.118 The composites of ordered mesoporous carbon (OMC) with CNTs: (A) CV curves for the composites with different CNT contents and (B) capacitances of the composite for different cations in 5 g/L aqueous solution at a scan rate of 10 mV/s [421].

Fig. 4.118A, revealing that the capacitance measured at a scan rate of 10 mV/s becomes the highest of 30 F/g at a CNT content of 10 wt.%. The capacitances of a composite in 0.5 g/L aqueous solution for various salts (Fig. 4.118B) exhibit that the capacitance for NaCl is larger than for other monovalent cations (LiCl and KCl), and that trivalent cation Fe^{3+} gives a higher capacitance than divalent Ca^{2+} and monovalent K^+ (comparable with Na^+), suggesting the predominant effect of the sizes of hydrated cations. The desalination capacity (electrosorption capacity) of the CNT/OMC composite measured with a voltage range of −0.4−1.2 V in an NaCl solution (93.5 μS/cm) with 25 mL/min flow was 10.7 μmol/g, larger than those of an AC and the OMC without CNTs (3.7 and 9.2 μmol/g, respectively). The KOH activation of the composite was effective in enhancing the desalination capacity to 11.8 μmol/g [422].

Microporous carbon spheres were prepared from the water/polyethylene glycol mixed solution of sucrose with sulfuric acid under microwave irradiation at 160°C, followed by carbonization at 1000°C in N_2 [423]. The resultant carbon was composed of spheres with an average diameter of about 800 nm and was microporous with S_{BET} of 1321 m^2/g and V_{total} of 0.59 cm^3/g. Its electrosorption capacity reached 5.81 mg/g in NaCl aqueous solution with an initial concentration of 0.5 g/L at a voltage of 1.6 V. N-doping of these microporous carbon spheres by carbonization in ammonia at 1000°C (N content of 6.7 at.%) increased S_{BET} and V_{total} to 1640 m^2/g and 0.79 cm^3/g, respectively, and electrosorption capacity to 14.91 mg/g [424].

The possibility of simultaneous removal of organic pollutants and some inorganic salts was investigated using a simple CDI cell with gas purging (O_2 and N_2) on a mixed solution of 1.1 mM phenol and 17.1 mM NaCl [425]. The carbon electrode was prepared by mixing an AC with PTFE in 80/20 by weight, of which S_{BET} was 300 m^2/g with S_{micro} of 98 m^2/g. Changes in phenol concentration, total organic carbon (TOC), and conductivity of the effluent with time are shown in Fig. 4.119A–C, respectively, as a function of the applied voltage. The application of 3.0 V under O_2 purging with a flow rate of 500 mL/min is effective in removing almost 100% of phenol, reducing TOC to 61%, accompanying a certain decrease in conductivity (i.e., decrease in NaCl concentration), where phenol was removed by electrochemical oxidation and NaCl by electrosorption in parallel.

Carbon aerogels were synthesized by the sol–gel polycondensation of RF in a slightly basic medium, followed by freeze-drying and carbonization at 1050°C, which were composed of interconnected colloid-like particles with diameters of about 10 nm [426]. The resultant carbons contained mesopores smaller than 50 nm among particles and micropores in the particles, resulting in relatively large S_{BET} of 400–1000 m^2/g. A CDI cell was constructed by stacking a number of carbon aerogel electrodes, of which the CDI performance was studied in aqueous solutions of different salts, NaCl and $NaNO_3$ [427], NH_4ClO_4 [428], and H_2CrO_4 [429]. Composite sheets were prepared by infiltrating the 70% RF solution into carbon fiber webs, followed by carbonization at 1050°C. Two sheets of the composite were glued onto both sides of a Ti metal plate (a current collector and a structural support) and used as the electrodes of a simple CDI cell. By using the stack of 192 pairs of these electrodes, NaC1 and $NaNO_3$ could effectively be removed from their solution at an applied voltage of

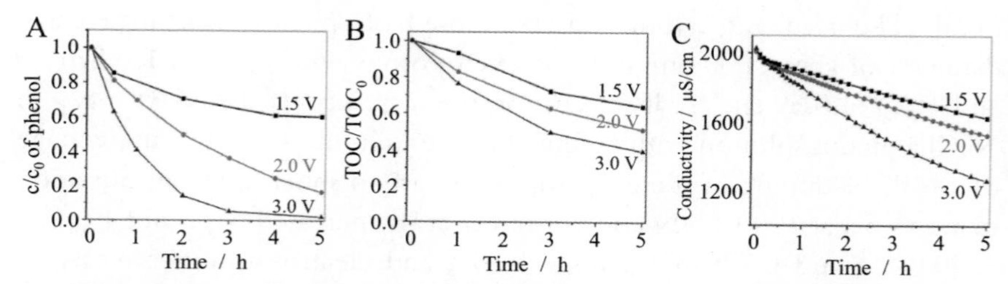

Figure 4.119 Effect of applied voltage on simultaneous removal of phenol and NaCl in aqueous solution under O_2 purging: (A) phenol concentration, (B) TOC, and (C) conductivity changes with time [425].

1.2 V [427]. By using a stack of 384 electrode pairs with an applied voltage of 1.2 V, approximately 95% of NH_4ClO_4 could be removed in a single pass of the solution [428]. A CDI unit composed of 144 electrode pairs could lower the concentration of Cr^{4+} from 35 to 2 ppb [429]. In Fig. 4.120, the concentrations of total chromium ions and Cr^{4+} in the effluents decreases quickly from 30-35 to 10 ppb by applying a voltage of 0.9 V and tends to increase during the next 7 h, but by increasing voltage to 1.2 V the concentrations can decrease and are kept at a low level of less than 10 ppb for up to 24 h.

The CDI performance (deionization and regeneration) was studied on a stack of six cells composed from a carbon aerogel, which was derived from RF at 800°C and had high porosity (about 80%), low bulk density (about 0.50 g/cm^3), and high conductivity (about 13.2 S/cm) [430]. The maximum removal efficiency from 50 mg/L NaCl solution with 400 mL/min flow rate at a cell voltage of 1.5 and 1.7 V reached 92.8% and 97.6%, respectively. The CDI performances of commercially available carbon aerogels were studied by focusing on the selectivity of cations [431] and deionization mechanisms, including electrosorption and physisorption [354]. N-doped carbon aerogels, which were prepared from a mixture of resorcinol (R), formaldehyde (F), and melamine (M) (a molar ratio M/R of 2 and 0.5) with or without prepolymerization, followed by carbonization/activation at 750°C in CO_2 flow, were used as electrodes of a simple CDI cell [432]. The aerogel without prepolymerization exhibited an electrosorption capacity of 8.2 mg/g, whereas the one with prepolymerization gave a capacity of 5.4 mg/g, at a cell voltage of 1.5 V with an NaCl solution having an initial concentration of 0.025 M, and the former having much a

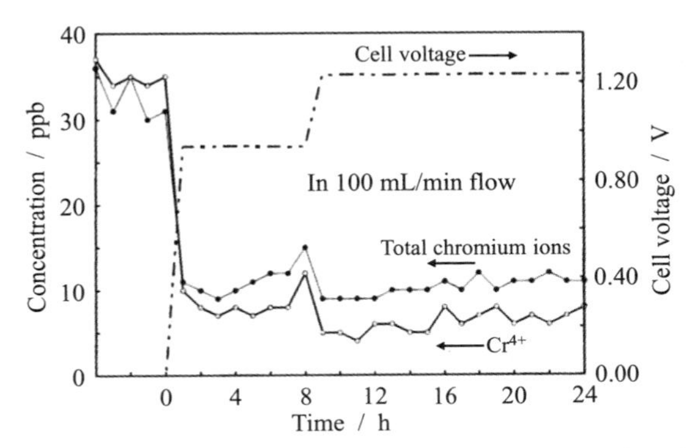

Figure 4.120 Changes in the concentrations of chromium ions (total and Cr^{4+}) and applied voltage with time for a CDI unit composed of 144 electrode pairs [429].

faster adsorption rate of 2.73 mg/cm^{3}·h than the latter of 2.12 mg/cm^{3}·h. The carbon aerogels derived from RF mixed with a carbon black (conductive additive, about 0.9% w/v) by CO_2 activation at 750°C delivered a maximum electrosorption capacity of 10.3 mg/g at 1.5 V with a NaCl solution having an initial concentration of 0.025 M [433]. Loading of MnO_2 and Fe_3O_4 onto RF-derived carbon aerogel was performed by immersing the gels into an acetone solution of $Mn(NO_3)_2$ and $Fe(NO_3)_2$, followed by freeze-drying with CO_2 and carbonization at 750°C [434]. MnO_2- and Fe_3O_4-loaded carbon aerogels possessed NaCl electrosorption capacities of 0.11 and 0.13 mmol/g at 1.5 V with 0.025 M NaCl solution, respectively.

4.3.3 Reduced graphene oxides and reduced graphite oxides

4.3.3.1 Flakes and foams

The CDI performance of reduced graphene- and/or graphite-oxides (here abbreviated as rGO) were studied in different forms, flakes, foams (aerogels), and mixtures with other active materials.

Electrosorption capacities of rGO before and after activation in a simple CDI cell with an NaCl solution (conductivity of 160 mS/cm, concentration of about 74 ppm) are shown as a function of the applied voltage in comparison with MWCNTs and an AC in Fig. 4.121 [435]. rGO flakes were prepared by reducing GO flakes with hydrazine at 80°C and their activation was performed by mixing with KOH at 800°C. The electrosorption capacity of rGO is markedly higher than for AC and MWCNTs, which is thought to be due to the interaction of Na^{+} ions with π-electrons

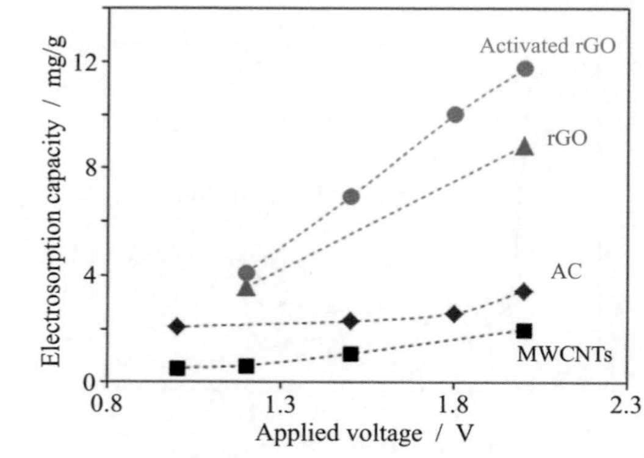

Figure 4.121 Electrosorption capacities of an rGO and an activated rGO in comparison with an AC and MWCNTs [435].

of rGO flakes (cation–π interaction). The electrosorption capacity of rGO at a voltage of 2 V is improved by KOH activation; from 8.86 mg/g before activation to 11.86 mg/g after activation, which is associated with a drastic increase in S_{BET} from 591 m^2/g to 3513 m^2/g. The deionization was greatly accelerated by activation, being completed within 20 min for the activated rGO, whereas the pristine rGO needed approximately 50 min for completion.

rGO nanoflakes prepared by a reduction of GO with hydrazine were formed into a sheet by mixing with PTFE binder, of which the CDI performance was measured using a simple cell with NaCl solution (55 μS/cm) [436]. The CDI performance improved by increasing the voltage up to 2.0 V. The electrosorption isotherm was well simulated by the Freundlich equation, giving an equilibrium electrosorption capacity of 73 μmol/g. The electrode of rGO nanoflakes was also prepared by dip-coating onto a graphite substrate, of which the electrosorption isotherm could also be approximated by the Freundlich equation [437]. rGO flakes prepared by vacuum-promoted low-temperature exfoliation (at 200°C), followed by annealing at 500 and 700°C delivered an electrosorption capacity for Pb^{2+} of 31–40 mg/g in aqueous solution with pH above 5 [438].

rGO foams were prepared by mixing a GO suspension with polystyrene (PS) microspheres (about 280 nm diameter) under ultrasonication, followed by vacuum filtration and then annealing at 900°C, where GO flakes were thermally reduced to rGO and PS was removed by leaving pores resulting in an rGO foam [439]. The CDI performance of these rGO foams was characterized by forming sheets mixed with 10 wt.% acetylene black and 10 wt.% PTFE as electrodes of a simple cell with NaCl solution (108.6 μS/cm) with a flow rate of 25 mL/min. Its electrosorption capacity and rate are improved by increasing the applied voltage up to 2.0 V and much better than for the pristine rGO flakes, as shown in Fig. 4.122. rGO foams prepared using a polyurethane sponge as a sacrificial template at 900°C delivered an electrosorption capacity of 4.95 mg/g at 1.5 V [440]. rGO foam prepared via direct freeze-drying of GO dispersion at −53°C followed by annealing and reducing at 800°C in N_2 flow delivered an electrosorption capacity of 14.9 mg/g in a flow of NaCl solution with about 500 mg/L concentration, which is much higher than the rGO prepared under the same conditions without freeze-drying (4.64 mg/g) [441]. Laminates were prepared by vacuum filtration of dispersed rGO flakes in deionized water after reduction of GO by hydrazine at 80°C, with which

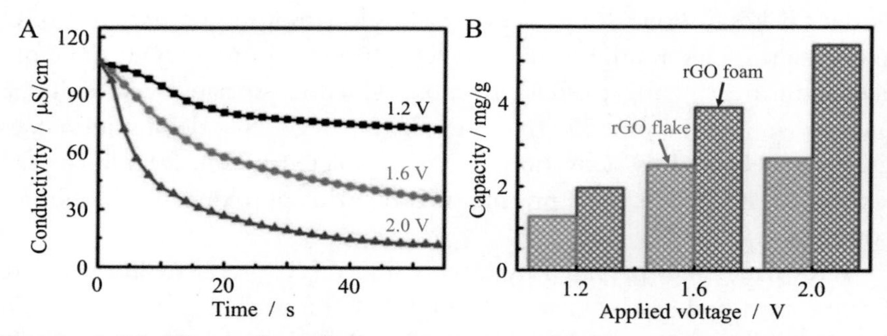

Figure 4.122 Effect of applied voltage on electrosorption of rGO foams: (A) electrosorption behaviors, and (B) comparison of capacities of rGO foam and flakes [439].

the CDI performance was characterized using the electrode sheets formed by mixing with graphite (20 mass%) and PTFE (8 mass%) in an MCDI cell [442].

As-prepared rGO flakes contain various functional groups gifted from GO, which cause pseudoadsorption in the observed electrosorption capacity of the CDI process, being observed as pseudo-capacitance as shown on the electrodes of supercapacitors. Nonlinear charge–discharge curves were observed as shown for both rGO flakes and foams prepared by thermal exfoliation and reduction in Fig. 4.123A [439], demonstrating the presence of marked pseudo-capacitance and the effectiveness of foaming of rGO to obtain improved capacity. However, it is interesting to note that the rGO flakes reduced with pyridine through ultrasonication show much

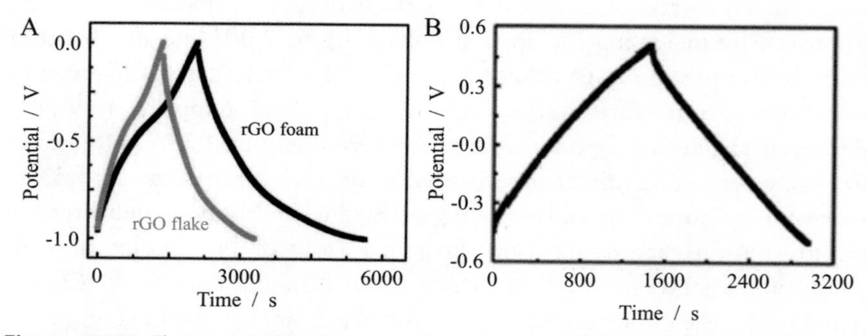

Figure 4.123 Charge–discharge curves for rGOs at a current density of 0.1 A/g in 0.5 M NaCl solution: (A) thermally reduced rGO flakes and foam, and (B) pyridine-reduced rGO [439,443].

better linearity in the curve, although a slight IR drop is observed, as shown in Fig. 4.123B [443]. The difference in electrosorption behaviors is due to the functional groups remaining on rGO after the reduction, but detailed analyses of functional groups have not been reported. An introduction of sulfonic groups into rGO nanoflakes was performed to control the hydrophilicity of rGO flakes [444]. The sulfonation of rGO flakes was confirmed by XPS, of which the improvement in wettability with water was proved by contact angle measurements. The sulfonated rGO exhibited a high NaCl removal efficiency of 83.4% and electrosorption capacity of 8.6 mg/g, whereas a nonsulfonated one showed 40% and 4.1 mg/g, respectively.

N-doping of rGO was also effective for improvement of CDI performance [445]. N-doping was performed by the thermal exfoliation of GO at 800°C in an NH_3 atmosphere and N content was determined to be 7.3 at.% by XPS. N-doped rGO showed S_{BET} of 359 m^2/g and V_{total} of 1.02 cm^3/g, though the rGO without N-doping gave S_{BET} of 155 m^2/g and V_{total} of 0.87 cm^3/g. The electrosorption behavior was compared between N-doped and nondoped rGOs as a function of applied voltage in a simple CDI cell with NaCl solution; the former had better CDI performance.

4.3.3.2 With metal oxides

Loading of nanoparticles of different metal oxides onto rGO flakes was reported to be effective for improving the CDI performance of rGO, loading of TiO_2, CeO_2, Fe_2O_3 and Mn_3O_4 [446], SiO_2 [447], SnO_2 [448], MnO_2 [449], and TiO_2 [450].

TiO_2 loading was performed by mixing GO with $TiCl_3$ and NH_4OH in water, followed by hydrothermal treatment at 180°C, and the resultant TiO_2/rGO composite was subjected to a CDI experiment [446]. In Fig. 4.124A, electrosorption behaviors were compared for the pristine rGO and an AC using a simple CDI cell with NaCl solution (500 mg/L) at an applied voltage of 1.2 V, showing that the composite has much higher electrosorption capacity and rate. The maximum capacity was calculated by applying the Langmuir equation to be 25.0, 15.8, and 3.6 mg/g for the TiO_2/rGO composite, the pristine rGO, and the AC, respectively. In Fig. 4.124B, cycles of electrosorption at 1.2 V and regeneration at 0 V are shown, suggesting the possibility of a fast response and high reversibility of electrosorption/regeneration of the composite, although the optimal

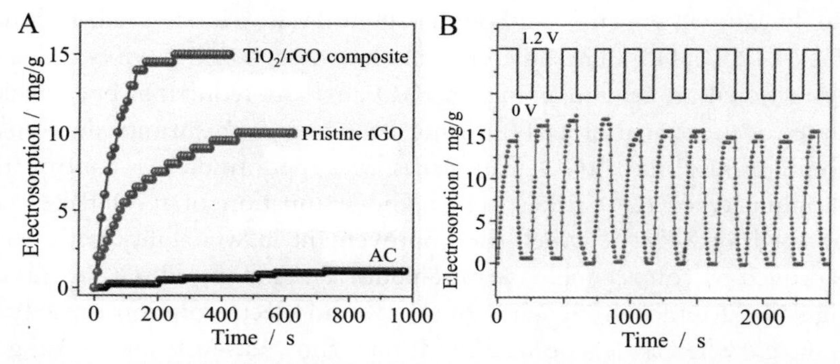

Figure 4.124 CDI performances of TiO_2/rGO composite with NaCl solution (500 mg/L) at 1.2 V: (A) electrosorption capacity in comparison with the pristine rGO and an AC, and (B) electrosorption/regeneration cycles of TiO_2/rGO composite [446].

conditions have to be searched. TiO_2 loading was also performed using commercially available TiO_2 nanoparticles (P25) through its dispersion with GO in 10 M KOH solution, followed by hydrothermal treatment at 120°C and then heat treatment at 650°C [450]. The composite showed much better performance than the pristine rGO in an MCDI cell; electrosorption capacity increased from 0.068 mmol/g for the pristine rGO to 0.281 mmol/g for 20 wt.% TiO_2-loaded rGO.

rGO loaded by MnO_2 nanoparticles was prepared by microwave irradiation of the mixture of graphite with $MnSO_4$, $(NH_4)_2S_2O_8$, and H_2O_2 [449]. Microwave irradiation was applied twice, at 500 W for 5 min and then at 800 W for either 15 or 30 min. The second irradiation time influenced the crystallinity of MnO_2 nanoparticles; those in the 30 min irradiated composite had higher crystallinity than those in the 15 min irradiated one, and as a consequence the former demonstrated improvements in both electrosorption capacity and rate.

Loading of metallic Ag on rGO flakes was performed by dispersing either $Ag(NO_3)_2$ or carbon-coated Ag nanoparticles in the GO solution, followed by a reduction with hexamethylene tetramine at 95°C [451]. C-coated Ag nanoparticles were prepared from Ag^+/β-cyclodextrin complexes by heating at 300°C. The CDI performance of these two composites was studied using eight simple cells in series and NaCl solution. NaCl removal efficiency was enhanced to 40% and 45% by loading Ag nanoparticles and carbon-coated Ag nanoparticles, respectively, from 18% for the pristine rGO (without Ag-loading). The antibacterial ability of these composite electrodes was also confirmed.

4.3.3.3 With other carbon materials

Composites of AC with rGO flakes were prepared by a reduction of GO flakes using hydrazine in their dispersion mixed with 10–50 wt.% AC granules after oxidation in HNO_3 [452]. They were formed into sheets with a thickness of 0.2 mm by mixing with graphite and PTFE in 70/20/10 by weight. The CDI performance of the composites was measured using a simple cell with NaCl solution with a flow rate of 25 mL/min at 25°C. The composite containing 20 wt.% rGO (GAC-20) showed the highest performance. In Fig. 4.125A and B, the electrosorption capacity and rate constant for GAC-20 are compared with an AC as a function of the applied voltage. Mixing 20 wt.% rGO is effective in improving the electrosorption capacity for NaCl, and is twice as large as AC, whereas the electrosorption rate became a little slower. rGO flakes were supposed to serve as a flexible bridge to form a "plane-to-point" (rGO-to-AC) conducting network.

The composites of mesoporous carbon with rGO flakes were prepared from a resol-type PF mixed with a thermally exfoliated rGO and a triblock-copolymer F127 by carbonization at 600°C in N_2 [453]. With the addition of 1–15 wt.% rGO, S_{BET} increases slightly, from 568 to 625–685 m^2/g, but electrosorption capacitance measured by a simple CDI cell with NaCl solution (40 mg/L) under a flow of 25 mL/min increased from 36 F/g for the pristine mesoporous carbon to 52 F/g for the composite containing 5 wt.% rGO. A mesoporous carbon composite with rGO was prepared from a mixture of GO with RF by carbonization at 900°C, which exhibited an electrosorption capacity of 3.2 mg/g for NaCl solution (65 mg/L), much higher than rGO, and a rapid regeneration at 0 V [454].

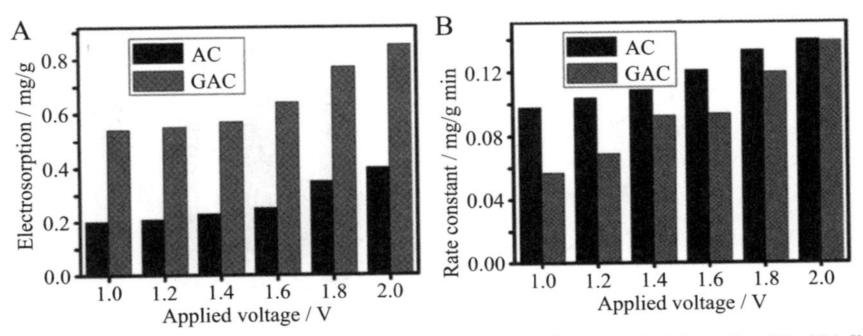

Figure 4.125 CDI performances of the composite of AC with 20 wt.% rGO (GAC): (A) electrosorption capacity for NaCl and (B) electrosorption rate constant versus applied voltage [452].

Hollow mesoporous carbon spheres coated by rGO flakes (GHMCSs) were prepared from the mixture of GO with PF-resin-coated polystyrene spheres by carbonization at 900°C [455]. By using a sheet formed by mixing with acetylene black and PTFE, the CDI performance was studied in a simple cell with NaCl solution (68.0 μS/cm) under a flow of 25 mL/min rate. Electrosorption behavior of GHMCS was compared with rGO and hollow mesoporous carbon sphere without rGO-coating (HMCS); GHMCS had better performance.

Porous carbon spheres were prepared from a mixture of GO, resolcinol, and formaldehyde with Na_2CO_3 (catalyst) in water by decanting in a solution of cyclohexane and span-80, followed by carbonization at 800°C [456]. The CDI performance of the resultant rGO-embedded porous carbon spheres was tested in a simple cell using an NaCl solution (800 mg/L) with a flow rate of 20 mL/min. The carbon spheres exhibit much better CDI performances than an AC, with much higher electrosorption capacity and rate (Fig. 4.126A) and higher charge efficiency (Fig. 4.126B).

A mixture of an auricularia with GO (30/1 by weight) was carbonized either under hydrothermal conditions at 180°C or at 400°C in Ar, and then activated at 850°C by mixing with KOH [457]. The carbon composite prepared from the mixture under hydrothermal conditions possessed electrosorption capacity of 7.74 mg/g, higher than the composite prepared from the same mixture in Ar (5.35 mg/g), although the former had lower S_{BET} and V_{micro} (1401 m^2/g and 0.716 cm^3/g, respectively) than the latter (1999 m^2/g and 1.016 cm^3/g, respectively). The improvement in CDI performance of the former was supposed to be due to the improvement in electrical conductivity.

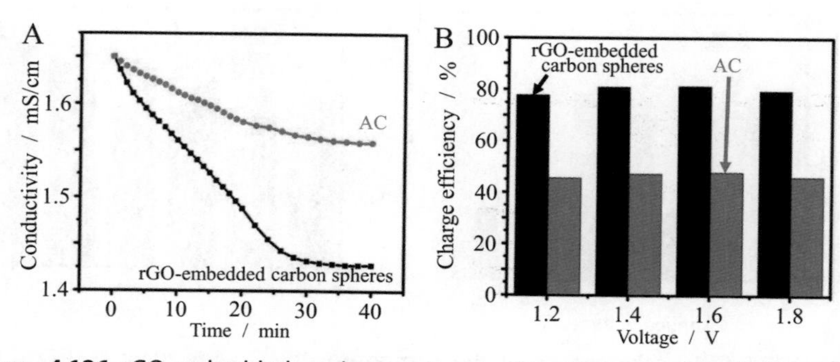

Figure 4.126 rGO-embedded carbon spheres: (A) electrosorption behavior and (B) charge efficiency at different applied voltages in comparison with an AC [456].

4.3.4 Carbon nanotubes and nanofibers

In order to use carbon nanotubes and nanofibers as electrodes of CDI cells, they have to be formed into the sheets (films) by using some binders. MWCNTs prepared by CVD of propylene using Ni catalyst at 750°C were mixed with PTFE (95/5 by weight) and formed into films [458]. The adsorption isotherms at 20°C with an applied voltage of 0−1.2 V were well approximated by Langmuir equation and the maximum electrosorption capacity at 1.2 V was calculated to be 9.35 mg/g, even though S_{BET} of the film was 153 m^2/g. Poly(vinyl alcohol) (PVA) was also used as a binder of MWCNTs for the electrodes of MCDI cells [459]. The electrode prepared by casting the mixture of MWCNTs oxidized in HNO_3 with PVA (1/1 by weight) showed better CDI performance than the electrode formed from a commercially available AC with polyvinylidene fluoride (PVDF) binder. MWCNTs, which were synthesized by CVD of methane using La_2NiO_4 as catalyst precursor followed by purification in HNO_3 and heat-treated at 850°C, were formed into sheets by compressing under 20 MPa with a phenolic resin binder, and their CDI performances were studied in a simple cell with NaCl solution (3 g/L) with a 10 mL/min flow rate at 1.2 V [460]. The MWCNTs after 850°C treatment exhibited better CDI performance than as-synthesized MWCNTs. MWCNTs with 1−2 nm diameter gave a capacity of 8.25 mg/g from NaCl solution (5 g/L) with a 10 mL/min flow rate at 1.0 V [461]. CNTs after air-plasma activation at 450 mV were coated onto a stainless steel with PTFE binder, which delivered enhanced CDI performance for Pb^{2+} [462].

Coating of CNTs by polypyrrole (PPy) was found to be effective in improving the CDI performance [463]. An asymmetric simple CDI cell composed of the negative electrode of PPy-coated CNTs and the positive electrode of the pristine CNTs gives better electrosorption performance, higher capacity and rate, than the symmetric cell composed of the pristine CNTs, as shown in Fig. 4.127.

MCDI cells were studied by employing various combinations of the negative and positive electrodes of MWCNTs with different amounts of cation- and anion-exchange polymers [464]. The mixture of MWCNTs with acetylene black and PVA (8/1/1 by weight) was converted to the slurries with deionized water by adding either PEI (cation-exchange polymer) or dimethyl diallyl ammonium chloride (DMDAAC) (anion-exchange polymer) and then cast on a graphite substrate. The mixing of either PEI or DMDAAC with CNTs improved the CDI performance;

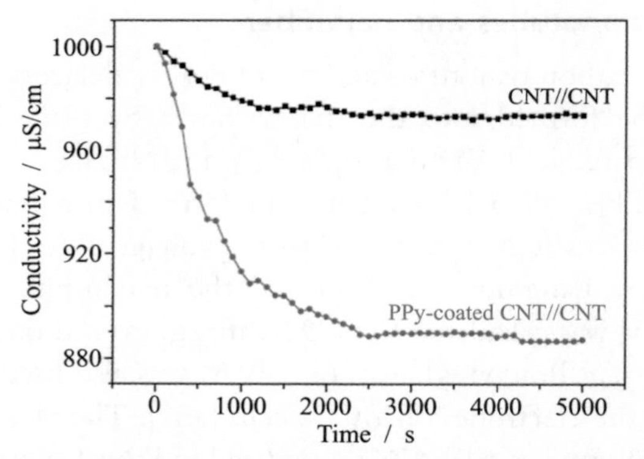

Figure 4.127 Changes in conductivity of the effluent with time for asymmetric PPy-coated CNT//CNT and symmetric CNT//CNT cells [463].

mixing DMDAAC into the positive electrode was more effective in comparison with mixing PEI into the negative electrode.

MWCNTs synthesized by catalytic CVD of methane at 680°C mixed with phenolic resin were pressed under a pressure of 25 MPa at 150°C, and then carbonized at 850°C to make a plate of 115 × 75 × 1 mm^3 [465–467]. The CDI performance of these MWCNT electrodes in an NaCl solution was studied using a simple CDI cell. In order to achieve high performance of CDI, carbonization of MWCNT/phenol resin composite is essential [465], as shown in Fig. 4.128A, and the pretreatment of MWCNTs before preparation of the composite is also important, as shown in Fig. 4.128B [466]. The pore structure in MWCNTs is thought to be

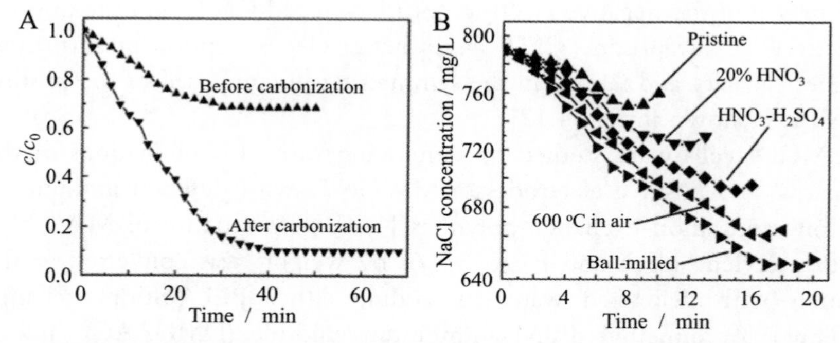

Figure 4.128 Desalination curves for a MWCNT/phenol resin composite electrode: (A) effect of carbonization and (B) effect of the pretreatments of MWCNTs before mixing with phenol resin [465,466].

changed more effectively by ball-milling, because S_{BET} increases to 128 m^2/g from 50 m^2/g for the pristine one.

The mixture of CNTs and CNFs was synthesized by low-pressure CVD of acetylene at 550°C on either Ni or graphite plate (CNTs/CNFs composite films) and their CDI performances were studied [362,468–471]. SEM and TEM images showed that CNTs and CNFs are mixed and entangled with each other [468]. The CDI performance depends greatly on the applied voltage [468] and also on the temperature of the NaCl solution [471]. CDI performances of these composites for NaCl, $ZnCl_2$, $CuCl_2$, and $FeCl_3$ were studied using a simple cell [470]. CDI performance of the cell composed from CNTs/CNFs electrodes with ion-exchange membranes was also studied [362], as mentioned in Section 4.3.1.

The electrodes were prepared by electrophoretic deposition (EPD) method in an acetone/ethanol mixed solution of commercial MWCNTs (10–20 nm diameters) with a small amount of $Al(NO_3)_3$ on graphite paper [472,473]. The CDI performance of these CNT electrodes in a simple cell depends greatly on the deposition time for EPD, as shown in Fig. 4.129 [472]. CNTs of about 120 mg were deposited during 30 min deposition and gave the highest electrosorption capacity and rate, whereas CNTs deposited over 60 min reached about 145 mg and gave a poor performance. The CNT electrode formed by adding polyacrylic acid in an EPD solution gave a much better performance when it was used as the positive electrode coupled with the negative electrode of CNTs (without PAA) than a simple symmetric cell CNT//CNT and an asymmetric cell of the CNTs//CNT

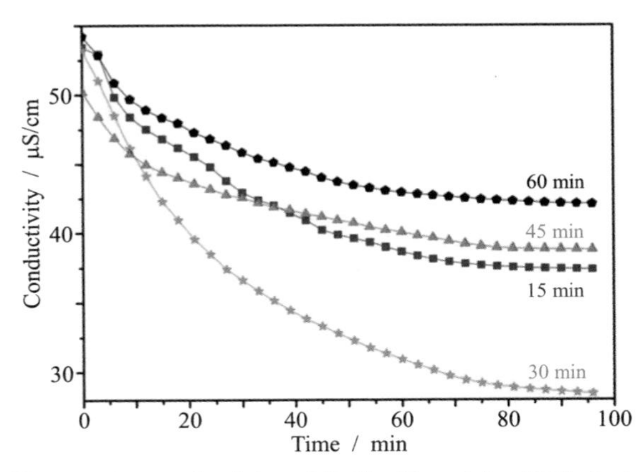

Figure 4.129 Changes in conductivity of NaCl effluent for CNT electrodes prepared with different EPD process times [472].

with cation-exchange polymer [473]. In the CDI cell, polyacrylic acid worked as an anion-exchange polymer, as well as a binder of CNTs.

CNT foams prepared by CVD of 1,2-dichlorobenzene containing ferrocene as catalyst precursor [474] showed an excellent electrosportion capacity for NaCl [475]. The foam was composed of CNTs with uniform diameters of 30–40 nm by forming a two-dimensional web as a uniform and flexible monolith. Its electrosorption capacity for NaCl was as high as 40.1 mg/g. By compression, the foams could be converted to a CNT membrane without any additives, which could be also used as electrodes of the CDI cell.

The composite of commercial SWCNTs after treatment in a concentrated mixed acid with rGO was prepared by suspending SWCNTs with GO in water, followed by reduction with hydrazine monohydrate [476]. On the composite prepared, electrosorption capacity as high as 26.42 mg/g, higher than 22.27 mg/g for rGO, was obtained [476], although a critical discussion on this result was presented [477,478]. The composites of CNTs after oxidation in HNO_3 with 10–50 wt.% rGO were prepared by dispersing CNTs with GO in water, followed by the reduction by hydrazine at 35°C [479]. Their CDI performance was studied by forming into a sheet with graphite powder and PTFE binder. For the composite, a much higher rate of electrosorption of NaCl was observed in a simple cell, in comparison with a CNT (without rGO) electrode. The rate constant was calculated to be 0.2885 for the composite electrode and 0.0459 for the CNT electrode by simulation of the pseudo-first-order adsorption equation. The composites of CNTs with rGO were prepared from a mixture of CNTs with GO in different ratios (5–20 wt.% GO) by thermal exfoliation at 300°C and made into a sheet by mixing with PTFE for CDI performance studies by a half cell with an NaCl solution (~57 μS/cm) flowing under a rate of 25 mL/min [480]. The composite electrode prepared from CNTs with 10 wt.% GO exhibited the highest capacity, more than twice as large as that from CNTs.

CDI performances of commercial SWCNTs and DWCNTs were measured by a simple cell with NaCl solution flowing at a 25 mL/min rate at a voltage of 2.0 V and compared with rGO nanoflakes [481]. The electrodes for CDI measurements were prepared by mixing with CNTs, graphite powder, and PTFE binder (7/2/1 by weight). As shown in Fig. 4.130, the CDI performance for DWCNTs is much better than for rGO, and slightly better than for SWCNTs. By approximation of the electrosorption isotherms by the Langmuir equation, maximum

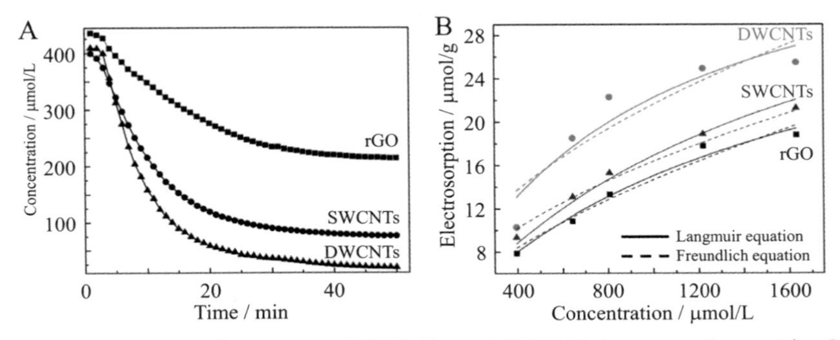

Figure 4.130 CDI performances of SWCNTs and DWCNTs in comparison with rGO nanoflakes in an NaCl solution flowing at a 25 mL/min rate at a voltage of 2.0 V: (A) electrosorption behaviors and (B) electrosorption isotherms [481].

electrosorption capacity was calculated to be 41−43 μmol/g for SWCNTs and DWSNTs, but 36 μmol/g for rGO. SWCNT electrodes together with cation- and anion-exchange membranes (MCDI cell) showed a high removal efficiency of about 97% at 1.2 V on the NaCl solution with an initial conductivity of 110 μS/cm, much higher than about 60% for the same electrodes without using ion-exchange membranes [482].

MWCNT/rGO composite aerogels were fabricated by supercritical CO_2 drying of their hydrogels obtained by heating the mixtures of GO and MWCNTs with vitamin C (reducing agent) [483]. MWCNTs were used either as-received or treated in an acid. The composite aerogels showed S_{BET} of 435 m^2/g, high conductivity (7.5 S/m), and an electrosorption capacity of 633 mg/g for 35 g/L NaCl solution. They showed high physical adsorption for not only NaCl but also for different metal ions (Pb^{2+}, Hg^{2+}, Ag^{+}, and Cu^{2+}) and dyes (methylene blue, rhodamine B, and fuchsin).

CNF webs were prepared from electrospun PAN nanofibers by stabilization at 280°C in air and carbonization at 800°C in N_2, followed by activation at 800°C in steam, and a simple CDI cell was constructed for characterization of the webs [484]. A marked micropore development by activation was observed; S_{BET} increased from 5 m^2/g for the as-carbonized web to 550 m^2/g for the activated web. The changes in NaCl concentration by the application of 1.2 V are shown for the CNF webs as-carbonized and activated in Fig. 4.131. Electrosorption performance is markedly improved by the activation, higher capacity and higher rate of electrosorption, in addition to the marked increase in physisorption before voltage application. CNFs prepared from PAN via electrospinning and following carbonization at 800°C were activated using $ZnCl_2$ at 500°C and

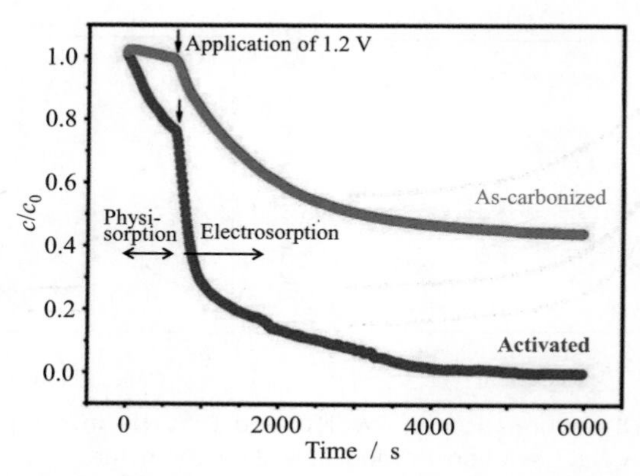

Figure 4.131 NaCl concentration changes with time for electrospun CNF webs as-carbonized and activated [484].

used as the electrodes of the CDI cell [485]. Activated CNFs were microporous and had improved hydrophilicity. Its electrosorption capacity was also improved to 20.52 mg/g from 1.8 mg/g for the pristine CNFs.

CNF webs were prepared by electrospinning of ethanol solution of resole-type PF with poly(vinyl butyral) (PVB) (0.6 wt.% to control the solution viscosity), followed by curing and then carbonization at 600, 800, and 1000°C, of which the CDI performance was studied using a simple cell with an NaCl solution [486]. The webs prepared were flexible and self-standing (Fig. 4.132A), of which the fiber diameters were relatively

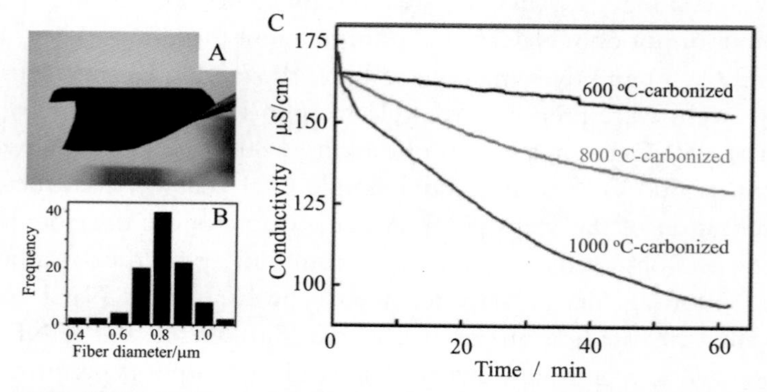

Figure 4.132 CNF web from phenol resin by electrospinning and carbonization at different temperatures: (A) image to show flexibility, (B) distribution of fiber diameter after 1000°C-carbonization, and (C) electrosorption behaviors with a flow rate of 6 mL/min at a voltage of 1.2 V on CNFs carbonized at different temperatures [486].

uniform with average diameters of 810 nm for 1000°C-carbonized webs (Fig. 4.132B). Although no marked development in pore structure was observed with an increase in carbonization temperature, a marked enhancement in CDI performance was obtained on 1000°C-carbonized webs, as compared to the electrosorption behaviors in Fig. 4.132C.

rGO-embedded porous CNF webs (G-PCNF) were prepared by electrospinning of the DMF solution of the mixture of PAN with GO, followed by stabilization at 280°C in air and carbonization at 800°C [487]. The CDI performance was measured in a simple cell with an NaCl solution flowing at a 6 mL/min rate at an applied voltage of 1.2 V by comparing G-PNCF with PAN-based CNF webs prepared without GO (PCNF) and a commercial ACF. G-PCNF has higher electrosorption capacity than PNCF and much higher than ACF, as shown in Fig. 4.133A. Both PCNF and G-PCNF have a certain amount of pseudo-sorption, as shown by the galvanostatic charge–discharge curves in Fig. 4.133B.

Commercial PAN-based carbon fibers (CFs) were treated with 5% HNO_3 solution at 80°C and then thin clusters of nano-sized MnO_2 particles were electrodeposited on them [488]. MnO_2-loaded CFs could electrosorb Cu^{2+}, of which the capacity was determined as 173 mg/g at 0.8 V and could be regenerated quickly by reversing the voltage. MnO_2-loading onto ordered mesoporous carbons (OMCs) also gave some improvement to CDI performance in a 1 M NaCl aqueous solution, although S_{BET} and V_{total} decreased by MnO_2-loading, from 720 m^2/g and 0.56 cm^3/g to 558 m^2/g and 0.35 cm^3/g, respectively [489].

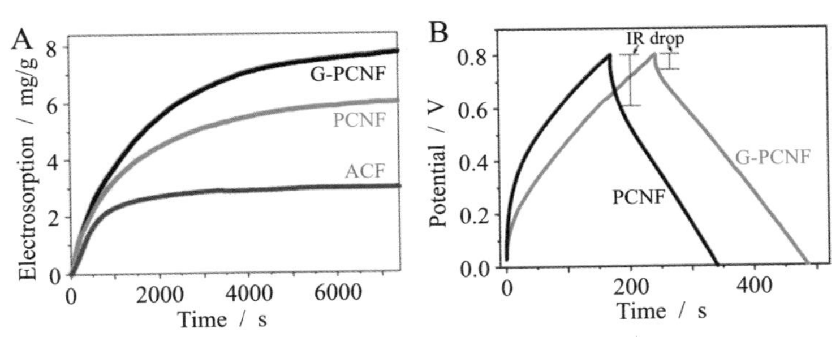

Figure 4.133 CDI performances of rGO-embedded PAN-based carbon nanofiber web (G-PCNF) in comparison with PAN-based carbon nanofiber web (without rGO) (PCNF) and ACF: (A) electrosorption behaviors in NaCl solution with an initial concentration of 1 g/L and (B) galvanostatic charge–discharge curves [487].

4.4 Electromagnetic interference shielding

Microwave technology has been under rapid development since the last century, covering extensive application areas, such as satellite communications, radar detections, information security, and microwave heating. Recently, the field of electromagnetic interference (EMI) shielding and electromagnetic (EM) absorption has become an important problem issue to serious EM wave pollution, and so the materials for these purposes, such as noise reduction and stealth technology, have received increased attention.

The EMI shielding material attempts to achieve high shielding efficiency together with several features, such as light weight, low density, cost efficiency, high thermal and chemical stabilities, mass productivity, and eco-friendliness. Most carbon materials show advantages of low density and low filler content to form composites with other materials, and so various carbon materials have been studied by aiming the applications for EMI shielding and EM absorption, i.e., porous carbons, including activated carbons, templated carbons, precursor-designed carbons, and fibrous carbon materials, including CNTs and CNFs. Among the carbon materials, graphene, including reduced graphene oxide (rGO), was recently shown to have high potentiality due to its high surface area, tunable electrical conductivity, low density, high stability, and good processability. EMI shielding was reviewed by focusing on graphene and rGO [490,491].

In this section, EMI shielding materials are explained by focusing on carbon-based materials after explanation of the fundamentals for EMI shielding by carbon materials. Carbon materials are classified into porous carbons, networks of CNTs and CNFs, graphene and its related materials, and carbon-based composite materials with inorganic materials.

4.4.1 Fundamentals

EMI shielding of the material is evaluated by shielding efficiency (SE) as defined and explained as the sum of the efficiencies due to the reflection (SE_R) and absorption (SE_A), as expressed by the following equation,

$$SE_{total} = 10 \log(P_I / P_T) = SE_R + SE_A,$$

where P_I is the power of incident wave and P_T that of transmitted wave, as illustrated in Fig. 4.134A. The powers of transmitted, reflected, and absorbed waves, denoted by P_T, P_R, and P_A, respectively, are usually normalized by the incident power P_I, and are called transmittance T,

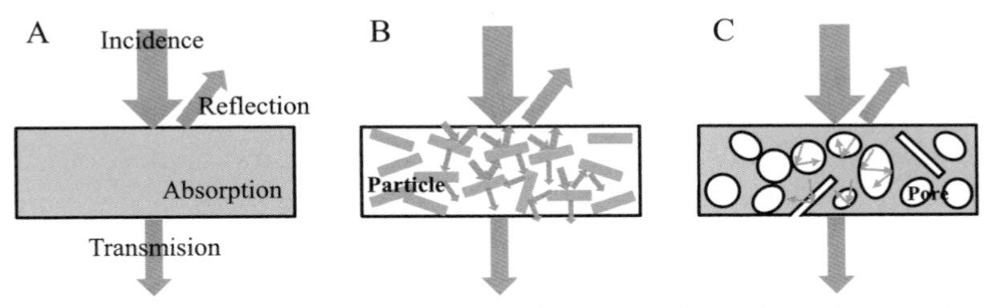

Figure 4.134 Schematic illustrations to explain reflection, absorption, and transmission with (A) matrix, (B) particles, and (C) pores.

reflectance R, and absorbance A, respectively. These parameters are related to each other as follows,

$$T + R + A = 1,$$

and so each shielding efficiency is expressed as follows;

$$SE_{\text{total}} = 10\log(1/T), \quad SE_{\text{R}} = -10\log(1-R), \quad \text{and} \quad SE_{\text{A}} = -10\log[T/(1-R)].$$

The dependences of SEs on frequency in the range of 2–18 GHz are shown, as an example, in Fig. 4.135. These data were measured on the composites of an ordered mesoporous carbon with a wax in different contents of carbon (1–20 wt.%) [492]. Transmitted power P_{T} markedly decreases with increasing carbon content and decreases gradually with increasing frequency, suggesting an increase in the shielding effect (Fig. 4.135A). Reflected power P_{R} increases with increasing carbon content (Fig. 4.135B), mainly because of an increase in reflection surfaces of the carbon particles dispersed in the matrix (wax), as illustrated in Fig. 4.134B.

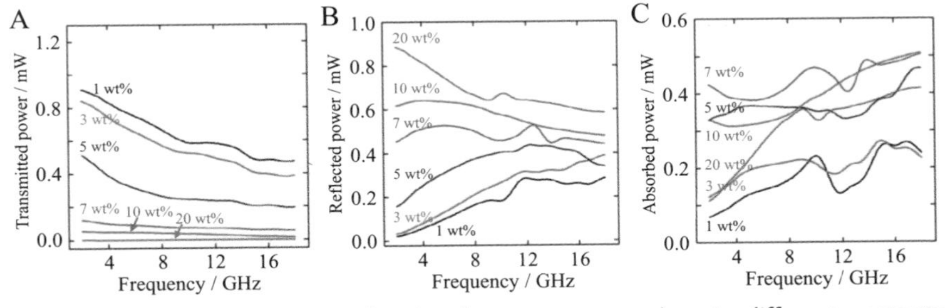

Figure 4.135 The composites of ordered mesoporous carbon in different contents with wax: (A) transmitted, (B) reflected, and (C) absorbed powers (P_{T}, P_{R}, and P_{A}, respectively) in the frequency range of 2–18 GHz [492].

The dependences of absorbed power P_A on the carbon content and frequency are slightly complicated, tending to increase with increasing carbon content up to 7 wt.%, but tending to decrease above 7 wt.%, and associated with some fluctuations. It has to be pointed out that multireflections at the walls of spaces between carbon particles and at the walls of pores in the particles (Fig. 4.134B and C) are mainly included in absorbed power, though partly in reflected power. This might cause the fluctuations observed in frequency dependences of these two powers (Fig. 4.135B and C). In Fig. 4.136A–C, the frequency dependences of shielding efficiencies, SE_{total}, SE_R, and SE_A, calculated on the same composites, are shown for different carbon contents, demonstrating a strong dependence on the carbon content.

The efficiency ES_R is caused by the reflection at the interface scattering during propagation in the material in addition to the reflection at the surfaces of the material, and increases with increasing conductivity of the material. The absorption refers to the dielectric permittivity, the real part ε' corresponding to the polarization capacity, and the imaginary part ε'' (dielectric loss factor) presents the attenuation toward the penetrating EM wave. Fig. 4.137A and B show the frequency dependences of the real and imaginary permittivities ε' and ε'' measured on the composites of ordered mesoporous carbon with wax, respectively.

To evaluate the EMI shielding performances, the reflection loss RL is often employed, which is calculated from the measured complex permittivity by the following equations,

$$RL(\mathrm{dB}) = 20\log|(Z_{in} - 1)/(Z_{in} + 1)|$$

$$Z_{in} = (\mu/\varepsilon)^{1/2}\tanh\left[j(2\pi/c)(\mu\varepsilon)^{1/2}fd\right],$$

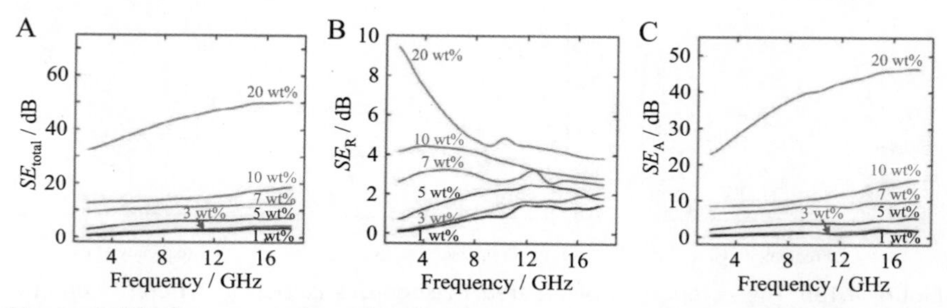

Figure 4.136 The composites of ordered mesoporous carbon in different contents with wax: (A) SE_{total}, (B) SE_R, and (C) SE_A in the frequency range of 2–18 GHz [492].

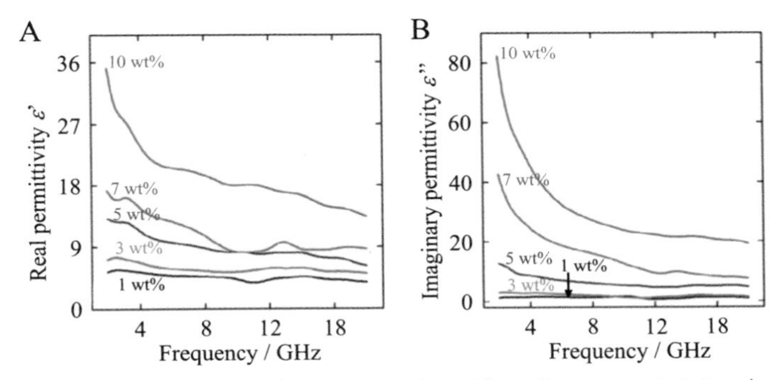

Figure 4.137 Frequency dependences of real and imaginary permittivity ε' and ε'' on the ordered mesoporous carbon/wax composites [492].

where ε and μ are the permittivity and permeability of the samples, f is the EM wave frequency, d is the thickness of the sample, and c is the velocity of light in a vacuum. When RL is less than -10 dB, around 90% of the EM wave power is absorbed.

The frequency range corresponding to RL lower than -10 dB (about 90% attenuation) is often defined as the efficient absorption bandwidth (EAB or simply bandwidth). However, the frequency range corresponding to RL of less than -20 dB is also used because an RL value of less than -20 dB (about 99.9% attenuation) is a measure of the practical application of the material as the EM absorber.

Graphite flakes were studied in the frequency range of 8–18 GHz using their composites with PF resin [493]. The RL of flaky graphite depended strongly on its milling time; its peak at 14.4 GHz changed from -10.4 dB for the pristine flake to -19.4 and -25.5 dB for the flakes milled for 8 and 12 h, respectively. Thin graphite flakes (TG) were prepared from commercial exfoliated graphite powder (EG) by further oxidation in nitric/sulfuric acid under sonication, TG and EG being confirmed to be composed of different thicknesses, 2–5 and 15–25 nm, respectively, and quite different broadness of 002 diffraction profiles [494]. In Fig. 4.138A and B, minimum RLs are plotted against the graphite content for two composites with 2 and 4 mm thicknesses. The minimum RL for the TG is lower than that of the EG at the same graphite content in the composites, for example, the minimum RL of the TG is close to -60 dB, whereas the EG gives a minimum RL of about -15 dB at the same graphite content of 14 wt.%. The experimental result reveals the effectiveness of thin graphite flakes (TG) for EMI shielding.

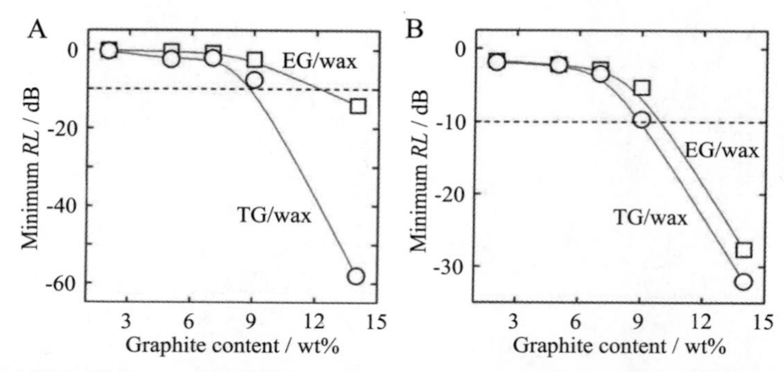

Figure 4.138 Minimum *RL* of EG/wax and TG/wax composites with thicknesses of (A) 2 mm and (B) 4 mm [494].

EMI shielding performance was occasionally evaluated using the product of the integral of reflection loss curve divided by the density and the thickness. Graphene foams prepared by hydrothermal treatment at 180°C, followed by annealing at 600°C in Ar (bulk density of 14 mg/cm^3) gives 2.2×10^5 dB cm^2/g [495], although conventional microwave absorption material gives more than two orders of magnitude lower, for example, about 3.1×10^2 dB cm^2/g for a carbonyl iron-based material.

In this section, reflection loss *RL* was mainly used to characterize the carbon materials for the application of EMI shielding.

4.4.2 Porous carbons

The effect of carbonization temperature on EMI shielding was studied using commercial polystyrene strong-base anion-exchange resin (PSAR) as a carbon precursor [496]. The carbons were prepared from a mixture of PSAR with a small amount of $K_3[Fe(CN)_6]$ at a temperature between 500 and 700°C in N_2, followed by refluxing in hot HCl solution. The carbon after mixing with 40 wt.% paraffin was pressed into a ring (outer diameter of 7 mm and inner diameter of 3 mm with different thicknesses) for shielding performance measurements in the frequency range of 2—18 GHz. Changes of *RL* with frequency are shown as functions of the carbonization temperature on the composites with 2 mm thickness and of the thickness of the composite prepared at 600°C in Fig. 4.139A and B, respectively. The carbons obtained at 500 and 550°C display very poor reflection in the whole frequency range measured, but that at 600°C delivers an intensive *RL* of −20.6 dB at 16 GHz (Fig. 4.139A). With higher carbonization temperature, the minimum of *RL* shifts to lower frequency. This marked

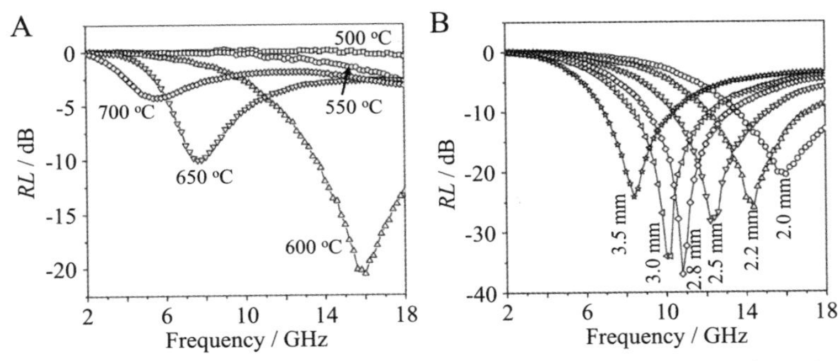

Figure 4.139 Reflection loss *RL* of the composite of the carbons derived from the resin with 60 wt.% paraffin: (A) with thickness of 2 mm on the composites carbonized at different temperatures and (B) with different thickness of the composite carbonized at 600°C [496].

dependence of *RL* on carbonization temperature was explained by crystallinity improvement on the basis of a small change in Raman parameter I_D/I_G. However, it has to be pointed out that the experimental data on I_D/I_G demonstrate a decrease with increasing carbonization temperature, from 0.65 for 550°C to 0.80 for 700°C, which suggests a decrease in crystallinity. It would be reasonable to give attention to the residual functional groups in the carbons and pore structure of the carbon, however, regrettably there is no information on pore structure on the carbons. With increasing thickness of the composite, the minimum of *RL* shifts to a lower frequency side and increases up to 2.8 mm, where *RL* reaches −37 dB at 10.8 GHz, and then decreases with a shift to the low-frequency side with a further increase in thickness above 2.8 mm, as shown in Fig. 4.139B.

Three kinds of carbon spheres with different pore structures were synthesized from RF resin mixed with tetraethyl orthosilicate (TEOS), i.e., solid spheres without TEOS (carbon solid microspheres, CSMs), hollow spheres using TEOS (carbon hollow spheres, CHMs), and hollow spheres using TEOS, of which the walls contain mesopores (porous carbon hollow spheres, PCHMs) [497]. These had characteristic pore structures as expressed by S_{BET} and V_{total}, 238 m^2/g and 0.06 cm^3/g for CSMs, 620 m^2/g and 0.28 cm^3/g for CHMs, and 1098 m^2/g and 1.41 cm^3/g for PCHMs, respectively. The *RL* of PCHMs reached −84 dB at 8.2 GHz, while CHMs gave −25 dB at 17.3 GHz and CSMs only −8.5 dB at 9.5 GHz.

Carbon fibers containing macropores were prepared by wet-spinning of the DMF solution of the mixture of PAN and PMMA, followed by stabilization at 180−280°C in air and carbonization at 1200°C in N_2

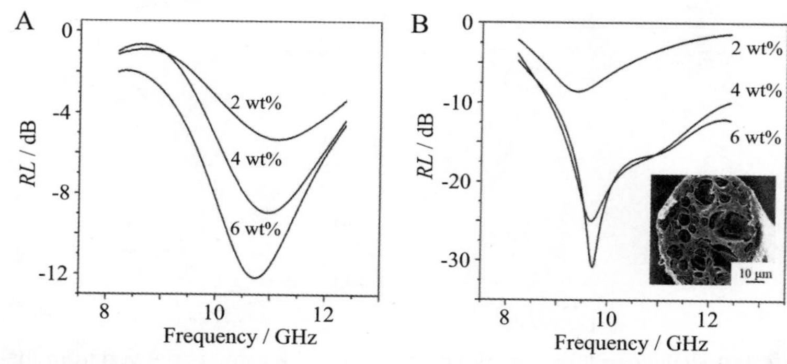

Figure 4.140 Reflection loss *RL* in 8.2–12.4 GHz: (A) CF/epoxy composite with a thickness of 3 mm and (B) porous CF/epoxy composite with 2.3 mm thickness with an SEM image of the fiber cross-section [498].

flow [498]. The fibers prepared from a PMMA/PAN mixture with a weight ratio of 3/7 contained many large pores (macropores), as shown in the SEM image of the cross-section as an inset in Fig. 4.140B (porous CF), whereas that from the mixture with 7/3 contained less macropores (CF). To determine the complex permittivity ε ($\varepsilon = \varepsilon' - j\varepsilon''$), the fibers after being ground up to micro- to millimeter lengths were mixed into epoxy in CF contents of 2, 4, and 6 wt.%, and then cured in a mold in air to form composite films with about 3 mm thickness. For CF (less macroporous), the real part ε' and imaginary part ε'' of complex permittivity increased with an increase in its content in the composite film and a small change in a frequency range of 8.2–12.4 GHz. For porous CF (macropore rich), ε' and ε'' increased with the increase in its content in the composite film, although ε' and ε'' decreased markedly with increasing frequency, particularly in the fiber contents of 4 and 6 wt.%. The microwave absorption properties (*RL*) of CFs and porous CFs are shown in Fig. 4.140A and B, respectively, with CF showing minimum loss at around 11 GHz and porous CFs giving an additional and higher loss at around 9.7 GHz.

Complex permittivity ε was determined on the composites of an ordered mesoporous carbon with a paraffin, the carbon being prepared from a phenol resin using a colloidal silica template at 800°C [492]. The carbon had S_{BET} and S_{micro} of 801 and 396 m^2/g, respectively, as a consequence S_{ext} was calculated to be about 405 m^2/g. Different contents of the carbon (1–20 wt.%) and different thicknesses of the composite (2–4.5 mm) were employed for the EMI shielding measurements. The powers of transmission, reflection, and absorption were measured and EMI shielding efficiencies (*SE*s) were evaluated as shown in the previous section as Figs. 4.135 and 4.136, respectively, revealing a marked increase of SE_{total} of the composite with 20 wt.% carbon, reaching 32−50 dB, and dominant contribution of SE_A on EMI shielding. By using 20 wt.% mesoporous carbon, SE_{total} of more than 20 dB, a commercially applicable level, could be obtained in the whole frequency region examined (2−18 GHz).

The composites of ACFs with epoxy resin in different ACF contents of 0.21−1.73 wt.% were prepared and their EMI shielding performances were measured [499]. ACFs were prepared from viscos fibers via carbonization and steam activation at 950°C, which had an average length of about 32 mm and diameter of 13−16 μm, and S_{BET} of 823 m^2/g. ACFs were mixed with a mixture of epoxy resin and polyamide (2/1 by weight) to form a sheet with a thickness of 3 mm by molding under 10 MPa at 80°C. As shown in Fig. 4.141A, the composite containing 0.43 wt.% ACF exhibits a strong peak at 8.2 GHz, of which the minimum *RL* reaches −26.8 dB. For the composites with other ACF contents, i.e., 0.21 wt.% and more than 0.76 wt.%, *RL* tends to decrease with increasing frequency and fluctuates around −12.5 dB in the frequency range of 5.8−18 GHz, whereas the composite containing 0.43 wt.% ACF also has similar fluctuations at the same frequency range. The marked peak in *RL* is characteristic for ACFs, whereas the composite containing nonactivated CFs with the same content shows only a broad peak at around 11 GHz, as shown in Fig. 4.141B.

The composites of polyaniline with carbon black (CB/PANI) were prepared by polymerization of the mixture of aniline with 5−10 wt.% CB (particle size of 50−80 nm (Degussa PHG-1P) at around 0°C using $(NH_4)_2S_2O_8$ catalyst, of which the EMI shielding characteristics were

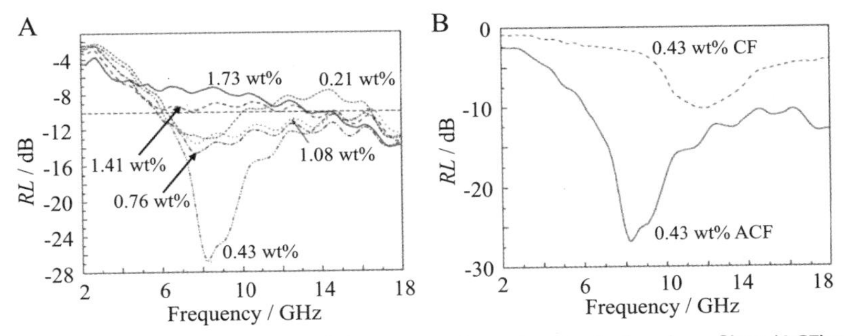

Figure 4.141 Reflection loss *RL* of the composites of activated carbon fiber (ACF) with epoxy/polyamide resin: (A) effect of ACF content and (B) effect of activation [499].

measured in frequency ranges of 2–18 and 18–40 GHz after forming into a film by mixing with epoxy resin [500]. The frequency dependences of the *RL* of the composites with different CB contents are shown for two frequency ranges in Fig. 4.142A and B, respectively. In the range of 2–18 GHz, the minimum *RL* peak shifts toward a lower frequency with increasing CB content, associating with marked intensity increase. The composite containing 30 wt.% CB exhibits a large loss, characterized by *RL* of less than −40 dB at around 11 GHz, although the composite containing 20 wt.% CB shows much a smaller loss at slightly higher frequency (Fig. 4.142A). In the range of 18–40 GHz, the composites with 30 and 20 wt.% CB exhibit much weaker loss at different frequencies, about −11 dB at 28 GHz and about −16 dB at 35 GHz, respectively (Fig. 4.142B).

Carbon spheres (diameters of about 5 μm) were synthesized from watermelon under hydrothermal conditions, of which the composite with a paraffin (25 wt.% carbon) delivered a high *RL* of −40 dB in the frequency range of 12–18 GHz [501]. Yolk-shell type carbon microspheres, which were synthesized by coating RF resin onto SiO_2 spheres, followed by carbonization at 700°C and dissolution of SiO_2 with HF, exhibited a high *RL* of −39.4 dB at 16.5 GHz on the composite film with a paraffin (50 wt.% carbon) [502]. This composite film with a thickness of 1.0–5.0 mm possessed the bandwidth (EAB below −20 dB) covering a wide frequency range from 4.5 to 18.0 GHz. The size of the hollow carbon spheres was shown to be an important factor in controlling the EMI shielding effect by using spheres synthesized by carbonization of hollow nanospheres of an aniline-pyrrole copolymer prepared by mixing TX-100 catalyst [503]. By increasing the amount of TX-100, the size of the carbon spheres became smaller, from about 100 nm without TX-100 to less than

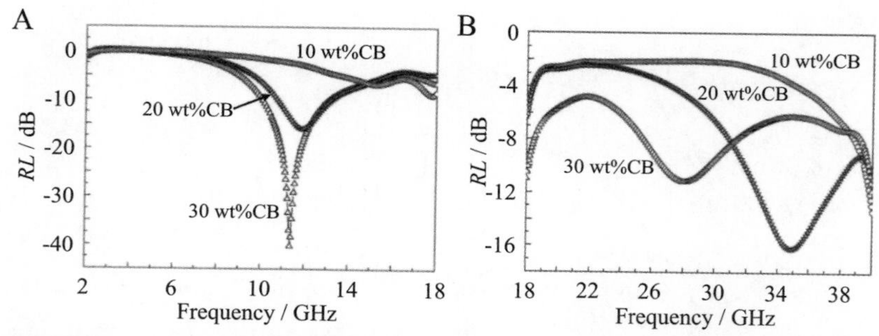

Figure 4.142 Reflection loss *RL* of PANI/CB composites with different CB contents: (A) in the frequency range of 2–18 GHz and (B) 18–40 GHz [500].

70 nm, with almost the same wall thickness of about 20 nm. As shown by the dependences of *RL* on frequency in the range of 2–18 GHz on the solid carbon spheres and hollow carbon spheres with 71.4 and 68.5 nm in Fig. 4.143A–C, respectively, large *RL* is obtained on the hollow carbon spheres with an average diameter of 71.4 nm, of which the minimum *RL* reaches −50.8 dB at 13.5 GHz. The carbonization temperature of the precursor spheres was optimized at 800°C. Microporous carbons were fabricated from cellulose-rich waste tissue papers mixed with PVA and formaldehyde by freeze-drying, stabilizing at 200°C in air, and carbonizing at 900°C, which delivered high SE_{total} as about 40 dB composed mainly of SE_A (about 35 dB) at 12.5 GHz [158].

Porous carbons were synthesized from RF gels by carbonization at 900°C [504]. By using the ratios of tert-butanol (solvent) to resorcinol (T/R) of 12, 7.5, and 5, different pore morphologies were obtained in the resultant carbon particles; T/R of 12 resulted in the aggregates of plate-like particles giving rise to slit-shaped pores (S_{BET} of 741 m^2/g with S_{meso} of 240 m^2/g) and T/R of 7.5 and 5 for uniform cage-like pores (S_{BET} of 571–513 m^2/g with S_{meso} of 158 m^2/g). The dielectric complex permittivity ε of these carbons was measured using composites with a paraffin in the frequency range of 8.2–12.4 GHz. The imaginary part ε'' of the permittivity of the carbon with T/R of 7.5 was much higher than that with T/R of 12, suggesting a much higher dielectric loss of the wave. This difference in ε, which certainly caused a difference in EMI shielding performance, was thought to be due to the difference in pore morphology.

Hollow PAN-based carbon fibers with dual walls, as shown in the SEM image of its cross-section in Fig. 4.144A, with a wide range of diameters from submicrometer to 200 μm, were prepared by either electrospinning or wet-spinning, followed by stabilization at 250°C in air and carbonization at

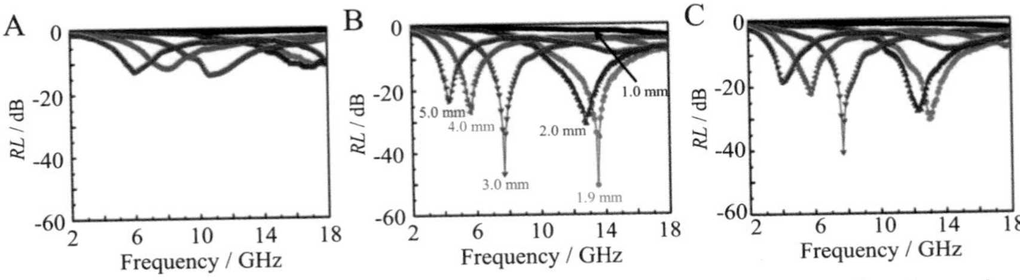

Figure 4.143 Reflection loss *RL* in the frequency range of 2–18 GHz for the carbon spheres derived from an aniline-pyrrole copolymer: (A) solid spheres, (B) hollow spheres with size of 71.4 nm, and (C) those of size 68.5 nm [503].

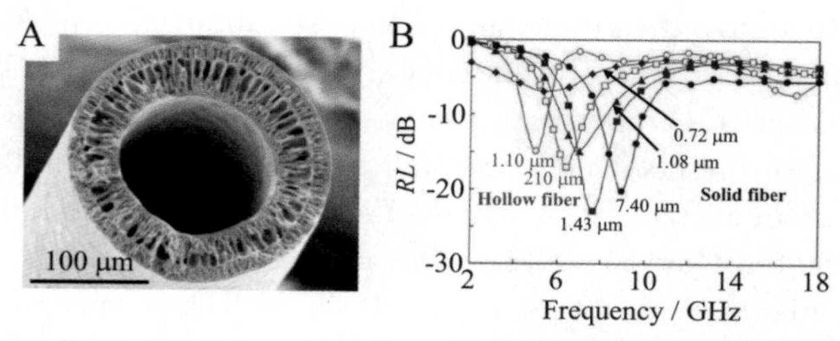

Figure 4.144 Hollow and solid carbon fibers: (A) SEM image of cross-section of hollow fibers and (B) dependences of *RL* on frequency for hollow and solid carbon fibers with different diameters measured using their composites with a paraffin having 3 mm thickness [505].

800°C [505]. EMI shielding performances were measured using the composite with a paraffin and compared with commercial PAN-based CFs (solid fibers). As shown in Fig. 4.144B, the *RL* peak shifts to lower frequencies with decreasing diameter of fibers, and the peak is located at a comparatively lower frequency for the hollow fibers than for solid fibers. The bandwidths at −10 dB for these fibers are not as wide at around 2 GHz, and the solid fibers with 0.72 μm diameter had zero bandwidth due to their weak absorption.

Carbon microcoils with diameters of 2−10 μm and lengths of 0.1−3 mm were shown to be effective for EMI shielding [506] and nanocoils with very short coil pitches with diameters and lengths of about 110 and 750 nm, respectively [507]. The former microcoils were studied in the frequency range of 75−105 GHz and exhibited an *RL* of less than −30 dB at 81. 91 and 102 GHz in the composite with PMMA (1 wt.% coil) [506]. The latter nanocoils delivered about −35 dB at about 17 GHz and about −27 dB at about 13 GHz in the composites with 15 wt.% coil and thickness of 2 and 2.5 mm, respectively, at a frequency range of 2−18 GHz [507].

Shielding of low-frequency EM was also studied on carbon materials. Shielding effectiveness *SE* was evaluated on carbon fiber mats (PAN-based CFs with a diameter of 7 μm), carbon nanofiber mats (CNFs with a diameter of 0.16 μm), and flexible graphite films in the frequency range of 0.1−1.5 GHz [508−510]. The CNF webs (2.9−5.4 mm thick, 0.13−0.22 g/cm^3 bulk density, 6.1−10 vol.% solid) provided a high SE_{total} of 52−81 dB at 1.5 GHz and high *SE*/density ratio (370−470 dB cm^3/g), though they had a low *SE*/thickness ratio of 14−18 dB/mm [508]. The addition of 0.5 wt.% graphite flakes into a cement containing 15 wt.% silica fume was effective in improving the EMI shielding in the frequency range of 10 Hz to 1 MHz [511]. Exfoliated graphite, which was prepared by thermal exfoliation and reduction of acid-intercalated graphite, exhibited a high real part of a relative dielectric constant at frequencies from 50 Hz to 2.0 MHz, which was roughly 10 times larger for the exfoliated graphite before washing (containing many residual acidic species) than that after washing [512].

4.4.3 Networks of carbon nanotubes and nanofibers

The composite films of SWCNTs with soluble cross-linked polyurethane (PU) were prepared by pouring DMF/toluene solutions of these mixtures with different SWCNT contents (1–25 wt.%) into a mold, followed by hot-pressing at 170°C under 15 MPa after solvent evaporation [513]. The changes in ε' and ε'' of complex permittivity with a microwave frequency in the range of 2–18 GHz are shown for the composites with different SWCNT contents and thickness of 2 mm in Fig. 4.145A and B, respectively. In Fig. 4.145C, the changes in *RL* with frequency are shown. Although the composite films containing 1 and 2 wt.% SWCNTs show weak wave-absorbing ability, i.e., low ε' and ε'' values and almost no dependencies on microwave frequency, a small increase in SWCNT content (from 1 to 2 wt.%) results in a marked decrease in *RL*; the composite containing 2 wt.% SWCNTs exhibits −13.3 dB at around 12 GHz. The *RL* decreases markedly and reaches the lowest value of −22 dB at 8.8 GHz with increasing SWCNT content up to 5 wt.%. However, a further increase in SWCNT content makes the *RL* larger and the frequency giving *RL* minimum shifting to the low-frequency side, whereas it makes ε' and ε'' markedly larger. The microwave absorption ability of these composites was thought to be attributed mainly to the dielectric loss.

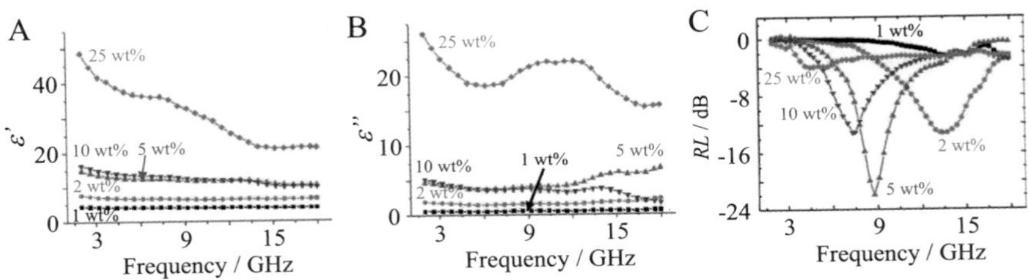

Figure 4.145 EMI shielding performances of SWCNT/PU composite films with different SWCNT contents: (A) real part (ε') and (B) imaginary part (ε'') of complex permittivity, and (C) reflection loss *RL* in the frequency range of 2–18 GHz [513].

The shielding effect of SWCNT-containing epoxy composite was studied in a wide range of frequencies of 10 MHz to 1.5 GHz [514]. SE_{total} values of 15–20 dB were obtained in 500 MHz to 1.5 GHz, with the highest value reaching 49 dB for 15 wt.% SWCNTs at 10 MHz. The long SWCNTs (average length of 1.43 μm) gave higher SE_{total} than short SWCNTs (0.74 μm long). Annealing of the short CNTs at a high temperature gave higher SE_{total} than the pristine CNTs, but was still inferior to the long CNTs.

Structural defects in CNTs were shown to influence strongly the EMI shielding performances. The permittivity was measured on the composite films of three kinds of MWCNTs with polystyrene [515]. The real part ε' of permittivity depends strongly on the CNTs, 8.5–9.3 for MWCNTs synthesized by arc discharge, fluctuation between 9.5 and 11.7 for the B-doped MWCNTs, and 25–30 for Fe-filled MWCNTs in the frequency range of 8–12 GHz. The imaginary part ε'', however, was very small and so could not be differentiated among the three MWCNTs. The Fe-filled CNTs were reasonably supposed to contain a large concentration of structural defects, the B-doped CNTs to contain the defects caused by B-doping, and the as-prepared CNTs to have the smallest amount of defects among these CNTs. CNTs with amorphous structure synthesized by arc discharge with Co-Ni alloy catalyst at 600°C (diameters in the range of 7–60 nm) delivered a slightly better *RL* (−13 dB at 13 GHz) than commercial MWCNTs (average diameter of 20 nm) [516]. Loading of 6 wt.% $La(NO_3)_3$ onto the CNTs with amorphous structure was effective in improving *RL* to −25 dB at 14 GHz. The shielding efficiency (*SE*) was compared between the virgin and recycled PAN-based CFs in the frequency range of 0.2–1.6 GHz using the composite films with a fiber areal density of 100 g/m^2 [517]. SE_{total} of about 50 dB for the virgin and a slightly lower 40 dB for the recycled CFs were observed. The recycled CFs were reasonably supposed to contain larger amounts of structural defects than the virgin CFs, owing to recycling processes. Porous CNFs prepared from PAN mixed with 5 wt.% $ZnCl_2$ by electrospinning and carbonization exhibited marked *RL*, −51 dB at 9.1 GHz with an effective bandwidth (less than −10 dB) ranging from 8 to 11 GHz for the film with a thickness of 3 mm [518].

A three-dimensional network of MWCNTs was fabricated by depositing MWCFTs (diameter of 8–15 nm and length of 30–50 μm) onto

PMMA microspheres (average diameter of 150 μm) via their dispersion solutions, followed by compression molding at 150°C under 10 MPa and then by supercritical CO_2-assisted foaming [519]. The resultant foams of MWCNT/PMMA composite with 5.0 wt.% MWCNT content, which had a low bulk density of 0.49 g/cm^3 and 2 mm thickness, exhibiting a good combination of high electrical conductivity of 3.19 S/m and high shielding effectiveness SE_{total} of 35.9 dB. In addition, they had good mechanical properties, and compressive strength and modulus of 5.12 and 36.43 MPa, respectively.

Aniline was polymerized in diluted HCl in the presence of dispersed colloidal graphite (43−90 wt.%) by adding ammonium persulfate [$(NH_4)_2S_2O_8$], of which the EMI shielding performances were measured in the frequency range of 8.2−12.4 GHz [520]. The resultant PANI/graphite composites showed much improved thermal stability and electrical conductivity, up to about 300°C and 67.4 S/cm, respectively, and delivered a maximum SE_{total} of 39.7 dB on the composite with 17.4 wt.% graphite, whereas pure PANI (without graphite) showed a conductivity of 1.1 S/cm and maximum SE_{total} of 31.4 dB. The further addition of MWCNTs (diameter of 10−70 nm and length of about 300 μm) into the PANI/graphite composites was effective in improving the shielding efficiency [521]. EMI shielding of the composites (MWCNT/PANI/graphite) depends greatly on the content of MWCNTs. The maximum SE_{total} reaches 98 dB for the composite with 10 wt.% MWCNTs and absorption is predominant in SE_{total}.

The flexible CNF networks were fabricated by electrospinning of the DMF solution of the mixture of PAN with GO to form webs, followed by stabilization of PAN at 270°C in air and carbonization of PAN accompanying with a reduction of GO at 1000°C in N_2 (rGO/CNF composite webs) [522]. The resultant composite web delivered a higher shielding efficiency than the neat CNF network (without rGO); SE_{total} of 25−28 dB in the frequency range of 8.2−12.5 GHz was obtained on a thin composite web (about 0.26 mm thick) containing 17.2 wt.% rGO, higher than the 17−18 dB for the neat CNFs. This enhancement was explained by a conduction improvement at the interfaces between CNFs by sandwiching rGO flakes. The resultant composite networks with a thickness of less than 0.3 mm with low density of less than 0.1 g/cm^3 were promising EMI shielding material.

4.4.4 Graphene and related materials

A single-layer graphene sheet was reported to have a very low shielding efficiency SE_{total} as about 2 dB [523], whereas the theoretical prediction was much higher, at about 35 dB up to the terahertz region [524]. Single-layer graphene was synthesized by an inductively coupled plasma CVD of C_2H_2 at 725°C and transferred onto PET substrate using PMMA, which was confirmed by strong G' and very weak D bands in Raman spectrum and low resistivity of 635 Ω/sq [523]. For comparison, a defective graphene sheet was also synthesized by modifying the CVD process parameters, which was confirmed by the appearance of strong D band in Raman spectrum and by high resistivity as about 2.3 kΩ/sq. The single-layer graphene delivered SE_{total} of 2.3 dB (corresponding to about 40% shielding) with SE_R and SE_A of −13.7 and −4.7 dB, respectively, but the defective graphene gave SE_{total} of only 0.01 dB.

Layer-by-layer composite of single-layer graphene with PET were prepared using different thicknesses of PET sheets, of which the EMI shielding performances were measured in the frequency ranges of 1−12 and 18−26.5 GHz [525]. A single-layer graphene film synthesized by CVD of CH_4 at 1000°C on Cu foil was transferred onto a PET sheet using PMMA, and the composite layer (graphene/PET-paired layer) thus prepared was stacked up to 20 layers. In Fig. 4.146A−C, changes in transmittance T, reflectance R, and absorbance A of the composites with the number of composite layers in the frequency range of 1−20 GHz are shown as a function of the thickness of PET sheet d, including without PET sheet ($d = 0$). With an increasing number of graphene layers stacked without PET layers, T decreases quickly and R increases rapidly, and A decreases.

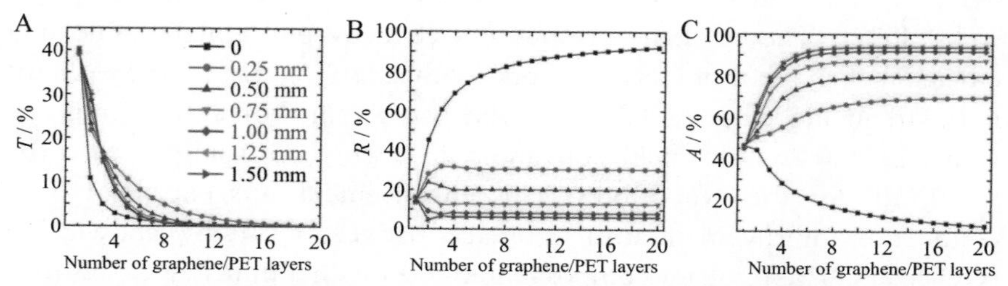

Figure 4.146 EMI shielding performances of layer-by-layer composites of single-layer graphene and PET sheet with different thicknesses in the frequency range of 1−20 GHz; dependences with number of graphene/PET-paired layers for (A) transmittance T, (B) reflectance R, and (C) absorbance A as a function of thickness d of PET layer [525].

The insertion of PET sheets between graphene layers influences markedly the *R* and *A* of the composites; *R* decreases markedly with insertion of a thin PET sheet, depending slightly on the thickness of the PET sheet *d* and losing the dependence on the number of paired layers, while *A* increases with increasing *d*, approaching 100% gradually with the increasing number of paired layers stacked. In Fig. 4.147, *A* and *T* of the composites in a frequency range of 18–16.5 GHz are shown as a function of the number of paired layers stacked; *A* increases and *T* decreases, *A* and *T* changing markedly up to four layers and then gradually in the whole range of frequencies employed. With increasing frequency from 18 GHz, both *A* and *T* tend to increase slightly.

An EMI shielding performance comparison was reported on two kinds of flakes; one denoted as GN (graphite nanosheet) was prepared from commercial exfoliated graphite by dispersion in alcohol/water under sonication and then oxidation in nitric/sulfuric acid, followed by washing and drying, and the other denoted as rGO was prepared by hydrazine reduction of GO synthesized by Hummers' method from natural graphite [526]. The flake GN had a much sharper 002 diffraction peak than the flake rGO, even though GN experienced oxidation in acid to form GO and thermal exfoliation/reduction at high temperature during production of the exfoliated graphite in an industry and further oxidation in acid before measurements. The rGO possessed almost 10 times higher SE_{total} than GN, which was thought to be caused by a higher number of structural defects due to the presence of more functional groups in the rGO flakes. EMI shielding performances were studied on thin graphite films prepared from

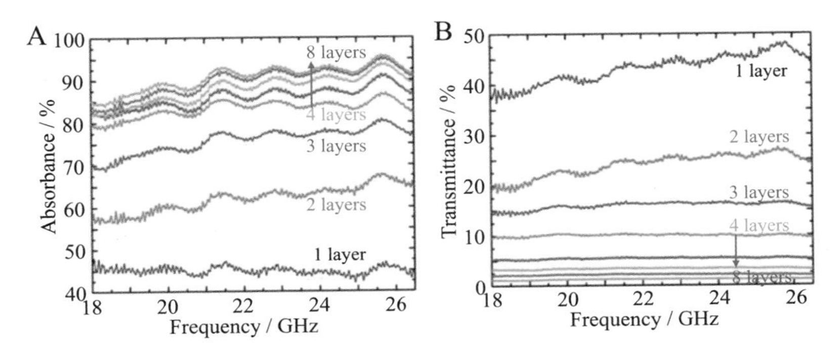

Figure 4.147 EMI shielding performances of the composites of single-layer graphene and PET ($d = 1.0$ mm) in the frequency range of 18–26.5 GHz as a function of the number of graphene/PET-paired layers stacked: (A) absorbance *A* and (B) transmittance *T* [525].

an exfoliated graphite by compression and rolling [527]. The exfoliated graphite was prepared using $HClO_4$ as intercalate, followed by exfoliation by microwave irradiation at 800 W for 50–60 s. The resultant flexible graphite sheet had a thickness of 0.1 mm and delivered SE_{total} of about 80 dB at a frequency of 12–18 GHz.

The composites of rGO with nitrile butadiene rubber (NBR) with different rGO contents were prepared by mixing in xylene, which was cast to a thin film on the surface of an Al panel [528]. rGO was prepared by thermal exfoliation and reduction of GO at 150°C. The dielectric properties were measured on these rGO/NBR composites over the frequency range of 4–12 GHz, as shown in Fig. 4.148. Both ε' and ε'' increase with increasing rGO content from 2 to 10 wt.% in the composites and decrease with increasing microwave frequency from 4 to 12 GHz. Tangent loss tan δ is quite high, increasing with increasing rGO content, 0.22–0.18 for 2 wt.% and 0.44–0.39 for 10 wt.%, and decreasing slightly with increasing frequency. The reflection loss *RL* calculated from these dielectric data reached 57 dB at 9.6 GHz for the composite film (3 mm thickness) containing 10 wt.% rGO.

The composites of rGO with poly(3,4-ethylenedioxythiophene) (PEDOT) were prepared by polymerization of EDOT monomers in a dispersion of rGO, where PEDOT nanofibers were thought to be grafted on the active sites of rGO flakes by electrostatic attraction [529]. The composite PEDOT/rGO exhibits enhanced EMI shielding performances, as shown by *RL* in a frequency range of 2–18 GHz measured on the films of 25 wt.% composite with a paraffin having different thicknesses, as shown in Fig. 4.149; the composite delivers a minimum *RL* of −48.1 dB at 10.5 GHz (Fig. 4.149C), while rGO gives only −8.9 dB at 9.5 GHz and PEDOT −14.5 dB at 7 GHz (Fig. 4.149A and B, respectively) on the films

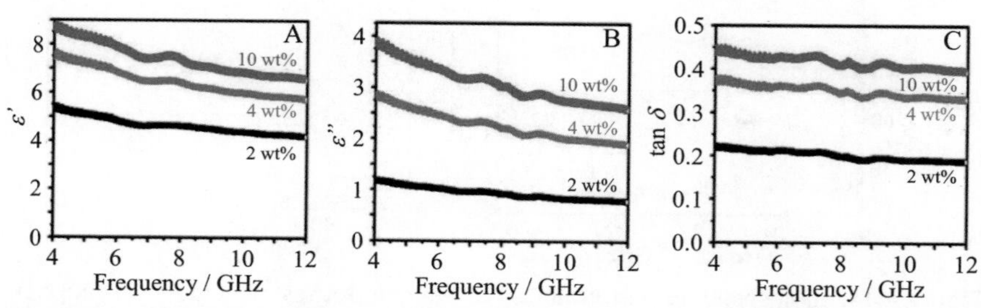

Figure 4.148 Dielectric properties of rGO/NBR composites with different contents of rGO; frequency dependences of (A) real part ε' and (B) imaginary part ε'' of permittivity, and (C) tangent loss tan δ [528].

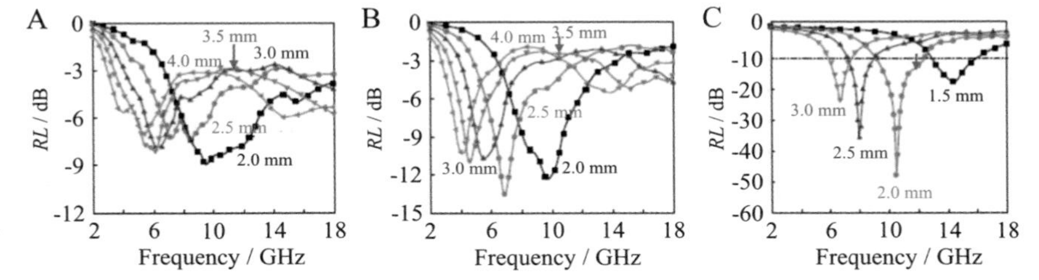

Figure 4.149 Reflection loss *RL* of (A) rGO, (B) PEDOT, and (C) rGO/PEDOT composite as a function of the thickness of the films [529].

with 2 mm thickness. The fibrous morphology of PEDOT on rGO may provide increased active sites for reflection and scattering of EM wave. The composites of rGO with PANI were proposed to achieve enhanced EMI shielding performances in different literatures [530–534]. The composites of PANI/rGO, where PANI nanorod arrays were deposited on the rGO flakes, exhibited enhanced *RL*, −45.1 dB at 12.9 GHz, whereas PANI nanorods (without rGO) gave −29.9 dB at 8.1 GHz [530]. The composites with PANI-covered rGO flakes delivered −41.4 dB at 13.8 GHz [532].

The preparation conditions, including annealing temperature, of rGO foams were found to govern their EMI shielding performances [535]. The foams were prepared by solvothermal treatment of the ethanol solutions of single-layer GO flakes in different concentrations (0.3, 0.6, and 0.9 mg/mL), followed by annealing at 200–800°C in Ar. In Fig. 4.150A, *RL* of the foams prepared from the 0.6 mg/mL solution at different annealing temperatures are shown in the frequency range of 2–18 GHz. With increasing

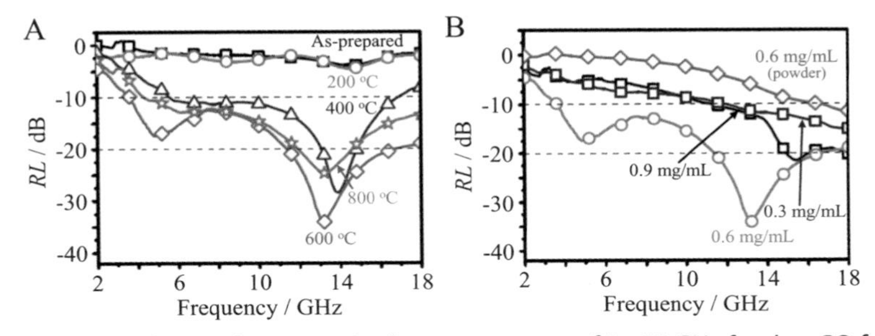

Figure 4.150 Reflection loss *RL* in the frequency range of 2–18 GHz for the rGO foams prepared by solvothermal process: (A) effect of annealing temperature for the foams prepared from 0.6 mg/mL GO concentration and (B) effect of the starting GO concentration for the foams annealed at 600°C in comparison with the powder of rGO obtained by mechanical pulverization of the foam [535].

annealing temperature, *RL* increases, associated with a marked decrease in bulk density; the as-prepared foam having a bulk density of 2.6 mg/cm^3 does almost no shielding, while the 600°C-annealed foam (bulk density of about 1.5 mg/cm^3) has a minimum loss of −34 dB at 13.1 GHz. In Fig. 4.150B, *RL* is shown as the function of the starting GO concentration to prepare rGO foams; the GO concentration of 0.6 mg/mL is optimal in achieving *RL* lower than −20 dB. In addition, the result on the rGO powder in Fig. 4.150B, which is obtained by mechanical pulverization of the foam prepared from 0.6 mg/mL GO and annealed at 600°C, demonstrates that foam morphology is an essential condition for achieving high *RL*.

The foams of rGO/PMMA composites with different rGO contents (0−1.8 vol%) were prepared from the mixture by compression at 0−25°C under 3.5−5.0 MPa CO_2 pressure, followed by rapid release of pressure in hot water at 70°C, of which the bulk density was 0.58−0.79 g/cm^3 [536]. EMI shielding performances were measured on the foam with thickness of about 4 mm, except the foam with 1.8 vol.% rGO (2.4 mm thick). SE_{total} at the frequency range of 8−12 GHz is shown on the foams with different rGO contents in Fig. 4.151A, revealing that the foam with 1.8 vol% rGO delivers SE_{total} greater than 10 dB in the whole frequency range examined. In Fig. 4.151B, *SE* values are plotted against rGO content by dividing the component SE_A and SE_R with SE_{total}, showing the predominant contribution of absorption for EMI shielding.

rGO foam was prepared by solvothermal treatment of ethanol solution of GO at 180°C, followed by freeze-drying after replacing ethanol by water and by annealing at 600°C in Ar [495]. As-prepared rGO foam is compressed by different strains (30%, 60%, and 90%) to obtain rGO foams with

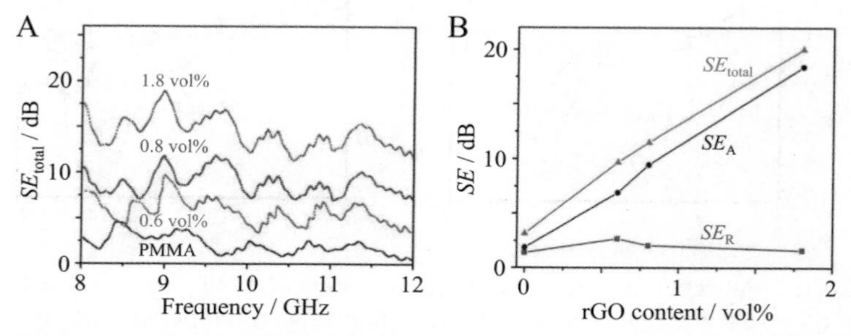

Figure 4.151 Foams of rGO/PMMA composite: (A) SE_{total} of the composites with different rGO contents in the frequency range of 8−12 GHz and (B) dependences of SE_{total}, SE_R, and SE_A at 9 GHz on rGO content in the composites [536].

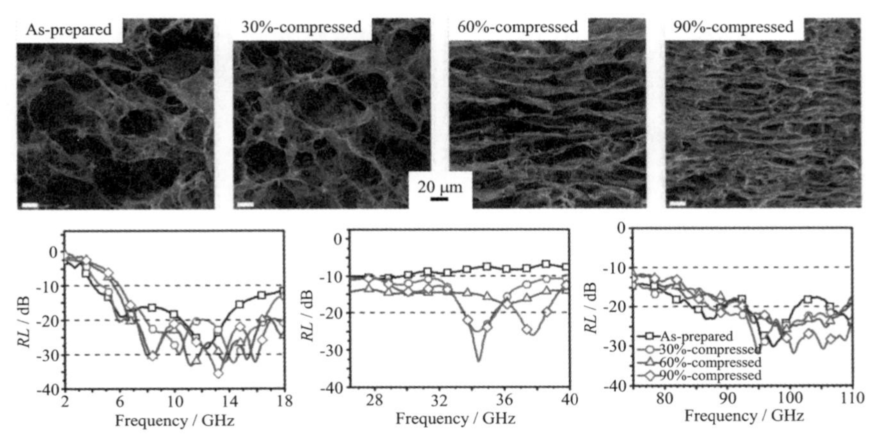

Figure 4.152 SEM images of the cross-sections of graphene foams compressed under different strains and reflection loss *RL* in three ranges of EM frequency [495].

different bulk densities, as shown by the SEM images of their cross-sections in Fig. 4.152. rGO flakes are randomly arranged in the as-prepared foam, and they tend to stack in parallel by making layers perpendicular to the compressing direction with the increase in applied compressive strain, and as a consequence void spaces shrink to flattened cross-sections and the solid matrix becomes denser. *RL* calculated on these foams is plotted against EM frequency in three ranges, 2–18, 26.5–40, and 75–110 GHz, in Fig. 4.152. In the whole frequency range, these rGO foams exhibit relatively high EMI shielding by having lower *RL* than −10 dB in most frequency ranges.

An aqueous suspension of commercial single-layer graphene oxide GO (lateral sizes of 40–50 μm) was spray-dried by a nebulizer using air at 140°C to form flower-shaped GO particles, which could be reduced by hydrazine at 80°C to be rGO [537]. The resultant rGO particles kept a flower-like morphology and consisted of crumpled and folded rGO thin flakes, as shown in Fig. 4.153A and B. EMI shielding characteristics of this flower-like rGO were measured on the composites with a paraffin in different rGO contents up to 12 wt.%. As shown in Fig. 4.153C, *RL* decreases with increasing thickness of the composites above 2 mm and its peak shifts to the low-frequency side with increasing thickness. The minimum *RL* reached −42.9 dB at around 7 GHz on the thickness of 4 mm, of which the efficient absorption bandwidth (EAB < -10 dB) was 5.59 GHz, whereas the rGO prepared from the same GO by thermal reduction at

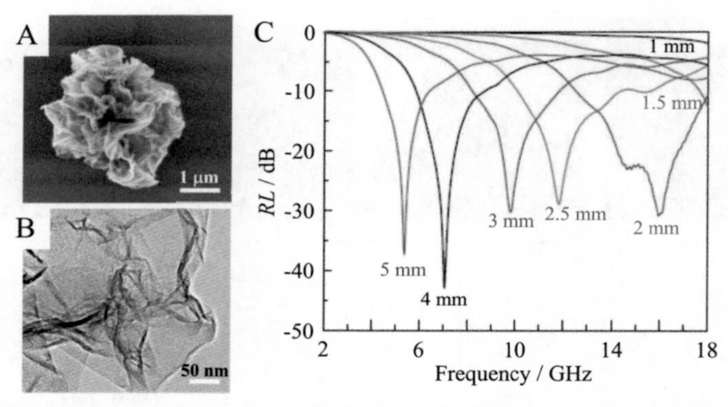

Figure 4.153 Flower-like rGO particles: (A) SEM and (B) TEM images, and (C) reflection loss *RL* for the 10 wt.% rGO/paraffin composites with different thicknesses [537].

1300°C (no flower-like morphology) gave a minimum *RL* of −30 dB and EAB of 4.24 GHz.

MWCNT/rGO composite foams were prepared by solvothermal treatment of an ethanol solution of MWCNTs and GO in different ratios (1/2, 1/3, 1/5, 1/7, 1/9, and 0/1 by weight) at 180°C, followed by freeze-drying after the replacement of ethanol by water, and annealing at 200−800°C in an Ar atmosphere, of which EMI shielding performances were measured in the frequency range of 2−18 GHz [538]. MWCNTs were commercially available ones having outer diameters of 20−40 nm and lengths of 5−15 μm and modified by treating in 8 M HNO_3/H_2SO_4 (1/3 v/v). GO was prepared from natural graphite flakes with lateral sizes above 10 μm. The *RL* of the MWCNT/rGO composites depends greatly on the annealing temperature and becomes smaller than rGO flakes without MWCNTs (denoted as 0/1) after annealing above 600°C, as shown on the composites annealed at 200 and 600°C in Fig. 4.154A and B, respectively. After annealing at 200°C, the composite foams with the ratios of 1/2, 1/3, and 1/5 exhibit clear minimums of about −30 dB at around 4.6−5.0 GHz and about −25 dB at around 10 GHz, although rGO foam without MWCNTs (composite 0/1) shows a shallow minimum of −7.6 dB. After 600°C annealing, however, the frequency giving the *RL* minimum looks likely to shift at high frequency above 18 GHz. The solvothermal treatment during the foaming process was shown to be effective in obtaining an improved shielding performance by comparing with the composite foams prepared by direct freeze-drying of aqueous solutions of MWCNTs and GO.

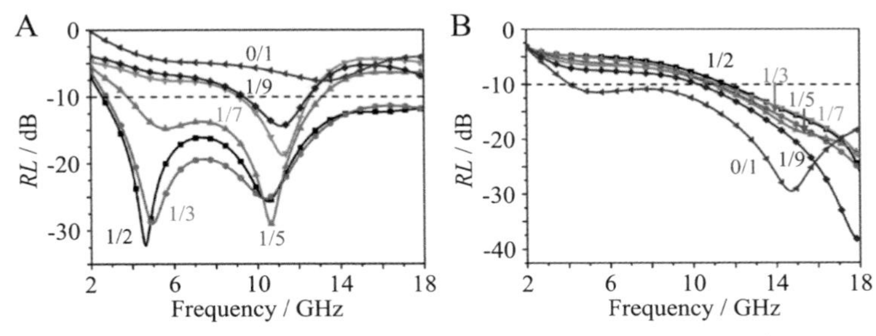

Figure 4.154 Reflection loss *RL* in the frequency range of 2–18 GHz for MWCNT/rGO composites with different MWCNT/GO ratios prepared via solvothermal treatment: (A) annealed at 200°C and (B) at 600°C [538].

The addition of a small amount of rGO to carbon spheres (1/10 by weight) was shown to be effective to improve EMI shielding performances of carbon spheres [539]. To form the composite, carbon spheres synthesized by solvothermal process of glucose were annealed at different temperatures of 500–800°C in N_2 and modified by aminopropyltrimethoxysilane, and then mixed with rGO treated in 1 M HCl in advance. The composites possessed markedly improved *RL*; the composite annealed at 700°C achieved a minimum *RL* of −27 dB at around 14 GHz, whereas the carbon spheres without rGO addition could not give *RL* lower than −10 dB in the frequency range of 4–18 GHz.

4.4.5 Composites with inorganic materials

Various metals and metal oxides were coupled with various carbon materials, mostly rGO, to achieve enhanced EMI shielding performances.

ZnO was often used by wrapping its hollow spheres by rGO flakes [540], depositing as its nanowires on rGO flakes [541], decorating MWCNTs by its nanocrystals [542], decorating rGO by its nanocrystals [543,544], mixing rGO with its tetrapod-like particles [545], loading it onto rGO foams under hydrothermal condition followed by UV irradiation [546]. rGO foams decorated with ZnO nanowires delivered high performance for EMI shielding [541]. rGO foams prepared by freeze-drying of a GO suspension with different GO concentrations and following reduction at 700°C were decorated with in situ grown ZnO nanowires via hydrothermal treatment at 90°C. ZnO nanowires, which were wurtzite-type hexagonal crystals with diameters of about 100 nm and lengths of about 3 μm, were grown vertically and densely on the rGO flakes by making their 002 crystal planes parallel to the surface of rGO flakes. EMI shielding

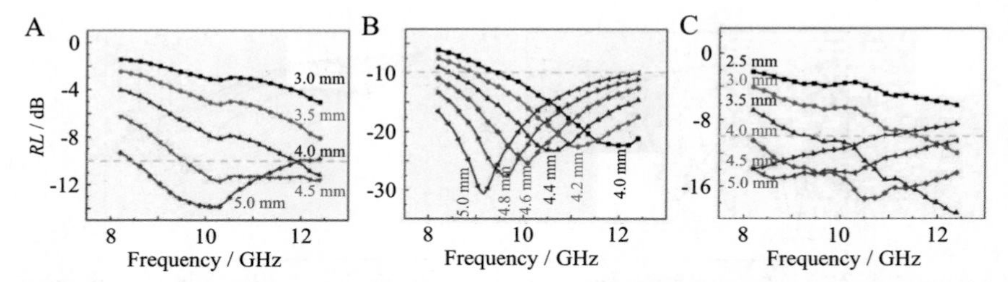

Figure 4.155 Reflection loss *RL* in the frequency range of 8.2–12.4 GHz for the composites of rGO loaded by ZnO nanowires and PDMS with different thicknesses; the composites prepared from the GO dispersions with (A) 0.6, (B) 0.8, and (C) 1.0 mg/mL GO concentrations [541].

performances of these ZnO-loaded rGO foams depend greatly on their preparation conditions, as shown by their composites with poly(-dimethylsiloxane) (PDMS) in Fig. 4.155. The rGO foam synthesized from 0.6 mg/L GO dispersion exhibits poor EM absorption (Fig. 4.155A), as well as that from 1.0 mg/L (Fig. 4.155C). However, the rGO foam from the 0.8 mg/mL GO dispersion exhibits high EMI shielding performances after loading ZnO nanorods, as shown in Fig. 4.155B, a minimum *RL* of −31.1 dB at 9.2 GHz and a wide bandwidth (EAB < −10 dB) covering the whole frequency range of 8.2–12.4 GHz). Uniformly deposited ZnO nanorods on rGO layers may interrupt the agglomeration of rGO flakes, which brings in more interfaces and improves the impedance matching of rGO foam, and as a consequence increases the reflection of the EM wave. rGO wrapped on hollow ZnO spheres delivered marked *RL* of −45 dB at 9.7 GHz [540]. The mixture of rGO with tetrapod-like ZnO (1/2 by weight) delivered a minimum *RL* of −59.5 dB at 14.43 GHz with a bandwidth (EAB < -20 dB) of 8.9 GHz (from 9.1 to 18.0 GHz) [545].

EMI shielding performances of rGO flakes were measured on SiO_2/rGO composites as a function of the temperature ranging from 323 to 473K in the frequency range of 8.5–11.5 GHz [547]. rGO flakes were prepared by a reduction of GO with HI and acetic acid, followed by freeze-drying, which were thought to be single-layered and have a lateral size of 1–2 μm. rGO flakes were mixed with SiO_2 xerogel nanoparticles in THF with different ratios. The shielding efficiency SE_{total} of the composite at a frequency range of 8.5–11.5 GHz is shown in Fig. 4.156 for the composites with different rGO contents by plotting against temperature, revealing marked dependences of SE_{total} on rGO content, EM frequency, and temperature. To achieve SE_{total} greater than 10 dB, the rGO content must

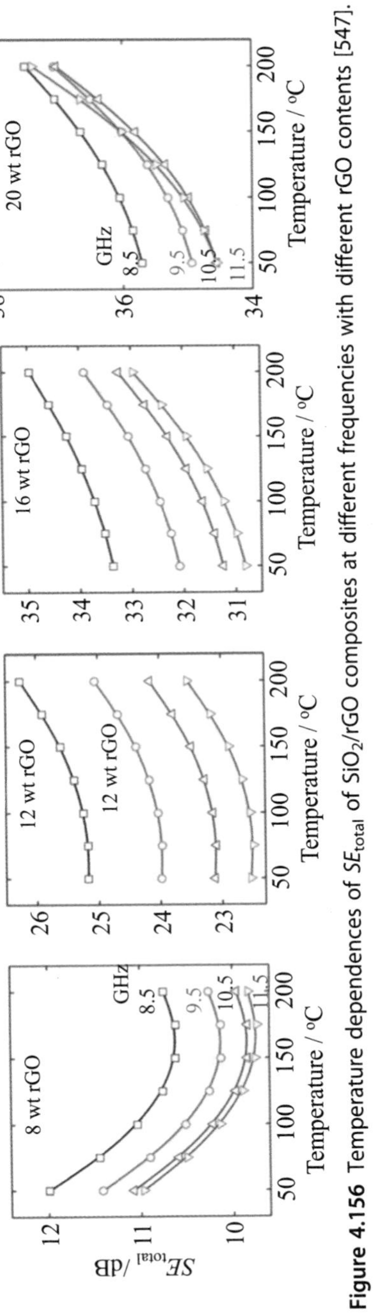

Figure 4.156 Temperature dependences of SE_{total} of SiO_2/rGO composites at different frequencies with different rGO contents [547].

be more than 8 wt.%. SE_{total} reaches about 38 dB at 200°C for the composite with 20 wt.% rGO content. The composites of rGO with SiO_2 textile were prepared by immersing a commercial textile into GO suspension containing hydroquinone, followed by heating at 100°C under a seal [548]. The composites could deliver *RL* minimum of −36 dB and effective absorption in the range of 8.5−11.5 GHz with a low rGO content (4.1 wt.%), as well as light weight (about 1 g/cm^3), strong tensile strength (40 MPa), and high thermal stability beyond 225°C.

ZrO_2-embedded carbon fibers were fabricated by electrospinning of the DMF solution of PAN with ZrO_2, followed by stabilization at 260°C in air and carbonization at 900, 1300, and 2100°C in Ar flow, where ZrO_2 clusters were embedded inside the fibers [549]. The composites could deliver an SE_{total} of 31.8 dB, which is 3.78 times higher than the carbon fibers (without ZrO_2), in which the contribution of absorption was dominant (over 88%). Fe_3O_4-embedded PAN-based carbon fibers were prepared by electrospinning with stabilization at 250°C and carbonization at 700 and 900°C [550]. The resultant composite fibers containing 3 and 5 wt.% Fe_3O_4 (particle size of 20−30 nm) delivered SE_{total} of about 60 and 67 dB in the range of 8.2−12.4 GHz, whereas the fibers without Fe_3O_4 gave 44 dB.

A ternary composite of PEDOT, Co_3O_4, and rGO delivers much enhanced EMI shielding performances compared with binary composites, PEDOT/rGO and Co_3O_4/rGO, as shown by *RLs* in the frequency range of 2−18 GHz in Fig. 4.157 [551]. The composites were obtained as the precipitates from mixtures of GO with 3,4-ethylenedioxythiophene monomer (EDOT) and/or $CoCl_2$ in water. The ternary composite PEDOT/Co_3O_4/rGO with a thickness of 2.0 mm delivers a minimum *RL* of −51.1 dB at 10.7 GHz with a bandwidth (EAB < -10 dB) of 3.1 GHz, with much higher *RLs* for binary composites, PEDOT/rGO and Co_3O_4/rGO.

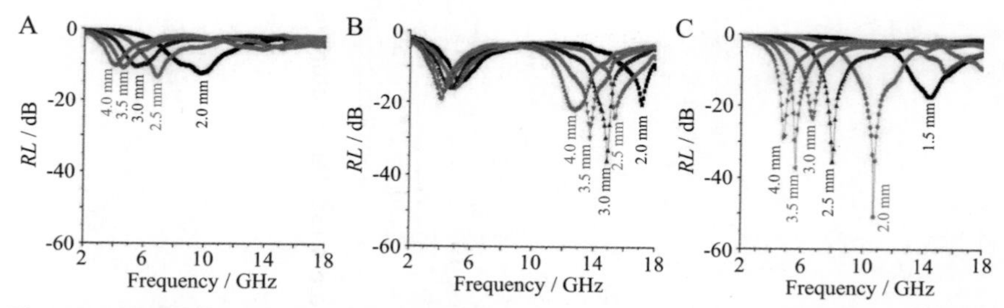

Figure 4.157 Reflection loss *RL* in the frequency range of 2−18 GHz: (A) PEDOT/rGO, (B) Co_3O_4/rGO, and (C) PEDOT/Co_3O_4/rGO composites with different thicknesses [551].

Loading of nanoparticles of ferromagnetic metal (Fe, Co, and Ni) onto CNFs was effective in achieving high EMI shielding performances [552]. The DMF solution of PAN with a metal acetylacetonate was electrospun to nanofibers which were stabilized at 250°C in air and then carbonized at a temperature of 1000–1200°C in Ar. The resultant carbon fibers contained 29.2 wt.% Ni, 29.8 wt.% Co, and Fe 16.4 wt.% Fe with some Fe_3O_4 had average fiber diameters of 475, 345, and 290 nm, respectively. These composites exhibited a well-defined peak of *RL* in the frequency range of 1–18 GHz by making films mixing with silicone (1.3 mm thick); Fe-loaded CNFs delivered a minimum of −67.5 dB at 16.6 GHz. Loading of metal nanoparticles onto MWCNTs was carried out by dispersing purified CNTs in metal–nitrate solutions, followed by reduction in H_2 at 900°C [553]. Fe-loaded MWCNTs exhibited a minimum *RL* of −39 dB at 2.68 GHz.

4.5 Sensing

Carbon-based sensors, which might be familiar as a glucose sensor composed of carbon paste to detect diabetes, were produced industrially by using porous carbons as sensing elements and are used in daily life by many. However, recent marked developments of carbon materials such as carbon nanotubes (CNTs) and graphene-related materials, have promoted rapid developments in materials science and engineering of carbon-based sensors. In most sensors, either adsorption of foreign ions and molecules or application of strain onto the materials are detected as a change in electrical resistivity and so the sensitivity (i.e., how much change is deduced by adsorption or straining) of the sensing materials is the most important parameter. In addition, small size, low weight, and low cost are often required qualities in sensors. Therefore, nano-sized carbon materials, such as CNTs and graphene-related materials, have attracted attention. Numerous reviews have been published on carbon-based sensors in various journals [554–571]. Strain sensors were reviewed by focusing on graphene foams (cellular graphene) [572]. Biological sensors based on graphene have recently been developed because of the biocompatibility of graphene, in addition to its extraordinary properties, such as high electrical conductivity, large surface area, high thermal conductivity, flexibility, etc., which have been reviewed in different journals [555,560,573–575].

In this section, various sensors based on carbon materials are summarized by dividing them into chemical sensors, strain sensors, and biosensors.

Chemical and strain sensors are described by subdividing the sensing carbon materials into porous carbons, such as activated carbons (ACs), templated carbons, etc., fibrous carbons, such as carbon nanotubes (CNTs) and nanofibers (CNFs), and graphene-related materials, including reduced graphene oxides (rGOs). For biosensors, however, the description is based on the materials which are detected by the sensor, such as small biomolecules, glucose, H_2O_2, proteins, etc.

4.5.1 Fundamentals

The resistivity of single-layer graphene prepared by the peeling technique was measured during the adsorption and desorption of strongly diluted NO_2 and NH_3 [576]. Micrometer-sized sensors made from graphene are capable of detecting individual gas molecules that attach to or detach from the graphene surface with a clear change in the local carrier concentration in graphene. As shown in Fig. 4.158A for NO_2 gas, each stepwise change in resistivity observed corresponds to a change in the local carrier concentration in graphene, which is a response to the addition and removal of a single electron during adsorption and desorption, respectively. The gas-induced changes in resistivity have different magnitudes for different gases, and the sign of the resistivity change indicates whether the gas is an electron acceptor (e.g., NO_2, H_2O, iodine) or an electron donor (e.g., NH_3, CO, ethanol), as shown in Fig. 4.158B. These findings gave scientific background to the chemical sensors based on various carbon materials, porous and fibrous carbons as well as graphene-related two-dimensional materials, due to the adsorption and desorption of ions and molecules.

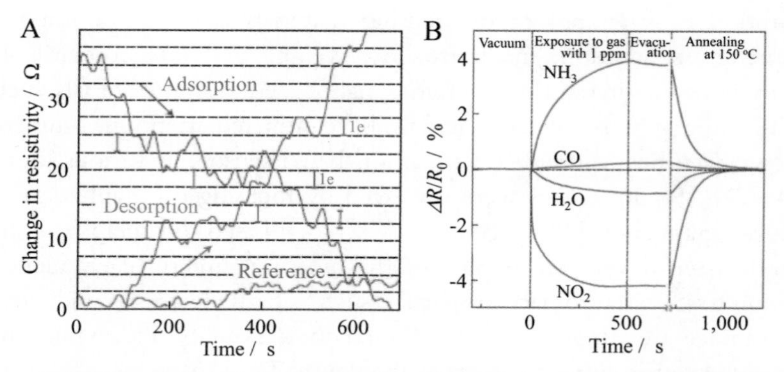

Figure 4.158 Sensitivity of single-layered graphene for gas adsorption/desorption: (A) change in resistivity in graphene with the adsorption and desorption of NO_2 and (B) change in relative resistivity $\Delta R/R_0$ with exposure to different gases, evacuation, and annealing [576].

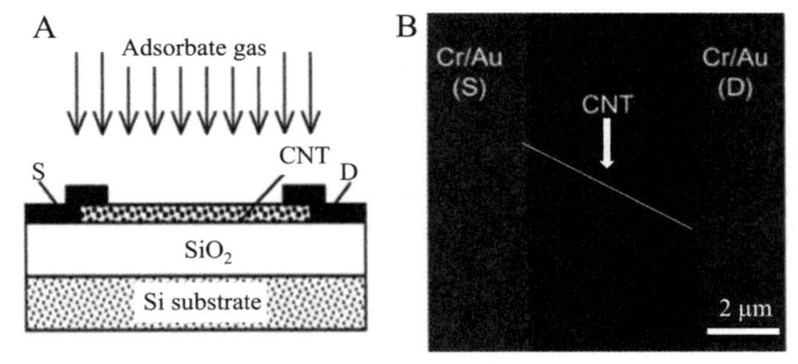

Figure 4.159 Field-effect transistor (FET) using CNTs: (A) schematic structure of the device and (B) SEM image of the device. The CNT in the electrode gap is exaggerated to make its presence clear [577].

Field-effect transistors (FETs) using different CNTs have been constructed, as shown by an illustration of the device and AFM image in Fig. 4.159A and B, respectively, where a single SWCNT bridges between the source and drain electrodes [577]. By using such a device, the sensing performances for NO_2 and NH_3 were studied on SWCNTs (1−3 nm diameter) synthesized by CVD of CH_4 at 1000°C using Fe catalyst [578]. On a single semiconducting SWCNT (s-SWCNT), rapid responses to the introduction of 200 ppm NO_2 and 1% NH_3 have been observed, as shown in Fig. 4.160A and B, respectively. The conductance of s-SWCNT increases substantially with exposure to NO_2, increasing by about 30 times during the initial 60 s but, in contrast, it decreases by about one order of magnitude in the initial 200 s with exposure to NH_3. On five s-SWCNTs, the response times (time for resistance change by one order of magnitude) and the sensitivity (R/R_0) were calculated to be in the ranges of 2−10 s and 100−1000, respectively, for 200 ppm of NO_2 and in the ranges

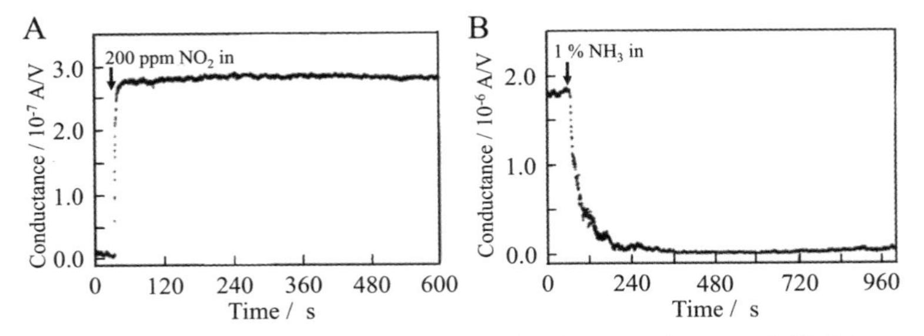

Figure 4.160 Responses to a single s-SWCNT for the introduction of (A) NO_2 and (B) NH_3 [578].

of 1−2 min and 10−100, respectively, for 1% NH_3. The responses for NO_2 and NH_3 molecules became somewhat slower on mats of SWCNTs, although the absolute values of responses were larger than for single SWCNTs. FET using SWCNT exhibited highly reversible and reproducible responses over many cycles of vapor exposure of various alcohols, including methanol, ethanol, 1-propanol, 2-propanol, 1-butanol, tertiary-butanol, 1-pentanol, and 1-octanol [577].

Three types of FET devices, nonpassivated, fully passivated by complete covering of a 70 nm thick SiO layer, and contact-passivated at the contacting parts between s-SWCNTs and leads (Ti/Au) at the drain and source electrodes, have been constructed [579]. The nonpassivated device exhibits large decreases in conductance with short exposures to NH_3, but it recovers only up to one-half after the removal of the gas (Fig. 4.161A). The fully passivated device shows a conductance change corresponding to the short exposures to gas but the response is small (Fig. 4.161B). In contrast, the contact-passivated device exhibits changes nearly as large as for the nonpassivated device and the changes are more reversible, as shown in Fig. 4.161C. Similar experimental results on the contributions from CNTs themselves and their contacts to metal leads have been reported [580]. These experimental results demonstrate that the interaction fo gases with the surface of SWCNTs, rather than with the contacts between CNTs and metallic leads, is responsible for the conductance (resistivity) change with the adsorption and desorption of gas molecules. The device with a single SWCNT and contact-passivation by Al_2O_3, was experimentally demonstrated to be able to detect concentrations of NO_2 as low as 100 ppb and was resettable after a short heat treatment at 110°C [581]. On the networks of CNTs, the junctions where two CNTs were crossing were also pointed out to be important sites for gas sensing, as well as the structural defects in CNT walls and the contacts between CNT and metallic electrodes [582].

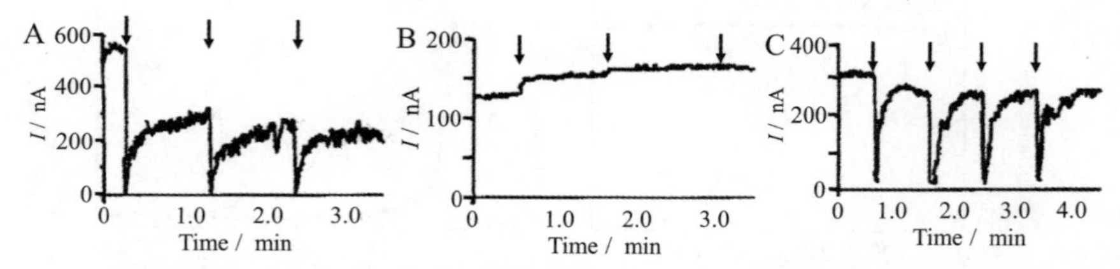

Figure 4.161 Electrical responses of SWCNT field-effect transistors at V_g of −10 V with short exposures to NH_3 (*arrows*): the responses of (A) the nonpassivated, (B) the fully passivated, and (C) the contact-passivated devices [579].

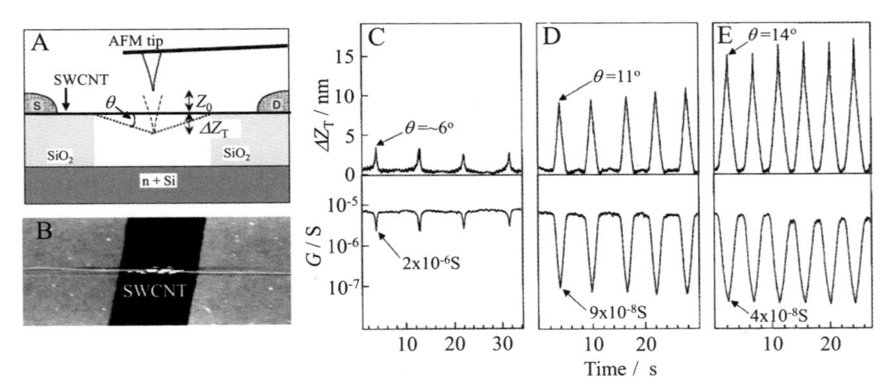

Figure 4.162 Strain sensor using SWCNTs: (A) scheme of the device and measurement, (B) AFM image of a SWCNT, (C)–(E) responses of deflection ΔZ_T to conductance G of SWCNTs at Z_0 of 8, 30, and 65, respectively [583].

A single SWCNT was experimentally shown to respond promptly to mechanical deformation with conductance reduction by two orders of magnitude [583]. The scheme of the device for the measurements and AFM image of the SWCNT are shown in Fig. 4.162A and B, respectively, and the conductance response of the device against repeated reflection ΔZ_T applied by three different tip-tube distances Z_0's of 8, 30, and 65 nm in Fig. 4.168C–E, respectively. The responses in conductance of the SWCNTs is spontaneous and both the mechanical deformation and electrical conductance of the SWCNTs are highly reversible.

4.5.2 Chemical sensors

4.5.2.1 Porous carbons

Sensing performances for various organic vapors, including toluene, benzene, ethyl acetate, methanol, ethanol, 2-propanol, hexane, chloroform, acetone, and tetrahydrofuran, have been studied on the composite films of a carbon black (CB, BP2000), with various organic polymers [584]. Resistance increased markedly during exposure for 15 s and returned to the initial value by discontinuing the vapor flow, as shown in Fig. 4.163A and B on the composite CB/PEVA [poly(ethylene)-co-vinyl acetate] for benzene of 1.1 ppt by volume in air and the composite CB/PVP [poly(*N*-vinylpyrrolidone)] for methanol of 1.5 ppt in air, respectively. As can be seen from Fig. 4.169, the baseline resistance value drifts by less than 0.02% for the CB/PEVA composite and less than 0.15% for the CB/PVP composite over 20 min. Resistivity changes for 60 s exposure to methanol

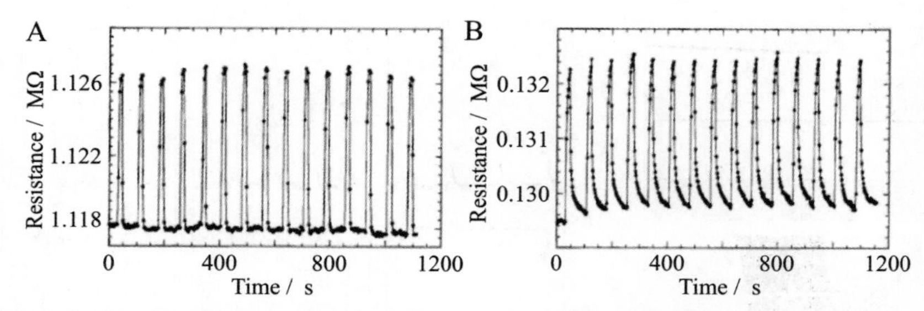

Figure 4.163 Changes in resistance of a CB/polymer composite sensor: (A) CB/PEVA composite for benzene (1.1 ppt) and (B) CB/PVP composite for methanol (1.5 ppt) [584].

(23.0 ppt), ethyl acetate (16.6 ppt), and benzene (17.1 ppt) were compared for 17 composites prepared from CB with different polymers, and the response of the sensors depended greatly on the polymer used.

A commercial activated carbon BAX-1500 (wood-based and H_3PO_4-activated) deposited on Au nanowires exhibited high sensitivity for methanol, chloroform, and toluene [585]. BAX-1500 has S_{BET} of 1903 m^2/g and V_{total} of 1.275 cm^3/g with V_{micro} of 0.543 cm^3/g and contained 8.7 at.% O. The cycling performances of the sensor for methanol and toluene are shown Fig. 4.164A and B, respectively. The responses for methanol, chloroform, and toluene are shown as a function of their concentration in Fig. 4.164C, demonstrating stable cyclic performance and linear relation between response $\Delta R/R_0$ and concentration for three VOC gases. The uptakes for methanol and chloroform were measured as 1.66 and 1 g/g, respectively, suggesting that some pores of BAX might be inaccessible to the larger chloroform molecule.

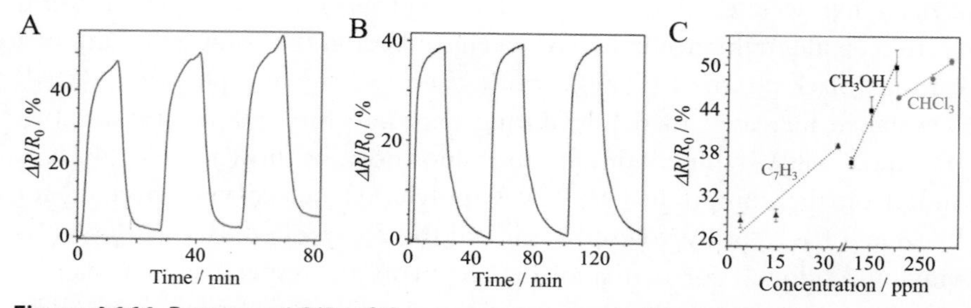

Figure 4.164 Response $\Delta R/R_0$ of the sensor composed of BAX coated on the Au chip: responses (A) for methanol (200 ppm) and (B) for toluene (34 ppm), and (C) its relation to the concentration of each adsorbate [585].

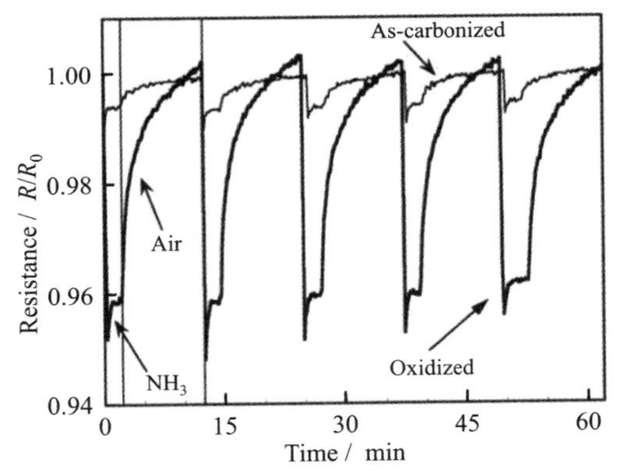

Figure 4.165 Resistance change with the repetition of exposure to NH_3 and purging with air for the porous carbon derived from poly(sodium 4-styrene sulfonate) before and after oxidation at 350°C [586].

The sensitivity for NH_3 of a porous carbon prepared from poly(sodium 4-styrene sulfonate) at 800°C was experimentally shown to be markedly improved by oxidation at 350°C in air [586]. As shown in Fig. 4.165, the response R/R_0 becomes much larger with oxidation. The response improvement by oxidation was thought to be due to the increase in S_{BET} and V_{micro}, from 330 to 756 m^2/g and from 0.107 to 0.275 cm^3/g, respectively. On the pristine carbon, complete recovery in the resistance by purging air could not be achieved in the initial few cycles (irreversible resistance changes), probably because of the reaction between NH_3 and the residual sulfonic groups on the carbon. On the oxidized carbon, however, complete recovery in resistance was achieved even at the first cycle due to the elimination of sulfonic groups by oxidation. Porous carbons were prepared from poly(4-ammonium styrene-sulfonic acid) and from a 1:1 mixture of poly(4-ammonium styrene-sulfonic acid) and poly(4-styrene-sulfonic acid-co-maleic acid) sodium salt at 800°C in N_2, followed by oxidation at 300°C in air [587]; the former carbon was simultaneously doped by S and N while the latter carbon was doped by S. The resultant carbons exhibit clear responses for different concentrations of NH_3 at room temperature, as shown in Fig. 4.166A. The sensitivity for NH_3 is quite different between two carbons, as shown by the dependences on NH_3 concentration in Fig. 4.166B, whereas pore structure parameters are not so different, S_{BET} and V_{total} are 727 m^2/g and 0.363 cm^3/g for S,N-codoped carbon and 847 m^2/g and 0.458 cm^3 g for the S-doped one, respectively, suggesting an important contribution of the dopants to sensing

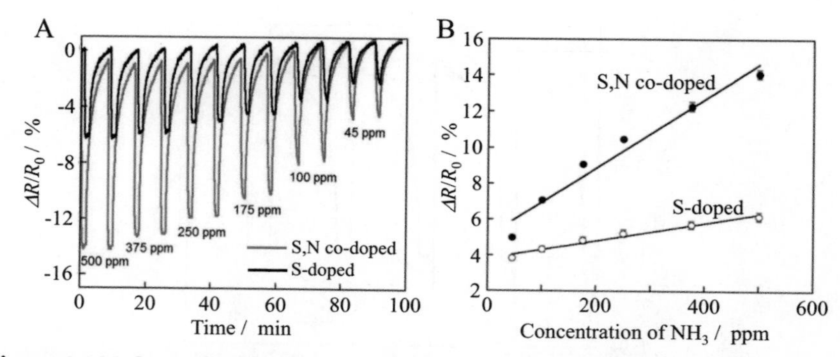

Figure 4.166 S,N-codoped and S-doped porous carbons: (A) responses for NH_3 with different concentrations, and (B) dependences of the sensitivity $\Delta R/R_0$ on NH_3 concentration [587].

performances. Sensitivity of $\Delta R/R_0$ of 14.1% for the S,N-codoped carbon is higher than that for the S-doped one.

Hollow carbon spheres were synthesized by CVD of toluene vapor at 900°C on the SiO_2 spheres with an average diameter of 210 nm, followed by dissolution of SiO_2 with HF. The resultant hollow carbon spheres exhibited high sensitivity to NH_3 after annealing at 300°C in air [588]. The response R/R_0 and sensibility $\Delta R/R_0$ for 74 ppm NH_3 at ambient humidity were 196% and 1.6, respectively, for the annealed spheres, although they were 6% and 0.08 for the pristine ones, respectively. The annealed spheres demonstrated a high sensitivity to NH_3 concentrations of 74–295 ppm over a broad range of relative humidity (10%–97%).

Mesoporous carbon nitride with a chemical composition of about $C_{2.6}N$ (*p*-type semiconductor) prepared from porous carbon nitride g-C_3N_4 with glucose at 700–900°C exhibited high sensitivity and selectivity for NO_2 at room temperature [589]. The composite of *g*-C_3N_4 with a CB (Ketjenblack) supported on a glassy carbon (GC) plate exhibited higher sensitivity for hydrazine, the sensitivity of 2.5 μA/μM and limit of detection (LOD) of 0.7 μM with linear response in the concentration range of 2–14 μM, was higher than for each component of the sensor, i.e., *g*-C_3N_4, Ketjenblack, and GC plate [590]. The composite was prepared by heating the mixture of melamine with the CB via dispersion in isopropyl alcohol at 550°C in N_2.

Microporous carbons synthesized from dried mango leaves by mixing with KOH and heating at 700–900°C in N_2 delivered high sensitivity for heavy metal ions, sensitivity of 0.025 mA/μM·cm^2, and LOD of 79.1 nM for Cu^{2+} [591].

4.5.2.2 Carbon nanotubes and nanofibers

4.5.2.2.1 Inorganic gases

CNT films (approximately 200-nm thick) were synthesized by plasma-enhanced CVD of CH_4 on Si_3N_4/Si substrates at 650°C, for which the sensitivity to NO_2 was studied under dynamic adsorption–desorption cycles (cyclic exposure to dry air, dry air containing 10–100 ppb NO_2, and dry air) at temperatures of 25–250°C [592,593]. In Fig. 4.167A and B, the resistance responses of the CNT thin film are shown at different temperatures to be exposed to NO_2 with 100 ppb concentration and at 165°C to be exposed to different concentrations of 10–100 ppb NO_2, respectively. The CNT film is sensitive to NO_2 at concentrations as low as 10 ppb at 165°C. When a high concentration of NO_2 is used, the baseline resistance (i.e., the resistance in dry air) is highly reproducible and stable (Fig. 4.167B). Thermal annealing of the CNT film at 250–320°C was effective in improving the sensitivity to NO_2.

Fluorination of MWCNTs with 2–15 nm inner diameter was reported to be effective in enhancing the sensitivity for NO [594]. Fluorination was performed at different temperatures of 100–1000°C for 10 min in F_2 gas under the pressure of 1 bar. The sensitivity $\Delta R/R_0$ for NO depends on the fluorination temperature; the CNTs fluorinated at a temperature below 400°C show n-type behavior, whereas the pristine CNTs and the CNTs fluorinated above 600°C behave as p-type, as shown in Fig. 4.168A. The 600°C-fluorinated CNTs exhibit an abrupt resistance change of about −15% but the 200°C-fluorinated ones show a slow increase in $\Delta R/R_0$ to about 20% with good cyclability, as shown in Fig. 4.168B. This change in sensing characteristics with fluorination temperature was discussed on the basis of the state analyses of fluorinated functional groups. UV irradiation

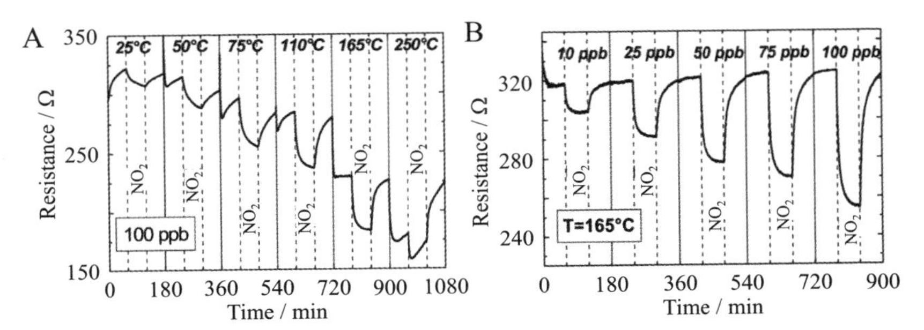

Figure 4.167 Resistance responses of CNT thin film to NO_2: (A) at different temperatures to be exposed to 100 ppb NO_2 and (B) at 165°C to different concentrations of NO_2 [593].

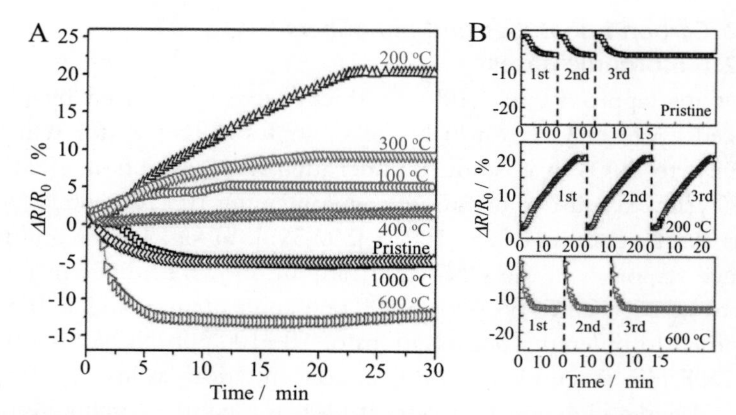

Figure 4.168 CNTs fluorinated at different temperatures: (A) resistivity changes with NO exposure time, and (B) cyclabilities for the pristine, 200°C-, and 600°C-fluorinated CNTs [594].

improved the sensing performances of CNT-based devices for NO, NO_2, and NH_3 [595]. UV irradiation during air flushing not only reduces the recovery time dramatically down to a few seconds but also enhances the response for NO from about 7% to 36%. LOD was improved from 27.1 ppm to 229 ppb before and after UV irradiation and LOD under UV irradiation reached 1.51 and 27.8 ppt for NO_2 and NH_3, respectively. Recovery of the single SWCNT sensor after exposure to NO_2 (adsorption of NO_2) was performed by self-heating in air to accelerate gas desorption [596]. Self-heating of the CNTs in the FET device was possible by increasing the bias pulse (V_{sd}) from 0.1 to 1.3 V, the power for this process being extremely low at 29 μW. The recovery process was repeatable and the sensor showed similar responses during repeated exposure to NO_2 and desorption (recovery).

The sensitivity of CNT-based sensors depends also on the forming process of the CNTs network. MWCNTs synthesized by CVD (diameters of 10−40 nm and lengths of 1−25 μm) were deposited on a metallic interdigitated electrode either by drop casting of their toluene dispersion or by self-assembled film formed at the toluene−water interface [597]. The self-assembled MWCNT film was more homogeneous and had better adhesion to the electrode (substrate), and consequently exhibited higher sensitivity for O_2 than the cast film, with 3.4 times higher sensitivity and a fast recovery time of 0.9 s at 160°C. Before drop casting of the SWCNT dispersion onto an interdigitated electrode, their debundling was shown to be important in achieving a high sensitivity for NH_3 using commercial SWCNTs [598]. Debundling was performed in a vortex mixer for a long time of 50 h in a solvent (DMF) and the subsequent sonication in DMF promoted CNT dispersion. The enhancement in sensitivity by debundling became marked with increasing NH_3 concentration, the sensitivity increasing from 8.4% for as-received SWCNTs to 27.7% for the debundled ones in 500 ppm NH_3.

An SWCNT sensor was fabricated by drop casting of an SWCNT aqueous dispersion onto a flexible plastic substrate [599]. Loading of indium-tin oxide (ITO) nanoparticles (average diameter of <50 nm) onto SWCNT bundles was also performed by dispersing ITO with SWCNTs before coating. The sensors showed high sensitivity to water vapor (humidity) and temperature. Responses for a momentary increase in relative humidity (*R*) from 50% to 60% are shown for SWCNT and ITO-loaded SWCNT sensors in Fig. 4.169. The sensor for the ITO-loaded SWCNTs was also sensitive to NH_3, of which the LOD reached 13 ppb.

Pd-loaded CNFs prepared by electrospinning showed high sensitivity for H_2 [600,601]. CNFs prepared from a pitch mixed with MWCNTs by electrospinning, followed by stabilization at 290°C in air, carbonization at 1650°C, and activation with KOH at 750°C exhibited high sensitivity for NO, 16% for 50 ppm NO in N_2 [602]. The sensor exhibited high reproducibility for adsorption–desorption cycles.

4.5.2.2.2 Organic gases and vapors

FET devices composed of random network of SWCNTs were fabricated and their performances as the sensors for various organic vapors were studied [603,604]. SWCNTs with diameters of about 1 nm were grown by CVD on a 250 nm thick SiO_2 layer on Si substrate and Pd electrode was deposited on the top of the CNTs network using photolithography, as

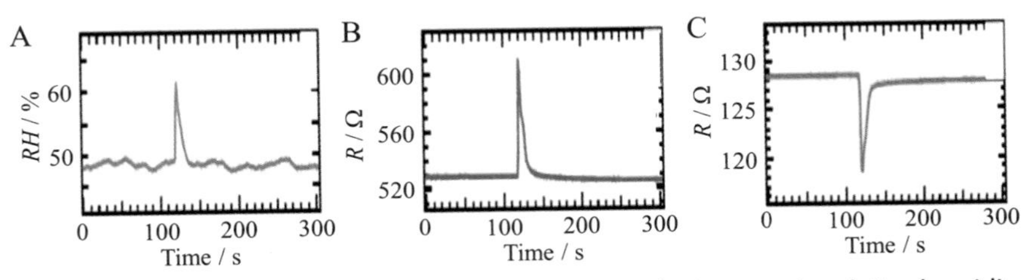

Figure 4.169 Responses of SWCNT-based sensors: (A) the increase in relative humidity (RH) due to human breath and the responses (R) of (B) the SWCNT sensor and (C) ITO-loaded SWCNT sensor [599].

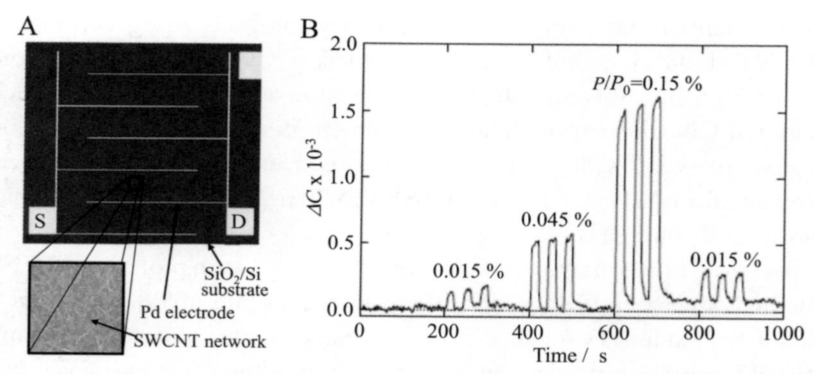

Figure 4.170 SWCNT random-network sensor (capacitor): (A) scheme of the capacitor with AFM image of the network and (B) sensing performance for DMF vapor with different concentrations [604].

shown schematically in Fig. 4.170A. Capacitance changes ΔC due to the repeated exposure of dimethylformamide (DMF) vapor with different concentrations for 20 s are shown in Fig. 4.170B, revealing that the responses are rapid, proportional to DMF concentration, and completely reversible [604]. These SWCNT random-network capacitors were possible to be prepared on flexible organic substrates, such as polyimide sheet [605]. On these capacitors, a change in conductance ΔG due to exposure to a vapor could be measured simultaneously with ΔC, the ratio of these responses $\Delta G/\Delta C$ being a concentration-independent intrinsic property of the vapor [606]. The responses ΔG and ΔC for diluted vapors of acetone [$(CH_3)_2CO$] and NH_3 are shown in Fig. 4.171A and B, respectively. Acetone gives a large capacitance change ΔC, with a very small conductance change ΔG (Fig. 4.171A), and as a consequence low $\Delta G/\Delta C$

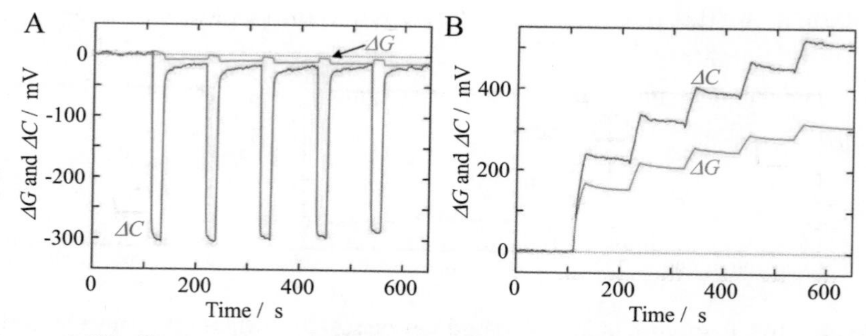

Figure 4.171 Changes in capacitance (ΔC) and conductance (ΔG) of the SWCNT random-network sensor (capacitor) with 20 s exposure to (A) acetone ($0.01P_0$) and (B) NH_3 ($0.0002P_0$) [606].

(−0.03), in contrast, NH_3 gives ΔG comparable to ΔC (Fig. 4.171B), and as a consequence $\Delta G/\Delta C$ is close to 1.

Metal oxide nanoparticles, ZnO and SnO_2, were loaded onto carbon nanofibers via electrospinning of the mixtures composed of the core of PAN and the shell of PVP containing various amounts of either zinc acetate $Zn(Ac)_2$, $SnCl_4$, or both, followed by stabilization at 400°C and carbonization at 800°C in N_2 [607]. Loading of these metal oxides was effective in enhancing the sensitivity for dimethyl methylphosphonate (DMMP, a warfare chemical) vapor, as shown in Fig. 4.172A, and the responses of the sensors increase linearly with increasing concentration of DMMP in a wide range of 0.1−1000 ppb, as shown in Fig. 4.172B. The sensor composed of CNF mats loaded by both ZnO and SnO_2 delivered the highest sensitivity in the whole concentration range of DMMP and LOD of less than 0.1 ppb. The same fabrication procedure was applied for the synthesis of WO_3-loaded CNFs, which showed high sensitivity for NO_2 with a LOD of 1 ppm at room temperature [608]. CNF-embedded Fe nanoparticles were prepared by electrospinning of the DMF solution of PAN and Fe(acetylacetonate)$_3$, followed by stabilization at 280°C in air, carbonization at 800 and 1000°C, and dissolution of Fe particles on the surface of the CNFs, with the resultant Fe-loaded CNFs exhibiting quick response and recovery for ethanol vapor [609]. Sn/SnO_2-loaded CNFs, which were prepared by electrospinning of DMF solution of PAN and $SnCl_4$ followed by partial oxidation of Sn metal on the surface of CNFs after carbonization at 800°C, delivered high sensitivity for ethanol vapor (1000 ppm) at a temperature range of 160−240°C, which is much higher than for Sn- and SnO_2-loaded CNFs [610].

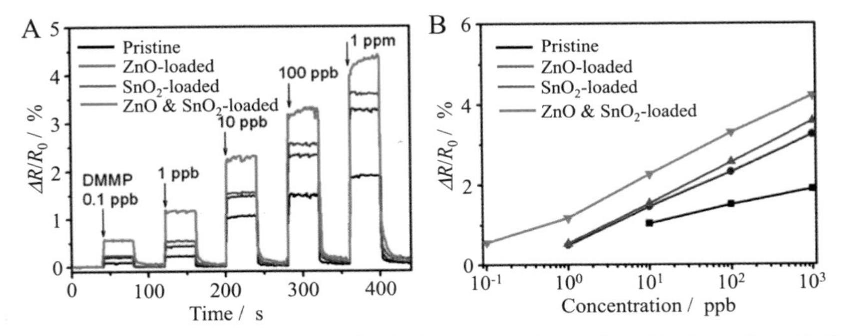

Figure 4.172 Electrospun CNFs loaded by metal oxides, ZnO and/or SnO_2: (A) responses of $\Delta R/R_0$ with different DMMP concentrations and (B) dependence of the response $\Delta R/R_0$ on DMMP concentration [607].

N-doped CNFs prepared from electrospun PAN nanofibers by stabilization at 300°C in N_2 and carbonization at 900°C in N_2 were shown to be electrochemically active for dihydroxybenzene isomers, i.e., hydroquinone (HQ), catechol (CC), and resorcinol (RC), with much higher sensitivity than CNFs prepared using the same procedure with stabilization in air [611]. Webs of N-doped CNFs were directly adhered to a glassy carbon electrode using 0.5 μL of chitosan solution (0.05 wt.%). The CV curve for each isomer showed a sharp oxidation peak at a specific potential, at about 136, 240, and 640 mV for HQ, CC, and RC, respectively. The current at this oxidation peak (response) increases linearly with increasing concentration of the isomer from 1 to 400 μM, as shown in Fig. 4.173A. The LODs for HQ, CC, and RC were estimated to be 0.3, 0.4, and 0.8 μM, respectively. By using N-doped CNFs as the electrode, three isomers can be detected with sufficient resolution in their mixture, as shown by the DPV curve for the mixture in Fig. 4.173B; the detection of each isomer in their mixture is much clearer using N-doped CNFs than other electrodes of CNF prepared via stabilization in air, chitosan, and glassy carbon. The electrode of electrospun CNFs cast on the mixture of graphite powder and mineral oil (7/3 by weight) packed into a pipette tube could detect CC and HQ separately with LODs of 0.2 and 0.4 μM, respectively [612]. The electrode composed from the mixture of CNFs with Nafion (sulfonated tetrafluoroethylene-based fluoropolymer-copolymer) and laccase cast onto GC substrate delivered a sensitivity of 41 μA/mM, LOD of 0.63 μM, linear range of 1−1310 μM, and response time of less than 2 s for CC [613].

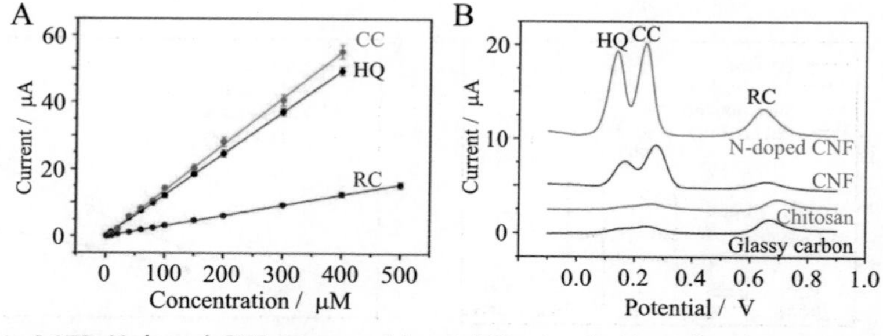

Figure 4.173 N-doped CNFs prepared by stabilization in N_2 and carbonization in N_2: (A) dependence of response (current) on the concentration of each isomer and (B) differential pulse voltammetry curves for the mixture of HQ, CC, and RC (50, 50, and 100 mM, respectively) in 0.1 M phosphate-buffered saline (PBS) measured by different electrodes, including N-doped CNFs [611].

4.5.2.3 *Graphene and related materials*

4.5.2.3.1 Inorganic gases

Patterning of single-layer graphene grown by CVD of ethanol onto a nanomesh by coupling with lithography and reactive ion etching resulted in enhancement of the sensitivity for NO_2 and NH_3 at room temperature [614]. Ozone (O_3) treatment of graphene was reported to be very effective in improving sensing performances, such as percentage response, LOD, and response time for NO_2 [615]. Single-layer graphene grown on Cu foil by CVD of CH_4 at 1000°C was transferred to the SiO_2/Si substrate using PMMA, as illustrated in the Raman spectrum in Fig. 4.174A, proving the formation of single-layer graphene, and then treated with O_3 for 60−90 s. With increasing exposure time for O_3, I_G/I_D decreases and resistance of the film increases, as shown in Fig. 4−174B, suggesting the introduction of structural defects in graphene film by O_3 exposure. Sensitivity for NO_2 of the sensor is improved markedly by increasing the O_3 exposure time up to

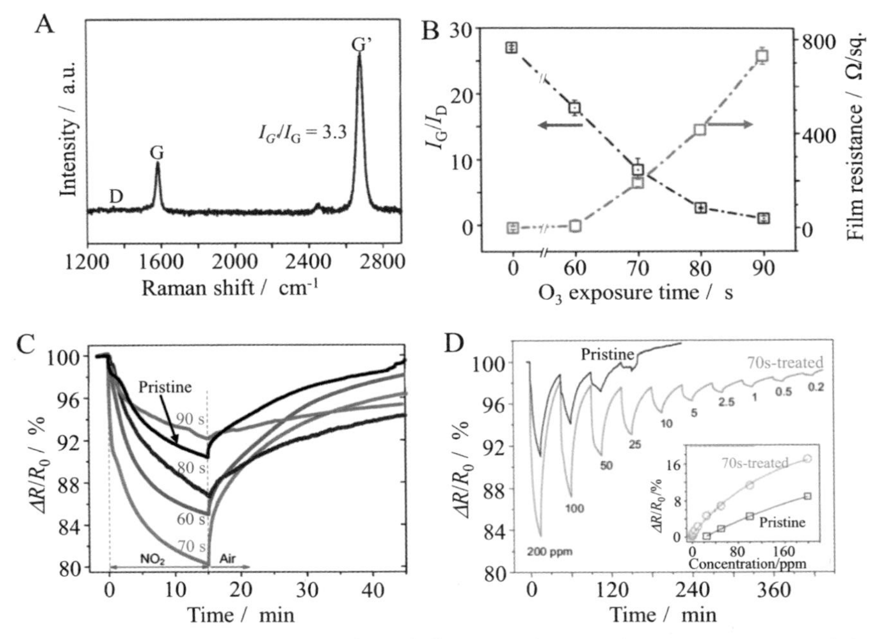

Figure 4.174 Ozone treatment of single-layer graphene: (A) Raman spectrum of the pristine graphene, (B) changes in I_G/I_D and film resistance with O_3 exposure, (C) response ($\Delta R/R_0$) for NO_2 exposure as a function of O_3 treatment time, and (D) change in response for the different concentrations of NO_2 of the pristine graphene and 70 s O_3-exposed one [615].

70 s and then starts to decrease, as shown in Fig. 4.174C. The graphene sensor after 70 s O_3 exposure delivered a much lower LOD of 0.2 ppm than the pristine graphene, as shown in Fig. 4.174D. B-doped single-layer graphene synthesized by CVD with 1 sccm Ar stream bubbled in a hexane solution of 0.5 M triethylborane (TEB), followed by heat treatment at 1000°C for 5 min exhibited high sensitivity for NO_2 and NH_3, with LODs of 95 and 60 ppt, respectively, although the graphene synthesized without TEB had LODs of 2.6 and 6.3 ppb, respectively [616]. Adsorption processes of various vapors, including NH_3 and NO_2, onto graphene and rGO have been discussed theoretically [617–620], demonstrating that the interaction of functional groups with the gas molecules is important, in addition to physisorption of gas molecules, particularly on rGO.

A single-layer graphene film prepared on SiO_2 substrate at 650°C by inductively coupled plasma CVD was patterned by laser interference lithography to the periodic array of graphene ribbons with a width of 200 nm at a pitch of 1 μm [621]. The sensing performance of the array for H_2 was studied after the deposition of Pd metal layer in 2 nm thickness. The array composed of 1000 graphene ribbons after Pd decoration exhibits an instant response to H_2, associated with quick and complete recovery in N_2 atmosphere, although the pristine film (before Pd decoration) shows slower responses to H_2 on/off, as shown in Fig. 4.175. In the Pd-decorated array, the graphene ribbon works as a support for homogeneous deposition of Pd nanoparticles and an electrically conductive path with less electrical noises. The resistance change $\Delta R/R_0$ (sensitivity) of the array (Fig. 4.175A) is much smaller than the pristine film (Fig. 4.175B).

FET based on graphene, which was epitaxially grown on 6H–SiC substrate, exhibited high sensitivity for pH change because both hydroxyl

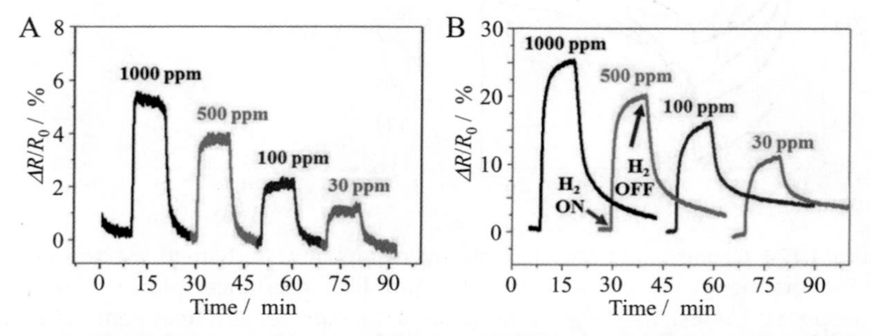

Figure 4.175 Hydrogen-sensing performances of (A) the Pd-decorated single-layer graphene ribbon array and (B) the pristine graphene film (without Pd) [621].

(OH^-) and hydroxonium (H_3O^+) ions were able to modulate the channel conductance by doping "holes" and "electrons," respectively [622]. The application of a negative gate potential, which induces the accumulation of OH^- ions on the surface, produces a larger increase in conductivity compared to the positive gate potential (accumulation of H_3O^+). The relation between threshold voltage and pH was approximated to be linear, with a slope of 98–99 mV/pH unit.

rGO nanoflakes contain many functional groups and defects, and so can offer great potential for practical development in rGO-based sensors because rGO is easy and cheap to produce on a large scale, readily functionalized, and able to tune its band-gap, in comparison with graphene nanosheets synthesized by the CVD process which is not reproducible on a large scale, and has no functional groups or zero band-gap [557]. By using an inkjet printing technique, a thin film of rGO was fabricated onto PET substrate with the reduction of GO with ascorbic acid, which exhibited high sensing ability for different gas molecules [623]. In Fig. 4.176, the response ($\Delta R/R_0$) for Cl_2 with different concentrations is shown, together with the appearance of a four-probe sensor and the responses for various vapors. The response for acetone vapor of rGO thin films was shown to depend on the degree of reduction with hydrazine hydrate vapor at 100°C [624]. On this rGO sensor, the accumulation of acetone vapor was observed during repetitive exposure to the vapor, which needed heating for complete desorption, whereas the sensor composed of SWCNTs showed almost no accumulation. The LOD of the rGO sensor was similar to that of the SWCNT sensor for warfare chemicals, chloroethylethyl sulfide (CEES) and DMMP, and an explosive (dinitrotoluene DNT). For HCN gas, however, the LOD of this sensor was as low as 70 ppb but that of the SWCNT sensor was more than 4000 ppb. The sensor based on rGO

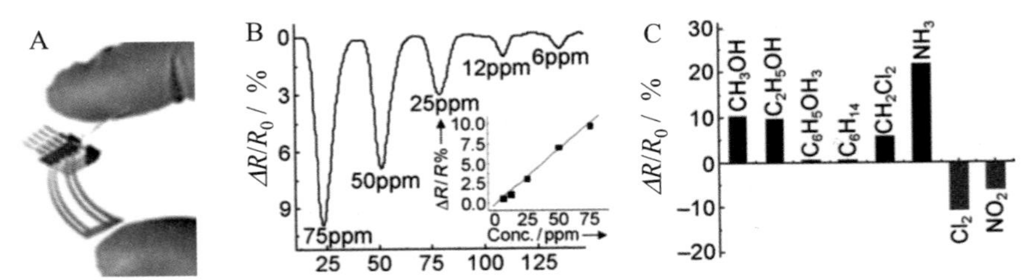

Figure 4.176 Inkjet-printed rGO/PET sensor: (A) a photograph of a four-probe device, (B) the response ($\Delta R/R_0$) to various concentrations of Cl_2 with the inset of vapor concentration dependence, and (C) the responses to various vapors [623].

reduced with p-phenylenediamine (PPD) could achieve high responses to DMMP with 30 ppm concentrations, which was much better than the sensor based on hydrazine-reduced rGO [625]. rGO reduced with pyrrole gave a detectable response even for the 1 ppb NH_3 gas [626].

The array of films with 10 μm width in 10 μm gap prepared from hydrazine-reduced rGO, of which flakes were mostly composed of single-layer flakes, by spin coating on SiO_2/Si substrate was studied for its sensing performance [627]. The sensing response was evaluated by a four-point resistance measurement. The performance of the array for 5 ppm NO_2 in dry N_2 is shown in Fig. 4.177, revealing the trend from the larger response with a longer response time at room temperature to the smaller response with a shorter response time at higher temperature. At 149°C, the sensor responds to 5 ppm NO_2 quickly and recovers the initial resistance quickly, but at low temperatures below 124°C the recovery cannot complete within the time span employed.

An rGO film deposited on Si substrate from the suspension of mostly single-layer flakes was settled in an FET device, bridging the source- and drain-electrodes of Au as the conducting channel [628]. The sensor exhibited an instantaneous decrease in resistance by flowing NH_3 and fast recovery in resistance by stopping NH_3 flow under a positive gate potential (40−10 V), far superior to the performance at zero and negative gate potentials. The film of rGO coated by porous conducting polymer PEDOT (porous-PEDOT/rGO composite) exhibited a high sensing performance for NH_3 gas [531]. As shown in Fig. 4.178, the sensor based on the

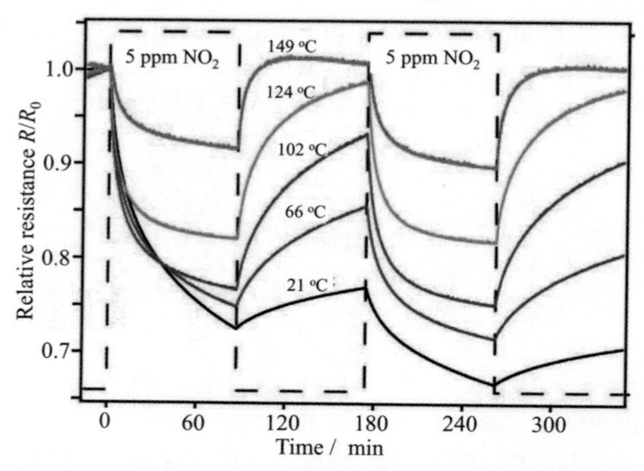

Figure 4.177 Sensing performance of the array of rGO film for NO_2 at different temperatures [627].

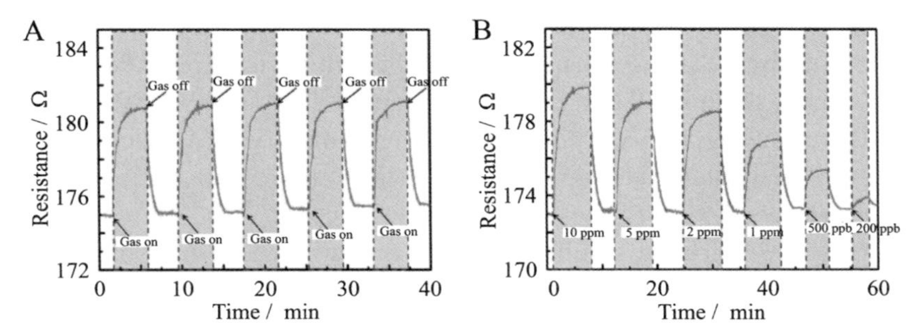

Figure 4.178 Porous-PEDOT/rGO composite film: (A) repeated sensing and recovery for 5 ppm NH_3 gas, and (B) sensitivities to different concentrations of NH_3 gas [629].

composite exhibits a high response, even after repeated cycles of exposure/recovery, and also a fast response and recovery performance even at the ppb level of NH_3 gas. This composite delivered high sensing performance for different gases, such as H_2S, SO_2, CH_2Cl_2, and CH_3OH, and was much better than rGO film without porous-PEDOT coating. The composites of rGO with poly(sodium 4-styrenesulfonate) (PSS) and poly(allylamine hydrochloride) (PAH), which were prepared from the mixed solutions of GO with either PSS or PAH by hydrogen reduction at 100°C, exhibited high sensitivity for NO_2 gas with concentrations of 1−15 ppm [630].

Surface acoustic wave (SAW) transducers composed from the hydrazine-reduced rGO nanosheets deposited on a 36 degrees YX $LiTaO_3$ surface were prepared, of which the gas sensing performance was assessed using H_2 and CO in air at 25 and 40°C [631]. The response of the composite is shown in Fig. 4.179, which is defined as the variation in operating frequency caused by the interaction with the gas. The responses for 1% H_2 and 1000 ppm CO are approximately 5.8 and 8.5 kHz at 25°C, but 1.7 and

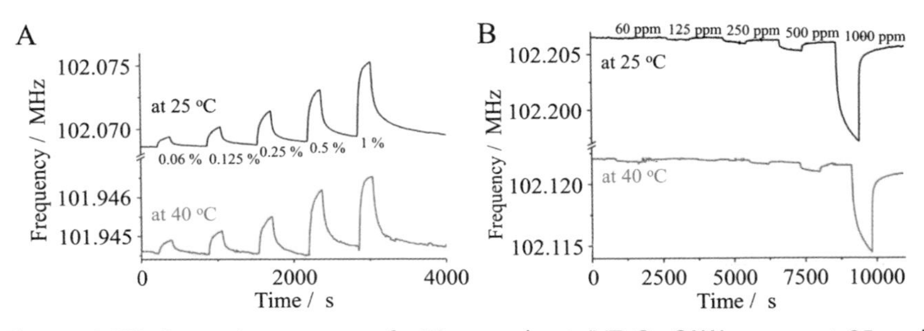

Figure 4.179 Dynamic response of rGO nanosheets/$LiTaO_3$ SAW sensor at 25 and 40°C: (A) H_2 and (B) CO with different concentrations [631].

7.0 kHz at 40°C, respectively. For the low concentration CO (60 and 125 ppm), the frequency change is almost insignificant. The 90% response times were 116 and 400 s at 25°C, but 12 and 300 s at 40°C for 0.125% H_2 and 250 ppm CO, respectively. The recovery times for CO were almost 30 and 20 min, while for H_2 they were less than 10 and 1 min at 25 and 40°C, respectively.

Metal-oxide loaded graphene and rGO were proposed for highly sensitive, selective, and cost-effective gas sensors, which could operate at room temperature and be applied in environmental protection for detecting toxic gases [557]. The composites of graphene and rGO with various metal oxides, including ZnO, Cu_2O, SnO_2, WO_3, Co_3O_4, NiO, In_2O_3, etc. were prepared for gas sensors [561]. ZnO-loaded rGO was prepared by simultaneous reduction of GO and zinc acetate $Zn(O_2CCH_3)_2$ by diluted LiOH in methanol at room temperature, which showed enhanced sensing for 1 ppm CO gas [632]. The composite of rGO loaded by highly anisotropic Cu_2O nanowires was fabricated under hydrothermal conditions [633]. Owing to the high surface area and improved conductivity, the Cu_2O/rGO composites achieved high sensitivity toward NO_2 at room temperature. The sensors based on SnO_2-loaded rGO had sensitivity for H_2 with a low concentration of 0.5% and additional loading of Pt on SnO_2/rGO composite by the reduction of H_2PtCl_6 in ethylene glycol at 150°C enhanced the H_2 sensitivity [634]. A Pt–SnO_2/rGO sensor was able to respond in 3–7 s and to be recovered within 2–6 s, depending on the hydrogen concentration. MoS_2-loaded rGO fibers were prepared by wet-spinning of GO dispersion containing Na_2MoO_4 into a coagulation bath of 5 wt.% $CaCl_2$ ethanol/water solution, followed by hydrothermal treatment at 240°C after mixing with L-cysteine ($C_3H_7NO_2S$) and annealing at 800°C, which exhibited high sensitivities for NO_2 and NH_3, its LOD reaching 53 ppb for NO_2 [635]. Vertically aligned SnO_2 nanorods on both surfaces of a CVD-grown graphene film were prepared by hydrothermal treatment, which exhibited improved sensing performances to various gases, particularly H_2S [636].

4.5.2.3.2 Organic gases and vapors

Oxygen functionalization was shown to be effective in increasing the response due to the adsorption of polar gas molecules by using double-layered graphene, which was epitaxially grown by CVD on 6H–SiC(0001) substrate [637]. The functionalization of graphene was performed using 5% O_2/Ar plasma at room temperature. The resistance

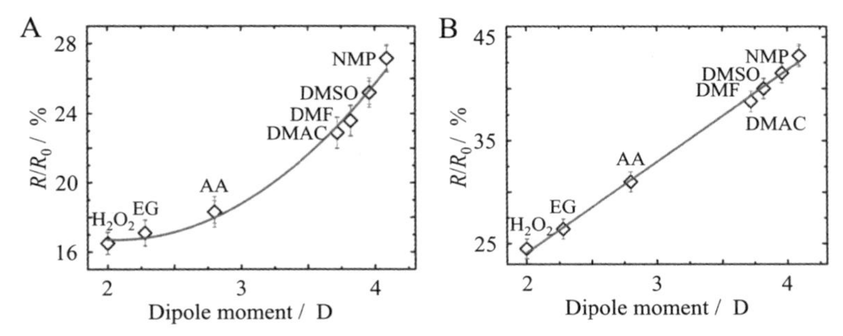

Figure 4.180 Relations between resistance change R/R_0 and dipole moment of the adsorbate gases for a double-layered graphene (A) before O_2-functionalization (as-prepared graphene) and (B) after O_2-functionalization [637].

change R/R_0 is observed more markedly on O_2-functionalized graphene than on an as-prepared one, as shown in Fig. 4.180A and B, and the response R/R_0 of O_2-functionalized graphene is linearly related to the dipole moment of adsorbed gas molecules. Although the average response time was about 10 s for both as-prepared and O_2-functionalized graphene sensors, the average recovery time was about 100 s for the latter, while it was 1.5−2 h for the former. Polyaniline-loaded rGO exhibited high sensitivity for NH_3 gas, which was much higher than for the pristine rGO and polyaniline itself [638]. The composite was prepared by mixing MnO_2-loaded GO with aniline monomers in solution, where MnO_2 worked as the template and oxidant for aniline monomers during polymerization and was washed out by a diluted HCl after the polymerization.

From GO flakes, hydrophilic and hydrophobic sensor elements could be fabricated [639]. On the tip of a polymer optical fiber bundle, a thin film of GO (hydrophilic) was deposited by drop-casting, and it was reduced by irradiation of sunlight, with the resultant rGO film being hydrophobic. The sensing element on the top of the polymer optical fibers could be half hydrophilic GO film and half hydrophobic rGO. This sensor provided the sensing ability to distinguish between tetrahydrofuran (THF) and dichloromethane. The sensing performances of these sensors, GO, rGO and GO/rGO, were studied using various VOC gases under an extreme environment of over 90% humidity.

The composite films of layer-by-layer assemblage of rGO and ionic liquids (IL), where the reduction of GO was performed by IL, were shown to work as a sensor for different organic gases, of which the response was evaluated as a frequency change on a quartz crystal microbalance [640]. The

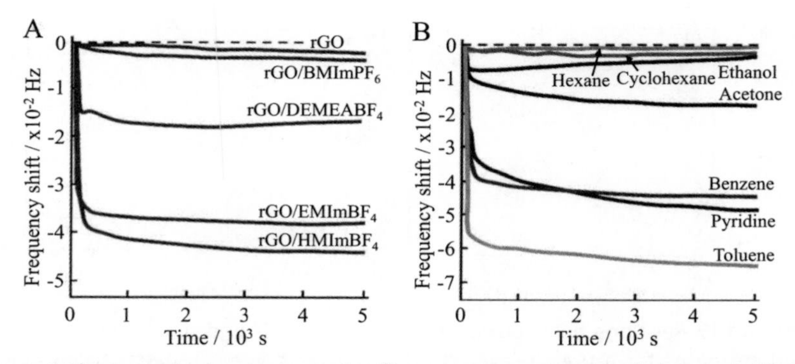

Figure 4.181 Layer-by-layer composite film sensor composed of rGO and different ionic liquids (ILs): (A) frequency shift of the films with different ILs measured by a quartz crystal microbalance due to the adsorption of benzene and (B) that due to the adsorption of various vapors on an rGO/$HMImBF_4$ composite film [640].

sensitivity of the composites depended greatly on the IL selected, $EMImBF_4$, $BMImPF_6$, $DEMEABF_4$, and $HMImBF_4$, as shown in Fig. 4.181A. The frequency shift depended greatly on aromatic compounds, as shown for the rGO/$HMImBF_4$ film in Fig. 4.181B, suggesting the possibility to detect aromatic compounds selectively.

By making an array of rGO-based gas sensors, the improvements in scattering, reproducibility, and selectivity of sensing responses were achieved owing to some inhomogeneity in structure and composition of rGO [641]. GO flakes, having lateral sizes in a wide range from a few hundred nm up to 200 μm and thickness of about 1 nm, were deposited on SiO_2/Si substrate by drop-casting of their water suspension to make the GO film with a thickness of about 10 nm. The GO film on the substrate was thermally reduced by heating at 150°C in a vacuum. By deposition of multiple Pt electrodes with about 90 μm gaps on the substrate in advance, an array of rGO-based sensors was possible to be fabricated. In Fig. 4.182A, resistance change by the cyclic exposure to isopropanol vapor (1000 ppm) and purging with dry air is shown for 20 individual sensors measured simultaneously, showing high cyclability of sensing, but each sensor showed quite different resistances to each other. In Fig. 4.182B, resistance changes averaged of 80 independent measurements by 20 individual sensors (four measurements by 20 sensors) (averaged responses, $\Delta R/R_0$) are plotted against the concentration of various vapors, revealing a high sensitivity for water vapor and suggesting a possibility to discriminate between different alcohols, such as methanol, ethanol, and isopropanol.

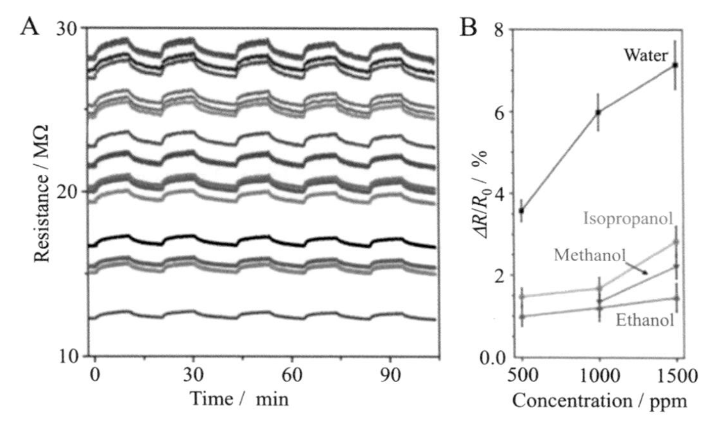

Figure 4.182 Gas sensor arrays based on rGO: (A) simultaneously measured resistances of 20 individual rGO devices for the cyclic exposure to isopropanol vapor (1000 ppm) and dry air, and (B) dependences of $\Delta R/R_0$ (response) averaged of 80 independent measurements (four times by 20 rGO devices) on the concentration of various organic vapors [641].

The composites were fabricated from vertically aligned ZnO nanorods grown on a ZnO conductive layer by covering a graphene film by CVD of CH_4 at 900–1000°C [642]. The composite is flexible and demonstrates high sensitivity for ethanol vapor at 300°C, as shown in Fig. 4.183, and it is possible to detect ethanol vapor at a ppm level concentration with high sensitivity.

A sensor composed of the film of rGO functionalized by octadecylamine (ODA) on a glass substrate was able to detect benzene, toluene, ethylbenzene, xylenes, and cyclohexane dissolved in water at low concentrations of 5–100 ppm [643].

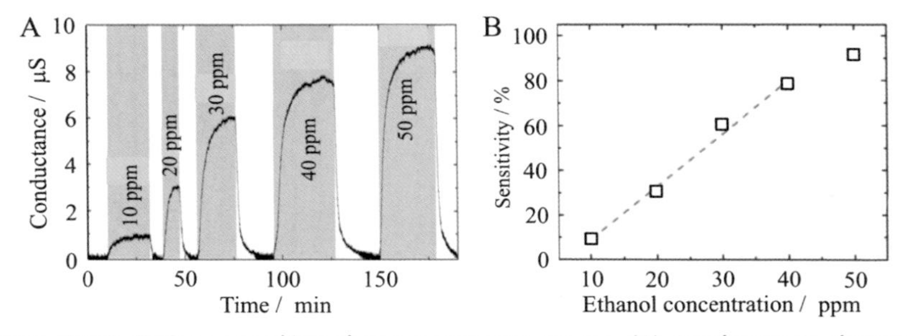

Figure 4.183 ZnO-nanorod/graphene composite sensor: (A) conductance change at different concentrations of ethanol vapor at 300°C and (B) change in sensitivity (ratio of resistance in air to that in ethanol) with ethanol concentration [642].

4.5.3 Strain sensors

4.5.3.1 Porous carbons

A three-layered composite sheet was fabricated from two electrode sheets of the mixture of B_4C-derived microporous carbon (CDC) powder (50 wt.%) with an ionic liquid (IL, $EMImBF_4$) (26 wt.%), and poly(vinylidene fluoride co-hexafluoropropylene) (PVF—HFP) (24 wt.%), and the separator layer of the mixture of PVF-HFP with IL (1/1 by weight), as shown in Fig. 4.184A, with three layers being hot-pressed at 75°C under 3.5 MPa for 10 s [644]. A change in the open-circuit voltage with bending deformation (r1—r3 in Fig. 4.184B) for 5 s was measured on a strip of the composite of 40-mm long and 20-mm wide, with the bending modes being illustrated in Fig. 4.184B, and the composite initially having a curvature of 20 m^{-1}. The peak voltage of 350 μV with a peak current of 22 μA is obtained under the maximum deformation (r3), as shown in Fig. 4.184C. Linear relations of the peak voltage and current to the final curvature were observed.

Electroconductive polydimethylsiloxane (PDMS)-based composite was prepared by mixing 15 wt.% carbon black (CB) with PDMS in toluene and printed in various substrates using stumping techniques; various patterns on a large area were easily obtained and the resistances of the sensors were possible to be controlled by changing the length [645]. Piezoelectric

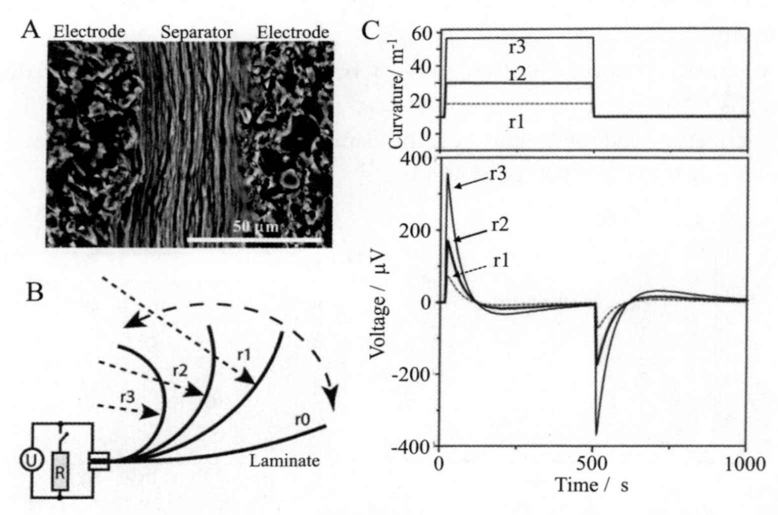

Figure 4.184 Carbide-derived carbon laminate sensor: (A) SEM image of the cross-section, (B) scheme of bending modes, (C) applied curvatures (r1—r3) with corresponding voltages [644].

performances of these electroconductive and flexible CB/PDMS composites were studied by mounting onto a glove to measure the motion of human fingers and by forming them into a rosette-type gauge to detect both the magnitude and direction of the principal strains.

Sensor arrays were fabricated by applying the direct laser writing technique (CO_2 laser with wavelength of 10.6 μm and beam size of about 100 μm) onto polyimide films (Kapton with 127 μm thickness) [646]. Various patterns of sensor arrays were possible to be prepared, including biaxial, shear, dual grid biaxial, diaphragm, etc. The sensitivity $\Delta R/R_0$ to the applied strain ε depends greatly on the power of the laser beam during the preparation process and high correspondences between $\Delta R/R_0$ and ε during cyclic application of strain is obtained, as shown in Fig. 4.185A. The gauge factor, *GF*, calculated as the ratio of $\Delta R/R_0$ to ε is improved markedly with increasing laser linear energy density (with proceeding carbonization), as shown in Fig. 4.185B, revealing the possibility of obtaining a high *GF*.

Carbonized patterns created by laser pyrolysis of a polyimide film were embedded into elastomeric substrates (PDMS), as shown in Fig. 4.186A, which worked as highly stretchable (up to 100% strain) and strain sensitive (*GF* of up to 20,000) sensors [647]. The sensing element obtained as a trace of laser beam was highly porous carbon nanoparticles with sheet resistances as low as 60 Ω/sq and very flexible. The device exhibits a very high change in resistance, R/R_0 increasing up to 10^4 (resistance up to 20 MΩ) for 100% longitudinal strain (graph in Fig. 4.186B), while it is negligibly low in

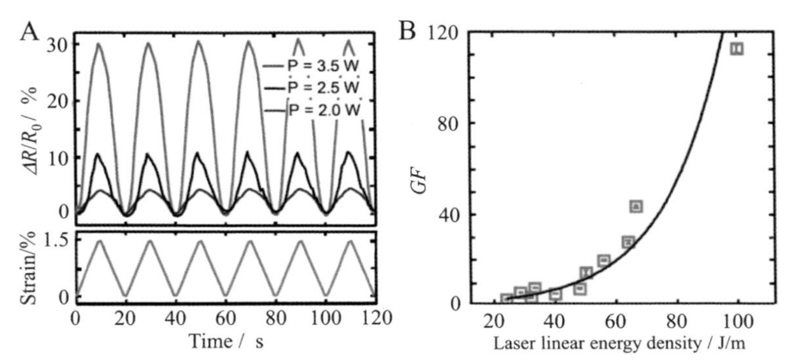

Figure 4.185 Sensors fabricated by direct laser writing on polyimide film: (A) correspondences between the response $\Delta R/R_0$ and strain on the sensors printed by different laser powers *P*, and (B) the dependence of gauge factor (*GF*) on laser linear energy density [646].

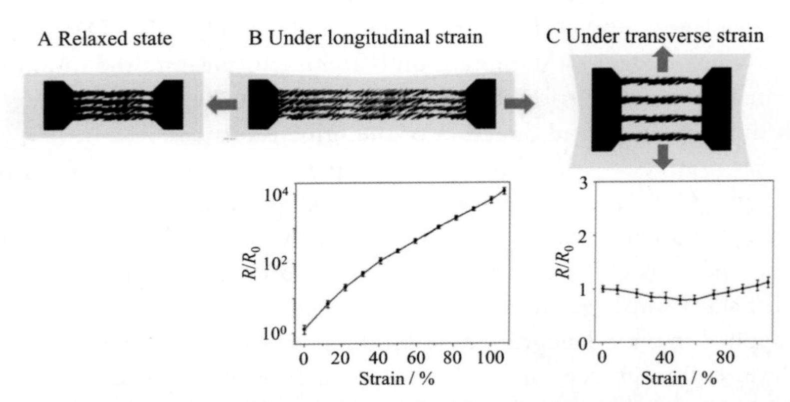

Figure 4.186 Strain sensor fabricated by embedding the laser-carbonized traces into PDMS: (A) appearance of the sensor, (B) the sensor under longitudinal strain accompanying with the curve of the response to strain, and (C) that under transverse strain with the response curve [647].

response to 100% transverse strain (Fig. 4.186C). The performance stability of the sensor was confirmed by subjecting it to 1000 stretch-and-release cycles (0%—100% strain). By embedding these sensors into a latex glove, the detailed motions of human fingers were possible to be monitored.

4.5.3.2 Carbon nanotubes and nanofibers

Electromechanical properties of SWCNTs are theoretically suggested to depend on chirality defined by the tube index (m, n) [648—651]. Armchair metallic SWCNTs (m-SWCNTs, $m = n$) retain their high symmetry even under tensile stress and the electrical properties are expected to be the least sensitive to the strain, but other types of SWCNTs with lower symmetries, i.e., semiconducting ones (s-SWCNTs with m-$n \neq 3$N where N is the integer) and quasimetallic ones (qm-SWCNTs with m-$n = 3$N, or small band-gap semiconducting ones), are more sensitive [652]. The experimental results reported, however, are slightly different from the theoretical prediction; qm-SWCNTs exhibit the largest $\Delta R/R_0$ and m-SWCNTs are the least sensitive under tensile strains, as shown in Fig. 4.187A—C for three types of SWCNTs. For qm-SWCNTs, the largest piezoresistive response ($\Delta R/R_0$) and very high GF (about 600—1000) were reported [653]. An increase and decrease in the band-gap of s- and qm-SWCNTs deposited on a Si_3N_4 substrate membranes were observed [654].

SWCNTs were sorted by chirality to improve the electromechanical properties. Enrichment of (6, 5) CNTs was done by suspending them in

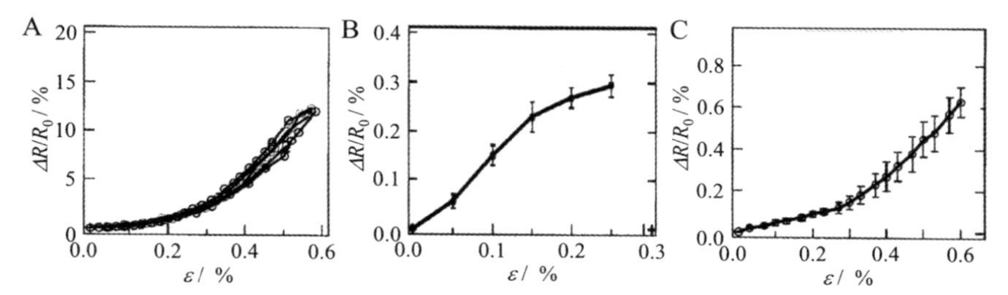

Figure 4.187 Responses ($\Delta R/R_0$) to tensile strain ε on SWCNTs: (A) qm-SWCNT, (B) s-SWCNT, and (C) m-SWCNT [652].

water with a sodium cholate surfactant [655], followed by separation using density gradient ultracentrifugation [656]. A device for sensing of mechanical strain was constructed: four electrodes were fixed on a flexure of which two inner electrodes were spaced 1 μm from each other where CNTs were deposited through a droplet of a 3 μg/mL aqueous solution and directed by applying AC voltage across these two electrodes [657]. Sensitivity was measured by applying the strain perpendicular to the flexure plane at the center and evaluated by the response as relative resistance change $\Delta R/R_0$ and gauge factor *GF*. The absorption spectrum measured over the ultraviolet (UV), visible (Vis), and near-infrared (nIR) spectra on enriched (6, 5) SWCNTs is shown in Fig. 4.188A, which is composed of 82.6% (6, 5) CNTs but still has a small amount of CNTs with other chirality. The resistance responses measured on an as-enriched CNT sensor in Fig. 4.188B suggest the necessity to eliminate the low resistance and low *GF* CNTs [i.e., (7, 5), (7, 6), (9, 1), 6, 4), etc.], because these CNTs limit the strain sensitivity of the sensor and so give negative *GF*. The electrical breakdown technique, applying 17.7 V under a strain, was performed on enriched (6, 5) CNTs. The resultant CNTs exhibits positive *GF* of 34.1,

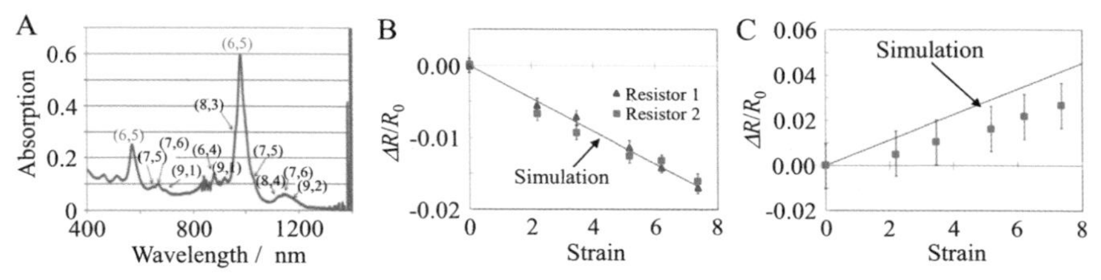

Figure 4.188 Enriched (6, 5) CNTs: (A) UV-Vis-nIR absorption spectrum for (6, 5) enriched CNTs, and resistance responses measured (B) on as-enriched CNT and (C) on enriched CNTs after high-strain electrical breakdown treatment with the line calculated by Monte Carlo simulation [657].

closer to the predicted value of 57 by Monte Carlo simulation, as shown in Fig. 4.188C.

PEG-grafted MWCNTs, which were fabricated by heat treatment of the mixture of PEG and CNTs in toluene solution at 70°C, were dispersed on a PDMS sheet with Au electrode (5×8 mm^2 with a gap of 0.6 mm) [658]. This sensor was implanted into the stomach of a rat to monitor the muscular contractions with the changes in the resistance of the sensor. Stomach contraction induced by the injection of acetylcholine was detected as the resistance change, with a larger change caused by a larger amount of injection, whereas physiological saline did not induce any stomach contractions (no resistance change).

Strain sensors were fabricated from the composites of commercial MWCNTs (diameters of about 100 nm) and vapor-grown carbon fibers (VGCFs) (diameters of 80–200 nm) with epoxy resin matrix [659]. The nanoparticles of various metals (Ni, Cu, and Ag) were loaded onto fibrous fillers by nonelectrolytic plating before composite fabrication. The resistance responses for the pristine MWCNTs and VGCFs increase nonlinearly with the increase in the strain and increase with decreasing filler content, as shown on the composites of the pristine MWCNTs and VGCFs in Fig. 4.189A and B, respectively. VGCFs are more sensitive to microstrain

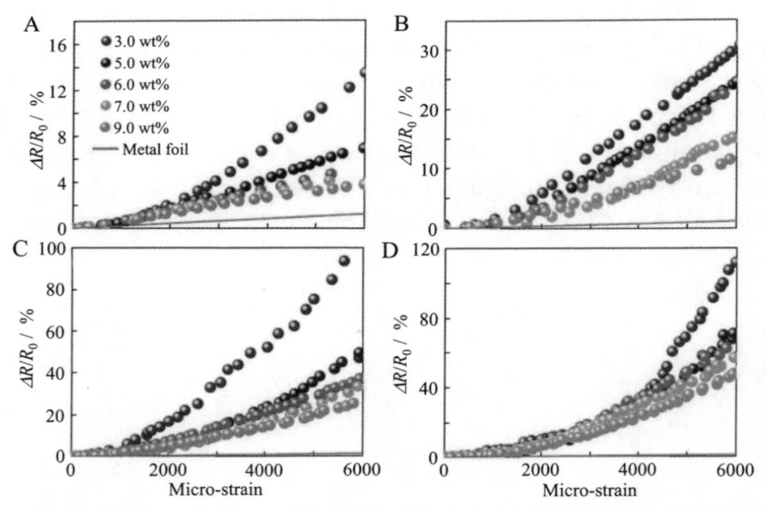

Figure 4.189 Strain responses of the MWCNTs/epoxy and VGCFs/epoxy composites in different filler contents with or without metal-loading: (A) the pristine MWCNTs, (B) the pristine VGCFs, (C) the V-loaded VGCFs, and (D) the Ag-loaded VGCFs. For comparison, the results of the metal-foil strain gauge are shown in each image [659].

than MWCNTs. Metal loading is very efficient for improving the resistance response, as shown for V-loaded and Ag-loaded VGCFs in Fig. 4.189C and D. The observed responses are much higher than the conventional metal-foil strain gauge. In particular, the composite using Ag/VGCF yielded the highest piezoresistivity among all the metal-loaded MWCNTs and VGCFs.

A composite of CNTs with graphite was fabricated by growing CNTs vertically on graphite nanoplatelets (4–5 μm in lateral size with hundreds of nanometers thickness) via floating-catalyst CVD of C_2H_2 with $C_6H_4(CH_3)_2$ at 650°C, and mass ratio CNT/graphite of approximately 1/1 [660]. The resultant composite CNT/graphite was mixed with PDMS and then formed into disks, which were flexible and exhibited piezoresistive sensitivity (*GF* of about 10^3 and pressure sensitivity around 0.6 kPa^{-1}). By compression into a thin film (150 mm thick), it was wearable for detecting slight motions of the finger.

CNTs could be used to construct a nanomechanical resonator with atomic mass resolution, with a mass sensitivity of 1.3×10^{-25} $kg/Hz^{1/2}$ [661,662].

4.5.3.3 Graphene and related materials

The electrical conductivity of a single-layer graphene, which was synthesized by CVD and suspended without any substrate, was found to be sensitive for strain, but its piezoresistive sensitivity represented by *GF* was not high, at about 1.9 with an applied strain of less than 3% [663]. Single-layer graphene prepared by mechanical peeling also gave *GF* of 1.9 [664]. A graphene sheet synthesized by CVD of CH_4 on Ni-coated SiO_2/Si substrate and transferred onto PDMS substrate, which had 95% transparency and was composed of one to three layers, showed a resistance increase from 492 to 522 kΩ with applied strain up to 1%, of which *GF* reaches about 6.1 [665]. Graphene films were grown on fluorophlogopite mica by a plasma-enhanced CVD of CH_4 without any catalyst (Fig. 4.190A), and were composed of nanograins of single-layer to few-layered graphene and exhibited high sensitivity to strain [666]. The $I-V$ relations are linear under strains in the range of −0.29%–0.37% and the resistance change $\Delta R/R_0$ with strain ε can be approximated to be linear, giving *GF* of 37, as shown in Fig. 4.190B. *GF* increases with increasing film resistance R_0, exceeding 300 as shown in Fig. 4.190C. A composite of sandwiched Ag-nanowires between graphene on a prestretched polyacrylate (PAC) was transparent (85% transparency), bendable up to 2 mm radius, and stretchable up to 200%, and could respond to a wide range of

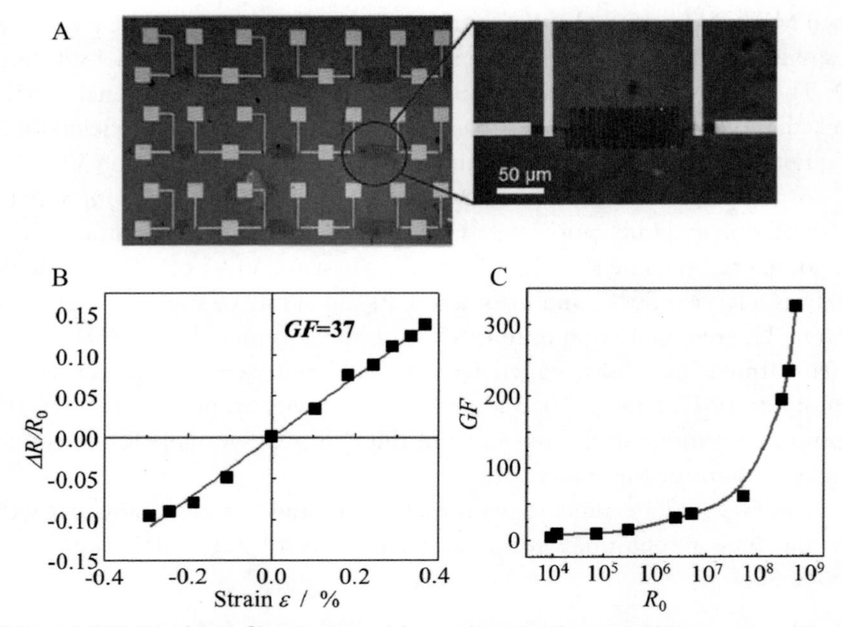

Figure 4.190 Graphene films prepared by a plasma-enhanced CVD on mica: (A) as-patterned devices, (B) resistance change $\Delta R/R_0$ with strain ε, and (C) change in *GF* with the film resistance R_0 [666].

strains from 5% to 200% with a short response time of less than 1 ms [667]. A composite was fabricated by impregnating PDMS into a graphene foam after dissolving out Ni foam used for CVD of CH_4 as a template, which displayed a high sensitivity of 8.5 in a range up to 2000 kPa [668]. A graphene mesh was prepared by CVD of CH_4 using Cu mesh, of which the composite with PDMS exhibited an exponential increase in electrical resistance with a tensile strain of 2%–6%, with *GF* reaching approximately 10^3 [669].

An all-graphene strain sensor was fabricated by transferring a single-layer graphene patterned by photolithography onto PDMS substrate and following pasting a dispersion of graphene flake (about three-layered and with a lateral size of about 10 μm) as the electrode [670]. The resultant sensor was fully flexible, stretchable up to 20% with a high *GF* of 42.2, and could detect strain as low as 0.1% with $\Delta R/R_0$ of about 0.005. As shown in Fig. 4.191A–C, the responses for a strain induced by stretching, bending, and twisting, respectively, are quick and highly reversible.

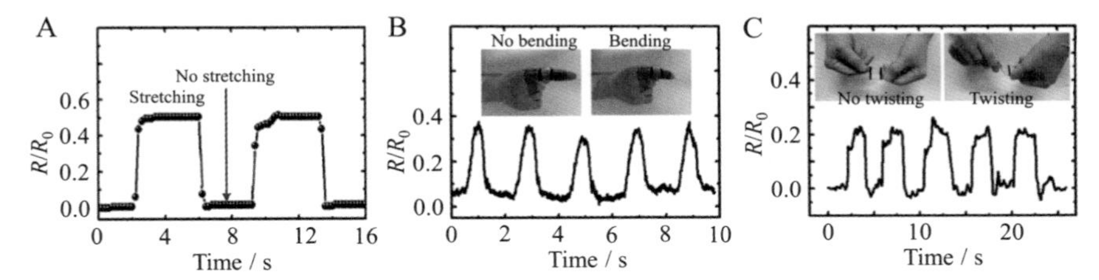

Figure 4.191 Responses to (A) stretching, (B) bending, and (C) twisting strains of all-graphene sensor [670].

Strain sensors were developed by the spray deposition of multilayered rGO nanoflakes suspended in 1-propanol on polycarbonate substrate [671]. rGO was prepared by thermal exfoliation of GO at 1150°C and sonication in 1-propanol. Quasistatic tests of this flexible sensor were performed by measuring the piezoresistive response using a universal tensile machine with a linearly increasing load. The resistance R of the sensor increased with increasing ε above 0.015%, and GF reached about 200 at 0.135% ε. rGO foams, which were prepared using an Ni foam template followed by infusion of PDMS via vacuum infiltration, worked as a flexible strain sensor able to record human blood pressure and heartbeat rate [672]. A composite of commercial graphene flakes (2–10 nm in size) with PDMS (graphene content of 20 wt.%) by mixing in solution could provide a capacitive strain sensor with GF of 3 [673].

A pressure sensor was proposed by using rGO film as the flexible positive electrode and CNTs as the negative electrode, as shown schematically in Fig. 4.192A [674]. A metallic Cr layer with a thickness of 100 nm was deposited on SiO_2/Si substrate by electron beam evaporation and then patterned to an array of microholes with a diameter of about 6 μm, followed by etching the substrate vertically to a depth of about

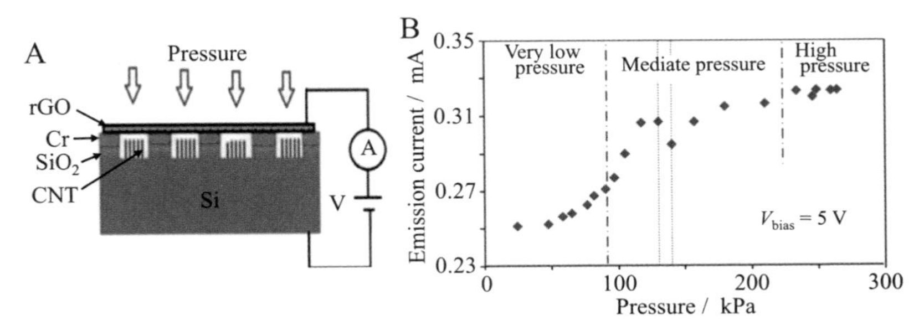

Figure 4.192 rGO-based pressure sensor: (A) scheme of the sensor fabricated, and (B) dependence of the field emission current on pressure [674].

1.5 μm, on which the deposition of Ni as a catalyst and then CNT growth by DC plasma-enhanced CVD were performed. Finally, a film of thermally exfoliated rGO was formed on the top of the substrate. The field emission occurs between rGO film and tips of CNTs, of which the current under a voltage of 5 V changes sensitively at low gas pressure, as shown in Fig. 4.192B. The change in emission current with applied pressure was discussed by dividing them into three pressure ranges: the highest output sensitivity of 21.9 μA/A·Pa in a pressure range of 100–170 kPa was obtained. The same device was used as a resonator, of which the resonance frequency reached about 10 MHz at an Ar pressure of about 100 kPa with a sensitivity of 11.32 Hz/Pa.

4.5.4 Biosensors

4.5.4.1 Small biomolecules

Monitoring of small biomolecules, such as ascorbic acid (AA), dopamine (DA), and uric acid (UA), in biological fluid is an important part of health care; AA is used for scurvy, mental illness, cancer, and AIDS prevention, a low level of DA in cerebral fluid is a sign of Parkinson and Alzheimer diseases, and an abnormal level of UA is related to gout. For electrochemical sensing of these molecules, a mixture of graphite powder (particle size of less than100 μm) with an IL (n-octylpyridinum hexafluorophosphate, OPFP) in 1/1 by weight was used as an electrode by packing into the cavity of a Teflon holder [675]. The electrode presented with high performance for the detection of different biomolecules, such as AA, as well as NADH (nicotinamide adenine dinucleotide disodium salt, a kind of coenzyme). PAN-based CNFs, which were prepared via electrospinning, followed by stabilization at 300°C in H_2/Ar and carbonization at 1100°C in Ar, exhibited sensitivity to AA and NADH by mixing with a mineral oil [676]. Replacing the graphite powder with CNFs and CNTs in the composite was effective in improving the sensitivity by only 10 wt.% CNF or CNT addition [677]. The sensing behavior of the composite of PAN-based CNFs with hexaflurophosphate (PFP) (CNF/PFP) was compared to those of the composites of the same CNFs with a mineral oil (CNF/MO) and of the CNTs with PFP (CNT/PFP), as shown by their CV curves measured on a mixture of 2 mM AA, 0.5 mM DA, and 0.5 mM UA in Fig. 4.193A–C, respectively [678]. The response of the CNF/MO electrode to these three analytes is observed as one broad oxidation peak.

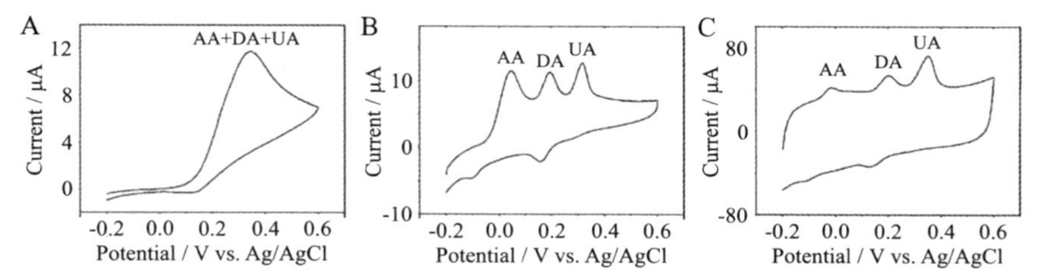

Figure 4.193 CV curves of the composites for the mixture of AA, DA, and UA: (A) PAN-based CNFs with a mineral oil (CNF/MO), (B) PAN-based CNFs with PFP (CNF/PFP), and (C) CNTs with PFP (CNT/PFP). The supporting electrolyte is 0.1 M PBS (pH 7.4) with a scan rate of 0.05 V/s [678].

At the CNF/PFP electrode, in contrast, three well-defined peaks, that are related to the oxidation of AA, DA, and UA, are clearly obtained with sufficient resolution. The CNT/PFP electrode can detect these three analytes separately, but their resolution is poor in comparison to the CNF/PFP composite. For guanine (G) and adenine (A), simultaneous detection was possible using the CNF/PFP composite with high sensitivity, which was much better than the CNF/MO and CNT/PFP composites. N-doping of CNFs was effective in improving the sensitivities and selectivity for AA, DA, and UA, as compared with CV and differential pulse voltammetry (DPV) curves for N-doped CNFs and nondoped CNFs on the glassy carbon (GC) electrode in Fig. 4.194A and B, respectively [679]. N-doped CNFs could detect AA, DA, and UA oxidation peaks clearly and separately. To improve the sensitivity for these biomolecules, loading of Pd [680] and Ag–Pt [681] onto electrospun PAN-based CNFs, and compositing of the

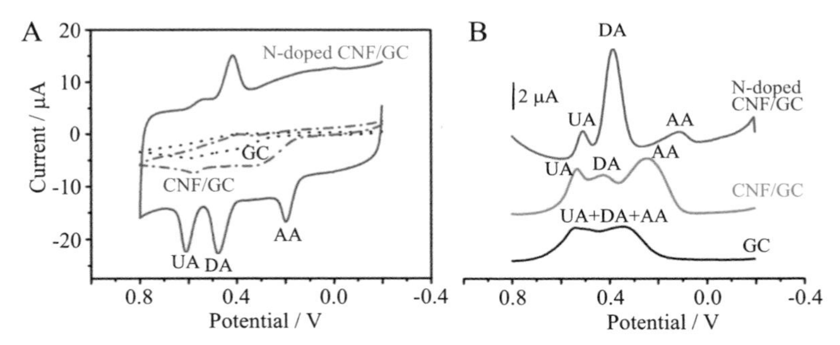

Figure 4.194 N-doped CNF/GC composite in comparison with a GC electrode (without CNF casting) and CNF/GC composite (without N-doping of CNFs): (A) CV curves and (B) differential pulse voltammetry (DPV) curves in a mixture of 1 mM AA, 0.2 mM DA, and 0.5 mM UA in 0.1 M PBS (pH 4.5) [679].

CNFs with a carbon paste [682] and a pyrolytic graphite [683], have been proposed.

A nanocomposite of rGO loaded with Ag nanoparticles, which was prepared by mixing rGO functionalized by poly(diallyldimethylammonium chloride) (PDDA) with Ag-nanoparticles through solution, was reported to work as an electrochemical aptasensor for the sensitive and selective determination of the antibiotic chloramphenicol [684]. Hollow rGO balls (about 100 nm diameter), which were prepared using Ni nanoparticles as a template, worked as an electrode for electrochemical determination of levodopa (enzymatically converted to DA) in the presence of uric acid [685]. It worked as a selective and reproducible electrode with a sensitivity of 0.69 $\mu A/\mu M \cdot cm^2$ and LOD of 1 μM.

4.5.4.2 Glucose

Glucose monitoring has attracted considerable attention in the diagnosis and management of diabetes mellitus as well as in the control of food processes. Mesoporous carbon spheres (S_{BET} of 490 m^2/g with V_{total} of 0.523 cm^3/g) were synthesized via electrospinning of the highly diluted DMF solution of PAN with SiO_2 spheres (weight ratio PAN/DMF/silica as 3/60/15), followed by stabilization at 250°C in air, carbonization at 1050°C in N_2 and dissolution of SiO_2 with HF [686]. Glucose oxidase (GOD, an enzyme) was immobilized onto the carbon spheres after the surface treatment by oxyfluorination at 200°C, and then drop-cast on the electrode of the sensor. Sensitivity to glucose on the carbon spheres reached 78 $\mu A/mM \cdot cm^2$. GOD immobilization was performed onto free-standing N-doped CNF webs fabricated via electrospinning of PAN by stabilization at 300°C in air and carbonization at 900°C in N_2 [687]. The response of the resultant web to glucose was linear in the concentration range of 0.05−3 mM at a voltage of 0.40 V. However, the selectivity toward glucose at 0.40 V became poor with the addition of 0.25 μM AA and UA into glucose solution. N-doped CNFs prepared by carbonization of electrospun PPy/PAN nanofibers at 900°C in N_2 displayed sensitivity to glucose after immobilization of GOD, giving a low LOD of 2 μM and a linear response in a concentration range of 12−1000 μM [688].

To satisfy the requirements of stable entrapment for a sufficient amount of enzyme and rapid mass transport of fuel (for example, glucose), a hierarchical pore structure of carbon having mesopores and macropores is required. The carbons synthesized using two MgO templates with sizes of 40 and 150 nm could satisfy these requirements [689,690]. For bilirubin oxidase (BOD), the optimum pore composition was demonstrated to be 33% macropores and 67% mesopores [689].

FET devices composed of graphene films after the immobilization of enzyme, GOD and glutamic dehydrogenase (GluD), in Na_2CO_3–$NaHCO_3$ buffer solution (pH 9.0) were successfully used to detect glucose and glutamate, respectively, by monitoring the current between the source and drain electrodes (I_{ds}) at a gate voltage V_g of 0 V [691]. Graphene films were synthesized by CVD of CH_4/H_2 mixed gas at 1000°C and composed of single-layer and few-layered domains. In Fig. 4.195, the current response I_{ds} of the sensor composed from GluD-modified graphene for glutamate is shown for different concentrations; the response seems to consist of two linear detection ranges, 5–50 μM and 50–1200 μM, respectively (lower inset in Fig. 4.195A). In contrast, glutamate (1 mM) did not lead to any signal from the device for graphene (without GluD-modification) (upper inset of Fig. 4.195A). The sensor composed of GluD-modified graphene did not respond to glucose, AA, UA, or acetaminophen. The sensor for GOD-modified graphene responded to the addition of 0.1–10 mM glucose (Fig. 4.195B).

Immobilization of GOD in N-doped rGO, which was prepared by nitrogen plasma treatment of hydrazine-reduced rGO, enhances the electrochemical response, as shown in Fig. 4.196 [692]. A marked response on N-doped rGO for successive addition of 0.1 mM glucose is obtained as shown in Fig. 4.196A. The clear responses to glucose ranging from 0.01 to 0.5 mM in the presence of 5 mM UA and 5 mM AA are observed for the GOD-modified N-doped rGO (Fig. 4.196B).

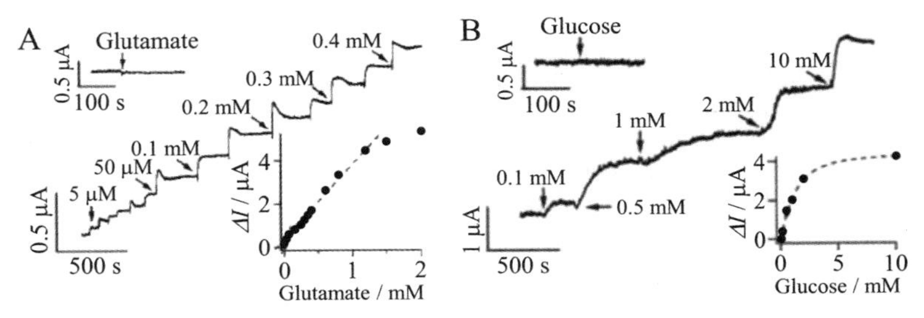

Figure 4.195 Current responses of graphene-based FETs: (A) GluD-modified graphene with the addition of glutamate with various concentrations, and (B) GOD-modified graphene with the addition of glucose at various concentrations. The lower insets show response curves of GluD- and GOD-modified graphene and the upper insets show the responses of graphene (without GluD- and GOD-modification) to 1 mM glutamate and 10 mM glucose, respectively [691].

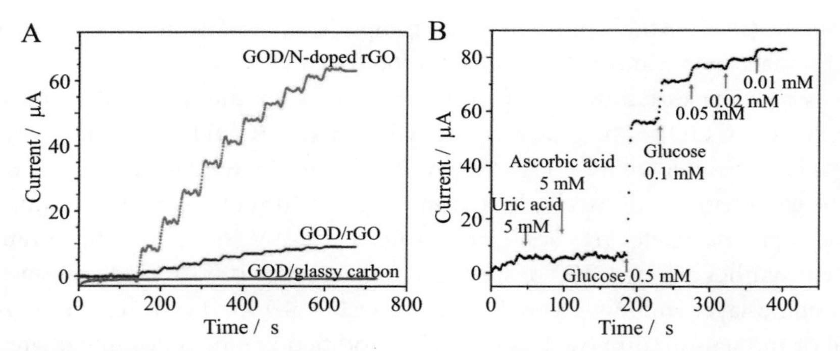

Figure 4.196 GOD-modified N-doped rGO: (A) electrochemical responses at −0.15 V in 0.1 M PBS (pH 7.0) with successive addition of 0.1 mM glucose, in comparison with GOD-modified rGO and GOD on the glassy carbon electrode, and (B) responses to successive addition of 5 mM uric acid, 5 mM ascorbic acid, and different amounts of glucose on GOD-modified N-doped rGO at −0.15 V in 0.1 M PBS (pH 7.0) [692].

4.5.4.3 Hydrogen peroxide

Hydrogen peroxide (H_2O_2) has widespread applications in the fields of foods and pharmaceuticals, but also can be involved in many pathological conditions including diabetes, neurodegeneration, Alzheimer disease, and cancer. Therefore, rapid, sensitive, and accurate H_2O_2 sensing is greatly demanded. Loading of Pt nanoparticles onto PAN-based CNFs synthesized by electrospinning was performed through the reaction of H_2PtCl_6 with HCOOH in an aqueous solution [693]. Amperometric responses were measured in 0.1 M phosphate buffer solution (PBS) at pH 7.4 under continuous stirring. The responses of the electrode upon injection of H_2O_2 are shown in Fig. 4.197A and B. The Pt-loaded CNFs exhibit a strong and

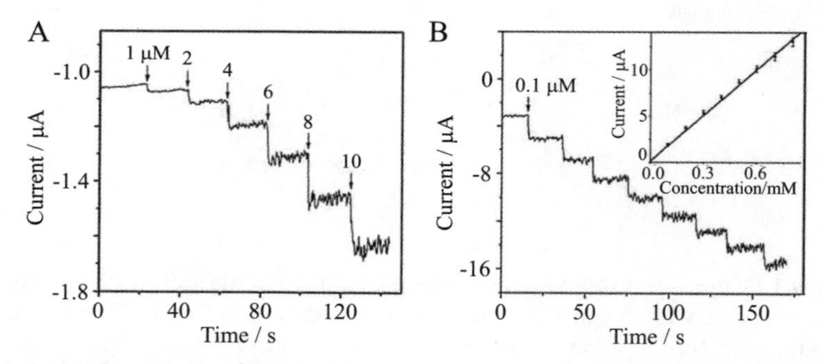

Figure 4.197 Responses of the Pt-loaded CNF electrode upon successive injection of H_2O_2 at (A) low concentrations and (B) repeated injection of 0.1 mM H_2O_2 at 0 V. The inset shows the calibration curve for H_2O_2 detection [693].

rapid response to each injection of the analyte, attributed to the high electrocatalytic efficiency of the well-dispersed Pt nanoparticles. The current signal of each injection of 0.1 mM H_2O_2 is linearly related to the concentration of H_2O_2 up to 0.8 mM, as shown by the inset in Fig. 4.197B, suggesting LOD of 0.6 μM. No interferences of AA, AP, or UA on the H_2O_2 sensitivity were experimentally confirmed.

N-doped CNFs were prepared by carbonization of electrospun PPy/PAN nanofibers at 900°C in N_2 after stabilization at 280°C, on which a high sensitivity to H_2O_2 was observed [694]. The CNFs derived from the precursor PPy/PAN nanofibers containing 12 wt.% PPy exhibited the highest electrochemical properties; stable and quick current responses were obtained in a wide range of H_2O_2 concentrations of 10 μM−15 mM with LOD of 3.4 μM, as shown in Fig. 4.198A. No interferences by AA, DA, and UA were experimentally confirmed. To testify to the feasibility of this CNF sensor in the practical analysis, a human serum was injected into the solution and then 20 mM H_2O_2 solution added. As shown by the amperometric responses in Fig. 4.198B, a stable and quick response to human serum is observed, as well as for H_2O_2. The H_2O_2 concentration in the serum was estimated to be 7.46 μM using this device, which was in accordance with the 8.14 μM determined by another method (permanganimetric method).

4.5.4.4 Detection of DNA

For DNA detection, the sensors were proposed using rGO enriched in −COOH functional groups [695], rGO modified by Au-nanoparticles [696], rGO functionalized with a conducting PPy graft copolymer [697],

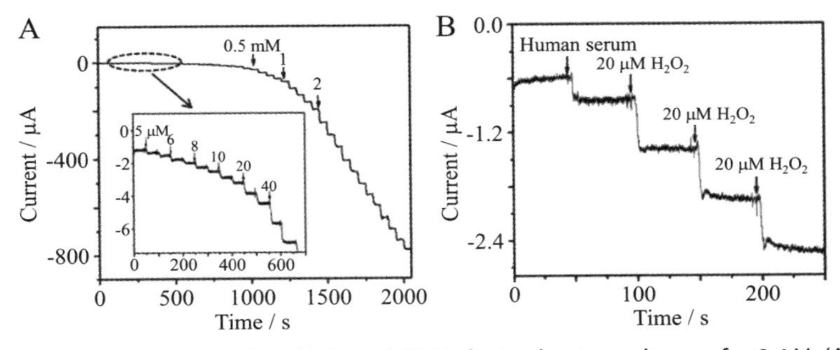

Figure 4.198 Responses of an N-doped CNF electrode at a voltage of −0.4 V: (A) the successive addition of H_2O_2 with different concentrations and (B) the addition of a human serum, followed by successive addition of H_2O_2 in N_2-saturated 0.1 M PBS (pH 7.4) [694].

and rGO hybridized with DNA [698]. The electrode was prepared by mounting a 0.15 mg/mL GO suspension on a screen-printed graphene electrode, followed by electrochemical reduction of GO, which exhibited high sensitivity for label-free detection of DNA [699]. The composite rGO/graphene formed on the surface of glassy carbon electrode (rGO/graphene double-layered electrode) gives a strong redox peak, which is much stronger than graphene, rGO, and GO/graphene, as shown in Fig. 4.199A. The hybridization of ssDNA (single-stranded DNA) immobilized on the electrode with cDNA (complementary DNA) gives dsDNA (double-stranded DNA), which shows a redox peak, as shown in Fig. 4.199B, which can be used for quantitative sensing of the DNA concentration. FET devices composed of graphene film, which was synthesized by CVD of CH_4 at 1000°C and consisted of about 60% single-layer and about 40% few-layered domains, were also shown to have high sensitivity to DNA [700,701]. These sensors had very low LOD of a few nM, even 100 fM.

PEG-grafted SWCNTs were reported to work as a biosensor [702–704]. In Fig. 4.200A, the responses of FET impedance to three kinds of oligonucleotides (1–3) are shown on the PEG-grafted SWCNTs after plasma ion irradiation, giving a sharp and quick decrease in the impedance, because the target oligo is negatively charged in a buffer solution. The PEG-grafted SWCNTs could be applied to detect taste materials, saltiness (NaCl and $MgCl_2$), sourness (acetic acid and citric acid), bitterness (caffeine and theophylline), sweetness (sucrose and saccharin sodium), etc., and also vitamins, as shown by the vitamins in Fig. 4.200B.

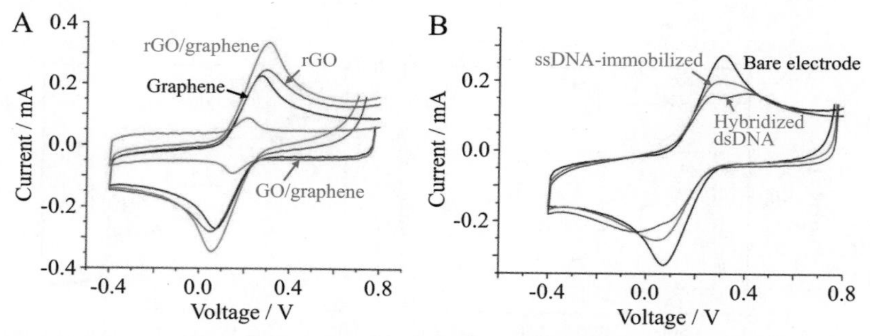

Figure 4.199 CV curves for an rGO/graphene double-layered electrode in 10 mM ferricyanide aqueous solution (1 M KCl as the supporting electrolyte) at room temperature: (A) comparison with rGO, graphene, and GO/graphene electrode, and (B) after immobilization of ssDNA on rGO/graphene double-layered electrode and hybridization of cDNA to ssDNA on the electrode to form dsDNA [699].

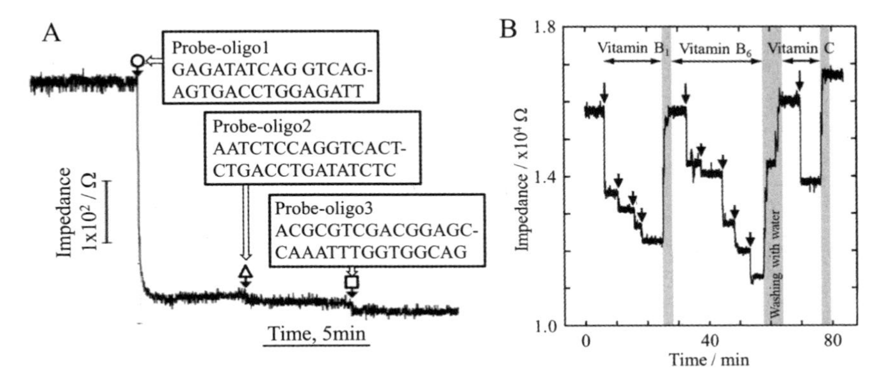

Figure 4.200 FET responses of a PEG-grafted SWCNT sensor (A) to three probe oligos and (B) to vitamins [703].

4.5.4.5 Proteins

A flexible and transparent rGO film with about 9 nm thickness was deposited on a PET substrate by reduction with hydrazine, which was settled in FET after patterning [705]. The resultant FET device exhibited sensitivity for proteins, fibronectin, and avidin. The electrodes were also prepared by casting the rGO/chitosan solution onto the surface of a thionine-modified glassy carbon disk, followed by dip-coating of Au nanoparticles, which exhibited high electrochemical sensitivity for α-fetoprotein (AFP) [706]. A linear decrease in the reduction current with a logarithmical increase in the concentration of AFP was obtained in the range of 1.0−10 ng/mL with a LOD of 0.7 ng/mL. The AFP concentration in human serum determined using this electrode (electrochemical immunosensor) agreed with that determined by electrochemiluminescent. The nanocomposites composed of GO, thionine, and Au nanoparticles were also proposed as a simple, label-free, and sensitive electrochemical immunosensor [707].

A FET device composed of thermally reduced rGO decorated by Au nanoparticles was applied for the detection of a specific protein after anchoring anti-immunoglobulin G (anti-IgG) on the surface of rGO through Au nanoparticles [708]. The response of the sensor ($\Delta R/R$) increased nonlinearly with increasing concentration of immunoglobulin (IgG) (a protein), from 10.8% to 65% with an increase in IgG concentration from 2 ng/mL to 0.02 mg/cm^3 and tended to saturate above 0.02 mg/cm^3.

4.5.4.6 Detection of bacteria

A FET of few-layered graphene synthesized by CVD could detect the bacterium *E. coli* with high sensitivity and specificity after loading

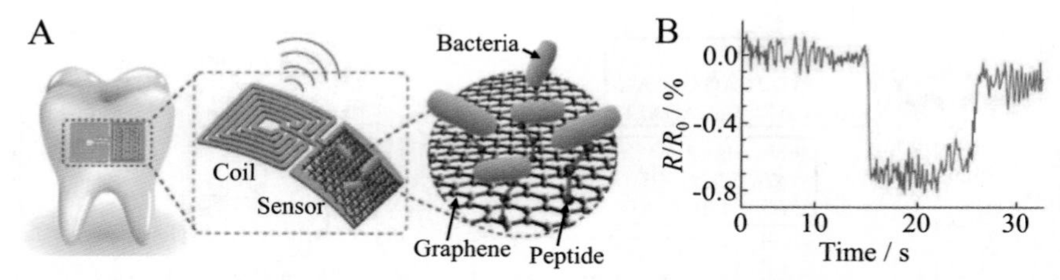

Figure 4.201 Graphene-based biosensor: (A) schematic illustration of the construction and setting of the sensor and (B) resistance response to touching/detouching of a single *E. coli* bacterium on the sensor [710].

anti-*E. coli* (antibody) on the surface of the graphene using a linker [709]. Through incubation with different concentrations of *E. coli* from 0 to 10^5 CFU/mL for 30 min, the conductance of the FET increases positively with increasing concentration, but no significant response of the FET is observed by incubation with another bacteria (*P. aeruginosa*), indicating high selectivity for the detection of *E. coli*.

A graphene-based nanosensor was fabricated by transfer of single-layer graphene synthesized by CVD onto a water-soluble silk film and a planar inductive and capacitive element was incorporated to enable wireless interrogation [710]. The graphene/silk sensor was functionalized by a peptide through its water solution to detect bacteria. This device could be transferred directly onto biomaterials, such as tooth enamel, and could have external connections through a resonant coil, as shown in Fig. 4.201A, and so the arrival and departure of a single bacterium can be monitored not only by the electrical response R/R_0 but also using a fluorescence image. The resistance response to arrival and departure of a single bacterium *E. coli* is shown in Fig. 4.201B, as an example.

References

[1] Mochida I, Hirayama T, Kisamori S, Kawano S, Fujitsu H. Marked increase of SO_2 removal ability of poly(acrylonitrile)-based active carbon fiber by heat treatment at elevated temperatures. Langmuir 1992;8(9):2290−4.

[2] Moreno-Castilla C, Carrasco-Marín F, Utrera-Hidalgo E, Rivera-Utrilla J. Activated carbons as adsorbents of SO_2 in flowing air. Effect of their pore texture and surface basicity. Langmiur 1993;9:1378−83.

[3] Kisamori S, Mochida I, Fujitsu H. Roles of surface oxygen groups on poly(acrylonitrile)-based active carbon fibers in SO_2 adsorption. Langmuir 1994;10:1241−5.

[4] Kisamori S, Kuroda K, Kawano S, Mochida I, Matsumura Y, Yoshikawa M. Oxidative removal of SO_2 and recovery of H_2SO_4 over poly(acrylonitrile)-based active carbon fiber. Energy Fuels 1994;8:1337−40.

[5] Mochida I, Miyamoto S, Kuroda K, Kawano S, Yatsunami S, Korai Y, Yasutake A, Yoshikawa M. Adsorption and adsorbed species of SO_2 during its oxidative removal over pitch-based activated carbon fibers. Energy Fuels 1999;13:369−73.
[6] Mochida I, Miyamoto S, Kuroda K, Kawano S, Yatsunami S, Korai Y, Yasutake A, Yoshikawa M. Oxidative fixation of SO_2 into aqueous H_2SO_4 over a pitch-based active carbon fiber above room temperature. Energy Fuels 1999;13:374−8.
[7] Ling L, Li K, Miyamoto S, Korai Y, Kawano S, Mochida I. Removal of SO_2 over ethylene tar pitch and cellulose based activated carbon fibers. Carbon 1999;37:499−504.
[8] Raymundo-Pinero E, Cazorla-Amoros D, Salinas-Martinez de Lecea C, Linares-Solano A. Factors controlling the SO_2 removal by porous carbons: relevance of the SO_2 oxidation step. Carbon 2000;38(3):335−44.
[9] Raymundo-Pinero E, Cazorla-Amoros D, Linares-Solano A. The role of different nitrogen functional groups on the removal of SO_2 from flue gases by N-doped activated carbon powders and fibres. Carbon 2003;41:1925−32.
[10] Raymundo-Pinero E, Cazorla-Amoros D, Linares-Solano A. Temperature programmed desorption study on the mechanism of SO_2 oxidation by activated carbon and activated carbon fibres. Carbon 2001;39(2):231−42.
[11] Daley MA, Mangun CL, DeBarr JA, Riha S, Lizzio AA, Donnals GL, Economy J. Adsorption of SO_2 onto oxidized and heat-treated activated carbon fibers (ACFs). Carbon 1997;35(3):411−7.
[12] Qiao WM, Yoon SH, Korai Y, Mochida I, Inoue S, Sakurai T, Shimohara T. Preparation of activated carbon fibers from polyvinyl chloride. Carbon 2004;42:1327−31.
[13] Lizzio AA, DeBarr JA. Mechanism of SO_2 removal by carbon. Energy Fuels 1997;11:284−91.
[14] Adib F, Bagreev A, Bandosz TJ. Analysis of the relationship between H_2S removal capacity and surface properties of unimpregnated activated carbons. Environ Sci Technol 2000;34:686−92.
[15] Lee WH, Reucroft PJ. Vapor adsorption on coal- and wood-based chemically activated carbons: (III) NH_3 and H_2S adsorption in the low relative pressure range. Carbon 1999;37(1):21−6.
[16] Adib F, Bagreev A, Bandosz TJ. Effect of pH and surface chemistry on the mechanism of H_2S removal by activated carbons. J Colloid Interface Sci 1999;216:360−9.
[17] Adib F, Bagreev A, Bandosz TJ. Effect of surface characteristics of wood-based activated carbons on adsorption of hydrogen sulfide. J Colloid Interface Sci 1999;214:407−15.
[18] Bandosz TJ. On the adsorption/oxidation of hydrogen sulfide on activated carbons at ambient temperatures. J Colloid Interface Sci 2002;246:1−20.
[19] Adib F, Bagreev A, Bandosz TJ. Adsorption/oxidation of hydrogen sulfide on nitrogen-containing activated carbons. Langmuir 2000;16:1980−6.
[20] Lu GQ, Lau DD. Characterisation of sewage sludge-derived adsorbents for H_2S removal. Part 2: surface and pore structural evolution in chemical activation. Gas Sep Purif 1996;10:103−11.
[21] Bagreev A, Bashkova S, Locke DC, Bandosz TJ. Sewage sludge-derived materials as efficient adsorbents for removal of hydrogen sulfide. Environ Sci Technol 2001;35:1537−43.
[22] Bagreev A, Locke D, Bandosz TJ. H2S adsorption/oxidation on adsorbents obtained from pyrolysis of sewage-sludge-derived fertilizer using zinc chloride activation. Ind Eng Chem Res 2001;40:3502−10.

[23] Bagreev A, Bandosz TJ. H_2S adsorption/oxidation on materials obtained using sulfuric acid activation of sewage sludge-derived fertilizer. J Colloid Interface Sci 2002;252:188−94.
[24] Bagreev A, Bandosz TJ. Efficient hydrogen sulfide adsorbents obtained by pyrolysis of sewage sludge derived fertilizer modified with spent mineral oil. Environ Sci Technol 2004;38:345−51.
[25] Bandosz TJ, Block KA. Municipal sludge-industrial sludge composite desulfurization adsorbents: synergy enhancing the catalytic properties. Environ Sci Technol 2006;40:3378−83.
[26] Ros A, Montes-Moran MA, Fuente E, Nevskaia DM, Martin MJ. Dried sludges and sludge-based chars for H_2S removal at low temperature: influence of sewage sludge characteristics. Environ Sci Technol 2006;40:302−9.
[27] Ros A, Lillo-Rodenas MA, Canals-Batlle C, Fuente E, Montes-Moran MA, Martin MJ, Linares- Solano A. A new generation of sludge-based adsorbents for H_2S abatement at room temperature. Environ Sci Technol 2007;41:4375−81.
[28] Ansari A, Bandosz TJ. Inorganic-organic phase arrangement as a factor affecting gas-phase desulfurization on catalytic carbonaceous adsorbents. Environ Sci Technol 2005;39:6217−24.
[29] Molina-Sabio M, Gonzalez JC, Rodriguez-Reinoso F. Adsorption of NH3 and H_2S on activated carbon and activated carbon-sepiolite pellets. Carbon 2004;42(2):448−50.
[30] Burchell TD, Judkins RR, Rogers MR, Williams AM. A novel process and material for the separation of carbon dioxide and hydrogen sulfide gas mixtures. Carbon 1997;35:1279−94.
[31] Przepiorskia J, Yoshida S, Oya A. Structure of K_2CO_3-loaded activated carbon fiber and its deodorization ability against H_2S gas. Carbon 1999;37:1881−90.
[32] Ellison MD, Crotty MJ, Koh DK, Spray RL, Tate KE. Adsorption of NH3 and NO_2 on single-walled carbon nanotubes. J Phys Chem B 2004;108:7938−43.
[33] Feng X, Irle S, Witek H, Morokuma K, Vidic R, Borguet E. Sensitivity of ammonia interaction with single-walled carbon nanotube bundles to the presence of defect sites and functionalities. J Am Chem Soc 2005;127(30):10533−8.
[34] Stoeckli F, Guillot A, Slasli AM. Specific and non-specific interactions between ammonia and activated carbons. Carbon 2004;42(8−9):1619−24.
[35] Domingo-Garcia M, Groszek AJ, Lopez-Garzon FJ, Perez-Mendoza M. Dynamic adsorption of ammonia on activated carbons measured by flow microcalorimetry. Appl Catal A 2002;233(1−2):141−50.
[36] Park SJ, Kim KD. Adsorption behaviors of CO_2 and NH_3 on chemically surface-treated activated carbons. J Colloid Interface Sci 1999;212(1):186−9.
[37] Guo J, Xu WS, Chena YL, Lua AC. Adsorption of NH_3 onto activated carbon prepared from palm shells impregnated with H_2SO_4. J Colloid Interface Sci 2005;281(2):285−90.
[38] Petit C, Kante K, Bandosz TJ. The role of sulfur-containing groups in ammonia retention on activated carbons. Carbon 2010;48:654−67.
[39] LeLeuch L, Bandosz TJ. Carbonaceous adsorbents for NH_3 removal at room temperature. Carbon 2007;45(3):568−78.
[40] Batllea C, Rosa A, Lillo-Rodenas MA, Fuente E, Montes-Moran MA, Martin MJ, Linares-Solano A. Carbonaceous adsorbents for NH_3 removal at room temperature. Carbon 2008;46(1):176−8.
[41] Petit C, Karwacki C, Peterson G, Bandosz TJ. Interactions of ammonia with the surface of microporous carbon impregnated with transition metal chlorides. J Phys Chem C 2007;111(34):12705−14.

[42] Petit C, Bandosz TJ. Activated carbons modified with aluminium-zirconium polycations as adsorbents for ammonia. Microp Mesop Mater 2008;114(1–3):137–47.
[43] Petit C, Bandosz TJ. Role of surface heterogeneity in the removal of ammonia from air on micro/mesoporous activated carbons modified with molybdenum and tungsten oxides. Microp Mesop Mater 2009;118(1–3):61–7.
[44] Zawadzki J, Wisniewski M. In situ characterization of interaction of ammonia with carbon surface in oxygen atmosphere. Carbon 2003;41(12):2257–67.
[45] Yadanaparthi SKR, Graybill D, von Wandruszka R. Adsorbents for the removal of arsenic, cadmium, and lead from contaminated waters. J Hazard Mater 2009;171:1–15.
[46] Luo X, Wang C, Wang L, Deng F, Luo S, Tu X, Au C. Nanocomposites of graphene oxide-hydrated zirconium oxide for simultaneous removal of As(III) and As(V) from water. Chem Eng J 2013;220:98–106.
[47] Yin Y, Zhou T, Luo H, Geng J, Yu W, Jiang Z. Adsorption of arsenic by activated charcoal coated zirconium-manganese nanocomposite; Performance and mechanism. Colloids Surf A 2019;575:318–28.
[48] Luo J, Luo X, Hu C, Crittenden JC, Qu J. Zirconia (ZrO_2) embedded in carbon nanowires via electrospinning for efficient arsenic removal from water combined with DFT studies. ACS Appl Mater Interfaces 2016;8:18912–21.
[49] Luo X, Wang C, Luo S, Dong R, Tu X, Zeng G. Adsorption of As(III) and As(V) from water using magnetite Fe_3O_4-reduced graphite oxide-MnO_2 nanocomposites. Chem Eng J 2012;187:45–52.
[50] Chandra V, Park J, Chun Y, Lee JW, Hwang I-C, Kim KS. Water-dispersible magnetite-reduced graphene oxide composites for arsenic removal. ACS Nano 2010;4(7):3979–86.
[51] Lou Z, Cao Z, Xu J, Zhou X, Zhu J, Liu X, Ali Baig S, Zhou J, Xu X. Enhanced removal of As(III)/(V) from water by simultaneously supported and stabilized Fe-Mn binary oxide nanohybrids. Chem Eng J 2017;322:710–21.
[52] Kanel SR, Manning B, Charlet L, Choi H. Removal of arsenic(III) from groundwater by nanoscale zero-valent iron. Environ Sci Technol 2005;39:1291–8.
[53] Zhu HJ, Jia Y, Wu X, Wang H. Removal of arsenic from water by supported nano zero-valent iron on activated carbon. J Hazard Mater 2009;172:1591–6.
[54] Gu ZM, Deng BL, Yang J. Synthesis and evaluation of iron-containing ordered mesoporous carbon (FeOMC) for arsenic adsorption. Microp Mesop Mater 2007;102:265–73.
[55] Wang C, Luo H, Zhang Z, Wu Y, Zhang J, Chen S. Removal of As(III) and As(V) from aqueous solutions using nanoscale zero valent iron-reduced graphite oxide modified composites. J Hazard Mater 2014;268:124–31.
[56] Vasudevan S, Lakshmi J. The adsorption of phosphate by graphene from aqueous solution. RSC Adv 2012;2:5234–42.
[57] Moreno-Castilla C. Adsorption of organic molecules from aqueous solutions on carbon materials. Carbon 2004;42:83–94.
[58] Adams LB, Boucher EA, Everett DH. Adsorption of organic vapors by Saran carbon fibers and powders. Carbon 1970;8:761–72.
[59] Silvestre-Albero A, Ramos-Fernandez JM, Martınez-Escandell M, Sepulveda-Escribano A, Silvestre- Albero J, Rodrıguez-Reinoso F. High saturation capacity of activated carbons prepared from mesophase pitch in the removal of volatile organic compounds. Carbon 2010;48:548–56.
[60] Das D, Gaur V, Verma N. Removal of volatile organic compound by activated carbon fiber. Carbon 2004;42:2949–62.

[61] Fuertes AB, Marbana G, Nevskaia DM. Adsorption of volatile organic compounds by means of activated carbon fibre-based monoliths. Carbon 2003;41:87–96.
[62] Zhang W, Chen L, Xu L, Dong H, Hu H, Xiao Y, Zheng M, Liu Y, Liang Y. Advanced nanonetwork- structured carbon materials for high-performance formaldehyde capture. J Colloid Interface Sci 2019;537:562–8.
[63] Wen Q, Li C, Cai Z, Zhang W, Gao H, Chen L, Zeng G, Shu X, Zhao Y. Study on activated carbon derived from sewage sludge for adsorption of gaseous formaldehyde. Bioresour Technol 2011;102(2):942–7.
[64] Rezaee A, Rangkooy H, Jonidi-Jafari A, Khavanin A. Surface modification of bone char for removal of formaldehyde from air. Appl Surf Sci 2013;286:235–9.
[65] Song Y, Qiao W, Yoon SH, Mochida I, Guo Q, Lang L. Removal of formaldehyde at low concentration using various activated carbon fibers. J Appl Polym Sci 2007;106:2151–7.
[66] Lee KJ, Jin M, Shiratori N, Yoon SH, Jang J. Toward an effective adsorbent for polar pollutants: formaldehyde adsorption by activated carbon. J Hazard Mater 2013;260(1):82–8.
[67] Lee KJ, Shiratori N, Gang HL, Jin M, Mochida I, Yoon SH, Jang J. Activated carbon nanofiber produced from electrospun polyacrylonitrile nanofiber as a highly efficient formaldehyde adsorbent. Carbon 2010;48(15):4248–55.
[68] Boonamnuayvitaya V, Saeung S, Tanthapanichakoon W. Preparation of activated carbons from coffee residue for the adsorption of formaldehyde. Sep Purif Technol 2005;42(2):159–68.
[69] Carter EM, Katz LE, Speitel Jr SG, Ramirez D. Gas-phase formaldehyde adsorption isotherm studies on activated carbon: correlations of adsorption capacity to surface functional group density. Environ Sci Technol 2011;45(15):6498–503.
[70] Asada T, Ishihara S, Yamane T, Toba A, Yamada A, Oikawa K. Science of bamboo charcoal: study on carbonizing temperature of bamboo charcoal and removal capability of harmful gases. J Health Sci 2002;48:473–9.
[71] Mizuta K, Matsumoto T, Hatate Y, Nishihara K, Nakanishi T. Removal of nitrate-nitrogen from drinking water using bamboo powder charcoal. Bioresour Technol 2004;95:255–7.
[72] Huang Z-H, Kang F, Zheng Y-P, Yang J-B, Liang K-M. Adsorption of trace polar methy-ethyl-ketone and non-polar benzene vapors on viscose rayon-based activated carbon fibers. Carbon 2002;40:1363–7.
[73] Foster KL, Fuerman RG, Economy J, Larson SM, Rood MJ. Adsorption characteristics of trace volatile organic compounds in gas streams onto activated carbon fibers. Chem Mater 1992;4(5):1068–73.
[74] Mangun CL, Daley MA, Braatz RD, Economy J. Effect pf pore size on adsorption of hydrocarbons in phenolic-based activated carbon fibers. Carbon 1998;36:123–31.
[75] Javed H, Luong DX, Lee C-G, Zhang D, Tour JM, Alvarez Pedro JJ. Efficient removal of bisphenol- A by ultra-high surface area porous activated carbon derived from asphalt. Carbon 2018;140:441–8.
[76] Moreno-Castilla C, Rivera-Utrilla J, Lopez-Ramon MV, Carrasco-Marin F. Adsorption of some substituted phenols on activated carbons from a bituminous coal. Carbon 1995;33:845–51.
[77] Singh S, Yenkie MKN. Competitive adsorption of some hazardous organic pollutants from their binary and ternary solutions onto granular activated carbon columns. Water Air Soil Pollut 2004;156:275–86.
[78] Liu Q-S, Zheng T, Wang P, Jiang J-P, Li N. Adsorption isotherm, kinetic and mechanism studies of some substituted phenols on activated carbon fibers. Chem Eng J 2010;157:348–56.

[79] Nabais JMV, Suhas JAG, Carrott PJM, Laginhas C, Roman S. Phenol removal onto novel activated carbons made from lignocellulosic precursors: influence of surface properties. J Hazard Mater 2009;167:904–10.
[80] Przepiorski J. Enhanced adsorption of phenol from water by ammonia-treated activated carbon. J Hazard Mater B 2006;135:453–6.
[81] Yuan M, Tong S, Zhao S, Jia CQ. Adsorption of polycyclic aromatic hydrocarbons from water using petroleum coke-derived porous carbon. J Hazard Mater 2010;181:1115–20.
[82] Valderrama C, Cortina JL, Farran A, Gamisans X, Lao C. Kinetics of sorption of polyaromatic hydrocarbons onto granular activated carbon and macronet hyper-cross-linked polymers (MN200). J Colloid Interface Sci 2007;310:35–46.
[83] Kong H, He J, Wu H, Wu H, Gao Y. Phenanthrene removal from aqueous solution on sesame stalk- based carbon. Clean Soil Air Water 2012;40:752–9.
[84] Xiao XM, Tian F, Yan YJ, Wu ZL, Wu Z, Cravotto G. Adsorption behavior of phenanthrene onto coal-based activated carbon prepared by microwave activation. Korean J Chem Eng 2015;32:1129–36.
[85] Gong Z, Alef K, Wilke B-M, Li P. Activated carbon adsorption of PAHs from vegetable oil used in soil remediation. J Hazard Mater 2007;143:372–8.
[86] Luna FMT, Pontes-Filho AA, Trindade ED, Silva Jr IJ, Azevedo DCS, Cavalcante Jr CL. Removal of aromatic compounds from mineral naphthenic oil by adsorption. Ind Eng Chem Res 2008;47:3207–12.
[87] Mastral AM, Garcia T, Murillo R, Callén MS, López JM, Navarro MV. PAH mixture removal from hot gas by porous carbons. From model compounds to real conditions. Ind Eng Chem Res 2003;42:5280–6.
[88] Tamai H, Kojima S, Ikeuchi M, Mondori J, Kanata T, Yasuda H. Preparation of mesoporous activated carbon fibers and their adsorption properties. TANSO 1996;175:243–8 [in Japanese)].
[89] Tamai H, Ikeuchi M, Kojima S, Yasuda H. Extremely large mesoporous carbon fibers synthesized by the addition of rare earth metal complexes and their unique adsorption behaviors. Adv Mater 1997;9:55–8.
[90] Karolczyk J, Janus M, Przepiorski J. Adsorption of humic acid on mesoporous carbons prepared from poly(ethylene terephthalate) templated with magnesium compounds. Pol J Chem Technol 2012;14:95–9.
[91] Sakoda A, Suzuki M, Hirai R, Kawazoe K. Trihalomethane adsorption on activated carbon fibers. Water Res 1991;25:219–25.
[92] Demirbas A. Agricultural based activated carbons for the removal of dyes from aqueous solutions: a review. J Hazard Mater 2009;167:1–9.
[93] Martin MJ, Artola A, Balaguer MD, Rigola M. Activated carbons developed from surplus sewage sludge for the removal of dyes from dilute aqueous solutions. Chem Eng J 2003;94:231–9.
[94] Otero M, Rozada F, Calvo LF, Garcia AI, Moran A. Elimination of organic water pollutants using adsorbents obtained from sewage sludge. Dyes Pigm 2003;57:55–65.
[95] Tan IAW, Hameed BH, Ahmed AL. Studies on the basic dye adsorption by palm fiber activated carbon. Chem Eng J 2007;127:111–9.
[96] Vinu A, Streb C, Murugesan V, Hartmann M. Adsorption of cytochrome C on new mesoporous carbon molecular sieves. J Phys Chem B 2003;107:8297–9.
[97] Kosonen H, Valkama S, Nykaenen A, Toivanen M, Brinke GT, Ruokolainen J, Ikkala O. Functional porous structures based on the pyrolysis of cured templates of block copolymer and phenolic resin. Adv Mater 2006;18:201–5.

[98] Parker HL, Hunt AJ, Budarin VL, Shuttleworth PS, Miller KL, Clark JH. The importance of being porous: polysaccharide-derived mesoporous materials for use in dye adsorption. RSC Adv 2012;2:8992−7.
[99] Han S, Sohn K, Hyeon T. Fabrication of new nanoporous carbons through silica templates and their application to the adsorption of bulky dyes. Chem Mater 2000;12:3337−41.
[100] Hoang NB, Nguyen TT, Nguyen TS, Bui TPQ, Bach LG. The application of expanded graphite fabricated by microwave method to eliminate organic dyes in aqueous solution. Cogent Eng 2019;6:1584939.
[101] Tryba B, Morawski AW, Kalenczuk RJ, Inagaki M. Exfoliated graphite as a new sorbent for removal of engine oils from wastewater. Spill Sci Technol Bull 2003;8:569−71.
[102] Carvallho MN, Da Silva KS, Sales DCS, Freire EMPL, Sobrinho MAM, Ghislandi MG. Dye removal from textile industrial effluents by adsorption on exfoliated graphite nanoplatelets: kinetic and equilibrium studies. Water Sci Technol 2016;73:2189−98.
[103] Liu T, Li Y, Du Q, Sun J, Jiao Y, Yang G, Wang Z, Xia Y, Zhang W, Wang K, Zhu H, Wu D. Adsorption of methylene blue from aqueous solution by graphene. Colloid Surf B 2012;90:197−203.
[104] Zhao J, Ren W, Cheng HM. Graphene sponge for efficient and repeatable adsorption and desorption of water contaminations. J Mater Chem 2012;22:20197−202.
[105] Ai L, Jiang J. Removal of methylene blue from aqueous solution with self-assembled cylindrical graphene−carbon nanotube hybrid. Chem Eng J 2012;192:156−63.
[106] Wang D-W, Li F, Lu GQ, Cheng H-M. Synthesis and dye separation performance of ferromagnetic hierarchical porous carbon. Carbon 2008;46:1593−9.
[107] Wang C, Feng C, Gao Y, Ma X, Wu Q, Wang Z. Preparation of a graphene-based magnetic nanocomposite for the removal of an organic dye from aqueous solution. Chem Eng J 2011;173:92−7.
[108] Dai H, Peng X, Yang W, et al. Synthesis and characterization of graphitic magnetic mesoporous nanocomposite and its application in dye adsorption. J Mol Liquids 2018;253:197−204.
[109] Krawiec P, Kockrick E, Borchardt L, Geiger D, Corma A, Kaskel S. Ordered mesoporous carbide derived carbons: novel materials for catalysis and adsorption. J Phys Chem C 2009;113(18):7755−61.
[110] Tao G, Zhang L, Hua Z, Chen Y, Guo L, Zhang J, Shu Z, Gao J, Chen H, Wu W, Liu Z, Shi J. Highly efficient adsorbents based on hierarchically macro/mesoporous carbon monoliths with strong hydrophobicity. Carbon 2014;66:547−59.
[111] Shen WE, Wen SZ, Cao NZ, Zheng L, Zhou W, Liu Y, Gu J. Expanded graphite A new kind of biomedical material. Carbon 1999;37:356−8.
[112] Kang F, Zheng Y, Zhao H, Wang H, Wang L, Shen W, Inagaki M. Sorption of heavy oils and biomedical liquids into exfoliated graphite − research in China. N Carbon Mater 2003;18:161−73.
[113] Yamaguchi NU, Bergamasco R, Hamoudi S. Magnetic $MnFe_2O_4$-graphene hybrid composite for efficient removal of glyphosate from water. Chem Eng J 2016;295:391−402.
[114] Cal MP, Rood MJ, Larson SM. Gas phase adsorption of volatile organic compounds and water vapor on activated carbon cloth. Energy Fuels 1997;11(2):311−5.
[115] Vagner C, Finqueneisel G, Zimny T, Weber JV. Water vapour adsorption on activated carbons: comparison and modelling of the isotherms in static and dynamic flow conditions. Fuel Process Technol 2002;77−78:409−14.

[116] Alcaniz-Monge J, Linares-Solano A, Rand B. Mechanism of adsorption of water in carbon micropores as revealed by a study of activated carbon fibers. J Phys Chem B 2002;106:3209–16.
[117] Inagaki M, Ohmura M, Tanaike O. Reversible water adsorption into carbonized polyimide films in ambient atmosphere. Carbon 2002;40:2502–5.
[118] Ohta N, Nishi Y, Morishita T, Tojo T, Inagaki M. Preparation of microporous carbon films from luorinated aromatic polyimides. Carbon 2008;40:1350–7.
[119] Ohta N, Nishi Y, Morishita T, Tojo T, Inagaki M. Water vapor adsorption of microporous carbon films prepared from fluorinated aromatic polyimides. Adsorpt Sci Technol 2008;26:373–82.
[120] Hou P-X, Orikasa H, Yamazaki T, Matsuoka K, Tomita A, Setoyama N, Fukushima Y, Kyotani T. Synthesis of nitrogen-containing microporous carbon with a highly ordered structure and effect of nitrogen doping on H_2O adsorption. Chem Mater 2005;17:5187–93.
[121] Horikawa T, Sekida T, Hayashi J, Katoh M, Do DD. A new adsorption-desorption model for water adsorption in porous carbons. Carbon 2011;49(2):416–24.
[122] Horikawa T, Sakao N, Sekida T, Hayashi J, Do DD, Katoh M. Preparation of nitrogen-doped porous carbon by ammonia gas treatment and the effects of N-doping on water adsorption. Carbon 2012;50:1833–42.
[123] Yang C-M, Kaneko K. Adsorption properties of nitrogen-alloyed activated carbon fiber. Carbon 2001;39:1075–82.
[124] Kockrick E, Schrage C, Borchardt L, Klein N, Rose M, Senkovska I, Kaskel S. Ordered mesoporous carbide derived carbons for high pressure gas storage. Carbon 2010;48:1707–17.
[125] Bandosz TJ, Jagiełło J, Schwarz JA. Effect of surface chemistry on sorption of water and methanol on activated carbons. Langmuir 1996;12:6480–6.
[126] Cossarutto L, Zimny T, Kaczmarczyk J, Siemieniewska T, Bimer J, Weber JV. Transport and sorption of water vapour in activated carbons. Carbon 2001;39:2339–46.
[127] Rudisill EN, Hacskaylo JJ, LeVan MD. Coadsorption of hydrocarbons and water on BPL activated carbon. Ind Eng Chem Res 1992;31(4):1122–30.
[128] Eissmann RN, LeVan MD. Coadsorption of organic compounds and water vapor on BPL activated carbon. 2. 1,1,2-Trichloro-1,2,2-trifluoroethane and dichloromethane. Ind Eng Chem Res 1993;32(11):2752–7.
[129] Inagaki M, Morishita T, Kuno A, Kito T, Hirano M, Suwa T, Kusakawa K. Carbon foams prepared from polyimide using urethane foam template. Carbon 2004;42:497–502.
[130] Ohta N, Nishi Y, Morishita T, Ieko Y, Ito A, Inagaki M. Preparation of microporous carbon foams for adsorption/desorption of water vapor in ambient air. New Carbon Mater 2008;23:216–20.
[131] Barton SS, Evans MJB. The adsorption of water vapor by porous carbon. Carbon 1991;29:1099–105.
[132] Lavanchy A, Stoeckli F. Dynamic adsorption, in active carbon beds, of vapour mixtures corresponding to miscible and immiscible liquids. Carbon 1999;37:315–21.
[133] Kaneko K, Hanzawa Y, Iiyama T, Kanda T, Suzuki T. Cluster-mediated water adsorption on carbon nanopores. Adsorption 1999;5:7–13.
[134] Do DD, Do HD. A model for water adsorption in activated carbon. Carbon 2000;38:767–73.
[135] Neitsch M, Heschel W, Suckow M. Water vapor adsorption by activated carbon: a modification to the isotherm model of Do and Do. Carbon 2001;39:1437–8.

[136] Sridhar S, Smitha B, Aminabhavi TM. Separation of carbon dioxide from natural gas mixtures through polymeric membranes - a review. Sep Purif Rev 2007;36:113–74.
[137] Balat H, Oz C. Technical and economic aspects of carbon capture and storage - a review. Energy Explor Exploit 2007;25:357–92.
[138] Yang H, Xu Z, Fan M, Gupta R, Slimane RB, Bland AE, Wright I. Progress in carbon dioxide separation and capture: a review. J Environ Sci China 2008;20:14–27.
[139] Morris RE, Wheatley PS. Gas storage in nanoporous materials. Angew Chem Int Ed 2008;47:4966–81.
[140] Yu KMK, Curcic I, Gabriel J, Tsang SCE. Recent advances in CO_2 capture and utilization. ChemSusChem 2008;1:893–9.
[141] Radosz M, Hu X, Krutkramelis K, Shen Y. Flue-gas carbon capture on carbonaceous sorbents: toward a low-cost multifunctional carbon filter for "Green" energy producers. Ind Eng Chem Res 2008;47:3783–94.
[142] Choi S, Drese JH, Jones CW. Adsorbent materials for carbon dioxide capture from large anthropogenic point sources. ChemSusChem 2009;2:796–854.
[143] D'Alessandro DM, Smit B, Long JR. Carbon dioxide capture: prospects for new materials. Angew Chem Int Ed 2010;49:6058–82.
[144] Bae YS, Snurr RQ. Development and evaluation of porous materials for carbon dioxide separation and capture. Angew Chem Int Ed 2011;50:11586–96.
[145] Olivares-Marın M, Maroto-Valer M. Development of adsorbents for CO_2 capture from waste materials: a review. Greenhouse Gases Sci Technol 2012;2:20–35.
[146] Lopez MAB, Martınez-Alonso A, Tascon JMD. N_2 and CO_2 adsorption on activated carbon fibres prepared from Nomex chars. Carbon 2000;38:1177–82.
[147] Lozano-Castello D, Cazorla-Amoros D, Linares-Solano A. Usefulness of CO_2 adsorption at 273 K for the characterization of porous carbons. Carbon 2004;42:1233–42.
[148] Ana H, Feng B, Sub S. CO_2 capture capacities of activated carbon fibre-phenolic resin composites. Carbon 2009;47:2396–405.
[149] Wahby A, Ramos-Fernandez JM, Martinez-Escandell M, Sepulveda-Escribano A, Silvestre-Albero J, Rodriguez-Reinoso F. High-surface-area carbon molecular sieves for selective CO_2 adsorption. ChemSusChem 2010;3:974–81.
[150] Maroto-Valer MM, Tang Z, Zhang Y. CO_2 capture by activated and impregnated anthracites. Fuel Process Technol 2005;86:1487–502.
[151] Wang L, Sun F, Hao F, Qu Z, Gao J, Liu M, Wang K, Zhao G, Qin Y. A green trace K_2CO_3 induced catalytic activation strategy for developing coal-converted activated carbon as advanced candidate for CO_2 adsorption and supercapacitors. Chem Eng J 2020;383:123205.
[152] Sevilla M, Al-Jumialy ASM, Fuertes AB, Mokaya R. Optimization of the pore structure of biomass- based carbons in relation to their use for CO_2 capture under low- and high- pressure regimes. ACS Appl Mater Interfaces 2018;10:1623–33.
[153] Narasimman R, Vijayan S, Prabhakaran K. Carbon foam with microporous cell wall and strut for CO_2 capture. RSC Adv 2014;4:578–82.
[154] Arami-Niya A, Rufford TE, Zhu Z. Activated carbon monoliths with hierarchical pore structure from tar pitch and coal powder for the adsorption of CO_2, CH_4 and N_2. Carbon 2016;103:115–24.
[155] Wang R, Wang P, Yan X, Lang J, Peng C, Xue Q. Promising porous carbon derived from celtuce leaves with outstanding supercapacitance and CO_2 capture performance. ACS Appl Mater Interfaces 2012;4:5800–6.
[156] Li D, Wang Y, Zhang X, Zhou J, Yang Y, Zhang Z, Wei L, Tian Y, Zhao X. Effects of compacting activated carbons on their volumetric CO_2 adsorption performance. Fuel 2020;262:116540.

[157] Hong L, Ju S, Liu X, Zhuang Q, Zhan G, Yu X. Highly selective CO_2 uptake in novel fishnet-like polybenzoxazine-based porous carbon. Energy Fuels 2019;33(11):11454−64.
[158] Vazhayal L, Wilson P, Prabhakaran K. Waste to wealth: lightweight, mechanically strong and conductive carbon aerogels from waste tissue paper for electromagnetic shielding and CO_2 adsorption. Chem Eng J 2020;381:122628.
[159] Ben T, Ren H, Ma S, Cao D, Lan J, Jing X, Wang W, Xu J, Deng F, Simmons JM, Qiu S, Zhu G. Targeted synthesis of a porous aromatic framework with high stability and exceptionally high surface area. Angew Chem Int Ed 2009;48:9457−60.
[160] Ben T, Li YQ, Zhu LK, Zhang DL, Cao DP, Xiang ZH, Yao XD, Qiu SL. Selective adsorption of carbon dioxide by carbonized porous aromatic framework (PAF). Energy Environ Sci 2012;5:8370−6.
[161] Zhang YM, Li BY, Williams K, Gao WY, Ma SM. A new microporous carbon material synthesized via thermolysis of a porous aromatic framework embedded with an extra carbon source for low- pressure CO_2 uptake. Chem Commun 2013;49:10269−71.
[162] Deng H-G, Jin S-L, Zhan L, Wang Y, Lu B, Qiao W, Ling L. Synthesis of porous carbons derived from metal-organic coordination polymers and their adsorption performance for carbon dioxide. New Carbon Mater 2012;27:194−9.
[163] Tian Z, Huang J, Zhang X, Shao G, He Q, Cao S, Yuan S. Ultra-microporous N-doped carbon from polycondensed framework precursor for CO_2 adsorption. Microp Mesop Mater 2018;257:19−26.
[164] Xing W, Liu C, Zhou Z, Zhou J, Wang G, Zhuo S, Xue Q, Song L, Yan Z. Oxygen-containing functional group-facilitated CO_2 capture by carbide-derived carbons. Nanoscale Res Lett 2014;9:189.
[165] Wang L, Yang RT. Significantly increased CO_2 adsorption performance of nano-structured templated carbon by tuning surface area and nitrogen doping. J Phys Chem C 2012;116:1099−106.
[166] Przepiorski J, Skrodzewicz M, Morawski AW. High temperature ammonia treatment of activated carbon for enhancement of CO_2 adsorption. Appl Surf Sci 2004;225:235−42.
[167] Sevilla M, Valle-Vigon P, Fuertes AB. N-doped polypyrrole based porous carbons for CO_2 capture. Adv Funct Mater 2011;21:2781−7.
[168] Yang H, Yuan Y, Tsang SCE. Nitrogen-enriched carbonaceous materials with hierarchical micro- mesopore structures for efficient CO_2 capture. Chem Eng J 2012;185−186:374−9.
[169] Nandi M, Okada K, Dutta A, Bhaumik A, Maruyama J, Derks D, Uyama H. Unprecedented CO_2 uptake over highly porous N-doped activated carbon monoliths prepared by physical activation. Chem Commun 2012;48:10283−5.
[170] Fan X, Zhang L, Zhang G, Shu Z, Shi J. Chitosan derived nitrogen-doped microporous carbons for high performance CO_2 capture. Carbon 2013;61:423−30.
[171] Sethia G, Sayari A. Nitrogen-doped carbons: remarkably stable materials for CO_2 capture. Energy Fuels 2014;28:2727−31.
[172] Fujiki J, Yogo K. The increased CO_2 adsorption performance of chitosan-derived activated carbons with nitrogen-doping. Chem Commun 2016;52:186−9.
[173] Chen T, Deng S, Wang B, Huang J, Wang Y, Yu G. CO_2 adsorption on crab shell derived activated carbons: contribution of micropores and nitrogen containing groups. RSC Adv 2015;5:48323−30.
[174] Wei HR, Deng SB, Hu BY, Chen ZH, Wang B, Huang J, et al. Granular bamboo-derived activated carbon for high CO_2 adsorption: the dominant role of narrow micropores. ChemSusChem 2012;5:2354−60.

[175] Chandra V, Yu SU, Kim SH, Yoon YS, Kim DY, Kwon AH, Meyyappan M, Kim KS. Highly selective CO_2 capture on N-doped carbon produced by chemical activation of polypyrrole functionalized graphene sheets. Chem Commun 2012;48:735–7.

[176] Sevilla M, Parra JB, Fuertes AB. Assessment of the role of micropore size and N-doping in CO_2 capture by porous carbons. ACS Appl Mater Interface Sci 2013;5:6360–8.

[177] Yang ZX, Xia YD, Zhu YQ. Preparation and gases storage capacities of N-doped porous activated carbon materials derived from mesoporous polymer. Mater Chem Phys 2013;141:318–23.

[178] Siriwardane RV, Shen M, Fisher EP, Poston JA. Adsorption of CO_2 on molecular sieves and activated carbon. Energy Fuels 2001;15:279–84.

[179] Chen C, Kim J, Ahn WS. Efficient carbon dioxide capture over a nitrogen-rich carbon having a hierarchical micro-mesopore structure. Fuel 2012;95:360–4.

[180] Zhao L, Bacsik Z, Hedin N, Wei W, Sun Y, Antonietti M, et al. Carbon dioxide capture on amine-rich carbonaceous materials derived from glucose. ChemSusChem 2010;3:840–5.

[181] Nakashima M, Shimada S, Inagaki M, Centeno TA. On the adsorption of CO_2 by molecular sieve carbons – volumetric and gravimetric studies. Carbon 1995;33:1301–6.

[182] Wang JD, Sentorun-Shalaby C, Ma X, Song C. High-capacity and low-cost carbon-based "Molecular Basket" sorbent for CO_2 capture from flue gas. Energy Fuels 2011;25:456–8.

[183] Arenillas A, Smith KM, Drage TC, Snape CE. CO_2 capture using some fly ash-derived carbon materials. Fuel 2005;84:2204–10.

[184] Plaza MG, Pevida C, Arenillas A, Rubiera F, Pis JJ. CO_2 capture by adsorption with nitrogen enriched carbons. Fuel 2007;86:2204–12.

[185] Maroto-Valer MM, Lu Z, Zhang YZ. Sorbents for CO_2 capture from high carbon fly ashes. Waste Manag 2008;28:2320–8.

[186] Xu W, Hussain A, Liu Y. A review on modification methods of adsorbents for elemental mercury from flue gas. Chem Eng J 2018;346:692–711.

[187] Li YH, Lee CW, Gullett BK. Importance of activated carbon's oxygen surface functional groups on elemental mercury adsorption. Fuel 2003;82:451–7.

[188] Manchester S, Wang X, Kulaots I, Gao Y, Hurt RH. High capacity mercury adsorption on freshly ozone-treated carbon surfaces. Carbon 2008;46:518–24.

[189] Pudasainee D, Gupta R, Khan A. Performance evaluation of functionalized biocarbon for mercury capture. Energy Fuels 2019;33:5867–74.

[190] Lee SS, Lee JY, Keener TC. Mercury oxidation and adsorption characteristics of chemically promoted activated carbon sorbents. Fuel Process Technol 2009;90:1314–8.

[191] Wang XQ, Wang P, Ning P, Ma YX, Wang F, Guo XL, Lan Y. Adsorption of gaseous elemental mercury with activated carbon impregnated with ferric chloride. RSC Adv 2015;5:24899–907.

[192] Zhang B, Xu P, Qiu Y, Yu Q, Ma J, Wu H, Luo G, Xu M, Yao H. Increasing oxygen functional groups of activated carbon with non-thermal plasma to enhance mercury removal efficiency for flue gases. Chem Eng J 2015;263:1–8.

[193] Zhang B, Zeng X, Xu P, Chen J, Xu Y, Luo G, Xu M, Yao H. Using the novel method of nonthermal plasma to add Cl active sites on activated carbon for removal of mercury from flue gas. Environ Sci Technol 2016;50:11837–43.

[194] Zhang J, Duan Y, Zhou Q, Zhu C, She M, Ding W. Adsorptive removal of gas phase mercury by oxygen non-thermal plasma modified activated carbon. Chem Eng J 2016;294:281–9.

[195] Hu P, Duan Y, Ding W, Zhang J, Bai L, Li N, Wei H. Enhancement of mercury removal efficiency by activated carbon treated with nonthermal plasma in different atmospheres. Energy Fuels 2017;31:13852−8.
[196] Shen F, Liu J, Wu D, Dong Y, Liu F, Huang H. Design of O_2/SO_2 dual-doped porous carbon as superior sorbent for elemental mercury removal from flue gas. J Hazard Mater 2019;366:321−8.
[197] Zabihi M, Ahmadpour A, Asl AH. Removal of mercury from water by carbonaceous sorbents derived from walnut shell. J Hazard Mater 2009;167:230−6.
[198] Goyal M, Bhagat M, Dhawan R. Removal of mercury from water by fixed bed activated carbon columns. J Hazard Mater 2009;171:1009−15.
[199] Chandra V, Kim KS. Highly selective adsorption of Hg^{2+} by a polypyrrole-reduced graphene oxide composite. Chem Commun 2011;47:3942−4.
[200] Ismaiel AA, Aroua MK, Yusoff R. Palm shell activated carbon impregnated with task-specific ionic- liquids as a novel adsorbent for the removal of mercury from contaminated water. Chem Eng J 2013;225:306−14.
[201] Meena AK, Mishra GK, Rai PK, Rajagopal C, Nagar PN. Removal of heavy metal ions from aqueous solutions using carbon aerogel as an adsorbent. J Hazard Mater 2005;122:161−70.
[202] Tian G, Geng JX, Jin YD, Wang CL, Li SQ, Chen Z, Wang H, Zhao YS, Li SJ. Sorption of uranium(VI) using oxime grafted ordered mesoporous carbon CMK-5. J Hazard Mater 2011;190:442−50.
[203] Wang YQ, Zhang ZB, Liu YH, Cao XH, Liu YT, Li Q. Adsorption of U(VI) from aqueous solution by the carboxyl mesoporous carbon. Chem Eng J 2012;198:246−53.
[204] Gorka J, Mayes RT, Baggetto L, Veith GM, Dai S. Sonochemical functionalization of mesoporous carbon for uranium extraction from seawater. J Mater Chem A 2013;1:3016−26.
[205] Carboni M, Abney CW, Taylor-Pashow KML, Vivero-Escoto JL, Lin WB. Uranium sorption with functionalized mesoporous carbon materials. Ind Eng Chem Res 2013;52:15187−97.
[206] Jana P, Ganesan V. Synthesis, characterization and radionuclide (137Cs) trapping properties of a carbon foam. Carbon 2009;47:3001−9.
[207] Babel S, Kurniawan TA. Low-cost adsorbents for heavy metals uptake from contaminated water: a review. J Hazard Mater 2003;97:219−43.
[208] Machida M, Mochimaru T, Tatsumoto H. Lead(II) adsorption onto the graphene layer of carbonaceous materials in aqueous solution. Carbon 2006;44:2681−8.
[209] Huang Z-H, Zhang F, Wang M-X, Lv R, Kang F. Growth of carbon nanotubes on low-cost bamboo charcoal for Pb(II) removal from aqueous solution. Chem Eng J 2012;184:193−7.
[210] Wang SG, Gong WX, Liu XW, Yao YW, Gao BY, Yue QY. Removal of lead(II) from aqueous solution by adsorption onto manganese oxide-coated carbon nanotubes. Sep Purif Technol 2007;58:17−23.
[211] Leyva-Ramos R, Berber-Mendoza MS, Salazar-Rabago J, Guerrero-Coronado RM, Mendoza- Barron J. Adsorption of lead(II) from aqueous solution onto several types of activated carbon fibers. Adsorption 2011;17:515−26.
[212] Cao X, Ma L, Gao B, Harris W. Dairy-manure derived biochar effectively sorbs lead and atrazine. Environ Sci Technol 2009;43:3285−91.
[213] Liu H, Dai P, Zhang J, Zhang C, Bao N, Cheng C, Ren L. Preparation and evaluation of activated carbons from lotus stalk with trimethyl phosphate and tributyl phosphate activation for lead removal. Chem Eng J 2013;228:425−34.

[214] Liu H, Zhang J, Ngo HH, Guo W, Wu H, Cheng C, Guo Z, Zhang C. Carbohydrate-based activated carbon with high surface acidity and basicity for nickel removal from synthetic wastewater. RSC Adv 2015;5:52048–56.
[215] Monser L, Adhoum N. Modified activated carbon for the removal of copper, zinc, chromium and cyanide from wastewater. Sep Purif Technol 2002;26:137–46.
[216] Jaramillo J, Gomez-Serrano V, Alvarez PM. Enhanced adsorption of metal ions onto functionalized granular activated carbons prepared from cherry stones. J Hazard Mater 2009;161:670–6.
[217] Liu W, Zhang J, Cheng C, Tian G, Zhang C. Ultrasonic-assisted sodium hypochlorite oxidation of activated carbons for enhanced removal of Co(II) from aqueous solutions. Chem Eng J 2011;175:24–32.
[218] Machida M, Fotoohi B, Amamo Y, Ohba T, Kanoh H, Mercier L. Cadmium(II) adsorption using functional mesoporous silica and activated carbon. J Hazard Mater 2012;221–222:220–7.
[219] Li Y-H, Wang S, Luan Z, Ding J, Xu C, Wu D. Adsorption of cadmium(II) from aqueous solution by surface oxidized carbon nanotubes. Carbon 2003;41:1057–62.
[220] Babic BM, Milonjic SK, Polovina MJ, Cupic S, Kaludjerovic BV. Adsorption of zinc, cadmium and mercury ions from aqueous solutions on an activated carbon cloth. Carbon 2002;40:1109–15.
[221] Sato S, Yoshihara K, Moriyama K, Machida M, Tatsumoto H. Influence of activated carbon surface acidity on adsorption of heavy metal ions and aromatics from aqueous solution. Appl Surf Sci 2007;253:8554–9.
[222] Gupta VK, Jain CK, Ali I, Sharma M, Saini VK. Removal of cadmium and nickel from wastewater using bagasse fly ash - a sugar industry waste. Water Res 2003;37:4038–44.
[223] Ulmanu M, Maranón E, Fernández Y, Castrillón L, Anger I, Dumitriu D. Removal of copper and cadmium ions from diluted aqueous solutions by low cost and waste material adsorbents. Water Air Soil Pollut 2003;142:357–73.
[224] Mohan D, Pittman Jr CU. Activated carbons and low cost adsorbents for remediation of tri- and hexavalent chromium from water. J Hazard Mater 2006;137:762–811.
[225] Aggarwal D, Goyal M, Bansal RC. Adsorption of chromium by activated carbon from aqueous solution. Carbon 1999;37(12):1989–97.
[226] Park S-J, Jang Y-S. Pore structure and surface properties of chemically modified activated carbons for adsorption mechanism and rate of Cr(VI). J Colloid Interface Sci 2002;249(2):458–63.
[227] Rivera-Utrilla J, Sanchez-Polo M. Adsorption of Cr(III) on ozonized activate carbon. Importance of Cπ–cation interactions. Water Res 2003;37:3335–40.
[228] Zhang Z, Luo H, Jiang X, Jiang Z, Yang C. Synthesis of reduced graphene oxide-montmorillonite nanocomposite and its application in hexavalent chromium removal from aqueous solutions. RSC Adv 2015;5:47408.
[229] Jiang X, Luo H, Yin Y, Zhou W. Facile synthesis of MoS_2/reduced graphene oxide composites for efficient removal of Cr(VI) from aqueous solutions. RSC Adv 2017;7:24149.
[230] Zhang L, Luo H, Liu P, Fang W, Geng J. A novel modified graphene oxide/chitosan composite used as an adsorbent for Cr(VI) in aqueous solutions. Int J Biol Macromol 2016;87:586–96.
[231] Song X, Gunawan P, Jiang R, Leong SSJ, Wang K, Xu R. Surface activated carbon nanospheres for fast adsorption of silver ions from aqueous solutions. J Hazard Mater 2011;194:162–8.
[232] Kadirvelu K, Goel J, Rajagopal C. Sorption of lead, mercury and cadmium ions in multi-component system using carbon aerogel as adsorbent. J Hazard Mater 2008;153:502–7.

[233] Faur-Brasquet C, Kadirvelu K, Le Cloirec P. Removal of metal ions from aqueous solution by adsorption onto activated carbon cloths: adsorption competition with organic matter. Carbon 2002;40:2387−92.
[234] Li YH, Ding J, Luan ZK, Di Z, Zhu Y, Xu C, Wu D, Wei B. Competitive adsorption of Pb^{2+}, Cu^{2+} and Cd^{2+} ions from aqueous solutions by multiwalled carbon nanotubes. Carbon 2003;41:2787−92.
[235] Shim J-W, Park S-J, Ryu S-K. Effect of modification with HNO_3 and NaOH on metal adsorption by pitch-based activated carbon fibers. Carbon 2001;39:1635−42.
[236] Adebajo MO, Frost RL, Kloprogge JT, Carmody O, Kokot S. Porous materials for oil spill cleanup: a review of synthesis and absorbing properties. J Porous Mater 2003;10:159−70.
[237] Toyoda M, Inagaki M. Heavy oil sorption using exfoliated graphite new application of exfoliated graphite to protect heavy oil pollution. Carbon 2000;38:199−210.
[238] Inagaki M, Nishi Y, Iwashita N, Toyoda M. Mechanism of heavy oil sorption into porous carbons. Fresenius Environ Bull 2004;13:183−9.
[239] Toyoda M, Iwashita N, Inagaki M. 4. Sorption of heavy oils into carbon materials. In: Radovic LR, editor. Chemistry and physics of carbon, vol. 30. Marcel Dekker; 2007. p. 177−234. 30.
[240] Inagaki M, Toyoda M, Iwashita N, Kang F. 27 Sorption of viscous organics by microporous carbons. In: Bottani EJ, Tascon JMD, editors. Adsorption by carbon. Elsevier; 2008. p. 711−34.
[241] Inagaki M, Kang F, Toyoda M, Konno H. 14. Carbon materials for spilled-oil recovery. In: Advanced materials science and engineering of carbon. Elsevier and Tsinghua Univ. Press; 2013.
[242] Takeuchi K, Fujishige M, Kitazawa H, Akuzawa N, Medina JO, Morelos-Gomez A, Cruz-Silva R, Araki T, Hayashi T, Terrones M, Endo M. Oil sorption by exfoliated graphite from dilute oil−water emulsion for practical applications in produced water treatments. J Water Process Eng 2015;8:91−8.
[243] Takeuchi K, Kitazawa H, Fujishige M, Akuzawa N, Medina JO, Morelos-Gomez A, Cruz-Silva R, Araki T, Hayashi T, Endo M. Oil removing properties of exfoliated graphite in actual produced water treatment. J Water Process Eng 2017;20:226−31.
[244] Inagaki M, Kawahara A, Konno H. Recovery of heavy oil from contaminated sand by using exfoliated graphite. Desalination 2004;170:77−82.
[245] Nishi Y, Dai G, Iwashita N, Iwashita N, Sawada Y, Inagaki M. Evaluation of sorption behavior of heavy oil into exfoliated graphite by wicking test. Mater Sci Res Intl 2002;8:243−8.
[246] Nishi Y, Iwashita N, Sawada Y, Inagaki M. Sorption kinetics of heavy oil into porous carbons. Water Res 2002;36:5029−36.
[247] Inagaki M, Nagata T, Suwa T, Toyoda M. Sorption kinetics of various oils onto exfoliated graphite. N Carbon Mater 2006;21:97−101.
[248] Toyoda M, Aizawa J, Inagaki M. Sorption and recovery of heavy oils by using exfoliated graphite. Desalination 1998;115:199−201.
[249] Toyoda M, Moriya K, Inagaki M. Temperature dependences of heavy oil sorption on exfoliated graphite. Sekiyu Gakkai shi 2001;44:169−72 [in Japanese].
[250] Toyoda M, Nishi Y, Iwashita N, Inagaki M. Sorption and recovery of heavy oils using exfoliated graphite Part IV: discussion of high oil sorption of exfoliated graphite. Desalination 2002;151:139−44.
[251] Toyoda M, Inagaki M. Sorption and recovery of heavy oils by using exfoliated graphite. Spill Sci Technol Bull 2003;8:467−74.
[252] Inagaki M, Kawahara A, Konno H. Sorption and recovery of heavy oils using carbonized fir fibers and recycling. Carbon 2002;40:105−11.

[253] Inagaki M, Kawahara A, Nishi Y, Iwashita N. Heavy oil sorption and recovery by using carbon fiber felts. Carbon 2002;40:1487–92.
[254] Yao T, Zhang Y, Xiao Y, Zhao P, Guo L, Yang H, Li F. The effect of environmental factors on the adsorption of lubricating oil onto expanded graphite. J Mol Liq 2016;218:611–4.
[255] Inagaki M, Konno H, Toyoda M, Moriya K, Kihara T. Sorption and recovery of heavy oils by using exfoliated graphite Part II: recovery of heavy oil and recycling of exfoliated graphite. Desalination 2000;128:213–8.
[256] Nishi Y, Iwashita N, Inagaki M. Evaluation of pore structure of exfoliated graphite by mercury porosimeter. TANSO 2002;201:31–4 [in Japanese].
[257] Zheng YP, Wang H-N, Kang F, Wang L-N, Inagaki M. Sorption capacity of exfoliated graphite for oils-sorption in and among worm-like particles. Carbon 2004;42:2603–7.
[258] Kumagai S, Noguchi Y, Kurimoto Y, Takeda K. Oil adsorbent produced by the carbonization of rice husks. Water Manag 2007;27:554–61.
[259] Angelova D, Uzunov I, Uzunova S, Gigova A, Minchev L. Kinetics of oil and oil products adsorption by carbonized rice husks. Chem Eng J 2011;172:306–11.
[260] Tryba B, Kalenczuk RJ, Kang F, Inagaki M, Morawski AW. Studies of exfoliated graphite (EG) for heavy oil sorption. Mol Cryst Liq Cryst 2000;340:113–9.
[261] Xu C, Jiao C, Yao R, Lin A, Jiao W. Adsorption and regeneration of expanded graphite modified by CTAB-KBr/H_3PO_4 for marine oil pollution. Environ Pollut 2018;233:194–200.
[262] He J, Song L, Yang H, Ren X, Xing L. Preparation of sulfur-free exfoliated graphite by a two-step intercalation process and its application for adsorption of oils. J Chem 2017:5824976.
[263] Zeng Y, Wang K, Yao J, Wang H. Hollow carbon beads fabricated by phase inversion method for efficient oil sorption. Carbon 2014;69:25–31.
[264] Bi H, Xie X, Yin K, Zhou Y, Wan S, He L, Xu F, Banhart F, Sun L, Ruoff RS. Spongy graphene as a highly efficient and recyclable sorbent for oils and organic solvents. Adv Funct Mater 2012;22(21):4421–5.
[265] Gui X, Wei J, Wang K, Cao A, Zhu H, Jia Y, Shu Q, Wu D. Carbon nanotube sponges. Adv Mater 2010;22:617–21.
[266] Lee C, Baik S. Vertically-aligned carbon nano-tube membrane filters with super-hydrophobicity and superoleophilicity. Carbon 2010;48:2192–7.
[267] Wang C, Yang S, Ma Q, Jia X, Ma P-C. Preparation of carbon nanotubes/graphene hybrid aerogel and its application for the adsorption of organic compounds. Carbon 2017;118:765–71.
[268] Du X, Liu HY, Mai YW. Ultrafast synthesis of multifunctional N-doped graphene foam in an ethanol flame. ACS Nano 2016;10(1):453–62.
[269] Pham VT, Nguyen TT, Bui TPQ, et al. Simple synthesis and characterization of cobalt ferrites on expanded graphite by the direct sol-gel chemistry for removal of oil leakage (fuel oil, diesel oil and crude oil). IOP Conf Series Mater Sci Eng 2019;479:012054.
[270] Saufi SM, Ismail AF. Fabrication of carbon membranes for gas separation - a review. Carbon 2004;42:241–59.
[271] Tin PS, Xiao Y, Chung T-S. Polyimide-carbonized membranes for gas separation: structural, composition, and morphological control of precursors. Sep Purif Rev 2006;35:285–318.
[272] Ismail AF, David LIB. A review on the latest development of carbon membranes for gas separation. J Membr Sci 2001;193:1–18.
[273] Ismail AF, Goh PS, Sanip SM, Aziz M. Transport and separation properties of carbon nanotube-mixed matrix membrane. Sep Purif Technol 2009;70:12–26.

[274] Tao Y, Endo M, Inagaki M, Kaneko K. Recent progress in the synthesis and applications of nanoporous carbon films. J Mater Chem 2011;21:313–23.
[275] Salleh WNW, Ismail AF, Matsuura T, Abdullah MS. Precursor selection and process conditions in the preparation of carbon membrane for gas separation: a review. Sep Purif Rev 2011;40:261–311.
[276] Ang EYM, Toh W, Yeo J, Lin R, Liu Z, Geethalakshmi KR, Ng TY. A review on low dimensional carbon desalination and gas separation membrane designs. J Membr Sci 2020;598:117785.
[277] Hatori H, Shiraishi M, Nakata H, Yoshitomi S. Carbon molecular sieve films from polyimide. Carbon 1992;30:719–20.
[278] Hatori H, Yamada Y, Shiraishi M, Nakata H, Yoshitomi S, Yoshihara M. Modification of pore structure in carbon molecular sieve films from polyimide. TANSO 1995;167:94–100.
[279] Sudo H, Haraya K. Gas permeation through micropores of carbon molecular sieve membranes derived from Kapton polyimide. J Phys Chem B 1997;101:3988–94.
[280] Suda H, Haraya K. Alkene/alkane permselectivities of a carbon molecular sieve membrane. Chem Commun 1997;1:93–4.
[281] Fuertes AB, Nevskaia DM, Centeno TA. Carbon composite membranes from Matrimid and Kapton polyimides for gas separation. Microp Mesop Mater 1999;33:115–25.
[282] Hatori H, Takagi H, Yamada Y. Gas separation properties of molecular sieving carbon membranes with nanopore channels. Carbon 2004;42:1169–73.
[283] Lua AC, Su J. Effects of carbonisation on pore evolution and gas permeation properties of carbon membranes from Kapton polyimide. Carbon 2006;44:2964–72.
[284] Tin PS, Chung TS, Hill AJ. Advanced fabrication of carbon molecular sieve membranes by nonsolvent pretreatment of precursor polymers. Ind Eng Chem Res 2004;43:6476–83.
[285] Favvas EP, Kapantaidakis GC, Nolan JW, Mitropoulos AC, Kanellopoulos NK. Preparation, characterization and gas permeation properties of carbon hollow fiber membranes based on Matrimid 5218 precursor. J Mater Process Tech 2007;186:102–10.
[286] Jones CW, Koros WJ. Carbon molecular sieve gas separation membranes I Preparation and characterization based on polyimide precursors. Carbon 1994;32:1419–25.
[287] Jones CW, Koros WJ. Characterization of ultramicroporous carbon membranes with humidified feeds. Ind Eng Chem Res 1995;34:158–63.
[288] Geiszler VC, Koros WJ. Effect of polyimide pyrolysis conditions on carbon molecular sieve membrane properties. Ind Eng Chem Res 1996;35:2999–3003.
[289] Kusuki Y, Shimazaki H, Tanihara N, Nakanishi S, Yoshinaga T. Gas permeation properties and characterization of asymmetric carbon membranes prepared by pyrolyzing asymmetric polyimide hollow fiber membrane. J Membr Sci 1997;134:245–53.
[290] Petersen J, Matsuda M, Haraya K. Capillary carbon molecular sieve membranes derived from Kapton for high temperature gas separation. J Membr Sci 1997;131:85–94.
[291] Ogawa M, Nakano Y. Gas permeation through carbonized hollow fiber membranes prepared by gel modification of polyamic acid. J Membr Sci 1999;162:189–98.
[292] Okamoto K, Kawamura S, Yoshino M, Kita H, Hirayama Y, Tanihara N, Kusuki Y. Olefin/paraffin separation through carbonized membranes derived from an asymmetric polyimide hollow fiber membrane. Ind Eng Chem Res 1999;38:4424–32.
[293] Singh-Ghosal A, Koros WJ. Air separation properties of flat sheet homogeneous pyrolytic carbon membranes. J Membr Sci 2000;174:177–88.

[294] Tin PS, Chung TS, Liu Y, Wang R. Separation of CO_2/CH_4 through carbon molecular sieve membranes derived from P84 polyimide. Carbon 2004;42:3123–31.
[295] Sim YH, Wang H, Li FY, Chua ML, Chung T-S, Toriida M, Tamai S. High performance carbon molecular sieve membranes derived from hyperbranched polyimide precursors for improved gas separation applications. Carbon 2013;53:101–11.
[296] Hayashi J, Yamamoto M, Kusakabe K, Morooka S. Simultaneous improvement of permeance and permselectivity of 3,3'-4,4'-biphenyltetracarboxylic dianhydride-4,4'-oxydianiline polyimide membrane by carbonization. Ind Eng Chem Res 1995;34:4364–70.
[297] Hayashi J, Mizuta H, Yamamoto M, Kusakabe K, Morooka S. Separation of ethane/ethyleneand propane/propylene systems with a carbonized BPDA-pp'ODA polyimide membrane. Ind Eng Chem Res 1996;35:4176–81.
[298] Hayashi J, Mizuta H, Yamamoto M, Kusakabe K, Morooka S. Pore size control of carbonized BPDA- pp'ODA polyimide membrane by chemical vapor deposition of carbon. J Membr Sci 1997;124:243–51.
[299] Hayashi J, Yamamoto M, Kusakabe K, Morooka S. Effect of oxidation on gas permeation of carbon molecular sieve membranes based on BPDA-pp'ODA polyimide. Ind Eng Chem Res 1997;36:2134–40.
[300] Kusakabe K, Yamamoto M, Morooka S. Gas permeation and micropore structure of carbon molecular sieving membranes modified by oxidation. J Membr Sci 1998;149:59–67.
[301] Ismail NH, Salleh WNW, Sazali N, Ismail AF, Yusof N, Aziz F. Disk supported carbon membrane via spray coating method: effect of carbonization temperature and atmosphere. Sep Purif Technol 2018;195:295–304.
[302] Kita H, Yoshino M, Tanaka K, Okamoto K. Gas permselectivity of carbonized polypyrrolone membrane. Chem Commun 1997:1051–2.
[303] Hosseini SS, Chung TS. Carbon membranes from blends of PBI and polyimides for N_2/CH_4 and CO_2/CH_4 separation and hydrogen purification. J Membr Sci 2009;328:174–85.
[304] Hosseini SS, Omidkhah MR, Moghaddam AZ, Pirouzfar V, Krantz WB, Tan NR. Enhancing the properties and gas separation performance of PBI–polyimides blend carbon molecular sieve membranes via optimization of the pyrolysis process. Sep Purif Technol 2014;122:278–89.
[305] Kim YK, Park HB, Lee YM. Carbon molecular sieve membranes derived from thermally labile polymer containing blend polymers and their gas separation properties. J Membr Sci 2004;243:9–17.
[306] Kim YK, Park HB, Lee YM. Gas separation properties of carbon molecular sieve membranes derived from polyimide/polyvinylpyrrolidone blends: effect of the molecular weight of polyvinylpyrrolidone. J Membr Sci 2005;251:159–67.
[307] Liu Q, Wang T, Liang C, Zhang B, Liu S, Cao Y, Qiu J. Zeolite married to carbon: a new family of membrane materials with excellent gas separation performance. Chem Mater 2006;18:6283–8.
[308] Tin PS, Chung T-S, Jiang L, Kulprathipanja S. Carbon–zeolite composite membranes for gas separation. Carbon 2005:2025–7.
[309] Park HB, Lee YM. Fabrication and characterization of nanoporous carbon/silica membranes. Adv Mater 2005;17:477–83.
[310] Fuertes AB, Centeno TA. Carbon molecular sieve membranes from polyetherimide. Microp Mesop Mater 1998;26:23–6.
[311] Sedigh MG, Xu L, Tsotsis TT, Sahimi M. Transport and morphological characteristics of polyetherimide-based carbon molecular sieve membranes. Ind Eng Chem Res 1999;38:3367–80.

[312] Sedigh MG, Jahangiri M, Liu PKT, Sahimi M, Tsotsis TT. Structural characterization of polyetherimide-based carbon molecular sieve membranes. AIChE J 2000;46:2245–55.
[313] Barbosa-Coutinho M, Salim VMM, Borgees CP. Preparation of carbon hollow fiber membranes by pyrolysis of polyetherimide. Carbon 2003;41:1707–14.
[314] Rao PS, Wey MY, Tseng HH, Kumar IA, Weng TH. A comparison of carbon/nanotube molecular sieve membranes with polymer blend carbon molecular sieve membranes for the gas permeation application. Microp Mesop Mater 2008;113:499–510.
[315] Salleh WNW, Ismail AF. Carbon hollow fiber membranes derived from PEI/PVP for gas separation. Sep Purif Technol 2011;80:541–8.
[316] Salleh WNW, Ismail AF. Effects of carbonization heating rate on CO_2 separation of derived carbon membranes. Sep Purif Technol 2012;88:174–83.
[317] Lee HJ, Yoshimune M, Suda H, Haraya K. Effects of oxidation curing on the permeation performances of polyphenylene oxide-derived carbon membranes. Desalination 2006;193:51–7.
[318] Lee HJ, Kim DP, Suda H, Haraya K. Gas permeation properties for the post-oxidized polyphenylene oxide (PPO) derived carbon membranes: effect of the oxidation temperature. J Membr Sci 2006;282:82–8.
[319] Lee HJ, Suda H, Haraya K. Characterization of the post-oxidized carbon membranes derived from poly (2,4-dimethyl-1,4-phenylene oxide) and their gas permeation properties. Sep Purif Technol 2008;59:190–6.
[320] Lee HJ, Suda H, Haraya K, Moon SH. Gas permeation properties of carbon molecular sieving membranes derived from the polymer blend of polyphenylene oxide (PPO)/polyvinylpyrrolidone (PVP). J Membr Sci 2007;296:139–46.
[321] Yoshimune M, Fujiwara I, Suda H, Haraya K. Novel carbon molecular sieve membranes derived from poly(phenylene oxide) and its derivatives for gas separation. Chem Lett 2005;34:958–9.
[322] Yoshimune M, Fujiwara I, Suda H, Haraya K. Gas transport properties of carbon molecular sieve membranes derived from metal containing sulfonated poly(phenylene oxide). Desalination 2006;193:66–72.
[323] Barsema JN, Balster J, Jordan V, van der Vegt NFA, Wessling M. Functionalized carbon molecular sieve membranes containing Ag-nanoclusters. J Membr Sci 2003;219:47–57.
[324] Barsema JN, van der Vegt NFA, Koops GH, Wessling M. Ag-functionalized carbon molecular-sieve membranes based on polyelecrolyte/polyimide blend precursors. Adv Funct Mater 2005;15:69–75.
[325] Kruczek B, Matsuura T. Effect of metal substitution of high molecular weight sulfonated polyphenylene oxide membranes on their gas separation performance. J Membr Sci 2000;167:203–16.
[326] Yoshimune M, Fujiwara I, Haraya K. Carbon molecular sieve membranes derived from trimethylsilyl substituted poly(phenylene oxide) for gas separation. Carbon 2007;45:553–60.
[327] Wei W, Hu H, You LB, Chen GH. Preparation of carbon molecular sieve membrane from phenol– formaldehyde novolac resin. Carbon 2002;40:465–7.
[328] Zhang X, Hu H, Zhu Y, Zhu S. Effect of carbon molecular sieve on phenol formaldehyde novolac resin based carbon membranes. Sep Purif Technol 2006;52:261–5.
[329] Zhang X, Hu H, Zhu Y, Zhu S. Carbon molecular sieve membranes derived from phenol formaldehyde novolac resin blended with poly (ethylene glycol). J Membr Sci 2007;289:86–91.

[330] Centeno TA, Vilas JL, Fuertes AB. Effects of phenolic resin pyrolysis conditions on carbon membrane performance for gas separation. J Membr Sci 2004;228:45–54.
[331] He X, Lie JA, Sheridan E, Hagg MB. Preparation and characterization of hollow fiber carbon membranes from cellulose acetate precursors. Ind Eng Chem Res 2011;50:2080–7.
[332] David LIB, Ismail AF. Influence of the thermal stabilization process and soak time during pyrolysis process on the polyacrylonitrile carbon membranes for O_2/N_2 separation. J Membr Sci 2003;213:285–91.
[333] Arnold BB, Murphy GW. Studies on the electrochemistry of carbon and chemically-modified carbon surfaces. J Phys Chem 1961;65:135–8.
[334] Johnson AM, Newman J. Desalting by means of porous carbon electrodes. J Electrochem Soc 1971;118:510–7.
[335] Oren Y, Soffer A. Electrochemical parametric pumping. J Electrochem Soc 1978;125:869–75.
[336] Oren Y, Soffer A. Water desalting by means of electrochemical parametric pumping. I. The equilibrium properties of a batch unit cell. J Appl Electrochem 1983;13:473–87.
[337] Oren Y, Soffer A. Water desalting by means of electrochemical parametric pumping. II. Separation properties of a multistage column. J Appl Electrochem 1983;13:489–505.
[338] Welgemoed TJ, Schutte CF. Capacitive Deionization TechnologyTM: an alternative desalination solution. Desalination 2005;183:327–40.
[339] Oren Y. Capacitive deionization (CDI) for desalination and water treatment - past, present and future (a review). Desalination 2008;228:10–29.
[340] Mauter MS, Elimelech M. Environmental applications of carbon-based nanomaterials. Environ Sci Technol 2008;26:5843–59.
[341] Foo KY, Hameed BH. A short review of activated carbon assisted electrosorption process: an overview, current stage and future prospects. J Hazard Mater 2009;170:552–9.
[342] Anderson MA, Cudero AL, Palma J. Capacitive deionization as an electrochemical means of saving energy and delivering clean water. Comparison to present desalination practices: will it compete? Electrochim Acta 2010;55:3845–56.
[343] Porada S, Zhao R, van der Wal A, Presser V, Biesheuvel PM. Review on the science and technology of water desalination by capacitive deionization. Prog Mater Sci 2013;58:1388–442.
[344] AlMarzooqi FA, Al Ghaferi AA, Saadat I, Hilal N. Application of capacitive deionisation in water desalination: a review. Desalination 2014;342:3–15.
[345] Aravindan V, Gnanaraj J, Lee Y-S, Madhavi S. Insertion-type electrodes for nonaqueous Li-ion capacitors. Chem Rev 2014;114:11619–35.
[346] Aghigh A, Alizadeh V, Wong HY, Islam MS, Amin N, Zaman M. Recent advances in utilization of graphene for filtration and desalination of water: a review. Desalination 2015;365:389–97.
[347] Subramani A, Jacangelo JG. Emerging desalination technologies for water treatment: a critical review. Water Res 2015;75:164–87.
[348] Lu K. Porous and high surface area silicon oxycarbide-based materials - a review. Mater Sci Eng R 2015;97:23–49.
[349] Liu Y, Nie C, Liu X, Xu X, Sun Z, Pan L. Review on carbon-based composite materials for capacitive deionization. RSC Adv 2015;5:15205–25.
[350] Gaikwad MS, Balomajumder C. Current progress of capacitive deionization for removal of pollutant ions. Electrochem Energy Technol 2016;2:17–23.
[351] Huang Z-H, Yang Z, Kang F, Inagaki M. Carbon electrodes for capacitive deionization. J Mater Chem A 2017;5:470–96.

[352] Oladunni J, Zain JH, Hai A, Banat F, Bharath G, Alhseinat E. A comprehensive review on recently developed carbon based nanocomposites for capacitive deionization: from theory to practice. Sep Purif Technol 2018;207:291–320.
[353] Inagaki M, Konno H, Tanaike O. Carbon materials for electrochemical capacitors. J Power Sources 2010;195:7880–903.
[354] Ying T-Y, Yang K-L, Yiacoumi S, Tsouris C. Electrosorption of ions from aqueous solutions by nanostructured carbon aerogel. J Colloid Interface Sci 2002;250:18–27.
[355] Avraham E, Bouhadana Y, Soffer A, Aurbach D. Limitation of charge efficiency in capacitive deionization I. On the behavior of single activated carbon. J Electrochem Soc 2009;156:P95–9.
[356] Avraham E, Noked M, Bouhadana Y, Soffer A, Aurbach D. The feasibility of boron removal from water by capacitive deionization. Electrochim Acta 2011;56:6312–7.
[357] Suss ME, Baumann TF, Bourcier WL, Spadaccini CM, Rose KA, Santiago JG, Stadermann M. Capacitive desalination with flow-through electrodes. Energy Environ Sci 2012;5:9511–9.
[358] Kim YJ, Choi JH. Enhanced desalination efficiency in capacitive deionization with an ion-selective membrane. Sep Purif Technol 2010;71:70–5.
[359] Liang P, Yuan L, Yang X, Zhou S, Huang X. Coupling ion-exchangers with inexpensive activated carbon fiber electrodes to enhance the performance of capacitive deionization cells for domestic wastewater desalination. Water Res 2013;47:2523–30.
[360] Zhao Y, Wang Y, Wang R, Wu Y, Xu S, Wang J. Performance comparison and energy consumption analysis of capacitive deionization and membrane capacitive deionization processes. Desalination 2013;324:127–33.
[361] Li HB, Gao Y, Pan LK. Electrosorptive desalination by carbon nanotubes and nanofibres electrodes and ion-exchange membranes. Water Res 2008;42:4923–8.
[362] Tian G, Liu L, Meng Q, Cao B. Preparation and characterization of cross-linked quaternised polyvinyl alcohol membrane/activated carbon composite electrode for membrane capacitive deionization. Desalination 2014;354:107–15.
[363] Lee J-Y, Seo S-J, Yun S-H, Moon S-H. Preparation of ion exchanger layered electrodes for advanced membrane capacitive deionization (MCDI). Water Res 2011;45:5375–80.
[364] Lee J, Kim S, Kim C, Yoon Y. Hybrid capacitive deionization to enhance the desalination performance of capacitive techniques. Energy Environ Sci 2014;7:3683–9.
[365] Biesheuvel PM, van der Wal A. Membrane capacitive deionization. J Membr Sci 2010;346:256–62.
[366] Jeon S-I, Park H-R, Yeo J-G, Yang S, Cho CH, Han MH, Kim DK. Desalination via a new membrane capacitive deionization process utilizing flow-electrodes energy. Environ Sci 2013;6:1471–5.
[367] Jeon SI, Yeo JG, Yang S, Choi J, Kim DK. Ion storage and energy recovery of a flow electrode capacitive deionization process. J Mater Chem A 2014;2:6378–83.
[368] Gendel Y, Rommerskirchen AKE, David O, Wessling M. Batch mode and continuous desalination of water using flowing carbon deionization (FCDI) technology. Electrochem Commun 2014;46:152–6.
[369] Hatzell KB, Iwama E, Ferris A, Daffos B, Urita K, Tzedakis T, Chauvet F, Taberna P-L, Gogotsi Y, Simon P. Capacitive deionization concept based on suspension electrodes without ion exchange membranes. Electrochem Commun 2014;43:18–21.
[370] Mitui N, Tomita T, Oda H. Removal of electrolytes from dilute aqueous solutions using activated carbon electrodes. TANSO 2000;194:243–7 [in Japanese].

[371] Oda H, Tada T, Nakagawa Y. Removal properties of ionic substances by electric double-layer adsorption using activated carbon electrodes. TANSO 2001;198:125–8 [in Japanese].
[372] Oda H, Nakagawa Y. Removal of ionic substances from dilute solution using activated carbon electrodes. Carbon 2003;41:1037–47.
[373] Wang G, Qian B, Dong Q, Yang J, Zhao Z, Qiu J. Highly mesoporous activated carbon electrode for capacitive deionization. Sep Purif Technol 2013;103:216–21.
[374] Hou C-H, Huang C-Y. A comparative study of electrosorption selectivity of ions by activated carbon electrodes in capacitive deionization. Desalination 2013;314:124–9.
[375] Huang SY, Fan CS, Hou CH. Electro-enhanced removal of copper ions from aqueous solutions by capacitive deionization. J Hazard Mater 2014;278:8–15.
[376] Huang C-C, He J-C. Electrosorptive removal of copper ions from wastewater by using ordered mesoporous carbon electrodes. Chem Eng J 2013;221:469–75.
[377] Choi J-H. Fabrication of a carbon electrode using activated carbon powder and application to the capacitive deionization process. Sep Purif Technol 2010;70:362–6.
[378] Hou C-H, Huang J-F, Lin H-R, Wang B-Y. Preparation of activated carbon sheet electrode assisted electrosorption process. J Taiwan Inst Chem Eng 2012;43:473–9.
[379] Mossad M, Zou LD. A study of the capacitive deionisation performance under various operational conditions. J Hazard Mater 2012;213:491–7.
[380] Zhao R, Biesheuvel PM, Miedema H, Bruning H, van der Wal A. Charge efficiency: a functional tool to probe the double-layer structure inside of porous electrodes and application in the modeling of capacitive deionization. J Phys Chem Lett 2010;1:205–10.
[381] Kang J, Kim T, Jo K, Yoon J. Comparison of salt adsorption capacity and energy consumption between constant current and constant voltage operation in capacitive deionization. Desalination 2014;352:52–7.
[382] Hu C-C, Wang C-C, Wu F-C, Tseng R-L. Characterization of pistachio shell-derived carbons activated by a combination of KOH and CO_2 for electric double-layer capacitors. Electrochim Acta 2007;52:2498–505.
[383] Yeh C-L, Hsi H-C, Li K-C, Hou C-H. Improved performance in capacitive deionization of activated carbon electrodes with a tunable mesopore and micropore ratio. Desalination 2015;367:60–8.
[384] Omosebi A, Gao X, Rentschler J, Landon J, Liu K. Continuous operation of membrane capacitive deionization cells assembled with dissimilar potential of zero charge electrode pairs. J Colloid Interface Sci 2015;446:345–51.
[385] Liu J, Lu M, Yang J, Cheng J, Cai W. Capacitive desalination of ZnO/activated carbon asymmetric capacitor and mechanism analysis. Electrochim Acta 2015;151:312–8.
[386] Ryoo MW, Kim JH, Seo G. Role of titania incorporated on activated carbon cloth for capacitive deionization of NaCl solution. J Colloid Interface Sci 2003;264:414–9.
[387] Zou L, Morris G, Qi D. Using activated carbon electrode in electrosorptive deionisation of brackish water. Desalination 2008;225:329–40.
[388] Kim C, Lee J, Kim S, Yoon J. TiO_2 sol–gel spray method for carbon electrode fabrication to enhance desalination efficiency of capacitive deionization. Desalination 2014;342:70–4.
[389] Liu P-I, Chunga L-C, Shaoa H, Liang T-M, Horng R-Y, Ma CCM, Chang M-C. Microwave-assisted ionothermal synthesis of nanostructured anatase titanium dioxide/activated carbon composite as electrode material for capacitive deionization. Electrochim Acta 2013;96:173–9.
[390] Han L, Karthikeyan KG, Anderson MA, Wouters JJ, Gregory KB. Mechanistic insights into the use of oxide nanoparticles coated asymmetric electrodes for capacitive deionization. Electrochim Acta 2013;90:573–81.

[391] Lado JJ, Pérez-Roa RE, Wouters JJ, Tejedor-Tejedor MI, Anderson MA. Evaluation of operational parameters for a capacitive deionization reactor employing asymmetric electrodes. Sep Purif Technol 2014;133:236–45.
[392] Kim Y-J, Kim J-H, Choi J-H. Selective removal of nitrate ions by controlling the applied current in membrane capacitive deionization (MCDI). J Membr Sci 2013;429:52–7.
[393] Yeo J-H, Choi J-H. Enhancement of nitrate removal from a solution of mixed nitrate, chloride and sulfate ions using a nitrate-selective carbon electrode. Desalination 2013;320:10–6.
[394] Huang ZH, Wang M, Wang L, Kang F. Relation between the charge efficiency of activated carbon fiber and its desalination performance. Langmuir 2012;28:5079–84.
[395] Li M, Chen Y, Huang Z-H, Kang F. Asymmetric electrodes constructed with PAN-Based activated carbon fiber in capacitive deionization. J Nanomater 2014. Article ID 204172.
[396] Laxman K, Myint MTZ, Al Abri M, Sathe P, Dobretsov S, Dutta J. Desalination and disinfection of inland brackish ground water in a capacitive deionization cell using nanoporous activated carbon cloth electrodes. Desalination 2015;362:126–32.
[397] Wang G, Pan C, Wang L, Dong Q, Yu C, Zhao Z, Qiu J. Activated carbon nanofiber webs made by electrospinning for capacitive deionization. Electrochim Acta 2012;69:65–70.
[398] Wang G, Dong Q, Ling Z, Pan C, Yua C, Qiu J. Hierarchical activated carbon nanofiber webs with tuned structure fabricated by electrospinning for capacitive deionization. J Mater Chem 2012;22:21819–23.
[399] Dong Q, Wang G, Qian B, Hu C, Wang Y, Qiu J. Electrospun composites made of reduced graphene oxide and activated carbon nanofibers for capacitive deionization. Electrochim Acta 2014;137:388–94.
[400] Dong Q, Wang G, Wua T, Peng S, Qiu J. Enhancing capacitive deionization performance of electrospun activated carbon nanofibers by coupling with carbon nanotubes. J Colloid Interface Sci 2015;446:373–8.
[401] Ahn HJ, Lee JH, Jeong Y, Lee J-H, Chi C-S, Oh H-J. Nanostructured carbon cloth electrode for desalination from aqueous solutions. Mater Sci Eng A 2007;449:841–5.
[402] Oh H-J, Lee J-H, Ahn H-J, Jeong Y, Kim Y-J, Chi C-S. Nanoporous activated carbon cloth for capacitive deionization of aqueous solution. Thin Solid Films 2006;515:220–5.
[403] Yamashita A, Minoura S, Miyake T, Ikenaga N, Oda H. Modification of functional groups on ACF surface and its application to electric double layer capacitor electrode. TANSO 2004;214:194–201.
[404] Noked M, Avraham E, Soffer A, Aurbach D. The rate-determining step of electroadsorption processes into nanoporous carbon electrodes related to water desalination. J Phys Chem C 2009;113:21319–27.
[405] Seo SJ, Jeon H, Lee JK, Kim GY, Park D, Nojima H, Lee J, Moon SH. Investigation on removal of hardness ions by capacitive deionization (CDI) for water softening applications. Water Res 2010;44:2267–75.
[406] Myint MTZ, Dutta J. Fabrication of zinc oxide nanorods modified activated carbon cloth electrode for desalination of brackish water using capacitive deionization approach. Desalination 2012;305:24–30.
[407] Laxman K, Myint MTZ, Bourdoucen H, Dutta J. Enhancement in ion adsorption rate and desalination efficiency in a capacitive deionization cell through improved electric field distribution using electrodes composed of activated carbon cloth coated with zinc oxide nanorods. ACS Appl Mater Interface 2014;6:10113–20.

[408] Laxman K, Myint MTZ, Khan R, Pervez T, Dutta J. Effect of a semiconductor dielectric coating on the salt adsorption capacity of a porous electrode in a capacitive deionization cell. Electrochim Acta 2015;166:329–37.
[409] Laxman K, Myint MTZ, Khan R, Pervez T, Dutta J. Improved desalination by zinc oxide nanorod induced electric field enhancement in capacitive deionization of brackish water. Desalination 2015;359:64–70.
[410] Ryoo MW, Seo G. Improvement in capacitive deionization function of activated carbon cloth by titania modification. Water Res 2003;37:1527–34.
[411] Porada S, Weinstein L, Dash R, van der Wal A, Bryjak M, Gogotsi Y, Biesheuvel PM. Water desalination using capacitive deionization with microporous carbon electrodes. ACS Appl Mater Interfaces 2012;4:1194–9.
[412] Porada S, Borchardt L, Oschatz M, Bryjak M, Atchison JS, Keesman KJ, Kaskel S, Biesheuvel PM, Presser V. Direct prediction of the desalination performance of porous carbon electrodes for capacitive deionization. Energy Environ Sci 2013;6:3700–12.
[413] Porada S, Sales BB, Hamelers HVM, Biesheuvel PM. Water desalination with wires. J Phys Chem Lett 2012;3:1613–8.
[414] Yang J, Zou LD. Recycle of calcium waste into mesoporous carbons as sustainable electrode materials for capacitive deionization. Microp Mesop Mater 2014;183:91–8.
[415] Zou L, Li L, Song H, Morris G. Using mesoporous carbon electrodes for brackish water desalination. Water Res 2008;42:2340–8.
[416] Li L, Zou L, Song H, Morris G. Ordered mesoporous carbons synthesized by a modified sol-gel process for electrosorptive removal of sodium chloride. Carbon 2009;47:775–81.
[417] Mayes RT, Tsouris C, Kiggans JO, Mahurin SM, DePaoli DW, Dai S. Hierarchical ordered mesoporous carbon from phloroglucinol-glyoxal and its application in capacitive deionization of brackish water. J Mater Chem 2010;20:8674–8.
[418] Tsouris C, Mayes R, Kiggans J, Sharma K, Yiacoumi S, DePaoli D, Dai S. Mesoporous carbon for capacitive deionization of saline water. Environ Sci Technol 2011;45:10243–9.
[419] Peng Z, Zhang D, Shi L, Yan T, Yuan S, Li H, Gao R, Fang J. Comparative electroadsorption study of mesoporous carbon electrodes with various pore structures. J Phys Chem C 2011;115:17068–76.
[420] Tsai Y-C, Doong R. Activation of hierarchically ordered mesoporous carbons for enhanced capacitive deionization application. Synth Met 2015;205:48–57.
[421] Peng Z, Zhang D, Shi L, Yan T. High performance ordered mesoporous carbon/carbon nanotube composite electrodes for capacitive deionization. J Mater Chem 2012;22:6603–12.
[422] Peng Z, Zhang D, Yan T, Zhanga J, Shia L. Three-dimensional micro/mesoporous carbon composites with carbon nanotube networks for capacitive deionization. Appl Surf Sci 2013;282:965–73.
[423] Liu Y, Pan L, Chen T, Xu X, Lu T, Sun Z, Chua DHC. Porous carbon spheres via microwave-assisted synthesis for capacitive deionization. Electrochim Acta 2015;151:489–96.
[424] Liu Y, Chen T, Lu T, Sun Z, Chua DHC, Pan L. Nitrogen-doped porous carbon spheres for highly efficient capacitive deionization. Electrochim Acta 2015;158:403–9.
[425] Duan F, Li Y, Cao H, Wang Y, Crittenden JC, Zhang Y. Activated carbon electrodes: electrochemical oxidation coupled with desalination for wastewater treatment. Chemosphere 2015;125:205–11.
[426] Wang J, Angnes J, Tobias H, Roesner R, Hong K, Glass R, Kong FM. Pekala R Carbon aerogel composite electrodes. Anal Chem 1993;65:2300–3.

[427] Farmer JC, Fix DV, Mack GV, Pekala RW, Poco JF. Capacitive deionization of NaCl and NaNO3 solutions with carbon aerogel electrodes. J Electrochem Soc 1996;143:159−69.
[428] Farmer JC, Fix DV, Mack GV, Pekala RW, Poco JF. Capacitive deionization of NH_4ClO_4 solutions with carbon aerogel electrodes. J Appl Electrochem 1996;26:1007−18.
[429] Farmer JC, Bahowick SM, Harrar JE, Fix DV, Martinelli RE, Vu AK, Carroll KL. Electrosorption of chromium ions on carbon aerogel electrodes as a means of remediating ground water. Energy Fuels 1997;11:337−47.
[430] Jung HH, Hwang SW, Hyun SH, Kang-Ho L, Kim GT. Capacitive deionization characteristics of nanostructured carbon aerogel electrodes synthesized via ambient drying. Desalination 2007;216:377−85.
[431] Gabelich CJ, Tran TD, Suffet IH. Electrosorption of inorganic salts from aqueous solution using carbon aerogels. Environ Sci Technol 2002;36:3010−9.
[432] Rasines G, Lavela P, Macıas C, Zafra MC, Tirado JL, Parra JB, Ania CO. N-doped monolithic carbon aerogel electrodes with optimized features for the electrosorption of ions. Carbon 2015;83:262−74.
[433] Rasines G, Lavela P, Macías C, Zafra MC, Tirado JL, Ania CO. Mesoporous carbon black-aerogel composites with optimized properties for the electro-assisted removal of sodium chloride from brackish water. J Electroanal Chem 2015;741:42−50.
[434] Zafra C, Lavela P, Rasines G, Macías C, Tiradoa JL, Ania CO. A novel method for metal oxide deposition on carbon aerogels with potential application in capacitive deionization of saline water. Electrochim Acta 2014;135:208−16.
[435] Li Z, Song B, Wu Z, Lin Z, Yao Y, Moon K-S, Wong CP. 3D porous graphene with ultrahigh surface area for microscale capacitive deionization. Nano Energy 2015;11:711−8.
[436] Li HB, Zou LD, Pan LK, Sun Z. Novel graphene-like electrodes for capacitive deionization. Environ Sci Technol 2010;44:8692−7.
[437] Li HB, Lu T, Pan LK, Zhang Y, Sun Z. Electrosorption behavior of graphene in NaCl solutions. J Mater Chem 2009;19:6773−9.
[438] Huang ZH, Zheng XY, Lv W, Wang M, Yang QH, Kang FY. Adsorption of lead(II) ions from aqueous solution on low-temperature exfoliated graphene nanosheets. Langmuir 2011;27:7558−62.
[439] Wang H, Zhang D, Yan T, Wen X, Zhang J, Shi L, Zhong Q. Three-dimensional macroporous graphene architectures as high performance electrodes for capacitive deionization. J Mater Chem A 2013;1:11778−89.
[440] Yang ZY, Jin LJ, Lu GQ, Xiao QQ, Zhang YX, Jing L, Zhang XX, Yan YM, Sun KN. Sponge- templated preparation of high surface area graphene with ultrahigh capacitive deionization performance. Adv Funct Mater 2014;24:3917−25.
[441] Xu X, Pan L, Liu Y, Lu T, Sun Z, Chua DHC. Facile synthesis of novel graphene sponge for high performance capacitive deionization. Sci Rep 2015;5:8458.
[442] Wimalasiri Y, Mossad M, Zou L. Thermodynamics and kinetics of adsorption of ammonium ions by graphene laminate electrodes in capacitive deionization. Desalination 2015;357:178−88.
[443] Wang H, Zhang D, Yan T, Wen X, Shi L, Zhang J. Graphene prepared via a novel pyridine−thermal strategy for capacitive deionization. J Mater Chem 2012;22:23745−8.
[444] Jia B, Zou LD. Wettability and its influence on graphene nanosheets as electrode material for capacitive deionization. Chem Phys Lett 2012;548:23−8.
[445] Xu X, Pan L, Liu Y, Lu T, Sun Z. Enhanced capacitive deionization performance of graphene by nitrogen doping. J Colloid Interface Sci 2015;445:143−50.

[446] Yin H, Zhao S, Wan J, Tang H, Chang L, He L, Zhao H, Gao Y, Tang Z. Three-dimensional graphene/metal oxide nanoparticle hybrids for high performance capacitive deionization of saline water. Adv Mater 2013;25:6270−6.
[447] Wen X, Zhang D, Yan T, Zhang J, Shi L. Three-dimensional graphene-based hierarchically porous carbon composites prepared by a dual-template strategy for capacitive deionization. J Mater Chem A 2013;1:12334−44.
[448] El-Deen AG, Barakat NAM, Khalil KA, Motlak M, Kim HY. Graphene/SnO_2 nanocomposite as an effective electrode material for saline water desalination using capacitive deionization. Ceram Int 2014;40:14627−34.
[449] El-Deen AG, Barakat NAM, Kim HY. Graphene wrapped MnO_2-nanostructures as effective and stable electrode materials for capacitive deionization desalination technology. Desalination 2014;344:289−98.
[450] El-Deen AG, Choi J-H, Kim CS, Khalil KA, Almajid AA, Barakat NAM. TiO_2 nanorod-intercalated reduced graphene oxide as high performance electrode material for membrane capacitive deionization. Desalination 2015;361:53−64.
[451] Cai P-F, Su C-J, Chang W-T, Chang F-C, Peng C-Y, Sun I-W, Wei Y-L, Jou C-J, Wang HP. Capacitive deionization of seawater effected by nano Ag and Ag@C on graphene. Marine Poll Bull 2014;85:733−7.
[452] Li HB, Pan LK, Nie CY, Liu Y, Sun Z. Reduced graphene oxide and activated carbon composites for capacitive deionization. J Mater Chem 2012;22:15556−61.
[453] Zhang D, Wen X, Shi L, Yan T, Zhang J. Enhanced capacitive deionization of graphene/mesoporous carbon composites. Nanoscale 2012;4:5440−6.
[454] Wang Z, Dou BJ, Zheng L, Zhang GN, Liu ZH, Hao ZP. Effective desalination by capacitive deionization with functional graphene nanocomposite as novel electrode material. Desalination 2012;299:96−102.
[455] Wang H, Shi L, Yan T, Zhang J, Zhong Q, Zhang D. Design of graphene-coated hollow mesoporous carbon spheres as high performance electrodes for capacitive deionization. J Mater Chem A 2014;2:4739−50.
[456] Liu L, Liao L, Meng Q, Cao B. High performance graphene composite microsphere electrodes for capacitive deionisation. Carbon 2015;90:75−84.
[457] Feng J, Yang Z, Hou S, Li M, Lv R, Kang F, Huang Z-H. GO/auricularia-derived hierarchical porous carbon used for capacitive deionization with high performance. Colloids Surf A 2018;547:134−40.
[458] Wang S, Wang DZ, Ji LJ, Gong Q, Zhu YF, Liang J. Equilibrium and kinetic studies on the removal of NaCl from aqueous solutions by electrosorption on carbon nanotube electrodes. Sep Purif Technol 2007;58:12−6.
[459] Hou C-H, Liu N-L, Hsu H-L, Den W. Development of multi-walled carbon nanotube/poly(vinyl alcohol) composite as electrode for capacitive deionization. Sep Pur Technol 2014;130:7−14.
[460] Dai K, Shi L, Fang J, Zhang D, Yu B. NaCl adsorption in multiwalled carbon nanotubes. Mater Lett 2005;59(16):1989−92.
[461] Zhang D, Shi L, Fang J, Dai K. Influence of diameter of carbon nanotubes mounted in flow-through capacitors on removal of NaCl from salt water. J Mater Sci 2007;42:2471−5.
[462] Yang L, Shi Z, Yang W. Enhanced capacitive deionization of lead ions using air-plasma treated carbon nanotube electrode. Surf Coatings Technol 2014;251:122−7.
[463] Wang Y, Zhang L, WuY, Xu S, Wang J. Polypyrrole/carbon nanotube composites as cathode material for performance enhancing of capacitive deionization technology. Desalination 2014;354:62−7.
[464] Liu Y, Pan L, Xu X, Lu T, Sun Z, Chua DHC. Enhanced desalination efficiency in modified membrane capacitive deionization by introducing ion-exchange polymers in carbon nanotubes electrodes. Electrochim Acta 2014;130:619−24.

[465] Zhang D, Shi L, Fang J, Dai K, Liu J. Influence of carbonization of hot-pressed carbon nanotube electrodes on removal of NaCl from saltwater solution. Mater Chem Phys 2006;96:140–4.
[466] Zhang D, Shi L, Fang J, Dai K, Li X. Preparation and desalination performance of multiwall carbon nanotubes. Mater Chem Phys 2006;97:415–9.
[467] Dai K, Shi L, Zhang D, Fang J. NaCl adsorption in multiwalled carbon nanotube/active carbon combination electrode. Chem Eng Sci 2006;61:428–33.
[468] Wang XZ, Li MG, Chen YW, Cheng RM, Huang SM, Pan LK, Sun Z. Electrosorption of ions from aqueous solutions with carbon nanotubes and nanofibers composite film electrodes. Appl Phys Lett 2006;89:53127.
[469] Wang X, Li M, Chen Y, Cheng R, Huang S, Pan L, Sun Z. Electrosorption of NaCl solutions with carbon nanotubes and nanofibers composite film electrodes. Electrochem Solid State Lett 2006;9:E23–6.
[470] Gao Y, Pan LK, Li HB, Zhang YP, Zhang ZJ, Chen YW, Sun Z. Electrosorption behavior of cations with carbon nanotubes and carbon nanofibres composite film electrodes. Thin Solid Films 2009;517:1616–9.
[471] Li H, Pan L, Zhang Y, Zou L, Sun C, Zhan Y, Sun Z. Kinetics and thermodynamics study for electrosorption of NaCl onto carbon nanotubes and carbon nanofibers electrodes. Chem Phys Lett 2010;485:161–6.
[472] Nie C, Pan L, Li H, Chen T, Lu T, Sun Z. Electrophoretic deposition of carbon nanotubes film electrodes for capacitive deionization. J Electroanal Chem 2012;666:85–8.
[473] Nie C, Pan L, Liu Y, Li H, Chen T, Lu T, Sun Z. Electrophoretic deposition of carbon nanotubes– polyacrylic acid composite film electrode for capacitive deionization. Electrochim Acta 2012;66:106–9.
[474] Gui XC, Wei JQ, Wang KL, Cao AY, Zhu HW, Jia Y, Shu QK, Wu DH. Carbon nanotube sponges. Adv Mater 2010;22:617–21.
[475] Wang L, Wang M, Huang Z-H, Cui T, Gui X, Kang F, Wang K, Wu D. Capacitive deionization of NaCl solutions using carbon nanotube sponge electrodes. J Mater Chem 2011;21:18295–9.
[476] Wimalasiri Y, Zou L. Carbon nanotube/graphene composite for enhanced capacitive deionization performance. Carbon 2013;59:464–71.
[477] Biesheuvel PM, Porada S, Presser V. Comment on "Carbon nanotube/graphene composite for enhanced capacitive deionization performance" by Y. Wimalasiri and L. Zou. Carbon 2013;63:574–5.
[478] Wimalasiri Y, Zou L. Response to "Comments on 'carbon nanotube/graphene composite for enhanced capacitive deionization performance' by Y. Wimalasiri and L. Zou". Carbon 2015;81:847–8.
[479] Li H, Liang S, Li J, He L. The capacitive deionization behaviour of a carbon nanotube and reduced graphene oxide composite. J Mater Chem A 2013;1:6335–41.
[480] Zhang D, Yan T, Shi L, Peng Z, Wen X, Zhang J. Enhanced capacitive deionization performance of graphene/carbon nanotube composites. J Mater Chem 2012;22:14696–704.
[481] Li H, Pan L, Lu T, Zhan Y, Nie C, Sun Z. A comparative study on electrosorptive behavior of carbon nanotubes and graphene for capacitive deionization. J Electroanal Chem 2011;653:40–4.
[482] Li H, Zou L. Ion-exchange membrane capacitive deionization: a new strategy for brackish water desalination. Desalination 2011;275:62–6.
[483] Sui Z, Meng Q, Zhang X, Ma R, Cao B. Green synthesis of carbon nanotube–graphene hybrid aerogels and their use as versatile agents for water purification. J Mater Chem 2012;22:8767–71.

[484] Wang M, Huang ZH, Wang L, Wang M-X, Kang F, Hou H. Electrospun ultrafine carbon fiber webs for electrochemical capacitive desalination. New J Chem 2010;34:1843–5.
[485] Liu J, Wang S, Yang J, Liao J, Lu M, Pan H. An L, ZnCl2 activated electrospun carbon nanofiber for capacitive desalination. Desalination 2014;344:446–53.
[486] Chen Y, Yue M, Huang Z-H, Kang F. Electrospun carbon nanofiber networks from phenolic resin for capacitive deionization. Chem Eng J 2014;252:30–7.
[487] Bai Y, Huang Z-H, Yu X-L, Kang F. Graphene oxide-embedded porous carbon nanofiber webs by electrospinning for capacitive deionization. Colloid Surf A 2014;444:153–8.
[488] Hu C, Liu F, Lan H, Liu H, Qu J. Preparation of a manganese dioxide/carbon fiber electrode for electrosorptive removal of copper ions from water. J Colloid Interface Sci 2015;446:359–65.
[489] Yang J, Zou LD, Song HH, Hao ZP. Development of novel MnO2/nanoporous carbon composite electrodes in capacitive deionization technology. Desalination 2011;276:199–206.
[490] Cao MS, Wang XX, Cao WQ, Yuan J. Ultrathin graphene electrical properties and highly efficient electromagnetic interference shielding. J Mater Chem C 2015;3:6589–99.
[491] Lv H, Guo Y, Yang Z, Cheng Y, Wang LP, Zhang B, Zhao Y, Xu ZJ, Ji G. A brief introduction to the fabrication and synthesis of graphene based composites for the realization of electromagnetic absorbing materials. J Mater Chem C 2017;5:491–512.
[492] Song WL, Cao MS, Fan LZ, Lu MM, Li Y, Wang CY, Ju HF. Highly ordered porous carbon/wax composites for effective electromagnetic attenuation and shielding. Carbon 2014;77:130–42.
[493] Fan YZ, Yang HB, Li MH, Zou GT. Evaluation of the microwave absorption property of flake graphite. Mater Chem Phys 2009;115:696–8.
[494] Song WL, Cao MS, Lu MM, Liu J, Yuan J, Fan LZ. Improved dielectric properties and highly efficient and broadened bandwidth electromagnetic attenuation of thickness-decreased carbon nanosheet/wax composites. J Mater Chem C 2013;1:1846–54.
[495] Zhang Y, Huang Y, Zhang T, Chang H, Xiao P, Chen H, Huang Z, Chen Y. Broadband and tunable high-performance microwave absorption of an ultralight and highly compressible graphene foam. Adv Mater 2015;27:2049–53.
[496] Du YC, Wang JY, Cui CK, Liu XR, Wang XH, Han XJ. Pure carbon microwave absorbers from anion- exchange resin pyrolysis. Synth Met 2010;160:2191–6.
[497] Xu HL, Yin XW, Zhu M, Han MK, Hou ZX, Li XL, Zhang LT, Cheng LF. Carbon hollow microspheres with a designable mesoporous shell for high-performance electromagnetic wave absorption. ACS Appl Mater Interfaces 2017;9:6332–41.
[498] Li G, Xie TS, Yang SL, Jin JH, Jiang JM. Microwave absorption enhancement of porous carbon fibers compared with carbon nanofibers. J Phys Chem C 2012;116:9196–201.
[499] Zou T, Zhao N, Shi C, Li J. Microwave absorbing properties of activated carbon fibre polymer composites. Bull Mater Sci 2011;34:75–9.
[500] Wu KH, Ting TH, Wang GP, Ho WD, Shih CC. Effect of carbon black content on electrical and microwave absorbing properties of polyaniline/carbon black nanocomposites. Polym Degrad Stab 2008;93:483–8.
[501] Wang NN, Wu F, Xie AM, Dai XQ, Sun MX, Qiu YY, Wang Y, Lv XL, Wang MY. One-pot synthesis of biomass derived carbonaceous spheres for excellent microwave absorption at the Ku band. RSC Adv 2015;5:40531–5.

[502] Qiang R, Du YC, Wang Y, Wang N, Tian CH, Ma J, Xu P, Han XJ. Rational design of yolk-shell C@C microspheres for the effective enhancement in microwave absorption. Carbon 2016;98:599−606.
[503] Zhou C, Geng S, Xu X, Wang T, Zhang L, Tian X, Yang F, Yang H, Li Y. Lightweight hollow carbon nanospheres with tunable sizes towards enhancement in microwave absorption. Carbon 2016;108:234−41.
[504] Huang YX, Wang Y, Li ZM, Yang Z, Shen CH, He CC. Effect of pore morphology on the dielectric properties of porous carbons for microwave absorption applications. J Phys Chem C 2014;118:26027−32.
[505] Chu Z, Cheng H, Xie W, Sun L. Effects of diameter and hollow structure on the microwave absorption properties of short carbon fibers. Ceram Int 2012;38:4867−73.
[506] Motojima S, Hoshiya S, Hishikawa Y. Electromagnetic wave absorption properties of carbon microcoils/PMMA composite beads in W bands. Carbon 2003;41:2658−66.
[507] Tang NJ, Zhong W, Au CT, Yang Y, Han MG, Lin KJ, Du YW, Synthesis. Microwave electromagnetic and microwave absorption properties of twin carbon nanocoils. J Phys Chem C 2008;112:19316−23.
[508] Hong X, Chung DDL. Carbon nanofiber mats for electromagnetic interference shielding. Carbon 2017;111:529−37.
[509] Eddib AA, Chung DDL. The importance of the electrical contact between specimen and testing fixture in evaluating the electromagnetic interference shielding effectiveness of carbon materials. Carbon 2017;117:427−36.
[510] Chung DDL, Eddib AA. Radio-frequency linear absorption coefficient of carbon materials, its dependence on the thickness and its independence on the carbon structure. Carbon 2017;124:473−8.
[511] Haddad AS, Chung DDL. Decreasing the electric permittivity of cement by graphite particle incorporation. Carbon 2017;122:702−9.
[512] Hong X, Chung DDL. Exfoliated graphite with relative dielectric constant reaching 360, obtained by exfoliation of acid-intercalated graphite flakes without subsequent removal of the residual acidity. Carbon 2015;91:1−10.
[513] Liu Z, Bai G, Huang Y, Li F, Ma Y, Guo T, He X, Lin X, Gao H, Chen Y. Microwave absorption of single-walled carbon nanotubes/soluble crosslinked polyurethane composites. J Phys Chem C 2007;111:13696−700.
[514] Li N, Huang Y, Du F, He XB, Lin X, Gao HJ, Ma YF, Li FF, Chen YS, Eklund PC. Electromagnetic interference (EMI) shielding of single-walled carbon nanotube epoxy composites. Nano Lett 2006;6:1141−5.
[515] Watts PCP, Hsu WK, Barnes A, Chambers B. High permittivity from defective multiwalled carbon nanotubes in the X-band. Adv Mater 2003;15:600−3.
[516] Zhao TK, Hou CL, Zhang HY, Zhu RX, She SF, Wang JG, Li TH, Liu ZF, Wei BQ. Electromagnetic wave absorbing properties of amorphous carbon nanotubes. Sci Rep 2014;4:1−7.
[517] Wong KH, Pickering SJ, Rudd CD. Recycled carbon fibre reinforced polymer composite for electromagnetic interference shielding. Compos Part A 2010;41:693−702.
[518] Zhen H, Wang H, Xu X. Preparation of porous carbon nanofibers with remarkable microwave absorption performance through electrospinning. Mater Lett 2019;249:210−3.
[519] Li T, Zhao G, Zhang L, Wang G, Li B, Gong J. Ultralow-threshold and efficient EMI shielding PMMA/MWCNTs composite foams with segregated conductive network and gradient cells. eXPRESS Polym Lett 2020;14:685−703.

[520] Saini P, Choudhary V, Dhawan SK. Electrical properties and EMI shielding behavior of highly thermally stable polyaniline/colloidal graphite composites. Polym Adv Technol 2009;20:355–61.
[521] Gupta TK, Singh BP, Mathur RB, Dhakate SR. Multiwalled carbon nanotube-graphene polyaniline multiphase nanocomposite with superior electromagnetic shielding effectiveness. Nanoscale 2014;6:842–51.
[522] Song W-L, Wang J, Fan L-Z, Li Y, Wang C-Y, Cao M-S. Interfacial engineering of carbon nanofiber– graphene–carbon nanofiber heterojunctions in flexible light weight electromagnetic shielding networks. ACS Appl Mater Interfaces 2014;6:10516–23.
[523] Hong SK, Kim KY, Kim TY, Kim JH, Park SW, Kim JH, Cho BJ. Electromagnetic interference shielding effectiveness of monolayer graphene. Nanotechnology 2012;23:455704.
[524] Lovat G. Equivalent circuit for electromagnetic interaction and transmission through graphene sheets. IEEE Trans Electromagn Compat 2012;54:101–9.
[525] Lu ZG, Ma LM, Tan JB, Wang HY, Ding XM. Transparent multi-layer graphene/polyethylene terephthalate structures with excellent microwave absorption and electromagnetic interference shielding performance. Nanoscale 2016;8:16684–93.
[526] Wen B, Wang XX, Cao WQ, Shi HL, Lu MM, Wang G, Jin HB, Wang WZ, Yuan J, Cao MS. Reduced graphene oxides: the thinnest and most lightweight materials with highly efficient microwave attenuation performances of the carbon world. Nanoscale 2014;6:5754–61.
[527] Sykam N, Rao GM. Lightweight flexible graphite sheet for high-performance electromagnetic interference shielding. Mater Lett 2018;233:59–62.
[528] Singh VK, Shukla A, Patra MK, Saini L, Jani RK, Vadera SR, Kumar N. Microwave absorbing properties of a thermally reduced graphene oxide/nitrile butadiene rubber composite. Carbon 2012;50:2202–8.
[529] Zhang X, Huang Y, Liu P. Enhanced electromagnetic wave absorption properties of poly(3,4- ethylenedioxythiophene) nanofiber-decorated graphene sheets by non-covalent interactions. Nano-Micro Lett 2016;8:131–6.
[530] Yu HL, Wang TS, Wen B, Lu MM, Xu Z, Zhu C, Chen Y, Xue X, Sun C, Cao M. Graphene/polyaniline nanorod arrays: synthesis and excellent electromagnetic absorption properties. J Mater Chem 2012;22:21679–85.
[531] Wang L, Huang Y, Huang H. N-doped graphene@polyaniline nanorod arrays hierarchical structures: synthesis and enhanced electromagnetic absorption properties. Mater Lett 2014;124:89–92.
[532] Liu P, Huang Y. Decoration of reduced graphene oxide with polyaniline film and their enhanced microwave absorption properties. J Polym Res 2014;21:430.
[533] Chen X, Meng F, Zhou Z, Tian X, Shan L, Zhu S, Xu X, Jiang M, Wang L, Hui D, Wang Y, Lua J, Gou J. One-step synthesis of graphene/polyaniline hybrids by in situ intercalation polymerization and their electromagnetic properties. Nanoscale 2014;6:8140–8.
[534] Luo J, Xu Y, Yao W, Jiang C, Xu J. Synthesis and microwave absorption properties of reduced graphene oxide-magnetic porous nanospheres-polyaniline composites. Compos Sci Technol 2015;117:315–21.
[535] Zhang Y, Huang Y, Chen HH, Huang ZY, Yang Y, Xiao PS, Zhou Y, Chen YS. Composite and structure control of ultralight graphene foam for high-performance microwave absorption. Carbon 2016;105:438–47.
[536] Zhang H-B, Yan Q, Zheng W-G, He Z, Yu Z-Z. Tough graphene-polymer microcellular foams for electromagnetic interference shielding. ACS Appl Mater Interfaces 2011;3:918–24.

[537] Chen C, Xi J, Zhou E, Peng L, Chen Z, Gao C. Porous graphene microflowers for high- performance microwave absorption. Nano-Micro Lett 2018;10:26.
[538] Chen H, Huang Z, Huang Y, Zhang Y, Ge Z, Qin B, Liu Z, Shi Q, Xiao P, Yang Y, Zhang T, Chen Y. Synergistically assembled MWCNT/graphene foam with highly efficient microwave absorption in both C and C bands. Carbon 2017;124:506–14.
[539] Lv H, Guo Y, Zhao Y, Zhang H, Zhang B, Ji G, Xu ZJ. Achieving tunable electromagnetic absorber via graphene/carbon sphere composites. Carbon 2016;110:130–7.
[540] Han MK, Yin XW, Kong L, Li M, Duan WY, Zhang LT, Cheng LF. Graphene-wrapped ZnO hollow spheres with enhanced electromagnetic wave absorption properties. J Mater Chem A 2014;2:16403–9.
[541] Song C, Yin X, Han M, Li X, Hou Z, Zhang L, Cheng L. Three-dimensional reduced graphene oxide foam modified with ZnO nanowires for enhanced microwave absorption properties. Carbon 2017;116:50–8.
[542] Lu M, Cao W, Shi H, Fang X, Yang J, Hou Z, Jin H, Wang W, Yuan J, Cao M. Multi-wall carbon nanotubes decorated with ZnO nanocrystals: mild solution process synthesis and highly efficient microwave absorption properties at elevated temperature. J Mater Chem A 2014;2:10540–7.
[543] Feng W, Wang Y, Chen J, Wang L, Guo L, Ouyang J, Jia D, Zhou Y. Reduced graphene oxide decorated with in-situ growing ZnO nanocrystals: facile synthesis and enhanced microwave absorption properties. Carbon 2016;108:52–60.
[544] Kim Y-J, Hadiyawarman, Yoon A, Kim M, Yi G-C, Liu C. Hydrothermally grown ZnO nanostructures on few-layer graphene sheets. Nanotechnology 2011;22:245603.
[545] Zhang L, Zhang X, Zhang G, Zhang Z, Liu S, Li P, Liao Q, Zhao Y, Zhang Y. Investigation on the optimization, design and microwave absorption properties of reduced graphene oxide/tetrapod-like ZnO composites. RSC Adv 2015;5:10197–203.
[546] Wu F, Xia Y, Wang Y, Wang M. Two-step reduction of self-assembled three dimensional (3D) reduced graphene oxide (RGO)/zinc oxide (ZnO) nanocomposites for electromagnetic absorption. J Mater Chem A 2014;2:20307–15.
[547] Wen B, Cao MS, Lu MM, Cao WQ, Shi HL, Liu J, Wang XX, Jin HB, Fang XY, Wang WZ, Yuan J. Reduced graphene oxides: light weight and high efficiency electromagnetic interference shielding at elevated temperatures. Adv Mater 2014;26:3484–9.
[548] Song WL, Guan XT, Fan LZ, Zhao YB, Cao WQ, Wang CY, Cao MS. Strong and thermostable polymeric graphene/silica textile for lightweight practical microwave absorption composites. Carbon 2016;100:109–17.
[549] Im JS, Kim JG, Bae T-S, Lee Y-S. Effect of heat treatment on ZrO_2-embedded electrospun carbon fibers used for efficient electromagnetic interference shielding. J Phys Chem Solids 2011;72:1175–9.
[550] Bayat M, Yang H, Ko FK, Michelson D, Mei A. Electromagnetic interference shielding effectiveness of hybrid multifunctional Fe_3O_4/carbon nanofiber composite. Polymer 2014;55:936–43.
[551] Liu PB, Huang Y, Sun X. Excellent electromagnetic absorption properties of poly (3, 4- ethylenedioxythiophene)-reduced graphene oxide-Co_3O_4 composites prepared by a hydrothermal method. ACS Appl Mater Interfaces 2013;5:12355–60.
[552] Xiang J, Li J, Zhang X, Ye Q, Xu J, Shen X. Magnetic carbon nanofibers containing uniformly dispersed Fe/Co/Ni nanoparticles as stable and high-performance electromagnetic wave absorbers. J Mater Chem A 2014;2:16905–14.

[553] Wen FS, Zhang F, Liu ZY. Investigation on microwave absorption properties for multiwalled carbon nanotubes/Fe/Co/Ni nanopowders as lightweight absorbers. J Phys Chem C 2011;115(29):14025−30.
[554] Bondavalli P, Legagneux P, Pribat D. Carbon nanotubes based transistors as gas sensors: state of the art and critical review. Sens Actuator B 2009;140:304−18.
[555] Pumera M, Ambrosi A, Bonanni A, Chng ELK, Poh HL. Graphene for electrochemical sensing and biosensing. Trends Anal Chem 2010;29:954−65.
[556] Kuila T, Bose S, Khanra P, Mishra AK, Kim NH, Lee JH. Recent advances in graphene-based biosensors. Biosens Bioelectron 2011;26:4637−48.
[557] Basua S, Bhattacharyya P. Recent developments on graphene and graphene oxide based solid state gas sensors. Sens Actuator B 2012;173:1−21.
[558] Chung C, Kim Y-K, Shin D, Ryoo S-R, Hong BH, Min D-H. Biomedical applications of graphene and graphene oxide. Acc Chem Res 2013;46:2211−24.
[559] Wu S, He Q, Tan C, Wang Y, Zhang H. Graphene-based electrochemical sensors. Small 2013;9:1160−72.
[560] Moldovan O, Iñiguez B, Jamal DM, Marsa LF. Graphene electronic sensors - review of recent developments and future challenges. IET Circuits, Devices Syst 2015;9:446−53.
[561] Chatterjee SG, Chatterjee S, Ray AK, Chakraborty AK. Graphene−metal oxide nanohybrids for toxic gas sensor: a review. Sensor Actuator B 2015;221:1170−81.
[562] Sajid MI, Jamshaid U, Jamshaid T, Zafar N, Fessi H, Elaissari A. Carbon nanotubes from synthesis to in vivo biomedical applications. Int J Pharm 2016;501:278−99.
[563] Yu X, Zhang W, Zhang P, Su Z. Fabrication technologies and sensing applications of graphene-based composite films: advances and challenges. Biosens Bioelectron 2017;89:72−84.
[564] Tian F, Lyu J, Shi J, Yang M. Graphene and graphene-like two-denominational materials based fluorescence resonance energy transfer (FRET) assays for biological applications. Biosens Bioelectron 2017;89:123−35.
[565] Bollella P, FuscoG, Tortolini C, Sanzò G, Favero G, Gorton L, Antiochia R. Beyond graphene: electrochemical sensors and biosensors for biomarkers detection. Biosens Bioelectron 2017;89:152−66.
[566] Bo X, Zhou M, Guo L. Electrochemical sensors and biosensors based on less aggregated graphene. Biosens Bioelectron 2017;89:167−86.
[567] Xiong M, Rong Q, Meng H, Zhang X. Two-dimensional graphitic carbon nitride nanosheets for biosensing applications. Biosens Bioelectron 2017;89:212−23.
[568] Janegitz BC, Silva TA, Wong A, Ribovski L, Vicentini FC, Sotomayor MPT, Fatibello-Filho O. The application of graphene for in vitro and in vivo electrochemical biosensing. Biosens Biuoelectron 2017;89:224−33.
[569] Wang H, Chen Q, Zhou S. Carbon-based hybrid nanogels: a synergistic nanoplatform for combined biosensing, bioimaging, and responsive drug delivery. Chem Soc Rev 2018;47:4198−232.
[570] Sinha A, Dhanjai JR, Zhao H, Karolia P, Jadon N. Voltammetric sensing based on the use of advanced carbonaceous nanomaterials: a review. Microchim Acta 2018;185:89.
[571] Baig N, Saleh TA. Electrodes modified with 3D graphene composites: a review on methods for preparation, properties and sensing applications. Microchim Acta 2018;185:283.
[572] Luo S, Samad YA, Chan V, Liao K. Cellular graphene: fabrication, mechanical properties, and strain- sensing applications. Matter 2019;1:1148−202.
[573] Liu Y, Dong X, Chen P. Biological and chemical sensors based on graphene materials. Chem Soc Rev 2012;41:2283−307.
[574] Wang F, Liu L, Li WJ. Graphene-based glucose sensors: a brief review. IEEE Trans Nanobiosci 2015;14:818−34.

[575] Luong JHT, Glennon JD, Gedanken A, Vashist SK. Achievement and assessment of direct electron transfer of glucose oxidase in electrochemical biosensing using carbon nanotubes, graphene, and their nanocomposites. Microchim Acta 2017;184:369–88.
[576] Schedin F, Geim AK, Morozov SV, Hill EW, Blake P, Katsnelson MI, Novoselov KS. Detection of individual gas molecules adsorbed on graphene. Nat Mater 2007;6:652–5.
[577] Someya T, Small J, Kim P, Nuckolls C, Yardley JT. Alcohol vapor sensors based on single-walled carbon nanotube field effect transistors. Nano Lett 2003;3:877–81.
[578] Kong J, Franklin NR, Zhou C, Chapline MG, Peng S, Cho K, Dai H. Nanotube molecular wires as chemical sensors. Science 2000;287:622–5.
[579] Bradley K, Gabriel JCP, Star A, Gruner G. Short-channel effects in contact-passivated nanotube chemical sensors. Appl Phys Lett 2003;83:3821–3.
[580] Liu X, Luo Z, Han S, Tang T, Zhang D, Zhou C. Band engineering of carbon nanotube field-effect transistors via selected area chemical gating. Appl Phys Lett 2005;86:243501.
[581] Mattmann M, Helbling T, Durrer L, Roman C, Hierold C, Pohle R, Fleischer M. Sub-ppm NO_2 detection by Al_2O_3 contact passivated carbon nanotube field effect transistors. Appl Phys Lett 2009;94:183502.
[582] Boyd A, Dube I, Fedorov G, Paranjape M, Barbara P. Gas sensing mechanism of carbon nanotubes: from single tubes to high-density networks. Carbon 2014;69:417–23.
[583] Tombler T, Zhou C, Alexseyev L, Kong J, Dai H, Liu L, Jayanthi CS, Tang M, Wu S. Reversible electromechanical characteristics of carbon nanotubes under local-probe manipulation. Nature 2000;405:769–72.
[584] Longergan MC, Severin EJ, Doleman BJ, Beaber SA, Grubbs RH, Lewis NS. Array-based vapor sensing using chemically sensitive, carbon black-polymer resistors. Chem Mater 1996;8:2298–312.
[585] Kante K, Florent M, Temirgaliyeva A, Leshayev B, Bandoz TJ. Exploring resistance changes of porous carbon upon physical adsorption of VOCs. Carbon 2019;146:568–71.
[586] Singh K, Travlou NA, Bashkova S, Rodrıguez-Castellon E, Bandosz TJ. Nanoporous carbons as gas sensors: exploring the surface sensitivity. Carbon 2014;80:183–92.
[587] Travlou NA, Seredych M, Rodríguez-Castellon E, Bandosz TJ. Insight into ammonia sensing on heterogeneous S- and N- co-doped nanoporous carbons. Carbon 2016;96:1014–21.
[588] Mutuma BK, Rodrigues R, Ranganathan K, Matsoso B, Wamwangi D, Hummelgen IA, Coville NJ. Hollow carbon spheres and a hollow carbon sphere/polyvinylpyrrolidone composite as ammonia sensors. J Mater Chem A 2017;5:2539–49.
[589] Wang D, Wen G, Zhang Y, Ying H, Zhang T, Tao X, Chen W. Novel C-rich carbon nitride for room temperature NO_2 gas sensors. RSC Adv 2014;4:18003–6.
[590] Thirupathi T, Ramanujam K. Carbon supported g-C_3N_4 for electrochemical sensing of hydrazine. Electrochem Energy Technol 2018;4:21–31.
[591] Madhu R, Sankar KV, Chen S-M, Selvan RK. Eco-friendly synthesis of activated carbon from dead mango leaves for the ultrahigh sensitive detection of toxic heavy metal ions and energy storage applications. RSC Adv 2014;4:1225–33.
[592] Valentini L, Armentano I, Kenny JM, Cantalini C, Lozzi L, Santuccia S. Sensors for sub-ppm NO_2 gas detection based on carbon nanotube thin films. Appl Phys Lett 2003;82:961–3.
[593] Valentini L, Cantalini C, Armentano I, Kenny JM, Lozzi L, Santucci S. Highly sensitive and selective sensors based on carbon nanotubes thin films for molecular detection. Diamond Relat Mater 2004;13:1301–5.

[594] Im JS, Kang SC, Bai BC, Bae T-S, In SJ, Jeong E, Lee S-H, Lee Y-S. Thermal fluorination effects on carbon nanotubes for preparation of a high-performance gas sensor. Carbon 2011;49:2235−44.
[595] Chen G, Paronyan TM, Pigos EM, Harutyunyan AR. Enhanced gas sensing in pristine carbon nanotubes under continuous ultraviolet light illumination. Sci Rep 2012;2:343.
[596] Chikkadi K, Muoth M, Maiwald V, Roman C, Hierold C. Ultralow power operation of self-heated, suspended carbon nanotube gas sensors. Appl Phys Lett 2013;103:223109.
[597] Cava CE, Salvatierra RV, Alves DCB, Ferlauto AS, Zarbin AJG, Roman LS. Self-assembled films of multi-wall carbon nanotubes used in gas sensors to increase the sensitivity limit for oxygen detection. Carbon 2012;50:1953−8.
[598] Teerapanich P, Myint MTZ, Joseph CM, Hornyak GL, Dutta J. Development and improvement of carbon nanotube-based ammonia gas sensors using ink-jet printed interdigitated electrodes. IEEE Trans Nanotechnol 2013;12:255−62.
[599] Rigoni F, Drera G, Pagliara S, Goldoni A, Sangaletti L. High sensitivity, moisture selective, ammonia gas sensors based on single-walled carbon nanotubes functionalized with indium tin oxide nanoparticles. Carbon 2014;80:356−63.
[600] Zhang L, Wang X, Zhao Y, Zhu Z, Fong H. Electrospun carbon nano-felt surface-attached with Pd nanoparticles for hydrogen sensing application. Mater Lett 2012;68:133−6.
[601] Zhao Y, Wang X, Lai C, He G, Zhang L, Fong H, Zhu Z. Electrospun carbon nanofibrous mats surface-decorated with Pd nanoparticles via the supercritical CO_2 method for sensing of H_2. RSC Adv 2012;2:10195−9.
[602] Lee S, Im J, Kang S, Lee S, Lee Y-S. Effects of improved porosity and electrical conductivity on pitch-based carbon nanofibers for high-performance gas sensors. J Porous Mater 2012;19:989−94.
[603] Snow ES, Novak JP, Campbell PM, Park D. Random networks of carbon nanotubes as an electronic material. Appl Phys Lett 2003;82:2145.
[604] Snow ES, Perkins FK, Houser EJ, Badescu SC, Reinecke TL. Chemical detection with a single-walled carbon nanotube capacitor. Science 2005;307:1942−5.
[605] Snow ES, Campbell PM, Novak JP. High-mobility carbon-nanotube thin-film transistors on a polymeric substrate. Appl Phys Lett 2005;86:033105.
[606] Snow ES, Perkins FK. Capacitance and conductance of single-walled carbon nanotubes in the presence of chemical vapors. Nano Lett 2005;5:2414−7.
[607] Lee JS, Kwon OS, Park SJ, Park EY, You SA, Yoon H, Jang J. Fabrication of ultrafine metal-oxide- decorated carbon nanofibers for DMMP sensor application. ACS Nano 2011;5:7992−8001.
[608] Lee JS, Kwon OS, Shin DH, Jang J. WO_3 nanonodule-decorated hybrid carbon nanofibers for NO_2 gas sensor application. J Mater Chem A 2013;1:9099−106.
[609] Li W, Zhang L-S, Wang Q, Yu Y, Chen Z, Cao C-Y, Song W-G. Low-cost synthesis of graphitic carbon nanofibers as excellent room temperature sensors for explosive gases. J Mater Chem 2012;22:15342−7.
[610] Yan S, Wu Q. Micropored Sn-SnO_2/carbon heterostructure nanofibers and their highly sensitive and selective C_2H_5OH gas sensing performance. Sensors Actuator B 2014;205:329−37.
[611] Huang J, Zhang X, Zhou L, You T. Simultaneous electrochemical determination of dihydroxybenzene isomers using electrospun nitrogen-doped carbon nanofiber film electrode. Sensors Actuator B 2016;224:568−76.
[612] Guo Q, Huang J, Chen P, Liu Y, Hou H, You T. Simultaneous determination of catechol and hydroquinone using electrospun carbon nanofibers modified electrode. Sensors Actuator B 2012;163:179−85.

[613] Li D, Pang Z, Chen X, Luo L, Cai Y, Wei Q. A catechol biosensor based on electrospun carbon nanofibers. Beilstein J Nanotechnol 2014;5:346−54.

[614] Paul RK, Badhulika S, Saucedo NM, Mulchandani A. Graphene nanomesh as highly sensitive chemiresistor gas sensor. Anal Chem 2012;84:8171−8.

[615] Chung MG, Kim DH, Lee HM, Kim T, Choi JH, Seo D, Yoo J-B, Hong S-H, Kang TJ, Kim YH. Highly sensitive NO_2 gas sensor based on ozone treated graphene. Sens Actuator B 2012;166−167:172−6.

[616] Lv R, Chen G, Li Q, McCreary A, Botello-Méndez A, Morozov SV, Liang L, Declerck X, Perea- López N, Cullen DA, Feng S, Elías AL, Cruz-Silva R, Fujisawa K, Endo M, Kang F, Charlier J-C, Meunier V, Pan M, Harutyunyan AR, Novoselov KS, Terrones M. Ultrasensitive gas detection of large-area boron-doped graphene. Proc Nat Acad Sci USA 2015;24:14527−32.

[617] Leenaerts O, Partoens B, Peeters FM. Adsorption of H_2O, NH_3, CO, NO_2, and NO on graphene: a first-principles study. Phys Rev B 2008;77:125416.

[618] Tang S, Cao Z. Adsorption of nitrogen oxides on graphene and graphene oxides: insights from density functional calculations. J Chem Phys 2011;134:044710.

[619] Tang S, Cao Z. Adsorption and dissociation of ammonia on graphene oxides: a first-principles study. J Phys Chem C 2012;116:8778−91.

[620] Mattson EC, Pande K, Unger M, Cui SM, Lu GH, Gaidardziska-Josifovska M, Weinert M, Chen J, Hirschmug CJ. Exploring adsorption and reactivity of NH_3 on reduced graphene oxide. J Phys Chem C 2013;117:10698−707.

[621] Pak Y, Kim S-M, Jeong H, Kang CG, Park JS, Song H, Lee R, Myoung N, Lee BH, Seo S, Kim JT, Jung G-Y. Palladium-decorated hydrogen-gas sensors using periodically aligned graphene nanoribbons. ACS Appl Mater Interfaces 2014;6:13293−8.

[622] Ang PK, Chen W, Wee ATS, Loh KP. Solution-gated epitaxial graphene as pH sensor. J Am Chem Soc 2008;130:14392−3.

[623] Dua V, Surwade SP, Ammu S, Agnihotra SR, Jain S, Roberts KE, Park S, Ruoff RS, Manohar SK. All-organic vapor sensor using inkjet-printed reduced graphene oxide. Angew Chem Int Ed 2010;49:2154−7.

[624] Robinson JT, Perkins FK, Snow ES, et al. Reduced graphene oxide molecular sensors. Nano Lett 2008;8:3137−40.

[625] Hu N, Wang Y, Chai J, Gao R, Yang Z, Kong ES, Zhang Y. Gas sensor based on p-phenylenediamine reduced graphene oxide. Sens Actuator B 2012;163:107−14.

[626] Hu N, Yang Z, Wang Y, Zhang L, Wang Y, Huang X, Wei H, Wei L, Zhang Y. Ultrafast and sensitive room temperature NH_3 gas sensors based on chemically reduced graphene oxide. Nanotechnology 2014;25:025502.

[627] Fowler JD, Allen MJ, Tung VC, Yang Y, Kaner RB, Weiller BH. Practical chemical sensors from chemically derived graphene. ACS Nano 2009;3:301−6.

[628] Lu G, Yu K, Ocola LE, Chen J. Ultrafast room temperature NH_3 sensing with positively gated reduced graphene oxide field-effect transistors. Chem Commun 2011;47:7761−3.

[629] Yang YJ, Yang XJ, Yang WY, Li SB, Xu JH, Jiang YD. Porous conducting polymer and reduced graphene oxide nanocomposites for room temperature gas detection. RSC Adv 2014;4:42546−53.

[630] Li X, UmarA, ChenZ, TianT, WangS, WangY. Supramolecular fabrication of polyelectrolyte- modified reduced graphene oxide for NO_2 sensing applications. Ceram Int 2015;41:12130−6.

[631] Arsat R, Breedon M, Shafiei M, Spizziri PG, Gilje S, Kaner RB, Kalantar-zadeh K, Wlodarski W. Graphen e-like nano-sheets for surface acoustic wave gas sensor applications. Chem Phys Lett 2009;467:344−7.

[632] Singh G, Choudhary A, Haranath D, Joshi AG, Singh N, Singh S, Pasricha R. ZnO decorated luminescent graphene as a potential gas sensor at room temperature. Carbon 2012;50:385–94.

[633] Deng S, Tjoa V, Fan HM, Tan HR, Sayle DC, Olivo M, Mhaisalkar S, Wei J, Sow CH. Reduced graphene oxide conjugated Cu_2O nanowire mesocrystals for high-performance NO_2 gas sensor. J Am Chem Soc 2012;134:4905–17.

[634] Russo PA, Donato N, Leonardi SG, Baek S, Conte DE, Neri G, Pinna N. Room-temperature hydrogen sensing with heteronanostructures based on reduced graphene oxide and tin oxide. Angew Chem Int Ed 2012;51:11053–7.

[635] Niu Y, Wang R, Jiao W, Ding G, Hao L, Yang F, He X. MoS2 graphene fiber based gas sensing devices. Carbon 2015;95:34–41.

[636] Zhang ZY, Zou RJ, Song GS, Yu L, Chen ZG, Hu JQ. Highly aligned SnO_2 nanorods on graphene sheets for gas sensors. J Mater Chem 2011;21:17360–5.

[637] Nagareddy VK, Chan HK, Hernández SC, Wheeler VD, Nyakiti LO, Myers-Ward RL, Eddy Jr CR, Goss JP, Wright NG, Walton SG, Gaskill DK, Horsfall AB. Improved chemical detection and ultra- fast recovery using oxygen functionalized epitaxial graphene sensors. IEEE Sens J 2013;13:2810–7.

[638] Huang X, Hu N, Gao R, Yu Y, Wang Y, Yang Z, Kong ES-W, Wei H, Zhang Y. Reduced graphene oxide–polyaniline hybrid: preparation, characterization and its applications for ammonia gas sensing. J Mater Chem 2012;22:22488.

[639] Some S, Xu Y, Kim Y, Yoon Y, Qin H, Kulkarni A, Kim T, Lee H. Highly sensitive and selective gas sensor using hydrophilic and hydrophobic graphenes. Sci Rep 2013;3:1–8.

[640] Ji Q, Honma I, Paek S-M, Akada M, Hill JP, Ajayan V, Ariga K. Layer-by-layer films of graphene and ionic liquids for highly selective gas sensing. Angew Chem Int Ed 2010;49:9737–9.

[641] Lipatov A, Varezhnikov A, Wilson P, Sysoev V, Kolmakov A, Sinitskii A. Highly selective gas sensor arrays based on thermally reduced graphene oxide. Nanoscale 2013;5:5426–34.

[642] Yi J, Lee JM, Park W. Vertically aligned ZnO nanorods and graphene hybrid architectures for high- sensitive flexible gas sensors. Sens Actuator B 2011;155:264–9.

[643] Myers M, Cooper J, Pejcic B, Baker M, Raguse B, Wieczorek L. Functionalized graphene as an aqueous phase chemiresistor sensing material. Sens Actuator B 2011;155:154–8.

[644] Must I, Kaasik F, Poldsalu I, Johanson U, Punning A, Aabloo A. A carbide-derived carbon laminate used as a mechanoelectrical sensor. Carbon 2012;50:535–41.

[645] Kong J-H, Jang N-S, Kim S-H, Kim J-M. Simple and rapid micropatterning of conductive carbon composites and its application to elastic strain sensors. Carbon 2014;77:199–207.

[646] Luo S, Hoang PT, Liu T. Direct laser writing for creating porous graphitic structures and their use for flexible and highly sensitive sensor and sensor arrays. Carbon 2016;96:522–31.

[647] Rahimi R, Ochoa M, Yu W, Ziaie B. Highly stretchable and sensitive unidirectional strain sensor via laser carbonization. ACS Appl Mater Interface 2015;7:4463–70.

[648] Yang L. Band-gap change of carbon nanotubes: effect of small uniaxial and torsional strain. Phys Rev B 1999;60:13874–8.

[649] Ito T, Nishidate K, Baba M, Hasegawa M. First principles calculations for electronic band structure of single-walled carbon nanotube under uniaxial strain. Surf Sci 2002;514:222–6.

[650] Chen Y, Weng C. Electronic properties of zigzag carbon nanotubes under uniaxial strain. Carbon 2007;45:1636–44.

[651] Cullinan M, Culpepper M. Carbon nanotubes as piezoresistive microelectromechanical sensors: theory and experiment. Phys Rev B 2010;82(11):115428.
[652] Cao J, Wang Q, Dai H. Electromechanical properties of metallic, quasimetallic, and semiconducting carbon nanotubes under stretching. Phys Rev Lett 2003;90:157601.
[653] Minot ED, Yaish Y, Sazonova V, Park JY, Brink M, McEuen PL. Tuning carbon nanotube band gaps with strain. Phys Rev Lett 2003;90:156401.
[654] Grow R, Wang Q, Cao J, Wang D, Dai H. Piezoresistance of carbon nanotubes on deformable thin- film membranes. Appl Phys Lett 2005;86:093104.
[655] Moore V, Strano M, Haroz E, Hauge R, Smalley R. Individually suspended single-walled carbon nanotubes in various surfactants. Nano Lett 2003;3:1379−82.
[656] Liu J, Hersam MC. Recent developments in carbon nanotube sorting and selective growth. MRS Bull 2010;35:315−22.
[657] Cullinan MA, Culpepper ML. Effects of chirality and impurities on the performance of carbon nanotube-based piezoresistive sensors. Carbon 2013;51:59−63.
[658] Hirata T, Takeda N, Tsutsui C, Koike K, Shimatani Y, Sakai T, Akiya M, Taguchi A. Measurement of contractile activity in small animal's digestive organ by carbon nanotube-based force transducer. Jpn J Appl Phys 2011;50:030210.
[659] Hu N, Itoi T, Akagi T, Kojima T, Xue J, Yan C, Atobe S, Fukunaga H, Yuan W, Ning H, Surina LY, Alamusi. Ultrasensitive strain sensors made from metal-coated carbon nanofiller/epoxy composites. Carbon 2013;51:202−12.
[660] Zhao H, Bai J. Highly sensitive piezo-resistive graphite nanoplatelet-carbon nanotube hybrids/polydimethylsilicone composites with improved conductive network construction. ACS Appl Mater Interfaces 2015;7:9652−9.
[661] Jensen K, Kim K, Zettl A. An atomic-resolution nanomechanical mass sensor. Nat Nanotech 2008;3:533−7.
[662] Chiu HY, Hung P, Postma HWC, Bockrath M. Atomic-scale mass sensing using carbon nanotube resonators. Nano Lett 2008;8:4342−6.
[663] Fu XW, Liao ZM, Zhou JX, Zhou -B, Wu H-C, Zhang R, Jing G, Xu J, Wu X, Guo W, Yu D. Strain dependent resistance in chemical vapor deposition grown graphene. Appl Phys Lett 2011;99:213107.
[664] Huang MY, Pascal TA, Kim H, Goddard WA, Greer JR. Simulations electronic−mechanical coupling in graphene from in situ nanoindentation experiments and multiscale atomistic. Nano Lett 2011;11:1241−6.
[665] Lee Y, Bae S, Jang H, Jang S, Zhu S-E, Sim SH, Song YI, Hong BH, Ahn J-H. Wafer-scale synthesis and transfer of graphene films. Nano Lett 2010;10:490−3.
[666] Zhao J, He CL, Yang R, Shi Z, Cheng M, Yang W, Xie G, Wang D, Shi D, Zhang G. Ultra-sensitive strain sensors based on piezoresistive nanographene films. Appl Phys Lett 2012;101:063112.
[667] Li Q, Ullah Z, Li W, Guo Y, Xu J, Wang R, Zeng Q, Chen M, Liu C, Liu L. Wide-range strain sensors based on highly transparent and supremely stretchable graphene/Ag-nanowires hybrid structures. Small 2016;12:5058−65.
[668] Pang Y, Tian H, Tao L, Li Y, Wang X, Deng N, Yang Y, Ren TL. Flexible, highly sensitive, and wearable pressure and strain sensors with graphene porous network structure. ACS Appl Mater Interfaces 2016;8:26458−62.
[669] Li X, Zhang R, Yu W, Wang K, Wei J, Wu D, Cao A, Li Z, Cheng Y, Zheng Q, Ruoff RS, Zhu H. Stretchable and highly sensitive graphene-on-polymer strain sensors. Sci Rep 2012;2:870.
[670] Chun S, Choi Y, Park W. All-graphene strain sensor on soft substrate. Carbon 2017;116:753−9.
[671] Rinaldi A, Proietti A, Tamburrano A, Ciminello M, Sarto MS. Graphene-based strain sensor array on carbon fiber composite laminate. IEEE Sens J 2015;15:7295−303.

[672] Samad YA, Li Y, Alhassan SM, Liao K. Novel graphene foam composite with adjustable sensitivity for sensor applications. ACS Appl Mater Interfaces 2015;7:9195−202.

[673] Filippidou MK, Tegou E, Tsouti V, Chatzandroulis S. A flexible strain sensor made of graphene nanoplatelets/polydimethylsiloxane nanocomposite. Microelectronic Eng 2015;142:7−11.

[674] Habibi M, Darbari S, Rajabali S, Ahmadi V. Fabrication of a graphene-based pressure sensor by utilising field emission behavior of carbon nanotubes. Carbon 2016;96:259−67.

[675] Maleki N, Safavi A, Tjabadi F. High-performance carbon composite electrode based on an ionic liquid as a binder. Anal Chem 2006;78:3820−6.

[676] Liu Y, Hou H, You T. Synthesis of carbon nanofibers for mediatorless sensitive detection of NADH. Electroanalysis 2008;20:1708−13.

[677] Kachoosangi RT, Musameh MM, Abu-Yousef I, Yousef JM, Kanan SM, Xiao L, Davies SG, Russell A, Compton RG. Carbon nanotube-ionic liquid composite sensors and biosensors. Anal Chem 2009;81:435−42.

[678] Liu Y, Wang D, Huang J, Hou H, You T. Highly sensitive composite electrode based on electrospun carbon nanofibers and ionic liquid. Electrochem Commun 2010;12:1108−11.

[679] Sun J, Li L, Zhang X, Liu D, Lv S, Zhu D, Wu T, You T. Simultaneous determination of ascorbic acid, dopamine and uric acid at a nitrogen-doped carbon nanofiber modified electrode. RSC Adv 2015;5:11925−32.

[680] Huang J, Liu Y, Hou H, You T. Simultaneous electrochemical determination of dopamine, uric acid and ascorbic acid using palladium nanoparticle-loaded carbon nanofibers modified electrode. Biosens Bioelectron 2008;24:632−7.

[681] Huang Y, Miao Y-E, Ji S, Tjiu WW, Liu T. Electrospun carbon nanofibers decorated with Ag-Pt bimetallic nanoparticles for selective detection of dopamine. ACS Appl Mater Interfaces 2014;6:12449−56.

[682] Liu Y, Huang J, Hou H, You T. Simultaneous determination of dopamine, ascorbic acid and uric acid with electrospun carbon nanofibers modified electrode. Electrochem Commun 2008;10:1431−4.

[683] Yue Y, Hu G, Zheng M, Guo Y, Cao J, Shao S. A mesoporous carbon nanofiber-modified pyrolytic graphite electrode used for the simultaneous determination of dopamine, uric acid, and ascorbic acid. Carbon 2012;50:107−14.

[684] Liu S, Lai G, Zhang H, Yu A. Amperometric aptasensing of chloramphenicol at a glassy carbon electrode modified with a nanocomposite consisting of graphene and silver nanoparticles. Microchim Acta 2017;184:1445−51.

[685] Gao X, Yue H, Song S, Huang S, Li B, Lin X, Guo E, Wang B, Guan E, Zhang H, Wu P. 3-Dimensional hollow graphene balls for voltammetric sensing of levodopa in the presence of uric acid. Microchim Acta 2018;185:91.

[686] Im JS, Kim JG, Bae T-S, Yu H-R, Lee Y-S. Surface modification of electrospun spherical activated carbon for a high-performance biosensor electrode. Sensors Actuator B 2011;158:151−8.

[687] Liu D, Zhang X, You T. Electrochemical performance of electrospun free-standing nitrogen-doped carbon nanofibers and their application for glucose biosensing. ACS Appl Mater Interface 2014;6:6275−80.

[688] Zhang X, Liu D, Li L, You T. Direct electrochemistry of glucose oxidase on novel free-standing nitrogen-doped carbon nanospheres@carbon nanofibers composite film. Sci Rep 2015;5:9885.

[689] Funabashi H, Takeuchi S, Tsujimura S. Hierarchical meso/macro-porous carbon fabricated from dual MgO templates for direct electron transfer enzymatic electrodes. Sci Rep 2017;7:45147.

[690] Mazurenko I, Clément R, Byrne-Kodjabachian D, de Poulpiquet A, Tsujimura S, Lojoua E. Pore size effect of MgO-templated carbon on enzymatic H_2 oxidation by the hyperthermophilic hydrogenase from *Aquifex aeolicus*. J Electroanal Chem 2018;812:221–6.
[691] Huang YX, Dong XC, Shi YM, Li CM, Li L-J, Chen P. Nanoelectronic biosensors based on CVD grown graphene. Nanoscale 2010;2:1485–8.
[692] Wang Y, Shao YY, Matson DW, Li J, Lin Y. Nitrogen-doped graphene and its application in electrochemical biosensing. ACS Nano 2010;4:1790–8.
[693] Liu Y, Wang D, Xu L, Hou H, You T. A novel and simple route to prepare a Pt nanoparticle-loaded carbon nanofiber electrode for hydrogen peroxide sensing. Biosens Bioelectron 2011;26:4585–90.
[694] Zhang X, Liu D, Yu B, You T. A novel nonenzymatic hydrogen peroxide sensor based on electrospun nitrogen-doped carbon nanoparticles-embedded carbon nanofibers film. Sensors Actuator B 2016;224:103–9.
[695] Huang KJ, Niu DJ, Sun JY, et al. Novel electrochemical sensor based on functionalized graphene for simultaneous determination of adenine and guanine in DNA. Colloids Surf B 2011;82:543–9.
[696] Du M, Yang T, Jiao K. Immobilization-free direct electrochemical detection for DNA specific sequences based on electrochemically converted gold nanoparticles/graphene composite film. J Mater Chem 2010;20:9253–60.
[697] Zhang J, Lei JP, Pan R, Xue Y, Ju H. Highly sensitive electrocatalytic biosensing of hypoxanthine based on functionalization of graphene sheets with water-soluble conducting graft copolymer. Biosens Bioelectron 2010;26:371–6.
[698] Wang Z, Zhang J, Chen P, Zhou X, Yang Y, Wu S, Niu L, Han Y, Wang L, Chen P, Boey F, Zhang Q. Label-free, electrochemical detection of methicillin-resistant *Staphylococcus aureus* DNA with reduced graphene oxide-modified electrodes. Biosens Bioelectron 2011;26:3881–6.
[699] Li B, Pan G, Avent ND, Lowry RB, Madgett TE, Waines PL. Graphene electrode modified with electrochemically reduced graphene oxide for label-free DNA detection. Biosens Bioelectron 2015;72:313–9.
[700] Dong X, Shi Y, Huang W, Chen P, Li L-J. Electrical detection of DNA hybridization with single-base specificity using transistors based on CVD-grown graphene sheets. Adv Mater 2010;22:1649–53.
[701] Stine R, Robinson JT, Sheehan PE, Tamanaha CR. Real-time DNA detection using reduced graphene oxide field effect transistors. Adv Mater 2010;22:5297–300.
[702] Hirata T, Takagi K, Akiya M. Development of a taste sensor based on a carbon nanotube-polymer composite material. Jpn J Appl Phys 2007;46:L314–6.
[703] Hirata T, Amiya S, Akiya M, Takei O, Sakai T, Hatakeyama R. Development of a vitamin-protein sensor based on carbon nanotube hybrid materials. Appl Phys Lett 2007;90:233106.
[704] Hirata T, Amiya S, Akiya M, Takei O, Sakai T, Nakamura T, Kawamura-Tsuzuku J, Yamamoto T, Hatakeyama R. Chemical modification of carbon nanotube based bionanosensor by plasma activation. Jpn J Appl Phys 2008;47:2068–71.
[705] He Q, Wu S, Gao S, Cao X, Yin Z, Li H, Chen P, Zhang H. Transparent, flexible, all-reduced graphene oxide thin film transistors. ACS Nano 2011;5:5038–44.
[706] Su BL, Tang JA, Huang JX, Yang H, Qiu B, Chen G, Tang D. Graphene and nanogold-functionalized immunosensing interface with enhanced sensitivity for one-step electrochemical immunoassay of alpha-fetoprotein in human serum. Electroanalysis 2010;22:2720–8.
[707] Han J, Ma J, Ma Z. One-step synthesis of graphene oxide–thionine–Au nanocomposites and its application for electrochemical immunosensing. Biosens Bioelectron 2013;47:243–7.

[708] Mao S, Lu G, Yu K, Bo Z, Chen J. Specific protein detection using thermally reduced graphene oxide sheet decorated with gold nanoparticle-antibody conjugates. Adv Mater 2010;22:3521–6.
[709] Huang Y, Dong X, Liu Y, Li L-J, Chen P. Graphene-based biosensors for detection of bacteria and their metabolic activities. J Mater Chem 2011;21:12358–62.
[710] Mannoor MS, Tao H, Clayton JD, Sengupta A, Kaplan DL, Naik RR, Verma N, Omenetto FG, McAlpine MC. Graphene-based wireless bacteria detection on tooth enamel. Nat Commun 2012;3:763.

CHAPTER 5

Concluding remarks and prospects

5.1 Concluding remarks

5.1.1 Syntheses of porous carbons

In this book, the synthesis processes of porous carbons are classified into activation, template-assisted carbonization, and precursor-designed carbonization, as summarized in Fig. 5.1. The process "activation" is explained by dividing it into chemical activation, physical activation, and activated carbon fibers. The process "template-assisted carbonization" is explained on the bases of the template materials employed, such as zeolites, mesoporous silicas, and MgO. The process "precursor-designed carbonization" consists of polymer blending, molecular design, and process design; molecular design includes the departure of labile functional (pending) groups from precursor molecules, defluorination, carbonization of different organic frameworks, carbon aerogels, and chlorination of metal carbides. For synthesizing macroporous carbons, the "process design" includes blowing of the precursors by adding blowing reagents and treatment under hydrothermal conditions, in addition to exfoliation through intercalation of acid molecules into graphite.

Porous carbons were represented by activated carbons (active carbons) until the end of the 20th century. The history of activated carbons goes back to the prehistoric era, when charcoals were used to purify drinking water, to control humidity in tombs, and to stop diarrhea as a medicine. Even now activated carbons are used in most water purification plants in large amounts and they are a principal component of diarrhea medicines. In order to respond to the requirements for the applications in modern industries, various activation techniques and reagents have been studied, which are classified into chemical and physical activations. Activated carbon fibers (ACFs), where micro- and/or mesopores are possible to be opened directly for adsorbate gases and liquids, present better performances in their

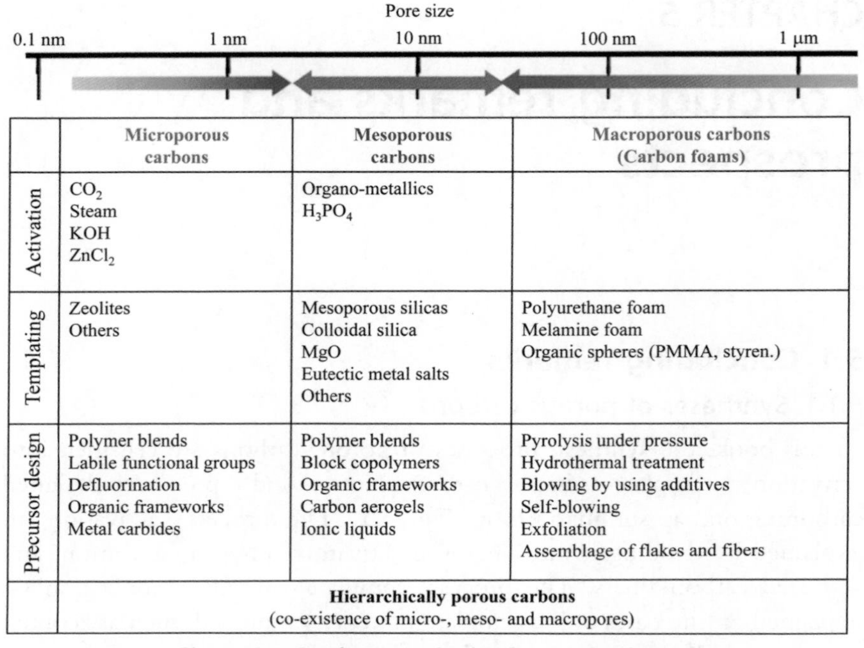

	Microporous carbons	Mesoporous carbons	Macroporous carbons (Carbon foams)
Activation	CO_2 Steam KOH $ZnCl_2$	Organo-metallics H_3PO_4	
Templating	Zeolites Others	Mesoporous silicas Colloidal silica MgO Eutectic metal salts Others	Polyurethane foam Melamine foam Organic spheres (PMMA, styren.)
Precursor design	Polymer blends Labile functional groups Defluorination Organic frameworks Metal carbides	Polymer blends Block copolymers Organic frameworks Carbon aerogels Ionic liquids	Pyrolysis under pressure Hydrothermal treatment Blowing by using additives Self-blowing Exfoliation Assemblage of flakes and fibers
	Hierarchically porous carbons (co-existence of micro-, meso- and macropores)		

Figure 5.1 Synthesis methods of porous carbons.

applications than granular activated carbons, in which the adsorbates have to diffuse through macro- and mesopores to reach micropores. The high performance of ACFs has created new applications for activated carbons.

In 1997, Kyotani reported the synthesis of microporous carbon by applying a template-assisted carbonization technique to zeolite [1], which gave a revolutionary impact for pore structure control in carbon materials. Other templates, mesoporous silicas, colloidal silicas, MgO, eutectic metal salts, etc., were successively proposed, most of them leading to higher carbonization yield than conventional activation processes. Some templates enable the formation of mesoporous carbons from different carbon precursors, which cannot give mesopores by simple carbonization. These template processes (templating) for the preparation of porous carbons exhibited various advantages in comparison with activation, e.g., homogeneous pore sizes, even-ordered pore arrangement, and high carbonization yield. In most cases, however, dissolution of template materials after the carbonization process is essential to recover porous carbons. These dissolution processes seemed to hamper the development of the industrial production of porous carbons through these templating processes. Just

MgO-templated carbon is industrially produced, because of a simple dissolution of MgO using diluted acid, and is on the market with the name CNovel [2]. Most templates used for the syntheses of porous carbons are hard templates, such as zeolite, silica, and MgO, while organic templates (soft templates) are also used for the synthesis of mesoporous carbons and macroporous carbons (carbon foams) without any additional process as template dissolution.

Just after the emergence of the zeolite-templating method, various techniques not using hard templates were proposed, which are summarized as the "precursor design" in this book, whereas some are discussed by including them in the template process. In this book, the processes for precursor design are classified into polymer blending, molecular design, and process design. Polymer blends consist mainly of the combinations of a polymer having a high carbonization yield with a labile polymer with low carbonization yields, some of the polymer combinations having been successfully applied to prepare porous carbons, particularly microporous carbon nanofibers. One of the polymer blending processes using block copolymers can produce mesoporous channels with ordered arrangements, although the coupled carbon precursors are limited. It has to be mentioned here that polymer blending is a shortened process to prepare mesoporous carbons, in comparison with templating processes using mesoporous silicas, because mesoporous silicas are produced using the same block copolymers. By designing the structure of precursor molecules (repeating units of polymers) to have labile pending groups (being classified by the molecular design), microporous carbons, even microporous carbon films, can be synthesized. Different organic frameworks (called porous-organic frameworks [POMs], covalent-organic frameworks [COFs], and metal-organic frameworks [MOFs]), and carbon aerogels can derive micro- and/or mesoporous carbons directly by carbonization. In the MOF containing Zn, the process for removing the residual metallic species is not needed because Zn vaporizes out during carbonization treatment. Different metal carbides result in microporous carbons directly by the heat treatment in Cl_2 gas flow (chlorination) because most metal chlorides can be vaporized out. Blowing of carbonaceous materials, such as pitches, either using blowing reagents or under hydrothermal conditions, followed by carbonization, leads to macroporous carbons, this process being classified as "process design." Exfoliation of graphite flakes via oxidation in acidic liquids is an efficient route to prepare macroporous graphite and reduced graphene oxide (rGO) foams, which is also classified as "process design."

To achieve a low-carbon society, the assessment of CO_2 emission over the whole life of a material (lifecycle assessment [LCA]) is required, from raw materials to its final disposal through its manufacturing and usage. For the production of carbon materials and their applications, the consideration on LCA for each carbon material is required. For example, the CO_2 emission is calculated to be 22.4 kg CO_2 for manufacturing 1 kg of polyacrylonitrile (PAN)-based carbon fibers, which is almost 10 times higher than that of iron, although carbon fibers have about 10 times higher mechanical strength than iron [3]. Now novel approaches for carbon fibers and the matrix plastics of carbon fiber-reinforced plastics (CFRPs) are strongly demanded by taking their recycling into consideration [4–6]. Therefore, the precursors for the production and applications of porous carbons have to be reconsidered by taking account of their LCA. It might be worth revisiting activated carbons because most were produced using various biomasses as precursors. Cycling uses of all materials for the production of porous carbons are also important. For example, the MgO template was experimentally shown to be used repeatedly just by adding a new carbon precursor to MgO-dissolved acid solution, such as acetic and citric acids [7].

5.1.2 Applications of porous carbons

The applications of porous carbons were explained by dividing into two parts, energy storage and conversion in Chapter 3, and environment remediation in Chapter 4, in this book. In Chapter 3, the storage of electric energy using rechargeable batteries, supercapacitors, and their hybrid cells are mainly discussed. Fuel cells and the storage of hydrogen and methane are discussed because of their importance for solving future energy problems. Thermal energy storage, in addition, is discussed where porous carbons play important roles as scaffolds and additives for improving thermal conductivity of the materials. In this field of energy problems, porous carbons are important structural components for devices. However, other carbon materials, such as carbon fibers, carbon nanotubes, graphene, and graphene-related materials, are included in the discussion because they are expected to give certain contributions for addressing energy problems. In Chapter 4, carbon materials contributing to environment remediation are discussed on the bases of functionalities of porous carbons. The applications of porous carbons as adsorbents are discussed by dividing them into adsorptive removals of various inorganic and organic pollutants. The adsorption of water vapor, CO_2, metal ions, and oils are separately discussed. The performances of porous carbons for gas separation, capacitive deionization for water desalination, electromagnetic interference shielding (microwave shielding), and sensing to chemicals, mechanical strains, and biomedicals, are discussed in comparison with carbon nanotubes and graphene-related materials.

In Table 5.1, the contributions of pores (the roles of pores) in carbon materials are summarized on each application described in Chapters 3 and 4. From Table 5.1, it can be pointed out that the pores in carbon materials have the following three roles in their applications; providing (1) large surface area from few hundreds to even more than 3000 m^2/g, (2) a wide range of pore sizes, from ultramicropores to macropores, and (3) a large amount of different functional groups on the pore surfaces. Role (3) is characteristic for carbon materials and includes a wide range of functional groups from acidic to basic ones.

Micropores in carbon materials are necessary for the storage of H_2, CH_4, and CO_2, smaller micropores being required for CO_2 adsorption in comparison with H_2 storage. In many applications, such as supercapacitors and adsorbents, however, these three roles of carbon pores work simultaneously, whereas in some applications one of these three becomes dominant. Roles (1) and (2) are governed by the synthesis conditions, i.e., the precursor and synthesis procedures including carbonization temperature and residence time. Role (3) also depends strongly on the synthesis conditions, but the functional groups can be modified by the treatment after carbonization. Post-treatment in HNO_3 is often employed to increase acidic functional groups.

For roles (1) and (2), the webs and crumbs of carbon nanotubes and nanofibers can fulfill the conditions for porous carbons, such as the nanosized spaces formed by entanglement of tubes and fibers.

For role (3), reduced graphene oxides (rGOs) can provide functional groups not only at the edges of the carbon layers but also by being associated with structural defects in the layer, as shown in a model for rGO flakes in Fig. 5.2A. On the graphene and rGO flakes, in addition, the π-electron clouds can have an interaction with π-electrons in organic molecules and some cations ($\pi-\pi$ and $\pi-$cation interactions), which promote their adsorption capabilities and so expand their application fields. For utilizing these functional groups and π-electron clouds efficiently, the stacking of graphene and rGO flakes in parallel needs to be avoided, as illustrated schematically in Fig. 5.2B. Bulky agglomeration of the flakes, however, is not appreciated in some applications, such as the electrode for supercapacitors.

Table 5.1 Contributions of pores in carbon materials.

Application			Contribution of pores	
			Principal	Subsidiary
Energy storage and conversion	Rechargeable batteries (negative electrode)	LIBs (intercalation type)	Intercalation between layers	Interaction with FGs (irreversible capacity)
		LIB (Redox-type)	Faradaic reaction with the material loaded on carbons	Interaction with FGs (irreversible capacity)
		Li–S	Holding sulfur in the micro- and mesopores of carbon electrode	Interaction of doped N in carbon electrode with Li-polysulfides
		SIBs	Trapping Na^+ in the pores and interaction with FGs	
	Supercapacitors		Electric double-layer formation on the large-area surfaces of electrodes	Faradaic reaction with FGs (pseudocapacitance)
	Hybrid cells		On one electrode: Electric double-layer formation On another electrode: Intercalation with FGs and/or redox reaction	
	Fuel cells		Holding of metal catalyst nanoparticles or holding of dopant atoms (N) on the surface to enhance oxygen reduction	
	Hydrogen storage		Micropore filling	
	Methane storage		Micropore filling with or without hydration	
	Thermal energy storage		Holding phase change materials in macropores	

Environment remediation	Adsorption	Inorganic pollutants	Interaction with FGs on the pore surfaces for oxidation and pore filling	
		Organic pollutants	Interaction with FGs on the pore surfaces and pore filling	
		H_2O adsorption	Micropore filling	Interaction with FGs at low P/P_0
		CO_2 capture	Filling of narrow micropores	Interaction with FGs
		Metal trapping	Interaction with FGs on the pore surfaces and oxidation	Pore filling
		Oil sorption	Filling of macropores, crevices on the surfaces, and void spaces between particles	Hydrophobicity (oleophilicity) of the pore surfaces
	Gas separation		Permeation through micro-/mesopores	
	Capacitive deionization		Micro- and mesopore filling and cation−π interaction on rGO	Interaction with FGs
	Electromagnetic interference shielding		Multireflections at surfaces of particles and pore walls	Absorption by carbon matrix
	Sensing	Chemical sensors	Pore filling and/or interaction with FGs	
		Strain sensors	Due to piezo-electrical properties of carbon	
		Biosensors	Modification of graphene and rGO (for example, immobilization of enzyme using FGs)	

FG, functional groups.

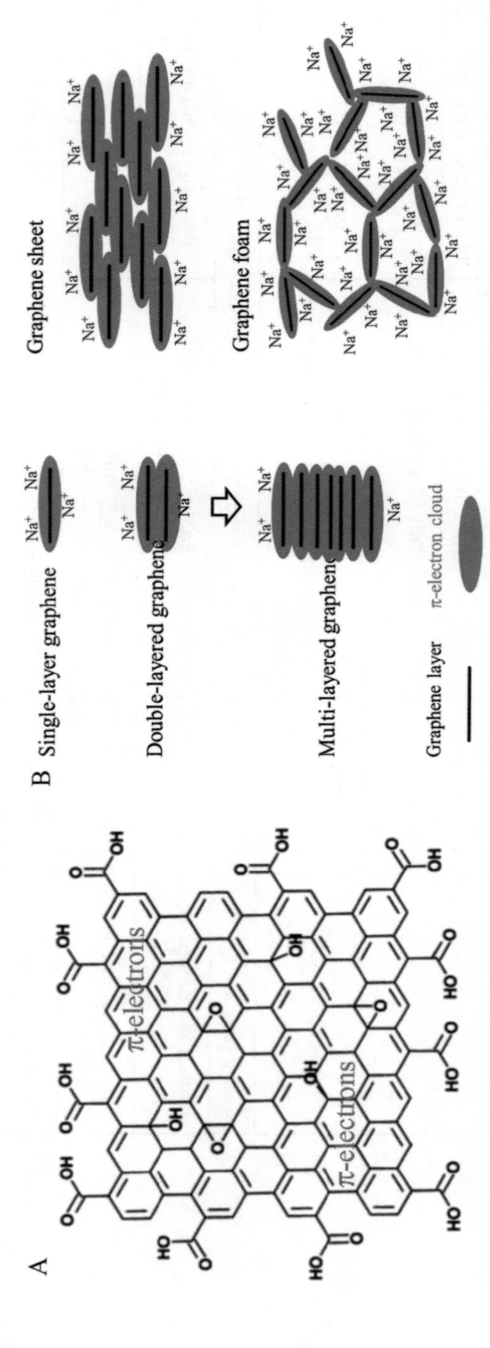

Figure 5.2 Schematic illustrations of (A) reduced graphene oxide flake and (B) stacking and agglomeration schema of graphene when Na^+ ions were adsorbed.

5.2 Constraint and reaction space in carbons, pores

The pores and channels of carbon materials have played important roles in their applications, as explained in Chapters 3 and 4 (and the summary in Table 5.1). The interlayer spaces in graphite crystals are also used to accommodate active materials, the products being called graphite intercalation compounds (e.g., lithium ions in rechargeable batteries). In these cases, the pores, channels, and interlayer spaces of carbon materials work as the accommodation spaces for the foreign materials. Recently, however, some research has demonstrated experimentally that the pores and channels of carbon materials work to endow the encapsulated functional materials with performances either improved or different from those of the bulk materials prepared by conventional processes (e.g., powders, films), as well as to enable the reactions of encapsulated materials, which are impossible in ambient conditions outside of the pores and channels. These pores and channels are reasonably supposed to give certain constraint to the materials by keeping them inside, accompanied by some changes in their structure, morphology, stability, properties, etc., and therefore they were called "constraint spaces" in carbon materials [8]. This might be prospective for the development of porous carbons, including carbon nanotubes, in the future.

Here, the experimental results demonstrating the constraint spaces in carbon materials are summarized, including porous carbons and carbon nanotubes, although some of them are duplicated with the presentation in previous chapters. Carbon-coated materials experience certain constraints from the carbon walls to the coated materials, which improves their performances. In addition, the pores in carbon materials provide also the reaction spaces to synthesize materials with new functionalities.

5.2.1 Porous carbons

Some of functional materials stored in the pores of carbon materials have been experimentally demonstrated to show unique properties, which are reasonably supposed to be caused by some constraints from nano-sized pores of carbon materials. Functional materials are stably supported in pores with constraint and have a large contact area with the conductive carbon surface, which endows them with new characteristics, different from those in the bulk state.

5.2.1.1 Electrochemically active materials

The confinement of electrochemically active organics, such as polyaniline (PANI) and polypyrene (PPY), into the micro- and mesopores of carbons was shown to be beneficial to enhance the volumetric capacitance of supercapacitors by retaining high power density, as reviewed [9] and explained by their electrochemical performance in Section 3.2.5.

Either aniline (ANI) or pyrene (PY) was adsorbed at 25 or 150°C, respectively, into a commercial AC (MSC30) having S_{BET} of 3160 m^2/g with V_{micro} of 0.99 and V_{meso} of 0.60 cm^3/g, followed by electrochemical polymerization in 1 M H_2SO_4 aqueous electrolyte [10]. A 2,2,6,6-tetramethylpiperidine-N-oxyl (TEMPO) derivative, 4-hydroxy-TEMPO benzoate (HTB), was also adsorbed on the same AC by heating a mixture of AC and HTB in different weight ratios (HTB/AC of 20/80–50/50) at 130°C in a vacuum-sealed glass ample [11]. The adsorption of 2,5-dichloro-1,4-benzoquinone (DCBQ) into Ketjenblack (KB) was performed by heating DCBQ and KB, which were separated in a glass ampule at different weight ratios (DCBQ/KB of 5/95–40/60), at 100°C, slightly higher than the sublimation temperature of DCBQ [12]. In Fig. 5.3, the change in the pore structure of the AC with HTB impregnation is shown by N_2 adsorption–desorption isotherms and pore-size distribution [11]. The result suggests that HTB molecules are impregnated into micropores and small mesopores of the AC. Occupation of DCBQ in mesopores of KB was also confirmed from the N_2 adsorption–desorption measurements of the composites [12]. Homogeneous distribution of these organics over the whole KB particles was confirmed from the EDS spectra of either N or Cl in organic molecules. The stable charge–discharge cycles of these composites as supercapacitor electrodes are explained in Section 3.2.5, suggesting that the electrochemical redox reactions of these organic molecules occur steadily and repeatedly in the pores of porous carbons.

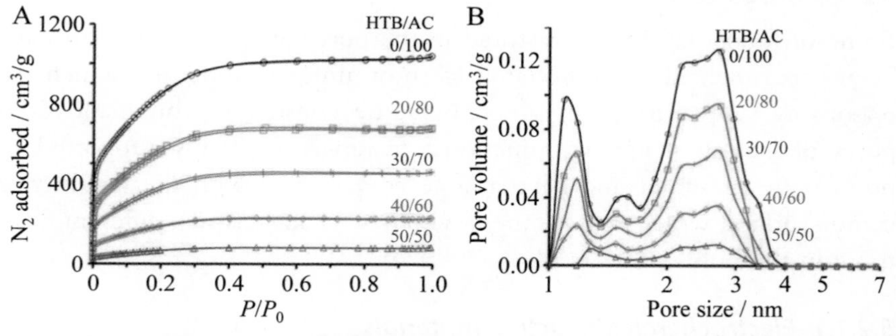

Figure 5.3 N_2 adsorption–desorption of HTB/AC composites: (A) isotherms collected at 77 K and (B) pore size distributions [11].

MnO_2 nanocrystals confined in micropores of porous carbon nanofibers (MnO_2/CNF composite) delivered high electrochemical capacitance in 1 M Na_2SO_4 aqueous electrolyte: 1282 F/g with the current density of 0.2 A/g [13]. Porous CNFs were synthesized from polymer blends of a polyimide (PMDA/ODA) and polyvinylpyrrolidone (PVP) via electrospinning, followed by imidization at 300°C and subsequent carbonization at 900°C in N_2. MnO_2 loading was performed by immersing porous CNFs into different concentrations of $KMnO_4$ aqueous solutions at room temperature, where MnO_4^- reacted with carbon to deposit MnO_2. For the preferential deposition of MnO_2 nanocrystals in nano-sized pores, the optimum concentration of $KMnO_4$ was determined to be 0.024 g/L. In Fig. 5.4A and B, the pore-size distributions are shown for the CNFs loaded by MnO_2 in $KMnO_4$ solution with different concentrations. The CNF loaded in 0.012 g/L $KMnO_4$ solution has almost the same pore-size distribution in the micropore range as that of the pristine CNF (no MnO_2 loading), while most of the mesopores are collapsed. In the CNF loaded in 0.024 g/L solution, the micropores with sizes of 0.6–1.2 nm are created by sacrificing small micropores. On the CNF obtained from the solution with a high concentration of 0.12 g/L, however, the deposition of MnO_2 occurs preferentially on the external surface of the CNF, because the micropores are retained and all mesopores are closed, as shown in Fig. 5.4A and B, respectively. The MnO_2 nanocrystals in the inner part of CNF were grown along with (200) crystal faces under nanopore constraint and thus formed ultrathin flakes, whereas the nanocrystals deposited on the surface of CNF were grown along (211) crystal faces. The former nanocrystals shortened the diffusion path of electrolyte and enlarged the active surface area.

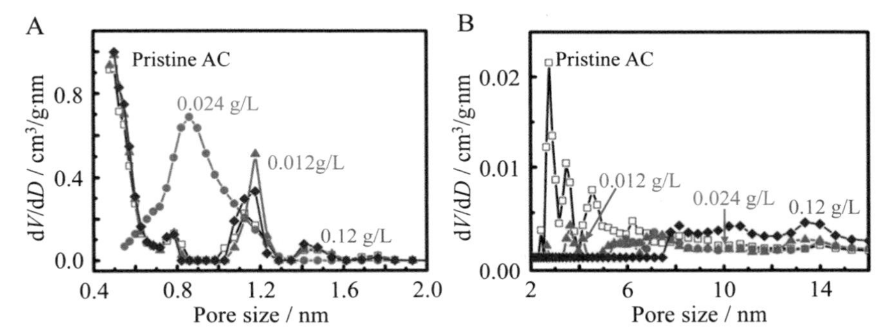

Figure 5.4 Pore size distributions of the porous CNFs loaded by MnO_2 nanocrystals in different concentrations: (A) micropore and (B) mesopore ranges [13].

A symmetrical supercapacitor of the MnO_2/CNF composite electrodes prepared from 0.024 g/L solution could deliver much better performance than that of the composites prepared from other concentrations and also than the pristine CNFs, with the energy density reaching 36 Wh/kg at a power density of 39 W/kg and 7.5 Wh/kg at 10.3 kW/kg.

The composite of mesoporous carbon with metallic Sn was fabricated by immersing the carbon into an ethanol solution of $SnCl_2$, followed by heat treatment at 700°C in Ar [14]. The mesoporous carbon was prepared from a mixture of melamine, phenol, and formalin with colloidal silica in water by polymerization at 450°C and carbonization at 900°C, followed by dissolution of SiO_2 with HF. The resulting mesoporous carbon has S_{BET} of 1560 m^2/g and V_{total} of 2.56 cm^3/g, together with a small amount of micropores and its N content is 14.51%. After Sn loading through $SnCl_2$ impregnation and decomposition at 700°C, most of the metallic Sn was confined into micro- and mesopores of the matrix carbon, as suggested from the changes in the N_2 adsorption–desorption isotherms in Fig. 5.5A, S_{BET} and V_{total} being reduced to 389 m^2/g and 0.84 cm^3/g, respectively. The resultant composite delivered highly reversible capacities of 1081 and 649 mAh/g with current densities of 0.2 and 4 A/g, respectively, and an excellent cycling performance in 1 M $LiPF_6$/(EC + DEC) electrolyte, as shown in Fig. 5.5B. The hybrid cell consisting of this composite after lithiation at the negative electrode and the microporous carbon prepared from biomass (pomelo peel) at the positive electrode exhibited a high energy density of 196 Wh/kg at a power density of 731 W/kg and 84.6 Wh/kg at 24.4 kW/kg.

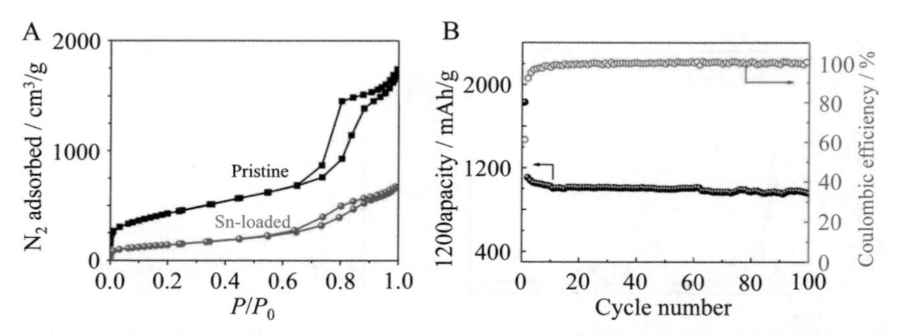

Figure 5.5 Composite of mesoporous carbon with metallic Sn nanoparticles: (A) N_2 adsorption–desorption isotherm in comparison with the pristine carbon and (B) cyclic performance with a current density of 0.2 A/g [14].

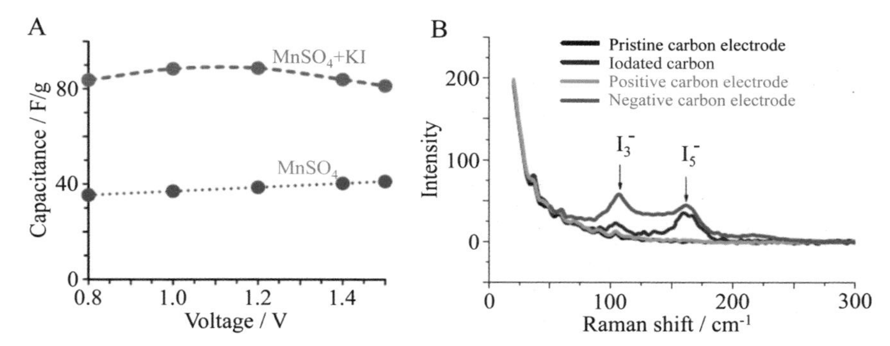

Figure 5.6 Symmetric cell of AC with aqueous electrolytes of ($MnSO_4$+KI) and $MnSO_4$: (A) capacitances of the cells against voltage and (B) Raman spectra comparison on the electrode carbons after aging in a cell with ($MnSO_4$+KI) electrolyte [15].

A symmetric cell of the same KOH-activated AC (S_{BET} of 2181 m^2/g) worked as a hybrid cell in (2 M $MnSO_4$+0.5 M KI) aqueous electrolyte, exhibiting battery-type behavior at the positive electrode and capacitor-type behavior at the negative electrode, although it worked as a supercapacitor in 2 M $MnSO_4$ electrolyte without KI [15], as mentioned in Section 3.3.1.1.2. The hybrid cell delivers much higher capacitance than the supercapacitor with $MnSO_4$ aqueous electrolyte, as shown in Fig. 5.6A. The symmetric cells of AC were also reported to work as hybrid cells in aqueous electrolytes of different alkali metal iodides [16]. This hybrid behavior in KI-based aqueous electrolyte was concluded to be due to the confinement of polyiodides in the micro- and mesopores of the carbon at the negative electrode, as shown in the Raman spectra in Fig. 5.6B [15]. The polyiodides contributed to add pseudo-capacitance to the cell capacitance through the interaction with electrolyte cations in the pores of negative electrode carbon.

5.2.1.2 Graphitic carbon nitride

Graphitic carbon nitride *g*-C_3N_4, which has high activity of oxygen reduction reaction (ORR) [17], was synthesized in a mesoporous carbon CMK-3 by impregnating cyanamide as a precursor, followed by calcination at 550°C in Ar [18]. As shown in Fig. 5.7A, the characteristic ordered mesoporous structure of the host CMK-3 was kept even after confinement of *g*-C_3N_4, reaching the *g*-C_3N_4 content of 27.1 wt.%. The ORR activity of the composite was comparable to a commercial Pt/C catalyst. However, its tolerance for methanol contamination was improved markedly, as shown in Fig. 5.7B. Since the CMK-3 has a high surface area and high electrical

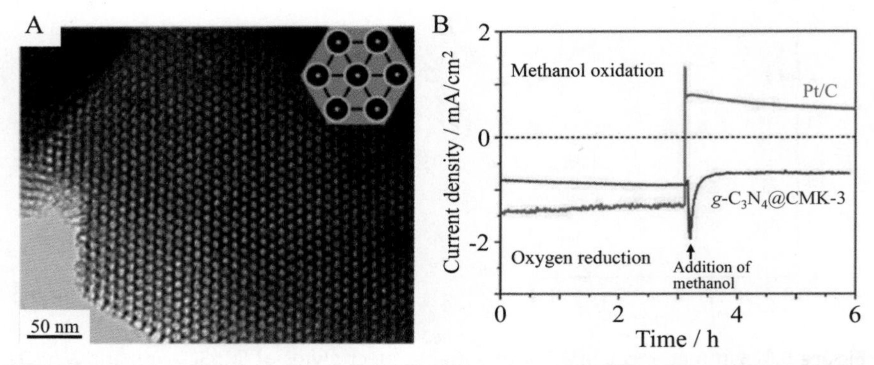

Figure 5.7 g-C_3N_4/CMK-3 composite: (A) TEM image with illustration, yellow: g-C_3N_4 and black: carbon (inset) and (B) chronoamperometric responses for methanol addition of the composite and a commercial Pt/C in O_2-saturated 0.1 M KOH solution [18].

conductivity, CMK-3 not only serves as a support for g-C_3N_4 but also confines the g-C_3N_4 catalyst in its nanospaces, which facilitates the electron transport necessary for an efficient $4e^-$ ORR pathway.

5.2.1.3 Sulfur

Sulfur can be confined to the pores of porous carbons for the application of negative electrodes of Li—S batteries (Section 3.1.3.1). At a negative electrode, anchoring of lithium polysulfides (Li_2S_x) is strongly demanded to inhibit their shuttling between electrodes during charge—discharge cycles. The shuttling would lead to the dissolution of Li_2S_x (x of 4—8) into electrolytes, resulting in short cycle lifetimes of the electrodes and contamination of the lithium positive electrode. N-doping and N/S-codoping of porous carbons were demonstrated to be effective in enhancing the affinity for Li_2S_x, preventing their dissolution into an electrolyte [19]. Doped mesoporous carbons were synthesized by mixing either N-allylurea or N-allylthiourea with colloidal silica in water and subsequent carbonization at 800°C (NMPC and NSMPC, respectively), while non-doped mesoporous carbon was synthesized from resorcinol-formaldehyde mixed with colloidal silica (MPC). As shown in Fig. 5.8, the codoped carbon NSMPC is capable of adsorbing one of the lithium polysulfides, Li_2S_6, much more than NMPC and MPC. It delivered a high capacity as 1267 mAh/g and very low capacity loss with charge—discharge cycling (0.041% per cycle), suggesting a firm anchoring of sulfurs on the pore wall of the NSMPC to make stable lithiation—delithiation cycles possible.

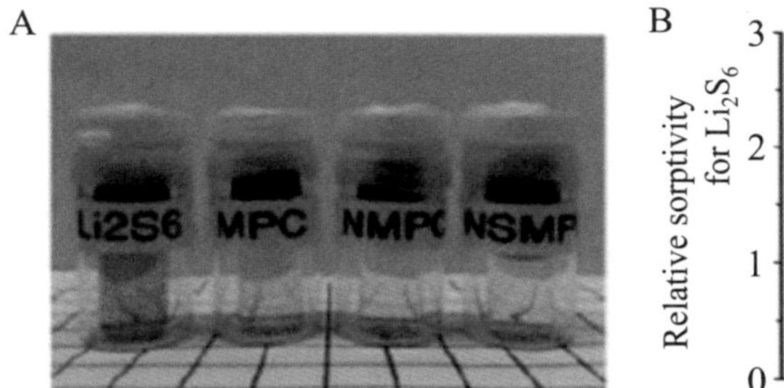

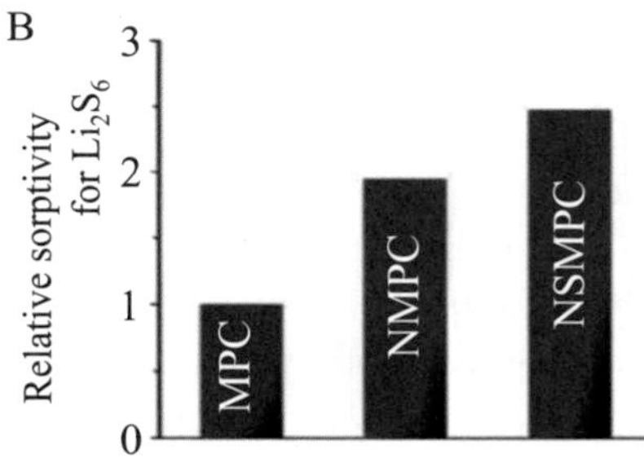

Figure 5.8 Relative adsorptivity of MPC, NMPC, and NSMPC for lithium polysulfide Li_2S_6: (A) photographs of lithium polysulfide solutions after adsorption for 20 h and (B) relative adsorptivity of MPC, NMPC, and NSMPC for Li_2S_6 [19].

5.2.1.4 Methane hydrate

Water molecules confined in micropores of AC could adsorb methane under pressure above 3.5 MPa at 2°C to form methane hydrate in the micropore [20]. The AC used was prepared from mesophase-pitch at 450°C, followed by activation using KOH at 800°C, and had S_{BET} of 3670 m^2/g with V_{micro} of 1.20 cm^3/g. In Fig. 5.9, methane adsorption–desorption isotherms at 2°C are shown for the AC with different pre-adsorbed water contents per 1 g of AC (R_w in g/g). Their methane adsorption capacity depended strongly on R_w and was much larger than the dry AC ($R_w = 0$). R_W of 1.8 corresponds to the saturation of water and R_W above 1.8 exceeds the saturation amount. At low methane pressure, the wet ACs show a lower adsorption capacity than the dry one, probably due to pore blocking by preadsorbed water. Above 3 MPa pressure, however, methane adsorption increases sigmoidally; the higher the R_W is, the more

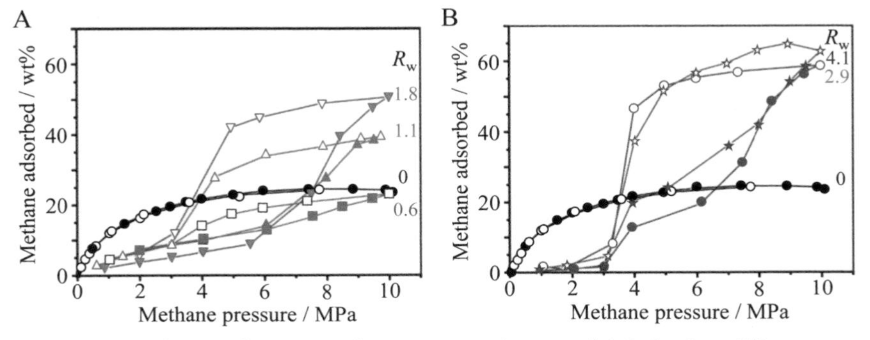

Figure 5.9 Methane adsorption–desorption isotherms of ACs having different pre-absorbed water contents (R_w) [20].

abruptly AC adsorbs methane. The formation of methane hydrate was confirmed by inelastic neutron scattering and synchrotron X-ray powder diffraction analyses. Hydration at 3.5 MPa and 2°C proceeds much milder and faster (finishing within minutes) than the occurrence of methane hydrate in nature (more than 6 MPa and slightly below room temperature). The ACs with high water contents, R_w of 2.9 and 4.1 (oversaturated), give two-step adsorption isotherms with pronounced hysteresis (Fig. 5.9B), suggesting that adsorption–desorption mechanisms of methane for wet ACs are different from those for the dry one, and the methane hydrate formed in carbon micropores is stable.

The storage of methane is one of most important strategies for energy storage in the future and hydrate storage is promising, as discussed in Section 3.6.2.

5.2.2 Carbon nanotubes

Carbon nanotubes (CNTs) provide long and straight channels with well-defined and uniform diameters. In these channels in CNTs, various materials, including CNTs, fullerenes, and different molecules, were reported to be encapsulated and to demonstrate specific structural changes in these constraint spaces. In this section, the structural changes of the encapsulated materials in CNT channels are summarized and discussed by classifying on the basis of encapsulated materials, i.e., CNTs themselves, iodine, sulfur, organic molecules (e.g., quinones), and water molecules.

5.2.2.1 Carbon nanotubes

Coalescence of double-walled carbon nanotubes was shown to occur by high-temperature treatment and to create novel structures [21]. In Fig. 5.10A, radial breathing mode (RBM) Raman spectra below 400 cm^{-1} are shown for DWCNTs heat-treated at different temperatures, where the Raman shift is inversely related to the CNT diameter and so the diameter values are shown for principal peaks. There are no significant changes in the spectrum up to 2100°C, except for the disappearance of the peak at 312 cm^{-1} (corresponding to 0.77 nm diameter), suggesting the transformation of SWCNTs to DWCNTs. After 2200°C treatment, an abrupt decrease in intensity of the peak at 270 cm^{-1} (0.90 nm) is observed. This decrease is interpreted as the destruction of the as-grown DWCNT structure which consists of the inner tubes with 0.9 nm diameter and the outer tubes with around 1.6 nm diameter. The pristine DWCNTs have a homogeneous size as shown in the high-resolution TEM image in

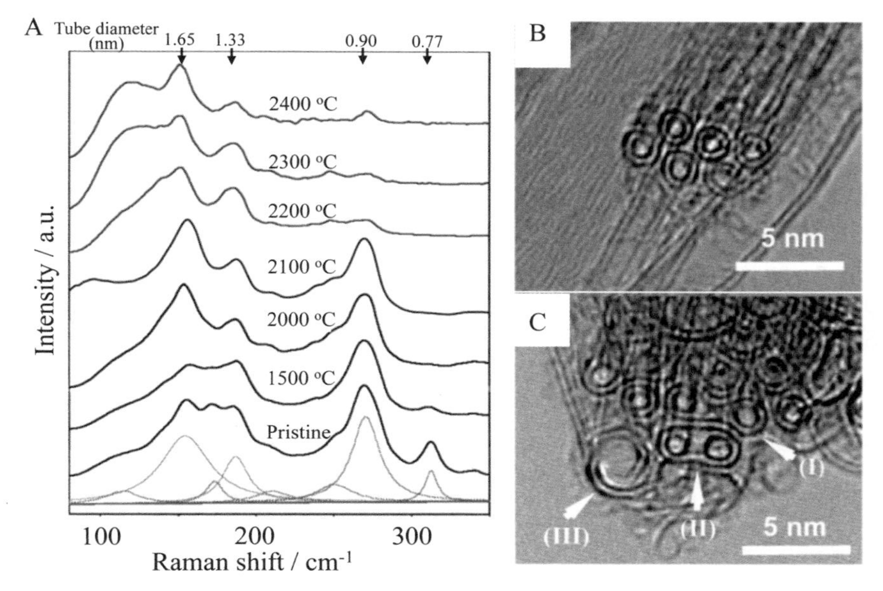

Figure 5.10 Heat-treatment of DWCNTs: (A) change in the radial breathing mode (RBM) Raman spectrum with heat-treatment temperature and (B) TEM image of the pristine DWCNTs and (C) that after 2100°C treatment [21].

Fig. 5.10B, whereas the TEM image of the 2100°C-treated DWCNTs (Fig. 5.10C) suggests the occurrence of structural reconstruction. The detailed analysis by TEM suggested that the reconstruction was supposed to occur sequentially; the original double-walled tube (I in Fig. 5.10C) coalesced into a bicable structure (II) and then changed to a large-diameter double-walled tube (III). The process of this reconstruction is shown in Fig. 5.11. Two adjacent DWCNTs (Fig. 5.11A) first merge into a bicable carbon nanotube by coalescence of the outer tubes (Fig. 5.11B), and the inner tubes coalesced to a large-diameter inner tube (II in Figs. 5.10C and 5.11C) [22]. Eventually, the off-centered inner tube interacted with the

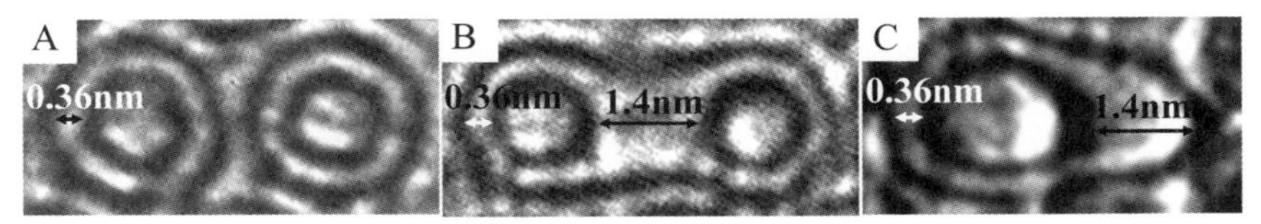

Figure 5.11 TEM images on the gradual changes from (A) adjacent DWCNTs, (B) bicable structure and to (C) off-centered DWCNT during heat treatment at 2100°C [22].

outer tube, leading to the off-centered DWCNT. Coalescence of SWCNTs to multiwalled CNTs was also reported [23].

Encapsulation of ferrocene ($FeCp_2$) into SWCNTs was performed by contacting SWCNTs with ferrocene vapor at 300°C under vacuum, followed by washing with diethyl ether to remove ferrocene deposited on the surface of tubes [24]. The resultant composite ($FeCp_2$@SWCNTs) retained the redox activity of $FeCp_2$ in 0.1 M Bu_4NClO_4/AN electrolyte. $FeCp_2$ encapsulated into SWCNTs was converted to metallic Fe nanoparticles by heating up to 700°C under vacuum to form Fe-encapsulated SWCNTs (Fe@SWCNT) [25]. Individual Fe particles formed in SWCNTs have a diameter of 0.5–0.7 nm, suggesting aggregation of several Fe atoms, and the filling yield (ratio of filled SWCNTs to empty SWCNTs) was estimated to be more than 80%. $FeCp_2$@SWCNTs converted to DWCNTs by high-temperature annealing [26]. Annealing of $FeCp_2$@SWCNTs at 600°C led to the formation of metastable iron carbide, which was supposed to be formed by absorbing $FeCp_2$-derived carbon atoms from one side of the tube and generated an inner tube to another side resulting in DWCNTs, as shown in the schematic illustration in Fig. 5.12. Once the inner tube had grown, the iron atoms diffused out of the tubes and aggregated into iron oxide nanoparticles on the surface of DWCNT. The intensity distributions of the RBM Raman spectra for the inner-tube of $FeCp_2$-derived DWCNTs were significantly different from those of C_{60}-peapod-derived DWCNTs (see the next section). Annealing

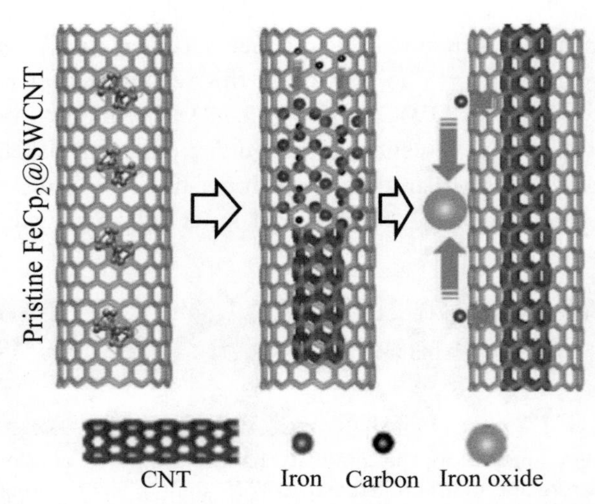

Figure 5.12 Schematic diagram for the growth process of the inner tube [26].

of $FeCp_2$@SWCNTs at 1150°C resulted in expelling of all Fe atoms from the tube. Encapsulation of $FeCp_2$ into DWCNTs ($FeCp_2$@DWCNT) resulted in a marked change in the electronic properties, probably due to the charge transfer between $FeCp_2$ molecules and DWCNTs, and Fe@DWCNT prepared from $FeCp_2$@DWCNT was unipolar n-type semiconductor [27]. Bis(cyclopentadienyl) cobalt (cobaltocene, $CoCp_2$) could be encapsulated only in the SWCNTs with a tube diameter of ~1 nm under vacuum at 100°C, and bis(ethylcyclopentadienyl) cobalt [$Co(EtCp)_2$] encapsulation was possible in a range of tube diameters greater than about 1 nm [28].

5.2.2.2 Fullerenes

Encapsulation of fullerenes, mostly C_{60} and C_{70}, into SWCNTs (C_{60}@SWCNT and C_{70}@SWCNT) was experimentally observed in the products of laser ablation of graphite impregnated by Co/Ni catalyst and in the products of carbon arc discharge using Ni/Y catalyst [29,30], the product being called "bucky-peapod" or simply "peapod." By using diameter-selected SWCNTs, encapsulations of C_{60} and C_{70} were performed by heat treatment at 650°C to yield high-density fullerene chains in each SWCNT, with filling factors of 85% and 72%, respectively [31]. The distance between C_{60} cages in the CNTs was 0.95 nm, which was shorter than that in the C_{60} crystal with face-centered cubic closest packing (1.00 nm) and longer than that in the polymer of C_{60} (0.92 nm). On the other hand, two different distances of 1.00 and 1.10 nm between C_{70} cages were observed for C_{70}@SWCNTs, which were explained by morphologies of C_{70} cages, standing and lying, respectively. Encapsulation of C_{60} and C_{70} into SWCNTs, DWCNTs, and triple-walled carbon nanotubes (TWCNTs), was facilitated by reacting CNTs with fullerene vapor [32,33]. The structural stability of C_{60}@SWCNTs was examined under high pressures and high temperatures by using in situ synchrotron X-ray diffraction [34]. The diffraction peak at 25.23 keV (corresponding to the spacing of 0.956 nm) due to one-dimensional C_{60} crystals is intensified with increasing pressure up to 10.7 GPa at room temperature, associated with weakening of hk diffraction peaks of SWCNT bundles as shown in Fig. 5.13A. The $C_{60}-C_{60}$ distance is gradually shortened from 0.956 nm at 0.1 MPa down to 0.892 nm at 10.7 GPa, as shown in Fig. 5.13B. By heating up to 1023°C under 4 GPa, the $C_{60}-C_{60}$ distance further reduced to 0.87 nm as a result of C_{60} polymerization [35].

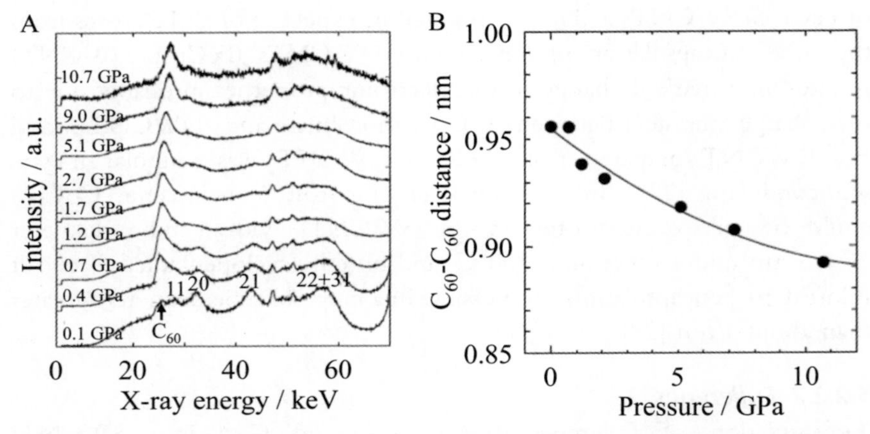

Figure 5.13 C_{60}-peapod: (A) change in XRD pattern with increasing pressure and (B) change in C_{60}—C_{60} distance with pressure [34].

Coalescence of C_{60} inside SWCNTs leads to a linear chain and finally to the inner tube wall, resulting in the formation of DWCNTs at high temperatures. C_{60}@SWCNTs were prepared by exposing SWCNTs to C_{60} vapor at 400°C in dry air, and then treated at high temperatures in vacuo (<10^{-6} Torr) [36]. Encapsulated C_{60} molecules are self-assembled to make a chain with nearly uniform center-to-center distances, as shown in the TEM image in Fig. 5.14A. C_{60} molecules remain mostly unchanged in CNTs up to 800°C and start to coalesce with adjacent ones to form linked beans and/or short nanotubes above 800°C, as shown on 1000°C-treated one in Fig. 5.14B. Around 1200°C, most C_{60} molecules coalesce to form a tube wall, making this part DWCNTs while retaining the empty

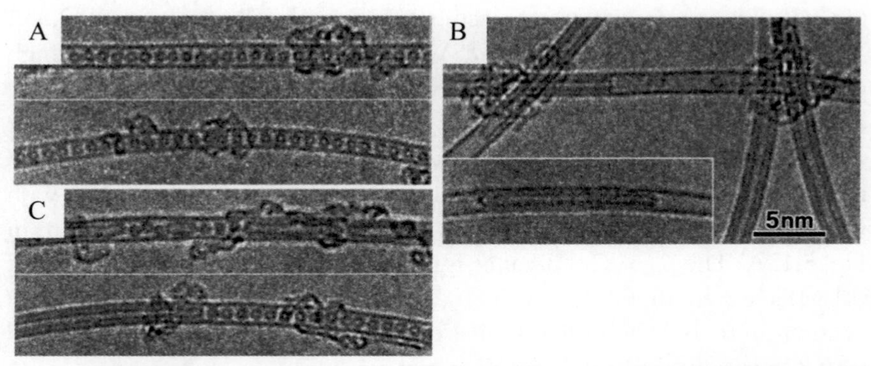

Figure 5.14 TEM images of C_{60}-encapsulated SWCNTs: (A) as-prepared, (B) after heat treatment at 1000°C, and (C) after 1200°C in vacuo [36].

single-wall tube partly, as shown in Fig. 5.14C. After the 1200°C treatment, no C_{60} molecules remain and most of the tubes are partly changed to double-walled. In some cases, the inner tubes are terminated by caps, shown as the inset in Fig. 5.14C. The thermal treatment of the peapods is the most effective way for fabricating catalyst-free and high-purity DWCNTs among the methods including arc-discharge and CVD methods using catalysts, particularly to form small-diameter inner tubes (0.4–0.7 nm). The diameter of the inner tube synthesized from the peapods of SWCNTs with filling factor of 60%–80% depended on the synthesis temperature [37]. Almost the same diameter distribution of the inner tubes is achieved by the coalescence of fullerenes at temperatures up to 1800°C, but it became broad after the heat treatment at 2000°C, including enlarged diameters due to coalescence of adjacent tubes. Coalescence of fullerenes in SWCNTs was induced also by electron irradiation to form DWCNTs, which was followed by in situ TEM observations [38]. DWCNTs are mechanically, thermally, and structurally more stable than SWCNTs, probably because of the buffer-like function of the outer tubes. The inner tube of DWCNTs exhibited unique transport and optical properties, such as extremely sharp Raman lines for the RBM Raman spectrum of the inner tube, suggesting its high crystallinity and highly unperturbed environment in the interior of the tubes [39]. Highly crystalline and uniform TWCNTs were also obtained from the heat treatment of DWCNT peapods at 2000°C in Ar [40]. The DWCNTs were prepared by catalytically grown method, and their diameters were enlarged to the sufficient interior spaces with 1.2–1.6 nm diameters for the encapsulation of C_{60} through the heat treatment at 2400°C in Ar atmosphere. The TWCNTs thus synthesized were uniform and highly crystalline, exhibiting new Raman and luminescent signals owing to the innermost tubes.

Cyclic electrochemical insertion–deinsertion of Li into C_{60}@SWCNT was possible in 1 M $LiClO_4$/(EC + DEC) with a potential range of 0–3.0 V [41]. As shown in Fig. 5.15, the cell gives a reversible capacity of 550–610 mAh/g, which is slightly larger than the pristine SWCNTs (460–490 mAh/g), whereas both C_{60}@SWCNTs and SWCNTs showed a large irreversible capacity. The enhancement in the capacity was explained by the following two possibilities: the change in the electronic structure of the SWCNTs and a steric effect from the encapsulated C_{60}. The encapsulated C_{60} is supposed to stabilize Li ions which are desolvated at the entrance of the tubes. K-insertion into C_{60}@SWCNTs was performed by exposing C_{60}@SWCNTs to potassium vapor at 473 K over 50 h [42].

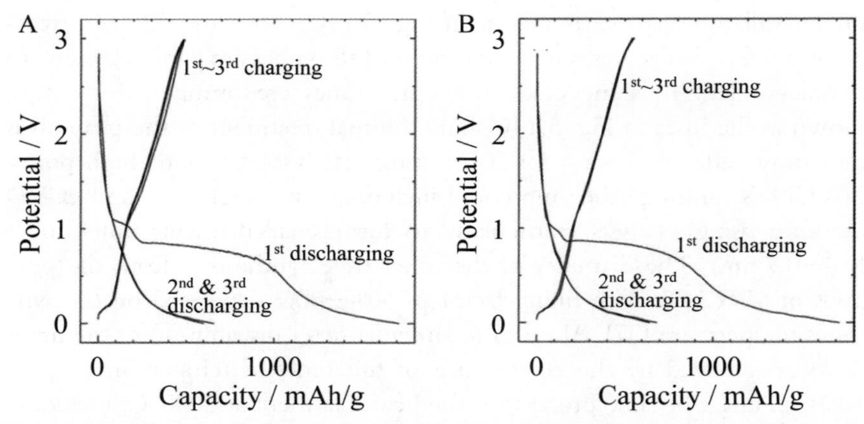

Figure 5.15 Charge–discharge curves of (A) SWCNTs and (B) C_{60}@SWCNTs in 1 M $LiClO_4$/(EC + DEC) [41].

K atoms in the tube were clearly detected as dark spots in TEM images and were allocated at the interfullerene sites. Gadolinium metallofullerenes Gd@C_{82} were encapsulated into SWCNTs, where the Gd@C_{82} molecules formed one-dimensional crystals [43].

5.2.2.3 Potassium iodide and iodine

Crystalline KI nanoparticles were grown in the nanospace of single-walled carbon nanohorns (SWCNHs) by heating the mixture at 800°C in a sealed quartz tube [44]. The nanospaces of SWCNHs, i.e., internal tubular spaces and interparticle micropores, were measured as volumes of 0.50 and 0.11 cm^3/g, respectively. Although KI is crystallized to NaCl-type structure (space group *Fm3m*) under ambient pressure in bulk, KI encapsulated in SWCNHs is crystallized into CsCl-type structure (space group *Pm3m*) with lattice constants of a = b = 0.35 nm and c = 0.38 nm, which is formed under pressure above 1.9 GPa in bulk crystal. The results suggest the highly constrained state of KI crystals in the nanospaces of SWCNHs. The c-axis of the KI nanocrystal is parallel and the a-axis is perpendicular to the nanohorn axis, as shown in Fig. 5.16.

One-dimensional KI crystal with two atomic thickness was formed in the channel of SWCNT with about 1.4 nm diameter by a capillary method and its structure was discussed on the basis of high-resolution TEM observation [45]. The structure of these biatomic crystal was observed to be distorted significantly. The distance between I ions was measured as 0.346 nm, whereas that in the NaCl-type bulk crystal was 0.353 nm.

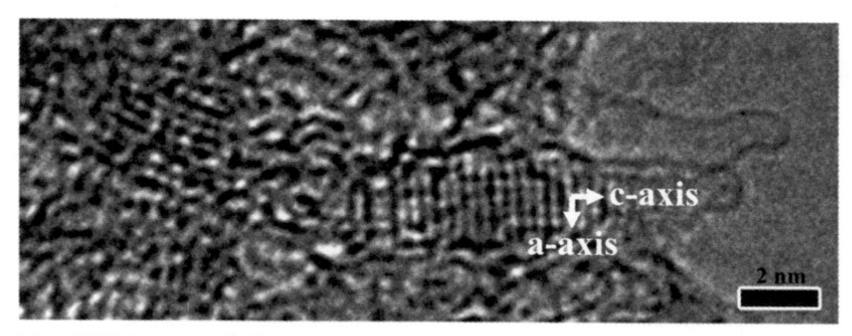

Figure 5.16 TEM image of the KI crystal in an SWCNH. The predominant contrasts are derived from I atoms in the image [44].

Encapsulation of I into SWCNTs (I@SWCNTs) was performed by heating the mixture of SWCNTs with iodine at 150°C in an evacuated glass tube and shown to improve their galvanomagnetic properties [46]. Electrochemical I-encapsulation in SWCNTs was performed in an NaI aqueous electrolyte, where its encapsulation level could be easily controlled by tuning the applied potential and time [47]. By reversing the polarity in an electrochemical cell, encapsulated-iodine molecules could be removed completely. To understand the structure of polyiodide ions encapsulated into CNTs and the charge-transfer from CNTs to iodine molecules, in situ Raman spectrum measurements were performed on SWCNTs with different tube diameters at low temperatures down to −100°C [48]. The amount of I encapsulated depended strongly on the diameter of SWCNTs; I contents estimated by TG analysis were 36.6, 49.0, and 50.0 wt.% for the SWCNTs having average diameters of 1.0, 1.5, and 2.5 nm, respectively. The Raman G-band peak position of the SWCNTs shifted toward the high wavenumber side with decreasing temperature, suggesting that the charge transfer from SWCNTs to encapsulated I molecules increased with decreasing temperature to form polyiodide ions, such as I_3^-, I_5^-, and I_7^-, as schematically shown in Fig. 5.17. A marked improvement in the electrical conductivity of SWCNTs by encapsulation of I may present the possibility of replacing the transparent conductive film, such as ITO, with these flexible transparent conductive films of SWCNTs.

5.2.2.4 Sulfur

Sulfur could be encapsulated into SWCNTs and DWCNTs as monoatomic chains with lengths up to 160 nm using a sealed glass tube at 873 K (S@SWCNT and S@DWCNT) [49]. High-resolution TEM images are shown for two straight chains in SWCNTs (Fig. 5.18A) and one zigzag

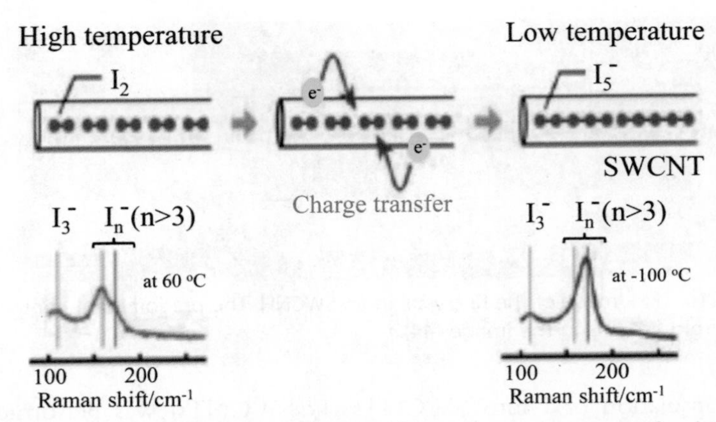

Figure 5.17 Schematic explanation for the change in the structure of polyiodide in SWCNTs [48].

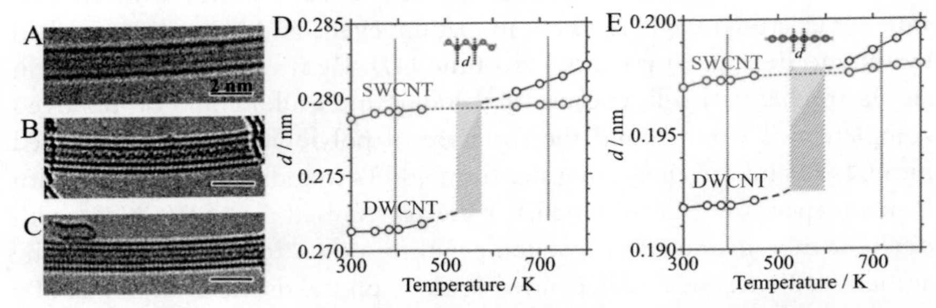

Figure 5.18 Sulfur encapsulation into CNTs: TEM images of (A) two straight S-chains in SWCNTs, (B) zigzag and (C) straight S-chains in DWCNTs, and temperature dependences of interatomic distance *d* in SWCNTs and DWCNTs of (D) zigzag and (E) straight S-chains. Melting and boiling points for bulk sulfur (about 393 and 718 K, respectively) are indicated by vertical lines for comparison [49].

chain and one straight chain in DWCNTs (Fig. 5.18B and C, respectively). As shown in Fig. 5.18D and E, the distance between two neighboring S atoms (*d*) in both straight and zigzag chains is slightly larger in SWCNTs than that in DWCNTs below 500 K, while *d* in DWCNTs approaches abruptly that in SWCNTs above 650 K. The S chains are immobilized in the host SWCNTs and DWCNTs up to 800 K (higher than the boiling point of bulk S), in other words, the channel in CNTs provides a suitable space for stabilizing the S chain. The sulfur atoms in CNTs were supposed to be bound with covalent character, forming a one-dimensional crystalline phase, as proved by XRD, and to be electro-conductive under ambient pressure and temperature. The electric resistivity at 300 K of SWCNTs

(1.1×10^{-3} Ωcm) decreased to 5.0×10^{-4} Ωcm by encapsulation of S chains. By taking into consideration the fact that the bulk S becomes metallic under ultrahigh pressure more than 90 GPa [50], encapsulated S chains in CNTs are strongly constrained, approaching the high-pressure phase of S in bulk.

5.2.2.5 Organic molecules

Various organic molecules, tetrakis(dimethylamino)ethylene (TDAE), tetramethyl-tetraselenafulvalene (TMTSF), tetrathiafulvalene (TTF), pentacene, anthracene, 3,5-dinitrobenzonitrile, tetracyano-*p*-quinodimethane (TCNQ), and tetrafluorotetracyano-*p*-quinodimethane (F_4TCNQ), were encapsulated into SWCNTs [51]. Encapsulation was performed simply by heating decapped SWCNTs with organic molecules just above their sublimation temperatures under vacuum. Encapsulation (confinement) of organic molecules resulted in the injection of amphoteric carriers into SWCNTs, which can be controlled by the ionization energy and electron affinity of the encapsulated molecules. As shown on TDAE@SWCNTs in Fig. 5.19A, the intensity of the optical absorption band at 0.68 eV decreases upon encapsulation. Since the absorption is changed by the encapsulation of organic molecules into SWCNTs, the intensity ratio of $I_{doped}/I_{pristine}$ is plotted against the ionization energy and the electron affinity of the molecules in Figs. 5.19B and C, respectively. The results reveal that TTF-, TMTSF-, and TDAE-confined SWCNTs show n-type behavior, while TCNQ- and F_4-TCNQ-confined SWCNTs show p-type behavior. These SWCNT semiconductors, by confinement of organic molecules, have advantages of air stability, controllable carrier doping, and simple process for preparation.

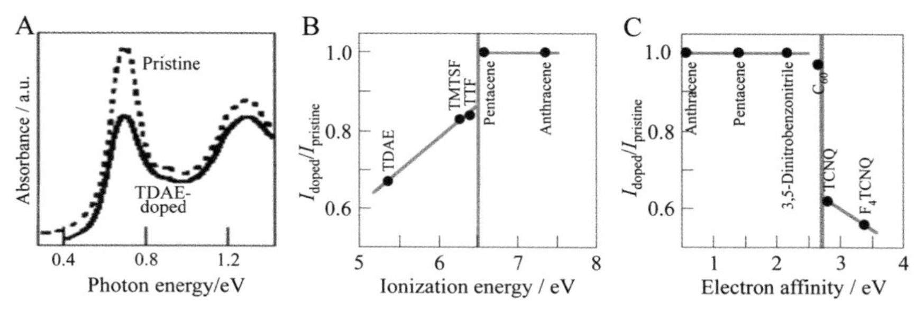

Figure 5.19 Optical absorption of organic molecule@SWCNTs: (A) absorption spectra of TDAE-doped and pristine SWCNTs, and the dependences of the absorption intensity ratio $I_{doped}/I_{pristine}$ on (B) ionization energy and (C) electron affinity of encapsulated molecules [51].

Encapsulation of quinone molecules, 9,10-anthraquinone (AQ) and 9,10-phenanthrene quinone (PhQ), into SWCNTs was performed by heating their mixture at 200°C in a glass tube [52,53]. After the heat treatment, AQ@SWCNTs and PhQ@SWCNTs were recovered by washing with organic solvents to remove the excess AQ and PhQ deposited on the outer surface of the SWCNTs. Self-supported films of AQ@SWCNTs and PhQ@SWCNTs were used as the electrodes for lithium-ion batteries (LIBs) and sodium-ion batteries (SIBs). The LIB capacities of AQ@SWCNTs and PhQ@SWCNTs suggested that almost all quinone molecules encapsulated in the SWCNTs were contributed to the reversible Li-ion storage, whereas the contribution of SWCNTs for Li-storage was negligible in the voltage range employed [52]. The effect of the tube diameter of SWCNTs on the Li- and Na-ion storage was studied using SWCNTs with different diameters of 1.5 and 2.5 nm (SWCNT-1.5 and SWCNT-2.5, respectively), of which encapsulated PhQ amounts were determined to be 22 and 38 wt.%, respectively [53]. Charge–discharge curves of PhQ@SWCNT-2.5 for Li^+ and Na^+ in the second to fifth cycles are shown in Fig. 5.20A and B, respectively. The charge–discharge profiles and the reversible capacities of PhQ@SWCNT-2.5 at room temperature are quite similar to those of the bulk PhQ for Li and Na ions, associating with two-step charging at 2.8 and 2.4 V for Li and at 2.3 and 1.9 V for Na. However, the reversible capacities of PhQ@SWCNT-1.5 and PhQ@SWCNT-2.5 are quite different. The former is almost as high as the theoretical capacity of PhQ (258 mAh/g), while the latter is about a half of the theoretical value. The capacity fading was observed on the bulk PhQ

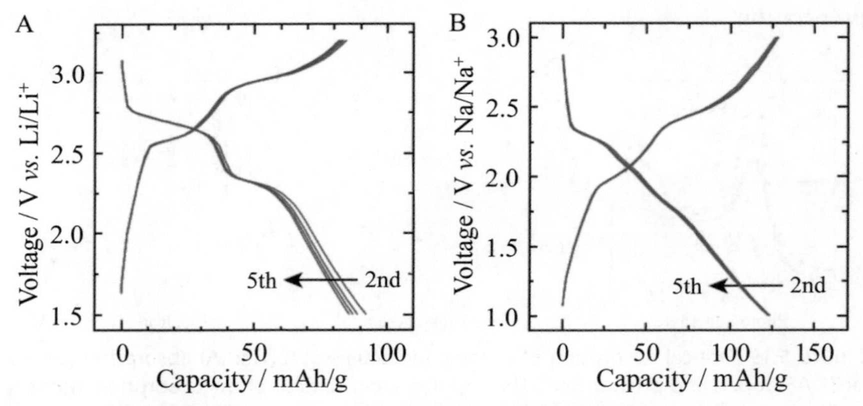

Figure 5.20 Charge–discharge curves of PhQ@SWCNT-2.5 with a current density of 100 mA/g at room temperature: (A) in 1 M $LiClO_4$/(EC + DEC) and (B) in 1 M $NaClO_4$/PC [53].

due to the dissolution of PhQ molecules into the electrolyte, but it was markedly suppressed by the encapsulation into SWCNTs [52]. The reversible capacities of LIBs using PhQ@SWCNT-1.5 and PhQ@SWCNT-2.5 at 0°C were much smaller than those at room temperature, probably due to kinetic problems. Meanwhile, the capacities of SIBs at 0°C for both electrodes slightly decreased in comparison with those at room temperature. Particularly, the reversible capacity of PhQ@SWCNT-2.5 at 0°C was almost the same as that at room temperature, probably because of facile Na ion diffusion in SWCNT-2.5 even at a low temperature. Encapsulation of β-carotene ($C_{40}H_{56}$), 9,10-dichloroanthracene, and coronene into SWCNTs resulted in the enhancement of LIB capacity [54]. Reversible capacity of coronene-encapsulated SWCNTs (7.2 wt.%) was calculated to be 793 mAh per CNT weight at 100 mA/g in 1 M $LiClO_4$/(EC + DEC), whereas pristine SWCNTs exhibited reversible capacity of 316 mAh/g. These organic-molecule-encapsulated SWCNTs showed highly irreversible capacities, but the enhanced reversible capacities and the high stability of these organic molecules in SWCNTs may be encouraging for practical applications as electrodes. Encapsulation of β-carotene was effective in suppressing its degradation under UV irradiation [55].

Encapsulation of large polycyclic aromatic hydrocarbons (PAHs), coronene and perylene, was performed by heating decapped SWCNTs (inner diameter of 1.7–2.0 nm) in their vapors with Ar (ambient pressure) at 438–530°C for coronene and 350–450°C for perylene [56]. High-resolution TEM images of SWCNTs encapsulated by coronene and perylene are shown in Fig. 5.21A and B, respectively. In SWCNTs,

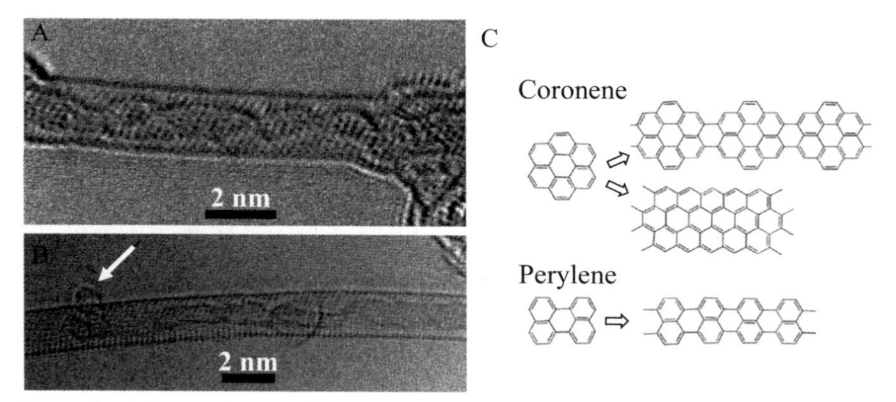

Figure 5.21 Encapsulation of coronene and perylene into SWCNT; TEM images of (A) coronene-encapsulated and (B) perylene-encapsulated SWCNTs, and (C) suggested structures of nanoribbons. Perylene molecule (0.8 nm in diameter) can be seen on the outside of SWCNT in (B) (marked by *arrow*) [56].

encapsulated coronene oligomers were found to be fused into ribbons, as shown schematically in Fig. 5.21C. An attempt to form nanoribbons in SWCNTs often resulted in nonuniform width in the range of 0.5–1.0 nm. Meanwhile, the maximal width was limited by the diameter of the CNTs and the geometrical shape was more complicated, appearing twisted. A columnar stacking of encapsulated coronene was also reported, where coronene molecules stacked with the molecular planes spacing of 0.35 nm and tilting by approximately 77° from the tube axis [57].

A linear-chain nano-diamond, where diamantane ($C_{14}H_{20}$) skeletons were connected together through C–C bonds, was formed in CNTs by the vapor-phase reaction of an apically dibromo-substituted diamantane, 4,9-dibromodiamantane, in the presence of either decapped SWCNTs or DWCNTs at 175°C under high vacuum [58]. A bis-hydroxylated diamantane, 1,6-bis(hydroxymethyl) diamantane, could be encapsulated at 230°C into DWCNTs, of which the inner diameter (<1.0 nm) was compatible with the diameter of diamantane (about 0.7 nm) [59].

5.2.2.6 Water molecules

Encapsulation of water into strongly hydrophobic CNTs had not been expected, but computer simulations predicted a certain possibility [60,61], even the formation of ice nanotubes inside SWCNTs [62]. Water molecules were experimentally shown to be adsorbed into decapped SWCNTs at room temperature and reversibly desorbed, and readsorbed above room temperature [63,64]. On water-encapsulated SWCNTs, a structural change of adsorbed water at low temperatures was studied using XRD [63]. Its XRD patterns of water-confined SWCNTs at different temperatures are shown in Fig. 5.22A. A new diffraction peak appears at reciprocal spacing Q of 22 nm^{-1} at the temperature below 235 K, and the intensity of the peaks at 22 nm^{-1} increases with decreasing temperature (Fig. 5.20B). A new peak was also observed at 7.2 nm^{-1}, of which the intensity behaves the same as the peak at 22 nm^{-1}. These results are explained by solidification of water molecules by forming nanotube (ice nanotube), as shown schematically in Fig. 5.22C. By using SWCNTs with different tube diameters (*D*) in the range of 1.17–2.40 nm, a temperature–diameter (*T*-*D*) phase diagram of water encapsulated into SWCNTs was determined on the bases of experimental results of XRD, ^{1}H and ^{2}D NMR, and electrical resistance measurements [64]. In thin SWCNTs ($1.17<D<\sim 1.3$ nm), water inside SWCNTs underwent a liquid–solid-like transition and formed hollow ice nanotube structures at low temperatures. For the large-diameter SWCNTs

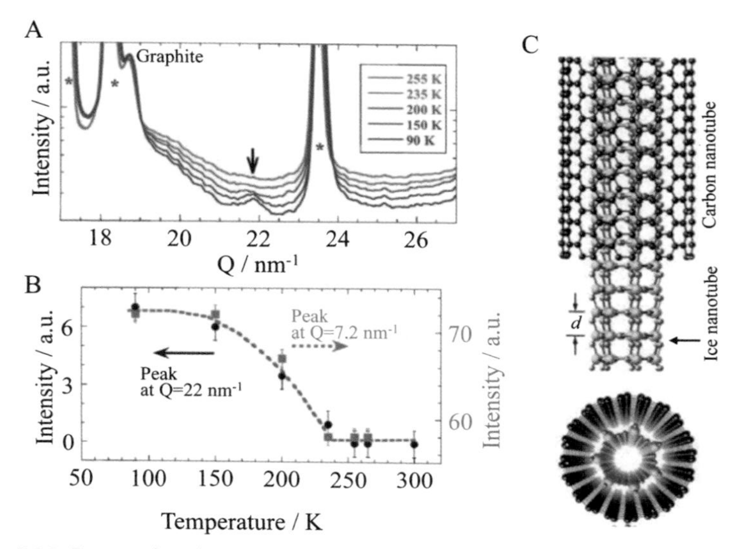

Figure 5.22 Encapsulated water in SWCNTs: (A) XRD profiles at different temperatures, (B) temperature dependence of the intensity of the peaks at 22 and 7.2 nm^{-1}, and (C) a schematic illustration of the ice nanotube inside an SWCNT [63].

($\sim 1.6 < D < 2.40$ nm), on the other hand, water could not be retained inside the SWCNTs. In the intermediate diameter range of about 1.4 nm, hollow ice nanotubes were formed at low temperatures, while a further increase in the water content led to filled structures like ice nanotubes containing a one-dimensional water chain inside.

A loading/releasing process of water molecules in SWCNTs was studied by XRD and hybrid reverse Monte Carlo simulations [65]. Water loading was performed at vapor pressures of 3.1 and then 3.8 kPa (corresponding to water-filling factors of 50% and 100%, respectively) and then water was released up to a vapor pressure of 2.3 kPa (corresponding to water filling of 50%). In the loading process, water molecules formed nanoclusters which were well stabilized in the channel of SWCNTs, while less stable water layers than nanoclusters were formed in the releasing process.

5.2.3 Creation of constrained spaces by carbon coating

Carbon coating has been performed on different functional materials to improve their performances, such as a marked improvement in photocatalytic activity of anatase-type TiO_2, a performance enhancement of electrode materials for LIBs and electrochemical capacitors, as reviewed earlier [66]. Recent experimental results on electrode materials also suggest that carbon coating can provide specific spaces for the structural changes of the matrix. In other words, the coated carbon may either give certain constraint to the functional materials or buffer physical changes in functional materials, such as volume change due to structural changes of the functional materials, similar to the pore walls of porous carbons and the tube walls of CNTs explained above.

5.2.3.1 Metallic silicon

Carbon coating is very effective for exerting the intrinsic functionality of metallic silicon as the negative electrode material for LIBs, such as high theoretical capacity (more than 3.5 Ah/g), by absorbing its large volume change (more than 300%) during lithiation–delithiation reactions and improving its relatively low electrical conductivity.

The particle size of Si exhibited marked influences on the rate and cycle performances as electrode materials for LIBs with 1 M $LiPF_6$/(EC + DEC) [67]. Si particles with a size of ca. 1 μm (micron-sized Si) showed an abrupt decrease in capacity with the cycle number, even at a small current density of 0.2 A/g due to pulverization of Si particles. The poor cycle lifetimes of the micron-sized Si particles were attributed to the typical bulk-like behavior. On the other hand, nano-sized Si particles with a size of 82 nm exhibited remarkably enhanced coulombic efficiency at the first and following cycles. The capacity retention of the nano-sized Si particles was enhanced by carbon coating with a thickness of ca. 10 nm through a pressure-pulsed chemical vapor deposition (P-CVD). The particle morphologies of the nano-sized Si particles and the carbon-coated ones changed from spherical particles to nano-sized wrinkled structures after 20 charge–discharge (lithiation–delithiation) cycles (Fig. 5.23A and B)

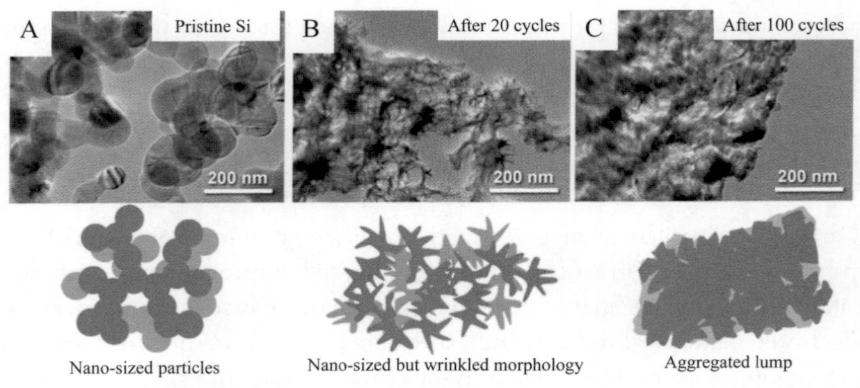

Figure 5.23 TEM images and morphology illustrations of nano-sized Si after cycling of lithiation–delithiation [68].

[67,68]. At the same time, the coated carbon was also deformed together with wrinkled Si, as evidenced by scanning transmission electron microscopy (STEM) imaging and energy-dispersive X-ray spectroscopy (EDS) analysis [67]. After 100 cycles, the wrinkled structure of C-coated Si is further transformed to an aggregated lump due to the repeated volumetric expansion and subsequent agglomeration of pulverized Si particles (Fig. 5.23C). The capacity fading accompanied by the structural deformation of the wrinkled Si was suppressed by limiting the capacity up to 1500 mAh/g during lithiation because of restricting the volumetric expansion to some extent, as described in Section 3.1.2.2. Similar structural transformation through wrinkled and improved battery performances was observed on ball-milled Si, which is less expensive than Si nanoparticles [68]. Flaky Si nanoparticles (thickness of ca. 16 nm and lateral size of 0.2–1 μm) were produced from Si sawdust by beads-milling in isopropyl alcohol. Since the Si sawdust contained about 4 wt.% graphite came from graphite substrate used for the cutting process of Si ingots, the graphite is homogeneously dispersed over the Si particles (carbon coating) after a beads-milling process [69]. The carbon layer coated on Si particles is supposed to work as a buffer between neighboring Si nanoparticles, even during marked morphology change from spherical or flaky to wrinkled.

Thin Si flakes were synthesized by CVD of silane (SiH_4) gas on the cubic crystals of NaCl at 550°C, followed by dissolution of NaCl [69]. Carbon coating of the Si flakes were performed by CVD of acetylene at 900°C, resulting in the carbon content of 7 wt.% and the thickness of about 10 nm. The nonporous structure of the pristine and carbon-coated Si flakes with a low surface area of ca. 10 m^2/g suppressed excessive SEI formation and reduced initial irreversible capacity, in comparison with that of the commercial Si nanoparticles. The initial reversible capacities of the pristine and the C-coated Si flakes reached 2255 and 2943 mAh/g, respectively, in 1.3 M $LiPF_6$/(EC + DEC). The C-coated Si flakes demonstrated a steady rate performance at around 2000 mAh/g capacity up to a 10 C rate, in marked contrast to the inferior rate capability of the pristine Si flakes.

5.2.3.2 Metallic germanium

Microporous and mesoporous carbons containing Ge nanoparticles were synthesized by hydrothermal treatment of a mixture of GeO_2 and glucose with NaOH at 180°C, followed by heating in H_2/Ar flow at 680°C [70]. The resultant carbon spheres with diameter from 50 to 100 nm contained Ge particles of about 16.6 wt.% and exhibited stable cycle performance at a

0.1 C rate in 1 M $LiPF_6$/(EC + DMC + DEC). The capacity at the first cycle exceeded 1000 mAh/g and was stabilized at about 400 mAh/g after the second cycle. In contrast, commercial Ge powder showed drastic capacity decay with the cycle number. The mixture of resol, tetramethoxygermane [$Ge(OEt)_4$] and a block copolymer PEO-*b*-PS was carbonized at 600°C in Ar, followed by heating at the same temperature in 4%H_2/Ar gas flow to reduce GeO_2 to partially metallic Ge [71]. The resultant composite was composed of mesoporous carbons encapsulating nanoparticles of 9.1 wt.% Ge and 71.1 wt.% GeO_2 and delivered highly reversible capacity (1631 mAh/g) and excellent cyclability in 1.3 M $LiPF_4$/(EC + EMC), which was much better performance than for the GeO_2/C composite without H_2 reduction. On the fabrication process of these composites, the starting GeO_2 particles were coated by carbonaceous materials derived from carbon precursors, either glucose or resol, and then carbonized at higher temperature to produce C-coated GeO_2, which was reduced by either carbon or H_2 to metallic Ge. Considering this process, reduction of GeO_2 resulted in the formation of Ge nanoparticles, leaving free spaces in the carbon cage to buffer the volumetric expansion by lithiation. The LIB performances of the Ge-containing carbon materials, including these composites, were explained and discussed previously (Section 3.1.2.3).

5.2.3.3 Metallic tin

Fine particles of metallic tin (Sn) could be dispersed in the mesopores of MgO-templated carbons [72]. MgO and SnO_2 (particle sizes of about 100 and 200 nm, respectively) were mixed with PVA in powder and carbonized at 900°C. Dissolution of MgO with diluted HCl left fine metallic Sn particles in the pores. On heating the mixture, SnO_2 was reduced to form molten Sn, which was supposed to wet with solid MgO particles and to be prevented from agglomeration during the carbonization of PVA. This carbon-coated Sn was successfully applied to the negative electrode of LIBs. The Sn/C composite containing 75 wt.% Sn delivered a reversible capacity of about 500 mAh/g after the second charge–discharge cycle, although it exhibited a relatively large irreversible capacity of about 550 mAh/g in the first cycle. The LIB performances of Sn and SnO_2 were described previously (Section 3.1.2.4).

5.2.3.4 Graphite

Carbon coating of graphite particles is effective to improve its charge–discharge performance in LIBs, as described in Section 3.1.1.1. Carbon coating was performed simply by heating the mixture of graphite powder with a carbon precursor, such as PVA, at around 900°C [73–75]. The disordered and porous carbons coated on the graphite particles worked as a buffer for a large volume change during lithiation–delithiation at the negative electrode. The optimum carbon coating conditions for delivering a small irreversible capacity with keeping reversible capacity more than 350 mAh/g are a coating of 5–10 wt.% carbon at 700–1000°C.

5.2.3.5 Lithium compounds

Carbon-coated $Li_3V_2(PO_4)_3$ (LVP) worked as either a negative or positive electrode by coupling with AC as the counter-electrodes in 1 M $LiPF_6$/(EC + DEC) electrolyte [76], because a monoclinic LVP was possible to serve as both the negative (V^{3+}/V^{2+}) and positive electrodes (V^{3+}/V^{5+}). In the cell I [C-coated LVP(anode)//AC], the principal reactions during charge–discharge are the insertion–deinsertion of Li^+ at the C-coated LVP electrode and the adsorption–desorption of PF_6^- at the AC electrode (Fig. 5.24A). In the cell II [AC//carbon-coated LVP (cathode)], however, the principal reactions during charge–discharge are the deinsertion–reinsertion of Li^+ at the C-coated LVP electrode and the adsorption–desorption of Li^+ at the AC electrode, as shown in the schematic insets in Fig. 5.24A and B, respectively. Charge–discharge curves exhibit the multiple plateaus, corresponding to the redox reactions on the LVP in the respective electrode. C-coating of LVP was effective to obtain steady cyclic and rate performances of both cells. Although the principal electrochemical reactions at the electrodes are different between two cells, the performances of these two hybrid cells were very similar, giving similar energy and power densities.

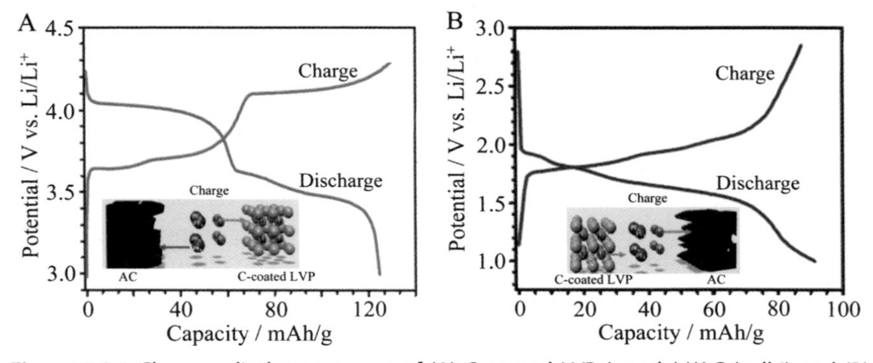

Figure 5.24 Charge–discharge curves of (A) C-coated LVP (anode)//AC (cell I) and (B) AC//C-coated LVP (cathode) (cell II) in 1 M $LiPF_6$/(EC + DEC) at a current density of 100 mA/g. The charging mechanisms in these cells are illustrated as insets. [76].

C-coating was successfully applied on different lithium compounds for electrodes in LIBs and hybrid cells. C-coated Li_2MnSiO_4 was synthesized by a solid-state reaction of the mixture of LiOH, $MnCO_3$, and SiO_2 with adipic acid ($C_6H_{10}O_4$) at 900°C [77], and carbon-coated Li_2FeSiO_4 was synthesized by a similar procedure also using adipic acid as a carbon precursor [78]. C-coating using adipic acid reduced the resistance between the particles of active materials, and as a consequence, achieved a high-power density for hybrid cells. Carbon coated $H_2Ti_{12}O_{25}$ was prepared from Na_2CO_3 and TiO_2 (molar ratio of 1/3) mixed with beta cyclodextrin (4.5 wt.%) by carbonization at 800°C for 2 h [79]. The thickness of the carbon layer coated was determined to be about 3 nm by TEM observation. The carbon coating was effective in improving the performance of a hybrid cell (lithium-ion capacitor [LIC]), giving a negligible capacitance decrease after 100 cycles and an energy density of 38.8 Wh/kg at a power density of 0.18 kW/kg in 1.5 M $LiBF_4$ (EC + DMC) [79]. A coated carbon layer was supposed to not only enhance the electrical conductivity but also suppress the swollen phenomenon due to the decomposition of $H_2Ti_{12}O_{25}$ in the electrolyte, generating C_2H_4, CO, and CH_4 gases.

5.2.3.6 Photocatalysts

C-coating of a photocatalyst, TiO_2 (anatase-type), was shown to be effective in enhancing an activity for the decomposition of some organic pollutants [80–84]. Carbon coating was performed by a simple heat-treatment of the mixed powders of either TiO_2 or its precursor with carbon precursor at 700–1100°C in an inert atmosphere [80]. By proper selection of the mixing ratio, the product obtained after heat-treatment was a black powder without any noticeable agglomeration. The phase transformation in TiO_2 from anatase-type to nonphotocatalytic rutile-type was found to be suppressed by the C-coating. In Fig. 5.25A, decomposition curves of methylene blue (MB) are shown for the C-coated TiO_2 prepared from commercial TiO_2 mixed with a carbon precursor (PVA) by heating at 900°C [81]. The experimental results revealed that the C-coating accelerates the decomposition of MB in aqueous solution, depending strongly on the amount of carbon coated and also on the temperature of the C-coating. In Fig. 5.25B, the rate constants of the decomposition of two different dyes, MB and reactive black 5 (RB5), are plotted against the amount of carbon coated on TiO_2 particles. Coating with more than

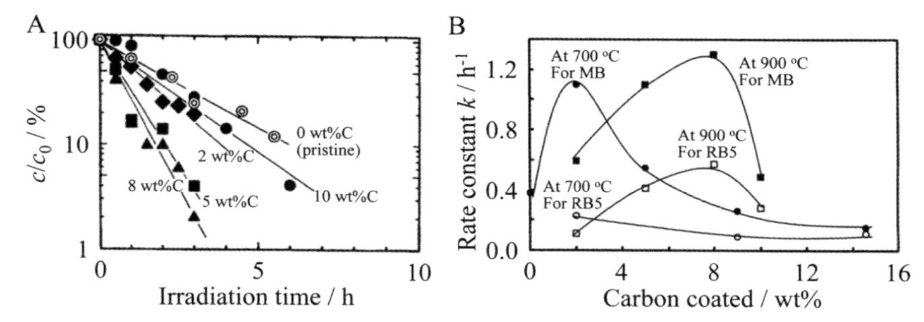

Figure 5.25 Effect of carbon-coating on photocatalytic activity of anatase-type TiO_2 under UV irradiation: (A) decomposition of methylene blue (MB) after adsorption saturation of the catalysts and (B) dependences of decomposition rate constants of MB and reactive black 5 (RB5) for TiO_2 coated by carbon at 700 and 900°C [81].

8 wt.%, rate constants for both dyes decrease abruptly, probably because of interference of UV rays by the coated carbon layer. Carbon coating temperature also affected the rate constants because of phase transition to rutile in TiO_2 particles. The catalyst prepared at 900°C with carbon coating of 8 wt.% delivered the highest rate constants for MB and RB5, where the crystallinity of anatase TiO_2 was improved. A carbon layer coated on the photocatalytic TiO_2 particles contributed to enhancing the photocatalytic reactions by suppressing phase transformation from anatase to non-photocatalytic rutile, enhancing the pollutant concentration near the catalyst particle by adsorption, and crystallinity improvement of the catalyst anatase.

In the course of carbon coatings of TiO_2, a partial reduction of TiO_2 to Ti_nO_{2n-1} was observed, most of which was photoactive in visible light [85,86]. Carbon-coated $W_{18}O_{49}$, which was also visible light active, was successfully synthesized from a mixture of para-ammonium tungstate [$(NH_4)_{10}W_{12}O_{41}$] with PVA by the heat treatment at 800°C in an inert atmosphere [87,88]. Carbon layers on these photocatalysts seemed to have various roles: to reduce TiO_2 to Ti_nO_{2n-1} and WO_3 to $W_{18}O_{49}$, to inhibit the sintering and crystal growth of catalyst particles keeping them small sized, and also to concentrate pollutants around catalyst particles due to pollutant adsorption by the carbon layer.

5.2.4 Syntheses of inorganic materials

5.2.4.1 Platinum

Subnanometer-sized Pt clusters (less than 6 Pt atoms) were obtained using porous carbon substrates by confining a Pt precursor (COD)$PtMe_2$

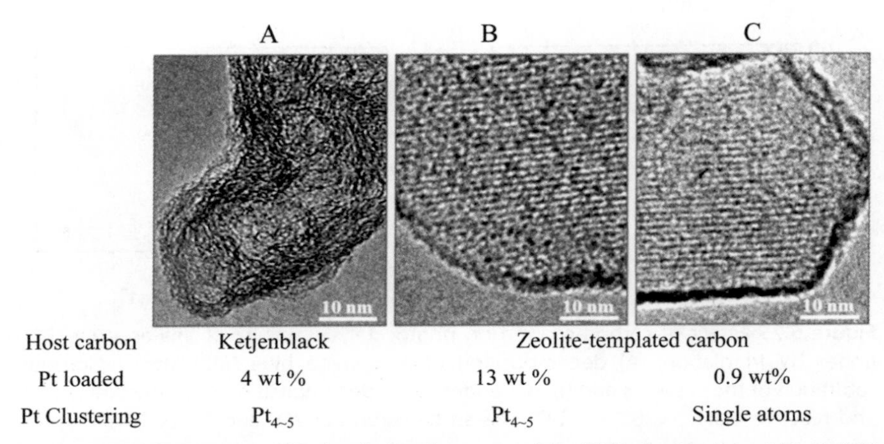

Figure 5.26 TEM images of Pt-loaded carbon materials: (A) Ketjenblack and (B, C) ZTC [89].

(COD = 1,5-cyclooctadiene) into their micropores, followed by thermal reduction at 300°C under vacuum [89]. Ketjenblack (KB; S_{BET}: 1290 m^2/g, V_{micro}: 0.46 cm^3/g) and zeolite-templated microporous carbon (ZTC, S_{BET}: 3680 m^2/g, V_{micro}: 1.55 cm^3/g) were employed as carbon substrates. The contents of the Pt-precursor after wet impregnation and that of metallic Pt after reduction were 6.6 and 3.96 wt.%, respectively, for KB, and 20.0 and 12.8 wt.%, respectively, for ZTC. The amount of the adsorbed Pt precursor proportionally increases with V_{micro} of the carbon substrates. The TEM images after Pt-loading on KB and ZTC are shown in Fig. 5.26A and B, respectively. Even after the thermal reduction at 300°C, no crystalline phase of Pt was detected by XRD, and the sizes of Pt particles were supposed to be less than 1 nm from TEM images. X-ray absorption spectroscopy (XAS) analysis revealed that subnanometer Pt_{4-5} clusters were supported on KB and ZTC, and the Pt—Pt distance was substantially shortened (2.5%—2.8%) in comparison with that in bulk Pt, because of the surface tension in the unsaturated coordination environment of the Pt atoms. The adsorption of the Pt-precursor into micropores was supposed to be crucial to prevent agglomeration and sintering of Pt nanoparticles. Reducing the adsorbed amount of the Pt precursor led to the dispersion of Pt as a single atom, as shown in Fig. 5.26C. The dispersion of Pt single atoms required a strong interaction between Pt atom and oxygen-containing functional group in ZTC, which was supported by the results of XAS analysis.

Pt nanoparticles were loaded on an Ar^+-ion irradiated glassy carbon plate by radiofrequency magnetron sputtering with plasma output of 20 W for 60 s at a fluence of up to 10^{16} ions/cm^2 [90]. Pt was supposed to be confined at the defect sites on the glassy carbon surface created by the irradiation at 380 keV Ar^+ with an average particle size of about 5 nm. A strong interfacial interaction between Pt nanoparticles and the glassy carbon substrate was estimated from its Pt $4f_{7/2}$ and C 1s XPS spectra.

5.2.4.2 Ceramics

Highly crystalline zeolites (ZSM-5 and Y) with uniform mesoporous channels between zeolite crystals were synthesized using a carbon aerogel as the template [91,92]. The precursors of ZSM-5 [tetraethyl orthosilicate (TEOS) and aluminum isopropoxide $Al(iPrO)_3$] were impregnated into mesopores of carbon aerogel, which was prepared from RF aerogels by freeze-drying and carbonization at 1050°C in N_2 flow, and then heat-treated in inert atmosphere to form zeolite ZSM-5 within the mesopores [91]. The zeolite/carbon composite was heated at 550°C in O_2/Ar flow to burn out the carbon and remove any organics that remained. The carbon aerogel had mesopores with a diameter of 23 nm and the resultant zeolite had mesopore channels with a diameter of 11 nm. The XRD patterns and N_2 adsorption–desorption isotherms of mesoporous ZSM-5 synthesized using carbon aerogel and ZSM-5 synthesized without carbon aerogel are compared in Fig. 5.27A and B, respectively. The pore structure parameters of mesoporous ZSM-5 and ZSM-5 are almost the same, except that the former had more mesopores with V_{meso} of 0.2 cm^3/g in addition to V_{micro}

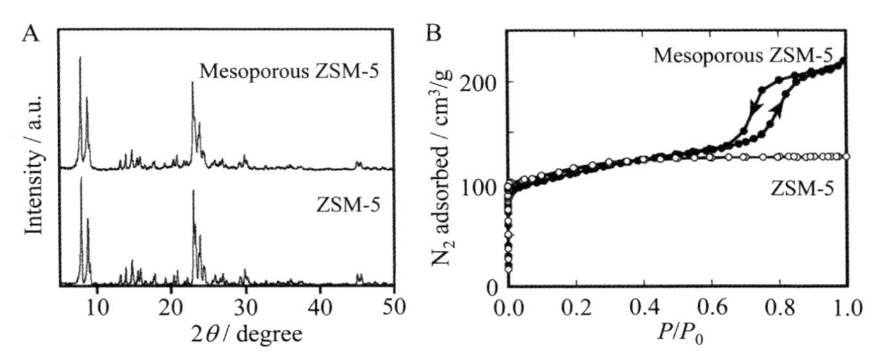

Figure 5.27 Mesoporous zeolite (ZSM-5) synthesized by templating carbon aerogel in comparison with ZSM-5 synthesized without carbon aerogel: (A) XRD and (B) N_2 adsorption–desorption isotherms [91].

of 0.15 cm^3/g. Mesoporous zeolite Y was successfully synthesized through the same procedure as in the case of ZSM-5 [92]. The mesoporous zeolite Y synthesized using carbon aerogel had a large V_{meso} (1.37 cm^3/g) but small V_{micro} (0.21 cm^3/g). Mesoporous carbons synthesized using colloidal silicas as templates were also used as the template for the synthesis of mesoporous ZSM-5 [93].

Reversible replication (templating) between carbon and silica was shown to be possible using highly ordered mesoporous silica SBA-15. The replication of SBA-15 with carbon to form ordered mesoporous carbon CMK-3 was conducted using furfuryl alcohol as a carbon precursor, and the replication of the resultant CMK-3 was performed using TEOS as a silica precursor [94]. The same procedures, i.e., the replication of SBA-15 with carbon and then reversible replication of the resultant CMK-3 with silica, were also examined using sucrose and Na_2SiO_3 [95] and also using propylene gas and TEOS vapor through the CVI process [96]. Well replications from SBA-15 to CMK-3 and from CMK-3 to silica replica in highly ordered alignments of mesopores as well as their sizes and volumes were proved by XRD patterns and N_2 adsorption–desorption isotherms shown in Fig. 5.28A and B, respectively. The parameters of superlattices and pore structures are listed for the aforementioned SBA-15, carbon replica CMK-3. and silica replica in Table 5.2.

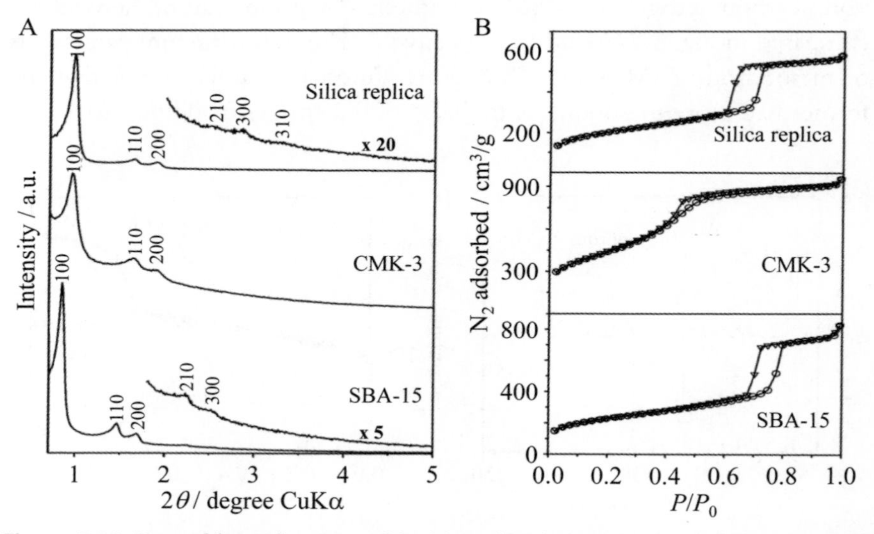

Figure 5.28 Reversible replications of highly ordered mesopore structures between silica (SBA-15) and carbon (CMK-3): (A) XRD patterns and (B) N_2 adsorption–desorption isotherms [95].

Table 5.2 Pore structure parameters on SBA-15, CMK-3, and silica replica shown in Fig. 5.28 [95].

Materials	Lattice parameter[a] (nm)	S_{BET} (m^2/g)	V_{total} (cm^3/g)	V_{meso}[b] (cm^3/g)	Pore size (nm)	Wall thickness (nm)
SBA-15	11.99	776	1.27	1.04	8.11	3.88
CMK-3	10.69	1603	1.48	1.02	3.43	—
Silica replica	10.20	685	0.90	0.69	6.47	3.73

[a] Parameter of superlattice calculated from 100 peak.
[b] Calculated by ($V_{total}-V_{micro}$).

By using the zeolite precursor with mesoporous carbons as templates, mesoporous aluminosilicate molecular sieves with zeolite framework were synthesized [97]. Mesoporous carbons employed as templates were CMK-1 and CMK-3 prepared using mesoporous silicas, MCM-48 and SBA-15, as templates, respectively. The carbon templates CMK-1 and CNK-3 after activation at 100°C in air were impregnated by zeolite precursors (TEOS + $Al(iPrO)_3$) in their ethanol solution and subjected to aging at 40°C for 48 h to form aluminosilicate/CMK composites. Tetrapropylammonium hydroxide was further added in the mixture for crystallization in an autoclave with saturated steam at 100°C, and then calcined at 600°C in air to obtain carbon-free aluminosilicate replicas, denoted RMM-1 and RMM-3, respectively. In Table 5.3, the pore structure parameters of the starting mesoporous silica and the resultant replicated aluminosilicates are listed, revealing a successful replication of mesopore structures in RMM-1 and RMM-3.

Different crystal morphologies of $BaTiO_3$, rod-like and spherical, were obtained in the nanopores of single-walled carbon nanohorns (SWCNHs) [98]. TiO_2 was first encapsulated into the nanopores of SWCNHs by

Table 5.3 Pore structure parameters of the starting mesoporous silicas (MCM-48 and SBA-15) and replicated aluminosilicates (RMM-1 and RMM-3) [97].

	S_{BET} (m^2/g)	V_{total} (cm^3/g)	V_{micro} (cm^3/g)	Mesopore diameter (nm)
MCM-48	1043	0.78	—	2.5–2.8
RMM-1	554	0.77	0.07	3–5
SBA-15	796	0.73	0.07	5–7
RMM-3	595	0.92	0.08	4–6

impregnating in an aqueous $TiCl_4$ solution and following annealing at 373 K. The formed TiO_2 was reacted with $Ba(OCH_2CH_3)_2$ at 400 K in an autoclave of N_2 atmosphere, and the resultant $BaTiO_3$ rods have high aspect ratio, i.e., 2–4 nm in diameter and 40–60 nm in length. Spherical $BaTiO_3$ particles with diameters of 2–4 nm were synthesized in the nanopores of SWCNHs by impregnating SWCNHs with a methanol/2-methoxyethanol mixed solution containing $Ba(OCH_2CH_3)_2$ and $Ti(OCH(CH_3)_2)_4$ and subsequent annealing at 400 K in an autoclave. Meanwhile, $BaTiO_3$ particles with diameters of 20–40 nm were synthesized via the same process without using SWCNHs.

A mesoporous anatase-type TiO_2 was synthesized from a mixture of a diblock copolymer, poly(ethylene oxide-*b*-acrylonitrile) (PEO-*b*-PAN), with TiO_2 sol [99]. A mixture of TiO_2 sol which was prepared in the polymer solution by reacting $TiCl_4$ with the corresponding Ti alkoxide was cast to form film. The film was heated at 250°C for cyclization of PAN in air and then heat-treated at 700°C in Ar to obtain a composite of highly crystalline anatase-type TiO_2 with amorphous carbon. During this process, the TiO_2 crystals nucleate, grow, and sinter into a wall material, while the PEO is thermally removed and the stabilized PAN is converted to a carbon material, resulting in a composite of nanostructured carbon with highly crystalline TiO_2. The TiO_2 exhibited an XRD pattern corresponding to an anatase structure with relative sharp diffraction peaks. A TEM image of the composite is shown in Fig. 5.29A. The composite gave mesoporous TiO_2

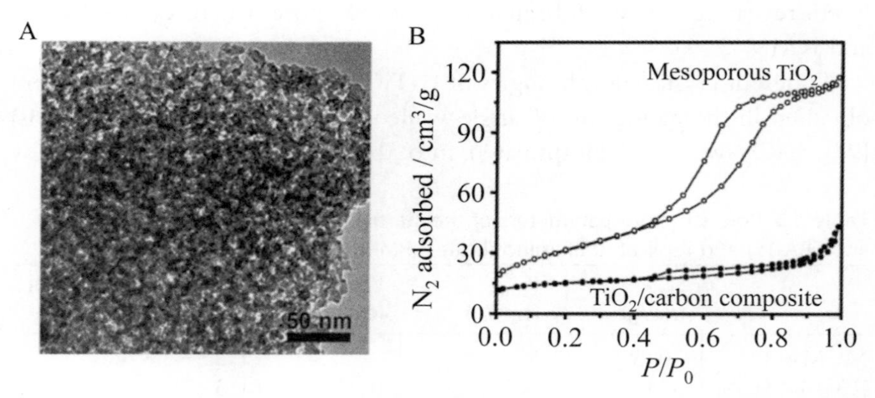

Figure 5.29 Mesoporous TiO_2: (A) TEM image of the composite of TiO_2 with carbon and (B) N_2 adsorption–desorption isotherms of the composite and the resultant mesoporous TiO_2 [99].

after heat treatment at 450°C in air to burn off the carbon. The N_2 adsorption–desorption isotherms are compared on the composites before and after air oxidation in Fig. 5.29B. The isotherm analysis on the mesoporous TiO_2 gave S_{BET} of 111 m^2/g, of which mesopores have relatively uniform sizes of around 7.4 nm.

References

[1] Kyotani T, Nagai T, Inoue S, Tomita A. Formation of new type of porous carbon by carbonization in zeolite nanochannels. Chem Mater 1997;9:609.

[2] n.d. http://www.toyotanso.com/Products/new_developed_products/cnovel.html.

[3] Zhang, X, Yamauchi, M, Takahashi, J. Life cycle assessment of CFRP in application of automobile, 18th international conference on composite materials, 21–26 August, 2011, Jeju, Korea; n.d.

[4] Pimenta S, Pinho ST. Recycling carbon fibre reinforced polymers for structural applications: technology review and market outlook. Waste Manag 2011;31:378–92.

[5] Asmatulu E, Twomey J, Overcash M. Recycling of fiber-reinforced composites and direct structural composite recycling concept. J Compos Mater 2013;48(5):593–608.

[6] La Rosa AD, Banatao DR, Pastine SJ, Latteri A, Cical G. Recycling treatment of carbon fibre/epoxy composites. Composer Part B 2016;104:17–25.

[7] Morishita T, Tsumura T, Toyoda M, Przepiórski J, Morawski AW, Konno H, Inagaki M. A review of the control of pore structure in MgO-templated nanoporous carbons. Carbon 2010;48:2690–707.

[8] Itoi H, Muramatsu H, Inagaki M. Constraint spaces in carbon materials. RSC Adv 2019;9:22823–40.

[9] Itoi H. Enhancing the performance of electrochemical capacitor electrodes by modifying their carbon nanopores with redox-active materials. TANSO 2019;288:103–13.

[10] Itoi H, Maki S, Ninomiya T, Hasegawa H, Matsufusa H, Hayashi S, Iwata H, Ohzawa Y. Electrochemical polymerization of pyrene and aniline exclusively inside the pores of activated carbon for high-performance asymmetric electrochemical capacitors. Nanoscale 2018;10:9760–72.

[11] Itoi H, Hasegawa H, Iwata H, Ohzawa Y. Non-polymeric hybridization of TEMPO derivatives with activated carbon for high-energy-density aqueous electrochemical capacitor electrodes. Sus Energy Fuels 2018;2:558–65.

[12] Itoi H, Yasue Y, Suda K, Katoh S, Hasegawa H, Hayashi S, Mitsuoka M, Iwata H, Ohzawa Y. Solvent-free preparation of electrochemical capacitor electrodes using metal-free redox organic compounds. ACS Sustainable Chem Eng 2017;5:556–62.

[13] Le TH, Yang Y, Yu L, Huang Z-H, Kang F. In-situ growth of MnO_2 crystals under nanopore- constraint in carbon nanofibers and their electrochemical performance. Sci Rep 2016;6:37368.

[14] Sun F, Gao J, Zhu Y, Pi X, Wang L, Liu X, Qin YA. High performance lithium ion capacitor achieved by the integration of a Sn-C anode and a biomass-derived microporous activated carbon cathode. Sci Rep 2017;7:40990.

[15] Przygocki P, Abbas Q, Babuchowska P, Beguin F. Confinement of iodides in carbon porosity to prevent from positive electrode oxidation in high voltage aqueous hybrid electrochemical capacitors. Carbon 2017;125:391–400.

[16] Lota G, Fic K, Frackowiak E. Alkali metal iodide/carbon interface as a source of pseudocapacitance. Electrochem Commun 2011;13:38–41.

[17] Inagaki M, Tsumura T, Kinumoto T. Toyoda M Graphitic carbon nitrides (*g*-C_3N_4) with comparative discussion to carbon materials. Carbon 2019;141:580−607.

[18] Zheng Y, Jiao Y, Chen J, Liu J, Liang J, Du A, Zhang W, Zhu Z, Smith SC, Jaroniec M, Lu GQ, Qiao SZ. Nanoporous graphitic-C_3N_4@carbon metal-free electrocatalysts for highly efficient oxygen reduction. J Am Chem Soc 2011;133:20116−9.

[19] Balach J, Singh HK, Gomoll S, Jaumann T, Klose M, Oswald S, Richter M, Eckert J, Giebeler L. Synergistically enhanced polysulfide chemisorption using a flexible hybrid separator with N and S dual-doped mesoporous carbon coating for advanced lithium-sulfur batteries. ACS Appl Mater Interfaces 2016;8:14586−95.

[20] Casco ME, Silvestre-Albero J, Ramırez-Cuesta AJ, Rey F, Jorda JL, Bansode A, Urakawa A, Peral I, Martınez-Escandell M, Kaneko K, Rodrıguez-Reinoso F. Methane hydrate formation in confined nanospace can surpass nature. Nat Commun 2015;6:6432.

[21] Endo M, Hayashi T, Muramatsu H, Kim YA, Terrones H, Terrones M, Dresselhaus MS. Coalescence of double-walled carbon nanotubes: formation of novel carbon bicables. Nano Lett 2004;4:1451−4.

[22] Muramatsu H, Hayashi T, Ahm KY, Terrones M, Endo M. Formation of off-centered double-walled carbon nanotubes exhibiting wide interlayer spacing from bi-cables. Chem Phys Lett 2006;432:240−4.

[23] Metenier K, Bonnamy S, Beguin F, Journet C, Bernier P, de La Chapelle ML, Chauvet O, Lefrant S. Coalescence of single-walled carbon nanotubes and formation of multi-walled carbon nanotubes under high-temperature treatment. Carbon 2002;40:1765−73.

[24] Guan L, Shi Z, Li M, Gu Z. Ferrocene-filled single-walled carbon nanotubes. Carbon 2005;43:2780−5.

[25] Li Y, Hatakeyama R, Kaneko T, Okada T. Nano sized magnetic particles with diameters less than 1 nm encapsulated in single-walled carbon nanotubes. Jpn J Appl Phys 2006;45:L428−31.

[26] Shiozawa H, Pichler T, Gruneis A, Pfeiffer R, Kuzmany H, Liu Z, Suenaga K, Kataura H. A catalytic reaction inside a single-walled carbon nanotube. Adv Mater 2008;20:1443−9.

[27] Li Y, Hatakeyama R, Kaneko T, Okada T, Kato T. Synthesis and electronic properties of ferrocene- filled double-walled carbon nanotubes. Nanotechnology 2006;17:4143.

[28] Li LJ, Khlobystov AN, Wiltshire JG, Briggs GAD, Nicholas RJ. Diameter-selective encapsulation of metallocenes in single-walled carbon nanotubes. Nat Mater 2005;4:481−5.

[29] Smith BW, Monthioux M, Luzzi DE. Encapsulated C_{60} in carbon nanotubes. Nature 1998;396:323.

[30] Smith BW, Luzzi DE. Formation mechanism of fullerene peapods and coaxial tubes: a path to large scale synthesis. Chem Phys Lett 2000;321:169−74.

[31] Kataura H, Maniwa Y, Abe M, Fujiwara A, Kodama T, Kikuchi K, Imahori H, Misaki Y, Suzuki S, Achiba Y. Optical properties of fullerene and non-fullerene peapods. Appl Phys A 2002;74:349−54.

[32] Kataura H, Maniwa Y, Kodama T, Kikuchi K, Hirahara K, Suenaga K, Iijima S, Suzuki S, Achiba Y, Krätschmer W. High-yield fullerene encapsulation in single-wall carbon nanotubes. Synth Met 2001;121:1195−6.

[33] Ning G, Kishi N, Okimoto H, Shiraishi M, Sugai T, Shinohara H. Structural stability and transformation of aligned C_{60} and C_{70} fullerenes in double-wall and triple-wall carbon nanotube- peapods. J Phys Chem C 2007;111:14652−7.

[34] Kawasaki S, Matsuoka Y, Yokomae T, Nojima Y, Okino F, Touhara H, Kataura H. XRD and TEM study of high pressure treated single-walled carbon nanotubes and C_{60}-peapods. Carbon 2005;43:37−45.

[35] Chorro M, Rols S, Cambedouzou J, Alverz L, Almairac R, Sauvajol J, Hodeau J-L, Marques L, Mezouar M, Kataura H. Structural properties of carbon peapods under extreme conditions studied using in situ x-ray diffraction. Phys Rev B 2006;74:205425.
[36] Bandow S, Takizawa M, Hirahara K, Yudasaka M, Iijima S. Raman scattering study of double-wall carbon nanotubes derived from the chains of fullerenes in single-wall carbon nanotubes. Chem Phys Lett 2001;337:48–54.
[37] Muramatsu H, Hayashi T, Kim YA, Shimamoto D, Endo M, Meunier V, Sumpter BG, Terrones M, Dresselhaus MS. Bright photoluminescence from the inner tubes of "Peapod"-derived double-walled carbon nanotubes. Small 2009;5:2678–82.
[38] Herandez E, Meunier V, Smith BW, Rurali R, Terrones H, Nardelli MB, Terrones M, Luzzi DE, Charlier J-C. Fullerene coalescence in nanopeapods: a path to novel tubular carbon. Nano Lett 2003;3:1037–42.
[39] Pfeiffer R, Kuzmany H, Kramberger C, Schaman C, Pichler T, Kataura H, Achiba Y, Kurti J, Zolyomi V. Unusual high degree of unperturbed environment in the interior of single-wall carbon nanotubes. Phys Rev Lett 2003;90:225501.
[40] Muramatsu H, Shimamoto D, Hayashi T, Kim YA, Terrones M, Endo M, Dresselhaus MS. Bulk synthesis of narrow diameter and highly crystalline triple-walled carbon nanotubes by coalescing fullerene peapods. Adv Mater 2011;23:1761–4.
[41] Kawasaki S, Iwai Y, Hirose M. Electrochemical lithium ion storage property of C_{60} encapsulated single-walled carbon nanotubes. Mater Res Bull 2009;44:415–7.
[42] Guan L, Suenaga K, Shi Z, Gu Z, Iijima S. Direct imaging of the alkali metal site in K-doped fullerene peapods. Phys Rev Lett 2005;94:045502.
[43] Hirahara K, Suenaga K, Bandow S, Kato H, Okazaki T, Shinohara H, Iijima S. One-dimensional metallofullerene crystal generated inside single-walled carbon nanotubes. Phys Rev Lett 2000;85:5384–7.
[44] Urita K, Shiga Y, Fujimori T, Iiyama T, Hattori Y, Kanoh H, Ohba T, Tanaka H, Yudasaka M, Iijima S, Moriguchi I, Okino F, Endo M, Kaneko K. Confinement in carbon nanospace-induced production of KI nanocrystals of high-pressure phase. J Am Chem Soc 2011;133:10344–7.
[45] Sloan J, Novotony MC, Bailey SR, Brown G, Xu C, Williams VC, Friedrichs S, Flahaut E, Callender RL, York APE, Coleman KS, Green MLH, Dunin-Borkowski RE, Hutchison JL. Two layer 4:4 co-ordinated KI crystals grown within single walled carbon nanotubes. Chem Phys Lett 2000;329:61–5.
[46] Ahn S, Kim Y, Nam Y, Yoo H, Park J, Park Y, Wang Z, Shi Z. Magnetotransport in iodine-doped single-walled carbon nanotubes. Phys Rev B 2009;80:165426.
[47] Song H, Ishii Y, Al-zubaidi A, Sakai T, Kawasaki S. Temperature-dependent water solubility of iodine-doped single-walled carbon nanotubes prepared using an electrochemical method. Phys Chem Chem Phys 2013;15:5767–70.
[48] Yoshida Y, Ishii Y, Kato N, Li C, Kawasaki S. Low-temperature phase transformation accompanied with charge-transfer reaction of polyiodide ions encapsulated in single-walled carbon nanotubes. J Phys Chem C 2016;120:20454–61.
[49] Fujimori T, Morelos-Gomez A, Zhu Z, Muramatsu H, Futamura R, Urita K, Terrones M, Hayashi T, Endo M, Hong SY, Choi YC, Tomanek D, Kaneko K. Conducting linear chains of sulphur inside carbon nanotubes. Nat Commun 2013;4:2162.
[50] Luo H, Desgreniers S, Vohra YK, Ruoff AL. High-pressure optical studies on sulfur to 121 GPa: optical evidence for metallization. Phys Rev Lett 1991;67:2998–3001.
[51] Takenobu T, Takano T, Shiraishi M, Murakami Y, Ata M, Kataura H, Achiba Y, Iwasa Y. Stable and controlled amphoteric doping by encapsulation of organic molecules inside carbon nanotubes. Nat Mater 2003;2:683–8.
[52] Ishii Y, Tashiro K, Hosoe K, Al-zubaidi A, Kawasaki S. Electrochemical lithium-ion storage properties of quinone molecules encapsulated in single-walled carbon nanotubes. Phys Chem Chem Phys 2016;18:10411–8.

[53] Li C, Ishii Y, Inayama S, Kawasaki S. Quinone molecules encapsulated in SWCNTs for low- temperature Na ion batteries. Nanotechnology 2017;28:355−401.
[54] Kawasaki S, Iwai Y, Hirose M. Electrochemical lithium ion storage properties of single-walled carbon nanotubes containing organic molecules. Carbon 2009;47:1081−6.
[55] Yanagi K, Miyata Y, Kataura H. Highly stabilized β-carotene in carbon nanotubes. Adv Mater 2006;18:437−41.
[56] Talyzin AV, Anoshkin IV, Krasheninnikov AV, Nieminen RM, Nasibulin AG, Jiang H, Kauppinen EI. Synthesis of graphene nanoribbons encapsulated in single-walled carbon nanotubes. Nano Lett 2011;11:4352−6.
[57] Okazaki T, Iizumi Y, Okubo S, Kataura H, Liu Z, Suenaga K, Tahara Y, Yudasaka M, Okada S, Iijima S. Coaxially stacked coronene columns inside single-walled carbon nanotubes. Angew Chem Int Ed 2011;50:4853−7.
[58] Nakanishi Y, Omachi H, Fokina NA, Schreiner PR, Kitaura R, Dahl JEP, Carlson RMK, Shinohara H. Template synthesis of linear-chain nanodiamonds inside carbon nanotubes from bridgehead- halogenated diamantane precursors. Angew Chem Int Ed 2015;54:10802−6.
[59] Nakanishi Y, Omachi H, Fokina NA, Schreiner PR, Becker J, Dahl JEP, Carlson RMK, Shinohara H. One-dimensional hydrogen bonding networks of bis-hydroxylated diamantane formed inside double- walled carbon nanotubes. Chem Commun 2018;54:3823−6.
[60] Sanson MSP, Biggin PC. Water at the nanoscale. Nature 2001;414:156−8.
[61] Hummer G, Rasaiah JC, Noworyta JP. Water conduction through the hydrophobic channel of a carbon nanotube. Nature 2001;414:188−90.
[62] Koga K, Gao GT, Tanaka H, Zeng XC. Formation of ordered ice nanotubes inside carbon nanotubes. Nature 2001;412:802−5.
[63] Maniwa Y, Kataura H, Abe M, Suzuki S, Achiba Y, Kira H, Matsuda K. Phase transition in confined water inside carbon nanotubes. J Phys Soc Jpn 2002;71:2863−6.
[64] Kyakuno H, Matsuda K, Yahiro H, Inami Y, Fukuoka T, Miyata Y, Yanagi K, Maniwa Y, Kataura H, Saito T, Yumura M, Iijima S. Confined water inside single-walled carbon nanotubes: global phase diagram and effect of finite length. J Chem Phys 2011;134:244501.
[65] Ohba T, Taira S, Hata K, Kanoh H. Mechanism of sequential water transportation by water loading and release in single-walled carbon nanotubes. J Phys Chem Lett 2013;4:1211−5.
[66] Inagaki M. Carbon coating for enhancing the functionalities of materials. Carbon 2012;50:3247−66.
[67] Iwamura S, Nishihara H, Kyotani T. Fast and reversible lithium storage in a wrinkled structure formed from Si nanoparticles during lithiation/delithiation cycling. J Power Sources 2013;222:400−9.
[68] Kasukabe T, Nishihara H, Iwamura S, Kyotani T. Remarkable performance improvement of inexpensive ball-milled Si nanoparticles by carbon-coating for Li-ion batteries. J Power Sources 2016;319:99−103.
[69] (a) Kasukabe T, Nishihara H, Kimura K, Matsumoto T, Kobayash H, Okai M, Kyotani T. Beads-milling of waste Si sawdust into high-performance nanoflakes for lithium-ion batteries. Sci Rep 2017;7:42734.(b) Ryu J, Chen T, Bok T, Song G, Ma J, Hwang C, Luo L, Song H-K, Cho J, Wang C, Zhang S, Park S. Mechanical mismatch-driven rippling in carbon-coated silicon sheets for stress-resilient battery anodes. Nat Commun 2018;9:2924.
[70] Xiao Y, Cao M, Ren L, Hu C. Hierarchically porous germanium-modified carbon materials with enhanced lithium storage performance. Nanoscale 2012;4(23):7469−74.
[71] Hwang J, Jo C, Kim MG, Chun J, Lim E, Kim S, Jeong S, Kim Y, Lee J. Mesoporous Ge/GeO_2/carbon lithium-ion battery anodes with high capacity and high reversibility. ACS Nano 2015;9:5299−309.

[72] Morishita T, Hirabayashi T, Okuni T, Ota N, Inagaki M. Preparation of carbon-coated Sn powders and their loading onto graphite flakes for lithium ion secondary battery. J Power Sources 2006;160:638–44.
[73] Nozaki H, Nagaoka K, Hoshi K, Ohta N, Inagaki M. Carbon-coated graphite for anode of lithium ion rechargeable batteries: carbon coating conditions and precursors. J Power Sources 2009;194:486–93.
[74] Ohta N, Nagaoka K, Hoshi K, Bitoh S, Inagaki M. Carbon-coated graphite for anode of lithium ion rechargeable batteries: graphite substrates for carbon coating. J Power Sources 2009;194:985–90.
[75] Hoshi K, Ohta N, Nagaoka K, Bitoh S, Yamanaka A, Nozaki H, Okuni T, Inagaki M. Production and advantages of carbon-coated graphite for the anode of lithium ion rechargeable batteries. TANSO 2009;2009:213–20.
[76] Satish R, Aravindan V, Ling WC, Madhavi S. Carbon-coated $Li_3V_2(PO_4)_3$ as insertion type electrode for lithium-ion hybrid electrochemical capacitors: an evaluation of anode and cathodic performance. J Power Sources 2015;281:310–7.
[77] Karthikeyan K, Aravindan V, Lee S, Jang I, Lim H, Park G, Yoshio M, Lee Y. Electrochemical performance of carbon-coated lithium manganese silicate for asymmetric hybrid supercapacitors. J Power Sources 2010;195:3761–4.
[78] Karthikeyan K, Aravindan V, Lee S, Jang I, Lim H, Park G, Yoshio M, Lee Y. A novel asymmetric hybrid supercapacitor based on Li_2FeSiO_4 and activated carbon electrodes. J Alloys Compd 2010;504:224–7.
[79] Yoon J-R, Baek E, Kim H-K, Pecht M, Lee S-H. Critical dual roles of carbon coating in $H_2Ti_{12}O_{25}$ for cylindrical hybrid supercapacitors. Carbon 2016;101:9–15.
[80] Tsumura T, Kojitani N, Izumi I, Iwashita N, Toyoda M, Inagaki M. Carbon coating of anatase-type TiO_2 and photoactivity. J Mater Chem 2002;12:1391–6.
[81] Tryba B, Morawski AW, Tsumura T, Toyoda M, Inagaki M. Hybridization of adsorptivity with photocatalytic activity - carbon-coated anatase. J Photochem Photobiol A 2004;167:127–35.
[82] Toyoda M, Yoshikawa Y, Tsumura T, Inagaki M. Photoactivity of carbon-coated anatase for decomposition of iminoctadine triacetate in water. J Photochem Photobiol A 2005;171:167–71.
[83] Inagaki M, Kojin F, Tryba B, Toyoda M. Carbon-coated anatase: the roles of the carbon layer for photocatalytic performance. Carbon 2005;43:1652–9.
[84] Toyoda M, Tsumura T, Tryba B, Mozia S, Janus M, Morawski AW, Inagaki M. Carbon materials in photocatalysts. Chem Phys Carbon 2012;31:171–267.
[85] Tsumura T, Hattori Y, Kaneko K, Inagaki M, Toyoda M. Formation of Ti_4O_7 phase through interaction between coated carbon and TiO_2. Desalination 2004;169:269–75.
[86] Toyoda M, Yano T, Tryba B, Mzia S, Tsumura T, Inagaki M. Preparation of carbon-coated Magneli phases Ti_nO_{2n-1} and their photocatalytic activity under visible light. Appl Catal B 2009;88:160–4.
[87] Kojin F, Mori M, Morishita T, Inagaki M. New visible light active photocatalyst, carbon-coated $W_{18}O_{49}$. Chem Lett 2006;35:388–9.
[88] Kojin F, Mori M, Noda Y, Inagaki M. Preparation of carbon-coated $W_{18}O_{49}$ and its photoactivity under visible light. Appl Catal B 2008;78:202–9.
[89] Itoi H, Nishihara H, Kobayashi S, Ittisanronnachai S, Ishii T, Berenguer R, Ito M, Matsumura D, Kyotani T. Fine dispersion of Pt_{4-5} subnanoclusters and Pt single atoms over porous carbon supports and their structural analyses with X-ray absorption spectroscopy. J Phys Chem C 2017;121:7892–902.
[90] Kimata T, Kato S, Yamaki T, Yamamoto S, Kobayashi T, Terai T. Platinum nanoparticles on the glassy carbon surface irradiated with argon ions. Surf Coating Technol 2016;306:123–6.

[91] Tao Y, Kanoh H, Kaneko K. ZSM-5 monolith of uniform mesoporous channels. J Am Chem Soc 2003;125:6044–5.
[92] Tao Y, Kanoh H, Kaneko K. Uniform mesopore-donated zeolite Y using carbon aerogel templating. J Phys Chem B 2003;107:10974–6.
[93] Kim S-S, Shah J, Pinnavaia TJ. Colloid-imprinted carbons as templates for the nanocasting synthesis of mesoporous ZSM-5 zeolite. Chem Mater 2003;15:1664–8.
[94] Lu A-H, Schmidt W, Taguchi A, Spliethoff B, Tesche B, Schueth F. Taking nanocasting one step further: replicating CMK-3 as a silica material. Angew Chem Int Ed 2002;41:3489–92.
[95] Kang M, Yi SH, Lee HI, Yie JE, Kim JM. Reversible replication between ordered mesoporous silica and mesoporous carbon. Chem Commun 2002:1944–5.
[96] Parmentier J, Vix-Guterl C, Saadallah S, Reda M, Illescu M, Werckmann J, Patarin J. Organised mesoporous silica synthesised by nanoscale duplication of an ordered mesoporous carbon material using a gas phase process. Chem Lett 2003;32:262–3.
[97] Sakthivel A, Huang S-J, Chen W-H, Lan Z-H, Chen K-H, Kim T-W, Ryoo R, Chiang AST, Liu S-B. Replication of mesoporous aluminosilicate molecular sieves (RMMs) with zeolite framework from mesoporous carbons (CMKs). Chem Mater 2004;16:3168–75.
[98] Hoshi D, Watanabe T, Ohba T. $BaTiO_3$ nanoparticles and nanorods synthesized in carbon nanohorns. TANSO 2017;280:198–202 [in Japanese].
[99] Stefik M, Lee J, Wiesner U. Nanostructured carbon-crystalline titania composites from microphase separation of poly(ethylene oxide-b-acrylonitrile) and titania sols. Chem Commun 2009;18:2532–4.

Index

Note: 'Page numbers followed by "*f*" indicate figures and "*t*" indicate tables.'

C

D

G

H

Q

R

T

U

V

W

X

Z